Boettcher

Marine
Chemical
Ecology

Marine Science Series

The CRC Marine Science Series is dedicated to providing state-of-the-art coverage of important topics in marine biology, marine chemistry, marine geology, and physical oceanography. The series includes volumes that focus on the synthesis of recent advances in marine science.

CRC MARINE SCIENCE SERIES

SERIES EDITOR

Michael J. Kennish, Ph.D.

PUBLISHED TITLES

Artificial Reef Evaluation with Application to Natural Marine Habitats, William Seaman, Jr.

Chemical Oceanography, Second Edition, Frank J. Millero

Coastal Ecosystem Processes, Daniel M. Alongi

Ecology of Estuaries: Anthropogenic Effects, Michael J. Kennish

Ecology of Marine Bivalves: An Ecosystem Approach, Richard F. Dame

Ecology of Marine Invertebrate Larvae, Larry McEdward

Ecology of Seashores, George A. Knox

Environmental Oceanography, Second Edition, Tom Beer

Estuary Restoration and Maintenance: The National Estuary Program, Michael J. Kennish

Eutrophication Processes in Coastal Systems: Origin and Succession of Plankton Blooms and Effects on Secondary Production in Gulf Coast Estuaries, Robert J. Livingston

Handbook of Marine Mineral Deposits, David S. Cronan

Handbook for Restoring Tidal Wetlands, Joy B. Zedler

Intertidal Deposits: River Mouths, Tidal Flats, and Coastal Lagoons, Doeke Eisma

Morphodynamics of Inner Continental Shelves, L. Donelson Wright

Ocean Pollution: Effects on Living Resources and Humans, Carl J. Sindermann

Physical Oceanographic Processes of the Great Barrier Reef, Eric Wolanski

The Physiology of Fishes, Second Edition, David H. Evans

Pollution Impacts on Marine Biotic Communities, Michael J. Kennish

Practical Handbook of Estuarine and Marine Pollution, Michael J. Kennish

Seagrasses: Monitoring, Ecology, Physiology, and Management, Stephen A. Bortone

Marine Chemical Ecology

Edited by

James B. McClintock

Professor of Biology
University of Alabama at Birmingham
Birmingham, Alabama

Bill J. Baker

Associate Professor of Chemistry
University of South Florida
Tampa, Florida

CRC Press
Boca Raton   London   New York   Washington, D.C.

Cover photograph © 1997 Norbert Wu/www.norbertwu.com. The antarctic red macroalga, *Phyllophora antarctica*, being held by the sea urchin, *Sterchinus neumayeri*, in a mutualistic relationship mediated by chemical and physical defenses. See Amsler, C.D., McClintock, J.B., and Baker, B.J., *Mar. Ecol. Prog. Ser.*, 1999, 183, 105, and Chapter 7 in this volume.

Library of Congress Cataloging-in-Publication Data

Marine chemical ecology / James B. McClintock, Bill J. Baker, editors.
    p. cm.—(Marine science series)
  Includes bibliographical references and index.
  ISBN 0-8493-9064-8 (alk. paper)
   1. Marine chemical ecology. I. McClintock, James B. II. Baker, Bill J. (Bill James),
  1958- III. Series.

QH541.5.S3 M254 2001
577.7′14—dc21   2001025040

Visit the CRC Press Web site at www.crcpress.com

Preface

As early as the middle of the 19th century, scientists were beginning to isolate and characterize organic compounds from nature. Modern studies of the nature of natural products, those compounds not involved in primary metabolic processes, began in the 1930s and 1940s. In the decades that followed, scientists began to focus on why organisms produce these compounds. At the same time, studies of factors that regulate the distribution and abundance of species were giving rise to the field of ecology. The subsequent combination of these two disciplines allowed for investigations of secondary metabolites and their functional roles in mediating reproductive processes, predator–prey interactions, and competition.

The inherently interdisciplinary field of chemical ecology has as its roots studies of chemical interactions in terrestrial microbial and plant systems. Terrestrial plant studies led to the logical extension of similar studies in marine plants, primarily macroalgae. The earliest studies of the chemical ecology of marine animals focused on invertebrates that were plant-like, lacking loco-motive ability or obvious physical means of protection. Surprisingly, such marine studies really only have been undertaken in earnest during the past 25 years, due largely to the lack of knowledge of the chemistry of marine organisms. It is probably no coincidence that the breadth of marine organisms studied expanded in concert with developments in underwater technology.

Today, the field of marine chemical ecology encompasses a broad interdisciplinary community of scientists. Among its many participants are those interested in understanding predator–prey interactions, competition, chemical communication, fouling, pathogen–host interactions, biosyn-thesis, reproduction, and the evolution of secondary metabolites. Bringing diverse disciplines to bear upon these fundamental questions has generated a body of knowledge that is truly synergistic and unequivocally greater than the sum of its parts.

The field of marine chemical ecology has been gaining momentum. Over the past decade, a number of excellent review articles and at least one book have been published on this topic. However, a field so diverse and interdisciplinary requires, from time to time, a conceptual synthesis. There has been, to date, no single source that provides this synthetic overview. This book is such an attempt. In four topical sections, this work spans aspects of marine chemical ecology from molecular to community levels, bridging these diverse disciplines. The authors have contributed their considerable experiences, resulting in a collective effort that will hopefully stimulate new ideas and inspire a new generation of researchers.

The introductory section, Chapters 1 through 3, provides a broad phylogenetic overview of marine organic chemistry. Emphasizing evolutionary, ecological, and biosynthetic considerations, the authors of these chapters set a foundation for the chapters that follow. Harper et al. provide the first balanced account of patterns of specific classes of secondary metabolites among marine organisms. Garson has provided an update of her periodic reviews of the field of marine biosynthesis with an emphasis on ecologically important compounds. Cimino and Ghiselin foray into theoretical analyses of evolutionary trends in marine biota in an effort to rationalize observed patterns of chemical defense in the marine realm.

The second section, Chapters 4 through 10, takes an organismal approach to understanding the role of secondary metabolites in mediating trophic interrelationships. While much of the content of this volume focuses on sessile organisms, Stachowicz reports on chemical ecology of mobile benthic invertebrates, and McClintock et al. review what is known to date about the chemical ecology of meroplankton and holoplankton and offer suggestions for future research. Trophic relationships in the oceanic water column have potential significance in influencing material and energy flux in planktonic food webs. Chemical ecology is ultimately driven by evolutionary pressures, which Cronin reviews from the point of view of resource allocation strategies. The macroalgae are perhaps most thoroughly studied with respect to their trophic chemical ecology. This fertile field is addressed by Paul et al. Marine chemical ecology has also been considered on broad biogeographic and temporal scales. Most notable among marine biogeographic analyses are those applied to the macroalgae, the resulting patterns of which are reviewed by Van Alstyne et al. Biogeographic patterns in higher organisms are less well defined; however, recent evidence from the south polar region, which Amsler et al. review from a chemical analytical perspective, is beginning to illuminate global patterns, at least with respect to trophic interactions. Furthering our understanding of the functional spatial distribution of metabolites, Steinberg et al. examines chemical interactions at organismal surfaces.

Section III, Chapters 11 through 15, reviews cellular and physiological aspects of marine chemical ecology. Targett and Arnold address the physiological consequences of ingesting secondary metabolites. Amsler and Iken review behavioral mechanisms by which algae and bacteria use and exploit chemical gradients. The important role of chemical cues in determining patterns of settlement and metamorphosis in many marine plants and invertebrates are well known. Hadfield and Paul take a comprehensive approach reviewing recent efforts to understand the chemical nature of settlement and metamorphic cues. Trapido-Rosenthal details the role of natural products in prey attraction and other aspects of chemoreception. Completing this section, Karentz sheds light on the important role of UV-absorbing compounds in defending marine organisms from damage due to harmful solar radiation.

Last, but by no means least, Chapters 16 through 18 review modern studies of marine chemical ecology, which, in effect if not in purpose, may lead to practical applications. Secondary metabolites long studied for the sake of knowledge have become commodities to the pharmaceutical industry; the means by which chemical ecological cues can be exploited for purposes of drug discovery are reviewed by Sennett in Chapter 16. Similarly, Rittschof takes a view of secondary metabolites which deter settlement as applied to the discovery of compounds which can be applied to marine surfaces to deter fouling, a major impediment to smooth sailing. Bernan reviews a relatively new area of research which focuses on the secondary metabolism of marine prokaryotic microbes. The success of drugs, agrochemicals, antifouling agents, and/or similar marine natural products in the marketplace will ultimately benefit the entire field due to increased public interest and heightened awareness of the resources available from the marine realm.

As we sit in our office here at Palmer Station, Antarctica, pursuing our own chemical ecological interests, our job of editing and organizing the contributed chapters is just beginning. We are especially grateful to the contributors to this volume who have taken considerable time from their busy schedules to produce, in every case, informative, well-written, and timely monographs; their dedication and thoroughness have made our jobs easy. Equally so, the efforts of the many reviewers of the individual chapters are gratefully acknowledged. The end result is now clear. The sum of this collection of topics creates a high-resolution snapshot of marine chemical ecology as we understand it at the beginning of the 21st century. We hope that readers will find much information within this book to interest them, some information to provoke them, and ultimately, at least one thing they are compelled to further investigate. Thus, the stage will be set for future developments in the field.

We are also indebted to the Office of Polar Programs and Division of Ocean Sciences at the National Science Foundation for their continued support which has, in part, made this book possible. J.B.M. would like to dedicate his input into this book to Ferne, Luke, and Jamie in acknowledgment of their patience and support. B.J.B. wishes to dedicate his contribution to Jill and Jeremy, whose enthusiastic interest and encouragement are a constant inspiration.

James B. McClintock
Bill J. Baker

Editors

James B. McClintock, Ph.D., is a professor in the Department of Biology and dean of the School of Natural Sciences and Mathematics at the University of Alabama at Birmingham. He received his B.S. degree from the University of California at Santa Cruz in 1978. Working in the laboratory of Dr. John Lawrence, he earned his M.S. and Ph.D. degrees from the University of South Florida in 1980 and 1984, respectively. He was a National Science Foundation (NSF) postdoctoral fellow in the laboratory of Dr. John Pearse at the University of California at Santa Cruz from 1984 to 1987. Dr. McClintock was appointed assistant professor of biology in the Department of Biology at the University of Alabama at Birmingham in 1987, and became an associate professor and then professor of biology in 1993 and 1997, respectively. In 1999 he was named dean of the School of Natural Sciences and Mathematics at the University of Alabama at Birmingham. Dr. McClintock has taught invertebrate zoology in the Department of Biology at the University of Alabama at Birmingham. For the past 8 years he has also taught field courses in Tropical Ecology at the Bahamian Field Station and Tropical Rainforest Ecology in Tortuguero National Park in Costa Rica.

Dr. McClintock has been the recipient of numerous grants from NSF and the National Oceanographic and Atmospheric Association (NOAA). He is a member of the Society for Integrative and Comparative Biology, the honorary society of Sigma Xi, and was elected a fellow of the American Association for the Advancement of Science in 1999. He has published over 150 papers and presented numerous invited lectures. His research interests are marine invertebrate chemical ecology, nutrition, reproduction, and evolution. He enjoys nature photography, hiking and camping, racquetball, and bluegrass and classical music.

Bill J. Baker, Ph.D., is associate professor of chemistry at the University of South Florida. He studied chemistry as an undergraduate at California Polytechnic State University, San Luis Obispo, receiving his B.S. degree in 1982. He studied bioorganic chemistry at the University of Hawaii at Manoa, under Paul J. Scheuer, earning his Ph.D. degree in 1986. After postdoctoral training in biosynthesis with Ron Parry at Rice University and Carl Djerassi at Stanford University, he was appointed assistant professor of chemistry at Florida Institute of Technology in 1990 where he advanced through the academic ranks before accepting his current position at the University of South Florida in 2001.

Dr. Baker has engaged in teaching and research in organic chemistry. He stays young by teaching sophomore organic chemistry and he stays current in his field by teaching graduate courses in the chemistry of natural products and in organic structure determination. He has published widely in the areas of marine natural products isolation, characterization, bioactivity, marine chemical ecology, synthesis and biosynthesis of marine metabolites, and marine biotechnology. His research has taken him to the underwater worlds of the warm South Pacific and Caribbean and to the cold polar waters of Antarctica, where he always finds the interface of field biology and organic chemistry fascinating. His research has been supported by a variety of sources, including the NSF, the National Institutes of Health, the NOAA, as well as private and commercial collaborators interested in potential uses of these products.

Dr. Baker has been active in the American Society of Pharmacognosy and is a member of the American Chemical Society, the American Association for the Advancement of Science, the International Society of Chemical Ecology, and Sigma Xi. He has served extensively as peer reviewer for granting agencies and journals and on advisory panels for several government agencies. He enjoys outdoor activities with his family, including camping, bicycling, and hiking, and he is an avid runner.

On July 21, 1998 the United States Board on Geographic Names recognized Drs. McClintock and Baker by naming geographic features after them on the north (McClintock Point) and south (Baker Point) sides of Explorers Cove, New Harbor, McMurdo Sound, Antarctica. This honor was bestowed upon them to recognize their extensive contributions to Antarctic marine biology and chemical ecology.

Contributors

Charles D. Amsler
Department of Biology
University of Alabama at Birmingham
Birmingham, Alabama

Thomas M. Arnold
Pesticide Research Laboratory
Pennsylvania State University
University Park, Pennsylvania

Bill J. Baker
Department of Chemistry
University of South Florida
Tampa, Florida

Valerie S. Bernan
Natural Products Microbiology Research
Wyeth-Ayerst Research
Pearl River, New York

Tim S. Bugni
Department of Medicinal Chemistry
University of Utah
Salt Lake City, Utah

Guido Cimino
Istituto per la Chimica di Molecole
  di Interesse Biologico
Napoli, Italy

Brent R. Copp
Department of Chemistry
University of Auckland
Auckland, New Zealand

Greg Cronin
Department of Biology
University of Colorado at Denver
Denver, Colorado

Edwin Cruz-Rivera
Marine Laboratory
University of Guam
Mangilao, Guam

Rocky de Nys
Schools of Biological Science
  and Microbiology and Immunology
University of New South Wales
Sydney, Australia

Megan N. Dethier
Friday Harbor Laboratories
University of Washington
Friday Harbor, Washington

David O. Duggins
Friday Harbor Laboratories
University of Washington
Friday Harbor, Washington

Mary J. Garson
Department of Chemistry
University of Queensland
Brisbane, Australia

Michael T. Ghiselin
Department of Invertebrate Zoology
California Academy of Sciences
San Francisco, California

Michael G. Hadfield
Department of Zoology
University of Hawaii at Manoa
Honolulu, Hawaii

Mary Kay Harper
Department of Medicinal Chemistry
University of Utah
Salt Lake City, Utah

Katrin B. Iken
Department of Biology
University of Alabama at Birmingham
Birmingham, Alabama

Chris M. Ireland
Department of Medicinal Chemistry
University of Utah
Salt Lake City, Utah

Robyn D. James
Department of Medicinal Chemistry
University of Utah
Salt Lake City, Utah

Deneb Karentz
Department of Biology
University of San Francisco
San Francisco, California

Staffan Kjelleberg
School of Microbiology and Immunology
University of New South Wales
Sydney, Australia

Brent S. Lindsay
Department of Medicinal Chemistry
University of Utah
Salt Lake City, Utah

James B. McClintock
Department of Biology
University of Alabama at Birmingham
Birmingham, Alabama

Valerie J. Paul
Marine Laboratory
University of Guam
Mangilao, Guam

Adam D. Richardson
Department of Medicinal Chemistry
University of Utah
Salt Lake City, Utah

Dan Rittschof
Marine Laboratory
Duke University
Beaufort, North Carolina

Peter C. Schnabel
Department of Medicinal Chemistry
University of Utah
Salt Lake City, Utah

Susan H. Sennett
Division of Biomedical Marine Research
Harbor Branch Oceanographic
  Institution, Inc.
Fort Pierce, Florida

John J. Stachowicz
Section of Evolution and Ecology
University of California
Davis, California

Deborah K. Steinberg
Bermuda Biological Station for Research, Inc.
Ferry Reach, Bermuda

Peter D. Steinberg
School of Biological Science
University of New South Wales
Sydney, Australia

Nancy M. Targett
Graduate College of Marine Studies
University of Delaware
Lewes, Delaware

Deniz Tasdemir
Department of Medicinal Chemistry
University of Utah
Salt Lake City, Utah

Robert W. Thacker
Marine Laboratory
University of Guam
Mangilao, Guam
and presently at
Department of Biology
University of Alabama at Birmingham
Birmingham, Alabama

Henry G. Trapido-Rosenthal
Bermuda Biological Station for Research, Inc.
Ferry Reach, Bermuda

Kathryn L. Van Alstyne
Shannon Point Marine Center
Western Washington University
Anacortes, Washington

Ryan M. VanWagoner
Department of Medicinal Chemistry
University of Utah
Salt Lake City, Utah

Sheryl M. Verbitski
Department of Medicinal Chemistry
University of Utah
Salt Lake City, Utah

Table of Contents

Section I

Background

1 Introduction to the Chemical Ecology of Marine Natural Products

*Mary Kay Harper, Tim S. Bugni, Brent R. Copp,
Robyn D. James, Brent S. Lindsay, Adam D. Richardson,
Peter C. Schnabel, Deniz Tasdemir, Ryan M. VanWagoner,
Sheryl M. Verbitski, and Chris M. Ireland**

CONTENTS

I. INTRODUCTION

Chemistry is the foundation of all life. Living organisms utilize chemistry for multiple purposes. Obvious examples include lipids utilizing chemistry for cell structure, DNA utilizing chemistry for genetic expression, and proteins utilizing chemistry for cell function and communication. The processes of chemical synthesis and degradation in living systems are termed metabolism, and, invariably, these processes are under enzymatic control. A well-defined set of biosynthetic pathways (see Table 1.1) is used by all organisms for the production of chemicals that are essential for the well-being and survival of the organism. Such chemicals are collectively known as primary metabolites and include lipids, DNA, and proteins. For many years now, chemists have isolated and characterized a diverse range of natural products from both terrestrial and marine organisms. The reason for the presence of such compounds in the organisms was not immediately apparent. It was obvious, however, that these compounds did not resemble the more classical primary metabolites, and they were therefore termed secondary metabolites or natural products. It is becoming increasingly apparent that this terminology is deceptive. Important functions, including ecological roles,

---

* Corresponding author.

are being attributed to secondary metabolites. Indeed, natural products may contribute as much as primary metabolites to the survival of the producing organism.

Theories abound as to why secondary metabolites are produced by marine organisms. Early theories suggested that such natural products represented chemical "ballast," waste or products of primary metabolic overflow. However, increasing evidence that many marine natural products play fundamental roles in ecology suggests a more premeditated scenario. In an authoritative account of the theories pertaining to secondary metabolite production, Williams et al. concluded "natural products have evolved under the pressure of natural selection to bind to specific receptors," and therefore represent ecological responses of organisms to their environment.[1] Marine organisms are under intense competitive pressure for space, light, and nutrients especially in such populous locations as tropical reefs. Thus, it is not surprising that these organisms have developed a range of defense mechanisms including behavioral (e.g., cryptic, nocturnal), physical (e.g., sclerites, tough external surfaces), and chemical strategies to ensure survival. Reputed ecological roles for marine natural products include anti-predation, mediation of spatial competition, prevention of fouling, facilitation of reproduction, and protection from ultraviolet radiation. Most studies to date have attempted to assess the more easily quantified feeding deterrence effects of purified compounds. The design, interpretation, and ecological relevance of such experiments, however, have recently come under close scrutiny.[2] Although difficult to implement, calls for a holistic approach to marine chemical ecology should not be ignored, as marine natural products do not act in isolation from one another or from physical defenses. For example, testing for synergy between chemical and physical defenses has only recently been addressed in any meaningful manner.[3-5]

In many cases, preliminary ecological studies have failed to demonstrate well-defined roles for natural products.[6] This may be interpreted as support for a chemical "baggage" model, but more likely reflects a present lack of understanding of the complexity of chemically mediated inter- and intra-specific interactions in the marine environment. Marine natural products have already proven their value as catalysts for the discovery and investigation of medicinally important biological systems (e.g., ion channels),[7] thus it is inevitable that similar developments will occur in marine biology. The factors that will drive such developments will most likely stem from studies in applied marine chemical ecology. For example, the quest for new nontoxic marine anti-fouling coatings will lead to the identification of natural products that could also be used to probe biochemical mechanisms associated with microbial adhesion, larval settlement, and the induction of metamorphosis. A recent example of such work is the report that furanone natural products interfere with the acylated homoserine lactone (AHL) signaling system in Gram-negative bacteria, a system which is implicated in bacteria bioluminescence, motility, and exoenzyme synthesis.[8]

Chemists typically classify secondary metabolites according to the metabolic pathways from which they were derived, also called their biosynthesis. The natural products presented in this chapter were classified according to their probable biosynthetic pathway, specifically isoprenoid, acetogenin, shikimate, amino acid, nucleic acid, or carbohydrate.[9] Glycosides were classified by the aglycone. Compounds arising from aromatic amino acids were listed as amino acid-derived rather than as shikimates. Compounds that appeared to originate from multiple pathways (i.e., mixed biosynthesis) were classified according to the probable source of the majority of carbons present. Table 1.1 lists representative chemical classes and structures for each biosynthetic pathway. No attempt has been made to second-guess the actual producing organism of secondary metabolites presented in the following sections. For example, natural products isolated from sponges are reported as sponge products, even though in some cases there is evidence that intrinsically associated symbionts are the actual producing organisms. In a similar fashion, metabolites isolated from nudibranchs, which are typically sequestered from dietary sources, are reported as products of the mollusc.

This chapter presents macroalgae first, followed by invertebrate phyla listed from the most primitive to the most advanced marine phyla. The seagrasses and several invertebrate phyla, which account for none or a very small number of natural products (<1% of the total from all sources),

TABLE 1.1
Biosynthetic Pathways and Representative Examples

Biosynthetic Pathway	Structural Classes	Chemical Example and Reference
Isoprenoid	Terpenes	Structure 1.1[10]
	Steroids	Structure 1.2[11]
	Carotenoids	Structure 1.3[12]
	Prenylated quinones and hydroquinones	Structure 1.4[13]
Acetogenin	Polyketides	Structure 1.5[14]
	Polyphenols	Structure 1.6[15]
	Fatty acids	Structure 1.7[16]
	Prostaglandins	Structure 1.8[17]
Amino Acid	Peptides (including cyclic)	Structure 1.9[18]
	Alkaloids	Structure 1.10[19]
Shikimate	Cinnamic acid derivatives	Structure 1.11[20]
	Flavonoids	Structure 1.12[21]
	Coumarins	Structure 1.13[22]
Nucleic Acid	Nucleic acids	Structure 1.14[23]
	Nucleo bases	Structure 1.15[24]
Carbohydrate	Sugars	Structure 1.16[25]
	Polysaccharides	Structure 1.17[26]

are listed in Table 1.2 but are not discussed in the text. Cultured organisms including fungi, bacteria, and dinoflagellates are also not discussed in this chapter.

The chemical distribution data presented in this chapter are based upon taxonomic classifications that were derived from several sources. The marine natural product database MarinLit[27] provided a general overview of the taxonomy, with more specialized works being consulted for several groups. The specific protocol used for each group is presented in its respective section. Compound totals in Table 1.2 and subsequent tables are based on searching taxonomy at the genus level in the *Chapman & Hall Dictionary of Natural Products* (CHDNP).[28] The phylogenetic scheme for each phylum shown in this chapter represents only those genera with chemistry reported in the literature as of June 1999.

The total number of compounds reported for each phylogenetic level (genus, family, etc.) appears in the "total" column. The number of compounds originating from each pathway lies to the right, along with the percentage of the level that subtotal represents (percentages are in parentheses). The taxonomy presented in each table includes only those genera represented in the natural product literature. The inherent limitations in such phylogenetic distributions include errors in taxonomy in the published literature, omissions in the natural product databases used, and that the literature usually only reports novel chemistry. This latter issue is particularly important in the current context as it may lead to distortion of the perceived phylogenetic distribution of secondary metabolites. As a consequence of these limitations, the resultant metabolite distribution tables presented within this chapter should not be used to evaluate taxonomic classifications or to define chemotaxonomic markers. Rather, they are intended to illustrate patterns or trends in metabolite distribution among marine organisms and to allow commentary on the evolution of secondary metabolite production and biosynthetic pathways.

As presented in Table 1.2, over half of reported marine natural products are derived from the isoprenoid biosynthetic pathway (56%), with the remainder split mainly between amino acid (19%) and acetogenin (20%) pathways. Secondary metabolites falling into the categories of nucleic acids and carbohydrates comprise only 1%. Such low levels are somewhat surprising given the fundamental importance of such classes of compounds as primary metabolites.

TABLE 1.2
Summary of Phylogenetic Distribution of Secondary Metabolites from Marine Organisms

Phylum	Total	Isoprenoid	Acetogenin	Amino Acid	Shikimate	Nucleic Acid	Carbohydrate
Plants							
Seagrasses	12	1 (8)	1 (8)	0	8 (67)	0	2 (17)
Chlorophyta	177	109 (62)	30 (17)	24 (14)	8 (4)	0	6 (3)
Rhodophyta	938	429 (46)	356 (38)	89 (9)	45 (5)	1 (<1)	19 (2)
Phaeophyta	759	513 (68)	116 (15)	15 (2)	91 (12)	0	24 (3)
Subtotal	1886	1052 (56)	503 (27)	128 (7)	152 (8)	1 (<1)	51 (3)
Invertebrates							
Porifera	2609	1295 (50)	574 (22)	681 (26)	37 (1)	19 (<1)	3 (<1)
Cnidaria	1499	1236 (82)	142 (9)	112 (7)	4 (<1)	4 (<1)	1 (<1)
Ctenophora	0	0	0	0	0	0	0
Platyhelminthes	12	0	2 (17)	10 (83)	0	0	0
Nemertea	0	0	0	0	0	0	0
Nematoda	2	1 (50)	0	1 (50)	0	0	0
Sipuncula	0	0	0	0	0	0	0
Echiura	0	0	0	0	0	0	0
Mollusca	575	310 (54)	138 (24)	104 (18)	13 (2)	2 (<1)	8 (1)
Annelida	28	0	6 (21)	16 (57)	5 (18)	0	1 (4)
Arthropoda	28	2 (7)	1 (4)	25 (89)	0	0	0
Echinodermata	464	317 (68)	73 (16)	25 (5)	31 (7)	0	18 (4)
Brachiopoda	0	0	0	0	0	0	0
Ectoprocta	102	5 (5)	29 (28)	65 (64)	0	3 (3)	0
Hemichordata	22	8 (36)	2 (9)	12 (55)	0	0	0
Urochordata	452	73 (16)	71 (16)	299 (66)	6 (1)	3 (<1)	0
Subtotal	5793	3247 (56)	1038 (18)	1350 (23)	96 (2)	31 (1)	31 (1)

Note: Numbers in parentheses represent percentages.

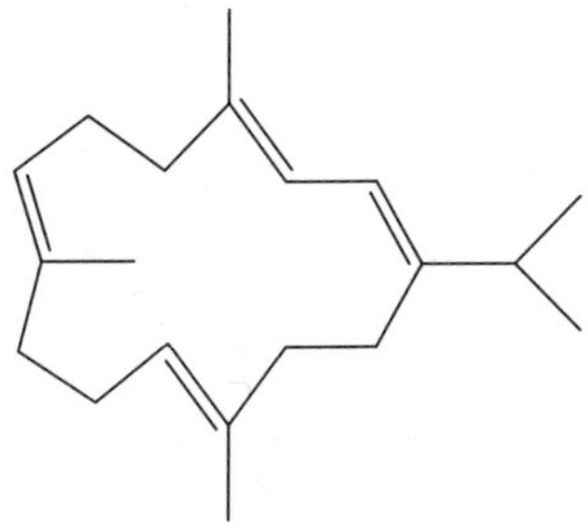

1.1

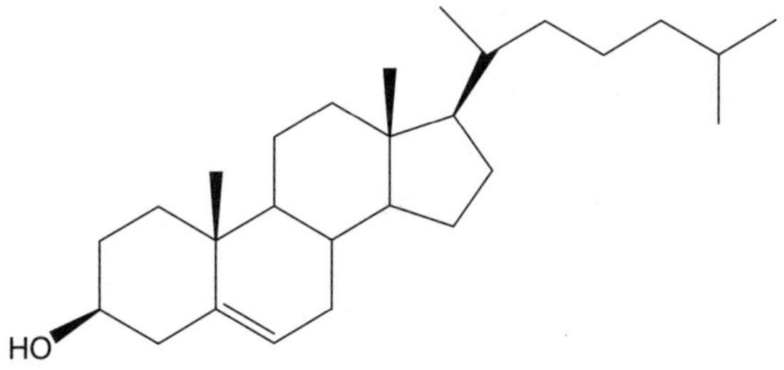

1.2

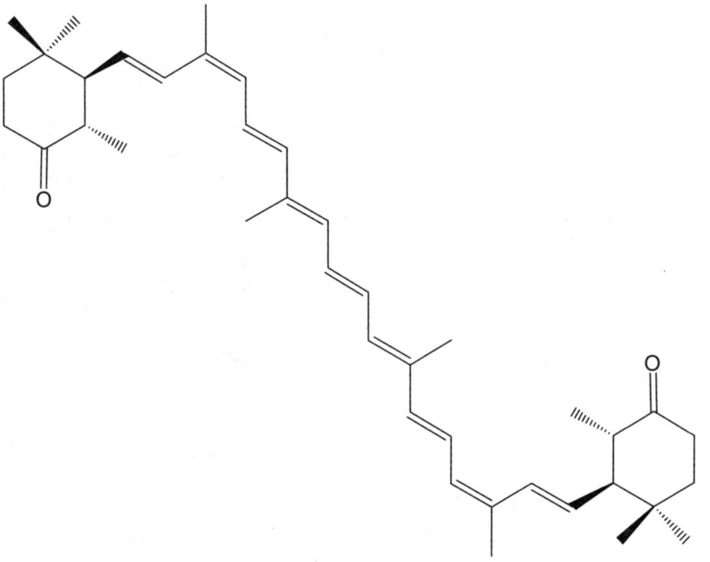

1.3

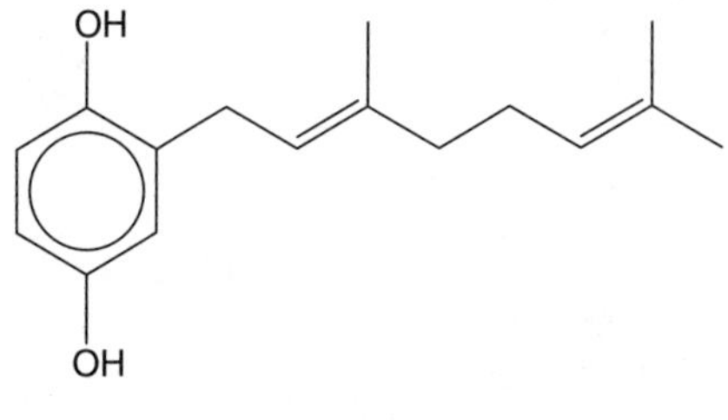

1.4

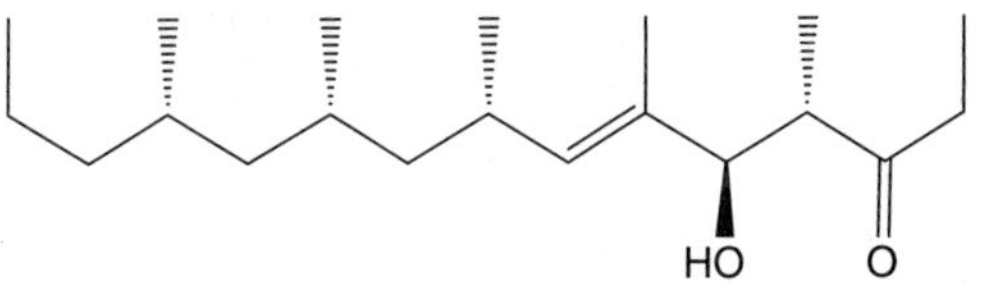

1.5

In an evolutionary context, biosynthetic pathways utilized for metabolite production seem to be conserved within each phylum. Thus in Table 1.2, no evidence can be found for organisms in a phylum using all available biosynthetic pathways to the same extent. This implies that secondary metabolite evolution has not been entropic in nature, which should result in the expression of a greater range of different types of chemistry utilized by a more advanced organism. Rather, this implies that with evolution has come a change in the dominant biosynthetic pathway utilized in the production of secondary metabolites. The reasons for such changes are obscure and must remain

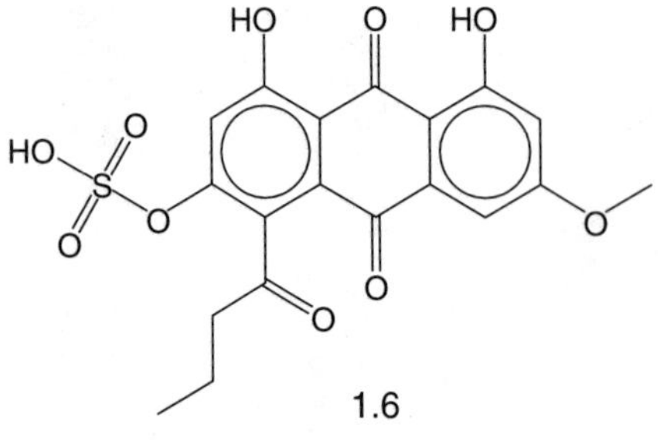

1.6

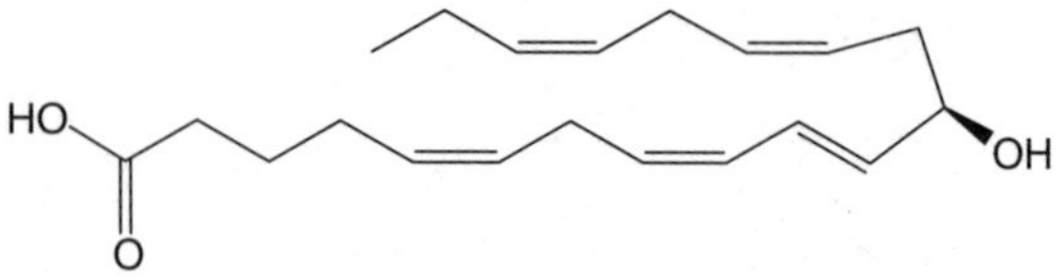

1.7

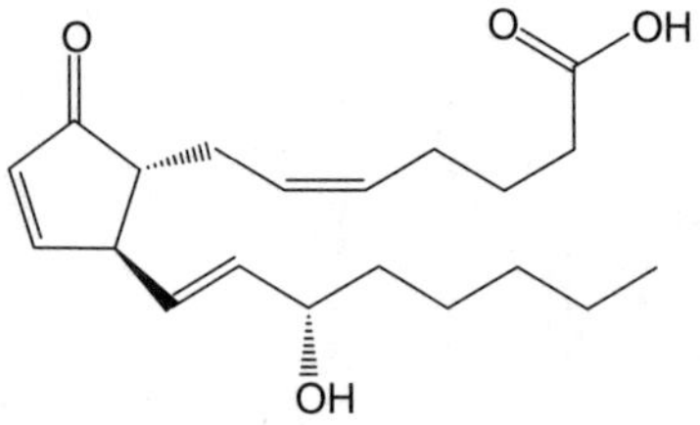

1.8

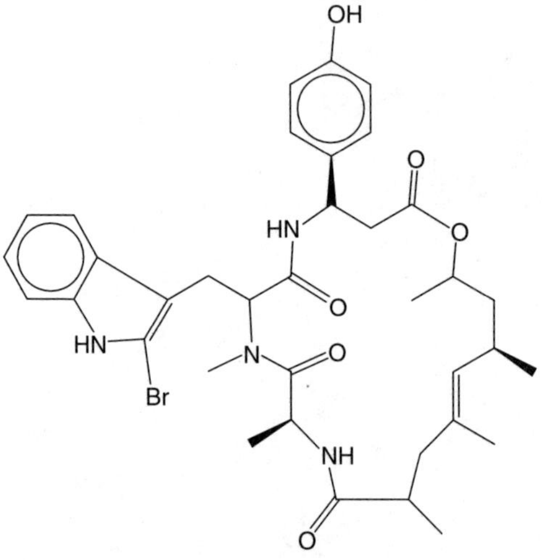

1.9

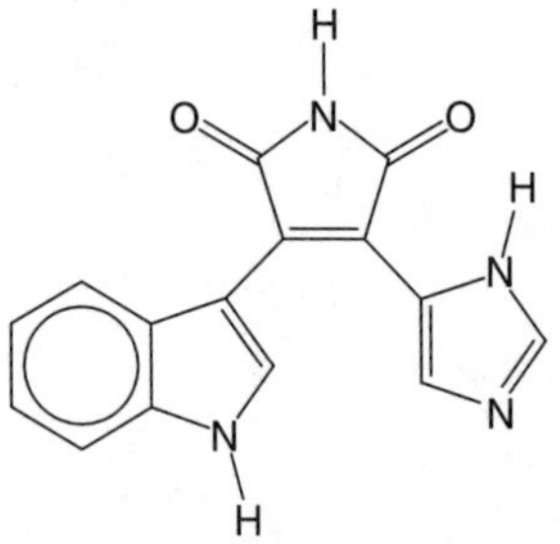

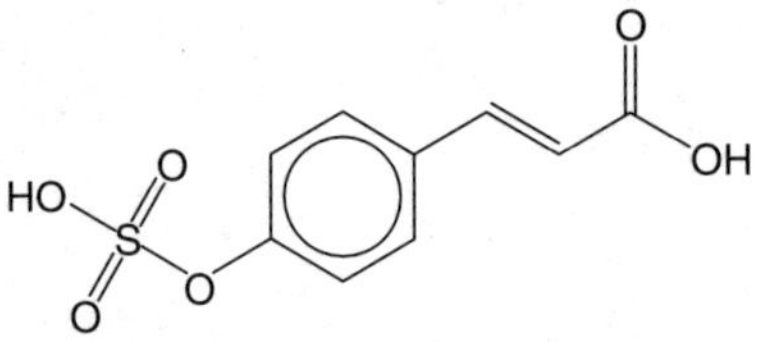

1.10

1.11

pure speculation. Clearly, however, change was not brought about by the poor survival enhancement properties of isoprenoid or acetogenin derived compounds, as organisms utilizing these compounds still flourish today.

II. ALGAE

Trainor described algae as "photosynthetic, nonvascular plants that contain chlorophyll *a* and have simple reproductive structures."[29] The fossil record indicates that algae may have existed before the Cambrian period.[30] This survey focuses on marine macroalgae, commonly referred to as seaweed. Seaweeds are the largest forms of algae and live attached to solid substrata between and

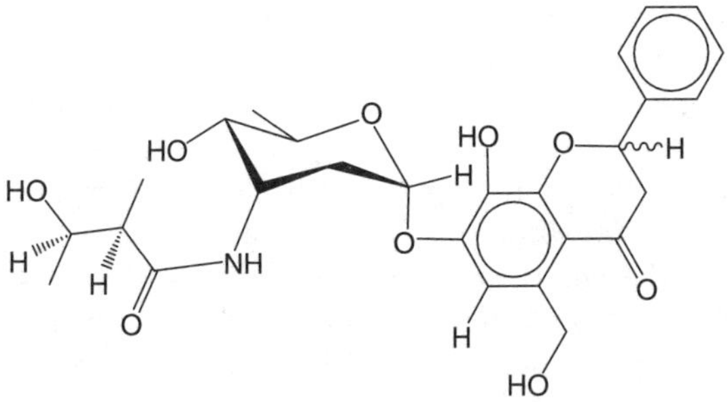

1.12

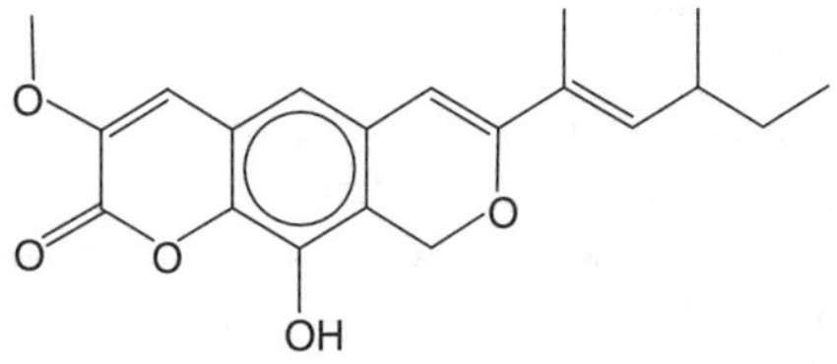

1.13

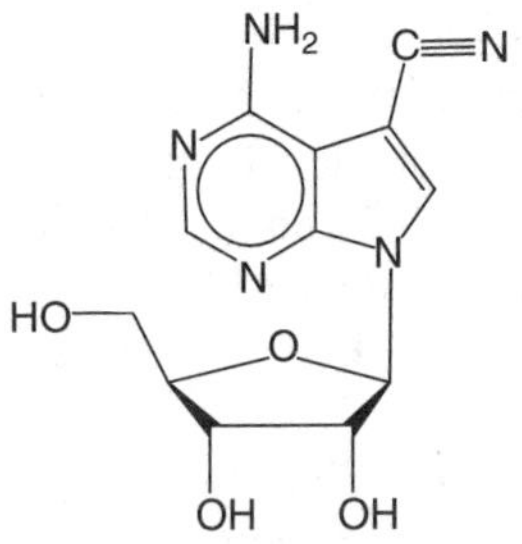

1.14

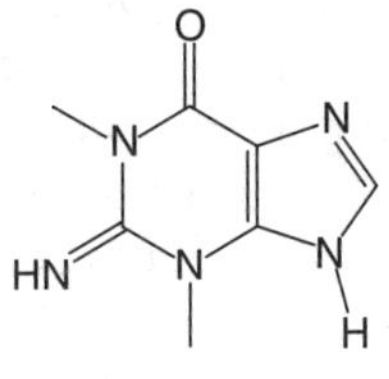

1.15

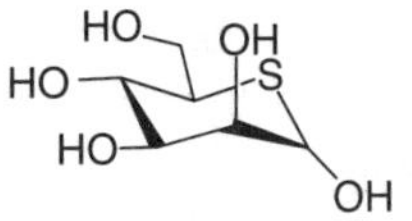

1.16

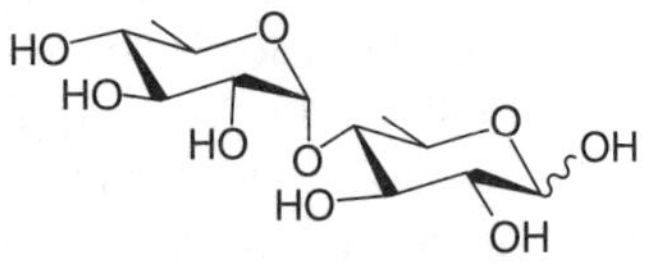

1.17

below tide marks. They are primarily found in three major habitats: rocky intertidal zones, tropical reefs, and kelp forests. Together with phytoplankton, seaweeds are the primary producers in oceans (i.e., they produce organic matter from inorganic starting materials via photosynthesis), and they serve as both food and habitat for a variety of herbivores.

Currently, most taxonomic systems classify algae in the Protist kingdom, although macroalgae may be more appropriately assigned to the Plantae Kingdom. Macroalgae are organized in three divisions: Chlorophyta (green algae, 13% marine), Rhodophyta (red algae, 98% marine), and Phaeophyta (brown algae, 99% marine).[30] Within these divisions, there are approximately 10,000 species of seaweed. Since 1841, the classification of algae has been based on the primary photosynthetic pigment chlorophyll (green), and the secondary photosynthetic pigments, carotenoids (brown or yellow) and phycobilins (red or blue). Today, there is wide variation in algal classification largely because different characteristics are emphasized. These characteristics are often biological, including photosynthetic pigments, storage products, chloroplast structure, cell structure, flagella, cell division, and life history.[30] The taxonomic classification scheme analyzed in this section is primarily based on Silva,[31] and is also based on Dawes,[30] Bold and Wynne,[32] and Van den Hoek[33] in cases where Silva makes no reference to particular genera.

Members of the division Chlorophyta, green algae, most resemble green plants and may be the ascendants of higher plants. The classification of Chlorophyta has been debated since the 1970s. Currently, Chlorophyta is categorized into three classes: Chlorophyceae, Prasinophyceae, and Charophyceae.[30] These classes contain approximately 500 genera and 8000 species that are largely freshwater.[33] Compounds isolated from marine representatives of the class Chlorophyceae are mainly isoprenoid derivatives, 91 (59%). Additionally, 29 acetogenins (19%), 21 amino acid derivatives (14%), and small percentages of carbohydrates and shikimates have been isolated. Within this class, certain orders appear to specialize in single biosynthetic pathways. Bryopsidales, for example, produce 62 (88%) isoprenoid derivatives while Acrosiphonales produce 6 (67%) aceto-genins. Compounds isolated from Prasinophyceae and Charophyceae are 100% isoprenoid and 100% amino acid derived, respectively. However, this statement should be judiciously interpreted, as these classes are not statistically well represented.

Rhodophyta is called the red algae due to its pigment, phycoerythrin. This division is predominantly marine and contains approximately 5000–5500 species in 500–600 genera.[33] One class, Rhodophyceae, and two subclasses, Florideophycidae and Bangiophycidae, are recognized in this division.[33] Compounds isolated from Rhodophyta are predominantly isoprenoid and acetogenin derivatives, 429 (46%) and 356 (38%), respectively. In addition, 89 (9%) amino acid, 45 (5%) shikimate, 19 (2%) carbohydrate, and 1 (<1%) nucleic acid derivatives have been isolated. In the subclass Florideophycidae, five orders contain mainly acetogenins: Bonnemaisoniales (124, 91%), Corallinales (7, 78%), Gelidiales (2, 33%), Gracilariales (14, 56%), and Nemaliales (9, 100%). The remaining orders contain predominantly isoprenoid derivatives: Ceramiales (285, 54%), Cryptonemiales (46, 66%), Gigartinales (25, 32%), Palmariales (52, 97%), and Plocamiales (4, 100%). The subclass Bangiophycidae contains three orders with marine species, yet only the order Bangiales is represented in the natural product literature. The compounds in this order are approximately equivalent percentages of isoprenoid, acetogenin, amino acid, and carbohydrate derivatives. Traditionally, the division Rhodophyta has been the primary source of halogenated compounds from algae. In fact, 93% of the 938 compounds known to be produced by Rhodophyta are halogenated, as compared to 7% halogenated compounds provided by Chlorophyta and 0.8% halogenated compounds produced by Phaeophyta.

For this discussion, Phaeophyta is classified as a distinct division. This division is almost exclusively marine and includes the common wracks and kelps. Phaeophyta is comprised of one class, Phaeophyceae, containing approximately 265 genera with 1500–2000 species.[30] The majority of compounds reported from Phaeophyta are isoprenoid (513, 68%). Interestingly, there are no reports of isoprenoids reported from the orders Ectocarpales and Scytosiphonales, and only one isoprenoid from the Laminariales.

Macroalgae are subject to biotic influences, such as grazers and their predators as well as competition from other algae, and abiotic factors such as nutrients and light; seaweeds actively respond to these biotic and abiotic influences. They produce a variety of secondary metabolites believed to be advantageous to the organism. For example, the brown alga *Dictyota menstrualis* has been proven less likely than other co-occurring seaweeds to be invaded by fouling agents. When larvae of the bryozoan *Bugula neritina* were exposed to extracts derived from this alga containing pachydictyol A and dictyol E (diterpene alcohols), larvae settlement was inhibited.[2] In addition, it has been hypothesized that phenolics might function as antibiotics. The green alga, *Chaetomorpha media,* doubled its production of phenolics when infected with fungi.[34]

III. PORIFERA

Members of the phylum Porifera (kingdom Animalia, subkingdom Metazoa) are aquatic sessile filter feeders. The Porifera are the most primitive multicellular organisms dating back to the Precambrian era. Sponge survival is critically dependent on the development of a stable aquiferous system through which water is pumped. Water currents generated by sponges are essential in feeding, waste excretion, respiration, and in many cases reproduction. Sponges feed upon bacteria and other microscopic nutrient sources from the water current. Sponges can be found in nearly all aquatic environments (marine and fresh water; shallow and deep; tropical and Antarctic), wherever there are substrata for sponge attachment and growth. There are over 9000 presently known species of sponges and as many species not yet described.[35]

The taxonomic study of sponges is fraught with difficulties. Only recently have authoritative taxonomic schemes for Porifera received general acceptance, though many details are still debated. There is still a considerable amount to be learned about sponges and the phylogenetic relationships among them. Identification of sponge specimens continues to be difficult in some cases, and there are examples in the literature in which original taxonomic identifications have required revision. For this study, the classification of Porifera was based on *Hooper's Sponge Guide*,[36] with recent amendments. Aside from difficulties in taxonomic identification, caution is required when searching for correlations between phylogeny and secondary metabolite production in sponges. For many metabolites, there is ambiguity as to whether the compounds are synthesized by sponge cells or by microorganisms associated with the sponge, since the mesohyl of sponges is often inhabited by microbes, and many poriferan natural products resemble metabolites produced by marine microbes.

Sponges are the most prolific marine invertebrate source of secondary metabolites. Porifera are attractive subjects for natural products chemists due to the sheer number of metabolites produced, the novelty of structure encountered, and the therapeutic potential of these compounds in the treatment of human diseases.[7] As sedentary filter feeders, sponges face a variety of ecological pressures including competition for space, predation, and fouling. Some of the proposed ecological roles for sponge metabolites include antifeedants, antifoulants, antibiotics, antisettlement cues, and photoprotective agents.[37]

Metabolites of the phylum Porifera account for almost 50% of the natural products reported from marine invertebrates. Of the 2609 poriferan metabolites, 98% are derived from amino acid, acetogenin, or isoprenoid pathways. Isoprenoids account for 50% of all sponge metabolites, while amino acid and polyketide pathways account for 26% and 22%, respectively. A significant number of sponge metabolites appear to be derived from mixed biosynthetic pathways. Most structures reported containing carbohydrate moieties were glycosides.

The phylum Porifera is comprised of three classes distinguished primarily by skeletal characteristics: Hexactinellida, Calcarea, and Demospongiae. Hexactinellida, deep water glass sponges, have recently been recognized as distinct at the subphylum level.[38] A few hexactinellid sponges have been investigated by natural products chemists, but there are no reports of metabolites from the 102 genera in this class. Sponges of the class Calcarea are exclusively marine and possess skeletons of free calcareous spicules. Within the Calcarea, only the genera *Leucetta* and *Clathrina* have reported chemistry, and all 31 calcarean metabolites are amino acid derived.

Demospongiae, the class which encompasses most living sponges, are a morphologically diverse group with fibrous protein skeletons sometimes supplemented by siliceous spicules having myriad possible configurations. There have been 2578 metabolites reported from Demospongiae; 50% are isoprenoid. The remainder are predominantly of acetogenin and amino acid biosynthetic pathways (22% and 25%, respectively). This distribution, however, is not uniform among the three subclasses within the Demospongiae. The subclass Homoscleromorpha represents a small, well-defined group of species and is believed to be an early stage in poriferan evolution. The dominant biosynthetic pathway within the Homoscleromorpha is acetogenin with 80 compounds (86%). The majority of these were reported from *Plakortis* species, exemplified by cyclic peroxide fatty acids.

The demosponge subclasses Tetractinomorpha and Ceractinomorpha are primarily defined by oviparous or viviparous larval development, respectively. This division is the subject of some dispute as there have been indications that the reproductive mode may be ecologically dependent and thus not a stable character for classification. There are 12 orders in the subclasses Tetractinomorpha and Ceractinomorpha, with the largest number of metabolites reported from the orders Halichondrida (424), Haplosclerida (489), and Dictyoceratida (663).

The sponges included in the order Halichondrida are quite varied with respect to morphology and chemistry. Some members of the Halichondrida produce compounds unique within the order, while others show chemical relatedness to sponges outside the order. For example, the curcuphenol-type sesquiterpenes have restricted distribution within the closely related halichondrid genera *Didiscus*,[39] *Myrmekioderma*,[40] and *Epipolasis*.[41] Some Axinellidae yield terpenoid isonitriles, formamides, and isothiocyanates which, with the exception of one report from a *Theonella* sp.,[42] are uniquely halichondrid metabolites. Other Axinellidae have been the source of pyrrole-2-carboxylic acid alkaloids such as oroidin,[43] which have also been reported from *Agelas*[44] and *Astrosclera*,[45] both of the order Agelasida, and the poecilosclerid sponge *Lissodendoryx* sp.[46]

Haplosclerida is a large, homogeneous order of sponges. Haplosclerid metabolites are relatively evenly distributed among the isoprenoid (158, 32%), acetogenin (190, 39%), and amino acid (140, 29%) biosynthetic pathways. While many of the isoprenoid metabolites reported from haplosclerid sponges are not unique to the order, there are some classes of compounds that are characteristic. Van Soest et al. studied the distribution of acetogenic straight chain acetylenes and determined that they are good chemotaxonomic markers for the order Haplosclerida, even though those authors proposed that these compounds are of symbiotic bacterial origin.[47] Due to their wide-ranging biological activities, 3-alkylpiperidine amino acid derivatives, ranging from halitoxins to the highly modified manzamines and sarains, have received a great deal of attention from natural products chemists. They are primarily restricted to the Haplosclerida, with a few reports from unrelated sponges such as *Theonella, Ircinia,* and *Stelletta*.[48] The localization of haliclonacyclamine to sponge cells supports a sponge origin for these 3-alkylpiperidine amino acid derivatives.[49] The genus *Xestospongia* has by far the greatest number (132) of reported compounds among haplosclerid genera. It is interesting to note that while haplosclerid sponges display diversity in their natural products, there does not appear to be overlap of biosynthetic pathways utilized within a given sponge.

Metabolites reported from the order Dictyoceratida are predominantly isoprenoid in origin (559, 84%). The distribution of isoprenoid classes is specific within the dictyoceratid families: linear furano sesterterpenes (Irciniidae), sesterterpenes with tetronic acid functional groups (Thorectidae), meroterpenoids (Spongiidae), and sesquiterpenes (Dysideidae). The isoprenoid pathway also dominates within the order Dendroceratida (74, 83%), but those metabolites are typically spongiane diterpenes. The largest number of sponge metabolites, 184, have been reported from the genus *Dysidea*. Several studies performed on *Dysidea* spp. have localized the isoprenoid terpenes avarol, dysidin, and spirodysin to sponge cells, while the shikimate-derived brominated diphenyl ethers and amino acid-derived chlorinated diketopiperazines were found in symbiotic cyanobacterial populations.[50-52]

Although a comprehensive treatment of every order within the Demospongiae is not the goal of this discussion, some deviations from the general trends of metabolite distribution within the subclass are noteworthy. Most compounds reported from the order Verongida are amino acid derivatives (138, 87%). In fact, one of the first chemotaxonomic observations noted within the Porifera was that brominated tyrosine metabolites were dominant for every genus in the order Verongida.[53] The first cellular localization study was performed on *Aplysina fistularis*. The metabolites aerothionin and homoaerothionin were localized to spherulous cells situated such that they could secrete the compounds into the exhalant water currents where they might serve defensive roles.[54]

Members of the order Agelasida produce isoprenoid metabolites (23, 32%), but amino acid metabolism is more prevalent (44, 62%). The diterpenes reported from some Agelasida are specific to *Agelas*, while the pyrrole-2-carboxylic acid derivatives are also reported from Axinellidae.[55] The Agelasida represent an interesting case study for chemotaxonomy. The two agelasid genera represented in the literature, *Agelas* and *Astrosclera,* are morphologically distinct. The former are fibrous and compressible, while the latter are extremely rigid and are living fossil relatives of the ancient reef building sponges. *Astrosclera* were originally placed in the now defunct class Sclerospongiae, which was comprised of sponges with siliceous and calcareous skeletons. However, spicule analysis showed a unique spicule type found otherwise only in *Agelas* sponges. Subsequent chemical analysis yielded pyrrole-2-carboxylic acid derivatives consistent with their current taxonomic placement in Agelasida.[45]

Members of the polyphyletic order Lithistida have been reported to contain 42 acetogenins (38%) and 49 amino acids (44%), with only 21 isoprenoid metabolites (19%). The order currently termed Lithistida probably arose from more than one divergence event in the past,[56] and may

internal space called the coelenteron, which serves for digestion and circulation, and a diploblastic body wall with a jelly-like middle layer. The major unique feature of Cnidaria is the possession of highly specialized cells called cnidae or nematocysts, located primarily on the tentacles. These cells serve to capture and paralyze food, protect against predators, and also settle to a solid substratum. Nematocysts contain toxic proteins that can be released upon direct stimulation.

Cnidarians may be planktonic or benthic and exhibit two different types of body form, medusa or polyp. Scyphozoa and Cubozoa exist only in the medusae form and the Anthozoa only in the polyp form, while Hydrozoa pass through both phases during their life cycle. Sexual and asexual reproduction occur in both body forms. Cnidarians are polytrophic feeders. They procure their food by predation, direct uptake of dissolved nutrients, feeding on bacteria, and through algal symbiosis. These endosymbiotic algae also help to build the calcareous skeleton of the reef building corals.

Taxonomy of the Cnidaria is often problematic as they usually lack a well-defined, consistent body form.[82] In the past, the classification of this phylum has been complicated since the medusae and polyp stages of the same organism were sometimes assigned different generic names. The environment has considerable effect on the morphology of cnidarians. Thus, many species, corals and gorgonians in particular, have been described more than once. The Cnidaria were formerly placed together with comb jellies in the phylum Coelenterata, but they are now universally recognized as two distinct phyla, the Cnidaria and the Ctenophora. The term coelenterate, however, is still widely used. The classical systematics of cnidarians is based mainly on life history and morphological characters; however, the genetic basis of these characters is unknown.[82] Furthermore, fossil records of most cnidarians are incomplete.[83] Chemotaxonomy and cladistic analysis have been used for the systematics of some colonial cnidarians.[82] The classification adopted here is based mainly on Brusca and Brusca,[81] with contributions from other sources.[83-86] Cnidarian taxonomy has been complicated by discrepancies concerning family level placement of some genera, particularly among soft corals and gorgonians.

Cnidarians have been a chemically prolific group of marine invertebrates yielding 1499 secondary metabolites. According to the Zoological Records database, the taxon Cnidaria is represented by 1,125 genera (10,737 sp. and subsp.), however, only approximately 12% of these genera have reports of chemistry. The most consistent aspect of the chemistry of these animals is that 1236 (82%) cnidarian metabolites are isoprenoid, 96% of which come from the subclass Octocorallia. Acetogenin (142, 9%) and amino acid pathways (112, 7%) are also significant, whereas shikimates, nucleic acids, and carbohydrates account for less than 1% of the total. As mentioned before, the phylum is divided into four classes: Hydrozoa, Scyphozoa, Cubozoa, and Anthozoa. The two primitive classes, Scyphozoa and Cubozoa, have no reports of natural products in the CHDNP. Of the eight orders in the class Hydrozoa, only Athecata and Thecata are represented in the CHDNP. Interestingly, the Athecata specialize in acetogenin metabolism, while the Thecata produce primarily amino acid metabolites. However, it must be noted that together they represent less than 3% of the natural products reported from cnidarians.

Almost 98% of the metabolites reported from cnidarians are from the largest class, Anthozoa. The subclass Octocorallia accounts for 87% of all cnidarian compounds. Terpenoids dominate across the Octocorallia, accounting for 92% of their reported metabolites. The order Alcyonacea accounts for half of the metabolites within the subclass. Particularly rich families include the Alcyoniidae (426, 65%) and Xeniidae (110, 17%). Gorgonacea (492, 38%) is the second richest order of Octocorallia, with two families, Gorgoniidae and Plexauridae, accounting for almost half of the metabolites within the order. Three of the remaining orders, Stolonifera, Pennatulacea, and Helioporacea, also show a preponderance of isoprenoid chemistry. However, the Telestacea deviate from the general trend. Of 24 (79%) telestacean metabolites reported, 19 are acetogenin in origin, represented primarily by a series of unusual halogenated prostanoids from the genus *Telesto*.

The chemistry of the subclass Zoantharia also shows significant deviation from the Octocorallia. Of 151 zoantharian metabolites reported, 88 (58%) are derived from the amino acid pathway. Zoanthids (order Zoanthidea) contain a high proportion of amino acid derivatives (46, 75%), the remaining metabolites being almost equal proportions of isoprenoids (7, 12%) and acetogenins (8, 13%). Sea anemones (order Actiniaria) are reported to produce 26 amino acid (60%), 13 isoprenoid (30%), 2 acetogenin (5%) and 2 nucleic acid (5%) derivatives. Most of these isolates originate from the family Actiniidae (36, 84%). Within the order Scleractinia (hard corals), the acetogenin pathway (19, 40%) appears to be slightly dominant over amino acid (16, 34%) and isoprenoid (9, 19%) pathways.

The distribution of terpenoids within the Cnidaria deserves a closer look. The order Helioporacea, which is represented by only one genus *Heliopora*, elaborates only diterpenes. True soft corals (order Alcyonacea) and gorgonians (Gorgonacea) are chemically similar — both produce a vast array of isoprenoids, namely sesqui- and diterpenoids, although cembrane diterpenes are the most common group. Alcyonacean and gorgonian cembranoids differ from each other by the fact that gorgonian cembranoids usually have 1*R* configurations, while alcyonacean cembranoids tend to possess 1*S* configurations.[87] The distribution of diterpenes and sesquiterpenes varies across families, yet volatile sesquiterpene hydrocarbons have been successfully used as chemotaxonomic markers for the Gorgonacea.[82]

Cnidarians produce an extensive range of structurally interesting compounds with high biomedical potential. The endogenous functions of some of these metabolites have also been investigated. Many cnidarian metabolites are assumed to be defense allemones, as predation is an important factor affecting the natural selection of marine organisms. Cnidarians are delicate animals and are relatively vulnerable to predators. Consequently, they have developed multiple defense mechanisms, both chemical and structural, probably due to the inability of a single defense to deter predators. Many colonial hydroids form a chitinous exoskeleton, whereas the medusae simply thicken their mesoglea to increase stiffness. A tough, horny, axial skeleton is found in several anthozoans, especially gorgonians, and is composed of gorgonin, a horn-like protein–polysaccharide complex. In most octocorals, mesenchymal cells secrete calcareous sclerites of various shapes and colors. A defensive role for sclerites against predation has been demonstrated.[88,89] Massive calcareous skeletons occur in stony corals and some hydrozoans, where the epidermal cells of polyps secrete a limestone skeletal case.

There is a strong negative correlation between the degree of physical defense and the diversity of natural products. The swimming cnidarians such as jellyfish, the relatively mobile sea anemones, and some solitary hydroids generally lack a rich complement of secondary metabolites and primarily produce neuropeptides and protein venoms. The most chemically prolific cnidarians are the soft-bodied colonial forms, such as corals, gorgonians, and zoanthids. Recent studies have concentrated on understanding the evolution of chemical defense and other chemical interactions among these organisms. As a result, many cnidarian metabolites were shown to have predator deterrent, antifouling, and overgrowth inhibitory activities. These marine allelochemicals are reported to play a role in the welfare of the organism by protecting it from UV radiation, mediating larval settlement and metamorphosis, regulating symbiotic relationships, and serving as pheromones.

Many cnidarians live in symbiotic relationships with intracellular dinoflagellates, which prompts questions about the origin of cnidarian metabolites. The controversy over the animal or plant origin of cnidarian terpenoids is still unresolved.[90] However, current evidence suggests an animal origin, as some aposymbiotic gorgonians also produce these complex terpenoids.[91] It is now known that marine invertebrate prostaglandin biosynthesis is different from the mammalian pathway and involves 8-lipoxygenase. It appears that the animals produce these metabolites; however, it has been proposed that algal symbionts provide arachidonic acid to the coral.[90] The origin and distribution of lipids between zooxanthellae and animal compartments in some symbiotic anthozoans have been studied and found to be completely different, quantitatively and qualitatively, in animals and zooxanthellae, indicating *de novo* synthesis of these metabolites by each organism. Fatty acid distribution patterns have been proposed as a good tool for taxonomic classifications of some cnidarians.[92]

The genus *Sinularia,* which is an important contributor to attached benthic communities on Indo-Pacific reefs, is probably the most studied cnidarian in relation to its chemistry and chemical ecology. Interference competition and fouling are very common phenomena in coral reefs, but octocorals are rarely fouled. The ability of octocorals to avoid fouling has been attributed to the production of toxic secondary metabolites, particularly diterpenoids. It has been shown that soft corals release large quantities of terpenes into the surrounding water to kill neighboring organisms.[93,94] Terpene concentration, distribution, and function in the colony appear to be very complex. In *Sinularia flexibilis,* 11,12-deoxyflexibilide was reported to serve as an ichthyotoxin, and sinulariolide as an algicide, while flexibilide was involved in interference competition.[95] Some diterpenes from the soft corals *Sinularia* and *Lobophytum* have been shown to have pheromone effects.[96,97]

Sea anemones are well known for their host–guest associations with anemone fish and hermit crabs. It has been shown that this species-specific recognition is chemically regulated.[98] A pyridinium compound, amphikuemin, secreted by the sea anemone *Radianthus* sp., induces the "attracted swimming" of the anemone fish.[99] The zoanthids specialize in amino acid metabolism. The zoanthoxanthins are the dominant class of the nitrogenous compounds from this order, and are responsible for the bioluminescence of this group.[100,101] Scleractinian corals employ acetylenic fatty alcohols and acids as sperm attractants,[97,102] and it has recently been reported that hard corals employ chemicals against fouling microbes.[103] Mycosporine-like amino acids provide natural protection to hard corals by absorbing harmful wavelengths of UV light.[104]

V. MOLLUSCA

The phylum Mollusca contains a highly diverse group of animals including 50,000 species (approximately half have been described) and 60,000 fossil molluscs.[81] Molluscs vary greatly in physical appearance, size, and feeding habits. The size range of molluscs is quite broad; the giant squid *Architeuthis* can grow to 20 m in length, but some bivalves are microscopic. Aside from the vastly diverse physical appearance of members of the phylum, molluscs have a common body plan. The bilaterally symmetrical body is composed of three major sections: a head, a foot, and a centralized visceral mass. Molluscs are covered with a sheet of skin, called the mantle, which is responsible for excreting a calcareous skeleton. The calcareous skeleton can vary from a hard external shell to small deposits within the mantle. One feature unique to the phylum Mollusca is the presence of a toothed apparatus, called the radula, which helps grind and process food. The extent of development within the nervous system and sensory structures varies among molluscs. Their dynamic nature has allowed them to live in almost all freshwater, marine, and terrestrial habitats. Some molluscs are sessile, while a large number are mobile, and feeding behavior varies as widely as habitat.

The phylum Mollusca contains eight classes: Caudofoveata, Scaphopoda, Aplacophora, Monoplacophora, Polyplacophora, Cephalopoda, Bivalvia, and Gastropoda. Caudofoveatans are shellless worm-like molluscs that live in deep sea floors. The ecology of the 70 known species has not been described to any extent. The Scaphopoda (tusk shells) include 300–500 species,[105] which have long narrow shells that are open at both ends. Aplacophorans are small (3–30 mm long) worm-like molluscs without shells, but instead covered with calcareous needle-like spicules. Monoplacophorans are comprised of 20 species,[105] and have a single shell, as evident from their class name. They are known largely from fossil records. Polyplachophora (chitons) represent a small class of Mollusca with a unique shell comprised of eight distinct plates. Cephalopoda (octopus, squids) are some of the most advanced invertebrate animals. Most cephalopods have fairly developed nervous systems allowing greater mobility than other molluscs. Interestingly, *Nautilus* is the only cephalopod genus to retain an external shell. The class Bivalvia (clams, oysters, mussels) contains all molluscs that have shells made from two calcareous plates or "valves." Gastropoda, with 40,000 species,[105] is the largest class within the phylum Mollusca. This class includes a variety of shelled and shellless animals including snails (orders Mesogastropoda and Neogastropoda), sea hares (order Anaspidea), and nudibranchs (order Nudibranchia).

The phylogenetic scheme used for classification of the phylum Mollusca was based primarily on that published by Brusca and Brusca.[81] However, most of the families and genera were obtained directly from the MarinLit database. Overall, the chemistry reported for the phylum Mollusca is not extensive with respect to the large number of genera. A total of 575 compounds have been reported for the phylum. The majority of these, 54%, are isoprenoid, while 24% are acetogenin and 18% are amino acids. The biosynthetic origin of the remaining compounds is 2% shikimate, 1% carbohydrate, and <1% nucleic acid. Chemotaxonomic trends can be seen among certain orders, families, and genera. For example, only acetogenins have been isolated from the bubble shells in the order Cephalaspidea. The order Anaspidea shows a strong trend with 99 (62%) of the isolated compounds derived from the isoprenoid pathway. While trends certainly exist, the molluscan ability to accumulate compounds from food sources makes this analysis somewhat misleading from a biosynthetic standpoint. However, if accumulation is selective, the trends could be important both ecologically and phylogenetically.

The gastropods account for the majority of chemistry reported for the phylum. Of the total reported for the phylum, 497 (86%) were isolated from the class Gastropoda. The subclass Opistho-branchia accounts for 378 compounds, 76% of the total reported from gastropods. More specifically, 317 compounds (55%) were isolated from nudibranchs and sea hares. For the order Nudibranchia, 159 compounds have been reported, of which 137 (86%) are isoprenoids. The subclass Prosobran-chia (snails) is one of the few groups that have yielded mainly amino acid derived compounds. A total of 72 compounds have been reported for prosobranchs, and 51% are derived from amino acids. A large number of groups within the phylum are underrepresented statistically with respect to reported chemistry.

Mollusc survival depends on several defensive mechanisms to escape or ward off predators. While some rely on hard shells for protection, a large number of molluscs are shell-less and rely on chemical defenses, mobility, or both. Some molluscs have been shown to selectively sequester compounds that have ecological roles or can be modified for ecological purposes. Furthermore, shell-less molluscs may have evolved from a shelled predecessor. Researchers have proposed that the evolutionary loss of a shell in nudibranchs was concomitant with the development of their ability to sequester and employ chemical defenses from their diet.[106] Studies of nudibranchs and sea hares provide good examples of molluscs selectively sequestering and utilizing compounds from food sources. Sea hares have been studied extensively, and their chemical defenses exhaus-tively reviewed.[107,108] The nudibranchs *Tambja eliora*, *T. abdere*, and *Roboastra tigris* were shown to contain the tambjamines A–D. The *Tambja* spp. obtained the tambjamines by feeding upon the bryozoan *Sessibugula translucens*, and the *R. tigris* from feeding upon *Tambja* spp. In the case of *T. abdere*, secretion of the tambjamines in its mucus was shown to deter predation by *R. tigris*. Interestingly, chemoreception studies with *R. tigris* suggested that the mollusc is attracted to its prey by low concentrations of tambjamines in their slime trail. The tambjamines have also been shown to be fish feeding deterrents.[109] While many molluscs have the ability to separate and sequester toxins, not all molluscs appear to use toxins for chemical defense. Some bivalves sequester toxins which are responsible for human shellfish poisoning but do not appear to protect against marine predators. The ability of some nudibranchs to acccomplish *de novo* synthesis of terpenes and other compounds has been reviewed.[90] Certainly, nudibranchs can synthesize secondary metab-olites, but utilizing food sources for chemical defense greatly reduces energy usage for secondary metabolite production.

The order Anaspidea (sea hares), specifically *Aplysia* spp., has been studied extensively and represents a substantial portion of molluscan chemistry. Of 158 compounds reported for the order Anaspidea, 116 can be attributed to *Aplysia* spp.; a large portion of the compounds appear to be from dietary sources. Of the 86 isoprenoid-derived compounds reported from *Aplysia*, 58% have similar halogenation patterns to compounds isolated mainly from the red alga *Laurencia* spp. The review by Carefoot[107] covers some of the ecological roles of diet-derived compounds isolated from *Aplysia* including the diet-dependent production of defensive ink. A pharmaceutically interesting group of peptides, the dolastatins, isolated from the Indian Ocean sea hare *Dolabella auricularia*,[110] have shown promising results as new anticancer agents.[111] The recent isolation of dolastatin 12 and related compounds from marine cyanobacterial sources[112–115] suggests that the herbivorous *D. auricularia* sequesters low levels of dolastatins from feeding.

Cone snails, *Conus* spp., have been investigated because of their production of conotoxin peptides. From an evolutionary standpoint, the production of conotoxins is quite interesting due to their wide range of neurophysiological activities. The conotoxins are small peptides, 10–30 amino acids, with conformations constrained by multiple disulfide bonds that target a number of receptors in vertebrate and invertebrate nervous systems. Cone snails use these toxins to immobilize prey, which allows the relatively slow-moving cone snails to feed on fish and worms. The wide variety of conotoxins isolated and the hypervariability within peptide sequences has led some to hypothesize a combinatorial biosynthetic approach for the production of conotoxins.[116,117]

The kidney of the giant red clam, *Tridacna maxima*, contains large amounts of arsenic-containing compounds. A wide variety of arsenic compounds have been isolated from red, brown, and green macroalgae. The source of arsenic may be unicellular green algal symbionts, or perhaps the source may be from organisms obtained through the filter feeding activities of *T. maxima*. Although the arsenic was probably derived from algal sources, the kidney of *Tridacna* was the source of several novel arsenic derivatives. Interestingly, 2′, 3′-dihydroxypropyl [5-deoxy-5-(demethylarsenyl)]ribofuranoside, which accounted for 30% of the arsenic found in the kidney of *T. maxima*, has been isolated as its phosphodiester derivative from filamentous blue-green algae.[118] Hydrolysis of the phosphodiester bond may represent a source of phosphorus for the clam. The giant clam may have the ability to utilize the phosphorus and store the remaining portion of the compound. Apparently, the arsenic compounds have no deleterious effects on the kidney of *T. maxima*.

A wide variety of ecological roles have been reported for compounds isolated from molluscs. Careful analysis of biosynthetic and phylogenetic relationships provides insight into the ecological stresses of many Mollusca. Further study of the chemical ecology of molluscs may also yield information about molluscan evolution.

VI. ECHINODERMATA

Echinoderms have existed since at least the early Cambrian period, with 7,000 extant and 13,000 extinct species.[81] Echinoderms are deuterostomes, a deuterostome being an "animal in which the anus is formed from the blastopore during development and the mouth arises as a secondary invagination."[119] Other characteristic Echinodermata features include a coelomic water vascular system, pentamerous radial symmetry, mutable connective tissue, and a spiny, calcareous endoskeleton.[81,119] Echinoderms range from suspension feeders to herbivores to predators. Many are scavengers. Echinoderms are widespread and often dominate the benthic ecosystem in terms of biomass.[81]

Echinodermata are divided into two subphyla, Pelmatozoa and Eleutherozoa, based upon variations in mobility and body plans. Organisms within the subphylum Pelmatozoa spend all or part of their life attached to substrate, while Eleutherozoa are sedentary. The members of Pelmatozoa also generally have the mouth and anus on the exposed surface, while Eleutherozoa have their oral opening either toward the substratum or horizontal.[120] The subphylum Pelmatozoa consists of a single class, Crinoidea. The subphylum Eleutherozoa contains five classes, including the recently discovered class Concentricycloidea.

Class Asteroidea, sea stars, is one of the most widely recognized groups of marine organisms due to their prevalence upon the shoreline and their characteristic shape. Asteroids are top predators in benthic ecosystems, often preying upon bivalves from the phylum Mollusca. The isoprenoid compound mytiloxanthin, for example, has been isolated from both the sea star *Asterias rubens* and the bivalve *Mytilus edulis*.[119] The class Concentricycloidea was first identified in 1986 and consists of two species in the genus *Xyloplax*. Class Ophiuroidea, brittle stars, contains the greatest number of echinoderm species. The remaining three classes within the phylum Echinodermata are Echinoidea (sea urchins and sand dollars), Holothuroidea (sea cucumbers), and Crinoidea (feather stars).

The phylogenetic tree used in this section was constructed as described in the introduction to this chapter. The genera were organized into the phylogenetic tree shown in Table 1.9 according to Brusca and Brusca.[81] Other sources were utilized as necessary, especially while organizing the table at the family level.[120–122] There are 464 compounds reported from the phylum Echinodermata. Isoprenoids dominate, accounting for 317 (68%). Acetogenins make up 73 (15%) of the reported compounds, and the remaining compounds are divided between shikimates (7%), amino acids (5%), and carbohydrates (4%).

The subphylum Pelmatozoa consists of one class, Crinoidea. The crinoids are divided into two orders, Isocrinida and Comatulida. Only the order Comatulida has any reported chemistry. Of the 29 compounds reported from Comatulida, 27 (93%) are acetogenins. This pattern differs from that of the phylum as a whole, as only 15% of the 464 compounds from Echinodermata are acetogenin derived. The subphylum Pelmatozoa is more primitive than the subphylum Eleutherozoa. The evolution of Eleutherozoa involves significant changes in echinoderm biology, including altered orientation of the mouth and anus, changes in feeding tactics, and increased mobility.[81] The concurrent evolution of the predominant secondary metabolite biosynthetic pathway from acetogenin (Pelmatozoa) to amino acid (Eleutherozoa) suggests a model for biosynthetic evolution.

The subphylum Eleutherozoa contains five classes. Class Concentricycloidea, containing only the genus *Xyloplax*, has no reported chemistry. Classes Asteroidea, Ophiuroidea, and Holothuroidea follow the pattern of the phylum as a whole, producing 222 (74%), 24 (100%), and 66 (94%) isoprenoid compounds, respectively. The majority of these isoprene compounds are sterols, many of which are cholestane, ergostadiene, or holostadiene derivatives.

The class Echinoidea, however, has yielded a variety of compounds: 3 isoprenoids (7%), 12 acetogenins (27%), 16 shikimates (36%), and 14 amino acids (31%). Most of these compounds (36, 80%) were reported from the order Echinoida. Many of the shikimate compounds are polyhydroxylated naphthoquinones and were isolated from multiple echinoid genera. The diversity of secondary metabolites isolated from the echinoids may arise from the algal diet. The order Spinulosida (Class Asteroidea) also breaks from the overall trends of the phylum pattern, with 3 isoprenoids, 10 shikimates, and 2 amino acids reported. The shikimates include a number of di- and triphenylated compounds.

Echinoderms have yielded a smaller range of secondary metabolites than other marine invertebrates such as sponges or ascidians, perhaps due to their ability to deter predators by other means. The ophiuroids are able to burrow into the substratum and reduce exposure to predators. Many echinoderms are nocturnal, moving and feeding when predator activity is at a minimum. The outer layer of most echinoderms is either calcified or composed of insoluble proteins, making the organism unfavorable as prey.[120]

The compounds most characteristic of the phylum are the saponins, glycosolated sterols, most of which are sulfated. The suite of secondary metabolites produced by the classes Asteroidea, Ophiuroidea, and Holothuroidea are dominated by these sterols. It is believed that echinoderms do not generally undertake *de novo* synthesis of these sterols, but create them by modifying precursors obtained through feeding.[90]

VII. ECTOPROCTA

The phylum Ectoprocta consists of approximately 4500 extant species[81] which are organized into three classes: Phylactolaemata, Stenolaemata, and Gymnolaemata. The class Phylactolaemata consists of freshwater ectoprocts and thus will not be considered here. Only a small fraction of the species in the class Stenolaemata are extant; these are grouped in one order, Cyclostomata. The majority of extant ectoproct species are found in the class Gymnolaemata, which is dominated by the order Cheilostomata.

The phylum Ectoprocta is grouped with Phoronida and Brachiopoda as the lophophorate phyla.[81] Neither of the latter two phyla are represented in the natural products literature. The lophophore consists of ciliated tentacles around the mouth of the organism. Ectoprocts are coelomates and have a U-shaped gut in which the anus is outside of the lophophore. Ectoprocts live in colonies, which arise asexually from a single zooid. The moss-like appearance of a colony of tiny zooids (a single zooid being less than 0.5 mm long)[123] was the source of the original phylum name, Bryozoa.[81] Within colonies, the zooids are polymorphic. The autozooids have lophophores and are responsible for feeding mainly upon diatoms. The heterozooids have functions other than feeding, such as attachment to the substratum or embryo incubation.[81,123] Ectoproct colonies nearly always grow upon a substratum. The phylogenetic tree used for the Ectoprocta was constructed as described in the introduction to this chapter. Genera were organized into the phylogenetic tree shown in Table 1.10 according to Brusca and Brusca.[81] Other sources were utilized as necessary, especially while organizing the table at the family level.[124,125] The phylum Ectoprocta has 102 reported compounds. Compounds originating from amino acids account for 64% of the 102 compounds. Twenty-nine (28%) acetogenin-derived structures have been reported; the remaining compounds consist of isoprenoids (5, 5%) and nucleosides (3, 3%).

The class Stenolaemata is represented in the literature by a single genus. The organism *Diaperoecia californica* was reported to produce desmethylphidolopin, a purine derivative. This compound was also isolated from the ectoproct *Phidolopora pacifica* (class Gymnolaemata, order Cheilostomata).[126]

The remaining 101 compounds arise from the orders Cheilostomata and Ctenostomata within the class Gymnolaemata. The 27 compounds from Ctenostomata originate from two genera, *Alcyonidium* and *Amathia,* and are almost exclusively amino acid derived. The amino acid compounds consist of brominated phenylalanine and tryptophan derivatives. Cheilostomata accounts for 74 (73%) of the compounds reported from the phylum. The majority of families and genera within Cheilostomata have yielded amino acid-derived compounds; the significant exception is the family Bugulidae. The Bugulidae genera *Dendrobeania* and *Sessibugula* have contributed 1 isoprenoid and 4 acetogenin compounds, respectively. The Bugulidae genus *Bugula* has yielded a single isoprenoid and 22 acetogenins. Sixteen of these acetogenins are members of the bryostatin family, isolated from *Bugula neritina*. The bryostatins are potent antineoplastic compounds currently in clinical trials.[127] The tambjamines and related compounds have also been isolated from *Bugula*. Independent reports of the tambjamines from *Sessibugula*, an ascidian, and marine bacteria suggest a possible microbial origin for the compounds. These pyrrole compounds have also been isolated from nudibranchs, members of the phylum Mollusca, which prey upon ectoprocts.[109]

The families Catenicellidae and Flustridae produce alkaloids derived from tryptophan. The majority of these contain bromine at carbon 6 of the indole ring, although some are more extensively brominated. The convolutamydines (brominated tryptophan derivates isolated from the genus *Amathia* in the order Ctenostomata)[128] are similarly brominated at carbon 6, suggesting a common biosynthetic pathway or bacterial symbiont.

Ectoprocts are able to sense their environment and respond to it accordingly. One potential role of secondary metabolites for these organisms is signaling or repelling other colonies or organisms. Ectoproct colonies compete for benthic space with other colonies and with other organisms, such as sponges and ascidians. Not only are ectoproct colonies able to sense the presence of other colonies, they are able to determine the size of the competing colony through chemical signals.[123] This inter-colony chemical signaling is a possible function of some ectoproct secondary metabolites as is deterring or inhibiting the growth of competing organisms such as sponges or ascidians.

VIII. UROCHORDATA

Urochordata is one of three subphyla in the phylum Chordata. Mammals, fish, birds, reptiles, and amphibians are all part of subphylum Vertebrata. Subphylum Cephalochordata is comprised of about 20 species of lancelets. Subphylum Urochordata contains about 3000 species in four classes: Ascidiacea (sea squirts or ascidians), Sorberacea (sorberaceans), Thaliacea (salps), and Appendicularia (larvaceans).[81] These four classes, and thus the subphylum Urochordata, are collectively known as the tunicates. Class Ascidiacea is the largest and most diverse of the four tunicate classes and the only one with reported chemistry. The number of extant ascidian species is not known, but is estimated between several hundred and several thousand.[129]

The tunicates differ in appearance from other chordates, especially the vertebrates. The larval stages of tunicates have nerve cords and notocords; however, these are lost in most adult forms. Tunicates have an exoskeleton called a tunic. This tunic is composed mostly of polysaccharides, along with proteins and blood cells. Tunicates possess a U-shaped gut with two openings called siphons. Tunicates generally feed upon plankton and small organic particles through non-selective filtering, although tentacles around the oral siphon exclude larger particles. Water flows in through the oral siphon, and food particles are trapped in a web of mucus. Eventually the mucal web is formed into strands and transported to the stomach for digestion. The waste water is ejected through the excurrent siphon.[81,130]

Ascidiacea is the largest and most diverse of the four tunicate classes. Ascidians may be solitary, social, or colonial (compound). Solitary ascidians live as individual animals, often camouflaged or in difficult to access environments. Social ascidians are attached to other members of a colony through a vascular system at their base. Colonial ascidians are composed of many individuals, or zooids, sharing a common tunic. The amount of each zooid embedded in the tunic varies, from only the lower abdomen to the entire individual. The oral siphon of each zooid is always exposed to allow water intake.[81,130] Ascidians are generally hermaphroditic with a single ovary and testis near the digestive tract. Solitary ascidians tend to produce many weakly yoked ova, which are fertilized externally by the sperm of another organism. Colonial ascidians tend to produce fewer eggs with higher yolk contents, which are released as larvae after development within the parent organism. The larval life span ranges from a few minutes to a few days. The attachment of the larva to the substratum and the beginning of metamorphosis appear to be linked.[81] Many ascidians are also able to replicate asexually, especially colonial ascidians. A single zooid may propagate into an entire colony through budding. There are three types of budding: stolonic, strobiliation, and esophageal-rectal.[130]

Ascidians play host to a range of organisms. These relationships may be commensal, symbiotic, or parasitic. Symbiotic relationships which involve exchange of metabolic products between the host and guest probably have the greatest influence on secondary metabolism and chemical ecology. Photosynthetic symbionts are the most common and consequential of these guests. Two unicellular symbionts frequently reported are the prokaryotes *Synechocystis* and *Prochloron*.[130] Ascidians pass symbionts to their larvae, and neither have been cultivated independently of one another.

Interestingly, only one class within the subphylum Urochordata, the Ascidiacea, is represented in the natural product literature. Within the class, only the orders Enterogona and Pleurogona are represented, with the Enterogona families Polyclinidae, Didemnidae, and Polycitoridae being the most widely studied. Members of the Ascidiacea display predominantly amino acid metabolism,[131,132] and account for 452 compounds in the CHDNP. The ascidian chemistry represents less than 8% of the total from marine invertebrates, probably reflecting difficulty in collecting and the complexity of their taxonomy. Nonetheless, the ascidians have been the source of a number of novel natural products with exciting structures and potent biological activity. Examples include the didemnins and ecteinascidins, with members from each of these classes of metabolites having been evaluated in human clinical trials as anticancer agents.[7]

The ecological roles played by ascidian metabolites are largely unknown. Some evidence exists for chemical defenses of both larvae and adults, ranging from feeding deterrence to protection from UV radiation. Ecologically relevant feeding deterrence has been reported for a number of ascidian-derived metabolites. These include the didemnimide series of alkaloids isolated from *Didemnum conchyliatum*,[133] extracts of the ascidian *Ecteinascidia turbinata*,[134] which contain the biologically active alkaloids the ecteinascidins, the tambjamine alkaloids (*Sigillina signifera*),[135] the didemnin depsipeptides (*Trididemnum solidum*),[6] and the polyandrocarpidine family of alkaloids (*Polyandrocarpa* sp.).[6] Other ecological roles attributed to ascidian metabolites include the antifouling potential of the eudistomin alkaloids (*Eudistoma olivaceum*)[136] and the UV protectant and anti-oxidant properties of mycosporine-like amino acids.[137,138]

IX. CONCLUDING REMARKS

An overall analysis of metabolite distributions points out a surprising trend. The distribution across the six biosynthetic categories is generally the same in both plants and animals. In fact, the major pathway in each is isoprenoid, representing 56% of the total in both cases. There are also interesting patterns observed within the animal kingdom. All phyla for which secondary metabolites are reported seem to express the acetogenin pathway much to the same extent, suggesting that acetogenins represent the generalist or default pathway for secondary metabolism. However, closer inspection shows a global perspective may be misleading as there are groups within each phylum that specialize in acetogenin metabolism. It is nonetheless intriguing to speculate that these apparent specialists are actually those that chose not to specialize, as the acetogenin pathway is the closest pathway to primary metabolism. In this same vein, isoprenoid metabolism which utilizes pathways that shunt intermediates involved in production of membrane sterols can also be viewed as closely allied to primary metabolism. It is interesting to note that isoprenoid metabolism is dominant among the less advanced phyla such as the Porifera and Cnidaria. It must also be acknowledged that these organisms are masters of isoprenoid metabolism and display both high degrees of specialization within taxonomic subgroups and at the same time unprecedented structural complexity and diversity. Conversely, the more advanced invertebrates such as the ectoprocts, hemichordates, and ascidians express amino acid metabolism at the expense of the isoprenoid pathway. They, too, are masters of their chosen pathway and express an exciting array of chemistry.

Marine natural products have evolved under the pressure of natural selection to bind to specific receptors in ecological targets. Indirect support for this comes from the observation of a wide range of receptor-specific interactions exhibited by marine secondary metabolites against mammalian receptor targets. In the same way that studies of such interactions have revolutionized our knowledge of human physiological processes, investigation of marine natural product–ecological receptor interactions will lead to a greater understanding of marine chemical ecology at the molecular level. This understanding will only come about when close collaborations are established between marine ecologists, for their field-based observations; pharmacologists, to purify appropriate receptor targets; molecular biologists, to decipher genetic codes; and chemists, to identify the active secondary metabolites. Evidence for the effectiveness of collaborations of this sort, for example, in the field of marine antifouling agent discovery, is already apparent. Only with such interactive collaborations will enough information be gathered in which to assess the true ecological roles played by marine secondary metabolites — a laudable goal for marine chemical ecologists of the 21st century.

REFERENCES

1. Williams, D. H., Stone, M. J., Hauck, P. R., and Rahman, S. K., Why are secondary metabolites (natural products) biosynthesized?, *J. Nat. Prod.*, 52, 1189, 1989.
2. Hay, M. E., Marine chemical ecology: what's known and what's next?, *J. Exp. Mar. Biol. Ecol.*, 200, 103, 1996.
3. Hay, M. E., Kappel, Q. E., and Fenical, W., Synergisms in plant defenses against herbivores: interactions of chemistry, calcification, and plant quality, *Ecology*, 75, 1714, 1994.
4. Hay, M. E., Defensive synergisms? Reply to Pennings, *Ecology*, 77, 1950, 1996.

5. Pennings, S. C., Testing for synergisms between chemical and mineral defenses — a comment, *Ecology*, 77, 1948, 1996.

6. Pawlik, J. R., Marine invertebrate chemical defenses, *Chem. Rev.*, 93, 1911, 1993.

7. Ireland, C. M., Copp, B. R., Foster, M. P., McDonald, L. A., Radisky, D. C., and Swersey, J. C., Biomedical potential of marine natural products, in *Pharmaceutical and Bioactive Natural Products, Marine Biotechnology*, Attaway, D. and Zaborsky, O. R., Eds., Plenum Publishing Corp., New York, 1993, 1.

8. Kjelleberg, S., Steinberg, P., Givskov, M., Gram, L., Manefield, M., and de Nys, R., Do marine natural products interfere with prokaryotic AHL regulatory systems?, *Aquat. Microb. Ecol.*, 13, 85, 1997.

9. Herbert, R. B., *The Biosynthesis of Secondary Metabolites*, 2nd edition, Chapman & Hall, New York, 1989, 1.

10. Bowden, B. F., Coll, J. C., Hicks, W., Kazlauskas, R., and Mitchell, S. J., Studies of Australian soft corals X: the isolation of epoxyisoneocembrene-A from *Sinularia grayi*; isoneocembrene-A from *Sarcophyton ehrenbergi*, *Aust. J. Chem.*, 31, 2702, 1978.

11. Slattery, M., Hamann, M. T., McClintock, J. B., Perry, T. L., Puglisi, M. P., and Yoshida, W. Y., Ecological roles for water-borne metabolites from Antarctic soft corals, *Mar. Ecol. Prog. Ser.*, 161, 133, 1997.

12. Tsushima, M., Fujiwara, Y., and Matsuno, T., Novel marine di-Z-carotenoids: cucumariaxanthins A, B, and C from the sea cucumber *Cucumaria japonica*, *J. Nat. Prod.*, 59, 30, 1996.

13. Aiello, A., Fattorusso, E., and Menna, M., Low molecular weight metabolites of three species of ascidians collected in the lagoon of Venice, *Biochem. Syst. Ecol.*, 24, 521, 1996.

14. Norte, M., Cataldo, F., and Gonzalez, A. G., Siphonarienedione and siphonarienolone, two new metabolites from *Siphonaria grisea* having a polypropionate skeleton, *Tetrahedron Lett.*, 29, 2879, 1988.

15. Rideout, J. A., Smith, N. B., and Sutherland, M. D., Chemical defense of crinoids by polyketide sulphates, *Experentia*, 35, 1273, 1979.

16. Bernart, M. and Gerwick, W. H., Isolation of 12-(*S*)-hepe from red alga *Murrayella periclados* and the structure of icosanoid from *Laurencia hybrida*. Implications to the synthesis of hybridolactone, *Tetrahedron Lett.*, 29, 2015, 1988.

17. Weinheimer, A. J. and Spraggins, R. L., The occurrence of two new prostaglandin derivatives (15-epi-PGA$_2$ and its acetate, methyl ester) in the gorgonian *Plexaura homomalla*. Chemistry of coelenterates XV, *Tetrahedron Lett.*, 5185, 1969.

18. Zabriskie, T. M., Klocke, J. A., Ireland, C. M., Marcus, A. H., Molinski, T. F., Faulkner, D. J., Xu, C., and Clardy, J. C., Jaspamide, a modified peptide from a *Jaspis* sponge with insecticidal and antifungal activity, *J. Am. Chem. Soc.*, 108, 3123, 1986.

19. Vervoort, H. C., Richards-Gross, S. E., Fenical, W., Lee, A. Y., and Clardy, J. C., Didemnimides A–D: novel, predator-deterrent alkaloids from the Caribbean mangrove ascidian *Didemnum conchyliatum*, *J. Org. Chem.*, 62, 1486, 1997.

20. Todd, J. S., Zimmerman, R. C., Crews, P., and Alberte, R. S., The antifouling activity of natural and synthetic phenolic acid sulphate esters, *Phytochemistry*, 34, 401, 1993.

21. Jiang, Z. D., Jensen, P. R., and Fenical, W., Actinoflavoside, a novel flavonoid-like glycoside produced by a marine bacterium of the genus *Streptomyces*, *Tetrahedron Lett.*, 38, 5065, 1997.

22. Guerriero, A., D'Ambrosio, M., Cuomo, V., and Pietra, F., A novel, degraded polyketidic lactone, leptosphaerolide, and its likely diketone precursor, leptosphaerodione. Isolation from cultures of the marine ascomycete *Leptosphaeria oraemaris* (Linder), *Helv. Chim. Acta*, 74, 1445, 1991.

23. Zabriskie, T. M. and Ireland, C. M., The isolation and structure of modified bioactive nucleosides from *Jaspis johnstoni*, *J. Nat. Prod.*, 52, 1353, 1989.

24. Lindsay, B. S., Battershill, C. N., and Copp, B. R., 1,3-Dimethylguanine, a new purine from the New Zealand ascidian *Botrylloides leachi*, *J. Nat. Prod.*, 62, 638, 1999.

25. Capon, R. J. and MacLeod, J. K., 5-Thio-D-mannose from sponge *Clathria pyramida* (Lendenfeld). First example of a naturally occurring 5-thiosugar., *J. Chem. Soc., Chem. Commun.*, 1200, 1987.

26. Ikegami, S., Hirose, Y., Kamiya, Y., and Tamura, S., Structure of carbohydrate moiety in asterosaponin A. Isolation of three new disaccharides, *Agric. Biol. Chem.*, 36, 1843, 1972.

27. Munro, M. H. G. and Blunt, J. W., MarinLit, a marine chemical literature database, version 10.4, Marine Chemistry Group, University of Canterbury, Christchurch, NZ, 1999.

28. Buckingham, J., *Dictionary of Natural Products*, Version 7:2, Chapman & Hall, London, 1999.
29. Trainor, F. R., *Introductory Phycology*, John Wiley & Sons, New York, 1978, 1.
30. Dawes, C. J., *Marine Botany*, 2nd edition, John Wiley & Sons, New York, 1998, 1.
31. Silva, P. C., Basson, P. W., and Moe, R. L., *Catalogue of the Benthic Marine Algae of the Indian Ocean*, University of California Press, Berkeley, 1996, 1.
32. Bold, H. C. and Wynne, M. J., *Introduction to the Algae*, 2nd edition, Prentice-Hall, Englewood Cliffs, NJ, 1985, 1.
33. Van den Hoek, C., Mann, D. G., and Jahns, H. M., *Algae: An Introduction to Phycology*, Cambridge University Press, Cambridge, 1995, 1.
34. Raghukumar, C. and Chandramohan, D., Changes in the marine green algae *Chaetomorpha media* on infection by a fungal pathogen, *Bot. Mar.*, 31, 311, 1988.
35. Bergquist, P. R., The Porifera, in *Invertebrate Zoology*, Anderson, D. T., Ed., Oxford University Press, Melbourne, 1998, 2.
36. Hooper, J. N. A., *Sponguide: Guide to Sponge Collection and Identification*, Queensland Museum, South Brisbane, 1995, 1, http://www.qmuseum.qld.gov.au.
37. Bergquist, P. R., *Sponges*, University of California Press, Berkeley, 1978, 143.
38. Levi, C., Ed., *Sponges of the New Caledonian Lagoon*, Orstom, Paris, 1998, 1.
39. Wright, A. E., Pomponi, S. A., McConnell, O. J., Kohmoto, S., and McCarthy, P. J., (+)-Curcuphenol and (+)-curcudiol, sesquiterpene phenols from shallow and deep water collections of the marine sponge *Didiscus flavus, J. Nat. Prod.*, 50, 976, 1987.
40. Harper, M. K., unpublished results, 1999.
41. Fusetani, N., Sugano, M., Matsunaga, S., and Hashimoto, K., (+)-Curcuphenol and dehydrocurcuphenol, novel sesquiterpenes which inhibit H, K-ATPase, from a marine sponge *Epipolasis* sp., *Experientia*, 43, 1234, 1987.
42. Nakamura, H., Kobayashi, J., and Ohizumi, Y., Novel bisabolene-type sesquiterpenoids with a conjugated diene isolated from the Okinawan sea sponge *Theonella* cf. *swinhoei, Tetrahedron Lett.*, 25, 5401, 1984.
43. Supriyono, A., Schwarz, B., Wray, V., Witte, L., Muller, W. E. G., van Soest, R., Sumaryono, W., and Proksch, P., Bioactive alkaloids from the tropical marine sponge *Axinella carteri, Z. Naturforsch. C*, 50, 669, 1995.
44. Forenza, S., Minale, L., Riccio, R., and Fattorusso, E., New bromo-pyrrole derivatives from the sponge *Agelas oroides, J. Chem. Soc., Chem. Commun.*, 1129, 1971.
45. Williams, D. H. and Faulkner, D. J., *N*-Methylated ageliferins from the sponge *Astrosclera willeyana* from Pohnpei, *Tetrahedron*, 52, 5381, 1996.
46. Schmitz, F. J., Gunasekera, S. P., Lakshmi, V., and Tillekeratne, L. M. V., Marine natural products: pyrrololactams from several sponges, *J. Nat. Prod.*, 48, 47, 1985.
47. van Soest, R. W. M., Fusetani, N., and Andersen, R. J., Straight-chain acetylenes as chemotaxonomic markers of the marine Haplosclerida, in *Sponge Sciences: Multidisciplinary Perspectives*, Watanabe, Y., and Fusetani, N., Eds., Springer-Verlag, Tokyo, 1998, 3.
48. Andersen, R. J., van Soest, R. W. M., and Kong, F., 3-Alkylpiperadine alkaloids isolated from marine sponges in the order Haplosclerida, in *Alkaloids: Chemical and Biological Perpectives*, Pelletier, S. W., Ed., Pergamon Press, Oxford, 1996, 301.
49. Garson, M. J., Flowers, A. E., Webb, R. I., Charan, R. D., and McCaffrey, E. J., A sponge/dinoflagellate association in the haplosclerid sponge *Haliclona* sp.: cellular origin of cytotoxic alkaloids by Percoll density gradient fractionation, *Cell Tissue Res.*, 293, 365, 1998.
50. Unson, M. D. and Faulkner, D. J., Cyanobacterial symbiont biosynthesis of chlorinated metabolites from *Dysidea herbacea* (Porifera), *Experientia*, 49, 349, 1993.
51. Unson, M. D., Holland, N. D., and Faulkner, D. J., A brominated secondary metabolite synthesized by the cyanobacterial symbiont of a marine sponge and accumulation of the crystalline metabolite in the sponge tissue, *Mar. Biol.*, 119, 1, 1994.
52. Uriz, M. J., Turon, X., Galera, J., and Tur, J. M., New light on the cell location of avarol within the sponge *Dysidea avara* (Dendroceratida), *Cell Tissue Res.*, 285, 519, 1996.
53. Bergquist, P. R. and Wells, R. J., Chemotaxonomy of the Porifera. The development and current status of the field, in *Marine Natural Products*, Scheuer, P. J., Ed., Academic Press, New York, 1983, 1.

54. Thompson, J. E., Barrows, K. D., and Faulkner, D. J., Localization of two brominated metabolites, aerothionin and homoaerothionin, in spherulous cells of the marine sponge *Aplysina fistularis* (= *Verongia thiona*), *Acta Zool.*, 64, 199, 1983.

55. Braekman, J. C., Daloze, D., Stoller, C., and van Soest, R. W. M., Chemotaxonomy of *Agelas* (Porifera, Demospongiae), *Biochem. Syst. Ecol.*, 20, 417, 1992.

56. Kelly-Borges, M. and Pomponi, S. A., Phylogeny and classification of lithistid sponges (Porifera: Demospongiae): a preliminary assessment using ribosomal DNA sequence comparisons, *Molec. Mar. Biol. Biotechnol.*, 3, 87, 1994.

57. Bewley, C. A. and Faulkner, D. J., Lithistid sponges: star performers or hosts to the stars?, *Angew. Chem. Int. Ed.*, 37, 2163, 1998.

58. Kobayashi, J., Murayama, T., and Ohizumi, Y., Theonelladins A–D, novel antineoplastic pyridine alkaloids from the Okinawan sponge *Theonella swinhoei*, *Tetrahedron Lett.*, 30, 4833, 1989.

59. Bewley, C. A., Holland, N. D., and Faulkner, D. J., Two classes of metabolites from *Theonella swinhoei* are localized in distinct populations of bacterial symbionts, *Experientia*, 52, 716, 1996.

60. Salomon, C. E., Deerinck, T., Ellisman, M., and Faulkner, D. J., The cellular localization of dercitamide in the Palauan sponge *Oceanapia sagittaria*, *Mar. Biol.*, in press.

61. Chan, W. R., Tinto, W. F., Manchand, P. S., and Todaro, L. J., Stereostructures of geodiamolides A and B, novel cyclodepsipeptides from the marine sponge *Geodia* sp., *J. Org. Chem.*, 52, 3091, 1987.

62. Talpir, R., Benayahu, Y., Kashman, Y., Pannell, L., and Schleyer, M., Hemiasterlin and geodiamolide TA; two new cytotoxic peptides from the marine sponge *Hemiasterella minor* (Kirkpatrick), *Tetrahedron Lett.*, 35, 4453, 1994.

63. de Silva, E. D., Andersen, R. J., and Allen, T. M., Geodiamolides C to F, new cytotoxic cyclodepsipeptides from the marine sponge *Pseudaxinyssa* sp., *Tetrahedron Lett.*, 31, 489, 1990.

64. Coleman, J. E., de Silva, E. D., Kong, F., and Andersen, R. J., Cytotoxic peptides from the marine sponge *Cymbastela* sp., *Tetrahedron*, 51, 10653, 1995.

65. D'Auria, M. V., Paloma, L. G., Minale, L., Zampella, A., Debitus, C., and Perez, J., Neosiphoniamolide A, a novel cyclodepsipeptide, with antifungal activity from the marine sponge *Neosiphonia superstes*, *J. Nat. Prod.*, 58, 121, 1995.

66. Pettit, G. R., Herald, C. L., Cichacz, Z. A., Gao, F., Schmidt, J. M., Boyd, M. R., Christie, N. D., and Boettner, F. E., Isolation and structure of the powerful human cancer cell growth inhibitors spongistatins 4 and 5 from an African *Spirastrella spinispirulifera* (Porifera), *J. Chem. Soc., Chem. Commun.*, 1805, 1993.

67. Fusetani, N., Shinoda, K., and Matsunaga, S., Cinachyrolide A: a potent cytotoxic macrolide possessing two spiro ketals from marine sponge *Cinachyra* sp., *J. Am. Chem. Soc.*, 115, 3977, 1993.

68. Kobayashi, M., Aoki, S., Sakai, H., Kihara, N., Sasaki, T., and Kitagawa, I., Altohyrtins B and C and 5-desacetylaltohyrtin A, potent cytotoxic macrolide congeners of altohyrtin A, from the Okinawan marine sponge *Hyrtios altum*, *Chem. Pharm. Bull.*, 41, 989, 1993.

69. Pettit, G. R., Cichacz, Z. A., Gao, F., Herald, C. L., and Boyd, M. R., Isolation and structure of the remarkable human cancer cell growth inhibitors spongistatins 2 and 3 from an eastern Indian Ocean *Spongia* sp., *J. Chem. Soc., Chem. Commun.*, 1166, 1993.

70. van Soest, R. W. M., Braekman, J. C., Hajdu, E., Faulkner, D. J., Harper, M. K., and Vacelet, J., The genus *Batzella*, a chemosystematic problem, *Bull. Inst. R. Sci. Nat. Belg.*, 66 (Suppl.), 89, 1996.

71. Bergquist, P. R., Cambie, R. C., and Kernan, M. R., Aaptamine, a taxonomic marker for sponges of the order Hadromerida, *Biochem. Syst. Ecol.*, 19, 289, 1991.

72. Pawlik, J. R., Chanas, B., Toonen, R. J., and Fenical, W., Defenses of Caribbean sponges against predatory reef fish. I. Chemical deterrency, *Mar. Ecol. Prog. Ser.*, 127, 183, 1995.

73. Lindquist, N. and Hay, M. E., Palatability and chemical defense of marine invertebrate larvae, *Ecol. Monogr.*, 66, 431, 1996.

74. Ebel, R., Brenzinger, M., Kunze, A., Gross, H. J., and Proksch, P., Wound activation of protoxins in marine sponge *Aplysina aerophoba*, *J. Chem. Ecol.*, 23, 1451, 1997.

75. Paul, V. J., Chemical defenses of benthic marine invertebrates, in *Ecological Roles of Marine Natural Products*, Paul, V. J., Ed., Cornell University Press, Ithaca, NY, 1992, 5.

76. Butler, A. J., van Altena, I. A., and Dunne, S. J., Antifouling activity of lyso-platelet-activating factor extracted from Australian sponge *Crella incrustans*, *J. Chem. Ecol.*, 22, 2041, 1996.

77. Thacker, R. W., Becerro, M. A., Lumbang, W. A., and Paul, V. J., Allelopathic interactions between sponges on a tropical reef, *Ecology*, 79, 1740, 1998.
78. Bandaranayake, W. M., Bemis, J. E., and Bourne, D. J., Ultraviolet absorbing pigments from the marine sponge *Dysidea herbacea*: isolation and structure of a new mycosporine, *Comp. Biochem. Physiol.*, 115C, 281, 1996.
79. Dunlap, W. C. and Shick, J. M., Ultraviolet radiation-absorbing mycosporine-like amino acids in coral reef organisms: a biochemical and environmental perspective, *J. Phycol.*, 34, 418, 1998.
80. Becerro, M. A., Uriz, M. J., and Turon, X., Chemically-mediated interactions in benthic organisms: the chemical ecology of *Crambe crambe* (Porifera, Poecilosclerida), *Hydrobiologia*, 356, 77, 1997.
81. Brusca, R. C. and Brusca, G. J., *Invertebrates*, Sinauer Associates, Sunderland, MD, 1990, 1.
82. Gerhart, D. J., The chemical systematics of colonial marine animals: an estimated phylogeny of the order Gorgonacea based on terpenoid characters, *Biol. Bull.*, 164, 71, 1983.
83. Veron, J. E. N., *Corals of Australia and the Indo-Pacific*, University of Hawaii Press, Honolulu, 1993, 61.
84. Cairns, S. D., Calder, D. R., Brinckmann-Voss, A., Castro, C. B., Pugh, P. R., Cutress, C. E., Jaap, W. C., Fautin, D. G., Larson, R. J., Harbison, G. R., Arai, M. N., and Opresko, D. M., *Common and Scientific Names of Aquatic Invertebrates from the United States and Canada: Cnidaria and Ctenophora*, American Fisheries Society, Bethesda, MD, 1991.
85. Cornelius, P. F. S., Manuel, R. L., and Ryland, L. S., *Handbook of the Marine Fauna of North-West Europe*, Oxford Press, New York, 1995, 1.
86. George, J. D. and George, J. J., *Marine Life, An Illustrated Encyclopedia of Invertebrates in the Sea*, John Wiley & Sons, New York, 1979, 18.
87. Coll, J. C., The chemistry and chemical ecology of octocorals (Coelenterata, Anthozoa, Octocorallia), *Chem. Rev.*, 92, 613, 1992.
88. Harvell, C. D., Fenical, W., and Greene, C. H., Chemical and structural defenses of Caribbean Gorgonians (*Pseudopterogorgia* spp.). I. Development of an *in situ* feeding assay, *Mar. Ecol. Prog. Ser.*, 49, 287, 1988.
89. Wylie, C. R. and Paul, V. J., Chemical defenses in three species of *Sinularia* (Coelenterata, Alcyonacea): effects against generalist predators and the butterflyfish *Chaetodon unimaculatus* Bloch, *J. Exp. Mar. Biol. Ecol.*, 129, 141, 1989.
90. Garson, M. J., The biosynthesis of marine natural products, *Chem. Rev.*, 93, 1699, 1993.
91. Bandurraga, M. M., McKittrick, B., Fenical, W., Arnold, E., and Clardy, J., Diketone cembrenolides from the Pacific gorgonian *Lophogorgia alba*, *Tetrahedron*, 38, 305, 1982.
92. Miralles, J., Diop, M., Ferrer, A., and Kornprobst, J.-M., Fatty acid composition of five Zoantharia from the Senegalese coast: *Palythoa dartevellei* Pax, *P. monodi* Pax, *P. variabilis* Duerden, *P. senegalensis* Pax and their associated Zooxanthellae, and *P. senegambiensis* Carter and its commensal, the decapoda *Diogenes ovatus* Miers, *Comp. Biochem. Physiol.*, 94B, 91, 1989.
93. Coll, J. C., Bowden, B. F., Tapiolas, D. M., and Dunlap, W. C., *In situ* isolation of allelochemicals released from soft corals (Coelenterata, Octocorallia): a totally submersible sampling apparatus, *J. Exp. Mar. Biol. Ecol.*, 60, 293, 1982.
94. Coll, J. C., Bowden, B. F., Tapiolas, D. M., Willis, R. H., Djura, P., Streamer, M., and Trott, L., Studies of Australian soft corals-XXXV: the terpenoid chemistry of soft corals and its implications, *Tetrahedron*, 41, 1085, 1985.
95. Maida, M., Carroll, A. R., and Coll, J. C., Variability of terpene content in the soft coral *Sinularia flexibilis* (Coelenterata: Octocorallia), and its ecological implications, *J. Chem. Ecol.*, 19, 2285, 1993.
96. Coll, J. C., Bowden, B. F., Heaton, A., Scheuer, P. J., Li, M. K. W., Clardy, J., Schulte, G. K., and Finer-Moore, J., Structures and possible functions of pukalide and epoxypukalide: diterpenes associated with eggs of sinularian soft corals (Cnidaria, Anthozoa, Octocorallia, Alcyonacea, Alcyoniidae), *J. Chem. Ecol.*, 15, 1177, 1989.
97. Coll, J. C., Leone, P. A., Bowden, B. F., Carroll, A. R., Koenig, G. M., Heaton, A., De Nys, R., Maida, M., Aliño, P. M., Willis, R. H., Babcock, R. C., Florian, Z., Clayton, M. N., Miller, R. L., and Alderslade, P. N., Chemical aspects of mass spawning in corals. II. (-)-*Epi*-thunbergol, the sperm attractant in the eggs of the soft coral *Lobophytum crassum* (Cnidaria: Octocorallia), *Mar. Biol.*, 123, 137, 1995.
98. Brooks, W. R., Chemical recognition by hermit crabs of their symbiotic sea anemones and a predatory octopus, *Hydrobiologia*, 216/217, 291, 1991.

99. Murata, M., Miyagawa-Kohshima, K., Nakanishi, K., and Naya, Y., Characterization of compounds that induce symbiosis between sea anemone and anemone fish, *Science*, 234, 585, 1986.

100. Cariello, L., Crescenzi, S., Prota, G. and Zanetti, L., Zoanthoxanthins of a new structural type from *Epizoanthus arenaceus* (Zoantharia), *Tetrahedron*, 30, 4194, 1974.

101. Cariello, L., Crescenzi, S., Zanetti, L., and Prota, G., A survey on the distribution of zoanthoxanthins in some marine invertebrates, *Comp. Biochem. Physiol.*, 63B, 77, 1979.

102. Coll, J. C., Bowden, B. F., Meehan, G. V., Koenig, G. M., Carroll, A. R., Tapiolas, D. M., Aliño, P. M., Heaton, A., de Nys, R., Leone, P. A., Maida, M., Aceret, T. L., Willis, R. H., Babcock, R. C., Willis, B. L., Florian, Z., Clayton, M. N., and Miller, R. L., Chemical aspects of mass spawning in corals. I. Sperm-attractant molecules in the eggs of the scleractinian coral *Montipora digitata*, *Mar. Biol.*, 118, 177, 1994.

103. Koh, E. G. L., Do scleractinian corals engage in chemical warfare against microbes?, *J. Chem. Ecol.*, 23, 379, 1997.

104. Wu Won, J. J., Rideout, J. A., and Chalker, B. E., Isolation and structure elucidation of a novel mycosporine-like amino acid from the reef-building corals *Pocillopora damicornis* and *Stylophora pistillata*, *Tetrahedron Lett.*, 36, 5255, 1995.

105. Healy, J. M., The Mollusca, in *Invertebrate Zoology*, Anderson, D. T., Ed., Oxford University Press, Melbourne, 1998, 122.

106. Faulkner, D. J. and Ghiselin, M. T., Chemical defense and evolutionary ecology of dorid nudibranchs and some other opisthobranch gastropods, *Mar. Ecol. Prog. Ser.*, 13, 295, 1983.

107. Carefoot, T. H., *Aplysia*: its biology and ecology, *Oceanogr. Mar. Biol. Annu. Rev.*, 25, 167, 1987.

108. Faulkner, D. J., Chemical defense of marine molluscs, in *Ecological Roles of Marine Natural Products*, Paul, V. J., Ed., Cornell University Press, Ithaca, NY, 1992, 119.

109. Carte, B. and Faulkner, D. J., Role of secondary metabolites in feeding associations between a predatory nudibranch, two grazing nudibranchs, and a bryozoan *J. Chem. Ecol.*, 12, 795, 1986.

110. Pettit, G. R., Kamano, Y., Herald, C. L., Fujii, Y., Kizu, H., Boyd, M. R., Boettner, F. E., Doubek, D. L., Schmidt, J. M., Chapuis, J. C., and Michel, C., Isolation of dolastatins 10–15 from the marine mollusc *Dolabella auricularia*, *Tetrahedron*, 49, 9151, 1993.

111. Pettit, G. R., The Dolastatins, in *Progress in the Chemistry of Organic Natural Products*, Herz, W., Kirby, G. W., Moore, R. E., Steglich, W., and Tamm, C., Eds., Springer-Verlag, Vienna, 1997, 1.

112. Harrigan, G. C., Luesch, H., Yoshida, W. Y., Moore, R. E., Nagle, D. G., Paul, V. J., Mooberry, S. L., Corbett, T. H., and Valeriote, F. A., Symplostatin 1: a dolastatin 10 analogue from the marine cyanobacterium *Symploca hydnoides*, *J. Nat. Prod.*, 61, 1075, 1998.

113. Harrigan, G. C., Yoshida, W. Y., Moore, R. E., Nagle, D. G., Park, P. U., Biggs, J., Paul, V. J., Mooberry, S. L., Corbett, T. H., and Valeriote, F. A., Isolation, structure determination, and biological activity of dolastatin 12 and lyngbyastatin 1 from *Lyngbya majuscula/Schizothrix calcicola* cyanobacterial assemblages, *J. Nat. Prod.*, 61, 1221, 1998.

114. Harrigan, G. C., Luesch, H., Yoshida, W. Y., Moore, R. E., Nagle, D. G., Biggs, J., Park, P. U., and Paul, V. J., Tumonic acids, novel metabolites from a cyanobacterial assemblage of *Lyngbya majuscula* and *Schizothrix calcicola*, *J. Nat. Prod.*, 62, 655, 1999.

115. Harrigan, G. C., Luesch, H., Yoshida, W. Y., Moore, R. E., Nagle, D. G., and Paul, V. J., Symplostatin 2: a dolastatin 13 analogue from the marine cyanobacterium *Symploca hydnoides*, *J. Nat. Prod.*, 62, 655, 1999.

116. Woodward, S. R., Cruz, L. J., Olivera, B. M., and Hillyard, D. R., Constant hypervariable regions in conotoxin propeptides, *EMBO*, 9, 1015, 1990.

117. Olivera, B. M., *Conus* venom peptides, receptor and ion channel targets, and drug design: 50 million years of neuropharmacology, *Mol. Biol. Cell*, 8, 2101, 1997.

118. Edmonds, J. S., Francesconi, K. A., and Stick, R. V., Arsenic compounds from marine organisms, *Nat. Prod. Rep.*, 10, 421, 1993.

119. Byrne, M., The Echinodermata, in *Invertebrate Zoology*, Anderson, D. T., Ed., Oxford University Press, Melbourne, 1998, 366.

120. Lawrence, J., *A Functional Biology of Echinoderms*, The Johns Hopkins University Press, Baltimore, 1987, 1.

121. Mah, C., Classification of the extant Echinodermata, World Wide Web, California Academy of Sciences, September 29, 1999, http://www.calacademy.org/research/izg/echinoderm/classify.htm.

122. Palagiano, E., De Marino, S., Minale, L., Riccio, R., and Zollo, F., Ptilomycalin A, Crambescidin 800 and related new highly cytotoxic guanidine alkaloids from the starfishes *Fromia monilis* and *Celerina heffernani*, *Tetrahedron*, 51, 3675, 1995.

123. Doherty, P. J., The Lophophorates: Phoronida, Brachiopoda, and Ectoprocta, in *Invertebrate Zoology*, Anderson, D. T., Ed., Oxford University Press, Melbourne, 1998, 343.

124. Woollacott, R. M. and Zimmer, R. L., Eds., *Biology of Bryozoans*, Academic Press, New York, 1977.

125. Bock, P., Systematic list of families of Bryozoa, World Wide Web, updated August 27, 1999, Department of Civil and Geological Engineering, RMIT University, http://www.civgeo.rmit.edu.au/bryozoa/famsys.html.

126. Tischler, M., Ayer, S. W., and Andersen, R. J., Nitrophenols from northeast Pacific bryozoans, *Comp. Biochem. Physiol.*, 84B, 43, 1986.

127. Propper, D. J., Macaulay, V., O'Byrne, K. J., Braybrooke, J. P., Wilner, S. M., Ganesan, T. S., Talbot, D. C., and Harris, A. L., A phase II study of bryostatin 1 in metastatic malignant melanoma, *Br. J. Cancer*, 78, 1337, 1998.

128. Kamano, Y., Zhang, H., Ichihara, Y., and Kizu, H., Convolutamydine A, a novel bioactive hydroxy-oxindole alkaloid from marine bryozoan *Amathia convoluta*, *Tetrahedron Lett.*, 36, 2783, 1995.

129. Colin, P. L. and Arneson, C., *Tropical and Pacific Invertebrates*, Coral Reef Press, Beverly Hills, CA, 1995, 267.

130. Monniot, C., Monniot, F., and Laboute, P., *Coral Reef Ascidians of New Caledonia*, Orstom, Paris, 1991, 1.

131. Davidson, B. S., Ascidians: producers of amino acid derived metabolites, *Chem. Rev.*, 93, 1771, 1993.

132. Molinski, T. F., Marine pyridoacridine alkaloids: structure, synthesis, and biological chemistry, *Chem. Rev.*, 93, 1825, 1993.

133. Vervoort, H. C., Pawlik, J. R., and Fenical, W., Chemical defense of the Caribbean ascidian *Didemnum conchyliatum*, *Mar. Ecol. Prog. Ser.*, 164, 221, 1998.

134. Young, C. M. and Bingham, B. L., Chemical defense and aposematic coloration in larvae of the ascidian *Ecteinascidia turbinata*, *Mar. Biol.*, 96, 539, 1987.

135. Paul, V. J., Lindquist, N., and Fenical, W., Chemical defense of the tropical ascidian *Atapazoa* sp. and its nudibranch predators *Nembrotha* spp., *Mar. Ecol. Prog. Ser.*, 59, 109, 1990.

136. Davis, A. R., Alkaloids and ascidian chemical defense: evidence for the ecological role of natural products from *Eudistoma olivaceum*, *Mar. Biol.*, 111, 375, 1991.

137. Dunlap, W. C. and Yamamoto, Y., Small-molecule antioxidants in marine organisms: antioxidant activity of mycosporine-glycine, *Comp. Biochem. Physiol.*, 112B, 105, 1995.

138. Dionisio-Sese, M. L., Ishikura, M., Maruyama, T., and Miyachi, S., UV-absorbing substances in the tunic of a colonial ascidian protect its symbiont, *Prochloron* sp., from damage by UV-B radiation, *Mar. Biol.*, 128, 455, 1997.

APPENDIX A: PHYLOGENETIC DISTRIBUTION OF SECONDARY METABOLITES (TABLES 1.3–1.11)

TABLE 1.3
Algae: Division Chlorophyta

Class	Order	Family	Genus	Total	Isoprenoid	Acetogenin	Amino Acid	Shikimate	Nucleic Acid	Carbohydrate
Charophyceae				177	109 (62)	30 (17)	24 (14)	8 (4)	0	6 (3)
	Charales			3			3 (100)			
		Characeae		3			3 (100)			
			Chara	3			3 (100)			
Chlorophyceae				154	91 (59)	29 (19)	21 (14)	7 (5)		6 (3)
	Acrosiphonales			9		6 (67)				3 (33)
		Acrosiphoniaceae		9		6 (67)				3 (33)
			Acrosiphonia	7		6 (86)				1 (14)
			Urospora	2						2 (100)
	Bryopsidales			71	62 (88)	1 (1)	2 (3)	5 (7)		1 (1)
		Bryopsidaceae		3	1 (33)		2 (67)			
			Bryopsis	3	1 (33)		2 (67)			
		Caulerpaceae		25	23 (92)	1 (4)		1 (4)		
			Caulerpa	25	23 (92)	1 (4)		1 (4)		
		Codiaceae		6	6 (100)					
			Codium	4	4 (100)					
			Rhipocephalus	2	2 (100)					
		Halimedaceae		9	9 (100)					
			Halimeda	9	9 (100)					
		Udoteaceae		28	23 (82)			4 (14)		1 (4)
			Avrainvillea	4				4 (100)		
			Chlorodesmis	2	2 (100)					
			Penicillus	7	6 (86)					1 (14)
			Pseudochlorodesmis	2	2 (100)					
			Tydemania	6	6 (100)					
			Udotea	7	7 (100)					

Order / Class	Family	Genus	n					
Cladophorales			6		4 (67)	2 (33)		
	Anadyomenaceae		1		1 (100)			
		Anadyomene	1		1 (100)			
	Cladophoraceae		3		1 (33)	2 (67)		
		Cladophora	3		1 (33)	2 (67)		
	Siphonocladaceae		2		2 (100)			
		Dictyosphaeria	2		2 (100)			
Chlorococcales			23	8 (35)	7 (31)	6 (26)	1 (4)	1 (4)
	Oocystaceae		16	6 (38)	4 (25)	5 (31)		1 (6)
		Chlorella	16	6 (38)	4 (25)	5 (31)		1 (6)
	Scenedesmaceae		7	2 (29)	3 (43)	1 (14)	1 (14)	
		Scenedesmus	7	2 (29)	3 (43)	1 (14)	1 (14)	
Dasycladales			16	15 (94)			1 (6)	
	Dasycladaceae		16	15 (94)			1 (6)	
		Cymopolia	11	11 (100)				
		Dasycladus	1				1 (100)	
		Neomeris	4	4 (100)				
Ulvales			20		10 (50)	9 (45)		1 (5)
	Monostromataceae		1			1 (100)		
		Monostroma	1			1 (100)		
	Ulvaceae		19		10 (53)	8 (42)		1 (5)
		Enteromorpha	11		6 (55)	5 (45)		
		Ulva	8	6 (78)	4 (50)	3 (38)		1 (12)
Volvocales			8					
	Chlamydomondaceae		1	1 (100)		1 (100)		
		Chlamydomonas	1	1 (100)	1 (11)	1 (11)		
	Polyblepharidaceae		7	5 (72)	1 (14)	1 (14)		
		Dunaliella	6	4 (66)	1 (16)	1 (16)		
		Pyramimonas	1	1 (100)				
Zygnematales								
	Desmidiaceae		1	1 (100)		1 (100)		
		Closterium	1	1 (100)		1 (100)		
Prasinophyceae			8	8 (100)				
Marriellales			8	8 (100)				
	Micromonadophyceae		8	8 (100)				
		Mantoniella	6	6 (100)				
		Micromonas	2	2 (100)				
Unknown								
	Unknown							
		Chlorophytum	12	10 (84)	1 (8)		1 (8)	

TABLE 1.4
Algae: Division Rhodophyta

Class	Subclass	Order	Family	Genus	Total	Isoprenoid	Acetogenin	Amino Acid	Shikimate	Nucleic Acid	Carbohydrate
Rhodophyceae					938	429 (46)	356 (38)	89 (9)	45 (5)	1 (<1)	19 (2)
	Bangiophycidae				14	4 (29)	4 (29)	3 (21)			3 (21)
		Bangiales			14	4 (29)	4 (29)	3 (21)			3 (21)
			Bangiaceae		6		2 (33)	3 (50)			1 (17)
				Porphyra	6		2 (33)	3 (50)			1 (17)
			Porphyridiaceae		8	4 (50)	2 (25)				2 (25)
				Porpyridium	7	4 (57)	2 (29)				1 (14)
				Rhodella	1						1 (100)
	Florideophycidae				924	425 (46)	351 (38)	86 (9)	45 (5)	1 (<1)	16 (2)
		Bonnemaisoniales			135	1 (1)	124 (91)	8 (6)	1 (1)		1 (1)
			Bonnemaisoniaceae		123	1 (1)	122 (99)				
				Asparagopsis	67	1 (1)	66 (99)				
				Bonnemaisonia	19		19 (100)				
				Delisea	33		33 (100)				
				Ptilonia	4		4 (100)				
			Halymeniaceae		12		2 (17)	8 (67)	1 (8)		1 (8)
				Grateloupia	10		1 (10)	8 (80)	1 (10)		
				Halymenia	1		1 (100)				
				Pachymenia	1						1 (100)
		Ceramiales			529	285 (54)	162 (31)	58 (11)	24 (4)		
			Ceramiaceae		19	5 (26)	11 (58)	3 (16)			
				Ceramium	2		1 (50)	1 (50)			
				Dasyphila	1		1 (100)				
				Griffithsia	1		1 (100)				
				Microcladia	10	5 (50)	5 (50)				
				Ptilota	5		3 (67)	2 (33)			

Order	Family	Genus					
	Dasyaceae		1			1 (100)	
		Dasya	1			1 (100)	
	Delesseriaceae		6	2 (33)		4 (67)	
		Martensia	4			4 (100)	
		Pantoneura	2	2 (100)			
	Rhodomelaceae		503	278 (55)	151 (30)	50 (10)	24 (5)
		Alsidium	2	1 (50)	1 (50)		
		Digenea	5		1 (20)	4 (80)	
		Chondria	29	2 (6)	8 (28)	19 (66)	
		Halopytis	2			2 (100)	
		Laurencia	431	274 (64)	138 (32)	18 (4)	1 (<1)
		Lophocladia	1			1 (100)	
		Maelanothamnus	1	1 (100)			
		Odonthalia	3			3 (100)	
		Osmundaria	3		1 (33)		2 (67)
		Polysiphonia	15		2 (13)	1 (7)	12 (80)
		Rhodomela	7				7 (100)
		Symphyocladia	2				2 (100)
		Vidalia	2			2 (100)	
Corallinales			9	1 (11)	7 (78)	1 (11)	
	Corallinaceae		9	1 (11)	7 (78)	1 (11)	
		Bossiella	2		2 (100)		
		Corallina	4		3 (75)	1 (25)	
		Lithophyllum	1	1 (100)			
		Lithothamnion	2		2 (100)		
Cryptonemiales			70	46 (66)	18 (26)	1 (1)	5 (7)
	Dumontiaceae		15		15 (100)		
		Constantinea	10		10 (100)		
		Farlowia	4		4 (100)		
		Neodilsea	1		1 (100)		
	Gloiosiphoniaceae		2		2 (100)		
		Gloiosiphonia	2		2 (200)		
	Peyssonneliaceae		5			1 (20)	4 (80)
		Prionitis	5			1 (20)	4 (80)
	Rhizophyllidacea		48	46 (96)	1 (2)		1 (2)
		Ochtodes	15	15 (100)			4 (80)
		Peyssonnelia	2	2 (100)			4 (80)
		Portieria	31	29 (94)	1 (3)		1 (3)

TABLE 1.4 (CONTINUED)
Algae: Division Rhodophyta

Class	Subclass	Order	Family	Genus	Total	Isoprenoid	Acetogenin	Amino Acid	Shikimate	Nucleic Acid	Carbohydrate
		Gelidiales	Gelidiaceae		6	1 (17)	2 (33)	1 (17)			2 (33)
				Beckerella	6	1 (17)	2 (33)	1 (17)			2 (33)
				Gelidium	3	1 (33)	2 (67)				1 (33)
		Gigartinales	Cystocloniaceae		75	25 (32)	14 (18)	10 (13)	15 (19)	1 (5)	10 (13)
				Pterocladia	11			2 (18)	9 (82)		
				Rhodophyllis	2			2 (100)			
					9				9 (100)	1 (100)	
			Furcellariaceae	Furcellaria	1					1 (100)	
			Gigartinaceae		8	5 (71)		5 (71)			3 (75)
				Chondrus	7			4 (57)			3 (43)
				Gigartina	3			1 (33)			2 (67)
			Hypneaceae		8			1 (14)			2 (29)
				Hypnea	6	5 (83)		1 (17)			1 (14)
			Phacelocarpaceae		34	20 (61)	13 (39)				1 (33)
				Ahnfeltia	3		2 (67)				
				Phacelocarpus	11		11 (100)				
				Sphaerococcus	20	20 (100)					
			Phyllophoraceae		3		1 (33)	2 (66)			
				Gymnogongrus	2			2 (100)			
				Phyllophora	1		1 (100)				
			Sarcodiaceae		1						1 (100)
				Sarcodia	1						1 (100)

Order	Family	Genus	N					
	Solieriaceae		9				6 (67)	3 (33)
		Agardhiella	1					1 (100)
		Eucheuma	1					1 (100)
		Kappaphycus	1					1 (100)
		Rhabdonia	6				6 (100)	
Gracilariales			25	8 (32)	14 (56)	2 (8)		1 (4)
	Gracilariaceae		25	8 (32)	14 (56)	2 (8)		1 (4)
		Gracilaria	15	8 (53)	5 (33)	1 (7)		1 (7)
		Polycavernosa	4	4 (100)				
		Gracilariopsis	6		5 (83)	1 (17)		
Palmariales			55	52 (97)				
	Palmariaceae		3		1 (33)	2 (66)		
		Palmaria	3		1 (33)	2 (67)		
Plocamiales			52	51 (98)	1 (2)			
	Plocamiaceae		52	51 (98)	1 (2)			
		Plocamium	4	4 (100)				
	Galaxauraceae		4	4 (100)				
		Galaxaura	4	4 (100)				
Nemaliales			9		9 (100)			
	Liagoraceae		4		4 (100)			
		Liagora	4		4 (100)			
	Lomentariaceae		5		5 (100)			
		Lomentaria	5		5 (100)			
Rhodymeniales	Rhodymeniaceae		7	2 (29)		3 (42)		2 (29)
		Botryocladia	7	2 (29)		3 (42)		2 (29)
		Rhodymenia	3			3 (100)		
Unknown	Unknown		4	2 (50)				2 (50)
		Coeloseira	1	1 (100)				

TABLE 1.5
Algae: Division Phaeophyta

Class	Order	Family	Genus	Total	Isoprenoid	Acetogenin	Amino Acid	Shikimate	Nucleic Acid	Carbohydrate
Phaeophyceae				759	513 (68)	116 (15)	15 (2)	91 (12)	0	24 (3)
	Cutleriales			2		2 (100)				
		Cutleriaceae		2		2 (100)				
			Cutleria	2		2 (100)				
	Desmarestiales			11	5 (46)	4 (36)	1 (9)			1 (9)
		Desmarestiaceae		11	5 (46)	4 (36)	1 (9)			1 (9)
			Desmarestia	11	5 (46)	4 (36)	1 (9)			1 (9)
	Dictyotales			387	346 (88)	36 (9)	1 (<1)	4 (3)		
		Dictyotaceae		387	346 (88)	36 (9)	1 (<1)	4 (3)		
			Dictyopteris	49	21 (43)	28 (57)				
			Dictyota	224	223 (100)	1 (<1)				
			Dilophus	44	44 (100)					
			Distromium	3				3 (100)		
			Glossophora	4	4 (100)					
			Lobophora	1				1 (100)		
			Pachydictyon	18	18 (100)					
			Padina	2	2 (100)					
			Spatoglossum	3	2 (67)	1 (33)				
			Stoechospermum	13	13 (100)					
			Stypopodium	8	8 (100)					
			Taonia	11	11 (100)					
			Zonaria	7		6 (86)	1 (14)			
	Ectocarpales			12		6 (50)		6 (50)		
		Ectocarpaceae		3		3 (100)				
			Ectocarpus	2		2 (100)				
			Giffordia	1		1 (100)				

Taxon						
Ralfsiaceae	7		1 (14)		6 (86)	
Analipus	7		1 (14)		6 (86)	
Spermatochnaceae	2		2 (100)			
Spermatochnus	2		2 (100)			
Fucales	282	160 (57)	45 (16)	9 (3)	57 (19)	11 (5)
Cystoseiraceae	149	124 (84)	11 (7)		14 (9)	
Acrocarpia	2		2 (100)			
Bifurcaria	17	14 (82)			3 (18)	
Caulocystis	6		6 (100)			
Cystophora	20	14 (70)	3 (15)		3 (15)	
Cystoseira	97	90 (93)			7 (7)	
Halidrys	7	6 (86)			1 (14)	
Fucaceae	23	6 (26)	3 (13)	2 (9)	5 (22)	7 (30)
Ascophyllum	2			1 (50)	1 (50)	
Fucus	17	6 (34)	3 (18)		4 (12)	4 (12)
Pelvetia	4			1 (25)		3 (75)
Himanthaliaceae	2				2 (100)	
Himanthalia	2				2 (100)	
Notheiaceae	24		24 (100)			
Notheia	24		24 (100)			
Sargassaceae	84	30 (36)	7 (8)	7 (8)	36 (43)	4 (5)
Carpophyllum	13				13 (100)	
Hizikia	3			1 (33)		2 (67)
Landsburgia	5				5 (100)	
Sargassum	61	28 (46)	7 (11)	6 (10)	18 (30)	2 (3)
Turbinaria	2	2 (100)				
Laminariales	50	1 (2)	17 (34)	2 (4)	18 (36)	12 (24)
Alariaceae	23	1 (4)	10 (44)	1 (4)	9 (39)	2 (9)
Alaria	1		1 (100)			
Egregia	3		3 (100)			
Eisenia	3	1 (33)		1 (33)	1 (33)	
Ecklonia	16		6 (38)		8 (50)	2 (12)
Chordaceae	3				3 (100)	
Chorda	3				3 (100)	
Laminariaceae	21		5 (24)	1 (5)	6 (29)	9 (42)
Agarum	1		1 (100)			

TABLE 1.5 (CONTINUED)
Algae: Division Phaeophyta

Class	Order	Family	Genus	Total	Isoprenoid	Acetogenin	Amino Acid	Shikimate	Nucleic Acid	Carbohydrate
			Cymathere	2		2 (100)				
			Laminaria	18		2 (11)	1 (1)	6 (33)		9 (50)
		Lessoniaceae		3		2 (67)				1 (33)
			Macrocystis	3		2 (67)				1 (33)
	Scytosiphonales			8		5 (63)	2 (25)	1 (12)		
		Scytosiphonaceae		8		5 (63)	2 (25)	1 (12)		
			Colpomenia	1				1 (100)		
			Petalonia	2			2 (100)			
			Scytosiphon	5		5 (100)				
	Sporochnales			7	1 (14)	1 (14)		5 (72)		
		Sporochnaceae		7	1 (14)	1 (14)		5 (72)		
			Encyothalia	1				1 (100)		
			Perithalia	5	1 (20)	1 (20)		3 (60)		
			Sporochnus	1				1 (100)		

TABLE 1.6
Phylum Porifera

Class	Subclass	Order	Family	Genus	Total	Isoprenoid	Acetogenin	Amino Acid	Shikimate	Nucleic Acid	Carbohydrate
					2609	1295 (50)	574 (22)	681 (26)	37 (1)	19 (<1)	3 (<1)
Hexactinellida					0						
Calcarea					31			31 (100)			
	Calcinea				31			31 (100)			
		Clathrinida			31			31 (100)			
			Clathrinidae		2			2 (100)			
				Clathrina	2			2 (100)			
			Leucettidae		29			29 (100)			
				Leucetta	29			29 (100)			
		Murrayonida			0						
	Calcaronea				0						
Demospongiae					2578	1295 (50)	574 (22)	650 (25)	37 (1)	19 (<1)	3 (<1)
	Homoscleromorpha				93	7 (8)	80 (86)	6 (6)			
		Homosclerophorida			93	7 (8)	80 (86)	6 (6)			
			Plakinidae		93	7 (8)	80 (86)	6 (6)			
				Corticium	3	2 (67)		1 (33)			
				Plakina	2	2 (100)					
				Plakinastrella	5		5 (100)				
				Plakortis	75	1 (1)	69 (92)	5 (7)			
				Oscarella	8	2 (25)	6 (75)				
	Tetractinomorpha				372	147 (40)	109 (29)	97 (26)	9 (2)	10 (3)	
		Spirophorida			13	8 (62)	3 (23)	2 (15)			
			Tetillidae		7	4 (57)	3 (43)				
				Cinachyra	4	2 (50)	2 (50)				
				Cinachyrella	3	2 (67)	1 (33)				

TABLE 1.6 (CONTINUED)
Phylum Porifera

Class	Subclass	Order	Family	Genus	Total	Isoprenoid	Acetogenin	Amino Acid	Shikimate	Nucleic Acid	Carbohydrate
			Sclerodermidae		6	4 (67)		2 (33)			
				Aciculites	1	1 (100)					
				Sclerotoderma	1			1 (100)			
				Microscleroderma	4	3 (75)		1 (25)			
		Astrophorida			116	55 (47)	28 (24)	18 (16)	8 (7)	7 (6)	
			Coppatiidae		55	25 (45)	19 (35)	1 (2)	8 (15)	2 (4)	
				Jaspis	55	25 (45)	19 (35)	1 (2)	8 (15)	2 (4)	
			Ancorinidae		31	18 (58)	5 (16)	8 (26)			
				Asteropus	7	7 (100)					
				Penares	8	3 (43)	4 (57)	1 (14)			
				Stelletta	16	8 (50)	1 (6)	7 (44)			
			Geodiidae		18	5 (28)	4 (22)	4 (22)		5 (28)	
				Erylus	6	3 (50)	3 (50)				
				Geodia	12	2 (17)	1 (8)	4 (33)		5 (42)	
			Pachastrellidae		12	7 (58)		5 (42)			
				Dercitus	5			5 (100)			
				Pachastrella	2	2 (100)					
				Poecillastra	5	5 (100)					
		Hadromerida			131	63 (48)	36 (27)	28 (21)	1 (<1)	3 (2)	
			Clionidae		35	24 (69)	2 (6)	9 (26)			
				Cliona	35	24 (69)	2 (6)	9 (26)			
			Chondrillidae		13	1 (8)	11 (85)		1 (8)		
				Chondrilla	5	1 (20)	3 (60)		1 (20)		
				Chondrosia	8		8 (100)				
			Hemiasterellidae		2			2 (100)			
				Hemiasterella	2			2 (100)			

Order	Family	Genus	n						
	Latrunculiidae		32	18 (56)	6 (19)	7 (22)			1 (3)
		Latrunculia	22	14 (64)		7 (32)			1 (4)
		Negombata	6		6 (100)				
		Sigmosceptrella	4	4 (100)					
	Polymastiidae		6	6 (100)					
		Polymastia	6	6 (100)					
	Spirastrellidae		12	3 (25)	6 (50)	3 (25)			
		Anthosigmella	3			3 (100)			
		Spirastrella	9	3 (33)	6 (67)				
	Suberitidae		22	7 (32)	8 (36)	7 (32)			
		Aaptos	7	1 (14)	2 (29)	4 (57)			
		Laxosuberites	6		6 (100)				
		Pseudosuberites	3			3 (100)			
		Suberites	5	5 (100)					
		Terpios	1	1 (100)					
	Tethyidae		7	4 (57)	3 (43)				
		Tethya	7	4 (57)	3 (43)				
	Trachycladidae		2					2 (100)	
		Trachycladus	2					2 (100)	
Lithistida			112	21 (19)	42 (38)	49 (44)			
	Theonellidae		95	14 (15)	42 (44)	39 (41)			
		Discodermia	20		10 (50)	10 (50)			
		Neosiphonia	8	1 (12)	6 (75)	1 (12)			
		Reidispongia	1		1 (100)				
		Theonella	66	13 (20)	25 (38)	28 (42)			
	Corallistidae		12	4 (33)		8 (67)			
		Callipelta	1			1 (100)			
		Corallistes	11	4 (36)		7 (64)			
	Pleromiidae		2			2 (100)			
		Pleroma	2			2 (100)			
	Azoricidae		3	3 (100)					
		Jereicopsis	3	3 (100)					
Ceractinomorpha			2113	1141 (54)	385 (18)	547 (26)	28 (1)	9 (<1)	3 (<1)
Agelasida	Agelasidae		71	23 (32)	3 (4)	44 (62)		1 (1)	
		Agelas	62	23 (37)	3 (5)	35 (56)		1 (2)	
			62	23 (37)	3 (5)	35 (56)		1 (2)	

TABLE 1.6 (CONTINUED)
Phylum Porifera

Class	Subclass	Order	Family	Genus	Total	Isoprenoid	Acetogenin	Amino Acid	Shikimate	Nucleic Acid	Carbohydrate
		Halichondrida	Astroscleridae		9			9 (100)			
				Astrosclera	9			9 (100)			
			Axinellidae		424	242 (57)	75 (18)	99 (23)	5 (1)	3 (1)	
				Acanthella	195	130 (67)	19 (10)	43 (22)	3 (1)		
				Auletta	70	63 (90)	2 (3)	5 (7)			
					1			1 (100)			
				Axinella	71	50 (70)	9 (13)	9 (13)	3 (4)		
				Cymbastela	14	11 (79)		3 (21)			
				Dragmacidon	1			1 (100)			
				Homaxinella	1	1 (100)					
				Phakellia	15		3 (20)	12 (80)			
				Phycopsis	2	1 (50)	1 (50)				
				Pseudaxinella	1			1 (100)			
				Ptilocaulis	7		4 (57)	3 (43)			
				Stylotella	9	2 (22)		7 (78)			
				Teichaxinella	3	2 (67)		1 (33)			
			Desmoxyiidae		18	14 (78)	4 (22)				
				Higginsia	6	5 (83)	1 (17)				
				Myrmekioderma	12	9 (75)	3 (25)				
			Dictyonellidae		12	6 (50)	3 (25)	3 (25)			
				Dictyonella	8	5 (62)	3 (38)				
				Ulosa	4	1 (25)		3 (75)			
			Halichondriidae		199	92 (46)	49 (25)	53 (27)	2 (1)	3 (2)	
				Axinyssa	13	12 (92)		1 (8)			
				Ciocalypta	4	4 (100)					
				Didiscus	8	8 (100)					

Taxon	n					
Epipolasis	20	20 (100)				
Halichondria	59	25 (42)	26 (44)	6 (10)	2 (3)	
Hymeniacidon	28	6 (21)	1 (4)	19 (68)		
Pseudaxinyssa	40	8 (20)	21 (52)	11 (28)		2 (7)
Spongosorites	8	8 (100)				
Topsentia	12	7 (58)	1 (8)	3 (25)		1 (8)
Prianos	7	2 (29)		5 (71)		
Poecilosclerida	218	69 (32)	80 (37)	60 (28)	5 (2)	4 (2)
Iophonidae						
Acarnus	23		2 (9)	21 (91)		
Damiria	2			2 (100)		
Zyzzya	19			19 (100)		
Microcionidae	34	15 (44)	18 (53)	1 (3)		
Pandaros	1		1 (100)			
Clathria	2		1 (50)	1 (50)		
Microciona	18	7 (39)	11 (61)			
Thalysias	9	8 (89)	1 (11)			
Echinochalina	4		4 (100)			
Raspailiidae						
Echinodictyum	35	20 (57)	14 (40)	1 (3)		
Ectyoplasia	1			1 (100)		
Eurypon	2		2 (100)			
Raspaciona	7	7 (100)				
Trikentrion	12	12 (100)				
Raspailia	9	1 (11)	8 (89)			
Raspaxilla	2		2 (100)			
Anchinoidae						
Hamigera	14	1 (7)	2 (14)	10 (71)	1 (7)	
Kirkpatrickia	3	1 (33)		2 (67)		
Phorbas	5			4 (80)	1 (20)	
Coelosphaeridae						
Forcepia	6		2 (33)	4 (67)		
Histodermella	8	1 (12)	3 (38)	4 (50)		
Lissodendoryx	6	1 (17)	2 (33)	3 (50)		
Crambeidae	12	1 (8)	10 (83)	1 (8)		
Crambe	10	1 (10)	9 (90)			
Monanchora	2		1 (50)	1 (50)		

TABLE 1.6 (CONTINUED)
Phylum Porifera

Class	Subclass	Order	Family	Genus	Total	Isoprenoid	Acetogenin	Amino Acid	Shikimate	Nucleic Acid	Carbohydrate
			Hymedesmiidae		1	1 (100)					
				Stylopus	1	1 (100)					
			Myxillidae		12	7 (58)		5 (42)			
				Iotrochota	4			4 (100)			
				Plocamissa	1			1 (100)			
				Stelodoryx	7	7 (100)					
			Phoriospongiidae		20	6 (30)	7 (35)	7 (35)			
				Batzella	13		6 (46)	7 (54)			
				Hemimycale	1		1 (100)				
				Psammoclemma	3	3 (100)					
				Strongylacidon	3	3 (100)					
			Tedaniidae		20	5 (25)	4 (20)	6 (30)	4 (20)	1 (5)	
				Tedania	20	5 (25)	4 (20)	6 (30)	4 (20)	1 (5)	
			Desmacellidae		4	2 (50)		2 (50)			
				Bienna	4	2 (50)		2 (50)			
			Hamacanthidae		2			2 (100)			
				Hamacantha	2			2 (100)			
			Mycalidae		33	10 (30)	20 (61)			3 (9)	
				Arenochalina	7	1 (14)	6 (86)				
				Esperiopsis	3	3 (100)					
				Mycale	23	6 (26)	14 (61)			3 (13)	
		Haplosclerida			489	158 (32)	190 (39)	140 (29)		1 (<1)	
			Chalinidae		110	40 (36)	26 (24)	43 (39)		1 (1)	
				Adocia	9	7 (78)	2 (22)				
				Cladocroce	2		2 (100)				
				Haliclona	41	12 (29)	13 (32)	15 (37)		1 (2)	

		Orina (& Gellius)	9			9 (100)		
		Reniera	39	15 (38)	5 (13)	19 (49)		
		Toxadocia	4		4 (100)			
		Toxiclona	6	6 (100)				
	Niphatidae		80	13 (16)	34 (43)	33 (41)		
		Aka (& Siphonodictyon)	12	12 (100)				
		Amphimedon	31		20 (65)	11 (35)		
		Cribrochalina	23	1 (4)	14 (61)	8 (35)		
		Niphates	14			14 (100)		
	Callyspongiidae		34	11 (32)	19 (56)	4 (12)		
		Callyspongia	17	2 (12)	11 (65)	4 (23)		
		Siphonochalina	17	9 (53)	8 (47)			
	Phloeodictyidae		39	16 (41)	21 (54)	2 (5)		
		Calyx	20	9 (45)	11 (55)			
		Oceanapia	3	1 (33)	2 (67)			
		Pachypellina	1			1 (100)		
		Pellina	15	6 (40)	8 (53)	1 (7)		
	Petrosiidae		226	78 (34)	90 (40)	58 (26)		
		Petrosia	75	11 (15)	54 (72)	10 (13)		
		Strongylophora	19	19 (100)				
		Xestospongia	132	48 (36)	36 (28)	48 (36)		
Dictyoceratida			663	559 (84)	33 (5)	51 (8)	18 (3)	2 (<1)
	Dysideidae		227	178 (78)	7 (3)	25 (11)	17 (8)	
		Dysidea	184	139 (75)	3 (2)	25 (14)	17 (9)	
		Euryspongia	16	12 (75)	4 (25)			
		Spongionella	25	25 (100)				
	Spongiidae		205	194 (94)	9 (4)		1 (1)	1 (1)
		Carteriospongia	10	10 (100)				
		Coscinoderma	2	2 (100)				
		Dactylospongia	11	11 (100)				
		Hippospongia	37	36 (97)	1 (3)			
		Hyatella	10	8 (80)	2 (20)			
		Lendenfeldia	2	2 (100)				
		Phyllospongia	25	24 (96)			1 (4)	
		Rhopaloeides	1	1 (100)				

TABLE 1.6 (CONTINUED)
Phylum Porifera

Class	Subclass	Order	Family	Genus	Total	Isoprenoid	Acetogenin	Amino Acid	Shikimate	Nucleic Acid	Carbohydrate
				Spongia	99	92 (93)	6 (6)				1 (1)
				Strepsichordaia	8	8 (100)					
			Irciniidae	Ircinia	47	40 (85)	2 (4)	5 (11)			
				Psammocinia	39	32 (82)	2 (5)	5 (13)			
				Sarcotragus	2	2 (100)					
					6	6 (100)					
			Thorectidae		184	147 (80)	15 (8)	21 (11)			1 (1)
				Aplysinopsis	3	3 (100)					
				Cacospongia	35	31 (89)	4 (11)				
				Collospongia	1	1 (100)					
				Fascaplysinopsis	10	1 (10)		9 (90)			
				Fasciospongia	19	16 (84)	1 (5)	2 (11)			
				Hyrtios	25	16 (64)	4 (16)	5 (20)			
				Luffariella	39	38 (97)					
				Petrosaspongia	17	17 (100)					
				Smenospongia	25	15 (60)	6 (24)	4 (16)			1 (3)
				Thorecta	7	6 (86)		1 (14)			
				Thorectandra	3	3 (100)					

Order	Family	Genus					
Dendroceratida			89	74 (83)		15 (17)	
	Darwinellidae		76	63 (83)		13 (17)	
		Aplysilla	27	27 (100)			
		Chelonaplysilla	19	15 (79)		4 (21)	
		Darwinella	2	1 (50)		1 (50)	
		Dendrilla	18	15 (83)		3 (17)	
		Hexadella	5			5 (100)	
		Pleraplysilla	5	5 (100)			
	Dictyodendrillidae		15	13 (87)		2 (13)	
		Dictyodendrilla	9	7 (78)		2 (22)	
		Igernella	6	6 (100)			
Halisarcida			0				
Verongida			159	16 (10)	4 (3)	138 (87)	1 (<1)
	Aplysinidae		72	7 (10)	3 (4)	62 (86)	
		Aplysina (= *Verongia*)	61	5 (8)	3 (5)	53 (87)	
		Verongula	11	2 (18)		9 (82)	
	Druinellidae		53	1 (2)	1 (2)	50 (94)	1 (2)
		Aplysinella	2			2 (100)	
		Druinella (= *Psammaplysilla*)	33			33 (100)	
		Pseudoceratina	18	1 (2)	1 (2)	15 (84)	1 (2)
	Ianthellidae		34	8 (24)		26 (76)	
		Anomoianthella	1			1 (100)	
		Ianthella	33	8 (24)		25 (76)	

TABLE 1.7
Phylum Cnidaria

Class	Subclass	Order	Suborder	Family	Genus	Total	Isoprenoid	Acetogenin	Amino Acid	Shikimate	Nucleic Acid	Carbohydrate
						1499	1236 (82)	142 (9)	112 (7)	4 (<1)	4 (<1)	1 (<1)
Hydrozoa												
		Actinulida				0						
		Athecata				37	8 (22)	18 (49)	11 (30)			
				Bougainvillidae		26	7 (27)	16 (62)	3 (12)			
				Eudendriidae	Garveia	16		16 (100)				
				Hydractiniidae	Eudendrium	16	6 (100)	16 (100)				
					Hydractinia	6	6 (100)		1 (100)			
				Polyorchidae		1			1 (100)			
				Tubulariidae	Polyorchis	1			1 (100)			
				Stylasteridae	Tubularia	1			1 (100)			
					Allopora	1	1 (100)		1 (100)			
						1	1 (100)					
		Limnomedusae				0						
		Narcomedusae				0						
		Pteromedusae				0						
		Siphonophora				0						
		Thecata				11	1 (9)	2 (18)	8 (73)			
				Aequoreidae	Aequoreae	2			2 (100)			
				Plumariidae		2			2 (100)			
					Aglaophenia	3			3 (100)			
				Sertulariidae		3			3 (100)			
					Abietinaria	6	1 (17)	2 (33)	3 (50)			
						2	2 (100)	2 (100)				

Class	Order	Family	Genus							
	Trachymedusae		Hydrallmania	1	1 (100)					
			Tridentata	3			3 (100)			
Scyphozoa				0						
Cubozoa				0						
Anthozoa				1462	1228 (84)	124 (8)	101 (7)	4 (<1)	4 (<1)	1 (<1)
	Cerianthipatharia			13	7 (54)	6 (46)				
	Anthipatharia	Antipathidae		13	7 (54)	6 (46)				
			Antipathes	7	7 (100)					
			Leiopathes	6		6 (100)				
	Ceriantharia			0						
Octocorallia				1298	1192 (92)	89 (7)	13 (1)	1 (<1)	2 (<1)	1 (<1)
	Stolonifera			73	56 (77)	17 (23)				
		Clavulariidae		71	56 (77)	17 (23)				
			Clavularia	60	43 (72)	17 (28)				
			Pachyclavularia	4	4 (100)					
			Sarcodictyon	7	7 (100)					
		Tubiporidae		2	2 (100)					
			Tubipora	2	2 (100)					
	Telestacea			24	5 (21)	19 (79)				
		Telestidae		24	5 (21)	19 (79)				
			Coelogorgia	3	3 (100)					
			Telesto	21	2 (10)	19 (90)				
	Helioporacea			11	10 (91)		1 (9)			
		Helioporidae		11	10 (91)		1 (9)			
			Heliopora	11	10 (91)		1 (9)			
	Alcyonacea			651	626 (96)	18 (3)	5 (1)			1 (<1)
		unknown		20	19 (95)	1 (5)				
			Maasella	3	3 (100)					
			Sclerophytum	17	16 (94)	1 (6)				
		Alcyoniidae		426	404 (95)	16 (4)	5 (1)			1 (<1)
			Alcyonium	39	39 (100)					
			Cladiella	15	15 (100)					
			Eleutherobia	3	3 (100)					
			Gersemia	15	15 (100)					

TABLE 1.7 (CONTINUED)
Phylum Cnidaria

Class	Subclass	Order	Suborder	Family	Genus	Total	Isoprenoid	Acetogenin	Amino Acid	Shikimate	Nucleic Acid	Carbohydrate
					Lithophyton	21	19 (90)	2 (10)				
					Lobophytum	79	71 (90)	7 (9)	1 (1)			
					Minabea	9	9 (100)					
					Paraerythropodium	10	10 (100)					
					Sarcophyton	91	89 (98)	2 (2)				
					Sinularia	142	134 (93)	5 (3)	4 (3)			1 (1)
				Nephtheidae		94	93 (99)			1 (1)		
					Capnella	21	21 (100)					
					Lemnalia	32	32 (100)					
					Nephthea	33	32 (97)			1 (3)		
					Paralemnia	8	8 (100)					
				Nidaliidae		1	1 (100)					
					Pieterfaurea	1	1 (100)					
				Xeniidae		110	109 (99)	1 (1)				
					Anthelia	10	10 (100)					
					Cespitularia	8	8 (100)					
					Efflatounaria	3	3 (100)					
					Heteroxenia	1		1 (100)				
					Xenia	88	88 (100)					
						492	457 (93)	29 (6)	4 (1)		2 (<1)	
		Gorgonacea		Anthothelidae		11	11 (100)					
					Erythropodium	11	11 (100)					
				Briareidae		136	135 (99)	1 (1)				
					Solenopodium	36	36 (100)					
					Briareum	100	99 (99)	1 (1)				
				Coralliidae		5	5 (100)					
					Corallium	5	5 (100)					

Family	Genus					
Melithaeidae		6	6 (100)			
	Melithaea	2	2 (100)			
	Acabaria	4	4 (100)			
Paragorgiidae		2	2 (100)			
	Paragorgia	2	2 (100)			
Subergorgiidae		4	4 (100)			
	Subergorgia	4	4 (100)			
Acanthogorgiidae		13	13 (100)			
	Acalycigorgia	13	13 (100)			
Ellisellidae		26	26 (100)			
	Ctenocella	4	4 (100)			
	Junceella	22	22 (100)			
Gorgoniidae		100	89 (89)	9 (9)	2 (2)	
	Gorgonia	2	2 (100)			
	Lasianthaea	2	2 (100)			
	Leptogorgia	8	8 (100)			
	Lophogorgia	4	4 (100)			
	Pacifigorgia	2	2 (100)			
	Pseudopterogorgia	73	71 (97)		2 (3)	
	Pterogorgia	9		9 (100)		
Isididae		15	15 (100)			
	Isis	15	15 (100)			
Paramuriceidae		21	16 (76)	3 (14)	2 (10)	
	Calicogorgia	9	6 (67)	3 (33)		
	Echinogorgia	4	4 (100)			
	Muricella	5	5 (100)			
	Paramuricea	1	1 (100)			
	Villogorgia	2			2 (100)	
Primnoidae		4	4 (100)			
	Primnoides	4	4 (100)			
Plexauridae		149	131 (88)	16 (11)		2 (1)
	Anthoplexaura	2	2 (100)			
	Astrogorgia	3	3 (100)			
	Eunicea	87	87 (100)			
	Eunicella	15	13 (87)			2 (13)
	Euplexaura	6	2 (33)	4 (67)		

TABLE 1.7 (CONTINUED)
Phylum Cnidaria

Class	Subclass	Order	Suborder	Family	Genus	Total	Isoprenoid	Acetogenin	Amino Acid	Shikimate	Nucleic Acid	Carbohydrate
					Muricea	5	5 (100)					
					Plexaura	18	7 (39)	11 (61)				
					Plexaureides	1	1 (100)					
					Plexaurella	2	2 (100)					
					Pseudoplexaura	8	7 (88)	1 (12)				
					Pseudothesia	1	1 (100)					
					Rumphella	1	1 (100)					
		Pennatulacea				47	38 (81)	6 (13)	3 (6)			
			Sessiflorae			11	9 (82)		2 (18)			
				Funiculinidae		6	6 (100)					
					Funiculina	6	6 (100)					
				Renillidae		5	3 (60)		2 (40)			
					Renilla	5	3 (60)		2 (40)			
			Subsessiflorae			36	29 (81)	6 (17)	1 (3)			
				Pennatulidae		8	8 (100)					
					Ptilosarcus	8	8 (100)					
				Pteroeidae		4	4 (100)					
					Pteroeides	4	4 (100)					
				Veretillidae		15	11 (73)	3 (20)	1 (7)			
					Cavernulina	3	3 (100)					
					Cavernularia	1			1 (100)			
					Veretillum	8	8 (100)					
					Lituaria	3		3 (100)				
				Virgulariidae		9	6 (67)	3 (33)				
					Scytalium	1	1 (100)					
					Stylatula	3	3 (100)					
					Virgularia	5	2 (40)	3 (60)				

Taxon				n					
Zoantharia	Actiniaria			151	29 (19)	29 (19)	88 (58)	3 (2)	2 (1)
		Actiniidae		43	13 (30)	2 (5)	26 (60)		2 (5)
			Actinia	36	13 (36)	2 (6)	20 (56)		1 (3)
			Anemonia	5	4 (80)		1 (20)		
			Anthopleura	5		1 (20)	4 (80)		
			Bunodosoma	24	9 (38)		15 (62)		
			Condylactis	1		1 (100)			
		Hormathidae		3			3 (100)		
			Calliactis	3			3 (100)		
		Metridiidae		1			1 (100)		
			Metridium	1			1 (100)		
		Sagartiidae		1					1 (100)
			Sagartia	1					1 (100)
		Stichodactylidae		2			2 (100)		
			Radianthus	2			2 (100)		
	Corallimorpharia			0					
	Zoanthidea			61	7 (12)	8 (13)	46 (75)		
		unknown		4	1 (25)		3 (75)		
			Gerardia	4	1 (25)		3 (75)		
		Zoanthidae		40	5 (13)	8 (20)	27 (68)		
			Palythoa	19		7 (37)	12 (63)		
			Zoanthus	21	5 (24)	1 (5)	15 (71)		
		Epizoanthidae		4			4 (100)		
			Epizoanthus	4			4 (100)		
		Parazoanthidae		13	1 (8)		12 (92)		
			Parazoanthus	13	1 (8)		12 (92)		
	Scleractinia			47	9 (19)	19 (40)	16 (34)	3 (6)	
		Astrocoeniina		12		10 (83)	2 (17)		
		Acroporidae		10		10 (100)			
			Acropora	4		4 (100)			
			Montipora	6		6 (100)			
		Pocilloporidae		2			2 (100)		
			Pocillopora	1					1 (100)
			Stylophora	1					1 (100)
		Fungiina		0					

TABLE 1.7 (CONTINUED)
Phylum Cnidaria

Class	Subclass	Order	Suborder	Family	Genus	Total	Isoprenoid	Acetogenin	Amino Acid	Shikimate	Nucleic Acid	Carbohydrate
			Faviina			6	3 (50)	2 (33)		1 (17)		
				Faviidae		4	3 (75)			1 (25)		
					Echinopora	4	3 (75)			1 (25)		
				Pectiniidae		2		2 (100)				
					Pectinia	2		2 (100)				
			Caryophylliina			6	6 (100)					
				Caryophylliidae		6	6 (100)					
					Deltocyathus	6	6 (100)					
			Dendrophyllina			23		7 (30)	14 (61)	2 (9)		
				Dendrophylliidae		23		7 (30)	14 (61)	2 (9)		
					Astroides	5			5 (100)			
					Dendrophyllia	1			1 (100)			
					Tubastrea	17		7 (41)	8 (47)	2 (12)		

TABLE 1.8
Phylum Mollusca

Class	Subclass	Order	Family	Genus	Total	Isoprenoid	Acetogenin	Amino Acid	Shikimate	Nucleic Acid	Carbohydrate
					575	310 (54)	138 (24)	104 (18)	13 (2)	2 (<1)	8 (1)
Caudofoveata					0						
Aplacophora					0						
Gastropoda	Prosobranchia				497	275 (55)	129 (26)	76 (15)	12 (2)	2 (<1)	3 (<1)
		Archaeogastropoda			72	19 (26)	3 (4)	37 (51)	11 (15)		2 (3)
			Trochidae		20		1 (5)	8 (40)	11 (55)		
				Monodonta	3			3 (100)			
					3			3 (100)			
			Turbinidae		4			4 (100)			
				Turbo	4			4 (100)			
			Neritidae		13		1 (8)	1 (8)	11 (84)		
				Nerita	13		1 (8)	1 (8)	11 (84)		
			Helicinidae		0						
		Mesogastropoda			11	4 (36)		7 (64)			
			Lamellariidae		4			4 (100)			
				Chelynotus	4			4 (100)			
			Planaxidae		4	4 (100)					
				Planaxis	4	4 (100)					
			Naticidae		3			3 (100)			
				Lamellaria	3			3 (100)			
		Neogastropoda			39	15 (38)	2 (5)	22 (56)			
			Buccinidae		6	1 (17)	2 (33)	3 (50)			
				Buccinium	2	1 (50)		1 (50)			
				Babylonia	4		2 (50)	2 (50)			
			Fasciolariidae		2	2 (100)					
				Fusinus	2	2 (100)					

TABLE 1.8 (CONTINUED)
Phylum Mollusca

Class	Subclass	Order	Family	Genus	Total	Isoprenoid	Acetogenin	Amino Acid	Shikimate	Nucleic Acid	Carbohydrate
			Muricidae		16	12 (75)		4 (25)			
				Ceratostoma	12	12 (100)					
				Murex	2			2 (100)			
				Nucella	2			2 (100)			
			Conidae		15			15 (100)			
				Conus	15			15 (100)			
			unknown	Kellitia	2						2 (100)
	Opisthobranchia				378	241 (64)	95 (25)	38 (10)	1 (<1)	2 (1)	1 (<1)
		Acochlidioidea			0						
		Cephalaspidea			30		30 (100)				
			Aglajidae		3		3 (100)				
				Aglaja	3		3 (100)				
			Haminoeidae		10		10 (100)				
				Haminoea	10		10 (100)				
			Scaphandridae		2		2 (100)				
				Scaphander	2		2 (100)				
			Smaragdinellidae		2		2 (100)				
				Smaragdinella	2		2 (100)				
			unknown		13		13 (100)				
				Bulla	4		4 (100)				
				Navanax	9		9 (100)				
		Ruuncinoidea			0						
		Sacoglossa			22	3 (14)	17 (77)	2 (9)			
			Elysiidae		9	3 (33)	4 (44)	2 (22)			
				Oxynoe	2	2 (100)					
				Elysia	7	1 (14)	4 (57)	2 (29)			
			unknown		13		13 (100)				
				Tridachiella	1		1 (100)				

Order	Family	Genus	N						
		Placobranchus	1						1 (100)
		Tridachia	11		11 (100)				
Anaspidea			158	99 (62)	33 (21)	25 (16)	1 (<1)		
	Aplysiidae		157	99 (63)	32 (20)	25 (16)	1 (<1)		
		Aplysia	116	86 (74)	17 (15)	12 (10)	1 (1)		
		Bursatella	2		1 (50)	1 (50)			
		Dolabella	35	13 (37)	10 (29)	12 (34)			
		Stylocheilus	4		4 (100)				
	Dolabriferidae		1		1 (100)				
		Dolabrifera	1		1 (100)				
Thecosomata			1						1 (100)
Gymnosomata			0						
Notaspidea			8	2 (25)	6 (75)				
	Pleurobranchaeidae		5	2 (40)	3 (60)				
		Pleurobranchus	3		3 (100)				
		Pleurobranchaea	2	2 (100)					
	Umbraculidae		3		3 (100)				
		Umbraculum	3		3 (100)				
Nudibranchia			159	137 (86)	9 (6)	11 (7)		2 (1)	
	Archidorididae		6	6 (100)					
		Archidoris	6	6 (100)					
	Arminidae		14	9 (64)	5 (36)				
		Armina	8	8 (100)					
		Hexabranchus	6	1 (17)	5 (83)				
	Cadlinidae		15	15 (100)					
		Cadlina	15	15 (100)					
	Chromodorididae		46	46 (100)					
		Chromodoris	38	38 (100)					
		Hypselodoris	8	8 (100)					
	Dendrodorididae		8	8 (100)					
		Dendrodoris	4	4 (100)					
		Doriopsilla	4	4 (100)					
	Discodorididae		9	3 (33)	4 (44)			2 (22)	
		Anisodoris	2	1 (50)				1 (50)	
		Diaulula	7	2 (29)	4 (57)			1 (14)	

TABLE 1.8 (CONTINUED)
Phylum Mollusca

Class	Subclass	Order	Family	Genus	Total	Isoprenoid	Acetogenin	Amino Acid	Shikimate	Nucleic Acid	Carbohydrate
			Dorididae	*Doris*	10	10 (100)					
				Austrodoris	4	4 (100)					
					6	6 (100)					
			Janolidae	*Janolus*	1			1 (100)			
			Kentrodorididae		1			1 (100)			
				Jorunna	4			4 (100)			
			Notodorididae		4			4 (100)			
				Notodoris	4			4 (100)			
			Onchidorididae	*Acanthodoris*	4	4 (100)					
				Adalaria	3	3 (100)					
					1	1 (100)					
			Phyllidiidae	*Phyllidiopsis*	12	12 (100)					
				Phyllidia	2	2 (100)					
					10	10 (100)					
			Tethyidae	*Melibe*	6	6 (100)					
				Tethys	2	2 (100)					
					4	4 (100)					
			Triophidae	*Triopha*	2	2 (100)					
					2	2 (100)					
			Tritoniidae	*Tochuina*	4	4 (100)					
			unknown	*Leminda*	14	12 (86)		2 (14)			
				Phestilla	4	4 (100)					
				Roboastra	2			2 (100)			
				Tambja	4	4 (100)					
					4	4 (100)					

Class	Subclass	Order	Family	Genus	n					
Pulmonata					47	15 (32)	31 (66)	1 (2)	1 (2)	
	Archaeopulmonata				0					
		Basommatophora			32	2 (6)	30 (94)			
			Siphonariidae		30		30 (100)			
				Siphonaria	30		30 (100)			
			Trumusculidae		2	2 (100)				
				Trimusculus	2	2 (100)				
		Stylommatophora			15	13 (87)	1 (7)	1 (7)		
			Onchidiidae		15	13 (87)	1 (7)	1 (7)		
				Onchidella	1	1 (100)				
				Onchidium	13	12 (92)		1 (8)		
				Peronia	1		1 (100)			
Bivalvia					65	35 (54)	9 (14)	15 (23)	1 (2)	5 (8)
	Protobranchia				0					
	Lamellibranchia				65	35 (54)	9 (14)	15 (23)	1 (2)	5 (8)
		Mytiloida			47	33 (70)	7 (15)	6 (13)		
			Mytilidae		21	15 (71)	2 (10)	3 (14)	1 (5)	
				Mytilus	20	15 (75)	1 (5)	3 (15)	1 (5)	
				Perna	1		1 (100)			
			Ostreidae		5	2 (40)	3 (60)			
				Crossostrea	4	1 (25)	3 (75)			
				Tiostrea	1	1 (100)				
			Pinnidae		3	3 (100)				
				Pinna	3	3 (100)				
			Pectinidae		17	13 (76)	2 (12)	2 (12)		
				Chlamys	1		1 (100)	1 (100)		
				Patinopecten	11	10 (91)		1 (9)		
				Pectinia	2		2 (100)			
				Placopecten	3	3 (100)				
			unknown		1			1 (100)		
				Saxidomus	1			1 (100)		
	Paleoheterodonta				0					
		Veneroida			17	1 (6)	2 (12)	9 (53)		5 (29)
			Tridacnidae		6			1 (17)		5 (83)
				Tridacna	6			1 (17)		5 (83)

TABLE 1.8 (CONTINUED)
Phylum Mollusca

Class	Subclass	Order	Family	Genus	Total	Isoprenoid	Acetogenin	Amino Acid	Shikimate	Nucleic Acid	Carbohydrate
			Mactridae		1	1 (100)					
				Mactra	1	1 (100)					
			Veneridae		10		2 (20)	8 (80)			
				Macrocallista	2			2 (100)			
				Ruditapes	6			6 (100)			
				Austrovenus	2		2 (100)				
		Myoida			1	1 (100)					
			Corbulidae		1	1 (100)					
				Mya	1	1 (100)					
	Anomalodesmata				0						
Scaphopoda					0						
Cephalopoda					13			13 (100)			
	Nautiloidea				0						
	Coleoidea				13			13 (100)			
		Sepioida			2			2 (100)			
				Sepia	2			2 (100)			
		Teuthoida			5			5 (100)			
				Loligo	5			5 (100)			
			Loliginidae		5			5 (100)			
		Octopoda			6			6 (100)			
			Octopodidae		6			6 (100)			
				Octopus	6			6 (100)			
		Vampyromorpha			0						

TABLE 1.9
Phylum Echinodermata

Subphylum	Class	Subclass	Superorder	Order	Family	Genus	Total	Isoprenoid	Acetogenin	Amino Acid	Shikimate	Nucleic Acid	Carbohydrate
Pelmatozoa	Crinoidea						464	317 (68)	73 (16)	25 (5)	31 (7)	0	18 (4)
				Isocrinida			29	2 (7)	27 (93)				
				Comatulida			29	2 (7)	27 (93)				
							0						
					Comasteridae		29	2 (7)	27 (93)				
						Comatula	5		5 (100)				
					Ptilometridae		5		5 (100)				
						Ptilometra	5		5 (100)				
					unknown		19	2 (11)	17 (89)				
						Comantheria	1		1 (100)				
						Comanthus	10		10 (100)				
						Gymnocrinus	2		2 (100)				
						Lamprometra	6	2 (33)	4 (67)				
Eleutherozoa							435	315 (72)	46 (11)	25 (6)	31 (7)		18 (4)
	Concentricycloidea						0						
	Asteroidea						296	222 (74)	33 (11)	10 (3)	14 (5)		17 (6)
				Platyasterida			0						
				Forcipulatida			87	68 (78)	8 (9)	3 (3)			8 (9)
					Asteriidae		86	68 (79)	8 (9)	2 (2)			8 (9)
						Aphelasterias	4	4 (100)					
						Asterias	53	37 (70)	6 (11)	2 (4)			8 (15)
						Coscinasterias	4	4 (100)					
						Cosmasterias	3	3 (100)					
						Dismolasterias	3	3 (100)					
						Distolasterias	2		2 (100)				
						Marthasterias	8	8 (100)					
						Pisaster	6	6 (100)					
						Pycnopodia	3	3 (100)					
					unknown								
						Evasterias	1			1 (100)			

TABLE 1.9 (CONTINUED)
Phylum Echinodermata

Subphylum	Class	Subclass	Superorder	Order	Family	Genus	Total	Isoprenoid	Acetogenin	Amino Acid	Shikimate	Nucleic Acid	Carbohydrate
				Paxillosida			46	42 (91)	3 (7)	1 (2)			
					Astropectinidae		28	24 (86)	3 (11)	1 (4)			
						Astropecten	12	9 (75)	3 (25)				
						Luidia	15	15 (94)		1 (6)			
					Porcellanasteridae		18	18 (100)					
						Clenodiscus	3	3 (100)					
						Styracaster	15	15 (100)					
				Valvatida			147	109 (74)	22 (15)	3 (2)	4 (3)		9 (6)
					Acanthasteridae		30	12 (40)	6 (20)		3 (10)		9 (30)
						Acanthaster	30	12 (40)	6 (20)		3 (10)		9 (30)
					Archasteridae		9	9 (100)					
						Archaster	9	9 (100)					
					Asterinidae		13	11 (85)		1 (8)	1 (8)		
						Asterina	6	4 (67)		1 (17)	1 (17)		
						Patiria	2	2 (100)					
						Tremaster	5	5 (100)					
					Asteropseidae		3	2 (67)		1 (33)			
						Dermasterias	3	2 (67)		1 (33)			
					Echinasteridae		36	30 (83)	6 (17)				
						Echinaster	11	6 (55)	5 (46)				
						Henricia	25	24 (96)	1 (4)				
					Goniasteridae		5	5 (100)					
						Rosaster	3	3 (100)					
						Sphaerodiscus	2	2 (100)					
					Mithrodiidae		1	1 (100)					
						Thromidia	1	1 (100)					
					Ophidiasteridae		23	13 (57)	10 (43)				
						Fromia	3	1 (33)	2 (67)				
						Gomophia	3	3 (100)					
						Hacelia	8	8 (100)					

Class	Order	Family	Genus	N	n (%)	additional
			Linckia	6	6 (100)	
			Ophidiaster	1	1 (100)	2 (100)
			Celerina	2	2 (100)	1 (4)
		Oreasteridae	Culcita	24	23 (96)	
			Halityle	3	3 (100)	
			Oreaster	1	1 (100)	
			Pentaceraster	13	13 (100)	1 (33)
			Poraster	3	2 (67)	
			Protoreaster	1	1 (100)	
		Solasteridae	Crossaster	3	3 (100)	
			Solaster	3	3 (100)	
	Spinulosida	unknown	Pteraster	2	2 (100)	
			Euretaster	1	1 (100)	
			Perknaster	15	3 (20)	2 (13); 10 (67)
			Neosmilaster	4	3 (75)	1 (25); 1 (100)
		unknown		2	2 (100)	
				1	1 (100)	
Ophiuroidea				24	24 (100)	
	Ophiurida			24	24 (100)	
		Ophiodermatidae	Ophioderma	9	9 (100)	
				9	9 (100)	
		Ophiuridae	Ophioplocus	2	2 (100)	
			Stegophiura	1	1 (100)	
		Ophiocomidae	Ophiocomina	1	1 (100)	
				1	1 (100)	
		unknown		12	12 (100)	
			Ophiomastix	1	1 (100)	
			Ophionereis	1	1 (100)	
			Ophionotus	3	3 (100)	
			Ophiopholis	6	6 (100)	
			Ophiura	1	1 (100)	
	Phrynophiurida			0		
Echinoidea				45	3 (7)	12 (27); 14 (31); 16 (36)
	Perischoechinoidea			0		
	Euechinoidea			45	3 (7)	12 (27); 14 (31); 16 (36)
		Diadematacea		5	5 (100)	
			Diadema	5	5 (100)	

TABLE 1.9 (CONTINUED)
Phylum Echinodermata

Subphylum	Class	Subclass	Superorder	Order	Family	Genus	Total	Isoprenoid	Acetogenin	Amino Acid	Shikimate	Nucleic Acid	Carbohydrate
			Atelostomata				0						
			Gnathostomata	Holectypoida			1			1 (100)			
				Clypeasteroida			0						
					unknown		1			1 (100)			
						Clypeaster	1			1 (100)			
			Echinacea	Echinoida			39	3 (8)	12 (31)	13 (33)	11 (28)		
							36	3 (8)	9 (25)	13 (36)	11 (31)		
					Toxopneustidae		5		3 (60)	1 (20)	1 (20)		
						Tripneustes	4		3 (75)	1 (25)			
						Sphaerechinus	1				1 (100)		
					Strongylocentrotidae		20	3 (15)	3 (15)	5 (25)	9 (45)		
						Strongylocentrotus	8				8 (100)		
						Paracentrotus	11	3 (27)	3 (27)	5 (46)			
						Pseudocentrotus	1				1 (100)		
					unknown		11		3 (27)	7 (64)	1 (9)		
						Hemicentrotus	5		1 (20)	4 (80)			
						Echinomerta	2		2 (100)				
						Anthocidaris	4			3 (75)	1 (25)		

Class	Subclass	Order	Family	Genus	N	n (%)				
		Arbacioida			3	3 (100)				
			unknown	Arbacia	3	3 (100)				
Holothuroidea					70	66 (94)	1 (1)	1 (1)	1 (1)	1 (1)
	Aspidochirotacea				33	31 (94)			1 (3)	1 (3)
		Aspidochirotida			33	31 (94)			1 (3)	1 (3)
			Holothuriidae		25	24 (96)			1 (4)	
				Bohadschia	9	9 (100)				
				Holothuria	16	15 (94)			1 (6)	
			Stichopodidae		6	5 (83)				1 (17)
				Stichopus	6	5 (83)				1 (17)
			unknown	Actinopyga	2	2 (100)				
	Dendrochirotacea				33	31 (94)	1 (3)	1 (3)		
		Dendrochirotida			33	31 (94)	1 (3)	1 (3)		
			Cucumariidae		32	30 (94)	1 (3)	1 (3)		
				Cucumaria	25	23 (92)	1 (4)	1 (4)		
				Eupentacta	3	3 (100)				
				Pentacta	1	1 (100)				
				Psolus	1	1 (100)				
				Thelenota	2	2 (100)				
			unknown	Neothyone	1	1 (100)				
	Apodacea	Apodida	Synaptidae		4	4 (100)				
				Synapta	4	4 (100)				

TABLE 1.10
Phylum Ectoprocta

Class	Order	Family	Genus	Total	Isoprenoid	Acetogenin	Amino Acid	Shikimate	Nucleic Acid	Carbohydrate
Gymnolaemata	Cheilostomata			102	5 (5)	29 (28)	65 (64)	0	3 (3)	0
				101	5 (5)	29 (29)	65 (65)		2 (2)	
				74	5 (7)	28 (38)	39 (53)		2 (3)	
		Bugulidae		28	2 (7)	26 (93)				
			Bugula	23	1 (4)	22 (96)				
			Dendrobeania	1	1 (100)					
			Sessibugula	4		4 (100)				
		Catenicellidae		4			4 (100)			
			Catenicella	1			1 (100)			
			Costaticella	1			1 (100)			
			Cribricellina	2			2 (100)			
		Flustridae		30			30 (100)			
			Chartella	7			7 (100)			
			Flustra	19			19 (100)			
			Securiflustra	4			4 (100)			

Class/Order	Family	Genus	n				
	Membraniporidae		2			2 (100)	
		Membranipora	2			2 (100)	
	Myriaporidae		3				3 (100)
		Myriapora	3				3 (100)
	Phidoloporidae		2	2 (100)			
		Phidolopora	2	2 (100)			
	Schizoporellidae		2		2 (100)		
		Dakaira	2		2 (100)		
	Vesiculariidae		3			3 (100)	
		Zoobotryon	3			3 (100)	
Ctenostomata			27		1 (4)	26 (96)	
	Alcyonidiidae		1			1 (100)	
		Alcyonidium	1			1 (100)	
	Vesiculariidae		26		1 (4)	25 (96)	
		Amathia	26		1 (4)	25 (96)	
Stenolaemata			1	1 (100)			
Cyclostomata			1	1 (100)			
	Diaperoeciidae		1	1 (100)			
		Diaperoecia	1	1 (100)			

TABLE 1.11
Subphylum Urochordata

Class	Order	Suborder	Family	Genus	Total	Isoprenoid	Acetogenin	Amino Acid	Shikimate	Nucleic Acid	Carbohydrate
					452	73 (16)	71 (16)	299 (66)	6 (1)	3 (<1)	0
Appendicularia					0						
Thaliacea					0						
Sorberacea					0						
Ascidiacea					452	73 (16)	71 (16)	299 (66)	6 (1)	3 (<1)	
	Enterogona				404	71 (18)	71 (18)	257 (64)	2 (1)	3 (1)	
		Aplousobranchiata			359	55 (15)	70 (19)	229 (64)	2 (1)	3 (1)	
			Polyclinidae		98	44 (45)	35 (36)	15 (15)	1 (1)	3 (3)	
				Aplidiopsis	1			1 (100)			
				Aplidium	33	10 (30)	17 (52)	3 (9)		3 (9)	
				Polyclinum	1				1 (100)		
				Pseudodistoma	15		10 (67)	5 (33)			
				Ritterella	46	32 (70)	8 (17)	6 (13)			
				Sidnyum	1	1 (100)					
				Synoicum	1	1 (100)					
			Didemnidae		151	5 (3)	28 (19)	118 (78)			
				Didemnum	55	3 (5)	16 (29)	36 (65)			
				Diplosoma	6		1 (17)	5 (83)			
				Leptoclinides	7			7 (100)			
				Lissoclinum	57	2 (4)	8 (14)	47 (82)			
				Polysyncraton	1		1 (100)				
				Trididemnum	25		2 (8)	23 (92)			
			Polycitoridae		110	6 (5)	7 (6)	96 (87)	1 (1)		
				Amaroucium	5	5 (100)					
				Atapozoa	2		2 (100)				
				Clavelina	21			21 (100)			

		Genus	N				
Phlebobranchiata		*Cystodytes*	48		4 (8)	43 (90)	1 (100)
		Eudistoma	8		1 (13)	7 (88)	
		Nephteis	1			1 (100)	
		Polycitor	1			1 (100)	
		Polycitorella	24	1 (4)		23 (96)	
	Ascidiidae	*Ascidia*	45	16 (36)	1 (4)	28 (62)	
		Phallusia	25	14 (56)	1 (4)	10 (40)	
	Cionidae	*Ciona*	20	12 (60)	1 (5)	7 (35)	
		Diazona	5	2 (40)		3 (60)	
		Rhopalaea	5	2 (40)		3 (60)	
			11			9 (82)	
			5			3 (60)	
			4			4 (100)	
			2			2 (100)	
	Perophoridae	*Ecteinascidia*	9			9 (100)	
			9			9 (100)	
Pleurogona			48	2 (4)		42 (88)	4 (8)
Stolidobranchiata			48	2 (4)		42 (88)	4 (8)
	Molgulidae	*Molgula*	2			2 (100)	
			2			2 (100)	
	Styellidae		43	1 (2)		38 (88)	4 (9)
		Amphicarpa	1			1 (100)	
		Botryllus	6			4 (67)	2 (33)
		Botrylloides	2			2 (100)	
		Cnemidocarpa	1			1 (100)	
		Dendrodoa	6			6 (100)	
		Eusynstyela	1			1 (100)	
		Polyandrocarpa	8			8 (100)	
		Polycarpa	17			15 (88)	2 (12)
		Styela	1	1 (100)			
	Pyuridae	*Microcosmus*	3	1 (33)		2 (67)	
		Pyura	2		1 (100)	2 (100)	

2 Ecological Perspectives on Marine Natural Product Biosynthesis

Mary J. Garson

CONTENTS

I. INTRODUCTION

Over the last four decades, the isolation and structure elucidation of biomedically important marine natural products has been a fruitful area of research for organic chemists. Compilations of marine structures reveal that by 1993 nearly 7000 different marine structures had been documented; these metabolites represent all the major structural classes of natural products, including terpenes, alkaloids, polyketides, peptides, and shikimate-derived metabolites, as well as compounds of mixed biogenesis.[1,2] Some phylogenetic trends have been recognized; for example, the majority of coelenterate metabolites are terpenoid in origin. Other phyla display broad structural diversity; metabolites of the polyketide, terpene, and alkaloid families, as well as compounds of mixed biosynthesis, have all been isolated from sponges (see Chapter 1, this volume).[2,3]

The extraordinary biomedical potential of marine natural products is well documented;[4–6] recent reviews provide details of their cytotoxic,[7,8] antiviral,[9] antiparasitic,[10] and antimalarial[11] activities. In addition, numerous marine metabolites show useful antimicrobial activity. Although some of these pharmacological effects are serendipitous, others may relate to the natural function of the metabolite *in situ*. It has long been assumed that bioactive secondary metabolites enhance the survival of marine organisms by protecting them from challenges such as microbial infection or predation, or by enabling them to compete for space, nutrients, and resources in the complex and competitive marine environment. Evolutionary pressures may have compelled marine organisms to produce antimicrobial or toxic substances causing growth inhibition or mortality in competitors. Additional biological roles that have been proposed for marine natural products include pheromones involved in chemical communication (molluscs) or in reproductive processes (molluscs, corals, algae); however, the organism pays a price for this ecological advantage. The biochemical pathways that generate marine natural products are complex, and significant amounts of metabolic energy are likely expended in their production that could otherwise have been directed to growth or reproduction (see Chapter 9, this volume).

This chapter addresses the hypothesis that marine organisms have evolved novel biosynthetic strategies in response to ecological pressures, and compares the current state of knowledge on the biosynthesis of marine metabolites with the ecological roles that have been evaluated for them. Benthic invertebrates and molluscs provide the major focus of the review, since these have been the most intensively studied by biosynthetic chemists and by ecologists; however, microorganisms and algae are also covered. Also considered are some ecological studies that have provided details of intra- or interspecific variations in metabolite content of marine organisms, and the genetic, environmental, or temporal factors responsible for this variation.

The reader is referred to articles by Barrow,[12] Garson,[13,14] and Moore[15] for background information on the field of marine secondary metabolite biosynthesis. Early work in this field used radiochemical tracers such as ^{14}C or ^{3}H; recently, studies with ^{13}C have proved popular because of the ease of detection of this stable isotope by nuclear magnetic resonance (NMR). In particular, our understanding of marine microbial biosynthesis and of *de novo* biosynthesis in marine molluscs has progressed significantly as a consequence of the use of ^{13}C-labeled precursors.

II. MARINE MICROORGANISMS

Marine microorganisms of interest to the natural products chemist include bacteria, dinoflagellates, protists, and cyanobacteria. Some of these microorganisms are free-living, while others are symbiotic in character, or found living in close association with algae or marine invertebrates.[16,17] The isolation of metabolites from marine microorganisms usually requires culturing of the isolated strains on nutrient-specific media, and is often guided by screening for antibiotic activity or cytotoxicity. Numerous chemical studies have now been reported on metabolites from cultured marine microorganisms.[1,16,18]

A. MARINE BACTERIA

The biosynthetic studies undertaken to date on microbial marine natural products well illustrate the diversity of metabolic pathways encountered in cultured marine bacteria. Examples include brominated alkaloids such as pentabromopseudiline (Structure 2.1),[19] polyketide or mixed polyketide metabolites such as oncorhyncholide (Structure 2.2),[20] aplasmomycin (Structure 2.3),[21] and andrimid (Structure 2.4),[22] or the cyclic depsipeptide salinamide A (Structure 2.5).[23] As researchers continue to define more specific culture media and a wider range of marine bacteria from diverse habitats are successfully placed into culture, the true biosynthetic potential of these prolific and adaptable microorganisms can be explored.

There has been interesting speculation on the ecological roles that marine bacterial metabolites play *in situ*.[17,24] The antibiotic activity of marine microbial metabolites is not surprising given the complex and mixed nature of marine microbial communities. Many species of marine bacteria are intimately associated with specific macroorganisms such as sponges, ascidians, or seaweeds and may provide protection to their host against predation or infection in exchange for a nutrient-rich and protected growth habitat. Salinamide A (Structure 2.5), for example, was isolated from a bacterium growing on the surface of a jellyfish,[25] while eggs of the shrimp *Palaemon macrodactylus* are protected from fungal attack by the bacterium *Alteromonas* sp., which produces isatin (Structure 2.6).[26] The adaptation of marine bacteria in response to diverse and often challenging marine habitats may have led to their rich secondary metabolism. There has been speculation about the indirect effects of marine bacterial products on ecosystem structure and nutrient recycling.[27]

Small organic molecules such as acylated homoserine lactones have been implicated in gene expression in marine bacteria and in signaling or sensing processes between bacterial communities.[28] Structurally related secondary metabolites, e.g., Structures 2.7 and 2.8, have been detected in marine bacterial isolates,[29,30] where they may be involved in signaling or alternately interfere with this process in competing bacterial strains.

A number of bioactive metabolites from marine bacteria have previously been ascribed to other marine species, either because of a close association between bacterium and host, or because the metabolic products accumulate as a consequence of passage through the food chain. The polyketide swinholide A (Structure 2.9) and the cyclic peptide theopalauamide (Structure 2.10), which were first isolated from the lithistid sponge *Theonella swinhoei*, are localized in unicellular heterotrophic bacteria and in filamentous bacteria, respectively.[31,32] Saxitoxin (Structure 2.11) and neosaxitoxin (Structure 2.12), which were originally thought to be metabolites from dinoflagellates or cyano-bacteria, are now known to be of bacterial origin. These two metabolites frequently accumulate in filter-feeding shellfish, thereby creating public health problems. Their biosynthesis from acetate, arginine, and methionine has been well reviewed by Shimizu.[33] Another marine bacterial metabolite suggested to be amino-acid derived is tetrodotoxin (Structure 2.13), isolated originally from a variety of sources including fish, octopus, and crabs. The ecological role of these highly toxic metabolites has not been explored in detail, but they may represent a storage form of amino acids

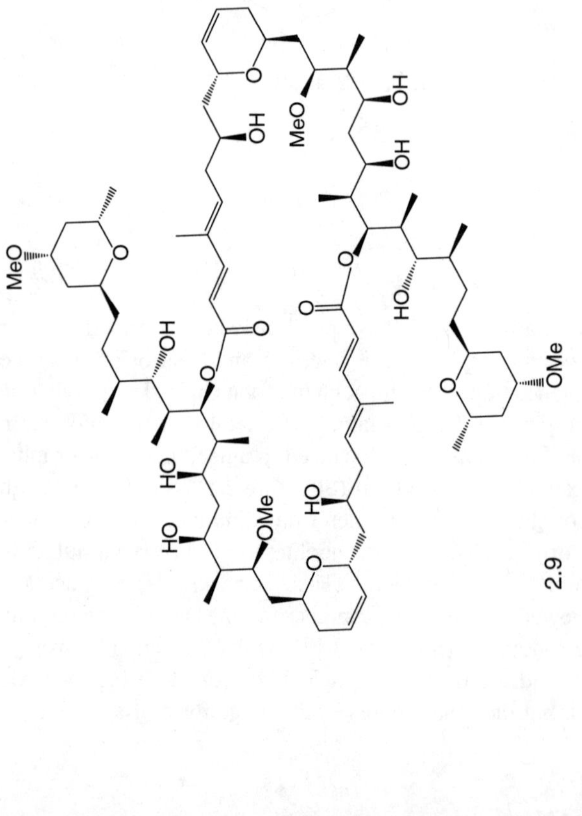

2.10

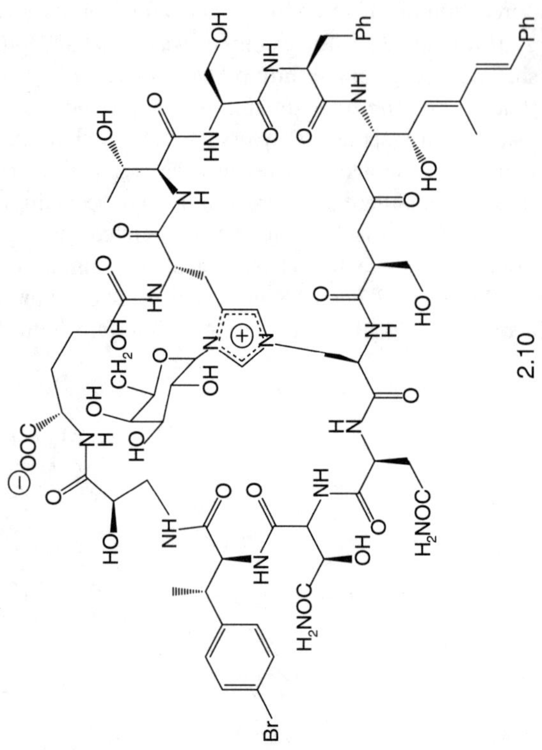

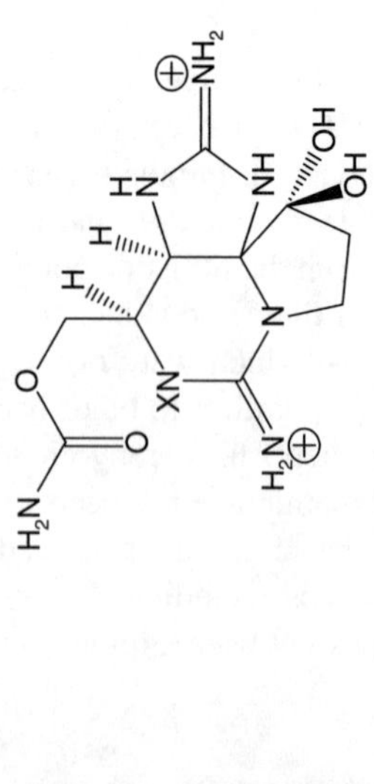

2.13

2.9

2.11 X = H
2.12 X = OH

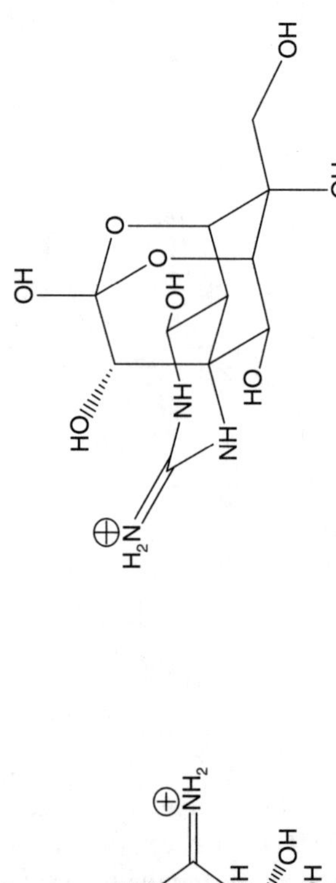

in the microorganism which can be used to enhance its survivability.[33] Both saxitoxin[34] and tetrodotoxin[27,35] afford chemical protection to the variety of marine animals in which they are found.

Knowledge of microbial diversity in the marine environment is currently being expanded by use of molecular biological techniques to study microorganisms in symbiosis with marine invertebrates, particularly in marine sponges[36] and in bryozoans such as *Bugula neritina*.[37,38] When linked to chemical studies[38] or to identification of biosynthetic gene clusters, this research may ultimately facilitate *ex situ* production of bioactive microbial marine natural products.[39] The biosynthesis of bryostatin 1, which may possibly occur in bacterial symbionts in the bryozoan *Bugula neritina*,[38] has been studied using tritiated or ^{14}C-labeled precursors in a fortified enzyme preparation prepared from host tissue. Acetate, *S*-adenosyl methionine, and glycerol were all used in bryostatin 1 (Structure 2.14) production, while propionate, succinate, *n*-butyrate, and isobutyrate were not incorporated.[40] *Bugula neritina* is a common fouling organism worldwide, whose ecological success may in part be explained by the presence of the suite of bryostatin metabolites; however, the invertebrate host must evolve mechanisms to survive the potentially lethal effects of these metabolites.

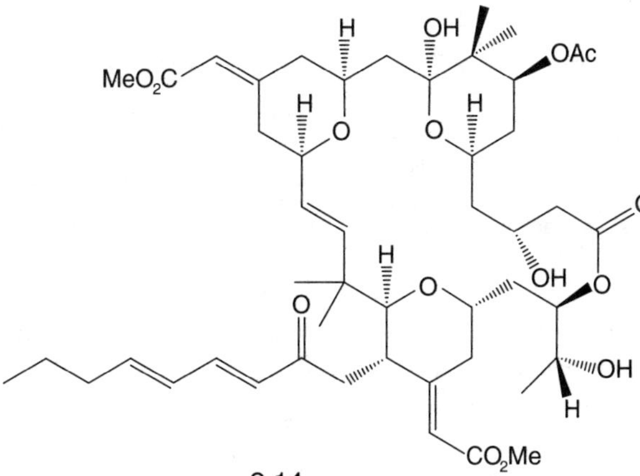

2.14

B. Marine Microalgae

Marine microalgae contain potent secondary metabolites such as okadaic acid (Structure 2.15) and dinophysistoxin-4 (Structure 2.16), and the neurotoxic brevetoxins, which are all complex cyclic polyethers; the severe toxicity of these compounds is in part due to the lipophilicity associated with their unique ladder-like skeletons. Wright et al. have used stable isotope labeling studies to demonstrate the acetate origin of dinophysistoxin-4 (Structure 2.16) and to deduce the mechanistic basis of carbon skeleton rearrangements (Scheme 2.1), which result in the loss of several carbon atoms derived from the carboxyl group of acetate.[41] Although at first sight metabolically uneconomical, this novel biosynthetic strategy sets up the correct mechanistic sequence for polyether ring formation via epoxide ring opening (Scheme 2.2). The amphidinolides, represented here by amphidinolide J (Structure 2.17), are toxic metabolites isolated from dinoflagellates present in flatworms; these metabolites are also acetate-derived[42] and probably represent additional examples of this single carbon atom deletion pathway. Perhaps this extraordinary biosynthetic capability has contributed to the ecological success of this unusual group of microorganisms. Their presence

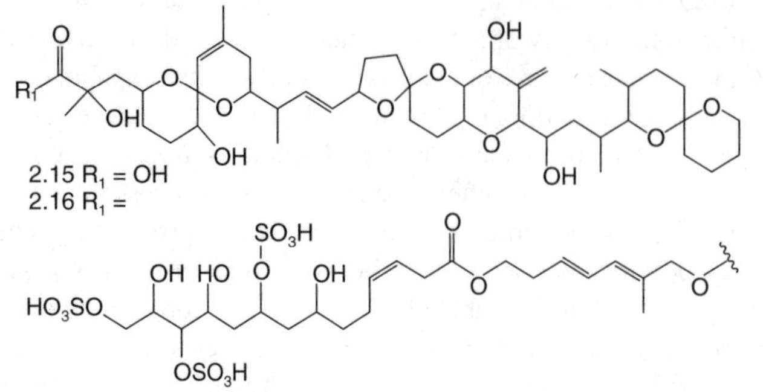

2.15 R$_1$ = OH
2.16 R$_1$ =

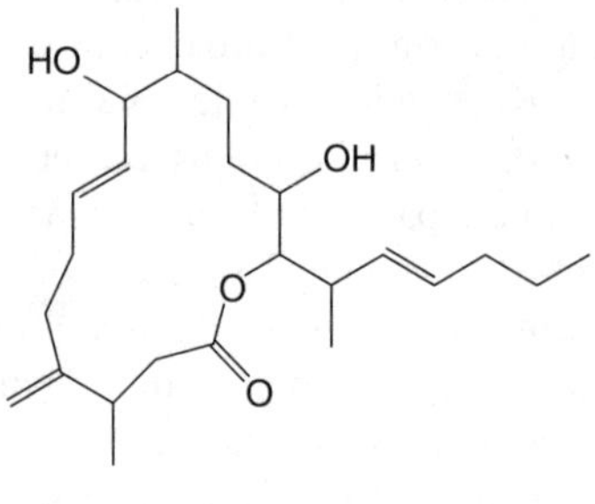

2.17

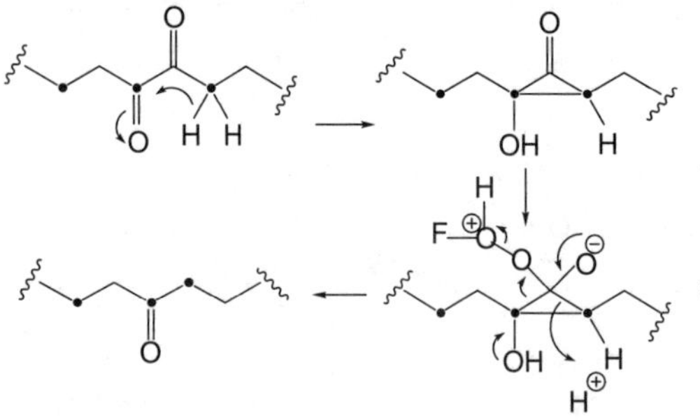

Scheme 2.1: F = flavin monooxygenase cofactor
• = methyl carbon of acetate precursor

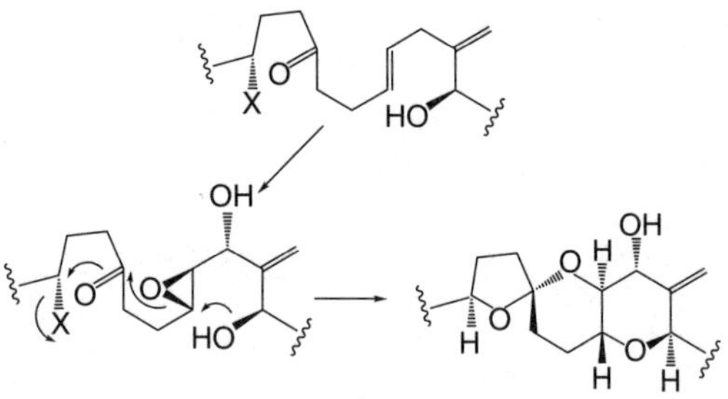

Scheme 2.2

would confer useful protection to an animal host that assimilates the toxins, via the marine food web, and can survive their potentially toxic effects.

C. Marine Cyanobacteria

Metabolites from cyanobacteria are generally of amino acid or polyketide origin and frequently show potent biological activity. The series of dolastatin metabolites, exemplified by dolastatin-10 (Structure 2.18), are linear peptides which show potent cytotoxic activity and are of clinical interest as anti-tumour agents. Originally isolated in very low yield from the Indian Ocean sea hare *Dolabella auricularia*, dolastatins are now known to be cyanobacterial products.[43,44] The discovery of a microbial source for these pharmaceutically important compounds will facilitate study of their biosynthesis and could potentially lead to the production of structural analogues by provision of modified biosynthetic precursors to the cultivar. As discussed below and in Section VI, toxic secondary metabolites from cyanobacteria have often been implicated in the chemical defenses of sea hares.[45–47]

It is believed that the presence of secondary metabolites impacts on the survivorship of a cyanobacterial strain in reef habitats that are subject to intense herbivory. Under suitable environmental conditions, cyanobacteria undergo rapid increases in population size, generating large cyanobacterial mats which are not calcified or of tough texture and which therefore present a potential source of food for herbivorous fish and other generalist predators. By restricting predation, potent cyanobacterial toxins facilitate the formation of cyanobacterial blooms.[48]

Lyngbya majuscula (syn. *Microcoleus lyngybaceus*) is a filamentous cyanobacterium which is widespread throughout the Indo-Pacific Ocean and which is implicated in large-scale fish kills as well as generating adverse dermatitic responses in humans. In view of the impact of *L. majuscula* on tourism and fishing activities, it is not surprising that extensive chemical and ecological studies have been undertaken. Various strains, either *in situ* or grown in culture, have been found to produce novel nitrogenous toxic metabolites exemplified by the malyngamides A and B (Structures 2.19 and 2.20), debromoaplysiatoxin (Structure 2.21), the lyngbyatoxins (e.g., Structure 2.22), curacin A (Structure 2.23), and barbamide (Structure 2.24), in addition to the dolastatins mentioned above. The biosynthesis of barbamide has been evaluated using stable isotope incorporation experiments; it involves chlorination of the pro-*S* methyl group of leucine, with chain extension by malonyl CoA.[49] Intact incorporation of [2-^{13}C, ^{15}N]glycine into barbamide suggested the cysteine origin of the thiazole ring.[50] Some speculative mechanistic hypotheses on the chlorination step have been presented.[51]

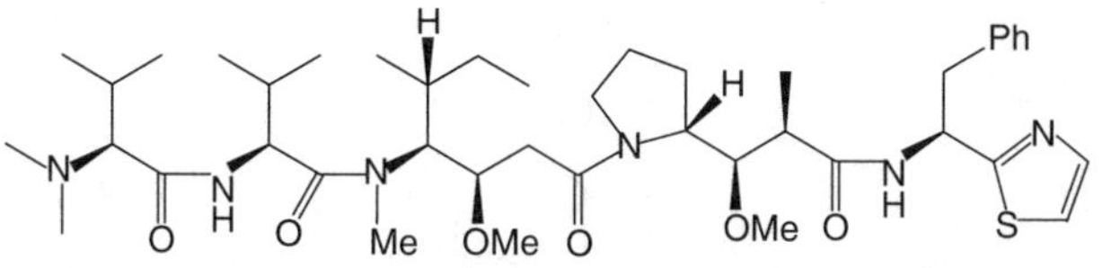

2.18

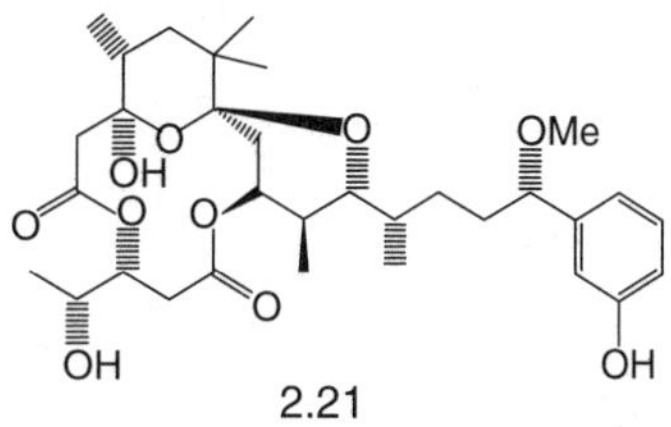

2.21

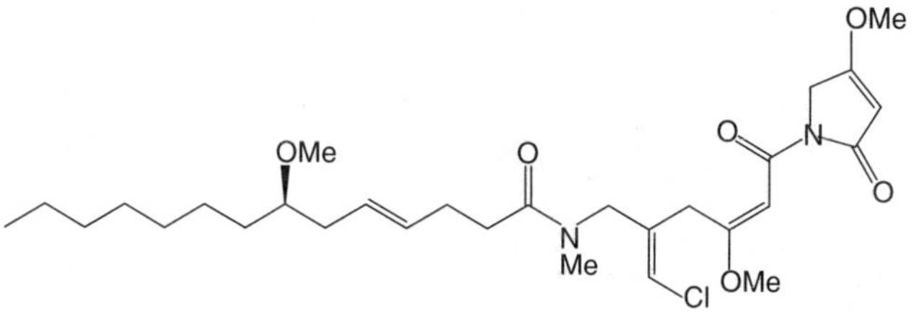

2.19

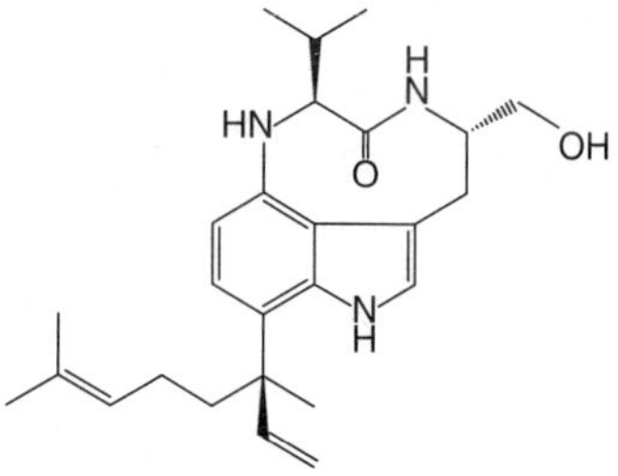

2.22

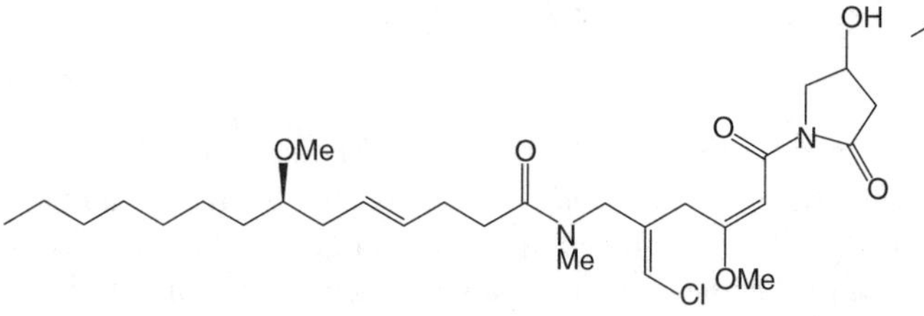

2.20

Extracts of *L. majuscula* at natural concentrations have been shown to deter feeding in surgeonfish,[52] rabbitfish,[53] and natural populations of reef fish,[45] as have some of its purified chemicals. Malyngamide A (Structure 2.19) was a deterrent at natural (1%) concentration for surgeonfish;[52] however, malyngamide B (Structure 2.20) did not deter juvenile rabbitfish when tested at that concentration.[53] When the effect of repeated metabolite exposure was tested, rabbitfish showed increased avoidance of both malyngamides, whereas parrotfish were more tolerant of malyngamide B.[54] Malyngamide A alone, or combinations of malyngamides A and B, were deterrents at 2% concentration when tested against natural populations of reef fish.[45] Both metabolites significantly deterred feeding by the pufferfish *Canthigaster solandri* and the crab *Leptodius* sp; although the toxicities of the metabolites differed, both were equally unpalatable to pufferfish.[55] The majusculamides (e.g., Structure 2.25), microcolin B (Structure 2.26), ypaomide (Structure 2.27), and malyngolide (Structure 2.28) have been found to be deterrent to pufferfish, parrotfish, and crabs.[55,56] Ypaomide also deterred feeding by rabbitfish and by the sea urchin *Echinometra mathaei*.[48]

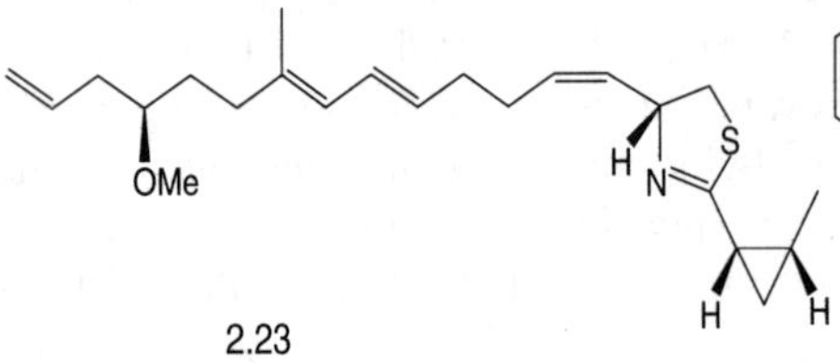

2.23

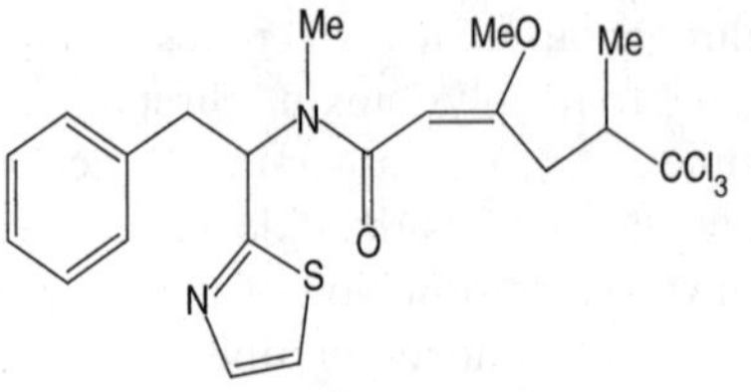

2.24

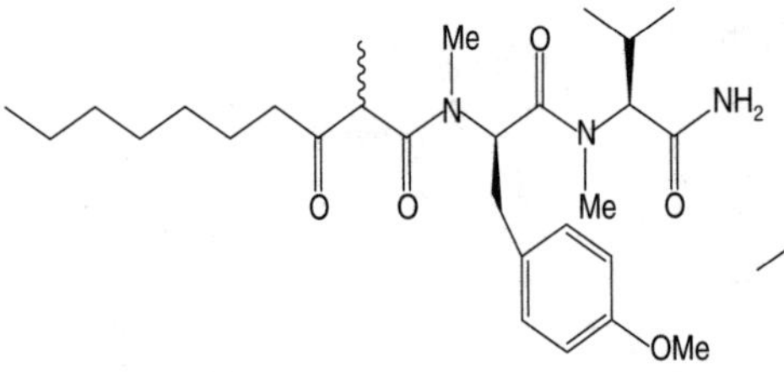

2.25

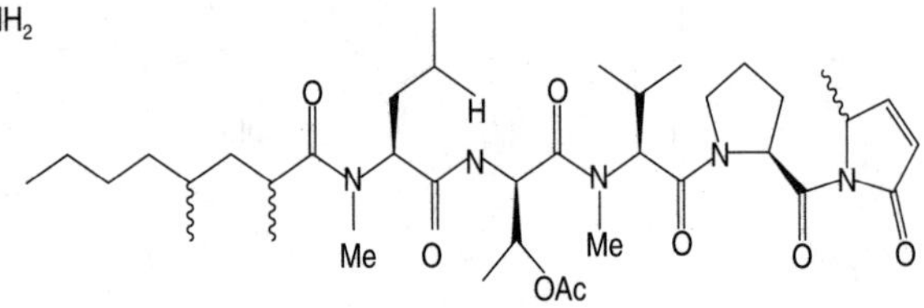

2.26

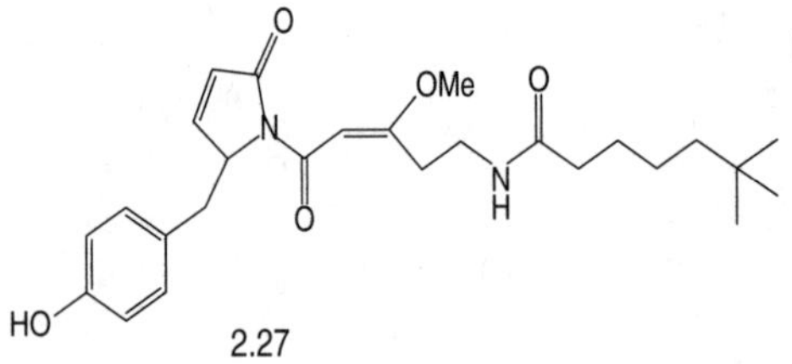

2.27

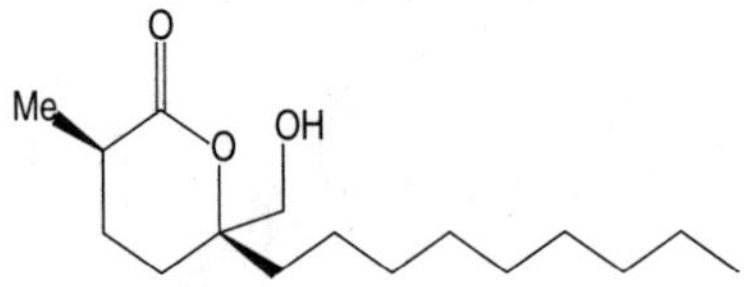

2.28

Some specialized herbivores select for *L. majuscula* in their diet. The sea hare *Stylocheilus longicauda* preferentially feeds on samples of *L. majuscula* which contain the malyngamides,[45] even though the malyngamides A and B individually deterred feeding of *S. longicauda* at natural concentrations.[47] Subsequently, it was found that the purified metabolites acted as feeding stimulants to *S. longicauda* at low concentrations, but were deterrents at higher concentrations.[57] Although many other nitrogenous secondary metabolites from *L. majuscula,* including the majusculamides, microcolin B, ypaomide, and malyngolide, have been shown to deter feeding by the sea hare *S. longicauda* at natural concentrations,[57] barbamide (Structure 2.24), in contrast, acts as a feeding stimulant at natural concentrations. Thus, closely related chemicals can have contrasting effects on the same consumer.[57] In contrast to *S. longicauda*, samples of *L. majuscula* deterred feeding by the sea hare *Dolabella auricularia*, but individual chemical deterrents have not been identified.[58]

Why does *L. majuscula* expend resources in manufacturing a range of metabolites when individual metabolites are clearly deterrent to a range of consumers? The chemical diversity detected in this cyanobacterium may reflect habitat adaptation or an induced response to herbivory,[56] or could result from the presence of closely related algal strains.[47] Slow-growing, chemically defended strains such as *Lyngbya* frequently co-occur with faster growing cyanobacterial strains which do not appear to be chemically defended.[48] The synthesis of a range of metabolites may enable the cyanobacteria to deter a broader range of consumers than if it manufactured a single compound alone. Additionally, it may be preferable for the host cyanobacterial strain to avoid the toxic effects of high levels of a single metabolite.[48,56,59]

The biosynthetic processes which lead to chemical diversity in other strains of cyanobacteria have been explored. The cyanobacterium *Microcystis aeruginosa* produces the potent hepatotoxin microcystin LR (Structure 2.29), which has been implicated in algal poisoning of water supplies and which accumulates in mussel tissue.[60] The toxin has been shown to be biosynthesized in the cyanobacterium by a mixed acetate/amino acid pathway; additional single carbon atoms are provided by methionine.[61] The non-ribosomal peptide synthetase genes for microcystin production have been reported.[62,63] Borophycin (Structure 2.30), structurally related to aplasmomycin (Structure 2.3), is a cytotoxic acetate-derived polyketide from *Nostoc linckia* whose additional methyl groups derive from methionine.[64] The biosynthesis of the toxins anatoxin-a(s) (Structure 2.31) and anatoxin-a (Structure 2.32) in the freshwater cyanophyte *Anabaena flos-aquae* has been investigated,[65,66] as has that of tolytoxin (Structure 2.33) in the freshwater *Scytonema mirabile*. Tolytoxin is a polyacetate with a glycine starter unit, and is structurally related to some sponge metabolites.[67] The hapalindoles, exemplified by hapalindole A (Structure 2.34) from the freshwater cyanobacterium *Hapalosiphon fontinalis*, contain the unusual isocyanide motif frequently seen in marine sponge metabolites; a tetrahydrofolate origin for the isocyanide moiety was inferred by incorporation of glycine, serine, methionine, and formate.[68]

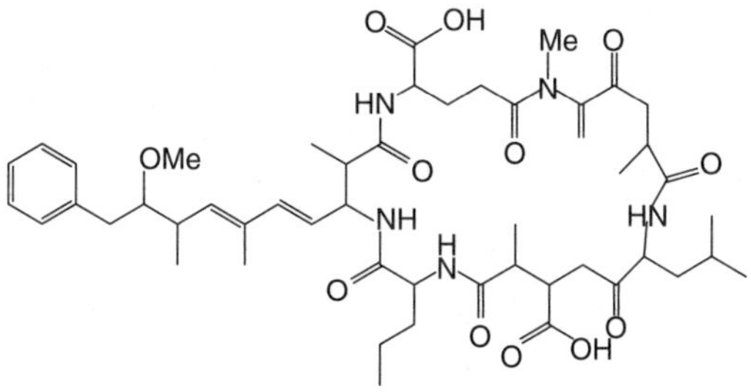

2.29 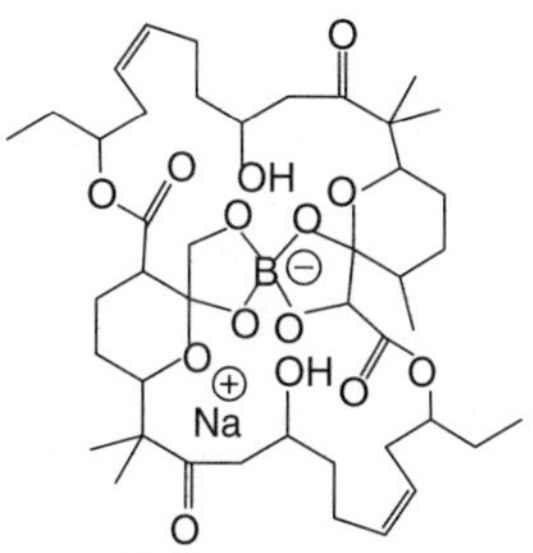   2.30 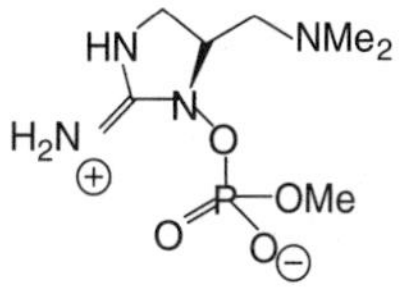   2.31

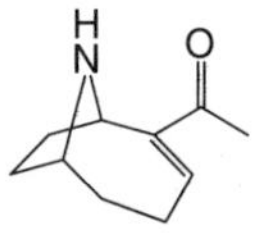

2.32 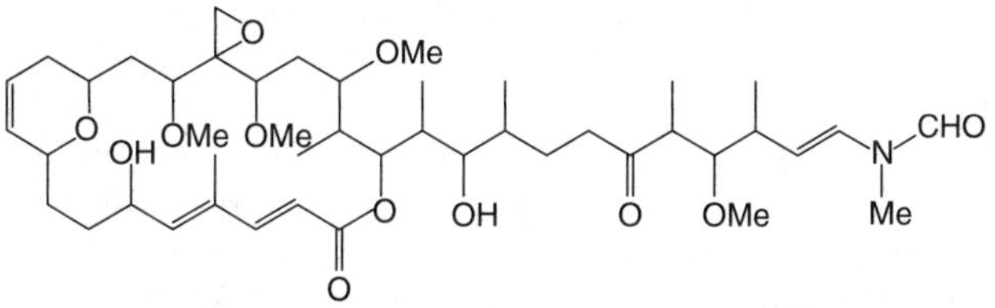   2.33 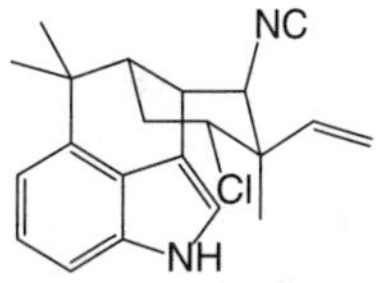   2.34

The structural resemblance between cyanobacterial metabolites, of both freshwater or marine origin, and some metabolites isolated from sponges has led to discussion of the symbiotic origin of the sponge compounds.[15,31,69] The biosynthesis and ecological role of cyanobacterial alkaloids has conveniently been studied in sponge tissue, since studies using isolated cyanobacterial preparations have been hampered by low incorporation levels.[70] The chlorinated *N,N*-dimethyldiketopiperazine (Structure 2.35) and the alkaloid 13-demethylisodysidenin (Structure 2.36) are both components of the cyanobacterial symbiont *Oscillatoria spongeliae* present in the tropical marine sponge *Dysidea herbacea*.[69,71,72] Some chlorinated alkaloids of *D. herbacea* are feeding deterrents and are believed to enhance the competitiveness of the sponge-cyanobacterial association.[71] The amino acid origin of the metabolites shown in Structures 2.35 and 2.36 has been demonstrated, with both *R*- and *S*-leucine incorporated, although the metabolites show a 5*R* configuration.[70,73] Valine, alanine, serine, and methionine were also incorporated into 13-demethylisodysidenin.[70] The sponge *D. herbacea* also produces brominated diphenyl ethers such as the antibacterial compound shown in Structure 2.37, which is present in the cells of the cyanobacterial symbiont as well as in crystalline form in sponge tissue where it may function as a physical defense.[74] An ecological role in preventing predation by generalist reef predators has been evaluated for the related bromophenol diphenyl ether (Structure 2.38);[47,75] however, this metabolite is sequestered by the sea hare *Stylocheilus longicauda*.[76] Labeling experiments using acetate and glucose to determine the biosynthetic origin of *Dysidea* bromophenols using sponge tissue have been inconclusive.[77]

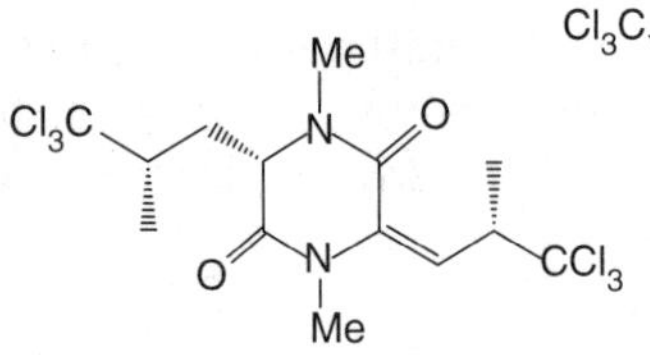

2.35 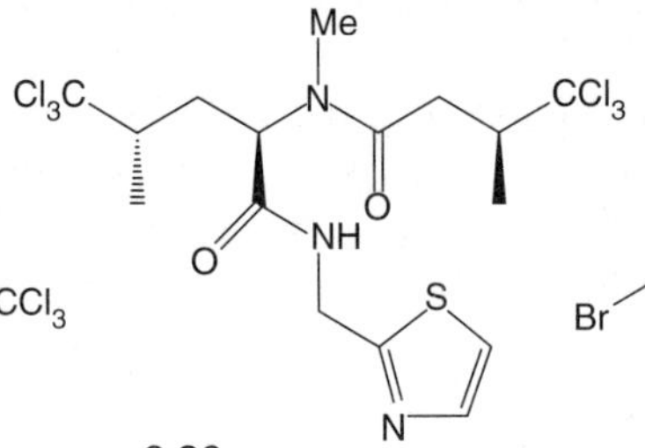   2.36 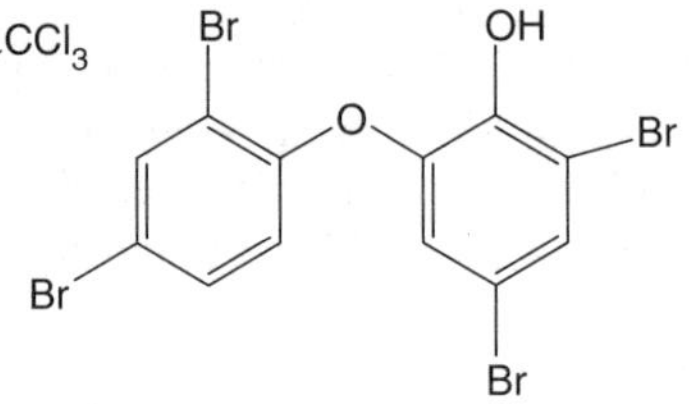   2.37

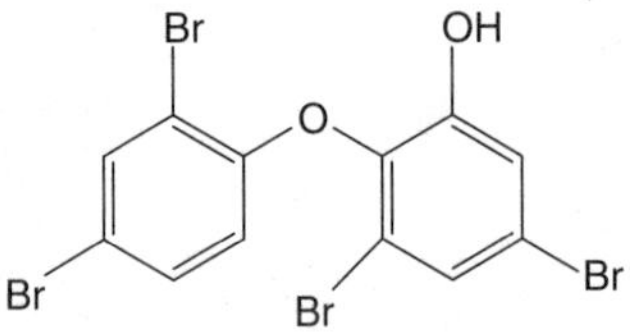

2.38

III. MARINE MACROALGAE

Because seaweeds and seagrasses could easily be collected, they were among the first marine specimens to be investigated chemically. Their secondary metabolite chemistry is typified by linear and cyclic halogenated sesquiterpenes and by diterpenes with diverse carbon skeletons.[1] Algal compounds are rarely nitrogenous;[2,3] this may be because seaweeds are often nitrogen-limited.[78] An understanding of the ecological roles of algal metabolites is becoming apparent. There are numerous reports of field and laboratory studies on antipredation, antifouling, and space competition roles for algal metabolites.[27,79,80] It is presumed that the high energetic cost of producing and maintaining novel biosynthetic capability, plus the need to safely store biologically potent, even toxic, compounds, is compensated for by a defensive benefit to the host plant. Plant chemical defenses are, therefore, often maximized in key plant parts such as reproductive or actively growing tissue.[79] Biosynthetic studies of macroalgae have utilized whole tissue samples,[81,82] enzyme preparations,[83] or female gametes of brown algae.[84,85]

A. GREEN ALGAE

Functionalized terpenes are the typical secondary metabolites of green alga, and their ecological roles have been well investigated. In principle, these metabolites could be derived either via the classical mevalonate pathway or the newly discovered mevalonate-independent pathway;[86] however, experimental verification is lacking. *Halimeda* spp. are calcified algae, commonly found on coral reefs, which produce diterpenoids (such as halimedatetraacetate (Structure 2.39), halimedatrial (Structure 2.40), or chlorodesmin (Structure 2.41)) that function as effective feeding deterrents against natural populations of reef fish (reviewed by Paul[80] and Hay[27]). The unusual trialdehyde structure of halimedatrial is reminiscent of other well-known antifeedants such as warburganal and polygodial.[79] Halimedatrial is present in newly produced uncalcified tissue, and the compound is synthesized at night when grazing pressure is lowest. The chemical defense shifts toward the less deterrent halimedatetraacetate in older, more calcified tissue.[78] The enol acetate moiety of halime-datetraacetate is a masked aldehyde group. Halimedatetraacetate is enzymatically converted into halimedatrial and its epimer epihalimedatrial (Structure 2.42) immediately upon plant damage. The algal protoxin and the enzyme may be separately compartmentalized within the *Halimeda* tissue in order to avoid autotoxicity.[87] A site-to-site variation in *Halimeda* chemistry suggests that the alga adjusts its metabolite composition in response to levels of herbivory in different habitats.[88] Other algal metabolites contain the same 1,4-diacetoxybutadiene unit as found in halimedatetraac-etate. The diterpene chlorodesmin (Structure 2.41) is an effective feeding deterrent in *Chlorodesmis* spp.[52,89] However, the sesquiterpene caulerpenyne (Structure 2.43) from *Caulerpa* spp. has no proven antifeedant activity against fish, despite being concentrated in plant parts prone to her-bivory;[90] it was effective against the gastropod mollusc *Dolabella auricularia*.[58]

The calcified green alga *Neomeris annulata* manufactures a range of brominated sesquiterpenes (Structures 2.44–2.46), which are concentrated in the fleshy tips of this alga[91] and which are individually deterrent to parrotfishes and the sea urchin. However, a mixture of the three metabolites did not increase deterrence, implying that other ecological roles may apply to these metabolites.[92]

Feeding deterrents in green algae are not exclusively terpenoid in character. Cymopol (Structure 2.47) and debromoisocymopol (Structure 2.48) from *Cymopolia barbata*, which may derive biosynthetically from alkylation of a bromophenol with geranyl pyrophosphate, both deter feeding by reef fishes.[93,94] However, cymopol stimulates feeding by sea urchins, which may explain why this alga manufactures a range of metabolites.[93]

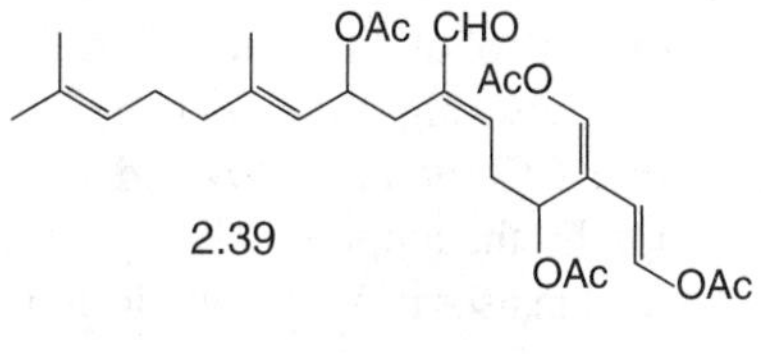

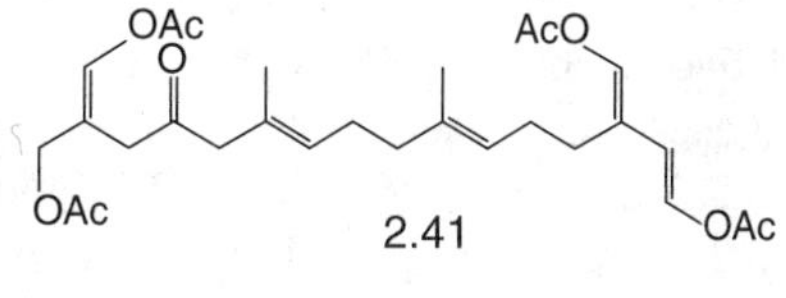

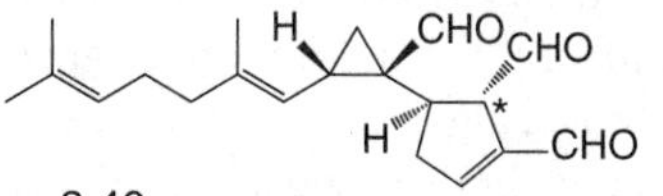

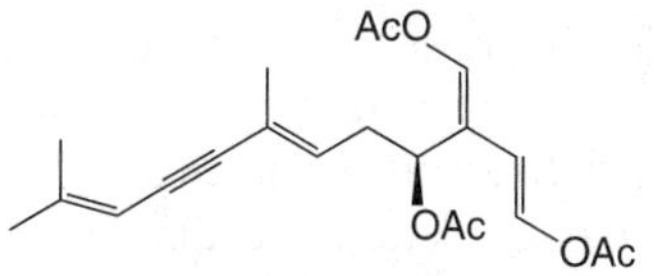

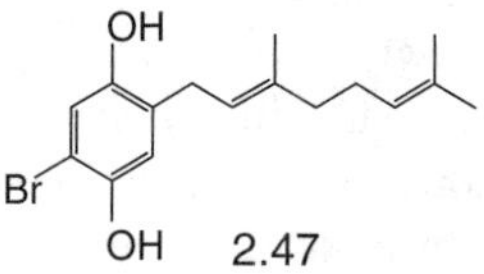

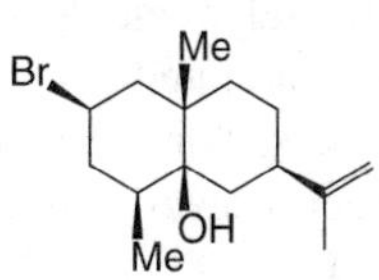

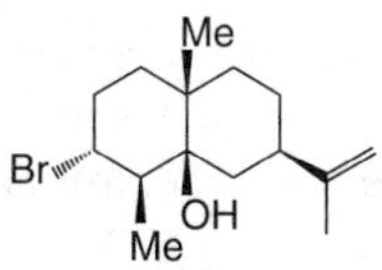

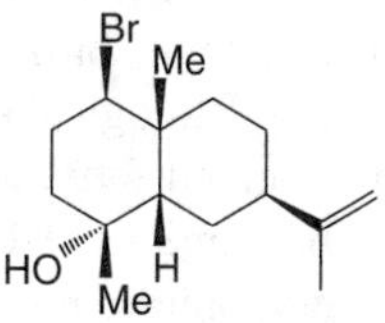

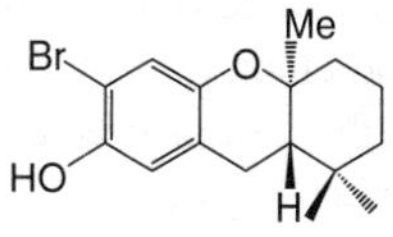

B. Brown Algae

The terpene biosynthetic pathway is prevalent throughout the marine brown algae, and frequently leads to biologically active metabolites. As with green algae, experimental information which confirms whether the biosynthetic precursor mevalonate is involved is lacking. However, additional secondary metabolites are afforded as a consequence of the involvement of enzymes which convert long-chain fatty acids into C_8 or C_{11} acyclic or cyclic hydrocarbons. These short-lived hydrocarbons have been shown to function as pheromones in sexual reproduction in brown algae,[85,95,96] while their decomposition products play a chemical defense role.[97,98] Experimental investigation of the pathways involved in pheromone production has been facilitated by the favorable rates of the biosynthetic steps involved; in many species of brown algae significant turnover of biosynthetic precursors can be detected in less than an hour.[85]

The experimental conversions of arachidonic acid into dictyotene (Structure 2.49) by female gametes of *Ectocarpus siliculosis* and of eicosapentaenic acid into various hydrocarbons, including ectocarpene (Structure 2.50), multifidene (Structure 2.51), hormosirene (Structure 2.52), finavarrene (Structure 2.53), and giffordene (Structure 2.54), by *E. siliculosis, Sphacelaria rigidula,* or *Giffordia mitchellae,* confirm that the pheromones derive from polyunsaturated C_{20} fatty acid metabolism.[84,99] A closely related biosynthetic pathway also operates in freshwater diatoms;[100,101] in contrast, the biosynthesis of C_{11} hydrocarbons in terrestrial plants of the Asteraceae involves the decarboxylation of unsaturated C_{12} fatty acids. Algal C_8 hydrocarbons, such as the triene fucoserratene (Structure 2.55) from *Fucus serratus,* are also derived by oxidation and cleavage of eicosapentaenic acid; again, this pathway has been tested in the freshwater diatom *Asterionella formosa.*[101] Precursor-directed biosynthesis has provided structural analogues of many of these pheromones,[84,99,101] and suggests the enzymatic systems involved have some plasticity.

The biosynthetic steps leading from arachidonic acid generate the divinylcyclopropane (Structure 2.56) which decomposes to ectocarpene (Structure 2.50) via a Cope rearrangement. Bioassay of ectocarpene (Structure 2.50) and of pre-ectocarpene (Structure 2.56) reveals that pre-ectocarpene is the actual male attractant and, hence, that the subsequent formation of ectocarpene represents a deactivation mechanism to prevent an undesirable build-up of chemical signal.[102,103] Cyclic-1,4-heptadienes such as dictyotene (Structure 2.49) and ectocarpene (Structure 2.50) are themselves unstable and undergo facile oxidation,[85,97] yielding alcohol and ketonic products. This oxidative mechanism may represent the deactivation process in those algae in which the cyclohepta-1,4-dienes are the actual pheromones. Interestingly, dictyotene (Structure 2.49) and its oxidative degradation products (Structures 2.57 and 2.58) deter feeding by the amphiphod *Amphithoe longimana,* but not by the sea urchin *Arbacia punctulata;* thus, this alga uses its biosynthetic end products as chemical defense agents.[98] The sulfur-functionalized C_{11} compound (Structure 2.59) of *Dictyopteris membranacea,* which is presumably also biosynthesized by a fatty-acid-related pathway, strongly deterred feeding by the mesograzer *Amphiphoe longimana,* but did not affect grazing by the sea urchin *A. punctulata.*[104] Synthetic samples of the three brown algal organosulfur metabolites shown in Structures 2.60–2.62 also deterred feeding by amphipods. In these algae, the biosynthetic pathway leading to C_{11} hydrocarbons, their oxidation products, or organosulfur compounds may have evolved as an additional specialist defense to protect susceptible female gametes from predation by amphipods. A consequence of generating a series of biologically potent compounds with diverse ecological roles from a single biosynthetic pathway is that the structural range of the pheromonal products is restricted. Biological control is achieved by using polyunsaturated acids as hydrocarbon precursors and saturated fatty acids to give organosulfur metabolites. For the many brown algae which produce the same suite of pheremones, control at the species level is likely achieved by use of discrete enantiomeric mixtures.[85]

The brown algal genus *Dictyopteris* shows extraordinary chemical diversity and contains examples both of fatty-acid-derived hydrocarbons and of products from terpene metabolism. The genus has been a favored target for chemical ecology studies, with varying results. A 1:2 ratio of hydrocarbons dictyopterene A (Structure 2.63) and B (Structure 2.64) from *D. delicatula* has been shown to deter grazing by reef fishes but has had no effect on grazing by amphipods,[105] a result which contrasts with the above data for Structures 2.57–2.62. Among the terpene metabolites, diverse patterns of deterrence against herbivores are apparent even for metabolites such as pachydictyol A (Structure 2.65), dictyol E (Structure 2.66), and dictyol B acetate (Structure 2.67), which represent subtle variations of the same basic biosynthetic pathway.[52,93,106–110] The terpenes acutilol A (Structure 2.68), acutilol A acetate (Structure 2.69), and acutilol B (Structure 2.70) from *Dictyota acutiloba* showed contrasting effects in feeding studies using tropical and temperate fishes or sea urchins.[111]

Small sedentary grazers such as amphipods appear to select chemically defended seaweeds as host plants since they would otherwise be subject to intense predation by reef fishes; however, they do not sequester metabolites as do other selective grazers such as sacoglossans and opistobranch molluscs.[105,109] Grazing by amphipods induces increased concentrations of acutilol A acetate and acutilol B in *Dictyopteris menstrualis* and makes the seaweed less susceptible to attack by other predators;[112] the same terpenes acted as antifoulants which prevented the settlement of bryozoan larvae.[107] This evidence for multiple roles for algal metabolites may provide an explanation of previously documented differences in chemical composition in *Dictyota*.[113]

2.49 R = $-(CH_2)_3CH_3$
2.50 R = $-CH{=}CH{-}CH_2CH_3$ *cis*

2.51

2.52

2.53

2.54

2.55

2.56

2.57

2.58

2.59

2.60

2.61

2.62

2.63 R = $CH_3(CH_2)_3$

2.64 R =

2.65 R' = H; R" = H
2.66 R' = OH; R" = H
2.67 R' = H; R" = OAc

2.68 R = H
2.69 R = Ac

2.70

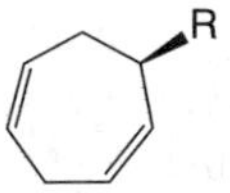

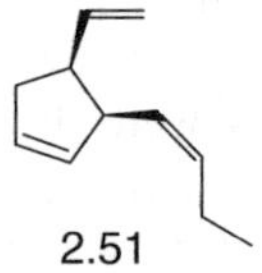

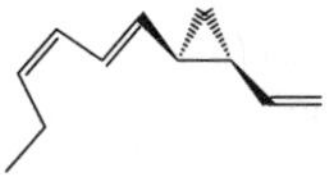

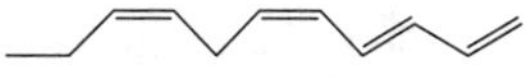

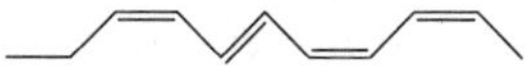

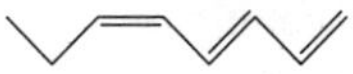

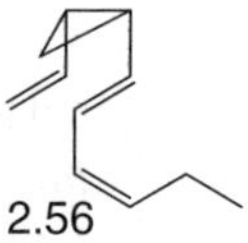

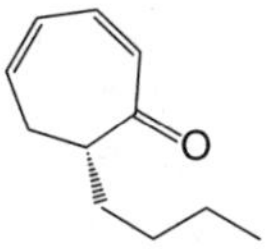

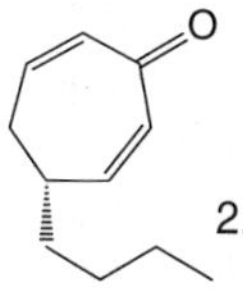

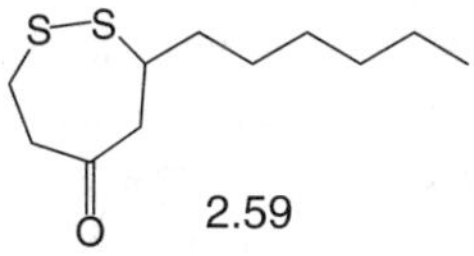

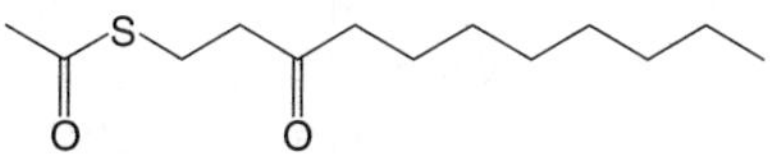

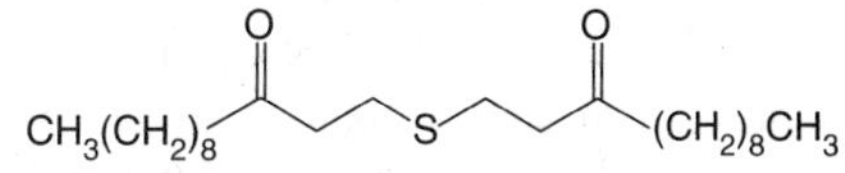

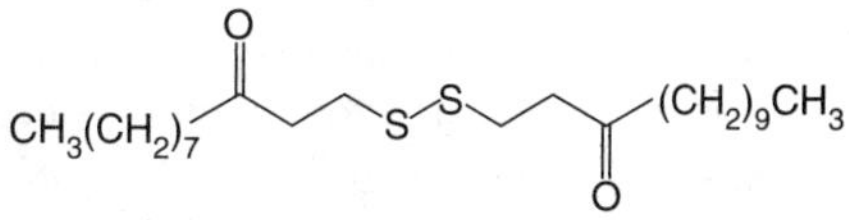

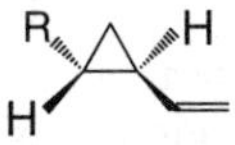

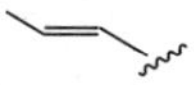

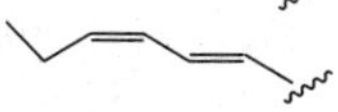

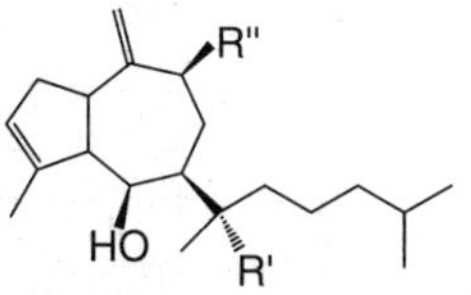

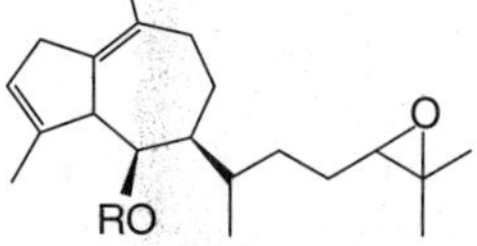

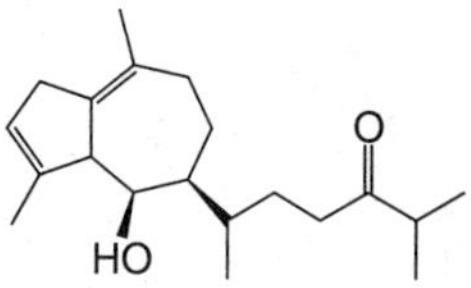

Studies of the nutrient requirements of algal biosynthetic pathways are highly relevant to these ecological studies. Cronin and Hay[114] report that increased nitrogen and phosphorus levels result in increased diterpene production in *Dictyota ciliolata*, whereas nutrient enhancement of *Sargassum filipendula* did not affect phlorotannin levels. Predator-induced chemical defense based on polyphenolics has been detected in many brown algae including *Fucus distichus*. These algae were shown to respond to injury by increasing concentrations of phenolic deterrents in its tissue. The damaged plants may also use the tannins to prevent attack by pathogens or as healing agents through interactions with proteins.[115,116]

C. Red Algae

Red algae produce a variety of polyketides and halogenated terpene metabolites exemplified by the polyacetate tribromoheptene oxide (Structure 2.71), fimbrolides such as that shown in Structure 2.72, and the cyclic monoterpenes ochtodene (Structure 2.73), chloromertensene (Structure 2.74) and chondrocole C (Structure 2.76). As with green and brown algae, metabolite diversity in red algae may provide protection against a wider range of consumers than if a single metabolite were produced; perhaps as partial compensation for the high metabolic cost involved, some plants appear not to be chemically defended.

The polyacetate origin of tribromoheptene oxide (Structure 2.71) has been confirmed by incorporation of acetate, and of palmitate (via β-oxidation),[81] while a low incorporation of acetate and bicarbonate has been reported into a fimbrolide (Structure 2.72) in *Delisea pulchra*.[12] Although there was little variation in secondary metabolite levels in different parts of this algae, there were large variations in individual metabolite levels; metabolites were also concentrated in the plant tips. Both crude extracts and selected furanone metabolites deterred feeding by generalist herbivore predators.[117] However, these metabolites have additional ecological roles. Recent work on the fimbrolides has led to the unravelling of their antifouling role which stems from their interference with bacterial signaling processes involving acylated homoserine lactones.[118] The alga is able to specifically control the abundance and composition of its epiphytic bacterial community.[28,119]

Low incorporations of bicarbonate, acetate, and mevalonate into the halogenated monoterpene metabolites (e.g., Structure 2.75) of *Plocamium cartilagineum* have been reported.[82] In contrast to *D. pulchra*, *Plocamium* spp. show significant intraspecimen variation in terpene chemical composition.[82,120,121] Life-cycle status, environmental factors, or the presence of mixed genetic populations may influence monoterpene composition and production.[82,121] Chloromertensene (Structure 2.74), isolated from the red algae *P. hamatum*, has been identified as the allelochemical responsible for contact-mediated necrosis in octocorals and sponges, and also has antifouling and antifeedant activity.[89,122] The chemical is not spontaneously released into the water column and therefore may be localized in cytoplasmic vesicles on the algal surface,[122] as has been demonstrated for the sesquiterpene β-snyderol (Structure 2.77) in *Laurencia snyderae*.[123] These chemicals are released when the cortical cells are disrupted by predators.[93] In this way, partitioning minimizes autotoxic effects.[79]

In *Portieria hornemannii*, ochtodene (Structure 2.73) production is independent of added nutrients, but light availability appears to influence its production.[121] Ochtodene is an effective feeding deterrent against adult and juvenile rabbitfish,[53] surgeonfish,[52] and natural populations of herbivorous fish; however, the structurally related chondrocole C (Structure 2.76) is not a deterrent against reef fish. The individual terpenes, Structures 2.73 and 2.76, were not effective deterrents of amphipod feeding, whereas an unresolved monoterpene mixture from this alga was a deterrent.[120] Reciprocal transplantation of *P. hornemannii* between two sites showing different secondary metabolite levels failed to show significant changes in monoterpene concentrations. Although this result is consistent with genetic rather than environmentally induced variation in chemistry, other factors such as interplant variation and seasonal effects may have influenced the secondary metabolites detected.[124]

The red algal chamigrene sesquiterpene elatol (Structure 2.78) has been shown to deter feeding by reef fishes.[93] Specimens of *Laurencia elata* from Southern Australia show a pronounced seasonal variation in elatol production. Incorporation studies using ^{14}C acetate failed to confirm an acetate-mevalonate path for elatol production.[125]

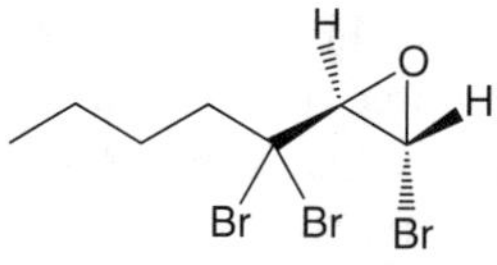

2.71 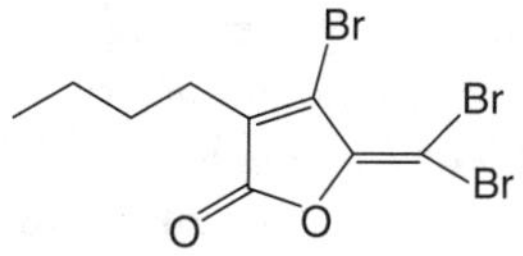   2.72 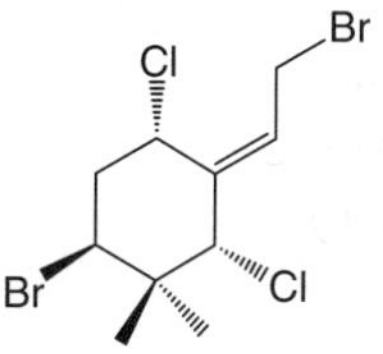   2.73 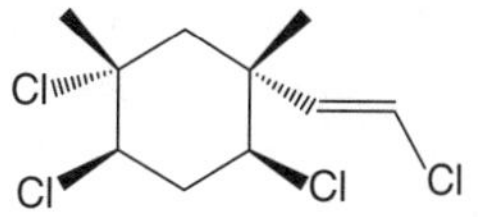   2.74

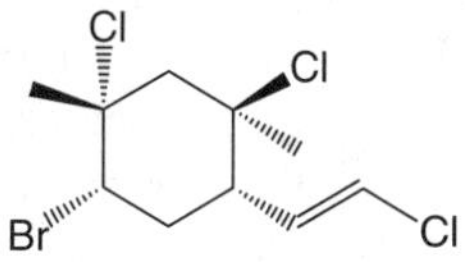

2.75 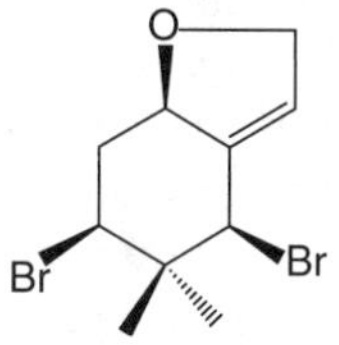   2.76 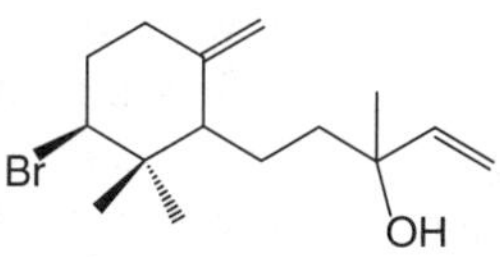   2.77 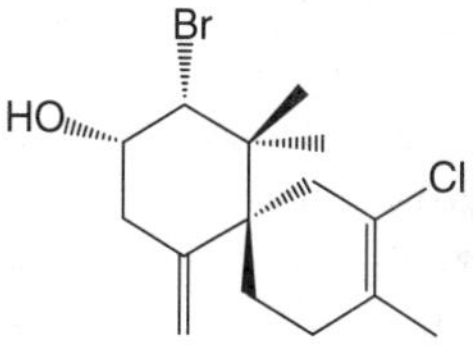   2.78

The distribution and biosynthesis of oxylipins, the bioactive oxidation products of fatty acids in many marine systems, particularly in red algae, has been reviewed.[126,127] In mammalian systems, oxylipins play key roles in cellular communication and in homeostasis, but their physiological roles in algae are not yet understood.

IV. MARINE SPONGES

Marine sponges are major components of benthic habitats. In contrast to ascidians and bryozoans,[80,128] predation on sponges appears restricted to a few specialized groups of consumers, notably turtles, some fish, and molluscs.[129] The ecological success of sponges is facilitated by the presence of bioactive secondary metabolites representing all the major compound classes[3] which have formed the basis of many detailed chemical and pharmacological studies.[1] Both laboratory and field assays have demonstrated the potency of sponge extracts and of individual sponge chemicals as feeding deterrents.[75,129,130] However, secondary metabolites alone may not determine feeding preferences of predators; toxic extracts do not always deter predators.[129,131] Spicules, tissue toughness, and nutritional quality are additional factors which need to be considered,[75,130,132] while some sponges may simply employ faster growth rates or early reproduction in order to sustain their populations.[132] Sponge metabolites may also play an important role in influencing marine invertebrate larval settlement and development.[133–135]

In sponges, a range of biosynthetic pathways appears to have evolved in response to ecological pressures. For example, sponges containing terpenes, polyketides, or alkaloids have been implicated in ecological interactions between sponges and corals,[136-138] while antifouling roles have been deduced for terpenes as well as alkaloids.[139–141] Fish feeding studies have identified diverse examples of brominated metabolites and terpenes as anti-predation chemicals.[75,142–144] Paul[80] has expressed the need for greater understanding of how secondary metabolite concentrations and types vary among individual populations of sponges, and the extent to which predators and environmental factors influence the production of secondary metabolites.

To date, biosynthetic studies in sponges have concentrated on nitrogenous metabolites and have generally used whole tissue specimens, either maintained *in situ* or in aquaria. Incorporation techniques have involved injection of precursors, liposomes, or the addition of water-soluble precursors directly into the aquarium water.[13,145] Two recent studies of alkaloids illustrate some specialized methodology for probing marine sponge biosynthesis. Production of the cytotoxic discorhabdin B (Structure 2.79) from phenylalanine has been demonstrated in thin slices of tissue maintained under tissue culture conditions and in the presence of antibiotics which were assumed to suppress any biosynthetic contribution from symbiotic bacteria.[146] Cultured archaeocyte cells prepared from the sponge *Axinella corrugata* (*Teichaxinella morchella*) utilized histidine, ornithine, and proline for the synthesis[147] of the feeding-deterrent stevensine (Structure 2.80).[148]

Verongiid sponges, characterized by the presence of novel, bioactive brominated alkaloids, are widespread in both tropical and temperate habitats and have been the focus of numerous chemical, biosynthetic, and ecological studies. Biosynthetically, these alkaloids derive from amino acids; of interest is the timing and manner of the bromination steps. Dibromoverongiaquinol (Structure 2.81), the rearranged dibromohomogentisamide (Structure 2.82), and aeroplysinin-1 (Structure 2.83) have all been shown to derive from phenylalanine and from tyrosine in *Aplysina fistularis*;[149,150] the sponge utilizes the tyrosine side-chain without significant deamination.[149] Although 3,5-dibromo-tyrosine and 3-bromotyrosine were confirmed as precursors to dibromoverongiaquinol and aeroplysinin-1, a range of labeled *O*-methyl nitriles or tyrosines were not utilized by the sponge, and so individual steps in the biosynthetic pathway could not be pinned down in detail. However methionine was specifically incorporated into the *O*-methyl group of aeroplysinin-1.[150] By examination of the range of natural product structures isolated from verongiid sponges, the timing of methylation was inferred. It was suggested that *O*-methylation occurs early in the biosynthetic pathway with an *O*-methyl nitrile undergoing epoxidation, eventually serving as precursor to aeroplysinin-1 and to a large group of oxazolidinone alkaloids, e.g., aerothionin (Structure 2.84); in contrast, non-methylated precursors provide α-oximino compounds such as the bastadins.

Experimental evidence supports a variety of ecological roles for sponge bromotyrosine metabolites. An analytical HPLC study of the chemistry of water or methanol-based extracts of *A. aerophoba* suggests that the biologically inactive isofistularin-3 (Structure 2.85), fistularin-1 (Structure 2.86), or aerophobin-2 (Structure 2.87) can be converted to aeroplysinin-1 (Structure 2.83) and then to dibromoverongiaquinol (Structure 2.81) by the action of enzymes.[151] Subsequently, proof of the enzymic nature of the biotransformation of oxazolidinone alkaloids such as isofistularin-3, fistularin-1, or aplysinamisin-1 (Structure 2.88) into aeroplysinin-1 (Structure 2.83) and dibromoverongiaquinol (Structure 2.81) were obtained from *in vitro* experiments using a cell-free extract prepared from sponge tissue. The bioconversions detected were shown to be extremely rapid.[144] The conversion was envisaged as (1) cleavage of the C8–C9 bond, ultimately generating aeroplysinin-1; (2) demethylation; and (3) hydration of the nitrile to give dibromoverongiaquinol (Structure 2.81). Both aeroplysinin-1 and dibromoverongiaquinol are algicides and show antimicrobial and cytotoxic activity; the metabolites also caused behavioral modification in the marine gastropod *Littorina littorea*,[151–153] while extracts from damaged sponge tissue strongly deterred the Caribbean wrasse *Thalassoma bifasciatum*.[144] The sponge *A. aerophoba* may represent an example of a wound-induced chemical defense that relies on enzymic conversion of inactive storage compounds into active defense compounds. Aeroplysinin-1 (Structure 2.83) may also be converted into dibromoverongiaquinol (Structure 2.81) by the slow action of alkaline sea water[153] or by storage.[150] An environmental influence on metabolite production is also apparent. Specimens of *Aplysina aerophoba* collected from shallow-water habitats in Yugoslavia contained both aeroplysinin-1 and dibromoverongiaquinol, whereas deep-water sponges contained only dibromoverongiaquinol. These deep water sponges quickly recovered their ability to manufacture aeroplysinin-1 when kept under illuminated aquarium conditions; the sponge surface tissue contained aeroplysinin-1, whereas inner cortical tissue contained only dibromoverongiaquinol.[154] This localization is consistent with a chemical defense role for the strongly cytotoxic aeroplysinin-1.

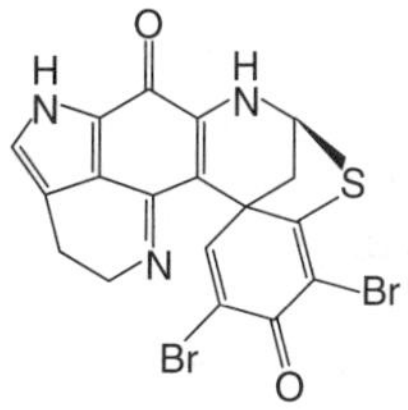

2.79 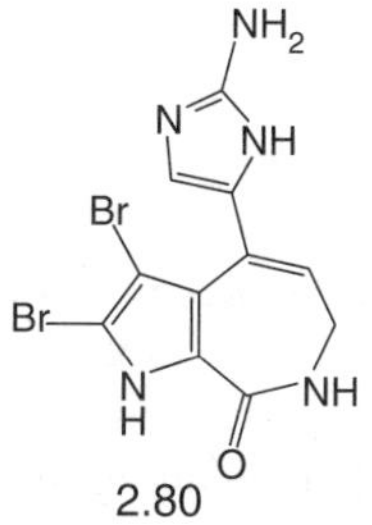 2.80 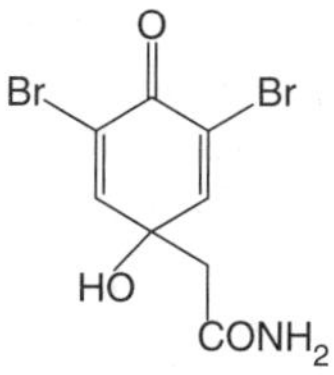 2.81 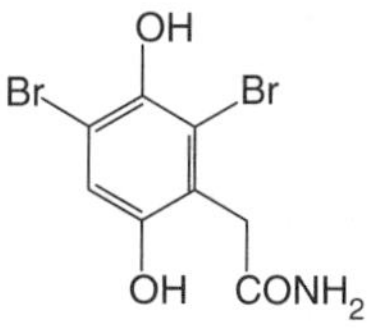 2.82 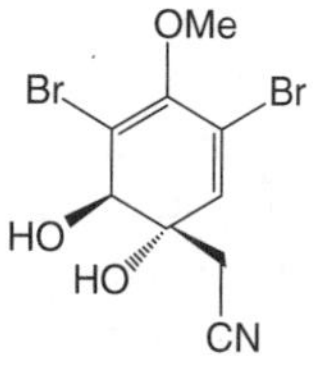 2.83

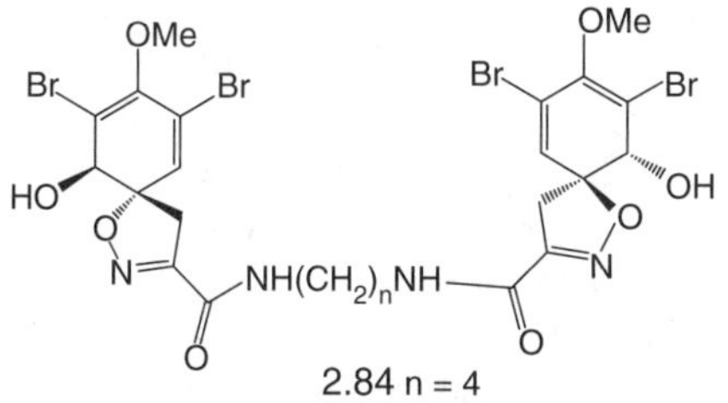

2.84 n = 4
2.89 n = 5

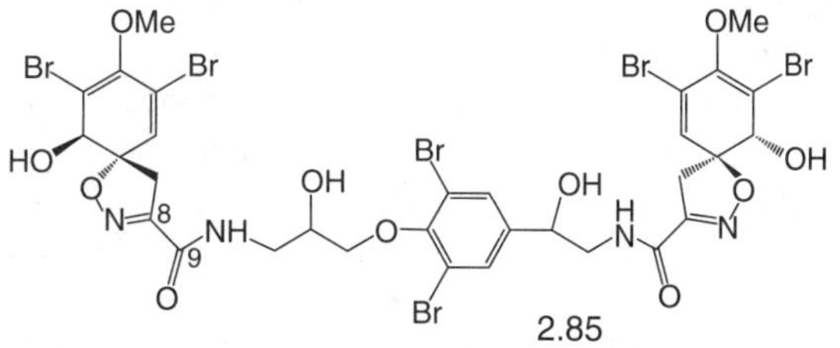

2.85

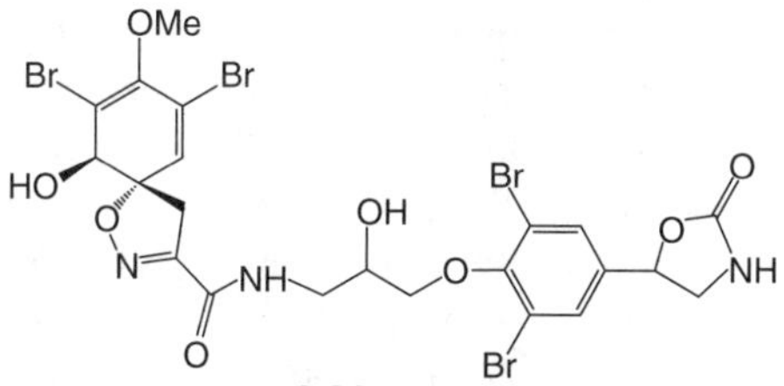

2.86

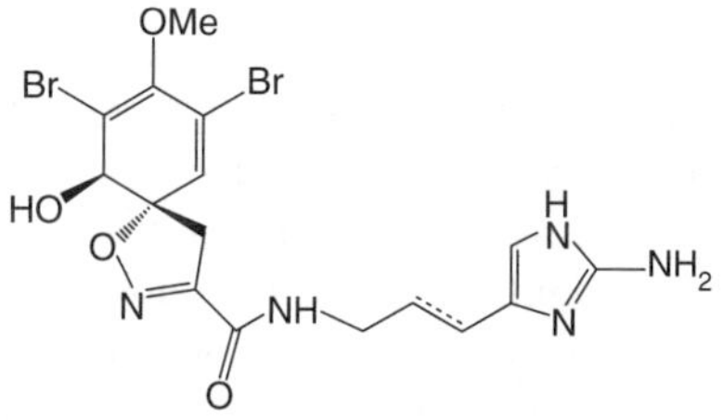

2.87 R = dihydro
2.88 R = dehydro

Thompson et al. have investigated the ecological role of the brominated isoxazoline alkaloids, aerothionin (Structure 2.84) and homoaerothionin (Structure 2.89), in Californian specimens of *A. fistularis*.[134,139] The alkaloids caused behavioral modification in marine invertebrates, were toxic to dorid nudibranchs other than one specialized feeder, inhibited the settlement and/or metamorphosis of invertebrate larvae, and were strongly antimicrobial and cytotoxic. Sponges exuded significantly more of these two alkaloids when wounded,[155] and, consistent with their defensive role, the metabolites were found to be localized in spherulous cells close to the aquiferous exhalant canals.[156]

A second group of bioactive sponge metabolites with an unusual structural motif are the terpene isocyanides, which often co-occur with structurally related isothiocyanates and formamides.[157–159] Less commonly encountered nitrogenous based substituents present in sponge terpene metabolites include isocyanates, thiocyanates, and dichloroimines (carbonimidic dichlorides). The biochemistry and ecology of these unique marine metabolites have been targeted for study by numerous researchers.[159,160]

Garson showed that the two isocyanide groups of the diterpene diisocyanoadociane (Structure 2.90) derived from inorganic cyanide.[161,162] Since then, numerous studies using classical precursor incorporation labeling techniques have shown that the N_1–C_1 substituents of marine sponge terpenes derive from inorganic cyanide,[163–166] and that these functional groups can also originate from thiocyanate.[164–167] ^{13}C–^{15}N labeling studies demonstrated that the carbon-nitrogen bond of cyanide remains intact during the biosynthesis of 2-isocyanoneopupukeanane

(Structure 2.91).[163] Advanced precursor studies have demonstrated metabolic conversions between isocyano- and isothiocyanato-metabolites.[167–169]

Although insects and higher plants use enzymatically generated cyanide as a chemical defense weapon and many terrestrial bacteria have the capacity to produce cyanide *in situ*,[170] inorganic cyanide has not been detected in the various sponges that produce N_1–C_1 metabolites. The question, then, is how do marine sponges utilize cyanide as a building block in the biosynthesis of antibacterial or cytotoxic metabolites? The biochemical origin of the cyanide or thiocyanate is not yet understood, and it is mechanistically plausible that the free anion is not generated *in situ*. The potential role of microbial symbionts in cyanide production in isocyanide-containing sponges has not yet been adequately explored. One marine sponge has clearly evolved mechanisms for handling inorganic cyanide; Hamann and Scheuer[171] have described a cyanide-smelling sponge which contains the terpene-quinone-metabolite (Structure 2.92). It has been proposed that Michael addition of cyanide ion generates this unusual adduct.

The diterpene isocyanide diisocyanoadociane (Structure 2.90) is localized in the sponge cells of *Amphimedon terpenensis*, where its synthesis is suggested.[172] The terpene components of several other sponge species, including spirodysin (Structure 2.93), avarol (Structure 2.94), and the fish feeding deterrent *ent*-furodysinin (Structure 2.95), have been shown to be localized in sponge cells rather than in any associated symbionts.[71,72,173,174] There is as yet no sound experimental evidence confirming the biosynthetic origin of the terpenoid skeleton of marine sponge isocyanides and associated N_1–C_1 metabolites. The classical terpene precursors, acetate and mevalonate, are not incorporated into sponge sesqui- and diterpenes,[161,162,175] At issue here may be the experimental constraints on the detection of terpene biosynthesis in marine systems,[13] and the fact that the biosynthetic precursors supplied may potentially be used preferentially for other metabolic processes.[176] The recently discovered mevalonate-independent route to terpenes has not yet been tested in marine sponges, but is known to operate in other marine phyla.[15,86]

There is an ecological benefit to the sponges that manufacture N_1–C_1 metabolites. Marine sponges containing isocyanides and isothiocyanates are rarely overgrown by other sessile invertebrates, a field observation which has been confirmed by laboratory assays.[140] Other isocyano-containing sponges inhibit the settlement of larvae of the ascidian *Herdmania curvata*.[159] A wide range of sesquiterpene and diterpene isocyanides and isothiocyanates inhibit settlement of the larvae of the barnacle *Balanus amphitrite* or of the worm *Halocynthia* sp.[177–179] In these assays, some of the isocyanoterpenes were nontoxic towards *B. amphitrite*.[177] A feeding-deterrence role has frequently been proposed for terpene isocyanide and isothiocyanate metabolites in sponges,[180] but this has yet to be experimentally demonstrated.[142,164]

The chemical strategies used by sponges in competition for space are further illustrated by a field study in which the sponge terpene metabolite 7-deacetoxyolepupuane (Structure 2.96) from *Dysidea* sp. caused necrosis when incorporated into agar strips and placed in contact with the sponge *Cacospongia* sp. In addition, crude extracts and the purified 7-deacetoxyolepupuane deterred feeding by the spongivorous fish *Promacanthus imperator*.[181] Although this bioactive metabolite has been isolated from nudibranchs, it is not sequestered from dietary sponges. The *de novo* biosynthesis of this metabolite in nudibranchs via the classical mevalonate pathway is discussed in Section VI. Thus, both sponges and molluscs appear to have evolved the same biosynthetic pathway.

Variations in sponge chemical defenses in response to environmental conditions have been examined, but disappointingly little is known yet of the underlying biosynthetic processes other than those for the bromotyrosine alkaloids described earlier. Thompson et al.[182] correlated the variable diterpene content in the sponge *Rhopaloeides odorabile* with environmental conditions, and suggested that the sponge modified diterpenes to avoid self-toxicity. Environmental conditions and specimen size influence the level of toxicity shown by the Mediterranean *Crambe crambe*,[183,184] which produces toxic guanidine alkaloids such as crambine A (Structure 2.97) and crambescidin 816 (Structure 2.98). These metabolites have been found localized in one type of spherulous cell located in the periphery of the sponge, consistent with a defense role.[184,185] Based on the range of bioactivities detected, Becerro et al.[141] proposed that individual alkaloids in the sponge extract may have acquired multiple ecological roles in response to diverse ecological pressures (such as predation or space competition), and that different alkaloids might share the same ecological function. Sponges such as *Crambe crambe* may use a range of metabolites to synergistically enhance biological activity, or to avoid self-toxicity through buildup of a single metabolite. In the sponge *Cacospongia* sp., the metabolite desacetylscalaradial (Structure 2.99) is concentrated in the tips rather than basal tissue where it may influence the distribution of the nudibranch predator *Glossodoris pallida*.[186] In the sponge *Oceanapia* sp., deterrent kuanoniamine metabolites (Structures 2.100–2.101) are localized in exposed body components involved in asexual reproduction.[187] It would be interesting to know whether intraspecimen chemical variation reflects different biosynthetic capabilities within individual sponge cell types.

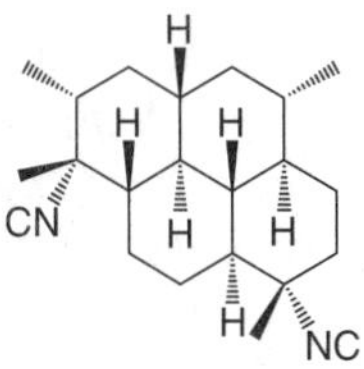

2.90 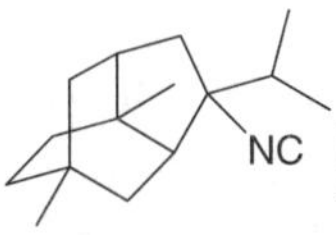   2.91 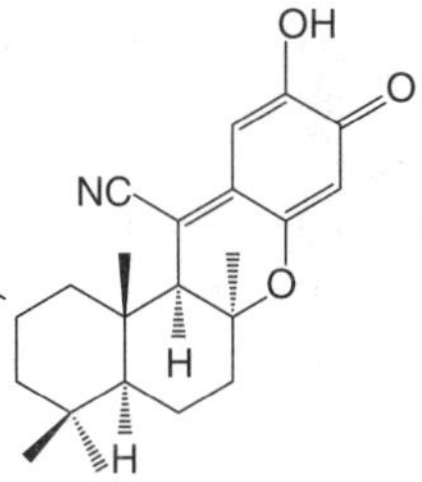   2.92 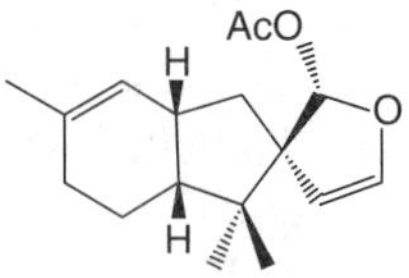   2.93 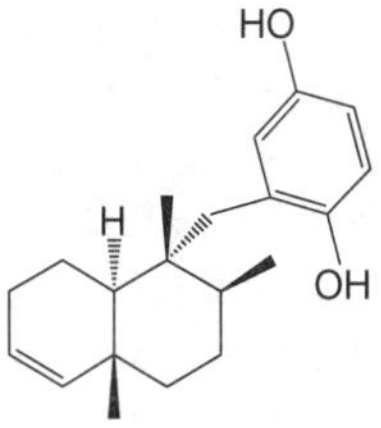   2.94

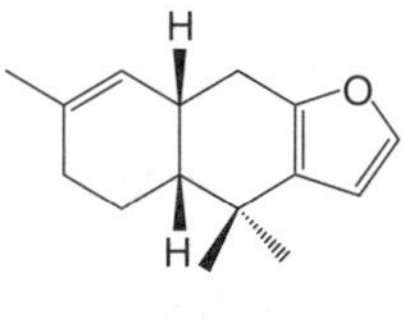

2.95 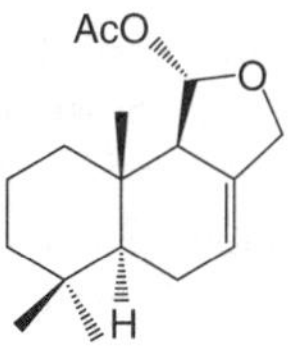   2.96 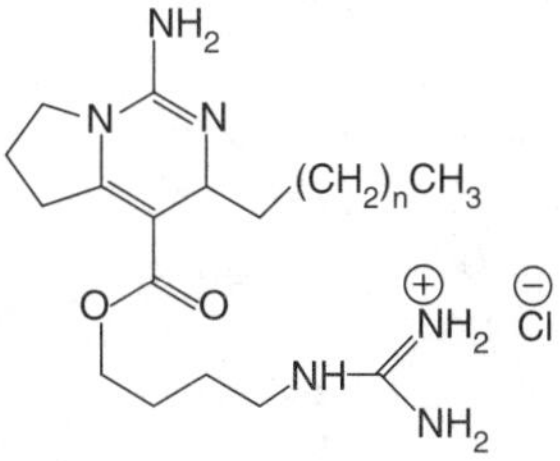   2.97

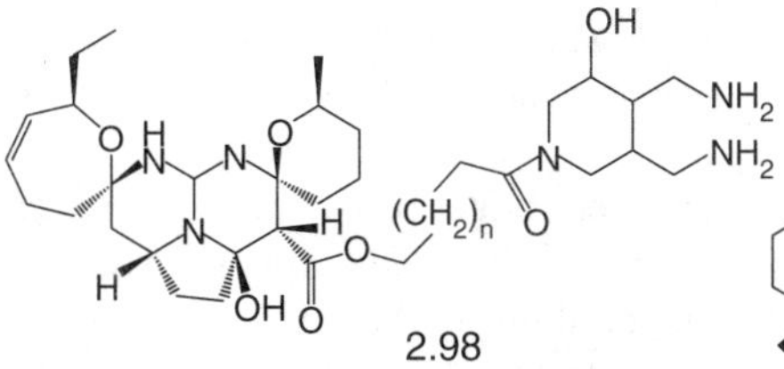

2.98 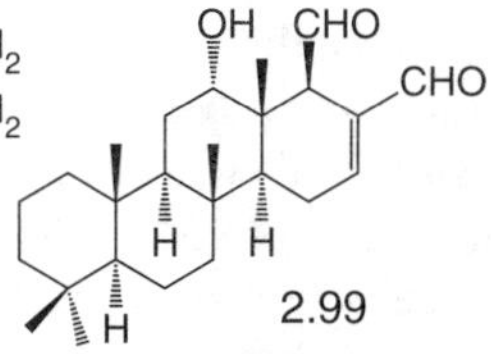   2.99 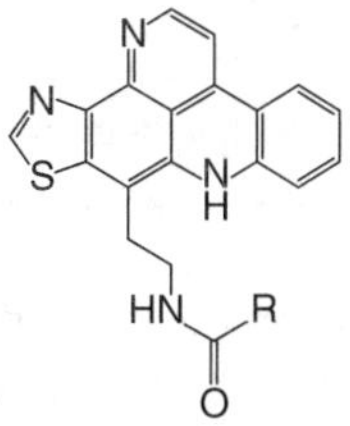   2.100 R = -CH₂CH₃   2.101 R = -CH₃

2.100 R = -CH$_2$CH$_3$
2.101 R = -CH$_3$

V. OTHER SESSILE MARINE ANIMALS

Tunicates, together with invertebrates such as soft corals, gorgonians, and bryozoans, all compete with marine sponges for space and resources, and may need to develop mechanisms to prevent fouling by larvae or bacteria. Additionally, they may suffer predation. Some potent pharmacologically active agents have been isolated from this group of marine animals, including the didemnins, ecteinascidins, pseudopterosins, and bryostatins. Biosynthetic experiments with this group of marine animals have utilized intact tissue[188] or cell-free extracts.[40,189,190]

A. SOFT CORALS, HYDROIDS, AND GORGONIANS

The typical secondary metabolites of soft corals are diterpenoids, although some species also produce sesquiterpenoids. The sesquiterpene portion of furoquinol (Structure 2.102) is labeled by [2-³H]mevalonolactone in *Sinularia capillosa*,[191] while *Heteroxenia* sp. converts acetate and mevalonate into cubebol (Structure 2.103) and clavukerin A (Structure 2.104).[188] In contrast, acetate is used for cetyl palmitate synthesis in *Alcyonium molle*, but not for *de novo* diterpene biosynthesis.[188]

The diterpene crassin acetate shown in Structure 2.105 is biosynthesized from acetate and bicarbonate in *Plexaura porosa*.[192] Its synthesis by a cell-free extract prepared from the intracellular dinoflagellate symbiont *Symbiodinium microadriaticum* was taken as evidence of the symbiont origin of the terpene. To account for the structural diversity of coelenterate terpenes, the animal host was assumed to control the biosynthetic process.[193] *De novo* squalene biosynthesis has been demonstrated in broken cell preparations from four species of Caribbean gorgonians, but the preparations did not convert mevalonate to diterpene metabolites such as crassin acetate.[194] However, it now seems unlikely that coelenterate terpenes are of dinoflagellate origin. Some soft corals lack algal symbionts but still contain terpenes, while dinoflagellates isolated from gorgonians produced different sterols in culture to those found in the intact symbiosis.[195] A stable isotope ratio study clearly implicates the coelenterate host as the source of terpenes in soft coral and gorgonian symbioses.[196]

Studies of the biosynthesis of the biomedically important pseudopterosin metabolites in the gorgonian *Pseudopterogorgia elisabethae* reveal the precursor roles of geranyl geranyl pyrophosphate and xylose. A cell-free extract converted tritiated geranyl geranyl pyrophosphate into pseudopterosin A (Structure 2.106); thus, acetylation of the xylose moiety, yielding other pseudopterosin metabolites, occurs at a late stage in biosynthesis.[190] Unlike many Caribbean gorgonians, *Pseudopterogorgia* spp. show antimicrobial activity.[197]

Terpenes have been implicated in the evolutionary success of the coelenterates. Soft coral terpenes play an important role in reproductive processes such as egg release or chemotaxis.[198] (-)-*Epi*-thunbergol (Structure 2.107) has been identified as the sperm attractant in *Lobophytum crassum*,[199] and the methyl ester of 15-*epi*-acetoxy-prostaglandin A₂ (Structure 2.108) has been implicated in egg release.[198] Some corals produce egg-specific defensive terpenes such as pukalide (Structure 2.109) or epoxypukalide (Structure 2.110).[200] Based on analyses which showed an increasing rather than decreasing level of metabolites, the larval stages of the soft coral *Sinularia polydactyla*[201] and the gorgonian coral *Briareum asbestinum*[202] appear capable of diterpene synthesis. Presumably, in each of these corals, the metabolic energy expended in biosynthesis is balanced by successful reproductive outcomes.

Despite the fleshy nature of soft corals relative to other invertebrates, predation is rare because the diterpene metabolites produced are potent feeding deterrents.[198,203] However, some specialist feeders feed exclusively on chemically protected species, generally from the tips where the terpene concentration is maximal, and have evolved mechanisms to detoxify the metabolites.[204,205] Mineralization appears ineffective as a means of defense against specialized predators.[204] Secondary chemicals are also used as a defense mechanism in gorgonians.[206] The concentrations of ichthyodeterrent terpenes is highest in the polyp-bearing region of *Pseudopterogorgia* spp. where they may deter vertebrate predators. In contrast to soft corals, basal spiculation in gorgonians may deter specialized predators such as nudibranchs.[207] Habitat-specific variation in gorgonian (*Briareum asbestinum*) chemistry has been reported, and is believed to reflect localized genetic variation.[208] In contrast, Maida et al. have considered environmental factors as the basis of variation in terpene chemistry in *Sinularia flexibilis*.[209]

Soft corals have been shown to release terpene chemicals into the water column.[210] Their interactions with other invertebrates,[211] or with algae[122,212,213] have been monitored. Some algal species cause localized tissue necrosis, or reduce terpene levels, in adjacent soft corals;[122,213] however, stressing soft corals by transplanting them to new sites stimulates terpene synthesis.[199] Some species of algae, notably the chemically protected *Halimeda* spp., grow in abundance at the base of soft coral colonies, but the soft coral chemical defense does not play a role in maintaining this association.[214]

Soft corals, gorgonians, and hydroids are often characterized by high endogenous levels of prostaglandins (e.g., PGA_2 in *Plexaura homomalla*), while some species produce biosynthetically related, but structurally specialized, secondary metabolites (e.g., punaglandins in *Telesto riisei* or clavulones in *Clavularia viridis*).[215] Although the mechanistic aspects of prostanoid biosynthesis in coelenterates are quite well understood,[127,216] the precise role played by these chemicals in the physiology and ecology of these animals is poorly understood. The high levels of prostanoids in coelenterates has led to speculation of additional ecological roles. A report that they may be antifeedants rather than allelochemicals[217] has been disputed.[218] Since metabolic processes leading to eicosanoid derivatives in algae are well documented, the host or symbiont origin of coelenterate prostaglandins is of interest.

B. TUNICATES

The biosynthesis of eudistomin H (Structure 2.111) and I (Structure 2.112), which are of clinical interest since eudistomin metabolites are cytotoxic and also show antimicrobial and antiviral activity, has been studied by Shen and Baker.[219,220] Proline and tryptophan were confirmed precursors, while ornithine and arginine were not significantly utilized.[220] The formation of eudistomin H from 5-bromotryptamine and 5-bromotryptophan, and of eudistomin I from tryptamine, suggests bromination as the key step which commits the amino acid to secondary metabolic processes.[219] The eudistomins are produced in very small quantities by the tunicate *Eudistoma olivaceum*, which is a major fouling organism commonly found on mangrove roots. An antifouling role for the eudistomins, suggested by the field observation that the surface of the tunicate is generally free from epibionts, was supported by testing either crude extract[221] or a mixture of eudistomins G (Structure 2.113) and H in laboratory settlement assays with four species of invertebrate larvae, and field settlement assays with the ascidian *Diplosoma glandulosum*. The eudistomin mixture did not inhibit feeding by the pinfish *Lagodon rhomboides* in laboratory assays.[222] Although these two eudistomins are not the most potent of the eudistomin family, these studies show they exert specific ecological roles. The tunicate metabolite shermilamine B (Structure 2.114), structurally related to the ichthyodeterrent sponge metabolites shown in Structures 2.100 and 2.101, is also derived from tryptophan.[223] The utilization of closely related biosynthetic pathways in two phyla could reflect the ecological significance of these natural products.

It has been anticipated that ascidian larvae would be chemically defended, since the larvae may settle in noncryptic habitats or in the vicinity of adults, where they may frequently be exposed to predators. Both larvae, as well as adults, of the Caribbean tunicate *Trididemnum solidum* contain cytotoxic didemnin metabolites which have been intensively studied as potential anticancer and immunosuppressive agents. A mixture of these didemnin metabolites, containing predominantly didemnin B (Structure 2.115), is now known to be a fish feeding deterrent.[224] Chemically defended larval mimics fed to the sea anemone *Aiptasia pallida* affected their growth and asexual reproduction and induced vomiting in the spotted pinfish *Lagodon rhomboides*, thus causing these fish to avoid feeding.[225] It is not yet clear whether these chemicals are synthesized within the larvae.

The tunichromes constitute a class of amino-acid-derived metabolites isolated from the blood cells of ascidians. The chemicals are involved in vanadium sequestration and reduction in the blood cells and may be involved in a primitive clotting mechanism to repair damaged tissue.[226,227] The *in vivo* incorporation of [^{14}C]phenylalanine into tunichrome An-1 (Structure 2.116) in *Ascidia ceratodes* has been reported.[228] Interestingly, vanadium is an inhibitor of larval settlement.[229]

VI. MARINE MOLLUSCS

Many species of marine molluscs are physically defended from predation by the presence of a hard shell or by the ingestion of undischarged nematocysts. The need for soft shelled or soft bodied molluscs to have an alternative, chemically-based defensive mechanism was first explored by Faulkner and Ghiselin.[230] When molested, many species of mollusc secrete a mucus that is deterrent to predators and which has been shown to be toxic;[131,180,231,232] many mollusc extracts are toxic.[233]

Molluscs can be divided into two biosynthetic categories based on the chemistry they exhibit relative to their diet; those that are dependent on dietary sources express chemistry that reflects their choice of diet. In some instances, molluscs have been shown to chemically modify the ingested compounds. These transformations may either enhance the deterrent nature of the metabolite or alternatively represent a detoxification mechanism; examples of both are documented below. The second category of molluscs are those which have the ability to biosynthesize metabolites *de novo* and, hence, may sometimes express a preference for a diet lacking secondary metabolites. The expense of maintaining secondary metabolic function is balanced against the lack of dietary constraints. Molluscs that can produce their own defensive allomones are likely to have an advantage over those dependent on a dietary source of metabolites.

Studies on the biosynthesis of mollusc natural products have contributed significantly to marine chemical ecology. To date, research has been conducted on herbivores such as sea hares, sacoglossans, and pulmonate limpets, and on carnivorous molluscs such as nudibranchs. Precursors used in labeling experiments can be injected into the mantle tissue or hepatopancreas of the molluscs or taken up by absorption through the skin.[234]

A. HERBIVORES

Sea hares are large, mobile herbivores that cannot restrict feeding to one plant species alone, but which show some specialization.[79] As described earlier (Sections IIC and III), sea hares acquire dietary metabolites from macroalgae and cyanobacteria. In some cases, sequestration appears to be selective; for example, the sea hare *Aplysia parvula* accumulates selected halogenated furanone metabolites from the red algae *Delisea pulchra*.[117] Dietary metabolites may be modified chemically, either to enhance their activity as feeding deterrents or as a detoxification mechanism. Stallard and Faulkner demonstrated the conversion of laurinterol (Structure 2.117) into aplysin (Structure 2.118) in *Aplysia californica* using a coated seaweed diet.[235] *Stylocheilus longicauda* feeds on the cyanobacterium *Lyngbya majuscula* (syn *Microcoleus lyngbyaceus*) and sequesters the malyngamides A (Structure 2.19) and B (Structure 2.20), but also modifies the ingested malyngamide B to its acetate derivative (Structure 2.119).[47,236] The cyanobacterial metabolites malyngamides A and B, and a mixture of majusculamides A and B, all deter feeding by pufferfish and crabs in assay, but the sea hare acetate does not deter feeding by either consumer. Since the sequestered metabolites are present in the digestive tract rather than in exterior parts or in the eggs, and thus are not optimally located for defense purposes, the sea hare may be feeding on a chemically defended algae simply to escape predation.[46,47] Acetylation may therefore represent a detoxification mechanism.

Sacoglossan molluscs feed suctorially on algae, notably on chemically protected species such as *Caulerpa* spp. and the calcified *Halimeda* spp., from which they may acquire terpene metabolites for defensive purposes. Extracts of *Elysia translucens* contain udoteal (Structure 2.120), the same metabolite as its algal diet,[237] whereas *E. halimedae* feeds on the lightly calcified tips of *H. macroloba*, which contain high levels of halimedatetraacetate (Structure 2.39), and converts this metabolite to an alcohol (Structure 2.121) for its own defensive purposes and to protect its egg masses.[236] Other *Elysia* spp. also contain the algal metabolites chlorodesmin (Structure 2.41),[89] or cyclic depsipep-

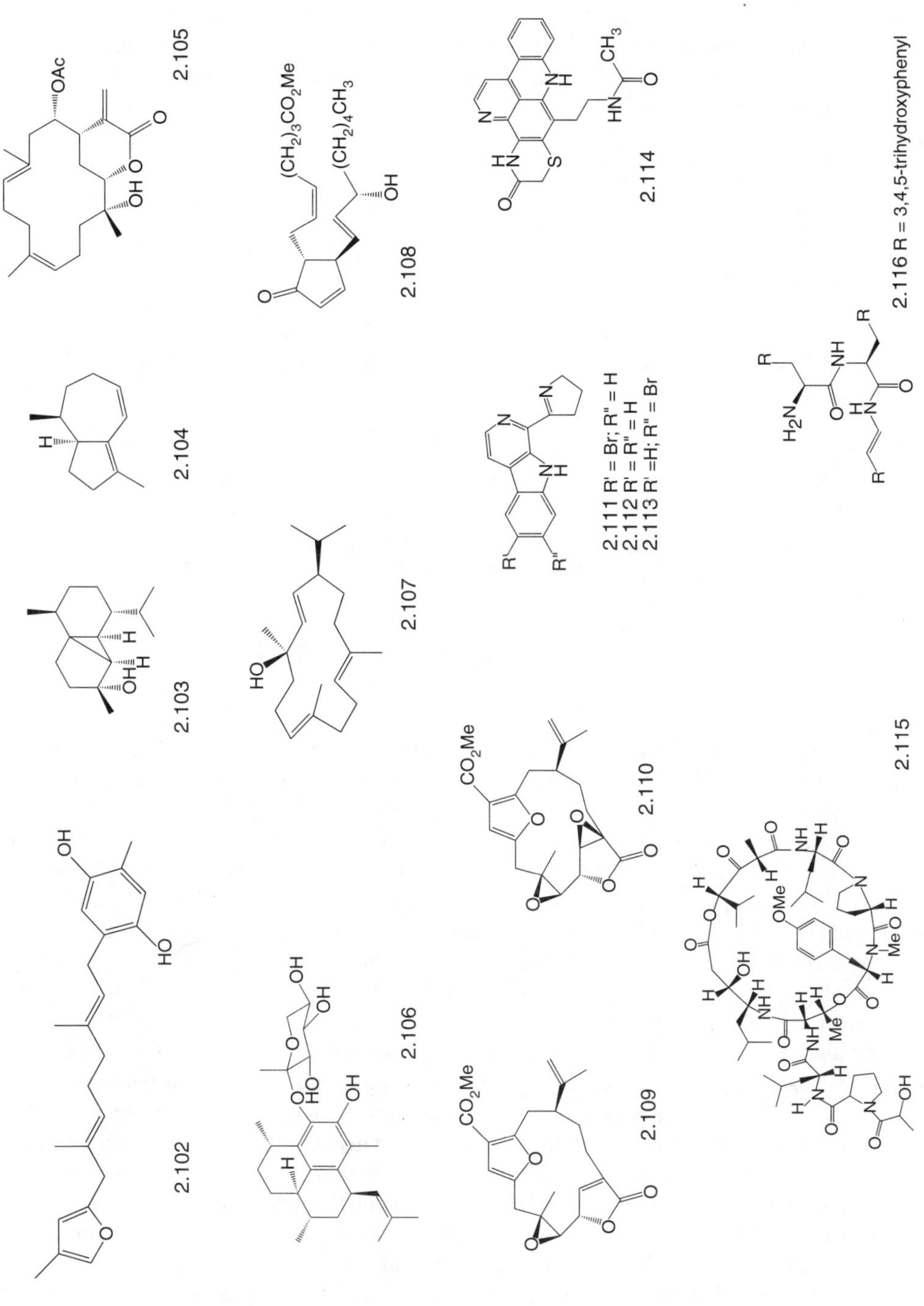

2.105

2.104

2.103

2.102

2.106

2.107

2.108

2.109

2.110

2.111 R' = Br; R" = H
2.112 R' = R" = H
2.113 R' = H; R" = Br

2.114

2.115

2.116 R = 3,4,5-trihydroxyphenyl

tides.[238,239] Oxynoid molluscs of the genera *Ascobulla*, *Oxynoe*, and *Lobiger* contain metabolites that are structurally related to caulerpenyne (Structure 2.43) present in their *Caulerpa prolifera* diet. The Mediterranean *Oxynoe olivacea* and *Lobiger serradifalci* modify caulerpenyne to the monoaldehyde oxytoxin-1 (Structure 2.122), which is compartmentalized in external parts that detach when the molluscs are molested. In *O. olivacea*, *L. serradifalci*, and *Ascobulla fragilis*, further modification to the potent 1,4-dialdehyde oxytoxin-2 (Structure 2.123) occurs. All three molluscs contain oxytoxin-1 in their mucus secretions, however, only *A. fragilis* retains traces of caulerpenyne in its body.[240]

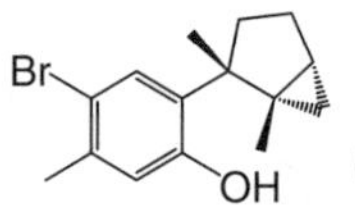

2.117 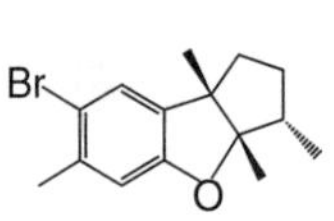   2.118 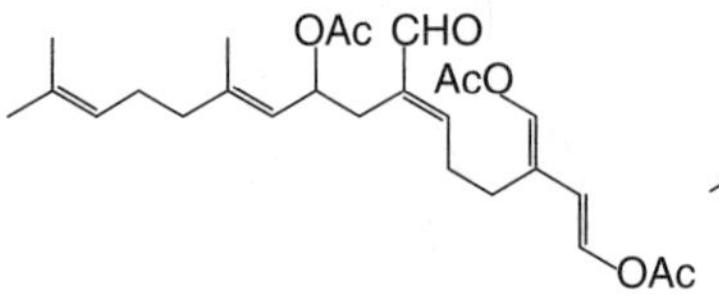   2.120 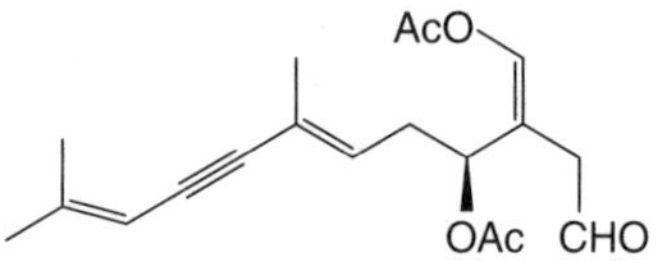   2.122

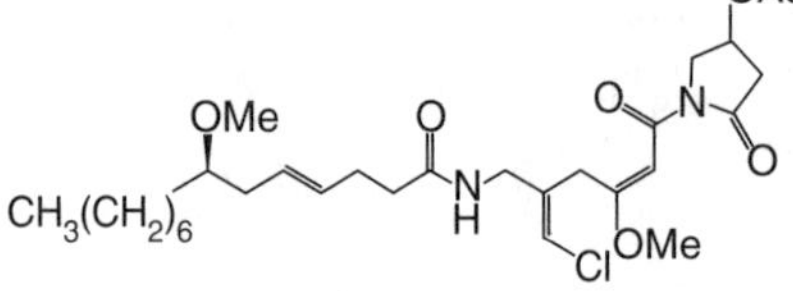

2.119 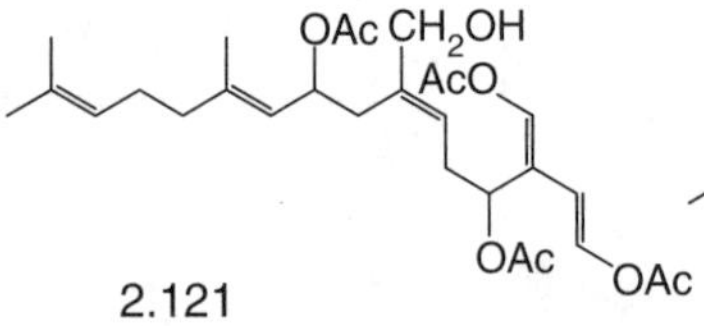   2.121 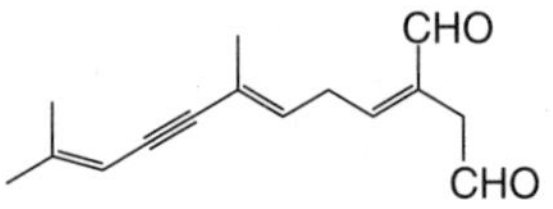   2.123

Many species of sacoglossans contain "polypropionate" metabolites, oxygenated compounds characterized by a highly branched carbon skeleton formally constructed from propionate units. Experimental evidence is accumulating that these metabolites are convincing markers of *de novo* biosynthesis in herbivorous molluscs: first, these polypropionate metabolites have not yet been found in algae, bryozoans, or ascidians, and their occurrence in sponges is restricted. Second, their biosynthesis from propionate or bicarbonate has been experimentally confirmed.[241] Chloroplasts, derived from an algal diet, are functional symbionts in many species of Sacoglossa; hence, carbon dioxide is another potential biosynthetic building block. Photosynthetically derived sugars may be converted to propionate via the tricarboxylic acid (TCA) cycle and succinate. This biosynthetic strategy reduces the dependence on an external food source.

When the sacoglossan *Placobranchus ocellatus* was immersed in seawater containing ^{14}C sodium hydrogen carbonate, ^{14}C label was incorporated into both 9,10-deoxytridachione (Structure 2.124) and the photorearrangement product photodeoxytridachione (Structure 2.125).[242] In *Elysia viridis*, sodium [1-^{14}C]propionate was incorporated into elysione (Structure 2.126), an ichthyotoxic metabolite found in the mollusc's mucus secretions.[237] Attempts to repeat this propionate labeling experiment in *Elysia timida* and to test the putative role of acetate in the biosynthesis of 15-norphotodeoxytridachione (Structure 2.127) were unsuccessful.[237,243] The metabolites shown in Structures 2.124, 2.125, 2.126, and 2.128, some of which were toxic, were present in the mucus secretion of *Elysia timida*, consistent with a defensive role.[243] The hypothesis that these complex pyrones function as suncreens[242] has been disputed.[243]

The brightly colored sacoglossan *Cyerce crystallina* contains toxic cyercene metabolites, which are variously localized in the mantle tissue, the detachable cerata, and the mucus extract produced by the animals when molested. It is therefore inferred that these metabolites serve defensive roles or act as inducers of cerata formation.[244,245] The freshly regenerated cerata from specimens of *C. crystallina* that had been kept in seawater containing [2-^{14}C]propionate yielded various radiolabeled cyercene metabolites. The highest incorporation was into cyercene A (Structure 2.129), consistent with a proposed growth-inducing role, and the lowest into cyercene 1 (Structure 2.130). It has not yet been determined whether the congeneric species *C. nigricans*, which contains polypropionates, manufactures these metabolites *de novo*.[246] None of the polypropionate metabolites, nor the traces of dietary chlorodesmin present, were found to be responsible for the repellant nature of the crude sacoglossan extract.[89,246]

The stiligerid sacoglossan *Ercolania funerea* contains several pyrone metabolites including cyercene B (Structure 2.131); preliminary evidence has been presented that these pyrones are propionate derived.[247] 7-Methylcyercene-2 (Structure 2.132) and -1 (Structure 2.133), both of which were growth inducers, were found mainly in the cerata, together with 7-methyl cyercene B (Structure 2.134). Mucus extracts from *E. funerea* extracts were modestly toxic.[245] In contrast, the polypropionate metabolites of *Placida dendritica* were not partitioned between various body parts, although slime extracts from *P. dendritica* were highly toxic to the mosquitofish *Gambusia affinis*. A probable *de novo* origin is inferred for the placidenes,[248] and also for the membrenones, which are feeding deterrents isolated from the skin of the pleurobrancoidean mollusc *Pleurobranchus membraneus*.[249]

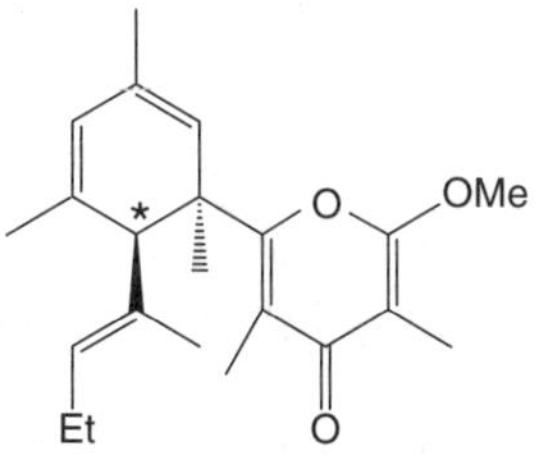

2.124
2.128 epimeric at *

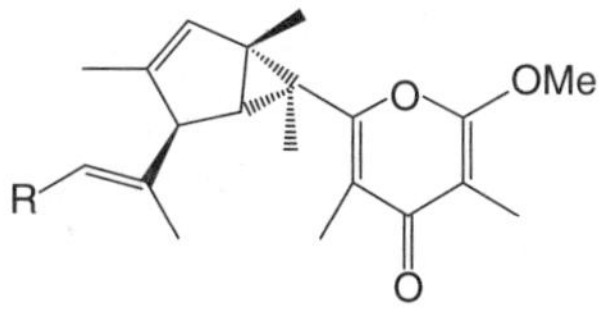

2.125 R = Et
2.127 R = Me

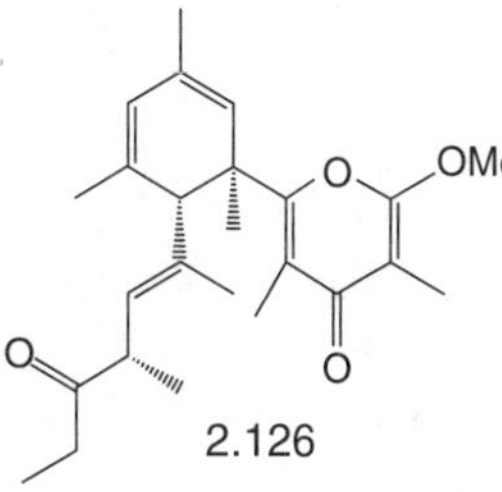

2.126

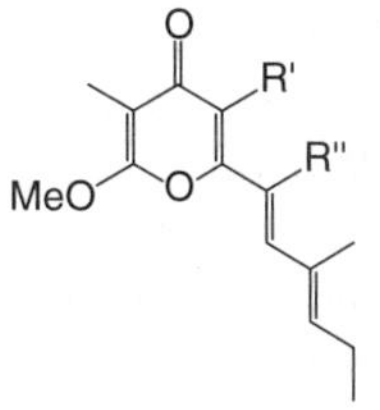

2.129 R' = R" = Me
2.131 R' = R" = H
2.134 R' = H; R" = Me

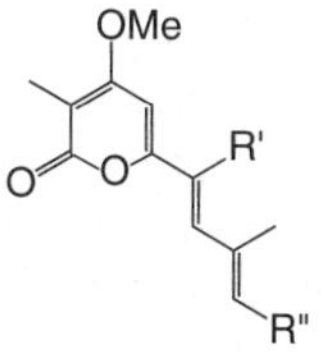

2.130 R' = H; R" = Me
2.132 R' = Me; R" = Et
2.133 R' = Me; R" = Me

A hypothetic evolutionary path has been traced for the Sacoglossa. In the families Oxynoidea and Polybranchiidea, the more ancestral species are chemically related to their algal diets, whereas more evolved species are capable of *de novo* biosynthesis.[237,250]

Like the Sacoglossa, some pulmonates contain propionate metabolites of *de novo* origin that have been implicated in their chemical protection. Molluscs of the genus *Siphonaria* are air breathing and roam the intertidal zone, feeding on algae, where they may be susceptible to predators despite the presence of a shell. The *de novo* origin of the denticulatins A and B (Structures 2.135 and 2.136) in *Siphonaria denticulata* and of the siphonarins A and B (Structures 2.137 and 2.138) has been confirmed. Incorporation of [1-^{14}C]acetate into *S. denticulata* did not provide significantly labeled denticulatins, but this precursor was utilized by *S. zelandica* for the synthesis of siphonarin A.[234,251] The role of succinate in furnishing propionate units is consistent with the presence of a functioning methylmalonyl mutase in these molluscs.[251] Siphonariid polypropionate compounds appear to play a limited role in chemical defense. Although the denticulatins are ichthyotoxic and have been shown to be located in the foot tissue and mucus of *S. denticulata*, the molluscs are commonly eaten by predators,[234] as are specimens of *S. maura* which produce the vallartanones. Vallartanone B (Structure 2.139) showed some fish-feeding deterrency, while vallartanone A (Structure 2.140) induced larval settlement in the tubeworm *Phragmatopoma californica*.[252] Siphonariid tetrahydropyran metabolites are chemically unstable, and may not be the actual natural products produced by the molluscs.[253,254]

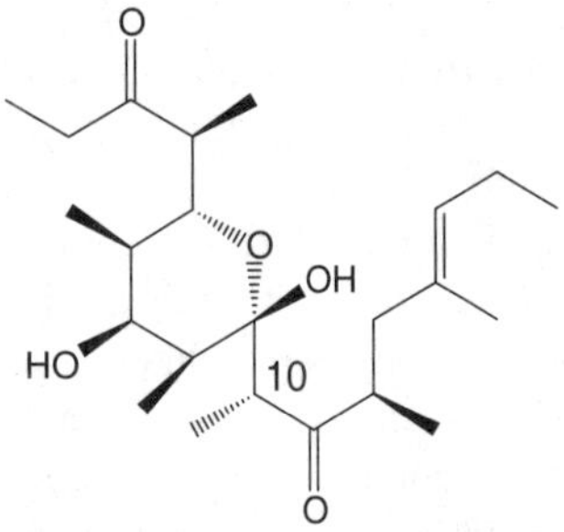

2.135
2.136 epimer at C-10

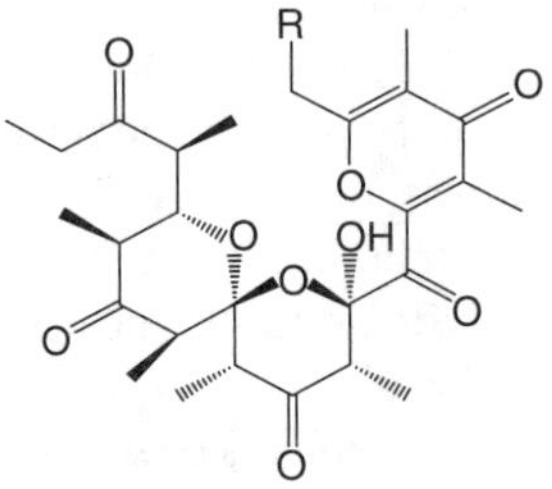

2.137 R = H
2.138 R = Me

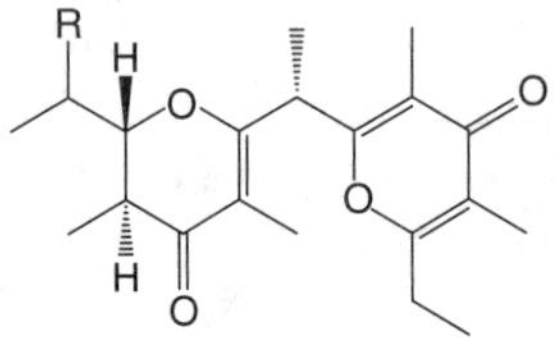

2.139 R = Me
2.140 R = H

B. CARNIVORES

Nudibranchs feed on invertebrates, or on other molluscs, and can be divided into two biosynthetic categories based on whether the composition of their secondary metabolites reflects their diet or *de novo* biosynthesis.[255] Nudibranchs that are dependent on dietary sources such as sponges, ascidians, or soft corals, contain chemicals that reflect their preferred diet (which may be a single sponge or a variety of sponges depending on availability). Such nudibranchs exhibit a range of chemical types and structures that reflect their choices of food, and their chemistry varies according to where they are collected.[131,180,232,256–259] The mechanisms of chemoreception involved are not understood,[142,260] although some progress has been made in understanding how metabolic selection confers an advantage on the mollusc. Although many of the sequestered metabolites are toxic, they are often localized in specialized spherical dorsal glands called mantle dermal formations (MDFs), or mantle border or in the gills, parts of the animals that may be expected to be prone to predation.[261–263] In this way, the benefits of utilizing a bioactive chemical as a defense weapon can be realized without toxic side effects.[142]

Two studies now provide convincing experimental support for dietary transfer. The MDFs of *Hypselodoris picta* (*webbi*) were shown to assimilate furodysinin (Structure 2.94) or the spiniferins-1 and 2 (Structures 2.141 and 2.142) in addition to their "regular" metabolite longifolin (Structure 2.143) when the molluscs were placed on sponges (*Dysidea fragilis*; *Pleraplysilla spinifera*) known to produce the components shown in Structures 2.94, 2.141, and 2.142.[264] Phyllidid nudibranchs are frequently associated with isocyanide/isothiocyanate-containing sponges from which they are inferred to sequester sesquiterpene metabolites.[260,262,265] Individual specimens of *Phyllidiella pustulosa* were allowed to feed on ^{14}C-cyanide- or thiocyanate-labeled specimens of the sponge *Acanthella cavernosa*, then starved for a short period to aid complete digestion of gut contents. On extraction, the nudibranchs contained radioactive axisonitrile-3 (Structure 2.144) and isothiocyanate-3 (Structure 2.145); thus, during the course of the experiment, the sponge synthesized ^{14}C-labeled quantities of the metabolites, which were then sequestered by the nudibranch. Control experiments involving injection of cyanide or thiocyanate into *P. pustulosa* demonstrated that these nudibranchs were unable to carry out *de novo* biosynthesis.[164]

Nudibranchs may selectively concentrate certain allelochemicals alone from their sponge diet, or they may chemically modify the ingested compounds.[142,261,262,266–268] The Mediterranean *Hypselodoris orsini* is able to transform dietary scalaradial (Structure 2.146) into deoxoscalarin (Structure 2.147), concentrated in the viscera, and into 6-keto-deoxoscalarin (Structure 2.148) found in MDFs.[261] The nudibranch *Glossodoris pallida* sequesters scalaradial (Structure 2.146) from the sponge *Cacospongia* sp. and converts it into deoxoscalarin (Structure 2.147), but does not sequester scalarin (Structure 2.149), the major sponge metabolite.[142] The nudibranchs also contain desacetylscalaradial (Structure 2.98). Deoxoscalarin is found in the reproductive system, eggs, and mantle border of *G. pallida,* while desacetylscalardial and scalaradial are predominantly present in the mantle border. Although removal of the nudibranch mantle increases susceptibility to predation by reef fish, the specific location of the diet-derived compounds was not significant.[269] Localization of compounds in the mantle tissue, however, may prevent autotoxicity or facilitate mucus production. The egg masses of *G. pallida*, which are eaten by a variety of fish, are low in the major sesterpenoid secondary metabolites.[142] Other glossodorid nudibranchs feed on sponges that contain heteronemin (Structure 2.150). Results of feeding-deterrence with *Cacospongia* or *Glossodoris* extracts in both field and laboratory assays were quite variable. The glossodorid nudibranchs may select to feed on *Cacospongia* sponges so that they are protected by hiding and feeding on a toxic sponge rather than the provision of a specific chemical defense.[142] Nudibranchs are typically found on the base of the sponge, rather than the tips, where they may be less accessible and where they are exposed to lower levels of terpene.[186]

Some nudibranchs have adapted to feed on soft corals,[270–273] ascidians,[274] bryozoans,[275] or on other molluscs rather than sponges. The mollusc *Ovula ovum* was reported to sequester the toxic sarcophytoxide (Structure 2.151) from *Sarcophyton* sp. and transform it into the less toxic terpene (Structure 2.152),[276] but the deoxy metabolite was subsequently found in the soft coral.[277]

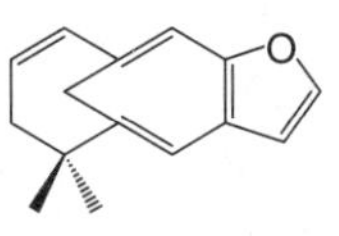

2.141 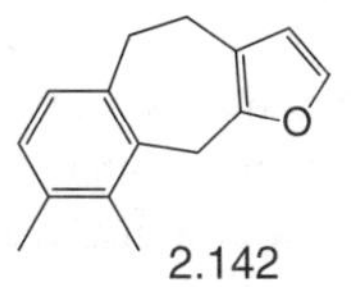   2.142 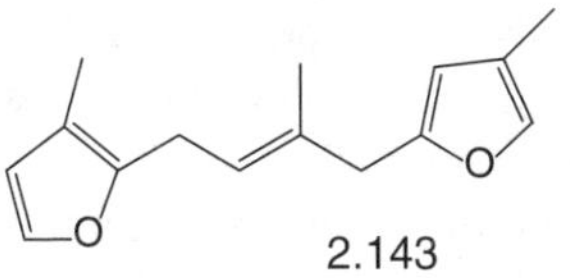   2.143 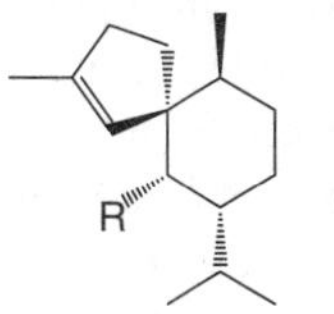   2.144 R = NC
                                    2.145 R = NCS

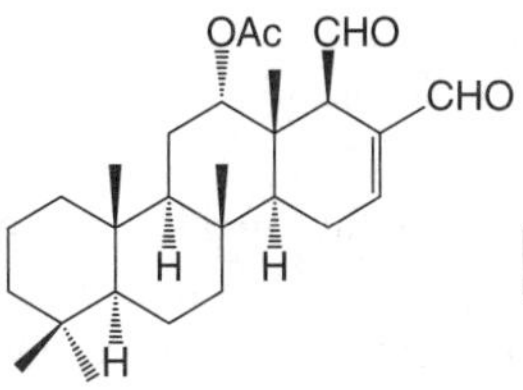

2.146 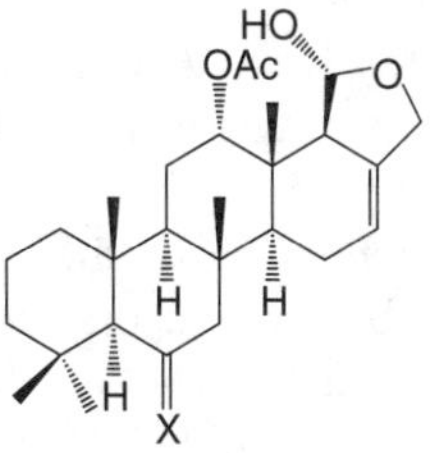   2.147 X = H$_2$ 2.149 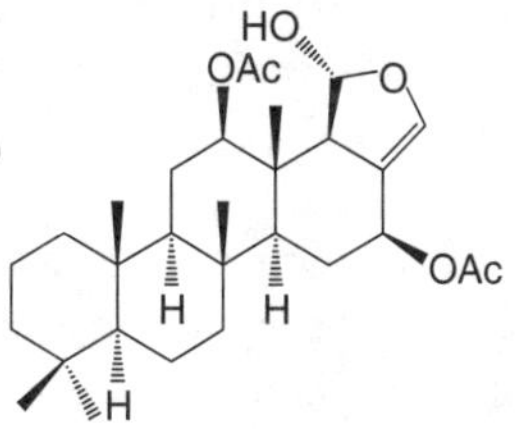   2.150
         2.148 X = O

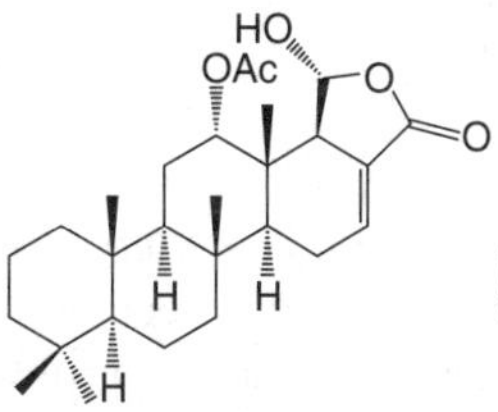

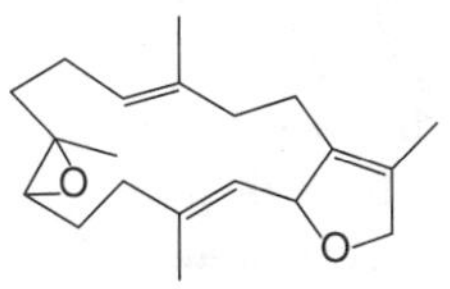

2.151 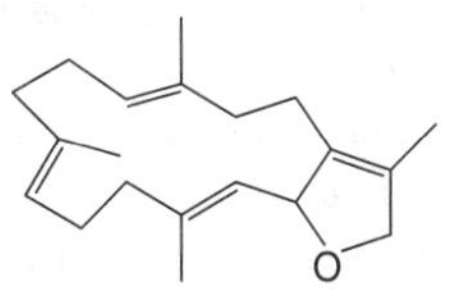   2.152

Cephalaspidean molluscs, notably the Aglajidae, eat other molluscs and thus acquire polypropionate metabolites indirectly, rather than by *de novo* biosynthesis. Thus *Philinopsis speciosa*, *Philinopsis depicta*, *Aglaja depicta*, and *Navanax inermis*, which all feed on *Bulla* spp., contain polypropionates of the aglajne and niuhinone classes.[278–280] Other Cephalaspideans contain alkylpyridine metabolites that are also considered to be indicators of dietary transfer. Molluscs of the Haminoeidae that contain pheremonal alkylpyridines[281–283] are frequently preyed upon by the Aglajidae.[283,284] The acetate-based biosynthesis of haminol metabolites in *Haminoea* spp. could not be demonstrated;[283] a preliminary report that the cephalaspidean mollusc *Navanax inermis* is capable of *de novo* alkylpyridine biosynthesis[285] is inconsistent with its predatory nature.[282] The Pacific *P. speciosa* also acquires cyanobacterial metabolites by feeding on sea hares such as *Stylocheilus longicaudus* and *Dolabella auricularia* which themselves feed on cyanobacteria.[286]

The second category of nudibranchs are those which have the ability to biosynthesize metabolites *de novo* and, hence, may express a preference for an invertebrate diet lacking secondary metabolites.[287] These molluscs show invariant chemistry independent of where they are collected.[255] Experiments conducted in the mid-1980s using ^{14}C-labeled precursors first tested the capacity of nudibranchs to undertake *de novo* biosynthesis.

Cimino et al. injected ^{14}C mevalonate into three species of dorid nudibranch (*Dendrodoris limbata*, *D. grandiflora*, and *D. arborescens*) and demonstrated incorporation of label into polygodial (Structure 2.153), 6β-acetoxyolepupuane (Structure 2.154), 7-deacetoxyolepupuane (Structure 2.95), and the sesquiterpene esters (Structure 2.155), all of which possess a drimane skeleton.[256,288–291] Incorporation of ^{13}C-glucose confirmed the terpenes were synthesized via the classical mevalonic acid route.[291] The tissue distribution of the various drimane sesquiterpenes was investigated, but the metabolites were not present in the mucus secretion where they might be expected to play a defensive role.[292] 7-Deacetoxyolepupuane (Structure 2.95), present in hermaphroditic tissue only, may be a masked form of the reactive and toxic polygodial which is a known feeding deterrent.[290,293] A low level of incorporation of mevalonic acid was reported into the sesquiterpenoid and diterpenoid acid glyceride metabolites, shown in Structures 2.156–2.158, of *Archidoris montereyensis* and *A. odhneri*.[294] However, Cimino et al. reported the inability of Mediterranean specimens of *Doris verrucosa* to manufacture the related glycerides verrucosins A and B (Structures 2.159 and 2.160),[295] but were able to demonstrate that this mollusc synthesizes xylosyl-methylthioadenine (Structure 2.161) *de novo*; although the metabolite is known to accumulate in the mollusc tissue, no biological role has been inferred.[296] The opistobranch mollusc *Tethys fimbrata* readily releases a defensive mucus secretion and sheds cerata when molested.[297] Precursor incorporation experiments have shown the mollusc manufactures novel prostaglandin 1,15-lactones of the A, E, and F series via the arachidonic acid pathway. These chemicals may function as defensive allomones and as precursors to standard prostaglandins in the cerata.[298,299] Their role in the reproductive and early developmental stages of this mollusc have also been probed.[300]

Convincing experimental confirmation of *de novo* biosynthesis in nudibranchs has been achieved using stable isotope-labeled precursors. Effective incorporation levels are achieved by taking advantage of the presumed ecological roles of the metabolites. Handling of the nudibranchs prior to injection leads to release of defensive substances in their mucus and activates the biosynthetic pathways by which the chemical defenses may be rejuvenated. This strategy led to impressive incorporations of [1,2-[13]C$_2$]acetate into the aldehydes nanaimoal (Structure 2.162), acanthodoral (Structure 2.163), and isoacanthodoral (Structure 2.164) in *Acanthodoris nanaimoensis*, and delineated an unexpected mode of biosynthesis of nanaimoal.[301] Subsequent ^{13}C studies by the Faulkner and Andersen groups established the mevalonate origin of the terpene portion of glyceride metabolites shown in Structures 2.156 and 2.158 and tanyolide B (Structure 2.165) in various nudibranchs,[302] and of the sesquiterpene metabolites albicanyl acetate (Structure 2.166), cadlinaldehyde (Structure 2.167), and luteone (Structure 2.168) by *Cadlina luteomarginata*.[303] The new cadlinaldehyde and luteane skeletons were shown to be formed by degradation of a sesterpenoid precursor. The acetyl residue of albicanyl acetate was labeled, but an incorporation into the terpene portion could only be demonstrated during the egg-laying period.[303] The nudibranch *Triopha catalinae* used acetate and butyrate in the *de novo* manufacture of polyketide metabolites such as triophamine (Structure 2.169); the labeling pattern demonstrated was consistent with the use of ethyl malonate units and hinted at a processive mode of biosynthesis.[304,305] The origin of the guanidino group in both triophamine and limaciamine (Structure 2.170) from the North Sea nudibranch *Limacia clavigera* is clearly of interest given the biosynthetic role of cyanide in N_1-C_1 metabolism in marine sponges. The 3-hydroxybutyrate fragment of diaulusterol (Structure 2.171) in the Northeastern Pacific nudibranch *Diaulula sandiegensis* is acetate derived, but stable isotope labeling experiments did not provide any evidence for the origin of the steroidal skeleton.[306]

Studies show that some nudibranchs utilize both biosynthetic strategies and are able to carry out *de novo* biosynthesis as well as ingesting metabolites. The North American *Cadlina luteomarginata* is characterized by an extensive repertoire of sesqui- and diterpenoid constituents, the majority of which can be traced to local sponges; a total of 37 terpenoids, representing 21 different carbon skeletons, have been isolated. However the nudibranch consistently yielded albicanyl acetate (Structure 2.166), a metabolite which deters fish feeding, from all collection sites, while the egg masses of this mollusc contain $1\alpha,2\alpha$–albicanyl acetate (Structure 2.172). The nudibranch may thus produce key defense compounds *de novo* in order to ensure reproductive success, but may take advantage of local sponges for additional defense purposes.[259,307] Likewise, drimane sesquiterpenes manufactured *de novo* by *Dendrodoris grandiflora* are implicated in the protection of egg masses,[292] although these molluscs additionally store metabolites such as microcionins 1 to 4 and fasciculatin, which are known sponge compounds.[256] In contrast to the Pacific *Cadlina luteomarginata*, Atlantic specimens of *Cadlina* acquire secondary metabolites from their diet;[308] it is not yet clear whether this species has dormant biosynthetic capability.

VII. CONCLUSIONS AND FUTURE DIRECTIONS

Species at all levels of the marine phylogenetic tree produce bioactive metabolites in response to the competitive marine environment. Marine microorganisms may need to produce a variety of metabolites because different compounds affect pathogens or host invertebrates in different ways. Any one compound may in itself be insufficient, especially as organisms become subject to intense ecological pressures; therefore, the need arises to produce a range of chemically based defenses combined with other protective mechanisms. The organisms most likely to survive are those with well-defined and novel biosynthetic capabilities that lead to ecologically potent metabolites. The macroalgae are characterized by terpenes or by pheremonal hydrocarbons that originate from eicosanoid metabolism, and frequently use subtle mixtures of these chemicals in response to ecological pressures. Herbivorous molluscs have adapted these algal chemical defenses to their own specialized needs. In invertebrate species, spatial competition necessitates production of compounds to prevent bacteria or the larvae of other invertebrate species from settling on or in the vicinity of the organism, while fleshy species utilize terpenes or nitrogenous metabolites as deterrents against predation. Specialized predators such as nudibranchs appear to overcome these chemical defenses and to sequester bioactive metabolites, either for their own defensive needs or other as yet unidentified functions. In some circumstances, modification of dietary chemicals occurs, either as a detoxification route or to enhance the biological activity of the sequestered metabolite. The ability of molluscs to adapt their diet or nutritional requirements in response to life cycle is an area worthy of further study.

An aspect of marine natural products research with potential significance to the biotechnology sector is the manner in which biosynthetic processes respond to ecological pressures, such as reproduction, predation, or infection, or are affected by habitat or resource needs. Studies of the variations in metabolite composition and concentration in cyanobacteria, algae, or sponges show that intraspecies differences can reflect genetic or environmental factors, or both. Marine ecologists need to work closely with taxonomists in utilizing the powerful new methods of molecular

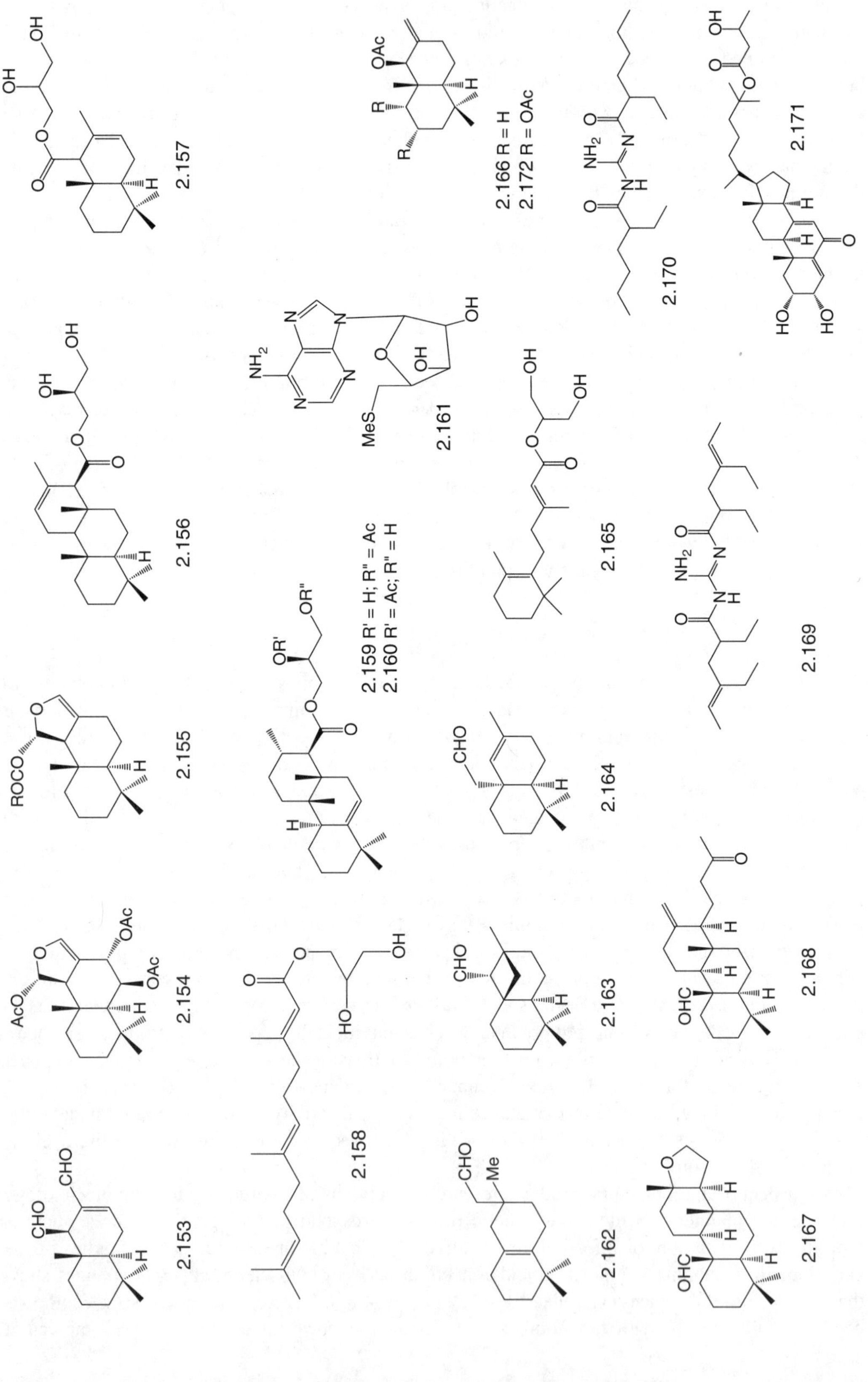

2.153
2.154
2.155
2.156
2.157
2.158
2.159 R' = H; R" = Ac
2.160 R' = Ac; R" = H
2.161
2.162
2.163
2.164
2.165
2.166 R = H
2.172 R = OAc
2.167
2.168
2.169
2.170
2.171

taxonomy. As demonstrated by recent work on the bryostatins,[38] the identification of high-yielding genetic strains can facilitate the biotechnological exploitation of marine natural products.

In cases where environmental factors appear to influence the biosynthetic outcomes in a marine organism, it becomes desirable to comprehend how the biosynthetic pathway is regulated at the molecular level. The biosynthetic switching "on" and "off" of genes that express individual enzymes, when better understood, will also aid environmentally sound production of marine metabolites for pharmaceutical or biotechnological use.

REFERENCES

1. Faulkner, D. J., Marine natural products, *Nat. Prod. Rep.*, 17, 7, 2000.
2. Dietzman, G. R., The marine environment as a discovery resource, in *High Throughput Screening: the Discovery of Bioactive Substances*, Devlin, J. P., Ed., Marcel Dekker, New York, 1997, 99.
3. Ireland, C. M., Molinski, T. F., Roll, D. M., Zabriskie, T. M., McKee, T. C., Swersey, J. F., and Foster, M. P., Uniqueness of the marine chemical environment: categories of marine natural products from invertebrates, in *Biomedical Roles of Marine Natural Products*, Fautin, D. G., Ed., California Academy of Sciences, San Francisco, 1988, 41.
4. Carté, B. K., Biomedical potential of marine natural products, *Bioscience*, 46, 271, 1996.
5. König, G. M. and Wright, A. D., Marine natural products. Current directions and future potential, *Planta Med.*, 62, 193, 1996.
6. Munro, M. H. G., Blunt, J. W., Dumdei, E. J., Hickford, S. J. H., Li, S., Battershill, C. N., and Duckworth, A. R., The discovery and development of marine compounds with pharmaceutical potential, *J. Biotech.*, 70, 15, 1999.
7. Munro, M. H. G., Luibrand, R. T., and Blunt, J. W., The search for antiviral and anticancer compounds from marine organisms, in *Bioorganic Marine Chemistry*, Vol. 1, Scheuer, P. J., Ed., Springer-Verlag, Berlin, 1988, 93.
8. Schmitz, F. J., Bowden, B. F., and Toth, S. I., Antitumour and cytotoxic compounds from marine organisms, in *Marine Biotechnology, Volume 1: Pharmaceutical and Bioactive Natural Products*, Attaway, D. H. and Zaborsky, O. R., Eds., Plenum Press, New York, 1993, 197.
9. Rinehart, K. L., Shield, L. S., and Cohen-Parsons, M., Antiviral substances, in *Marine Biotechnology, Volume 1: Pharmaceutical and Bioactive Natural Products*, Attaway, D. H. and Zaborsky, O. R., Eds., Plenum Press, New York, 1993, 309.
10. Crews, P., and Hunter, L., The search for antiparasitic agents from marine animals, in *Marine Biotechnology, Volume 1: Pharmaceutical and Bioactive Natural Products*, Attaway, D. H. and Zaborsky, O. R., Eds., Plenum Press, New York, 1993, 343.
11. Wright, A. D., König, G. M., Angerhofer, C. K., Greenidge, P., Linden, A., and Desqueyroux-Fáundez, R., Antimalarial activity: the search for marine derived natural products with selective antimalarial activity, *J. Nat. Prod.*, 59, 710, 1996.
12. Barrow, K. D., Biosynthesis of marine metabolites, in *Marine Natural Products: Chemical and Biological Perspectives*, Vol. 5, Scheuer, P. J., Ed., Academic Press, New York, 1983, 51.
13. Garson, M. J., Biosynthetic studies on marine natural products, *Nat. Prod. Rep.*, 6, 143, 1989.
14. Garson, M. J., The biosynthesis of marine natural products, *Chem. Rev.*, 93, 1699, 1993.
15. Moore, B. S., Biosynthesis of marine natural products: microorganisms and macroalgae, *Nat. Prod. Rep.* 16, 653, 1999.
16. Fenical, W. and Jensen, P. R., Marine microorganisms: a new biomedical resource, in *Marine Biotechnology, Volume 1: Pharmaceutical and Bioactive Natural Products*, Attaway, D. H. and Zaborsky, O. R., Eds., Plenum Press, New York, 1993, 419.
17. Jensen, P. R. and Fenical, W., Strategies for the discovery of secondary metabolites from marine bacteria: ecological perspectives, *Annu. Rev. Microbiol.*, 48, 559, 1994.
18. Fenical, W., Chemical studies of marine bacteria: developing a new resource, *Chem. Rev.*, 93, 1673, 1993.

19. Hanefield, U., Floss, H. G., and Laatsch, H., Biosynthesis of the marine antibiotic pentabromopseud-ilin.1. The benzene ring, *J. Org. Chem.*, 59, 3604, 1994.

20. Needham, J., Andersen, R. J., and Kelly, M. T., Biosynthesis of oncorhyncolide, a metabolite of the bacterial seawater isolate MK157, *J. Chem. Soc. Chem. Commun.*, 1367, 1992.

21. Lee, J. J., Dewick, P. M., Gorst-Allman, C. P., Spreafico, F., Kowal, C., Chang, C.-J., McInnes, A. G., Walter, J. A., Keller, P. J., and Floss, H. G., Further studies on the biosynthesis of the boron-containing antibiotic aplasmomycin, *J. Am. Chem. Soc.*, 109, 5426, 1987.

22. Needham, J., Kelly, M. T., Ishige, M., and Andersen, R. J., Andrimid and moiramides A–C, metabolites produced in culture by a marine isolate of the bacterium *Pseudomonas fluorescens*: structure eluci-dation and biosynthesis, *J. Org. Chem.*, 59, 2058, 1994.

23. Moore, B. S. and Seng, D., Biosynthesis of the bicyclic depsipeptide salinamide A in *Streptomyces* sp. CNB-091: origin of the carbons, *Tetrahedron Lett.*, 39, 3915, 1998.

24. Jensen, P. R. and Fenical, W., Marine bacterial diversity as a resource for novel microbial products, *J. Ind. Microbiology*, 17, 346, 1996.

25. Trischmann, J. A., Tapiolas, D. M., Jensen, P. R., Fenical, W., McKee, T. C., Ireland, C. M., Stout, T. J., and Clardy, J., Salinamides A and B: anti-inflammatory depsipeptides from a marine strepto-mycete, *J. Am. Chem. Soc.*, 116, 757, 1994.

26. Gil-Turnes, M. S., Hay, M. E., and Fenical, W., Symbiotic marine bacteria chemically defend crus-tacean embryos from a pathogenic fungus, *Science*, 246, 116, 1989.

27. Hay, M. E., Marine chemical ecology: what's known and what's next?, *J. Exp. Mar. Biol. Ecol.*, 200, 103, 1996.

28. Steinberg, P. D., Schneider, R., and Kjelleberg, S., Chemical defenses of seaweeds against microbial colonization, *Biodegradation*, 8, 211, 1997.

29. Pathirana, C., Dwight, R., Jensen, P. R., Fenical, W., Delgado, A., Brinen, L. S., and Clardy, J., Structure and synthesis of a new butanolide from a marine actinomycete, *Tetrahedron Lett.*, 32, 7001, 1991.

30. Hernandez, I. L. C., Godinho, M. J. L., Magalhães, A., Schefer, A. B., Ferreira, A. G., and Berlinck, R. G. S., *N*-acetyl-γ-hydroxyvaline lactone, an unusual amino acid derivative from a marine strepto-mycete, *J. Nat. Prod.*, 63, 664, 2000.

31. Bewley, C. A., Holland, N. D., and Faulkner, D. J., Two classes of symbionts from *Theonella swinhoei* are localised in distinct populations of bacterial symbionts, *Experientia*, 52, 716, 1996.

32. Bewley, C. A. and Faulkner, D. J., Lithistid sponges: Star performers or hosts to the stars?, *Angew. Chem. Int. Ed.*, 37, 2163, 1998.

33. Shimizu, Y., Microalgal metabolites, *Chem. Rev.*, 93, 1685, 1993.

34. Kvitek, R., DeGange, A. R., and Beitler, M. K., Paralytic shellfish poisoning toxins mediate feeding behaviours of sea otters, *Limnol. Oceanog.*, 36, 393, 1991.

35. Thuesen, E. V., The tetrodotoxin venom of chaetognaths, in *The Biology of Chaetognaths*, Bone, Q., Kapp, H., and Pierrot-Bults, A. C., Eds., Oxford University Press, Oxford, 1991, 55.

36. Friedrich, A. B., Merkert, H., Fendert, T., Hacker, J., Proksch, P., and Hentschel, U., Microbial diversity in the marine sponge *Aplysina cavernicola* (formerly *Verongia cavernicola*) analysed by fluorescence in situ hybridization (FISH), *Mar. Biol.*, 134, 461, 1999.

37. Haygood, M. G. and Davidson, S. K., Small-subunit rRNA genes and in situ hybridisation with oligonucleotides specific for the bacterial symbionts in the larvae of the bryozoan *Bugula neritina* and proposal of "*Candidatus endobugula sertula*," *Appl. Env. Microbiol.*, 63, 4612, 1997.

38. Davidson, S. K. and Haygood, M. G., Identification of sibling species of the bryozoan *Bugula neritina* that produce different anticancer bryostatins and harbour distinct strains of its bacterial symbiont *Candidatus endobugula sertula*, *Biol. Bull.*, 196, 273, 1999.

39. Haygood, M. H., Schmidt, E. W., Davidson, S. K., and Faulkner, D. J., Microbial symbionts of marine invertebrates: opportunities for microbial biotechnology, *J. Mol. Microbiol. Biotech.*, 1, 33, 1999.

40. Kerr, R. G., Lawry, J., and Gush, K. A., *In vitro* biosynthetic studies of the bryostatins, anticancer agents from the marine bryozoan *Bugula neritina*, *Tetrahedron Lett.*, 37, 8305, 1996.

41. Wright, J. L. C., Hu, T., McLachlan, J. L., Needham, J., and Walter, J. A., Biosynthesis of DTX-4: confirmation of a polyketide pathway, proof of a Baeyer-Villiger oxidation step, and evidence for an unusual carbon deletion process, *J. Am. Chem. Soc.*, 118, 8757, 1996.

42. Kobayashi, J., Takahashi, M., and Ishibashi, M., Biosynthetic studies of amphidinolide J: explanation of the generation of the unusual odd-numbered macrocyclic lactone, *J. Chem. Soc. Chem. Commun.*, 1639, 1995.

43. Harrigan, G. G., Yoshida, W. Y., Moore, R. E., Nagle, D. G., Park, P. U., Biggs, J., Paul, V. J., Mooberry, S. L., Corbett, T. H., and Valeriote, F. A., Isolation, structure determination and biological activity of dolastatin 12 and lyngbyastatin 1 from *Lyngbya majuscula/Schizothrix calcicola* cyanobacterial assemblages, *J. Nat. Prod.*, 61, 1221, 1998.

44. Mitchell, S. S., Faulkner, D. J., Rubins, K., and Bushman, F. D., Dolastatin 3 and two novel cyclic peptides from a Palauan collection of *Lyngbya majuscula*, *J. Nat. Prod.*, 63, 279, 2000.

45. Paul, V. J. and Pennings, S. C., Diet-derived chemical defenses in the sea hare *Stylocheilus longicauda* (Quoy et Gaimard 1824), *J. Exp. Mar. Biol. Ecol.*, 151, 227, 1991.

46. Pennings, S. C. and Paul, V. J., Sequestration of dietary secondary metabolites by three species of sea hares: location, specificity and dynamics, *Mar. Biol.*, 117, 535, 1993.

47. Pennings, S. C. and Paul, V. J., Secondary chemistry does not limit dietary range of the specialist sea hare *Stylocheilus longicauda* (Quoy and Gaimard 1824), *J. Exp. Mar. Biol. Ecol.*, 174, 97, 1993.

48. Nagle, D. and Paul, V. J., Chemical defenses of a marine cyanobacterial bloom, *J. Exp. Mar. Biol. Ecol.*, 225, 29, 1998.

49. Sitachitta, N., Rossi, J., Roberts, M., Gerwick, W. H., Fletcher, M. D., and Willis, C. L., Biosynthesis of the marine cyanobacterial metabolite barbamide. 1. Origin of the trichloromethyl group, *J. Am. Chem. Soc.*, 120, 7131, 1998.

50. Williamson, R. T., Sitachitta, N., and Gerwick, W. H., Biosynthesis of the marine cyanobacterial metabolite barbamide. 2: Elucidation of the origin of the thiazole ring by application of a new GHNMBC experiment, *Tetrahedron Lett.*, 40, 5175, 1999.

51. Hartung, J., The biosynthesis of barbamide - A radical pathway for "biohalogenation"?, *Angew. Chem. Int. Ed.*, 38, 1209, 1999.

52. Wylie, C. R. and Paul, V. J., Feeding preferences of the surgeonfish *Zebrasoma flavescens* in relation to chemical defenses of tropical algae, *Mar. Ecol. Prog. Ser.*, 45, 23, 1988.

53. Paul, V. J., Nelson, S. G., and Sanger, H. R., Feeding preferences of adult and juvenile rabbitfish *Siganeus argenteus* in relation to chemical defenses of tropical seaweeds, *Mar. Ecol. Prog. Ser.*, 60, 23, 1990.

54. Thacker, R. W., Nagle, D. G., and Paul, V. J., Effects of repeated exposure to marine cyanobacterial secondary metabolites on feeding by juvenile rabbitfish and parrotfish, *Mar. Ecol. Prog. Ser.*, 147, 21, 1997.

55. Pennings, S. C., Weiss, A. M., and Paul, V. J., Secondary metabolites of the cyanobacterium *Microcoleus lyngbyaceus* and the sea hare *Stylocheilus longicauda*, *Mar. Biol.: Palatability and Toxicity*, 126, 735, 1996.

56. Nagle, D. G. and Paul, V. J., Production of secondary metabolites by filamentous tropical marine cyanobacteria: ecological functions of the compounds, *J. Phycol.*, 35, 1412, 1999.

57. Nagle, D. G., Camacho, F. T., and Paul, V. J., Dietary preferences of the opistobranch mollusc *Stylocheilus longicauda* for secondary metabolites produced by the tropical cyanobacterium *Lyngbya majuscula*, *Mar. Biol.*, 132, 267, 1998.

58. Pennings, S. C. and Paul, V. J., Effect of plant toughness, calcification and chemistry on herbivory by *Dolabella auricularia*, *Ecology*, 73, 1606, 1992.

59. Pennings, S. C., Pablo, S. R., and Paul, V., J,, Chemical defenses of the tropical benthic marine cyanobacterium *Hormothamnion entereomorphoides*: diverse consumers and synergisms, *Limnol. Ocean.*, 42, 911, 1997.

60. Williams, D. E., Dawe, S. C., Kent, M. L., Andersen, R. J., Craig, M., and Holmes, C. F. B., Bioaccumulation and clearance of microcystins from salt water mussels *Mytilus edulis* and *in vivo* evidence for covalently-bound microcystins in mussel tissue, *Toxicon*, 35, 1617, 1997.

61. Moore, R. E., Chen, J. L., Moore, B. S., Patterson, G. M. L., and Carmichael, W. W., Biosynthesis of microcystin-LR. Origin of the carbons in the Adda and Masp units, *J. Am. Chem. Soc.*, 113, 5083, 1991.

62. Dittmann, E., Neilan, B., M., E., Von Dohren, H., and Borner, T., Insertional mutagenesis of a peptide synthetase gene that is responsible for hepatotoxin production in the cyanobacterium *Microcystis aeruginosa* PCC 7806, *Mol. Microbiol.*, 26, 779, 1997.

63. Neilan, B. A., Dittman, E., Rouhiainen, L., Bass, A. R., Schaub, V., Sivonen, K., and Borner, T., Nonribosomal peptide synthesis and toxigenicity of cyanobacteria, *J. Bacteriol.*, 181, 4089, 1999.

64. Hemscheidt, T., Puglisi, M. P., Larsen, L. K., Patterson, G. M. L., Moore, R. E., Rios, J. L., and Clardy, J., Structure and biosynthesis of borophycin, a new boeseken complex of boric acid from a marine strain of the blue green alga *Nostoc linckia*, *J. Org. Chem.*, 59, 3467, 1994.

65. Hemscheidt, T., Burgoyne, D. L., and Moore, R. E., Biosynthesis of anatoxin-A(S) - (2S, 4S)-4-hydroxyarginine as an intermediate, *J. Chem. Soc. Chem. Commun.*, 205, 1995.

66. Hemscheidt, T., Rapala, J., Sivonen, K., and Skulberg, O. M., Biosynthesis of anatoxin-A in *Anabaena flos-aquae* and homoanatoxin-A in *Oscillatoria formosa*, *J. Chem. Soc. Chem. Commun.*, 1361, 1995.

67. Carmeli, S., Moore, R. E., Patterson, G. M. L., and Yoshida, W. Y., Biosynthesis of tolytoxin origin of the carbons and heteroatoms, *Tetrahedron Lett.*, 34, 5571, 1993.

68. Bornemann, V., Patterson, G. M. L., and Moore, R. E., Isonitrile biosynthesis in the cyanophyte *Hapalosiphon fontinalis*, *J. Am. Chem. Soc.*, 110, 2339, 1988.

69. Faulkner, D. J., He, H.-Y., Unson, M. D., Bewley, C. A., and Garson, M. J., New metabolites from marine sponges: are symbionts important?, *Gazz. Chim. Ital.*, 123, 301, 1993.

70. Dumdei, E. J., Dexter, A. L., Field, K. L., Flowers, A. E., Garson, M. J., Gehrmann, J., and Molinski, T. F., unpublished data, 1995.

71. Unson, M. D. and Faulkner, D. J., Cyanobacterial biosynthesis of chlorinated metabolites from *Dysidea herbacea* (Porifera), *Experientia*, 49, 349, 1993.

72. Flowers, A. E., Dumdei, E. J., Charan, R. D., Garson, M. J., and Webb, R. I., Cellular origin of chlorinated diketopiperazines in the dictyoceratid sponge *Dysidea herbacea* (Keller), *Cell Tiss. Res.*, 292, 597, 1998.

73. Garson, M. J., The biosynthesis of marine natural products: why it is important, in *Sponges in Time and Space*, Van Soest, R. W. M., Van Kempen, T. M. G., and Braekman, J. C., Eds., Balkema, Amsterdam, 1994, 427.

74. Unson, M. D., Holland, N. D., and Faulkner, D. J., A brominated secondary metabolite synthesized by the cyanobacterial symbiont of a marine sponge and accumulation of the crystalline metabolite in the sponge tissue, *Mar. Biol.*, 119, 1, 1994.

75. Duffy, J. E. and Paul, V. J., Prey nutritional quality and the effectiveness of chemical defenses against tropical reef fish, *Oecologia*, 90, 333, 1992.

76. Pennings, S. C., Puglisi, M., Pitlik, T. J., Himaya, A. C., and Paul, V. J., Effects of secondary metabolites and $CaCO_3$ on feeding by surgeonfish and parrotfishes: within-plant comparisons, *Mar. Ecol. Prog. Ser.*, 134, 49, 1996.

77. Flowers, A. E. and Garson, M. J., unpublished data, 1994.

78. Hay, M. E., Paul, V. J., Lewis, S. M., Gustafson, K., Tucker, J., and Trindell, R. N., Can tropical seaweeds reduce herbivory by growing at night? Diel patterns of growth, nitrogen content, herbivory, and chemical versus morphological defenses, *Oecologia*, 75, 233, 1988.

79. Hay, M. E. and Fenical, W., Marine plant herbivore interactions: the ecology of chemical defense, *Ann. Rev. Ecol. Syst.*, 19, 111, 1988.

80. Paul, V. J., *Ecological Roles of Marine Natural Products*, Comstock Publishing Associates, Ithaca, NY, 1992.

81. Young, D. N., McConnell, O. J., and Fenical, W., *In vivo* biosynthesis of tribromoheptene oxide in *Bonnemaisonia nootkana*, *Phytochemistry*, 20, 2335, 1981.

82. Barrow, K. D. and Temple, C. A., Biosynthesis of halogenated monoterpenes in *Plocamium cartilagineum*, *Phytochemistry*, 24, 1697, 1985.

83. Moghaddam, M. F. and Gerwick, W. H., 12-lipoxygenase activity in the red marine algae *Gracilariopsis lemaneiformis*, *Phytochemistry*, 29, 2457, 1990.

84. Stratmann, K., Boland, W., and Muller, D. G., Pheromones of marine brown algae: a new branch of eicosanoid metabolism, *Angew. Chem. Int. Ed.*, 31, 1246, 1992.

85. Boland, W., The chemistry of gamete attraction; chemical structures, biosynthesis and (a)biotic degradation of algal pheremones, *Proc. Natl. Acad. Sci.*, 92, 37, 1995.

86. Rohmer, M., The discovery of a mevalonate-independent pathway for isoprenoid biosynthesis in bacteria, algae and higher plants, *Nat. Prod. Rep.*, 16, 565, 1999.

87. Paul, V. J. and Van Alstyne, K. L., Activation of chemical defenses in the tropical green algae *Halimeda* spp., *J. Exp. Mar. Biol. Ecol.*, 160, 191, 1992.

88. Paul, V. J. and Van Alstyne, K. L., Chemical defense and chemical variation in some tropical Pacific species of *Halimeda* (*Halimedaceae: chlorophyta*), *Coral Reefs*, 6, 263, 1988.

89. Hay, M. E., Pawlik, J. R., Duffy, J. E., and Fenical, W., Seaweed–herbivore–predator interaction: host-plant specialisation reduces predation on small herbivores, *Oecologia*, 81, 411, 1989.

90. Meyer, K. D. and Paul, V. J., Intraplant variation in secondary metabolite concentration in three species of *Caulerpa* (Chlorophyta: Caulerpales) and its effects on herbivorous fishes, *Mar. Ecol. Prog. Ser.*, 82, 249, 1992.

91. Meyer, K. D. and Paul, V. J., Variations in secondary metabolite and aragonite concentrations in the tropical green seaweed *Neomeris annulata*: effects on herbivory by fishes, *Mar. Biol.*, 122, 537, 1995.

92. Lumbang, W. A. and Paul, V. J., Chemical defense of the tropical green seaweed *Neomeris annulata* Dickie: effects of multiple compounds on feeding by herbivores, *J. Exp. Mar. Biol. Ecol.*, 201, 185, 1996.

93. Hay, M. E., Fenical, W., and Gustafson, K., Chemical defense against diverse coral reef herbivores, *Ecology*, 68, 1581, 1987.

94. Park, M., Fenical, W., and Hay, M. E., Debromoisocymobarbatol, a new chromanol feeding deterrent from the marine alga *Cymopolia barbata*, *Phytochemistry*, 31, 4115, 1992.

95. Maier, I. and Müller, D. G., Sexual pheromones in algae, *Biol. Bull.*, 170, 145, 1986.

96. Maier, I., Brown algal pheromones, *Prop. Phycol. Res.*, 11, 51, 1995.

97. Oldham, N. J. and Boland, W., Chemical ecology: multifunctional compounds and multitrophic interactions, *Naturwissenschaften*, 83, 248, 1996.

98. Hay, M. E., Piel, J., Boland, W., and Schnitzler, I., Seaweed sex pheromones and their degradation products frequently suppress amphipod grazing but rarely suppress sea urchin feeding, *Chemoecology*, 8, 91, 1998.

99. Stratmann, K., Boland, W., and Müller, D. G., Biosynthesis of pheromones in female gametes of marine brown algae (Phaeophyceae), *Tetrahedron*, 49, 3755, 1993.

100. Pohnert, G. and Boland, W., Biosynthesis of the algal pheromone hormosirene by the freshwater diatom *Gomphonema parvulum* (Bacillariophyceae), *Tetrahedron*, 52, 10073, 1996.

101. Hombeck, M. and Boland, W., Biosynthesis of the algal pheromone fucoserratene by the freshwater diatom *Asterionella formosa* (Bacillariophyceae), *Tetrahedron*, 11033, 1998.

102. Boland, W., Pohnert, G., and Maier, I., Pericyclic reactions in nature: spontaneous Cope rearrangement inactivates algal pheromones, *Angew. Chem. Int. Ed.*, 34, 1602, 1995.

103. Pohnert, G. and Boland, W., Pericyclic reactions in nature: synthesis and Cope rearrangement of thermolabile bis-alkenylcyclopropanes from female gametes of marine brown algae (Phaeophyceae), *Tetrahedron*, 53, 13681, 1997.

104. Schnitzler, I., Boland, W., and Hay, M. E., Organic sulphur compounds from *Dictyopteris* spp. deter feeding by an herbivorous amphipod (*Ampithoe longimana*) but not by an herbivorous sea urchin (*Arbacia punctulata*), *J. Chem. Ecol.*, 24, 1715, 1998.

105. Hay, M. E., Duffy, J. E., Fenical, W., and Gustafson, K., Chemical defense in the seaweed *Dictyopteris delicatula*. Differential effects against reef fishes and amphipods, *Mar. Ecol. Prog. Ser.*, 48, 185, 1988.

106. Duffy, J. E. and Hay, M. E., Herbivore resistance to seaweed defenses: the role of mobility and predator risk, *Ecology*, 75, 1304, 1994.

107. Schmitt, T. M., Hay, M. E., and Lindquist, N., Constraints on chemically mediated coevolution: multiple functions for seaweed secondary metabolites, *Ecology*, 76, 107, 1995.

108. Cronin, G. and Hay, M. E., Susceptibility to herbivores depends on recent history of both the plant and animal, *Ecology*, 77, 1531, 1996.

109. Hay, M. E., Duffy, J. E., Pfister, C. A., and Fenical, W., Chemical defense against different marine herbivores: are amphipods insect equivalents?, *Ecology*, 68, 1567, 1987.

110. Hay, M. E., Renaud, P. E., and Fenical, W., Large mobile versus small sedentary herbivores and their resistance to seaweed chemical defenses, *Oecologia*, 75, 246, 1988.

111. Cronin, G., Paul, V. J., Hay, M. E., and Fenical, W., Are tropical herbivores more resistant than temperate herbivores to seaweed chemical defenses? Diterpenoid metabolites from *Dictyota acutiloba* as feeding deterrents for tropical versus temperate fishes and urchins, *J. Chem. Ecol.*, 23, 289, 1997.

112. Cronin, G. and Hay, M. E., Induction of seaweed chemical defenses by amphipod grazing, *Ecology*, 77, 2287, 1996.

113. Boland, W. and Muller, D. G., On the odour of the Mediterranean seaweed *Dictyopteris membranacea*: new C_{11} hydrocarbons from the marine brown algae, *Tetrahedron Lett.*, 28, 307, 1987.

114. Cronin, G. and Hay, M. E., Effects of light and nutrient availability on the growth, secondary chemistry, and resistance to herbivory of two brown seaweeds, *Oikos*, 77, 93, 1996.

115. Van Alstyne, K. L., Herbivore grazing increases polyphenolic defenses in the intertidal brown alga *Fucus distichus*, *Ecology*, 69, 655, 1988.

116. Yates, J. L. and Peckol, P., Effects of nutrient availability and herbivory on polyphenolics in the seaweed *Fucus vesiculosus*, *Ecology*, 74, 1757, 1993.

117. De Nys, R., Steinberg, P. D., Rogers, C. N., Charlton, T. S., and Duncan, M. W., Quantitative variation of secondary metabolites in the sea hare *Aplysia parvula* and its host plant *Delisea pulchra*, *Mar. Ecol. Prog. Ser.*, 130, 135, 1996.

118. De Nys, R., Steinberg, P. D., Willemsen, P., Dworjanyn, S. A., Gabelish, C. L., and King, R. J., Broad spectrum effects of secondary metabolites from the red alga *Delisea pulchra* in antifouling assays, *Biofouling*, 8, 259, 1995.

119. Maximilien, R., De Nys, R., Holmström, C., Gram, L., Kjelleberg, S., and Steinberg, P. D., Bacterial fouling is regulated by secondary metabolites from the red alga *Delisea pulchra*, *Aq. Micr. Ecol.*, 15, 233, 1998.

120. Paul, V. J., Hay, M. E., Duffy, J. E., Fenical, W., and Gustafson, K., Chemical defense in the seaweed *Ochtodes Secundiramea* (Montagne) Howe (Rhodophyta); effects of its monoterpenoid components upon diverse coral reef herbivores, *J. Exp. Mar. Biol. Ecol.*, 114, 249, 1987.

121. Puglisi, M. P. and Paul, V. J., Intraspecific variation in the red algae *Portieria hornemannii*: monoterpene concentrations are not influenced by nitrogen or phosphorus enrichment, *Mar. Biol.*, 128, 161, 1997.

122. De Nys, R., Coll, J. C., and Price, I. R., Chemically mediated interactions between the red alga *Plocamium hamatum* (Rhodophyta) and the octocoral *Sinularia cruciata* (Alcyonacea), *Mar. Biol.*, 108, 315, 1991.

123. Young, D. N., Howard, B. M., and Fenical, W., Subcellular localization of brominated secondary metabolites in the red algae *Laurencia snyderae*, *J. Phycol.*, 16, 182, 1980.

124. Matlock, D. B., Ginsburg, D. W., and Paul, V. J., Spatial variability in secondary metabolite production by the tropical red algae *Portieria hornemannii*, *Hydrobiologia*, 398/399, 267, 1999.

125. David, D. M. and Garson, M. J., unpublished data, 1987.

126. Gerwick, W. H., Structure and biosynthesis of marine algal oxylipins, *Biochim. Biophys. Acta.*, 1211, 243, 1994.

127. Gerwick, W. H., Epoxy allylic carbocations as conceptual intermediates in the biogenesis of diverse marine oxylipins, *Lipids*, 31, 1215, 1996.

128. Proksch, P., Defensive roles for secondary metabolites from marine sponges and sponge-feeding nudibranchs, *Toxicon*, 32, 639, 1994.

129. Pawlik, J. R., Chanas, B., Toonen, R. T., and Fenical, W., Defenses of Caribbean sponges against predatory reef fish. I. Chemical deterrency, *Mar. Ecol. Prog. Ser.*, 127, 183, 1995.

130. Pennings, S. C., Pablo, S. R., Paul, V. J., and Duffy, J. E., Effects of sponge secondary metabolites in different diets on feeding by three groups of consumers, *J. Exp. Mar. Biol. Ecol.*, 180, 137, 1994.

131. Schulte, G. R. and Scheuer, P. J., Defense allomones of some marine mollusks, *Tetrahedron*, 38, 1857, 1982.

132. Chanas, B. and Pawlik, J. R., Defense of Caribbean sponges against predatory reef fishes. II. Spicules, tissue toughness and nutritional quality, *Mar. Ecol. Prog. Ser.*, 127, 195, 1995.

133. Bingham, B. L. and Young, C. M., Influence of sponges on invertebrate recruitment: a field test of allelopathy, *Mar. Biol.*, 109, 19, 1991.

134. Thompson, J. E., Walker, R. P., and Faulkner, D. J., Screening and bioassays for biologically-active substances from forty marine sponge species from San Diego, California, USA, *Mar. Biol.*, 88, 11, 1985.

135. Martin, D. and Uriz, M. J., Chemical bioactivity of Mediterranean benthic organisms against embryos and larvae of marine invertebrates, *J. Exp. Mar. Biol. Ecol.*, 173, 11, 1993.

136. Porter, J. W. and Targett, N. M., Allelochemical interactions between sponges and corals, *Biol. Bull.*, 175, 230. 1988.

137. Sullivan, B., Faulkner, D. J., and Webb, L., Siphonodictidine, a metabolite of the burrowing sponge *Siphonodictyon* sp. that inhibits coral growth, *Science*, 221, 1175, 1983.

138. Garson, M. J., Clark, R. J., Webb, R. I., Field, K. L., Charan, R. D., and McCaffrey, E. J., The ecological role of cytotoxic alkaloids: *Haliclona* n.sp., an unusual sponge/dinoflagellate association, *Mem. Qld. Queensl. Mus.*, 44, 205, 1999.

139. Thompson, J. E., Exudation of biologically-active metabolites in the sponge *Aplysina fistularis*. I. Biological evidence, *Mar. Biol.*, 88, 23, 1985.

140. Fusetani, N., Marine natural products influencing larval settlement and metamorphosis of benthic invertebrates, *Curr. Org. Chem.*, 1, 127, 1997.

141. Becerro, M. A., Turon, X., and Uriz, M. J., Multiple functions for secondary metabolites in encrusting marine invertebrates, *J. Chem. Ecol.*, 23, 1527, 1997.

142. Rogers, S. D. and Paul, V. J., Chemical defenses of three *Glossodoris* nudibranchs and their dietary *Hyrtios* sponges, *Mar. Ecol. Prog. Ser.*, 77, 221, 1991.

143. Chanas, B., Pawlik, J. R., Lindel, T., and Fenical, W., Chemical defense of the Caribbean sponge *Agelas clathrodes* (Schmidt), *J. Exp. Mar. Biol. Ecol.*, 208, 185, 1996.

144. Ebel, R., Brenzinger, M., Kunze, A., Gross, H. J., and Proksch, P., Wound activation of protoxins in marine sponge *Aplysina aerophoba*, *J. Chem. Ecol.*, 23, 1451, 1997.

145. Djerassi, C. and Silva, C. J., Biosynthetic studies of marine lipids. 41. Sponge sterols: origin and biosynthesis, *Acc. Chem. Res.*, 24, 371, 1991.

146. Lill, R. E., Major, D. A., Blunt, J. W., Munro, M. H. G., Battershill, C. N., McLean, M. G., and Baxter, R. L., Studies on the biosynthesis of discorhabdin B in the New Zealand sponge *Latrunculia* sp. B, *J. Nat. Prod.*, 58, 306, 1995.

147. Andrade, P., Willoughby, R., Pomponi, S., and Kerr, R. G., Biosynthetic studies of the alkaloid, stevensine, in a cell culture of the marine sponge *Teichaxinella morchella*, *Tetrahedron Lett.*, 40, 4775, 1999.

148. Wilson, D. M., Puyana, M., Fenical, W., and Pawlik, J. R., Chemical defense of the Caribbean sponge *Axinella corrugata* against predatory fishes, *J. Chem. Ecol.*, 25, 2811, 1999.

149. Tymiak, A. A., and Rinehart, K. L., Jr., Biosynthesis of dibromotyrosine-derived antimicrobial compounds by the marine sponge *Aplysina fistularis* (*Verongia aurea*), *J. Am. Chem. Soc.*, 103, 6763, 1981.

150. Carney, J. R. and Rinehart, K. L., Jr., Biosynthesis of brominated tyrosine metabolites by *Aplysina fistularis*, *J. Nat. Prod.*, 58, 971, 1995.

151. Teeyapant, R. and Proksch, P., Biotransformation of brominated compounds in the marine sponge *Verongia aerophoba* — evidence for an induced chemical defense, *Naturwissenschaften*, 80, 369, 1993.

152. Koulman, A., Proksch, P., Ebel, R., Beekman, A. C., Van Uden, W., Konings, A. W. T., Pedersen, J. A., Pras, N., and Woerdenbag, H. J., Cytotoxicity and mode of action of aeroplysinin-1 and a related dienone from the sponge *Aplysina aerophoba*, *J. Nat. Prod.*, 59, 591, 1996.

153. Weiss, B., Elbrächter, M., Kirchner, M., and Proksch, P., Defense metabolites from the marine sponge *Verongia aerophoba*, *Biochem. Syst. Ecol.*, 24, 1, 1996.

154. Kreuter, M. H., Robitzki, A., Chang, S., Steffen, R., Michaelis, M., Kljajíc, Z., M., B., Schröder, H. C., and Müller, W. E. G., Production of the cytostatic agent aeroplysinin by the sponge *Verongia aerophoba* in *in vitro* culture, *Comp. Biochem. Physiol.*, 101C, 183, 1992.

155. Walker, R. P., Thompson, J. E., and Faulkner, D. J., Exudation of biologically-active metabolites in the sponge *Aplysina fistularis*. II. Chemical evidence, *Mar. Biol.*, 88, 27, 1985.

156. Thompson, J. E., Barrow, K. D., and Faulkner, D. J., Localization of two brominated metabolites, aerothionin and homoaerothionin in spherulous cells of the marine sponge *Aplysina fistularis* (= *verongia thiona*), *Acta Zool. (Stockh.)*, 64, 199, 1983.

157. Scheuer, P. J., Isocyanides and cyanides as natural products, *Acc. Chem. Res.*, 25, 433, 1992.

158. Chang, C. W. J. and Scheuer, P. J., Marine isocyano compounds, in *Topics in Current Chemistry*, Vol. 167, Scheuer, P. J., Ed., Springer-Verlag, Berlin, 1993, 33.

159. Garson, M. J., Simpson, J. S., Flowers, A. E., and Dumdei, E. J., Cyanide and thiocyanate derived functionality in marine organisms-Structures, biosynthesis and ecology, in *Bioactive Natural Products (Part B)*, Vol. 21, Rahman, A., Ed., Elsevier, Amsterdam, 2000, 329.

160. Chang, C. W. J. and Scheuer, P. J., Biosynthesis of marine isocyanoterpenoids in sponges, *Comp. Biochem. Physiol.*, 97B, 227, 1990.

161. Garson, M. J., Biosynthesis of the novel diterpene isonitrile diisocyanoadociane by a marine sponge of the *Amphimedon* genus: incorporation studies with [^{14}C]cyanide and sodium [2-^{14}C]acetate, *J. Chem. Soc. Chem. Commun.*, 35, 1986.

162. Fookes, C. J. R., Garson, M. J., MacLeod, J. K., Skelton, B. W., and White, A. H., Biosynthesis of diisocyanoadociane, a novel diterpene from the marine sponge *Amphimedon* sp. Crystal structure of a monoamide derivative, *J. Chem. Soc. Perkin Trans. 1*, 1003, 1988.

163. Karuso, P. and Scheuer, P. J., Biosynthesis of isocyanoterpenes in sponges, *J. Org. Chem.*, 54, 2092, 1989.

164. Dumdei, E. J., Flowers, A. E., Garson, M. J., and Moore, C. J., Biosynthetic evidence for transfer of secondary chemicals between the marine sponge *Phakellia cavernosa* (Dendy) and the dorid nudibranch *Phyllidiella pustulosa*, *Comp. Biochem. Physiol.*, 118A, 1385, 1997.

165. Simpson, J. S., Raniga, P., and Garson, M. J., Biosynthesis of dichloroimines in the tropical marine sponge *Stylotella aurantium*, *Tetrahedron Lett.*, 38, 7947, 1997.

166. Simpson, J. S., and Garson, M. J., Thiocyanate biosynthesis in the tropical marine sponge *Axinyssa* n.sp., *Tetrahedron Lett.*, 39, 5819, 1998.

167. Simpson, J. S. and Garson, M. J., Advanced precursors in marine biosynthetic study: The biosynthesis of diisocyanoadociane in *Amphimedon terpenensis*, *Tetrahedron Lett.*, 40, 3909, 1999.

168. Hagadone, M. R., Scheuer, P. J., and Holm, A., On the origin of the isocyano function in marine sponges, *J. Am. Chem. Soc.*, 106, 2447, 1984.

169. Simpson, J. S. and Garson, M. J., unpublished data, 2000.

170. Vennesland, B., Conn, E. E., Knowles, C. J., Westley, J., and Wissing, J., *Cyanide in Biology*, Academic Press, London, 1981.

171. Hamann, M. T. and Scheuer, P. J., Cyanopuupehenol, an antiviral metabolite of a sponge of the order Verongida, *Tetrahedron Lett.*, 32, 5671, 1991.

172. Garson, M. J., Thompson, J. E., Larsen, R. M., Battershill, C. N., Murphy, P. T., and Berquist, P. R., Terpenes in sponge cell membranes: cell separation and membrane fractionation studies with the tropical marine sponge *Amphimedon* sp., *Lipids*, 27, 378, 1992.

173. Uriz, M. J., Turon, X., Galera, J., and Tur, J. M., New light on the cell location of avarol within the sponge *Dysidea avara*, *Cell Tiss. Res.*, 285, 519, 1996.

174. Marin, A., Lopez, M. D., Esteban, M. A., Meseguer, J., Muñoz, J., and Fontana, A., Anatomical and ultrastructural studies of chemical defense in the sponge *Dysidea fragilis*, *Mar. Biol.*, 131, 639, 1998.

175. De Rosa, M., Minale, L., and Sodano, G., Metabolism in porifera. I. Some studies on the biosynthesis of fatty acids, sterols and bromo compounds by the sponge *Verongia aerophoba*, *Comp. Biochem. Physiol.*, 45B, 883, 1973.

176. Garson, M. J., Partali, V., Liaaen-Jensen, S., and Stoilov, I. L., Isoprenoid biosynthesis in a marine sponge of the *Amphimedon* genus: incorporation studies with [1-^{14}C]cholesterol and [2-^{14}C]mevalonate, *Comp. Biochem. Physiol.*, 91B, 293, 1988.

177. Fusetani, N., Hirota, H., Okino, T., Tomono, Y., and Yoshimura, E., Antifouling activity of isocyanoterpenoids and related compounds isolated from a marine sponge and nudibranchs, *J. Nat. Toxins*, 5, 249, 1996.

178. Hirota, H., Tomono, Y., and Fusetani, N., Terpenoids with antifouling activity against barnacle larvae from the marine sponge *Acanthella cavernosa*, *Tetrahedron*, 52, 2359, 1996.

179. Hirota, H., Okino, T., Yoshimura, E., and Fusetani, N., Five new antifouling sesquiterpenes from two marine sponges of the genus *Axinyssa* and the nudibranch *Phyllidia pustulosa*, *Tetrahedron*, 54, 13971, 1998.

180. Thompson, J. E., Walker, R. P., Wratten, S. J., and Faulkner, D. J., A chemical defense mechanism for the nudibranch *Cadlina luteomarginata*, *Tetrahedron*, 38, 1865, 1982.

181. Thacker, R. W., Becerro, M. A., Lumbang, W. A., and Paul, V. J., Allelopathic interactions between sponges on a tropical reef, *Ecology*, 79, 1740, 1998.

182. Thompson, J. E., Murphy, P. T., Bergquist, P. R., and Evans, E. A., Environmentally induced variation in diterpene composition of the marine sponge *Rhopaloeides odorabile*, *Biochem. Syst. Ecol.*, 15, 595, 1987.

183. Becerro, M. A., Turon, X., and Uriz, M. J., Natural variation of toxicity in encrusting sponge *Crambe crambe* (Schmidt) in relation to size and environment, *J. Chem. Ecol.*, 21, 1931, 1995.

184. Becerro, M. A., Uriz, M. J., and Turon, X., Chemically-mediated interactions in benthic organisms: the chemical ecology of *Crambe crambe* (Porifera: Poecilosclerida), *Hydrobiologia*, 356, 77, 1997.

185. Uriz, M. J., Becerro, M. A., Tur, J. M., and Turon, X., Location of toxicity within the Mediterranean sponge *Crambe crambe* (Demospongiae: Poecilosclerida), *Mar. Biol.*, 124, 583, 1996.

186. Becerro, M. A., Paul, V. J., and Starmer, J., Intracolonial variation in chemical defenses of the sponge *Cacospongia* sp. and its consequences on generalist fish predators and the specialist nudibranch predator *Glossodoris pallida*, *Mar. Ecol. Prog. Ser.*, 168, 187, 1998.

187. Schupp, P., Eder, C., Paul, V. J., and Proksch, P., Distribution of secondary metabolites in the sponge *Oceanapia* sp. and its ecological implications, *Mar. Biol.*, 135, 573, 1999.

188. Dai, M. C., Garson, M. J., and Coll, J. C., Biosynthetic processes in soft corals. I. A comparison of terpene biosynthesis in *Alcyonium molle* (Alcyoniidae) and *Heteroxenia* sp. (Xeniidae), *Comp. Biochem. Physiol.*, 99B, 775, 1991.

189. Kerr, R. G. and Miranda, N. L., Biosynthetic studies of ecteinascidins in the marine tunicate *Ecteinascidia turbinata*, *J. Nat. Prod.*, 58, 1618, 1995.

190. Coleman, A. C., Mydlarz, L. D., and Kerr, R. G., *In vivo* and *in vitro* investigations into the biosynthetic relatedness of the pseudopterosins, *Org. Lett.*, 1, 2173, 1999.

191. Coll, J. C., Bowden, B. F., Tapiolas, D. M., Willis, R. H., Djura, P., Streamer, M., and Trott, L., Studies of Australian soft corals-XXXV. The terpenoid chemistry of soft corals and its implications, *Tetrahedron*, 41, 1085, 1985.

192. Rice, J. R., Papastephanou, C., and Anderson, D. G., Isolation, localisation and biosynthesis of crassin acetate in *Pseudoplexaura porosa* (Houttyun), *Biol. Bull.*, 138, 334, 1970.

193. Papastephanou, C. and Anderson, D. G., Crassin acetate biosynthesis in a cell-free homogenate of zooxanthellae from *Plexauara porosa* (Houttyun), *Comp. Biochem. Physiol.*, 73B, 617, 1982.

194. Anderson, D. G., Gorgosterol biosynthesis: localisation of squalene formation in the zooxanthellar component of various gorgonians, *Comp. Biochem. Physiol.*, 81B, 423, 1985.

195. Kokke, W. C. M. C., Fenical, W., Bohlin, L., and Djerassi, C., Sterol synthesis by cultured zooxanthellae; implications concerning sterol metabolism in the host-symbiont association in Caribbean gorgonians, *Comp. Biochem. Physiol.*, 68B, 281, 1981.

196. Kokke, W. C. M. C., Epstein, S., Look, S. A., Rau, G. H., Fenical, W., and Djerassi, C., On the origin of terpenes in symbiotic associations between marine invertebrates and algae (zooxanthellae), *J. Biol. Chem.*, 259, 8168, 1984.

197. Jensen, P. R., Harvell, C. D., Wirz, K., and Fenical, W., Antimicrobial activity of extracts of Caribbean gorgonian corals, *Mar. Biol.*, 125, 411, 1996.

198. Coll, J. C., Bowden, B. F., Alino, P. M., Heaton, A., König, G. M., De Nys, R., Willis, R. H., Sammarco, P. W., and Clayton, M. N., Chemically mediated interactions between marine organisms, *Chem. Scripta*, 29, 383, 1989.

199. Coll, J. C., Leone, P. A., Bowden, B. F., Carroll, A. R., König, G. M., Heaton, A., De Nys, R., Maida, M., Alino, P. M., Willis, R. H., Babcock, R. C., Florian, Z., Clayton, M. N., Miller, R. L., and Alderslade, P. N., Chemical aspects of mass spawning in corals. II. (-)-*epi*-thunbergol, the sperm attractant in the eggs of the soft coral *Lobophytum crassum* (Cnidaria: Octocorallia), *Mar. Biol.*, 123, 137, 1995.

200. Coll, J. C., Bowden, B. F., Heaton, A., Scheuer, P. J., Li, M. K. W., Clardy, J., Schulte, G. K., and Finer-Moore, J., Structures and possible functions of epoxypukalide and pukalide, *J. Chem. Ecol.*, 15, 1177, 1989.

201. Slattery, M., Hines, G. A., Starmer, J., and Paul, V. J., Chemical signals in gametogenesis, spawning and larval settlement and defense of the soft coral *Sinularia polydactyla*, *Coral Reefs*, 18, 75, 1999.

202. Harvell, C. D., West, J. M., and Griggs, C., Chemical defense of embryos and larvae of a West Indian gorgonian coral, *Invert. Reprod. Dev.*, 30, 239, 1996.

203. Coll, J. C., The chemistry and chemical ecology of octocorals (Coelenterata, Anthozoa, Octocorallia), *Chem. Rev.*, 90, 613, 1992.

204. Wylie, C. R. and Paul, V. J., Chemical defenses in three species of *Sinularia* (Coelenterata, Alcyonacea): effects against generalist predators and the butterfly fish *Chaetodon unimaculatus* Bloch, *J. Exp. Mar. Biol. Ecol.*, 129, 141, 1989.

205. Alino, P. M., Sammarco, P. W., and Coll, J. C., Toxic prey discrimination in highly specialised predators: visual v. chemical cues as determinants, *J. Exp. Mar. Biol. Ecol.*, 164, 209, 1992.

206. Harvell, C. D., Fenical, W., and Greene, C. H., Chemical and structural defenses of Caribbean gorgonians (*Pseudopterogorgia* spp.). I. Development of an *in situ* feeding assay, *Mar. Ecol. Prog. Ser.*, 49, 287, 1988.

207. Harvell, C. D. and Fenical, W., Chemical and structural defenses of Caribbean gorgonians (*Pseudopterogorgia* sp.): Intracolony localisation of defense, *Limn. Oceanogr.*, 34, 382, 1989.

208. Harvell, C. D., Fenical, W., Roussis, V., Ruesink, J. L., Griggs, C. C., and Greene, C. H., Local and geographic variation in the defensive chemistry of a West Indian gorgonian coral (*Briareum asbestinum*), *Mar. Ecol. Prog. Ser.*, 93, 165, 1993.

209. Maida, M., Carroll, A. R., and Coll, J. C., Variability of terpene content in the soft coral *Sinularia flexibilis* (Coelenterata: Octocorallia) and its ecological implications, *J. Chem. Ecol.*, 19, 2285, 1993.

210. Coll, J. C., Bowden, B. F., Tapiolas, D. M., and Dunlap, W. C., *In situ* isolation of allelochemicals released from soft corals (Coelenterata: Octocorallia): a totally-submersible sampling apparatus, *J. Exp. Mar. Biol. Ecol.*, 60, 293, 1982.

211. Aceret, T. L., Sammarco, P. W., and Coll, J. C., Toxic effects of alcyonacean diterpenes on scleractinian corals, *J. Exp. Mar. Biol. Ecol.*, 188, 63, 1995.

212. Coll, J. C., Price, I. R., König, G. M., and Bowden, B. F., Algal overgrowth of alcyonacean soft corals, *Mar. Biol.*, 96, 129, 1987.

213. Leone, P. A., Bowden, B. F., Carroll, A. R., and Coll, J. C., Chemical consequences of relocation of the soft coral *Lobophytum compactum* and its placement in contact with the red alga *Plocamium hamatum*, *Mar. Biol.*, 122, 675, 1995.

214. Kerr, J. N. Q. and Paul, V. J., Animal-plant defense association: the soft coral *Sinularia* sp. (Cnidaria, Alcyonacea) protects *Halimeda* spp. from herbivory, *J. Exp. Mar. Biol. Ecol.*, 186, 183, 1995.

215. Gerwick, W. H., Carbocyclic oxylipins of marine origin, *Chem. Rev.*, 93, 1807, 1993.

216. Brash, A. R., Baertschi, S. W., and Harris, T. M., Formation of prostaglandin A analogues via an allene oxide, *J. Biol. Chem.*, 265, 6705, 1990.

217. Gerhart, D. J., Prostaglandin A$_2$ in the Caribbean gorgonian *Plexaura homomalla*: evidence against allelopathic and antifouling roles, *Biochem. Syst. Ecol.*, 14, 417, 1986.

218. Pawlik, J. and Fenical, W., A re-evaluation of the ichthyodeterrent role of prostaglandins in the Caribbean gorgonian coral *Plexaura homomalla*, *Mar. Ecol. Prog. Ser.*, 52, 95, 1989.

219. Shen, G. Q. and Baker, B. J., Biosynthetic studies of eudistomin H in the tunicate *Eudistoma olivaceum*, *Tetrahedron Lett.*, 35, 4923, 1994.

220. Shen, G. Q. and Baker, B. J., Biosynthetic studies of the eudistomins in the tunicate *Eudistoma olivaceum*, *Tetrahedron Lett.*, 35, 1141, 1994.

221. Davis, A. R. and Wright, A. E., Interspecific differences in fouling of two congeneric ascidians (*Eudistoma olivaceum* and *E. capsulatum*): Is surface acidity an effective defense?, *Mar. Biol.*, 102, 491, 1989.

222. Davis, A. R., Alkaloids and ascidian chemical defense: evidence for the ecological role of natural products from *Eudistoma olivaceum*, *Mar. Biol.*, 111, 375, 1991.

223. Steffan, B., Brix, K., and Pütz, W., Biosynthesis of shermilamine B, *Tetrahedron*, 49, 6223, 1993.

224. Lindquist, N., Hay, M. E., and Fenical, W., Defenses of ascidians and their conspicuous larvae: adult versus larval chemical defenses, *Ecol. Monogr.*, 62, 547, 1992.

225. Lindquist, N. and Hay, M. E., Can small rare prey be chemically defended? The case for marine larvae, *Ecology*, 76, 1347, 1995.

226. Taylor, S. W., Kammerer, B., and Bayer, E., New perspectives in the chemistry and biochemistry of the tunichromes and related compounds, *Chem. Rev.*, 97, 333, 1997.

227. Smith, M. J., Kim, D., Horenstein, B., Nakanishi, K., and Kustin, K., Unravelling the chemistry of tunichrome, *Acc. Chem. Res.*, 24, 117, 1991.

228. He, X., Kustin, K., Parry, D., Robinson, W. E., Ruberto, G., and Nakanishi, K., *In vivo* incorporation of ^{14}C phenylalanine into ascidian tunichrome, *Experientia*, 48, 367, 1992.

229. Stoecker, D., Resistance of a tunicate to fouling, *Biol. Bull.*, 155, 615, 1978.

230. Faulkner, D. J. and Ghiselin, M. T., Chemical defense and evolutionary ecology of dorid nudibranchs and some other opistobranch gastropods, *Mar. Ecol. Prog. Ser.*, 13, 295, 1983.

231. Burreson, B. J., Scheuer, P. J., Finer, J., and Clardy, J., 9-Isocyanopupukeanane, a marine invertebrate allomone with a new sesquiterpene skeleton, *J. Am. Chem. Soc.*, 97, 4763, 1975.

232. Cimino, G., De Rosa, S., De Stefano, S., and Sodano, G., The chemical defense of four Mediterranean nudibranchs, *Comp. Biochem. Physiol.*, 73B, 471, 1982.

233. Gunthorpe, L. and Cameron, A. M., Bioactive properties of extracts from Australian dorid nudibranchs, *Mar. Biol.*, 94, 39, 1987.

234. Manker, D. C., Garson, M. J., and Faulkner, D. J., *De novo* biosynthesis of polypropionate metabolites in the marine pulmonate *Siphonaria denticulata*, *J. Chem. Soc. Chem. Commun.*, 1061, 1988.

235. Stallard, M. O. and Faulkner, D. J., Chemical constituents of the digestive gland of the sea hare *Aplysia californica*. II. Chemical transformations, *Comp. Biochem. Physiol.*, 49B, 37, 1974.

236. Paul, V. J. and Van Alstyne, K. L., Use of ingested algal diterpenoids by *Elysia halimedae* Macnae as antipredator defenses, *J. Exp. Mar. Biol. Ecol.*, 119, 15, 1988.

237. Gavagnin, M., Marin, A., Mollo, E., Crispino, A., Villani, G., and Cimino, G., Secondary metabolites from Mediterranean Elysioidea: origin and biological role, *Comp. Biochem. Physiol.*, 108B, 107, 1994.

238. Hamann, M. T. and Scheuer, P. J., Kahalide F: a bioactive depsipeptide from the sacoglossan mollusk *Elysia rufescens* and the green alga *Bryopsis* sp., *J. Am. Chem. Soc.*, 115, 5825, 1993.

239. Hamann, M. T., Otto, C. S., Scheuer, P. J., and Dunbar, D. C., Kahalides, bioactive peptides from a marine mollusc *Elysia rufescens* and its algal diet *Bryopsis* sp., *J. Org. Chem.*, 61, 6594, 1996.

240. Gavagnin, M., Marin, A., Castellucio, F., Villani, G., and Cimino, G., Defensive relationships between *Caulerpa prolifera* and its shelled sacoglossan predators, *J. Exp. Mar. Biol. Ecol.*, 175, 197, 1994.

241. Davies-Coleman, M. T. and Garson, M. J., Marine polypropionates, *Nat. Prod. Rep.*, 15, 477, 1998.

242. Ireland, C. M. and Scheuer, P. J., Photosynthetic marine mollusks: *in vivo* [14]C incorporation into metabolites of the sacoglossan *Placobranchus ocellatus*, *Science*, 205, 922, 1979.

243. Gavagnin, M., Spinella, A., Castellucio, F., Cimino, G., and Marin, A., Polypropionates from the Mediterranean mollusk *Elysia timida*, *J. Nat. Prod.*, 57, 298, 1994.

244. Di Marzo, V., Vardaro, R. R., De Petrocellis, L., Villani, G., Minei, R., and Cimino, G., Cyercenes, novel pyrones from the ascoglossan mollusc *Cyerce cristallina*. Tissue distribution, biosynthesis and possible involvement in defense, *Experientia*, 47, 1221, 1991.

245. Di Marzo, V., Marin, A., Vardaro, R. R., De Petrocellis, L., Villani, G., and Cimino, G., Histological and biochemical bases of defense mechanisms in four species of polybranchioidea ascoglossan molluscs., *Mar. Biol.*, 117, 367, 1993.

246. Roussis, V., Pawlik, J. R., Hay, M. E., and Fenical, W., Secondary metabolites of the chemically rich ascoglossan *Cyerce nigricans*, *Experientia*, 46, 327, 1990.

247. Vardaro, R. R., Di Marzo, V., Marin, A., and Cimino, G., α and γ pyrone polypropionates from the Mediterranean ascoglossan mollusc *Ercolania funerea*, *Tetrahedron*, 48, 9561, 1992.

248. Vardaro, R. R., Di Marzo, V., and Cimino, G., Placidenes: cyercene-like polypropionate γ pyrones from the Mediterranean ascoglossan mollusc *Placida dendritica*, *Tetrahedron Lett.*, 33, 2875, 1992.

249. Ciavatta, M. L., Trivellone, E., Villani, G., and Cimino, G., Membrenones – new polypropionates from the skin of the Mediterranean mollusc *Pleurobranchus membranaceus*, *Tetrahedron Lett.*, 34, 6791, 1993.

250. Cimino, G. and Ghiselin, M. T., Chemical defense and evolution in the Sacoglossa (Molluscs: Gastropoda: Opistobranchia), *Chemoecology*, 8, 51, 1998.

251. Garson, M. J., Jones, D. D., Small, C. J., Liang, J., and Clardy, J., Biosynthetic studies on polypropionates: a stereochemical model for siphonarins A and B from the pulmonate limpet *Siphonaria zelandica*, *Tetrahedron Lett.*, 35, 6921, 1994.

252. Manker, D. C. and Faulkner, D. J., Vallartenones A and B, polypropionate metabolites of *Siphonaria maura* from Mexico, *J. Org. Chem.*, 54, 5734, 1989.

253. Blanchfield, J. T., Brecknell, D. D., Brereton, I. M., Garson, M. J., and Jones, D. D., Caloundrin B and funiculatin A: new polypropionates from siphonariid limpets, *Aust. J. Chem.*, 47, 2255, 1994.

254. Brecknell, D. D., Collett, L., Davies-Coleman, M. T., Jones, D. D., and Garson, M. J., New non-contiguous polypropionates from marine molluscs: a comment on their natural product status, *Tetrahedron*, 56, 2497, 2000.

255. Faulkner, D. J., Molinski, T. F., Andersen, R. J., Dumdei, E. J., and de Silva, E. D., Geographic variation in defensive chemicals from Pacific coast nudibranchs and some related marine molluscs, *Comp. Biochem. Physiol.*, 97C, 233, 1990.

256. Cimino, G., De Rosa, S., De Stefano, S., Morrone, R., and Sodano, G., The chemical defense of nudibranch molluscs. Structure, biosynthetic origin and defensive properties of terpenoids from the dorid nudibranch *Dendrodoris grandiflora*, *Tetrahedron*, 41, 1093, 1985.

257. Kassühlke, K. E., Potts, B. C. M., and Faulkner, D. J., New nitrogenous sesquiterpenes from two Philippine nudibranchs, *Phyllidia pustulosa* and *P. varicosa,* and from a Palauan sponge, *Halichondria* cf. *lendenfeldi*, *J. Org. Chem.*, 56, 3747, 1991.

258. Fusetani, N., Wolstenholme, H. J., Matsunaga, S., and Hirota, H., Two new sesquiterpene isonitriles from the nudibranch *Phyllidia pustulosa*, *Tetrahedron Letters*, 32, 7291, 1991.

259. Dumdei, E. J., Kubanek, J., Coleman, J. E., Pika, J., Andersen, R. J., Steiner, J. R., and Clardy, J., New terpenoid metabolites from the skin extracts, an egg mass, and dietary sponges of the North Eastern Pacific dorid nudibranch *Cadlina luteomarginata*, *Can. J. Chem.*, 75, 773, 1997.

260. Karuso, P., Chemical ecology of the nudibranchs, in *Bioorganic Marine Chemistry*, Vol. 1, Scheuer, P. J., Ed., Springer-Verlag, Berlin, 1987, 31.

261. Cimino, G., Fontana, A., Gimenez, F., Marin, A., Mollo, E., Trivellone, E., and Zubia, E., Biotrans-formation of a dietary sesterpenoid in the Mediterranean nudibranch *Hypselodoris orsini*, *Experientia*, 49, 582, 1993.

262. Cimino, G. and Sodano, G., Transfer of sponge secondary metabolites to predators, in *Sponges in Time and Space*, Van Soest, R. W. M., Van Kempen, T. M. G., and Braekman, J. C., Eds., Balkema, Amsterdam, 1994, 459.

263. Avila, C. and Durfort, M., Histology of epithelia and mantle glands of selected species of doridacean molluscs with chemical defensive strategies, *Veliger*, 39, 148, 1996.

264. Fontana, A., Gimenez, F., Marin, A., Mollo, E., and Cimino, G., Transfer of secondary metabolites from the sponges *Dysidea fragilis* and *Pleraplysilla spinifera* to the mantle dermal formations (MDF's) of the nudibranch *Hypserlodoris webbi*, *Experientia*, 50, 510, 1994.

265. Cimino, G., Fontana, A., and Gavagnin, M., Marine opisthobranch molluscs: chemistry and ecology in sacoglossans and dorids, *Curr. Org. Chem.*, 3, 327, 1999.

266. Carte, B., Kernan, M. R., Barrabee, E. B., and Faulkner, D. J., Metabolites of the nudibranch *Chromodoris funerea* and the singlet oxygen oxidation products of furodysin and furodysinin, *J. Org. Chem.*, 51, 3528, 1986.

267. Pawlik, J. R., Kernan, M. R., Molinski, T. F., Harper, M. K., and Faulkner, D. J., Defensive chemicals of the Spanish dancer nudibranch *Hexabranchus sanguineus* and its egg ribbons: macrolides derived from a sponge diet, *J. Exp. Mar. Biol. Ecol.*, 119, 99, 1988.

268. Cimino, G. and Sodano, G., Biosynthesis of secondary metabolites in marine molluscs, in *Marine Natural Products – Diversity and Biosynthesis*, Vol. 167, Scheuer, P. J., Ed., Springer-Verlag, Berlin, 1993, 77.

269. Avila, C. and Paul, V. J., Chemical ecology of the nudibranch *Glossodoris pallida*: is the location of diet-derived metabolites important for defense?, *Mar. Ecol. Prog. Ser.*, 150, 171, 1997.

270. Rudman, W. B., The anatomy and biology of alcyonarian-feeding aeolid opistobranch mollusks and their development of symbiosis with zooxanthellae, *Zool. J. Linn. Soc.*, 72, 219, 1981.

271. Baker, B. J. and Scheuer, P. J., The punaglandins: 10-chloroprostanoids from the octocoral *Telesto riisei*, *J. Nat. Prod.*, 57, 1346, 1994.

272. McClintock, J. B., Baker, B. J., Slattery, M., Heine, J. N., Bryan, P. J., Yoshida, W., Davies-Coleman, M. T., and Faulkner, D. J., Chemical defense of common Antarctic shallow-water nudibranch *Tritoniella belli* Eliot (Molluscs: Tritonidae) and its prey *Clavularia frankliniana* Rouel (Cnidaria: Octocorallia), *J. Chem. Ecol.*, 20, 3361, 1994.

273. Slattery, M., Avila, C., Starmer, J., and Paul, V. J., A sequestered soft coral diterpene in the aeolid nudibranch *Phyllodesmiun guamensis* Avila, Ballesteros, Slattery, Starmer and Paul, *J. Exp. Mar. Biol. Ecol.*, 226, 33, 1998.

274. Paul, V. J., Lindquist, N., and Fenical, W., Chemical defenses of the tropical ascidian *Atapozoa* sp. and its nudibranch predators *Nembrotha* spp., *Mar. Ecol. Prog. Ser.*, 59, 109, 1990.

275. Carte, B. and Faulkner, D. J., Role of secondary metabolites in feeding associations between a predatory nudibranch, two grazing nudibranchs and a bryozoan, *J. Chem. Ecol.*, 12, 795, 1986.

276. Coll, J. C., Tapiolas, D. M., Bowden, B. F., Webb, L., and Marsh, H., Transformation of soft coral (Coelenterata: Octocorallia) terpenes by *Ovula ovum*, *Mar. Biol.*, 74, 35, 1983.

277. Webb, L., Studies on the apparent detoxification of sarcophytoxide by the prosobranch mollusc *Ovula ovum, Comp. Biochem. Physiol.*, 90C, 155, 1988.

278. Coval, S. J., Schulte, G. R., Matsumoto, G. K., Roll, D. M., and Scheuer, P. J., Two polypropionate metabolites from the Cephalaspidean mollusk *Philinopsis speciosa, Tetrahedron Lett.*, 5359, 1985.

279. Cimino, G., Sodano, G., and Spinella, A., New propionate-derived metabolites from *Aglaja depicta* and from its prey *Bulla striata* (Opisthobranch molluscs), *J. Org. Chem.*, 52, 5326, 1987.

280. Spinella, A., Alvarez, L. A., and Cimino, G., Predator–prey relationship between *Navanax inermis* and *Bulla gouldiana*: a chemical approach, *Tetrahedron*, 49, 3203, 1993.

281. Cimino, G., Spinella, A., and Sodano, G., Potential alarm pheremones from the Mediterranean opistobranch *Scaphander lignarius, Tetrahedron Lett.*, 30, 5003, 1989.

282. Cimino, G., Passeggio, A., Sodano, G., Spinella, A., and Villani, G., Alarm pheremones from the Mediterranean opistobranch *Haminoea navicula, Experientia*, 47, 61, 1991.

283. Spinella, A., Alvarez, L. A., Passeggio, A., and Cimino, G., New 3-alkylpyridines from three Cephalaspidean molluscs: structure, ecological role and taxonomic relevance, *Tetrahedron*, 49, 1307, 1993.

284. Marin, A., Alvarez, L. A., Cimino, G., and Spinella, A., Chemical defence in cephalaspidean gastropods: origin, anatomical location and ecological role, *J. Mollusc. Stud.*, 65, 121, 1999.

285. Fenical, W., Sleeper, H. L., Paul, V. J., Stallard, M. O., and Sun, H. H., Defensive chemistry of *Navanax* and related Opistobranch molluscs, *Pure Appl. Chem.*, 51, 1865, 1979.

286. Nakao, Y., Yoshida, W. Y., Szabo, C. M., Baker, B. J., and Scheuer, P. J., More peptides and other diverse constituents of the marine mollusk *Phylinopsis speciosa, J. Org. Chem.*, 63, 3272, 1998.

287. Davies-Coleman, M. T., and Faulkner, D. J., New diterpenoic acid glycerides from the Antarctic nudibranch *Austrodoris kerguelensis, Tetrahedron*, 47, 9743, 1991.

288. Cimino, G., de Rosa, S., De Stefano, S., Sodano, G., and Villani, G., Dorid nudibranch elaborates its own chemical defense, *Science*, 219, 1237, 1983.

289. Cimino, G., De Rosa, S., De Stefano, S., and Sodano, G., Marine natural products: new results from Mediterranean invertebrates, *Pure Appl. Chem.*, 58, 375, 1986.

290. Fontana, A., Ciavatta, M. L., Miyamoto, T., Spinella, A., and Cimino, G., Biosynthesis of drimane terpenoids in dorid molluscs: Pivotal role of 7-deacetoxyolepupuane in two species of *Dendrodoris* nudibranchs, *Tetrahedron*, 55, 5937, 1999.

291. Fontana, A., Villani, G., and Cimino, G., Terpene biosynthesis in marine molluscs: incorporation of glucose in drimane esters of *Dendrodoris* nudibranchs via classical mevalonate pathway, *Tetrahedron Lett.*, 41, 2429, 2000.

292. Avila, C., Cimino, G., Crispino, A., and Spinella, A., Drimane sesquiterpenoids in Mediterranean *Dendrodoris* nudibranchs: anatomical distribution and biological role, *Experientia*, 47, 306, 1991.

293. Cimino, G., Sodano, G., and Spinella, A., Occurrence of olepupuane in two Mediterranean nudibranchs: a protected form of polygodial, *J. Nat. Prod.*, 51, 1010, 1988.

294. Gustafson, K. and Andersen, R. J., Chemical studies of British Columbian nudibranchs, *Tetrahedron*, 41, 1101, 1985.

295. Avila, C., Ballesteros, M., Cimino, G., Crispino, A., Gavagnin, M., and Sodano, G., Biosynthetic origin and anatomical distribution of the main secondary metabolites in the nudibranch mollusc *Doris verrucosa, Comp. Biochem. Physiol.*, 97B, 363, 1990.

296. Porcelli, M., Cacciapuoti, G., Cimino, G., Gavagnin, M., Sodano, G., and Zappia, V., Biosynthesis and metabolism of 9-[5′deoxy-5′-(methylthio)-β-D-xylofuranosyl]adenine, a novel natural analogue of methylthioadenosine, *Biochem. J.*, 263, 635, 1989.

297. Marin, A., Di Marzo, V., and Cimino, G., A histological and chemical study of the cerata of the opistobranch mollusc *Tethys fimbria, Mar. Biol.*, 111, 353, 1991.

298. Di Marzo, V., Cimino, G., Crispino, A., Minardi, C., Sodano, G., and Spinella, A., A novel multifunctional metabolic pathway in a marine mollusc leads to unprecedented prostaglandin derivatives (prostaglandin-1,15-lactones), *Biochem J.*, 273, 593, 1991.

299. Cimino, G., Crispino, A., Di Marzo, V., Sodano, G., Spinella, A., and Villani, G., A marine mollusc provides the first example of *in vivo* storage of prostaglandins: prostaglandins-1,5-lactones, *Experientia*, 47, 56, 1991.

300. Di Marzo, V., Minardi, C., Vardaro, R. R., Mollo, E., and Cimino, G., Prostaglandin F-1,15-lactone fatty acyl esters: a prostaglandin lactone pathway branch developed during the reproduction and early larval stages of a marine mollusc, *Comp. Biochem. Physiol.*, 101B, 99, 1992.

301. Graziani, E. I. and Andersen, R. J., Investigations of sesquiterpenoid biosynthesis by the dorid nudibranch *Acanthodoris nanaimoensis*, *J. Am. Chem. Soc.*, 118, 4701, 1996.

302. Graziani, E. I., Andersen, R. J., Krug, P. J., and Faulkner, D. J., Stable isotope incorporation evidence for the *de novo* biosynthesis of terpenoic acid glycerides by dorid nudibranchs, *Tetrahedron*, 52, 6869, 1996.

303. Kubanek, J., Graziani, E. I., and Andersen, R. J., Investigations of terpenoid biosynthesis by the dorid nudibranch *Cadlina luteomarginata*, *J. Org. Chem.*, 62, 7239, 1997.

304. Graziani, E. I. and Andersen, R. J., Stable isotope incorporation evidence for a polyacetate origin of the acyl residues in triophamine, a diacylguanidine metabolite obtained from the dorid nudibranch *Triopha catalinae*, *J. Chem. Soc. Chem. Commun.*, 2377, 1996.

305. Kubanek, J. and Andersen, R. J., Evidence for the incorporation of intact butyrate units in the biosynthesis of triophamine, *Tetrahedron Lett.*, 38, 6327, 1997.

306. Kubanek, J. and Andersen, R. J., Evidence for *de novo* biosynthesis of the polyketide fragment of diaulusterol A by the Northeastern Pacific dorid nudibranch *Diaulula sandiegensis*, *J. Nat. Prod.*, 62, 777, 1999.

307. Kubanek, J., Faulkner, D. J., and Andersen, R. J., Geographic variation and tissue distribution of endogenous terpenes in the North Eastern Pacific dorid nudibranch *Cadlina luteomarginata*: implications for the regulation of *de novo* biosynthesis, *J. Chem. Ecol.*, 26, 377, 2000.

308. Fontana, A., Gavagnin, M., Mollo, E., Trivellone, E., Ortea, J., and Cimino, G., Chemical studies of *Cadlina* molluscs from the Cantabrian Sea (Atlantic Ocean), *Comp. Biochem. Physiol.*, 111B, 283, 1995.

3 Marine Natural Products Chemistry as an Evolutionary Narrative

Guido Cimino* and Michael T. Ghiselin

CONTENTS

I. INTRODUCTION

Chemistry is generally looked upon as one of the "nomothetic" sciences, i.e., one that seeks to establish the laws of nature and does not concern itself with particular objects or events. Natural

---

* Corresponding author.

0-8493-9064-8/01/$0.00+$1.50

products chemistry is an exception, being very much concerned with what are called "individuals" in a broad metaphysical sense.[1] Like geology, paleontology, and systematic zoology, it is very much a natural history discipline. The various natural kinds of secondary metabolites are classes of molecules, and, being classes, they do not evolve any more than does the calcium carbonate that forms the shells of molluscs. What do evolve are individual populations and lineages, which change with respect to the properties of the organisms and their parts, including the enzymes that produce secondary metabolites. Metabolism has a history and we ought to be able to reconstruct that history just like the chemical and physical aspects of defense. There seems to have been an arms race between shelled molluscs and the crabs that have preyed upon them.[2] A chemical arms race involving molluscs in which the shell has become reduced is a straightforward extrapolation.

It is no longer fashionable to dismiss secondary metabolites as mere waste products or as substances that no longer play an important role in the lives of organisms. They are distributed in the body very much like other features that have obvious value in the struggle for existence.[3] Even today, however, much of the discussion of the supposed function of natural products continues to treat natural selection as little more than background material. Indeed, historical situations are often invoked to explain away anomalies where efforts to find adaptation fail. Adaptation can be treated as if it were nothing more than a condition or state. At least implicitly, however, the product is defined in terms of the process. In other words, when we claim that something is an adaptation, we presuppose a historical narrative, even if the narrative is concerned only with the very recent past. If we really want to understand the adaptive significance of secondary metabolites, we need to ask some truly historical questions.

These authors' contributions in this area have mainly dealt with the evolution of chemical defense in opisthobranch gastropods.[4,5] Faulkner and Ghiselin[6] addressed the question of whether the reduction of the shell in these animals preceded the evolution of chemical defense (a post-adaptive scenario) or whether the loss of the shell was made possible by the presence of chemical defense (a pre-adaptive scenario). The latter hypothesis was preferred on the grounds that in groups in which the shell is relatively well developed, chemical defense is already present. The reasoning is basically a matter of plotting features on the branches of a phylogenetic tree and inferring the sequence of events. But the biological plausibility of the sequence in question may provide an additional line of evidence. This is a traditional mode of reasoning that goes back to Darwin and his follower Anton Dohrn, who founded the Zoological Station at Naples.[7,8] Evolution proceeds by steps; in each step the functioning of the organism as a whole is conserved, but particular functions often succeed one another over time.

Various patterns in the evolution of chemical defense have been documented, including detoxification, modification and sequestration of metabolites, and their positioning in places where they will more effectively repel predators. Of particular interest is the evolution of *de novo* synthesis. The work of these authors has suggested how this might happen. It also suggests that asking questions about the evolution of biosynthetic capacity might provide a unifying theme for the study of natural products chemistry.

This chapter first discusses how chemical defense might be acquired. The authors then suggest reasons why it should evolve differently in various kinds of organisms such as autotrophs and heterotrophs. The chapter then gives some examples, presented in an order that is not, strictly speaking phylogenetic, although the taxonomic groups discussed are generally thought to be natural ones in the sense that they represent genealogical wholes (clades). The opisthobranchs are discussed in more than one place because they derive metabolites from various sources, including *de novo* synthesis. The acquisition and use of some metabolites rather than others provides evidence that they are, as suspected, defensive chemicals. However, they may be defensive in a broad sense that includes dealing with fouling organisms, spatial competitors, and various other things.

II. THE EVOLUTION OF BIOSYNTHETIC CAPACITY

It is well understood that the synthesis and modification of metabolites is under enzymatic control. The enzymes may function as catalysts, and the reactions themselves are not restricted only to living systems. So the evolution of biosynthetic capacity is largely the result of changes in enzymes by mutation, gene duplication, and other familiar processes. The organisms synthesize and modify secondary metabolites in a stepwise fashion, much as organic chemists do, and in neither case are the laws of nature violated.

The term "secondary metabolite" is generally understood to mean that the chemicals in question are not directly involved in the basic maintenance of the organism. Secondary metabolites are produced from a remarkably limited range of starting materials known as primary metabolites. There are three major classes of secondary metabolites of interest here (Figure 3.1). Acetogenins are produced by head-to-tail condensation of acetic acid into linear units starting with acetyl-CoA. Terpenes are made from acetate units that are turned into isoprene units and then put together into larger units by head-to-tail condensation via isopentenyl diphosphate, often followed by cyclization. Finally, alkaloids are usually formed by the Mannich reaction in which amino acids are transformed into amines and aldehydes. There are, of course, less common metabolites, sometimes very interesting ones. These include the polypropionates, discussed later in this chapter. As shown in Figure 3.1, they form similar to acetogenins, but the starting compound is propionyl CoA. Acetogenins and polypropionates are often referred to in the literature as polyketides. One should bear in mind that much of the diversity of metabolites can be explained as a result of stepwise synthesis of larger and larger units, with some divergent variants in skeletal structure and a lot of rearrangements and other modifications of the basic structures. Such patterns of synthesis can be explained historically, and stepwise modification of biosynthetic pathways through time is a basic phylogenetic theme. The fact that the same compound may be synthesized by different pathways is not an impediment to such historical analysis, but rather an opportunity. Different pathways often reflect separate historical origins.

Before discussing how organisms might evolve such pathways, it is convenient to consider how they might acquire pathways from other organisms. One such possibility is through symbiosis, especially mutualism. Such mutualism is well documented in the phylum Porifera, or sponges, which often contain bacteria within their tissues that produce some of the metabolites that defend the sponge and presumably the bacteria as well. The sponges did not have to evolve the chemicals that defend them. Another possibility is lateral gene transfer. The well-known spread of antibiotic resistance between lineages of bacteria makes such a transfer seem highly plausible. Lateral gene transfer may be quite common among marine microorganisms.[9] It can enable them to acquire the capacity for biosynthesis without having to evolve it through the modification of pre-existing metabolic apparatus. Such capacity means that the organisms are not constrained by the necessity of obtaining the metabolites from symbionts or food. But whether such a transfer is not just possible, but has in fact occurred, has to be established on the basis of empirical evidence. For multicellular animals it is mere conjecture.

We have suggested elsewhere that there are two modes by which the capacity for biosynthesis of secondary metabolites might evolve.[5] The first of these is the straightforward and well-documented anasynthetic mode in which more and more steps are added and, perhaps, a molecule of increasing size is produced. Such evolution has been well documented in terrestrial plants, and some marine examples are discussed below. In some cases it has been shown that the end product of the most derived evolutionary stage is accumulated in the tissues, but that the intermediates are present in lesser concentrations. These intermediates, however, may be the metabolites that are concentrated in the tissues of related forms that represent ancestral conditions in a historical sequence. This case presents a chemical analogue of the traditional notion that ontogeny recapitulates phylogeny.[10] Such recapitulation occurs only under restricted conditions, namely, where there has been terminal addition

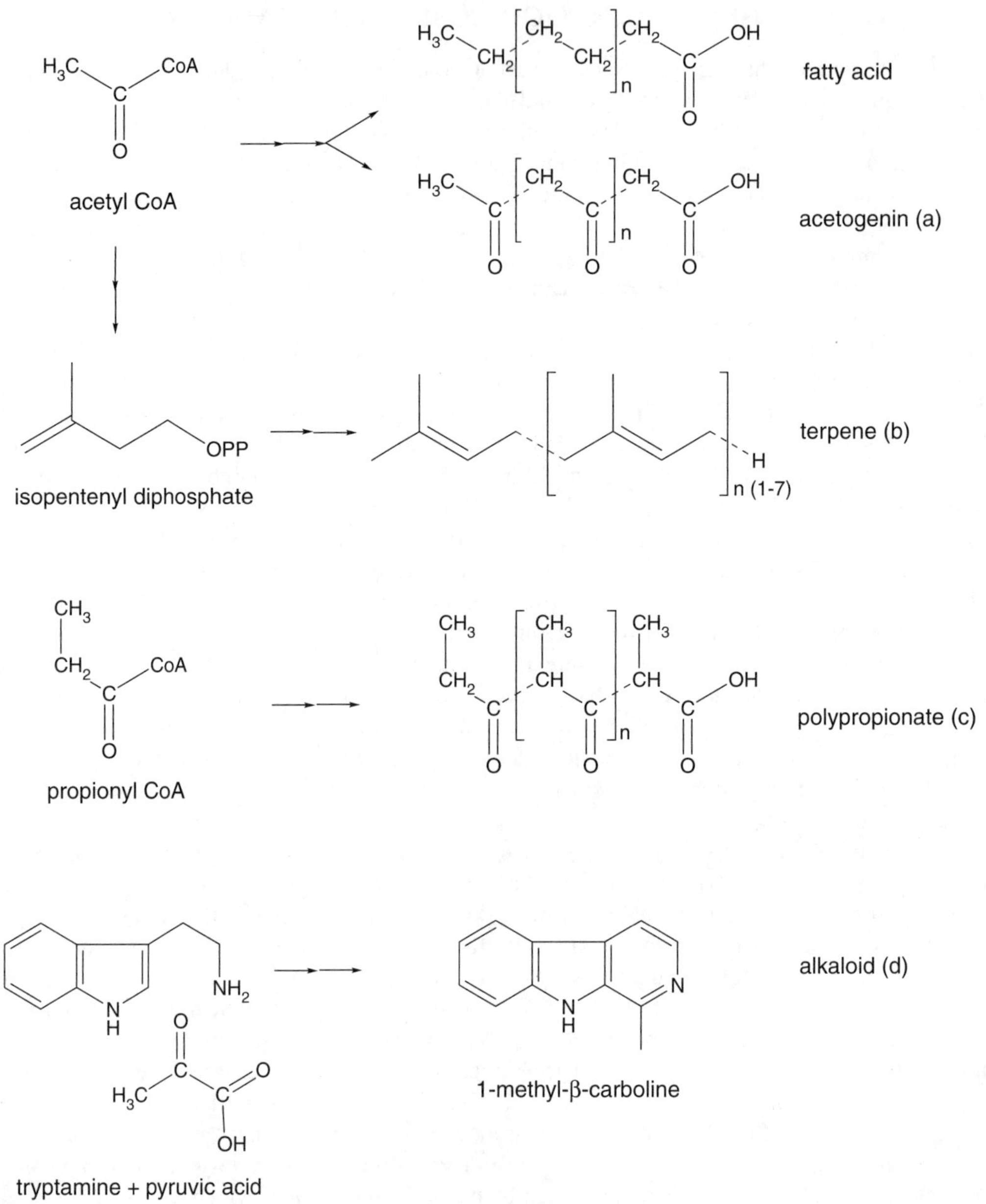

FIGURE 3.1 Main classes of secondary metabolites. Fatty acids and acetogenins are derived from acetyl Coenzyme A, which forms the isopentyl diphosphate that forms terpenes. The (unusual) polypropionates derive from propionyl CoA. Alkaloids are typically modified amino acids.

of new stages. There is nothing to prevent the secondary loss of developmental or biosynthetic stages, and it is possible that earlier steps in a pathway might be affected.

Another possibility is called the retrosynthetic mode. We recognized this possibility because opisthobranchs have evolved the capacity for *de novo* synthesis of metabolites similar to those that they originally derived from food. How could they evolve an entire pathway that supposedly had not been part of their evolutionary heritage, a process that usually takes a long time and a lot of unusual events? One possibility was lateral gene transfer. However, the metabolites in question sometimes did not have the same chirality as those in the food organisms, suggesting that different enzymes of different historical origins are involved. The chirality is just one example of the fact that when nudibranchs have evolved *de novo* synthesis, the metabolites that they produce, although similar, are almost never identical to the originals. So, elaborating on some earlier ideas,[11,12] we proposed that the predators first evolved the ability to modify the last stages of a biosynthetic pathway while still relying upon intermediates that are available in food. Given that the intermediates are present in the food, any mutation that increases their concentration would be selectively advantageous. Steps in the synthesis could be added backward until replacing a point in which a precursor was available that the predator could synthesize on its own. This would clearly not be a situation in which ontogeny recapitulates phylogeny.

III. EVOLUTIONARY PATTERNS IN DIFFERENT CLASSES OF ORGANISMS

Classes of organisms refers here to abstract kinds, as opposed to taxonomic groups, which are concrete historical units. Classes here means groups of organisms that share such properties as the manner in which food is obtained, whether they are sessile or motile, large or small, etc. Large, sessile autotrophs have evolved separately and independently upon numerous occasions. So have filter-feeders, grazers, and predators. These groups of unrelated organisms have many of the same ecological requirements and functional constraints. They often display some evolutionary convergences that help us infer what the important selection pressures have been.

Microorganisms include prokaryotes (bacteria such as Cyanobacteria) and unicellular eukaryotes (Protozoa and algae of various taxa). Because of their small size and considerable biosynthetic versatility, they are predisposed to assume the position of mutualistic symbionts within the bodies of other organisms. Lateral gene transfer is, of course, particularly well documented in bacteria. Unicellular organisms that remain together as a group and form clones can defend the group as a whole in the same way that an entire multicellular organism does. Part of the unit can be sacrificed, leaving the rest still able to survive and reproduce.

Sessile autotrophs in marine environments are almost all algae. Many algae remain unicellular, but there are several lineages which are multicellular organisms crudely convergent with terrestrial plants. Because the source of nutriment is photosynthesis, multicellular plants have no dietary source of secondary metabolites, and, furthermore, the opportunities for evolving symbiotic relationships that might provide defense are quite limited. So, we would expect the anasynthetic mode to be virtually the only way of acquiring chemical defense in such organisms. On the other hand, the sessile life style, where organisms cannot move away from either grazers or spatial competitors, makes chemical defense a particularly important mechanism. There are, to be sure, examples of terrestrial plants with defensive symbiotic relationships, for example, with ants, and there may be some marine examples as well.

Sessile animals (and ones that are virtually so) have much the same problems in defending themselves and warding off predators as terrestrial plants. In fact, many of these animals supplement their food supply by means of symbiotic unicellular organisms, as do a few animals that move from place to place, but slowly. Most sessile marine animals feed upon material that is either suspended in the water or deposited on the substrate. This material sometimes contains metabolites that might be put to use defensively. However, the food is usually heterogeneous and relatively unpredictable as to content; therefore, it does not supply a reliable source of metabolites. It is generally not used defensively, and such use is facultative. Some bivalves, however, do concentrate saxitoxin, derived from dinoflagellates, in tissues that are exposed to predators.[13]

A large number of sessile filter feeders do have defensive metabolites that are not derived from food. Some of these metabolites are synthesized by the animals themselves, and some are synthesized by symbionts. In the latter case, there must be a reliable way of providing a supply of symbionts for the next generation. This is especially important because sessile and sedentary marine animals generally disperse as larvae or as in even earlier developmental stages. Special adaptations that transmit the guest symbionts from host to host across generations are presumptive evidence that the host species actually receives a benefit, i.e., that the host is either a mutualist or a parasite. Of course, unwelcome guests often do succeed in getting transmitted from parent to offspring, but this occurs in spite of the noncooperation of the host. In that case the parasite is the exploiting organism, and is not necessarily contained within the other organism, so an ectoparasite could actually surround its host.

Slow-moving grazers are also in a poor position to flee from predators and require some sort of suitable protection, either mechanical, chemical, or both. Spatial competitors and fouling organisms are less of a problem. Depending on what they eat, it may or may not be easy for them to obtain defensive metabolites from food. If they do obtain such metabolites from food, one would expect them to be specialists that feed upon organisms of those taxa that contain a copious and reliable supply of such metabolites. Obtaining defensive metabolites from symbionts is a distinct possibility, but there is little evidence of that in these organisms. The difficulty of overcoming the food organism's chemical (and mechanical) defenses combined with relying upon that same organism as a source of protective metabolites tends to produce specialization, yet it constrains the animal to a narrower range of food items. Hence one might predict shifts of host and the evolution of a capacity for *de novo* synthesis.

IV. TAXONOMIC SURVEY

A. BACTERIA

The bacteria of interest to us here are restricted to one major clade of Eubacteria: the Actinobacteria and the closely related Cyanobacteria. They are particularly significant as sources of metabolites that are used by other organisms, either because the other organisms are their symbionts or feed upon them. A few fungi are also significant for the same reason. The possible origin of their use of such compounds has been subject to some speculation. A reasonable possibility is that the metabolites were used in competing with other prokaryotes, perhaps in or on the surface of sediments. The bacterial metabolites of interest here are macrolides, polyketides, cyclic lactones, cyclic peptides, and alkaloids (Figure 3.2). Macrolides of sponges are produced by symbiotic bacteria in their tissues. Polyketides of urochordates are also produced by symbiotic bacteria, and these are concentrated by the opisthobranch gastropods that feed upon them. Figure 3.2 shows typical metabolites from bacteria and fungi.

1. Cyanobacteria

Some characteristic metabolites of Cyanobacteria are macrolides and chlorinated amides (Figure 3.3). Cyanobacteria were formerly called blue-green algae, and, as the name suggests, some of them are convergent with the prokaryotic algae in general structure and also in sometimes containing secondary metabolites. In the opisthobranch order Anaspidea, there are isolated cases of shifts from feeding on true algae to including Cyanobacteria in the diet. The availability of metabolites that could be used defensively in food organisms with suitable physical properties can be viewed as allowing an opportunistic shift in niche. However, the anaspideans are remarkably adept at coping with a broad range of metabolites, and an alternative scenario in which Cyanobacteria were present in the diet from a much earlier period cannot presently be excluded.

B. ALGAE

This section discusses the three groups of algae within which multicellular, mainly sessile lineages have evolved. All three groups have members rich in secondary metabolites that are pressed into service defensively by opisthobranchs that feed upon them. Presumably, they are also used

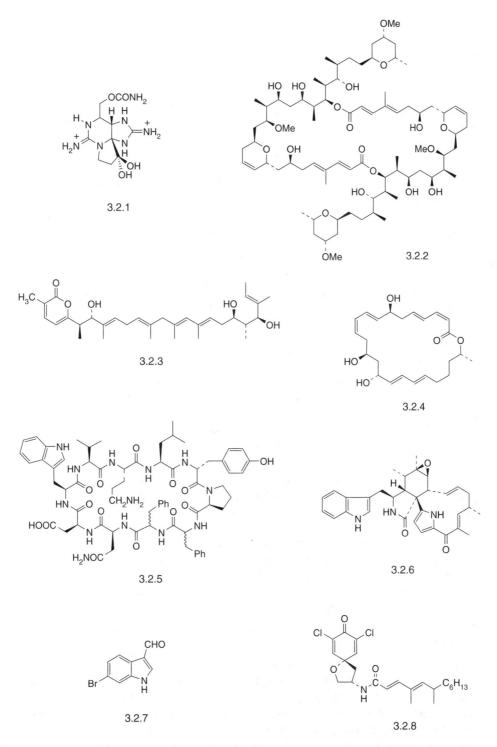

FIGURE 3.2   Selected metabolites after marine bacteria and fungi. 3.2.1 Saxitoxin.[173] 3.2.2 Swinholide from *Theonella swinhoei*.[174] 3.2.3 Lagunapyrone A from an actinomycete.[175] 3.2.4 (-)Macrolactin A from an unidentified deep-sea bacterium.[176] 3.2.5 Loloatin B from *Bacillus* sp. isolated from an unidentified tube worm.[177] 3.2.6 Penochalasin A, from *Penicillum* sp., symbiotic with the green alga *Enteromorpha intestinalis*.[178] 3.2.7 6-Bromindole-3-carbaldehyde from the ascidian *Stomozoa murrayi* and an associated *Acinetobacter* sp.[179] 3.2.8 Gymnastatin A from a strain of *Gymnasella dankaliensis* isolated from the sponge *Halichondria japonica*.[180]

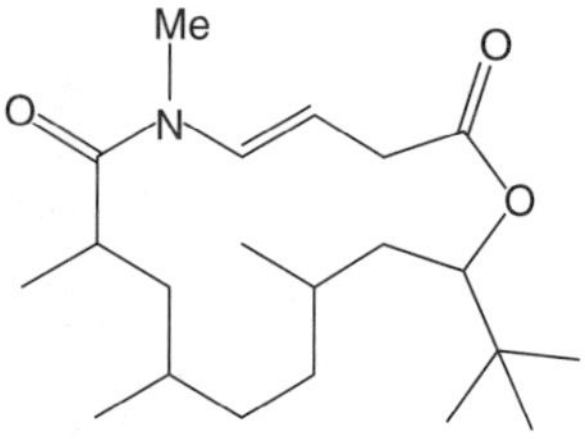

3.3.1 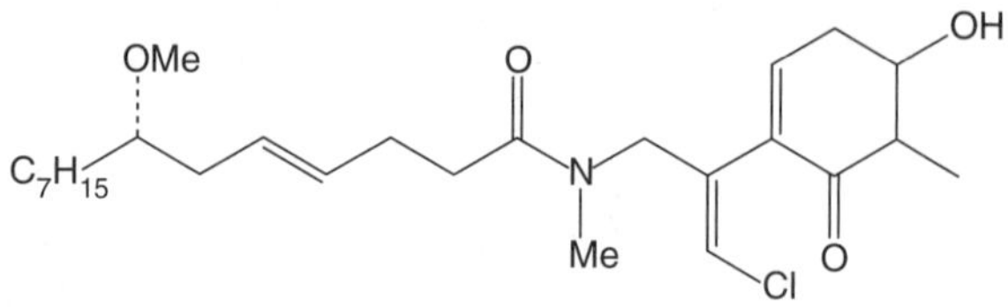   3.3.2

FIGURE 3.3   Selected metabolites from cyanobacteria. 3.3.1 Laingolide from *Lyngbya bouillonii*.[181] 3.3.2 Malyngamide from *Lyngbya majuscula*.[182]

defensively by plants, though, again, the wider range of effects upon other organisms under that rubric is included. The differences between these three groups in the kinds of metabolites that they elaborate suggest separate historical origins. The possibility that these metabolites could be traced back to an earlier common ancestry cannot be ruled out, but there seems to be no evidence for that theory. What has been documented is the elaboration and diversification of metabolites within each of these major algal groups.

1.   Phaeophyta

The first algal lineage considered here is the brown algae, or Phaeophyta, which include the familiar, macroscopic kelps. This chapter does not go into the fascinating and highly controversial issue of whether polyphenolics in brown algae are used defensively (see Chapter 6 in this volume). These compounds are not expropriated by algivorous animals, and indeed, there would seem to be no defensive use of these compounds by any animals. Terpenoids are another matter altogether. Dolabellanes (Figure 3.4.7), for example, are highly toxic compounds. They are named after the genus *Dolabella*, which are opisthobranchs of the order Anaspidea (see below).

Amico[14] studied the secondary metabolites in brown algae of the genus *Cystoseira* (family Cystoseiraceae, order Fucales) from a phylogenetic point of view. He was able to arrange metabolites in series, such that the more evolutionarily derived algae have metabolites that require more steps in their biosynthesis (Figure 3.4). Linear diterpenoids in his scheme are succeeded by what he calls a pool of open-chain meroditerpenoids, and these in turn are elaborated into several groups of derived and divergent ring systems. His arrangement agreed well with morphological classifications and made sense in terms of the biogeography of the group. Because the genus originated in the late Cretaceous period (ca. 80 MYBP), it seems clear that the elaboration and diversification of the metabolites occurred during the Cenozoic Period.

2.   Rhodophyta

Important metabolites of red algae are shown in Figure 3.5. Red algae have traditionally been divided into Bangiaceae and Floridiophycidae. The former are morphologically simpler, and efforts to find defensive metabolites in them have failed. On the other hand, C-15 halogenated acetogenins and halogenated monoterpenoids are widespread among the latter. Within the Floridiophycidae, halogenated monoterpenes predominate in the Gigartinales, whereas the Ceramiales emphasize brominated sesquiterpenes. Other significant metabolites of red algae include sesquiterpenoids, diterpenoids, polyethers, and dipeptides.

Opisthobranchs of the order Anaspidea (sea hares) feed upon a diversity of algae. That they sometimes obtain metabolites from blue-green algae (Cyanobacteria) and brown algae (Phaeophyta) has already been mentioned. They also feed upon green algae (Chlorophyta, see below), but these

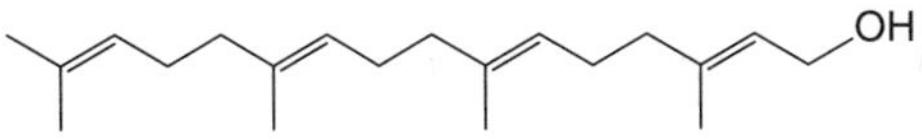

3.4.1

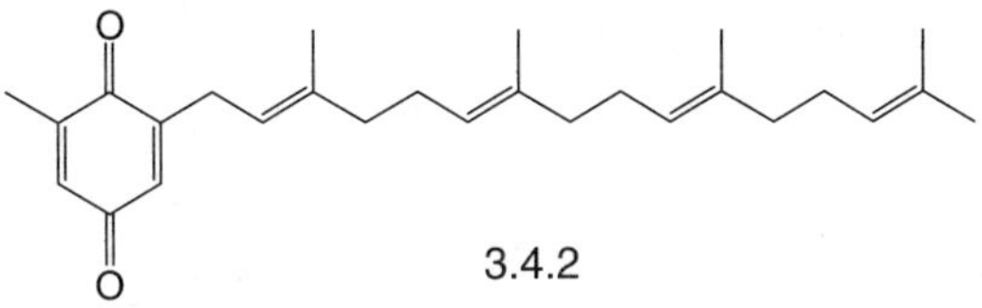

3.4.2

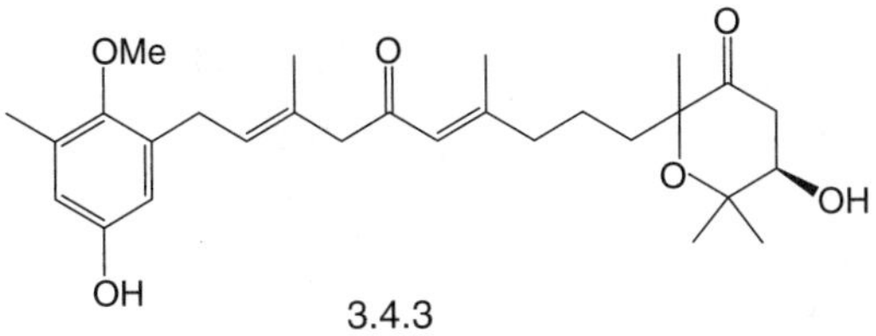

3.4.3

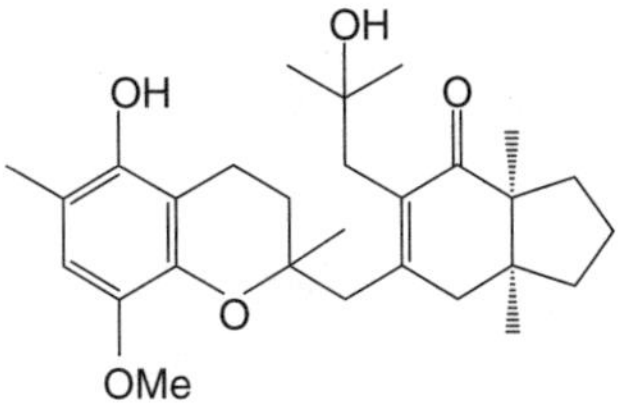

3.4.4

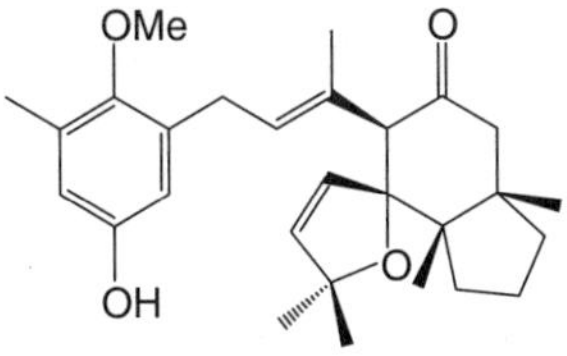

3.4.5

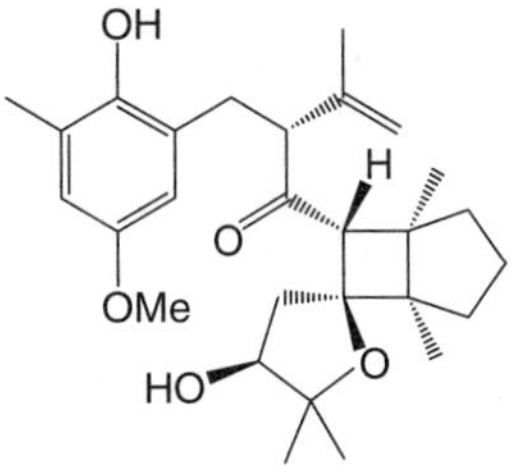

3.4.6

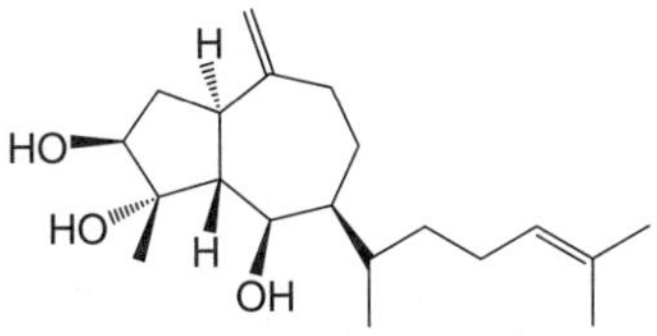

3.4.7

FIGURE 3.4 Metabolites of brown algae. 3.4.1–3.4.6 Selected metabolites from the *Cystoseira* showing simple and derived conditions.[183] 3.4.7 Dictyotatriol A from *Dictyota dichotoma*.[184]

are not known to provide them with an important source of metabolites. A remarkably wide range of chlorinated terpenoids has been recovered from red algae, especially from the genus *Laurencia* and from sea hares of the genus *Aplysia* that feed upon them. The ability to feed upon algae that are rich in halogenated terpenoids seems to have been a major innovation within the Anaspidea, but the group in general is adept at feeding upon plants with a broad range of algal metabolites.

To what extent the anaspideans are expropriating the metabolites and using them defensively, and to what extent they are merely obtaining them from food and disposing of them, has been a topic of controversy. It has been pointed out that the metabolites are mostly concentrated in the

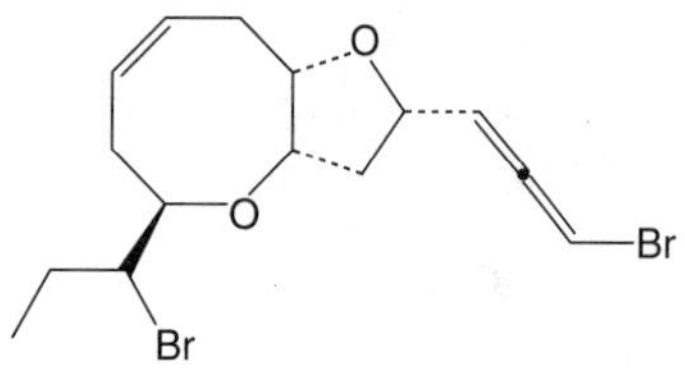

3.5.1

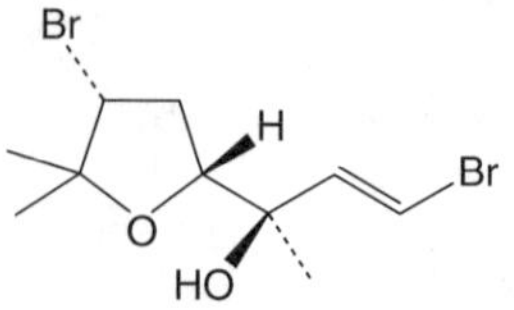

3.5.2

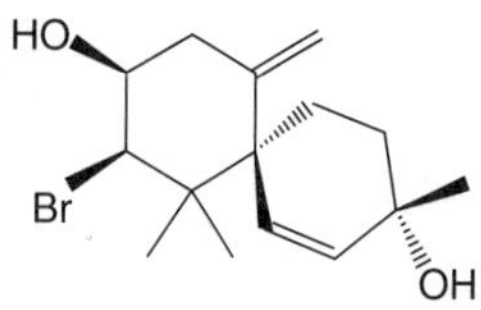

3.5.3

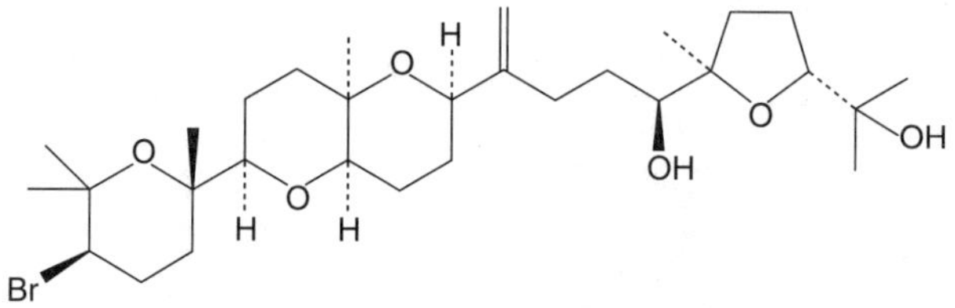

3.5.4

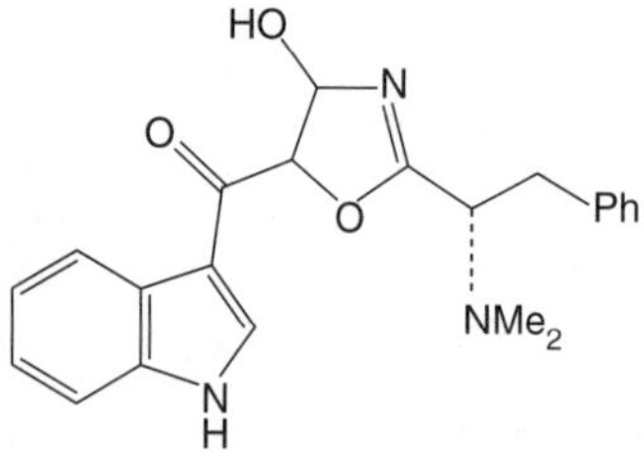

3.5.5

FIGURE 3.5   Selected metabolites from red algae. 3.5.1 Pannosallene from *Laurencia pannosa*.[185] 3.5.2 Pantafuranoid A from *Pantoneura plocamioidea*.[186] 3.5.3 Rigidol from *Laurencia rigida*.[187] 3.5.4 10-Epide-hydroxythyrsiferol from *Laurencia viridis*.[188] 3.5.5 Almazole D from *Haraldiophyllum* sp.[189]

digestive gland, which is not where they would most effectively deal with predators.[15] However, evolution from a relatively ineffective mechanism of defense to a more effective one is what is expected and has precedents in other groups of opisthobranchs. Some oversimplifications about selective mechanisms may be implicit in such discussions. It is generally assumed that a defensive adaptation cannot evolve if the organism that possesses it is killed and therefore fails to reproduce more than conspecifics without it. This stricture would apply to the anaspideans in question were it a scenario with metabolites becoming present in the digestive gland because of their defensive role. It does not apply, however, to a scenario in which the metabolites were initially present in the food and, therefore, in the digestive gland. An animal containing such metabolites in its digestive gland would not be protected from predators that devoured it. But if the predators were sickened or killed, they would tend to leave the survivors alone. Selection could then favor putting the metabolites in a more effective position. Kin selection might also operate in spite of some theoretical considerations first put forth by Faulkner and Ghiselin,[6] and subsequently by other authors.[16,17] The problem with kin selection in opisthobranchs is that most of them have larval dispersal, and, therefore, the juveniles and adults do not live with close relatives. Under such circumstances, kin selection does not seem to be a good explanation for the aposematic coloration that is so common in the group. One important point, however, has been overlooked. Opisthobranchs are internally fertilizing hermaphrodites that store sperm, and they often occur in aggregations of conspecifics.

Killing such an organism eliminates its ability to lay eggs, but its sperm may survive and fertilize another organism's eggs.

Furthermore, the present-day situation in anaspideans turns out to be a bit more complicated than expected. According to some recent field studies, *Aplysia juliana* ate only green algae and had no defensive metabolites, whereas *A. parvula*, which fed upon red algae, used diet-derived metabolites quite effectively in defense.[18,19] The latter species is considered the most primitive representative of the genus,[20] and it seems likely that the former has ceased to accumulate metabolites or was not doing so when the study was conducted. It is noteworthy that metabolites typical of Rhodophyta have been recorded from what has been identified, at least, as *A. juliana*.[21] The prevailing pattern of feeding mainly upon taxa with metabolites, coupled with occasional shifts to taxa that also contain secondary metabolites, likewise points toward a defensive role.

3. Chlorophyta

Green algae, or Chlorophyta, which are genealogically closer to vascular plants than they are to other algae, have well-documented chemical defense in the order Bryopsidales. In this order (Figure 3.6), the main defensive metabolites are diterpenoids and sesquiterpenoids. Minor metabolites of interest include cyclic peptides (Figure 3.6.1), and bicyclic lipids (Figure 3.6.2). Of particular interest are the closely related families Udoteaceae and Caulerpaceae.[22,23] The Udoteaceae include several genera (*Udotea*, *Halimeda*, *Chlorodesmis*) which have a combination of chemical defense and mechanical defense based on calcification. The combination has been shown to be more effective against the grazing animals that occur in the largely tropical or subtropical habitats of the algae.[24] Caulerpaceae consists only of the genus *Caulerpa*, and its separation evidently does not have a genealogical basis, so it seems likely that the absence of calcification is secondary. The metabolites of these algae have much in common. Their biological activity has been shown to correlate with the structure of the molecules, which generally have a protected 1,4-conjugated aldehyde moiety which would react with primary amines. This was established by comparing a series of natural and synthetic analogs.[25]

The green algae are fed upon by opisthobranchs of the order Sacoglossa, which pierce the cells and suck out their contents. The evolution of this relationship has recently been treated in considerable detail by these authors.[4] Only those aspects that are germane to the theme of this review are included here. Although it is not clear which genus among these algae was the ancestral food, shifts to other groups of algae are obviously secondary. Thus, although most species of *Elysia* continue to feed upon and utilize algae of the family Udoteaceae, with the typical sesquiterpenoid metabolites, *E. rufescens* feeds upon *Bryopsis* (family Bryopsidaceae) and obtains from it toxic peptides.[26] These kahalalides, as they are called, are suspected of coming from epibiotic cyanobacteria rather than from the alga.[27] Even within the Udoteaceae, *Avrainvillea longicaulis* is the source of a brominated diphenylmethane derivative found in *Costasiella ocellifera* (= *C. liliane*), which is in a different family of sacoglossans; this compound is an effective deterrent to feeding by fish.[28]

Also secondary are shifts to other kinds of food, abandonment of chemical defense, and *de novo* synthesis. In any event, the animals almost (but not quite) always continue to specialize upon food organisms that contain such metabolites. When the animals evolve the capacity for *de novo* synthesis, they may produce variants of the ancestral compound, but these variants are more physiologically active than the originals. Finally, like some other opisthobranchs, they sometimes use polypropionates synthesized *de novo* (see below).

C. Metazoa

1. Porifera

Porifera, or sponges, are rich in secondary metabolites; a representative sample is shown in Figure 3.7. Some of these are biosynthesized by the sponges themselves, but others are biosynthesized by bacterial

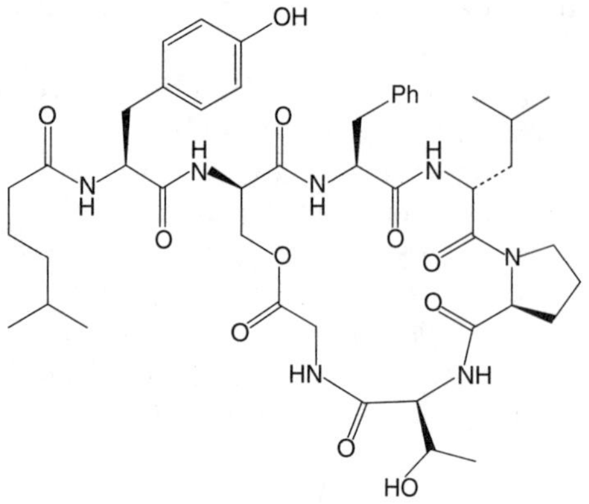

FIGURE 3.6 Selected metabolites from green algae. 3.6.1 Kahalalide A from *Bryopsis* sp. and the sacoglossan *Elysia rufescens*.[190] 3.6.2 Dictyospharerin from *Dictyospheria sericea*.[191] 3.6.3 Halitunal from *Halimeda tuna*.[192] 3.6.4 Caulerpenyne from *Caulerpa prolifera*.[193] 3.6.5 Avrainvilleol from *Avrainvillea longicaulis* and the sacoglossan *Costasiella ocellifera*.[194]

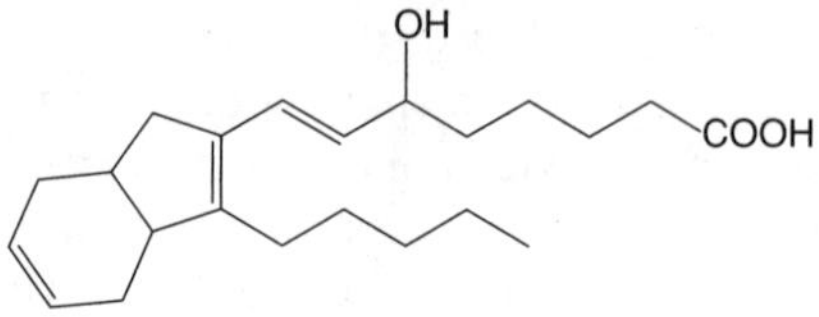

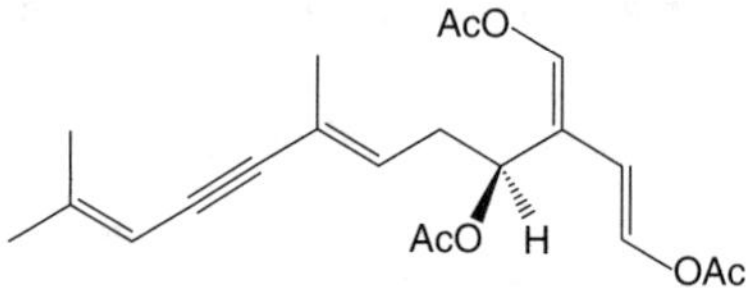

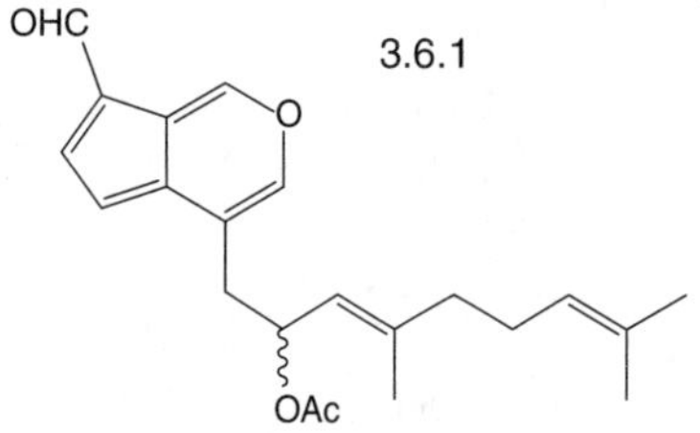

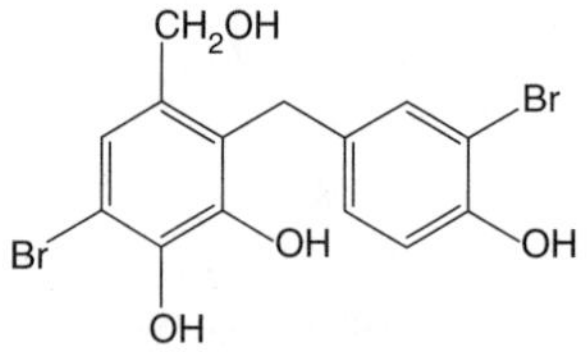

symbionts.[29–31] The metabolites of interest here may conveniently be listed under two categories, according to whether they are probably synthesized by the sponges or their symbionts. Metabolites produced by the symbionts are often identified as such by their presence in bacteria that are not symbionts or at least are symbionts of organisms other than sponges.

Sponge systematics is a difficult area of research, and, partly for this reason, secondary metabolites have been pressed into service in efforts to create a system that more nearly reflects the phylogeny. Secondary metabolites make good taxonomic markers or, better, diagnostic characters, for lineages of sponges. For example, brominated metabolites derived from tyrosine (Figure 3.7.1) are characteristic of the order Verongida in the subclass Ceractinomorpha,[32] and isonitriles (Figure 3.7.2) are mainly known from the order Halichondrida including the Axinellida.[33] Developmental characters have also been used, but the emphasis has been upon anatomy, especially the morphology of the spicules that help support the organism and may also provide the organism some protection from getting eaten. It seems somewhat enigmatic that Chanas and Pawlik[34] did not get positive results in efforts to corroborate the defensive role of spicules against fish grazing. Similar experiments had given positive results with other organisms. Laboratory tests with artificial food are hard to interpret, and even field studies may not relate to actual selection pressures. Evidence to the contrary is that the Keratosa, a group in which the spicules have regressed, are particularly rich in biologically active secondary metabolites.[32] That fits the notion of an arms race with a shift from mechanical to chemical defense.

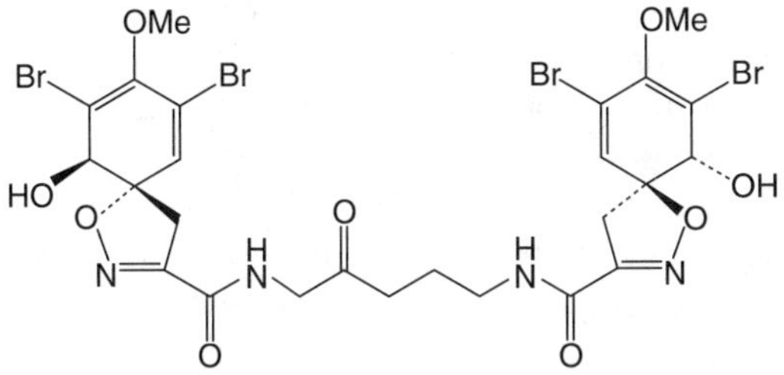

3.7.1

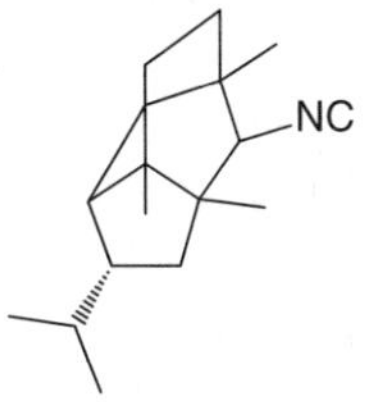

3.7.2

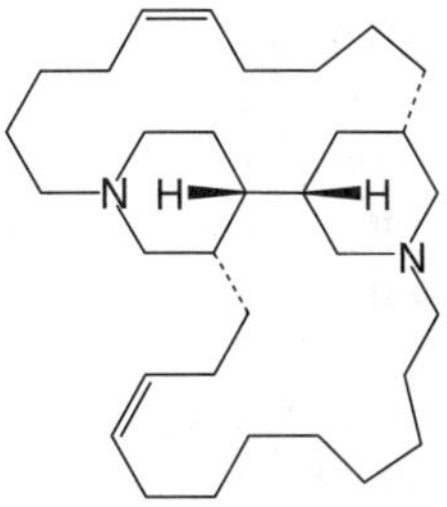

3.7.3

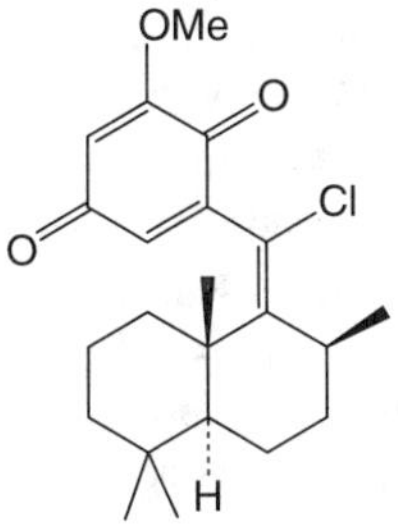

3.7.4

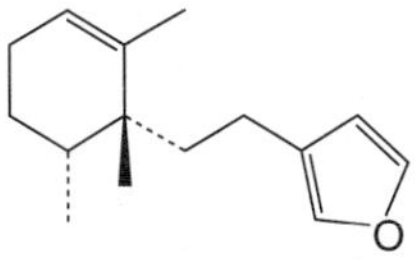

3.7.5

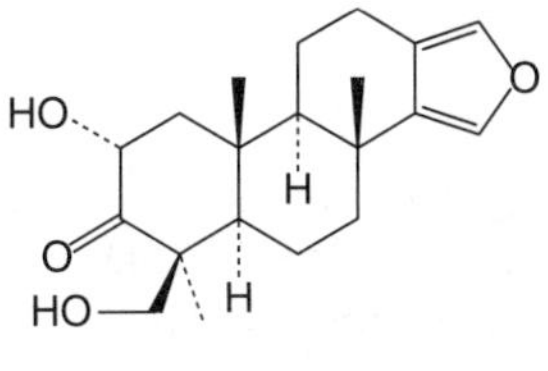

3.7.6

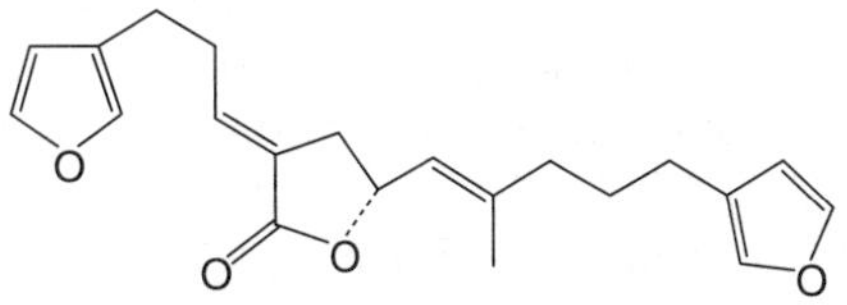

3.7.7

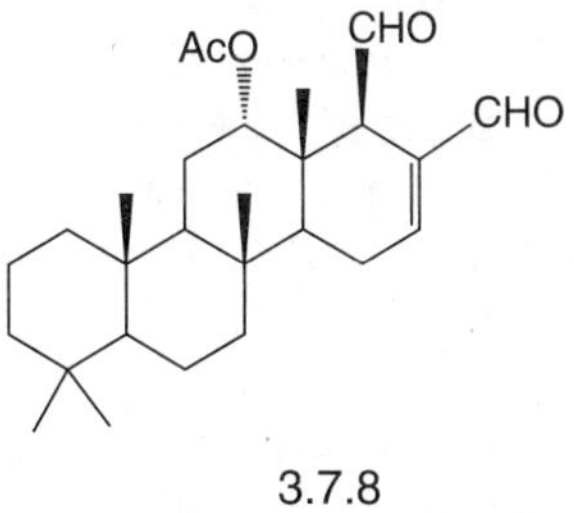

3.7.8

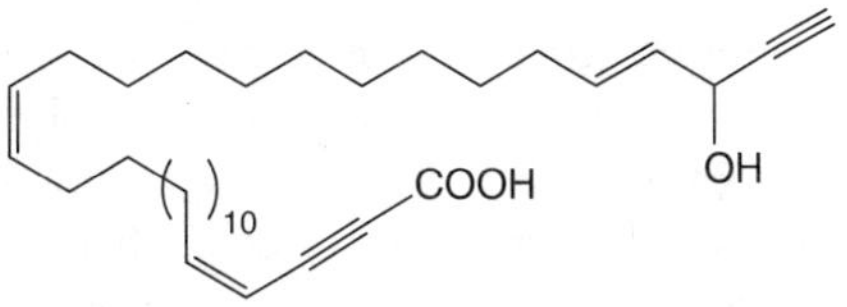

3.7.9

FIGURE 3.7 Selected metabolites from sponges. 3.7.1 Oxohomoaerothionin from *Aplysina cavernicola*.[195] 3.7.2 9-Isocyanopupukeanane from *Hymeniacidon* sp. and the dorid nudibranch *Phyllidia varicosa*.[196] 3.7.3 Haliclonacyclamine B from *Haliclona* sp.[197] 3.7.4 *E*-chlorodeoxyspongiaquinone from *Euryspongia* sp.[198] 3.7.5 (-)-Microcionin from *Microciona toxystila*.[199] 3.7.6 Isospongiadiol from *Spongia* sp.[200] 3.7.7 Isonitenin from *Spongia officinalis*.[201] 3.7.8 Scalaradial from *Cacospongia mollior*.[202] 3.7.9 Pellynic acid from *Pellina triangulata*.[203]

The role of symbionts in producing metabolites that occur in sponges has received increasing documentation. The relationship may reasonably be considered a mutualistic one, i.e., one that benefits both parties, rather than parasitism or commensalism. There is a good possibility of coevolution between the two partners. Symbiotic bacteria are passed from one generation of sponge to the next via maternal cells that surround the egg, even when the animals are oviparous,[35] indicating that the sponges have evolved adaptations that ensure the presence of the bacteria. The sponges and their associated bacterial symbionts are known to produce quite different metabolites. The sponge *Dysidea herbacea* produces sesquiterpenoids, whereas its symbiotic cyanobacteria produce polychlorinated amino acids.[36] Likewise, the sponge *Suberea creba* produces quinolines, whereas its symbionts produce tyrosine metabolites.[37]

For the order Haplosclerida in the subclass Ceractinomorpha and their close relatives, there have been two particularly interesting phylogenetic studies on metabolites produced by symbionts. Andersen, Van Soest, and Kong[38] studied the 3-alkylpiperidene alkaloids from this assemblage. They placed the molecules in series beginning with a monomeric unit leading to dimers and derived and rearranged skeletons. The resulting tree agreed quite well with more traditional skeletal characters. It has two branches, one consisting of the viviparous families Callyspongiidae and Niphatidae with Chalanidae as the sister group, and one consisting of the oviparous families Phloeodictyidae and Petrosiidae.

More recently, Van Soest, Fusetani and Andersen[39] studied the straight-chain acetylenes that occur in the same group (including close relatives of Haplosclerida) with broadly similar results. Again, they produced a phylogeny based on derived and divergent biosynthetic pathways, a phylogeny that agrees well with morphological data. The preliminary results also suggest that symbiotic bacteria have occasionally moved from one lineage to another, much as one might predict. It must be mentioned that the Haplosclerida contain quite a variety of metabolites other than these two examples. Some of these make their way into nudibranchs that feed upon the sponges; for example, both sterols and high molecular weight polyacetylenes have been recovered from *Peltodoris atro-maculata* and the sponge *Petrosia* upon which it feeds.[40,41]

The isonitriles mentioned above as characteristic of Halichondrida (in some classifications this group includes the Axinellidae) are interesting as they provide the basis for a minor radiation in dorid nudibranchs. Namely, the Porostomata include a lineage of slugs, the family Phylidiidae, that feed exclusively on sponges that are rich in such isonitriles. The biosynthetic pathways whereby these isonitriles are produced have been studied, and it is noteworthy that cyanide is incorporated into the molecule.[42] The nudibranchs have not evolved the capacity for *de novo* synthesis of these compounds.[43] Some authors have questioned whether they are used defensively by the nudibranchs. It is true that rigorous experiments to this effect have not been done. However, the animals themselves provide a sort of natural experiment or assay. The species that live in the most exposed positions, readily visible to predators in the tropics, are the ones with the most obvious smell.[44] Furthermore, there are mimetic complexes consisting of these nudibranchs, other nudibranchs, polyclad flatworms and sea cucumbers, all with similar color patterns.[45]

Sponges themselves apparently do biosynthesize a variety of secondary compounds. Some examples from the Keratosa (mentioned above as having lost the spicules) are of particular interest.[32] There are three orders in subclass Ceractinomorpha of the class Demospongiae. The order Verongida consists of the families Aplysinidae, Aplysinellidae, and Ianthellidae, and is distinct from the other two by being oviparous and containing brominated tyrosine derivatives, rather than being viviparous and containing terpenes. The order Dendroceratida, with families Darwinellidae and Dictyoden-dillidae, has diterpenes. Finally in the order Dictyoceratida, there are three families with sesterterpenes — Spongiidae, Thorectidae, and Ircinidae — and one family, Dysideidae, with sesquiterpenes. Although the phylogeny is perhaps debatable, the taxonomic pattern suggests a diversification in the metabolites that are synthesized by the sponges.

Among these keratose sponges, *Aplysina aerophoba* (Verongida) is said to be at least capable of biotransformation of the characteristic brominated tyrosine derivatives.[46] They are held, evidently as inactive precursors, and modified by enzymatic action at the time of release.[47] However, there is some question as to how well the experiments represent *in vivo* conditions.

There is a significant correlation between the phylogeny of sponges and that of the dorid nudibranchs of the Family Chromodorididae that feed upon them.[5] The early branches of this group feed on a wide range of sponges, and some of them are quite euryphagous. In general, *Chromodoris* and *Glossodoris* have diterpenes and sesterterpenes of the kind found in most Dictyoceratida, whereas *Hypselodoris* is characterized by sesquiterpenes characteristic of the family Dysideidae. There is at least one exception to the rule, *H. orsini*, with a sesterterpene scalaradial (Figure 3.7.8) derived from a sponge of the family Thorectidae.[48] The scalaradial is modified by the nudibranch by selective reduction into deoxoscalarin (Figure 3.14.1), which is less toxic, then by selective oxidation into 6-ketodeoxoscalarin (Figure 3.14.2).

Various dorid nudibranchs are thought to have evolved the capacity for *de novo* synthesis of metabolites originally derived from sponges, but the evidence is not always compelling. The following cases have been well established experimentally. *Sclerodoris tanya* synthesizes tanyolide (Figure 3.14.3), a terpenoic acid glyceride.[49] *Archidoris odhneri* synthesizes an unnamed farnesic acid glyceride (Figure 3.14.4).[50] *Dendrodoris limbata* synthesizes 7-deacetoxyolepupuane (Figure 3.14.5).[51] This is the only example we have of a nudibranch biosynthesizing a defensive metabolite that is identical to one found in potential prey. *D. grandiflora* and some other species of the same genus synthesize drimane terpenoids.[52–54] The aforementioned are all Porostomata or their supposed sister group. In the Chromodorididae (discussed above), *Cadlina luteomarginata* synthesizes albicanyl acetate (Figure 3.14.6), cadlinaldehyde (Figure 3.14.7), and luteone (Figure 3.14.8).[55]

2. Cnidaria

Representative metabolites from Cnidaria (= Coelenterata s.s.) are shown in Figure 3.8. The Cnidaria have venomous stinging capsules that are beyond the scope of this study. Also excluded from this discussion are some occasional secondary metabolites in various groups that are not sufficiently well enough known to generalize about. This discussion is restricted to a single order, Gorgonacea, in subclass Octocorallia (Alcyonaria) and class Anthozoa. These gorgonians, or sea fans and sea whips, are sessile animals that feed on small planktonic organisms. Their stinging capsules are small relative to those of some other cnidarians, and are less important for defense. They have proven to be a rich source of secondary metabolites, and there are a few interesting connections with nudibranchs and other gastropods.

Gorgonians are particularly rich in terpenoids, notably sesquiterpenoids and diterpenes. Gerhardt[56] produced a biosynthetic tree for these compounds and found that it agreed well with the existing, morphological system of classification. Although rooting the tree under the assumption that the ancestral form had no terpenoids at all would seem to be unrealistic, he got two distinct lineages. In the first branch, with the genera *Gorgonia*, *Pseudopterogorgia*, *Plexaurella*, and *Muricea*, sesquiterpenes were present and diterpenes were absent. In the second branch, with the genera *Briareum*, *Eunicella*, *Eunicea*, and *Pseudoplexaura*, diterpenes are always present, but in one subclade there are also sesquiterpenes. Given that *Briareum* and *Eunicella* occupy a more basal position in the tree, Gerhardt reasoned that the ability to produce the sesquiterpenes had evolved twice within this lineage. It is a case of parallel evolution, but not the rampant variety. The presence of terpenes in other Octocoralia may perhaps represent parallel evolution as well. The terpene skeletons of *Alcyonium* (Octocorallia: Alcyonacea) are somewhat different from those of Porifera.[57] Another intriguing class of metabolites that occur in Gorgonacea is the prostanoids. These compounds are ubiquitous among metazoans, and have a wide range of physiological functions.[58–61] However, a defensive function, involving a much increased level of concentration and some modification in structure, is unusual. Claims that the compounds are allelopathic rather than defensive have been rejected.[62,63] Negative evidence about feeding deterrence by fish turned out to be the result of the fish not responding immediately but vomiting after a delay.[64] The delay, coupled with

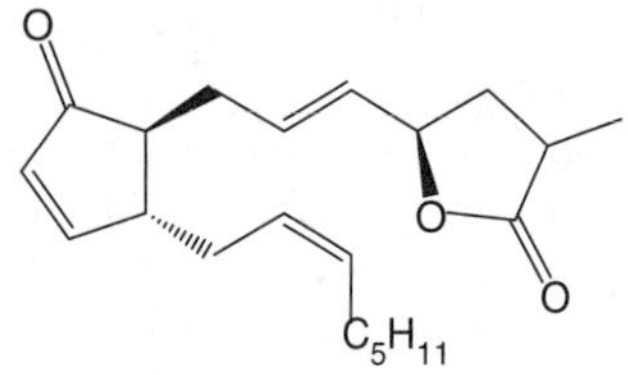

3.8.1

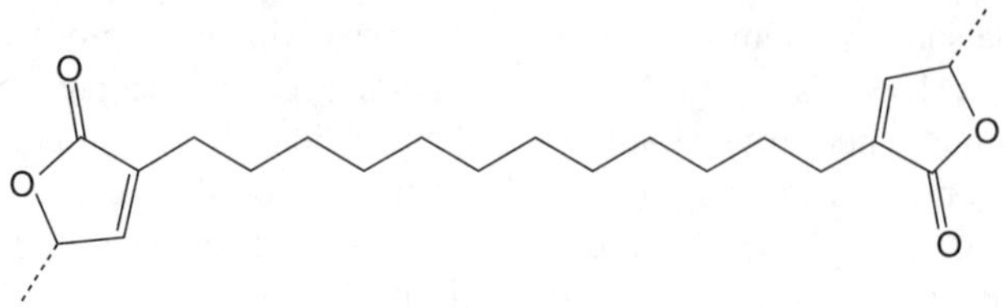

3.8.2

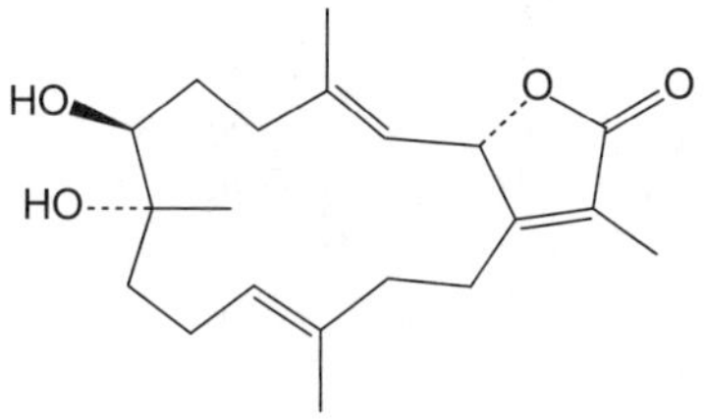

3.8.3

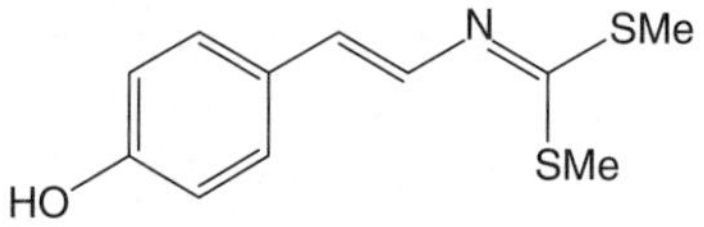

3.8.4

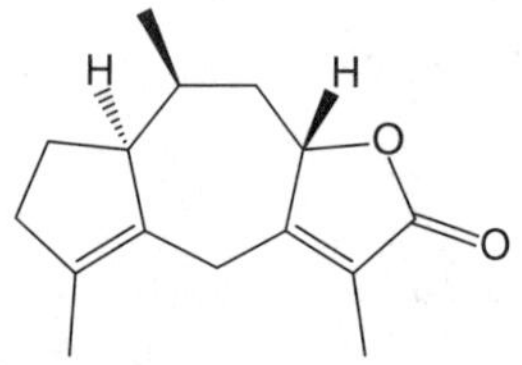

3.8.5

FIGURE 3.8 Selected metabolites from anthozoans. 3.8.1 Clavulone 1 from *Clavularia viridis*.[204] 3.8.2 (+)-Ancepsenolide from *Pterogorgia anceps*.[205] 3.8.3 7β, 8α-dihydroxydeepoxysarcophine from *Sarcophiton trocheliophorum*.[206] 3.8.4 Tridentatol A from *Tridentata marginata*.[207] 3.8.5 Americanolide F from *Pseudopterogorgia americana*.[208]

aversive learning, is effective in deterring predators that take a bite and then return to feed again. The prosobranch gastropod *Cyphoma gibbosum* feeds upon gorgonians that contain high levels of Prostaglandin A$_2$. The snail feeds preferentially upon gorgonian colonies that have not previously been attacked, apparently because the defense is inducible and more of a deterrent at higher levels.[63] It feeds upon a range of gorgonians, and has been found not to be immune to terpenoids from *Gorgonia ventalina*.[65] *Cyphoma* belongs to a group of cowrie-like animals that commonly feed upon Octocorallia. Another of these, *Ovula ovum*, was thought to detoxify terpenoids from a soft coral,[66] but further investigation showed that this was not the case.[67] The snails are generally believed to use the metabolites defensively.

A few nudibranchs feed upon gorgonians, which may represent the original food of the branch (Cladohepatica) that switched from sponges to cnidarians as food. Unfortunately, the lineage of nudibranchs (Dendronotacea) that are the sister group of the other Cladohepatica and include many of the animals that feed upon gorgonians are among the less common and less studied nudibranchs, especially from the point of view of natural products. When Gosliner and Ghiselin[68] described *Tritonia hamnerorum*, their editor deleted the remark that the animals smelled like camphor. Despite that, a furan was recovered both from this nudibranch and a gorgonian upon which it feeds.[69] Within the Dendronotacea there is one lineage, the family Tethyidae, that has ceased to feed upon cnidarians and eats small arthropods instead. *Tethys* has been found to defend itself with prostaglandins.[70–75] We have suggested that this may be no coincidence, and that perhaps the use of prostaglandins that were derived from food was followed by the use of prostaglandins synthesized by the nudibranchs themselves.[5] Because prostaglandins are so widespread in metazoan tissues, it is not necessary to invoke the evolution of *de novo* synthesis in the retrosynthetic mode for this particular case.

3. Sessile Filter Feeders with Symbionts: Ectoprocta and Urochordata

The next two groups to be discussed contain defensive metabolites that are generally acknowledged to be produced by symbionts, and not by the animals themselves. These are the phylum Ectoprocta (= Bryozoa s.s.) and the chordate subphylum Urochordata. Although Ectoprocta are lophophorates, and lophophorates have previously been considered related, albeit remotely, to the Chordata and other Deuterostomes, molecular evidence has made this relationship seem quite dubious; rather, lophophorates are closer to annelids and molluscs.[76–80] At any rate, the ancestry of lophophorates is so remote that cross-lineage transfer of symbionts seems much more likely than their having been present since the Precambrian era. We treat Ectoprocta and Urochordata together because of their feeding relationships with opisthobranchs as well as their chemical similarities. The metabolites in question are mainly alkaloids.

a.   Tentaculata: Phoronida and Brachiopoda

The phylum Ectoprocta (= Bryozoa s.s.) is generally thought to be most closely related to the phyla Phoronida and Brachiopoda. The three are often placed together into a single phylum or superphylum Tentaculata. The phoronid *Phoronopsis viridis* is a tube-dwelling worm that is abundant in the mudflats where it was studied. It contains the halogenated phenolics 2,6-dibromophenol and 2,4,6-tribromophenol.[81] These compounds are similar to those secreted by hemichordates of the class Enteropneusta, and in both groups they are thought to suppress the growth of bacteria. Otherwise, little is known about their natural products chemistry. Brachiopods are said to be repellent,[82,83] but the only marine natural products reported from them are four long-chained glycerol enol ethers (Figure 3.9.3) from the terebratulid *Gryphus vitreus*.[84]

b.   Tentaculata: Ectoprocta = Bryozoa s.s.

Secondary metabolites of Ectoprocta occur in distantly related lineages within the group, indicating that they have been acquired since diverging. With trivial exceptions, marine ectoprocts are colonial animals with good mechanical defense in the form of calcareous shells that surround the zooids. There is an excellent fossil record showing a sustained tendency over 100 million years for the cheilostomes to out-compete ctenostomes by overgrowth, though apparently there has not been an arms race.[85] Mechanical defense by means of spines has been shown to be inducible.[86,87]

Although, for example, tetracyclic terpenoid lactones (Figure 3.9.1) have been found,[88] the main bryozoan metabolites of interest are alkaloids and macrocyclic ethers.[89] The tambjamines are a good example of alkaloids. These are bipyrroles that occur in ascidians as well as in nudibranchs that derive them from eating ectoprocts, ascidians, or even other nudibranchs (see section below on Urochordata). The bryostatins (Figure 3.9.2) are the important polyethers from ectoprocts.[90] They are macrocyclic lactones.

A microbial origin for both the alkaloids and the polyethers has been maintained.[91] Dietary derivation seems dubious, since ectoproct food is not likely to provide a rich and reliable source of metabolites. Ectoprocts are suspension feeders that take up small particles by means of tentacles. Furthermore, most ectoprocts have a reproductive biology that would facilitate transfer of bacteria across generations: they brood the developing embryos. Bacteria suspected of producing bryostatins were found in the larvae in some, but not all, of the species surveyed by Woollacott.[92] What was formerly thought to be a single species (*Bugula neritina*) of ectoproct that varied in its bryostatin content is actually a complex of cryptic species each with characteristic bacterial symbionts.[93]

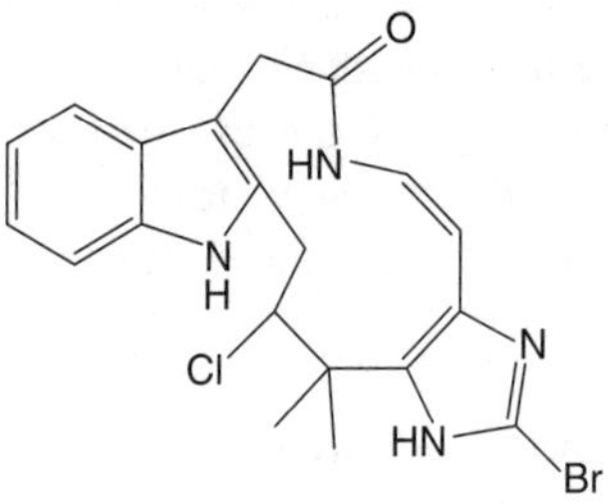

FIGURE 3.9 Selected metabolites from bryozoans and a brachiopod. 3.9.1 Securine A from *Securiflustra securirons*.[209] 3.9.2 Bryostatin 18 from *Bugula neritina*.[210] 3.9.3 Glycerol ethers from the brachiopod *Gryphus vitreus*.[211]

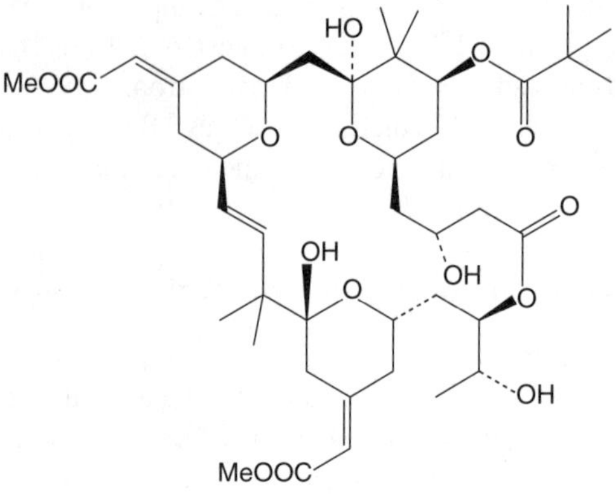

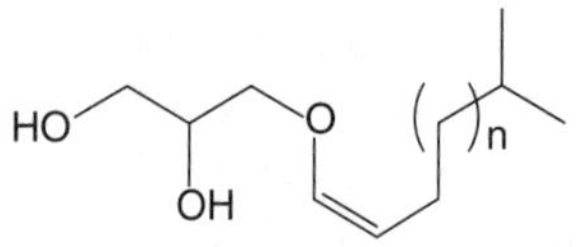

c.   Urochordata

The phylum Chordata consists of three subphyla: Urochordata, Cephalochordata, and Vertebrata. Among the Urochordata (tunicates), chemical defense has not yet been recorded in the two planktonic classes (Appendicularia and Thaliacea), perhaps because of inadequate sampling. However, it does occur quite commonly among the Ascidiacea, which are benthic organisms. Ascidians are mainly sessile, and their asexual reproduction has repeatedly led to the formation of highly organized colonies. The sessile habitus is conducive to the evolution of both mechanical and chemical modes of defense. Ascidians are well defended by a tough but flexible tunic that contains cellulose. The ability of the colony as a whole to resist predators even though some members of the colony are destroyed may further the utilization of particular kinds of defensive metabolites. Some (compound) ascidians of the genus *Didemnum* produce eicosanoids, evidently of non-symbiotic origin, providing an interesting parallel with gorgonians (see above).[94,95]

Symbiosis with algae and bacteria (including Cyanobacteria) is widespread in ascidians and is responsible for coloration as well as defensive metabolites, which may themselves be colorful. Symbiotic bacteria are also responsible for the (probably defensive) bioluminescence of their pelagic relatives, the Thaliacea.[96] Brooding of the young "tadpole" larva may facilitate the transfer of symbionts from one generation to the next. Many of the metabolites in ascidians are quite similar to those known to be produced by bacteria, and the taxonomic distribution of the metabolites in the animals tends to confirm this.[97] Others, however, are evidently produced by the tunicates themselves.

Examples of secondary metabolites from Ascidians are shown in Figure 3.10. Ascidians are noteworthy for the presence of vanadium (and related metals) together with high concentrations of sulphuric acid and tunichromes, which are unstable hydroquinoid compounds. A defensive role for these has been suspected,[98] and it is noteworthy that gastropod molluscs (both prosobranch and opisthobranch) that feed upon tunicates often secrete large amounts of sulfuric acid, used in both feeding and defense.[99] A fair range of chemicals may be characterized as occasionally present, but not of major significance, for example, a few terpenoids, isoprenoid hydroquinones, fatty-acid derivatives, and polyethers. The predominant metabolites are amino-acid-derived metabolites, including both linear and cyclic peptides and various alkaloids.[100]

A few groups of gastropods feed upon ascidians. The presence of metabolites in the molluscs provides a fair indication of which ones are actually used defensively. In the subclass Prosobranchia, the Lamellariidae are a good, if perhaps understudied, example. An unidentified species of Lamellaria from Palau was found to contain alkaloids called lamellarins (Figure 3.10.9).[101] They occur in the ascidian genus *Didemnum,* and about 30 of them have been described.[102] Another lamellariid, *Chelynotus semperi*, contains a series of pyridoacridines, kuanoniamines (Figure 3.10.10), derived from an otherwise unidentified colonial tunicate.[103] Homarine has been found in the lamellariid *Marseniopsis mollis*; however, even though it has a defensive role, this metabolite evidently derives from epibionts, not from the ascidian upon which the snail feeds.[104]

In the subclass Opisthobranchia, several lineages have switched to feeding upon ascidians. Within the order Notaspidea, which probably fed originally upon sponges, *Pleurobranchus* feeds upon ascidians, and at least one of them, *P. forskalli*, contains a cyclic peptide.[105] Within the order Nudibranchia, originally sponge-feeders, there is one group of "phanerobranch" dorids, the Suctoria, which evidently switched to bryozoans and (subsequently, it would seem) to ascidians. The pattern here is particularly instructive. Tambjamines (Figure 3.10.5) occur in both ectoprocts and ascidians. Nudibranchs of the genus *Tambja* obtain tambjamines from the bryozoans upon which they feed. *Roboastra*, in the same family, feeds upon *Tambja* and uses the metabolites defensively.[106] Another nudibranch of the same family, *Nembrotha*, obtains tambjamines by feeding upon ascidians.[107] Here, the metabolites are obviously a resource, over and above the nutritional content of the food, that provides the basis for adaptive changes in diet. The nudibranchs in question are remarkably conspicuous and are often encountered in the open by divers, unlike many tropical nudibranchs that are cryptic or keep out of sight.

Some tunicates are fed upon by flatworms. Not much information is available, but there are some intriguing parallels between the flatworms and the nudibranchs, which resemble each other superficially and for which cases of mimicry are known. Five alkaloids (heterocycles), including lepadin B (Figure 3.10.8), have been described from the marine turbellarian *Prostheceraeus villatus;* they derive from the tunicate *Clavelina lepadiformis* upon which it feeds.[108] Attention has been drawn to the convergence with nudibranchs: the flatworm is aposematically colored. Another example is staurosporine derivatives, which have been recorded from the ascidian *Eudistoma toealensis*, and a turbellarian flatworm of the genus *Pseudoceros* which feeds upon it.[109] These indolocarbazole alkaloids (Figure 3.10.9) resemble alkaloids known from fungi rather than the usual ascidian alkaloids that are suspected of having bacterial origin.

4. Sessile Filter or Deposit Feeders Evidently without Symbionts

a. Annelida: Polychaeta

Marine annelids of the class Polychaeta have traditionally been divided into the free-living Errantia and the sedentary-to-sessile Sedentaria, and although the division is somewhat artificial, it suffices for the present discussion. The natural products literature is not very extensive, and the literature on the Errantia mainly deals with their toxins. Among Sedentaria, there is a report of brominated phenolics (Figure 3.11.1) in *Thelepus setosus* (family Terebellidae).[110] These compounds are of some interest as a case of convergent evolution with unrelated marine worms having somewhat similar habits: Phoronida and Enteropneusta. Some other marine annelids are known to be toxic, and some opisthobranchs of the order Cephalaspidea feed upon them. They form a distinct clade that is not recognized in formal classifications, but consists of such genera as *Actaeon* and *Hydatina*. Natural products chemists have not yet investigated this relationship.

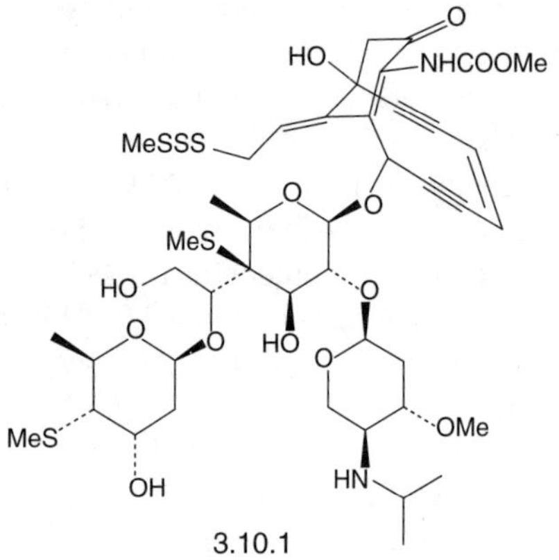

3.10.1

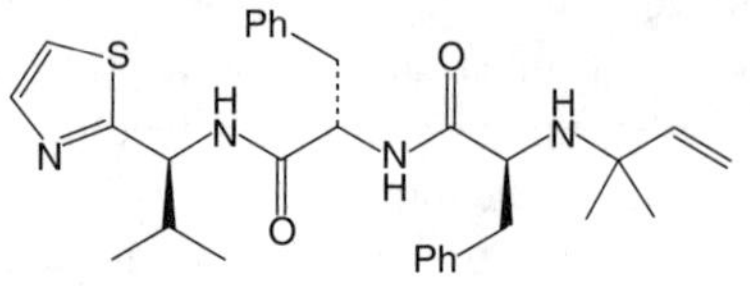

3.10.2

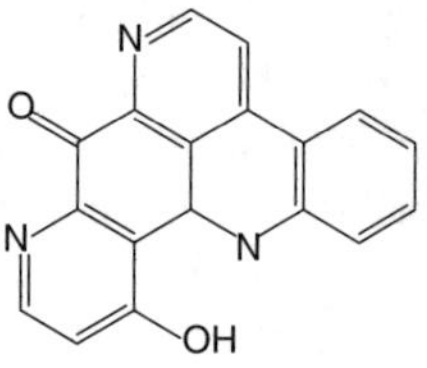

3.10.4

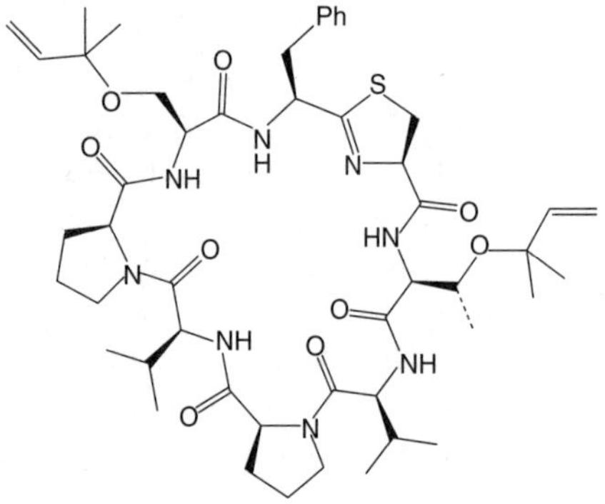

3.10.3

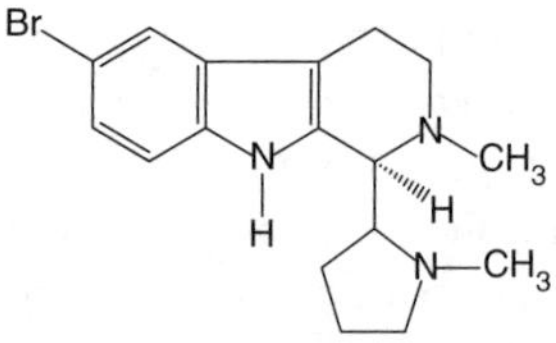

3.10.5

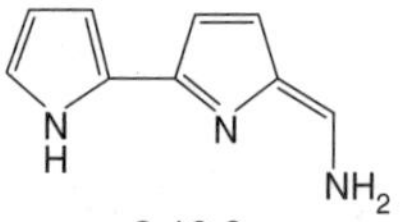

3.10.6

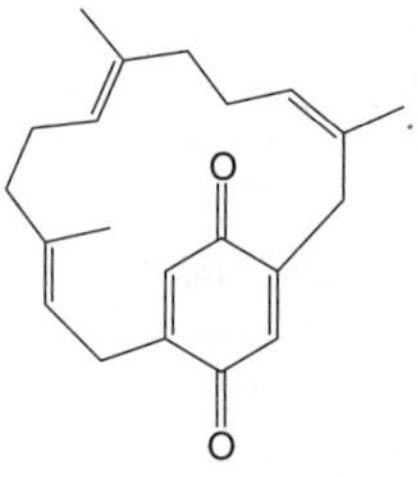

3.10.7

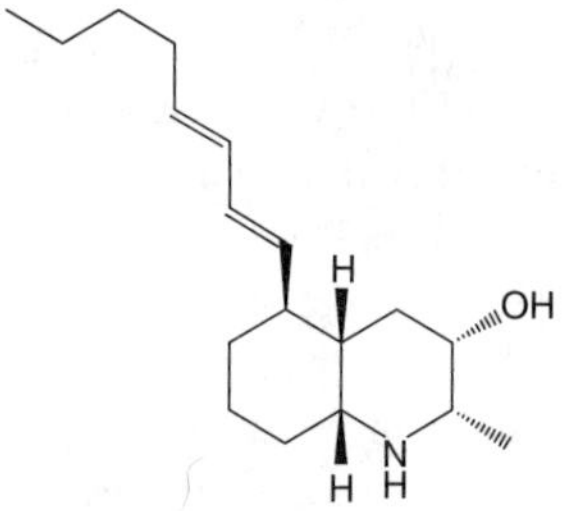

3.10.8

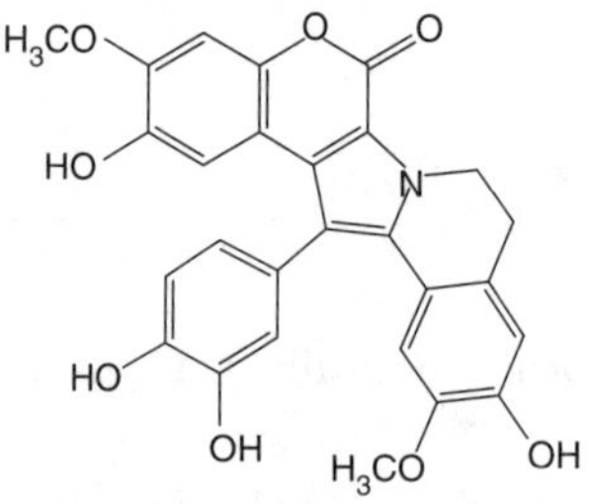

3.10.9

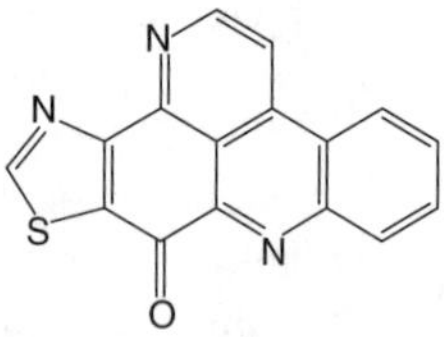

3.10.10

FIGURE 3.10 Selected metabolites from tunicates. 3.10.1 Namenamicin from *Polysyneraton lithostratum*.[212] 3.10.2 Virenamide B from *Diplosoma virescens*.[213] 3.10.3 Meridine from *Amphicarpa meridiana*.[214] 3.10.4 Patellin 5 from *Lissoclinum* sp.[215] 3.10.5 Tambjamine A from *Atapozoa* sp.[216] 3.10.6 Woodinine from *Eudistoma fragum*.[217] 3.10.7 Longithorone B from *Aplidium longithorax*.[218] 3.10.8 Lepadin B from *Clavelina lepadiformis* and the flatworm *Prostheceraeus villatus*.[219] 3.10.9 Lamellarin Z from *Didemnum chartceum*.[220] 3.10.10 Kuanoniamine A from an unidentified tunicate and the prosobranch gastropod *Chelynotus semperi*.[221]

b. Hemichordata: Enteropneusta and Pterobranchia

The Hemichordata are a phylum of lower deuterostomes and, therefore, are more closely related to echinoderms and chordates than to annelids. They are discussed here because they are separate lineages that have evolved distinctive metabolites and also because of their feeding mechanisms. Very little work has been done on their natural products, however.

The Pterobranchia are uncommon, colonial animals that feed upon plankton with an assemblage of tentacles in a fashion comparable to that of bryozoans. The cephalostatins (Figure 3.11.2) of *Cephalodiscus ritteri* are cell-growth inhibitors.[111,112] Although they are nitrogenous compounds and therefore alkaloids, they are steroid dimers, not derivatives of amino acids,[113] and the type of compound that the animals should be able to synthesize on their own. A single species of ascidian, *Ritterella tokioka*, has yielded a diversity of similar compounds, the ritterazines.[114–117] Given that this class of compounds is so unusual, one is tempted to speculate that it was present in the common ancestor of Chordata and Hemichordata. This possibility would be more credible if the metabolites were found in a wider range of chordates.

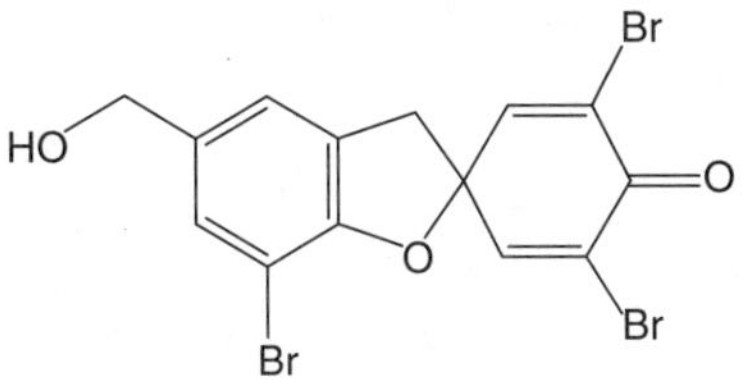

3.11.1

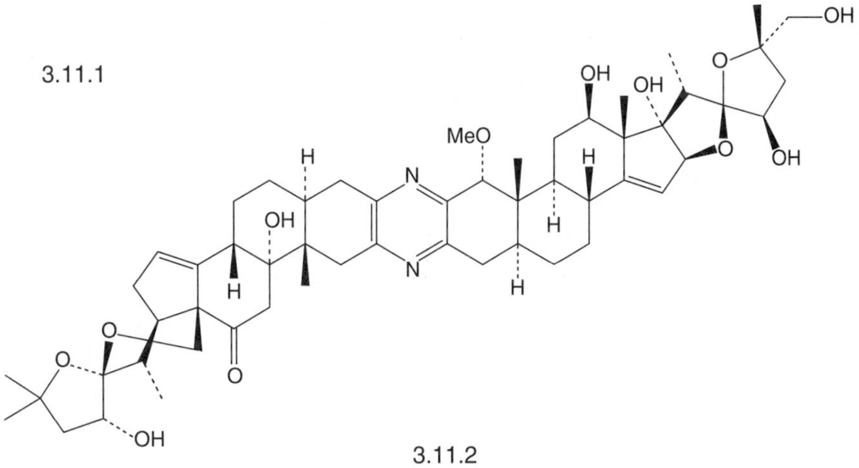

3.11.2

FIGURE 3.11 Selected metabolites from a polychaete and a hemichordate 3.11.1 Thelepin, from the polychaete *Thelepus setosus*.[222] 3.11.2 Cephalostatin 10, from *Cephalodiscus gilchristi*.[223]

The Enteropneusta are soft-bodied worms that burrow in sediments, where they surround themselves with mucus and feed by taking up deposited material and straining it with a pharyngeal apparatus that is similar to that of primitive chordates. The nature of the food is such that derivation of metabolites seems unlikely, and symbiotic origin is only somewhat more plausible. A wide variety of metabolites are involved, mainly found in the mucus secretion that lines the animals' burrows.[118–120] They include bromophenols, halogenated indoles, and indigotin pigments. The iodoform-like smell that has been widely reported by naturalists turns out to be due to dibromophenols having antimicrobial activity.[121–122] They are suspected, however, of playing an important role in deterring predators. The presence of similar compounds in *Phoronida* and an annelid has already been mentioned. These unrelated worms are all infaunal suspension and deposit feeders. The ones that have been studied occur at high population densities.

5. Slow-Moving Grazers and Predators

a.   Echinodermata

The echinoderms are distinguished by a distinct morphological gap, and relationships to other Deuterostomia including chordates and hemichordates are evident only on the basis of embryology and molecular biology. What the adult ancestor was like — whether sessile like a pterobranch or perhaps more vagile — cannot be decided on the basis of presently available evidence. Deuterostomes do not share defensive metabolites that might be traced back to a common ancestor in the Precambrian era. The ancestral echinoderm was, in all probability, a sessile organism, but only one of the five extant classes remains sessile in the adult stage. The spiny endoskeleton provides some protection, but it tends to be reduced when the animals evolve in the direction of greater motility, sometimes associated with feeding upon larger food items. Primitive echinoderms were probably nonspecific feeders upon material suspended in the water or deposited on the substrate. This would give them little opportunity for specializing upon food from which they might obtain defensive metabolites; however, their way of life would make them vulnerable and give an advantage to metabolites synthesized by the echinoderms themselves. (Examples of metabolites from echinoderms are given in Figure 3.12.)

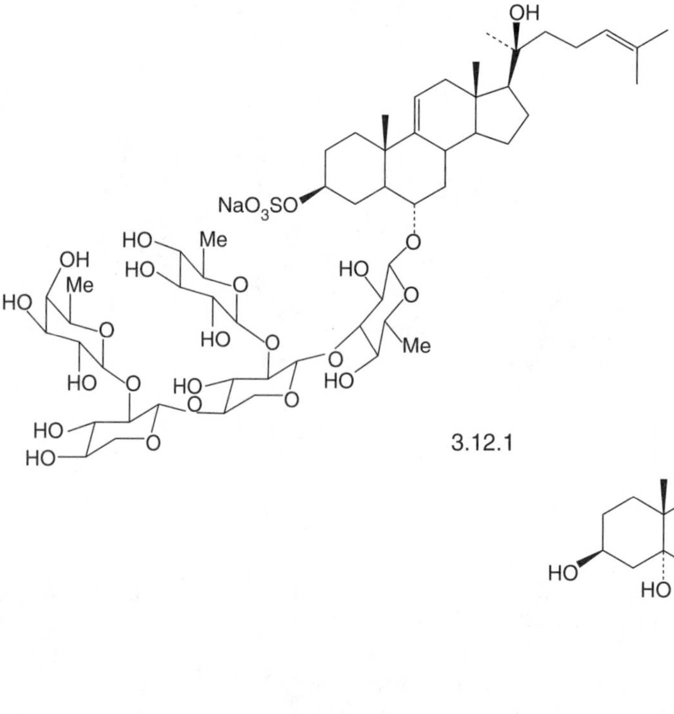

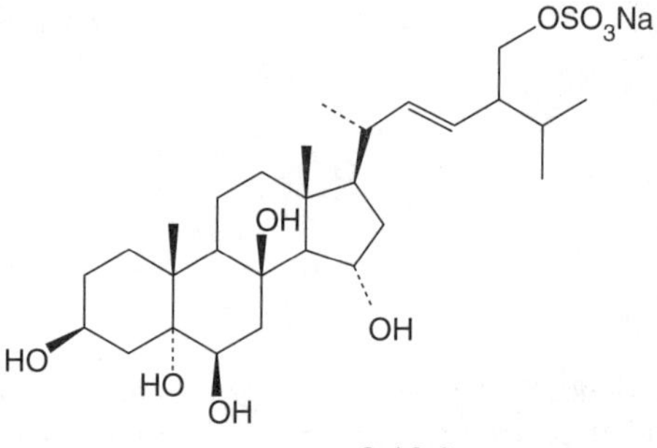

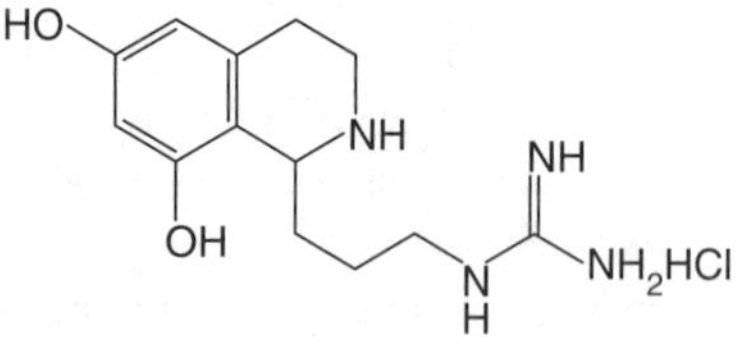

FIGURE 3.12 Selected metabolites from echinoderms. 3.12.1 Comasteride A from *Comasterias lurida*.[224] 3.12.2 A polyhydroxylated sterol from *Stylaster caroli*.[225] 3.12.3 Fuscusine from *Perknaster fuscus*.[226]

Molecular evidence is decisive for Crinoidea being the sister group of the Eleutherozoa, which are the four legitimate, extant, free-living classes: Asteroidea (sea stars), Ophiuroidea (brittle stars), Holothuroidea (sea cucumbers), and Echinoidea (sea urchins). The evidence is less compelling for relationships among these four classes, but a consensus among specialists has lately become increasingly popular for Asteroidea plus Ophiuroidea being the sister group of Holothuroidea plus Echinoidea. This is particularly interesting from the point of view of defensive metabolites, because some biochemical similarities between Asteroidea and Holothuroidea indicated a common ancestry. The presence of Δ^7-sterols instead of the ancestral Δ^5-sterols was interpreted as a synapomorphy for Asteroidea plus Holothuroidea. That and a few other biochemical similarities have to be treated, therefore, as convergences in the scenario that follows, which suggests how the apparent conflicts might be reconciled.

Among animals, quinoid pigments are essentially limited to echinoderms. They are of polyketide origin. In crinoids, quinoid pigments have an important role in defending the animals from predators. Since they may also play a role in repelling fouling organisms, their precise ancestral function is not obvious, but it is known that at least the surface of the body was initially defended by such pigments. In other echinoderms these pigments are present, but they are somewhat different both in chemistry and in function. The main pigments of crinoids are anthraquinoids, and some asteroids have them as well. Napthaquinoids occur occasionally in all five classes, but in only one of them, Echinoidea, are they well developed. Many of them are internal and evidently used in defense against bacteria. Echinoids are defended externally by spines and pedicellariae, which are modified groups of spines; both can deliver venoms.

Echinoderms are also noteworthy for the presence of saponins, which are steroid (Figure 3.12.1) and triterpenoid oligoglycosides. Similar compounds (with a defensive function) occur in plants, but they are otherwise not found in animals. The presence of saponins in both Asteroidea and Holothuroidea was used as evidence for a relationship between these two classes. However, it seems more reasonable to treat this as a convergence between animals that have evolved softer bodies, although something intermediate may have been present in the common ancestor of all four classes. The steroid glycosides are characteristic of the Asteroidea and the triterpenoid oligoglycosides are characteristic of the Holothuroidea. The discovery of steroid glycosides in Ophiuroidea provides an additional line of evidence that Asteroidea and Holothuroidea are sister groups.[123-125] On the other hand, the triterpenoid oligoglycosides are characteristic of the Holothuroidea. Both kinds of saponins are useful for diagnosing taxonomic groups within each class,[126] and it is obvious that the defensive use of these compounds has been greatly elaborated since the time of their common ancestry. It is a good candidate for an instance of parallel evolution. We should mention that there are metabolites with a possible defensive function other than those discussed above. For example, alkaloids have been recorded from sea stars (Figure 3.12.3).[127] An alkaloid in *Dermasterias imbricata* has been shown to evoke defensive behavior in the anthozoan *Stomphia* upon which it feeds.[128]

b. Platyhelminthes and Nemertea

Two phyla of marine invertebrates are mentioned here because they have free-living representatives that move around slowly in pursuit of prey; in that respect they are somewhat similar to gastropods. In both of these phyla there are reports of defensive metabolites and aposematic color patterns. Again, the relationships of these animals to other phyla are very remote, and in all probability they evolved within the lineages in question.

The toxicity of both marine and terrestrial flatworms and its association with aposematic coloration has long been known. Arendt and Walther provide a useful survey of the older literature.[129] A few flatworms have been studied with respect to metabolites produced by symbionts. The section about Urochordata discussed flatworms that derive metabolites from food.

Nemerteans are slow-moving predators that can subdue prey of considerable size by means of a proboscis. A nicotine-like alkaloid from the integument has been described; it was observed that the metabolite is not used to subdue prey.[130] It would appear that the animals have evolved a chemical defense based on the biosynthesis of alkaloids. The literature on toxins of this group is not very extensive, but it has been well reviewed.[131]

c. Mollusca: Gastropoda

The ancestral gastropod, like primitive molluscs of other classes, is considered a slow-moving organism that is protected by a shell and feeds by means of a rasp-like feeding organ (radula). This combination of features implies an animal that is adept at dealing with mechanically resistant, but not highly motile, food items. Material such as algae and deposits could be taken up from the substrate in the form of small particles. Subsequent elaborations on this same basic theme have involved a variety of shifts in diet and elaborations of the feeding apparatus, sometimes accompanied by a reduction in the shell. Some examples have already been provided of how opisthobranchs and other gastropods have shifted to various kinds of food in correlation with the presence of metabolites, and also of how the use of defensive metabolites has been further elaborated by the molluscs. Representative metabolites from gastropods and the bivalve *Pinna* are shown in Figures 3.13 and 3.14.

Opisthobranchs also have the ability to biosynthesize polypropionates, which have been the topic of a recent review.[132] This ability is also possessed by their closest relatives, the pulmonates, but not by any other animals. It does occur in bacteria and fungi, and some of the products are important antibiotics. Given the amount of research that has been done, the negative evidence cannot be dismissed lightly as an artifact of sampling. One might speculate that gene transfer or symbiosis has been involved. But the taxonomic distribution makes either of these hypotheses seem unlikely. The propionates have often appeared when there has been a shift from food that is chemically defended to food that is not chemically defended. The food items are quite heterogeneous. But there is a real puzzle, simply because the defensive use of polypropionates seems to have evolved repeatedly within a single clade. This would, perhaps, be a case of parallel evolution. The possibility of secondary losses should likewise be considered. The data must be examined in a bit more detail.

Opisthobranchia and Pulmonata are generally considered sister groups, though there is a strong possibility that some groups of opisthobranchs are on a lineage that leads to the pulmonates.[133] Perhaps the Acoela (Notaspidea plus Nudibrancha) branched off first.[134] Be this as it may, pulmonates are a single lineage derived from something anatomically close to a primitive opisthobranch in which the mantle cavity (or part of it) got converted into a lung. Some of the descendants of the common ancestor of Pulmonata gave rise to the air-breathing pulmonates of land and freshwater, but many have remained in the sea, where they are aquatic organisms that breathe mainly air. Defensive metabolites, including polypropionates, have been recorded from two distinct lineages of marine pulmonates. Like most pulmonates, they remain herbivorous.

The first such pulmonate lineage (Basommatophora) includes the genera *Siphonaria* and *Trimusculus*. They are cap-shaped snails (limpets) that are abundant on intertidal rocks, especially in the tropics. Metabolites are secreted in the mucus and at the edge of the shell, and their absence in the gut is one reason to assume that they are not of dietary origin. The defensive use of the metabolites has been questioned, purely on the basis that some predation seems to occur. The diversity of polypropionates in the limpets has evidently been overestimated as a result of laboratory artifacts. Nonetheless, there is a fair range of polypropionates including both acyclic and cyclic compounds in *Siphonaria* (Figure 3.13.1).[135,136] The related genus *Trimusculus* contains diterpenes, not polypropionates.[137]

The second pulmonate lineage having polypropionates is the Onchidiacea, which are pulmonate slugs. They, of course, have no shell as adults. The dorsal surface, especially the edge of the mantle, contains specialized repugnatorial glands that contain various secondary compounds. In particular, there are propionate esters in the genera *Onchidium*[138,139] and *Peronia*.[140] In the genus *Onchidella*, the repugnatorial secretion consists largely of the enol acetate sesquiterpenoid onchidial (Figure 3.13.2).[141,142]

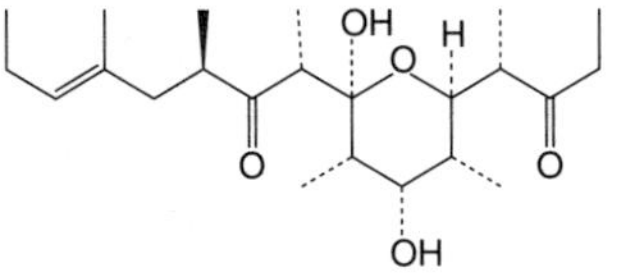

3.13.1

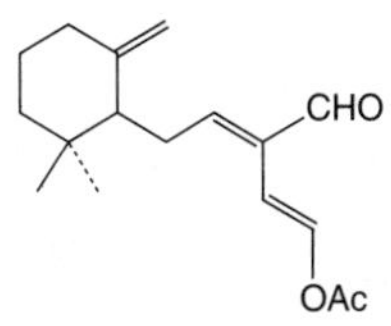

3.13.2

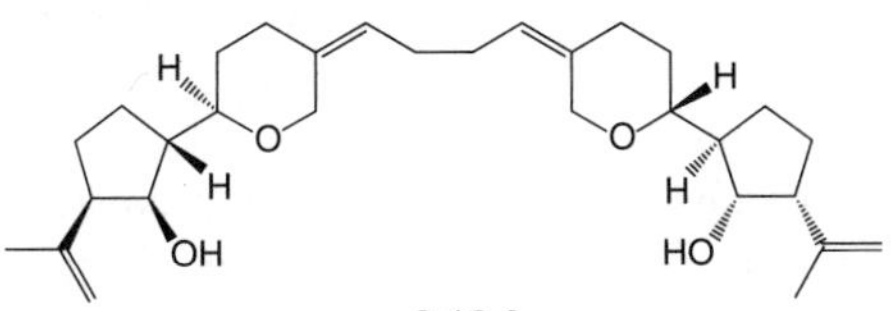

3.13.3

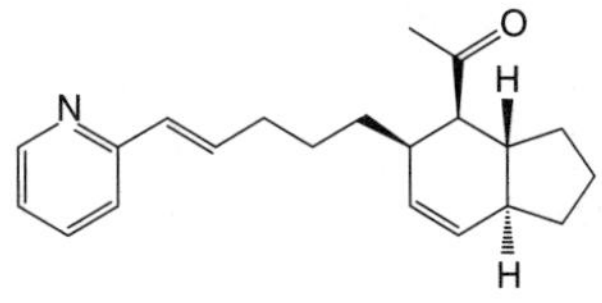

3.13.4

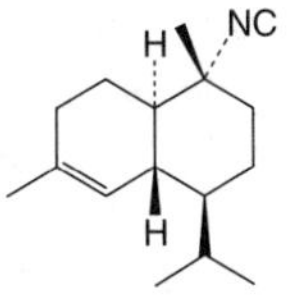

3.13.5

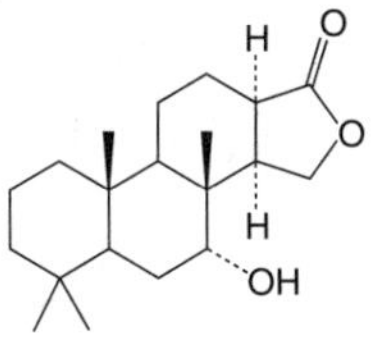

3.13.6

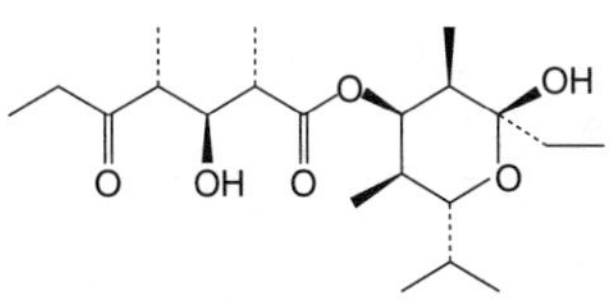

3.13.7

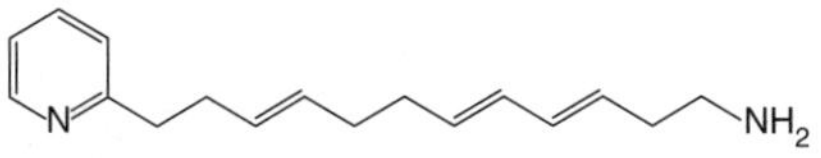

3.13.8

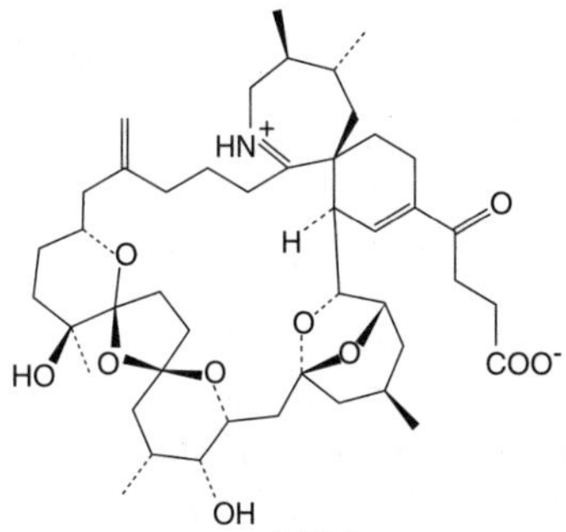

3.13.9

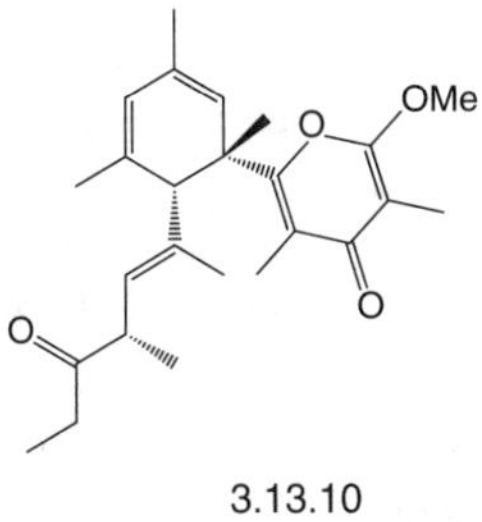

3.13.10

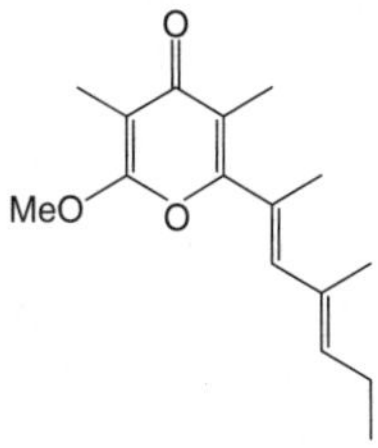

3.13.11

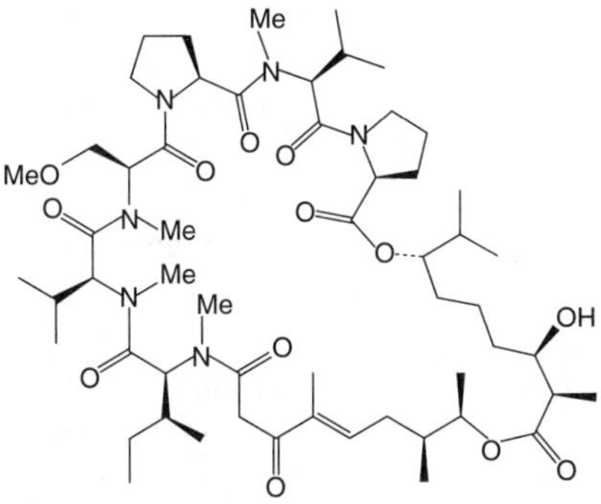

3.13.12

FIGURE 3.13 Selected metabolites from molluscs. 3.13.1 Dentaculatin A, from *Siphonaria denticulata*.[227] 3.13.2 Onchidal from *Onchidella binneyi*.[228] 3.13.3 Testudinariol A from *Pleurobranchus testudinarius*.[229] 3.13.4 (-)Isopulo'upone from *Bulla gouldiana*.[230] 3.13.5 10-Isocyano-4-cadinene from *Phyllidia* spp.[231] 3.13.6 7α-Hydroxyspongian-16-one from *Chromodoris obsoleta*.[232] 3.13.7 Dolabriferol from *Dolabrifera dolabrifera*.[233] 3.13.8 Noloamine from *Smaragdinella calyculata*.[234] 3.13.9 Pinnatoxin D, from *Pinna muricata*.[235] 3.13.10 Elysione from *Elysia chlorotica*.[236] 3.13.11 Cyercene A from *Cyerce crystallina*.[237] 3.13.12 Nordolastin G from *Dolabella auricularia*.[238]

The "pteropods" are a presumably artificial group of holopelagic opisthobranchs that are classified as separate orders: Thecosomata and Gymnosomata. Chemical defense has only been recorded in the latter group, but little work has been done so far on either of them. The Gymnosomata have been interpreted as close relatives of the Anaspidea by some authors,[143] but recent, and highly provisional, molecular results do not support this suggestion.[133] Be this as it may, the gymnosomes are pelagic predators, and their use of polypropionates in correlation with a shift in both habitat and diet fits the overall pattern that we have observed in other groups of opisthobranchs. The gymnosome *Clione antarctica* contains the polypropionate pteroenone.[144] Its effectiveness in repelling predators is evident from a kind of parasitism: the pteropods are abducted by hyperiid amphipods.[145] *Clione* feeds upon *Limacina*, which are herbivorous pteropods of the order Thecosomata. The polypropionates are absent from the prey, and, therefore, dietary derivation seems unlikely.[146]

The orders Nudibranchia and Notaspidea form a monophyletic assemblage in which chemical defense is extensive and well documented. It seems peculiar that propionates have been found only isolated within the latter group. The skin of *Pleurobranchus membranaceus* contains membrenones.[147] Membrenone-C is remarkably similar to vallartanone-B from the basommatophoran pulmonate *Siphonaria maura*.[136] A closely related notaspidean, *P. testudinarius*, has been found to contain triterpenoids (Figure 3.13.3).[148]

Polypropionates also occur sporadically in three other orders of opisthobranchs: Cephalaspidea, Anaspidea, and Sacoglossa. The latter two have already been discussed with respect to their basic feeding pattern and major metabolites. Cephalaspidea is generally considered a paraphyletic grade, in other words, a primitive stock from which some of its descendants have been subtracted. The three orders in question have the characteristic head-shield from which the name Cephalaspidea is derived, and the other two orders have primitive members with that feature, among others, which indicates a very close relationship. The exact branching sequences remain highly debatable, but the natural groups are obvious enough for our purposes. The common ancestor of these animals was herbivorous. A few lineages of Cephalaspidea have switched to feeding upon animals, including other opisthobranchs.

Among the Cephalaspidea, there are several herbivorous forms with defensive metabolites.[149] Two species of the generalist herbivore *Bulla* probably biosynthesize polypropionates that resemble those of pulmonates. *B. striata* produces a series of such metabolites called aglajanes. It is fed upon by the carnivorous cephalaspidean *Aglaja depicta*, which uses the polypropionates defensively.[150] A similar relationship exists between *B. gouldiana* and *Navanax inermis*.[151] Another herbivorous cephalaspidean, one that lives in the upper littoral, is *Smaragdinella calyculata*. It contains a polypropionate and an α-substituted pyridine (Figure 3.13.8).[152] The specialized gizzard plates and other characters indicate that this animal is genealogically close to the genus *Haminoea*, which has somewhat different feeding habits. These animals are recorded as sources of various metabolites, including alkaloids[153] and sesquiterpenes.[154] The alkaloids are suggestive of ones found in Cyanobacteria, and it is evident that there has been a switch from true algae to blue-green "algae," as has happened in some Anaspidea. Tripeptides from the carnivorous cephalaspidean *Philinopsis speciosa* are evidently derived from cyanobacteria via other opisthobranchs.[155]

The order Anaspidea (sea hares) has already been discussed in connection with the derivation of metabolites from algae and cyanobacteria upon which these animals feed. There has been some question as to how well defended they are by metabolites in the skin. Part of the answer is that defensive metabolites do occur in the integument. *Dolabella auricularia* contains cyclopeptides called auripyrones (Figure 3.13.12).[156] *Dolabrifera dolabrifera* and *Petalifera petalifera* contain the polypropionate dolabriferol (Figure 3.13.7).[157]

Finally, propionates are very common in the order Sacoglossa (Figures 3.13.10 and 3.13.11).[4] As discussed above, these animals primitively, and with few isolated exceptions, feed upon algae by piercing the cell wall and sucking out the contents. Transitional forms are morphologically very close to cephalaspideans in their external anatomy. Exactly which alga was the ancestral food is debatable, but this is a point of minor significance, for the algae are closely related and have similar metabolites such as udoteal and caulerpenyne. The sacoglossans have diversified and specialized, and changes in the food supply have been accompanied by changes in defensive metabolites. Often, the metabolites that appear under such circumstances are polypropionates. Those which are primitive, insofar as they retain shells as adults, all feed upon algae with defensive metabolites, but none of those are known to contain polypropionates. Those in which the shell has been lost in the adult are divided by phylogeneticists into two separate groups. The first is the family Elysiidae in which the rim of the foot forms a continuous flap. The second consists of the families Caliphyllidae and Stiligeridae, which have separate dorsal processes called cerata. The Elysiidae are discussed first.

One noteworthy nutritive adaptation that has evolved within the Sacoglossa is a kind of ectoparasitism in which the mollusc utilizes chloroplasts derived from the algae upon which it feeds by keeping them alive and exploiting their photosynthetic ability. This is rather common among the Elysiidae. It is particularly well developed in the genus *Plakobranchus*, which produces polypropionates.[158] These are rather similar in structure to those of animals known in the literature as *Tridachia* and *Tridachiella*.[159–162] However, *Plakobranchus* is not closely related to them. In the same family, *Elysia viridis* has shifted to the alga *Codium* and is now protected by the polypropionate elysione (Figure 3.13.10).[163] The same correlation of a new algal host with production of polypropionates is characteristic of *Elysia timida*, which eats *Acetabularia*,[164] and *Elysia chlorotica*, which eats *Cladophora*.[165]

The phylogeny of the Elysiidae is sufficiently well known for one to infer with some confidence that such shifts to polypropionates have occurred more than once, but one cannot be sure how often this has happened. Similar changes have occurred, however, in the lineage with cerata, to which we now turn our attention. *Placida dendritica*, which feeds upon *Bryopsis*, biosynthesizes gamma pyrones.[166] In another family, *Cyerce*, which are known to feed upon *Chlorodesmus* among other algae, the cyercenes (Figure 3.13.11) are likewise known to be synthesized *de novo*.[167,168]

What might we conclude from the pattern just described? Given that polypropionate biosynthesis occurs only in the higher gastropods and in no other Metazoa, it seems clear that the capacity to do so does not evolve readily. And yet, within one monophyletic group, it would seem to have evolved repeatedly. One alternative that comes to mind is that the ability to synthesize polypropionates has been secondarily lost. This is certainly a possibility for some cases, but it does not fit the overall pattern. The polypropionates do not disappear upon the occasion that some new dietary source of defensive metabolites becomes available. On the contrary, they seem to appear within groups in which dietary metabolites are the original condition, but where the diet-derived metabolites are no longer available or, in a few cases, not particularly effectual. Like biotransformation, biosynthesis appears to be a derived condition. It evolves as a kind of *ad hoc* adaptation where it is needed, in a few groups of organisms.

A second possibility is that the polypropionates are produced by symbionts rather than by the molluscs themselves, and that these symbionts have been independently acquired. The spread of symbionts that produce secondary metabolites across lineages is well documented, and on that account this alternative explanation seems superficially attractive. But several other considerations make that seem much less plausible. First, the symbionts that are known to have spread across lineages often have gone to quite distantly related lineages rather than to a group of closely related ones such as the Euthyneura. Second, there is no plausible source from which the hypothetical symbionts could have been derived, except, perhaps, from other gastropods. Third, the polypropionates of *Siphonaria* and a few others are often quite simple, straight-chain compounds such as one would expect if biosynthesis had begun within the group. Fourth, the polypropionates are restricted to the integument of the animals, and there is no place where a reservoir of symbionts producing those polypropionates might be maintained. Indeed, in the shelled cephalaspideans *Scaphander* and *Bulla*, the polypropionates are secreted only along the rim of the mantle. Fifth, the propionates display a remarkable facility for appearing when other defensive metabolites have ceased to be available.

Given the present state of knowledge, a third possibility seems a bit more plausible than the previous two. This possibility is that the ability to synthesize polypropionates was present in the common ancestor of Opisthobranchia and Pulmonata. The compounds that were synthesized may have had some function other than, or in addition to, chemical defense. If the initial function (primary function) was something other than chemical defense, then the polypropionates could then subsequently be pressed into defensive service as a secondary function. The evolution of the secondary function would not require the capacity to biosynthesize polypropionates, but rather the secretion of polypropionates in large quantities; of course, they would subsequently be modified and delivered more effectively. The secondary function would now be the primary function. Parallel evolution of propionates within the group would result from repeated changes of this kind (Funktionswechsel). Loss of the primary function would preclude any such Funktionswechsel, which may explain why polypropionates have never been detected in the Nudibranchia.

A minor variation on this theme would be that the initial function was indeed chemical defense, but that the propionates acquired some minor secondary function that has been retained in many lineages of opisthobranchs after the defensive function was lost. The re-evolution of a defensive function would then be possible, and it would be a curious example of atavism. At any rate, the situation with respect to prostaglandins in *Tethys* suggests that not all apparent absence of metabolites necessarily means an inability to synthesize them. Further research as to the validity of such scenarios is obviously necessary. In particular, there are obvious opportunities for seeking propionates at lower concentrations and in as yet unexamined taxa. Other possibilities for new research include comparative biosynthesis and comparative enzymology.

V. GENERAL DISCUSSION

To the beginner, both chemistry and biology appear bewilderingly complex. Yet, in spite of the wealth of detail, everything seems much simpler when viewed in light of the fundamental concepts. For chemistry, Atkins reduces the basics to a list of nine great ideas, all of them quite simple.[169] Among these is the idea that matter consists of about 100 elements. These are enough to produce a vast range of chemical diversity. Although not part of Atkins' list, the same kind of simplicity can be found in natural products chemistry. As previously mentioned, secondary metabolites are constructed from a remarkably limited number of starting materials. It might not be too difficult to construct a similar list of great ideas for evolutionary biology. In terms of how to begin, all living beings are linked together by relationships of common ancestry of a greater or lesser degree. That provides the basis for systematics as an historical science in which everything has a place within the context of a single whole. Adding one more item, evolutionary change occurs primarily as the result of reproductive competition between individuals of the same species. That tells us, among other things, that adaptation as a state is only intelligible in terms of adaptation as a process. By implication, any claim about adaptive significance logically entails and presupposes some kind of historical narrative.

Organisms do what they do because they repeat whatever it was that contributed to their ancestors' reproductive success. What they do in some other context, whether it be the artificial conditions of the laboratory or the more natural ones of field studies, may be quite irrelevant to the issue of adaptive significance. It is not enough to replace freshwater fish with marine fish when doing a bioassay. There is no such thing as a "rigorous" experimental test of a hypothesized raison d'etre for the presence of a secondary compound. Comparative biology is the only thing we have to go on when we deal with the past. Natural products chemistry is an integral part of such comparative biology. Its role, however, is not to "reduce" the organisms to chemicals. Rather, the chemicals are placed within the context of a narrative about the organisms. The organisms synthesize and modify the metabolites and use them as resources in the struggle for existence. The two sciences come together because the narrative must be plausible both as chemistry and as biology. There can be no contradiction of the laws and principles of either.

The general picture that emerges from our scenario is of gradual elaboration of metabolites, mainly by sedentary or sessile organisms and their symbionts and predators. It is not clear to what extent the original function was "defensive" in a broader sense. Perhaps dealing with spatial competitors, fouling organisms, and parasites was more important. That issue remains unresolved for contemporary organisms as well. However, groups of animals that feed upon metabolite-rich marine organisms have specialized in doing so. They have tracked the metabolites through evolutionary time, often switching from one source of metabolite-rich food to another. The classes of metabolites that are tracked in this fashion are ones for which a role in reducing predation is manifest. The animals that specialize in metabolite-rich food have also evolved the capacity to detoxify the metabolites and otherwise to modify them. The biotransformation of metabolites is often accompanied by more effective delivery systems and by mechanisms that render the metabolites less harmful to the animal that uses them defensively. At later stages in evolution, dependency upon the original food source is reduced by *de novo* synthesis of the original metabolite or of one that is similar to the original. Indeed, the metabolites that are synthesized *de novo*, although similar to those derived from food, are usually not identical to them, and a further stage of adaptation seems to have been attained. In some cases (the propionates), chemicals of a different class are used, but these are often convergent with respect to the reactive part of the molecule.

"Arms race" is a figurative term that has been applied in different contexts, sometimes with unfortunate results. It has generally been linked to the notion of coevolution in a strict sense, in which predator and prey species exert a mutual influence upon each other as individual populations. This is not exactly what Vermeij[2] seemingly had in mind for the long-term interaction of shelled molluscs and crabs. Rather the oceans became increasingly populated by heavily shelled molluscs that belonged to very independent lineages, and by similarly independent lineages of crabs and other predators that were able to overcome the molluscs' defenses. Defensive chemicals would seem to be a more recent variant on the same basic theme with innovations in the use of metabolites accumulating in separate lineages. There is some correspondence between the systematics of opisthobranchs and that of the organisms upon which they feed, but that happens because the metabolites that the opisthobranchs exploit are good taxonomic markers. Thus, our work on opisthobranchs supports Steneck's basic conclusions with respect to sacoglossan gastropods and their algal food.[170] We would only emphasize that the extent to which species coevolve is probably a matter of degree. One would expect to find more coevolution between sponges and their symbiotic bacteria than between sponges and the nudibranchs that feed upon them.

The macroevolutionary pattern of opisthobranch gastropods can be described as an adaptive radiation. It began with the group exploiting a fairly narrow range of food items that were also a source of defensive metabolites. The group diversified so as to become specialists upon a wider range of such food items. Diversification does not seem to have been accompanied by a loss of evolutionary plasticity, whether physiological or morphological. Reduction of the shell set the stage for the evolution of a broader range of morphologies. The ability to exploit a wider range of food and metabolites, and, hence, expansion into new niches, reflects continued physiological innovation. The physiological innovations have evolved repeatedly, and the trends are good examples of progressive evolution.[171,172] There has been remarkably little extinction of the sort that wipes out the intermediates between higher taxa and thereby gives the false impression that evolution has occurred through saltation. Someone looking for alternatives to Darwinism should not waste his time on opisthobranchs.

ACKNOWLEDGMENTS

For their advice regarding this manuscript, we thank Julia Kubanek, Ernesto Mollo, C. C. Naik, and P. S. Parameswaran. Support for travel was provided by the Short Term Mobility Program of the CNR (Italy).

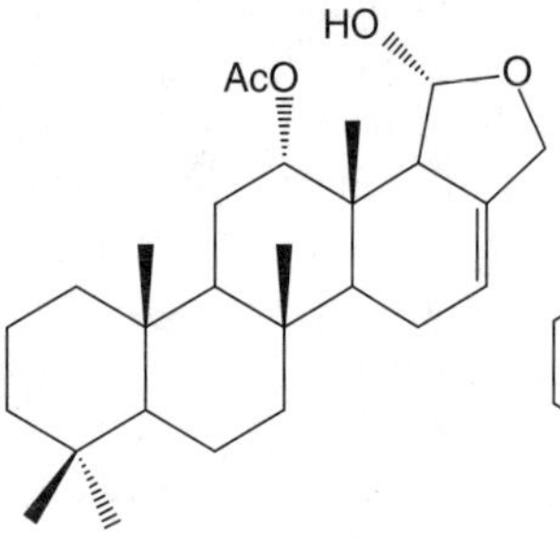

3.14.1 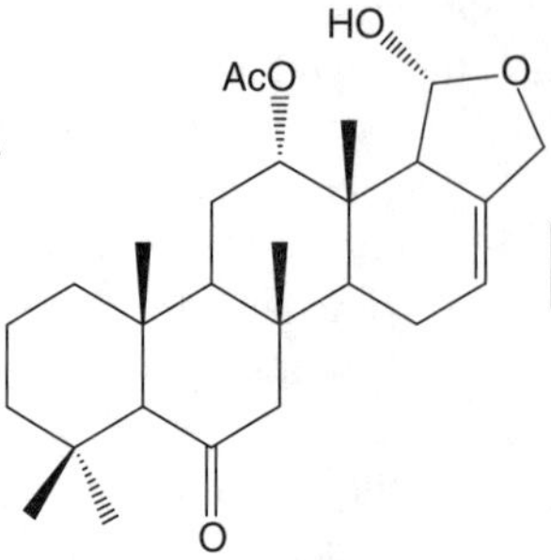   3.14.2 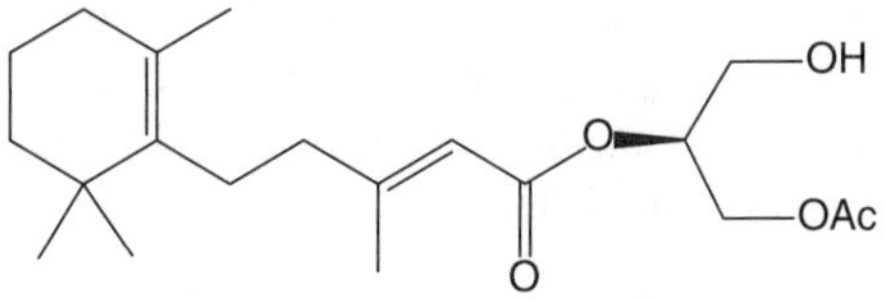   3.14.3

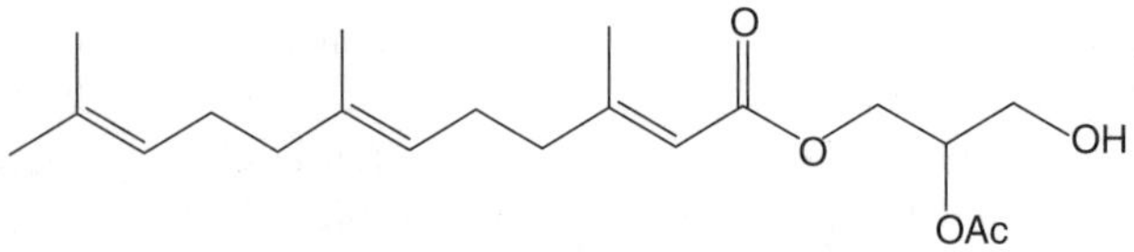

3.14.4 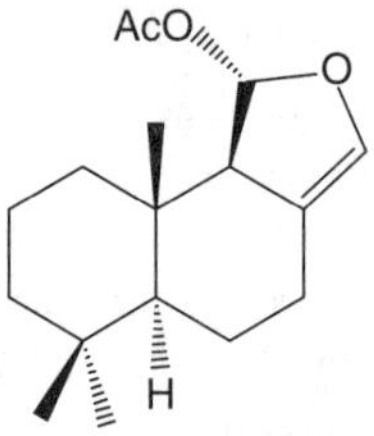   3.14.5

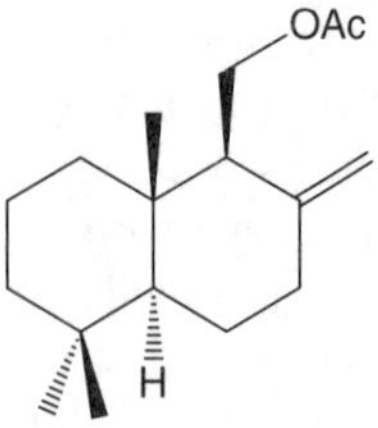

3.14.6 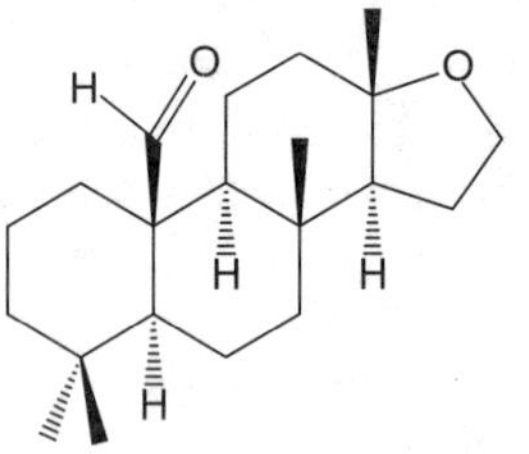   3.14.7 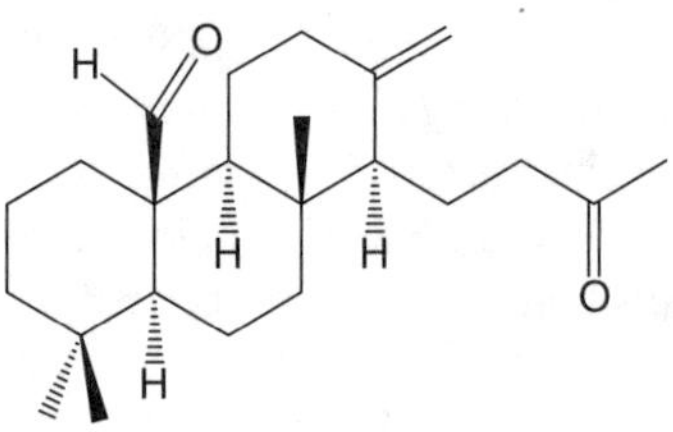   3.14.8

FIGURE 3.14 Selected metabolites biotransformed or biosynthesized by opisthobranch gastropods. 3.14.1 Deoxoscalarin from *Hypselodoris orsini*.[239] 3.14.2 6-Ketodeoxoscalarin from *Hypselodoris orsini*.[240] 3.14.3 Tanyolide from *Sclerodoris tanya*.[241] 3.14.4 Farnesic acid glyceride.[242] 3.14.5 7-Deacetoxyolepupuane from *Dendrodoris*.[243] 3.14.6 Albicanyl acetate from *Cadlina luteomarginata*.[243] 3.14.7 Cadlinaldehyde from *Cadlina luteomarginata*.[243] 3.14.8 Luteone from *Cadlina luteomarginata*.[244]

REFERENCES

1. Ghiselin, M. T., *Metaphysics and the Origin of Species*, State University of New York Press, Albany, NY, 1997.
2. Vermeij, G. J., *Evolution and Escalation: An Ecological History of Life*, Princeton University Press, Princeton, NJ, 1987.
3. Williams, D. H., Stone, M. J., Hauck, P. R., and Rahman, S. K., Why are secondary metabolites (natural products) biosynthesized?, *J. Nat. Prod.*, 52, 1189, 1989.
4. Cimino, G. and Ghiselin, M. T., Chemical defense and evolution in the Sacoglossa (Mollusca: Gastropoda: Opisthobranchia), *Chemoecology*, 8, 51, 1998.

5. Cimino, G. and Ghiselin, M. T., Chemical defense and evolutionary trends in biosynthetic capacity among dorid nudibranchs (Mollusca: Gastropoda: Opisthobranchia), *Chemoecology*, 9, 187, 1999.

6. Faulkner, D. J. and Ghiselin, M. T., Chemical defense and the evolutionary ecology of dorid nudibranchs and some other opisthobranch gastropods, *Mar. Ecol. Prog. Ser.*, 13, 295, 1983.

7. Dohrn, A., *Der Ursprung der Wirbelthiere und das Princip des Functionswechsels. — Genealogische Skizzen von Anton Dohrn*, Wilhelm Engelmann, Leipzig, 1875.

8. Ghiselin, M. T., The origin of vertebrates and the principle of succession of functions. Genealogical sketches by Anton Dohrn. 1875. An English translation from the German, introduction and bibliography, *History and Philosophy of the Life Sciences*, 16, 5, 1994.

9. Pietra, F., Structurally similar natural products in phylogenetically distant marine organisms, and a comparison with terrestrial species, *Chem. Soc. Rev.*, 24, 65, 1995.

10. Muller, F., *Für Darwin*, Wilhelm Engelmann, Leipzig, 1864.

11. Horowitz, N. H., On the evolution of biochemical syntheses, *Proc. Nat. Acad. Sci. USA*, 31, 153, 1945.

12. Horowitz, N. H., The evolution of biochemical synthesis: retrospect and prospect, in *Evolving Genes and Proteins*, Bryson, V. and Vogel, H. J., Eds., Academic Press, New York, 1965, 15.

13. Kvitek, R. G. and Beitler, M. K., Relative insensitivity of butter clam neurons to saxitoxin: a pre-adaptation for sequestering paralytic shellfish poisoning toxins as a chemical defense, *Mar. Ecol. Prog. Ser.*, 69, 47, 1991.

14. Amico, V., Marine brown algae of family Cystoseiraceae: chemistry and chemotaxonomy, *Phytochemistry*, 39, 1257, 1995.

15. Pennings, S. C. and Paul, V. J., Sequestration of dietary secondary metabolites by three species of sea hares: location, specificity and dynamics, *Mar. Bio.*, 117, 535, 1993.

16. Rudman, W. B., Purpose in pattern: the evolution of colour in chromodorid nudibranchs, *J. Molluscan Stud.*, 57, 5, 1991.

17. Tullrot, A. and Sundberg, P., The conspicuous nudibranch *Polycera quadrilineata*: aposematic coloration and individual selection, *Ann. Behav.*, 41, 175, 1991.

18. Rogers, C. N., Steinberg, P. D., and De Nys, R., Factors associated with oligophagy in two species of sea hares (Mollusca: Anaspidea), *J. Exp. Mar. Biol. Ecol.*, 192, 47, 1995.

19. De Nys, R., Steinberg, P. D., Rogers, C. N., Charlton, T. S., and Duncan, M. W., Quantitative variation of secondary metabolites in the sea hare *Aplysia parvula* and its host plant, *Delisea pulchra*, *Mar. Ecol. Prog. Ser.*, 130, 135, 1996.

20. Eales, N. B., Revision of the world species of *Aplysia*, *Bull. Br. Mus. (Natural History)*, 5, 267, 1960.

21. Atta-ur-Rahman, K. A., Alvi, K. A., Abbas, S. A., Sultana, T., Shameel, M., Choudhary, M. I., and Clardy, J. C., A diterpenoid lactone from *Aplysia juliana*, *J. Nat. Prod.*, 54, 886, 1991.

22. Paul, V. J. and Fenical, W., Natural products chemistry and chemical defense in tropical marine algae of the phylum Chlorophyta, in *Bioorganic Marine Chemistry*, Vol. 1, Scheuer, P. J., Ed., Springer-Verlag, Berlin, 1987, 1.

23. Paul, V. J., Chemical defenses of benthic marine invertebrates, in *Ecological Roles of Marine Natural Products*, Paul V. J., Ed., Cornell University Press, Ithaca, NY, 1992, 164.

24. Hay, M. E., Kappel, Q. E., and Fenical, W., Synergisms in plant defenses against herbivores: interactions of chemistry, calcification, and plant quality, *Ecology*, 75, 1714, 1994.

25. Caprioli, V., Cimino, G., Colle, R., Gavagnin, M., Sodano, G., and Spinella, A., Insect antifeedant activity and hot taste for humans of selected natural and synthetic 1,4-dialdehydes, *J. Nat. Prod.*, 50, 146, 1987.

26. Hamann, M. T., Otto, C. S., Scheuer, P. J., and Dunbar, D. C., Kahalalides: bioactive peptides from a marine mollusk *Elysia rufescens* and its algal diet *Bryopsis* sp., *J. Org. Chem.*, 61, 6594, 1996.

27. Kan, Y., Fujita, T., Sakamoto, B., Hokama, Y., and Nagai, H., Kahalalide K: a new cyclic depsipeptide from the Hawaiian green alga *Bryopsis*, *J. Nat. Prod.*, 62, 1169, 1999.

28. Hay, M. E., Duffy, J. E., Paul, V. J., Renaud, P. E., and Fenical, W., Specialist herbivores reduce their susceptibility to predation by feeding on the chemically defended seaweed *Avrainvillea longicaulis*, *Limn. Ocean.*, 35, 1734, 1990.

29. Fusetani, N. and Matsunaga, S., Bioactive sponge peptides, *Chem. Rev.*, 93, 1793, 1993.

30. Kobayashi, M., Aoki, S., Gato, K., Matsunami, K., Kurosu, M., and Kitagawa, I., Trisdoline, a new antibiotic indole trimer, produced by a bacterium of *Vibrio* sp. separated from the marine sponge *Hyrtios altum*, *Chem. Pharm. Bull.*, 42, 2449, 1994.

31. Oclarit, J. M., Yaaoka, Y., Kamimura, K., Ohta, S., and Ikegami, S., Andrimid, an antimicrobial substance in the marine sponge *Hyatella*, produced by an associated *Vibrio* bacterium, in *Sponge Sciences: Multidisciplinary Perspectives*, Watanabe, Y. and Fusetani, N., Eds., Springer-Verlag, Tokyo, 1998, 391.

32. Bergquist, P. R., Walsh, D., and Gray, R. D., Relationships within and between the orders of Demospongiae that lack a mineral skeleton, in *Sponge Sciences: Multidisciplinary Perspectives*, Watanabe, Y. and Fusetani, N., Eds., Springer-Verlag, Tokyo, 1998, 31.

33. Van Soest, R. W. M., Demosponge higher taxa classification re-examined, in *Fossil and Recent Sponges*, Reitner, J. and Keupp, H., Eds., Springer-Verlag, Berlin, 1991, 54.

34. Chanas, B. and Pawlik, J. R., Defenses of Caribbean sponges against predatory reef fish: II. Spicules, tissue toughness, and nutritional quality, *Mar. Ecol. Prog. Ser.*, 127, 195, 1995.

35. Levi, C. and Levi, P., Embryogenese de *Chondrosia reniformis* (Nardo), demosponge ovipaire, et transmission des bacteries symbiotiques, *Ann. Sci. Nat. Zoologie*, (12) 18, 367, 1976.

36. Unson, M. D. and Faulkner, D. J., Cyanobacterial symbiont biosynthesis of chlorinated metabolites from *Dysidea herbacea*, *Experientia*, 49, 349, 1993.

37. Debitus, C., Guella, G., Mancini, I., Waikedre, J., Guemas, J. P., Nicolas, J. L., and Pietra, F., Quinolines from a bacterium and tyrosine metabolites from its host sponge, *Suberea creba* from the Coral Sea, *J. Mar. Biotech.*, 6, 136, 1998.

38. Andersen, R. J., Van Soest, R. W. M., and Kong, F., 3-Alkylpiperidine alkaloids isolated from marine sponges in the order Haplosclerida, in *Alkaloids: Chemical and Biological Perspectives*, Pelletier, S. W., Ed., Pergamon Press, New York, 1996, 10, 301.

39. Van Soest, R. W. M., Fusetani, N., and Andersen, R. J., Straight-chain acetylenes as chemotaxonomic markers of the marine Haplosclerida, in *Sponge Sciences: Multidisciplinary Perspectives*, Watanabe, Y. and Fusetani, N., Eds., Springer-Verlag, Tokyo, 1998, 3.

40. Castiello, D., Cimino, G., De Rosa, S., De Stefano, S., Izzo, G., and Sodano, G., Studies on the chemistry of the relationship between the opisthobranch *Peltodoris atromaculata* and the sponge *Petrosia ficiformis*, in *Biologie des Spongiaires*, Levi, C. and Boury-Esnault, N., Eds., Centre Nationale de la Recherche Scientifique, Paris, 1979, 413.

41. Castiello, D., Cimino, G., De Rosa, S., De Stefano, S., and Sondo, G., High molecular weight polyacetylenes from the nudibranch *Peltodoris atromaculata* and the sponge *Petrosia ficiformis*, *Tetrahedron Lett.*, 21, 5047, 1980.

42. Garson, M. J., Biosynthesis of the novel diterpene isonitrile diisocyanodociane by a marine sponge of the *Amphimedon* genus; incorporation studies with sodium [^{14}C]cyanide and sodium [2-^{14}C]acetate, *J. Chem. Soc., Chem. Commun.*, 35, 1986.

43. Dumdei, E. J., Flowers, A. E., Garson, M. J., and Moore, C. J., The biosynthesis of sesquiterpene isocyanides and isothiocyanates in the marine sponge *Acanthella cavernosa* (Dendy); evidence for dietary tranfer to the dorid nudibranch *Phyllidiella pustulosa*, *Comp. Biochem. Physiol.*, 113A, 1385, 1997.

44. Brunckhorst, D. J., Do phyllidiid nudibranchs demonstrate behavior consistent with their apparent warning coloration? Some field observations, *J. Molluscan Stud.*, 57, 481, 1991.

45. Gosliner, T. M. and Behrens, D. W., Special resemblance, aposematic coloration and mimicry in opisthobranch gastropods, in *Adaptive Coloration in Invertebrates*, Wicksten, M., Ed., Texas A&M University Sea Grant College Program, College Station, TX, 1990, 127.

46. Teeyapant, R. and Proksch, P., Biotransformation of brominated compounds in the marine sponge *Verongia aerophoba* — evidence for an induced chemical defense, *Naturwissenschaften*, 80, 369, 1993.

47. Ebel, R., Brenzinger, M., Kunze, A., Gross, H., and Proksch, P., Wound activation of prototoxins in the marine sponge *Aplysina aerophoba*, *J. Chem. Ecol.*, 23, 1451, 1997.

48. Fontana, A., Avila, C., Martinez, E., Ortea, J., Trivellone, E., and Cimino, G., Defensive allomones in three species of *Hypselodoris* (Gastropoda, Nudibranchia), from the Cantabrian Sea, *J. Chem. Ecol.*, 19, 339, 1993.

49. Graziani, E. I., Andersen, R. J., Krug, P. J., and Faulkner, D. J., Stable isotope incorporation evidence for *de novo* biosynthesis of terpenoic acid glycerides by dorid nudibranchs, *Tetrahedron*, 52, 6869, 1996.

50. Gustafson, K. and Andersen, R. J., Chemical studies of British Columbia nudibranchs, *Tetrahedron*, 41, 1101, 1985.

51. Fontana, A., Ciavatta, M. L., Miyamoto, T., Spinella, A., and Cimino, G., Biosynthesis of drimane terpenoids in dorid molluscs: pivotal role of 7-deacetoxyolepupuane in two species of *Dendrodoris* nudibranchs, *Tetrahedron*, 55, 5937, 1999.

52. Cimino, G., De Rosa, S., De Stefano, S., Morrone, R., and Sodano, G., The chemical defense of nudibranch molluscs. Structure, biosynthetic origin and defensive properties of terpenoids from the dorid nudibranch *Dendrodoris grandiflora*, *Tetrahedron*, 41, 1093, 1985.

53. Cimino, G. and Sodano, G., Transfer of sponge secondary metabolites to predators, in *Sponges in Time and Space: Biology, Chemistry, Paleontology*, Van Soest, R. W. M., Van Kempen, T. M. G., and Braekman, J.C., Eds., A.A. Balkema, Rotterdam, 1994, 459.

54. Avila, C., Cimino, G., Crispino, A., and Spinella, A., Drimane sesquiterpenoids in Mediterranean *Dendrodoris* nudibranchs: anatomical distribution and biological role, *Experientia*, 47, 306, 1991.

55. Kubanek, J., Graziani, E. I., and Andersen, R. J., Investigations of terpenoid biosynthesis by the dorid nudibranch *Cadlina luteomarginata*, *J. Org. Chem.* 62, 7239, 1997.

56. Gerhart, D. J., The chemical systematics of colonial marine animals: an estimated phylogeny of the order Gorgonacea based on terpenoid characters, *Biol. Bull.*, 164, 71, 1983.

57. Dai, M. C., Garson, M. J., and Coll, J. C., Biosynthetic processes in soft corals. I. A comparison of terpene biosynthesis in *Alcyonium molle* (Alcyoniidae) and *Heteroxenia* sp. (Xeniidae), *Comp. Biochem. Physiol.*, 99B, 775, 1991.

58. Stanley-Samuelson, D. W., Comparative eicosanoid physiology in invertebrate animals, *Am. J. Physiol.*, 260, R849, 1991.

59. Stanley-Samuelson, D., The biological significance of prostaglandins and related eicosanoids in invertebrates, *Am. Zool.*, 34, 589, 1994.

60. De Petrocellis, L. and Di Marzo, V., Aquatic invertebrates open up new perspectives in eicosanoid research: biosynthesis and bioactivity, *Prostaglandins, Leukotrienes and Essential Fatty Acids*, 51, 215, 1994.

61. Stanley-Samuelson, D. and Howard, R. W., The biology of prostaglandins and related eicosanoids in invertebrates: cellular, organismal and ecological actions, *Am. Zool.*, 38, 369, 1998.

62. Gerhart, D. J., Prostglandin A_2: an agent of chemical defense in the Caribbean gorgonian *Plexaura homomalla*, *Mar. Ecol. Prog. Ser.*, 19, 181, 1984.

63. Gerhart, D. J., Prostaglandin A_2 in the Caribbean gorgonian *Plexaura homomalla*: evidence against allelopathic and antifouling roles, *Biochem. Syst. Ecol.*, 14, 417, 1986.

64. Gerhart, D. J., Emesis, learned aversion and chemical defense in octocorals: a central role for prostaglandins?, *Am. J. Phys.*, 260, R839, 1991.

65. Van Alstyne, K. L. and Paul, V. J., Chemical and structural defenses in the sea fan *Gorgonia ventalina*: effects against generalist and specialist predators, *Coral Reefs*, 11, 155, 1992.

66. Coll, J. C., Tapiolas, B. F., Bowden, B. F., Webb, L., and Marsh, H., Transformation of soft coral (Coelenterata: Octocorallia) terpenes by *Ovula ovum* (Mollusca: Prosobranchia), *Mar. Biol*, 74, 35, 1983.

67. Webb, L., Studies on the apparent detoxification of sarcophytoxide by the prosobranch mollusc *Ovula ovum*, *Comp. Biochem. Physiol.*, 90C, 155, 1988.

68. Gosliner, T. M. and Ghiselin, M. T., A new species of *Tritonia* from the Caribbean Sea, *Bull. Mar. Sci.*, 40, 428, 1987.

69. Cronin, G., Hay, M. E., Fenical, W., and Lindquist, N., Distribution, density, and sequestration of host chemical defenses by the specialist nudibranch *Tritonia hamnerorum* found at high densities on the sea fan *Gorgonia ventalina*, *Mar. Ecol. Prog. Ser.*, 119, 177, 1995.

70. Cimino, G., Spinella, A., and Sodano, G., Naturally occurring prostaglandin-1,15-lactones, *Tetrahedron Lett.*, 30, 3589, 1989.

71. Cimino, G., Crispino, A., Di Marzo, V., Sodano, G., Spinella, A., and Villani, G., A marine mollusc provides the first example of in vivo storage of prostaglandins: Prostaglandin-1,15-lactones, *Experientia*, 47, 56, 1991.

72. Cimino, G., Crispino, A., Di Marzo, V., Spinella, A., and Sodano, G., Prostaglandin 1,15-lactones of the F series from the nudibranch mollusc *Tethys fimbria*, *J. Org. Chem.*, 56, 2907, 1991.

73. Di Marzo, V., Cimino, G., Sodano, G., Spinella, G., and Villani, G., A novel prostaglandin metabolic pathway from a marine mollusc: prostaglandin-1,15-lactones, *Adv. Prostaglandin Thromboxane Leukotriene Res.*, 21, 129, 1990.

74. Di Marzo, V., Cimino, G., Crispino, A., Minardi, C., Sodano, G., and Spinella, A., A novel multifunctional metabolic pathway in a marine mollusc leads to unprecedented prostaglandin derivatives: prostaglandin 1,15-lactones, *Biochem. J.*, 273, 593, 1991.

75. Di Marzo, V., Minardi, C., Vardaro, R. R., Mollo, E., and Cimino, G., Prostaglandin F-1,15-lactone fatty acyl esters: a prostaglandin lactone pathway branch developed during the reproduction and early larval stages of a marine mollusc, *Comp. Biochem. Physiol.*, 101B, 99, 1992.

76. Field, K. G., Olsen, G. J., Lane, D. J., Giovannoni, S., Ghiselin, M. T., Raff, E. C., Pace, N. R., and Raff, R. A., Molecular phylogeny of the animal kingdom, *Science*, 239, 748, 1988.

77. Ghiselin, M. T., Summary of our current knowledge of metazoan phylogeny, in *The hierarchy of life: molecules and morphology in phylogenetic analysis*, Fernholm, B., Bremer, K., and Jornvall, H., Eds., Exerpta Medica, Amsterdam, 1989, 261.

78. Halanych, K. M., Bacheller, J. D., Aguinaldo, A. M. A., Liva, S. M., Hillis, D. M., and Lake, J. A., Evidence from 18S ribosomal DNA that the lophophorates are protostome animals, *Science*, 267, 1641, 1995.

79. Halanych, K. M., Convergence in the feeding apparatus of lophophorates and pterobranch hemichordates revealed by 18S rDNA: an interpretation, *Biol. Bull.*, 190, 1, 1996.

80. Williams, A., Carlson, S. J., Brunton, C. H. C., Holmer, L. E., and Popov, L., A supra-ordinal classification of the Brachiopoda, *Philos. Trans. R. Soc. Lond.* (B), 351, 1171, 1996.

81. Sheikh, Y. M. and Djerassi, C., 2,6-Dibromophenol and 2,4,6-tribromophenol — antiseptic secondary metabolites of *Phoronopsis viridis*, *Experientia*, 31, 265, 1975.

82. Thayer, C. W., Brachiopods versus mussels: competition, predation, and palatability, *Science*, 228, 1527, 1985.

83. McClintock, J. B., Slattery, M., and Thayer, C. W., Energy content and chemical defense of the articulate brachiopod *Liothyrella uva* from the Antarctic Peninsula, *J. Exp. Mar. Biol. Ecol.*, 169, 103, 1993.

84. D'Ambrosio, M., Guerriero, A., and Pietra, F., Glycerol enol ethers of the brachiopod *Gryphus vitreus* from the Tuscan archipelago, *Experientia*, 52, 624, 1996.

85. Mckinney, F. K., One hundred million years of competitive interactions between bryozoan clades: asymmetrical but not escalating, *Biol. J. Linnean Soc.*, 56, 465, 1995.

86. Yoshioka, P. M., Predator-induced polymorphism in the bryozoan *Membranipora membranacea* (L.), *J. Exp. Mar. Biol. Ecol.*, 61, 233, 1982.

87. Harvell, C. D., Genetic variation and polymorphism in the inducible spines of a marine bryozoan, *Evolution*, 52, 80, 1998.

88. Yu, C. M. and Wright, J. L. C., Murrayanolide, an unusual C_{21} tetracyclic terpenoid lactone from the marine bryozoan *Dendrobeania murrayana*, *J. Nat. Prod.*, 58, 1978, 1995.

89. Christophersen, C., Secondary metabolites from marine bryozoans. A review, *Acta Chem. Scand.*, B, 39, 517, 1985.

90. Pettit, G. R., The bryostatins, *Fortschritte der Chemie Organischer Naturstoffe*, 57, 153, 1991.

91. Anthoni, U., Nielsen, P. H., Pereira, M., and Christophersen, C., Bryozoan secondary metabolites: a chemotaxonomical challenge, *Comp. Biochem. Physiol.*, 96B, 431, 1990.

92. Woollacott, R. M., Association of bacteria with bryozoan larvae, *Mar. Biol.*, 65, 155, 1981.

93. Davidson, S. K. and Haygood, M. G., Identification of sibling species of the bryozoan *Bugula neritina* that produce different anticancer bryostatins and harbor distinct strains of the bacterial symbiont "*Candidatus endobugula sertula*," *Biol. Bull.*, 196, 273, 1999.

94. Lindquist, N. and Fenical, W., Ascidiatrienolides A-C, novel lactonized eicosanoids from the marine ascidian *Didemnum candidum*, *Tetrahedron Lett.*, 30, 2735, 1989.

95. Niwa, H., Inagaki, H., and Yamada, K., Didemnilactone and neodidemnilactone, two new fatty acid metabolites possessing a 10-membered lactone from the tunicate *Didemnum mosleyi* (Herdman), *Tetrahedron Lett.*, 32, 5127, 1991.

96. Mackie, G. O. and Bone, Q., Luminescence and associated effector activity in *Pyrosoma* (Tunicata: Pyrosomida), *Proc. Royal Soc. London (B)*, 202, 483, 1978.

97. Kang, H., Jensen, P. R., and Fenical, W., Isolation of microbial antibiotics from a marine ascidian of the genus *Didemnum*, *J. Org. Chem.*, 61, 1543, 1996.

98. Stoecker, D., Distribution of acid and vanadium in *Rhopalaea birkelandi* Tokioka, *J. Exp. Mar. Biol. Ecol.*, 48, 277, 1980.

99. Thompson, T. E., Defensive acid-secretion in marine gastropods, *J. Exp. Mar. Biol. Assoc. U.K.*, 39, 115, 1960.

100. Davidson, B. S., Ascidians: producers of amino acid derived metabolites, *Chem. Rev.*, 93, 1771, 1993.

101. Andersen, R. J., Faulkner, D. J., He, C. H., Van Duyne, G. D., and Clardy, J., Metabolites of the marine prosobranch mollusc *Lamellaria* sp., *J. Am. Chem. Soc.*, 107, 5492, 1985.

102. Davis, R. A., Carroll, A. R., Pierens, G. K., and Quinn, R. J., New lamellarin alkaloids from the Australian ascidian, *Didemnum chartaceum*, *J. Nat. Prod.*, 62, 419, 1999.

103. Carroll, A. R. and Scheuer, P. J., Kuanoniamines A, B, C, and D: pentacyclic alkaloids from a tunicate and its prosobranch mollusk predator *Chelynotus semperi*, *J. Org. Chem.*, 55, 4426, 1990.

104. McClintock, J. B., Baker, B. J., Hamann, M. T., Yoshida, W., Slattery, M., Heine, J. N., Bryan, P. J., Jayatilake, G. S., and Moon, B. H., Homarine as a feeding deterrent in common shallow-water Antarctic lamellarian gastropod *Marseniopsis mollis*: a rare example of chemical defense in a marine prosobranch, *J. Chem. Ecol.*, 20, 2539, 1994.

105. Wesson, K. J. and Hamann, M. T., Keenamide A, A bioactive cyclic peptide from the marine mollusk *Pleurobranchus forskalii*, *J. Nat. Prod.*, 59, 629, 1996.

106. Carte, B. and Faulkner, D. J., Role of secondary metabolites in feeding associations between a predatory nudibranch, two grazing nudibranchs, and a bryozoan, *J. Chem. Ecol.*, 12, 795, 1986.

107. Paul, V. J., Lindquist, N., and Fenical, W., Chemical defenses of the tropical ascidian *Atapozoa* sp. and its nudibranch predators *Nembrotha* spp., *Mar. Ecol. Prog. Ser.*, 59, 109, 1990.

108. Kubanek, J., Williams, D. E., de Silva, E. D., Allen, T., and Andersen, R. J., Cytotoxic alkaloids from the flatworm *Prostheceraeus villatus* and its tunicate prey *Clavelina lepadiformis*, *Tetrahedron Lett.*, 36, 6189, 1995.

109. Schupp, P., Eder, C., Proksch, P., Wray, V., Schneider, B., Herderich, M., and Paul, V., Staurosporine derivatives from the ascidian *Eudistoma toealensis* and its predatory flatworm *Pseudoceros* sp., *J. Nat. Prod.*, 62, 959, 1999.

110. Higa, T. and Scheuer, P. J., Thelepin, a new metabolite from the marine annelid *Thelepus setosus*, *J. Am. Chem. Soc.*, 96, 2246, 1974.

111. Pettit, G. R., Inoue, M., Kamano, Y., Herald, D. L., Arm, C., Dufresne, C., Christie, N. D., Schmidt, J. M., Doubek, D. L., and Krupa, T. S., Isolation and structure of the powerful cell growth inhibitor Cephalostatin 1, *J. Am. Chem. Soc.*, 110, 2006, 1988.

112. Pettit, G. R., Kamano, Y., Inoue, M., Dufresne, C., Boyd, M. R., Herald, C. L., Schmidt, J. M., Doubek, D. L., and Christie, N. D., Antineoplastic agents. 214. Isolation and structure of cephalostatins 7-9, *J. Org. Chem.*, 57, 429, 1992.

113. Winterfeldt, E., Marine natural products — synthetic exercises and biological activity, *Pure Appl. Chem.*, 71, 1095, 1999.

114. Fukuzawa, S., Matsunaga, S., and Fusetani, N., Ritterazine A, a highly cytotoxic dimeric steroidal alkaloid, from the tunicate *Ritterella tokioka*, *J. Org. Chem.*, 59, 6164, 1994.

115. Fukuzawa, S., Matsunaga, S., and Fusetani, N., Isolation and structure elucidation of ritterazines B and C, highly cytotoxic dimeric steroidal alkaloids, from the tunicate *Ritterella tokioka*, *J. Org. Chem.*, 60, 608, 1995.

116. Fukuzawa, S., Matsunaga, S., and Fusetani, N., Ten more ritterazines, cytotoxic steroidal alkaloids from the tunicate *Ritterella tokioka*, *Tetrahedron*, 51, 6707, 1995.

117. Fukuzawa, S., Matsunaga, S., and Fusetani, N., Isolation of 13 new ritterazines from the tunicate *Ritterella tokioka* and chemical transformation of ritterazine B., *J. Org. Chem.*, 62, 4484, 1997.

118. Higa, T., Fujiyama, T., and Scheuer, P. J., Halogenated phenol and indole constituents of acorn worms, *Comp. Biochem. Physiol.*, 65B, 525, 1980.

119. Higa, T., Ichiba, T., and Okuda, R. K., Marine indoles of novel substitution pattern from the acorn worm *Glossobalanus* sp. *Experientia*, 41, 1487, 1985.

120. Higa, T., Okuda, R. K., Severns, R. M., Scheuer, P. J., He, C.-H., Xu, C., and Clardy, J., Unprecedented constituents of a new species of acorn worm, *Tetrahedron*, 43, 1063, 1987.

121. King, G. M., Inhibition of microbial activity in marine sediments by a bromophenol from a hemichordate, *Nature*, 323, 257, 1986.

122. Jensen, P., Emrich, R., and Weber, K., Brominated metabolites and reduced numbers of meiofauna organisms in the burrow wall lining of the deep-sea enteropneust *Stereobalanus canadensis*, *Deep-Sea Res.*, A, 39, 1247, 1992.

123. Riccio, R., D'Auria, M. V., and Minale, L., Two new steroidal glycoside sulfates, longicaudoside-A and -B from the Mediterranean ophiuroid *Ophioderma longicaudum*, *J. Org. Chem.*, 51, 533, 1986.

124. D'Auria, M. V., Minale, L., and Riccio, R., Polyoxygenated steroids of marine origin, *Chem. Rev.*, 93, 1839, 1993.

125. Minale, L., Riccio, R., and Zollo, F., Steroidal oligoglycosides and polyhydroxysteroids from echinoderms, *Fortschritte der Chemie Organischer Naturstoffe*, 62, 74, 1993.

126. Stonik, V. A. and Elyakov, G. B., Secondary metabolites from echinoderms as chemotaxonomic markers, in *Bioorganic Marine Chemistry*, Scheuer, P. J., Ed., Springer-Verlag, Berlin, 2, 43, 1988.

127. Kong, F., Harper, M. K., and Faulkner, D. J., Fuscusine, a tetrahydroisoquinoline alkaloid from the sea star *Perknaster fuscus antarcticus*, *Nat. Prod. Lett.*, 1, 71, 1992.

128. Burgoyne, D. L., Miao, S., Pathirana, C., Andersen, R. J., Ayer, W. A., Singer, P. P., Kokke, W. C. M. C., and Ros, D. M., The structure and partial synthesis of imbricatine, a benzyltetrahydroisoquinoline alkaloid from the starfish *Dermasterias imbricata*, *Can. J. Chem.*, 69, 20, 1991.

129. Arndt, W. and Manteuffel, P., Die Turbellarien als Trager von Giften, *Zeitschrift fur Morphologie und Okologie der Tiere*, 11, 344, 1925.

130. Bacq, Z. M., Les poisons des Nemertiens, *Bulletin de l'Academie Royale de Belgique*, (5), 32, 1072, 1936.

131. Kem, W. R., Structure and action of nemertean toxins, *Am. Zool.*, 25, 99, 1985.

132. Davies-Coleman, M. T. and Garson, M. J., Marine polypropionates, *Nat. Prod. Rep.*, 15, 477, 1998.

133. Thollesson, M., Phylogenetic analysis of Euthyneura (Gastropoda) by means of the 16S rRNA gene: use of a 'fast' gene for 'higher-level' phylogenies, *Proc. R. Soc. Lond. B*, 266, 75, 1999.

134. Wollscheid, E. and Wagele, H., Initial results on the molecular phylogeny of the Nudibranchia (Gastropoda, Opisthobranchia) based on 18S rDNA data, *Mol. Phylogenetics Evol.*, 13, 215, 1999.

135. Manker, D. C. and Faulkner, D. J., The baconipyrones: novel polypropionates from the pulmonate *Siphonaria baconi*, *J. Org. Chem.*, 54, 5371, 1989.

136. Manker, D. C. and Faulkner, D. J., Vallartanones A and B, polypropionate metabolites of *Siphonaria maura* from Mexico, *J. Org. Chem.*, 54, 5374, 1989.

137. Manker, D. C. and Faulkner, D. J., Diterpenes from the marine pulmonate *Trimusculus reticulatus*, *Tetrahedron*, 43, 3677, 1987.

138. Ireland, C. M., Biskupiak, J. E., Hite, G. J., Rapposch, M., Scheuer, P. J., and Ruble, J. R., Ilikonapyrone esters, likely defense allomones of the mollusc *Onchidium verruculatum*, *J. Org. Chem.*, 49, 559, 1984.

139. Rodiguez, J., Riguera, R., and Debitus, C., The natural polypropionate-derived esters of the mollusc *Onchidium* sp., *J. Org. Chem.*, 57, 4624, 1992.

140. Biskupiak, J. E. and Ireland, C. M., Cytotoxic metabolites from the mollusc *Peronia peroni*, *Tetrahedron Lett.*, 26, 4307, 1985.

141. Ireland, C. and Faulkner, D. J., The defensive secretion of the opisthobranch mollusc *Onchidella binneyi*, *Bioorg. Chem.*, 7, 125, 1978.

142. Abramson, S. N., Radic, Z., Manker, D., Faulkner, D. J., and Taylor, P., Onchidal: a naturally occurring irreversible inhibitor of acetylcholinesterase with a novel mechanism of action, *Mol. Pharm.*, 36, 349, 1989.

143. Pelseneer, P., Report on the Pteropoda collected by H.M.S. Challenger during the years 1873–1876. Part III. Anatomy, *Challenger Rep.*, 23, 1, 1888.

144. Yoshida, W. Y., Bryan, P. J., Baker, B. J., and McClintock, J. B., Pteroenone: a defensive metabolite of the abducted Antarctic pteropod *Clione antarctica*, *J. Org. Chem.*, 60, 780, 1995.

145. McClintock, J. B., and Janssen, J., Pteropod abduction as a chemical defense in a pelagic antarctic amphipod, *Nature*, 346, 462, 1990.

146. Bryan, P. J., Yoshida, W. Y., McClintock, J. B., and Baker, B. J., Ecological role for pteroenone, a novel antifeedant produced by the conspicuous antarctic pteropod *Clione antarctica* (Gymnosomatia: Pteropoda), *Mar. Biol.*, 122, 271, 1995.

147. Ciavatta, M. L., Trivellone, E., Villani, G., and Cimino, G., Membrenones: new polypropionates from the skin of the Mediterranean mollusc *Pleurobranchus membranaceus*, *Tetrahedron Lett.*, 34, 6791, 1993.

148. Spinella, A., Mollo, E., Trivellone, E., and Cimino, G., Testudinariol A and B, two unusual triterpenoids from the skin and mucus of the marine mollusc *Pleurobranchus testudinarius*, *Tetrahedron*, 53, 16891, 1997.

149. Marin, A., Alvarez, L. A., Cimino, G., and Spinella, A., Chemical defense in cephalaspidea gastropods: origin, anatomical location and ecological roles, *J. Molluscan Stud.*, 65, 121, 1999.

150. Cimino, G., Sodano, G., and Spinella, A., New propionate-derived metabolites from *Aglaja depicta* and from its prey *Bulla striata* (opisthobranch molluscs), *J. Org. Chem.*, 52, 5326, 1987.

151. Spinella, A., Alvarez, L. A., and Cimino, G., Predator-prey relationship between *Navanax inermis* and *Bulla gouldiana*: a chemical approach, *Tetrahedron*, 49, 3203, 1993.

152. Szabo, C. M., Nakao, Y., Yoshida, W. Y., and Scheuer, P. J., Two diverse constituents of the cephalaspidean mollusk *Smaragdinella calyculata*, *Tetrahedron*, 52, 9681, 1996.

153. Matikainen, J., Kaltia, S., Hase, T., and Kuronen, P., The synthesis of haminols A and B, *J. Nat. Prod.*, 58, 1622, 1995.

154. Poiner, A., Paul, V. J., and Scheuer, P. J., Kumepaloxane, a rearranged trisnor sesquiterpene from the bubble shell *Haminoea cymbalum*, *Tetrahedron*, 45, 617, 1989.

155. Nakano, Y., Yoshida, W., and Scheuer, P. J., Pupukeamide, a linear tetrapeptide from a cephalaspidean mollusk *Philinopsis speciosa*, *Tetrahedron Lett.*, 37, 8993, 1996.

156. Suenaga, K., Kigoshi, H., and Yamada, K., Auripyrones A and B, cytotoxic polypropionates from the sea hare *Dolabella auricularia*: isolation and structures, *Tetrahedron Lett.*, 39, 5151, 1996.

157. Ciavatta, M. L., Gavagnin, M., Puliti, R., and Cimino, G., Dolabriferol: a new polypropionate from the skin of the anaspidean mollusc *Dolbarifera dolabrifera*, *Tetrahedron*, 52, 12831, 1996.

158. Ireland, C. and Scheuer, P. J., Photosynthetic marine mollusks: in vivo ^{14}C incorporation into metabolites of the sacoglossan *Placobranchus ocellatus*, *Science*, 205, 922, 1979.

159. Ireland, C., Faulkner, D. J., Solheim, B. A., and Clardy, J., Tridachione, a propionate-derived metabolite of the opisthobranch mollusc *Tridachiella diomedea*, *J. Am. Chem. Soc.*, 100, 1002, 1978.

160. Ireland, C. and Faulkner, D. J., The metabolites of the marine molluscs *Tridachiella diomedea* and *Tridachia crispata*, *Tetrahedron*, 37 (Suppl. 1), 233, 1981.

161. Ksebati, M. B. and Schmitz, F. J., Tridachiapyrones: propionate-derived metabolites from the sacoglossan mollusc *Tridachia crispata*, *J. Org. Chem.*, 50, 5637, 1985.

162. Gavagnin, M., Mollo, E., and Cimino, G., A new γ-dihydropyrone-propionate from the Caribbean sacoglossan *Tridachia crispata*, *Tetrahedron Lett.*, 37, 4259, 1996.

163. Gavagnin, M., Marin, A., Mollo, E., Crispino, A., Villani, G., and Cimino, G., Secondary metabolites from Mediterranean Elysioidea: origin and biological role, *Comp. Biochem. Physiol.*, 108B, 107, 1994.

164. Gavagnin, M., Spinella, A., Castelluccio, F., Cimino, G., and Marin, A., Polypropionates from the Mediterranean mollusk *Elysia timida*, *J. Nat. Prod.*, 57, 298, 1994.

165. Dawe, R. D. and Wright, J. L. C., The major polypropionate metabolites from the sacoglossan mollusc *Elysia chlorotica*, *Tetrahedron Lett.*, 27, 2559, 1986.

166. Vardaro, R. R., Di Marzo, V., and Cimino, G., Placidenes: cyercene-like polypropionate γ-pyrones from the Mediterranean sacoglossan mollusc *Placidia dendritica*, *Tetrahedron Lett.*, 33, 2875, 1992.

167. Di Marzo, V., Vardaro, R. R., De Petrocellis, L., Villani, G., Minei, R., and Cimino, G., Cyercenes, novel pyrones from the sacoglossan mollusc *Cyerce cristallina*. Tissue distribution, biosynthesis and possible involvement in defense and regenerative processes, *Experientia*, 47, 1221, 1991.

168. Vardaro, R. R., Di Marzo, V., Crispino, A., and Cimino, G., Cyercenes, novel polypropionate pyrones from the autotomizing Mediterranean mollusc *Cyerce cristallina*, *Tetrahedron*, 47, 5569, 1991.

169. Atkins, P. Chemistry: the great ideas, *Pure Appl. Chem.*, 71, 927, 1999.

170. Steneck, R. S., Plant-herbivore coevolution: a reappraisal from the marine realm and its fossil record, in *Plant–Animal Interactions in the Marine Benthos*, John, D. M., Hawkins, S. S., and Price, J. H., Eds., Clarendon Press, Oxford, 1992, 477.

171. Ghiselin, M. T., Darwin, progress, and economic principles, *Evolution*, 49, 1029, 1995.

172. Ghiselin, M. T., Progress and the economy of nature, *J. Bioecon.*, 1, 35, 1999.

173. Shimizu, Y., Recent progress in marine toxin research, *Pure Appl. Chem.*, 54, 1973, 1982.

174. Kitagawa, I., Kobayashi, M., Katori, T., Yamashita, M., Tanaka, J., Doi, M., and Ishida, T., Absolute stereostructure of swinholide A, a potent cytotoxic macrolide from the Okinawan marine sponge *Theonella swinhoei*, *J. Am. Chem. Soc.*, 112, 3710, 1990.

175. Lindel, T., Jensen, P. R., and Fenical, W., Lagunapyrones A-C: cytotoxic acetogenins of a new skeletal class from a marine sediment bacterium, *Tetrahedron Lett.*, 37, 1327, 1996.

176. Gustafson, K., Roman, M., and Fenical, W., The macrolactins, a novel class of antiviral and cytotoxic macrolides from a deep-sea marine bacterium, *J. Am. Chem. Soc.*, 111, 7519, 1989.

177. Yamada, K., Iijumi, Y., Tsuji, T., Yamada, K., Tomono, Y., and Uemura, D., Aburatobolactam, a novel inhibitor of superoxide anion generation from a marine microorganism, *Heterocyclic Comm.*, 2, 315, 1996.

178. Numata, A., Takahashi, C., Ito, Y., Minoura, K., Yamada, T., Matsuda, C., and Nomoto, K., Phenochalasins, a novel class of cytotoxic cytochalasans from a *Penicillium* species separated from a marine alga: structure determination and solution information, *J. Chem. Soc. Perk. Trans. I*, 239, 1995.

179. Olguin-Uribe, G., Abou-Mansour, E., Boulander, A., Debard, H., Francisco, C., and Combaut, G., 6-Bromoindole-3-carbaldehyde, from an *Acinetobacter* sp. bacterium associated with the ascidian *Stomozoa murrayi*, *J. Chem. Ecol.*, 23, 2507, 1997.

180. Numata, A., Amagata, T., Minoura, K., and Ito, T., Gymnastatins, novel cytoxic metabolites produced by a fungal strain from a sponge, *Tetrahedron Lett.*, 38, 5675, 1997.

181. Nagle, D. G., Paul, V. J., and Roberts, M. A., Ypaoamide, a noo broadly acting feeding deterrent from the marine cyanobacteriaum *Lyngbya majuscula*, *Tetrahedron Lett.*, 37, 7107, 1996.

182. Wu, M., Milligan, K.E., and Gerwick, W. H., Three new Malyngamides from the marine cyanobacterium *Lyngbya majuscula*, *Tetrahedron*, 53, 15983, 1997.

183. Amico, V., Marine brown algae of family Cystoseiraceae: chemistry and chemotaxonomy, *Phytochem.*, 39, 1257, 1995.

184. Durán, R., Zubia, E., Ortega, M. J., and Salvá, J., New diterpenoids from the alga *Dictyota dichotoma*, *Tetrahedron*, 53, 8675, 1997.

185. Suzuki, M., Takahashi, Y., Matsuo, Y., and Masuda, M., Pannosallene, a brominated C_{15} norterpenoid from *Laurencia pannosa*, *Phytochem.*, 41, 1101, 1996.

186. Cueto, M. and Darias, J., Uncommon tetrahydrofuran monoterpenes from Antarctic *Pantoneura plocamioides*, *Tetrahedron*, 52, 5899, 1997.

187. Konig, G. M. and Wright, A. D., *Laurencia ridida*: chemical investigations of its antifouling dichloromethane extract, *J. Nat. Prod.*, 60, 967, 1997.

188. Norte, M., Fernandez, J. J., Souto, M. L., and Garcia-Gravalos, M. D., Two new antitumor polyether squalene derivatives, *Tetrahedron Lett.*, 37, 2671, 1996.

189. N'Diaye, I., Guella, G., Mancini, I., and Pietra, F., Almazole D, a new type of antibacterial 2,5 disubstituted oxazolic dipeptide from a red alga of the coast of Senigal, *Tetrahedron Lett.*, 37, 3049, 1996.

190. Hamann, M. T. and Scheuer, P. J., Kahalalide F: a bioactive depsipeptide from the sacoglossan mollusk *Elysia rufescens* and the green alga *Bryopsis* sp., *J. Am. Chem. Soc.*, 115, 5825, 1993.

191. Rochfort, S. J., Watson, R., and Capon, R. J., Dictyosphaerin: a novel bicyclic lipid from a southern Australian marine green alga, *Dictyosphaeria seriacea*, *J. Nat. Prod.*, 59, 1154, 1996.

192. Koehn, F. E., Gunaserka, S. P., Niel, D. N., and Cross, S. S., Halitunal, an unusual diterpene aldehyde from the marine alga *Halimeda tuna*, *Tetrahedron Lett.*, 32, 169, 1991.

193. Amico, V., Oriente, G., Piattelli, M., and Tringali, C., Caulerpenyne, an unusual sesquiterpenoid from the green alga *Caulerpa prolifera*, *Tetrahedron Lett.*, 19, 3593, 1978.

194. Hay, M. E., Duffey, J. E., Paul, V. J., Renaud, P. E., and Fenical, W., Specialist herbivores reduce their susceptibility to predation by feeding on the chemically defended seaweed *Avrainvillea longicaulis*, *Limnol. Oceanogr.*, 35, 1734, 1990.

195. Ciminiello, P., Fattorusso, E., Forino, M., Magno, S., and Pansini, M., Chemistry of Verongida sponges VIII — bromocompounds from the Mediterranean sponges *Aplysina aerophoba* and *Aplysina cavernicola*, *Tetrahedron*, 53, 6565, 1997.

196. Burreson, B. J., Scheuer, P. J., Finer, J., and Clardy, J., 9-Isocyanopupukeanane, a marine invertebrate allomone with a new sesquiterpene skeleton, *J. Am. Chem. Soc.*, 97, 4763, 1975.

197. Charan, R. D., Garson, M. J., Brereton, I. M., Willis, A. C., and Hooper, J. N. A., Haliclonacyclamines A and B, cytotoxic alkaloids from the tropical marine sponge *Haliclona* sp., *Tetrahedron*, 52, 9111, 1996.

198. Urban, S. and Capon, R. J., Deoxyspongiaquinones: new sesquiterpene quoinones and hydroquinones from a southern Australian marine sponge *Euryspongia* sp., *Aust. J. Chem.*, 49, 611, 1996.

199. Cimino, G., De Stefano, S., Guerriero, A., and Minale, L., Furanosesquiterpenoids in sponges IV. Microcionins from *Microciona toxystila*, *Tetrahedron Lett.*, 3723, 1975.

200. Kohmoto, S., McConnell, O. J., Wright, A., and Cross, S., Isospongiadiol, a cytotoxic and antiviral diterpene from sponge, *Spongia* sp., *Chem. Lett.*, 1687, 1987.

201. Lenis, L., Nunez, L., Jimenez, C., and Riguera, R., Isonitenin and acetylhomoagmatine, new metabolites from the sponges *Spongia officinalis* and *Cliona celata* collected at the Galician coast (N.W. Spain), *Nat. Prod. Lett.*, 8, 15, 1996.

202. Cimino, G., De Stefano, S., and Minale, L., Scalaradial, a third sesterterpene with the tetracarbocyclic skeleton of scalarin, from the sponge *Cacospongia mollior*, *Experientia*, 30, 846, 1974.

203. Fu, X., Abbas, A. S., Schmitz, F. J., Vidavsky, I., Gross, M. L., Laney, M., Schatzman, M., and Cabuslay, R. D., New acetylinic metabolites from the marine sponge *Pellina triangulata*, *Tetrahedron*, 53, 799, 1997.

204. Iwashima, M., Watanabe, K., and Iguche, K., New marine prostanoids, preclavulone lactones, from the Okinawan soft coral *Clavularia viridis*, *Tetrahedron Lett.*, 38, 8319, 1997.

205. Schmitz, F. J., Kraus, K. W., Ciereszko, L. S., Sifford, D. H., and Weinheimer, A. J., Ancepsenolide, a novel bisbutenolide from the gorgonian *Pterogorgia anceps* (Pallas) (synonymous with *Xiphogorgia anceps*), *Tetrahedron Lett.*, 97, 1966.

206. Duh, C. Y. and Hou, R. S., Cytotoxic cembranoids from the soft corals *Sinularia gibberosa* and *Sarcophyton trocheilophorum*, *J. Nat. Prod.*, 59, 595, 1996.

207. Lindquist, N., Lobkovsky, E., and Clardy, J., Tridentatols A-C, novel natural products of the marine hydroid *Tridentata marginata*, *Tetrahedron Lett.*, 37, 9131, 1996.

208. Rodriguez, A. D. and Boulanger, A., New guaiane metabolites from the Caribbean gorgonian coral, *Pseudopterogorgia americana*, *J. Nat. Prod.*, 60, 207, 1997.

209. Rahbaek, L., Anthoni, U., Christophersen, C., Nielsen, P. H., and Petersen, B. O., Marine alkaloids. 18. Securamines and securines, halogenatd indole-imidazole alkaloids from the marine bryozoan *Securiflustra securifrons*, *J. Org. Chem.*, 61, 887, 1996.

210. Pettit, G. R., Gao, F., Blumberg, P. M., Herald, C. L., Coll, J. C., Kamano, Y., Lewin, N. E., Schmidt, J. M., and Chapuis, J. C., Antineoplastic agents. 340. Isolation and structural elucidation of bryostatins 16-18, *J. Nat. Prod.*, 59, 286, 1996.

211. D'Ambrosio, M., Guerriero, A., and Pietra, F., Glycerol enol ethers of the brachiopod *Gryphus vitreus* from the Tuscan archipelago, *Experientia*, 52, 624, 1996.

212. McDonald, L. A., Capson, T. L., Krishnamurthy, G., Ding, W. D., Ellestad, G. A., Bernan, V. S., Maiese, W. M., Lassota, P., Discafani, C., Kramer, R. A., and Ireland, C. M., Namenamicin, a new enediyne antitumor antibiotic from the marine ascidian *Polysyncraton lithostrotum*, *J. Am. Chem. Soc.*, 118, 10898, 1996.

213. Carroll, A. R., Feng, Y., Bowden, B. F., and Coll, J. C., Studies of Australian ascidians. 5. Virenamides A-C, new cytotoxic linear peptides from the colonial didemnid ascidian *Diplosoma virens*, *J. Org. Chem.*, 61, 4059, 1996.

214. Schmitz, F. J., De Guzman, F. S., Hossain, M. B., and van der Helm, D., Cytotoxic aromatic alkaloids from the ascidian *Amphicarpa meridiana* and *Leptoclinides* sp.: meridine and 11-hydroxyascididemnin, *J. Org. Chem.*, 56, 804, 1991.

215. Carroll, A. R., Coll, J. C., Bourne, D. J., MacLeod, J. K., Zabriskie, T. M., Ireland, C. M, and Bowden, B. F., Patellins 1-6 and trunkamide A: novel hexa-, hepta- and octa-peptides from colonial ascidians, *Lissoclinum* sp., *Aust. J. Chem.*, 49, 659, 1996.

216. Paul, V. J., Lindquist, N., and Fenical, W., Chemical defenses of the tropical ascidian *Atapozoa* sp. and its nudibranch predators *Nembrotha* spp., *Mar. Ecol. Prog. Ser.*, 59, 109, 1990.

217. Debitus, C., Laurent, D., and Pais, M., Alcaloïdes d'une ascidie neocaledonienne, *Eudistoma fragum*, *J. Nat. Prod.*, 51, 799, 1988.

218. Fu, X., Hossain, M. B., Schmitz, F. J., and van der Helm, D., Longithorones, unique prenylated para- and meta-cyclophane type quinones from the tunicate *Aplidium longithorax*, *J. Org. Chem.*, 62, 3810, 1997.

219. Kubanek, J., Williams D. E., de Silva, E. D., Allen, T., and Andersen, R. J., Cytotoxic alkaloids from the flatworm *Prostheceraeus villatus* and its tunicate prey *Clavelina lepadiformis*, *Tetrahedron Lett.*, 36, 6189, 1995.

220. Davis, R. A., Carroll, A. R., Pierens, G. K., and Quinn, R. J., New lamellarin alkaloids from the Australian ascidian, *Didemnum chartaceum*, *J. Nat. Prod.*, 62, 419, 1999.

221. Carroll, A. R. and Scheuer, P. J., Kuanoniamines A, B, C, and D: pentacyclic alkaloids from a tunicate and its prosobranch mollusk predator *Chelynotus semperi*, *J. Org. Chem.*, 55, 4426, 1990.

222. Higa, T. and Scheuer, P. J., Constituents of the marine annelid *Thelepus setosus*, *Tetrahedron*, 31, 2379, 1975.

223. Pettit, G. R., Xu J., Ichihara, Y., Williams, M. D., and Boyd, M. R., Antineoplastic agents 285. Isolation and structures of cephalostatins 14 and 15, *Can. J. Chem.*, 72, 2260, 1994.

224. Roccatagliata, A. J., Maier, M. S., Seldes, A. M., Iorizzi, M., and Minale, L., Starfish saponins, part 2. Steroidal oligoglycosides from the starfish *Comasterias lurida*, *J. Nat. Prod.*, 57, 747, 1994.

225. Iorizzi, M., de Riccardis, F., Minale, L., Palagiano, E., Riccio, R., Debitus C., and Duhet, D., Polyoxygenated marine steroids from the deep water starfish *Styracaster caroli*, *J. Nat. Prod.*, 57, 1361, 1994.

226. Kong, F., Harper, M. K., and Faulkner, D. J., Fuscusine, a tetrahydroisoquinoline alkaloid from the sea star *Perknaster fuscus antarcticus*, *Nat. Prod. Lett.*, 1, 71, 1992.

227. Hochlowski, J. E., Faulkner, D. J., Matsumoto, G. K., and Clardy, J., The denticulatins, two propionate metabolites from the pulmonate *Siphonaria denticulata*, *J. Am. Chem. Soc.*, 105, 7413, 1983.

228. Ireland, C. and Faulkner, D. J., The defensive secretion of the opisthobranch mollusc *Onchidella binneyi*, *Bioorg. Chem.*, 7, 125, 1978.

229. Spinella, A., Mollo, E., Trivellone, E., and Cimino, G., Testudinariol A and B two unusual triterpenoids from the skin and mucus of the marine mollusc *Pleurobranchus testudinarius*, *Tetrahedron*, 53, 16891, 1997.

230. Spinella, A., Alvarez, L. A., and Cimino, G., Predator-prey relationship between *Navanax inermis* and *Bulla gouldiana*: a chemical approach, *Tetrahedron*, 49, 3203, 1993.

231. Okino, T., Yoshimura, E., Hirota, H., and Fusetani, N., New antifouling sesquiterpenes from four nudibranchs of the family Phyllidiidae, *Tetrahedron*, 52, 9447, 1996.

232. Miyamoto, T., Sakamoto, K., Arao, K, Komori, T., Higuchi, R., and Sasaki, T., Dorisenones, cytotoxic spongian diterpenoids, from the nudibranch *Chromodoris obsoleta*, *Tetrahedron*, 52, 8187, 1996.

233. Ciavatta, M. L., Gavagnin, M., Puliti, R., Cimino, G., Martinez, E., Ortega, J., and Mattia, C. A., Dolabriferol: a new polypropionate from the skin of the anaspidean mollusc *Dolabrifera dolabrifera*, *Tetrahedron*, 52, 12831, 1996.

234. Szabo, C. M., Nakao, Y., Yoshida, W. Y., and Scheuer, P. J., Two diverse constituents of the cephalaspidean mollusk *Smaragdinella calyculata*, *Tetrahedron*, 52, 9681, 1996.

235. Chou, T., Haino, T., Kuramoto, M., and Uemura, D., Isolation and structure of pinnatoxin D, a new shellfish poison from the Okinawan bivalve *Pinna muricata*, *Tetrahedron Lett.*, 37, 4027, 1996.

236. Dawe, R. D. and Wright, J. L. C., The major polypropionate metabolites from the sacoglossan mollusc *Elysia chlorotica*, *Tetrahedron Lett.*, 27, 2559, 1986.

237. Vardaro, R. R., Di Marzo, V., Crispino, A., and Cimino, G., Cyercenes, novel polypropionate pyrones from the autotomizing Metiterranean mollusc *Cyerce cristallina*, *Tetrahedron*, 47, 5569, 1991.

238. Mutou, T., Kondo, T., Ojika, M., and Yamada, K., Isolation and stereostructures of dolastatin G and nordolastatin G, cytotoxic 35-membered cyclodepsipeptides from the Japanese sea hare *Dolabella auricularia*, *J. Org. Chem.*, 61, 6340, 1996.

239. Cimino, G., Fontana, A., Giminez, F., Marin, A., Mollo, E., Trivellone, E., and Zubia, E., Biotransformation of a dietary sesterterpenoid in the Mediterranean nudibranch *Hypselodoris orsini*, *Experientia*, 49, 582, 1993.

240. Graziani, E. I., Andersen, R. J., Krug, P. J., and Faulkner, D. J., Stable isotope incorporation evidence for *de novo* biosynthesis of terpenoic acid glycerides by dorid nudibranchs, *Tetrahedron*, 52, 6869, 1996.

241. Fontana, A., Civatta, M. L., Miamoto, T., Spinella, A., and Cimino, G., Biosynthesis of drimane terpenoids in dorid molluscs: pivotal role of 7-deacetoxyolepupuane in two species of *Dendrodoris* nudibranchs, *Tetrahedron*, 55, 5937, 1999.

242. Kubanek, J., Graziani, E. I., and Andersen, R. J., Investigations of terpenoid biosynthesis by the dorid nudibranch *Cadlina luteomarginata*, *J. Org. Chem.*, 62, 7239, 1997.

Section II

Organismal Patterns
in Marine Chemical Ecology

4 Chemical Ecology of Mobile Benthic Invertebrates: Predators and Prey, Allies and Competitors

John J. Stachowicz

CONTENTS

I. INTRODUCTION

The diversity of topics addressed in this volume attests to the fact that marine chemical ecology is more than just animals and plants producing chemicals that deter predation. Chemicals are involved in mediating a diverse array of inter- and intraspecific interactions including predation, competition, mutualism, and reproductive processes, as well as interactions between organisms and their physical environment. This diversity is best exemplified in the mobile invertebrates. Mobile invertebrates are the dominant predators and herbivores in many marine systems and serve as "keystone" species in several of these systems. Thus, factors (including chemistry) that determine their distribution, abundance, and impact on communities and ecosystems should be of broad interest to marine biologists and ecologists. Straightforward production of predator-deterrent chemicals is rare in this group as compared to sessile invertebrates and seaweeds, and this has led the ecologists and chemists studying these organisms to diversify in terms of the types of interactions they study. Waterborne chemicals help mobile invertebrates locate food, mates, and appropriate habitats or symbiotic partners; they also help regulate and synchronize reproductive cycles and alert organisms to the danger of nearby predators. Nevertheless, the bulk of research on chemically mediated interactions has focused on predator–prey interactions, so much of the chapter is necessarily devoted to these interactions. In areas where rigorous studies involving mobile benthic invertebrates are rare (e.g., antifouling and allelopathy), examples from other groups (plants, sessile invertebrates, or verte- brates) or habitats (open water marine, freshwater, or terrestrial) are provided to identify areas deserving increased attention. More detailed treatments of particular types of interactions or habitats can be found in the other chapters of this volume.

Several excellent reviews currently exist on particular aspects of marine chemical ecology,[1–6] so this chapter does not attempt to provide a comprehensive or historic overview, but rather tries to provide a sound conceptual discussion of the diversity and importance of chemically mediated interactions involving mobile invertebrates. Due to space constraints, not all relevant studies can be included, and recent studies are sometimes cited in favor of more classical work, as these provide similar conceptual information but often use more advanced methodologies and provide greater access to other literature on the topic. Where possible, this chapter highlights studies that assess the importance of chemically mediated interactions within the broader context of ecology and evolutionary biology.

II. CHEMICAL MEDIATION OF PREDATOR–PREY INTERACTIONS

Both primary and secondary metabolites from marine organisms play an important role in mediating all phases of predator–prey interactions, from defending prey against detection and attack to helping predators locate prey from a distance and subdue it once it is captured.

A. PREY DEFENSES AGAINST PREDATORS

Although relatively few mobile invertebrates produce their own defensive compounds, many more use the defensive compounds produced by other organisms, either by physiologically sequestering them from their prey, or by developing commensal or mutualistic associations with other chem- ically unpalatable organisms (see Section IV.B). Additionally, some animals use waterborne cues to detect the presence of predators and adjust their behavior and use of refuges to minimize the risk of detection.

1. *De Novo* Production

As with sessile animals and plants (see other chapters, this volume), the chemical deterrence of mobile invertebrates is best assessed using an approach in which ecologically relevant consumers are offered palatable food items with chemical extracts coated on, or embedded within, them.[7] Assays in which the toxicity of compounds is assessed by dissolving them in the water containing the assay organisms have been repeatedly shown to bear no relation to the effects of compounds when ingested with prey.[1,8,9] Most feeding deterrents of mobile invertebrates appear to be lipid-soluble, thus these assays should not encounter problems with compounds dissolving or leaching into the water, and extract or compound concentration can be carefully controlled. Several investigators have found minimal loss of lipophilic extracts from test foods during the duration of a bioassay.[1,10] Given the long retention time of the few compounds or extracts that have been evaluated and the similar solubility characteristics of many marine secondary metabolites, this general methodology can probably be used with most nonpolar, lipid-soluble metabolites. Methods for assaying the feeding deterrent properties of marine organisms have recently been critically reviewed,[7] and interested readers should consult that paper.

Using these methodologies, chemical defenses against predation have been reported from sea spiders, echinoderms, and molluscs. However, compared to sessile invertebrates[4] and seaweeds (see Chapter 6 in this volume), relatively few mobile invertebrates appear to produce their own chemical feeding deterrents. Although this may be due in part to phylogenetic constraints, mobile invertebrates also have a broader array of behavioral defenses, including flight, aggression, and avoidance of predators by restricting activity to periods when predators are less active. Not surprisingly, then, chemical defenses among the mobile invertebrates appear most common among groups that lack obvious morphological or behavioral mechanisms of defenses. For example, shell-less gastropods, including nudibranchs, sea hares, and ascoglossans (sacoglossans), are often supposed to elaborate some form of chemical defense.[4,11] Although many of these animals obtain their chemical defense from their prey either directly or in an altered form (Section II.B.2), a few dorid nudibranchs and sacoglossans are known to produce the deterrent chemicals *de novo*.[12–18] In the first example of *de novo* synthesis of chemical defenses by a dorid nudibranch, Cimino et al.[12] noted that polygodial (Structure 4.1), a defensive compound isolated from the dorid nudibranch *Dendrodoris limbata*, was not present in the sponges on which the animal fed. Using radiolabeling techniques, the authors demonstrated that the nudibranch produces deterrent chemicals not directly derived from its diet. Several other dorid nudibranchs appear to be capable of synthesizing sesquiterpenoids, diterpenoids, and sesterterpenoids that are effective feeding deterrents, but only a few have been demonstrated to employ both sequestration and *de novo* synthesis.[16,19] Species with *de novo* synthesis are freed from the constraints of specialization on a chemically defended food in order to obtain defensive compounds and are thus able to exploit a broader taxonomic range of food items.[18] Cimino and Ghiselin[18] have suggested that in some cases, *de novo* synthesis may evolve retrospectively from sequestration rather than independently, as enzymes and biochemical pathways originally employed in detoxification and sequestration are modified to synthesize compounds originally derived from the diet. The exciting possibility of unraveling the evolutionary history of chemical defenses in this group (and other groups) may benefit from collaborations with the emerging field of molecular phylogenetics.

Some shelled gastropods do produce chemical defenses, although this is far less common. One South African limpet, *Siphonaria capensis*, occurs at very high densities on rocky shores, apparently protected from predators by chemical feeding deterrents. These animals are rarely consumed relative to *Patella granularis* (a similar limpet that lacks defensive chemistry) and exude a repellent mucus onto the surface of their shell when attacked. Nonpolar extracts from *Siphonaria* confer resistance from predation to *Patella* when they are coated on its shell.[20] Because the metabolites responsible for the chemical defense have not been fully isolated and characterized, it is still unclear whether the compounds that confer resistance to predation in *Siphonaria* are diet derived or synthesized *de novo*.

Chemical defenses are less commonly reported in other groups of mobile marine invertebrates, but they may exist. Heine et al.[21] showed that a common Antarctic nemertean worm is rejected as prey by co-occurring fishes despite the lack of obvious structural defenses. The unpalatability has been attributed to a highly acidic mucus coating (pH 3.5), although toxic peptides were also present[22] and are thought to serve a defensive function in other nemertean worms.[23] However, rigorous experimental data in support of a defensive function for these peptides are generally lacking.

Despite their diversity, and in contrast to their terrestrial counterparts, examples of the presence of either diet-derived or *de novo* production of defensive chemicals among marine arthropods are rare. However, several studies provide evidence that suggests no chemical defense in this group. A pinnotherid pea crab has been shown to be unpalatable to mummichogs that consumed similar-sized blue crabs, although it is unclear whether this defense is chemical or structural in nature.[24] Several marine amphipods have bright coloration that has been thought to function as warning coloration,[25] but rigorous bioassays to determine whether these species are chemically unpalatable have yet to be reported. In the one example where chemical components of a marine arthropod have been shown to deter predation by ecologically realistic predators at natural concentrations, ecdysteroids (Structures 4.2 and 4.3) protected a pycnogonid sea spider from predation by green crabs.[26] These compounds serve a normal function as a molting hormone,[27] but were present in all developmental stages, including nonmolting stages. Additionally, concentrations were much higher than normally required for the induction of molting, suggesting their alternative function of predator deterrence. In general, secondary metabolites isolated from marine arthropods have not been shown to deter feeding by ecologically relevant predators.

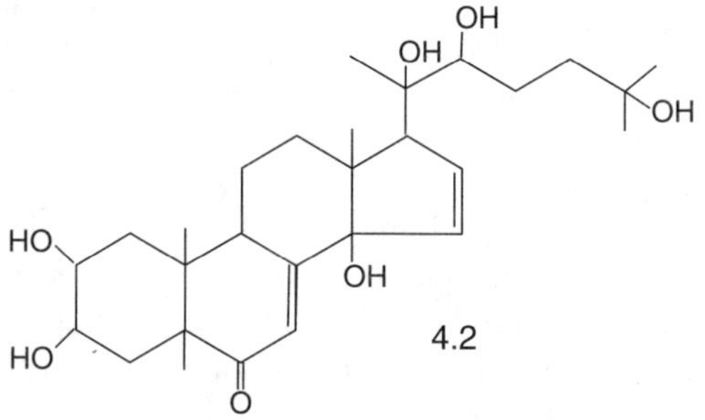

4.2

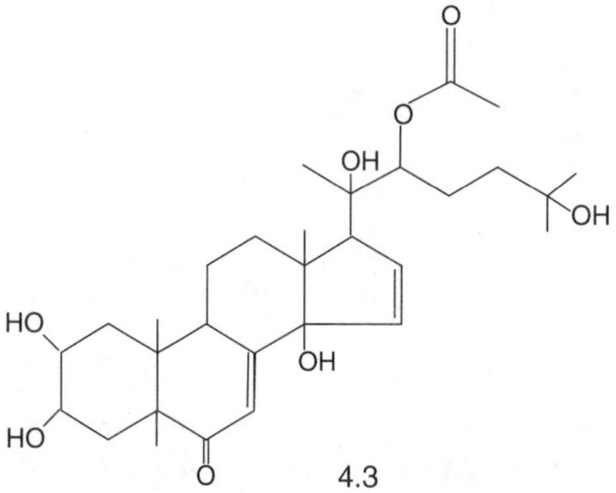

4.3

A survey of the frequency of chemical defense in echinoderms from the Gulf of Mexico found that a number of asteroids (10/12 species examined) and ophiuroids (3/3 species) echinoderms contained deterrent chemicals within their body walls.[28] Although the specific chemicals responsible for deterrence among the echinoderms have only rarely been isolated and characterized, crude chemical extracts varied in their effectiveness against different predators. Many extracts deterred feeding by the pinfish (*Lagodon rhomboides*), while fewer extracts were effective against predation by a majid crab (*Stenorhynchus seticornis*), mirroring the differences in susceptibility to algal chemical defenses observed in large, mobile (fishes) vs. small, sedentary (amphipods, crabs) herbivores (Section II.B.1).[29,30]

2. *Sequestration of Diet-Derived Defensive Compounds*

Although relatively few mobile invertebrates produce their own defensive chemicals, many more are able to physiologically sequester defensive compounds from their prey. The opisthobranch molluscs, in particular, offer a diversity of examples in which species with highly specialized diets obtain chemical defenses directly from their prey; for example, dorid nudibranchs from sponges and sea hares and sacoglossans from red algae. The evolutionary progression of shell loss in this group has been hypothesized to be the result of the deployment of diet-derived defensive compounds that rendered a hard shell obsolete for defense.[31] Several reviews[4,11,17,18,32] describe the mechanisms behind physiological sequestration. Described here are several taxonomically diverse examples in which both the ecological *and* chemical aspects of the interaction have been particularly well characterized.

Dorid nudibranchs feed almost exclusively on sponges and commonly sequester sponge-produced defensive compounds. For example, the Spanish dancer nudibranch, *Hexabranchus sanguineus*, feeds on sponges in the genus *Halichondria* which produce oxazole-containing macrolides that deter feeding by fishes.[33] The nudibranch sequesters halichondramide (Structure 4.4), alters it slightly (Structure 4.5), and concentrates these compounds in its dorsal mantle and egg masses where they serve as a potent defense against consumers. Concentrations of the defensive compounds are lowest in the sponge, higher in the nudibranch, and highest in the nudibranch egg masses, but even the lowest natural concentrations strongly deter feeding by fishes.

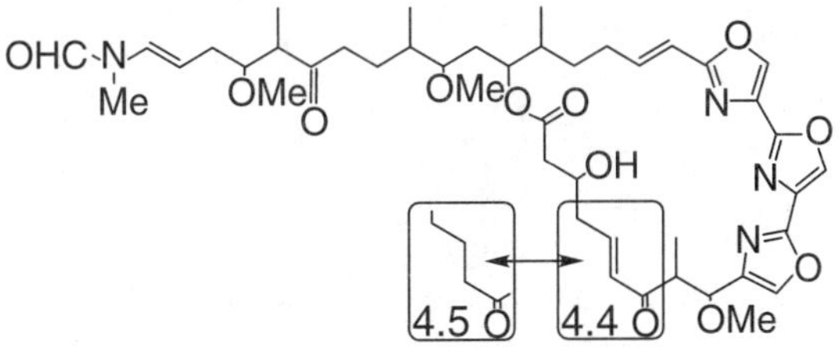

As mentioned previously, compounds are often not sequestered uniformly throughout the body tissue. For example, many dorid nudibranchs accumulate sequestered compounds along the mantle border.[33–35] In some cases, to avoid autotoxicity, inactive precursor compounds are stored in the digestive gland and are converted to the toxic form and transferred to the mantle border where they may be more effective deterrents.[34] Although it has been hypothesized that such localization of compounds is important for chemical defense, there is little experimental evidence in support of this. On Guam, the nudibranch *Glossodoris pallida* sequesters defensive compounds from its sponge prey, localizing them in mantle dermal formations (MDF) on the surface of the animal.[35] In the most direct test available to date, removal of these tissues of locally high concentration of defensive compounds increased the palatability of these animals to predation by fishes and crabs, but assays with artificial foods showed no difference in palatability of foods with localized vs. uniform concentrations of metabolites.[35] Thus, a high, localized concentration of chemicals in the MDFs was no more effective at reducing predation than lower, uniform levels. However, localization of compounds in the surface tissues of the mantle may facilitate excretion of compounds into mucus on the surface of the animal, enhancing the effectiveness of the defense. Alternatively, such localization may serve nondefensive functions such as sequestration of noxious compounds away from vital internal organs and avoidance of autotoxicity.[35] The causes and consequences of within-individual variation in the concentration of defensive compounds should provide an area of research worthy of further consideration.

Ascoglossan sea slugs (Sacoglossa) feed suctorially on marine algae and sequester functional chloroplasts from their prey in the tissues of their mantle.[14,36] Additionally, these animals often store sequester seaweed secondary metabolites for defense against predation.[37–39] In some instances, precursors to defensive compounds that are obtained from prey and converted to more deterrent compounds prior to deployment. For example, *Elysia halimedae* obtains halimedatetraacetate (Structure 4.6) from *Halimeda macroloba*, reduces the aldehyde group on this compound into the corresponding alcohol (Structure 4.7), and uses this compound in its own defense.[37] Some ascoglossans use fixed carbon from sequestered chloroplasts to produce their own defensive compounds.[14,15,17]

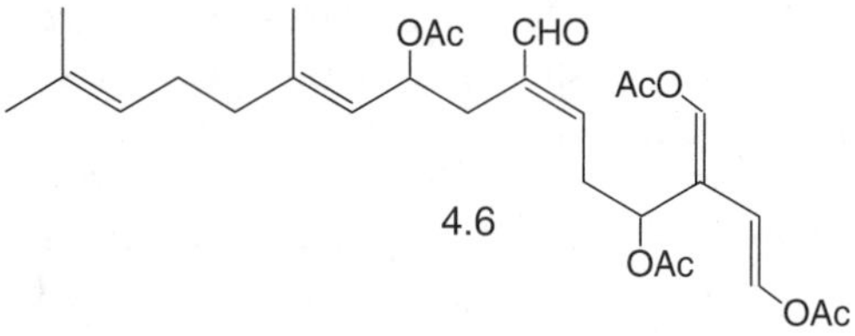

4.6

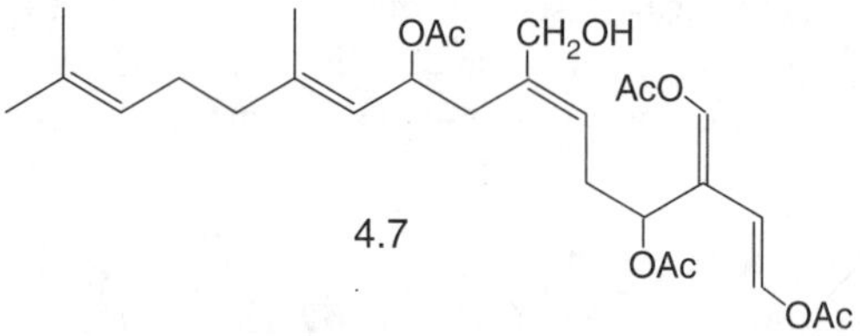

4.7

Sea hares (order Anaspidea) have been repeatedly shown to sequester metabolites that defend seaweeds from generalist herbivores[11,40–43] and may use a combination of diet-derived and *de-novo*-produced compounds for defense.[44] In contrast to the strategic location of compounds in the mantle border by nudibranchs and sacoglossans, sequestered compounds appear most concentrated internally, in the digestive gland of sea hares. This suggests that the accumulation of these compounds in sea hares may be a simple consequence of the detoxification of ingested metabolites rather than an adaptation to reduce predation.[42] In *Dolabella auricularia*, for example, whole body extracts deter predators, but this pattern is due almost entirely to the unpalatability of the digestive gland, as feeding assays with other body tissues and their extracts showed no effect on palatability.[42] However, some sea hares do contain sequestered algal metabolites in the skin and surface tissues at concentrations that are deterrent to predators.[43] Many opisthobranchs also secrete copious amounts of diet-derived compounds into their egg masses, which is often thought to render them unpalatable to generalist predators, although rigorous evidence for this is rare. For example, egg masses from the sea hare *Aplysia juliana* are chemically unpalatable to reef fishes, but diet-derived metabolites do not appear to be the cause of this unpalatability.[42]

Sea hares not only sequester compounds from their algal prey into their body tissues, but also produce copious amounts of "ink" that has (largely through anecdotal evidence) been postulated to serve a defensive function.[44] These animals are generally too slow moving to use the ink cloud as a "smoke screen" to escape all but the most sedentary predators (e.g., anemones), but noxious chemicals in the ink cloud could stun or repel more mobile predators. These hypotheses have been tested by manipulating ink production by the sea hare *Aplysia californica* by altering the diet; *Aplysia* fed red algae (*Gracilaria* sp.) produce copious amounts of ink, whereas individuals fed green algae (*Ulva* sp.) do not.[45] When ensnared in the tentacles of sea anemones (*Anthopleura xanthogrammica*, a natural predator of *Aplysia* in Pacific coast tide pools), red-algal fed sea hares released ink, causing the anemone to release it unharmed; green-algal fed individuals did not release ink and were readily consumed. Similar amounts of ink applied to inkless (green-algal fed) *Aplysia* when trapped by sea anemones caused these otherwise palatable animals to be rejected as prey.[45]

Aplysia with chemical defenses in their tissues but without ink were consumed at a similar rate to those without toxins (20% vs. 12%), whereas those without toxins in their tissues, but with ink exhibited much greater survival (71%), suggesting that the excretion of ink may be the primary defense of these sea hares when being consumed by slow-moving predators like sea anemones.[45] Pennings[42] also found that ink from some (but not all) sea hares could deter predators, however he found no evidence that the metabolites responsible for defense were diet derived. Sea hares also secrete opaline when attacked by predators, although the function of this secretion has yet to be unambiguously determined.[44]

As a final example of the physiological sequestration of defensive chemicals from prey items, some authors have argued that the enhanced concentration of toxins from marine phytoplankton that accumulate in filter feeding bivalves should be considered a form of sequestration of chemical defenses. However, many, if not most, filter feeders are harmed by the ingestion of these toxins,[46] so any benefit of reduced predation levels may be outweighed by costs. Yet some bivalves are particularly resistant to phytoplankton toxins like those that cause paralytic shellfish poisoning (PSP). For example, some species of butter clam (*Saxidomus*) are 1 to 2 orders of magnitude more resistant to the effects of saxitoxin (Structure 4.8) produced by red-tide forming dinoflagellates than other co-occurring bivalves.[47] These clams sequester the toxins for up to two years in their siphon, the most exposed part of the animal and thus the most vulnerable tissue to predation, and use these sequestered toxins as a chemical defense against predation by siphon-nipping fishes.[48] Because they were less susceptible to siphon nipping, clams containing saxitoxin consistently extended their siphons further into the water column, presumably increasing their access to food.

4.8

These sequestered toxins were effective deterrents against a range of potential clam predators, including sea otters.[49] Otters are historically rare in areas where toxic phytoplankton blooms are common, but are present where these blooms have been rare, so the sequestration of phytoplankton defenses by bivalves may limit the distribution of this important predator. Otters are also voracious predators of sea urchins, which can reach high numbers in the absence of otters, and devastate kelp beds through their grazing activities.[50,51] Thus, in the absence of otters, ecosystem structure and function are altered dramatically, so the sequestration of toxins from phytoplankton may dramatically alter nearshore communities like kelp beds through indirect effects on keystone species such as sea otters.[49] However, because areas prone to blooms of toxic phytoplankton may also be more subject to degradation by humans, including loading of nutrients and pollutants, a causal link between red tides and kelp forest health may be difficult to conclusively demonstrate.

3. Predator Detection and Avoidance

There are three main types of chemical cues that prey use as warnings of the threat of predation: (1) those actively released by conspecifics that can serve as warning signals, (2) those released passively when prey tissue is damaged, and (3) odors released directly by predators. Much of this work has involved aquatic vertebrates (fishes,[52–56] amphibians,[57] and also freshwater algae[58]), which often use chemical cues released by conspecifics injured by predators as an alarm signal and take appropriate predator avoidance measures such as hiding or reducing movement. However, a diverse array of mobile marine invertebrates appear to exhibit similar responses to the presence of injured or stressed conspecifics.[59–62]

Chemicals released when organisms are attacked can serve to mark a location as dangerous to conspecifics. Many gastropods leave a slime trail behind them as they move, which can allow for easier location by conspecifics in search of mates. However, when sufficiently molested (as by a predator), *Navanax inermis* secretes a mixture of bright yellow chemicals (navenones A–C; Structures 4.9–4.11) into its slime trail, which causes an avoidance response in trail-following conspecifics.[60] Bioassay-guided fractionation of snail slime indicated that the navenones were responsible for the trail-breaking behavior and may be used as a warning cue by conspecifics. Other opisthobranchs can take advantage of species-specific chemicals in the slime trails and use them to track prey. The nudibranch *Tambja abdere* sequesters compounds (the tambjamines; see Structures 4.12–4.15) from its bryozoan prey that serve as deterrents against predation by fishes, and secretes these compounds in low amounts into the slime trail.[63] The predatory nudibranch *Roboastra tigris* preys on *Tambja* and uses the low concentrations of tambjamines in the slime trail to locate its prey. However, when attacked by *Roboastra*, *Tambja* secretes a mucus containing a higher concentration of tambjamines, causing *Roboastra* to break off the attack. This particular study highlights how the function of compounds can be altered not only by changes in structure, but also by changes in concentration: the tambjamines attracted predators at low concentrations, but repelled them at higher levels.[63]

4.9

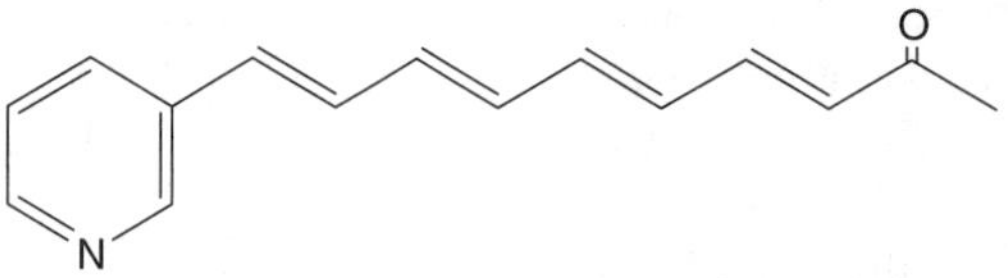

4.10

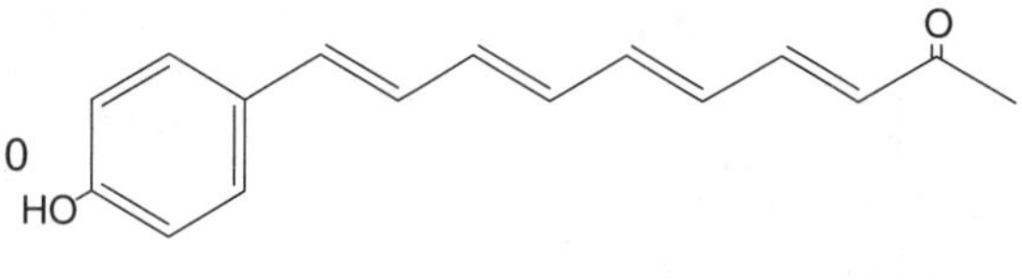

4.11

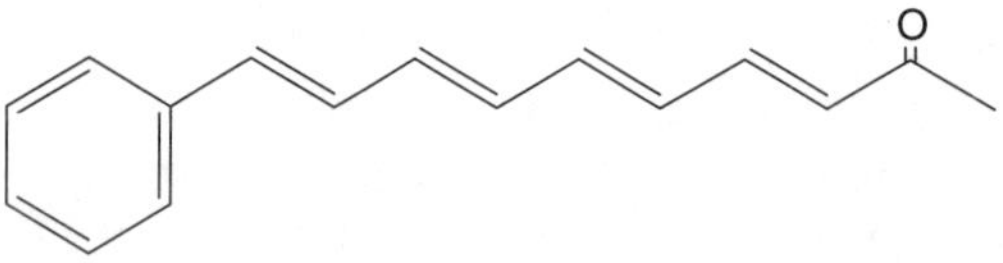

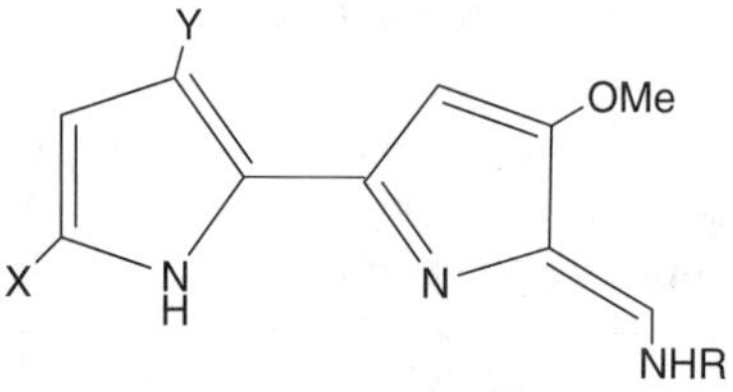

4.12 X = H Y = H R = H
4.13 X = Br Y = H R = H
4.14 X = H Y = H R = *i*-Bu
4.15 X = H Y = Br R = *i*-Bu

The slime trail examples of alarm pheromones offer systems that are relatively tractable experimentally, since chemical cues are bound to the substrate in the mucus. More frequently, chemicals involved in detection of danger are waterborne, posing significant challenges to investigators, including accurate reproduction and characterization of the stimulus and the effects of moving water on the dispersal of chemical signals.[7,64,65] This is not a trivial point given that even moderate turbulence can have a substantial effect on chemical concentrations and the spatial distribution of an odor plume,[66] and thus an organism's ability to locate an odor source.[65,67] Investigators in the lab have attempted to mimic natural field conditions using flumes (see Section II.C.1 for a more complete description of these methodological issues). As an example, an experiment with an intertidal marine gastropod used a flume with some vertical drop to mimic the organism's intertidal habitat and tested the effects of the chemical scent of both predators and injured conspecifics on foraging behavior.[62] Gastropod activity was reduced by odors from crushed conspecifics or from crushed conspecifics and crabs (predators). Additionally, more snails sought refuge out of the water in the presence of these cues. However, when gastropods were starved, the predator and injured conspecific cues had no effect on snail behavior, suggesting that physiological state and diet history of the prey organisms may alter their willingness to take risks. In addition to their ecological importance, these findings also suggest that the common procedure of starving test organisms before use in bioassays may significantly alter results, as has been demonstrated in some feeding bioassays.[68]

Specific chemical substances associated with flight responses have rarely been isolated, but this has apparently not been necessary for the adaptation of this phenomenon to applied problems. As one example, spider crabs in the genus *Libinia* are a nuisance for lobster fishermen in the northeastern United States because they have little market value, consume bait, and increase the number of person-hours required to process traps. When crushed spider crabs were placed in lobster traps, catches of spider crabs decreased markedly, while catches of commercially valuable rock crabs and lobsters were unaffected.[69] Spider crabs (*Libinia dubia*) also decrease feeding rate in response to predator odor.[70] Although the mechanistic details of a species-specific alarm cue are unresolved in this case, it seems unlikely that this would concern lobstermen who utilize this "technology" to increase their livelihood.

Flight responses are unlikely to be effective when predators are more mobile than prey, and in these cases the presence of predators can induce morphological or chemical changes designed to reduce their susceptibility to predation. Not surprisingly, "inducible defenses" appear to be particularly common among sessile marine organisms including seaweeds,[71,72] bryozoans,[73] and cnidarians,[74] phytoplankton,[75] and among terrestrial plants.[76] However, the phenomenon is by no means restricted to sessile organisms, as inducible defenses have been extensively studied in freshwater rotifers[77] and cladocerans.[78] Marine mobile invertebrates such as snails[79–81] and mussels[82] have also been shown to exhibit morphological shifts in the presence of highly mobile predators such as crabs. Three species of intertidal snail, *Nucella lamellosa*, *Nucella lapillus*, and *Littorina obtusata*, all produce thicker shells when subjected to water containing effluent from decapod crabs that commonly prey on snails. *N. lamellosa* exhibits even greater induction of shell thickness when exposed to water in which crabs were consuming conspecifics.[79] A combination of predator and killed prey appears to be the most effective stimulus for eliciting a range of antipredator behaviors.[62,79] However, none of these studies demonstrate that the measured increase in the defensive trait results in a decrease in susceptibility to predation, although Leonard et al.[82] showed that the increased shell thickness of mussels exposed to green crabs and injured conspecifics increased the force required to break the shells.

This type of correlative approach is widespread, as only a few marine studies involving inducible defenses (and none with mobile invertebrates) have directly demonstrated that the induction results in a decrease in the susceptibility of the organism to predation.[71,72] Statistically significant differences in shell thickness or concentrations of defensive chemicals may or may not meaningfully affect predator preferences in ecologically relevant field situations. For chemical defenses, compound dose–response relationships may be nonlinear, and threshold levels of defense could be sufficient to deter predators so that further induction has little additional benefit. Thus, future studies should focus on directly demonstrating whether an induced response reduces predation on prey organisms.

Implicit in any evolutionary argument for inducible defenses is the idea that defenses are costly to deploy, and, thus, in situations where attack is predictable, they can be selectively deployed during periods of maximum predator pressure.[83] However, unambiguous demonstrations of the fitness costs of inducible defenses for marine organisms are rare. Many advances in measuring the costs of induced defense have been made in those systems in which the organism induces defensive characteristics after being exposed to a chemical cue indicative of predator presence, as this allows quantification of the costs of induction without the confounding influence of tissue loss due to consumption.[73] The ability of waterborne cues from predators and damaged prey to induce a morphological change in gastropods and bivalves[79–82] suggests that these animals may be useful study organisms for addressing theoretical issues surrounding inducible defenses.

B. Consequences of Feeding Deterrents for Predators

1. Susceptibility of Consumers to Defensive Chemicals

An emerging generalization from studies of the susceptibility of consumers to prey chemical defense is that many small, low-mobility invertebrates such as amphipods, polychaetes, shell-less gastropods, and crabs readily consume seaweeds that produce chemicals that deter feeding by larger, mobile grazers like fishes and urchins.[30,38,39,84–87] From most of these studies it is unclear whether body size or mobility is more important in selecting for resistance to defensive chemicals. Smaller herbivores sometimes specialize on a single, chemically defended host species because they either physiologically or behaviorally sequester the defensive metabolites from that host (see Sections II.A.2 and IV.B). Such specialization is far less common among larger consumers, in part because they may be too large to benefit from an associational refuge.[70,87,88] However, this pattern may be generalizable to closely related, similar-sized species that differ in mobility. Among amphipods, for example, low-mobility species are more tolerant of seaweed chemical defenses than higher-mobility species,[89] and are also better able to employ compensatory feeding to substitute food quantity for quality.[90] Among similarly sized brachyuran crabs, those with reduced mobility are less selective feeders and are unaffected by algal chemical defenses.[30] Among the brachyuran crabs tested, the relationship between low mobility and resistance to algal chemical defenses held, both among species within a family as well as between families, suggesting that the pattern may be robust.[30] However, additional data from different taxonomic groups, particularly outside the Crustacea, are needed to test this hypothesis rigorously. Additionally, it is not yet clear whether low mobility drives resistance to chemical defenses or whether resistance to chemical defenses facilitates a low-mobility lifestyle. The resolution of this question may be aided by the application of phylogenetic methods.

For most marine invertebrates that readily consume chemically defended seaweeds, it is not known whether they are actually resistant to, or simply tolerant of, algal secondary metabolites. In the case of specialist consumers (e.g., nudibranchs, ascoglossans, some amphipods or crabs; see Section IV.B), a means of resistance to specific chemicals seems likely. However, for marine invertebrates that consume a diverse array of prey that produce different chemical defenses against a broad suite of predators,[85,86] perhaps tolerance or less-specific mechanisms of resistance (i.e., gut pH) become more important. The actual mechanisms by which marine consumers avoid harmful effects of consuming chemical defenses (detoxification or dietary mixing) are even less well understood (see Section II.B.2).

Although feeding by most small, specialist predators and herbivores is either unaffected or stimulated by the chemical defenses produced by their hosts, this is not always the case. The amphipod *Ampithoe longimana* readily feeds on the chemically defended brown alga *Dictyota menstrualis*, however, high concentrations of the diterpene alcohol pachydictyol A (Structure 4.16) found in some plants deter feeding by these herbivores.[72] These higher levels of defensive chemicals in *Dictyota* appear to be an induced response to attack by small herbivores like *A. longimana*.[72] Such intraspecific variation in concentration of defensive chemicals has the potential to significantly impact the distribution and abundance of these small, specialist predators. As one example, on Guam, the nudibranch *Glossodoris pallida* feeds exclusively on the branching sponge *Cacospongia*, from which it sequesters the defensive compounds scalaradial (Structure 4.17) and desacetylsca-laradial (Structure 4.18).[91] Concentrations of these metabolites are highest in the growing tips of the sponge and lowest at the base, but even the lowest concentrations strongly deter predation by generalist fishes. However, the higher concentrations typical of sponge tips deter feeding by *Glossodoris*, whereas the lower concentrations at the base do not. *Glossodoris* are equally suscep-tible to predators at the base and tips of sponges, yet are found almost exclusively near sponge bases, suggesting that intra-individual differences in concentrations of defensive metabolites drives the distribution of these specialist predators.[91]

Knowledge of the variability in the susceptibility of different guilds and species of mobile invertebrates to chemical defenses produced by sessile invertebrates and seaweeds is critical for a mechanistic understanding of the distribution of the sessile benthos in the sea. Large mobile invertebrates like sea urchins commonly alter benthic community composition from palatable to unpalatable species.[92,93] Most notably, chemical defenses produced by tropical seaweeds have been widely implicated in the persistence of these species in areas of intense herbivory like coral reefs[1,3] (also see Chapter 6 in this volume).

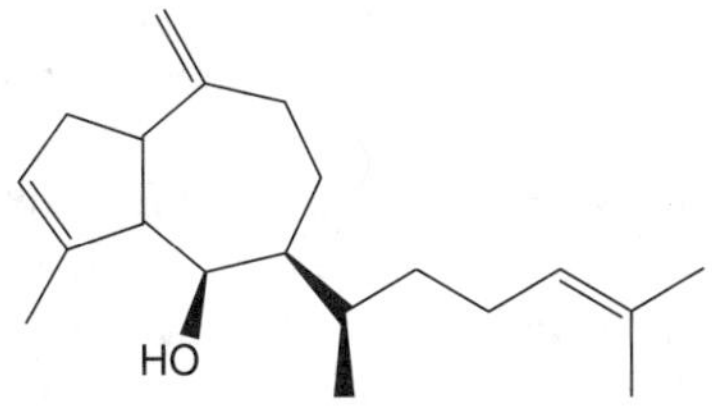

In contrast to the well-known effects of large mobile grazers on benthic communities, the community and ecosystem-level consequences of small consumers that are resistant to chemical defenses are still poorly understood and probably currently underestimated. In coastal North Carolina, *Ampithoe longimana*, an amphipod that readily feeds on several chemically defended brown algae, is capable of shifting the seaweed community from a brown- to a red-algal dominated community, although this effect only occurs in the absence of fishes, which prey heavily on amphipods and directly consume red algae that lack chemical defenses.[94] Because many small consumers that are resistant to chemical defenses exhibit limited mobility, their impact may be spatially restricted. Low-mobility, nonselective grazers like some crabs and amphipods create small patches of intense, nonselective grazing which are superimposed on the background of selective grazing by more mobile herbivores like fishes and urchins.[86] On a landscape level, the combined effect of both types of grazing should result in a mosaic of patches with high and low algal density, with important consequences for species diversity at both the local and regional scale.[95–99] Additionally, local reductions in density can alter the nature of inter- and intraspecific interactions among seaweeds,[100–103] reduce the density of seaweed-associated invertebrates,[99,104] and decrease the abundance and recruitment of fishes.[105–107] Thus, although the impact of small, nonselective grazers may be spatially restricted, this type of grazing clearly merits consideration in models of herbivore impact on marine community dynamics and ecosystem function.

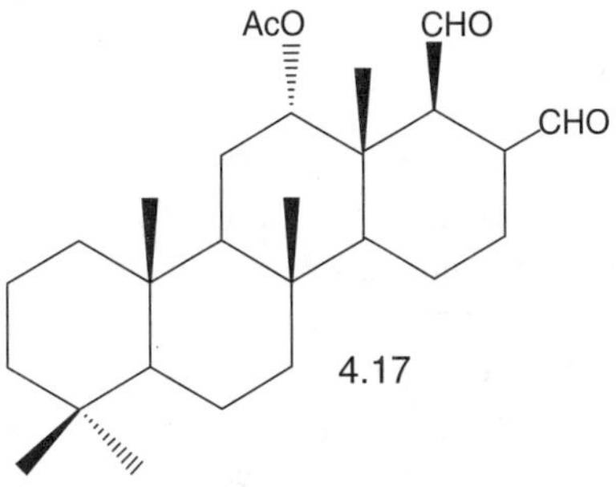

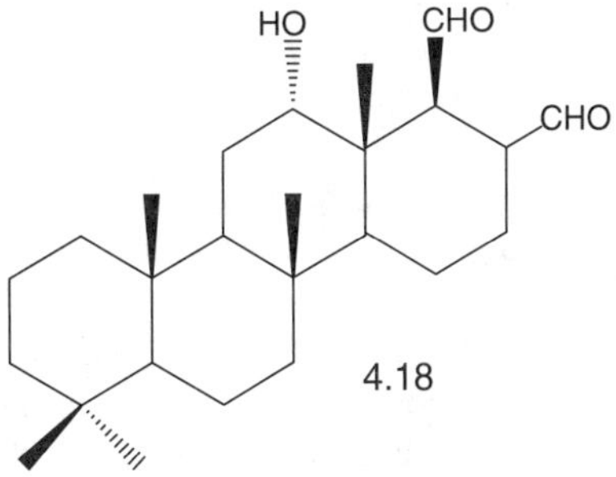

2. Consequences of Consuming Defensive Metabolites

Although there is a considerable amount known about the effects of prey chemicals on predator feeding preferences, much less is known about the proximate or ultimate reasons why marine invertebrates avoid certain compounds. Even when compounds cause behavioral avoidance of a food, few studies have assessed how consumption of prey secondary metabolites affects the physiology (and ultimately the fitness) of invertebrate consumers. Two basic approaches have been used: (1) comparing effects of natural prey items which naturally contain or lack various secondary metabolites, or (2) comparing the effects of artificially prepared diets with and without metabolites. Studies of the first group[108,109] have been able to correlate metabolite presence with certain effects on consumers, but the effects of secondary metabolites are confounded by other traits (e.g., protein, caloric content, morphology) that may also vary among the different foods. The second approach allows a more direct test of the effects of metabolites on consumer performance.

Few experiments of any kind have rigorously examined the long-term effects of chemical defenses on mobile invertebrates. This is, in part, due to the difficulty of getting consumers to eat foods that contain chemical defenses. Co-occurring generalist predators either avoid consuming foods with noxious chemicals or rapidly learn to avoid them,[87,110,111] and the effects of compounds on specialist predators that readily eat chemically defended prey may have limited applicability to most consumers. Although some studies do demonstrate reduced growth, survival, or fitness of consumers on foods with chemical defenses added,[87,111–114] it is sometimes unclear whether the reduced growth rate observed on chemically defended foods is due to behavioral (reduced consumption rates) or physiological (toxic or digestibility-reducing) effects.

The effects of ingested metabolites can occur relatively quickly, by altering assimilation rates of food, or they can be more chronic. Phlorotannins produced by brown algae are analogs of the condensed tannins produced by terrestrial plants and, thus, are thought to function by complexing with proteins in the guts of herbivores, reducing the ability of animals to assimilate ingested material. However, in general, phlorotannins seem to have little measurable effect on the assimilation or conversion efficiency of the crabs, gastropods, isopods, and echinoids for which that has been measured.[109,115–117] Furthermore, even if secondary metabolites did decrease assimilation efficiencies, herbivores might compensate for this by increasing feeding rates, as has been observed for crabs and amphipods feeding on low-quality plants or artificial diets.[30,85,90] Phlorotannins did reduce the digestion rate of algal protein by the gut fluids of two limpets in vitro, but protein digestion in two species of isopod was unaffected.[115] Gut surfactants produced by these isopods appeared to inhibit the binding of polyphenolics to proteins in the gut, and the occurrence of these gut surfactants is widespread among marine consumers.[115] Assimilation rates of the tropical crab *Mithrax sculptus* and of several temperate gastropods were also not well correlated with the phenolic content of plants being eaten,[116,117] but a lack of effect of a compound on assimilation does not preclude the possibility of effects of growth and fitness.

The effects of phlorotannins on the growth rates of herbivores are also not clear cut. Although there are strong differences in herbivore growth rates and fecundities among different algal species,[90,108,109,118,119] manipulation of the phenolic content of artificial foods has little effect in the few species for which data are available.[109] Some studies do show reduced growth rates of herbivores feeding on artificial diets to which phlorotannins have been added;[113] however, unnaturally high compound concentration and relatively low quality of artificial diets complicate the interpretation of these results.[120] Steinberg and Van Altena[109] found no direct effect of phlorotannins and suggested that the differences they observed in herbivore growth across algal species in Australia correlated better with the presence of smaller, nonpolar metabolites, as herbivores generally exhibited lowest growth and survival when fed monospecific diets of species containing these metabolites.[119]

Relatively few studies have directly examined the long-term fitness consequences to mobile marine invertebrates of consuming lipid-soluble chemical defenses, and clear generalizations have not emerged from the data that are available. For example, diterpene alcohols from seaweeds in the genus *Dictyota* have been assayed for effects on several different consumers, often with dramatically varying results. These compounds are well known to deter feeding by a variety of urchins, fishes, and crustaceans.[68,86,87,108,114] Fishes (the pinfish, *Lagodon rhomboides*) fed fish food laced with pachydictyol A (Structure 4.16) grew more slowly than those fed control diets,[87] but two species of sea hare were apparently unaffected by ingesting this metabolite at identical concentrations.[121] A mixture of dictyols [pachydictyol A and dictyol E (Structure 4.19)] incorporated into an artificial algal diet at natural concentrations did not affect survivorship, growth, or fecundity of the gammarid amphipod *Ampithoe longimana*, but growth, survival, and fecundity of a congener (*A. valida*) was strongly suppressed, and the fitness of a

4.19

distantly related isopod (*Paracerceis caudata*) was actually enhanced by the presence of these "feeding-deterrent" compounds.[114] Although the dictyols deterred feeding by all these consumers, there was no consistent relation between behavioral deterrence and the long-term effects of the compounds on consumer fitness. In some cases, compounds may function defensively just because they taste bad or because they mimic the taste of compounds that do have deleterious fitness effects.

For small consumers like amphipods that may have relatively limited mobility, the effects of confining animals to a monospecific, chemically noxious diet may be ecologically relevant, but this may be less acceptable for invertebrates with greater mobility or a more varied diet. In recognition of this, Lindquist and Hay[111] assessed the effects of occasional consumption of chemically defended prey on consumer fitness. They fed sea anemones (*Aiptasia pallida*) large meals of high-quality food pellets followed several hours later by small meals of food pellets either containing or lacking several structurally related defensive compounds [the didemnins, e.g., didemnin B (Structure 4.20)] from larvae of the ascidian *Trididemnum solidum*. This mimicked anemones getting the bulk of their food from palatable prey, but feeding at low levels (1.8% of total diet) on defended foods; thus, all anemones consumed the same total amount of prey. Even this low level of feeding on chemically defended prey dramatically decreased growth (by 75%) and vegetative propagation (by 50%) of anemones.[111]

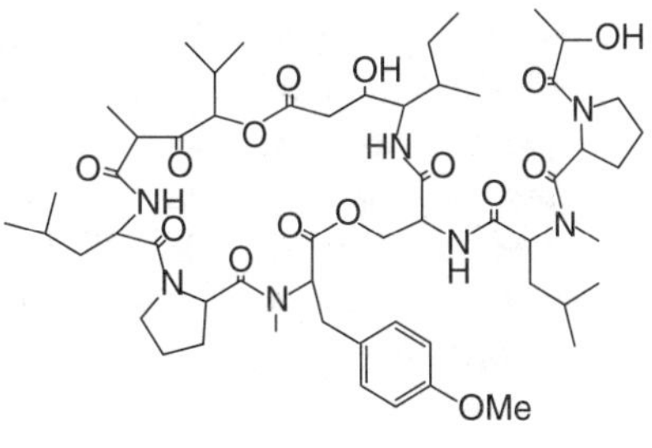

4.20

C. CHEMICALLY AIDED PREDATION

Thus far, this section has focused primarily on ways in which mobile invertebrates use chemicals to defend themselves against predators and the consequences of these defenses for the behavior and fitness of predators. However, predators also employ chemicals in all phases of their search for prey, including prey location, capture, and initiation of feeding.

1. Foraging and Prey Detection

In marine systems, the ability to detect and orient to food from a distance is potentially of considerable advantage, allowing consumers to detect food over an area larger than they can profitably physically search. Mobile invertebrates from such diverse groups as amphipods,[122] lobsters,[123,124] crabs,[65] shrimp,[125] nudibranchs,[126] bivalves,[127] snails,[128] cephalopods,[129] and polychaetes[130] have been reported to detect chemical signals from prey items at a distance. This ability may be particularly advantageous in marine systems where vision can be severely restricted due to attenuation of light in deep or turbid waters. In contrast to the spatial limitation of visual cues, chemical odors can be carried over considerable distances, although variation in currents and bottom characteristics alter the strength and quality of the signal.[65,66,131,132] In addition to the information contained in a chemical signal that reaches an animal, an animal's response can vary depending on the animal's activity state,[133,134] hunger level,[68,135] feeding history,[62,136] and the presence of conspecifics[137] or predators.[62] Other methodological issues surrounding studies of chemically mediated foraging are addressed in detail elsewhere.[7,64]

The isolation of specific compound(s) responsible for the attraction of predators to prey has been elusive. Although it is well known that specific amino acids are contained in prey items and do attract predators, field measurements have shown that fluxes of amino acids from carrion are only occasionally above the threshold concentrations required for detection by scavengers, and that fluxes from live organisms are often well below levels detectable by predators.[138] Additionally, amino acids are rapidly degraded by bacteria when they are released into the water column[139] and thus are probably more likely to serve as feeding stimulants once prey is encountered than as cues for locating more distant prey (see Section II.C.3). In contrast, peptides appear to be much less susceptible to bacterial degradation[139] and are known to serve as chemical cues in a wide variety of communication systems (see also Sections IV.A and V). Many of these peptides consist of low molecular weight compounds (ca. 500–5000 Da) with a basic amino acid residue (often arginine) at the carboxy terminus.[140] Particularly in foraging, natural stimuli can be complex mixtures of attractants, repellents, and neutral chemicals from multiple individuals and species whose combined activity may differ considerably from that of any component in isolation.[125,141–143] Rather than detail the specific compounds or mixtures that have been shown to attract predators to prey, this chapter emphasizes studies that attempt to identify conditions under which chemically mediated location of intact prey or prey exudates are likely to be important in the field.

Foragers are able to locate prey from great distances without visual signals in the field,[122,133,144–146] although the mechanisms by which they do this are often unclear. Most of the environments in which marine organisms typically forage are characterized by at least some water flow (currents or waves) and topographic complexity, both of which can increase turbulence. Turbulence transforms an easily discernible gradient of odor from prey to predator into a patchy assortment of odor pulses that vary in strength and frequency with distance from the source.[66,147,148] Although different mobile invertebrates are known to use different mechanisms to respond and orient to odor sources in the field, the flow speed and the bottom characteristics are critical to the ability of most organisms to efficiently track an odor to its source.[65,132,148]

As one example, Weissburg and Zimmer-Faust[65,67] tested the ability of blue crabs (*Callinectes sapidus*) to locate a live and intact clam (*Mercenaria mercenaria*) or whole clam extract at flow speeds from 0 to 14.4 cm s^{-1} on sandy and gravel bottoms. In the absence of flow, predators were unable to locate prey, regardless of substrate type. Slow flow (~3 cm s^{-1}) resulted in efficient search paths and tracking success of nearly 100%, while fast flow (14 cm s^{-1}) resulted in convoluted search paths and low to moderate tracking success. Turbulence at high flow rates effectively dispersed the odor plume, apparently making it more difficult to track. When bottom composition was altered from sand to gravel (with flow speed held constant), turbulence also increased, markedly decreasing tracking success.[65] Blue crabs appear to track odor to its source by a combination of chemotaxis and rheotaxis, moving upstream when chemical signals from the prey are detected (orienting into the direction of the current). Such chemically mediated rheotaxis has also been demonstrated for gastropods.[128] Flow also affects the success of chemically mediated prey location by blue crabs indirectly by causing a shift in crab orientation from cross-stream to along stream. This behavior is probably intended to reduce drag on the organism, but also has the indirect effect of decreasing cross-stream movements and thus decreases the probability of a forager encountering an odor plume.[65]

The effects of flow speed and turbulence on chemolocation suggest that the use of chemical cues by blue crabs to locate distant prey may only function under a narrow range of conditions in nature. However, in shallow estuaries where blue crabs are abundant, slow unidirectional flow over a period of several hours may commonly occur; thus, chemoreception could be an important foraging tool in these habitats. Recent field experiments have shown that over 80% of crabs tested were successful in following odor plumes emanating from injured bivalve prey or artificial plumes of clam bivalve mantle fluid to their source.[149] This is one of the few studies to demonstrate that foragers can use chemical signals alone to locate natural prey items in the field. Additionally, even within high flow areas, slower flow near slack water could allow temporary establishment of conditions conducive for chemically mediated foraging. If this is the case, predation rates could vary as a function of tidal stage, even in subtidal locations. The implications of spatial variability in the success of chemically mediated foraging on prey populations and communities are largely unknown and should provide an interesting area for future study.

Few studies have rigorously assessed the roles of flow and turbulence on chemically mediated foraging, so generalization of results obtained using blue crabs is unclear. Experiments designed to test the effects of turbulence on chemoreception by crayfishes in artificial freshwater streams have produced different results.[132] When bottom characteristics were altered to increase roughness of artificial stream beds, foraging success did not change, and the time required for crayfishes to locate artificial food sources actually decreased. This may be because turbulence caused an increase in the frequency of the signal fluctuation, reducing the time interval between odor bursts, thereby facilitating odor tracking. Similar results have been obtained for moths locating mates via phero-mones.[148] Although the effects of flow speed and turbulence on the success of chemically mediated foraging behavior may vary among species, it does seem clear that these parameters are critical in determining the distribution of chemical signals and, therefore, should play an important role in the tracking of waterborne signals in the marine environment.

One large expanse of marine benthos in which turbulence and high flow may be less important is the deep sea. The relatively mild hydrodynamic conditions in the deep sea should enhance the persistence of chemical gradients and simplify tracking of odor plumes by predators or scaven-gers.[150] However, the logistical difficulties associated with experimentation on animals in this habitat have precluded extensive testing of this hypothesis. To date, location of carrion by deep sea scavengers using odor plumes has been documented in highly mobile amphipods and hagfish, but not in less-mobile gastropods or echinoderms.[122,146] These results have led some to suggest that sensitivity to chemical cues may be positively correlated with mobility.[122] Because they would have little chance of locating and obtaining a distant food source, it seems intuitive that animals with low mobility should require greater concentrations of stimuli before responding,[151] but additional data are needed to rigorously evaluate this hypothesis.

2. *Toxin-Mediated Prey Capture*

Toxins delivered in the bites or stings of aquatic organisms have been subjects of intense interest from a medicinal and natural history perspective for centuries. Despite improved understanding of the chemistry, toxicology, and pharmacology of many of these substances, studies on their ecological roles are still relatively uncommon. It is often assumed that venomous mouthparts or stinging tentacles play a role in prey capture or defense; however, the ecological mechanisms underlying these hypotheses have rarely been tested directly. The role of toxins in prey capture has generally been inferred from: (1) observations of foraging behavior and reactions of the attacked prey, (2) the existence of structures apparently adapted to deliver toxins, (3) isolation of toxins or venoms associated with these structures, and (4) assessment of their toxicity by injecting the chemicals into standard lab animals (often of little ecological relevance) such as mice, crayfishes, or insects. These studies have uncovered a great deal of taxon-specificity in the effects of toxins on laboratory animals,[23,152] highlighting the need to assess the effects of realistic doses of toxins delivered in an ecologically meaningful way to relevant prey organisms.

As a consequence, many ecological studies of toxin-mediated prey capture have been descriptive or have not established clear relations between the occurrence of suspected toxins and prey capture. Nevertheless, available evidence suggests that mobile benthic invertebrates as diverse as nemertean worms,[23] gastropods,[152,153] cephalopods,[154,155] and chaetognaths[156] can inject toxins into their prey to facilitate prey capture. These toxins are diverse in structure (ranging from hydrocarbon- to peptide-based) and mode of action both within and among taxa.[157] Cnidarians, including hydroids and jellyfish, provide probably the most well-studied example of toxin-mediated prey capture in the marine environment, as they produce a diverse array of subcellular structures called nematocysts, some of which are able to penetrate even calcified exoskeletons or fish scales and inject proteina-ceous toxins.[158,159] Different structural types of nematocysts appear to have different functions, such as prey capture,[159] intra- or interspecific aggression,[160,161] or defense against predators.[162]

The toxins identified from cone snails, nemerteans, cephalopod molluscs, etc. undoubtedly play some role in prey capture, but their importance relative to other predatory behaviors is generally unknown. For some predators, toxins may serve as a secondary rather than primary mode of capturing prey. For example, Olivera et al.[153] note that certain cone snails capture fish by stinging them with their poisonous proboscis and then engulfing them, while others distend their rostrum and sting the fish only after it has been at least partially trapped. Perhaps the most unambiguous demonstration of the role that injected toxins play in prey capture is in the octopus (*Eledone cirrhosa*). It is well known that octopuses bore holes into the shells of their molluscan or crustacean prey, but shell penetration requires at least 10–20 minutes for crabs and longer for molluscs, yet crabs removed from the grasp of an octopus within two minutes of attack do not recover.[154] *Eledone* injects saliva into its crustacean prey by puncturing a hole through the eye, and this can occur within the first few minutes of an attack.[155] Experimental injection of saliva showed that a bite through the eye was the most rapid and effective means of toxin entry and accounted for the rapid subdual of prey.[155] The saliva contains a complex mix of both toxins and enzymes and not only paralyzes and kills the prey, but also begins digestion from within the shell, allowing tissue to be more easily and thoroughly removed from the carapace.

It is currently unclear whether toxin-mediated prey capture by mobile invertebrates has a significant impact on prey population size or community composition. In freshwater systems, chemically mediated prey capture by flatworms has been demonstrated to significantly impact prey populations in the laboratory. Neurotoxic chemicals released from the mucus webs of the flatworm *Mesostoma* can drive entire populations of the cladoceran *Daphnia magna* to extinction in culture, but the concentration these chemicals normally attain under realistic field conditions is unknown. Nevertheless, because the mucus webs these flatworms build function to trap prey, Dumont and Carels[163] likened these flatworms to spiders with toxic webs. Similar impacts may occur in open water marine systems where organisms that employ toxin-mediated prey capture are abundant, or even dominant, predators (e.g., chaetognaths and cnidarians).

3. Feeding Stimulants

Most ecological research on the role of secondary chemicals in food selection has focused on identifying compounds that serve as defenses against consumers.[1–4] Feeding stimulants (compounds that promote ingestion and continuation of feeding) may be of equal importance but are less thoroughly studied within an ecological context. These differ from feeding attractants (see Section II.C.1) which are waterborne chemicals that predators use to locate prey from a distance. Sakata[164] reviews feeding stimulants for marine gastropods and shows that specific amino acids, sugars, carbohydrates, glycerolipids, etc. can induce gastropods to begin feeding on otherwise inert materials such as filter paper or crystalline cellulose (Avicel SF). These procedures are likely to work best at identifying lipid-soluble feeding stimulants because it is unlikely that water-soluble compounds would remain in the Avicel for any significant amount of time once the plate is placed in seawater. Other investigators have focused on water-soluble chemicals using different methodologies, often in the context of the attraction of predators to prey from a distance (see Section II.C.1). Interest in feeding stimulants has been driven more by the economic benefits of maximizing feeding rates of cultured gastropods and crustaceans than by ecological and evolutionary implications, and most investigations in this area have tested commercially available compounds rather than conducted bioassay-guided discovery of the metabolites that affect feeding in nature.

Natural feeding stimulants may be present in a variety of marine organisms, but they are often discovered only accidentally (see, however, Sakata[164]). As studies of chemical deterrence apply crude extracts to already palatable foods, many assays designed to assess feeding deterrence would not detect stimulants even if they were present, because palatable control foods are regularly completely consumed. Nevertheless, in surveys of the activity of chemical extracts of seaweeds and marine invertebrates for defense against predators, extracts occasionally exhibit stimulatory properties,[40,86,120] although the specific metabolites are rarely identified and the ecological implications of these stimulants are not addressed. However, Sakata and co-workers[164–166] have isolated and characterized several natural feeding stimulants from algae that are highly preferred by abalone and other molluscan grazers.

Compounds that serve as chemical feeding deterrents to large generalist consumers also can serve as feeding stimulants to a variety of smaller, specialist grazers that behaviorally or physiologically sequester these compounds for use in their own defense[33,37–39,84,167] (see Sections II.A.2 and IV.B). Feeding stimulants may also mediate mutualistic interactions in which one of the partners gains a nutritional supply from the other (see Section IV.C).

III. CHEMICAL MEDIATION OF COMPETITION AMONG MOBILE INVERTEBRATES

Intense competition for limiting resources (space, light, food, etc.) has been well documented among marine benthic organisms and can play an important role in determining the overall structure of marine communities. Secondary chemicals may impact the outcome of competitive interactions when chemicals released from, or bound on the surface of, one organism reduce the growth, survival, or reproductive success of another organism. Antifoulants remove or prevent the settlement of organisms directly on the surface of another, whereas allelopathic chemicals mediate interactions among two organisms growing directly on a primary substrate. Thus, both processes potentially alter the outcome of competitive/overgrowth interactions. Although little research on chemical mediation of competitive processes among mobile invertebrates is available, the few available examples, combined with a greater number of examples involving sessile animals and plants, highlight the broader, community-level importance of secondary metabolites in marine systems.

A. ANTIFOULANTS

Fouling by micro- and macroorganisms is usually thought to be the bane of sessile species, and considerable effort has been focused on determining the ways in which these plants and animals employ chemicals to deter colonization and growth of fouling organisms.[168–170] However, slow-moving marine invertebrates with rigid exteriors are often subject to colonization and overgrowth by sessile macroinvertebrates or potentially pathogenic fouling bacteria. Even relatively fast-moving organisms (e.g., whales and sea turtles) can be colonized by sessile fouling organisms. Fouling can negatively impact mobile benthic invertebrates by increasing drag, which can increase the probability of dislodgment from the substrate during storms and increase the amount of energy required for movement, thereby decreasing growth rates.[171] Many mobile animals have behavioral mechanisms such

4.21

as frequent burial to remove fouling organisms,[172] but some do apparently use chemical antifoulants. For example, the intertidal limpet *Trimusculus reticulatus* concentrates a diterpenoid (Structure 4.21) in its mantle, foot, and mucus that kills settling larvae of the reef-building tube worm *Phragmatopoma californica*.[173]

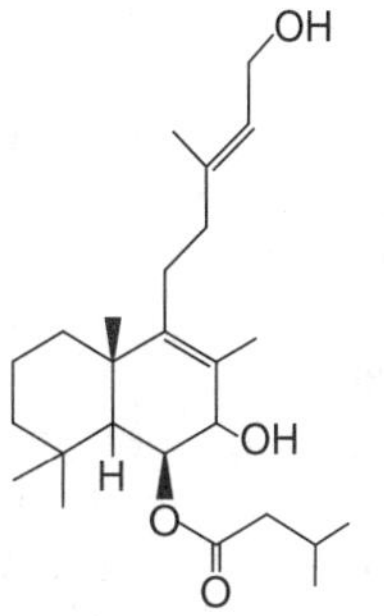

Studies of chemical antifoulants have been driven largely by the search for nontoxic alternatives to the fouling paints applied to the bottom of ships and other manmade structures (see Chapter 17 in this volume). Consequently, relatively few studies have successfully addressed the role of chemical compounds in deterring fouling in an ecologically meaningful way. It is generally unknown whether most proposed antifoulants are bound to the surface of organisms or released gradually into the water column, and very different methods will be appropriate to assessing the effectiveness of each.[169,170] Where surface chemistry appears to be important, extracts in bioassays will be too concentrated if the extract from a three-dimensional organism is applied to a two-dimensional surface; here surface extraction techniques will be most relevant.[169,174] Slow release of organic extracts from sessile invertebrates or seaweeds can be achieved by placing them into

Phytagel discs which can be outplanted in the field for up to 6 weeks without being degraded.[170] Fouling organisms colonize untreated Phytagel surfaces at about the same rate as they colonize Plexiglas® plates, but Phytagel discs with certain extracts will either facilitate or deter colonization. The concentration of sponge crude extract placed in Phytagel declined steadily over a 21 day period in running seawater, at which point 56% of the initial extract was still present, suggesting that the extract was slowly released from this gel. For organisms that can be shown to naturally release metabolites, this method appears to offer an efficient and ecologically realistic method of conducting field assays on potential allelopathic interactions, although differences in surface texture between experimental surfaces and intact organisms could confound the application of results.

A broad survey of echinoderms from the Gulf of Mexico determined that while few species possessed body wall extracts that deterred settlement by bacteria, many deterred settlement by barnacles and bryozoans.[175] This study represents an excellent initial effort in the search for chemical antifoulants among mobile marine invertebrates and should be commended in particular for the large number of species and higher taxonomic groupings assayed. However, several important methodological issues warrant mentioning so that future studies might make further progress. Extracts in this particular study were dissolved in seawater and tested for effects on settlement on plastic dishes, not echinoderm surfaces, thus, the relevance of these results to patterns of fouling on echinoderms in the field is unclear. Further, assays were performed in the lab in still water, and it is unclear whether assayed concentrations might ever occur in nature where flow can rapidly dissipate compounds and decrease the realized concentration of metabolites experienced by settling fouling organisms (see Hay et al.[7] for a discussion of methodological concerns and suggestions regarding still water assays). These methodological problems are by no means confined to this study, and solutions are not always readily available. However, data on the relative degree of fouling found on organisms in the field from species with and without extracts that deter settlement by sessile invertebrate larvae would allow an assessment of the relative importance of chemical vs. other means of avoiding fouling (e.g., sloughing of surface tissues, abrasion, or burial in sediment).

B. ALLELOPATHY AND COMMUNITY STRUCTURE

Although allelopathy is usually the domain of sessile organisms fighting for limiting resources such as space or light, mobile invertebrates are also known to employ this method of excluding competitors. Organisms living in soft sediments often physically modify their environment both by burrowing through and disrupting sediment structure and also by creating tubes that increase the biogenic complexity of the environment. A wide variety of soft sediment polychaete and hemichordate worms also produce halogenated organic compounds (Structures 4.22–4.24),[176] but the ecological relevance of these compounds is still relatively poorly understood. Bromophenols (Structure 4.22) secreted into the sediment by capitellid polychaetes deter colonization of these sediments by juvenile bivalves,[177] and similar compounds released into the sediment by a terebellid polychaete deter colonization of that sediment by nereid worms.[178] These compounds have relatively low solubility in water and are thus likely to have long residence times in the sediment, so they may potentially impact community composition on large temporal and spatial scales. Recent work by Woodin and co-workers suggests that by secreting these chemicals into the sediment, large infauna not only deter settlement by other large organisms that are potential competitors, but they also create refuges for species that are tolerant of these chemicals.[179] The increase in local diversity due to this refuge effect is similar to that seen as a result of physical refuges such as those provided by the tubes of large polychaete worms.

Allelopathy has been studied in greater detail for sessile invertebrates and plants. In several instances, the mechanistic details of these interactions have been elucidated, so these examples are mentioned here to guide future work on allelopathy among mobile invertebrates. The most convincing studies of allelopathy use manipulative field experiments to demonstrate direct inhibition

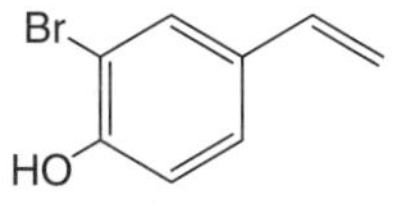

4.22

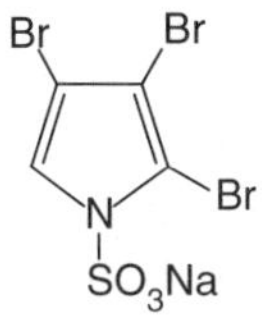

4.23

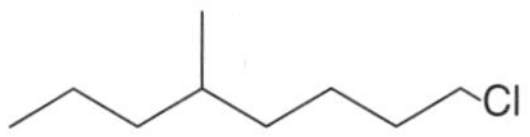

4.24

of one species by another and also assay extracts in the field to show a chemical mechanism for the inhibition. As one example, De Nys et al.[168] found that when the red alga *Plocamium hamatum* was located near the soft coral *Sinularia cruciata*, the soft coral showed signs of tissue necrosis where the alga had contacted it. Reciprocal transplants demonstrated that this necrosis was a result of contact of polyps with the alga. Natural concentrations of the monoterpene chloromertensene (Structure 4.25) from *Plocamium* coated onto artificial plants caused similar necrosis, demonstrating that the negative effect of *Plocamium* on *Sinularia* was chemically mediated.

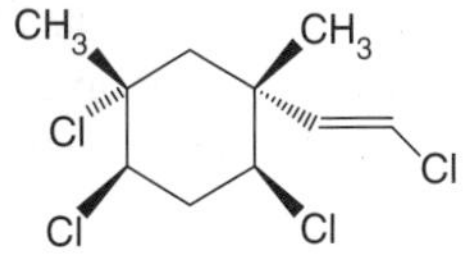

4.25

Since allelopathy usually involves surface chemistry or the release of compounds into the water, future studies should take the difficult step of attempting to target the surface of the organism in their extraction procedures or quantify natural release of metabolites into the water column. Schmitt et al.[169] made a first attempt at this by swabbing the surface of an apparently deterrent organism (a brown alga) in an attempt to extract only the surface chemistry, showing that levels of diterpene alcohols (Structures 4.16 and 4.19) on the plant surface were sufficient to deter settlement by arborescent bryozoans. A more quantitative approach has been developed to remove lipid soluble secondary metabolites from the surface of two Australian seaweeds; submerging intact plants in hexane for 20 to 40 s extracted surface bound chemicals without damaging surface cells or extracting metabolites from the interior of the plants.[174] However, the applicability of this method to animals, other plants, and other classes of metabolites remains untested. In cases where the relevant metabolites are exuded into the water column, careful quantification of the spatial distribution of these metabolites will be necessary to design bioassays that deliver ecologically realistic concentrations. Coll et al.[180] have developed a submersible sampling apparatus capable of *in situ* isolation of exuded allelopathic chemicals that should prove useful in such endeavors.

IV. CHEMICAL MEDIATION OF MUTUALISTIC AND COMMENSAL ASSOCIATIONS

Population and community ecologists studying species interactions in both marine and terrestrial systems have often focused on predator–prey and competitive interactions[181,182] at the expense of mutualisms and commensalisms. Given the growing appreciation in the ecological community for the importance of positive interactions (facilitation, commensalism, mutualism) in both marine and terrestrial communities,[181–183] it seems appropriate to discuss here the importance of chemistry in directly and indirectly mediating these interactions. While chemical mediation of location of mutualistic or commensal partners are well known, the importance of chemically mediated positive interactions as a mechanism driving ecological and evolutionary patterns is just beginning to be understood.

A. Host Location

Because associations involving mobile invertebrates depend at least initially upon two partners locating each other, chemical signals that allow mobile organisms to track and recognize appropriate host organisms are likely to be critical for the maintenance of these associations. Much of the early literature on the topic offered evidence for chemical-mediation of host location between symbiotic species using a Y-maze. In a Y-maze, or Y-tube, design, water flows into a central chamber from two arms (thus the apparatus is shaped like a "Y"). Water entering one arm is laden with the chemical that is the hypothesized stimulus, while the other arm serves as a control. In most experiments, the choice offered is between an intact host organism (or water conditioned by the presence of the host) and a seawater control or a nonhost organism. This early work demonstrated that potential associates could orient toward and follow the odor coming from a host organism, and in some cases the chemical signals involved were highly host specific (reviewed by Ache[184]).

However, the use of Y-mazes may overemphasize the importance of chemical cues in host location. The density of host organisms and the flow rate through the chamber in which they are held will dramatically affect the concentration of the stimulus. Nearly all studies stock host animals at higher than natural densities (or do not report host density in the field relative to the experimental conditions), and use slow rates of water flow. Thus, these assays should be regarded only as demonstrations of the potential of animals to locate mutualistic partners using chemoreception. Flow rate and turbulence associated with changing bottom roughness can dramatically alter the concentration and structure of odor plumes,[66,131] and the ability of organisms to locate a target using distance chemoreception[65,131] (see Section II.C.1). Thus, the ability of chemistry to mediate host–symbiont interactions may be restricted to relatively small-scale interactions or habitats with relatively slow flow.

Nevertheless, numerous studies offer convincing evidence that mobile invertebrates have the potential to use waterborne cues to locate hosts. One of the more well-studied examples involves hermit crabs (e.g., *Dardanus venosus* and *Pagurus pollicaris*) that place sea anemones (e.g., *Calliactis tricolor*) on their shell as a defense against predators. *Calliactis* "willingly" releases its grip on the substrate when mechanically stimulated by the crab, and gains a refuge from its own predators (e.g., sea stars and polychaete worms) as a result of its location on hermit crab shells.[185] Crabs actively search for and remove anemones from the substrate and attach them to their shells and appear to be aided in the location of anemones by chemical cues exuded from live, intact individuals.[186] In Y-maze experiments, over 80% (16/19 and 13/16 crabs in two separate experiments) of *Dardanus venosus* chose the arm of the "Y" that held *Calliactis* at its end. In contrast, *Pagurus pollicaris* and *Dardanus* that already had anemones on their shells did not discriminate between the arms of the Y-maze. In all experiments, flow rate into each arm was 3 cm s^{-1}, thus total flow in the central portion of the "Y" was 6 cm s^{-1}. No mention was made of the flow rate at the 30-m-deep site where these crabs were collected, so it is difficult to assess the applicability of these results to a field setting. However, crabs collected from this site did commonly have anemones on their shells, so they were able to locate anemones, and chemoreception provides a plausible mechanism for how this might occur.

Interestingly, crabs from shallower depths (where currents or wave-induced turbulence might be expected to be stronger and result in dilution and dispersion of chemical signals) rarely had anemones on their shells,[186] suggesting that water motion might disrupt the ability of crabs to locate symbionts. Clearly, the role of hydrodynamics in the ability of organisms to locate symbionts deserves attention similar to that of the role of chemoreception in locating prey[65,131,132] (see Section II.C.1). It may be that chemically mediated host location occurs primarily under low flow conditions, such as might occur at slack high or low tides in coastal locations. The impacts of how spatial and temporal patterns of water movement might impact the establishment and maintenance of symbioses is currently poorly understood.

Other species facultatively exploit the hermit crab–anemone mutualisms and use chemical cues produced by one or more of the partners to locate appropriate habitat. For example, the porcellanid crab *Porcellana sayana* is commonly found within the shells of hermit crabs with *Calliactis* anemones. Laboratory assays demonstrated that these crabs were attracted to odor from *Calliactis*, but not hermit crabs, and not another species of anemone, *Aiptasia pallida*, that is not found symbiotically with hermit crabs,[187] although flow rates in these experiments were very low (10 ml min⁻¹). Interestingly, when these porcellanids did contact *Aiptasia*, they were captured and killed in its tentacles, whereas porcellanids were often found moving freely amongst the tentacles of *Calliactis*, despite the fact that this species produces similar nematocysts. Although the mechanism by which porcellanids avoid being stung by *Calliactis* is unclear, a mechanism has been elucidated for a similar interaction among tropical Pacific sea anemones and "anemone fishes." These fishes do not appear to depend on obtaining masking compounds from the mucus of their host anemones, but rather have innate protection because they produce a mucus layer on the surface of their scales that is 3 to 4 times thicker than that of other fishes.[188,189]

Unlike studies of anti-predator defenses, rarely is an attempt made in studies of host location to identify specific chemical(s) responsible for eliciting host-location behavior. The lone exception occurs for the interactions between sea anemones and anemone fish that often occur in species-specific pairs. Naïve juvenile fish are attracted to their host anemones by chemicals released into the water and do not appear to use visual cues to locate hosts.[189] Murata et al.[190] demonstrated that young anemone fish (*Amphiprion perideraion*) were attracted to crude extracts of the mucus collected from their normal host anemone *Radianthus kuekenthali*. They also extracted intact anemones and partitioned the extract into aqueous and lipophilic fractions, and then further partitioned these fractions to isolate several compounds that appear to be responsible for attracting fish to anemones. In *Radianthus kuekenthali*, a single aqueous compound, amphikuemin (Structure 4.26), appears to be primarily responsible for the attraction of young anemone fish (*Amphiprion perideraion*). Various other lipophilic compounds were isolated from *Radianthus* crude extract, but these appeared to have less of an impact on fish behavior and required higher concentrations (10^{-6} M vs. 10^{-10} M for amphikuemin) to elicit a response. The species-specific nature of many anemone–anemone fish symbioses has been puzzling given that most anemone fishes do not illicit nematocyst discharge in a number of anemone species and thus have available to them a wide array of potential hosts. Species-specific associations appear to be maintained in some cases by species-specific host location cues,[189] although the ultimate reasons for the specificity remain unclear.

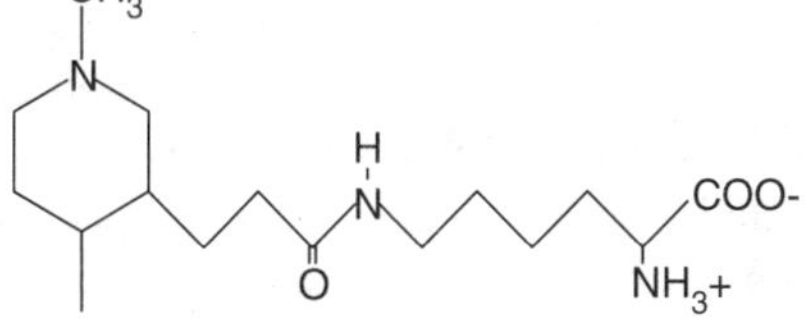

4.26

Numerous small invertebrates use chemical cues to initiate and maintain associations with sessile organisms that provide a structural or chemical refuge from predators. Some of these also benefit their hosts. Structurally complex organisms like branching corals and coralline algae that provide these refuges are often slow growing relative to chemically defended fleshy seaweeds, and become overgrown in the absence of herbivorous symbionts capable of consuming these seaweeds. Several species of small crab in the genus *Mithrax* readily consume chemically defended seaweeds like *Dictyota* and *Halimeda*, and thereby indirectly benefit the corals and calcified seaweeds in which they reside.[85,86,191] Because a diverse suite of other animals use these corals and coralline algae as habitat, these crabs may be thought of as "keystone" species in that without them, the critical habitat on which so many other species depend might disappear due to competition from more weedy species. Host chemistry plays a direct role in the maintenance of at least some of these associations. Crude lipophilic extracts of the coral *Oculina arbuscula* attract and enhance feeding rates of the crab *Mithrax forceps* on artificial diets. In the field, crabs do not feed directly on the coral, but consume lipid-rich mucus from the surface of the coral.[86] Chemical mediation of the establishment and maintenance of this type of cleaning mutualism could be widespread given that several other corals reward mutualistic crabs or shrimp with nutritional supplements.[192–194]

B. Associational Refuges

An associational refuge occurs when one organism that is resistant to some form of biotic or abiotic stress locally ameliorates that stress, facilitating the persistence of less tolerant species.[183] For many (if not most) chemically mediated associational refuges, that stress is predation. Those organisms that cannot physiologically sequester defensive compounds from chemically defended prey items (see Section II.A.2) may still gain refuges from predation by associating with chemically defended plants or invertebrates, in some cases "behaviorally" sequestering defensive metabolites. Reliance on an associational defense occurs in a wide range of both sessile and mobile taxa in terrestrial,[195] marine benthic,[38,39,70,84–86,95,108,167,196,197] and pelagic[198] environments. Palatable species may be at less risk of predation when growing near, or living on, unpalatable species because they are less likely to be discovered by predators. For example, the Caribbean crab *Thersandrus compressus* lives on and eats primarily the green alga *Avrainvillea longicaulis*, which produces a brominated diphenylmethane derivative called avrainvilleol (Structure 4.27) that deters feeding by reef fishes. Although the crab does not sequester avrainvilleol within its tissues, it is highly cryptic on its host alga and rarely preyed upon because omnivorous consumers avoid foraging in patches of chemically defended seaweeds like *Avrain-villea*.[39] The portunid crab *Caphyra rotundifrons* gains a similar refuge from predators by closely associating with the chemically defended alga, *Chlorodesmis fastigiata*, on the Great Barrier Reef.[38] Because mobile associates often resemble their host plant, many such refuges were initially interpreted as arising from visual crypsis, although it is increasingly apparent that host chemistry plays an important role in these interactions.[70,84,167]

4.27

A second, perhaps more intimate, form of associational defense involves the use of chemically defended seaweeds to construct unpalatable shelters or "domiciles" in which animals "behaviorally sequester" chemical defenses. The Caribbean amphipod *Pseudampithoides incurvaria* builds domiciles out of chemically defended seaweeds in the genus *Dictyota*. Amphipods within domiciles constructed of *Dictyota* are rejected by fishes as prey, but amphipods that have been forced to create domiciles of palatable seaweeds are readily consumed.[84] While plant secondary metabolites are most often thought of as having negative effects on herbivores (see Section II.B.2 and also Chapter 6 in this volume), the growing appreciation for associational defense highlights the importance of the positive indirect effects of plant defensive chemistry on higher trophic levels. Biogenic chemical complexity, in the form of predator-deterrent secondary metabolites, can thus be thought of as analogous to the structural complexity produced by coral reefs, kelp forests, and seagrass beds in that complexity generally enhances ecosystem diversity by providing refuges from predation.[199]

An increasing number of specialized interactions between small mobile invertebrates and chemically defended algae or sessile invertebrates have been reported in the marine environment,[6,167,183] and such interactions are also well known between terrestrial plants and insects.[200] However, there is some debate as to the ultimate factors driving this ecological specialization. Several authors have suggested that a plant's value as a refuge from predators (i.e., "enemy-free space"[84,201]) may be more important than the value of the plant as a food resource,[202] but this assertion has been controversial.[203–209] Herbivores in both marine and terrestrial systems specialize on particular hosts that provide both food and shelter from natural enemies, thus it is often difficult to assess the relative importance of these factors in selecting for ecological specialization. Decorator crabs (brachyuran crabs in the family Majidae) in the marine environment offer a novel opportunity for the study of specialization on chemically defended plants because these crabs place seaweeds on their backs as camouflage but do not necessarily feed on these same plants; thus, the choice of plants as food and shelter may be decoupled. One species in the southeastern United States, *Libinia dubia*, decorates almost exclusively with the brown alga *Dictyota menstrualis*, which produces the diterpene alcohol dictyol E (Structure 4.19) that makes it unpalatable to fishes; crabs use this chemical as a cue in selecting *Dictyota* for use as decoration.[70] Because these fishes also consume small invertebrates like crabs, *Dictyota* makes an ideal camouflage, and crabs decorated in this way experience much less predation than crabs decorated with seaweeds that are palatable to fishes.[70] In contrast to their specialized predator-avoidance and deterrence behaviors, these crabs are very generalized consumers and avoid consuming *Dictyota*, suggesting that predation pressure, rather than diet selection, drives specialization in this crab. Other studies involving specialization on chemically defended plants by nonherbivorous species have supported the conclusion that predation pressure alone can drive ecological specialization by small invertebrates.[210]

Although all of the chemically mediated associational defenses discussed thus far involve macroorganisms, there are likely to be undiscovered defenses involving microbes. As advances in microbial ecology have percolated through to chemical ecology, it has become apparent that many marine microbes produce bioactive secondary metabolites[211] that may be exploited by mobile marine invertebrates. As an example, embryos of the shrimp *Palaemon macrodactylus* are covered by a strain of *Alteromonas* sp. that produces 2,3-indolinedione (Structure 4.28) that chemically defends the embryos from attack by a pathogenic fungus.[212] It may turn out that many of the ecologically important compounds supposedly produced by marine seaweeds and macroinvertebrates are actually produced by microbial symbionts. Such is the case with the symbiotic cyanobacteria that live within the sponge *Dysidea herbacea*, as the brominated diphenyl ether (Structure 4.29) that defends the sponge from predation is located exclusively within (and is presumably produced by) the microbial symbionts.[213] Further research in this rapidly moving field will likely reveal that chemical ecologists have traditionally underemphasized the importance of microbes and microbial symbioses in producing the secondary metabolites that are important to defense of macroorganisms from predators, competitors, and physical stresses such as ultraviolet radiation.

4.28   4.29

It is increasingly recognized that mutualisms and commensalisms, like other forms of species interactions, can be highly conditional in nature.[181,183] The bulk of mutualisms may be facultative, and the strength and nature of their outcome may vary in space and time depending on prevailing environmental conditions and the presence of other species. One of these conditions can be the presence or concentration of particular secondary metabolites. For example, the refuge provided to small invertebrates by associating with a chemically defended seaweed should become less effective if the concentration of the defensive chemical decreases due to environmental stress,[88,214–216] or more effective if it increases due to induction,[71–73] although no studies have tested this directly. Because the small herbivores that use chemically defended plants as both food and shelter appear most likely to cause induction,[5] there is the possibility that these herbivores could indirectly control the refuge value of their host plant. Furthermore, because secondary metabolites often play multiple ecological roles (e.g., deterring predation and colonization by fouling organisms[169] or protection from predators and UV radiation[217]), compound concentrations (and thus the value of a plant as a refuge) may change due to factors other than herbivory. How these complex interdependencies alter the net outcome of plant–herbivore associations have yet to be examined experimentally and should provide a stimulating area for both empirical and theoretical research.

C. Local Specialization and Population Subdivision

Chemical mediation of symbiotic interactions in species-specific pairs has the potential to contribute to the isolation of populations within a species, and possibly speciaton. When individuals within a population exhibit dramatically different host choices, they may rarely contact each other, leading to differentiation and, eventually, isolation.[218] For example, nearly all pinnotherid crabs live symbiotically (generally as parasites or commensals) within the mantle cavities of bivalves, the tubes of polychaete worms, in the respiratory trees of holothurians, and among the spines of other echinoderms. Because crabs spend most of their benthic life associated with their hosts, strong host preferences among populations within a species have the potential to subdivide a species into reproductively isolated populations. Numerous pinnotherids have used chemoreception to locate appropriate hosts,[219–221] although many of these studies suffer from some of the methodological concerns discussed for Y-mazes (Section IV.A).

Many of these crabs are thought to be generalists on one group of hosts (e.g., sand dollars, mussels, etc.), but there is some evidence that different populations within a single pinnotherid species exhibit strikingly different preferences for individual host species, mediated by chemoreception. Thus, as is the case for some insects, species thought to be generalist in host choice may, in fact, be comprised of populations of specialists.[222,223] In some cases, these preferences are induced by contact with the host and may facilitate colonization of new hosts of the same species after the crab leaves the host to mate or in the event of the death of the host.[221] However, in other cases, preferences are not altered by exposure to new hosts. For example, Stevens[224,225] presents evidence for two genetically distinct host races of the pinnotherid *Pinnotheres novaezelandiae*: one that inhabits the mussel *Mytilus edulis aoteanus* and one that inhabits the mussel *Perna canaliculus*. Field-collected *Pinnotheres* were attracted to chemical cues from the host species in which they were found, but not from other host species, and 4 weeks of acclimation with novel hosts failed to induce attraction.[224] An electrophoretic analysis of the population structure of this species indicated that individuals collected sympatrically from *Perna* and *Mytilus* exhibited striking levels of genetic differentiation, suggesting the possibility of cryptic sibling species formation.[225,226] Another subpopulation of *Pinnotheres novaezelandiae* was found exclusively in another bivalve *Mactra ovata ovata*; although chemolocation assays were not performed with these crabs, they are genetically and morphologically distinct from other *P. novaezelandiae* and probably represent a different species.[225,227]

In all the chemolocation assays with pea crabs, the attraction of crabs from each host race to their respective host was substantially less than 100%, and some crabs chose "incorrect" hosts, suggesting that there may still be at least the potential for gene flow among populations. These experiments were conducted with adult crabs, and host location experiments with larval or post-larval forms that are normally responsible for initial host selection might be particularly informative. Nevertheless, this series of studies suggests that within-species differences in chemically mediated host location may lead to population subdivision and reproductive isolation among marine species.

Other evidence for chemical mediation of local specialization in marine invertebrates comes from several studies of associational refuges between seaweeds and herbivores (discussed in Section IV.B.). Dictyol E (Structure 4.19) from the brown alga *Dictyota menstrualis* deters feeding by omnivorous fishes and stimulates the decorator crab *Libinia dubia* to cover its carapace with this alga as a refuge from predators.[70] Similarly, pachydictyol A (Structure 4.16) from the related alga *Dictyota bartraysii* stimulates domicile construction by the amphipod *Pseudampithoides incurvaria*.[84] Interestingly, in both cases, these specialized recognition systems appear to be localized adaptations. *P. incurvaria* from Belize recognize appropriate hosts using pachydictyol A, but populations of the same amphipod from the Bahamas do not.[84] Similarly, the decorator crab and its preferred decoration material do not have completely overlapping ranges, and crabs from outside the range of *D. menstrualis* are unable to recognize this alga as an appropriate decoration material.[228] Thus, there seems to be differentiation among populations of these small herbivores with respect to chemically mediated host recognition. The study of the ecological and evolutionary consequences of such variation is in its infancy in the marine environment, but terrestrial studies have shown that such variability in chemically mediated host recognition can play an important role in coevolutionary processes and speciation.[229]

V. CHEMICAL MEDIATION OF REPRODUCTIVE PROCESSES

A wide array of reproductive processes, from intraspecific contests to mate location to induction of spawning, are mediated by chemistry. Many of these interactions involve chemical mediation of mate recognition during spawning. These "sex pheromones" may play an important role in the reproductive isolation of species, as differences in the structure or amounts of these pheromones that are released, or differences in sensitivities to these compounds, can result in assortative mating. Chemical mediation of sperm–egg attraction and fusion can also erect barriers to reproduction. In the few cases where these chemical cues have been identified, important evolutionary insights into mechanisms of reproductive isolation have been gleaned.

A. Sex Pheromones

Sex pheromones have been widely reported among crustaceans, however many of these reports are anecdotal in nature or, if experimental, lack appropriate controls.[230,231] Nevertheless, there is solid, mounting evidence for the importance of chemicals in controlling mating behavior in a diversity of crustaceans including amphipods, anomurans, and brachyurans. Bioassays using a Y-maze design (see Section IV.A) to deliver chemical stimuli from live organisms have been useful in demonstrating chemically mediated mate attraction. For example, Borowsky[232] showed that male amphipods (*Microdeutopus gryllotalpa*) are attracted to receptive females by a waterborne substance that is not released by other males or nonreceptive females. The compound(s) responsible are known to be polar, and their effectiveness at eliciting a response is reduced when combined with exudates from the green alga, *Ulva*.[233] This suggests that pheromone detection and interpretation by organisms within the context of the diversity of odors found in the field may be more complicated than in the filtered-seawater environment typical of most laboratory assays.

These types of assays[232,233] have been successfully combined with amphipod culture techniques[108] to investigate the effects of diet on pheromone composition in the context of a potential mechanism promoting reproductive isolation.[234] Genetically distinct populations of the estuarine amphipod *Eogammarus confervicolus* from British Columbia differed in their ability to modify their pheromone secretions with diet-derived chemicals. Females living on the fucoid algae *Fucus* and *Pelvetia* produced pheromones that attracted males living on fucoids but did not attract males collected from salt marshes or woody debris. Even geographically distant (several hundred Km) populations of amphipods living on fucoids were mutually attractive. Restriction fragment length polymorphism (RFLP) analysis of population structure indicated that all fucoid-dwelling populations were genetically similar and distinct from other populations of *E. confervicolus*, but all *E. confervicolus* were more closely related to each other than to a congener.[235] This suggests that individuals living on fucoids represent a distinct host race, at least partially reproductively isolated from other populations by the diet-modified pheromone they produce.

B. SYNCHRONIZATION OF REPRODUCTION

Although gametes of free-spawning organisms can potentially be transported great distances, the probability of finding conspecific gametes with which to fuse decreases exponentially with distance from release.[236,237] In most free-spawning organisms, the probability of successful gamete fusion approaches zero when spawning organisms are not located in close proximity.[238–241] Thus, there can be a strong advantage not only to living in a high-density aggregation of conspecifics, but also to having reproduction among those conspecifics highly synchronized to enhance fertilization success.

Although many reproductive cycles appear to be on lunar, circadian, or circatidal rhythms,[242,243] chemical cues are important in fine-tuning spawning synchrony. For example, reproduction in the lugworm *Arenicola marina* occurs only once a year, and local populations exhibit highly synchronous spawning. However, populations from neighboring beaches spawn at different times, so concordant environmental cues are not the proximate factor synchronizing reproduction in this species; well-synchronized reproduction in this species is achieved largely by pheromonal control.[244] Gamete release in both sexes is stimulated by exposure to water in which other individuals had previously spawned, and release is triggered by a waterborne cue rather than by direct contact with gametes. However, dilution of sperm in the water is so rapid,[237] that active transport of sperm into burrows where females have laid their eggs is required to ensure fertilization.[245] Another waterborne chemical cue associated with spawning induces females to pump seawater through their burrow, increasing the probability of contact between sperm and egg.[244] Similar chemical synchronization of spawning may occur in other polychaetes,[246,247] sponges,[248] and echinoderms.[249–252]

Synchronization of reproduction may be particularly important for the persistence of species that have semelparous reproductive strategies. For many semelparous polychaetes, exogenous environmental cues such as photoperiod or lunar phase serve to synchronize the timing of spawning.[243] However, mate location, sex recognition, swarming behavior, and gamete release are controlled by a variety of chemical cues.[247,253,254] Few of the proposed pheromones involved in the synchronization of reproduction in mobile marine invertebrates have been purified or structurally characterized, with the exception of those produced by some nereid polychaetes.[253–255] Studies that separate and isolate these compounds have led to the discovery that differences in the structure or concentration of pheromones used to coordinate reproduction may serve to facilitate reproductive isolation in closely related species. For example, the polychaetes *Nereis succinea* and *Platynereis dumerilii* co-occur in Danish fjords and spawn at the same time of year. Male swarming behavior and sperm release is enhanced in both species by the pheromone 5-methyl-3-heptanone (Structure 4.30). However, these species apparently do not form heterospecifc mating swarms, because induction of spawning in *N. succinea* requires 25 times the concentration of sex pheromone as *P. dumerilii*; concentrations of the pheromone that are high enough to induce spawning in *N. succinea* inhibit swarming behavior in *P. dumerilii*.[246] In *P. dumerilii*, different optical isomers of the same compound (Structure 4.30) control male and female gamete release.[253]

In most, if not all, of these cases, environmental cues or endogenous rhythms cue the initiation of gametogenesis, and chemical cues only impact the final, coordinated phases of spawning (i.e., courtship, gamete release). One exception is the holothurian *Cucumaria frondosa*, which does not exhibit any sort of aggregative behavior, but emits a waterborne pheromone that synchronizes the development of gametes among individuals within a population.[252]

Fertilization success could also be enhanced by chemical attraction of sperm to eggs. This has been reported for a variety of free-spawning mobile invertebrates including holothurians, ophiuroids, asteroids, and molluscs,[256–258] as well as sessile invertebrates,[259,260] and seaweeds.[261,262] In general, the distance over which attraction occurs is usually only a few hundred μm, and long-distance attraction seems unlikely to occur due to the small amounts of attractant released and the weak swimming ability of sperm relative to ocean currents or wave motion. Attraction over short distances, however, may increase the effective target size of the egg and result in an increase in fertilization success due to a higher probability of sperm–egg collision.[218] The importance of sperm chemotaxis under realistic flow conditions remains to be seen as turbulence can markedly decrease fertilization success.[263] Sperm chemotaxis may also serve a gamete recognition function, similar to the species-specific sperm–egg binding found in some sea urchins.[264] Minor changes in the sperm acrosomal protein (bindin) and the sperm receptor on the egg may play a major role in the formation of reproductively isolated echinoid species that have long-range larval dispersal.[265] However, sperm chemotaxis among many other mobile invertebrates is more often specific at only the family or genus level,[256–258] so it is unclear how broadly important this mechanism might be in the formation of new species for the bulk of marine taxa.

C. Timing of Larval Release

For species that brood larvae, chemical cues may induce larval release. The timing of larval release in these species is often broadly synchronized to environmental cues such as light–dark cycles, tidal amplitude, and lunar phase, in order to minimize predation, temperature stress, desiccation, photodamage, and salinity stress.[266,267] As with synchronized spawning, chemical cues may act to narrow the window within which larval release occurs. For example, in the estuarine crab *Rhithropanopeus harrisii*, larval release appears to be synchronized on a circadian or circatidal rhythm, depending on local environmental conditions.[268] However, when pregnant female crabs with mature embryos are exposed to water containing recently expelled larvae, they pump their abdomen in a manner normally associated with the expulsion of larvae.[269] Exposure to conspecific larvae themselves had no effect on crab behavior, while crushed eggs did elicit the pumping behavior. Additionally, eggs close to hatching (<1 day) elicited a pumping response in more females than eggs 8 to 10 days from hatching, suggesting that a cue is released from eggs at or near the time of hatching to signal expulsion by the mother. If this cue reaches other individuals within the population, it could synchronize larval release among individuals.

A similar release pheromone occurs in a related xanthid crab *Neopanope sayi*.[270] This pheromone has been partially characterized as a peptide rich in cysteine, glycine, methionine, and isoleucine less than 1000 Da in size. Interestingly, *N. sayi* was also induced to spawn by water conditioned with newly hatched eggs of several other decapod crabs (*Rhithropanopeus harrisii* and *Uca pugilator*), although higher concentrations were required.[270] Nevertheless, waterborne chemicals have the potential to help synchronize larval release across a number of distantly related brachyuran crabs, potentially leading to the local swamping of predators.

VI. CONCLUSION

Chemical mediation of interactions among species in the marine environment goes well beyond anti-predator defense, and in perhaps no group is the diversity of chemically mediated interactions better represented than in mobile invertebrates. Mobile invertebrates use chemicals to track prey, detect and repel predators, find mates or mutualistic partners, and compete for space or other limited resources. Of course, these research areas transcend taxonomic boundaries, and active research is ongoing with a wide variety of marine plants (both macroscopic and microscopic) and mobile and sessile animals in both benthic and pelagic systems (see other chapters in this volume). Some subfields of marine chemical ecology have progressed rapidly due to innovative and highly successful collaborations between chemists and marine ecologists. The combination has undoubtedly been mutualistic, with ecologists gaining an appreciation of the chemical complexity involved in their systems and the need for careful quantification of compounds and concentrations and chemists gaining an appreciation for the inherently "noisy" behavioral and ecological world within which chemicals act as signals. The resulting blend of ecologically realistic bioassays and rigorous chemical separation, quantification, and structural elucidation has yielded quantum leaps in our understanding of chemical mediation of interactions among organisms. Similar collaborations involving chemists, ecologists, and microbial ecologists are beginning to assess how commonly macroorganisms rely on microbial symbionts for chemical defense against predators, pathogens, competitors, and other environmental hazards.

At this point, we know chemistry mediates a tremendous diversity of behavioral interactions among individual organisms. However, much remains to be learned about both the broader implications of these behaviors for community ecology and evolutionary biology and the mechanistic details underlying these behaviors. New collaborations combining ecology, genetics, and chemistry have made important strides in understanding processes leading to reproductive isolation and speciation and should be encouraged. On the mechanistic side, our understanding of the actual basis for how most chemically mediated interactions work is rudimentary. Although the sensory physiology behind the reception of waterborne chemical cues has been investigated in some detail for particular species,[271] it is much less clear why particular compounds in natural foods serve as feeding deterrents, either from a physiological or an ecological perspective. Increased collaboration among not only ecologists and chemists but also physiologists, behaviorists, and evolutionary biologists will provide the framework to answer the larger questions in marine ecology that increasingly cross traditional disciplinary boundaries.

ACKNOWLEDGMENTS

References and access to unpublished data were provided by E. Cruz-Rivera, N. Lindquist, E. Sotka, and S. Woodin. Comments from S. Bullard, C. Kicklighter, E. Cruz-Rivera, E. Sotka, and an anonymous reviewer improved earlier drafts of the manuscript.

REFERENCES

1. Hay, M.E. and Fenical, W., Marine plant–herbivore interactions: the ecology of chemical defense, *Ann. Rev. Ecol. Syst.* 19, 111, 1988.
2. Hay, M.E. and Steinberg, P.D., The chemical ecology of plant–herbivore interactions in marine versus terrestrial communities, in *Herbivores: Their Interaction with Secondary Plant Metabolites: Ecological and Evolutionary Processes*, Vol. 2, Rosenthal, G.A. and Berenbaum, M.R., Eds., Academic Press, San Diego, 1992, 371.
3. Paul, V.J., *Ecological Roles of Marine Natural Products*, Comstock, Ithaca, NY, 1992.
4. Pawlik, J.R., Marine invertebrate chemical defenses, *Chem. Rev.*, 93, 1911, 1993.
5. Hay, M.E., Marine chemical ecology: what's known and what's next?, *J. Exp. Mar. Biol. Ecol,* 200, 103, 1996.
6. McClintock, J.B. and Baker, B.J., A review of the chemical ecology of Antarctic marine invertebrates, *Am. Zool.*, 37, 329, 1997.
7. Hay, M.E., Stachowicz, J.J., Cruz-Rivera, E., Bullard, S., Deal, M., and Lindquist, N., Bioassays with marine and freshwater macro-organisms, in *Methods in Chemical Ecology*, Millar, J.G. and Haynes, K.F., Eds., Chapman and Hall, New York, 1998, 39.
8. Sammarco, P.W. and Coll, J.C., Chemical adaptations in the Octocorallia: evolutionary considerations, *Mar. Ecol. Prog. Ser.*, 88, 93, 1992.
9. Pawlik, J.R., Chanas, B., Toonen, R.J., and Fenical, W., Defenses of Caribbean sponges against predatory reef fish. I. chemical deterrency, *Mar. Ecol. Prog. Ser.*, 127, 183, 1995.
10. McConnell, O.J., Hughes, P.A., Targett, N.M., and Daley, J., Effects of secondary metabolites on feeding by the sea urchin, *Lytechinus variegatus, J. Chem. Ecol.*, 8, 1437, 1982.
11. Faulkner, D.J., Chemical defenses of marine molluscs, in *Ecological Roles of Marine Natural Products*, Paul, V.J., Ed., Comstock, Ithaca, NY, 1992, 119.
12. Cimino, G., De Rosa, S., De Stefano, S., Sodano, G., and Villani, G., Dorid nudibranch elaborates its own chemical defense, *Science*, 219, 1237, 1983.

13. Gustafson, K. and Anderson, R.J., Chemical studies of British Columbia nudibranchs, *Tetrahedron*, 41, 1101, 1985.

14. Jensen, K.R., Evolution of the Sacoglossa (Mollusca, Opisthobranchia) and the ecological associations with their food plants, *Evol. Ecol.*, 11, 301, 1997.

15. Ireland, C. and Scheuer, P.J., Photosynthetic marine mollusks: In vivo ^{14}C incorporation into metabolites of the sacoglossan *Placobranchus ocellatus*, *Science*, 205, 922, 1979

16. Kubanek, J., Graziani, E.I., and Andersen, R.J., Investigations of terpenoid biosynthesis by the dorid nudibranch *Cadlina luteomarginata*, *J. Org. Chem.*, 62, 7239, 1997.

17. Cimino, G. and Ghiselin, M.T., Chemical defense and evolution in the sacoglossa (Mollusca: Gastropoda: Opisthobranchia), *Chemoecology* 8, 51, 1998.

18. Cimino, G. and Ghiselin, M.T., Chemical defense and evolutionary trends in biosynthetic capacity among dorid nudibranchs, *Chemoecology*, 9, 187, 1999.

19. Avila, C., Cimino, G., Crispino, A., and Spinella, A., Drimane sesquiterpenoids in Mediterranean *Dendrodoris* nudibranchs: anatomical distribution and biological role, *Experientia*, 47, 306, 1991.

20. McQuaid, C.D., Cretchley, R., and Rayner, J.L., Chemical defence of the intertidal pulmonate limpet *Siphonaria capensis* (Quoy & Gaimard) against natural predators, *J. Exp. Mar. Biol. Ecol*, 237, 141, 1999.

21. Heine, J.N., McClintock, J.B., Slattery, M., and Weston, J., Energetic composition, biomass, and chemical defense in the common Antarctic nemertean *Parborlasia corrugatus* McIntosh, *J. Exp. Mar. Biol. Ecol.*, 153, 15, 1991.

22. Kem, W.R., unpublished data cited in ref. 6.

23. Kem, W.R., Structure and action of nemertine toxins, *Am. Zool.*, 25, 99, 1985.

24. Luckenbach, M.W. and Orth, R.J., A chemical defense in Crustacea? *J. Exp. Mar. Biol. Ecol.*, 137, 79, 1990.

25. Norton, S.F. and Stallings, C.D., The distribution and abundance of three aposematic, chemically defended gammarid amphipods off Bell Island, WA, *Am. Zool.*, 39, 119A, 1999.

26. Tomaschko, K.H., Ecdysteroids from *Pycnogonum litorale* (Arthropoda, Pantopoda) act as a chemical defense against *Carcinus maenas* (Crustacea, Decapoda), *J. Chem. Ecol.*, 20, 1445, 1994.

27. Buckmann, D. and Tomaschko, K.H., 20-Hydroxyecdysone stimulates molting in pycnogonid larvae (Arthropoda, Pantopoda), *Gen. Comp. Endocrinol.*, 88, 261, 1992.

28. Bryan, P.J., McClintock, J.B., and Hopkins, T.S., Structural and chemical defenses of echinoderms from the northern Gulf of Mexico, *J. Exp. Mar. Biol. Ecol.*, 210, 173, 1997.

29. Hay, M.E., Renaud, P.E., and Fenical, W., Large mobile versus small sedentary herbivores and their resistance to seaweed chemical defenses, *Oecologia*, 75, 246, 1988.

30. Stachowicz, J.J. and Hay, M.E., Reduced mobility is associated with compensatory feeding and increased diet breadth of marine crabs, *Mar. Ecol. Prog. Ser.*, 188, 169, 1999.

31. Faulkner, D.J. and Ghiselin, M.T., Chemical defense and evolutionary ecology of dorid nudibranchs and some other opisthobranch gastropods, *Mar. Ecol. Prog. Ser.*, 13, 295, 1983.

32. Cimino, G. and Sodano, G., Transfer of sponge secondary metabolites to predators, in *Sponges in Time and Space. Biology, Chemistry, Paleontology*, van Kempen, T.M.G. and Braekman, J.C., Eds, A.A. Balkema, Rotterdam, 1994, 459.

33. Pawlik, J.R., Kernan, M.R., Molinski, T.F., Harper, M.K., and Faulkner, D.J., Defensive chemicals of the Spanish dancer nudibranch *Hexabranchus sanguineus* and its egg ribbons: Macrolides derived from a sponge diet, *J. Exp. Mar. Biol. Ecol.*, 119, 99, 1988.

34. Avila, C., Cimino, G., Fontana, A., Gavagnin, M., Ortea, J., and Trivellone, E., Defensive strategy of two *Hypselodoris* nudibranchs from Italian and Spanish coasts, *J. Chem. Ecol.*, 17, 625, 1991.

35. Avila, C. and Paul, V.J., Chemical ecology of the nudibranch *Glossodoris pallida*: is the location of diet-derived metabolites important for defense? *Mar. Ecol. Prog. Ser.*, 150, 171, 1997.

36. Ireland, C. and Faulkner, D.J., The metabolites of the marine molluscs *Tridachiella diomedea* and *Tridachia crispata*, *Tetrahedron*, 37, 233, 1981.

37. Paul, V.J. and Van Alstyne, K.L., Use of ingested algal diterpenoids by *Elysia halimedae* Macnae (Opisthobranchia: Ascoglossa) as anti-predator defenses, *J. Exp. Mar. Biol. Ecol.*, 119, 15, 1988.

38. Hay, M.E., Pawlik, J.R., Duffy, J.E., and Fenical, W., Seaweed–herbivore–predator interactions: host-plant specialization reduces predation on small herbivores, *Oecologia*, 81, 418, 1989.

39. Hay, M.E., Duffy, J.E., Paul, V.J., Renaud, P.E., and Fenical, W., Specialist herbivores reduce their susceptibility to predation by feeding on the chemically-defended seaweed *Avrainvillea longicaulis*, *Limnol. Oceanogr.*, 35, 1734, 1990.

40. Hay, M.E., Fenical, W., and Gustafson, K., Chemical defense against diverse coral reef herbivores, *Ecology*, 68, 1581, 1987.

41. Kinnel, R.B., Dieter, R.K., Meinwald, J., Van Engen, D., Clardy, J., Eisner, T., Stallard, M.O., and Fenical, W., Brasilenyne and cis-dihydrorhodophytin: antifeedant medium-ring haloethers from a sea hare (*Aplysia brasiliana*), *Proc. Nat. Acad. Sci. USA*, 76, 3576, 1979.

42. Pennings, S.C., Interspecific variation in chemical defenses in the sea hares (Opisthobranchia: Anaspidea), *J. Exp. Mar. Biol. Ecol.*, 180, 203, 1994.

43. de Nys, R., Steinberg, P.D., Rogers, C.N., and Charleton, T., Quantitative variation of secondary metabolites in the sea hare *Aplysia parvula* and its host plant, *Mar. Ecol. Prog. Ser.*, 130, 135, 1996.

44. Johnson, P.M. and Willows, A.O.D., Defense in sea hares (gastropoda, opisthobranchia, anaspidea): multiple layers of protection from egg to adult, *Mar. Fresh. Behav. Physiol.*, 32, 147, 1999.

45. Nolen, T.G., Johnson, P.M., Kicklighter, C.E., and Capo, T., Ink secretion by the marine snail *Aplysia californica* enhances its ability to escape from a natural predator, *J. Comp. Physiol. A*, 176, 239, 1995.

46. Shumway, S.E., A review of the effects of algal blooms on shellfish and aquaculture, *J. World Aquaculture Soc.*, 21, 65, 1990.

47. Kvitek, R.G. and Beitler, M.K., Relative insensitivity of butter clam neurons to saxitoxin: a preadaptation for sequestering paralytic shellfish poisoning toxins as a chemical defense, *Mar. Ecol. Prog. Ser.*, 69, 47, 1991.

48. Kvitek, R.G., Sequestered paralytic shellfish poisoning toxins mediate glaucous-winged gull predation on bivalve prey, *Auk*, 108, 381, 1991.

49. Kvitek, R.G., DeGange, A.R., and Beitler, M.K., Paralytic shellfish poisoning toxins mediate feeding behavior of sea otters, *Limnol. Oceanogr.*, 36, 393, 1991.

50. Simenstad, C.A., Estes, J.A., and Kenyon, K.W., Aleuts, sea otters, and alternate stable states, *Science*, 200, 403, 1978.

51. Duggins, D.O., Simenstad, C.A., and Estes, J.A., Magnification of secondary production by kelp detritus in coastal marine ecosystems, *Science*, 245, 170, 1989.

52. Pfeiffer, W., The distribution of fright reaction and alarm substance cells in fishes, *Copeia*, 1977, 653.

53. Smith, R.J.F., Alarm signals in fishes, *Rev. Fish Biol. Fishes*, 2, 33, 1992.

54. Mathis, A. and Smith, R.J.F., Chemical alarm signals increase the survival time of fathead minnows (*Pimephales promelas*) during encounters with northern pike (*Esox lucius*), *Behav. Ecol.*, 4, 260, 1993.

55. Mathis, A. and Smith, R.J.F., Intraspecific and cross-superorder responses to chemical alarm signals by brook stickleback, *Ecology*, 74, 2395, 1993.

56. Mathis, A., Chivers, D.P., and Smith, R.J.F., Chemical alarm signals: predator deterrents or predator attractants?, *Am. Nat.*, 145, 994, 1995.

57. Hews, D.K., Alarm response in larval western toads, *Bufo boreas*: release of larval chemicals by a natural predator and its effect on predator capture efficiency, *Anim. Behav.*, 36, 125, 1988.

58. Hannson, L.A., Behavioural response in plants: adjustment in algal recruitment induced by herbivores, *Proc. R. Soc. Lond. B*, 263, 1241, 1996.

59. Snyder, N. and Snyder, H., Alarm response of *Diadema antillarum*, *Science*, 168, 276, 1970.

60. Sleeper, H.L., Paul, V.J., and Fenical, W., Alarm pheromones from the marine opisthobranch *Navanax inermis*, *J. Chem. Ecol.*, 6, 57, 1980.

61. Lawrence, J.M., A chemical alarm response in *Pycnopodia helianthoides* (Echinodermata: Asteroidea), *Mar. Behav. Physiol.*, 19, 39, 1991.

62. Vadas, R.L., Burrows, M.T., and Hughes, R.N., Foraging strategies of dogwhelks, *Nucella lapillus* (L.): interacting effects of age, diet and chemical cues to the threat of predation, *Oecologia*, 100, 439, 1994.

63. Carte, B. and Faulkner, D.J., Role of secondary metabolites in feeding associations between a predatory nudibranch, two grazing nudibranchs, and a bryozoan, *J. Chem. Ecol.*, 12, 795, 1986.

64. Zimmer-Faust, R.K., The relationship between chemoreception and foraging behavior in crustaceans, *Limnol. Oceanogr.*, 34, 1367, 1989.

65. Weissburg, M.J. and Zimmer-Faust, R.K., Life and death in moving fluids: hydrodynamic effects on chemosensory-mediated predation, *Ecology*, 74, 1428, 1993.

66. Moore, P.A., Weissburg, M.J., Parrish, J.M., Zimmer-Faust, R.K., and Gerhardt, G.A., Spatial distribution of odors in simulated benthic boundary layer flows, *J. Chem. Ecol.*, 20, 255, 1994.

67. Weissburg, M.J. and Zimmer-Faust, R.K., Odor plumes and how blue crabs use them in finding prey, *J. Exp. Biol.*, 197, 349, 1994.

68. Cronin, G. and Hay, M.E., Susceptibility to herbivores depends on recent history of both plant and animal, *Ecology*, 77, 1531, 1996.

69. Richards, R.A. and Cobb, J.S., Use of avoidance responses to keep spider crabs out of traps for American lobsters, *Trans. Am. Fish. Soc.*, 116, 282, 1987.

70. Stachowicz, J.J. and Hay, M.E., Reducing predation through chemically-mediated camouflage: indirect effects of plant defenses on herbivores, *Ecology*, 80, 495, 1999.

71. Van Alstyne, K.L., Herbivore grazing increases polyphenolic defenses in an intertidal brown alga *Fucus distichus*, *Ecology*, 69, 655, 1988.

72. Cronin, G. and Hay, M.E., Induction of seaweed chemical defenses by amphipod grazing, *Ecology*, 77, 2287, 1996.

73. Harvell, C.D., The ecology and evolution of inducible defenses in a marine bryozoan: cues, costs and consequences, *Am. Nat.*, 128, 810, 1986.

74. Ayre, D.J. and Grosberg, R.K., Aggression, habituation, and clonal co-existance in the sea anemone *Anthopleura elegantissima*, *Am. Nat.*, 146, 427, 1995.

75. Van Donk, E., Lurling, M., and Lampert, W., Consumer-induced changes in phytoplankton: inducibility, costs, benefits, and the impact on grazers, in *The Ecology and Evolution of Inducible Defenses*, Tollrian, R. and Harvell, C.D., Eds., Princeton University Press, Princeton, NJ, 1999, 89.

76. Karban, R. and Baldwin, I.T., *Induced responses to herbivory*, University of Chicago Press, Chicago, IL, 1997.

77. Gilbert, J.J., Rotifer ecology and embryological induction, *Science,* 151, 1234, 1966.

78. Dodson, S.I., The ecological role of chemical stimuli for the zooplankton: predator induced morphology in *Daphnia, Oecologia,* 78, 361, 1989.

79. Appleton, R.D. and Palmer, A.R., Water-borne stimuli released by predatory crabs and damaged prey induce more predator-resistant shells in a marine gastropod, *Proc. Nat. Acad. Sci. USA*, 85, 4387, 1988.

80. Palmer, A.R., Effect of crab effluent and scent of damaged conspecifics on feeding, growth, and shell morphology of the Atlantic dogwhelk *Nucella lapillus* (L.), *Hydrobiologia*, 193, 155, 1990.

81. Trussell, G.C., Phenotypic plasticity in an intertidal snail: the role of a common crab predator, *Evolution*, 50, 448, 1996.

82. Leonard, G.H., Bertness, M.D., and Yund, P.O., Crab predation, waterborne cues, and inducible defenses in the blue mussel *Mytilus edulis*, *Ecology*, 80, 1, 1999.

83. Tollrian, R. and Harvell, C.D., The evolution of inducible defenses: current ideas, in *The Ecology and Evolution of Inducible Defenses*, Tollrian, R. and Harvell, C.D., Eds., Princeton University Press, Princeton, NJ, 1999, 157.

84. Hay, M.E., Duffy, J.E., and Fenical, W., Host-plant specialization decreases predation on a marine amphipod: an herbivore in plant's clothing, *Ecology*, 71, 733, 1990.

85. Stachowicz, J.J. and Hay, M.E., Facultative mutualism between an herbivorous crab and a coralline alga: advantages of eating noxious seaweeds, *Oecologia*, 105, 377, 1996.

86. Stachowicz, J.J. and Hay, M.E., Mutualism and coral persistence: the role of herbivore resistance to algal chemical defense, *Ecology*, 80, 2085, 1999.

87. Hay, M.E., Duffy, J.E., and Pfister, C.A., Chemical defense against different marine herbivores: are amphipods insect equivalents?, *Ecology*, 68, 1567, 1987.

88. Pennings, S.C., Multiple factors promoting narrow host range in the sea hare, *Aplysia californica*, *Oecologia*, 82, 192, 1990.

89. Duffy, J.E. and Hay, M.E., Herbivore resistance to seaweed chemical defense: the role of mobility and predation risk, *Ecology*, 75, 1304, 1994.

90. Cruz-Rivera, E. and Hay, M.E., Can quantity replace quality? Food choice, compensatory feeding, and fitness of marine mesograzers, *Ecology*, 81, 201, 2000.

91. Becerro, M.A., Paul, V.J., and Starmer, J., Intracolonial variation in chemical defenses of the sponge *Cacospongia* sp. and its consequences on generalist fish predators and the specialist nudibranch predator *Glossodoris pallida*, *Mar. Ecol. Prog. Ser.*, 168, 187, 1998.

92. Hay, M.E., The ecology of seaweed-herbivore interactions on coral reefs, *Coral Reefs*, 16, S67, 1997.

93. Miller, M.W. and Hay, M.E., Coral-seaweed-grazer-nutrient interactions on temperate reefs, *Ecol. Monogr.*, 66, 323, 1996.

94. Duffy, J.E. and Hay, M.E., Strong impacts of grazing amphipods on the organization of a benthic community, *Ecol. Monogr.*, 70, 237, 2000.

95. Hay, M.E., Associational plant defenses and the maintenance of species diversity: turning competitors into accomplices, *Am. Nat.*, 128, 617, 1986.

96. Abele, L.G., Species diversity of decapod crustaceans in marine habitats, *Ecology*, 55, 156, 1974.

97. Heck, K.L. and Wetstone, G.S., Habitat complexity and invertebrate species richness and abundance in tropical seagrass meadows, *J. Biogeog.*, 4, 135, 1977.

98. Hay, M.E., Spatial patterns of herbivore impact and their importance in maintaining algal species richness, *Proc. Fifth Int. Coral Reef Congr.*, 4, 29, 1985.

99. Stoner, A.W. and Lewis, F.G., The influence of quantitative and qualitative aspects of habitat complexity in tropical sea-grass meadows, *J. Exp. Mar. Biol. Ecol.*, 94, 19, 1985.

100. Schiel, D.R., Growth, survival and reproduction of two species of marine algae at different densities in natural stands, *J. Ecol.*, 73, 199, 1985.

101. Paine, R.T., Benthic macroalgal competition: complications and consequences, *J. Phycol.*, 26, 12, 1990.

102. Steneck, R.S., Hacker, S.D., and Dethier, M.N., Mechanisms of competitive dominance among crustose coralline algae: an herbivore mediated competitive reversal, *Ecology*, 72, 938, 1991.

103. Bertness, M.D. and Leonard, G.H., The role of positive interactions in communities: lessons from intertidal environments, *Ecology*, 78, 1976, 1997.

104. Dean, R.L. and Connell, J.H., Marine invertebrates in algal succession. II. Tests of hypotheses to explain changes in diversity with succession, *J. Exp. Mar. Biol. Ecol.*, 109, 217, 1987.

105. Anderson, T.W., Role of macroalgal structure in the distribution and abundance of a temperate reef fish, *Mar. Ecol. Prog. Ser.*, 113, 279, 1994.

106. Carr, M.H., Effects of macroalgal dynamics on recruitment of a temperate reef fish, *Ecology*, 75, 1320, 1994.

107. Levin, P.S. and Hay, M.E., Responses of temperate reef fishes to alterations in algal structure and species composition, *Mar. Ecol. Prog. Ser.*, 134, 37, 1996.

108. Duffy, J.E. and Hay, M.E., Food and shelter as determinants of food choice by an herbivorous marine amphipod, *Ecology*, 72, 1286, 1991.

109. Steinberg, P.D. and van Altena, I., Tolerance of marine invertebrate herbivores to brown algal phlorotannins in temperate Australasia, *Ecol. Monogr.*, 62, 189, 1992.

110. Gerhart, D.J., Prostaglandin A_2: an agent of chemical defense in the Caribbean gorgonian *Plexaura homomalla*, *Mar. Ecol. Prog. Ser.*, 19, 181, 1984.

111. Lindquist, N. and Hay, M.E., Can small rare prey be chemically defended? The case for marine larvae, *Ecology*, 76, 1347, 1995.

112. Paul, V.J. and Fenical, W. Chemical defense in tropical green algae, order Caulerpales, *Mar. Ecol. Prog. Ser.*, 34, 157, 1986.

113. Winter, F.C. and Estes, J.A., Experimental evidence for the effects of polyphenolic compounds from *Dictyoneurum californicum* Ruprecht (Phaeophyta: Laminariales) on feeding rate and growth in the red abalone *Haliotus rufescens* Swainson, *J. Exp. Mar. Biol. Ecol.*, 155, 263, 1992.

114. Cruz-Rivera, E., Effects of diet on temperate mesograzers: consequences of prey traits for consumer feeding and fitness, Ph. D. dissertation, University of North Carolina, Chapel Hill, 1999.

115. Tugwell, S. and Branch, G.M., Effects of herbivore gut surfactants on kelp polyphenol defenses, *Ecology*, 73, 205, 1992.

116. Targett, N.M., Boettcher, A.A., Targett, T.E., and Vrolijk, N.H., Tropical marine herbivore assimilation of phenolic-rich plants, *Oecologia*, 103, 170, 1995.

117. Targett, N.M. and Arnold, T.M., Predicting the effects of brown algal phlorotannins on marine herbivores in tropical and temperate oceans, *J. Phycol.*, 34, 195, 1998.

118. Duggins, D.O. and Eckman, J.E., Is kelp detritus a good food for suspension feeders? Effects of kelp species, age, and secondary metabolites, *Mar. Biol.*, 128, 489, 1997.

119. Poore, A.G.B. and Steinberg, P.D., Preference–performance relationships and effects of host plant choice in an herbivorous marine amphipod, *Ecol. Monogr.*, 69, 443, 1999.

120. Duffy, J.E. and Paul, V.J., Prey nutritional quality and the effectiveness of chemical defenses against tropical reef fishes, *Oecologia*, 90, 333, 1992.

121. Pennings, S.C. and Carefoot, T.H., Post-ingestive consequences of consuming secondary metabolites in sea hares (Gastropoda: Opisthobranchia), *Comp. Biochem. Physiol.*, 111C, 249, 1995.

122. Tamburri, M.N. and Barry, J.P., Adaptations for scavenging by three diverse bathyal species, *Eptatretus stouti*, *Neptunea amianta* and *Orchomene obtusus*, *Deep Sea Res.* I, 46, 2079, 1999.

123. Zimmer-Faust, R.K., ATP: A potent prey attractant evoking carnivory, *Limnol. Oceanogr.*, 38, 1271, 1993.

124. Atema, J., Eddy chemotaxis and odour landscapes: exploration of nature with animal sensors, *Biol. Bull.*, 191, 129, 1996.

125. Carr, W.E.S. and Derby, C.D., Chemoreception in the shrimp, *Palaemonetes pugio*, identification of active compounds in food extracts and evidence of synergistic mixture interactions, *Chem. Senses*, 11, 49, 1986.

126. Allmon, R.A. and Sebens, K. P., Feeding biology and ecological impact of an introduced nudibranch, *Tritonia plebia*, New England, USA, *Mar. Biol.*, 99, 375, 1988.

127. Ward, J.E., Cassell, H.K., and MacDonald, B.A., Chemoreception in the sea scallop *Placopecten magellanicus* (Gmelin). I. Stimulatory effects of phytoplankton metabolites on clearance and ingestion rates, *J. Exp. Mar. Biol. Ecol.*, 163, 1992.

128. Rittschof, D., Williams, L.G., Brown, B., and Carriker, M.R., Chemical attraction of newly hatched oyster drills, *Biol. Bull.*, 164, 493, 1983.

129. Boyle, P.R., Responses to water-borne chemicals by the octopus *Eledone cirrhosa* (Lamarck 1798), *J. Exp. Mar. Biol. Ecol.*, 104, 23, 1986

130. Glynn, P.W., An amphinomid worm predator of the crown-of-thorns sea star and general predation on asteroids in eastern and western Pacific coral reefs, *Bull. Mar. Sci.*, 35, 54, 1984.

131. Finelli, C.M., Pentcheff, N.D., Zimmer-Faust, R.K., and Wethey, D.S., Odor transport in turbulent flows: constraints on animal navigation, *Limnol. Oceanogr.*, 44, 1056, 1999.

132. Moore, P.A. and Grills, J.L., Chemical orientation to food by the crayfish *Orconectes rusticus*: influence of hydrodynamics, *Anim. Behav.*, 58, 953, 1999.

133. Zimmer-Faust, R.K. and Case, J.F., A proposed dual role of odor in foraging by the California spiny lobster, *Panulirus interruptus* (Randall), *Biol. Bull.*, 164, 341, 1983.

134. Zimmer-Faust, R.K., O'Neill, P.B., and Schar, D.W., The relationship between predator activity state and sensitivity to prey odor, *Biol. Bull.*, 190, 82, 1996.

135. Costero, M. and Meyers, S.P., Evaluation of chemoreception by *Pennaeus vannamei* under experimental conditions, *Progressive Fish-Culturist*, 55, 157, 1993.

136. Daniel, P.C. and Bayer, R.C., Development of chemically mediated prey-search response in postlarval lobsters (*Homarus americanus*) through feeding experience, *J. Chem. Ecol.*, 13, 1217, 1987.

137. Steele, C.W., Scarfe, A.D., and Owens, D.W., Effects of group size on the responsiveness of zebrafish, *Brachydanio verio* (Hamilton Buchanan), to alanine, a chemical attractant, *J. Fish. Biol.*, 38, 553, 1991.

138. Zimmer-Faust, R.K., Commins, J.E., Schar, D.W., Browne, K.A., Wethey, D.S., Pentcheff, N.D., and Finelli, C.M., Mechanisms regulating predation: the role of free amino acids as prey attractants, *Eos, Trans. Amer. Geophys. Union*, 77(3) (supplement), OS70, 1996.

139. Decho, A.W., Browne, K.A., and Zimmer-Faust, R.K., Chemical cues: why basic peptides are signal molecules in marine environments, *Limnol. Oceanogr.*, 43, 1410, 1998.

140. Rittschof, D., Peptide-mediated behaviors in marine organisms, *J. Chem. Ecol.*, 16, 261, 1990.

141. Zimmer-Faust, R.K., Tyre, J.E., Michel, W.C., and Case, J.F., Chemical mediation of adaptive feeding in a marine decapod crustacean: the importance of suppression and synergism, *Biol. Bull.*, 167, 339, 1984.

142. Carr, W.E.S. and Derby, C.D., Chemically stimulated feeding behavior in marine animals: implications of chemical mixtures and involvement of mixture interactions, *J. Chem. Ecol.*, 12, 989, 1986.

143. Zimmer-Faust, R.K., Towards a theory on optimal chemoreception, *Biol. Bull.*, 172, 10, 1987.

144. Himmelman, J.H., Movement of whelks (*Buccinum undulatum*) toward a baited trap, *Mar. Biol.*, 97, 521, 1988.

145. Wellins, C.A., Rittschof, D., and Wachowiak, M., Location of volatile odor sources by ghost crab *Ocypode quadrata* (Fabricius), *J. Chem. Ecol.*, 15, 1161, 1989.

146. Young, C.M., Tyler, P.A., Emson, R.H., and Gage, J.D., Perception and selection of macrophyte detrital falls by the bathyal echinoid *Stylocidaris lineata*, *Deep Sea Res.* I, 40, 1475, 1993.

147. Moore, P.A. and Atema, J., Spatial information in the three-dimensional fine structure of an aquatic odor plume, *Biol. Bull.*, 181, 408, 1991.

148. Mafra-Neto, A. and Carde, R.T., Fine-scale structure of pheromone plumes modulates upwind orientation of flying moths, *Nature*, 369, 142, 1994.

149. Zimmer-Faust, R.K., Finelli, C.M., Pentcheff, N.D., and Wethey, D.S., Odor plumes and animal navigation in turbulent water flow: a field study, *Biol. Bull.*, 188, 111, 1995.

150. Jumars, P.A. and Gallagher, E.D., Deep-sea community structure: three plays on the benthic proscenium, in *The Environment of the Deep Sea*, Ernst, W.G., Ed., Prentice-Hall, Englewood Cliffs, NJ, 1982, 217.

151. Stockton, W.L. and DeLaca, T.E., Food falls in the deep sea: occurrence, quality, and significance, *Deep Sea Res.*, 29A, 157, 1982.

152. Kobayashi, M., Kobayashi, J., and Ohizumi, Y., Cone shell toxins and the mechanisms of their pharmacological action, in *Bioorganic Marine Chemistry*, vol. 3, Schever, P.J., Ed., Springer-Verlag, New York, 1989, 71.

153. Olivera, B.M., Gray, W.R., and Cruz, L.J., Marine snail venoms, in *Handbook of Natural Toxins, Marine Toxins and Venoms, Vol. 3*, Tu, A.T., Ed., Marcel Dekker, Inc., New York, 1988, 327.

154. Boyle, P.R., Prey handling and salivary secretions in octopi, in *Trophic Relationships in the Marine Environment*, Barnes, M. and Gibson, R.N., Eds., *Proceedings of the 24th European Marine Biology Symposium*, Aberdeen University Press, Aberdeen, 1990, 541.

155. Grisley, M.S., Boyle, P.R., and Key, L.N., Eye puncture as a route of entry for saliva during predation on crabs by the octopus *Eledone cirrhosa* (Lamarck), *J. Exp. Mar. Biol. Ecol.*, 202, 225, 1996.

156. Thuesen, E.V., Kogure, K., Hashimoto, K., and Nemoto, T., Poison arrowworms: a tetrodotoxin venom in the marine phylum Chaetognatha, *J. Exp. Mar. Biol. Ecol.*, 116, 249, 1988.

157. Hal, S. and Strichartz, G., *Marine Toxins: Origin, Structure and Molecular Pharmacology*, ACS Symposium Series 418, American Chemical Society, Washington DC, 1990.

158. Mariscal, R.N., Nematocysts, in *Coelenterate Biology*, Muscatine, L. and Lenhoff, H.M., Eds., Academic Press, New York, 1974, 129.

159. Purcell, J.E. and Mills, C.E., The correlation between nematocyst types and diets in pelagic hydrozoa, in *The Biology of Nematocysts*, Hessinger, D.A. and Lenhoff, H.M., Eds., Academic Press, San Diego, 1988, 463.

160. Lang, J.C., Interspecific aggression by scleractinian corals. II. Why the race is not only to the swift, *Bull. Mar. Sci.*, 21, 952, 1973.

161. Grosberg, R.K., The evolution of allorecognition specificity in marine invertebrates, *Quart. Rev. Biol.*, 63, 377, 1988.

162. Stachowicz, J.J. and Lindquist, N., Hydroid defenses against predators: the importance of secondary metabolites vs. nematocysts, *Oecologia*, 124, 280, 2000.

163. Dumont, H.J. and Carels, I., Flatworm predator (*Mesostoma* cf. *lingua*) releases a toxin to catch planktonic prey (*Daphnia magna*), *Limnol. Oceanogr.*, 32, 699, 1987.

164. Sakata, K., Feeding attractants and stimulant for marine gastropods, in *Bioorganic Marine Chemistry, Vol. 3*, Scheuer, P.J., Ed., Springer-Verlag, New York, 1989, 115.

165. Sakata, K., Tsuge, M., and Ina, K., A simple bioassay for feeding stimulants for the young sea hare *Aplysia juliana*, *Mar. Biol.*, 91, 509, 1986.

166. Sakata, K., Itoh, T., and Ina, K., A new bioassay method for phagostimulants for a young abalone, *Halitosis discus* Reeve, *Agric. Biol. Chem.*, 48, 425, 1984.

167. Hay, M.E., Seaweed chemical defenses: their role in the evolution of feeding specialization and in mediating complex interactions, in *Ecological Roles of Marine Natural Products*, Paul, V.J., Ed., Comstock, Ithaca, NY, 1992, 93.

168. de Nys, R., Coll, J.C., and Price, I.R., Chemically mediated interactions between the red alga *Plocamium hamatum* (Rhodophyta) and the octocoral *Sinularia cruciata* (Alcyonacea), *Mar. Biol.*, 108, 315, 1991.

169. Schmitt, T.M., Hay, M.E., and Lindquist, N., Constraints on chemically mediated coevolution: multiple functions for seaweed secondary metabolites, *Ecology*, 76, 107, 1995.

170. Henrikson, A.A. and Pawlik, J.R., A new antifouling assay method: results from field experiments using extracts of four marine organisms, *J. Exp. Mar. Biol. Ecol.*, 194, 157, 1995.

171. Wahl, M., Increased drag reduces growth of snails: comparison of flume and in situ experiments, *Mar. Ecol. Prog. Ser.*, 151, 291, 1997.

172. Becker, K. and Wahl, M., Behaviour patterns as natural antifouling mechanisms of tropical marine crabs, *J. Exp. Mar. Biol. Ecol.*, 203, 245, 1996.

173. Manker, D.C. and Faulkner, D.J., Investigation of the role of diterpenes produced by marine pulmonates *Trimusculus reticulatus* and *T. conica*, *J. Chem. Ecol.*, 22, 23, 1996.

174. de Nys, R., Dworjanyn, S.A., and Steinberg, P.D., A new method for determining surface concentrations of marine natural products on seaweeds, *Mar. Ecol. Prog. Ser.*, 162, 79, 1998.

175. Bryan, P.J., Rittschof, D., and McClintock, J.B., Bioactivity of echinoderm ethanolic body wall extracts: an assessment of marine bacterial attachment and macroinvertebrate larval settlement, *J. Exp. Mar. Biol. Ecol.*, 196, 79, 1996.

176. Fielman, K.T., Woodin, S.A., Walla, M.D., and Lincoln, D.E., Widespread occurrence of natural halogenated organics among temperate marine infauna, *Mar. Ecol. Prog. Ser.*, 181, 1, 1999.

177. Woodin, S.A., Lindsay, S.M., and Lincoln, D.E., Biogenic bromophenols as negative recruitment cues, *Mar. Ecol. Prog. Ser.*, 157, 303, 1997.

178. Woodin, S.A., Marinelli, R.L., and Lincoln, D.E., Allelochemical inhibition of recruitment in a sedimentary assemblage, *J. Chem. Ecol.*, 19, 517, 1993.

179. Woodin, S.A., personal communication, 2000.

180. Coll, J.C., Bowden, B.F., and Tapiolas, D.M., In situ isolation of allelochemicals released from soft corals (Coelenterata: Octocorallia): a totally submersible sampling apparatus, *J. Exp. Mar. Biol. Ecol.*, 60, 293, 1982.

181. Bronstein, J.L., Our current understanding of mutualism, *Q. Rev. Biol.*, 69, 31, 1994.

182. Bruno J.F. and Bertness, M.D, Habitat modification and facilitation in benthic marine communities, in *Marine Benthic Community Ecology*, Bertness, M.D., Hay, M.E., and Gaines, S.D., Eds., Sinauer Associates, Massachusetts, 2000, 201.

183. Stachowicz, J.J., Mutualism, facilitation and the structure of ecological communities, *BioScience*, 51, 2001.

184. Ache, B.W., The experimental analysis of host location in symbiotic marine invertebrates, in *Symbiosis in the Sea*, Vernberg, W.B., Ed., University of South Carolina Press, Columbia, South Carolina, 1972, 45.

185. Brooks, W.R. and Gwaltney, C.L., Protection of symbiotic cnidarians by their hermit crab hosts: evidence for mutualism, *Symbiosis*, 15, 1, 1993.

186. Brooks, W.R., Chemical recognition by hermit crabs of their symbiotic sea anemones and a predatory octopus, *Hydrobiologia*, 216/217, 291, 1991.

187. Brooks, W.R. and Rittschof, D., Chemical detection and host selection by the symbiotic crab *Porcellana sayana*, *Invert. Biol.*, 114, 180, 1995.

188. Lubbock, R., Why are clownfishes not stung by anemones?, *Proc. R. Soc. Lond. B*, 207, 35, 1980.

189. Miyagawa, K., Experimental analysis of the symbiosis between anemonefish and sea anemones, *Ethology*, 80, 19, 1989.

190. Murata, M., Miyagawa-Kohshima, K., Nakanishi, K., and Naya, Y., Characterization of compounds that induce symbiosis between sea anemone and anemone fish, *Science*, 234, 585, 1986.

191. Coen, L.D., Herbivory by crabs and the control of algal epibionts on Caribbean host corals, *Oecologia*, 75, 198, 1988.

192. Castro, P., Brachyuran crabs symbiotic with scleractinian corals: a review of their biology, *Micronesica*, 12, 99, 1976.

193. Glynn, P.W., Some ecological consequences of coral-crustacean guard mutualisms in the Indian and Pacific Oceans, *Symbiosis*, 4, 301, 1987.

194. Stimson, J., Stimulation of fat-body production in the polyps of the coral *Pocillopora damicornis* by the presence of mutualistic crabs in the genus *Trapezia*, *Mar. Biol.*, 106, 211, 1990.

195. Atsatt, P.R. and O'Dowd, D.J., Plant defense guilds, *Science*, 193, 24, 1976.

196. Littler, M.M., Taylor, P.R., and Littler, D.S., Plant defense associations in the marine environment, *Coral Reefs*, 5, 63, 1986.

197. Pfister, C.A. and Hay, M.E., Associational plant refuges: convergent patterns in marine and terrestrial communities result from differing mechanisms, *Oecologia*, 77, 118, 1988.

198. McClintock, J.B. and Janssen, J., Pteropod abduction as a chemical defense in a pelagic Antarctic amphipod, *Nature*, 346, 462, 1990.

199. Hay, M.E., and Fenical, W., Chemical ecology and marine biodiversity: insights and products from the sea, *Oceanogr.*, 9, 10, 1996.

200. Strong, D.R., Lawton, J.H., and Southwood, R., *Insects on Plants*, Harvard University Press, Cambridge, Massachusetts, 1984.

201. Bernays, E.A. and Graham, M., On the evolution of host specificity in phytophagous insects, *Ecology*, 69, 886, 1988.

202. Ehrlich, P.R. and Raven, P.H., Butterflies and plants: a study in coevolution, *Evolution*, 18, 586, 1964.

203. Barbosa, P., Some thoughts on "the evolution of host range", *Ecology*, 69, 912, 1988.

204. Courtney, S., If it's not coevolution it must be predation?, *Ecology*, 69, 910, 1988.

205. Ehrlich, P.R. and Murphy, D.D., Plant chemistry and host range in insect herbivores, *Ecology*, 69, 908, 1988.

206. Fox, L.R., Diffuse coevolution within complex communities, *Ecology*, 69, 906, 1988.

207. Jermy, T., Can predation lead to narrow food specialization in phytophagous insects?, *Ecology*, 69, 902, 1988.

208. Rausher, M.D., Is coevolution dead?, *Ecology*, 69, 898, 1988.

209. Thompson, J.N., Coevolution and alternative hypotheses on insect/plant interactions, *Ecology*, 69, 893, 1988.

210. Sotka, E.E., Hay, M.E., and Thomas, J.D., Host-plant specialization by a non-herbivorous amphipod: advantages for the amphipod and costs for the seaweed, *Oecologia*, 118, 471, 1999.

211. Fenical, W., Chemical studies of marine bacteria: developing a new resource, *Chem. Rev.*, 93, 1673, 1993.

212. Gil-Turnes, M.S., Hay, M.E., and Fenical, W., Symbiotic marine bacteria chemically defend crustacean embryos from a pathogenic fungus, *Science*, 246, 116, 1989.

213. Faulkner, D.J, Unson, M.D., and Bewley, C.A., The chemistry of some sponges and their symbionts, *Pure Appl. Chem.*, 66, 1983, 1994.

214. Renaud, P.E., Hay, M.E., and Schmitt, T.M., Interactions of plant stress and herbivory: intraspecific variation in the susceptibility of a palatable versus an unpalatable seaweed to sea urchin grazing, *Oecologia*, 82, 217, 1990.

215. Yates, J.L. and Peckol, P., Effects of nutrient availability and herbivory on polyphenolics in the seaweed *Fucus vesiculosus*, *Ecology*, 74, 1757, 1993.

216. Van Donk, E. and Hessen, D.O., Grazing resistance in nutrient-stressed phytoplankton, *Oecologia*, 93, 508, 1993.

217. Stachowicz, J.J. and Lindquist, N., Chemical defense among hydroids on pelagic *Sargassum*: predator deterrence and absorption of solar UV radiation by secondary metabolites, *Mar. Ecol. Prog. Ser.*, 155, 115, 1997.

218. Duffy, J.E., Resource-associated population subdivision in a symbiotic coral-reef shrimp, *Evolution*, 50, 360, 1996.

219. Davenport, D., Camougis, G., and Hickok, J., Analyses of the behavior of commensals in host-factor. 1. A hesionid polychaete and pinnotherid crab, *Anim. Behav.*, 8, 209, 1960.

220. Gray, I.E., McCloskey, L.R., and Weihe, S.C., The commensal crab *Dissodactylus mellitae* and its reaction to sand dollar host factor, *J. Elisha Mitchell Soc.*, 84, 472, 1968.

221. Derby, C.D. and Atema, J., Induced host odor attraction in the pea crab *Pinnotheres maculatus*, *Biol. Bull.*, 158, 26, 1980.

222. Fox, L.R. and Morrow, P.A., Specialization: species property or local phenomenon?, *Science*, 211, 887, 1981.

223. Thompson, J.N. and Pellemyr, O., Evolution of oviposition behavior and host preference in lepidoptera, *Ann. Rev. Entomol.*, 36, 65, 1991.

224. Stevens, P.M., Specificity of host-recognition of individuals from different host races of symbiotic pea crabs (Decapoda: Pinnotheridae), *J. Exp. Mar. Biol. Ecol.*, 143, 193, 1990.

225. Stevens, P.M., A genetic analysis of the pea crabs (Decapoda: Pinnotheridae) of New Zealand. I. Patterns of spatial and host-associated genetic structuring in *Pinnotheres novaezelandiae* Filhol, *J. Exp. Mar. Biol. Ecol.*, 141, 195, 1990.

226. Stevens, P.M., A genetic analysis of the pea crabs (Decapoda: Pinnotheridae) of New Zealand. II. Patterns and intensity of spatial population structure in *Pinnotheres atrinicola*, *Mar. Biol.*, 108, 403, 1990.

227. Stevens, P.M., New host record for pea crabs (*Pinnotheres* spp.) symbiotic with bivalve mollusks in New Zealand (Decapoda, Brachyura), *Crustaceana*, 63, 216, 1992.

228. Stachowicz, J.J. and Hay, M.E., Geographic variation in camouflaging behavior by the decorator crab *Libinia dubia*, *Am. Nat.*, 156, 59, 2000.

229. Thompson, J.N., *The Coevolutionary Process*, University of Chicago Press, Chicago, 1994.

230. Dunham, P.J., Pheromones and social behavior in Crustacea, in *Endocrinology of Selected Invertebrate Types*, Laufer, H., and Downer, R., Eds., Alan R. Liss, New York, 1988, 375.

231. Dunham, P.J., Sex pheromones in crustacea, *Biol. Rev.*, 53, 555, 1978.

232. Borowsky, B., Effects of receptive females' secretions on some male reproductive behavior in the amphipod crustacean *Microdeutopus gryllotalpa*, *Mar. Biol.*, 84, 183, 1984.

233. Borowsky, B., Augelli, C.E., and Wilson, S.R., Towards chemical characterization of waterborne pheromone of amphipod crustacean *Microdeutopus gryllotalpa*, *J. Chem. Ecol.*, 13, 1673, 1987.

234. Stanhope, M.J., Connelly, M.M., and Hartwick, B., Evolution of a crustacean chemical communication channel: behavioral and ecological genetic evidence for a habitat-modified, race-specific pheromone, *J. Chem. Ecol.*, 18, 1871, 1992.

235. Stanhope, M.J., Hartwick, B., and Baillie, D., Molecular phylogeographic evidence for multiple shifts in habitat preference in the diversification of an amphipod species, *Mol. Ecol.,* 2, 99, 1993.

236. Denny, M.W., *Biology and Mechanics of Wave Swept Environments*, Princeton University Press, Princeton, New Jersey, 1988.

237. Denny, M.W. and Shibata, M.F., Consequences of surf-zone turbulence for settlement and external fertilization, *Am. Nat.*, 134, 859, 1989.

238. Pennington, T., The ecology of fertilization of echinoid eggs: the consequences of sperm dilution, adult aggregation, and synchronous spawning, *Biol. Bull.*, 169, 417, 1985.

239. Yund, P.O., An in situ measurement of sperm dispersal in a colonial marine hydroid, *J. Exp. Zool.*, 253, 102, 1990.

240. Levitan, D.R., Influence of body size and population density on fertilization success and reproductive output in a free-spawning invertebrate, *Biol. Bull.*, 181, 261, 1991.

241. Levitan, D.R., The ecology of fertilization in free-spawning invertebrates, in *Marine Invertebrate Larvae*, McEdward, L., Ed., CRC Press, Boca Raton, FL, 1995, 123.

242. Forward, R.B., Jr., Larval release rhythms of decapod crustaceans: an overview, *Bull. Mar. Sci.*, 41, 165, 1987.

243. Bentley, M.G. and Pacey, A.A., Physiological and environmental control of reproduction in polychaetes, *Oceanogr. Mar. Biol. Annu. Rev.*, 30, 443, 1992.

244. Hardege, J.D. and Bentley, M.G., Spawning synchrony in *Arenicola marina*: evidence for sex pheromonal control, *Proc. R. Soc. Lond. B Biol.*, 264, 1041, 1997.

245. Williams, M.E., Bentley, M.G., and Hardege, J.D., Assessment of field fertilization success in the infaunal polychaete *Arenicola marina* (L.): intermittent irrigation of the tube and the intermittent aerial respiration, *J. Mar. Biol. Assoc. UK*, 33, 51, 1997.

246. Zeeck, E., Hardege, J.D., Bartels-Hardege, H., Sex pheromones and reproductive isolation in two nereid species, *Nereis succinea* and *Platynereis dumerlii*, *Mar. Ecol. Prog. Ser.*, 67, 183, 1990.

247. Bartels-Hardege, H.D., Hardege, J.D., Zeeck, E., Muller, C., Wu, B.L., and Zhu, M.Y., Sex pheromones in marine polychaetes V: a biologically active volatile compound from the coelomic fluid of a female *Nereis* (*Neanthes*) *japonica* (Annelida Polychaeta), *J. Exp. Mar. Biol. Ecol.*, 201, 275, 1996.

248. Reiswig, H.M., Porifera: sudden sperm release by tropical demospongiae, *Science*, 170, 538, 1970.

249. Ormond, R.F.G., Campbell, A.C., Head, S.H., Moore, R.J., Rainbow, P.R., and Saunders, A.P., Formation and breakdown of aggregations of the crown-of-thorns starfish *Acanthaster planci* (L.), *Nature*, 246, 167, 1973.

250. Young, C.M., Tyler, P.A., Cameron, J.L., and Rumrill, S.G., Seasonal breeding aggregations in low-density populations of the bathyal echinoid *Stylocidaris lineata*, *Mar. Biol.*, 113, 603, 1992.

251. Hamel, J.F. and Mercier, A., Prespawning behavior, spawning, and development of brooding starfish *Leptasterias polaris*, *Biol. Bull.*, 188, 32, 1995.

252. Hamel, J.F. and Mercier, A., Evidence of chemical communication during the gametogenesis of holthuroids, *Ecology*, 77, 1600, 1996.

253. Zeeck, E., Hardege, J.D., Bartels-Hardege, H., and Wesselmann, G., Sex pheromone in a marine polychaete: determination of the chemical structure, *J. Exp. Zool.*, 246, 285, 1988.

254. Zeeck E., Harder, T., Beckmann, M., and Muller, C.T., Marine gamete-release pheromones, *Nature*, 382, 214, 1996.

255. Zeeck, E., Hardege, J.D., Bartels-Hardege, H., Willig, A., and Wesselmann, G., Sex pheromones in a marine polychaete: biologically active compounds from female *Platynereis dumerlii*, *J. Exp. Zool.*, 260, 93, 1991.

256. Miller, R.L., Chemotactic behavior of the sperm of chitons (Mollusca: Polyplacophora), *J. Exp. Zool.*, 202, 203, 1977.

257. Miller, R.L., Demonstration of sperm chemotaxis in Echinodermata: Asteroidea, Holothuroidea, Ophiuroidea, *J. Exp. Zool.*, 234, 283, 1985.

258. Miller, R.L., Specificity of sperm chemotaxis among Great Barrier Reef shallow-water holothurians and ophiuroids, *J. Exp. Zool.*, 279, 189, 1997.

259. Miller, R.L., Chemotaxis during fertilization in the hydroid Campanularia, *J. Exp. Zool.*, 162, 22, 1966.

260. Miller, R.L., Sperm chemotaxis in ascidians, *Am. Zool.*, 22, 827, 1982.

261. Boland, W., The chemistry of gamete attraction; chemical structures, biosynthesis, and (a)biotic degradation of algal pheromones, *Proc. Nat. Acad. Sci. USA*, 92, 31, 1995.

262. Maier, I., Brown algal pheromones, *Prog. Phycological Res.*, 11, 51, 1995.

263. Mead, K.S. and Denny, M.W., The effect of hydrodynamic sheer stress on fertilization and early development of the purple sea urchin *Stongylocentrotus purpuratus*, *Biol. Bull.*, 188, 46, 1995.

264. Metz, E.C., Kane, R.E., Tanagimachi, H., and Palumbi, S.R., Fertilization between closely related sea urchins is blocked by incompatibilities during sperm-egg attachment and early stages of fusion, *Biol. Bull.*, 187, 23, 1994.

265. Palumbi, S.R., Marine speciation on a small planet, *Trends Ecol. Evol.*, 7, 114, 1992.

266. Morgan, S.G., The timing of larval release, in *Marine Invertebrate Larvae*, McEdward, L., Ed., CRC Press, Boca Raton, FL, 1995, 157.

267. Morgan, S.G. and Christy, J.H., Adaptive significance of the timing of larval release, *Am. Nat.*, 145, 457, 1995.

268. Forward, R.B., Jr., Lohmann, K.J., and Cronin, T.W., Rhythms in larval release by an estuarine crab (*Rhithropanopeus harrisii*), *Biol. Bull.*, 163, 287, 1982.

269. Forward, R.B., Jr. and Lohmann, K.J., Control of egg hatching in the crab *Rhithropanopeus harrisii* (Gould), *Biol. Bull.*, 165, 154, 1983.

270. De Vries, M.C., Rittschof, D., and Forward, R.B., Jr., Chemical mediation of larval release behaviors in the crab *Neopanope sayi*, *Biol. Bull.*, 180, 1, 1991.

271. Atema, J., Fay, R.R., Popper, A.N., and Tavolga, W.N., *Sensory Biology of Aquatic Animals*, Springer-Verlag, New York, 1988.

5 The Chemical Ecology of Invertebrate Meroplankton and Holoplankton

James B. McClintock,* Bill J. Baker, and Deborah K. Steinberg

CONTENTS

---

* Corresponding author.

I. INTRODUCTION

A. The Problem

Meroplankton is comprised of organisms that spend some component of their life history in the plankton, usually the eggs and larvae of benthic or nektonic adults. There are a few examples of adult benthic marine invertebrates that spend brief periods of time in the plankton. These include the adult reproductive phase (epitoke) of some marine polychaetes, whose benthic life history is interrupted at reproductive maturity by a dramatic ontogeny of swimming appendages followed by swimming behaviors that result in the swarming and bursting of a pelagic reproductive phase. Some deep sea holothuroids (sea cucumbers) spend some period of their adult life swimming in the plankton.[1] Moreover, there are a number of demersal marine invertebrates such as copepods and amphipods that migrate up into the water column at night.[2,3] Nonetheless, the vast majority of meroplankton in the world's oceans are comprised of the propagules of algae or the eggs and larvae of benthic invertebrates and fish. Little is known of the chemical defenses of the propagules of algae or the eggs and larvae of fish. Therefore, the present review and discussion of the chemical ecology of meroplankton will focus primarily on feeding-deterrent properties of marine invertebrate eggs and larvae. While all marine invertebrate groups have the potential of producing defensive chemistry in their larval offspring, studies to date have focused specifically on the eggs and larvae of sponges, cnidarians, molluscs, echinoderms, and ascidians, groups of organisms that are well known to possess chemical defenses in their adult stages.

Benthic marine invertebrates possess a relatively discrete repertoire of reproductive modes.[4-7] Typically, they fall into two categories: those that broadcast large numbers of small eggs that are fertilized and develop into small feeding planktotrophic larvae, or those that produce small numbers of large eggs that are fertilized and develop into large, often conspicuous, nonfeeding lecithotrophic larvae that are subsequently brooded (protected by the parent) or released into the plankton.[5] Eggs or larvae that are released into the plankton, while in some cases having limited mobility generated by ciliary beating, are extremely sluggish and generally lack protective skeletization, and evidence would suggest that they are exposed to considerable predatory pressure from planktivores.[8,9] Moreover, an extensive literature indicates that eggs and larvae face a formidable array of predators (reviewed by Rumrill[10]). One example of planktivory was aptly described by Emery[11] for planktivorous pomacentrid fish when he offered that they constituted a "wall of mouths" facing the plankton. Therefore, one might expect strong evolutionary pressure for selection of defensive chemicals that would decrease the likelihood of predation. This is particularly the case for those benthic marine invertebrates that produce very small numbers of large, conspicuously colored, nutrient-rich lecithotrophic larvae that are released into the plankton, where loss of even small numbers of larval progeny would have strong negative effects on the probability of successful recruitment.[6,12] In summary, information on the chemical defenses of eggs, embryos, and larvae of marine invertebrates is important because models of evolutionary selection of life history patterns make assumptions about patterns of mortality of offspring.[5,13-18] These models generally assume that eggs, embryos, and larvae are vulnerable to predators, and have primarily considered marine invertebrates with planktotrophic modes of development.

Unlike meroplankton, holoplankton is comprised of organisms that spend their entire life cycles in the plankton. The holoplankton contain representatives of nearly every algal and animal group. Over 10,000 species of copepods (crustacea) alone are known, and these can reach abundances of 70,000 per square meter of water in the surface waters of the North Sea.[19] A large variety of gelatinous zooplankton inhabit the sea, with prominent members including medusae, siphonophores, ctenophores, pelagic molluscs, and pelagic ascidians (e.g., salps, larvaceans). The ubiquitous salps, for example, are periodically encountered in swarms extending hundreds of kilometers,[20,21] and, although patchy, can reach densities of 1000 animals per cubic meter.[21] The planktonic protozoa, unicellular, and colonial animals such as acantharia, foraminifera, and radiolaria, are also numerous and widely distributed.[22,23] Similar to meroplankton, marine holoplanktonic algae and invertebrates are likely subject to intense predation pressure, primarily from crustaceans and fish. Cyanobacteria (formerly known as blue-green algae), radiolarians, foraminiferans, and larvaceans move passively with the currents, while some gelatinous zooplankton such as salps, cnidarians, ctenophores, pteropods, and heteropods are generally sluggish swimmers. It is unlikely that they can swim rapidly enough to avoid predatory crustaceans and fish. One might expect that holoplanktonic marine organisms have evolved secondary metabolites to deal with the problem of predation, particularly those species that are conspicuous. Moreover, organisms symbiotically associated with chemically defended holoplanktonic organisms may derive some protection by simply associating with holoplankton.

B. ROLE IN REGULATION OF MATERIAL AND ENERGY FLUX

Holoplankton, and to some extent meroplankton, are responsible for the regulation of material and energy flow in oceanic food webs.[24] Zooplankton grazing plays a key role in the recycling of all biogenic elements, and the community structure of the pelagic food web determines the export of elements from the upper water column. The abundance of particular taxa as influenced by ecological processes like chemical defense provides a mechanism to affect this structure. The size distribution of pelagic producers (phytoplankton) and trophic position of consumers (zooplankton and micronekton) determines the proportion of primary production that is lost and the composition and sedimentation rate of sinking particles from surface communities, and has a significant impact on nutrient cycling.[24] For example, the production in most oceanic food webs tends to be dominated by microbial processes, with protozoan and small crustacean grazers in complex food webs. Much of the carbon and nutrients are recycled in the surface waters with little export, due to loss of energy at each of the many trophic levels. Alternatively, copepods or other large grazers feeding directly on large diatoms in coastal upwelling areas may contribute directly to flux, and a larger fraction of the phytoplankton production is exported in this short food web. There are also generalist consumers, such as the pelagic tunicates that feed with mucous food webs with which they can filter the smallest size particles.[25,26] When abundant, pelagic tunicates can account for large exports of material from the surface waters to the deep sea[21,27–29] through the flux of their fecal pellets or discarded feeding webs. Since they feed at the base of the microbial food web, even in an oceanic ecosystem, they will short circuit the normal paradigm for material and energy flow within these communities. Clearly, there is a need to begin to understand how chemical deterrents may mediate these important patterns of material and energy flow in oceanic systems.

C. THE PARADOX OF THE PLANKTON

Chemically mediated defenses among holoplankton and meroplankton may help resolve why there is a great diversity of co-existing species, all competing for the same resources, in a seemingly homogeneous habitat such as the open sea — "the paradox of the plankton."[30] According to the theory of competitive exclusion, one species should out-compete them all. However, this uniform environment is characterized by small-scale spatial and temporal heterogeneity, such as development of microhabitats in low-turbulence situations,[31] and, thus, offers a variety of niches. Factors such as selective zooplankton grazing are important as well. For example, if one species is consumed by a predator, while another species is chemically defended, this will result in diversification of both predator and prey and make co-existence possible. An excellent example of niche diversification in the pelagic environment is the association of crustaceans with gelatinous zooplankton, such as copepods associated with salps[32] and the mucus feeding webs of "houses" of larvaceans.[33,34] Interestingly, virtually all hyperiid amphipods are associates of gelatinous zooplankton such as salps, ctenophores, siphonophores, medusae, or radiolarian colonies for at least part of their lives.[35–38] Amphipods may use gelatinous zooplankton as a feeding platform, food source, or brooding chamber.[35-37] Hyperiids are thought to be descendants of benthic amphipods that have evolved to live in a benthic-like habitat in midwater that is provided by the gelatinous zooplankton.[37] Virtually nothing is known about the use of gelatinous zooplankton as a refuge from predation in the pelagic zone. However, a benthic marine amphipod has been shown to build a "domicile" from algal material in which it resides.[39] The amphipods selectively chose for their domiciles algal species with secondary metabolites that deter predation by reef fishes, and are thus chemically defended against predation. Whether pelagic amphipods might reduce their chance of being consumed by associating with chemically defended gelatinous zooplankton in the pelagic zone is unknown.

II. LABORATORY STUDIES OF CHEMICAL DEFENSES

The vast majority of work conducted to date on the chemical ecology of meroplankton and holoplankton has employed laboratory-based approaches. There are both pros and cons associated with laboratory studies. Bioassays conducted in the laboratory can be carefully controlled, and small organisms or their eggs and larvae can be observed directly, whereas field observations would be much more difficult or even impossible. Nonetheless, it is fair to say that the field of marine chemical ecology has been moving ideologically towards increasingly ecologically relevant approaches to hypothesis testing.[40] The goldfish toxicity assays employed to evaluate chemical defenses in the earlier years[41-43] have given way to sympatric marine predator models, the use of extracts rather than homogenates, employment of ecologically relevant concentrations of extracts or pure compounds, and increasing numbers of studies that couple laboratory and field assays.

A. MEROPLANKTON

To date, we are aware of no laboratory studies of the chemical feeding-deterrent properties of meroplankton that have an adult planktonic phase, nor of meroplankton comprised of the propagules of marine macroalgae. Lucas et al.[44] were the first known researchers to experimentally examine the feeding-deterrent properties of the eggs and larvae of a marine invertebrate. The focus of their study was on the coralivorous sea star *Acanthaster planci,* a species known to produce copious numbers of small eggs that develop as planktotrophic larvae. Lucas and others had noted earlier[45,46] that some species of fish discriminated against the larvae of *A. planci.* Knowing that saponins occurred in the adult body wall and eggs of *A. planci,*[47] Lucas et al.[44] examined the potential role of saponins as feeding deterrents in eggs and larvae. Gelatin pellets were prepared with yeast extract as a feeding stimulant and contained ecologically conservative concentrations of crude saponin extracts. Four species of sympatric pomacentrid fish were employed as potential predators. In almost all cases, fish rejected experimental gelatin pellets containing saponins while readily consuming control pellets. Interestingly, Lucas et al.[44] noted that with decreasing hunger level, fish increased their discrimination against pellets containing saponins, indicating that the nutritional condition or hunger level of fish predators can influence the ability of chemicals to effectively deter predators. Moreover, Lucas et al.[44] noted that the amount of feeding stimulant added to the experimental pellets influenced acceptability. While their study did not involve an accurate modeling of the nutritional or energetic content of eggs or larvae, they extended this observation to a comparison of planktotrophic and lecithotrophic eggs and larvae, arguing that planktotrophic larvae may be nutritionally less acceptable to fish than yolky eggs and yolky lecithotrophic larvae. Lucas et al.[44] concluded that saponins sequestered in eggs and larvae appear to be effective deterrents against fish, and they offered some preliminary qualitative observations that the larvae of *A. planci* may also be rejected by planktivorous invertebrate carnivores and benthic polychaetes. The logical extension of this work, posed in a question by Lucas et al.[44] — "Do larvae of other sea stars contain saponins as chemical defenses?" — still remains unanswered.

While no bioassays were conducted on discrete gametes, embryos, or larvae, De Vore and Brodie[48] examined the palatability of gravid ovaries of the temperate sea cucumber *Thyone briareus* to the common killifish *Fundulus diaphanus*. Fish were held on a maintenance diet and presented pieces of ovaries or control food over an 11 day period, with the first day consisting of a feeding trial on control food (mussel tissue) and the subsequent 10 days consisting of a randomized presentation of either experimental gravid ovary or control food. Fish demonstrated a strong and significant rejection of the gravid ovaries as compared to the controls, and investigators suggest that this may reflect the concentration of a toxin within the ova to protect the progeny. As this species possesses a vitellarian larva that develops in the water column, it is possible that these larvae may possess a chemical defense. However, further work is needed to verify that the deterrent properties of the gravid ovary are indeed related to feeding deterrents sequestered in the ova and not simply in the supporting ovarian tissues.

Young and Bingham[49] addressed both issues of chemical defenses and aposematism (warning coloration) in their study of the large, brightly colored larvae of the subtropical colonial ascidian *Ecteinascidia turbinata*. Larvae were presented to sympatric pinfish (*Lagodon rhomboides*) and acceptance or rejection noted. Importantly, these investigators examined whether rejected larvae were still capable of normal swimming behaviors, an indication that fish mouthing did not cause subsequent mortality. Swimming larvae were rejected by pinfish. The researchers further demonstrated that rejection was chemically based by observing rejection of agar pellets containing larval homogenates; defensive chemicals were effective at a concentration approximately 170 times lower than they occur in the larva. The identity of the defensive chemicals was not determined but shown to have a molecular weight of less than 14,000 Da, and to unlikely be proteins since boiling did not inhibit bioactivity. Employing a rather ingenious approach, Young and Bingham[49] further examined the question of larval aposematism in *E. turbinata*. Utilizing the palatable larvae of the ascidian *Clavulina oblonga*, they dyed these larvae a color similar to that of the unpalatable larvae of *E. turbinata*. Fish that had been conditioned on larvae of *E. turbinata* generally ignored the dyed larvae of *C. oblonga*. Indeed, these fish appeared to learn how to avoid colored larvae rapidly, whereas fish that had not been conditioned on the larvae of *E. turbinata* consumed the larvae of dyed or undyed larvae of *C. oblonga* in equal frequency. Their observations provide a unique test of whether warning coloration operates at the group or kin level, since larval prey die if sampled during the learning process.[50] Their results argue against group selection since the majority of larvae survive the learning process, and fish retain their recognition of unpalatable larvae for only a very short period of time (Young and Bingham, unpublished). Instead, they argue that if larval survival decreases with successive strikes, then individual selection[51] should be invoked as an explanation for the evolution of aposematic coloration in the larvae of *E. turbinata*, even given the short memory of the pinfish. This study raises intriguing questions about how widespread aposematism might be among marine invertebrates or even fish eggs, embryos, or larvae. Certainly, visual predators such as fish can comprise a significant component of the planktonic predator population, and other investigators have indicated that colored and thus conspicuous larvae may be more likely to possess chemical defenses.[52,53] A larger data base from carefully controlled studies of aposematism in marine invertebrate larvae is needed before a general pattern can be adequately evaluated.

Although the chemical deterrent properties of the eggs and planktotrophic larvae of the tropical nudibranch *Hexabranchus sanguineus* were not investigated, Pawlik et al.[54] demonstrated that the egg ribbons of this nudibranch are defended from fish predation by macrolides (see kabiramide, Figure 5.2). Sequestering these compounds from its sponge diet, *H. sanguineus* incorporates the compounds into the mantle tissues and egg cases after chemical modification. The presence of defensive macrolides in the egg cases raises intriguing questions about whether these defensive compounds might also be provisioned in the eggs and subsequently serve a defensive role in meroplanktonic larvae.

McClintock and Vernon[52] furthered the study of chemical defenses of echinoderm offspring by examining feeding-deterrent properties of the eggs and embryos of Antarctic sea stars, sea urchins, and a sea cucumber. They chose to use an allopatric fish model as a predator (the marine killifish *Fundulus grandis*), arguing that by using a temperate/subtropical fish, this model predator has had no exposure to Antarctic echinoderm eggs or embryos over evolutionary time and would not be expected to have co-evolved adaptations of resistance to toxins. Other chemical ecologists have argued that it is more important to employ sympatric predators in order to effectively evaluate whether chemical deterrents are truly effective against ecologically relevant predators.[40] Both sides of this argument likely have merit, and perhaps a dualistic approach is best if possible, with both sympatric and allopatric models tested. Because many Antarctic echinoderms have been shown to release their eggs into the water column, in contrast to Thorson's Rule,[7] and have extremely slow rates of development that expose them to long periods of predation, McClintock and Vernon[52] predicted that chemical defenses may be relatively common in Antarctic echinoderm eggs and embryos. They found that lyophilized egg and embryo tissues of four species of sea stars embedded at ecologically relevant concentrations in agar pellets deterred feeding in killifish. Three of these species produced large yolky eggs or embryos (lecithotrophs), and one of these, *Perknaster fuscus*, did not brood but rather released large yolky eggs into the water column where they developed as pelagic lecithotrophic larvae. A fourth species, *Porania antarctica*, produced intermediate-sized eggs that developed as relatively large planktotrophic larvae. The nature of the chemical compound(s) was not determined, but they are likely to be saponins[44] such as steroid oligoglycosides and polyhydroxysteroids.[55] The results indicate that Antarctic echinoderms that produce small numbers of lecithotrophic or large numbers of planktotrophic eggs and embryos can employ chemicals to defend their offspring, but that chemical defenses may be somewhat more common in lecithotrophic species. Interestingly, both lecithotrophic species that brood and broadcast their eggs and embryos were found to be chemically defended. One obvious question that arises from these observations is why brooding species should invest defensive chemistry in their offspring when they are presumably protected from predation by the adult. The answer may be associated with the especially vulnerable period of time when the juvenile leaves the protection of the adult, presumably provisioned with defensive chemicals until some refuge in size is attained. This may be particularly important in polar marine environments where slow growth rates in sea stars and other marine invertebrates may result in periods of years spent in the vulnerable juvenile phase.[56]

McClintock et al.[57] extended the analysis of potential chemical defenses to another phylum of Antarctic marine invertebrates in their analysis of the biochemical and energetic composition and chemical defenses of the common Antarctic ascidian *Cnemidocarpa verrucosa*. Their chemical studies were limited to an examination of the palatability of the gonad to a sympatric planktonic fish (*Pagothenia borchgrevinki*). Unable to trigger a spawning response in order to collect eggs or embryos or raise larvae, they were only able to test small pieces of intact ovitestes using similarly sized pieces of cod muscle as controls. They found significant rejection of ovitestes by Antarctic fish. While homogenates or extracts were not tested in feeding pellets, it is unlikely that deterrence was related to structural defenses (no skeletal material) or low nutritional content (ovitestes were found to be very high in energy[57]). Their findings, nonetheless, must be interpreted with caution since feeding deterrence could be attributable to chemicals in the sperm (although unlikely) or the nongametic gonadal tissues. The data do suggest that the bright orange planktotrophic eggs, embryos, and larvae of *C. verrucosa* may possess a chemical defense. The nature of a potential chemical deterrent was not investigated, but the ovitestes was determined to be mildly acidic (pH = 5.86), a factor that may have contributed to their rejection by fish. Feeding deterrence in the outer tunic of some ascidians has been attributed to sulfuric acid sequestered in small bladders.[58]

Using the eggs and larvae of marine ascidians as a model system, Lindquist et al.[53] examined the question, "Why are embryos so tasty?" posed by Orians and Janzen,[59] who had pointed out that birds, reptiles, amphibians, fish, and insects all seem to lose large proportions of their eggs and larvae to predators and that evolution should strongly favor those organisms that produce distasteful eggs. Orians and Janzen[59] speculated that (1) actively developing tissues such as those in eggs and embryos are incompatible with toxic chemicals (autotoxicity), (2) there are energetic constraints that limit the ability of eggs and embryos to produce toxins, and (3) there may be tradeoffs between the production of deterrents and potential development rates. Lindquist et al.[53] noted that the focus of much of this speculation revolved around vertebrates, and that there was a need to extend the evaluation to marine invertebrates before generalizing about any apparent lack of chemical defenses in eggs and embryos.

Lindquist et al.[53] selected ascidians as a model group of organisms for such a study because they have large conspicuous eggs and embryos that are amenable to chemical evaluation, and their larval ecology is generally well known.[49,60–62] Shipboard laboratory-based bioassays included an investigation of chemical defenses of both living larvae and larval crude extracts of the Caribbean ascidian *Trididemnum solidum* exposed to predation by the bluehead wrasse *Thalassoma bifasciatum* found in sympatry with this ascidian. Groups of wrasses were held in small aquaria and maintained on a diet sufficient to prevent starvation. Using a rather innovative approach, Lindquist et al.[53] employed the eyes of freeze-dried krill as a larval mimic since they were similar in size and color to larvae and were readily consumed by these fish. Replicate groups of fish were presented a krill eye, then a *T. bifasciatum* larva, and then yet another krill eye to ensure that a rejection response to the larva was not due to satiation. Crude extracts of single larvae were impregnated into single krill eyes and were presented to fish along with controls and fish ingestion experiments conducted in a similar fashion. The researchers found the tadpole larvae and the krill eyes treated with crude extract of *T. bifasciatum* to be highly unpalatable to the groups of bluehead wrasse. Coupled with field observations of ascidian larval chemical defenses (see below) and a review of the literature on the unpalatability of ascidian larvae conducted to date (Table 5.1), Lindquist et al.[53] argue that brooding ascidians that produce large conspicuous larvae, and often release their larvae over short durations and distances during daylight hours to ensure that larvae can employ photic cues to enhance settlement, are under strong selection pressure to evolve chemical defenses. Lindquist et al. also point out that many predatory reef fish have limited home ranges,[63] and that clumping of unpalatable larval prey may increase the likelihood that fish will learn to avoid ingesting chemically defended eggs and larvae (also see Young and Bingham[49]). They also propose that chemical defenses among larvae that require several weeks or more to develop in the plankton may be less common because pelagic eggs and larvae are likely to be transported offshore where predation levels are lower.[63,64] However, it would seem that while predation may indeed be lower in these pelagic offshore habitats, the longer duration of exposure may offset any benefit attributable to habitat-specific differences in predation level.

Importantly, Lindquist et al.[53] also document that ascidians can exhibit chemical differences between defensive secondary metabolites among adults and larvae. For example, larvae from colonies of *Sigillina* cf. *signifera* contained more tambjamine C, less tambjamine E, and no tambjamine F as compared to adults.[65] Moreover, larvae of *Trididemnum solidum* contain only four of the six didemnins found in adults.[53] This could be the result of different selective pressures during planktonic vs. benthic life history phases. In contrast, Lucas et al.[44] found no differences in the saponin chemical defenses of the embryos, larvae, and adults of the sea star *Acanthaster planci*. Clearly, additional studies are needed to expand the evaluation of ontogenetic shifts in defensive chemistry in marine organisms.

Based on their findings as well as those of others for ascidians, Lindquist et al.[53] question the adequacy of the autotoxicity, energetic, or developmental constraints suggested by Orians and Janzen[59] to explain a presumed lack of chemical defenses in the eggs and embryos of animals. Coupled with other reports of chemical defenses in the eggs and embryos of amphibians,[66] insects,[67] and additional marine invertebrates,[44,48,52,54,68–70] there appears to be ample evidence to question the validity of these presumed constraints. However, Slattery et al.[70] recently suggested that the lack of chemical defenses in the larvae of the soft coral *Sinularia polydactyla* may be attributable to autotoxicity constraints.

In yet another study focusing on the larvae of a colonial ascidian, Lindquist and Hay[71] evaluated not only whether secondary metabolites in the large brooded larvae of *Trididemnum solidum* cause

TABLE 5.1
Chemically Defended or Distasteful Eggs, Embryos, and Larvae of Benthic Marine Invertebrates

Taxon	Reproductive Phase	Reproductive Mode	Predator(s) Deterred	References
Sponges				
Callyspongia vaginalis	Larva	Lecithotrophic	Coral, fish	68, 72
Monanchora unguifera	Larva	Lecithotrophic	Coral, fish	68, 72
Niphates digitalis	Larva	Lecithotrophic	Coral, fish	68, 72
Pseudoceratina crassa	Larva	Lecithotrophic	Coral, fish	68, 72
Ptilocaulis spiculifera	Larva	Lecithotrophic	Coral, fish	68, 72
Tedania ignis	Larva	Lecithotrophic	Coral, fish	68, 72
Calyx pedatypa	Larva	Lecithotrophic	Fish	72
Mycale laxissima	Larva	Lecithotrophic	Fish	72
Ulosa ruetzleri	Larva	Lecithotrophic	Fish	72
Ectyoplasia ferox	Larva	Lecithotrophic	Fish	72
Xestospongia muta	Larva	Lecithotrophic	Fish	72
Isodictya setifera	Egg	Lecithotrophic	Sea anemone, sea star, amphipod	113
Soft Corals				
Briareum asbestinum	Larva	Lecithotrophic	Coral, fish	68, 72
Eunicea mammosa	Larva	Lecithotrophic	Coral, fish	68, 72
Erythropodium caribaeorum	Larva	Lecithotrophic	Coral, fish	68, 72
Plexaura flexuosa	Larva	Lecithotrophic	Coral	68, 72
Pseudoplexaura porosa	Larva	Lecithotrophic	Coral, fish	68, 72
Sinularia polydactyla	Larva	Lecithotrophic	Fish	70
Hard Corals				
Agaricia agaricites	Larva	Lecithotrophic	Coral	68
Siderastrea radians	Larva	Lecithotrophic	Coral	68
Porites asteroides	Larva	Lecithotrophic	Coral	68
Hydroids				
Eudendrium carneum	Larva	Lecithotrophic	Sea anemone, fish	68, 72
Corydendrium parasticuml	Larva	Lecithotrophic	Sea anemone, fish	68, 72
Bryozoan				
Bugula neritina	Larva	Lecithotrophic	Sea anemone, fish	68, 72
Polychaetes				
Streblospio benedicti	Larva	Planktotrophic, lecithotrophic	Crab, fish	78
Capitella sp.	Larva	Lecithotrophic	Crab, fish	78
Nudibranchs				
Hexabranchus sanguineus	Egg ribbon	—	Fish	54
Tritoniella belli	Egg ribbon	—	Sea star	113
Echinoderms — Sea Stars				
Porania antarctica	Egg	Lecithotrophic	Fish	52
Diplasteria brucei	Embryo	Lecithotrophic	Fish	52
Notasterias armata	Embryo	Lecithotrophic	Fish	52

TABLE 5.1 (CONTINUED)
Chemically Defended or Distasteful Eggs, Embryos, and Larvae of Benthic Marine Invertebrates

Taxon	Reproductive Phase	Reproductive Mode	Predator(s) Deterred	References
Diplasteria brucei	Embryo	Lecithotrophic	Sea star, sea anemone	113
Perknaster fuscus	Larva	Lecithotrophic	Sea star, sea anemone, amphipod, fish	52, 113
Psilaster charcoti	Larva	Lecithotrophic	Sea star, sea anemone, amphipod	113
Acanthaster planci	Larva	Planktotrophic	Fish	44
Ascidians				
Ecteinascidia turbinata	Larva	Lecithotrophic	Fish	49
Trididemnum solidum	Larva	Lecithotrophic	Fish	53
Sigillina cf. signifera	Larva	Lecithotrophic	Fish	53

feeding deterrence, but rather extended their evaluation to measure whether changes in consumer growth or survivorship result from physiological effects of the ingested noxious compounds. Such an evaluation is appropriate considering that many consumers ingest small amounts of noxious compounds when sampling prey, often with minimal apparent detrimental effects. First, they presented the pinfish *Lagodon rhomboides* with alginate pellets (larval mimics) containing ecologically relevant concentrations of larval didemnin cyclic peptides[53] and squid puree as a feeding stimulant, or, alternatively, control alginate pellets containing only squid. They found that fish rapidly learned to avoid the larval mimics and consequently found it impossible to evaluate long-term impacts of the ingestion of noxious secondary metabolites on fitness. Nonetheless, Lindquist and Hay[71] found that the sea anemone *Aptasia pallida* provided an excellent model to evaluate long-term effects on fitness. Sea anemones were presented experimental and control pellets daily for a period of 32 days. At the end of this time period, it became clear that while didemnins were capable of causing an emetic response, sea anemones did not become conditioned to avoid ingestion of the experimental pellets. Therefore, the sea anemones consumed some minimal level of didemnins over the experimental period. The results of this minimal consumption of noxious compounds were profound. Growth of the adult sea anemones was reduced by 82% in the experimental group. Moreover, the production of asexual clones was reduced by 44%, and the average mass of a clonal offspring was reduced by 41% as compared to clones produced by control sea anemones that had not ingested didemnins. These findings clearly demonstrate that ingestion of ecologically relevant concentrations of noxious secondary metabolites can cause a significant reduction in consumer fitness. This was the first demonstration of an effect on fitness resulting from the consumption of secondary metabolites from larval meroplankton. Lindquist and Hay[71] argue that such dramatic decreases in fitness could clearly select for consumers that recognize and reject prey containing defensive chemicals, even when they form a very small portion of a generalist's diet. In the case of the ascidian *Trididemnum solidum*, larvae are released year round, during daylight hours, and remain in the vicinity of the adult. It is likely that predators would repeatedly encounter such larvae. As hypothesized by Young and Bingham,[49] because rejected ascidian larvae are not killed by predators, group or kin selection need not be invoked to explain the evolution of chemical defenses in these larvae. In summary, Lindquist and Hay[71] provide additional support for the hypothesis that there should be strong evolutionary selection of chemical defenses in the eggs, embryos, and larvae of meroplanktonic marine invertebrates, particularly those that have lecithotrophic reproductive modes.[12,49,52,53]

Lindquist[68] extended the analysis of the palatability of lecithotrophic marine invertebrate larvae in his comparative investigation of the larvae of a variety of temperate and tropical sponges (nine species), gorgonians (nine species), corals (three species), hydroids (two species), and bryozoans (one species). Noting that while larvae of benthic invertebrates search for an appropriate settlement site they encounter a variety of sessile invertebrate predators, he selected three species of corals and a species of sea anemone as model larval predators. The methodological approach involved placing predators held on a maintenance diet individually in containers and presenting them first with a palatable control food, comprised of a sodium alginate pellet containing squid mantle flesh to ensure they were feeding, and then subsequently presenting them a larva. All larvae presented to corals or sea anemones were monitored to see if they were rejected, ingested, or ingested and regurgitated. Larvae that were regurgitated were followed to ensure that they developed and metamorphosed to the juvenile stage normally. Feeding assays were done across a period of several days to prevent predators from becoming satiated during feeding trials. Lindquist found that many of the species tested were unpalatable to these predators; indeed, only the larvae of three of nine species of sponges and two of nine species of gorgonians were consumed. Importantly, both larval survival and metamorphosis were not significantly different in regurgitated larvae and control larvae that were never attacked. Although Lindquist[68] did not evaluate the basis of rejection, it is likely that this was chemically based, since the larvae had no potential skeletal or behavioral defenses.

In yet another survey that focused on fish rather than invertebrate predators, Lindquist and Hay[72] demonstrated that the brooded larvae of 11 species of Caribbean sponges and 3 species of gorgonians, in addition to the brooded larvae of 2 species of temperate hydroids and a bryozoan, were unpalatable to fish. In contrast, brooded larvae of three species of temperate ascidians, a temperate sponge, and three species of Caribbean hard corals were consumed. Larval laboratory assays were conducted by first presenting a single brine shrimp to a species of sympatric fish that had been held on a maintenance diet. Only fish that consumed the brine shrimp were presented larvae, and only larvae that had been sampled by fish and either ingested or rejected were considered in the experimental design. As seen by Lindquist,[68] larvae that had been mouthed and rejected showed no significant decrease in metamorphic competence. Of the species with unpalatable larvae, five were further examined to determine whether noxious chemicals were responsible for deterrence. In all five cases, fish rejected alginate pellets containing a feeding stimulant and ecologically relevant concentrations of larvae extracts, while consuming control pellets containing only feeding stimulant, indicating a chemically based deterrence. While not directly tested, it is likely that deterrent chemistry is responsible for the unpalatable nature of many, if not all, of the larvae tested. Interestingly, brooded larvae were most likely to be unpalatable, while broadcasted larvae (both lecithotrophic and planktotrophic) were generally consumed. Providing additional and more broadly based evidence to their conjecture[71] that larvae of the tropical ascidian *Trididemnum solidum* are chemically defended in part because they release conspicuous larvae during daylight hours, Lindquist and Hay later found[72] that unpalatable larvae were almost always released during the day (89% of total species investigated), while palatable larvae were seldom found to do so (23% of total species investigated). Many of the unpalatable larvae were brightly colored (60% of total species investigated), while all palatable larvae lacked coloration, supporting earlier predictions that aposematism may operate in chemically defended lecithotrophic marine invertebrate larvae.[49]

Extending the analysis of the palatability of marine invertebrate lecithotrophic and planktotrophic eggs, embryos, and larvae to the polar regions, McClintock and Baker[12] examined a suite of Antarctic marine invertebrates with contrasting modes of reproduction. These included the spawned eggs and larvae of a sea urchin and the intraovarian eggs of a sea star, both with planktotrophic larvae, and the lecithotrophic embryos and larvae of three sea stars with either brooding or broadcasting modes of reproduction. Moreover, a nudibranch and sponge with egg ribbons and brooded lecithotrophic embryos, respectively, were examined. Gravid ovaries, spawned eggs, and developing embryos and larvae were tested for palatability against three common sympatric predators with very different feeding patterns: the antarctic sea star *Odontaster validus*, a benthic omnivorous scavenger suggested by Dayton et al.[73] to be sufficiently abundant to act as a larval filter for settling marine invertebrate larvae; the large, abundant, and voracious sea anemone *Isotealia antarctica* that feeds benthically by scavenging organisms that drift, swim, or crawl near its tentacles; and the swarming amphipod *Paramoera walkeri* that scavenges seasonally on the benthos and in the water column. Where sufficient amounts of material were available for extraction, crude lipophilic and hydrophilic extracts were prepared and imbedded at ecologically relevant concentrations in alginate pellets containing a feeding stimulant and tested against predators using pellets containing only feeding stimulant and the appropriate solvent carrier as a control. Alginate pellets containing lipophilic and hydrophilic extracts of spawned eggs of the sea urchin *Sterechinus neumayeri* and gravid ovaries of the sea star *Odontaster validus*, both with broadcasting planktotrophic modes of reproduction, were readily consumed by all three predators. Pellets containing lipophilic and hydrophilic extracts of the four-armed plutei of *S. neumayeri* were also readily consumed by the sea anemone predator, indicating a lack of chemical defense. In contrast, at least one of the three predators displayed significant feeding deterrence for eggs, embryos, larvae, or their lipophilic or hydrophilic extracts among the remaining five lecithotrophic benthic marine invertebrates tested. The basis of rejection was demonstrated to be chemically derived in the brooded embryos of the sea star *Diplasteria brucei*, and it is very likely that defense is also chemically based in the remaining four lecithotrophic species which possess conspicuous eggs, embryos, or larvae that are high in energy content (and thus attractive prey), lack morphological defenses, and are immobile or sluggish swimmers. This study demonstrates that feeding deterrence is species-specific among predators, and supports observations that lecithotrophic embryos or larvae may be particularly well suited to chemical defenses.[49,53,71,72] Moreover, in Antarctica, where both broadcasting and brooding modes of lecithotrophy are particularly common, rates of larval development are several orders of magnitude longer than comparable species at temperate and especially tropical latitudes. Polar marine invertebrate larvae may literally spend 2 to 6 months on the benthos or in the water column during development, thereby substantially increasing predation risk prior to juvenile recruitment.

Another recent study of chemical defenses of the lecithotrophic progeny of a marine invertebrate is that of Slattery et al.,[70] who examined the developing embryos and larvae of the tropical soft coral *Sinularia polydactyla*. Whole blastulae, early planula larvae, and competent planula larvae, along with their extracts imbedded in alginate pellets containing a krill feeding stimulant, were presented to the pufferfish *Canthigaster solandri*. Rejected larvae were followed to observe survival, and fish were presented a food pellet after an experimental pellet to make sure they were not satiated during the trial. All whole developmental stages and their respective extracts were found to be deterrent against pufferfish. The defensive metabolites pukalide and 11β-acetoxypukalide were detected in blastulae and larvae and are likely to be responsible for feeding deterrence,[70,74,75] although Coll et al.[76] concluded pukalide was an unlikely feeding deterrent, since a number of common tropical reef fish will consume the eggs of this species of coral. It is noteworthy that Slattery et al.[70] detected increasing concentrations of defensive metabolites during the ontogeny of the larvae, although even blastulae contained concentrations sufficient to prevent predation by pufferfish (see also Harvell et al.[77]). This suggests that developing progeny are capable of secondary metabolite synthesis and that defensive compounds need not be derived directly from adults.

The most recent study of defensive metabolites in marine invertebrate planktotrophic and lecithotrophic larvae is that of Cowart et al.,[78] who examined the presence of halogenated metabolites with known feeding deterrent properties[79] (also Woodin, Richmond, Lincoln, and Lewis, unpublished data) in both pre- and post-release larvae of the benthic polychaetes *Streblospio benedicti* (known to possess both lecithotrophic and planktotrophic development as a genetic polymorphism) and *Capitella* sp. (lecithotrophic development). Researchers found that pre-release larvae of *S. benedicti* with planktotrophic development had the lowest levels of halogenated metabolites, while post-release larvae had intermediate concentrations, suggesting that the plank-totrophic larvae are synthesizing these defensive metabolites during their development. Even higher levels of halometabolites were found in post-release lecithotrophic larvae of *S. benedicti*, suggesting that lecithotrophic females may be expending more energy on chemical defenses than their plank-totrophic counterparts by supplying their embryos with greater amounts of compounds, their precursors, or sufficient energy for their biosynthesis.[78] Pre-release larvae of the lecithotrophic *Capitella* sp. contained the highest concentrations of halogenated metabolites when comparing both species. Levels of halogenated metabolites in the larvae of both species were greater than those known to deter predation. However, it should be noted that these feeding-deterrent assays were conducted using epibenthic predators and there is a need to evaluate their effectiveness at deterring potential larval predators.

The breadth of the surveys by Lindquist,[68] Lindquist and Hay,[72] and McClintock and Baker,[12] across a variety of phyla and from temperate, tropical, and polar geographic regions, coupled with the data from studies reviewed above,[49,53,71] indicates that previous hypotheses, predicting that predation on marine larvae is relatively ubiquitous[10] should be reconsidered for lecithotrophic larvae. These studies also add significant new information to a growing database on the palatability of marine invertebrate species with lecithotrophic larvae (Table 5.1). A similar assessment for plank-totrophic larvae must await broadening of the database for species with this mode of development. To date there have been comparatively few studies of the palatability of eggs, embryos, and larvae of species of marine invertebrates with planktotrophic larvae,[12,44] although one recent study has, for the first time, examined the palatability of a suite of temperate planktotrophic larvae.[80] These investigators presented three species of predatory fish and a hard coral with a selection of live planktotrophic larvae of benthic marine invertebrates collected from the plankton. Those that were rejected were crushed to render potential morphological defenses useless and were then re-offered to predators in order to assess whether defense was morphologically or chemically based. Research-ers found that for at least one fish predator, a significant number of gastropod veligers, barnacle cyprids, crab zoeae, and stomatopod larvae were likely morphologically defended (34% of meroplankton tested; these were rejected whole and then consumed once crushed). Others, including polychaete larvae, barnacle nauplii, bivalve veligers, shrimp zoeae, crab megalopae, phoronid actinotrochs, and hemichordate tornaria, were readily consumed (65% of meroplankton tested), apparently lacking chemical means of defense. A number of nemertean pilidia, asteroid bipinnaria, and cnidaria planulae were rejected both whole and crushed, suggesting that they were chemically defended. These findings suggest that while select taxa of meroplankton have planktotrophic larvae that are chemically defended, a considerable proportion of species with planktotrophic larvae may rely on the production of copious numbers of small feeding larvae to offset a lack of a morphological or chemical defense.

B. HOLOPLANKTON

The first experimental documentation of a potential chemical defense in a holoplanktonic marine organism was conducted by Shanks and Graham[81] on the scyphozoan jellyfish *Stomolopus melea-gris,* a common inhabitant of bays along the Atlantic coast of North America. The study was stimulated by observations that when physically disturbed, this jellyfish produced a sticky mucus that appeared to have toxic effects on fish held in the same collecting container. Laboratory bioassays were conducted by collecting mucus produced over a discrete time period from an individual jellyfish of known volume. This mucus was then combined with seawater, and two experimental treatments were prepared: one by leaving the mucus-seawater mixture undisturbed and the other by centrifuging the mucus-seawater mixture such that the particulates and pieces of mucus were removed (mucus-free). A control consisted of seawater alone. Three sympatric species were tested for toxic effects including two fish, the juvenile planefish and the Atlantic bumper, and spider crabs. A fourth test animal was the allopatric pinfish *Logodon rhomboides.* Experimental fish placed into either the mucus-free or mucus-seawater mixture reacted by displaying gaping behaviors or laying on their sides, and a number died within a 24-h period. Spider crabs were unaffected by the treatments. While no direct measurements of feeding deterrence were conducted, the investigators proposed that the toxic effects observed were evidence of a chemical defense. While these immersion toxicity assays are not directly ecologically relevant, one can speculate that fish predators that bite into *S. meleagris* and become exposed to mucus might find the toxins distasteful (see also the field experiments below). Shanks and Graham[81] did not examine the nature of the toxin but suggested that it may be derived from nematocysts trapped in the mucus. Other species of gelatinous zooplankton including jellyfish and ctenophores are known to produce mucus upon contact,[82–84] but it is currently unknown if this indicates widespread mucus-bound chemical defenses among gelatinous holoplankton.

McClintock and Janssen[85] examined the feeding deterrent properties of the shell-less Antarctic pteropod *Clione antarctica*. This circumpolar sea butterfly occurs in vast swarms (up to 300 individuals per cubic meter[86]) and is likely, therefore, to play an important role in energy transfer in planktonic ecosystems.[87] Ranging from 1–3 cm in body length, it has a conspicuous orange coloration and is a sluggish swimmer.[88] Individuals caught in plankton nets in the field were subjected to laboratory feeding bioassays using an Antarctic fish as an ecologically relevant predator. The common circumpolar fish *Pagothenia borchgrevinki* feeds in the water column on zooplankton but does not include *C. antarctica* in its diet.[87] First, fish were presented either a live intact sea butterfly or a control piece of cod tissue of similar size and shape. Fish consistently consumed the cod while always rejecting the sea butterflies. In a second experiment, fish were presented either agar pellets containing sea butterfly homogenates or agar pellets with fish meal alone as a control. Again, fish always consumed agar pellets with fish meal while always rejecting the pellets with sea butterfly homogenates, indicating the butterflies' chemical defense against fish predation. Employing flash and high-pressure liquid chromatographic (HPLC) techniques, pteroenone (Figure 5.2), a linear β-hydroxyketone and the first example of a defensive secondary metabolite from a pelagic gastropod, was isolated from tissues of *C. antarctica*.[89] When embedded in alginate food pellets at ecologically relevant concentrations, pteroenone caused significant feeding deterrence in *P. borchgrevinki* and *Pseudotrematomus bernachii*, two Antarctic fish known to feed on planktonic organisms.[86,87] Concentrations of pteroenone were variable among pteropods, but even those individuals with the lowest recorded natural concentrations were effectively protected from fish predation. Further chemical analysis indicated that the primary dietary item of this carnivorous sea butterfly, the shelled pteropod *Limacina helicina*, does not contain pteroenone, evidence that points to *de novo* production of this fish antifeedant by *C. antarctica*.[86]

The cyanobacterium *Trichodesmium* is one of the most abundant phytoplankters in tropical seas, contributing a major fraction of new nitrogen and fixed carbon to surface waters.[90–92] To date, however, there are anecdotal and qualitative observations of grazing by crabs and fish on *Trichodesmium*.[93,94] Interestingly, some pelagic copepods such as *Macrosetella gracilis* not only feed on *Triochodesmium*, but use it as a substrate for juvenile development.[95] Hawser et al.[96] suggest that these copepods gain protection by associating with toxic algae. However, *Trichodesmium* is lethal to other copepods,[96] and McCarthy and Carpenter[97] concluded that *Trichodesmium* must be subject to little grazing pressure to account for its high standing stocks. O'Neil and Roman[93] indicated the necessity of quantitative experiments to determine whether *Trichodesmium* is chemically defended against potential crustacean and fish predators (see preliminary studies described below). Other phytoplankton that produce chemical feeding deterrents include the dinoflagellates.[98,99] Copepods such as *Calanus pacificus* will reject dinoflagellate prey containing toxins, subsequently regurgitating cells and failing to maintain a full gut, or are killed by the toxins.[100–102] Huntley et al.[101] suggest that production of feeding deterrents and resultant release from grazing pressure allow the slow-growing dinoflagellates to form significant blooms.

McClintock et al.[103] examined the feeding deterrent properties of *Trichodesmium* and eight other species of common oceanic holoplankton representing five phyla from Bermudian waters. Holoplankton were collected at sites 5–20 km southeast of Bermuda using plankton nets or by conducting blue-water dives and capturing individuals in handheld jars. Live holoplankton were immediately returned to the laboratory and subjected to feeding trials using the common planktivorous fish *Abudefduf saxatilis* (sergeant major) as a model predator. Large zooplankton (salps, heteropods, ctenophores) were cut into small pieces, while smaller zooplankton (cyanobacteria colonies, radiolarians, and foraminiferans) were presented intact to fish, in both cases along with equivalent numbers of squid mantle tissue controls. Significant feeding deterrence was detected in fish presented with eight of the nine holoplankton species including the colonial cyanobacterium *Trichodesmium,* three species of radiolarians, a foraminiferan, ctenophore, and heteropod, and one of two species of salps. These findings indicate that feeding deterrence occurs across a wide diversity of holoplankton. While the basis of the feeding deterrent responses observed were not determined in this study, these organisms are sluggish and generally depauperate in apparent structural defenses, and, therefore, it seems likely that defense is chemically based. Additional studies were recently conducted by McClintock, Baker, and Steinberg in Bermuda to address this question (see below). Chemical defenses, as documented here for conspicuous members of the holoplanktonic community, may allow these organisms to play a prominent role in structuring marine food webs. For example, short food webs of large organisms allow a large proportion of the primary production to be exported.[24] Complex food webs based on microbial populations recycle most of the primary production with little subsequent export.[24] The size of the grazers in each community determines the production rate of large sinking detrital particles; large animals generally produce large, rapidly sinking waste products. In addition, some of the pelagic ascidians, such as salps and larvaceans, feed with fine-mesh mucus webs that allow them to feed directly on the smallest phytoplankton, thus creating short food webs in microbe-dominated waters.[26] These same unique grazers are ubiquitous and can form massive blooms which dominate export when they occur.[24] Chemical defense may reduce predation sufficiently to allow these large bloom populations to occur. Large planktonic protozoa (radiolaria, acantharia, and foraminifera), with associated symbiotic algae, are also important in elemental cycles because they create a direct link between primary production (via their associated symbionts) and export (they sink during reproduction[24,104]). Feeding deterrent properties, as documented here for many of these species, may allow them to play these prominent roles in elemental cycles.

McClintock, Baker, and Steinberg (unpublished) conducted further bioassays to examine the role of chemical defenses in oceanic holoplankton from Bermuda. Bioassays consisted of presenting groups of the planktivorous fish *Harengula humeralis* (red-ear sardine) with pieces of or whole holoplankton. Also tested were alginate/agar pellets containing whole organism homogenates or lipophilic and hydrophilic extracts of holoplankton plus a feeding stimulant, along with appropriate solvent/feeding stimulant controls. McClintock and colleagues found that both whole individuals and alginate pellets containing homogenates of the large black pelagic copepod *Candacia ethiopica* were significantly deterrent to fish (Figure 5.1). Pieces of whole colonies of two species of radiolarians were also significantly deterrent toward fish predators, while both intact colonies and extracts of the cyanobacterium *Trichodesmium* embedded in agar were also highly deterrent (Figure 5.1). Moreover, extracts of the ctenophore *Mnemiopsis macrydi* embedded in agar pellets were essentially deterrent ($P < 0.06$) to fish predators (Figure 5.1). These findings underscore the need to continue rigorous studies to evaluate the incidence and importance of chemical defenses among marine holoplankton.

The epiphytic hydroid *Tridentata turbinata* commonly occurs in association with the pelagic *Sargassum* community. As such, this hydroid, although not exclusively pelagic (it can occur benthically as well), can conditionally be considered a component of the holoplankton. Stachowicz and Lindquist[105] examined the palatability of this hydroid and its extracts to the sympatric pelagic filefish *Monocanthus hispidus*. Individual fish were presented bite size portions of hydroid followed by a

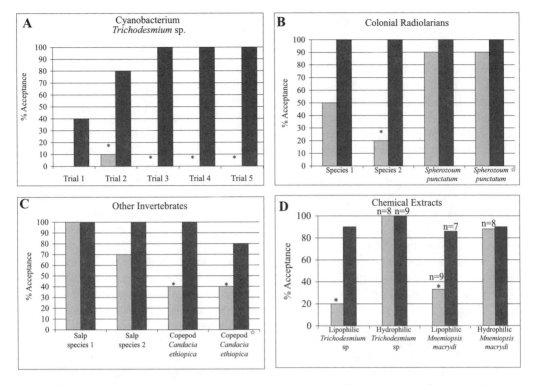

FIGURE 5.1 Histograms showing the feeding-deterrent properties of holoplankton collected near Bermuda and presented to the common planktivorous fish *Harengula humeralis* (red-ear sardine). A. Multiple trials used whole colonies of the colonial cyanobacterium *Trichodesmium* sp. B. Three species of colonial radiolaria. C. Other invertebrates including two species of salps and the copepod *Candacia ethiopica* were presented randomly to five groups of three fish in separate aquaria, along with equal numbers of controls (equivalent sized pieces of squid mantle tissue). Shown is percent acceptance for holoplankton presented intact (*Trichodesmium* sp., radiolarian species 1 and 2, and *C. ethiopica*) or cut into small pieces (salp species 1 and 2 and *Spherozoum punctatum*). Note: in B and C, stars indicate experiments that were conducted with alginate pellets containing whole-organism homogenates embedded in alginate pellets containing dried fish powder, as a feeding stimulant vs. control alginate pellets with feeding stimulant only. D. Chemical extracts of the colonial cyanobacterium *Trichodesmium* sp., and the ctenophore *Mnemiopsis macrydi*, embedded in agar pellets containing dried fish powder were presented to a group of 60 fish in a single, large aquarium. Control agar pellets containing feeding stimulant only were presented in equal numbers. For all experiments, holoplankton or pellets that were taken into the mouth and then spit out within a 1-min period were considered rejected. Light bars are experimental, dark bars are controls. Asterisks indicate experimental treatments that differed significantly from controls (Fisher's Exact Test; $P < 0.05$). Unless otherwise noted, n = 10.

brine shrimp in order to ensure that fish were not satiated if they rejected a hydroid piece. Control group of fish were individually presented only brine shrimp. Consistent rejection responses for *T. marginata* were observed and bioassay-guided fractionation pursued to determine the nature of the deterrent. An assay food consisting of pureed squid mantle tissue (control) or mantle tissue with hydroid extract, both treated with a calcium chloride solution as a hardener, was spread into a thin paste and cut into small pellets (2–3 microliter volume). Pellets were presented to file fish. The lipophilic extract proved deterrent, and HPLC revealed three novel compounds (tridentatol A–C). Subsequent feeding bioassays revealed that only tridentatol A was deterrent to fish. Interestingly, benthic populations of *T. marginata* contained another novel metabolite, tridentatol D (nondeterrent), but contained neither tridentatol B nor C. Benthic populations did contain the active feeding deterrent tridentatol A, suggesting that there may be intraspecific differences in chemical defenses in response to selective pressures associated with holoplanktonic and benthic environments.

III. FIELD STUDIES OF CHEMICAL DEFENSES

A. MEROPLANKTON

Very few studies have been conducted whereby the chemical defenses of meroplankton have been examined in a field setting. This is likely testament to both the youth of this field of enquiry and the inherent difficulties involved in conducting carefully controlled field experiments with plankton. Nonetheless, there is an intrinsic value in conducting field bioassays, and when linked with controlled laboratory assays, this is likely the strongest experimental approach to hypothesis testing. Lindquist et al.[53] performed field fish feeding bioassays on the large larvae (5.2 micoliter volume) of the ascidian *Sigillina* cf. *signifera*. They dissected intact larvae from adults and then released an individual larva or a similar-sized piece of squid in a haphazard sequence on a patch reef in the Bahamas where this ascidian was abundant. By the time the larvae were released on the reef, they were no longer capable of swimming. Twenty-one larvae and controls were released and followed until they had drifted 3 m from the experimental test site. In almost all cases, these larvae were found to be avoided by coral reef fish.

In another *in situ* study, Lindquist[68] offered larvae of the coral *Agaricia agaricites* to the Caribbean corals *Dichocoenia stokesii* and *Montastrea cavernosa*. A palatable alginate food control was prepared with pureed squid-mantle flesh cut into pieces of the same approximate size as the larvae. Lindquist argues convincingly that these corals are potentially important predators of marine invertebrate larvae because they are known to supplement their zooxanthellae-based nutrition with small planktonic organisms,[106] and they are among the most abundant particle feeders on reef systems. *D. stokesii* did not consume any of the larvae, while only 30% of the larvae were consumed by *M. cavernosa*, indicating that both corals displayed statistically significant rejection of *A. agaricites* larvae. The rejected larvae of *D. stokesii* were collected and held for 3 to 5 days, and larval survivorship was 100% after this time period.

B. HOLOPLANKTON

Only two studies have investigated potential chemical defenses of holoplankton in the field. Shanks and Graham,[107] after conducting laboratory experiments to demonstrate that mucus produced by the scyphomedusa *Stromolopus meleagris* was toxic to sympatric fish, tested the effectiveness of this presumed chemical defense in the field. A diver approached an *S. meleagris*, frightening the associated fish (planehead filefish and Atlantic bumpers) into the bell of the jellyfish. The diver then simulated an attack on the jellyfish by pinching the bell margin with forceps. Within one minute of the simulated attack, the fish abandoned the host jellyfish and swam rapidly away. This was repeated ten times with identical results each time. Further manipulation of the jellyfish in captivity revealed that when disturbed, the jellyfish releases mucus from the underside of the bell, suggesting that fish hiding in the host bell may have been repelled by the release of toxic mucus, and likely would be deterred from feeding on the jellyfish, although the possible role of nematocysts must also be considered.

In an investigation of the feeding deterrent properties of the Antarctic pteropod *Clione antarctica*, Bryan et al.[86] conducted field feeding assays presenting live pteropods to the Antarctic fish *Pagothenia borchgrevinki* and *Pseudotrematomus bernacchii*. Both species of fish displayed strong rejection, suggesting that they were tasting a defensive compound. Subsequent laboratory bioassays revealed that a linear β-hydroxyketone (pteroenone) was responsible for fish feeding deterrence (*vide supra*).

IV. CHEMISTRY OF MEROPLANKTON AND HOLOPLANKTON

We know very little about the secondary metabolic chemistry of the plankton. Studies of the meroplankton, specifically of the pelagic eggs, embryos, and larvae of benthic invertebrates, have benefitted from the knowledge of chemical compositions of the adults, since the analysis of a known compound is considerably easier than that of an unknown compound. Holoplankton, on the other hand, are often difficult to collect in quantities sufficient for chemical analysis. Coupled with the common perception that the vastness of the open ocean provides a haven for plankton and thus reduces selective pressure for the evolution of chemical defenses, pelagic organisms have received little attention from chemists.

Among studies of the meroplankton, extracts prepared from the eggs and larvae of the sea star *Acanthaster planci* were shown to deter feeding by several pomacentrid fish.[44] The extract used in this study was a purified fraction rich in saponins, rather than discrete chemical compounds. However, the saponin content of adult *A. planci* is well documented, represented by saponins such as thornasterol and its retro-aldol 3-sulfate analogue sapogenol (Figure 5.2); whether the saponin content of the eggs and larvae mirror that found in the adults is an area in need of further study. More detailed investigations of the chemical nature of echinoderm secondary metabolites in eggs and larvae and their role in UV protection have been carried out.[108,109] Ascidians appear adept at provisioning their larvae with predator deterrents.[53,72] The Indo-Pacific ascidian *Sigillina signifera* produces larvae which contain the pigment tambjamine E (Figure 5.2) at concentrations sufficient to deter predation by coral reef fishes. Didemnin B (Figure 5.2) and nordidenmin B similarly protect Caribbean *Trididemnum solidum* larvae from Atlantic reef fishes. Pukalide (Figure 5.2) and 11β-acetoxypukalide are present in the eggs and larvae of the Pacific soft coral *Sinularia polydactyla* at levels sufficient to deter an omnivorous fish predator and a sympatric microbe.[70] A recent investigation of the polychaete worm *Streblospio benedicti* and *Capitella* sp. has documented halogenated alkanes and aromatics (respectively) in four life history stages;[78] halogenated alkanes were found to be highest in post-release larvae of *S. benedicti* and in pre-release larvae of *Capitella* sp., and in the latter case, the concentrations identified are above those known to deter predation of adults, albeit from adult (benthic) predators.

Only three studies of organisms spending their entire life history in the water column have succeeded in associating discrete chemical substances with deterrence of predation. Pteroenone (Figure 5.2) is the fish feeding deterrent of the pteropod *Clione antarctica*.[86,89] In addition to protecting the pteropod, the amphipod *Hyperiella dilatata* gains protection of pteroenone by abducting the mollusc and positioning it advantageously on its dorsum.[85] Tridentatols A and B (Figure 5.2) are produced by the hydroid *Tridentata marginata* found among pelagic *Sargassum*.[110] *T. marginata* and its eggs, which may occur on the benthos, are protected from predators found among the *Sargassum* by these unusual *N*-[(dialkylthio)methylene]tyramine derivatives, which may also serve as natural sunscreens. Embryos of the shrimp *Palaemon macrodactylus* are protected from infection by the fungal crustacean pathogen *Lagenidium callinectes* by isatin (Figure 5.2), a compound not produced by the shrimp itself but rather by a commensal *Alteromonas* sp. bacterium.[111]

In addition to the well-documented cases described above, there is a wealth of chemical and biological data suggestive of chemical deterrence in the plankton, although the data are incomplete. Consider, for example, that a number of pelagic organisms known to elaborate secondary metabolites have yet to be subjected to ecological evaluation. Microalgae are excellent examples since they produce toxins, some of which bloom and result in massive ecological disasters (see Shimizu[99]), yet we know little of the ecology behind the toxins. Could these toxins be produced during non-blooming periods as chemical deterrents? Ulapualide is found in the egg ribbons of a nudibranch, the Spanish Dancer *Hexabranchus sanguineus*,[54,112] whose larvae are pelagic; does the chemical protection afforded the egg ribbons follow the larvae into the water column? In addition to the lack of data on the ecological role of many secondary metabolites from pelagic organisms, there are notable cases of deterrence in which active metabolites have yet to be identified.[44,103,113] The question of whether eggs, embryos, and/or larvae are capable of *de novo* biosynthesis, or whether adults provision them, has yet to be addressed. Clearly, further study of the chemical nature of feeding deterrence in the plankton is in order.

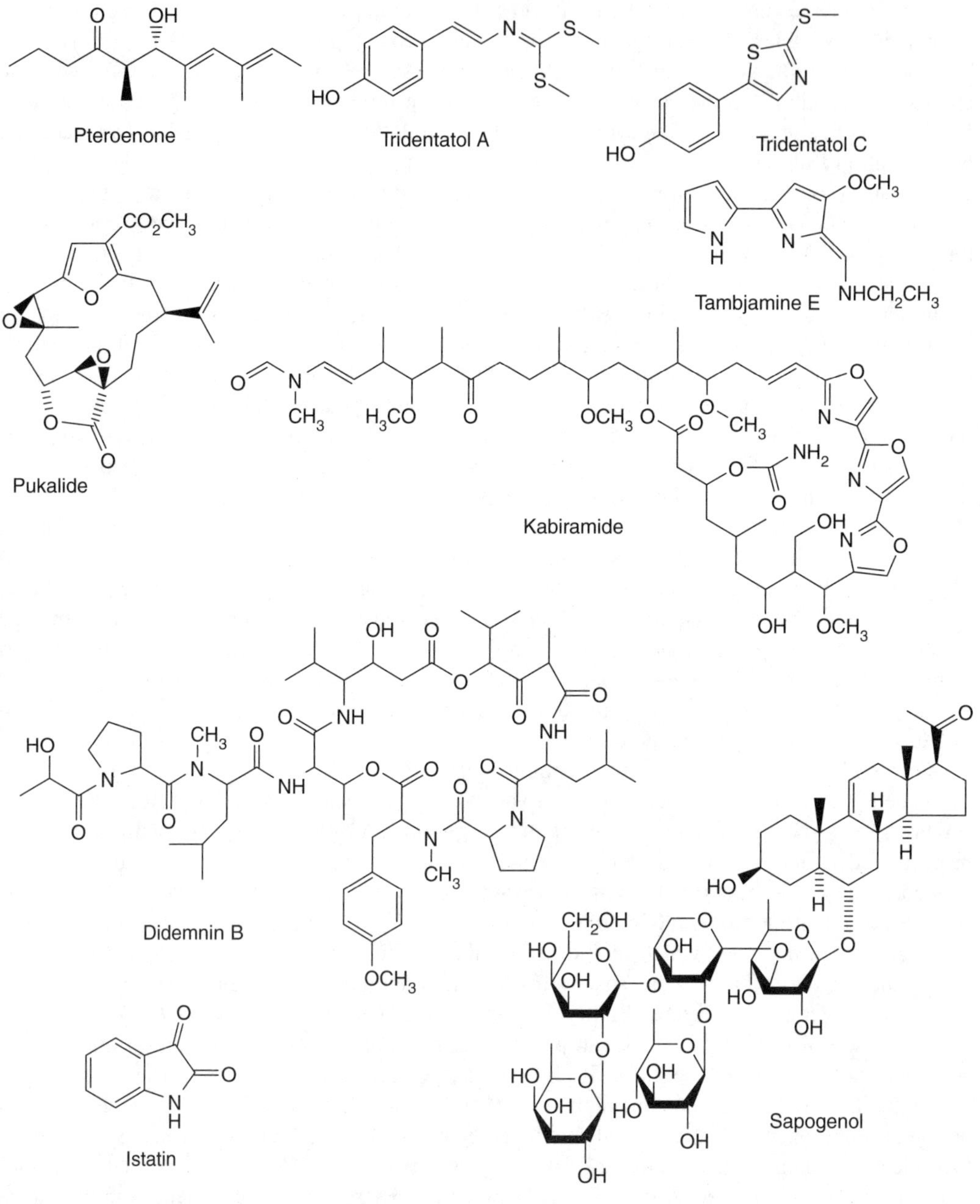

FIGURE 5.2 Chemical structures of deterrents from planktonic organisms.

V. OTHER MODES OF PREDATOR AVOIDANCE

The absence of cover in the pelagic realm and the inability of plankton to hide from visually orienting predators in the surface, sunlit zone has led to a number of adaptations including: (1) small size, (2) transparency and other forms of crypsis, (3) diel vertical migration, (4) exploitation of the sea surface or surfaces of other organisms and particles, (5) structural defenses, and (6) aposematism (warning coloration). All these adaptations must be considered in predator–prey interactions in an ecosystem nearly devoid of cover. Different taxa of plankton may exhibit one or more of these adaptations, but under what circumstances might it be advantageous to be chemically defended?

A. Size

Although categorization schemes of plankton size vary, plankton are typically categorized into the picoplankton (<2 μm; e.g., viruses and bacteria), nanoplankton (2–20 μm; e.g., many of the phytoplankton and protozoa), microplankton (20–200 μm; e.g., larger phytoplankton and protozoa and some invertebrate larvae), mesoplankton (0.2–2 μm; e.g., most species of copepods), macroplankton (2–20 mm; e.g., larger copepods, chaetognaths, and many gelatinous zooplankton such as ctenophores and pteropods), micronekton (2–20 cm; e.g., euphausiids and shrimps), and the megaplankton (>20 cm; e.g., the large gelatinous zooplankton such as the scyphozoan medusae, siphonophores, or colonial tunicates).[114,115] The majority of the "net zooplankton" (those caught in a standard 200 μm mesh net) in the surface waters during the day are under 2 mm in size (especially in open ocean environments[116,117] because larger zooplankton are easily seen and consumed by visual predators[118]). Exceptions include the larger, transparent gelatinous zooplankton and larger migrating taxa that can venture into the surface waters at night (see below). One might expect likely candidates for harboring defensive chemistry to be larger plankton living in the surface waters. Among meroplanktonic organisms, one might expect those that are larger and more conspicuous to be more frequently chemically defended. While data are still relatively rare for small feeding planktotrophic larvae, it appears likely that the larger yolk-laden lecithotrophic larvae of marine invertebrates more often possess chemical means of defense (studies reviewed above). There are no comparative studies of size effects on defensive chemistry in the holoplankton, but certainly small phytoplankton, such as some species of dinoflagellates and diatoms, produce chemical feeding deterrents against copepod predators,[100–102,119] indicating that even some of the minute plankton are candidates for chemical defenses.

B. Transparency and Other Forms of Crypsis

Convergent evolution in response to environmental light has produced three main forms of crypsis in the pelagic environment including transparency (the most prevalent), reflection of most (or all) visible wavelengths of light, and counter-illumination by bioluminescence.[120] Many zooplankton living in the euphotic zone of the sea are transparent, making them difficult for a potential predator to see. Examples include gelatinous zooplankton such as chaetognaths, ctenophores, hydromedusae, siphonophores, pteropods, salps, doliolids, larvaceans, and some species of copepods and mysids. Reflection of available light is a common defense used by some zooplankton and many pelagic fishes. A common strategy for many transparent plankton is to make reflective that part of the body where transparency cannot be maintained, such as the gut.[120] The bright, blue coloration of many open-ocean, near-surface-living plankton is another example of reflected light used as cover (and may protect against harmful UV radiation as well). Many pelagic fishes are counter-shaded, where body surfaces that are directed downward are lighter in color than those directed toward the surface. Marine predators are, therefore, unable to distinguish the prey's silhouette from above or below. The red and black coloration of deep-sea plankton and fish is cryptic because red light attenuates rapidly with depth in the water, so a red organism appears black at depth, and black blends in, rendering these organisms invisible to their predators.[120] Analogous to countershading, ventral luminescence or "counter-illumination" is used in some zooplankton and fishes below the euphotic zone to mimic down-welling light to camouflage their silhouettes.[121,122] The luminescence can be similar in intensity, color, and direction to that of down-welling light in the mesopelagic zone.[123,124]

Do transparent or other cryptic zooplankton need chemical defenses as well? Shanks and Graham[107] found that the mucus secreted from the scyphozoan jellyfish *Stomolopus meleagris* was toxic to fish [this species can be pigmented in varying degrees among individuals, and although they are not transparent, they are translucent (M. Graham, personal communication)]. McClintock et al.[103] found significant feeding deterrence in fish presented with transparent plankton including radiolarians, a ctenophore, a heteropod, and a salp. Moreover, McClintock, Baker, and Steinberg (unpublished) found that extracts of the transparent ctenophore *Mnemiopsis macrydi* embedded in agar pellets were essentially deterrent (P < 0.06) to fish predators. There are so few studies to date, that it is difficult to conclude if there are additional advantages for planktonic organisms to use defensive chemistry to ward off predators in addition to being transparent or cryptically colored. However, current evidence suggests that more than one form of defense is used by many species.

C. VERTICAL MIGRATION

Diel vertical migration, whereby animals feed in surface waters at night and return to deeper waters at dawn, is a widespread phenomenon in both freshwater and marine systems. It is thought that between 20 and 50% of zooplankton vertically migrate in the sea.[125] One of the several hypotheses suggested to explain the adaptive significance of vertical migration is predator avoidance.[126,127] Evidence has accumulated to support this hypothesis that prey are safer from visually orienting predators by feeding in the food-rich surface waters under the cover of darkness and spending their days at depth below the sun-lit zone.[128–130] One might expect that zooplankton that do not undergo vertical migrations are more likely to be chemically defended. Organisms that are not strong swimmers or cannot afford the considerable energetic costs of vertical migration may have evolved a chemical defense rather than a behavioral, migratory defense. The inventory of chemically defended plankton is still too small to compare migrators vs. nonmigrators to test this idea. However, it is known that chemical exudates produced by predators can influence vertical migration behavior in several species of freshwater zooplankton.[129,131,132] As studies of chemical defenses in the plankton progress, we should be able to determine if a "defense trade-off" exists between migratory and chemically defended plankton. Interestingly, a defense trade-off does exist in the timing of spawning of some invertebrate larvae, where more unpalatable invertebrate larvae are spawned during daylight hours, while more palatable larvae are spawned during the night.[72]

D. EXPLOITATION OF SEA SURFACE OR SURFACES OF OTHER ORGANISMS AND PARTICLES

The sea surface may provide a refuge for some plankton and larval fish as the "shimmering, rippling surface provides an optically complex habitat."[133] The neuston and pleuston (swimming and floating organisms living at the air–sea interface) are a unique group of organisms including *Halobates* (the only open ocean insect), *Physalia* and *Velella* (Cnidaria), and *Janthina* (Gastropoda). Some plankton may also take refuge from predators on the surface of other organisms or substrates. Mentioned above are the associations of hyperiid amphipods with gelatinous zooplankton.[35–36] The ubiquitous, nonliving, organic aggregates that are easily visible by eye in the sea (marine snow) harbor rich communities of associated organisms. Marine snow typically consists of a detrital or mucus matrix with associated dinoflagellates, ciliates and other protozoa, and bacteria, and these microorganisms are often enriched over those in the surrounding water (reviewed in Alldredge and Silver[134]). A number of zooplankton also use marine snow as a habitat and food source.[33,34] Steinberg et al.[34] suggest that some species may also reside on marine snow to take refuge from predators, as some particles are considerably larger than the individual zooplankters.

We are aware of two examples where holoplankton may benefit by associating with chemically defended organisms. The colonial cyanobacterium *Trichodesmium* spp. harbors a number of micro-organims and zooplankton.[135] Interestingly, some pelagic copepods such as *Macrosetella gracilis* not only feed on *Trichodesmium*, but use it as a substrate for juvenile development.[95,135] However, *Trichodesmium* is lethal to other copepods.[96] Hawser et al.[96] suggest that these copepods gain protection by associating with toxic algae. The spider crab *Libinia dubia* is a common associate and predator of the apparently chemically defended medusa *Stomolopus meleagris*.[81] Discharged mucus from *S. meleagris* killed potential fish predators but did not kill or change the behavior of *L. dubia*. Thus, the crab may gain some additional protection from its predators by associating with a chemically defended host (in addition to protection already provided by the nematocysts of the jellyfish). Whether other plankton can benefit by associations with other organisms that are chemically defended remains to be studied.

E. STRUCTURAL DEFENSE

Morphological defenses appear to be generally more common among meroplanktonic organisms than holoplanktonic organisms. Morgan[136] summarizes proposed morphological defenses in marine invertebrate larvae which include spicules, setae, nematocysts, spines, mucus, and shells of sponge, coral, polychaete, crustacean, brachiopod, mollusc, and echinoderm larvae. Several specific examples include the spiny protuberances that defend polychaete and crustacean larvae, actually used to pierce the tissues of predators that try to feed on them.[136,137] Some echinoderm larvae are provisioned with spicules that can effectively deter predators.[138] Among the holoplankton, many diatoms (e.g., *Chaetoceros* spp.) are equipped with spines that may serve to both slow sinking (by increasing surface to volume ratio) and deter predation, in addition to their hard silica frustules (shells). Many marine protozoa (e.g., foraminifera, acantharia, and radiolaria) also are equipped with spines and hard shells. Large spines are uncommon in most holoplanktonic crustacea (e.g., copepods, euphausiids), although setae are widespread. Few studies have directly tested the effectiveness of spines, setae, and hard shells at defending predators. Gelatinous zooplankton lack morphological defenses with the notable exception of those that possess nematocysts.

F. APOSEMATISM

Young and Bingham[49] demonstrated that among the meroplankton, the bright orange larvae of the ascidian *Ecteinascida turbinata* contained defensive chemistry against the juvenile pinfish *Lagodon rhomboides*. Young and Bingham[49] suggested that warning coloration (aposematism) may be common in larvae that are chemically defended. Subsequently, Lindquist and Hay[72] showed that the frequency of bright coloration (red, orange, or yellow) of unpalatable larvae (tested against co-occurring fishes) of a variety of benthic invertebrates was high (12 out of 20 species). Among the holoplankton, the occurrence of bioluminescence in toxic dinoflagellates has been suggested as a form of aposematism. McClintock and Janssen[85] examined the feeding deterrent properties of the holoplanktonic Antarctic pteropod *Clione antarctica*. Ranging from 1 to 3 cm in body length, it has a conspicuous orange coloration and is a sluggish swimmer.[88] That this chemically defended Antarctic sea butterfly is brightly colored raises the intriguing question of whether aposematism is operating. In McMurdo Sound, Antarctica, there is little or no light for up to 6 months of the year, and even when light is present, snow covered sea ice greatly reduces the levels of irradiance which penetrate the water. Antarctic fish are often considered to function near their visual threshold, relying on visual cues in combination with their lateral lines to detect mechanical stimulation from prey.[139] Pankhurst and Montgomery[140] demonstrated that the Antarctic fish *Pagothenia borchgrevinki* is most sensitive to visible light at 500 nm, with sensitivity dropping drastically past 550 nm. The color orange has a wavelength of 600 nm, a wavelength these fish are unable to detect. If this is generally true for all Antarctic fish, then coloration in *C. antarctica* would not appear to be an example of aposematism. Observations of the conspicuous black copepod *Candacia ethiopica* in the surface waters off Bermuda (Steinberg, personal observation) led to the hypothesis that it may be chemically defended. Preliminary results from subsequent experiments indicated that both whole individuals and alginate pellets containing homogenates of *C. ethiopica* were significantly deterrent to fish (Figure 5.1). In freshwater, Kerfoot[141] suggests that visual predation by fishes upon pigmented but mildly unpalatable groups of plankton (water mites) has selected for enhanced conspicuousness (bright coloration) and unpalatability. It is possible that the same mechanism may be operating in the marine environment.

G. OTHER CONSIDERATIONS

1. Speed/Swimming Behaviors

Although the word "plankton" comes from the Greek word *planktos*, meaning that which is passively drifting or wandering, many plankton are quite capable of locomotion. Some of the best examples of the capability for speed come from estimates of swimming speeds of vertical migrators. Estimates of migratory swimming speeds in the field for a variety of individual copepod species range from about 42 to 122 m h^{-1} (mean absolute upward and downward rates[142]) and 68 to 186 m h^{-1} (median depth migration speeds[143]). Migration speeds reported for a variety of euphausiid species range from 120 to 191 m h^{-1}.[143] Even gelatinous zooplankton can move at impressive speeds, with mean *in situ* swimming speeds ranging from about 1 to 9 cm s^{-1} (36 to 324 m h^{-1}) for a variety of salps (reviewed in Bone[144]), and a mean speed of 15 cm s^{-1} (540 m h^{-1}, but only short distances, 5 to 14 m, were swum at a time) in the jellyfish *Stomolopus meleagris*.[81] Many plankton have evasive predator escape behaviors other than speed. Schooling, seen in many micronektonic organisms such as the euphausiids,[145] provides a means of protection other than speed by other members of the school.[146] Other escape behaviors may include the hop-and-sink swimming behavior of some copepods,[147] or the sinking behavior of pteropods[148] and colonial radiolarians[149] when disturbed. One would expect the more sluggish plankton, or those without escape maneuverability, to be chemically defended, but as yet there are no studies to address this.

2. Nutritional Content

It is also unlikely that meroplankton and holoplankton are afforded any protection due to their low nutritional content. Pelagic pteropods can be relatively rich in protein and lipids.[86] While some gelatinous zooplankton have been considered to have relatively low carbon contents,[150,151] the low body carbon–nitrogen ratio in gelatinous zooplankton such as salps[152] indicates that they are potentially quite nutritious. Moreover, the energetic value of the internal organs can be very high.[153] Indeed, some predators of gelatinous zooplankton feed almost exclusively on these internal organs, avoiding the ingestion of gelatinous body parts.[35,36,154] Some gelatinous zooplankton attain large body sizes or occur in vast numbers or even swarms, making them energetically attractive prey items even if low in energy on an individual basis.

3. Time in the Plankton

Length of time in the plankton is yet another consideration when evaluating predation risk. While holoplankton spend their entire lives in the plankton, different groups of meroplankton can exhibit highly variable amounts of time among the plankton. For example, pelagic lecithotrophic larvae of most marine invertebrates spend comparatively less time in the plankton than planktotrophic larvae.[7] As noted above, the brooded larvae of ascidians or sponges may be released and swim for only a few hours or days before settlement. In contrast, most planktotrophic marine invertebrate larvae spend weeks in the plankton, and in polar systems they may spend up to 6 months in the plankton.[7] Even pelagic lecithotrophic larvae in polar environments may spend several months in the water column prior to settlement.[7] Clearly, such meroplankton that spend long periods in the plankton will be subject to increased levels of predation pressure and should be under strong selective pressure to evolve defenses, including those of a chemical nature.

VI. SYMBIOSES

While there are likely a myriad of symbiotic relationships between chemically defended and non-chemically defended organisms in the plankton awaiting discovery, only two documented cases of such symbiotic interactions are known. Both examples are comprised of holoplanktonic organisms, and perhaps it is likely that future work will yield a greater proportion of species among the holoplankton rather than the meroplankton possessing symbiotic relationships involving chemical defenses. Our prediction is based on the assumption that meroplankton generally (with the polar exceptions discussed above) spend a much shorter period of their life history in the plankton where opportunities for the evolution of symbiotic relationships among planktonic organisms are greatest.

Gil-Turnes et al.[111] examined the brooded embryos of the shrimp *Palaemon macrodactylus* and, in contrast to the juvenile and adult life phase, found them to be remarkably resistant to infection by the fungus *Lagenidium callinectes* (a common and deadly pathogen among crustaceans). Suspecting that there might be a microbial basis to the lack of fungal growth on embryos, those researchers cultured bacteria from the surface of the embryos and discovered a common strain of *Alteromonas* sp. Further work revealed that when this particular strain of *Alteromonas* sp. was grown in pure culture, it produced, and released into the culture medium, large quantities of an antifungal compound subsequently identified as 2,3-indolinedione (istatin, see Figure 5.2). In subsequent experiments, Gil-Turner et al.[111] demonstrated that: (1) embryos treated with penicillin to remove bacteria died, (2) embryos treated with penicillin and reinfected with *Alteromonas* sp. displayed about 60% survival, (3) embryos treated with penicillin and then dipped periodically in 2,3-indolinedione displayed similar high levels of survival, and (4) control embryos not treated with penicillin had the highest survival (80%). These results clearly indicate that the commensalistic bacteria *Alteromonas* sp. is capable of deterring fungal infection of shrimp embryos by producing a potent chemical defense in the form of the antifungal compound 2,3-indolinedione. There are important evolutionary and ecological ramifications of this symbiotic interaction. For one, it is evident that marine organisms are likely to be subject to widespread pressure from pathogenic microorganisms, and that through the evolution of such symbiotic relationships they may have evolved the capacity to successfully reproduce and ultimately survive. As evidenced by this study, it is likely that marine microbial chemical ecology is perhaps the most understudied, and potentially fruitful, avenue of research in marine chemical ecology.

A second study documenting a symbiotic relationship involving chemical defenses among marine organisms is that of McClintock and Janssen,[85] who worked with the Antarctic pteropod (sea butterfly) *Clione antarctica* and the hyperiid amphipod *Hyperiella dilatata*. Observations of sea butterflies and amphipods in the field under annual sea ice revealed that large numbers of amphipods were carrying an individual sea butterfly on their back using their sixth and seventh pair of swimming appendages (pereiopods). Laboratory feeding assays employing the common antarctic planktivorous fish *Pagothenia borchgrevinki* demonstrated that live amphipods are consumed by fish, while amphipod-sea butterfly pairs are consistently rejected. Subsequent studies revealed that the sea butterflies contained a potent fish antifeedant compound which was given the name pteroenone.[86,89]

Clearly, this relationship between the sea butterfly and the amphipod is a type of symbiosis, because it essentially involves two dissimilar species that live together in an intimate association. However, none of the currently accepted relationships defined within the context of symbiosis — mutualism, commensalism, or parasitism — is suitable to describe this specific interaction. In this association, the antagonist (amphipod) benefits greatly (although there are some negatives including slowed swimming speeds and restricted mobility[69,85]), while the sea butterfly is essentially at the mercy of the amphipod and cannot feed and sustain its energy requirements. Sea butterflies are apparently released after some period of time and replaced with a new individual, as McClintock and Janssen[85] never found an amphipod carrying a dead sea butterfly, despite examining hundreds of amphipod–sea butterfly pairs.

The relationship described above might lead one to suspect that the reason that the hyperiid amphipod *Hyperiella dilatata* abducts and carries a chemically defended sea butterfly (*Clione antarctica*) is that it lacks the ability to synthesize or produce its own chemical defenses. Indeed, Hay et al.[39] have also shown that amphipods may associate with chemically defended algae to provide defense against fish predators, rather than produce defensive compounds themselves.

However, recently there has been evidence that the production of chemical defenses by benthic marine amphipods is indeed possible, as Norton and Stallings[155] report on the distribution and abundance of three aposematic, chemically defended gammarid amphipods in the northwest Pacific Ocean. There is no reason to suspect that planktonic amphipods cannot similarly evolve their own chemical defenses. Hyperiid amphipods, which are known to associate extensively with gelatinous zooplankton, are perhaps uniquely situated to exploit the chemical defenses of other planktonic organisms.

VII. POTENTIAL ANTIFOULANTS

Marine organisms, both benthic and planktonic, are subjected to intense fouling pressure from settling bacteria, diatoms, algal spores, and marine invertebrate larvae. Secondary metabolites have been shown to function as inhibitors of fouling in both benthic marine algae[156,157] and invertebrates.[158,159] To our knowledge, there is no information available on the antifoulant properties of secondary metabolites from meroplankton or holoplankton. There is no *a priori* reason to believe antifoulants may not occur widely in meroplankton, and especially in holoplankton.

VIII. SUMMARY AND FUTURE DIRECTIONS

The field of chemical ecology of plankton is relatively young. The ecological roles of bioactive secondary metabolites derived from marine algae (reviewed by Hay and Fenical[160]) and marine invertebrates (reviewed by Bakus et al.,[161] Paul,[162] Pawlik,[158] McClintock and Baker,[113] and Amsler et al.[163]) have almost exclusively focused on benthic marine organisms.[113,159-162] A few studies have investigated toxins in freshwater zooplankton,[141,164] but comparatively little information is available on the chemical ecology of planktonic marine organisms,[165] arguably one of the most significant biotic components of the world's oceans. Thus, there exists a wide range of future directions for investigation.

Many of the studies reviewed in this chapter have focused on the meroplankton. However, little is known about ontogenetic shifts in concentrations and patterns of defense in marine invertebrate larval forms.[40] Further work is needed to determine if, for a wider range of species, developing larvae are capable of secondary metabolite synthesis or if defensive compounds are derived directly from adults. While a number of studies have been conducted on chemical defenses in lecithotrophic larvae of benthic invertebrates, the database is still quite small for planktotrophic larvae. Additional carefully controlled studies of aposematism in marine invertebrate larvae are also needed to determine if there is indeed a general pattern of chemical defenses in conspicuously colored larvae.

The chemical ecology of holoplankton is clearly vastly understudied. Do defense trade-offs exist between chemically defended plankton and plankton with other kinds of defenses such as small size, transparency, or cryptic coloration, or behavioral defenses such as vertical migration? Aposematism, which has been demonstrated in benthic invertebrate larvae, has not been explicitly demonstrated in any conspicuously colored holoplankton. There are many unique organism associations in the plankton. Do some of these organisms, such as copepods on *Trichodesmium*, derive protection or defensive chemistry from their hosts? Studies need to be conducted with ecologically relevant predators, which may be experimentally challenging, since many planktivores may be difficult to keep in laboratory settings.

Further directions for the study of both meroplankton and holoplankton chemical ecology include development of field bioassays to couple with laboratory studies. Studies are needed to determine how consumers may perceive secondary metabolites produced by plankton.[166] The ability of planktonic organisms to sequester defensive chemistry in specific tissues or mucus, as seen in benthic invertebrates, is unknown. Very little is known about the specific compounds responsible for chemical defenses in plankton, and we expect the library of secondary metabolites produced by plankton to grow as interest and resources are focused on the major difficulty inherent in this research; only small amounts of biomass are often available for analysis. Analytical techniques, such as LC/MS, LC/NMR, and micro- and/or nano-bore NMR spectroscopy, for isolation and structure determination of secondary metabolites continue to push the frontier of detectable limits, and these should continue to improve. Application of these cutting-edge techniques to projects with resources such as ship time and personnel available for the tedious collection and sorting efforts will be required before we can begin to understand the role of secondary metabolism in the plankton. Future investigations of the chemical ecology of plankton, while often experimentally challenging, are certain to change the way we view planktonic food webs and material and energy cycling in the sea.

ACKNOWLEDGMENTS

This chapter resulted from synergy facilitated by an SGER grant from the National Science Foundation to McClintock (OCE 9714402), Baker (OCE 9725040), and Steinberg (OCE 9725041) to initiate studies on the chemical ecology of gelatinous zooplankton in Bermuda. Funds provided by NSF OPP-9814538 to J.B. McClintock and C.D. Amsler were of assistance in preparing this chapter. We are grateful to the undergraduate and graduate students that contributed to these studies including Tom Barlow, Vanessa Voss, Jason Stanko, Dan Swenson, Jessica Bohonowych, and Toby Jarvis. We thank Chuck Amsler and Katrin Iken for providing comments and suggestions on our chapter.

REFERENCES

1. Pawson, D.L., Some aspects of the biology of deep-sea echinoderms, *Thalassia Jugoslav.*, 12, 287, 1976.
2. Alldredge, A.L. and King, J.M., Distribution, abundance, and substrate preference of demersal reef zooplankton at Lizard Island Lagoon, Great Barrier Reef, *Mar. Biol.*, 41, 317, 1977.
3. Alldredge, A.L. and King, J.M., The distance demersal zooplankton migrate above the benthos: implications for predation, *Mar. Biol.*, 84, 253, 1985.
4. Thorson, G., Reproductive and larval ecology of marine bottom invertebrates, *Biol. Rev.*, 25, 1, 1950.
5. Vance, R.R., On reproductive strategies in marine benthic invertebrates, *Am. Nat.*, 107, 339, 1973.
6. Strathmann, R.R., Feeding and nonfeeding larval development and life-history evolution in marine invertebrates, *Ann. Rev. Ecol. Syst.*, 16, 339, 1985.
7. Pearse, J.S., McClintock, J.B., and Bosch, I., Reproduction of antarctic marine benthic invertebrates: tempos, modes and timing, *Am. Zool.*, 31, 65, 1991.
8. Young, C.M. and Chia., F.S., Abundance and distribution of pelagic larvae as influenced by predation, behavior, and hydrographic features, in *Reproduction of Marine Invertebrates*, Vol. 9, Giese, A. and Pearse, J.S., Eds., Aberdeen University Press, Aberdeen, Scotland, 1987, 385.
9. Morgan, S.G., Life and death in the plankton, in *Ecology of Marine Invertebrate Larvae,* McEdward, L., Ed., CRC Press, New York, 1995, 279.
10. Rumrill, S.S., Natural mortality of marine invertebrate larvae, *Ophelia,* 32, 163, 1990.
11. Emery, A.R., Comparative ecology and functional osteology of fourteen species of damselfish (Pisces: Pomacentridae) at Alligator Reef, Florida Keys, *Bull. Mar. Sci.*, 23, 649, 1973.
12. McClintock, J.B. and Baker, B.J., Palatability and chemical defense of eggs, embryos and larvae of shallow-water antarctic marine invertebrates, *Mar. Ecol. Prog. Ser.,* 154, 121, 1997.
13. Mileikovsky, S.A., Types of larval development in marine bottom invertebrates, their distribution and ecological significance: a re-evaluation, *Mar. Biol.,* 10, 193, 1971.
14. Strathmann, R.R. and Strathmann, M.F., The relation between adult size and brooding in marine invertebrates, *Am. Nat.*, 119, 91, 1985.

15. Emlet, R.R., McEdward, L.R., and Strathmann, R.R., Echinoderm larval biology viewed from the egg, in *Echinoderm Studies,* Jangoux, M. and Lawrence, J.M., Eds., A.A. Balkema Press, Rotterdam, 1987, 55.

16. Roughgarden, J., Gaines, S., and Possingham, H., Recruitment dynamics in complex life-cycles, *Science,* 241, 1460, 1988.

17. Roughgarden, J., The evolution of marine life cycles, in *Mathematical Evolutionary Theory,* Feldmans, M.W., Ed., Princeton University Press, Princeton, New Jersey, 1989, 270.

18. Alexander, S.E. and Roughgarden, J., Larval transport and population dynamics of intertidal barnacles: a coupled benthic/oceanic model, *Ecol. Monogr.,* 66, 259, 1996.

19. Huys, R. and Bosshall, G.A., Copepod evolution, Ray Society, 1991, 9.

20. Berner, L., Distributional atlas of Thalacia in the California Current region, *CalCOFI Atlas,* 8, 1967.

21. Andersen, V., Salp and pyrosomid blooms and their importance in biogeochemical cycles, in *The Biology of Pelagic Tunicates,* Bone, Q., Ed., Oxford University Press, New York, 1998, 340.

22. Lee, J.J. and Capriulo, G.M., The ecology of marine protozoa: an overview, in *The Ecology of Marine Protozoa,* Capriulo, G.M., Ed.., Oxford University Press, New York, 1990, 3.

23. Caron, D.A. and Swanberg, N.R., The ecology of planktonic sarcodines, *Rev. Aquatic Sci.,* 3, 147, 1990.

24. Michaels, A.F. and Silver, M.W., Primary production, sinking fluxes and the microbial food web, *Deep Sea Res.,* 35, 473, 1988.

25. Alldredge, A.L. and Madin, L.P., Pelagic tunicates: unique herbivores in the marine plankton, *Bio-Science,* 32, 655, 1982.

26. Kremer, P. and Madin, L.P., Particle retention efficiency in salps, *J. Plank. Res.,* 4, 1009, 1992.

27. Silver, M.W. and Bruland, K.W., Differential feeding and fecal pellet consumption of salps and pteropods, and the possible origin of the deep water flora and olive green "cells," *Mar. Biol.,* 62, 263, 1981.

28. Madin, L.P., The production, composition and sedimentation of salp fecal pellets in oceanic waters, *Mar. Biol.,* 67, 39, 1982.

29. Caron, D.A., Madin, L.P., and Cole, L.L., Composition and degradation of salp fecal pellets: implications for vertical flux in oceanic environments, *J. Mar. Res.,* 47, 829, 1989.

30. Hutchinson, G.E., The paradox of the plankton, *Am. Nat.,* 95, 137, 1961.

31. Richerson, P., Armstong, R., and Goldman, C.R., Contemporaneous disequilibrium, a new hypothesis to explain "the paradox of the plankton," *Proc. Nat. Acad. Sci.,* 67, 1710, 1970.

32. Heron, A.C., A specialized predator–prey relationship between the copepod *Sapphirina augusta* and the pelagic tunicate *Thalia democratia, J. Mar. Biol. Assoc. UK,* 53, 429, 1973.

33. Alldredge, A.L., Discarded appendicularian houses as sources of food, surface habitats, and particulate organic matter in planktonic environments, *Limnol. Oceanogr.,* 21, 14, 1976.

34. Steinberg, D.K., Silver, M.W., Pilskaln, C.H., Coale, S.L., and Paduan, J.B., Midwater zooplankton communities on pelagic detritus (giant larvacean houses) in Monterey Bay, California, *Limnol. Oceanogr.,* 39, 1606, 1994.

35. Harbison, G.R., Biggs, D.C., and Madin, L.P., The association of amphipoda Hyperiida with gelatinous zooplankton. II. Associations with cnidaria, ctenophora and radiolaria, *Deep Sea Res.,* 24, 465, 1977.

36. Madin, L.P. and Harbison, G.R., The association of amphipoda Hyperiida with gelatinous zooplankton. I. Association with salpidae, *Deep Sea Res.,* 24, 449, 1977.

37. Laval, P., Hyperiid amphipods as crustacean parasitoids associated with gelatinous zooplankton, *Oceanogr. Mar. Biol. Annu. Rev.,* 18, 11, 1980.

38. Diebel, C.E., Observations on the anatomy and behavior of *Phronima sedentaria, J. Crust. Biol.,* 8, 79, 1988.

39. Hay, M.E., Duffy, J.E., and Fenical, W., Host-plant specialization decreases predation on marine amphipods: an herbivore in plant's clothing, *Ecology,* 71, 733, 1990.

40. Hay, M.E., Marine chemical ecology: what's known and what's next?, *J. Exp. Mar. Biol. Ecol.,* 200, 103, 1996.

41. Bakus, G.J. and Green, G., Toxicity in sponges and holothurians: a geographic pattern, *Science,* 185, 951, 1974.

42. Bakus,G.J., Chemical defense mechanisms on the Great Barrier Reef, *Science,* 211, 497, 1981.

43. McClintock, J.B., Investigation of the relationship between invertebrate predation and biochemical composition, energy content, spicule armament and toxicity of benthic sponges at McMurdo Sound, Antarctica, *Mar. Biol.*, 94, 479, 1987.

44. Lucas, J.S., Hart, R.J., Howden, M.E., and Salathe R., Saponins in eggs and larvae of *Acanthaster planci* (L.) (Asteroidea) as chemical defenses against planktivorous fish, *J. Exp. Mar. Biol. Ecol.*, 40, 155, 1979.

45. Yamaguchi, M., Coral reef asteroids of Guam, *Biotropica*, 7, 12, 1975.

46. Lucas, J.S., Environmental influences on the early development of *Acanthaster planci* (L.), in *Crown-of-Thorns Starfish Seminar Proceedings*, Brisbane, September 6, 1974, Australian Government Public Service, Canberra, 1975, 109.

47. Howden, M.E.H., Lucas, J.S., McDuff, M., and Salathe, R., Chemical defenses of *Acanthaster planci*, in *Crown-of-Thorns Starfish Seminar Proceedings*, Brisbane, September 6, 1974, Australian Government Public Service, Canberra, 1975, 67.

48. De Vore, D.D. and Brodie, E.D., Jr., Palatability of the tissues of the holothurian *Thyone briareus* (Lesueur) to fish, *J. Exp. Mar. Biol. Ecol.*, 61, 279, 1982.

49. Young, C.M. and Bingham, B.L., Chemical defense and aposematic coloration in larvae of the ascidian *Ecteinascidia turbinata*, *Mar. Biol.*, 96, 539, 1987.

50. Harvey, P.H. and Paxton, R.J., The evolution of aposematic coloration, *Oikos,* 37, 391, 1981.

51. Jarvi, T., Sillen-Tullberg, B., and Wiklund, C., Individual versus kin selection for aposematic coloration: a reply to Harvey and Paxton, *Oikos*, 37, 393, 1981.

52. McClintock, J.B. and Vernon, J.D., Chemical defense in the eggs and embryos of antarctic sea stars (Echinodermata), *Mar. Biol.*, 105, 491, 1990.

53. Lindquist, N., Hay, M.E., and Fenical, W., Defenses of ascidians and their conspicuous larvae: adult vs. larval chemical defenses, *Ecol. Monogr.*, 62, 547, 1992.

54. Pawlik, J.R., Kernan, M.R., Molinski, T.F., Harper, M.K., and Faulkner, D.J., Defensive chemicals of the Spanish dancer nudibranch *Hexabranchus sanguineus* and its egg ribbons: macrolides derived from a sponge diet, *J. Exp. Mar. Biol. Ecol.*, 119, 99, 1988.

55. De Marino, S., Iorizzi, M., Zollo, F., Minale, L., Amsler, C.D., Baker, B.J., and McClintock, J.B., Isolation, structure elucidation, and biological activity of the steroid oligoglycosides and polyhydroxysteroids from the Antarctic starfish *Acodontaster conspicuus*, *J. Nat. Prod.*, 60, 959, 1997.

56. McClintock, J.B., Pearse, J.S., and Bosch, I., Population structure and energetics of the shallow-water antarctic sea star *Odontaster validus* in contrasting habitats, *Mar. Biol.*, 99, 235, 1988.

57. McClintock, J.B., Heine, J., Slattery, M., and Weston, J., Biochemical and energetic composition, population biology, and chemical defense of the antarctic ascidian *Cnemidocarpa verrucosa* Lesson, *J. Exp. Mar. Biol. Ecol.,* 147, 163, 1991.

58. Stoecker, D., Chemical defenses of ascidians against predators, *Ecology,* 61, 1327, 1980.

59. Orians, G.H. and Janzen, D.H., Why are embryos so tasty? *Am. Nat.,* 108, 581, 1974.

60. Davis, A.D. and Butler, A.L., Direct observations of larval dispersal in the colonial ascidian *Padoclavella moluccensis* Sluiter: evidence for closed populations, *J. Exp. Mar. Biol. Ecol.,* 127, 189, 1989.

61. Svane, I. and Young, C.M., The ecology and behavior of ascidian larvae, *Oceanogr. Mar. Biol. Ann. Rev.*, 27, 45, 1989.

62. Bingham, B.L. and Young, C.M., Larval behavior of the ascidian *Ecteinascidia turbinata* Herdman; an in situ experimental study of the effects of swimming on dispersal, *J. Exp. Mar. Biol. Ecol.,* 145, 189, 1991.

63. Sale, P.F., The ecology of fishes on coral reefs, *Oceanogr. Mar. Biol. Ann. Rev.*, 18, 367, 1980.

64. Strathmann, R.R., What controls the type of larval development? Summary statement for the evolution session, *Bull. Mar. Sci.*, 39, 616, 1986.

65. Lindquist, N. and Fenical, W., New tambjamine class alkaloids from the marine ascidian *Atapozoa* sp. and its nudibranch predators — origin of the tambjamines in *Atapozoa, Experientia,* 47, 504, 1991.

66. Brodie, E.D., Jr. and Formanowicz, D.R., Jr., Antipredator mechanisms of larval anurans: protection of palatable individuals, *Herpetologia,* 43, 369, 1987.

67. Brower, L.P., Chemical defenses in butterflies, *Symposium of the Royal Entomological Society of London*, 11, 109, 1984.

68. Lindquist, N., Palatability of invertebrate larvae to corals and sea anemones, *Mar. Biol.*, 126, 745, 1996.

69. McClintock, J.B. and Baker, B.J., Chemical ecology in Antarctic seas, *Am. Sci.*, 86, 254, 1998.

70. Slattery, M., Hines, G.A., Starmer, J., and Paul, V.J., Chemical signals in gametogenesis, spawning, and larval settlement and defense of the soft coral *Sinularia polydactyla*, *Coral Reefs*, 18, 75, 1999.

71. Lindquist, N. and Hay, M.E., Can small rare prey be chemically defended? The case for marine larvae, *Ecology*, 76, 1347, 1995.

72. Lindquist, N. and Hay, M.E., Palatability and chemical defense of marine invertebrate larvae, *Ecol. Monogr.*, 66, 431, 1996.

73. Dayton, P.K., Robilliard, G.A., Paine, R.T., and Dayton, L.B., Biological accommodation in the benthic community at McMurdo Sound, Antarctica, *Ecol. Monogr.*, 44, 105, 1974.

74. Wylie, C.R. and Paul, V.J., Chemical defenses in three species of *Sinularia* (Coelenterata: Alcyonacea) — effects against generalist predators and the butterflyfish *Chaetodon unimaculatus* Bloch, *J. Exp. Mar. Biol. Ecol.*, 129, 141, 1989.

75. Slattery, M., Paul, V.J., Van Alstyne, K.L., and Wylie, C.R., Ecological roles for secondary metabolites in tropical Pacific soft corals and their eggs, *J. Cell Biochem.*, 19B, 34, 1995.

76. Coll, J.C., Bowden, B.F., Alino, P.F., Heaton, A., Konig, G.M., De Nys, R., and Willis, R.H., Chemically mediated interactions between marine organisms, *Chem. Sci.*, 29, 383, 1989.

77. Harvell, C.D., West, J.M., and Griggs, C., Chemical defense of embryos and larvae of a West Indian gorgonian coral, *Briareum asbestinum, Invertebrate Reprod. Dev.*, 30, 239, 1996.

78. Cowart, J.D., Fielman, K.T., Woodin, S.A., and Lincoln, D.E., Halogenated metabolites in two marine polychaetes and their planktotrophic and lecithotrophic larvae, *Mar. Biol.*, 136, 993, 2000.

79. Woodin, S.A., Walla, M.D., and Lincoln, D.E.., Occurrence of brominated compounds in soft-bottom benthic organisms, *J. Exp. Mar. Biol. Ecol.*, 107, 209, 1987.

80. Bullard, S.G., Lindquist, N.L., and Hay, M.E., Susceptibility of invertebrate larvae to predators: how common is predator resistance?, *Mar. Ecol. Prog. Ser.*, 191, 153, 1999.

81. Shanks, A.L. and Graham, W.M., Chemical defense in a scyphomedusa, *Mar. Biol.*, 45, 81, 1988.

82. Greve, W., Okologische Untgersucchungen an *Pleurobrachia pileus* II. Laboratoriumsuntersuchungen, *Helogolander wiss. Meeresunters,* 23, 141, 1972.

83. Horridge, G.A., Non-motile sensory cilia and neuormuscular functions in a ctenophore independent effector organ, *Proc. R. Soc. (Ser. B)*, 162, 335, 1965.

84. Caron, D.A., Davis, P.G., Madin, L.P., and Sieburth, J.McN., Heterotrophic bacteria and bacterivorous protozoa in oceanic macroaggregates, *Science,* 218, 795, 1982.

85. McClintock, J.B. and Janssen, J., Pteropod abduction as a chemical defence in a pelagic antarctic amphipod, *Nature,* 346, 462, 1990.

86. Bryan, P.J., Yoshida, W.Y., McClintock, J.B., and Baker, B.J., Ecological role for pteroenone, a novel antifeedant from the conspicuous antarctic pteropod *Clione antarctica* (Gymnosomata: Gastropoda), *Mar. Biol.*, 122, 271, 1995.

87. Foster, B.A., Carghill, J.M., and Montgomery, J.C., Planktivory in *Pagothenia borchgrevinki* (Pisces: Notheniidae) in McMurdo Sound, Antarctica, *Polar Biol.*, 8, 49, 1987.

88. Gilmer, R.W. and Lalli, C.M., Bipolar variation in *Clione*, a gymnosomatous pteropod, *Am. Malac. Bull.*, 8, 67, 1990.

89. Yoshida, W.Y, Bryan, P.J., Baker, B.J., and McClintock, J.B., Pteroenone: a defensive metabolite of the abducted antarctic pteropod *Clione antarctica*, *J. Org. Chem.*, 60, 780, 1995.

90. Carpenter, E.J. and Romans, K., Major role of the cyanobacterium *Trichodesmium* in the nutrient cycling in the North Atlantic Ocean, *Science,* 254, 1356, 1991.

91. Carpenter, E.J. and Roenneberg, T., The marine planktonic cyanobacteria *Trichodesmium* spp.: photosynthetic rate measurements in the SW Atlantic Ocean, *Mar. Ecol. Prog. Ser.*, 118, 267, 1995.

92. Capone, D.G., Zehr, J.P., Pearl, H.W., Bergman, B, and Carpenter, E.J., *Trichodesmium*, a globally significant marine cyanobacterium, *Science,* 276, 1221, 1997.

93. O'Neil, J.M. and Roman, M.R., Grazers and associated organisms of *Trichodesmium*, in *Marine Pelagic Cyanobacteria: Trichodesmium and other Diazotrophs*, Carpenter, E.J., Ed., Kluwer Academic Press, Netherlands, 1992, 61.

94. Sellner, K.G., Trophodynamics of marine cyanobacteria blooms, in *Marine Pelagic Cyanobacteria: Trichodesmium and other Diazotrophs*, Carpenter, E.J., Ed., Kluwer Academic Press, Netherlands, 1992, 75.

95. Roman, M.R., Ingestion of blue-green algae *Trichodesmium* by the harpacticoid copepod, *Macrosetella gracilis*, *Limnol. Oceanogr.*, 23, 1245, 1978.

96. Hawser, S.P., O'Neil, J.M., Roman, M.R., and Gould, G.A., Toxicity of blooms of cyanobacterium *Trichodesmium* to zooplankton, *Appl. Phycol.*, 4, 79, 1992.

97. McCarthy, J.J. and Carpenter, E.J., *Osciallatoria* (*Trichodesmium*) *theubautii* (Ctanophyta) in the central North Atlantic Ocean, *J. Phycol.*, 15, 75, 1979.

98. White, A.W., Dinoflagellate toxins as probable cause of an Atlantic Herring (*Clupea harengus)* kill, and pteropods as an apparent vector, *J. Fish. Res. Board Can.*, 34, 2421, 1977.

99. Shimizu, Y., Microalgal metabolites, *Chem. Rev.*, 93, 1685, 1993.

100. Huntley, M.P., Sykes, S., and Marin, V., Chemically-mediated rejection of dinoflagellate prey by the copepods *Calanus pacificus* and *Paracalanus parvus*: mechanism, occurrence and significance, *Mar. Ecol. Prog. Ser.*, 28, 105, 1986.

101. Huntley, M.E., Ciminiello, P., and Lopez, M.D.G., Importance of food quality in determining development and survival of *Calanus pacificus* (Copepoda: Calanoida), *Mar. Biol.*, 95, 103, 1987.

102. Sykes, P.F. and Huntley, M.E., Acute physiological reactions of *Calanus pacificus* to selected dinoflagellates: direct observations, *Mar. Biol.*, 94, 19, 1987.

103. McClintock, J.B., Swenson, D.P., Steinberg, D.K., and Michaels, A.F., Feeding-deterrent properties of common oceanic holoplankton from Bermudian waters, *Limnol. Oceanogr.*, 41, 798, 1996.

104. Michaels, A.F., Caron, D.A., Swanberg, N.R., Howse, F.A., and Michaels, C.M., Planktonic sarcodines (Acantharia: Radiolaria and Foraminifera) in surface waters near Bermuda: abundance, biomass and vertical flux, *J. Plankton Res.*, 17, 131, 1995.

105. Stachowicz, J.J. and Lindquist, N., Susceptibility to predation and chemical defense among hydroids on pelagic *Sargassum*, *Mar. Ecol. Prog. Ser.*, in press.

106. Wellington, G.M., Depth zonation of corals in the Gulf of Panama: control and facilitation by resident reef fishes, *Ecol. Monogr.*, 52, 223, 1982.

107. Shanks, A.L. and Graham, W.M., Oriented swimming in the jellyfish *Stomolopus meleagris* L.Agassiz (Scyphozoan: Rhizostomomida), *J. Exp. Mar. Biol. Ecol.*, 108, 159, 1987.

108. Chioccara, F.L., Zeuli, L., and Novellino, E., Occurrence of mycosporine related compounds in sea urchin eggs, *Comp. Biochem. Physiol.*, 85B, 459, 1986.

109. Bandaranayake, W.M. and Des Rocher, A., Role of secondary metabolites and pigments in the epidermal tissues, ripe ovaries, viscera, gut contents and diet of the sea cucumber *Holothuria atra*, *Mar. Biol.*, 133, 163, 1999.

110. Stachowicz, J.J. and Lindquist, N., Chemical defense among hydroids on pelagic *Sargassum*: predator deterrence and absorption of solar UV radiation by secondary metabolites, *Mar. Ecol. Prog. Ser.*, 155, 115, 1997.

111. Gil-Turnes, M.S., Hay, M.E., and Fenical, W., Symbiotic marine bacteria chemically defend crustacean embryos from a pathogenic fungus, *Science*, 246, 116, 1989.

112. Roesner, J.A. and Scheuer, P.J., Ulapualide A and B; extraordinary antitumor macrolides from nudibranch eggmasses, *J. Am. Chem. Soc.*, 108, 846, 1986.

113. McClintock, J.B. and Baker, B.J., A review of the chemical ecology of Antarctic marine invertebrates, *Am. Zool.*, 37, 329, 1997.

114. Omori, M. and Ikeda, T., *Methods in Marine Zooplankton Ecology*, John Wiley and Sons, New York, 1984.

115. Parsons, T.R., Takahashi, M., and Hargrave, B., *Biological Oceanographic Processes*, Pergamon Press, New York, 1984.

116. Madin, L.P., Horgan, E.F., and Steinberg, D.K., Zooplankton at the Bermuda Atlantic Time-Series (BATS) station: diel, seasonal and interannual variation in biomass, 1994-1998, *Deep Sea Res. II*, in press.

117. Landry, M.R., Al-Mutairi, H., Selph, K.E., Christensen, S., and Nunnery, S., Seasonal patterns of mesozooplankton abundance and biomass at Station ALOHA, *Deep Sea Res. II* , in press.

118. Hamner, W.M., Hamner, P.P., Strand, S.W., and Gilmer, R.W., Behavior of Antarctic krill, *Euphausia superba*: chemoreception, feeding, schooling, and molting, *Science,* 220, 433, 1983.

119. Shaw, B.A., Harrison, P.J., and Anderson, R.J., Feeding deterrent properties of apo-fucoxanthinoids from marine diatoms. II. Physiology of production of apo-fucoxanthinoids by the marine diatoms *Phaeodactylum tricornutum* and *Thalssiosira pseudonana*, and their feeding deterrent effects on the copepod *Tigriopus californicus, Mar. Biol.*, 124, 473, 1995.

120. McFall-Ngai, M.J., Crypsis in the pelagic environment, *Am. Zool.*, 30, 175, 1990.

121. Clarke, W.D., Function of bioluminescence in mesopelagic organisms, *Nature*, 265, 244, 1963.

122. McAllister, D.E., The significance of ventral bioluminescence in fishes, *J. Fish. Res. Bd. Can.*, 24, 537, 1967.

123. Young, R.E., Ventral bioluminescent countershading in midwater cephalopods, *Symp. Zool. Soc. Lond.*, 38, 161, 1977.

124. Latz, M.I., Physiological mechanisms in control of bioluminescent countershading in a midwater shrimp, *Mar. Fresh. Behav. Physiol.*, 26, 207, 1995.

125. Longhurst, A.R. and Harrison, W.G., The biological pump: profiles of plankton production and consumption in the upper ocean, *Prog. Oceanogr.*, 22, 47, 1989.

126. Angel, M.V., Vertical migration in the oceanic realm: possible causes and probable effects, in *Migration: Mechanisms and Adaptive Significance*, Rankin, M.A., Ed., Port Aransas, Marine Science Institute (Contributions in Marine Science Suppl. to vol. 27), 1986, 45.

127. Lampert, W., The adaptive significance of diel vertical migration of zooplankton, *Func. Ecol.*, 3, 21, 1989.

128. Bollens, S.M. and Frost, B.W., Predator-induced diel vertical migration in a planktonic copepod, *J. Plankton Res.*, 11, 1047, 1989.

129. Neil, W.E., Induced vertical migration in copepods as a defense against invertebrate predation, *Nature*, 345, 524, 1990.

130. Bollens, S.M., Frost, B.W., Thoreson, D.S., and Watts, S.J., Diel vertical migration in zooplankton: field evidence in support of the predator avoidance hypothesis, *Hydrobiologia*, 234, 33, 1992.

131. Dobson, S.I., The ecological role of chemical stimuli for the zooplankton: predator avoidance behavior in *Daphnia, Limnol. Oceanogr.*, 33, 1431, 1988.

132. Loose, C.J., von Elert, E., and Dawidoxicz, P., Chemically-induced diel vertical migration in *Daphnia*: a new bioassay for kairomones exuded by fish, *Arch. Hydrobiol.*, 126, 329, 1993.

133. Hamner, W.M., Predation, cover, and convergent evolution in epipelagic oceans, *Mar. Fresh. Behav. Physiol.*, 26, 71, 1995.

134. Alldredge, A.L. and Silver, M.W., Characteristics, dynamics, and significance of marine snow, *Prog. Oceanogr.*, 20, 41, 1988.

135. O'Neil, J.M. and Roman, M.R., Ingestion of the cyanobacterium *Trichodesmium* spp. by pelagic harpacticoid copepods *Macrosetella, Miracia*, and *Oculosetella, Hydrobiologia*, 292/293, 235, 1994.

136. Morgan, S.G., Adaptive significance of spination in estuarine crab zoeae, *Ecology*, 70, 464, 1989.

137. Wilson, D.P., On the mirtraria larvae of *Oweniua fusiformis* delle Chiaje, *Phil. Trans. Roy. Soc. Lond. B*, 221, 231, 1932.

138. Pennington, J.T. and Chia, F.S., Morphological and behavioral defenses of trochophore larvae of *Sabellaria cementarium* (Polychaeta) against four planktonic predators, *Biol. Bull.*, 167, 168, 1984.

139. Montgomery, J.C., Pankhurst, N.W., and Foster, B.A., Limitations on visual food-location in the planktivorous antarctic fish *Pagothenia borchgrevinki, Experentia*, 45, 395, 1989.

140. Pankhurst, N.W. and Montgomery, J.C., Visual function in four antarctic nototheniid fishes, *J. exp. Biol.*, 142, 311, 1989.

141. Kerfoot, W.C., A question of taste: crypsis and warning coloration in freshwater zooplankton communities, *Ecology*, 63, 538, 1982.

142. Roe, H.S.J., The diel migrations and distributions within a mesopelagic community in the North East Atlantic. 4. The copepods, *Prog. Oceanogr.*, 13, 353, 1984.

143. Wiebe, P.H., Copley, N.J., and Boyd, S.H., Coarse-scale horizontal patchiness and vertical migration of zooplankton in Gulf Stream warm-core ring 82-H, *Deep Sea Res.*, 39, S247, 1992.

144. Bone, Q., Locomotion, locomotor muscles and buoyancy, in *Biology of Pelagic Tunicates*, Bone, Q., Ed. Oxford University Press, New York, 1998, 35.

145. Mauchline, J., The biology of mysids and euphausiids, *Adv. Mar. Biol.*, 18, 3, 1980.

146. Pitcher, T.J. and Parrish, J.K., Functions of shoaling behavior in teleosts, in *Behavior of Teleost Fishes*, 2nd ed., Pitcher, T.J., Ed., Chapman and Hall, New York, 1993, 363.

147. Peterson, W.T. and Dam, H.G., The influence of copepod swimmers on pigment fluxes in brine-filled vs. ambient seawater-filled sediment traps, *Limnol. Oceanogr.*, 35, 448, 1990.

148. Gilmer, R.W. and Harbison, G.R., Morphology and field behavior of pteropod molluscs in the families Cavoliniidae, Limnacinidae and Peraclididae (Gastropoda: Thecosomata), *Mar. Biol.*, 91, 47, 1986.

149. Swanberg, N.R., The ecology of colonial radiolarians: their colony morphology, trophic interactions and associations, behavior, distribution, and the photosynthesis of their symbionts, Ph.D. thesis, Woods Hole Oceanographic Institution, 1979.

150. Curl, H., Standing crops of carbon, nitrogen, phosphorus and transfer between trophic levels in continental shelf waters south of New York, *Rapp. Proc.-Verb. Cons. Int. Explor. Mer.*, 153, 183, 1962.

151. Beers, J.R., Studies on the chemical composition of the major zooplankton groups in the Sargasso Sea off Bermuda, *Limnol. Oceanogr.*, 11, 520, 1966.

152. Madin, L.P., Cetta, C.M., and McAlister, V.L., Elemental and biochemical composition of salps (Tunicata: Thalacia), *Mar. Biol.*, 63, 217, 1981.

153. Shenker, J.M., Carbon content of the neritic scyphomedusae *Chrysaora fuscenscens, J. Plankton Res.*, 7, 169, 1985.

154. Jansson, J. and Harbison, G.R., Fish in salps: the association of square tails (*Tetragbonurus* spp.) with pelagic tunicates, *J. Mar. Biol. Ass. U.K.*, 61, 917, 1981.

155. Norton, S.F. and Stallings, C.D., The distribution and abundance of three aposematic, chemically defended gammarid amphipods off Bell Island, WA *Proc. Am. Acad. Underwater Sci.*, in press.

156. Davis, A.R., Targett, N.M., McConnell, O.J., and Young, C.M., Epibiosis of marine algae and benthic invertebrates: natural products chemistry and other mechanisms inhibiting settlement and overgrowth, in *Bioorganic Marine Chemistry*, Vol. 3, Scheuer, P.J., Ed., Springer-Verlag, Berlin, 1989, 85.

157. Paul, V.J. and Fenical, W., Natural products chemistry and chemical defense in tropical marine algae of the order Chlorophyta, in *Bioorganic Marine Chemistry*, Scheuer, P.J., Ed., Springer-Verlag, Berlin, 1987, 1.

158. Pawlik, J.R., Chemical ecology of the settlement of benthic marine invertebrates, *Oceanogr. Mar. Biol. Ann. Rev.*, 30, 273, 1992.

159. Pawlik, J.R., Marine invertebrate chemical defense, *Chem. Rev.*, 93, 1911, 1993.

160. Hay, M.E. and Fenical, W., Marine plant-herbivore-predator interactions: the ecology of chemical defense, *Ann. Rev. Ecol. Syst.*, 19, 111, 1988.

161. Bakus, G.J., Targett, N.M., and Schulte, B., Chemical ecology of marine organisms: an overview, *J. Chem. Ecol.*, 12, 951, 1986.

162. Paul, V.J., Ed., *Ecological Roles of Marine Natural Products*, Comstock Press, Ithaca, NY, 1992.

163. Amsler, C.D., Iken, K.B., McClintock, J.B., and Baker, B.J., Secondary metabolites from Antarctic marine organisms and their ecological implications, in *Marine Chemical Ecology*, McClintock, J.B. and Baker, B.J., Eds., CRC Press, Boca Raton, FL, 2001.

164. Conde-Porcuna, J.M., Chemical interference by *Daphnia* on *Keratella*: a lifetable experiment, *J. Plankton Res.*, 20, 1637, 1998.

165. Wolfe, G.V., The chemical defense ecology of marine unicellular plankton: constraints, mechanisms and impacts, *Biol. Bull.*, 198, 225, 2000.

166. Zimmer, R.K. and Butman, C.A., Chemical signaling processes in the marine environment, *Biol. Bull.*, 198, 168, 2000.

6 Chemical Mediation of Macroalgal–Herbivore Interactions: Ecological and Evolutionary Perspectives

Valerie J. Paul,* Edwin Cruz-Rivera, and Robert W. Thacker

CONTENTS

I. DIVERSITY AND NATURAL PRODUCTS CHEMISTRY OF MARINE MACROALGAE

This chapter considers interactions between herbivores and benthic marine algae and seagrasses. It also discusses benthic cyanobacteria because many species are large, conspicuous, and common components of tropical communities, and recent studies suggest that their role is similar to that of true algae in coral reefs. The term "seaweeds" traditionally includes the macroscopic, multicellular, marine green, brown, and red algae; the benthic, filamentous blue-green algae are also sometimes considered. All seaweeds are unicellular at some stage of their life cycle (usually as spores or zygotes), and they are viewed as "primitive" photosynthetic organisms because of their simple construction and reproduction. There is a wide variation in algal classification among systematists, but the traditional divisions for algae are the Cyanobacteria (prokaryotic blue-green algae, sometimes termed Cyanophyta), Chlorophyta (green algae), Phaeophyta (brown algae), and

---

* Corresponding author.

Rhodophyta (red algae).[1-3] The seagrasses are the only truly submerged angiosperms in the marine environment.

Benthic algae are influenced by diverse biological and environmental factors such as herbivory, competition, light, temperature, salinity, water motion, and nutrient availability.[1,2] Variations in these parameters directly influence algal distribution and growth[1,2] and can indirectly affect algal susceptibility to herbivores by altering the nutrient or defensive chemical content of algae.[4] Furthermore, algae acquire nutrients from the water column rather than from roots, and they rely on the buoyancy of water, rather than stiff structural material or woody tissues, to float toward the surface and light. Thus, virtually all of the algal thallus is exposed to fouling by epiphytic microorganisms, algae, and sessile invertebrates, and consumption by herbivores and other consumers (including humans).[1]

Seaweeds have several mechanisms for tolerating or resisting herbivory, and these defensive strategies have been discussed previously.[4,5] Many seaweeds can deter herbivores by morphological, structural, and chemical defenses[6-9] or by associating with deterrent seaweeds or other benthic organisms that reduce herbivore foraging.[10-15] Structural defenses such as calcification and toughness are common in certain groups of green and red seaweeds and have been previously discussed.[6,7,16-19] Chemical defenses of seaweeds have been recently reviewed[8,9,20-24] and will be discussed more thoroughly in this chapter. Often, several defensive mechanisms may be functioning simultaneously,[16,23,25-31] and the importance of multiple defenses may be very significant in herbivore-rich tropical waters. The common co-occurrence of $CaCO_3$ and secondary metabolites in tropical seaweeds has been suggested to be adaptive because the high diversity of tropical herbivores limits the effectiveness of any single defense.[4,23-31] Multiple defenses could act additively or synergistically to reduce the ability of herbivores to adapt to seaweed defenses. For example, combinations of $CaCO_3$ and secondary metabolites have been tested as feeding deterrents, and both additive[28-31] and synergistic[27] effects of these combined defenses have been observed.

Over 2400 natural products have been isolated from marine red, brown, and green algae, and the majority of these have come from tropical algae.[32] In general, these compounds occur in relatively low concentrations, ranging from 0.2% to 2% of algal dry mass, although compounds such as the polyphenolics in brown algae can occur at concentrations as high as 15% of algal dry mass.[8,33] Except for metabolites from phytoplankton and cyanobacteria (blue-green algae), very few nitrogenous compounds have been isolated from algae.[32,34] Some red algae, blue-green algae, and a few green algae incorporate halides from seawater into the organic compounds they produce.[8,35,36] Bromine and chlorine are the most common halides found in marine algae. Halogenating enzymes such as bromoperoxidases and chloroperoxidases function in the biosynthesis of these halogenated compounds.[37] The majority of macroalgal compounds are terpenoids, especially sesqui- and diterpenoids. Acetogenins (acetate-derived metabolites) including unusual fatty acids constitute another common class of algal secondary metabolites.[34] Most of the remaining metabolites result from mixed biosynthesis and are often composed of terpenoid and aromatic portions.

Blue-green algae (cyanobacteria) occur in a variety of marine benthic habitats, including rocky shores, sandy shores, and salt marshes.[38-40] Cyanobacteria are important in marine environments because they play a major role in nitrogen fixation, and they are important primary producers.[2] Their ability to fix nitrogen may explain their production of many nitrogen-containing secondary metabolites including peptides and lipopeptides.[41-43] In general, the ecological roles for cyanobacterial compounds are poorly known; most studies have focused on their biomedical potential.[41-43]

High abundances of *Lyngbya* spp. and *Oscillatoria* spp. have been observed on coral reefs, where benthic mats of these blue-green algae can cover thousands of square meters.[38,40,44] These benthic, filamentous cyanobacteria produce a wide variety of secondary metabolites, many of which are toxic or pharmacologically active.[41-45] Cyclic peptides and depsipeptides isolated from blue-green algae have recently been reviewed.[43] Members of the Oscillatoriaceae, especially strains of *Lyngbya majuscula*, have proven to be rich sources of natural products. Metabolites isolated from *L. majuscula* include (see Figure 6.1) malyngolide, a lipid metabolite with antibiotic activity;[46]

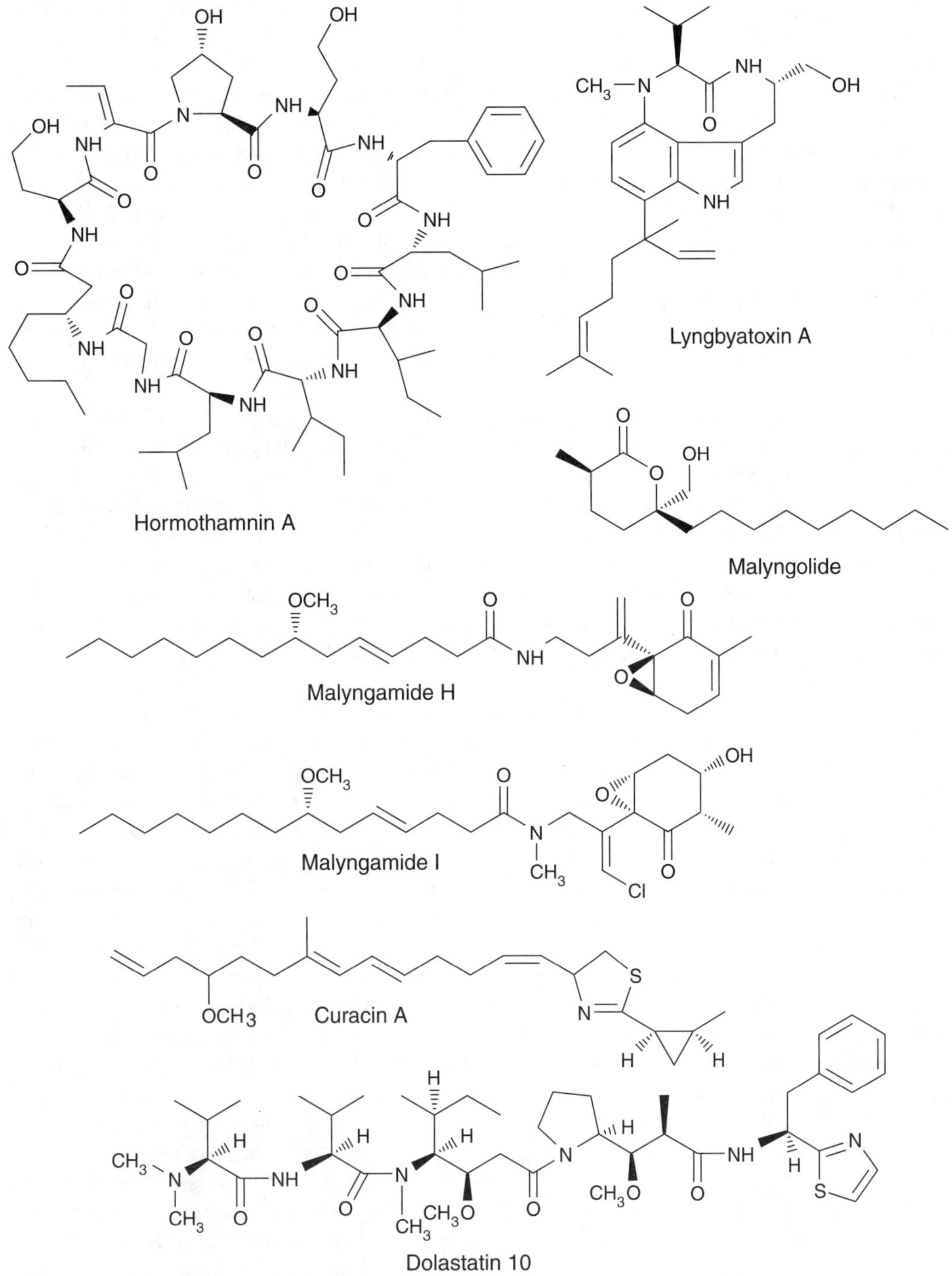

FIGURE 6.1 Examples of secondary metabolites from marine cyanobacteria.

lyngbyatoxin A, a cyclic depsipeptide that is a potent phorbol-ester-type tumor promoter;[47–49] curacin A, an antimitotic agent with potent brine shrimp toxicity;[50] and the lipopeptides malynga-mide H,[51] I,[52] J, K, and L,[53] which are toxic to brine shrimp and goldfish.

Other species of marine blue-green algae also produce unusual secondary metabolites. Hor-mothamnin A, a cyclic peptide isolated from *Hormothamnion enteromorphoides*, (Nostocaceae) has both antibiotic and cytotoxic activity.[54,55] Several dolastatins and dolastatin analogs have recently been isolated from the marine cyanobacteria *Symploca hydnoides*,[56,57] *Lyngbya majuscula*,[58–60] and mixed cyanobacterial assemblages.[61] Dolastatins and dolastatin analogs are potent cytotoxins, three of which are currently in clinical evaluation as potential anticancer drugs.[62] Dolastatins were originally isolated from collections of the herbivorous sea hare *Dolabella auricularia*.[63–65] The discovery of dolastatins in blue-green algae suggests that their occurrence in sea hares is a result of dietary consumption. Recent investigations have also demonstrated that *Dolabella auricularia* readily consumes certain species of cyanobacteria and is capable of growing on a strictly cyano-bacterial diet.[66]

The marine green algae (Chlorophyta) range from cold temperate to tropical waters; several families are exclusively tropical.[2] Most of the compounds isolated from the green algae (229 compounds)[32] are terpenes; sesquiterpenes and diterpenes are particularly common (see Figure 6.2).[8] Tropical green algae of the order Caulerpales have been especially well studied; members of this group, including species of *Caulerpa* and *Halimeda*, contain acyclic or monocyclic sesqui- and diterpenoids.[67,68] Triterpenes, which are not very common in marine algae, have been reported from *Tydemania expeditionis*.[69,70] Only a few green algae produce halogenated compounds. *Neomeris annulata* (Dasycladaceae) produces brominated sesquiterpenes,[71,72] *Cymopolia barbata* (Dasycla-daceae) produces brominated prenylquinones of mixed terpenoid and aromatic biosynthesis,[73] and *Avrainvillea* spp. produce brominated aromatic compounds.[74,75]

The brown algae are almost exclusively marine and primarily dominant in temperate waters. They range in size from small filamentous forms to subtidal kelps, which are the largest and most abundant benthic marine algae in temperate seas.[2,76] Brown algae are the only seaweeds that produce polyphenolic compounds. Although these compounds may function like terrestrial tannins by binding proteins or other macromolecules, they are structurally different compounds that are complex polymers derived from a simple aromatic precursor, phloroglucinol (1,3,5-trihydroxyben-zene).[33,77,78] These metabolites are often termed phlorotannins to distinguish them from the terrestrial tannins. Tannins found in higher plants are divided into two classes: condensed tannins (including lignins) and hydrolyzable tannins. Tannins are polymeric, consisting of multiple structural units containing phenolic groups. Lignins are large phenolic polymers bound to polysaccharides in plant cell walls. Hydrolyzable tannins are formed from polymerization of esters of glucose with gallic acid or related compounds.[79,80] None of these types of tannins occur in marine algae. Polyphenolics in brown algae may function as defenses against herbivores,[33,78] as antifoulants[81,82] (but see Jennings and Steinberg[83]), as chelators of metal ions,[77] and in UV absorption.[84]

Over 980 secondary metabolites have been isolated from the brown algae.[32] In addition to polyphenolics, brown algae in the order Dictyotales produce nonpolar metabolites such as terpe-noids, acetogenins, and compounds of mixed terpenoid-aromatic biosynthesis (see Figure 6.3). For example, over 230 compounds have been isolated from *Dictyota* spp.[32] *Sargassum* species also produce acetogenins and compounds of mixed terpenoid-aromatic biosynthesis. Brown algae of the genus *Desmarestia* are also known to concentrate high amounts of sulfuric acid, which may be used in defense.[1,8,85,86]

The greatest variety of secondary metabolites is probably found among the red algae, where all classes of compounds except phlorotannins are represented and many metabolites are haloge-nated.[35,87] Over 1240 compounds have been reported from the red algae.[32] Red seaweeds from the families Bonnemaisoniaceae, Rhizophyllidaceae, and Rhodomelaceae are rich in halogenated com-pounds that range from halogenated methanes, haloketones, and phenolics to more complex terpenes

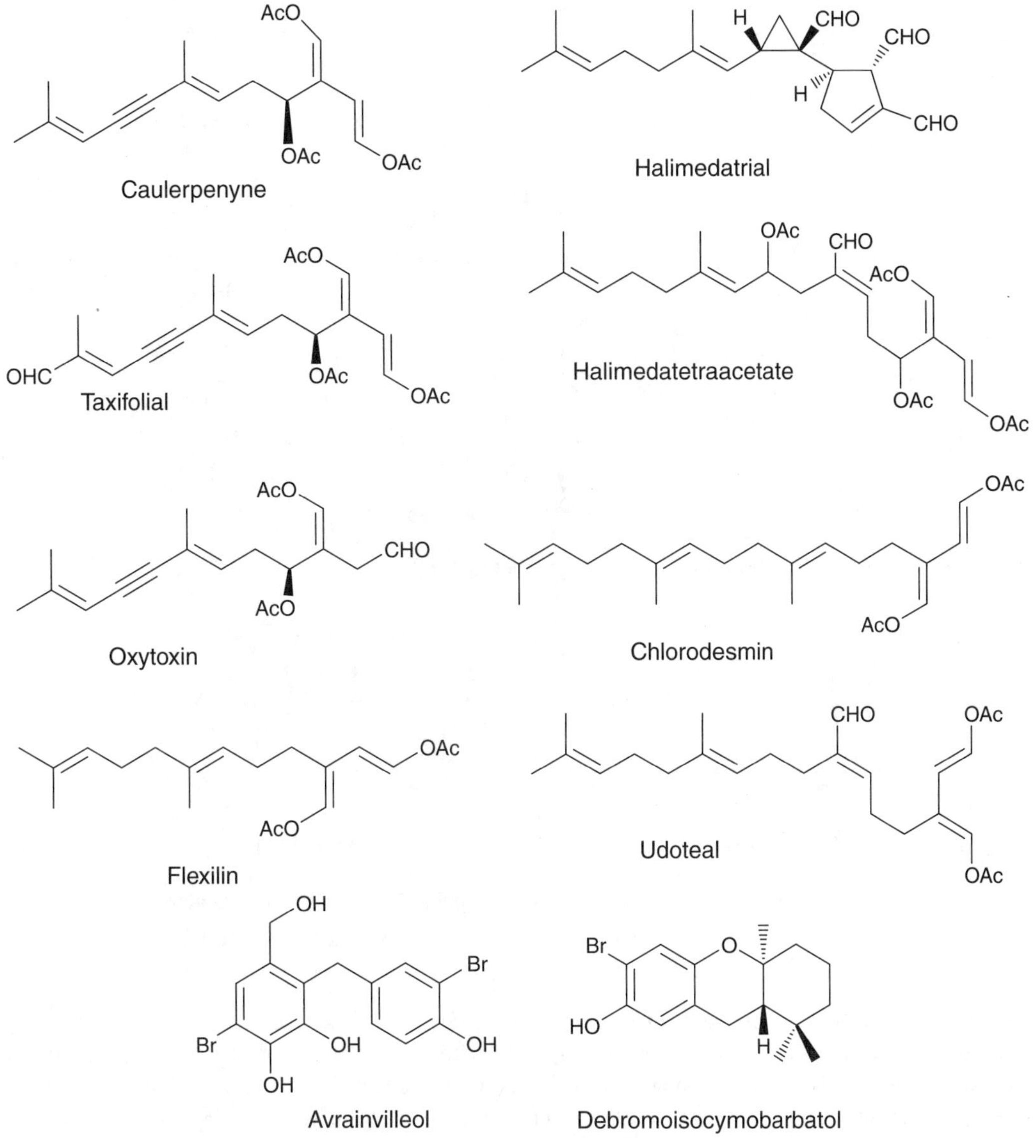

FIGURE 6.2 Examples of secondary metabolites from marine green algae.

(see Figure 6.4).[35,36,87,88] The red algal genus *Laurencia*, the subject of extensive investigations, produces over 570 compounds,[32] many of which are halogenated and of unique structural types.[32,87,89]

Seagrasses are known to produce phenolic acids,[90] phenolic acid sulfate esters,[91] and sulfated flavonoids.[92] While the natural functions of most of these compounds are not known, some of the phenolic acids and the sulfate esters inhibit growth of microorganisms and fouling organisms such as barnacles[91,93] and deter grazing by amphipods.[93] Herbivory on seagrasses may be important for the structuring of seagrass communities in nearshore environments,[94] but very little is known about the role of chemical defenses in the ecology of seagrasses.

Many possible defensive functions for algal secondary metabolites have been proposed including antimicrobial, antifouling, and antifeedant activities. To date, the role of these compounds as

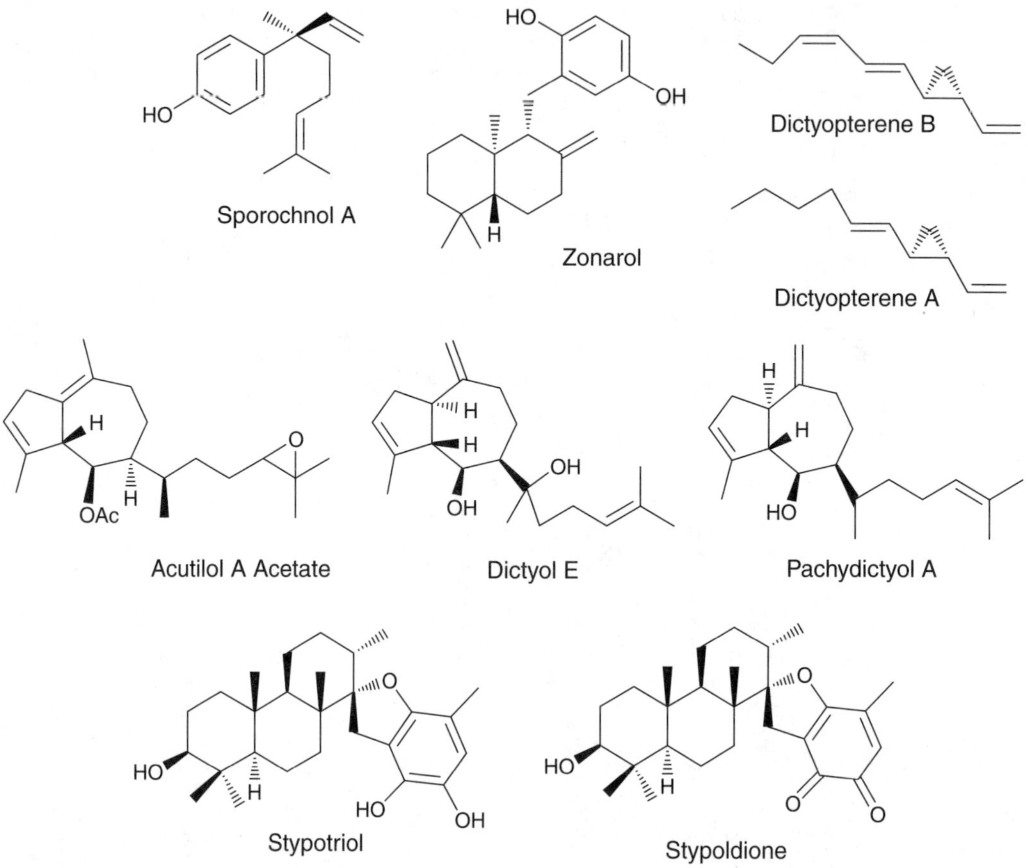

FIGURE 6.3 Examples of secondary metabolites from marine brown algae.

defenses against herbivores has been studied most. Recent studies have clearly shown that many seaweed natural products function as feeding deterrents toward herbivores.[8,21-24] However, many compounds may also have other roles or may function simultaneously as defenses against pathogens, fouling organisms, and herbivores, thereby increasing the adaptive value of these metabolites.[67,95] A consideration of multiple functions for secondary metabolites is important, because even though algal secondary metabolites may function as defenses against herbivores, they may have evolved for other reasons such as resistance to pathogens or competitors. Some algal secondary metabolites do show antimicrobial[96-98] or antifouling[81,95,99-101] effects. Chapters 10 and 17 in this volume cover the topic of natural antifoulants in more detail.

II. MARINE HERBIVORES

A. MARINE BENTHIC HERBIVORY AND HERBIVORE NUTRITION

Herbivory in benthic marine systems is intense. For example, on coral reefs, herbivores can remove almost 100% of the biomass produced daily by marine macroalgae, whereas in the most intensely grazed terrestrial systems — African grasslands — herbivores only consume about 66% of the above-ground plant biomass.[9,21,102] While terrestrial plants produce subterranean structures such as roots, bulbs, and tubers that are generally inaccessible to most animals, most marine algae do not

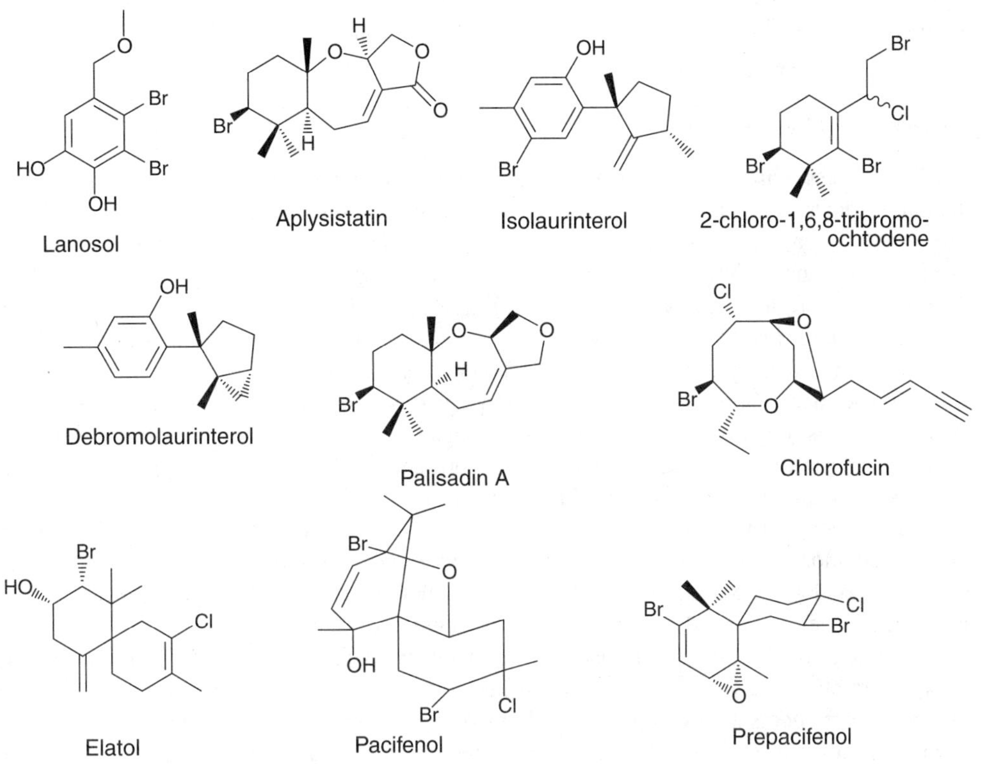

FIGURE 6.4 Examples of secondary metabolites from marine red algae.

produce equivalent underground parts (although at least one species of *Caulerpa* is known to absorb nutrients through its underground rhizomes[103]). Therefore, virtually all algal biomass in the ocean is potentially exposed to consumers, epibionts, and pathogens.[104]

The feeding activities of herbivores constitute an important ecological force controlling the structure and dynamics of plant communities; however, from an herbivore's point of view, feeding is only a means of gaining adequate nutrition for survival, growth, and reproduction. Herbivores make a living consuming food that is much lower in nutritional value and much higher in indigestible structural material than their own tissues.[105–107] As a result, herbivores must process large quantities of food[108–110] or rely on alternative strategies that enhance nutrient uptake per unit of food.[105,106] Although marine algae do not have the large quantities of nondigestible structural material that terrestrial plants have,[9,104] nutrient uptake can nevertheless be constrained for marine herbivores. For example, coral reef fishes may eat many times their required energetic needs in order to gain enough nitrogen from seaweeds.[108]

The nutritional component of marine plant–herbivore interactions has often been overlooked by many workers, particularly chemical ecologists.[22,111,112] Chemical defenses, in essence, keep herbivores from effectively exploiting potential nutrient sources and, therefore, may affect herbivore nutrition both directly and indirectly.[27,112–115] For example, prey nutritional value can alter the effectiveness of marine chemical defenses against diverse consumers,[27,113,115] and interactions between nutritional value and algal chemistry can affect consumer fitness.[112] Prey quality can affect the perception of a chemical defense[27,113,115] and the consumer's ability to digest or detoxify the compound,[116,117] or may indirectly alter digestive associations with gut symbionts (see Slansky[114] for terrestrial examples). Although digestive associations with microbes are known for a number of herbivorous fishes,[111,118,119] the effects of algal defenses on gut microbes and consequences for the host have not been demonstrated.

B. HERBIVORE DIVERSITY

Herbivores in the sea are phylogenetically more diverse than on land. Seaweeds are eaten by diverse vertebrate ectotherms and an array of invertebrate consumers that vary both in their selectivity and in their impact on algae. Vertebrate herbivores are comprised of various fish families and species, as well as some turtles.[24,111,120–125] Invertebrate herbivores span at least four different phyla and include an array of gastropods (such as snails, limpets, sacoglossans, sea hares, cephalaspideans, and chitons), urchins, crabs, amphipods, isopods, shrimps, polychaete worms, copepods, and a few insect species that feed on algae living in the upper littoral zone.[24,110,123,126–135]

The importance of different herbivore groups varies geographically, and herbivore species diversity increases towards the tropics in most groups.[111,122,136] This contrasts with the patterns of speciation in algae (and in some particular invertebrate groups) for which the number of species increases with latitude.[137,138] However, differences in the latitudinal effects of herbivorous fishes could arise from higher fish abundances, rather than species diversity, in the tropics.[139] Nonetheless, herbivore radiation in the tropics likely has had important consequences for the evolution of seaweeds, as evidenced by the increased diversity of defenses and higher chemical deterrence of tropical algae when compared to their temperate counterparts.[24,140–144]

The biogeography of herbivorous fishes has been reviewed elsewhere.[111,122,139,145] For invertebrate grazers, information is available for some gastropods[146] and urchins.[147] Although there are obvious limitations in our understanding of herbivore biogeography, some patterns do emerge. Latitudinally, herbivory by gastropods is more important in temperate systems than in the tropics, fish herbivory shows the opposite pattern, and urchin herbivory appears to be intense in both temperate and tropical zones.[136] Although amphipod diversity increases with latitude,[137] little is known about the feeding habits of most species, precluding biogeographical comparisons in herbivory.

Despite latitudinal variation in the importance of particular groups of consumers, herbivory and predation increase in intensity and more developed prey defenses occur in the tropics than in temperate regions.[140,141, 148–154] Likewise, studies of seaweed–herbivore interactions have suggested that a higher diversity and higher tissue concentrations of small, lipophilic secondary metabolites are found in tropical seaweeds than in temperate seaweeds.[8,32] However, few studies have rigorously tested this hypothesis.[143,144] In contrast, larger water-soluble phlorotannins, namely, the polyphenolic compounds produced by brown seaweeds, have been hypothesized to be more abundant in temperate species than in tropical species.[33] Recent investigations by Targett and co-workers[78,155,156] have challenged this view, finding that tropical brown seaweeds can have high concentrations of phlorotannins. However, these phlorotannins appear to have little effect on common tropical herbivores.[156]

C. HERBIVORE SIZE AND GUILDS

Marine herbivores span a smaller size range than their terrestrial counterparts.[102] For algal consumers, marine equivalents of terrestrial megaherbivores (such as elephants, rhinoceri, hippopotami, and a variety of large extinct grazers)[102,157–159] are not known, nor is there fossil evidence that they have existed. Although seagrasses are consumed by large mammals such as manatees and dugongs,[94] algae are consumed by these animals only under severe food limitation,[160] and, thus, the largest grazers on algae are fishes,[111,122] turtles,[121] and the larger sea hares (e.g., *Aplysia vaccaria* can reach more than 15 kg and are the largest gastropods in the world).[161]

Because size can impose constraints on mobility, predation risk, and per capita impact on food plants, it is often useful to divide marine herbivores into functional groups or guilds sharing certain general characteristics. Although such categories are useful for describing patterns and processes of plant–herbivore interaction, they should be used with caution; closely related consumers can sometimes exploit quite different food sources, and broad categorization may potentially overlook important complexities in the interaction of animals with plants.[110,111,135,162–164] In terms of size, marine herbivores can be roughly divided into macro-, meso-, and micrograzers. Macrograzers or macroherbivores include larger, more mobile herbivores that can cover considerable areas while foraging, whose feeding activities often have strong impact on seaweeds, and that generally do not live on the algae they consume (see Steinberg[165] for an exception). Most herbivorous fishes, turtles, sea urchins, and large gastropods (e.g., the sea hare *Dolabella auricularia*)[30,129] qualify as macrograzers, and tend to have broad, generalized diets.

Mesograzers, in contrast, often feed and live on plants larger than themselves. In terms of size, mesograzers include herbivores between 0.2 and 20 mm, of potentially limited mobility, that live on the plants they eat but generally cause limited damage, and for which predation risk is often large.[134,166–168] Aside from smaller consumers, many larger animals may spend part of their lives as functional mesograzers.

A number of mesoherbivores are trophic specialists that consume one or a few algal species,[9,167,169–177] often chemically defended ones.[9,167,175,176] Other, more generalized feeders may also have strong preferences for, and form close associations with, chemically defended algae.[135,166,178–180] However, in comparison with terrestrial environments,[181,182] feeding specialization among marine herbivores is rare.[9,104,167]

Mesograzers find shelter in the seaweeds they consume and are often more resistant to algal chemical defenses than larger herbivores.[131,133,167] Because predation on mesograzers can be intense, associating with noxious or unpalatable hosts can have clear selective advantages.[9,22,167] Association with particular seaweeds can reduce predation risk of these smaller herbivores either via crypsis, reduced mobility coupled with selection of unpalatable algae as habitats, adaptations that reduce the risk of being removed from their host plants, behavioral or physiological sequestration of chemical defenses from the algae they eat, or a combination of these[9,167] (also reviewed in Chapter 4 in this volume).

Metabolites that are toxic or deterrent toward generalist herbivores such as fishes and sea urchins are often ineffective against mesograzers and may even function as feeding stimulants. Specialist herbivorous molluscs such as sea hares and sacoglossans (ascoglossans) selectively consume chemically rich seaweeds and often concentrate seaweed secondary metabolites[8,9,183] as defenses against their own natural enemies. These metabolites can be concentrated on the skin or actively exuded when the molluscs are attacked.[8,87,170,173,184–186] Some mesograzers do not obtain secondary metabolites from their algal diets but appear to biosynthesize their own metabolites.[187] Sacoglossans feed suctorially and some can retain photosynthetically functional chloroplasts from their host algae.[177] Thus, certain sacoglossans may be capable of synthesizing their chemical defenses from primary metabolites derived from photosynthesis.[188]

Other small mesograzers such as amphipods, crabs, and polychaetes may also preferentially consume seaweeds that are chemically defended.[8,9,166,167,189] Amphipods, crabs, and polychaetes do not appear to sequester algal secondary metabolites. These small marine herbivores sometimes feed selectively or exclusively on seaweeds that are chemically defended from fish. Hay and coworkers[9,22,132,167] have shown that the association between mesograzers and chemically defended seaweeds reduces predation on the herbivores. They have hypothesized that escape from predation may be a dominant factor selecting for dietary specialization among these herbivores (see also Chapter 4 in this volume). However, Poore and Steinberg[134] showed that the nest-building amphipod *Peramphithoe parmerong* seemed to rely primarily on intrinsic host plant qualities in determining its seaweed diet, rather than on extrinsic properties, such as predation, in determining food choice.

Although the role of micrograzers (e.g., copepods, cladocerans, ostracods, etc.) in pelagic communities is well studied, extremely little is known about micrograzers on the marine benthos. This is probably due to the assumption that such small consumers are constrained by size to feeding on microalgae. At least some species of harpacticoid copepods are known to consume macroalgae, and one species can become a pest for seaweed aquaculture.[190–193] In Israel, a copepod tentatively identified as *Amphiascus minutus*[193] infests ponds containing *Gracilaria conferta* and tunnels through the alga. During their summer reproductive peak, copepod infestations can become severe and lead to the collapse of algal cultivation ponds.[192,193] Other benthic copepods form associations with algae[194] and cyanobacterial mats,[195] but their feeding habits have not been addressed. However, copepods of the genus *Diarthrodes* are known to bore into algae in a fashion similar to the example described above.[190] More information on the ecology of benthic marine micrograzers is clearly needed.

Marine herbivores are a diverse group, and strict compartmentalization into guilds should not be interpreted as an all-encompassing scheme. In particular, the relation between size and mobility of mesograzers has been criticized by some authors and has led to debate.[162,163,168] Larger consumers such as urchins may have reduced mobility and live closely associated with their food plants, essentially acting as "large" mesograzers,[165] while small consumers may perform diel migrations of several meters.[168,196] It is also clear now that related mesograzers can vary in their relative mobility.[110,135,164,168,180] In the case of amphipods[110] and crabs,[164] lower mobility appears to be associated with a higher ability for compensatory feeding, and, therefore, an increased ability to exploit lower quality algal foods.

Given their large impact on algal communities, macroherbivores appear to be more important selective agents for the evolution of chemical defenses than meso- or microherbivores.[9,21,167] Although mesograzers can have strong effects on algal communities,[197–202] they are rarely acknowledged as definitive forces shaping the evolution of seaweed defensive chemistry. The grazing pressure of meso- and micrograzers could be more important for early algal stages and gametes.[201–204] Although information on the chemical defense of early life stages is available for marine invertebrates,[205,206] similar studies on algae are lacking (see Deal,[207] Hay et al.,[203] and Schnitzler et al.[204]). Evaluating marine herbivore selective pressures at different scales needs more attention. Despite its shortcomings, the use of guilds has proven useful in explaining ecological and evolutionary processes between seaweeds and their consumers[9,167] and should be refined rather than abandoned.

III. CHEMICALLY MEDIATED INTERACTIONS BETWEEN MARINE MACROALGAE AND HERBIVORES

A. FEEDING-DETERRENT EFFECTS OF ALGAL SECONDARY METABOLITES

The common method of testing for feeding-deterrent effects against herbivores has been to incorporate seaweed extracts or isolated metabolites at natural concentrations into a palatable diet, either a preferred seaweed or an artificial diet, and then to compare feeding rates of the grazers on treated foods with those on appropriate controls.[208] Deterrent effects observed in these assays appear to be based primarily on the taste of the treated food. If a compound is deterrent toward an herbivore, the degree of avoidance is often related to the concentration of the extract or metabolite in the diet. Higher concentrations often result in a more pronounced deterrent effect.[8,209–211] These methods do not assess toxicity or other physiological effects on the consumers or possible detoxification methods by herbivores.

Compounds produced by marine cyanobacteria deterred feeding by several species of herbivorous fishes,[44,55,176,212–214] sea urchins,[55,214] and crabs.[55] In contrast, the opisthobranch sea hare *Stylocheilus longicauda* specializes on *Lyngbya majuscula* and prefers artificial diets containing compounds produced by blue-green algae; however, high concentrations of some of these metabolites can still deter feeding by *S. longicauda*.[176,211] The sea hares *S. longicauda* and *Dolabella auricularia* sequester cyanobacterial compounds from their diets, gaining protection from fish and invertebrate predators.[66,176,212,215,216]

Green algae produce a variety of terpenoid compounds which have been shown to defend against herbivores. The tropical Atlantic seaweed *Cymopolia barbata* was one of the first calcified green algae recognized to be chemically rich, and it has been studied from a variety of locations around the Caribbean, Bermuda, and the Florida Keys.[73,217, 218] The alga contains complex mixtures of bromine-containing prenylated hydroquinones. The major metabolite, cymopol, deterred feeding by the sea urchin *Lytechinus variegatus*[219] and natural assemblages of herbivorous fishes on Caribbean coral reefs,[220] but not by the rabbitfish *Siganus doliatus* on the Great Barrier Reef.[221] Cymopol stimulated feeding by the Caribbean sea urchin *Diadema antillarum*.[220] Another minor metabolite found in Florida Keys collections, bromoisocymobarbatol, significantly reduced feeding by parrotfishes in field assays and by the omnivorous pinfish *Lagodon rhomboides* and a Caribbean amphipod, *Hyale macrodactyla*, in laboratory assays.[73]

Halimeda spp. are among the most common algae in coral reef communities and are some of the best studied chemically defended green algae. Their ability to persist in areas of intense herbivory likely results from the production of both chemical and structural defenses. Levels of calcification in this genus can range from 70 to 90%.[222,223] Several structurally similar diterpenes, including halimedatetraacetate (Figure 6.2) and the epimers halimedatrial and epihalimedatrial, are produced.[223–225] Both halimedatetraacetate and halimedatrial are feeding deterrents toward herbivorous fishes on Guam; however, halimedatrial significantly reduces grazing by 24% more than does halimedatetraacetate.[226] Epihalimedatrial is an unstable metabolite that decomposes rapidly within minutes after isolation, making its feeding deterrent properties difficult to determine.[223,226]

Several species of *Halimeda* can rapidly convert the less toxic and deterrent major metabolite halimedatetraacetate to the more toxic and deterrent halimedatrial.[227] This process, termed activation[227] or biotransformation,[228] occurs within seconds of tissue injury and appears to be enzymatically mediated. Only the portion of the organism in the immediate vicinity of the injury is affected. Extracts from damaged algae contain higher amounts of halimedatrial and are more deterrent toward herbivorous fishes than extracts from undamaged algae.[227] These conversions occur after any mechanical injury and could occur when a fish bites or chews on the algae. Activation is also common in some terrestrial plants[229–231] and has been reported to occur in some marine sponges.[228,232] Examples include plants that produce HCN from organic precursors,[229,230] plants that convert glucosinolates to thiocyanates and isothiocyanates after tissue damage,[233,234] and the activation of oleuropein, a phenolic secoiridoid glycoside with strong protein denaturing activity.[231] In these cases, precursor compounds are compartmentalized and physically separated from the enzymes that activate them.

Chemical defenses in calcified seaweeds may be particularly important against herbivores such as parrotfishes and sea urchins that can readily consume calcified foods.[19,31,235] In a series of laboratory and field experiments designed to examine differences among fish species in their responses to both chemical and mineral defenses in *Halimeda*, Schupp and Paul[31] found that *Halimeda* diterpenes can limit feeding by the parrotfish *Scarus sordidus*, which is not affected by the levels of $CaCO_3$ found in *Halimeda* spp. In contrast, the rabbitfish *Siganus spinus* and the surgeonfish *Naso lituratus* were deterred by $CaCO_3$ in their diets but were unaffected by *Halimeda* diterpenes.[31] In general, combined defenses ($CaCO_3$ and terpenes) increased the number of fish species that were deterred relative to either single defense. However, for the parrotfish *S. sordidus*, the deterrent effect of the diterpenes seemed to be diluted by the addition of aragonite to the food, making combinations of $CaCO_3$ and chemical defenses less effective as a defense than secondary metabolites alone because these fish preferred eating calcified foods.

Hay et al.[27] examined the effects of secondary metabolites and $CaCO_3$ from three tropical green seaweeds, *Rhipocephalus phoenix, Udotea cyathiformis*, and *Halimeda goreauii*, on feeding by three different types of herbivores (mixed species groups of parrotfishes, the amphipod *Cymadusa filosa*, and the sea urchin *Diadema antillarum*). Addition of finely powdered commercial grade calcite as 69% of food dry mass had no effect on feeding by parrotfishes, deterred feeding by *Cymadusa*, and deterred *Diadema* when food organic content was low, but not when it was higher.

Natural concentrations of semipurified secondary metabolites from *Rhipocephalus* or *Udotea* deterred feeding by all three herbivores, and appeared to be more effective than $CaCO_3$ at depressing feeding. The major metabolite from *H. goreauii* did not deter feeding by any of the herbivores. In two of nine feeding assays, a synergistic effect between $CaCO_3$ and chemical defenses was observed.[27] The synergistic effects were apparent for some combinations of algal metabolites and herbivores but not others, and the mechanisms producing these effects were not clear. These observations of synergism between chemical and mineral defenses clearly warrant further studies to determine how often synergisms occur and what mechanisms may be involved.

Neomeris annulata, another calcified tropical green alga, produced brominated sesquiterpenes[71] that deterred natural assemblages of herbivorous fishes.[72] The fleshy, apical portions of the thalli were high in secondary metabolites (1.5% of dry mass) and relatively low in calcium carbonate (aragonite) (65% ash), whereas the basal portions of the thalli were low in sesquiterpenes (0.2% of dry mass) and high in calcium carbonate (90% ash).[28] When a naturally occurring combination of extract and aragonite found in the tips (10% crude extract and 65% aragonite) was paired with the combination found in the bases (0.8% crude extract and 90% aragonite) and offered to natural assemblages of reef fishes, no significant difference in deterrence was observed. Combinations of secondary metabolites and aragonite were tested against one or the other single defenses in similar feeding assays; the combinations of defenses always proved a more effective deterrent than either secondary metabolites or aragonite alone.[28] In laboratory assays, the three brominated sesquiterpenes found in *N. annulata* on Guam were tested for their ability to deter three species of herbivores that are not strongly deterred by calcium carbonate in their diets, the parrotfishes *Scarus schlegeli* and *Chlorurus sordidus* and the sea urchin *Diadema savignyi*.[236] All three compounds deterred feeding by all three herbivores at or below their natural concentrations. Natural combinations of the three compounds were not more effective than either of the major compounds alone, suggesting no enhanced or synergistic effect of multiple secondary metabolites.[236] Sesquiterpenes from *N. annulata* also deterred feeding by the surgeonfish *Naso lituratus*.[237] This surgeonfish was also deterred by compounds found in other green algae including avrainvilleol from *Avrainvillea obscura* and diterpenes from *Tydemania expeditionis*.[237]

Pennings et al.[238] used within-individual variation in defenses in the alga *Neomeris annulata* to test the hypothesis proposed by Schupp and Paul[31] that surgeonfishes are deterred from feeding by $CaCO_3$ but not by most chemical defenses, whereas parrotfishes are deterred from feeding by chemical defenses but not by $CaCO_3$. In assays using whole algal thalli, two species of parrotfishes fed primarily on the metabolite-poor, $CaCO_3$-rich basal portions of *N. annulata*, whereas two species of surgeonfishes fed primarily on the metabolite-rich, $CaCO_3$-poor tips. These differences in grazing location by different types of herbivores had important consequences for *N. annulata*. Most of the photosynthesis likely takes place in the green, actively growing, apical portions of *N. annulata*, and individuals grazed at the tips grew in length more slowly than individuals grazed around the base.[238] In contrast, thalli grazed at the base might be more likely to be lost in heavy wave action.[239]

Brown algal compounds, especially compounds from *Dictyota* spp., have been shown to deter a variety of herbivores in temperate and tropical waters. Pachydictyol A (Figure 6.3), which is produced by tropical and warm temperate species of brown algae, deterred herbivores in field and laboratory assays,[166,220,221,240] but its effects varied depending on the type of herbivore.[21] Fishes were generally deterred by the compound,[21,24,166,220,240] but some amphipods, polychaetes, and sea urchins were not affected.[220] A specialist amphipod, *Pseudoamphithoides incurvaria*, which builds bivalved domiciles from *Dictyota bartayresii*, actually cued on pachydictyol A for building domiciles.[172] Dictyol E, from *Dictyota dichotoma*, deterred both temperate fish and urchins but not small grazers such as amphipods or polychaetes.[220,221] The same compound, however, stimulated decorating behavior of the majid crab *Libinia dubia*. The crab exploits the deterrence of the alga by "behaviorally sequestering" the chemistry and using it as a defense against omnivorous fishes.[132] On the Great Barrier Reef, dictyol E was not a deterrent against the rabbitfish *Siganus doliatus*, but pachydictyol A and related dictyols B and H did deter feeding by this rabbitfish.[221] Acutilols, from a Hawaiian collection of *Dictyota acutiloba,*[241] deterred the temperate pinfish more strongly than tropical herbivorous fishes.[144] Only the major metabolite acutilol A acetate deterred two species of parrotfishes and the tropical sea urchin *Diadema savignyi* at natural concentrations, but the temperate herbivores were deterred by all of the compounds at or below natural concentrations.[144]

Stypotriol (Figure 6.3), from the brown seaweed *Stypopodium zonale,* deterred feeding by Caribbean reef fishes and the urchin *Diadema antillarum,*[220] but not *Siganus doliatus* on the Great Barrier Reef.[221] Instead, stypoldione was much more effective than stypotriol in assays against *S. doliatus* conducted on the Great Barrier Reef.[221] Other brown algal compounds including spatane, and dolastane class diterpenes and zonarol, did not affect feeding by *S. doliatus* on the Great Barrier Reef.[221] Sporochnol A, a prenylated phenol from the Caribbean brown alga *Sporochnus bolleanus*, deterred feeding by parrotfish but not by the tropical sea urchin *Diadema antillarum* nor the amphipod *Cymadusa filosa.*[242] Two prenylated phenols were isolated from a related tropical alga, *Encyothalia cliftonii*, collected near Exmouth in Western Australia. The crude extract of *E. cliftonii* and the purified major phenol showed significant feeding deterrence toward the herbivorous sea urchin *Tripneustes esculentus*, but the minor phenol was not active at its natural concentration.[243]

Dictyopterenes A and B (Figure 6.3) are simple C_{11} hydrocarbons from the Caribbean brown alga *Dictyopteris deliculata*. Similar C_{11} hydrocarbons function as sexual pheromones in brown algae, but pheromonal activity has not been confirmed for dictyopterenes A and B.[244] In field assays, the alga was of intermediate preference to herbivorous reef fishes, and the dictyopterenes were significant feeding deterrents toward herbivorous fishes. However, the compounds did not deter feeding by a natural mixed-species assemblage of amphipods found in a turf of *D. deliculata* and associated algae.[189] Organic C_{11} sulfur compounds from *Dictyopteris* spp., which are biosynthetically related to the C_{11} sex pheromones, deterred feeding by an herbivorous amphipod, *Ampithoe longimana*, but not by the sea urchin *Arbacia punctulata*,[204] a pattern opposite to that seen for most seaweed chemical defenses that are often active against larger generalist herbivores but not smaller, less mobile mesograzers. Hay et al.[203] tested a variety of naturally occurring C_{11} hydrocarbons and their degradation products toward the amphipod *Ampithoe longimana* and the sea urchin *Arbacia punctulata*. Seven of twelve metabolites or mixtures deterred the amphipod, but only two of eleven deterred the sea urchin. The C_{11} hydrocarbon and sulfur compounds were more frequently and more strongly deterrent to the amphipod than the sea urchin and appeared to be more effective against the mesograzer.[203,204] Because mesograzers are more likely to consume gametes, zygotes or young sporelings, the high concentrations of C_{11} compounds may be especially important for defending these early life history stages.[203]

A variety of red algal compounds have also been shown to deter feeding by marine herbivores. Elatol and isolaurinterol (Figure 6.4), halogenated terpenes from the red algal genus *Laurencia*, deterred feeding by reef fishes and the sea urchin *Diadema antillarum* in the Caribbean,[220] but elatol was not as effective in field and laboratory assays with reef fishes conducted on Guam.[24,240] The halogenated terpenes from *Laurencia*, including elatol, debromolaurinterol, and chlorofucin, also deterred the rabbitfish *Siganus doliatus* in field assays on the Great Barrier Reef, but other halogenated terpenes such as pacifenol, prepacifenol, palisadin A, and a *Laurencia* chamigrene had no affect on feeding by this herbivorous fish.[221] Palisadin A was deterrent only at twice the natural concentrations in field assays on Guam, and its related oxidation product aplysistatin had no effect on feeding even at the higher concentrations.[240] Several diterpenes from Japanese collections of *Laurencia saitoi* deterred feeding by sympatric young sea urchins and abalone.[245]

The Caribbean red alga *Ochtodes secundiramea* produces halogenated monoterpenes; major metabolites include ochtodene (Figure 6.4) and chondrocole C. Ochtodene deterred feeding by herbivorous reef fishes in the Grenadine Islands and on Guam while chondrocole C did not.[246] The two compounds and an unresolved mixture of monoterpenes from *O. secundiramea* were also tested for their effects on feeding by amphipods collected in the Grenadine Islands. Only the unresolved monoterpene mixture deterred feeding by the amphipods. A variety of secondary metabolites, such as the mixture of monoterpenes synthesized by *O. secundiramea*, may be necessary to deter the diverse herbivores found in coral reef habitats. Ochtodene and apakaochtodenes A and B[247] found in *Portieria hornemannii* on Guam also deterred feeding by different herbivorous reef fishes[237,248,249] and the oligophagous sea hare *Aplysia parvula*.[175]

Usually the presence or absence of deterrent secondary metabolites in seaweeds correlates well with the susceptibility of seaweeds toward herbivores. Seaweeds that are least susceptible to grazing fishes often employ chemical and structural defenses;[23,25–29,31] however, some herbivorous fishes consume chemically rich seaweeds. For instance, the rabbitfish *Siganus argenteus* readily consumes the green seaweed *Chlorodesmis fastigiata*,[249,251] which is avoided by another rabbitfish *S. spinus* and many other herbivorous fishes.[248,250] It is not known whether *S. argenteus* detoxifies or simply tolerates the diterpenoid metabolites produced by *C. fastigiata*. In field assays conducted on the Great Barrier Reef, the rabbitfish *S. doliatus* was also not deterred by chlorodesmin (Figure 6.2), the major diterpene from *C. fastigiata*.[221] Another green seaweed, *Caulerpa racemosa*, is readily consumed by many herbivorous fishes,[26,252] although the alga produces sesquiterpenoid metabolites in relatively high concentrations (1 to 2 % of dry weight). Neither the extract of *Caulerpa* nor the major metabolite caulerpenyne (Figure 6.2) are deterrent toward many of the herbivorous fishes on Guam;[248–250,252] however, the compound may influence other types of herbivores such as molluscs more than fishes.[30,67]

As we have discussed, a variety of compounds from all classes of marine algae have now been tested for their effects on feeding by many different temperate and tropical herbivores. Many of these compounds effectively deter feeding by herbivores. However, there is considerable variance in the responses of different types of herbivores to even very similar compounds. Some metabolites inhibit feeding by most herbivores, whereas other compounds deter only a few herbivores.[21,24] There is also considerable variation among different herbivores, even closely related species, in their responses to secondary metabolites from seaweeds. Thus, as the diversity of herbivore species increases, the probability of having herbivores that are not affected by any particular chemical defense undoubtedly increases. In these cases, complex mixtures of secondary metabolites[253] or multiple defenses may be particularly important.[27,29,31]

It is not clear what chemical features determine the toxicity or antifeedant effects of any particular secondary metabolite. Toxicity and deterrency are not intrinsic properties of any compound; they result from physiological interactions between a metabolite and its consumer. The same metabolite may show pronounced differences in its effects even toward closely related species of herbivores.[78,112,180] Some metabolites inhibit feeding by most herbivores, whereas other compounds deter a few herbivores but not others.[9,21,24] Compounds that differ only slightly in chemical structure can vary greatly in their deterrent effects.[9,24,220,240,246,249] For example, closely related green algal metabolites including udoteal, flexilin, chlorodesmin, an epoxylactone from *Pseudochlorodesmis furcellata*, and caulerpenyne (Figure 6.2) differ in their feeding deterrent effects, although all contain the same major bis-enol acetate functional group.[24,240,248,254] Similarly, halogenation in red algal metabolites does not necessarily enhance their feeding deterrent effects. Of three halogenated sesquiterpenoids from *Laurencia*, elatol is broadly effective as a feeding deterrent, palisadin A (Figure 6.4) is deterrent toward some herbivores but not others, and aplysistatin does not appear to have feeding deterrent effects.[21,24,221,240,253] Thus, predictions about the toxic or deterrent effects of particular secondary metabolites toward natural predators may be difficult to make based upon chemical structures or results of pharmacological assays. Field and laboratory assays with natural herbivores are necessary to examine these ecological interactions.[208]

Most studies in marine chemical ecology examine the effects of secondary metabolites on consumers as the initial palatability of whole organisms, crude extracts, or isolated metabolites. Although the effectiveness of chemical defenses can depend on the experiences of the consumer, especially when the consumer has repeated experience with chemical defenses,[210,213,255–257] only a few studies in marine chemical ecology have examined the ability of consumers to learn about chemical defenses.[206,210,213,258–260] In addition to constantly rejecting foods of low palatability, consumers may increase their acceptance of initially unpalatable foods that have positive post-ingestive consequences or decrease their acceptance of initially palatable foods that have negative post-ingestive consequences.[206,210,213,256–259] Clearly, additional studies are needed to determine how foragers respond to multiple encounters with chemical defenses and to variation in chemical defenses among prey.

B. Physiological Effects of Algal Chemical Defenses

Ecologists have often assumed that defensive metabolites are avoided because consuming them negatively affects consumer fitness. Furthermore, different types of plant defenses have often been qualified as either digestibility reducers ("quantitative" defenses such as tannins, lignins, and other polyphenolics) or toxins ("qualitative" defenses such as alkaloids, terpenes, acetogenins, and other smaller molecules).[261,262] An alternative view is that defensive metabolites do not need to cause negative effects on consumers to be avoided. Chemicals may simply taste bad or mimic the taste of other metabolites that do carry a physiological cost if ingested.[22] Taking into account studies in both terrestrial and aquatic environments, there is very limited support for dividing defensive plant chemicals as either toxins or digestibility reducers. For example, terrestrial polyphenolics can act both as toxins or digestibility reducers,[263–265] and many nonphenolic terrestrial and marine compounds that act as effective chemical defenses are not overtly toxic.[112,220,266,267]

Many early studies addressing the physiological or fitness costs of marine secondary metabolites on herbivores lacked rigorous design or presented compounds to consumers in an unrealistic fashion (e.g., injecting or dissolving compounds in the water as opposed to feeding them to the animals) and sometimes led to erroneous conclusions (see examples in Hay[22] and Hay et al.[208]). In the case of macroalgal metabolites, a few studies have provided important insights through better designed experiments. Marine workers have generally concentrated on two approaches to study post-ingestive effects of algal chemical defenses: (1) measuring the effects of secondary metabolites on consumer assimilation of nutrients or other digestive processes (see Chapter 11 in this volume), and (2) quantifying the long-term effects of ingesting secondary metabolites on survivorship, growth, and reproduction (fitness). Some studies have also sought insights about the physiological effects of algal metabolites by correlating food metabolite concentrations with consumer growth.[268] Pennings and Carefoot[269] introduced a novel technique to the study of such interactions by measuring blood glucose levels in sea hares as a proxy for stress induced by metabolite ingestion. Although blood glucose has been used to measure stress due to environmental factors in some marine invertebrates,[269–272] their study showed no correlation between the toxicity of secondary metabolites and blood glucose concentrations, limiting the usefulness of the technique.

Given the emphasis on polyphenolics as digestibility reducers, many authors have equated reductions in digestive processes with decreased fitness. This may not always be the case. No effect on assimilation, in turn, does not preclude other effects on consumer fitness. Animals may also counteract reduced assimilation of nutrients by compensatory feeding[109,110,164] or by mixing diets that may both balance nutritional needs and dilute noxious metabolites.[114,129,135,257] Compensatory feeding on chemically defended prey increases both the ingestion of nutrients and of secondary metabolites, which can have strong effects on herbivore fitness.[112,114] However, gut-level interactions between nutrients and secondary chemicals can ameliorate the negative effects of ingested compounds.[114] Furthermore, animals deal with ingested metabolites through a variety of physiological, metabolic, and behavioral adaptations that likely act in concert. Examples include sequestration of compounds, enzymatic degradation of metabolites, target-site insensitivity, gut pH, the production of tannin-binding substances, surfactants, peritrophic membranes, geophagy, and others.[114,263,273] Although little is known about these adaptations in marine consumers, some marine herbivores produce gut surfactants that could counteract the effects of phlorotannins,[274] and detoxification enzymes have been monitored in animals that feed on noxious prey.[275]

Measurements of growth (which often correlate with reproductive potential) or of production of eggs and gonads have often been used to quantify how diet affects fitness.[110,129,135,276] Because fitness is a measurement of the genetic contribution of an individual to the next generation via its offspring, this approach provides a more accurate description of how food traits may influence consumers through their life cycle. It should be noted, however, that secondary metabolites may not directly affect egg production but rather the viability of eggs. For example, diatom metabolites potentially can cause abnormal embryonic development and suppress egg viability in copepods without necessarily reducing female fecundity.[277–279]

Several marine studies have focused on the physiological effects of secondary metabolites produced by diatoms,[277–280] planktonic dinoflagellates,[281–288] benthic dinoflagellates,[289–291] salt marsh macrophytes,[292] or invertebrates[206,258] and will not be treated in detail here. Interested readers should consult the references above and also reviews by Targett and Ward,[293] Hay,[22] and Hay et al.[208] Nevertheless, a few marine studies have measured the post-ingestive effects of macroalgal metabolites in ecologically relevant settings.

Boettcher and Targett[294] measured the effects of algal polyphenolics on the assimilation efficiency of the temperate fish *Xiphister mucosus*. In their study, extracted polyphenolics were separated by molecular weight, coated onto palatable algal discs, and fed to the fish. Assimilation efficiency was reduced by the larger (>10 kDa) phlorotannins, whereas smaller phlorotannins had little or no effect. Similar results were also found when phlorotannins of different size were fed to the fish *Girella nigricans*.[295] Irelan and Horn[296] measured assimilation efficiencies of nitrogen and carbon for the fish *Cebidichthys violaceous* fed on polyphenolics from two different brown algae. Polyphenolics from the temperate alga *Fucus gardneri* reduced assimilation of nitrogen but not carbon. However, researchers found no effects on assimilation of either nutrient when fish were fed on phlorotannins from the alga *Macrocystis integrifolia*.

In contrast with the above studies, long-term ingestion of algal phlorotannins did not affect assimilation, survivorship, or growth of a snail and an urchin,[142] even though both temperate herbivores consumed phlorotannin extracts for several months. Similarly, lipophilic metabolites from green algae did not affect assimilation efficiency of a parrotfish.[116] Although Winter and Estes[297] reported a decrease in growth of abalone when the animals were fed on diets containing polyphenolic extracts from the alga *Dictyonerum californicum*, both control and treatment animals lost mass, suggesting that the basic artificial diet was inadequate for the animals. Also, food disks were changed every 12 to 14 days, so it is difficult to assess whether the abalone were affected by the chemicals, or whether the patterns observed were actually driven by differences in microbial degradation of treatment vs. control foods.[110] Increasing polyphenolic concentrations can be inversely correlated with consumer growth for filter feeders ingesting kelp detritus[268] and for snails grazing on decaying salt marsh macrophytes.[292] However, correlation should not be accepted as causality without more direct testing, as multiple food traits may covary or correlate with observed patterns of growth.[292]

Although some of the above studies provide partial support for the role of polyphenolics as digestibility reducers,[261,262] they also show that assuming all polyphenolics have similar modes of action is misleading. Terrestrial studies have found similar variability in the physiological and fitness effects of tannins and other polyphenolics,[263,265,298] and assimilation of phlorotannin-rich algae by marine herbivores can be high.[78,142,156] These studies argue against the simple structure–activity relations originally proposed for such "quantitative" defenses.[261]

A few studies have analyzed the effects of lipophilic algal metabolites on consumer fitness. Paul and Fenical[67] showed that survivorship of a snail (*Strombus costatus*) was decreased when any one of three green algal metabolites were coated onto a palatable alga. However, food consumption was not quantified.

The effects of any one compound varies depending on the consumer in question. For example, ingestion of the terpene alcohol pachydictyol A (Figure 6.3) reduced growth of a temperate sparid fish[166] but not of two species of sea hares.[269] The sea hares sequestered pachydictyol A in their digestive glands, an adaptation often interpreted as a means of overcoming the negative effects of ingested metabolites.[114] However, the sea hare *Aplysia juliana* also readily sequestered malyngamide B from the benthic blue-green alga *Lyngbya majuscula*, even though this compound reduced growth of this sea hare by 70%.[269]

Mixtures of the terpene alcohols pachydictyol A and dictyol E (Figure 6.3), the two principal feeding deterrents of the alga *Dictyota menstrualis*, also affected consumers differentially. In a recent study,[112] five amphipods and one isopod were fed low and high quality artificial diets either containing or lacking natural concentrations of the two metabolites. The consumers represented a continuum of more closely to more distantly related crustaceans, beginning with the amphipod *Ampithoe longimana*, which prefers to feed on *D. menstrualis* and is a relatively sedentary tube builder which lives in close association with the alga.[166,179,180,209] Effects of the metabolites on survivorship, growth, and reproductive output varied considerably among species, were unrelated to phylogenetic relations among grazers, and were strongly influenced by food nutritional quality in some cases. For example, dictyols did not significantly affect survivorship, growth, or fecundity (although there was a non-significant trend towards lower egg production) of *Ampithoe longimana*, which feeds normally on *D. menstrualis*. In contrast, the congeneric *A. valida* experienced reduced survivorship, growth, and a dramatic supression in egg output when fed on the compounds. *Cymadusa compta*, which belongs to the same family as these two amphipods, was also generally unaffected by the dictyols, and for the isopod *Paracerceis caudata*, the chemicals enhanced survivorship. For the other two amphipods studied, *Gammarus mucronatus* and *Elasmopus levis*, food nutritional quality interacted with secondary chemistry to ameliorate the negative effects of the dictyols in the high quality diet. Although the dictyols deterred feeding for all six consumers in both low and high quality foods, there was no consistent relation between deterrence and the long-term effects of the compounds on consumer fitness. Furthermore, the compounds acted indirectly on consumers (at least in part) by reducing overall food intake rather than by post-ingestive toxicity.[112] Standard toxicity bioassays also support this conclusion as dictyols do not appear to be overtly toxic.[8]

Other physiological effects of defensive chemicals have not been observed or studied for plant–herbivore interactions. For example, studies of chemically defended sessile invertebrates have shown that some compounds have emetic properties against fish predators, and that fishes can learn quickly to avoid these chemicals after regurgitating ingested food.[206,258] Macroalgal compounds could potentially act in similar ways, but such effects have never been documented.

Secondary metabolites may act by directly affecting consumer physiological processes or by indirectly reducing overall food consumption, thus compromising nutrient acquisition.[112,114] Studies have not always decoupled these two components, making interpretations of the results difficult. The "ecological realism" of some investigations has also been compromised by feeding consumers amounts of compounds they would probably never eat in nature. These problems have been noted before,[22,206,208,220] and the need for more research on both methodological approaches and the ecology of deterrence and fitness effects of defensive metabolites has been discussed.[22,208] More work on these issues will provide important insights on the factors constraining or favoring the evolution of particular prey chemical defenses, metabolite concentrations, and consumer adaptations to cope with prey natural products.

C. Intraspecific Variation in Secondary Metabolite Production

Given the variability in responses of herbivores toward seaweed chemical defenses, it is not surprising that a great deal of variation occurs in the production of secondary metabolites by seaweeds. Different herbivores have different methods of foraging and digestive physiologies; thus, the effectiveness of any single defense should decrease as the diversity of herbivore types within a community increases.[5,299] It is possible that seaweeds produce a variety of secondary metabolites because different compounds deter different herbivores. This may be especially true for tropical seaweeds that are exposed to a diverse assemblage of herbivores. Metabolite types and concentrations vary within and among individuals in a population, with the age of the individuals, and among conspecific populations of seaweeds.[8,9,207]

Variation in secondary metabolites occurs at a number of levels. Differences in concentrations and types of chemical defenses within algae occurs in *Halimeda* spp. and *Neomeris annulata*, in which more toxic and deterrent compounds are allocated to new growth.[28,226,300] In several species of *Caulerpa*, crude organic extract and the major metabolite caulerpenyne were up to five times higher in the upright portions of the thalli as compared to the runners (stolons). While this distribution of compounds seems ideal for chemical defense against herbivores, neither the extracts nor pure compounds were defensive against herbivory by reef fishes.[252] The temperate red alga *Neorhodomela larix* showed considerable variation within individual thalli in production of bromophenols (lanosol and its 1,4-disulfate ester).[301,302] Adjacent portions of thalli varied by as much as an order of magnitude in bromophenol content. Younger, growing areas of thalli and some reproductive structures produced more lanosol than interior and basal vegetative regions of thalli.[302] The temperate red alga *Delisea pulchra* showed highest levels of brominated furanones in the apical meristematic tips of thalli.[303] Further study of the location of these compounds within thalli indicated that they are stored in the central vesicle of gland cells. These cells release the furanones onto the surface of the alga; furanone levels are highest near the apical tips and decrease toward the base of the alga.[304] The location of these compounds on the surface of the alga is consistent with their function as antifoulants.[100] Release of defensive metabolites to the surface of thalli is not common to all red seaweeds. *Laurencia obtusa*, which produces halogenated sesquiterpenes, did not release substantial amounts of these compounds onto the surfaces of thalli.[101] The terpenoid metabolites produced by *Laurencia* species are stored in membrane-bound vesicles (*corps en cerise*),[305] which are not typically released to the external environment except following damage to the plant.[101,305]

In contrast to the pattern of higher levels of compounds located in younger tissues observed in many seaweeds, the brown seaweed *Dictyota ciliolata* showed an opposite distribution of dictyol terpenes.[306] Higher levels of chemical defenses were found in older tissues; these parts of the thalli were less palatable than young apices to an herbivorous amphipod and a sea urchin. The authors suggest that because *D. ciliolata* does not have translocation systems like coenocytic algae do, secondary metabolites cannot be transported from older to younger portions of the thalli, and, therefore, are in greatest concentrations in the older growth.[306]

Chemical variation also occurs among individuals within a population. This has been observed for *Halimeda* spp.[223,226] and red algae including *Portieria hornemannii*[307,308] and *Laurencia palisada*.[253] Some of this variation may be related to the age of individual thalli.[223,253,300] The preferences of four common intertidal and subtidal invertebrates from the Pacific northwest for juvenile and adult brown algae showed that the susceptibility of these algae to herbivores depended on both the algal stage and species of herbivore.[309] Differences in chemical defenses in these different algal stages were not examined in this study, so it is not known if chemical defenses or other characteristics of the algae were responsible for this variation in susceptibility. Diploid sporophytes and male and female haploid gametophytes of the isomorphic brown alga *Dictyota ciliolata* were examined for differences in their chemical defenses and susceptibility to herbivory. The different life stages had similar levels of chemical defenses and were equally susceptible to herbivorous amphipods and sea urchins.[310]

Variation in types and concentrations of secondary metabolites also occurs among populations of seaweeds growing in different habitats. Populations of *Halimeda* from habitats where herbivory is intense tend to contain higher levels of the more potent deterrent halimedatrial (Figure 6.2) than do populations from areas of low herbivory.[226] Other green algae such as *Penicillus, Udotea, Rhipocephalus,* and *Caulerpa* also often produce higher concentrations or different types of secondary metabolites in populations from herbivore-rich reef habitats than in populations from herbivore-poor areas such as reef flats or seagrass beds.[67,68,311] Shallow and deep water populations of the brown alga *Stypopodium zonale* produced different secondary metabolites.[312] The red alga *Portieria hornemanni* is well known to vary in its composition of halogenated monoterpenes among different collection sites in the tropical Pacific.[247] Puglisi and Paul[307] found that populations of the red alga *P. hornemannii* from different reef sites around Guam varied considerably in concentrations of monoterpenes. The among-site variation in *P. hornemannii* was not environmentally mediated because transplantation of algae between sites did not have a significant effect on monoterpene concentrations.[308] Different chemotypes of *Laurencia nipponica* were recently shown to be genetically distinct based on results of intra- and interpopulational crosses between female and male gametophytes in laboratory culture studies.[313] In most cases, however, we do not know if variation in concentrations and types of secondary metabolites results from herbivore-induced chemical defenses, localized selection resulting in high levels of defense, genetic differences, or other factors not related to herbivory.

Studies have pointed to the influences that physical environmental factors have on algal chemistry and susceptibility to herbivory. Interactions among environmental stresses, plant defenses, and nutrient status can affect herbivore preferences. For example, after dessication, the palatable seaweed *Gracilaria tikvahiae* was less susceptible to grazing by the sea urchin *Arbacia punctulata*, while the unpalatable seaweed *Padina gymnospora* became more susceptible.[314] The change in palatability in *P. gymnospora* seemed to result from loss of chemical defenses, although the particular metabolites involved could not be isolated. Mild dessication of *Dictyota ciliolata* reduced concentrations of dictyol terpenes (see Figure 6.3) by 7 to 38%, and the seaweed became much more susceptible to herbivores as a consequence.[210]

Both light and nutrients can sometimes affect secondary metabolite concentrations in seaweeds as they have been shown to do in terrestrial plants. The carbon/nutrient balance (CNB) hypothesis[315] predicts that resource availability, especially carbon and nitrogen availability, influences the phenotypic expression of chemical defenses. When resources exceed the necessary levels for growth, excess nutrients are available for chemical defense. Several marine studies have tested the CNB hypothesis. Puglisi and Paul[307] found that levels of monoterpenes produced by the red seaweed *Portieria hornemannii* did not follow the predictions of the CNB hypothesis. Nitrogen and phosphorus enrichment of the alga in laboratory and field experiments had no effect on monoterpene concentrations.[307] Similarly, Cronin and Hay[316] found that terpenes produced by *Dictyota ciliolata* often did not follow the predictions of the CNB hypothesis. Findings for seaweeds parallel those of terrestrial plants that suggest that the CNB hypothesis often works well for predicting the responses of polyphenolic compounds to nutrients and light, but terpenes do not appear to follow the same patterns.[316]

Variation in production of polyphenolics by brown algae has received the most attention of any group of algal secondary metabolites.[33,78] This is largely because polyphenolics, as a class of compounds, can be assayed by colorimetric techniques such as the Folin–Denis assay,[78,317,318] which quickly yields information about the intraspecific and interspecific patterns of variation of this group of compounds (including monomeric and polymeric phenols). Unfortunately, the assays do not provide any information about levels of individual compounds or even about changes in relative concentrations of compounds of different size classes. Nonetheless, studies with different species of brown algae have shown that phlorotannins largely follow the predictions of the CNB hypothesis, and polyphenolics tend to accumulate under nitrogen deficiency.[316,319–322] Cronin and Hay[316] caution that while the CNB hypothesis provides a good theoretical framework for testing how environmental factors influence chemical defenses and susceptibility to herbivory, other factors may be equally important in predicting how environmental factors influence chemical variation. For example, the brown alga *Ascophyllum nodosum* increased polyphenolic concentrations in response to increased UV-B radiation over a 2-week period, and researchers suggested that phlorotannins may function as a screen against harmful UV radiation.[84]

Polyphenolics have been found to vary within individual thalli depending on tissue type,[323–328] among individual thalli within and between populations,[329,330] and seasonally.[331] Simulated herbivory (mechanical wounding) has been shown to induce phlorotannin production in several studies[320,322,327,332–334] but not in others.[84,335] Phlorotannin induction can occur rapidly, within 1 to 3 days of wounding.[322,334] Pavia and Toth[336] found that a few weeks of grazing by the gastropod *Littorina obtusata* could induce the production of phlorotannins in *Ascophyllum nodosum*, but grazing by the isopod *Idotea granulosa* and simulated herbivory caused no significant changes in phlorotannin levels. They proposed that patterns of grazing by *L. obtusata*, which lives and feeds on a few species of fucoid algae, could be an important factor in explaining natural variation in the levels of phlorotannins in *A. nodosum*.[336]

There has been surprisingly little evidence showing that inducible defenses occur in seaweeds other than brown algae. In addition to the examples of phlorotannin induction in brown algae, only one other example of an inducible chemical defense has been reported in seaweeds. *Dictyota menstrualis* induced increased levels of dictyol terpenes in response to grazing by the amphipod *Ampithoe longimana*, which made the seaweed less susceptible to further grazing by the amphipod.[209] Among-site differences in amphipod densities, grazing scars, seaweed defensive chemistry, and palatability were clearly documented in one year, but patterns were less clear in two subsequent years, illustrating the variability and complexity among patterns of grazing and algal response.[209]

IV. INDIRECT EFFECTS OF ALGAL CHEMISTRY: ASSOCIATIONAL DEFENSES, MUTUALISMS, AND SHARED DOOM

Algal chemical defenses may have indirect positive and negative effects on associated species. The term "associational defense" has been used to describe interactions in which one species gains protection from natural enemies by living spatially close to a deterrent species.[11,12] While most discussions of this phenomenon have focused on sessile organisms living in close proximity to each other,[11,12,15] motile species often benefit indirectly from associating with deterrent organisms.[132,172,173] A species may also indirectly decrease the deterrence or increase the susceptibility of another. The end result would be an enhanced risk to one or both organisms involved, a phenomenon termed "shared doom."[15]

Palatable algae can find partial protection by associating with unpalatable species. Littler et al.[12] showed that the deterrent alga *Stypopodium zonale* provides a refuge to palatable algae growing at the base of it. Plastic mimics also provided some refuge to palatable species, but to a lesser degree than real algae, suggesting that physical as well as chemical traits of *Stypopodium* are responsible for these patterns. In both temperate and tropical systems, palatable algae can reduce herbivore pressure by associating with noxious seaweeds, thus turning "competitors into accomplices."[10–12,14] Similar spatial escapes can be achieved by algae living in proximity of sessile invertebrates such as gorgonians, soft corals, and fire corals that have structural as well as chemical deterrents.[13,337] However, chemistry may not always be the primary deterrent component of the association.[337]

Epiphytes can also alter the palatability of either deterrent or palatable species. Wahl and Hay[15] showed that the susceptibility of algae to herbivory by sea urchins depended on the chemical traits of both host and epiphyte. Decreased susceptibility of the host by a deterrent epiphyte was considered an associational defense, whereas increased susceptibility of a host by a palatable epiphyte was termed "shared doom."

Although associational defense has been mostly discussed for sessile organisms,[11,12] motile organisms can indirectly benefit from algal chemical defenses. For example, the amphipod *Ampithoe longimana* and the polychaete *Platynereis dumerilii* live and feed on the brown alga *Dictyota menstrualis* (previously *D. dichotoma*).[166,179,180,209] This alga produces dictyols that deter feeding by omnivorous fishes and urchins.[166,210] Because it is defended, the alga is visited less often by omnivorous fish, which in turn enhances survival of the mesograzers inhabiting it. This strategy is important for a number of mesograzers that do not physiologically sequester defensive chemicals from their algal hosts. Examples include amphipods within the genera *Hyale* and *Ampithoe* as well as small crabs (e.g., *Thersandrus compressus* and *Caphyra rotundifrons*) that develop similar associations with chemically defended algae.[135,167,171,173,178,207]

Algal chemistry has likely promoted the evolution of more specific associations as well. For example, the tropical amphipod *Pseudamphithoides incurvaria* carries around a domicile constructed from the chemically defended alga *Dictyota bartayresii*.[172,338] Amphipods within the domiciles are rejected by fish, whereas amphipods without domiciles or amphipods forced to build domiciles out of palatable algae are readily consumed.[172] Pachydictyol A (Figure 6.3), the main defensive metabolite in *D. bartayresii*, stimulates domiciles construction in a dose-dependent fashion.[172]

The role of chemical defenses in inquilinistic relationships is likely to be more common than currently appreciated. For example, the filter-feeding amphipod *Ericthonius brasiliensis* gains protection by constructing domiciles from the chemically defended alga *Halimeda tuna*.[339] Similarly, algal domiciles are constructed by other amphipods within the genus *Pseudamphithoides*,[340] and various species of alpheid shrimp build tubes out of algae and cyanobacteria, some of which are known to be chemically defended.[66]

Although decorator crabs have been assumed to cover themselves to match their surroundings, Stachowicz and Hay[132] showed that the crab *Libinia dubia* selectively decorated with the alga *Dictyota menstrualis* in North Carolina. Crabs covered with palatable algae suffered higher predation than those decorated with *Dictyota,* and the chemical that deterred omnivorous fish in this alga (dictyol E) stimulated decoration behavior in *Libinia*. Interestingly, intraspecific variation in decoration behavior occurs in *Libinia*; crabs collected from Connecticut, where *Dictyota* does not occur, did not selectively decorate with the alga.[132] Other decorator crabs also appear to be selective rather than "match the background." In Spain, the small decorator crab *Macropodia rostrata* decorates almost exclusively with the brown alga *Dictyota linearis*.[195] This alga contains a number of dictyol-class metabolites similar to those of *D. menstrualis*. Other sympatric decorator crabs do not show this preference.[195]

Positive associations may arise from herbivore resistance to algal chemical defenses. Two recent studies of herbivorous crabs from the genus *Mithrax* have demonstrated that these animals will readily consume various chemically defended seaweeds.[131,133] *Mithrax sculptus* associates with the calcified alga *Neogoniolithon* in Florida, and *M. forceps* associates with the temperate coral *Oculina arbuscula*. In both cases, the crabs keep their hosts clean from encroaching competitors and epiphytes, many of which are chemically defended algae. Thus, the coral or algal host is able to persist in the environments because it shelters mesograzers that can cope with a broad range of algal secondary metabolites.

V. ALGAL CHEMICAL DEFENSES
AND BENTHIC COMMUNITY STRUCTURE

Chemically mediated interactions have important direct and indirect effects on communities from both ecological and evolutionary standpoints.[22,341] Chemical defense or communication cannot be properly understood unless it is viewed through the lenses of population, community, and ecosystem processes, and this requires consideration of both the biotic and abiotic components of the natural environment.[342] For example, chemically mediated foraging is affected by water flow because it relies on water-soluble cues that are carried away from prey.[343–345] Similar constraints likewise modify the effectiveness of other waterborne cues, such as alarm signals, sexual pheromones, and settlement cues, in both mobile and sessile organisms.[244,345–350]

Benthic community structure is strongly influenced by chemical defenses. In the tropics, most algae living on reef slopes, where herbivory is most intense, contain structural or chemical defenses that allow them to establish populations in the presence of abundant and diverse herbivores.[21,24] Chemically defended algae, in turn, serve as safe habitats against predation for small herbivores and as defensive allies to algae susceptible to herbivory.[167] Also, algal chemical defenses could potentially indirectly affect ecosystem-wide processes. Seagrass metabolites can influence decomposition and (concomitantly) nutrient remineralization rates in marine systems.[351] Given the stability, potential toxicity, and antimicrobial properties of a number of algal metabolites, it is possible that these compounds act in a similar fashion in benthic food web processes. These interactions have not been properly studied.

Abiotic variables also influence the outcome of algal–herbivore interactions with potential consequences for community and ecosystem processes. Dessication and UV exposure can decrease algal defense, thus increasing susceptibility to grazing.[210,314] In one study, Pavia et al.[84] showed that UV-B treated algae increased in polyphenolic content and that the isopod *Idotea granulosa* preferred to feed on these more chemically rich algae. Furthermore, plant allocation to growth or defense can be strongly affected by environmental variables such as light and nutrient availability.[261,262,315]

The influence of nutrients and herbivory in algal community structure has received increased attention in recent years with the documentation of macroalgal blooms and phase shifts in benthic systems.[352–354] Phase shifts on coral reefs from coral-dominated to macroalgal-dominated communities have been attributed to increased nutrient availability due to eutrophication and reduced herbivore abundance due to overfishing and disease.[352,355–357] Studies of nutrient uptake by macroalgae[358,359] and herbivore-exclusion experiments (reviewed by Hay,[21] Carpenter,[360] and Hixon[361]) have documented the mechanisms underlying changes in macroalgal abundance and community structure. Algae that dominate communities following the reduction of herbivory are often chemically defended species such as *Halimeda* and *Dictyota* spp.,[353,356,362–365] which are relatively unpalatable to herbivores.

Studies of phase shifts from coral-dominated to macroalgal-dominated communities have often overlooked filamentous blue-green algae (cyanobacteria), which generally have been grouped with turf algae in ecological research (e.g., Steneck and Dethier[366]). Benthic cyanobacteria may play an important role in these phase shifts, as they can be early colonizers of dead coral and disturbed substrates.[367–369] Littler and Littler[370] found that cyanobacterial turf grew abundantly after the death of crustose coralline algae by coralline lethal orange disease (CLOD) on a reef in Fiji. The cyanobacterial turf was observed to overgrow coral branches and kill coral polyps.[370] Although rapid increases in the abundance of cyanobacteria may be caused by factors similar to those that influence macroalgal blooms, the effects of eutrophication and reduced herbivory on cyanobacterial populations are not well understood. Many cyanobacteria can fix nitrogen, and their abundance is often not positively correlated with either nitrogen or phosphorus availability,[40,371] thereby reducing the impact of eutrophication on the formation of blooms. Tsuda and Kami[367] suggested that selective browsing by herbivorous fishes on macroalgae removed potential competitors and favored the establishment of unpalatable blue-green algae. Crude extracts and isolated secondary metabolites of several strains of blue-green algae have been tested in assays for feeding deterrence and are usually deterrent to generalist herbivores.[44,55,176,212–214] Thus, the chemical defenses of cyanobacteria may play a critical role in bloom formation and persistence.

The chemically rich seaweed, *Caulerpa taxifolia,* has invaded coastal areas of the northwestern Mediterranean. Since first being detected near Monaco, it has spread to many areas, in some cases hundreds of kilometers away from the site of its accidental introduction.[372,373] The alga is known to produce caulerpenyne, a sesquiterpene found in many species of *Caulerpa*, at relatively high concentrations, as well as oxytoxin, taxifolials (see Figure 6.2), and other terpenes.[374,375] *C. taxifolia* is unpalatable to generalist herbivores in the Mediterranean[376] and can affect the physiology of sympatric fishes.[377] The chemical defenses of this alga appear to have facilitated this biological invasion, which is affecting the benthic community structure in areas where it occurs.[378,379]

VI. CONCLUSIONS

Over the past 20 years, chemical ecologists have proposed hypotheses regarding the evolution of chemical defenses in plants and animals.[9,158,261,262,315,380–383] Many of these hypotheses suggest that the evolution of plant defense mechanisms is responsive to factors such as the plant's risk of discovery by herbivores, the cost of defense, and the relative value of various plant parts. These predictions are based on preliminary data, and in many cases still have not been rigorously tested.

Thus, although there is currently general acceptance of the defensive roles of many secondary metabolites, there is considerable speculation regarding how herbivores and the physical environment interact to affect plant chemistry.[22,78,84,262,315,383,384] The diversity and ubiquity of secondary metabolites produced by plants has generated debate regarding their costs and benefits and the selective forces influencing their biosynthesis. Studies of marine plant–herbivore interactions have contributed greatly to our knowledge of how chemical defenses influence the ecology, physiology, and behavior of herbivores and to our understanding of specialization, chemical variation, and biogeographic patterns of chemical defense.[9,22,78]

The importance of algal chemical defenses is evidenced by the direct effects these have on species survival and the dominance of chemically defended algae in areas of high grazing intensity. Defended algae also serve as microhabitats to small consumers and, thus, indirectly contribute to enhance local biodiversity.[385] Because defended algae persist in environments where most other seaweeds do not and produce compounds that can be deterrent and toxic, chemical defenses could indirectly affect local patterns of productivity, decomposition, and nutrient remineralization. However, little is known about these complex phenomena, and more direct links between chemical defenses and ecosystem-wide processes need to be established.[22] Research is also needed for understanding the ultimate causes of chemical variation in algal populations and its consequences for ecological and evolutionary processes.[9,22,78] This will require better integration among ecological and molecular approaches from both taxonomical and physiological standpoints. Similarly, direct tests of latitudinal evolutionary patterns in algal chemical defense are few and sometimes opposed to predictions.[78,142,143]

Recent and ongoing research (Paul and co-workers[44,55]) has focused on the role of benthic cyanobacteria in benthic communities and has unraveled interesting parallels between consumer-prey interactions and the chemical ecology of prokaryotic and eukaryotic algae. Cyanobacteria in Pacific coral reefs contain chemical defenses that are deterrent to generalist grazers, but not to smaller mesograzers, and the cyanobacteria serve as habitats to a suite of smaller organisms, many of which have yet to be described.[66] Thus, these cyanobacteria may constitute an understudied source of marine biodiversity. This approach is novel in that cyanobacteria are often treated as negative signs of disturbance in marine environments rather than important components of the community. Although bloom-forming cyanobacteria can be detrimental to reefs, numerous species form part of the normal biota of marine communities.[3]

Finally, the role of chemical defenses in algae may aid our understanding of phase shifts and biological invasions in marine environments. An increasing number of studies has shown chemically defended algae to be responsible for macroalgal blooms (see above). Chemically defended algae, such as *Caulerpa taxifolia*, can also be aggressive invaders in marine communities, potentially because control by herbivores is constrained by algal defenses. Knowledge about the chemical ecology of algae will aid management decisions when options for biocontrol are evaluated.

ACKNOWLEDGMENTS

We are grateful to A. A. Boettcher, R. C. Bolser, M. Deal, M. Friedlander, M. E. Hay, R. Kennish, A. Miralto, H. Pavia, A. G. B. Poore, D. I. Walker, and B. Worm for providing us with reprints, manuscripts in press, and personal communications of ongoing research. Research on macroalgal–herbivore interactions in V. J. Paul's laboratory has been supported by NSF grant HRD-9023311 and EPA grant R82-6220 through the U.S. ECOHAB program. Although the research described has been funded in part by the United States Environmental Protection Agency, it has not been subjected to the Agency's peer and policy review and, therefore, does not necessarily reflect the views of the Agency and no official endorsement should be inferred. E. Cruz-Rivera has been supported by the NIH Research Supplements for Under-Represented Minorities Program with a postdoctoral supplement to grant CA53001 from the National Cancer Institute.

REFERENCES

1. Lobban, C. S. and Harrison, P. J., *Seaweed Ecology and Physiology*, Cambridge University Press, Cambridge, 1994.
2. Dawes, C. J., *Marine Botany*, 2nd Ed., John Wiley & Sons, Inc., New York, 1998.
3. Littler, D. S. and Littler, M. M., *Caribbean Reef Plants,* OffShore Graphics, Inc., Washington D.C., 2000.
4. Duffy, J. E. and Hay, M. E., Seaweed adaptations to herbivory, *BioScience,* 40, 368, 1990.
5. Lubchenco, J. and Gaines, S. D., A unified approach to marine plant–herbivore interactions. I. Populations and communities, *Annu. Rev. Ecol. Syst.,*12, 405, 1981.
6. Steneck, R. S., The ecology of coralline algal crusts: convergent patterns and adaptive strategies, *Annu. Rev. Ecol. Syst.*, 17, 273, 1986.
7. Steneck, R. S., Herbivory on coral reefs: a synthesis, *Proc. 6th Int. Coral Reef Sym.*, 1, 37, 1988.
8. Hay, M. E. and Fenical, W., Marine plant–herbivore interactions: the ecology of chemical defense, *Annu. Rev. Ecol. Syst.*, 19, 111, 1988.
9. Hay, M. E. and Steinberg, P. D., The chemical ecology of plant–herbivore interactions in marine versus terrestrial communities, in *Herbivores: Their Interactions with Secondary Plant Metabolites: Ecological and Evolutionary Processes*, Vol. 2, Rosenthal, G.A. and Berenbaum, M.R., Eds., Academic Press, San Diego, 1992, chap. 10.
10. Hay, M. E., Spatial patterns of herbivore impact and their importance in maintaining algal species richness, *Proc. 5th Int. Coral Reef Cong.*, 4, 29, 1985.
11. Hay, M. E., Associational plant defenses and the maintenance of species diversity: turning competitors into accomplices, *Am. Nat.*, 128, 617, 1986.
12. Littler, M. M., Taylor, P. R., and Littler, D. S., Plant defense associations in the marine environment, *Coral Reefs*, 5, 63, 1986.
13. Littler, M. M., Littler, D. S., and Taylor, P. R., Animal–plant defense associations: effects on the distribution and abundance of tropical reef macrophytes, *J. Exp. Mar. Ecol.*, 105, 107, 1987.
14. Pfister, C. A. and Hay, M. E., Associational plant refuges convergent patterns in marine and terrestrial communities result from differing mechanisms, *Oecologia*, 77, 118, 1988.
15. Wahl, M. and Hay, M. E., Associational resistance and shared doom: effects of epibiosis on herbivory, *Oecologia*, 102, 329, 1995.
16. Littler, M. M. and Littler, D. S., The evolution of thallus form and survival strategies in benthic marine macroalgae: field and laboratory tests of a functional form model, *Am. Nat.*, 116, 25, 1980.
17. Littler, M. M., Taylor, P. R., and Littler, D. S., Algal resistance to herbivory on a Caribbean barrier reef, *Coral Reefs*, 2, 111, 1983.
18. Steneck, R.S. and Watling, L., Feeding capabilities and limitations of herbivorous molluscs: a functional group approach, *Mar. Biol.*, 68, 299, 1982.
19. Pitlik, T. J. and Paul, V. J., Effects of toughness, calcite level, and chemistry of crustose coralline algae (Rhodophyta, Corallinales) on grazing by the parrotfish *Chlorurus sordidus*, *Proc. 8th Int. Coral Reef Sym.*, 1, 701, 1997.
20. Van Alstyne, K. L. and Paul, V. J., The role of secondary metabolites in marine ecological interactions, *Proc. 6th Int. Coral Reef Sym.*, 1, 175, 1988.
21. Hay, M. E., Fish–seaweed interactions on coral reefs: effects of herbivorous fishes and adaptations of their prey, in *The Ecology of Fishes on Coral Reefs*, Sale, P. F., Ed., Academic Press, New York, 1991, chap. 5.
22. Hay, M. E., Marine chemical ecology: what's known and what's next?, *J. Exp. Mar. Biol. Ecol.*, 200, 103, 1996.
23. Hay, M. E., Calcified seaweeds on coral reefs: complex defenses, trophic relationships, and value as habitats, *Proc. 8th Int. Coral Reef Sym.*, 1, 713, 1997.
24. Paul, V. J., Ed., *Ecological Roles of Marine Natural Products*, Cornell University Press, Ithaca, NY, 1992.
25. Hay, M. E., Predictable spatial escapes from herbivory: how do these affect the evolution of herbivore resistance in tropical marine communities?, *Oecologia*, 64, 396,1984.
26. Paul, V. J. and Hay, M.E., Seaweed susceptibility to herbivory: chemical and morphological correlates, *Mar. Ecol. Prog. Ser.*, 33, 255, 1986.

27. Hay, M. E., Kappel, Q. E., and Fenical, W. Synergisms in plant defenses against herbivores: interactions of chemistry, calcification and plant quality, *Ecology*, 75, 1714, 1994.

28. Meyer, K. D. and Paul, V. J., Variation in aragonite and secondary metabolite concentrations in the tropical green seaweed *Neomeris annulata*: effects on herbivory by fishes, *Mar. Biol.*, 122, 537, 1995.

29. Paul, V. J., Secondary metabolites and calcium carbonate as defenses of calcareous algae on coral reefs, *Proc. 8th Int. Coral Reef Symp.*, 1, 707, 1997.

30. Pennings, S. C. and Paul, V. J., Plant defenses against *Dolabella* herbivory: the role of chemistry, calcification and toughness, *Ecology*, 73, 1606, 1992.

31. Schupp, P. J. and Paul, V. J., Calcium carbonate and secondary metabolites in tropical seaweeds: variable effects on herbivorous fishes, *Ecology*, 75, 1172, 1994.

32. Munro, M.H.G. and Blunt, J.W., *MarinLit*, version 10.4, Marine Chemistry Group, University of Canterbury, Christchurch, New Zealand, 1999.

33. Steinberg, P. D., Geographical variation in the interaction between marine herbivores and brown algal secondary metabolites, in *Ecological Roles of Marine Natural Products*, Paul, V. J., Ed., Comstock Publishing Associates, Ithaca, NY, 1992, chap. 2.

34. Ireland, C. M., Roll, D. M., Molinski, T. F., Mckee, T.C ., Zabriskie, T. M., and Swersey, J. C., Uniqueness of the marine chemical environment: categories of marine natural products from inverte- brates, in *Biomedical Importance of Marine Organisms*, Fautin, D.G., Ed., California Academy of Sciences, San Francisco, 1988, 41.

35. Fenical, W., Halogenation in the Rhodophyta: a review, *J. Phycol.*, 11, 245, 1975.

36. Fenical, W., Natural products chemistry in the marine environment, *Science,* 215, 923, 1982.

37. Butler, A. and Walker, J. V., Marine haloperoxidases, *Chem. Rev.*, 93, 1937, 1993.

38. Whitton, B. A. and Potts, M., Marine littoral, in *The Biology of Cyanobacteria*, Carr, N. G. and Whitton, B. A., Eds., University of California Press, Berkeley, 1982, chap. 20.

39. Whitton, B. A. and Potts, M., Eds., *The Ecology of Cyanobacteria: Their Diversity in Time and Space,* Kluwer Academic Publishers, Dordrecht, 2000.

40. Thacker, R. W. and Paul, V. J., Are benthic cyanobacteria indicators of nutrient enrichment? Rela- tionships between cyanobacterial abundance and environmental factors on the reef flats of Guam, *Bull. Mar. Sci.*, in press.

41. Moore, R. E., Constituents of blue-green algae, in *Marine Natural Products*, Vol. 3, Scheuer, P. J., Ed., Academic Press, New York, 1981, chap. 1.

42. Moore, R. E., Toxins, anitcancer agents, and tumor promoters from marine prokaryotes, *Pure Appl. Chem.*, 54, 1919, 1982.

43. Moore, R. E., Cyclic peptides and depsipeptides from cyanobacteria: a review, *J. Ind. Microbiol.*, 16, 134, 1996.

44. Nagle, D. G. and Paul, V. J., Production of secondary metabolites by filamentous tropical marine cyanobacteria: ecological functions of the compounds, *J. Phycol.*, 35, 1412, 1999.

45. Gerwick, W. H., Roberts, M. A., Proteau, P. J., and Chen, J. L., Screening cultured marine microalgae for anticancer-type activity, *J. Appl. Phycol.*, 6, 143, 1994.

46. Cardellina, J. H., II, Moore, R. E., Arnold, E. V., and Clardy, J., Structure and absolute configuration of malyngolide, an antibiotic from the marine blue-green alga *Lyngbya majuscula* Gomont, *J. Org. Chem.*, 44, 4039, 1979.

47. Cardellina, J. H., II, Marner, F. J., and Moore, R. E., Seaweed dermatitis: structure of lyngbyatoxin A, *Science*, 204, 193, 1979.

48. Sims, J. K. and Zandee Van Rilland, R. D., Escharotic stomatitis caused by the "stinging seaweed" *Microcoleus lyngbyaceus* (formerly *Lyngbya majuscula*): a case report and review of literature, *Hawaii Med. J.*, 40, 243, 1981.

49. Fujiki, H., Suganuma, M., Hakiii, H., Bartolini, G., Moore, R. E., Takayama, S., and Sugimura, T., A two-stage mouse skin carcinogenesis study of lyngbyatoxin A, *J. Cancer Res. Clin. Oncol.*, 108, 174, 1984.

50. Gerwick, W. H., Proteau, P. J., Nagle, D. G., Hamel, E., Blokhin, A., and Slate, D. L., Structure of curacin A, a novel antimitotic, antiproliferative, and brine shrimp toxic natural product from the marine cyanobacterium *Lyngbya majuscula*, *J. Org. Chem.*, 59, 1243, 1994.

51. Orjala, J., Nagle, D. G., and Gerwick, W. H., Malyngamide H, an ichthyotoxic amide possessing a new carbon skeleton from the Caribbean cyanobacterium *Lyngbya majuscula*, *J. Nat. Prod.*, 58, 764, 1995.

52. Todd, J. T. and Gerwick, W. H., Malyngamide I from the tropical marine cyanobacterium *Lyngbya majuscula* and the probable structure revision of stylocheilamide, *Tetrahedron Lett.*, 36, 7837, 1995.

53. Wu, M., Milligan, K. E., and Gerwick, W. H., Three new malyngamides from the marine cyanobacterium *Lyngbya majuscula*, *Tetrahedron*, 53, 15983, 1997.

54. Gerwick, W. H., Jiang, Z. D., Agarwal, S. K., and Farmer, B. T., Total structure of hormothamnin A, a toxic cyclic undecapeptide from the tropical marine cyanobacterium *Hormothamnion enteromorphoides*, *Tetrahedron*, 48, 2313, 1992.

55. Pennings, S. C., Pablo, S. R., and Paul, V. J., Chemical defenses of the tropical benthic marine cyanobacterium *Hormothamnion enteromorphoides*: diverse consumers and synergisms, *Limnol. Oceanogr.*, 42, 911, 1997.

56. Harrigan, G. G., Luesch, H., Yoshida, W. Y., Moore, R. E., Nagle, D. G., Paul, V. J., Mooberry, S. L., Corbett, T. H. and Valeriote, F. A., Symplostatin 1: a dolastatin 10 analog from the marine cyanobacterium *Symploca hydnoides*, *J. Nat. Prod.*, 61, 1075, 1998.

57. Harrigan, G. G., Luesch, H., Yoshida, W. Y., Moore, R. E., Nagle, D. G., and Paul, V. J., Symplostatin 2, a dolastatin 13 analogue from the marine cyanobacterium *Symploca hydnoides*, *J. Nat. Prod.*, 62, 655, 1999.

58. Luesch, H., Yoshida, W. Y., Moore, R. E., and Paul, V. J., Lyngbyastatin 2 and norlyngbyastatin 2, analogues of dolastatin G and nordolastatin G from the marine cyanobacterium *Lyngbya majuscula*. *J. Nat. Prod.*, 62, 1702, 1999.

59. Luesch, H., Yoshida, W. Y., Moore, R. E., Paul, V. J., and Mooberry, S. L., Isolation, structure determination and biological activity of lyngbyabellin A from the marine cyanobacterium *Lyngbya majuscula*, *J. Nat. Prod.*, 63, 611, 2000.

60. Mitchell, S. S., Faulkner, D. J., Rubins, K., and Bushman, F. D., Dolastatin 3 and two novel cyclic peptides from a Palauan collection of *Lyngbya majuscula*, *J. Nat. Prod.*, 63, 279, 2000.

61. Harrigan, G. G., Yoshida, W., Moore, R. E., Nagle, D. G., Park, P. U., Biggs, J., Paul, V. J., Mooberry, S. L., Corbett, T. H, and Valeriote, F. A., Isolation, structure determination, and biological activity of dolastatin 12 and lyngbyastatin 1 from *Lyngbya majuscula/Schizothrix calcicola* cyanobacterial assemblages, *J. Nat. Prod.*, 61, 1221, 1998.

62. Persinos, G., Update: promising new compounds, *Wash. Insight*, 11, 6, 1998.

63. Pettit, G. R., Kamano, Y., Herald, C. L., Tuinman, A. A., Boettner, F. E., Kizu, H., Schmidt, J. M., Baczynskyj, L., Tomer, K. B., and Bontems, R. J., The isolation and structure of a remarkable marine animal antineoplastic constituent: dolastatin 10, *J. Am. Chem. Soc.*, 109, 6883, 1987.

64. Pettit, G. R., Progress in the discovery of biosynthetic anticancer drugs, *J. Nat. Prod.*, 59, 812, 1996.

65. Pettit, G. R. Dolastatins, in *Progress in the Chemistry of Organic Natural Products*, Herz, W., Kirby, G. W., Moore, R. E., Steglich, W., and Tamm, C., Eds., Springer-Verlag, New York, 70, 1997, 1.

66. Cruz-Rivera, E. and Paul, V. J., work in progress.

67. Paul, V. J. and Fenical, W., Chemical defense in tropical green algae, order Caulerpales, *Mar. Ecol. Prog. Ser.*, 34, 157, 1986.

68. Paul, V. J. and Fenical, W., Natural products chemistry and chemical defense in tropical marine algae of the phylum Chlorophyta, in *Bioorganic Marine Chemistry*, vol. 1, Scheuer, P. J., Ed., Springer-Verlag, Berlin, 1987, 1.

69. Paul, V. J., Fenical, W., Raffii, S., and Clardy, J., The isolation of new norcycloartene triterpenoids from the tropical marine alga *Tydemania expeditionis* (Chlorophyta), *Tetrahedron Lett.*, 23, 3459, 1982.

70. Govindan, M., Abbas, S. A., and Schmitz, F. J., New cycloartanol sulfates from the alga *Tydemania expeditionis*: inhibitors of the protein tyrosine kinase pp60[v-src], *J. Nat. Prod.*, 57, 74, 1994.

71. Barnekow, D. E., Cardellina, J. H., II, Zekter, A. S., and Martin, G. E., Novel cytotoxic and phytotoxic halogenated sesquiterpenes from the green alga *Neomeris annulata*, *J. Am. Chem. Soc.*, 111, 3511, 1989.

72. Paul, V. J., Cronan, J. M., Jr., Cardellina, J. H., II, Isolation of new brominated sesquiterpene feeding deterrents from the tropical green alga *Neomeris annulata* (Dasycladaceae: Chlorophyta), *J. Chem. Ecol.*, 19, 1847, 1993.

73. Park, M., Fenical, W., and Hay, M. E., Debromoisocymobarbatol, a new chromanol feeding deterrent from the marine alga *Cymopolia barbata*, *Phytochemistry*, 31, 4115, 1992.

74. Sun, H. H., Paul, V. J., and Fenical, W., Avrainvilleol, a brominated diphenylmethane derivative with feeding deterrent properties from the tropical green alga *Avrainvillea longicaulis*, *Phytochemistry*, 22, 743, 1983.

75. Chen, J. L., Gerwick, W. H., Schatzman, R., and Laney, M., Isorawsonol and related IMP dehydrogenase inhibitors from the tropical green alga *Avrainvillea rawsonii*, *J. Nat. Prod.*, 57, 947, 1994.

76. Dayton, P. K., Ecology of kelp communities, *Annu. Rev. Ecol. Syst.*, 16, 215, 1985.

77. Ragan, M. A. and Glombitzka, K. W., Phlorotannins, brown algal polyphenols, *Prog. Phycol. Res.*, 4, 130, 1986.

78. Targett, N. M. and Arnold, T. M., Predicting the effects of brown algal phlorotannins on marine herbivores in tropical and temperate oceans, *J. Phycol.*, 34, 195, 1998.

79. Swain, T., Tannins and lignins, in *Herbivores: Their Interactions with Secondary Plant Metabolites*, Rosenthal, G. A. and Janzen, D. H., Eds., Academic Press, New York, 1979, chap. 19.

80. Hagerman, A. E. and Butler, L. G., Tannins and lignins, in *Herbivores: Their Interactions with Secondary Plant Metabolites: The Chemical Participants*, Vol. 1, Rosenthal, G. A. and Berenbaum, M. R., Eds., Academic Press, San Diego, 1991, chap. 10.

81. Sieburth, J. M. and Conover, J. T., *Sargassum* tannin, an antibiotic which retards fouling, *Nature*, 208, 52, 1965.

82. Lau, S. C. K. and Qian, P. Y., Phlorotannins and related compounds as larval settlement inhibitors of the tube-building polychaete *Hydroides elegans*, *Mar. Ecol. Prog. Ser.*, 159, 219, 1997.

83. Jennings, J. G. and Steinberg, P. D., Phlorotannins versus other factors affecting epiphyte abundance on the kelp *Ecklonia radiata*, *Oecologia*, 109, 461, 1997.

84. Pavia, H., Cervin, G., Lindgren, A., and Aberg, P., Effects of UV-B radiation and simulated herbivory on phlorotannins in the brown alga *Ascophyllum nodosum*, *Mar. Ecol. Prog. Ser.*, 157, 139, 1997.

85. Dayton, P. K., The structure and regulation of some South American kelp communities, *Ecol. Monogr.*, 55, 447, 1985.

86. McClintock, M., Higinbotham, N., Uribe, E. G., and Cleland, R. E., Active, irreversible accumulation of extreme levels of H_2SO_4 in the brown alga *Desmarestia*, *Plant Physiol.*, 70, 771, 1982.

87. Faulkner, D. J., Marine natural products: metabolites of marine algae and herbivorous marine molluscs, *Nat. Prod. Rep.*, 1, 251, 1984.

88. Marshall, R. A., Harper, D. B., McRoberts, W. C., and Dring, M. J., Volatile bromocarbons produced by *Falkenbergia* stages of *Asparagopsis* sp. (Rhodophyta), *Limnol. Oceanogr.*, 44, 1348, 1999.

89. Erickson, K. L., Constituents of *Laurencia*, in *Marine Natural Products: Chemical and Biological Perspectives, Vol. 5*, Scheuer, P. J., Ed., Academic Press, New York, 1983, chap. 4.

90. Zapata, O. and McMillan, C., Phenolic acids in seagrasses, *Aq. Bot.*, 7, 307, 1979.

91. Todd, J. S., Zimmerman, R. C., Crews, P., and Alberte, R. S., The antifouling activity of natural and synthetic phenolic acid sulphate esters, *Phytochemistry*, 34, 401, 1993.

92. McMillan, C., Sulfated flavonoids and leaf morphology of the *Halophila ovalis–H. minor* complex (Hydrocharitaceae) in the Pacific Islands and Australia, *Aq. Bot.*, 16, 337, 1983.

93. Harrison, P. G., Control of microbial growth and amphipod grazing by water-soluble compounds from leaves of *Zostera marina*, *Mar. Biol.*, 67, 225, 1982.

94. Valentine, J. F. and Heck, K. L., Jr., Seagrass herbivory: evidence for the continued grazing of marine grasses, *Mar. Ecol. Prog. Ser.*, 176, 291, 1999.

95. Schmitt, T. M., Hay, M. E., and Lindquist, N., Antifouling and herbivore deterrent roles of seaweed secondary metabolites: constraints on chemically-mediated coevolution, *Ecology*, 76, 107, 1995.

96. Almodovar, L. R., Ecological aspects of some antibiotic algae in Puerto Rico, *Bot. Mar.*, 6, 143, 1964.

97. Caccamese, S., Azzolina, R., Furnari, G., Cormaci, M., and Grasso, S., Antimicrobial and antiviral activities of extracts from Mediterranean algae, *Bot. Mar.*, 23, 285, 1980.

98. Hodgson, L. M., Antimicrobial and antineoplastic activity in some south Florida seaweeds, *Bot. Mar.*, 27, 387, 1984.

99. Schmitt, T. M., Lindquist, N., and Hay, M. E., Seaweed secondary metabolites as antifoulants: effects of *Dictyota* spp. diterpenes on survivorship, settlement, and development of marine invertebrate larvae, *Chemoecology*, 8, 125, 1998.

100. De Nys, R., Steinberg, P. D., Willemsen, P., Dworjanyn, S. A., Gabelish, C. L., and King, R. J., Broad spectrum effects of secondary metabolites from the red alga *Delisea pulchra* in antifouling assays, *Biofouling*, 8, 259, 1995.

101. De Nys, R., Dworjanyn, S. A., and Steinberg, P. D., A new method for determining surface concentrations of marine natural products on seaweeds, *Mar. Ecol. Prog. Ser.*, 162, 79, 1998.

102. Choat, J. H. and Clements, K. D., Vertebrate herbivores in marine and terrestrial environments: a nutritional ecology perspective, *Annu. Rev. Ecol. Syst.*, 29, 375, 1998.

103. Williams, S. L., Uptake of sediment ammonium and translocation in a marine green macroalga *Caulerpa cupressoides*, *Limnol. Oceanogr.*, 29, 374, 1984.

104. Hay, M. E., Marine–terrestrial contrasts in the ecology of plant chemical defenses against herbivores, *Tr. Ecol. Evol.*, 6, 362, 1991.

105. Mattson, W. J., Jr., Herbivory in relation to plant nitrogen content, *Annu. Rev. Ecol. Syst.*, 11, 119, 1980.

106. White, T. C. R., *The Inadequate Environment: Nitrogen and the Abundance of Animals*, Springer-Verlag, New York, 1993.

107. Bowen, S. H., Lutz, E. V., and Ahlgren, M. O, Dietary protein and energy as determinants of food quality: trophic strategies compared, *Ecology*, 76, 899, 1995.

108. Hatcher, B. G., The interaction between grazing organisms and the epilithic algal community of a coral reef: a quantitative assessment, *Proc. 4th Int. Coral Reef Symp.*, 2, 515, 1981.

109. Simpson, S. J. and Simpson, C. L., The mechanisms of nutritional compensation by phytophagous insects, in *Insect–Plant Interactions, Vol. 2*, Bernays, E. A., Ed., CRC Press, Boca Raton, FL, 1990, 111.

110. Cruz-Rivera, E. and Hay, M. E., Can quantity replace quality? Food choice, compensatory feeding, and fitness of marine mesograzers, *Ecology*, 81, 201, 2000.

111. Choat, J. H., The biology of herbivorous fishes on coral reefs, in *The Ecology of Fishes on Coral Reefs*, Sale, P. F., Ed., Academic Press, San Diego, 1991, chap. 6.

112. Cruz-Rivera, E., Effects of diet on temperate mesograzers: consequences of prey traits for consumer feeding and fitness, Ph.D. thesis, University of North Carolina at Chapel Hill, Chapel Hill, North Carolina, 1999.

113. Duffy, J. E. and Paul, V. J., Prey nutritional quality and the effectiveness of chemical defenses against tropical reef fishes, *Oecologia*, 90, 333, 1992.

114. Slansky, F., Jr., Allelochemical–nutrient interactions in herbivore nutritional ecology, in *Herbivores: Their Interactions with Secondary Plant Metabolites: Ecological and Evolutionary Processes*, Vol. 2, Rosenthal, G. A., and Berenbaum, M. R., Eds., Academic Press, San Diego, 1992, chap. 4.

115. Pennings, S. C., Pablo, S. R., Paul, V. J., and Duffy, J. E., Effects of sponge secondary metabolites in different diets on feeding by three groups of consumers, *J. Exp. Mar. Biol. Ecol.*, 180, 137, 1994.

116. Targett, T. E. and Targett, N. M., Energetics of food selection by the herbivorous parrotfish *Sparisoma radians*: roles of assimilation efficiency, gut evacuation rate, and algal secondary metabolites, *Mar. Ecol. Prog. Ser.*, 66, 13, 1990.

117. Yang, C. S., Brady, J. F., and Hong, J., Dietary effects on cytochromes P450, xenobiotic metabolism, and toxicity, *FASEB J.*, 6, 737, 1992.

118. Rimmer, D. W. and Wiebe, W. J., Fermentative microbial digestion in herbivorous fishes, *J. Fish Biol.*, 31, 229, 1987.

119. Luczkovich, J. J. and Stellwag, E. J., Isolation of cellulolytic microbes from the intestinal tract of the pinfish *Lagodon rhomboides*: size-related changes in diet and microbial abundance, *Mar. Biol.*, 116, 381, 1993.

120. Randall, J. E., Food habits of reef fishes of the West Indies, *Stud. Trop. Oceanogr.*, 5, 655, 1967.

121. Bjorndal, K. A., Nutrition and grazing behavior of the green turtle *Chelonia mydas*, *Mar. Biol.*, 56, 147, 1980.

122. Horn, M. H., Biology of marine herbivorous fishes, *Oceanogr. Mar. Biol. Annu. Rev.*, 27, 167, 1989.

123. John, D. M., Hawkins, S. J., and Price, J., Eds., *Plant–Animal Interactions in the Marine Benthos*, Clarendon Press, Oxford, 1992.

124. Gerking, S. D., *Feeding Ecology of Fish*, Academic Press, San Diego, 1994.

125. Myers, R. F., *Micronesian Reef Fishes: A Comprehensive Guide to the Fishes of Micronesia*, 3rd ed., Coral Graphics, Guam, 1999.

126. Fauchald, K. and Jumars, P. A., The diet of worms: a study of polychaete feeding guilds, *Oceanogr. Mar. Biol. Ann. Rev.*, 17, 193, 1979.

127. Jangoux, M. and Lawrence, J. M., Eds., *Echinoderm Nutrition*, A. A. Balkema, Rotterdam, 1982.

128. Hawkins S. J. and Hartnoll, R. G., Grazing of intertidal algae by marine invertebrates, *Oceanogr. Mar. Biol. Annu. Rev.*, 21, 195, 1983.

129. Pennings, S. C., Masatomo, T. M., and Paul V. J., Selectivity and growth of the generalist herbivore *Dolabella auricularia* feeding upon complimentary resources, *Ecology*, 74, 879, 1993.

130. Littler, M. M., Littler, D. S., and Taylor, P. R., Selective herbivore increases the biomass of its prey: a chiton-coralline reef building association, *Ecology*, 76, 1666, 1995.

131. Stachowicz, J. J. and Hay, M. E., Facultative mutualism between an herbivorous crab and a coralline alga: advantages of eating noxious seaweeds, *Oecologia*, 105, 377, 1996.

132. Stachowicz, J. J. and Hay, M. E., Reducing predation through chemically mediated camouflage; indirect effects of plant defenses on herbivores, *Ecology*, 80, 495, 1999.

133. Stachowicz, J. J. and Hay, M. E., Mutualism and coral persistence: the role of herbivore resistance to algal chemical defense, *Ecology*, 80, 2085, 1999.

134. Poore, A. G. B. and Steinberg P. D., Preference–performance relationships and effects of host plant choice in a herbivorous marine amphipod, *Ecol. Monogr.*, 69, 443, 1999.

135. Cruz-Rivera, E. and Hay, M. E., The effects of diet mixing on consumer fitness: macroalgae, epiphytes, and animal matter as food for marine amphipods, *Oecologia*, 123, 252, 2000.

136. Gaines, S. D. and Lubchenco, J., A unified approach to marine plant–herbivore interactions. II. Biogeography, *Ann. Rev. Ecol. Syst.*, 13, 111, 1982.

137. Barnard, J. L., Amphipodological agreement with Platnick, *J. Nat. Hist.*, 20, 1675, 1991.

138. Santelices, B. and Marquet, P. A., Seaweeds, latitudinal diversity patterns, and Rapoport's rule, *Diversity and Distributions*, 4, 71, 1998.

139. Meekan, M. G. and Choat, J. H., Latitudinal variation in abundance of herbivorous fishes: a comparison of temperate and tropical reefs, *Mar. Biol.*, 128, 373, 1997.

140. Vermeij, G.J., *Biogeography and Adaptation*, Harvard University Press, Cambridge, 1978.

141. Vermeij, G. J., *Evolution and Escalation: An Ecological History of Life*, Princeton University Press, New Jersey, 1987.

142. Steinberg, P. D. and van Altena, I., Tolerance of marine invertebrate herbivores to brown algal phlorotannins in temperate Australasia, *Ecol. Monogr.*, 62, 189, 1992.

143. Bolser, R. C. and Hay, M. E., Are tropical plants better defended? Palatability and defenses of temperate vs. tropical seaweeds, *Ecology*, 77, 2269, 1996.

144. Cronin, G., Paul, V. J., Hay, M. E., and Fenical, W., Are tropical herbivores more resistant than temperate herbivores to seaweed chemical defenses? Diterpenoid metabolites from *Dictyota acutiloba* as feeding deterrents for tropical versus temperate fishes and urchins, *J. Chem. Ecol.*, 23, 289, 1997.

145. Bellwood, D. R., Reef fish biogeography: habitat associations, fossils and phylogenies. *Proc. 8th Int. Coral Reef Symp.*, 1, 379, 1997.

146. Spight, T. M., Diversity of shallow-water gastropod communities on temperate and tropical beaches, *Am. Nat.*, 111, 1088, 1977.

147. Lawrence, J. M. and Sammarco, P. W., Effects of feeding on the environment: Echinoidea, in *Echinoderm Nutrition*, Jangoux, M. and Lawrence, J. M., Eds., A.A. Balkema, Rotterdam, 1982, chap. 24.

148. MacArthur, R. H., *Geographical Ecology: Patterns in the Distribution of Species*, Harper and Row, New York, 1972.

149. Bakus, G. J. and Green, G., Toxicity in sponges and holothurians: a geographical perspective, *Science*, 185, 951, 1974.

150. Bertness, M. D., Garrity, S. D., and Levings, S. C., Predation pressure and gastropod foraging: a tropical-temperate comparison, *Evolution*, 35, 995, 1981.

151. Menge, B. A. and Lubchenco, J., Community organization in temperate and tropical rocky intertidal habitats: prey refuges in relation to consumer pressure gradients, *Ecol. Monogr.*, 51, 429, 1981.

152. Fawcett, M. H., Local and latitudinal variation in predation on an herbivorous marine snail, *Ecology*, 65, 1214, 1984.

153. Coley, P. D. and Aide, T. M., Comparison of herbivory and plant defenses in temperate and tropical broad-leaved forests, in *Plant–Animal Interactions: Evolutionary Ecology in Tropical and Temperate Regions*, Price, P. W., Ed., John Wiley & Sons, Ithaca, NY, 1991.

154. Pennings, S. C., Siska, E. L., and Bertness, M. D., Latitudinal differences in plant palatability in Atlantic coast salt marshes, *Ecology*, in press.
155. Targett, N. M., Coen, L. D., Boettcher, A. A., and Tanner, C. E., Biogeographic comparisons of marine algal polyphenolics: evidence against a latitudinal trend, *Oecologia*, 89, 464, 1992.
156. Targett, N. M., Boettcher, A. A., Targett, T. E., Vrolijk, N. H., Tropical marine herbivore assimilation of phenolic-rich plants, *Oecologia*, 103, 170, 1995.
157. Owen-Smith, R. N., *Megaherbivores: The Influence of Very Large Body Size on Ecology*, Cambridge University Press, Cambridge, UK, 1988.
158. Crawley, M. J., *Herbivory: the Dynamics of Plant–Animal Interactions. Studies in Ecology*, Vol. 10, University of California Press, Berkeley, 1983.
159. Van Soest, P., *Nutritional ecology of the ruminant*, Cornell University Press, Ithaca, NY, 1994.
160. Spain, A. V. and Heinsohn, G. E., Cyclone-associated feeding changes in the dugong, *Mammalia*, 37, 678, 1973.
161. Beeman, R. D., The order Anaspidea, *Veliger*, 3, 87, 1968.
162. Bell, S. S., Amphipods as insect equivalents? An alternative view, *Ecology*, 72, 350, 1991.
163. Duffy, J. E. and Hay, M. E., Amphipods are not all created equal: a reply to Bell, *Ecology*, 72, 354, 1991.
164. Stachowicz, J. J. and Hay, M. E., Reduced mobility is associated with compensatory feeding and increased diet breadth of marine crabs, *Mar. Ecol. Prog. Ser.*, 188, 169, 1999.
165. Steinberg, P. D., Interaction between the canopy dwelling echinoid *Holopneustes purpurescens* and its host kelp *Ecklonia radiata*, *Mar. Ecol. Prog. Ser.*, 127, 169, 1995.
166. Hay, M. E., Duffy J. E., Pfister C. A., and Fenical W., Chemical defense against different marine herbivores: are amphipods insect equivalents?, *Ecology*, 68, 1567, 1987.
167. Hay, M. E., The role of seaweed chemical defenses in the evolution of feeding specialization and in the mediation of complex interactions, in *Ecological Roles of Marine Natural Products*, Paul, V. J., Ed., Cornell University Press, Ithaca, NY, 1992, chap. 3.
168. Brawley, S. H., Mesoherbivores, in *Plant–Animal Interactions in the Marine Benthos*, John, D. M., Hawkins, S. J., and Price, J., Eds., Clarendon Press, Oxford, UK, 1992, chap. 11.
169. Jensen, K. R., A review of sacoglossan diets, with comparative notes on radular and buccal anatomy, *Malac. Rev.*, 13, 55, 1980.
170. Paul, V. J. and Van Alstyne, K. L., Use of ingested algal diterpenoids by *Elysia halimedae* Macnae (Opisthobranchia: Ascoglossa) as antipredator defenses, *J. Exp. Mar. Biol. Ecol.*, 119, 15, 1988.
171. Hay, M. E., Pawlik, J. R., Duffy J. E., and Fenical W., Seaweed–herbivore–predator interactions: host-plant specialization reduces predation on small herbivores, *Oecologia*, 81, 418, 1989.
172. Hay, M. E., Duffy, J. E., and Fenical W., Host-plant specialization decreases predation on a marine amphipod: an herbivore in plant's clothing, *Ecology*, 71, 733, 1990.
173. Hay, M. E., Duffy, J. E., Paul, V. J., Renaud, P. E., and Fenical W., Specialist herbivores reduce their susceptibility to predation by feeding on the chemically defended seaweed *Avrainvillea longicaulis*, *Limnol. Oceanogr.*, 35, 1734, 1990.
174. Trowbridge, C. D., Diet specialization limits herbivorous sea slug's capacity to switch among food species, *Ecology*, 72, 1880, 1991.
175. Ginsburg, D. W. and Paul, V. J., Chemical defenses in the sea hare *Aplysia parvula*: importance of diet and sequestration of algal secondary metabolites, *Mar. Ecol. Prog. Ser.*, in press.
176. Paul, V. J. and Pennings, S. C., Diet-derived chemical defenses in the sea hare *Stylocheilus longicauda* (Quoy and Gaimard 1824), *J. Exp. Mar. Biol. Ecol.*, 151, 227, 1991.
177. Williams, S. I. and Walker, D. I., Mesoherbivore-macroalgal interactions: feeding ecology of sacoglossan sea slugs (Mollusca, Opisthobranchia) and their effects on their food algae, *Oceanogr. Mar. Biol. Annu. Rev.*, 37, 87, 1999.
178. Hay, M. E., Renaud, P. E., and Fenical, W., Large mobile versus small sedentary herbivores and their resistance to seaweed chemical defenses, *Oecologia*, 75, 246, 1988.
179. Duffy, J. E. and Hay, M. E., Food and shelter as determinants of food choice by an herbivorous marine amphipod, *Ecology*, 72, 1286, 1991.
180. Duffy, J. E. and Hay, M. E., Herbivore resistance to seaweed chemical defense: the roles of mobility and predation risk, *Ecology*, 75, 1304, 1994.

181. Strong, D. R., Lawton, J. H., and Southwood, R., *Insects on Plants*, Harvard University Press, Cambridge, Massachusetts, 1984.

182. Futuyma, D. J. and Moreno, G., The evolution of ecological specialization, *Annu. Rev. Ecol. Syst.*, 19, 207, 1988.

183. Avila, C., Natural products of opisthobranch molluscs: a biological review, *Oceanogr. Mar. Biol. Annu. Rev.*, 33, 487, 1995.

184. Faulkner, D. J., Chemical defenses of marine molluscs, in *Ecological Roles of Marine Natural Products*, Paul, V. J., Ed., Cornell University Press, Ithaca, NY, 1992, chap. 4.

185. Di Marzo, V., Marin, A., Vardaro, R. R., De Petrocellis, L., Villani, G., Cimino, G., Histological and biochemical bases of defense mechanisms in four species of Polybranchioidea ascoglossan molluscs, *Mar. Biol.*, 117, 367, 1993.

186. Johnson, P. M. and Willows, A. O. D., Defense in sea hares (Gastropoda, Opisthobranchia, Anaspidea): multiple layers of protection from egg to adult, *Mar. Fresh. Behav. Physiol.*, 32, 147, 1999.

187. Cimino, G. and Ghiselin, M. T., Chemical defense and evolution in the Sacoglossa (Mollusca: Gastropoda: Opisthobranchia), *Chemoecology*, 8, 51, 1998.

188. Ireland, C. and Faulkner, D. J., The metabolites of the marine molluscs *Tridachiella diomedea* and *Tridachia crispata*, *Tetrahedron*, 37, 233, 1981.

189. Hay, M. E., Duffy, J. E., Fenical W., and Gustafson, K., Chemical defense in the seaweed *Dictyopteris deliculata*: differential effects against reef fishes and amphipods, *Mar. Ecol. Prog. Ser.*, 48, 185, 1988.

190. Fahrenbach, W. F., The biology of a harpacticoid copepod, *La Cellule*, 3, 304, 1962.

191. Friedlander, M., Weintraub, N., Freedman, A., Sheer, J., Snovsky, Z., Shapiro, J., and Kissil, G. W., Fish as potential biocontrollers of *Gracilaria* (Rhodophyta) culture *Aquaculture*, 145, 113, 1996.

192. Friedlander, M. and Levy, I., Cultivation of *Gracilaria* in outdoor tanks and ponds, *J. Appl. Phycol.*, 7, 315, 1995.

193. Friedlander, M., personal communication, 1999.

194. Gunill, F. C., Macroalgae as habitat patch islands for *Scutellidium lamellipes* (Copepoda: Harpacticoida) and *Ampithoe tea* (Amphipoda: Gammaridae), *Mar. Biol.*, 69, 103, 1982.

195. Cruz-Rivera, E., work in progress.

196. Rogers, C. N., Williamson, J. E., Carson, D. G., and Steinberg, P. D., Diel vertical movement by mesograzers on seaweeds, *Mar. Ecol. Prog. Ser.*, 166, 301, 1998.

197. Tegner, M. J. and Dayton, P. K., El Nino effects on southern California kelp forest communities, *Adv. Ecol. Res.*, 17, 243, 1987.

198. Trowbridge, C. D., Mesoherbivory: the ascoglossan sea slug *Placida dendritica* may contribute to the restricted distribution of its algal host, *Mar. Ecol. Prog. Ser.*, 83, 207, 1992.

199. Parker, T., Johnson, C., and Chapman, A. R. O., Gammarid amphipods and littorinid snails have significant but different effects on algal succession in littoral fringe tidepools, *Ophelia*, 38, 69, 1993.

200. Chess, J. R., Effects of the stipe-boring amphipod *Peramphithoe stypotrupetes* (Corophioidea: Amphithoidae) and grazing gastropods on the kelp *Laminaria setchellii*, *J. Crust. Biol.* 13, 638, 1993.

201. Duffy, J. E. and Hay, M. E., Strong impacts of grazing amphipods on the organization of a benthic community, *Ecol. Monog.*, 70, 237, 2000.

202. Lotze, H. K., Worm, B., and Sommer, U., Propagule banks, herbivory and nutrient supply control population development and dominance patterns in macroalgal blooms, *Oikos*, 89, 46, 2000.

203. Hay, M. E., Piel, J., Boland, W., and Schnitzler, I., Seaweed sex pheromones and their degradation products frequently suppress amphipod feeding but rarely suppress sea urchin feeding, *Chemoecology*, 8, 91, 1998.

204. Schnitzler, I., Boland, W., and Hay, M. E., Organic sulfur compounds from *Dictyopteris* spp. (Phaeophyceae) deter feeding by an herbivorous amphipod (*Ampithoe longimana*) but not by an herbivorous sea urchin (*Arbacia punctulata*), *J. Chem. Ecol.*, 24, 1715, 1998.

205. Lindquist, N., Hay, M. E., and Fenical, W., Defense of ascidians and their conspicuous larvae: adult vs. larval chemical defenses, *Ecol. Monogr.*, 62, 547, 1992.

206. Lindquist, N. and Hay, M. E., Can small rare prey be chemically defended? The case for marine larvae, *Ecology*, 76, 1347, 1995.

207. Deal, M. S., The causes and consequences of within-species variation in seaweed chemical defenses, Ph.D. dissertation, University of North Carolina at Chapel Hill, Chapel Hill, North Carolina, 1997.

208. Hay, M. E., Stachowicz, J. J., Cruz-Rivera, E., Bullard, S., Deal, M. S., Lindquist, N., Bioassays with marine and freshwater macroorganisms, in *Methods in Chemical Ecology, Vol. 2, Bioassay Methods,* Haynes, K. F. and Millar, J. G., Eds., Chapman & Hall, Norwell, MA, 1998, chap. 2.

209. Cronin, G. and Hay, M. E., Induction of seaweed chemical defenses by amphipod grazing, *Ecology,* 77, 2287, 1996.

210. Cronin, G. and Hay, M. E., Susceptibility to herbivores depends on recent history of both the plant and animal, *Ecology,* 77, 1531, 1996.

211. Nagle, D. G., Camacho, F. T., and Paul, V. J., Dietary preferences of the opisthobranch mollusc *Stylocheilus longicauda* for secondary metabolites produced by the tropical cyanobacterium *Lyngyba majuscula, Mar. Biol.,* 132, 267, 1998.

212. Pennings, S. C., Weiss, A. M., and Paul, V. J., Secondary metabolites of the cyanobacterium *Microcoleus lyngbyaceus* and the sea hare *Stylocheilus longicauda*: palatability and toxicity, *Mar. Biol.,* 123, 735, 1996.

213. Thacker, R. W., Nagle, D. G., and Paul, V. J., Effects of repeated exposures to marine cyanobacterial secondary metabolites on feeding by juvenile rabbitfish and parrotfish, *Mar. Ecol. Prog. Ser.,* 147, 21, 1997.

214. Nagle, D. G. and Paul, V. J., Chemical defense of a marine cyanobacterial bloom, *J. Exp. Mar. Biol. Ecol.,* 225, 29, 1998.

215. Pennings, S. C. and Paul, V. J., Sequestration of dietary metabolites by three species of sea hares: location, specificity, and dynamics, *Mar. Biol.,* 117, 535, 1993.

216. Pennings, S. C., Paul, V. J., Dunbar, D. C., Hamann, M. T., Lumbang, W. A., Novack, B., Jacobs, R. S., Unpalatable compounds in the marine gastropod *Dolabella auricularia*: distribution and effect of diet, *J. Chem. Ecol.,* 25, 735, 1999.

217. Hogberg, H. E., Thompson, R. H., and King, T. J., The cymopols, a group of prenylated bromohydroquinones from the green calcareous alga *Cymopolia barbata, J. Chem. Soc. Perkins I,* 1696, 1976.

218. Wall, M. E., Wani, M. C., Manikumar, G., Taylor, H., Hughes, T. J., Gaetano, K., Gerwick, W. H., McPhail, A. T., and McPhail, D. R., Plant antimutagenic agents, 7. Structure and antimutagenic properties of cymobarbatol and 4-isocymobarbatol, new cymopols from green alga (*Cymopolia barbata*), *J. Nat. Prod.,* 52, 1092, 1989.

219. McConnell, O. J., Hughes, P. A., Targett, N. M., and Daley, J., Effects of secondary metabolites on feeding by the sea urchin *Lytechinus variegatus, J. Chem. Ecol.,* 8, 1427, 1982.

220. Hay, M. E., Fenical, W., and Gustafson, K., Chemical defense against diverse coral-reef herbivores, *Ecology,* 68, 1581, 1987.

221. Hay, M. E., Duffy, J. E., and Fenical W., Seaweed chemical defenses: among-compound and among-herbivore variance, *Proc. 6th Int. Coral Reef Sym.,* 3, 43, 1988.

222. Hillis-Colinvaux, L. Ecology and taxonomy of *Halimeda*: primary producer of coral reefs, *Adv. Mar. Biol.,* 17, 1, 1980.

223. Paul, V. J. and Van Alstyne, K. L., Antiherbivore defenses in *Halimeda, Proc. 6th Int. Coral Reef Sym.,* 3, 133, 1988.

224. Paul, V. J. and Fenical, W., Isolation of halimedatrial: chemical defense adaptation in the calcareous reef-building alga *Halimeda, Science,* 221, 747, 1983.

225. Paul, V. J. and Fenical, W., Bioactive diterpenoids from tropical marine algae of the genus *Halimeda* (Chlorophyta), *Tetrahedron,* 40, 3053, 1984.

226. Paul, V. J. and Van Alstyne, K. L., Chemical defense and chemical variation in some tropical Pacific species of *Halimeda* (Chlorophyta, Halimedaceae), *Coral Reefs,* 6, 263, 1988.

227. Paul, V. J. and Van Alstyne, K.L. Activation of chemical defenses in the tropical green algae *Halimeda* spp., *J. Exp. Mar. Biol. Ecol.,* 160, 191, 1992.

228. Teeyapant, R. and Proksch, P., Biotransformation of brominated compounds in the marine sponge *Verongia aerophoba* — Evidence for an induced chemical defense? *Naturwissenschaften,* 80, 369, 1993.

229. Conn, E. E., Cyanide and cyanogenic glycosides, in *Herbivores: Their Interaction with Secondary Plant Metabolites,* Rosenthal, G. A. and Janzen, D. H., Eds., Academic Press, New York, 1979, 387.

230. Seigler, D. S., Cyanide and cyanogenic glycosides, in *Herbivores: Their Interaction with Secondary Plant Metabolites, Vol. I, The Chemical Participants,* 2nd ed., Rosenthal, G. A. and Berenbaum, M. R., Eds., Academic Press, San Diego, 1991, chap. 2.

231. Konno, K., Hirayama, C., Yasui, H., and Nakamura, M., Enzymatic activation of oleuropein: A protein crosslinker used as a chemical defense in the privet tree, *Proc. Nat. Acad. Sci. USA*, 96, 9159, 1999.

232. Ebel, R., Brenzinger, M., Kunze, A., Gross, H. J., and Proksch, P., Wound activation of protoxins in marine sponge *Aplysina aerophoba*, *J. Chem. Ecol.*, 23, 1451, 1997.

233. Chew, F. S., Biological effects of glucosinolates, in *Biologically Active Natural Products: Potential Use in Agriculture*, Cutler, E. D., Ed., American Chemical Society Symposium Series No. 380, 1988, 155.

234. Louda, S. and Mole, S., Glucosinolates: chemistry and ecology, in *Herbivores: Their Interaction with Secondary Plant Metabolites, Vol. I, The Chemical Participants*, 2nd ed., Rosenthal, G. A. and Berenbaum, M. R., Academic Press, San Diego, 1991, chap. 4.

235. Pennings, S. C. and Svedberg, J. M., Does $CaCO_3$ in food deter feeding by sea urchins? *Mar. Ecol. Prog. Ser.*, 101,163, 1993.

236. Lumbang, W. A. and Paul, V. J., Chemical defenses of the tropical green seaweed *Neomeris annulata* Dickie: effects of multiple compounds on feeding by herbivores, *J. Exp. Mar. Biol. Ecol.*, 201, 185, 1996.

237. Meyer, K. D., Paul, V. J., Sanger, H. R., and Nelson, S. G., Effects of seaweed extracts and secondary metabolites on feeding by the herbivorous surgeonfish *Naso lituratus*, *Coral Reefs*, 13, 105, 1994.

238. Pennings, S. C., Puglisi, M. P., Pitlik, T. J., Himaya, A. C., and Paul, V. J., Effects of secondary metabolites and $CaCO_3$ on feeding by surgeonfishes and parrotfishes: within-plant comparisons, *Mar. Ecol. Prog. Ser.*, 134, 49, 1996.

239. Padilla, D. K., Rip stop in marine algae: minimizing the consequences of herbivore damage, *Evol. Ecol.*, 7, 634, 1993.

240. Paul, V. J., Wylie, C., and Sanger, H., Effects of algal chemical defenses toward different coral-reef herbivorous fishes: a preliminary study, *Proc. 6th Int. Coral Reef Sym.*, 3, 73, 1988.

241. Hardt, I. H., Fenical, W., Cronin, G., and Hay, M. E., Acutilols, potent herbivore feeding deterrents from the tropical brown alga, *Dictyota acutiloba*, *Phytochemistry*, 43, 71, 1996.

242. Shen, Y. C., Tsai, P. I., Fenical, W., and Hay, M. E., Secondary metabolite chemistry of the Caribbean marine alga *Sporochnus bolleanus*: a basis for herbivore chemical defence, *Phytochemistry*, 32, 71, 1993.

243. Roussis, V., King, R. L., and Fenical, W., Secondary metabolite chemistry of the Australian brown alga *Encyothalia cliftonii*: evidence for herbivore chemical defence, *Phytochemistry*, 34, 107, 1993.

244. Maier, I. and Muller, D.G., Sexual pheromones in algae, *Biol. Bull.*, 170, 145, 1986.

245. Kurata, K., Taniguchi, K., Agatsuma, Y., and Suzuki, M., Diterpenoid feeding-deterrents from *Laurencia saitoi*, *Phytochemistry*, 47, 363, 1998.

246. Paul, V. J., Hay, M. E., Duffy, J. E., Fenical, W., and Gustafson, K., Chemical defense in the seaweed *Ochtodes secundiramea* (Rhodophyta): effects of its monoterpenoid components upon diverse coral-reef herbivores, *J. Exp. Mar. Biol. Ecol.*, 114, 249, 1987.

247. Gunatilaka, A. A. L., Paul, V. J., Park, P. U., Puglisi, M. P., Gitler, A. D., Eggleston, D. S., Haltiwanger, R. C., and Kingston, D. G. I., Apakaochtodenes A and B: two tetrahalogenated monoterpenes from the red marine alga *Portieria hornemannii*, *J. Nat. Prod.*, 62, 1376, 1999.

248. Wylie, C. R. and Paul, V. J., Feeding preferences of the surgeonfish *Zebrasoma flavescens* in relation to chemical defenses of tropical algae, *Mar. Ecol. Prog. Ser.*, 45, 23, 1988.

249. Paul, V. J., Nelson, S. G., and Sanger, H. R., Feeding preferences of adult and juvenile rabbitfish *Siganus argenteus* in relation to chemical defenses of tropical seaweeds, *Mar. Ecol. Prog. Ser.*, 60, 23, 1990.

250. Paul, V. J., Meyer, K. D., Nelson, S. G., and Sanger, H. R., Deterrent effects of seaweed extracts and secondary metabolites on feeding by the rabbitfish *Siganus spinus*, *Proc. 7th Int. Coral Reef Sym.*, 2, 867, 1993.

251. Tsuda, R. T. and Bryan, P. G., Food preferences of juvenile *Siganus rostratus* and *S. spinus* in Guam, *Copeia*, 604, 1973.

252. Meyer, K. D. and Paul, V. J., Intraplant variation in secondary metabolite concentration in three species of *Caulerpa* (Chlorophyta: Caulerpales) and its effects on herbivorous fishes, *Mar. Ecol. Prog. Ser.*, 82, 249, 1992.

253. Biggs, J. S., The role of secondary metabolite complexity in the red alga *Laurencia palisada* as a defense against diverse consumers, Master's thesis, University of Guam, Mangilao, Guam, 2000.

254. Paul, V. J., Ciminiello, P., and Fenical, W., New diterpenoid feeding deterrents from the Pacific green alga *Pseudochlorodesmis furcellata*, *Phytochemistry,* 27, 1011, 1988.

255. Lee, J. C. and Bernays, E. A., Food tastes and toxic effects: associative learning by the polyphagous grasshopper *Schistocerca americana* (Drury) (Orthoptera: Acrididae), *Anim. Behav.*, 39, 163, 1990.

256. Provenza, F. D. and Cincotta, R. P., Foraging as a self-organizational learning process: accepting adaptability at the expense of predictability, in *Diet Selection*, Hughes, R.N., Ed., Blackwell Scientific, London, 1993, 78.

257. Freeland, W. J. and Janzen, D. H., Strategies in herbivory by mammals: the role of plant secondary compounds, *Am. Nat.,*108, 269, 1974.

258. Gerhart, D. J., Prostaglandin A_2: an agent of chemical defense in the Caribbean gorgonian *Plexaura homomalla*, *Mar. Ecol. Prog. Ser.*, 19, 181, 1984.

259. Gerhart, D. J., Emesis, learned aversion, and chemical defense in octocorals: a central role for prostaglandins?, *Am. J. Physiol.*, 260, R839, 1991.

260. Giménez-Casalduero, F., Thacker, R. W., and Paul, V. J., Association of color and feeding deterrence by tropical reef fishes, *Chemoecology*, 9, 33, 1999.

261. Feeney, P., Plant apparency and chemical defense, *Rec. Adv. Phytochem.*, 10, 1, 1976.

262. Coley, P. D., Bryant, J. P., and Chapin, F. S., III, Resource availability and plant antiherbivore defense, *Science*, 230, 895, 1985.

263. Bernays, E. A., Cooper Driver, G., and Bilgener, M., Herbivores and plant tannins, *Adv. Ecol. Res.*, 19, 263, 1989.

264. Guglielmo, C. G., Karasov, W. H., and Jakubas, W. J., Nutritional costs of a plant secondary metabolite explain selective foraging by ruffed grouse, *Ecology*, 77, 1103, 1996.

265. Dearing, M. D., Effects of *Acomastylis rosii* tannins on a mammalian herbivore, the North American pika, *Ochotona princeps*, *Oecologia*, 109, 122, 1997.

266. Bernays, E. A., Relationship between deterrence and toxicity of plant secondary compounds for the grasshopper *Schistocerca americana*, *J. Chem. Ecol.*, 17, 2519, 1991.

267. Bernays, E. A. and Cornelius, M., Relationship between deterrence and toxicity of plant secondary compounds for the alfalfa weevil *Hypera brunneipennis*, *Entomol. Exp. Appl.*, 64, 289, 1992.

268. Duggins, D. O. and Eckman, J. E., Is kelp detritus a good food for suspension feeders? Effects of kelp species, age and secondary metabolites, *Mar. Biol.*, 128, 489, 1997.

269. Pennings, S. C. and Carefoot, T. H., Post-ingestive consequences of consuming secondary metabolites in sea hares (Gastropoda: Opisthobranchia), *Comp. Biochem. Physiol.*, 111C, 249, 1995.

270. Taylor, A.C. and Spicer, J. I., Metabolic responses of the prawns *Palaemon elegans* and *P. serratus* (Crustacea: Decapoda) to acute hypoxia and anoxia, *Mar. Biol.*, 95, 521, 1987.

271. Spicer, J. I., Hill, A. D., Taylor, A. C., and Strang, R. H. C., Effect of aerial exposure on concentrations of selected metabolites in blood of the Norwegian lobster *Nephrops norvegicus* (Crustacea: Nephropidae), *Mar. Biol.*, 105, 129, 1990.

272. Carefoot, T. H., Effects of environmental stressors on blood glucose levels in sea hares, *Aplysia dactylomela*, *Mar. Biol.*, 118, 579, 1994.

273. Brattsten, L. B. Metabolic defenses against plant allelochemicals, in *Herbivores: Their Interaction with Secondary Plant Metabolites: Ecological and Evolutionary Processes,* Vol. 2, Rosenthal, G. A. and Berenbaum, M. R., Eds., Academic Press, San Diego, 1992, 175.

274. Tugwell, S. and Branch, G. M., Effects of gut surfactants on kelp polyphenol defenses, *Ecology*, 73, 205, 1992.

275. Vrolijk, N. H., Targett, N. M., Woodin, B. R., and Stegeman, J. J., Toxicological and ecological implications of biotransformation enzymes in the tropical teleost *Chaetodon capistratus*, *Mar. Biol.*, 119, 151, 1994.

276. Kennish, R., Diet composition influences the fitness of the herbivorous crab *Grapsus albolineatus*, *Oecologia*, 105, 22, 1996.

277. Poulet, S. A., Laabir, M., Ianora, A., and Miralto, A., Reproductive response of *Calanus helgolandicus*. I. Abnormal embryonic and naupliar development, *Mar. Ecol. Prog. Ser.*, 129, 85, 1995.

278. Chaudron, Y., Poulet, S. A., Laabir, M., Ianora, A., and Miralto, A., Is hatching success of copepod eggs diatom density-dependent?, *Mar. Ecol. Prog. Ser.*, 144, 185, 1996.

279. Miralto, A., Barone, G., Romano, G., Poulet, S. A., Ianora, A., Russo, G. L., Buttino, I., Mazzarella, G., Laabir, M., Cabrini, M., and Glacobbe, M. G., The insidious effect of diatoms on copepod reproduction, *Nature*, 402, 173, 1999.

280. White, A. W., Sensitivity of marine fishes to toxins from the red-tide dinoflagellate *Gonyaulax excavata* and implications for fish kills, *Mar. Biol.*, 65, 255, 1981.

281. Huntley, M., Sykes, P., Rohan, S., and Martin, V., Chemically-mediated rejection of dinoflagellate prey by the copepods *Calanus pacificus* and *Paracalanus parvus*: mechanism, occurrence and significance, *Mar. Ecol. Prog. Ser.*, 28, 105, 1986.

282. Ives, J. D., Possible mechanisms underlying copepod grazing responses to levels of toxicity in red tide dinoflagellates, *J. Exp. Mar. Biol. Ecol.*, 112, 131, 1987.

283. Sykes, P. F. and Huntley, M. E., Acute physiological reactions of *Calanus pacificus* to selected dinoflagellates: direct observations, *Mar. Biol.*, 94, 19, 1987.

284. Kvitek, R. G. and Beitler, M. K., A case for sequestering of paralytic shellfish toxins as a chemical defense, *J. Shellfish Res.*, 7, 629, 1988.

285. Gosselin, S., Fortier, L., and Gagné, J. A., Vulnerability of marine fish larvae to the toxic dinoflagellate *Protogonyaulax tamarensis*, *Mar. Ecol. Prog. Ser.*, 57, 1, 1989.

286. Uye, S. and Takamatsu, K., Feeding interactions between planktonic copepods and red-tide dinoflagellates from Japanese coastal waters, *Mar. Ecol. Prog. Ser.*, 59, 97, 1990.

287. Kvitek, R. G., Sequestered paralytic shellfish poisoning toxins mediate glaucous-winged gull predation on bivalve prey, *Auk*, 108, 381, 1991.

288. Kvitek, R. G., DeGange, A. R., and Beitler, M. K., Paralytic shellfish poisoning toxins mediate feeding behavior of sea otters, *Limnol. Oceanogr.*, 36, 393, 1991.

289. Davin, W. T., Jr., Kohler, C. C., and Tindall, D. R., Effects of ciguatera toxins on the bluehead, *Trans. Am. Fish. Soc.*, 115, 908, 1986.

290. Davin, W. T., Jr., Kohler, C. C., and Tindall, D. R., Ciguatera toxins adversely affect piscivorous fishes, *Trans. Am. Fish. Soc.*, 117, 374, 1988.

291. Magnelia, S. J., Kohler, C. C., and Tindall, D. R., Acanthurids do not avoid consuming cultured toxic dinoflagellates yet do not become ciguatoxic, *Trans. Am. Fish. Soc.*, 121, 737, 1992.

292. Barlöcher, F. and Newell, S. Y., Growth of the salt marsh periwinkle *Littoraria irrorata* on fungal and cordgrass diets, *Mar. Biol.*, 118, 109, 1994.

293. Targett, N. M. and Ward, J. E., Bioactive microalgal metabolites: mediation of subtle ecological interactions in phytophagous suspension feeding marine invertebrates, in *Bioorganic Marine Chemistry*, Vol. 4, Scheuer, P. J., Ed., Springer-Verlag, Berlin, 1991, 91.

294. Boettcher, A. A. and Targett, N. M., Role of polyphenolic molecular size in reduction of assimilation efficiency in *Xiphister mucosus*, *Ecology*, 74, 891, 1993.

295. Boettcher, A. A., The role of polyphenolic molecular size in the reduction of assimilation efficiency in some marine herbivores. Master's thesis, University of Delaware, Newark, Delaware, 1992.

296. Irelan, C. D. and Horn, M. H., Effects of macrophyte secondary chemicals on food choice and digestive efficiency of *Cebidichthys violaceus* (Girard), an herbivorous fish of temperate marine waters, *J. Exp. Mar. Biol. Ecol.*, 153, 179, 1991.

297. Winter, F. C. and Estes, J. A., Experimental evidence for the effects of polyphenolic compounds from *Dictyoneurum californicum* Ruprecht (Phaeophyta: Laminariales) on feeding rate and growth in the red abalone *Haliotis rufescens* Swainson, *J. Exp. Mar. Biol. Ecol.*, 155, 263, 1992.

298. Ayres, M. P., Clausen, T. P., Maclean, S. F., Jr., Redman, A. A. M., and Reichardt, P. B., Diversity of structure and antiherbivore activity in condensed tannins, *Ecology*, 78, 1696, 1997.

299. Gaines, S. D., Herbivory and between-habitat diversity: the differential effectiveness of defenses in a marine plant, *Ecology*, 66, 473, 1985.

300. Hay, M. E., Paul, V. J., Lewis, S. M., Gustafson, K., Tucker, J., and Trindell, R. N., Can tropical seaweeds reduce herbivory by growing at night?: diel patterns of growth, nitrogen content, herbivory, and chemical versus morphological defenses, *Oecologia*, 75, 233, 1988.

301. Phillips, D. W. and Towers, G. H. N., Chemical ecology of red algal bromophenols. I. Temporal, interpopulational and within-thallus measurements of lanosol levels in *Rhodomela larix* (Turner) C. Agardh, *J. Exp. Mar. Biol. Ecol.*, 58, 285, 1982.

302. Carlson, D. J., Lubchenco, J., Sparrow, M. A., and Trowbridge, C. D., Fine-scale variability of lanosol and its disulfate ester in the temperate red algae, *J. Chem. Ecol.*, 15, 1321, 1989.

303. De Nys, R., Steinberg, P. D., Rogers, C. N., Charlton, T. S., and Duncan, M. W., Quantitative variation of secondary metabolites in the sea hare *Aplysia parvula* and its host plant, *Delisea pulchra*, *Mar. Ecol. Prog. Ser.*, 130, 135, 1996.

304. Dworjanyn, S. A., De Nys, R., and Steinberg, P. D., Localisation and surface quantification of secondary metabolites in the red alga *Delisea pulchra*, *Mar. Biol.*, 133, 727, 1999.

305. Young, D. N., Howard, B. M., and Fenical, W., Subcellular localization of brominated secondary metabolites in the red alga *Laurencia snyderae*, *J. Phycol.*, 16, 182, 1980.

306. Cronin, G. and Hay, M. E., Within-plant variation in seaweed palatability and chemical defenses: optimal defense theory versus the growth-differentiation balance hypothesis, *Oecologia*, 105, 361, 1996.

307. Puglisi, M. P. and Paul, V. J., Intraspecific variation in secondary metabolite production in the red alga *Portieria hornemannii*: monoterpene concentrations are not influenced by nitrogen and phosphorus enrichment, *Mar. Biol.*, 128, 161, 1997.

308. Matlock, D. B., Ginsburg, D. W., and Paul, V. J., Spatial variability in secondary metabolite production by the tropical red alga *Portieria hornemannii*, *Hydrobiologia*, 398/399, 267, 1999.

309. Van Alstyne, K. L., Ehlig, J. M., and Whitman, S. L., Feeding preferences for juvenile and adult algae depend on algal stage and herbivore species, *Mar. Ecol. Prog. Ser.*, 180, 179, 1999.

310. Cronin, G. and Hay, M. E., Chemical defenses, protein content, susceptibility to herbivory of diploid vs. haploid stages of the isomorphic brown alga *Dictyota ciliolata* (Phaeophyta), *Bot. Mar.*, 39, 395, 1996.

311. Paul, V. J., Littler, M. M, Littler, D. S., and Fenical, W., Evidence for chemical defense in the tropical green alga *Caulerpa ashmeadii* (Caulerpaceae: Chlorophyta): isolation of new bioactive sesquiterpenoids, *J. Chem. Ecol.*, 13, 1171, 1987.

312. Gerwick, W. H., Fenical, W., and Norris, J. N., Chemical variation in the tropical seaweed *Stypopodium zonale* (Dictyotaceae), *Phytochemistry*, 24, 1279, 1985.

313. Masuda, M., Abe, T., Sato, S., Suzuki, T., and Suzuki, M., Diversity of halogenated secondary metabolites in the red alga *Laurencia nipponica* (Rhodomelaceae, Ceramiales), *J. Phycol.*, 33, 196, 1997.

314. Renaud, P. E., Hay, M. E., and Schmitt, T. M., Interactions of plant stress and herbivory: intraspecific variation in the susceptibility of a palatable versus an unpalatable seaweed to sea urchin grazing, *Oecologia*, 82, 217, 1990.

315. Bryant, J. P., Chapin, F. S., and Klein, D. R., Carbon/nutrient balance of boreal plants in relation to vertebrate herbivory, *Oikos*, 40, 357, 1983.

316. Cronin, G. and Hay, M. E., Effects of light and nutrient availability on the growth, secondary chemistry, and resistance to herbivory of two brown seaweeds, *Oikos*, 77, 93, 1996.

317. Van Alstyne, K. L., Comparison of three methods for quantifying brown algal polyphenolic compounds., *J. Chem. Ecol.*, 21, 45, 1995.

318. Stern, J. L., Hagerman, A. E., Steinberg, P., Winter, F. C., and Estes, J. A., A new assay for quantifying brown algal phlorotannins and comparisons to previous methods, *J. Chem. Ecol.*, 22, 1273, 1996.

319. Ilvessalo, H. and Tuomi, J., Nutrient availability and accumulation of phenolic compounds in the brown alga *Fucus vesiculosus*, *Mar. Biol.*, 101, 115, 1989.

320. Yates, J. L. and Peckol, P., Effects of nutrient availability and herbivory on polyphenolics in the seaweed *Fucus vesiculosus*, *Ecology*, 74, 1757, 1993.

321. Arnold, T. M., Tanner, C. E., and Hatch, W. I., Phenotypic variation in polyphenolic content of the tropical brown alga *Lobophora variegata* as a function of nitrogen availability, *Mar. Ecol. Prog. Ser.*, 123, 177, 1995.

322. Peckol, P., Krane, J. M., and Yates, J. L., Interactive effects of inducible defense and resource availability on phlorotannins in the North Atlantic brown algae *Fucus vesiculosus*, *Mar. Ecol. Prog. Ser.*, 138, 209, 1996.

323. Poore, A. G. B., Selective herbivory by amphipods inhabiting the brown alga *Zonaria angustata*, *Mar. Ecol. Prog. Ser.*, 107, 113, 1994.

324. Steinberg, P. D., Algal chemical defenses against herbivores: allocation of phenolic compounds in the kelp *Alaria marginata*, *Science*, 223, 405, 1984.

325. Tugwell, S. and Branch, G. M., Differential polyphenolic distribution among tissues in the kelps *Ecklonia maxima*, *Laminaria pallida* and *Macrocystis angustifolia* in relation to plant-defense theory, *J. Exp. Mar. Biol. Ecol.*, 129, 219, 1989.

326. Tuomi, J., Ilvessalo, H., Niemela, P., Siren, S., and Jormalainen, V., Within-plant variation in phenolic content and toughness of the brown alga *Fucus vesiculosus* L., *Bot. Mar.*, 32, 505, 1989.

327. Van Alstyne, K. L., Adventitous branching as a herbivore-induced defense in the intertidal brown alga *Fucus distichus*, *Mar. Ecol. Prog. Ser.*, 56, 169, 1989.

328. Van Alstyne, K. L., McCarthy, J. J., III, Hustead, C. L., and Kearns, L. J., Phlorotannin allocation among tissues of Northeastern Pacific kelps and rockweeds, *J. Phycol.*, 35, 483, 1999.

329. Pavia, H. and Aberg, P., Spatial variation in polyphenolic content of *Ascophyllum nodosum* (Fucales, Phaeophyta), *Hydrobiologia*, 326/327, 199, 1996.

330. Van Alstyne, K. L., McCarthy, J. J., III, Hustead, C. L., and Duggins, D. O., Geographic variation in polyphenolic levels of Northeastern Pacific kelps and rockweeds, *Mar. Biol.*, 133, 371, 1999.

331. Steinberg, P. D., Seasonal variation in the relationships between growth rate and phlorotannin production in the kelp *Ecklonia radiata*, *Oecologia*, 102, 169, 1995.

332. Van Alstyne, K. L., Herbivore grazing increases polyphenolic defenses in the intertidal brown alga *Fucus distichus*, *Ecology*, 69, 655, 1988.

333. Peckol, P. and Yates, J. L, Inducible phlorotannin levels in brown algae from backreef sites, *Proc. 8th Int. Coral Reef Sym.*, 2, 1259, 1997.

334. Hammerstrom, K., Dethier, M. N., and Duggins, D. O., Rapid phlorotannin induction and relaxation in five Washington kelps, *Mar. Ecol. Prog. Ser.*, 165, 293, 1998.

335. Steinberg, P. D., Lack of short-term induction of phlorotannins in the Australasian brown algae *Ecklonia radiata* and *Sargassum vestitum*, *Mar. Ecol. Prog. Ser.*, 112, 129, 1994.

336. Pavia, H. and Toth, G., Inducible chemical resistance to herbivory in the brown seaweed *Ascophyllum nodosum*, *Ecology*, 81, 3212, 2000.

337. Kerr, J. Q. and Paul, V. J., Animal-plant defense association: the soft coral *Sinularia* sp. (Cnidaria, Alcyonacea) protects *Halimeda* spp. from herbivory, *J. Exp. Mar. Biol. Ecol.*, 186, 183, 1995.

338. Lewis, S. M. and Kensley, B., Notes on the ecology and behavior of *Pseudamphithoides incurvaria* (Just) (Crustacea, Amphipoda, Amphithoidae), *J. Nat. Hist.*, 16, 267, 1982.

339. Sotka, E. E., Hay, M. E., and Thomas, J. D., Host-plant specialization by a non-herbivorous amphipod: advantages for the amphipod and costs for the seaweed, *Oecologia*, 118, 471, 1999.

340. Barnard, J. L. and Karaman, G. S., The families and genera of marine gammaridean Amphipoda (except marine gammaroids), *Rec. Aust. Mus.*, Suppl. 13, 1991.

341. Vet, L. E. M., From chemical to population ecology: infochemical use in an evolutionary context, *J. Chem. Ecol.*, 25, 31, 1996.

342. Zimmer-Faust, R. K., Towards a theory of optimal chemoreception, *Biol. Bull.*, 172, 10, 1987.

343. Weissburg, M. J. and Zimmer-Faust, R. K., Life and death in moving fluids: hydrodynamic effects on chemosensory-mediated predation, *Ecology*, 74, 1428, 1993.

344. Zimmer, R. K., Commins, J. E., and Browne, K. A., Regulatory effects of environmental chemical signals on search behavior and foraging success, *Ecology*, 80, 1432, 1999.

345. Sweatman, H., Field evidence that settling coral reef fish larvae detect resident fishes using dissolved chemical cues, *J. Exp. Mar. Bol. Ecol.*, 14, 163, 1988.

346. Pawlik, J. R., Chemical ecology of the settlement of benthic marine invertebrates, in *Oceanogr. Mar. Biol. Annu. Rev.*, 30, 273, 1992.

347. Rittschof, D., Tsai, D. W., Massey, P. G., Blanco, L., Keuber, G. L., Jr., and Haas, R. J., Jr., Chemical mediation of behavior in hermit crabs: alarm and aggregation cues, *J. Chem. Ecol.*, 18, 959, 1992.

348. Smith, R. J. F., Alarm signals in fishes, *Rev. Fish. Biol. Fishes*, 2, 33, 1992.

349. Tamburri, M. N., Finelli, C. M., Wethey, D. S., and Zimmer-Faust, R. K., Chemical induction of larval settlement behavior in flow, *Biol. Bull.*, 191, 367, 1996.

350. Zimmer, R. K. and Butman, C. A., Chemical signaling processes in the marine environment, *Biol. Bull.*, 198, 168, 2000.

351. Valiela, I., Koumjian, L., Swain, T., Teal, J. M., and Hobbie, J. E., Cinnamic acid inhibition of detritus feeding, *Nature*, 280, 55, 1979.

352. Done, T. J., Phase shifts in coral reef communities and their ecological significance, *Hydrobiologia*, 247, 121, 1992.

353. Hughes, T. P., Catastrophes, phase shifts, and large-scale degradation of a Caribbean coral reef, *Science*, 265, 1547, 1994.

354. Valiela, I., McClelland, J., Hauxwell, J., Behr, P. J., Hersh, D., and Foreman, K., Macroalgal blooms in shallow estuaries: controls and ecophysiological and ecosystem consequences, *Limnol. Oceanogr.*, 42, 1105, 1997.

355. Littler, M. M. and Littler, D. S., Models of tropical reef biogenesis: the contribution of algae, *Prog. Phycol. Res.*, 3, 323, 1984.

356. Lapointe, B. E., Nutrient thresholds for bottom-up control of macroalgal blooms on coral reefs in Jamaica and southeast Florida, *Limnol. Oceanogr.*, 42, 1119, 1997.

357. McCook, L. J., Macroalgae, nutrients, and phase shifts on coral reefs: scientific issues and management consequences for the Great Barrier Reef, *Coral Reefs*, 18, 357, 1999.

358. Larned, S. T., Nitrogen- versus phosphorous-limited growth and sources of nutrients for coral reef macroalgae, *Mar. Biol.*, 132, 409, 1998.

359. Schaffelke, B. and Klumpp, D. W., Short-term nutrient pluses enhance growth and photosynthesis of the coral reef macroalga *Sargassum baccularia*, *Mar. Ecol. Prog. Ser.*, 170, 95, 1998.

360. Carpenter, R. C., Invertebrate predators and grazers, in *Life and Death of Coral Reefs*, Birkeland, C., Ed., Chapman & Hall, New York, 1997, chap. 9.

361. Hixon, M. A., Effects of reef fishes on corals and algae, in *Life and Death of Coral Reefs*, Birkeland, C., Ed., Chapman & Hall, New York, 1997, chap. 10.

362. Rogers, C. S., Garrison, V., and Grober-Dunsmore, R., A fishy story about hurricanes and herbivory: seven years of research on a reef in St. John, U.S. Virgin Islands., *Proc. 8th Int. Coral Reef Sym.*, 1, 555, 1997.

363. Woodley, J. D., Sea-urchins exert top-down control of macroalgae on Jamaican coral reefs (1), *Coral Reefs*, 18, 192, 1999.

364. McClanahan, T. R., Aronson, R. B., Precht, W. F., and Muthiga, N. A., Fleshy algae dominate remote coral reefs of Belize, *Coral Reefs*, 18, 61, 1999.

365. Lirman, D. and Biber, P., Seasonal dynamics of macroalgal communities of the Northern Florida Reef Tract, *Bot. Mar.*, 43, 305, 2000.

366. Steneck, R. S. and Dethier, M. N., A functional group approach to the structure of algal-dominated communities, *Oikos*, 69, 476, 1994.

367. Tsuda, R. T. and Kami, H. T., Algal succession on artificial reefs in a marine lagoon environment in Guam, *J. Phycol.*, 9, 260, 1973.

368. Borowitzka, M. A., Larkum, A. W. D., and Borowitzka, L. J., A preliminary study of algal turf communities of a shallow coral reef lagoon using an artificial substratum, *Aq. Bot.*, 5, 365, 1978.

369. Larkum, A. W. D., High rates of nitrogen fixation on coral skeletons after predation by the crown of thorns starfish *Acanthaster planci*, *Mar. Biol.*, 97, 503, 1988.

370. Littler, M. M. and Littler, D. S., Disease-induced mass mortality of crustose coralline algae on coral reefs provides rationale for the conservation of herbivorous fish stocks, *Proc. 8th Int. Coral Reef Sym.*, 1, 719, 1997.

371. Cowell, B. C. and Botts, P. S., Factors influencing the distribution, abundance and growth of *Lyngbya wollei* in central Florida, *Aq. Bot.*, 49, 1, 1994.

372. Meinesz, A. and Hesse, B., Introduction et invasion de l'algue tropicale *Caulerpa taxifolia* en Mediterranee nord-occidentale, *Oceanologica acta*, 14, 415, 1991.

373. Meinesz, A., de Vaugelas, J., Hesse, B., and Mari, X., Spread of the introduced tropical green alga *Caulerpa taxifolia* in northern Mediterranean waters, *J. Appl. Phycol.*, 5, 141, 1993.

374. Guerriero, A., Meinesz, A., D'Ambrosia, M., and Pietra, F., Isolation of toxic and potentially toxic sesqui- and monoterpenes from the tropical green seaweed *Caulerpa taxifolia* which has invaded the Region of Cap Martin and Monaco, *Helv. Chim. Acta*, 75, 689, 1992.

375. Guerriero, A., Marchetti, F., D'Ambrosia, M., Senesi, S., Dini, F., and Pietra, F., New ecotoxicologically and biogenetically relevant terpenes of the tropical green seaweed *Caulerpa taxifolia* which is invading the Mediterranean, *Helv. Chim. Acta*, 76, 855, 1993.

376. Boudouresque, C. F., Lemee, R., Mari, X., Meinesz, A., The invasive alga *Caulerpa taxifolia* is not a suitable diet for the sea urchin *Paracentrotus lividus*, *Aq. Bot.*, 53, 245, 1996.

377. Uchimura, M., Sandeauz, R., and Larroque, C., The enzymatic detoxifying system of a native Mediterranean scorpio fish is affected by *Caulerpa taxifolia* in its environment, *Environ. Sci. Technol.*, 33, 1671, 1999.

378. Francour, P., Harmelin-Vivien, M., Harmelin, J. G., and Duclerc, J., Impact of *Caulerpa taxifolia* colonization on the littoral ichthyofauna of North-western Mediterranean: preliminary results, *Hydrobiologia*, 300–301, 345, 1995.

379. Bellan-Santini, D., Arnaud, P. M., Bellan, G., and Verlaque, M., The influence of the introduced tropical alga *Caulerpa taxifolia* on the biodiversity of the Mediterranean marine biota, *J. Mar. Biol. Assoc. UK*, 76, 235, 1996.

380. Fox, L. R., Defense and dynamics in plant–herbivore systems, *Am. Zool.*, 21, 853, 1981.

381. Rhoades, D. F. and Cates, R. G., Toward a general theory of plant antiherbivore chemistry, *Rec. Adv. Phytochem.*, 10, 168, 1976.

382. Rhoades, D. F., Evolution of plant chemical defense against herbivores, in *Herbivores: Their Interaction with Secondary Plant Metabolites*, Rosenthal, G. A. and Janzen, D. H., Eds., Academic Press, New York, 1979, 3.

383. Denno, R. F. and McClure, M. S., Eds., *Variable plants and herbivores in natural and managed systems*, Academic Press, New York, 1983.

384. Bryant, J. P., Kuropat, P. J., Cooper, S. M., Frisby, K, and Owen-Smith, N., Resource availability hypothesis of plant antiherbivore defence tested in a South African savanna ecosystem, *Nature*, 340, 227, 1989.

385. Hay, M. E. and Fenical, W., Chemical ecology and marine biodiversity: insights and products from the sea, *Oceanogr.*, 9, 10, 1996.

7 Secondary Metabolites from Antarctic Marine Organisms and Their Ecological Implications

Charles D. Amsler, Katrin B. Iken, James B. McClintock, and Bill J. Baker*

CONTENTS

I. INTRODUCTION

The Antarctic continent evokes images of extreme environments, rugged explorers, unusual animal life, and, especially for scientists, discovery. Science has, in fact, been the driving force for much of the effort to explore this harshest of continents. Perhaps it was the efforts of Robert Scott and members of his expedition, who studied the wintering habits of the Emperor Penguin, recounted

---

* Corresponding author.

in 1922 by Cherry-Garrard in his epic novel *The Worst Journey in the World*,[1] that set the tone for future research endeavors. First sighted less than two centuries ago and fully mapped only in the latter half of the twentieth century, there still remains much to be discovered in Antarctica.

Today, Antarctic research is governed largely by the Antarctic Treaty, which is an agreement stemming from the efforts of the International Geophysical Year (IGY) that designates the continent to be used primarily for research purposes.[2] Forty-four countries are signatories to the Antarctic Treaty or have acceded to its provisions and currently conduct research in the Antarctic, many of them operating year-round stations. The largest of these laboratory facilities is McMurdo Station (US) in the Ross Sea. Most research stations are located along the coastline of the continent and can be supported by ship, but some, such as the Admundson-Scott base at the South Pole, are hundreds of miles inland and can only be supported by fixed-wing aircraft. A large part of the Antarctic research effort also takes place on research vessels, most of which are highly capable icebreakers.

The availability of Antarctic marine laboratories and research vessels has provided the facilities and support to undertake study of the unique environment. The breakup of Gondwanaland in the Cretaceous Period resulted in movement of the Antarctic continent toward the South Pole and facilitated the development of a circumpolar current, effectively isolating the continent and its coastal marine biota. The Antarctic climate remained temperate to subtropical until 22 million years ago.[3] The biota, therefore, has its origin in warmer climates, and its marine biota is generally considered very old.[4,5] According to Dayton et al.[3] and Brey et al.,[6] the shallow-water Antarctic marine fauna is derived from a relict autochthonous fauna, a eurybathic fauna from deeper water, and cool-temperate species, mostly from South America. The isolation of Antarctic waters leads to an extremely high proportion of endemism.[4,7] Because of this unique biogeographical evolution, studies of Antarctic species and their ecology provide opportunities to study phenomena which are very difficult or impossible to study in other ecosystems.

The study of interactions of Antarctic marine invertebrates that are mediated by secondary metabolites is one such area of opportunity. Early studies in other geographical regions examining the toxicity of marine invertebrates (primarily sponges and holothurians) concluded that the incidence of defensive chemicals was much higher at tropical than temperate latitudes.[8–11] These early studies indicated that sessile and sluggish marine invertebrates commonly contain defensive chemicals, a phenomenon which has now been well established in marine organisms[12–15] as well as terrestrial plants.[16–19] Tropical–temperate comparisons led to a latitudinal hypothesis, suggesting an inverse correlation between the incidence of chemical defense in marine invertebrates and latitude.[8] This "latitudinal gradient" in chemical defense is thought to be driven by predation pressure, with tropical sessile and sluggish marine invertebrates under proportionately higher levels of predation by browsing fish.[8,20] Therefore, in an evolutionary context, selective pressure for defensive compounds would be expected to be highest at low latitudes and lowest at high latitudes. Nonetheless, recent toxicity[21,22] and feeding deterrence[23,24] bioassays conducted with sessile or sluggish Antarctic marine invertebrates have detected an incidence of bioactivity commensurate with temperate, and perhaps even tropical, marine environments.[25]

In tropical marine benthic systems, which are thought to be predominated by biological interactions,[26,27] there is an abundance of marine invertebrates that use chemicals as a means of defense from predators, inhibiting settling and fouling invertebrates, and preventing overgrowth.[8,11,27–30] Therefore, it is not unexpected that in the Antarctic marine benthos, where biological interactions also seem to dominate,[31,32] a large number of chemically mediated interactions also occur. Indeed, recent laboratory studies, in many cases using sympatric bioassay species, have indicated that chemical activity occurs in Antarctic sponges,[21,23,24,33–42] bryozoans,[43] coelenterates,[44,45] nemerteans,[46,47] opisthobranchs,[44,48–50] prosobranchs,[51] pteropods,[52–54] embryonic and adult echinoderms,[55–59] tunicates,[60] macroalgae,[61–66] and brachiopods.[67] It is likely that bioactive compounds in Antarctic species serve a variety of functions and are important in structuring interactions at the individual, population, and community levels.[68,69]

The following review is a comprehensive account of chemical investigations of Antarctic species with an emphasis on known ecological roles of specific secondary metabolites. It differs from prior accounts of Antarctic chemical ecology[68,69] in that ecological roles mediated by secondary metabolites which have not been isolated and characterized are not treated. It is organized in a partially redundant fashion so that characterized secondary metabolites are presented from a phylogenetic point of view and their known ecological activity discussed. Following the phylogenetic presentation are two sections that treat ecological functions, referring back to specific secondary metabolites which are known to mediate the function. This redundancy allows a broader picture of our knowledge of Antarctic chemical ecology to develop, highlights the advances that have been made in recent years, and sets the stage for establishing priorities for future endeavors.

II. PHYLOGENETIC DISTRIBUTION OF MARINE NATURAL PRODUCTS IN ANTARCTIC ORGANISMS

The community of the shallow continental shelf waters of Antarctica is known for its high diversity and abundance of many large species.[21,31,70–74] Alcyonians, actinians, bryozoans, and hydroids are common sessile marine invertebrates associated with the dominant sponge community, while sluggish nemerteans, opisthobranchs, and echinoderms are abundant. [5,31] Although annual ice scour and anchor ice prevent the establishment of a community dominated by sessile macroinvertebrates above 30 m depth around parts of the continent,[75,76] below this depth the diversity and biomass of sponges and other sessile marine invertebrates are high.[77] In some places, the sponge spicule mat may even attain a depth of several meters, [72,78] while sponges in some locales cover more than 55% of the benthic surface area.[31,32]

Below the zone of ice scour, the physical environment of the Antarctic marine community is very stable. Such stability has, in part, led to the conclusion of Dayton et al.[31] that the Antarctic benthic community is structured in large part by biological factors such as predation and competition, that is, the community is "biologically accommodated." For example, Dayton et al.[31] found that the sea star *Perknaster fuscus* preys predominately on the fast-growing and potentially space-dominating sponge *Mycale acerata*. This predation on *M. acerata* prevents competitive exclusion of more slow-growing sponge species.

To date, there are nearly 200 different secondary metabolites described from Antarctic organisms, excluding fatty acids and sterols, many of which are not found in congeners from temperate and tropical regions. The distribution of known secondary metabolites from these phyla (Table 7.1) differs considerably from that of the general marine realm (see Table 1.2 in this volume[79]). For example, while sponges are one of the major targets of chemical investigations in both Antarctica (36% of articles in MarineLit[80] that studied organisms from the Antarctic geographic region focused on sponges) and elsewhere (29% of all MarineLit articles), they comprise only 28 of 130 Antarctic invertebrate-derived compounds (22%), while 45% of non-Antarctic invertebrate secondary metabolites are derived from sponges.[79] Echinoderms, on the other hand, comprise 61% of described Antarctic invertebrate compounds, but only 8% of non-Antarctic invertebrate-derived compounds. Cnidaria are the second richest source of non-Antarctic invertebrate compounds (26%), but Antarctic Cnidaria are very poor sources so far (<1%). Whether this reflects how little is known about secondary metabolism in Antarctic macrofauna or whether it notes an evolutionary trend toward different defensive strategies can only be answered by continued study.

A. Sponges

Sponges are the dominant macroinvertebrate found on the Antarctic benthos.[31,32,70,71] Dayton et al.[31] reported that sponges occupy 55% of the shallow-water McMurdo Sound benthos, and Burton[70] drew an analogy between the Antarctic benthos and the benthos of the Cretaceous era, which was

Table 7.1 Summary of Phylogenetic Distribution of Secondary Metabolites in Antarctic Organisms

Biosynthetic pathway	Total Compounds	Isoprenoid	Acetogenin	Amino Acid	Shikimate	Nucleic Acid	Carbohydrate[†]
Isolated Metabolites (%)	212*	131 (62)	46 (22)	21 (10)	8 (4)	6 (3)	0
Invertebrate Phylum	130 (61)						
Porifera	28 (13)	8 (29)	1 (4)	11 (39)	2 (8)	6 (21)	
Cnidaria	1 (<1)		1 (100)				
Arthropoda	2 (1)		2 (100)				
Mollusca	14 (7)	9 (64)	4 (29)	1 (7)			
Echinodermata	79 (37)	78 (99)		1 (1)			
Ectoprocta[†]	0						
Chordata[†]	0						
Polychaeta	6 (3)				6 (100)		
Algal Class	72 (34)						
Rhodophyceae	47 (22)	28 (60)	19 (40)				
Phaeophyceae	18 (9)	6 (33)	12 (66)				
Ulvophyceae	4 (2)		4 (100)				
Chrysomerophyceae	2 (1)		2 (100)				
Microorganisms	10 (5)	2 (20)		8 (80)			

*Some duplication is reflected in this number since it tallies the total number of compounds reported from each phylum/class; the same compound found in two different organisms, therefore, gets counted twice.
[†]Reported for comparison to Table 1.2 in this volume.

the period during which the Porifera was most abundant and widespread. Although Gutt and Koltun[81] have more recently demonstrated that sponge distribution can be patchy, there can be little doubt as to the central position the Porifera play in the ecology of the Antarctic ecosystem.[3,31]

Besides the high biomass of sponges on the Antarctic benthos, the well-documented ability of sponges to elaborate secondary metabolites has led researchers to focus investigations of Antarctic natural products on this phylum. One of the first such studies was an investigation of *Dendrilla membranosa*. This bright yellow sponge is one of several conspicuous sponges on the McMurdo Sound benthos and shallow-water benthos along the Antarctic Peninsula, and yet it is devoid of spicule armament. Dayton et al.[31] suggested, prior to any chemical investigations of Antarctic invertebrates, that *D. membranosa* was likely to be chemically defended since it avoids predation despite its lack of physical defense. Molinski and Faulkner characterized membranolide (Structure 7.1) and 9,11-dihydrogracilin (Structure 7.2), and suggested these compounds may act as defensive metabolites.[82] The relative stereochemistry of 9,11-dihydrogracilin has been established by x-ray analysis of the C-8 ketone obtained by ozonolysis of the natural product.[83] Other *D. membranosa* terpenes include dendrillin (Structure 7.3)[37] and dendrinolide (Structure 7.4),[84] reported from different collections of the sponge. Two alkaloids have been isolated from *D. membranosa*; picolinic acid (Structure 7.5)[37] and a quinoline pigment (Structure 7.6)[85] are likely responsible for much of the anti-predator and antibiotic activity reported for the sponge.[31,37,82] The terpenes found in this sponge are not unlike those found in temperate and tropical sponges: 9,11-dihydrogracilin is the saturated analogue of gracilin, from the Mediterranean sponge *Spongionella gracilis*,[86] and membranolide is closely related to aplysulphurin from the Australian sponge *Aplysilla sulphurea*.[87] Stereochemical analysis of membranolide has secured the relative stereochemistry,[88] and a chemical synthesis has been carried out.[89]

Other terpenes from Antarctic sponges display unusual chemical features. Originally found in *Suberites* sp. from the King George Island[90] and subsequently in McMurdo Sound[42] collections, suberitenones A (Structure 7.7) and B (Structure 7.8) are sesterterpenes with a cyclohexenone substituent on a phenanthrene tricyclic skeleton. These sesterterpenes have defensive properties toward a major Antarctic spongivore, the sea star *Perknaster fuscus*.[42] The sponge *Lissodendoryx flabellata* from McMurdo Sound elaborates the antiproliferative cembrane flabellatene (Structure 7.9).[91] This is the first report of a cembrane diterpene from a sponge; cembranes are usually characteristic of soft corals.[92] An unidentified Antarctic sponge collected near the Western MacRobertson Shelf yielded an antibiotic sesquiterpene alcohol (Structure 7.10).[93]

Discorhabdin and related pigments have been isolated from temperate and tropical sponges of the genera *Latrunculia*, *Prianos*, *Zyzzya*, and *Batzella*;[94] discorhabdin C (Structure 7.11)[95] and G (Structure 7.12)[38] were recently found in *Latrunculia apicalis* collected from McMurdo Sound.[96] Discorhabdins often bear significant cytotoxicity. Discorhabdin C is perhaps the strongest sea star deterrent studied to date and displays broad spectrum antibiotic activity toward sympatric bacteria.[38,96]

One of the most striking sponges on the Antarctic benthos is the bright red sponge *Kirkpatrickia variolosa*. The pigments which bestow this bright coloration are themselves various shades of orange, yellow and purple. Among pigments characterized to date, variolins A (Structure 7.13), B (Structure 7.14), and C (Structure 7.15) have the very unusual pyridopyrrolopyrimidine ring system and represent the first aminopyrimidine marine natural products; variolin A is zwitterionic, and all three variolins are potent cytotoxins toward P388 cell line.[97,98] *K. variolosa* also elaborates resveritrol triacetate (Structure 7.16).[99] Stilbenes such as seen in Structure 7.16 are found in many plant genera, and some are considered to be phytoalexins,[100,101] and resveritrol has been shown to display cancer preventative effects.[102] The psammopemmins A–C (Structures 7.17–19), from *Psammopemma* sp., share the pyrimidine ring system of the variolins, substituted on the hydroxyindole ring, rather than the pyridopyrrolopyrimidine system found in the variolins.[103]

The Antarctic polychaete sponge, *Isodictya erinacea*, produces a number of isolable metabolites, some of which may or may not be secondary metabolites. Among the ecologically relevant natural products, the yellow pigment eribusinone (Structure 7.20) appears to be a tryptophan catabolite and behaves much as other such catabolites in regulating crustacean molting (*vide infra*).[41] Among a host of purine and nucleoside metabolites found in this sponge,[40] erinacine (Structure 7.21) is an unusual alkylated purine with no currently understood functional role. Excess inosine, uridine, 2′-deoxycytidine, 1,9-dimethylguanine, and 7-methyladenine are also found in *I. erinacea*. Ecologically relevant concentrations of *p*-hydroxybenzaldehyde from *I. erinacea* display feeding inhibitory properties toward *Perknaster fuscus*.[40]

Leucetta leptorhapsis, the Antarctic rubber sponge, so called due to its stretched appearance, produces the acetogenin rhapsamine (Structure 7.22), which bears the unusual bis(1,3-diaminoglyc-erol) group.[39] Rhapsamine has antipredatory and cytotoxic activity and is structurally related to the coriacenins, which are produced by the Mediterranean sponge *Clathria coriacea*.[104]

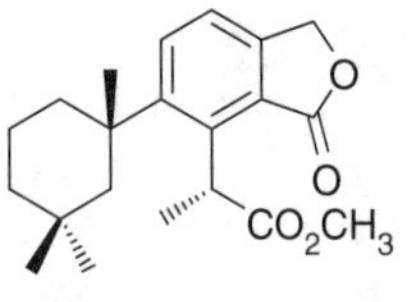

7.1

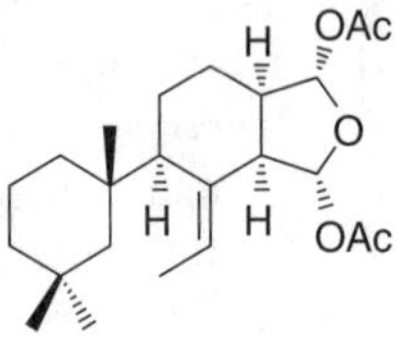

7.2

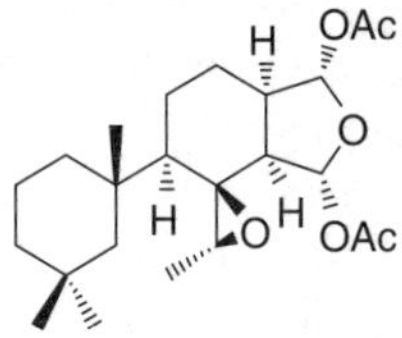

7.3

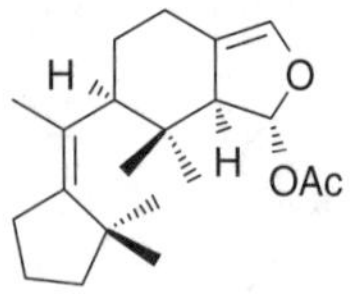

7.4

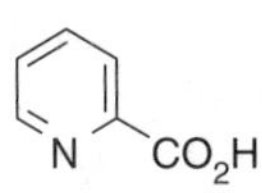

7.5

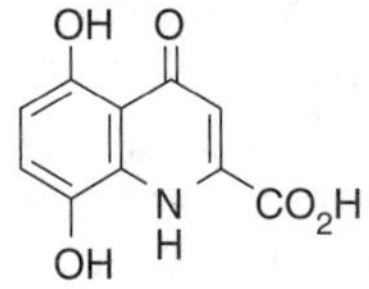

7.6

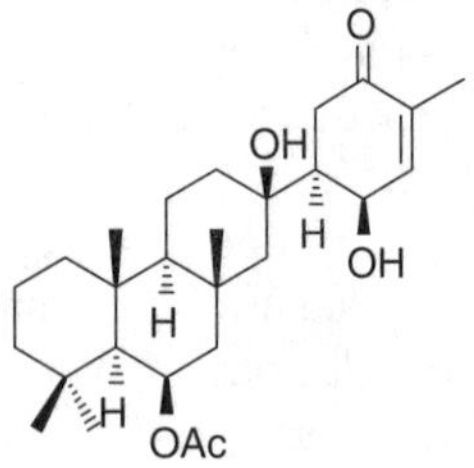

7.7

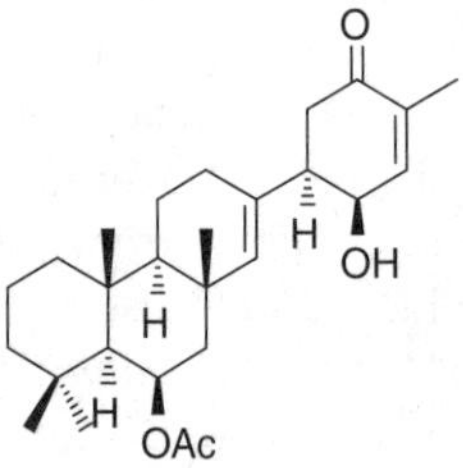

7.8

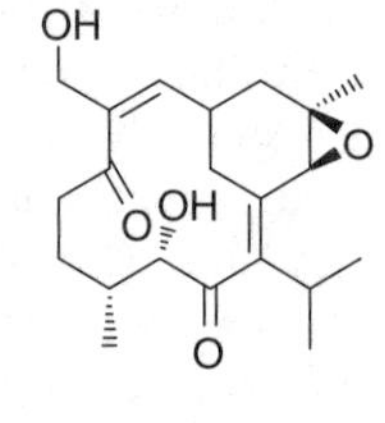

7.9

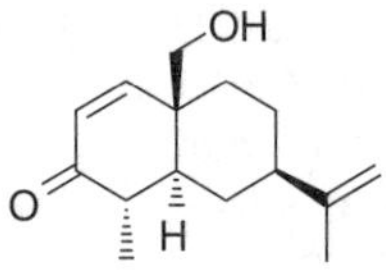

7.10

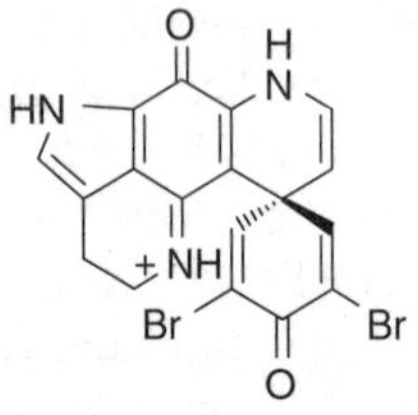

7.11

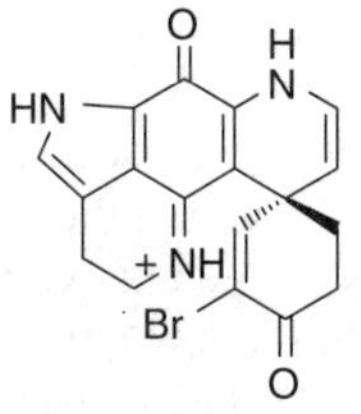

7.12

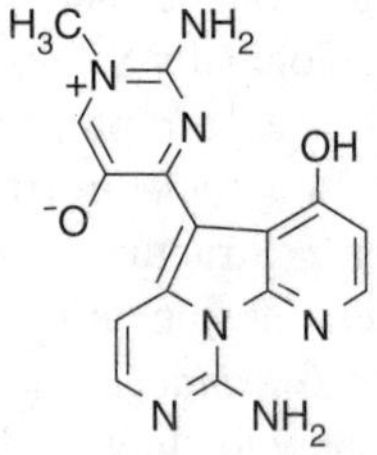

7.13

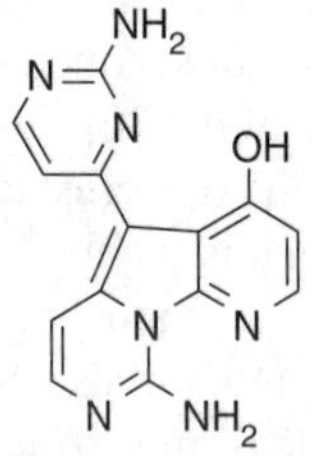

7.14

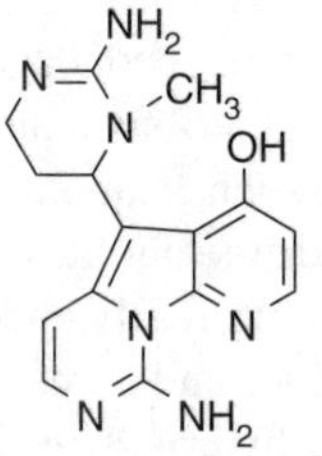

7.15

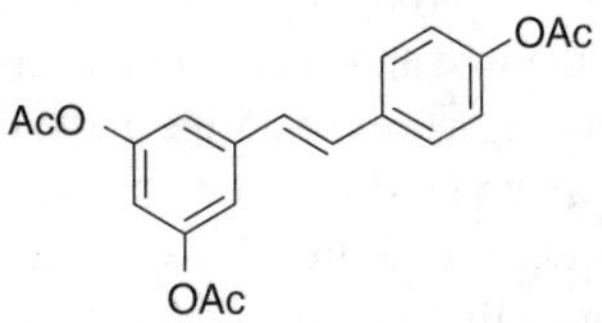

7.16

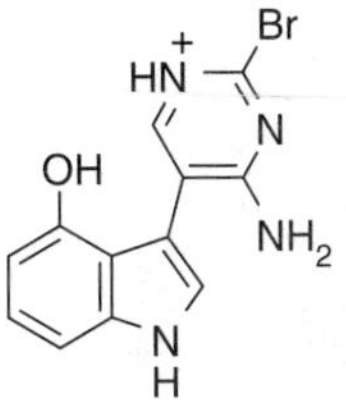

7.17

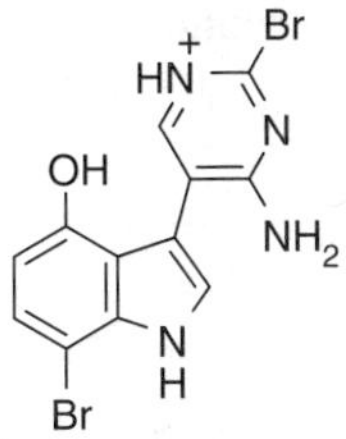

7.18

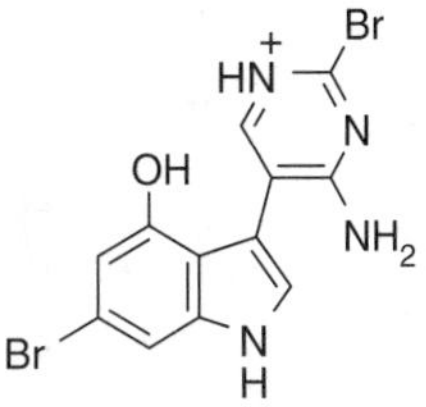

7.19

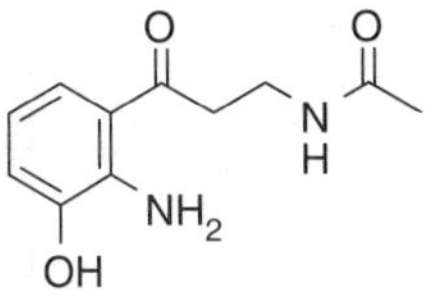

7.20

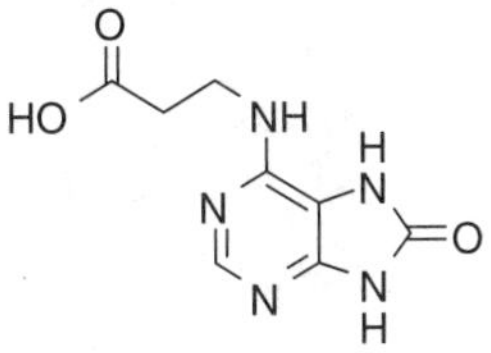

7.21

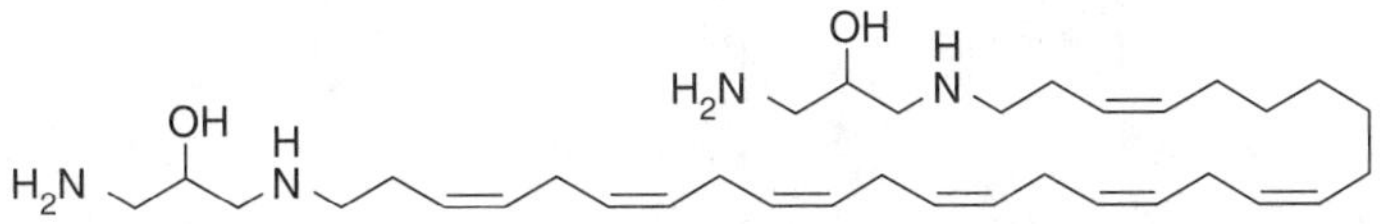

7.22

B. Algae

Macroalgae dominate shallow marine communities on hard substrates along the Antarctic Peninsula, often covering over 80% of the bottom and with standing biomass levels comparable to temperate kelp forests.[105–108] Although these Antarctic algal forests provide cover to many fish and invertebrates,[66,109] there is little evidence of significant macroalgal grazing in the field.[110–112] A number of Antarctic marine invertebrates and fish can consume macroalgae.[65,66,113] However, with the exception of amphipods,[66] one fish,[113] and two gastropods,[65,66] macroalgae have been reported only rarely in the guts of potential herbivores and then only in small amounts.[44,66,114–116] In general, macroalgae are known to defend themselves against herbivory by physical mechanisms[117–120] and by chemical means such as production and sequestration of metabolites which deter feeding.[121–125] Interestingly, we have recently shown that some Antarctic macroalgae, when used as cover, physically protect sea urchins from their major predator, large sea anemones.[62]

Many red macroalgae (Class: Rhodophyceae), especially those belonging to the genera *Laurencia*, *Plocamium*, and *Asperigopsis*, are prolific producers of halogenated terpenes and C_{15} acetogenins.[92,126] To date, 47 compounds have been reported from Antarctic red macroalgae.[127–132] In the very first investigation of the natural product chemistry of Antarctic biota, *Plocamium* sp. from the Antarctic Peninsula was reported to elaborate halogenated monoterpenes, including acyclic (Table 7.2) and carbocyclic (Structures 7.33–7.37) derivatives;[129–131] acyclic halogenated monoterpenes have recently generated considerable interest due to their often differential cytotoxicity.[133,134] With the exception of the monohalogenated derivatives from *Pantoneura plocamioides*, acyclic monoterpenes in Antarctic red macroalgae studied to date are limited in distribution to *Plocamium* sp. Oxygenated acyclic monoterpenes, such as those from *P. plocamioides*, are largely limited to southern latitudes.[131,135–140] Of Antarctic red macroalgae, *Plocamium* sp. is the sole source of carbocyclic halogenated monoterpenes (Structures 7.33–7.37),[128,138] which distinguishes this red macroalga in chemical diversity from other Antarctic red macroalgae such as *Pantoneura plocamioides*[141,142] and *Delisea fimbriata*,[127,143] which elaborate primarily furano- (Table 7.3) or pyrano-terpenes (Table 7.4) and/or γ lactone acetogenins (Table 7.5).

Table 7.2 Halogenated Monoterpenes from Antarctic
Red Macroalgae

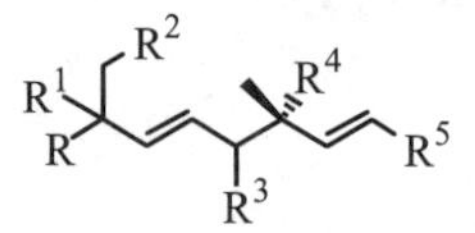

Monoterpene scaffold

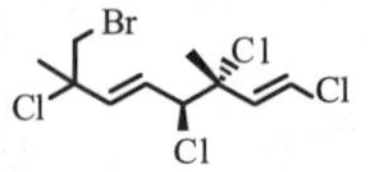

Example halogenated
monoterpene: **7.24**

	R	R¹	R²	R³	R⁴	R⁵	Citation
7.23:	Cl	CH_2Cl	Br	βCl	Cl	Br	129
7.24:	Cl	CH_3	Br	βCl	Cl	Cl	129
7.25:	Cl	CH_3	Br	βCl	βCl	Cl	129
7.26:	Cl	CH_2Cl	Br	βCl	Cl	Cl	129
7.27:	Cl	CH_2Cl	Cl	βCl	Cl	Br	138
7.28:	Cl	CH_3	Br	βCl	Cl	Br	138
7.29:	Cl	CH_2Cl	Br	βCl	OH	Cl	131, 140
7.30:	Cl	CH_2Cl	Br	αCl	OH	Cl	130
7.31:	OH	CH_3	H	αOH	OH	Br	130
7.32:	OH	CH_3	H	βOH	OH	Br	130

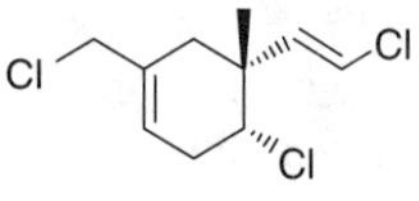

7.33 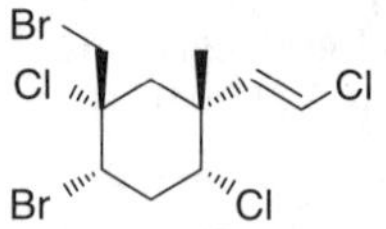   7.34 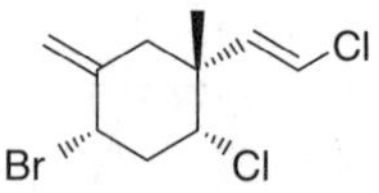   7.35 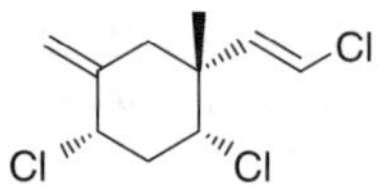   7.36 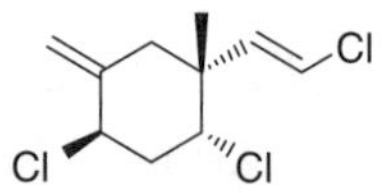   7.37

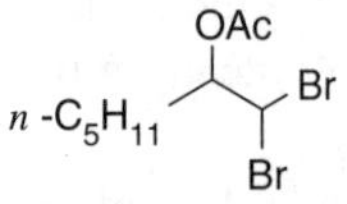

7.60 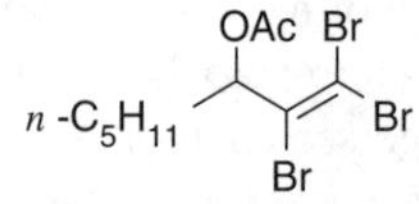   7.61 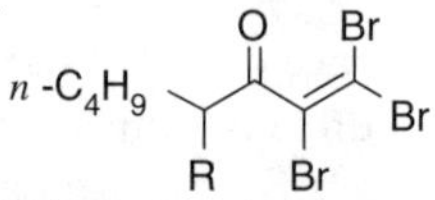   7.62 (R = H or Cl)

Antarctic Rhodophyceae, like their temperate and tropical congeners,[144–146] produce a variety of halogenated C_1 to C_8 alkanes.[147,148] The C_1 and C_2 alkyl halides are of interest for their ozone-depleting properties, though the brown macroalgae appear to be the major sources of such volatile halogenated organic compounds (VHOC).[147] The rhodophyte *Delisea fimbriata* contains, in addition to γ-lactones (Table 7.5), four linear C_7 or C_8 halogenated ketones (Structures 7.60–7.62), two of which (Structures 7.61 and 7.62) are unique to Antarctic red macroalga.[127,143]

Brown macroalgae (Class: Phaeophyceae) elaborate, in addition to the VHOC mentioned above, prenylated terpenes and C_{11} hydrocarbons.[92] Antarctic brown macroalgae have been subjected to very little chemical investigation. Seven prenylated terpenes (Structures 7.63–7.68) of the plasto-quinone family have been reported[149,150] from *Desmarestia menziesii* collected on the Antarctic Peninsula. Such prenylated terpenes have been suggested as defensive agents for brown macroalgae, though no ecological investigations of Antarctic predators have yet been undertaken.

Table 7.3 Halogenated Monoterpenes from Antarctic Red Macroalgae:
Furanoses

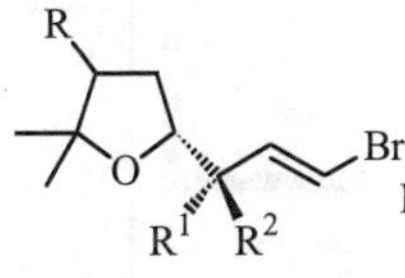

Scaffold A 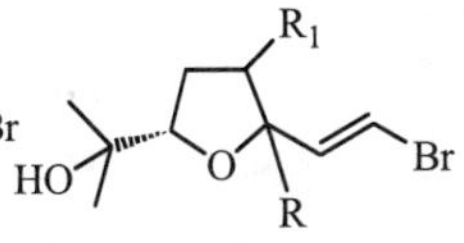   Scaffold B

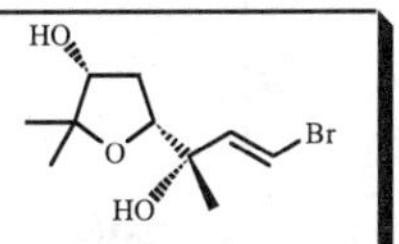

Example halogenated
furanose monoterpene: 7.38

	Scaffold	R	R^1	R^2	Citation
7.38	A	αOH	OH	CH₃	141
7.39	A	αOH	CH₃	OH	141
7.40	A	βOH	OH	CH₃	141
7.41	A	βBr	OH	CH₃	141
7.42	A	αBr	OH	CH₃	141
7.43	A	αBr	CH₃	OH	141
7.44	B	αOH	αCH₃		132
7.45	B	βOH	αCH₃		132
7.46	B	βOH	βCH₃		132

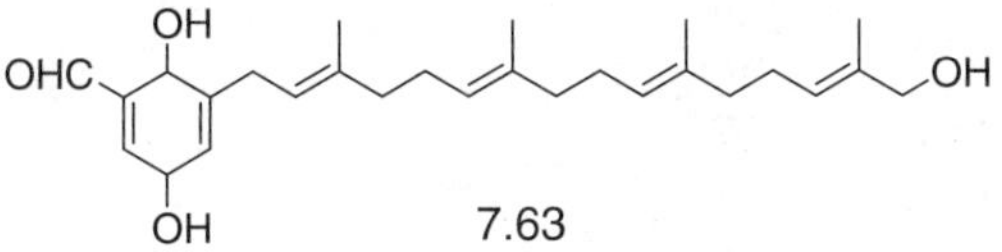

7.63

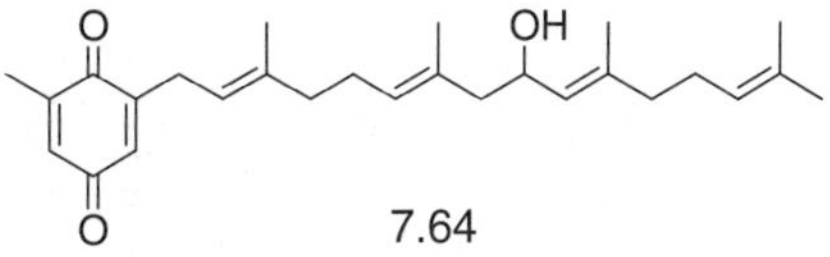

7.64

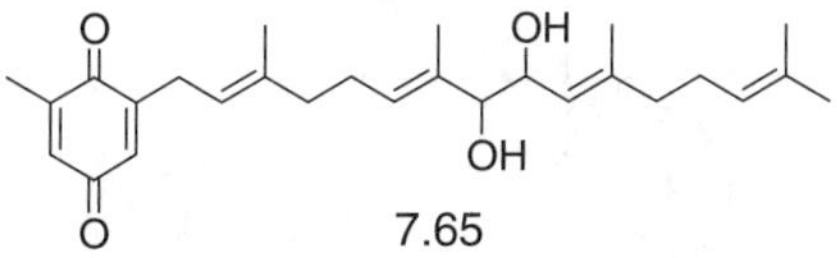

7.65

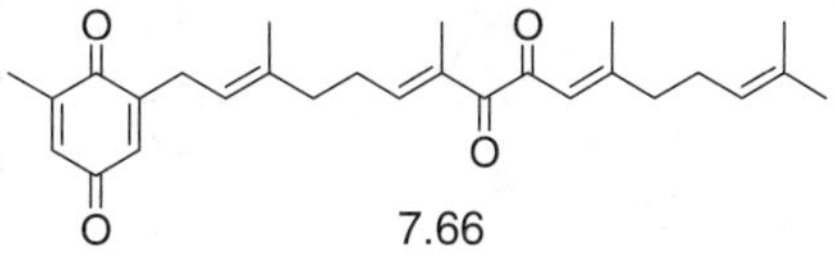

7.66

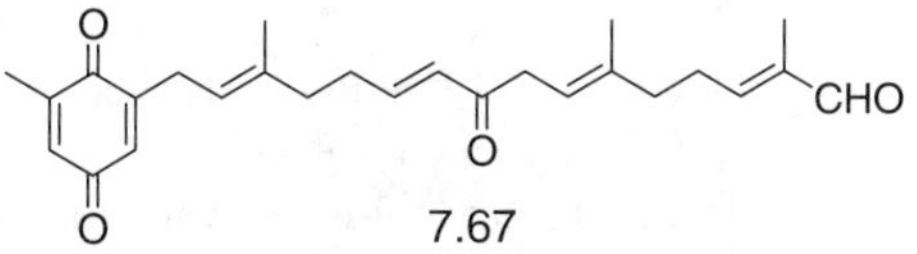

7.67

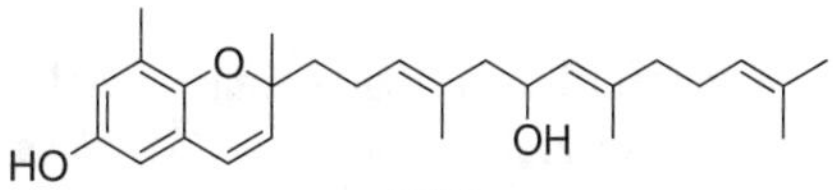

7.68

With regard to VHOC, *Desmarestia menziesii* and several other species of Antarctic brown macroalgae are significant producers of these ozone-depleting chemicals. This is particularly noteworthy since *D. menziesii* is a dominant macroalga;[106–108,110,151] a single plant (up to 10 kg weight) can produce 14.3 g of bromoform per year.[147] While VHOC are implicated as defensive in function,[152] and Laturnus et al.[147] provide evidence that some species produce their highest levels in tissues important for survival, *D. menziesii* is nonetheless a potential food source for a variety of herbivores.[66]

Table 7.4 Halogenated Monoterpenes from
Antarctic Red Macroalgae: Pyranoses

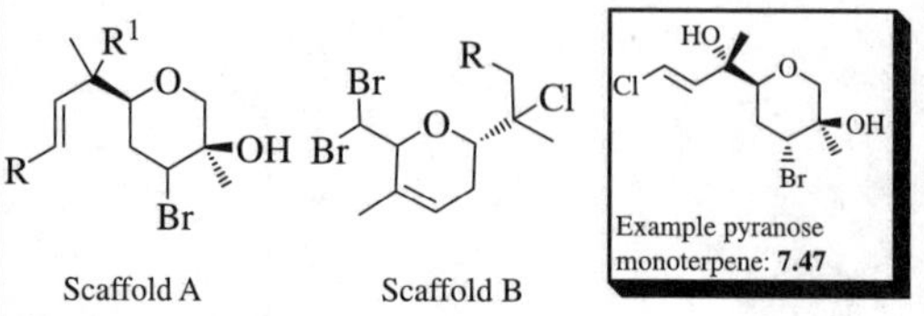

Scaffold A   Scaffold B

Example pyranose
monoterpene: **7.47**

	Scaffold	R	R^1	Other	Citation
7.47	A	Cl	αOH	6α	132
7.48	A	Cl	βOH	6α	132
7.49	A	Br	βOH	6α	132
7.50	B	H		2β	142
7.51	B	H		2α	142

Table 7.5 Halogenated
Compounds from Antarctic Red
Macroalgae: γ-lactones

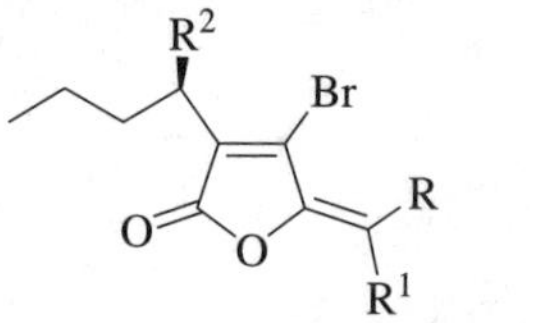

	R	R^1	R^2	Citation
7.52	H	Cl	OAc	127
7.53	H	I	OAc	127
7.54	I	H	OAc	127
7.55	Br	Br	OAc	127
7.56	H	Br	OAc	127
7.57	Br	H	OAc	127
7.58	H	Br	H	143
7.59	H	Br	H	143

Antarctic brown macroalgae are also reported to elaborate phlorotannins (polyphenolics)[65,66] which, in several temperate and tropical phaeophytes,[13] are known to have antifeedant effects. However, the chemical structure and the ecological role of phlorotannins in Antarctic brown macroalgae still remain to be established.

Terpenes are the predominant natural product in green macroalgae and diatoms and related groups.[92] With the exception of an evaluation of VHOC production, for which these algae are only minor producers,[147,148] there have been no published investigations of their secondary metabolite chemistry from Antarctic regions.

C. MOLLUSCS

The secondary metabolite chemistry of molluscs often reflects their diet since many molluscs have evolved the ability to sequester dietary metabolites for their own defense. This led to the suggestion that the sequestration of secondary metabolites from their diet has provided molluscs the protection they needed to evolve a shell-less morphology.[153–156] A number of mollusc species are capable of biosynthesizing their own defensive chemistry;[157] in Antarctica, this appears to be the dominant strategy.

The pteropod *Clione antarctica*, a pelagic mollusc, produces the polyketide pteroenone (Structure 7.69).[53] Pteroenone is distasteful to sympatric fish.[54] *C. antarctica* are subject to capture by the amphipod *Hyperiella dilatata*, which holds them in place on its dorsum and is similarly avoided by the same fish predators (see Section III.A, below).

Considerable work has been carried out on the secondary metabolite composition of the nudibranch *Austrodoris kerguelenensis* (*A. kerguelensis*, *A. mcmurdensis*). Diterpenoic acid glycerides (Structures 7.70–7.77) have been reported from several different collections,[158–162] and tissue distributions of the secondary metabolites have been studied (see Section III.A).[163]

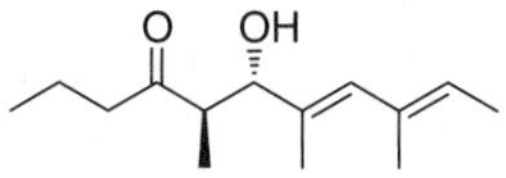

7.69

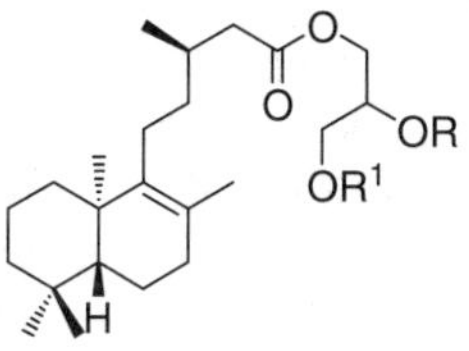

7.70: R = Ac, R¹ = H
7.71: R = H, R¹ = Ac
7.72: R = R¹ = H

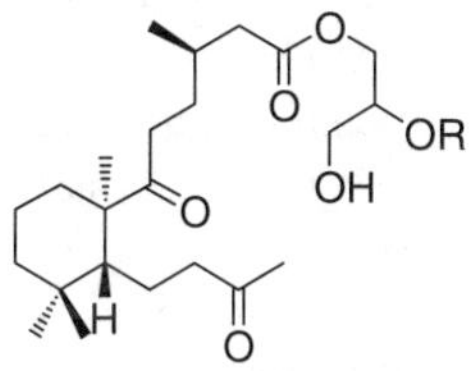

7.73: R = Ac
7.74: R = H

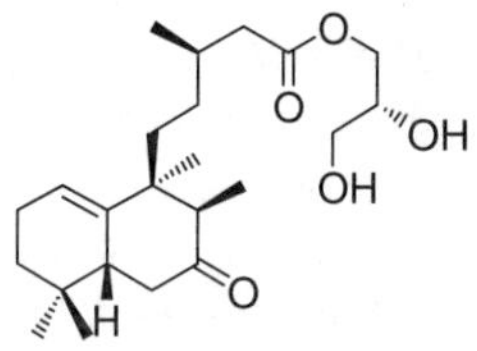

7.75

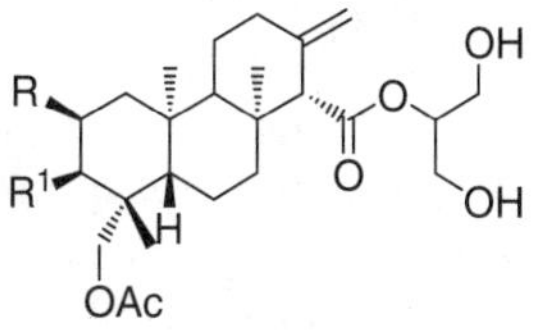

7.76: R = OAc, R¹ = H
7.77: R = H, R¹ = OAc

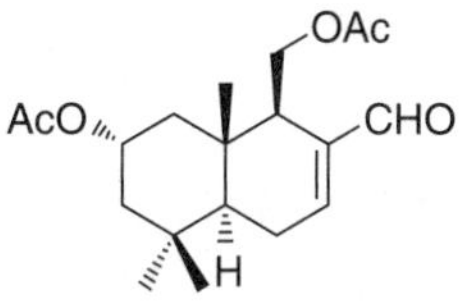

7.78

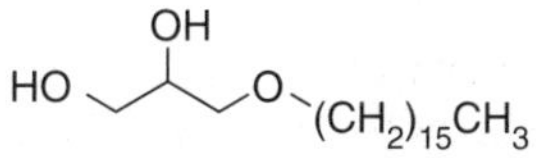

7.79

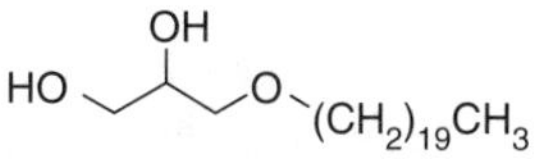

7.80

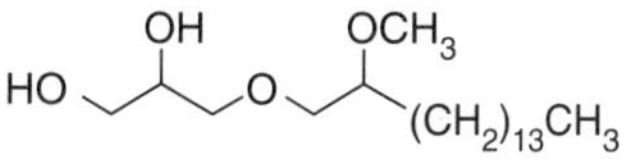

7.81

Bathydoris hodgsoni can be found in the deep waters of the Weddell Sea. This nudibranch elaborates the drimane sesquiterpene hodgsonal (Structure 7.78) and sequesters it in its mantle tissues.[50,164] The nudibranch *Tritoniella belli* collected from McMurdo Sound is the only documented example of an Antarctic mollusc sequestering defensive chemistry from its diet.[44] Among the glyceride esters (Structures 7.79–7.81) isolated from this species, chimyl alcohol (Structure 7.79) can also be found in one of its prey items, the stoloniferan coral *Clavularia frankliniana*. Sequestration of these defensive chemicals is opportunistic, however, and other, still undescribed metabolites are more commonly associated with defense in this mollusc.[49]

The bright yellow lamellarian gastropod *Marseniopsis mollis* collected at McMurdo Sound produces the feeding deterrent homarine.[51] Homarine is widely distributed in marine phyla and has been suggested to serve as an osmolyte.[165]

D. ECHINODERMS

Antarctic sea stars (Class: Asteroidea) and brittle stars (Class: Ophiuroidea) have been extensively studied, yielding 79 compounds to date, making Echinodermata the phylum with the largest number of known secondary metabolites among Antarctic organisms. Although high in absolute numbers, the diversity of Antarctic echinoderm chemical structures and bioactivity is quite limited (Tables 7.6–7.10). With the exception of the isoquinoline alkaloid fuscusine (Structure 7.82), Antarctic sea stars elaborate steroid glycosides, polyhydroxysteroid glycosides, and their sulfates. Sulfated glycosides, or saponins, generally regarded as toxins, are abundant in sea stars (asterosaponins) and sea cucumbers (holothurins) from more northern climes, often accompanied by sulfated and nonsulfated polyhydroxysteroids;[166,167] no reports from Antarctic sea cucumbers have yet appeared. Polyhydroxysteroids and their sulfates are the predominant compounds reported to date from ophiuroids, regardless of geographic origin.

Fuscusine (Structure 7.82) is an unusual echinoderm metabolite and only the second alkaloid reported from sea stars.[168] Isolated from the common asteroid *Perknaster fuscus*, fuscusine appears to be a condensation product derived from 3,5-dihydroxyphenylalanine and arginine and is thus biosynthetically related to the only other known sea star alkaloid, imbricatine (Structure 7.83),[169,170] which bears a tyrosine in the place of fuscusine's arginine and includes a methylhistidine substituent. Isolated from the temperate sea star *Dermasterias imbricata*, imbricatine causes an unusual swimming escape response in the anemones *Stomphia coccinea* and *S. didemon*, which are common prey items of *D. imbricata*.[171] A similar ecological role for fuscusine, however, has not been established yet, though *P. fuscus* is spongivorous and not known to include anemones in its diet.[31]

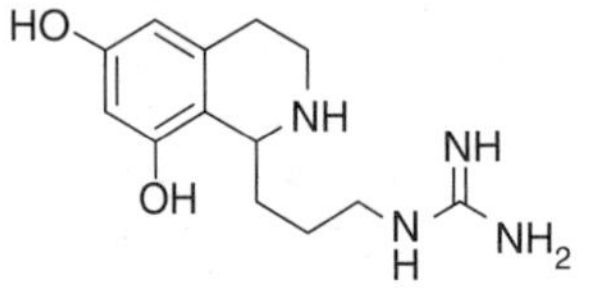

7.82 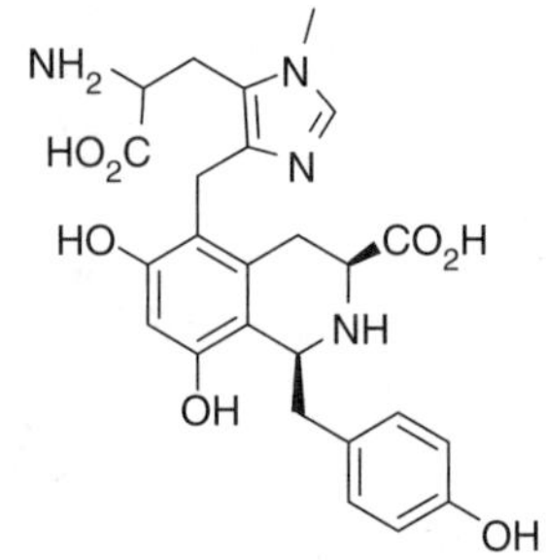   7.83

Three species of Antarctic ophiuroids have been investigated, leading to the isolation and characterization of seventeen polyhydroxysteroid sulfates (Table 7.6). *Ophioderma longicaudum* is a cosmopolitan ophiuroid which was collected from Gondwana, Antarctica, at 70 m depth. The acetone extract from *O. longicaudum* contained three new sulfates of polyhydroxysteroids (Structures 7.84–7.86).[172] *Ophionotus victoriae* contained, in addition to Structure 7.86, five related compounds (Structures 7.87–7.91).[173] More recently, *Astrotoma agassizii*, from the sub-Antarctic South Georgia Islands, was similarly reported with *O. longicaudum* chemistry (Structures 7.84 and 7.85) as well as new polyhydroxysteroid sulfates (Structures 7.92–7.94).[174] Biogeographic considerations suggest that not only are Structures 7.84–7.86 common among Antarctic ophiuroids studied to date, but Structure 7.86, first reported[175] from Pacific brittle stars, is quite common in non-Polar ophiuroids;[173] sulfates shown in Structures 7.87–7.91 can be found in Mediterranean collections of *Ophiotrix fragilis* and/or *Ophiura texturata*.[173] Polyhydroxysteroids are of biomedical interest due

Table 7.6 AntarcticOphioroid Polyhydroxysteroid Sulfates

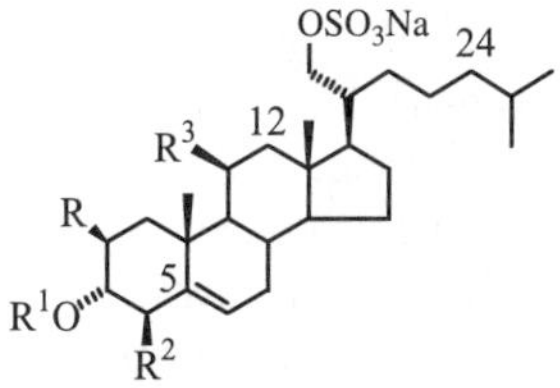

Scaffold A
Scaffold B: 5α,6-H$_2$

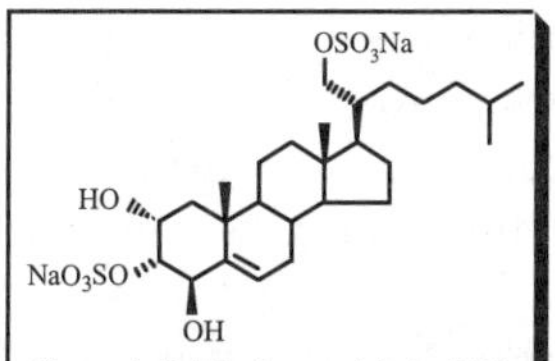

Example Polyhydroxysulphate: **7.84**

	Scaffold	R	R^1	R^2	R^3	Δ^X	Citation
7.84	A	α-OH	SO$_3$Na	OH	H		172
7.85	A	OSO$_3$Na	H	H	H		172
7.86	B	H	SO$_3$Na	α-OH	β-OH		172
7.87	A	H	SO$_3$Na	OH	H		173
7.88	A	OH	SO$_3$Na	H	H		173
7.89	A	OH	SO$_3$Na	H	H	Δ^{22}	173
7.90	A	H	SO$_3$Na	H	H		173
7.91	B	H	SO$_3$Na	H	H		173
7.92	A	OSO$_3$Na	H	H	H	Δ^{24}	174
7.93	B	OH	SO$_3$Na	H	H	Δ^{24}	174
7.94	A	α-OH	SO$_3$Na	OH	H	Δ^{24}	174

to their antiviral activity, as demonstrated by halistanol sulfate from the tropical sponge *Halichon-dria moorei*.[176]

Steroidal sulfates, which are typically characteristic of asteroids from other geographical regions, are thus far uncommon in Antarctic collections. Two reports have identified four such steroids.[177,178] The steroid sulfates represented by Structures 7.95–7.97 were isolated from a member of the family Asteriidae.[177] Collected from deep water at Tethis Bay, the single animal yielded only the second known example of the ring-B oxidized lactone (Structure 7.95), as well as two 6-keto derivatives (Structures 7.96–7.97). Oxidation at C-25 of the steroid side chain had previously been encountered,[179] although the 25-sulfate group has only been reported from an ophiuroid.[175] The 7-membered lactone has only been previously reported from the terrestrial realm.[180] Thornasterol B 3-sulfate (Structure 7.98) was found among more than two dozen steroid glycosides from a sea star belonging to the family Echinasteridae (*vide infra*).[178]

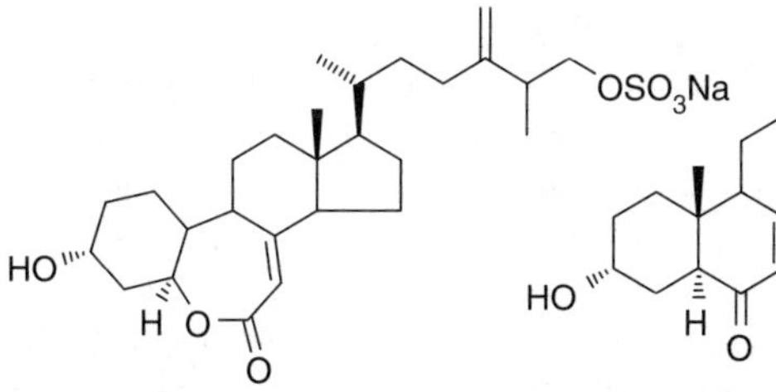

7.95

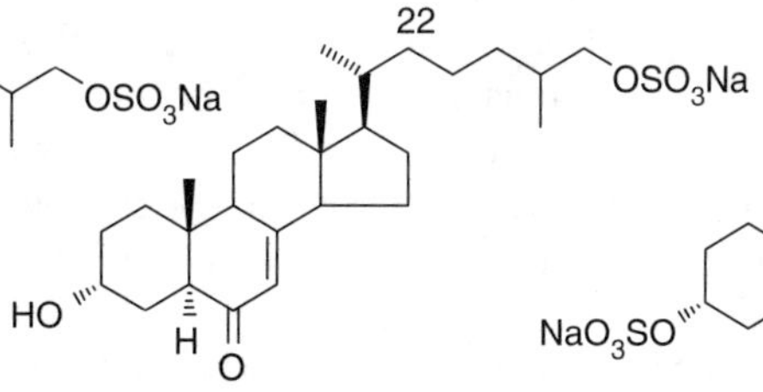

7.96
7.97: Δ^{22}

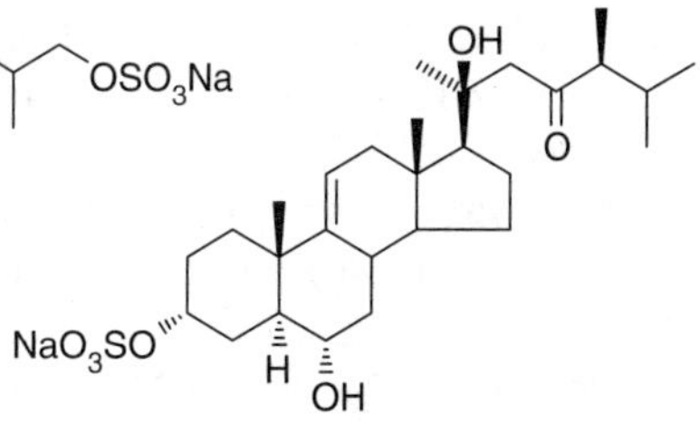

7.98

Steroidal glycosides are predominant among the Antarctic sea star natural products. Chemical investigations of 5 species of Antarctic sea stars have resulted in the isolation of 16 steroid oligoglycosides (asterosaponins), 18 polyhydroxysteroid glycosides (PHG), and 12 sulfated polyhydroxysteroid glycosides (SPG). These glycosides primarily bear conventional saccharides,* such as β-xylo- and β-quinovopyroanoses, though the occasional saccharide or substitution pattern is rare.

Santiagoside (Structure 7.99, Table 7.7), from *Neosmilaster georgianus*, was among the first compounds reported from Antarctic asteroids, and its tetrasaccharide was the first four-sugar oligoglycoside reported among sea star asterosaponins.[181] The aforementioned Echinasteridae sea star,[178] which was said to resemble the genus *Henricia*, was collected at Tethis Bay; a single specimen yielded seven asterosaponins (Structures 7.100–7.104),[182] including two which were previously reported from Caribbean[183] or Japanese[184,185] sea stars. *Acodontaster conspicuus*, from McMurdo Sound, was shown to contain three asterosaponins (Structures 7.105–7.107) which had previously been described from *Asterias amurensis* collected from Japanese waters.[186,187] The same individual from the Asteriidae family, which yielded the steroidal sulfates seen in Structures 7.95–7.97 (see above), also contained five previously uncharacterized asterosaponins, the asteriidosides A–E (Structures 7.110–7.114).[177] Several of the aforementioned asterosaponins displayed cytotoxicity; antarcticoside C (Structure 7.102) was the most active ($IC_{50} < 3.3$ μg/mL toward NSCLC-N6 cell line).[182] The *A. conspicuus* asterosaponins were inactive toward sympatric microorganisms.[57]

Asteroid polyhydroxysteroid glycosides (PHG) differ from the asterosaponins described above in several ways. Besides the lack of a C-6 sulfate group and greater oxidation of the aglycone (see scaffolds A–C, Table 7.8), glycosidation is found on the side chain and, occasionally, at the C-3 position of the steroid nucleus (scaffold C), and Antarctic PHG are limited to mono- and disaccharides. PHG are accompanied by similarly glycosylated sulfates (SPG, Table 7.9) which bear the C-6 sulfate found on the asterosaponins. Antarcticosides D–P (Structures 7.115–7.124, and 7.135–7.137), isolated from the Echinasteridae family asteroid which yielded the asterosaponins represented by Structures 7.100–7.106,[178] share the same pentahydroxyl aglycone (scaffold A, Table 7.8) and bear variously methylated xylopyranose and arabanofuranose sugars. The exceptional structures are those glycosides (Structures 7.120, 7.122, and 7.137) which bear galactofuranose, which is found only rarely in asteroids.[178] *Acodontaster conspicuus*, in addition to asterosaponins described above, was found to contain PHG, the acodontasterosides D–I (Structures 7.125–7.130), and SPG, including halityloside I (Structure 7.138) and acodontasterosides A–C (Structures 7.139–7.141).[57] Halityloside I was originally reported from the Pacific sea star *Halityle regularis*.[188] The validosides (Structures 7.134–7.135) are similarly substituted SPG and are isolated from *Odontaster validus* collected in shallow water at King George Island.[189] Asteriidosides H, I, and L (Structures 7.142–7.144) are unusual SPG in that they bear glycosidation at C-3.[190] All acodontasteroside PHG showed modest to strong inhibition of sympatric microorganisms,[57] suggesting a possible role in regulation of surface flora.

The remaining group of sea star compounds are the polyhydroxysteroids (Table 7.10). The aforementioned Echinasteridae family[178] elaborates the polyhydroxysteroids shown in Structures 7.145–7.157, one of which (Structure 7.154)[191] was previously reported from a Mediterranean sea star *Sphaerodiscus placenta*. The Antarctic *Acodontaster conspicuus* produces polyols (Structures 7.157–7.162);[57] hexol (Structure 7.159) can be found in the Mediterranean *Coscinasterias tenuispina*.[192] Polyhydroxysteroid (Structure 7.152) has modest cytotoxicity and the polyols shown in Structures 7.158, 7.161, and 7.162 were inhibitory toward sympatric microorganisms; none of the others tested displayed notable bioactivity.

---

* Examples of the sugars found in Antarctic asteroid glycosides can be found at the bottom of tables containing glycosides. Note that the designation of a particular glycosidic linkage in the table identifies the linkage from the anomeric position to a designated position, indicated by superscript, on the attached glycoside. Thus, the santiagoside (Structure 7.99) tetrasaccharide, abbreviated in Table 7.7 as Glc⁴-Qui[²-Qui]⁴-Fuc, specifies that the glucopyranose (Glc) attached to the aglycone has a quinovopyranose (Qui) in position C-4, and that quinovopyranose itself has another quinovopyranose in position C-2 and a fucopyranose (Fuc) in position C-4. See santiagoside as an example asterosaponin at the top of Table 7.7.

Table 7.7 Antarctic Sea Star Asterosaponins

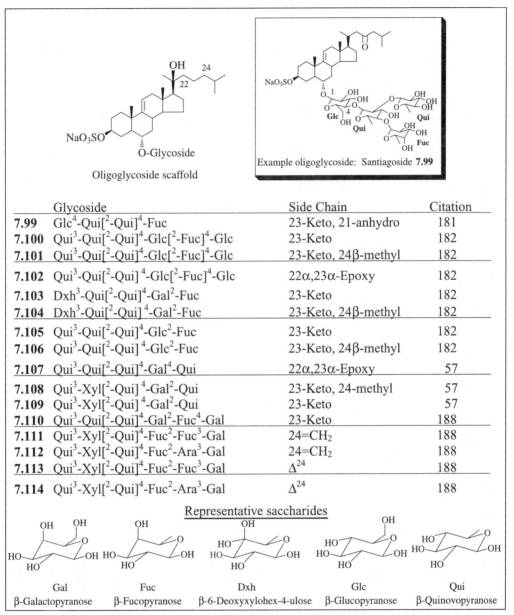

	Glycoside	Side Chain	Citation
7.99	Glc^4-Qui[^{2}-Qui]^4-Fuc	23-Keto, 21-anhydro	181
7.100	Qui^3-Qui[^{2}-Qui]^4-Glc[^{2}-Fuc]^4-Glc	23-Keto	182
7.101	Qui^3-Qui[^{2}-Qui]^4-Glc[^{2}-Fuc]^4-Glc	23-Keto, 24β-methyl	182
7.102	Qui^3-Qui[^{2}-Qui]^4-Glc[^{2}-Fuc]^4-Glc	22α,23α-Epoxy	182
7.103	Dxh^3-Qui[^{2}-Qui]^4-Gal^2-Fuc	23-Keto	182
7.104	Dxh^3-Qui[^{2}-Qui]^4-Gal^2-Fuc	23-Keto, 24β-methyl	182
7.105	Qui^3-Qui[^{2}-Qui]^4-Glc^2-Fuc	23-Keto	182
7.106	Qui^3-Qui[^{2}-Qui]^4-Glc^2-Fuc	23-Keto, 24β-methyl	182
7.107	Qui^3-Qui[^{2}-Qui]^4-Gal^4-Qui	22α,23α-Epoxy	57
7.108	Qui^3-Xyl[^{2}-Qui]^4-Gal^2-Qui	23-Keto, 24-methyl	57
7.109	Qui^3-Xyl[^{2}-Qui]^4-Gal^2-Qui	23-Keto	57
7.110	Qui^3-Qui[^{2}-Qui]^4-Gal^2-Fuc^4-Gal	23-Keto	188
7.111	Qui^3-Xyl[^{2}-Qui]^4-Fuc^2-Fuc^3-Gal	24=CH$_2$	188
7.112	Qui^3-Xyl[^{2}-Qui]^4-Fuc^2-Ara^3-Gal	24=CH$_2$	188
7.113	Qui^3-Xyl[^{2}-Qui]^4-Fuc^2-Fuc^3-Gal	Δ^{24}	188
7.114	Qui^3-Xyl[^{2}-Qui]^4-Fuc^2-Ara^3-Gal	Δ^{24}	188

E. SOFT CORALS

Only one report of Antarctic soft coral secondary metabolite chemistry has appeared in the literature. As described above, the stoloniferan coral *Clavularia frankliniana* has been demonstrated to produce fatty glyceride esters, chimyl alcohol (Structure 7.79) in particular, that are also found in its predator, the nudibranch *Tritoniella belli*.[44] Chimyl alcohol has been investigated in ecological bioassays and has been implicated in deterring predation by the omnivorous sea star *Odontaster validus*.[44]

Table 7.8 Antarctic Sea Star Polyhydroxysteroid Glycosides

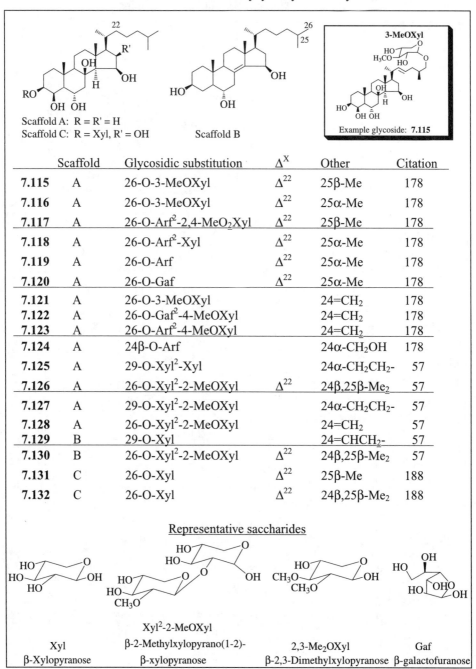

	Scaffold	Glycosidic substitution	Δ^X	Other	Citation
7.115	A	26-O-3-MeOXyl	Δ^{22}	25β-Me	178
7.116	A	26-O-3-MeOXyl	Δ^{22}	25α-Me	178
7.117	A	26-O-Arf^2-2,4-MeO$_2$Xyl	Δ^{22}	25β-Me	178
7.118	A	26-O-Arf^2-Xyl	Δ^{22}	25α-Me	178
7.119	A	26-O-Arf	Δ^{22}	25α-Me	178
7.120	A	26-O-Gaf	Δ^{22}	25α-Me	178
7.121	A	26-O-3-MeOXyl		24=CH$_2$	178
7.122	A	26-O-Gaf^2-4-MeOXyl		24=CH$_2$	178
7.123	A	26-O-Arf^2-4-MeOXyl		24=CH$_2$	178
7.124	A	24β-O-Arf		24α-CH$_2$OH	178
7.125	A	29-O-Xyl^2-Xyl		24α-CH$_2$CH$_2$-	57
7.126	A	26-O-Xyl^2-2-MeOXyl	Δ^{22}	24β,25β-Me$_2$	57
7.127	A	29-O-Xyl^2-2-MeOXyl		24α-CH$_2$CH$_2$-	57
7.128	A	26-O-Xyl^2-2-MeOXyl		24=CH$_2$	57
7.129	B	29-O-Xyl		24=CHCH$_2$-	57
7.130	B	26-O-Xyl^2-2-MeOXyl	Δ^{22}	24β,25β-Me$_2$	57
7.131	C	26-O-Xyl	Δ^{22}	25β-Me	188
7.132	C	26-O-Xyl	Δ^{22}	24β,25β-Me$_2$	188

Representative saccharides

Xyl
β-Xylopyranose

Xyl^2-2-MeOXyl
β-2-Methylxylopyrano(1-2)-
β-xylopyranose

2,3-Me$_2$OXyl
β-2,3-Dimethylxylopyranose

Gaf
β-galactofuranose

Table 7.9 Antarctic Sea Star Polyhydroxyglycoside Sulfates

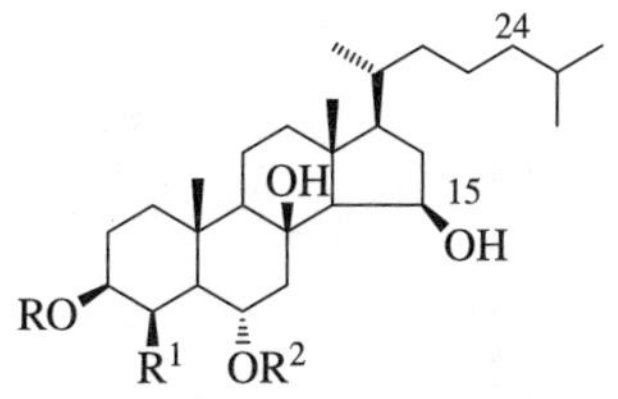

Polyhydroxyglycoside
sulfate scaffold

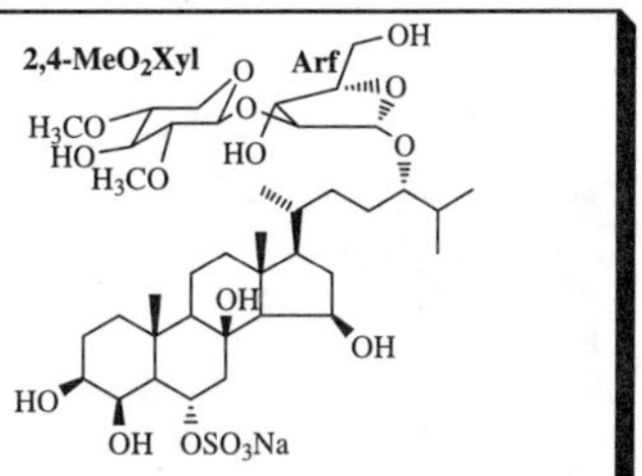

Example polyhydroxyglycoside sulfate: **7.133**

	R	R^1	R^2	Glycoside	Δ^X	Other	Citation
7.133	H	OH	SO$_3$Na	24α-O-Arf^2-2,4-MeO$_2$Xyl			190
7.134	H	H	H	24α-O-2-SulO-3-MeOXyl			190
7.135	H	OH	H	26-O-3-MeO-4-SulOXyl		24=CH$_2$	178
7.136	H	OH	SO$_3$Na	26-O-3-MeOGlc	Δ^{22}	25β, 24β-Me	178
7.137	H	OH	SO$_3$Na	26-O-Gaf^2-2,4-Me$_2$OXyl	Δ^{22}	25β, 24β-Me	178
7.138	H	OH	SO$_3$Na	26-O-Xyl^2-2-MeOXyl	Δ^{22}	25β, 24β-Me	57
7.139	H	OH	SO$_3$Na	28-O-Gaf^2-2-MeOXyl	Δ^{22}	24α-CH$_2$-	57
7.140	H	OH	SO$_3$Na	28-O-Xyl^2-2-MeOXyl	Δ^{22}	24α-CH$_2$-	57
7.141	H	OH	SO$_3$Na	29-O-Xyl^2-2-MeOXyl		24=CHCH$_2$-	57
7.142	Xyl	H	H	24α-O-3-SulOXyl	Δ^{22}		188
7.143	Xyl	H	H	24α-O-3-SulOXyl	Δ^{22}	7α-OH	188
7.144	Xyl	H	H	28-OSO$_3$Na	Δ^{22}	24β-CH$_2$-, 15α	188

Representative saccharides

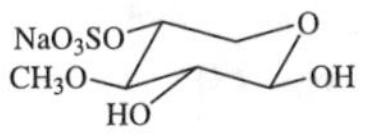

4-SulO-3-MeOXyl
β-2-Sulfonyl-3-
methylxylopyranose

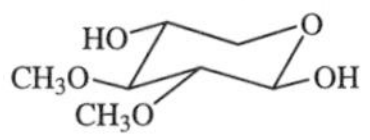

2,3-Me$_2$OXyl
β-2,3-Dimethylxylo-
pyranose

Table 7.10 Antarctic Sea Star Polyhydroxysteroids

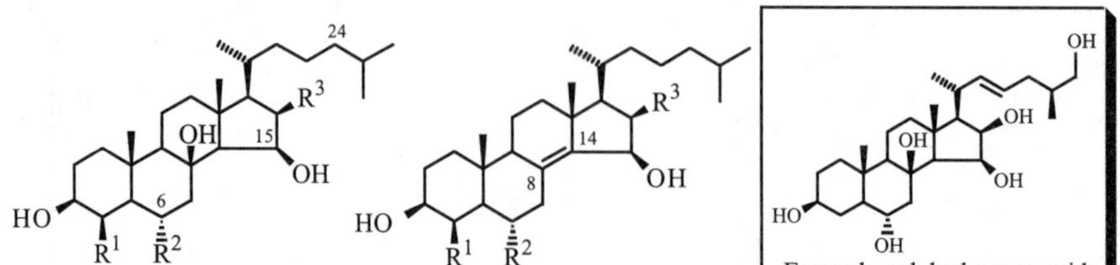

Scaffold A

Scaffold B

Scaffold C: 8β,14α-H$_2$, 16α

Example polyhydroxysteroid: **7.145**

	R^1	R^2	R^3	Scaffold	Side Chain	Other	Citation
7.145	H	OH	OH	A	Δ^{22}, 25β, 26-OH		178
7.146	H	OH	OH	A	Δ^{22}, 25α, 26-OH		178
7.147	H	OH	OH	A	25α, 26-OH		178
7.148	H	OH	OH	A	25α-Me, 25β,26-(OH)$_2$		178
7.149	H	OH	OH	A	Δ^{22}, 21,26-(OH)$_2$, 25β		178
7.150	H	OH	OH	B	Δ^{22}, 25β, 26-OH,	15α	178
7.151	H	OH	OH	B	Δ^{22}, 25α, 26-OH	15α	178
7.152	H	OH	OH	B	25α, 26-OH	15α	178
7.153	H	OH	OH	B	25α-Me, 25β,26-(OH)$_2$	15α	178
7.154	H	OH	OH	B	24=CH$_2$, 25α, 26-OH	15α	178
7.155	H	O	OH	B	26-OH	6-keto	178
7.156	H	β-OH	H	B	Δ^{22}, 24α-OH	15α	178
7.157	OH	OH	H	B	Δ^{22}, 24α-OH		178
7.158	H	OH	H	B	Δ^{22}, 24α-OH		57
7.159	OH	OH	H	B	Δ^{22}, 24-nor, 26-OH		57
7.160	OH	OH	OH	B	24α-ethyl, 25β, 26-OH		57
7.161	H	OH	H	C	Δ^{22},24β-Me, 25β, 26-OH		57
7.162	H	OH	H	C	Δ^{22}, 24-nor, 26-OH		57

F. MICROBIOTA

Chemical investigations of Antarctic marine microbiota have only recently begun to appear in the literature. One of the first such reports described six diketopiperazines (Structures 7.163–7.168) and two phenazine alkaloids (Structures 7.169–7.170) from an Antarctic sponge-associated bacterium, *Pseudomonas aeruginosa*.[193] The diketopiperazines were inactive in antibiotic and/or cytotoxicity bioassays, though the phenazine alkaloids, which are not uncommon bacterial biosynthetic products,[194] had significant antibiotic activity. No similar alkaloids were detected in the sponge *Isodictya setifera*, collected at McMurdo Sound, from which *P. aeruginosa* was isolated.

Two diterpenoids, focardin (Structure 7.171) and an epoxide derivative (Structure 7.172), are produced by the marine ciliate *Euplotes focardii*, collected from Ross Sea coastal waters.[195] Focardin is autotoxic and inhibits other ciliates, including one sympatric species, suggesting they are present

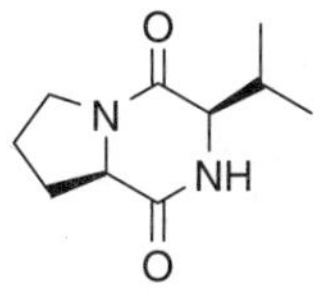

7.163: *Cyclo*-(L-Pro-L-Val) 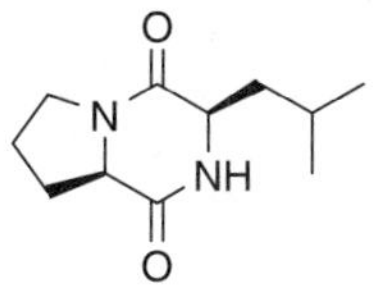 7.164: *Cyclo*-(L-Pro-L-Leu) 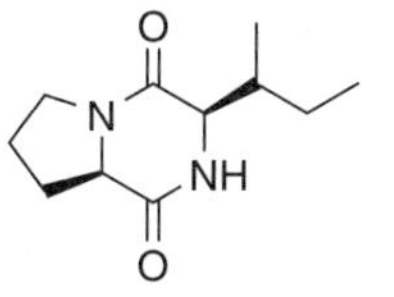 7.165: *Cyclo*-(L-Pro-L-Ile) 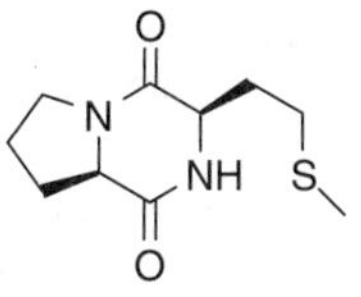 7.166: *Cyclo*-(L-Pro-L-Met)

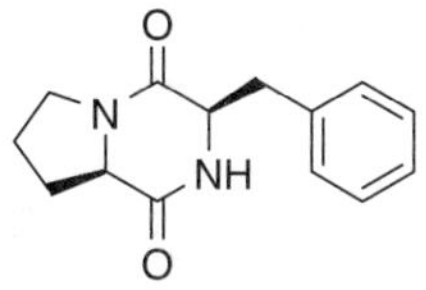

7.167: *Cyclo*-(L-Pro-L-Phe) 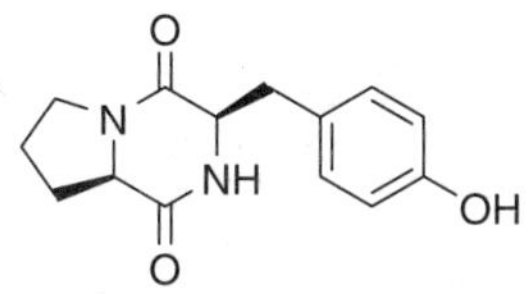 7.168: *Cyclo*-(L-Pro-L-Tyr) 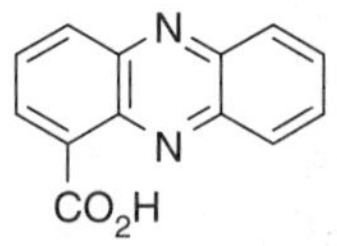 7.169 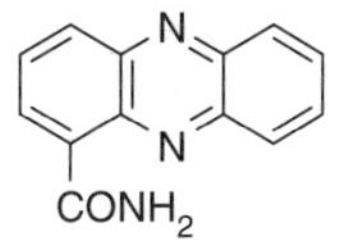 7.170

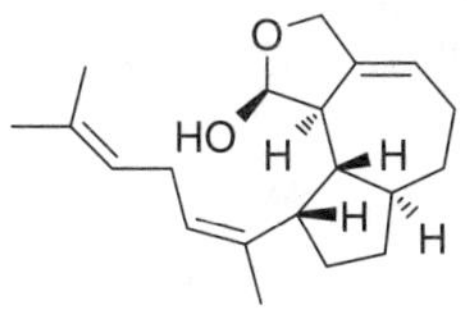

7.171: Focardin 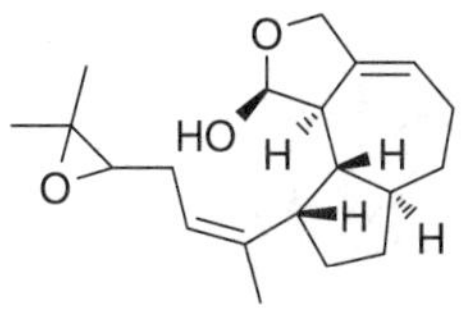 7.172

in a defensive role. Because little research on ciliates[92] appears in the chemical literature, it is not possible to discuss the uniqueness of focardin.

G. Other Organisms

Large individuals of the polychaete *Thelepus extensus* are found in the bays of the sub-Antarctic Kerguelen Islands. Antibiotic bromophenols (Structures 7.173–7.178) are present predominantly in the distal regions of the worm, suggesting a role in fouling and/or infection control.[196] The same compounds were distributed similarly in the tissues of five other geographically distinct collections of terebellid worms, though the polar species *Amphitrite kerguelensis* and ten other temperate or tropical species were devoid of bromophenols.

Salps are a significant component of the Antarctic ocean food web, occupying a trophic niche parallel with krill.[197] To date, chemical evaluations of Antarctic salps have been limited to fatty acid[198] and sterol analyses.[199] While these types of metabolites are arguably primary metabolites or otherwise of little consequence to chemical ecology, it is interesting to note the hemolytic activity of several polyunsaturated acids found in *Salpa thompsoni*.[198]

Antarctic krill have been characterized as having high levels of prostaglandins, including the physiologically active $PGF_{2\alpha}$ (Structure 7.179) and PGE_2 (Structure 7.180).[200,201] Mezykowski and Ignatowska-Switalska found prostaglandins present in krill at levels considerably higher than hormonal levels typically present in mammalian tissues,[200] although Pawlowicz later used different techniques to revise the concentrations downward to levels commensurate with mammalian tissues.[201] Pawlowicz suggests that the earlier determination, which used a radioimmunological method of detection, may not have sufficiently distinguished among other prostaglandins. From a chemical ecology viewpoint, high levels of prostanoids found in the tissues of the tropical soft coral *Plexaura homomalla*[202] are known to deter fish predation,[203] which raises the question of whether the higher levels of prostanoids detected by Mezykowski and Ignatowska-Switalska may act in concert or individually to deter fish predation on krill.

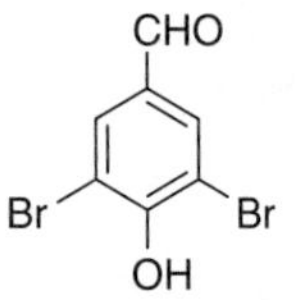

7.173

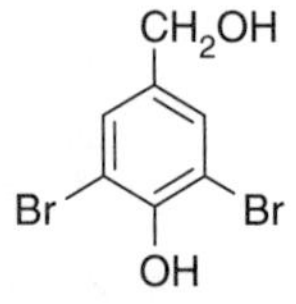

7.174

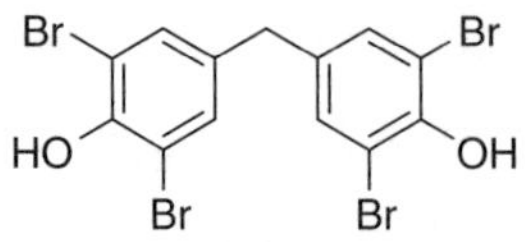

7.175

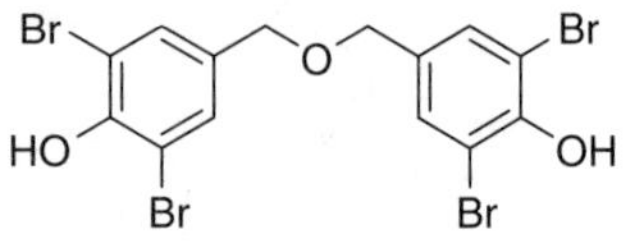

7.176

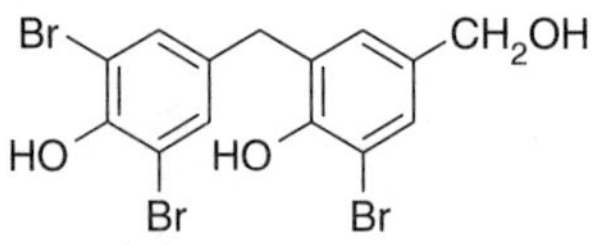

7.177: Thelephenol

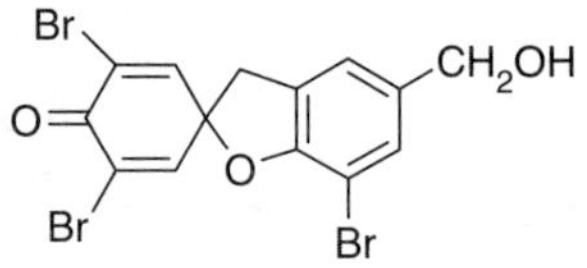

7.178: Thelepin

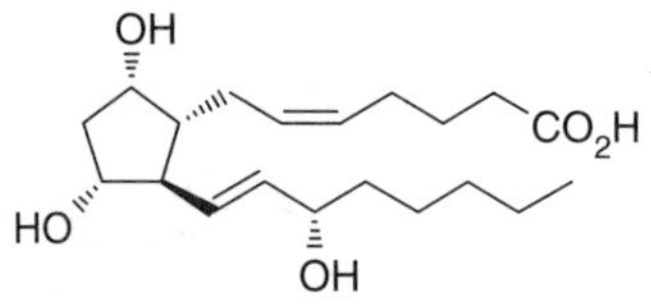

7.179: PGF$_{2\alpha}$

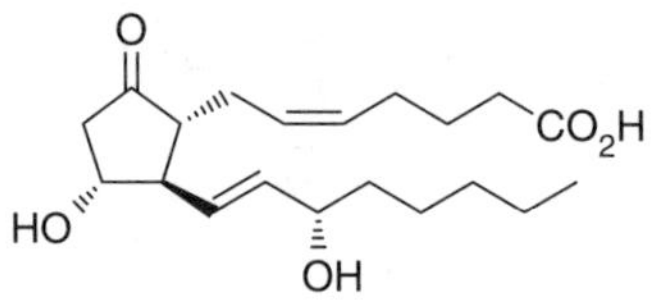

7.180: PGE$_2$

III. FUNCTIONAL ROLES
OF ANTARCTIC MARINE NATURAL PRODUCTS

Secondary metabolites produced by organisms increase their fitness[204–207] and differ from primary metabolites most notably in having a taxonomically distinct distribution. In order to assess the functional role of secondary metabolites, bioassays which model ecological interactions, such as predator/prey and other survival-related relationships, must be employed. Feeding deterrence assays have been at the forefront of studies focusing on the ecological role of secondary metabolites in the marine realm, due to the prominent role that natural products play in structuring trophic relationships.[11,30] Inhibition of fouling and/or other surface phenomena has also been studied extensively as a functional role of marine natural products. In addition, several other likely functions include defense against pathogens, protection from ultraviolet radiation, attraction by pheromones, and roles in competitive interactions.

A. FEEDING DETERRENTS

The role of secondary metabolites in mediating feeding deterrence of Antarctic invertebrate predators has mainly been investigated in sponges and molluscs. Sponges largely dominate the benthic fauna in Antarctica, and sea stars are known to be major predators on sponges, yielding ecologically relevant predator–prey pairs for bioassays. Sea stars have chemosensory tube feet which they use to detect their prey and then feed extraorally by extruding their cardiac stomachs over their prey,[208] digesting sponge tissue from its skeletal structures. A commonly used behavioral assay is based on the response of chemosensory tube feet in a major Antarctic spongivor, *Perknaster fuscus*. This assay has been employed to assess the palatability of sponge extracts, and results from this assay have correlated well with observed feeding patterns of the sea star.[23,24] For example, *P. fuscus* feeds predominately on two sponges, *Mycale acerata* and *Homaxinella balfourensis*, two of the sponges whose extracts elicited among the lowest levels of tube foot retraction response. In addition, five sponges displaying high levels of retraction have been subject to chemical investigation, yielding compounds which themselves caused significant responses in the assay. These include discorhabdin C (Structure 7.11) from *Latrunculia apicalis*,[38,96] rhapsamine (Structure 7.22) from *Leucetta leptorhapsis*,[34,39] suberitenones A and B (Structures 7.7 and 7.8) from *Suberites* sp.,[42] picolinic acid from *Dendrilla membranosa*,[34,37] and p-hydroxybenzaldehyde from *Isodictya erinacea*.[40] Neither the norditerpenes (Structures 7.2 and 7.3) nor the pigment (Structure 7.6) isolated from *Dendrilla membranosa* showed deterrent activity in the assay,[37] and variolin A (Structure 7.13) from *Kirkpatrickia variolosa* similarly failed to elicit a response from *P. fuscus*.

Other than in tropical and most temperate regions where browsing fish are the main omnivorous predators on invertebrates, the Antarctic benthos is conspicuously depauperate in fish that graze on invertebrates. Here, mainly invertebrate predators such as omnivorous sea stars occupy this niche.[31] Especially in shallow waters, *Odontaster validus* is the most common predator, and this species has been widely used in feeding deterrence bioassays, especially to test chemical defense in opisthobranch molluscs. With the evolutionary loss of a shell in opisthobranchs, feeding-deterrent natural products are considered especially important in this group.[153,154]

Chimyl alcohol (Structure 7.79), isolated from the mantle tissues of the nudibranch *Tritoniella belli*,[44] caused rejection by *O. validus* of shrimp-treated disks at natural tissue-level concentrations. *T. belli* sequesters this defensive chemistry from its diet, the stoloniferan coral *Clavularia frankliniana*. Sequestration of chimyl alcohol from the diet, however, seems to be opportunistic, and *T. belli* also provisions its mucus with other yet undescribed deterrent natural products.[49] Chimyl alcohol is reported from other non-Antarctic molluscs[209] and has been demonstrated to function as an antibacterial and a fish antifeedant at levels found in the tissues of the dorid nudibranchs *Archidoris montereyensis* and *Aldisa sanguinea cooperi*.[210]

Mantle tissue, as well as ecologically relevant concentrations of the isolated sesquiterpene hodgsonal (Structure 7.78) from the nudibranch *Bathydoris hodgsoni*, showed a strong feeding deterrent effect on *O. validus*.[50] As is consistent with the *de novo* biosynthetic conjecture of Faulkner and Ghiselin,[153] hodgsonal found in *B. hodgsoni* is quantitatively and qualitatively indistinguishable regardless of collection site. Also, *B. hodgsoni* is an omnivorous feeder which makes sequestration of defensive chemistry from a particular diet highly unlikely.[50]

It was shown earlier that mantle tissue and crude extracts of the most common Antarctic nudibranch, *Austrodoris kerguelenensis*, exhibit deterrent activity against predators such as the sea star *O. validus*.[48] But only recently, feeding-deterrent activity of the dominant secondary metabolites, diterpene diacyglycerides (Structures 7.70–7.77) and related monoacylglycerides, was established in bioassays with *O. validus*.[163] These terpenes are similar to those found in *Archidoris* spp., leading to speculation[158] that, like *Archidoris* spp., *Austrodoris kerguelenensis* biosynthesizes these compounds *de novo*. In support of this argument, Dayton et al.[31] found that *A. kerguelenensis* feeds selectively on hexactinellid sponges, which are especially poor producers of natural products (none are known[92]), and hexactinellid extracts have displayed little bioactivity in bioassays.[23,68] Consistently, no precursors of the active diacylglycerides in *A. kerguelenensis* could be found in its main diet, the hexaxtinellid sponge *Rosella racovitzae*.[163]

Among Antarctic shell-less molluscs, the pteropod *Clione antarctica* produces the polyketide pteroenone (Structure 7.69).[53] Pteroenone, but no other compounds from *C. antarctica*, proved to be ichthyodeterrent against sympatric fish species such as *Pagothenia borchgrevinki* and *Pseudotrematomus bernacchii*.[54] In what is perhaps one of the most uniquely Antarctic chemical ecological relationships, the hyperiid amphipod *Hyperiella dilatata* abducts *C. antarctica* for its own defensive purposes.[52,69] The amphipod uses its pereiopods to hold and carry live pteropods on its back in order to benefit from their chemical defense. Laboratory experiments showed that *H. dilatata* carrying *C. antarctica* were rejected by fish while amphipods alone were readily eaten.[52,69]

In the bright yellow lamellarian *Marseniopsis mollis*, soft mantle tissue covers a reduced shell, which makes this species, similar to opisthobranchs, susceptible to predation. In the first demonstration of chemical defense in a prosobranch mollusc, ecologically relevant concentrations of homarine, which is present in high concentrations in the tissues of *Marseniopsis mollis*, caused

O. validus to reject shrimp-treated disks which were otherwise readily fed upon in the absence of the feeding deterrent.[51] Homarine is widely distributed in marine phyla and has otherwise been characterized as an osmolyte and an antifoulant.[165,211] McClintock et al.[51] were unable to detect the compound in the tunicate *Cnemidocarpa verrucosa*, the primary diet of *M. mollis*, but rather the compound is present in epibionts found on the surface of *C. verrucosa*.

B. Surface Defenses: Inhibition of Fouling and Infectious Agents

Microbiota such as bacteria, diatoms, and marine invertebrate larvae exert considerable fouling pressure on benthic invertebrates. While some organisms employ physical means of fouling prevention, such as mucus production, secondary metabolites have been shown to function as antifoulants in a number of notable cases.[15,57,212–215] Few studies of Antarctic organisms have investigated activity toward sympatric microorganisms and/or other fouling organisms. Among Antarctic sponge-derived compounds, discorhabdin C (Structure 7.11) from *Latrunculia apicalis*, and the pigment (Structure 7.6) from *Dendrilla membranosa*, are potent antimicrobial compounds active toward both antibiotic tester strains as well as sympatric microorganisms isolated from the water column and/or the surfaces of other benthic invertebrates.[34,37,38,85,216] Other compounds isolated from *Dendrilla membranosa*, membranolide (Structure 7.1) and 9,11-dehydrogracilin (Structure 7.2), have antibiotic activity toward antibiotic tester strains (*Bacillus subtilis* and/or *Staphylococcus aureus*), though the ecological relevance of that activity is unclear. Acodontasterosides (Structures 7.125–7.141) from the sea star *Acodontaster conspicuus* are similarly antimicrobial toward sympatric microorganisms.[57]

Sponges are also subject to fouling by diatoms, which becomes especially severe during annual microalgal blooms.[217] Fouling diatoms may be a potential food source for sponges, but surface fouling can become a vital problem for sponges when microalgae start to clog up pores and thus significantly reduce filter feeding and respiration. Recent studies of water-soluble fractions of lipophilic and hydrophilic extracts from sponges collected at McMurdo Sound suggest a likely role of secondary metabolism in controlling diatom fouling.[218] Some species (e.g., *Leucetta leptorhapsis* and *Kirkpatrickia variolosa*) showed low natural densities of diatom fouling, high activity against sympatric diatoms in bioassays, and activity in other chemical defenses. Extracts of *Calyx acerata*, a species which is naturally heavily fouled by diatoms, did not show significant activity against diatoms in bioassays, however, extracts are strongly active against sea star predators.[23,68] Among sponge species which are observed to be naturally fouled by diatoms, extracts of *Mycale acerata* and *Homaxinella balfourensis* showed moderate to high levels of antifouling activity against sympatric diatoms in laboratory bioassays.[218] Both sponge species, however, are not or only weakly chemically defended against predation by sea stars[23,24] (see above, Section III.A). This shows a certain level of specificity of chemical defense against diatom fouling.[217,218]

C. Other Roles

Scarce resources, such as food or space, in some marine environments can lead to production of compounds for the offensive role of reducing competition. Focardin (Structure 7.171) and its 11,12-epoxide (Structure 7.172), produced by the Antarctic ciliate *Euplotes focardii*, are autotoxic and cytotoxic toward a sympatric ciliate and a selection of temperate and tropical predatory ciliates.[195]

Functional roles treated above are strictly organismal responses to biotic stresses. The use of chemical means to address physical stresses is largely nonexistent. A notable exception to this is the response many organisms make to harmful ultraviolet (UV) radiation. Sunscreens are small organic molecules which absorb UV radiation so that damage to DNA via cross-linking[219–225] is prevented. Production of micosporin-like amino acids (MAA) appears to be widespread in shallow-water Antarctic marine invertebrates and macroalgae,[223,224] which may be particularly adaptive at present because of the Antarctic ozone hole.

IV. ECOLOGICAL IMPLICATIONS OF FEEDING DETERRENTS

The benthic environment of Antarctica is unique in many ways. There are several implications that have come to light as studies of chemical defense mechanisms in Antarctic invertebrates have advanced. The isolation and characterization of chemical defenses from all highly pigmented sponges on the McMurdo Sound benthos, for example, raises questions of the role of coloration in an ecosystem largely devoid of visually oriented predators. Similarly, the nature of predation on Antarctic sponges, e.g., sea star extraoral feeding behavior, leads to the conjecture that defenses in these organisms would be most useful if concentrated on the outermost surface tissues, as would be predicted by the optimal defense theory.[226]

A. The Role of Pigments in Antarctic Sponges

A number of Antarctic sponges, as well as other Antarctic benthic marine invertebrates, contain brightly colored pigments. The presence of this bright coloration is intriguing in a marine system where grazing pressure by visually oriented predators such as fish is generally lacking. In such a system, one would anticipate that pigmentation could not be driven by aposematism.[227,228] Recent studies of the defensive chemistry of three brightly colored Antarctic sponges has indicated that these pigments themselves are bioactive and/or cause feeding deterrence. For example, the green alkaloid discorhabdin C (Structure 7.11), from the *Latrunculia apicalis*, is active in the chemotactic tube-foot retraction assay,[38] as are purple variolin pigments found in *Kirkpatrickia variolosa* (Baker and McClintock, unpublished data). The *Dendrilla membranosa* pigment (Structure 7.6) is inhibitory towards a water-column bacterium. These bioactive pigments raise the intriguing question of whether evolutionary selection for, or retention of, pigmentation has been driven by predation pressure in brightly colored benthic Antarctic marine invertebrates. In other sessile organisms such as marine and terrestrial plants, pigments are almost exclusively employed in energy capture or serve as antioxidants.[229] These are unlikely roles for pigments of benthic marine invertebrates. There is little information on whether pigments from temperate or tropical benthic marine invertebrates may serve as antifeedants. Brightly colored pigments have been invoked as warning coloration in temperate and tropical marine invertebrates,[230] though at least one study disputed the role of pigments in aposematic coloration of tropical sponges.[231]

B. Sequestration

The optimal defense theory predicts that the distribution of defenses within individual organisms should be directly correlated with risk of attack.[226] Two Antarctic sponge species proved amenable to test this theory as it applies to sponges. The roughly spherical *Latrunculia apicalis* and *Suberites* sp. both contain natural products which influence the feeding behavior of the spongivorous sea star *Perknaster fuscus*. Since *P. fuscus* feeds by extruding its cardiac stomach over its prey, a sponge would best be protected by concentrating defensive chemistry in its surface tissues. Both *L. apicalis* and *Suberites* sp. concentrated as much as 90% of their defensive chemistry within 2 mm of their surface.[232] Similar to the results of Thompson et al.,[233] who found potential defensive compounds aerothionin and homoaerothionin limited in cellular distribution to the spherulous cells of the mesohyl, and that of Schupp et al.,[234] who found defensive compounds localized in *Oceanapia* sp. in a manner that selectively protected the capitum, this result suggests that sponges are capable of sequestering defensive chemistry in tissue distribution patterns that addresses the nature of the predatory threat, in agreement with the optimal defense theory.

Sequestration of defensive chemicals in discrete tissues of an Antarctic mollusc has also been investigated. *Marseniopsis mollis* contains the defensive metabolite homarine throughout its tissues (mantle, foot, viscera), with the highest concentration in the foot.[51] In the nudibranchs *Austrodoris kerguelenensis*[162] and *Bathydoris hodgsoni*,[50] defensive chemicals are concentrated only in the outer tissues such as mantle, foot, and gills, but not in the internal viscera. In *B. hodgsoni*, the defensive compound hodgsonal is also present in dorsal papillae which were observed to detach easily during physical contact, and a combined morphological–chemical defense strategy is discussed.[50]

C. ORIGIN OF ANTARCTIC SECONDARY METABOLITES

Marine organisms often obtain defensive chemistry from their diet, sequestering the secondary metabolites of their prey in their own tissues.[14,15] The Antarctic nudibranch *Tritoniella belli* feeds predominantly on the soft coral *Clavularia frankliniana*, and both contain the sea star deterrent chimyl alcohol, suggesting that the mollusc sequesters the metabolite from its diet.[44] *Marseniopsis mollis* contains a sea star defensive metabolite, homarine. Homarine occurs in many different marine invertebrates, and its functional role in many of those organisms is obscure. *M. mollis* was reported to feed on the tunic of the solitary tunicate *Cnemidocarpa verrucosa*, which does not contain homarine. High concentrations of homarine in epizooites fouling the tunic of *C. verrucosa* suggest that *M. mollis* may be feeding on these animals and deriving its protective chemistry from them.[51] Pteroenone, from the pteropod *Clione antarctica*, is likely produced *de novo* since it is not detected in the primary prey of *C. antarctica*, the shelled pteropod *Limacina helicina*.[54] It is also assumed that the nudibranchs *Austrodoris kerguelenensis* and *Bathydoris hodgsoni* biosynthesize their defensive chemicals *de novo*, since no precursors are found in their diet and levels of active compounds are similar regardless of collection site.[50, 163]

V. FUTURE DIRECTIONS

There remains considerable interest in Antarctic chemical ecology. Issues in need of clarification include the functional role of many of the natural products which have been described as well as the chemical nature of bioactivity in a number of Antarctic organisms. Continued chemical investigations are necessary in order to establish how secondary metabolic pathways of Antarctic organisms compare with organisms from more northern latitudes. Unique physical features of Antarctic coastal environments that may interact with secondary metabolite biochemistry can also make it an ideal natural laboratory for testing models concerning the evolution of chemical defenses. For example, coastal Antarctica is unique among the world's marine environments in that light, rather than nutrients, is growth limiting for macroalgae a very broad biogeographic region (and likely has been for millions of years). This makes coastal Antarctica the best, and perhaps only, place to test carbon limitation predictions of the Resource Allocation Model[235] and the carbon nutrient balance hypothesis[236] in a marine system.

ACKNOWLEDGMENTS

The authors would like to thank their students and colleagues who contributed much of the described work in this chapter. Work on this manuscript was supported by grants OPP98-14538 (JBM, CDA) and OPP99-01076 (BJB) from the National Science Foundation and by R/LR-MB-10 (BJB, Co-PI) by Florida Sea Grant.

REFERENCES

1. Cherry-Garrard, A., *The Worst Journey in the World*, Carroll and Graf Publishers, 1997 (reprint of the original 1922 version).
2. U.S. Government document NSF 92-134, www.nsf.gov/od/opp/antarct/usaphist.htm.
3. Dayton, P. K., Mordida, B. J., and Bacon, F., Polar marine communities, *Am. Zool.*, 34, 90, 1994.
4. Dell, R. K., Antarctic benthos, *Adv. Mar. Biol.*, 10, 1, 1972.

5. Arntz, W. E., Brey, T., and Gallardo, V. A., Antarctic zoobenthos, *Oceanogr. Mar. Biol. Ann. Rev.*, 32, 241, 1994.
6. Brey, T., Dahm, C., Gorny, M., Klages, M., Stiller, M., and Arntz, W. E., Do Antarctic benthic invertebrates show an extended level of eurybathy?, *Antarctic Sci.*, 8, 3, 1996.
7. Brandt, A., Origin of Antarctic Isopoda (Crustacea, Malacostraca), *Mar. Biol.*, 113, 415, 1992.
8. Bakus, G. J. and Green, G., Toxicity in sponges and holothurians: a geographic pattern, *Science*, 185, 51, 1974.
9. Green, G., Ecology of toxicity in marine sponges, *Mar. Biol.*, 40, 207, 1977.
10. Bakus, G. J. and Thun, M. A., Bioassays on the toxicity of Caribbean sponges, in *Sponge Biology*, Levi, C. and Bourny-Esnault, N., Eds., Centre National de la Recherche Scientifique (C. N. R. S.), Paris, 1979, 417.
11. Bakus, G. J., Chemical defense mechanisms on the Great Barrier Reef, Australia, *Science*, 211, 497, 1981.
12. Steinberg, P. D. and Paul, V., Fish feeding and chemical defense of tropical brown algae in Western Australia, *Mar. Ecol. Prog. Ser.*, 58, 253, 1990.
13. Van Alstyne, K. L. and Paul, V. J., The biogeography of polyphenolic compounds in marine macroalgae: temperate brown algal defenses deter feeding by tropical herbivorous fishes, *Oecologia*, 84, 158, 1990.
14. Paul, V. J., *Ecological Roles of Marine Natural Products*, Comstock Publishing Association, Ithaca, NY, 1992.
15. Pawlik, J. R., Marine invertebrate chemical defenses, *Chem. Rev.*, 93, 1911, 1993.
16. Rosenthal, G. A. and Janzen, D. H., *Herbivores: Their Interaction with Secondary Plant Metabolites*, Academic Press, New York, 1979.
17. Berenbaum, M. R. and Rosenthal, G. A., Eds., *Herbivores: Their Interactions with Secondary Plant Metabolites: The Chemical Participants*, 2nd ed., Academic Press, Orlando, FL, 1991.
18. Harborne, J. B., *Introduction to Ecological Biochemistry*, 4th ed., Academic Press, New York, 1994.
19. Harborne, J. B., Recent advances in chemical ecology, *Nat. Prod. Rep.*, 16, 509, 1999.
20. Vermeij, G. J., *Biogeography and Adaptation*, Harvard University Press, Boston, MA, 1978.
21. Blunt, J. W., Munro, M. H. G., Battershill, C. N., Copp, B. R., McCombs, J. D., Perry, N. B., Prinsep, M., and Thompson, A. M., From the antarctic to the antipodes; 45° of marine chemistry, *New Zealand J. Chem.*, 14, 751, 1990.
22. McClintock, J. B., Slattery, M., Heine, J., and Weston, J., The chemical ecology of the antarctic spongivorous sea star *Perknaster fuscus*, *Antarctic J. U. S.*, 27, 129, 1992.
23. McClintock, J. B., Baker, B. J., Slattery, M., Hamann, M., Kopitzke, R., and Heine, J., Chemotactic tube-foot responses of a spongivorous sea star *Perknaster fuscus* to organic extracts from antarctic sponges, *J. Chem. Ecol.*, 20, 859, 1994.
24. McClintock, J. B., Baker, B. J., Amsler, C. D., and Barlow, T. L., Chemotactic tube-foot responses of the spongivorous sea star *Perknaster fuscus* to organic extracts of sponges from McMurdo Sound, Antarctica, *Antarctic Sci.*, 12, 41, 2000.
25. Amsler, C. D., McClintock, J. B., and Baker, B. J., Chemical defenses of antarctic marine organisms: a reevaluation of the latitudinal hypothesis, in *Antarctic Ecosystems: Models for Wider Ecological Understanding*, Proceedings of the Seventh SCAR International Biology Symposium, Davison, W., Howard-Williams, C., Broady, P., Eds., New Zealand Natural Sciences, Christchurch, New Zealand, 158, 2000.
26. Sanders, H. L., Marine benthic diversity: a comparative study, *Amer. Nat.*, 102, 243, 1968.
27. Jackson, J. B. C. and Buss, L., Allelopathy and spatial competition among coral reef invertebrates, *Proc. Nat. Acad. Sci. U. S. A.*, 72, 5160, 1975.
28. Coll, J. C., La Barre, S., Sammarco, P. W., Williams, W. T., and Bakus, G. J., Chemical defenses in soft corals (Coelenterata: Octocorallia) of the Great Barrier Reef: a study of comparative toxicities, *Mar. Ecol. Prog. Ser.*, 8, 271, 1982.
29. La Barre, S. C., Coll, J. C., and Sammarco, P. W., Defensive strategies of soft corals (Coelenterata: Octocorallia) of the Great Barrier Reef. II. The relationship between toxicity and feeding deterrence, *Biol. Bull.*, 171, 565, 1986.
30. Bakus, G. J., Targett, N. M., and Schulte, B., Chemical ecology of marine organisms: an overview, *J. Chem. Ecol.*, 12, 951, 1986.

31. Dayton, P. K., Robilliard, G. A., Paine, R. T., and Dayton, L. B., Biological accommodation in the benthic community at McMurdo Sound, Antarctica, *Ecol. Monogr.*, 44, 105, 1974.

32. Barthel, D. and Gutt, J., Sponge associations in the eastern Weddell Sea, *Antarctic. Sci.*, 4, 137, 1992.

33. McClintock, J. B., Investigation of the relationship between invertebrate predation and biochemical composition, energy content, spicule armament and toxicity of benthic sponges at McMurdo Sound, Antarctica, *Mar. Biol.*, 94, 479, 1987.

34. Baker, B. J., Kopitzke, W., Hamann, M., and McClintock, J. B., Chemical ecology of antarctic sponges from McMurdo Sound, Antarctica: chemical aspects, *Antarctic J. U. S.*, 28, 132, 1993.

35. McClintock, J. B., Slattery, M., Baker, B. J., and Heine, J. N., Chemical ecology of antarctic sponges from McMurdo Sound, Antarctica: Ecological aspects, *Antarctic J. U. S.*, 28, 134, 1993.

36. Baker, B. J., Yoshida, W. Y., and McClintock, J. B., Chemical constituents of four antarctic sponges in McMurdo Sound, Antarctica, *Antarctic J. U. S.*, 24, 153, 1994.

37. Baker, B. J., Kopitzke, R. W., Yoshida, W. Y., and McClintock, J. B., Chemical and ecological studies of the antarctic sponge *Dendrilla membranosa*, *J. Nat. Prod.*, 58, 1459, 1995.

38. Yang, A., Baker, B. J., Grimwade, J. E., Leonard, A. C., and McClintock, J. B., Discorhabdin alkaloids from the Antarctic sponge *Latrunculia apicalis*, *J. Nat. Prod.*, 58, 1596, 1995.

39. Jayatilake, G., Baker, B. J., and McClintock J. B., Rhapsamine, a cytotoxin from the Antarctic sponge *Leucetta leptorhapsis*, *Tetrahedron Lett.*, 38, 7507, 1997.

40. Moon, B. H., Baker, B. J., and McClintock, J. B., Purine and nucleoside metabolites from the Antarctic sponge *Isodictya erinacea*, *J. Nat. Prod.*, 61, 116, 1998.

41. Moon, B. H., Park, Y. C., McClintock, J. B., and Baker, B. J., Structure and bioactivity of erebusinone, a pigment from the Antarctic sponge *Isodictya erinacea*, *Tetrahedron*, 56, 9057, 2000.

42. Baker, B. J., Barlow, T. L., and McClintock, J. B., Evaluation of the functional role of suberitenones A and B from the sponge *Suberites* sp. found in McMurdo Sound, Antarctica, *Antarctic J. U. S.*, 32, 90, 1997.

43. Winston, J. E. and Bernheimer, A. W., Haemolytic activity in an antarctic bryozoan, *J. Nat. Hist.*, 20, 369, 1986.

44. McClintock, J. B., Baker, B. J., Slattery, M., Heine, J. N., Bryan, P. J., Yoshida, W., Davies-Coleman, M. T., and Faulkner, D. J., Chemical defense of the common Antarctic shallow-water nudibranch *Tritoniella belli* (Mollusca: Tritonidae) and its prey, *Clavularia frankliniana* Rouel (Cnidaria: Octocorallia), *J. Chem. Ecol.*, 20, 3361, 1994.

45. Slattery, M. and McClintock, J. B., An overview of the population biology and chemical ecology of three species of Antarctic soft corals, in *Antarctic Communities: Species, Structure and Survival*, Battaglia, B., Valencia, J., and Walton, D. W. H., Eds., Cambridge University Press, Cambridge, U.K., 1997, 309.

46. McClintock, J. B., Heine, J., Slattery, M., and Weston, J., Chemical bioactivity in shallow-water antarctic marine invertebrates, *Antarctic J. U. S.*, 25, 260, 1990.

47. Heine, J. N., McClintock, J. B., Slattery, M., and Weston, J., Energetic composition, biomass and chemical defense in the common antarctic nemertean *Parborlasia corrugatus*, *J. Exp. Mar. Biol. Ecol.*, 153, 15, 1991.

48. McClintock, J. B., Slattery, M., Heine, J., and Weston J., Chemical defense, biochemical composition and energy content of three shallow-water antarctic gastropods, *Polar Biol.*, 11, 623, 1992.

49. Bryan P. J., McClintock, J. B., and Baker, B. J., Population biology and antipredator defenses of the shallow water antarctic nudibranch *Tritoniella belli*, *Mar. Biol.*, 132, 259, 1998.

50. Avila, C., Iken, K., Fontana, A., and Cimino, G., Chemical ecology of the Antarctic nudibranch *Bathydoris hodgsoni* Eliot, 1907: defensive role and origin of its natural products, *J. Exp. Mar. Biol. Ecol.*, 252, 27, 2000.

51. McClintock, J. B., Baker, B. J., Hamann, M., Yoshida, W., Slattery, M., Heine, J., Jayatilake, G., Moon, B., and Bryan, P. J., Homarine as a feeding deterrent in the common shallow-water Antarctic lamellarian gastropod *Marseniopsis mollis*: a rare example of chemical defense in a marine proso-branch, *J. Chem. Ecol.*, 20, 2539, 1994.

52. McClintock, J. B. and Janssen, J., Pteropod abduction as a chemical defense in a pelagic antarctic amphipod, *Nature*, 346, 424, 1990.

53. Yoshida, W., Bryan, P., Baker, B. J., and McClintock, J. B., Pteroenone, the feeding deterrent substance from the Antarctic pteropod *Clione antarctica*, *J. Org. Chem.*, 60, 780, 1995.

54. Bryan, P., Yoshida, W., McClintock, J. B., and Baker, B. J., Ecological role for pteroenone, a novel antifeedant from the conspicuous Antarctic pteropod *Clione antarctica*, *Mar. Biol.*, 122, 272, 1995.
55. McClintock, J. B., Toxicity of shallow-water antarctic echinoderms, *Polar Biol.*, 9, 461, 1989.
56. McClintock, J. B. and Vernon, J., Chemical defense in the eggs and the embryos of antarctic sea stars (Echinodermata), *Mar. Biol.*, 105, 491, 1990.
57. De Marino, S., Iorizzi, M., Zollo, F., Amsler, C. D., Baker, B. J., McClintock, J. B., and Minale, L., Isolation, structure elucidation and biological activity of the steroid oligoglycosides and polyhydroxysteroids from the Antarctic sea star *Acodontaster conspicuus*, *J. Nat. Prod.*, 60, 959, 1997.
58. McClintock, J. B., Trophic biology of antarctic echinoderms, *Mar. Ecol. Prog. Ser.*, 111, 191, 1994.
59. McClintock, J. B. and Baker, B. J., Palatability and chemical defense in the eggs, embryos and larvae of shallow-water antarctic marine invertebrates, *Mar. Ecol. Prog. Ser.*, 154, 121, 1997.
60. McClintock, J. B., Heine, J., Slattery, M., and Weston, J., Biochemical and energetic composition, population biology, and chemical defense of the antarctic ascidian *Cnemidocarpa verrucosa* Lesson, *J. Exp. Mar. Biol. Ecol.*, 147, 163, 1991.
61. Amsler, C. D., McClintock, J. B., and Baker, B. J., Chemical defense against herbivory in the antarctic marine macroalgae *Iridaea cordata* and *Phyllophora antarctica* (Rhodophyceae), *J. Phycol.*, 34, 53, 1998.
62. Amsler, C. D., McClintock, J. B., and Baker, B. J., An Antarctic feeding triangle: defensive interactions between macroalgae, sea urchins, and sea anemones, *Mar. Ecol. Prog. Ser.*, 183, 105, 1999.
63. Amsler, C. D., McClintock, J. B., Tedeschi, D., Dunton, K. H., and Baker, B. J., Quantitative estimation of the phlorotannin content of three antarctic brown macroalgae, *Antarctic J. U. S.*, 32, 84, 1997.
64. McClintock, J. B. and Baker, B. J., Chemical feeding deterrent properties of the benthic algae *Phyllophora antarctica* and *Iridaea cordata* from McMurdo Sound, Antarctica, *Antarctic J. US*, 30, 155, 1995.
65. Iken, K., Feeding ecology of the Antarctic herbivorous gastropod *Laevilacunaria antarctica* Martens, *J. Exp. Mar. Biol. Ecol.*, 236, 133, 1999.
66. Iken, K., Trophische Beziehungen zwischen Makroalgen und Herbivoren in der Potter Cove (King-George-Insel, Antarktis), *Ber. Polarforsch.*, 201, 1, 1996.
67. McClintock, J. B., Slattery, M., and Thayer, C. W., Energy content and chemical defense of the articulate brachiopod *Liothyrella uva* from the Antarctic Peninsula, *J. Exp. Mar. Biol. Ecol.*, 169, 103, 1993.
68. McClintock, J. B. and Baker, B. J., A review of the chemical ecology of shallow-water antarctic marine invertebrates, *Am. Zool.*, 37, 329, 1997.
69. McClintock, J. B. and Baker, B. J., Chemical ecology in Antarctic seas, *Am. Sci.*, 86, 254, 1998.
70. Burton, M., Sponges, *Discovery Rep.*, 6, 237, 1932.
71. Koltun, V. M., Sponges of the arctic and antarctic: a faunistic review, *Symp. Zool. Soc. Lond.*, 25, 285, 1970.
72. Dayton, P. K., Observations of growth, dispersal and population dynamics of some sponges in McMurdo Sound, Antarctica, in *Sponge Biology*, Levi, C. and Bourny-Esnault, N., Eds., Centre National de la Recherche Scientifique (C. N. R. S.), Paris, 1979, 271.
73. Dayton, P. K., Interdecadal variation in an antarctic sponge and its predators from oceanographic climate shifts, *Science*, 245, 1484, 1989.
74. Arntz, W. E., Gutt, J., Klages, M., Antarctic marine biodiversity: an overview, in *Antarctic Communities: Species, Structure and Survival*, Battaglia, B., Valencia, J., and Walton, D. W. H., Eds., Cambridge University Press, Cambridge, U.K., 1997, 3.
75. Dayton, P. K., Robilliard, G. A., and DeVries, A. L., Anchor ice formation in McMurdo Sound, Antarctica and its biological effects, *Science*, 163, 173, 1969.
76. Dayton, P. K., Robilliard, G. A., and Paine, R. T., Benthic faunal zonation as a result of anchor ice at McMurdo Sound, Antarctica, in *Antarctic Ecology*, Holgate, M. W., Ed., Academic Press, London, 1970, 244.
77. Brey, T. and Gerdes, D., Is Antarctic benthic biomass really higher than elsewhere?, *Antarctic Sci.*, 9, 266, 1997.
78. Koltun, V. M., Spicules of sponges as an element of the bottom sediments of the Antarctic, in *Symposium on Antarctic Oceanography, Santiago de Chile 1966*, Scott Polar Research Institute, Cambridge, U.K., 1968, 1.

79. Harper, M. K., Bugni, T. S., Copp, B. R., James, R. D., Lindsay, B. S., Richardson, A. D., Schnabel, P. C., Tasdemir, D., VanWagoner, R. M., Verbitski, S. M., and Ireland, C. M., Introduction to the chemical ecology of marine natural products, in *Marine Chemical Ecology*, McClintock, J. B. and Baker, B. J., Eds., CRC Press, Boca Raton, FL, 2001.

80. Munro, M. H. G. and Blunt, J. W., MarinLit, version 10.4. Marine Chemistry Group, University of Canterbury, Christchurch, NZ, 1999.

81. Gutt, J. and Koltun, V. M., Sponges of the Lazarev and Weddell Sea, Antarctica: explanations for their patchy occurrence, *Antarctic Sci.*, 7, 227, 1995.

82. Molinski, T. F. and Faulkner, D. J., Metabolites of the Antarctic sponge *Dendrilla membranosa*, *J. Org. Chem.*, 52, 296, 1987.

83. Puliti, R., Fontana, A., Cimino, G., Mattia, C. A., and Mazzarella L., Structure of a keto derivative of 9,11-dihydrogracilin-A, *Acta Cryst. Sect. C*, 49, 1373, 1993.

84. Fontana, A., Scognamiglio, G., and Cimino, G., Dendrinolide, a new degraded diterpenoid from the Antarctic sponge *Dendrilla membranosa*, *J. Nat. Prod.*, 60, 475, 1997.

85. Molinski, T. F. and Faulkner, D. J., An antibacterial pigment from the sponge *Dendrilla membranosa*, *Tetrahedron Lett.*, 29, 2137, 1988.

86. Mayol, L., Piccialli, V., and Sica, D., Gracilin A, a unique nor-diterpene metabolite from the marine sponge *Spongionella gracilis*, *Tetrahedron*, 26, 1357, 1985.

87. Karuso, P., Skelton, B. W., Taylor, W. C., and White, A. H., The isolation from *Aplysilla sulphurea* of 1-acetoxy-4-ethyl-5-(1,3,3-trimethylcyclohexyl)-1;3-dihydroisobenzofuran-1′(4),3-carbolactone and determination of the crystal structure, *Aust. J. Chem.*, 37, 1081, 1984.

88. Manriquez, V., San-Martin, A., Rovirosa, J., Darias, J., and Peters, K., Structure of membranolide, a diterpene from the antarctic sponge *Dendrilla membranosa*, *Acta Cryst. Sect. C*, 46, 2486, 1990.

89. Yoo, S.-E. and Yi, K. Y., Total synthesis of (±)-membranolide, *Synlett*, 697, 1990.

90. Shin, J., Seo, Y., Rho, J. R., Baek, E., Kwon, B. M., Jeong, T. S., and Bok, S. H., Suberitenones A and B: sesterterpenoids of an unprecedented skeletal class from the Antarctic sponge *Suberites* sp., *J. Org. Chem.*, 60, 7582, 1995.

91. Fontana, A., Ciavatta, M. L., Amodeo, P., and Cimino, G., Single solution phase conformation of new antiproliferative cembranes, *Tetrahedron*, 55, 1143, 1999.

92. Faulkner, D. J., Marine natural products, *Nat. Prod. Rep.*, 16, 155, 1999.

93. Urban, S., Wilton, H., Lu, C. C., and Capon, R. J., A new sesquiterpene alcohol from an Antarctic sponge, *Nat. Prod. Lett.*, 6, 187, 1995.

94. Ding, Q. Z., Chichak, K., and Lown, J. W., Pyrroloquinoline and pyridoacridine alkaloids from marine sources, *Curr. Med. Chem.*, 6, 1, 1999.

95. Perry, N. B., Blunt, J. W., McCombs, J. D., and Munro, M. H. G., Discorhabdin C; a highly cytotoxic pigment from a sponge of the genus *Latrunculia*, *J. Org. Chem.*, 51, 5476, 1986.

96. Yang, A., Chemical and biological studies of the antarctic sponge *Latrunculia apicalis*, Master's thesis, Florida Institute of Technology, Melbourne, FL, 1994.

97. Perry, N. B., Ettouati, L., Litaudon, M., Blunt, J. W., Munro, M. H. G., Parkin, S., and Hope, H., Alkaloids from the Antarctic sponge *Kirkpatrickia varialosa*. Part 1: Variolin B, a new antitumor and antiviral compound, *Tetrahedron*, 50, 3987, 1994.

98. Trimurtulu, G., Faulkner, D. J., Perry, N. B., Ettouati, L., Litaudon, M., Blunt, J. W., Munro, M. H. G., and Jameson, G. B., Alkaloids from the Antarctic sponge *Kirkpatrickia varialosa*. Part 2: Variolin A and N(3′)-methyl tetrahydrovariolin B, *Tetrahedron*, 50, 3993, 1994.

99. Jayatilake, G. S., Baker, B. J., and McClintock, J. B., 1,3,5′-Triacetoxystilbene, a plant growth regulator, from the Antarctic sponge *Kirkpatrickia variolosa*, *J. Nat. Prod.*, 58, 1958, 1995.

100. Brinker, A. M. and Seigler, D. S., Isolation and identification of piceatannol as a phytoalexin from sugarcane, *Phytochemistry*, 30, 3229, 1991.

101. Bezhuashvili, M. G., Mudzhiri, L. A., Kurkin, V. A., and Zapesochnya, G. G., Stilbene from grapevine, *Khim. Drev.*, 75, 1991.

102. Jang, M., Cai, L., Udeani, G. O., Slowing, K. V., Thomas, C. F., Beecher, C. W. W., Fong, H. H. S., Farnsworth, N. R., Kinghorn, A. D., Mehta, R. G., Moon, R. C., and Pezzuto, J. M., Cancer chemo-preventive activity of resveratrol, a natural product derived from grapes, *Science*, 275, 218, 1997.

103. Butler, M. S., Capon, R. J., and Lu, C. C., Psammopemmins (A–C), novel brominated 4-hydroxyindole alkaloids from an Antarctic sponge, *Psammopemma* sp., *Aust. J. Chem.*, 45, 1871, 1992.

104. Casapullo, A., Fontana, A., and Cimino, G., Coriacenins: a new class of long alkyl chain amino alcohols from the Mediterranean sponge *Clathrina coriacea*, *J. Org. Chem.*, 61, 7415, 1996.

105. Amsler, C. D., Rowley, R. J., Laur, D. R., Quetin, L. B., and Ross, R. M., Vertical distribution of antarctic peninsular macroalgae: cover, biomass and species composition, *Phycologia*, 34, 424, 1995.

106. Brouwer, P. E. M., Geilen, E. F. M., Gremmen, N. J. M., and van Lent, F., Biomass, cover and zonation pattern of sublittoral macroalgae at Signy Island, South Orkeny Islands, Antarctica, *Bot. Mar.*, 38, 259, 1995.

107. Klöser, H., Quartino, M. L., and Wiencke, C., Distribution of macroalgae and macroalgal communities in gradients of physical conditions in Potter Cove, King George Island, Antarctica, *Hydrobiology*, 333, 1, 1996.

108. Klöser, H., Mercuri, G., Laturnus, F., Quartino, M. L., and Wiencke, C., On the competitive balance of macroalgae at Potter Cove (King George Island, South Shetlands), *Pol. Biol.*, 14, 11, 1994.

109. Richardson, M. D., The ecology and reproduction of the brooding antarctic bivalve *Lissarca miliaris*, *Br. Antarctic Surv. Bull.*, 45, 19, 1979.

110. DeLaca, T. E. and Lipps, J. H., Shallow-water marine associations, Antarctic Peninsula, *Antarctic J. U. S.*, 11, 12, 1976.

111. Gilbert, N. S., Microphytobenthic seasonality in near-shore marine sediments at Signy Island, South Orkney Islands, Antarctica, *Estuarine Coastal Shelf Sci.*, 33, 89, 1991.

112. Miller, K. A. and Pearse, J. S., Ecological studies of seaweeds in McMurdo Sound, Antarctica, *Am. Zool.*, 31, 35, 1991.

113. Iken, K., Barrera-Oro, E. R., Quartino, M. L., Casaux, R. J., and Brey, T., Grazing in the Antarctic fish *Notothenia coriiceps*: evidence for selective feeding on macroalgae, *Antarctic Sci.*, 9, 386, 1997.

114. Shabica, S. V., The general ecology of the antarctic limpet *Patinigera polaris*, *Antarctic J. U. S.*, 6, 160, 1971.

115. Pearse, J. S. and Giese, A. C., Food, reproduction and organic constitution of the common antarctic echinoid *Sterechinus neumayeri* (Meissner), *Biol. Bull.*, 130, 387, 1966.

116. Daniels, R. A., Feeding ecology of some fishes of the Antarctic Peninsula, *Fish. Bull.*, 80, 575, 1981.

117. Steneck, R. S. and Watling, L., Feeding capabilities and limitation of herbivorous molluscs: a functional approach, *Mar. Biol.*, 68, 299, 1982.

118. Schupp, P. J. and Paul, V. J., Calcification and secondary metabolites in tropical seaweeds: variable effects on herbivorous fishes, *Ecology*, 75, 1172, 1994.

119. Littler, M. M. and Littler, D. S., The evolution of thallus form and survival strategies in benthic marine macroalgae: field and laboratory tests of a functional form model, *Amer. Nat.*, 116, 25, 1980.

120. Padilla, D. K., Algal structural defenses: form and calcification in resistance to tropical limpets, *Ecology*, 70, 835, 1989.

121. Hay, M. E. and Fenical, W., Marine plant–herbivore interactions: the ecology of chemical defense, *Ann. Rev. Ecol. Syst.*, 19, 111, 1988.

122. Hay, M. E. and Fenical, W., Chemical mediation of seaweed–herbivore interactions, in *Plant–Animal Interactions in the Marine Benthos*, John, V. M., Hawkins, S. J., and Price, J. H., Eds., Systematic Association, Clarendon Press, Oxford, UK, 1992, 317.

123. Hay, M. E., Marine chemical ecology: what's known and what's next?, *J. Exp. Mar. Biol. Ecol.*, 200, 103, 1996.

124. Paul, V. J., Seaweed chemical defenses on coral reefs, in *Ecological Roles of Marine Natural Products*, Paul, V. J., Ed., Cornell University Press, Ithaca, NY, 1992, 24.

125. Steinberg, P. D., Geographical variation in the interaction between marine herbivores and brown algal secondary metabolites, in *Ecological Roles of Marine Natural Products*, Paul, V. J., Ed., Cornell University Press, Ithaca, NY, 1992, 51.

126. Erickson, K., Constituents of *Laurencia*, in *Marine Natural Products, Chemical and Biological Perspectives*, Vol. 5, Scheuer P. J., Ed., Academic Press, New York, 1984, 131.

127. Pettus, J. A., Wing, R. M., and Sims, J. J., Marine natural products XII. Isolation of a family of multihalogenated γ-methylene lactones from the red seaweed *Delisea fimbriata*, *Tetrahedron Lett.*, 41, 1977.

128. Stierle, D. B. and Sims, J. J., Polyhalogenated cyclic monoterpenes from the red alga *Plocamium cartilagineum* of Antarctica, *Tetrahedron*, 35, 1261, 1979.

129. Stierle, D. B., Wing, R. M., and Sims, J. J., Marine natural products XVI. Polyhalogenated acyclic monoterpenes from red alga *Plocamium* of Antarctica, *Tetrahedron*, 35, 2855, 1979.

130. Cueto, M., Darias, J., Rovirosa, J., and San-Martin, A., Pantoneurotriols: probable biogenic precursors of oxygenated monoterpenes from Antarctic *Pantoneura plocamioides*, *Tetrahedron*, 54, 3575, 1998.

131. Rovirosa, J., Sanchez, I., Palacios, Y., Darias, J., and San-Martin, A., Antimicrobial activity of a new monoterpene from *Plocamium cartilagineum* from Antarctic peninsula, *Bol. Soc. Chil. Quim.*, 35, 131, 1990.

132. Cueto, M., Darias, J., Rovirosa, J., and San-Martin, A., Unusual polyoxygenated monoterpenes from the Antarctic alga *Pantoneura plocamioides*, *J. Nat. Prod.*, 61, 17, 1998.

133. Fuller, R. W., Cardellina, J. H., II, Jurek, J., Scheuer, P. J., Alvarado-Lindner, B., McGuire, M., Gray, G. N., Steiner, J. R., Clardy, J., Menez, E., Shoemaker, R. H., Newman, D. J., Snader, K. M., and Boyd, M. R., Isolation and structure/activity features of halomon-related antitumor monoterpenes from the red alga *Portieria hornemannii*, *J. Med. Chem.*, 37, 4407, 1994.

134. König, G. M. and Wright, A. D., Marine natural products research. Current directions and future potential, *Planta Med.*, 62, 193, 1996.

135. Blunt, J. W., Bowman, N. J., Munro, M. H. G., Parsons, M. J., Wright, G. J., and Kon, Y. K., Polyhalogenated monoterpenes of the New Zealand marine red alga *Plocamium cartilagineum*, *Aust. J. Chem.*, 38, 519, 1985.

136. Coll, J. C. and Wright, A. D., Tropical marine algae. VI. New monoterpenes from several collections of *Chondrococcus hornemanni* (Rhodophyta, Gigartinales, Rizophyllidaceae), *Aust. J. Chem.*, 42, 1983, 1989.

137. Wright, A. D., König, G. M., Sticher, O., and de Nys, R., Five new monoterpenes from the marine red alga *Portieria hornemannii*, *Tetrahedron*, 47, 5717, 1991.

138. Darias, J., Rovirosa, J., and San-Martin, A., Estudio Quimiotaxonomico de organismos de la antartida, in Proceedings Actas del II Symposium Espanol de Estudios Antarcticos, Madrid, 1987, 89.

139. Cueto, M., Darias, J., San-Martin, A., Rovirosa, J., Seldes, A., Metabolitos secundarios de organismos marinos antarticos, in Proceedings Actas del IV Symposium Espanol de Estudios Antarcticos, Madrid, 1991, 95.

140. Rovirosa, J., Sanchez, I., Palacios, Y., Darias, J., and San-Martin, A., Actividad antimicrobiana de un nuevo monoterpeno del *Plocamium cartilagineum* de la peninsula antarctica, in Proceedings Actas del II Symposium Espanol de Estudios Antarcticos, Gredos, 1987, 128.

141. Cueto, M. and Darias, J., Uncommon tetrahydrofuran monoterpenes from Antarctic *Pantoneura plocamioides*, *Tetrahedron*, 52, 5899, 1996.

142. Cueto, M., Darias, J., Rovirosa, J., and San-Martin, A., Tetrahydropyran monoterpenes from *Plocamium cartilagineum* and *Pantoneura plocamioides*, *J. Nat. Prod.*, 61, 1466, 1998.

143. Cueto, M., Darias, J., San-Martin, A., and Rovirosa, J., New acetyl derivatives from Antarctic *Delisea fimbriata*, *J. Nat. Prod.*, 60, 279, 1997.

144. Burreson, B. J., Moore, R. E., and Roller, P. P., Volatile halogen compounds in the alga *Asparagopsis taxiformis* (Rhodophyta), *J. Agricul. Food Chem.*, 24, 856, 1976.

145. Gribble, G., Naturally occurring organohalogen compounds — a survey, *J. Nat. Prod.*, 55, 1353, 1992.

146. Moore, R. E., Algal nonisoprenoids, in *Marine Natural Products, Chemical and Biological Perspectives*, Vol. 1, Scheuer P. J., Ed., Academic Press, New York, 1978, 44.

147. Laturnus, F., Wiencke, C., and Klöser, H., Antarctic macroalgae — sources of volatile halogenated organic compounds, *Mar. Env. Res.*, 41, 169, 1996.

148. Giese, B., Laturnus, F., Adams, F. C., and Wiencke, C., Release of volatile iodinated C_1–C_4 hydrocarbons by marine macroalgae from various climate zones, *Environ. Sci. Technol.*, 33, 2432, 1999.

149. Rivera, P. L., Plastoquinones and a chromene isolated from the antarctic brown alga *Desmarestia menziesii*, *Bol. Soc. Chil. Quim.*, 41, 103, 1996.

150. Rivera, P. L., Podesta, F., Norte, M., Cataldo, F., and Gonzalez, A. G., New plastoquinones from the brown alga *Desmarestia menziesii*, *Can. J. Chem.*, 68, 1399, 1990.

151. Neushul, M., Diving observation of subtidal antarctic marine vegetation, *Bot. Mar.*, 8, 234, 1965.

152. Fenical, W., Halogenation in the Rhodophyta, *J. Phycol.*, 11, 245, 1975.

153. Faulkner, D. J. and Ghiselin, M. T., Chemical defense and evolutionary ecology of dorid nudibranchs and some other opisthobranch gastropods, *Mar. Ecol. Prog. Ser.*, 13, 295, 1983.

154. Avila, C., Natural products of opisthobranch molluscs: a biological review, *Oceanogr. Mar. Biol. Ann. Rev.*, 33, 487, 1995.

155. Karuso, P., Chemical ecology of the nudibranchs, in *Bioorganic Marine Chemistry*, Vol. 1, Scheuer, P. J., Ed., Springer-Verlag, New York, 1987, 31.

156. Faulkner, D. J. Chemical defenses of marine molluscs, in *Ecological Roles of Marine Natural Products*, Paul, V. J., Ed., Comstock Publishing Associates, Ithaca, New York, 1992, 119.

157. Cimino, G., De Rosa, S., De Stefano, S., Sodano, G., and Villani, G., Dorid nudibranch elaborates its own chemical defense, *Science*, 219, 1237, 1983.

158. Davies-Coleman, M. T. and Faulkner, D. J., New diterpenoic acid glycerides from the antarctic nudibranch *Austrodoris kerguelensis*, *Tetrahedron*, 47, 9743, 1991.

159. Gavagnin, M., Trivellone, E., Castelluccio, F., Cimino, G., and Cattaneo-Vietti, R., Glyceryl ester of a new halimane diterpenoic acid from the skin of the Antarctic nudibranch *Austrodoris kerguelensis*, *Tetrahedron Lett.*, 36, 7319, 1995.

160. Gavagnin, M., Ungur, N., Castelluccio, F., Muniain, C., and Cimino G., New minor diterpenoid diacylglycerols from the skin of the nudibranch *Anisodoris fontaini*, *J. Nat. Prod.*, 62, 269, 1999.

161. Ungur, N., Gavagnin, M., Fontana, A., and Cimino, G., Absolute stereochemistry of natural sesquiterpenoid diacylglycerols, *Tetrahedron Asym.*, 10, 1263, 1999.

162. Gavagnin, M., De Napoli, A., Cimino, G., Iken, K., Avila, C., and Garcia, F. J., Absolute stereochemistry of diterpenoid diacylglycerols from the Antarctic nudibranch *Austrodoris kerguelenensis*, *Tetrahedron Asym.*, 10, 2647, 1999.

163. Iken, K., Avila, C., Fontana, A., and Gavagnin, M., in preparation.

164. Iken, K., Avila, C., Ciavatta, M. L., Fontana, A., and Cimino, G., Hodgsonal, a new drimane sesquiterpene from the mantle of the antarctic nudibranch *Bathydoris hodgsoni*, *Tetrahedron Lett.*, 39, 5635, 1998.

165. Gasteiger, E. L., Gergen, J., and Haake, P., A study of the distribution of homarine (*N*-methyl picolinic acid), *Biol. Bull.*, 109, 345, 1955.

166. Kerr, R. G. and Baker, B. J., Marine sterols, *Nat. Prod. Reports*, 8, 465, 1991.

167. D'Auria, M. V., Minale, L., and Riccio, R., Polyoxygenated steroids of marine origin, *Chem. Rev.*, 93, 1939, 1993.

168. Kong F., Harper, M. K., and Faulkner, D. J., Fuscusine, a tetrahydroisoquinoline alkaloid from the seastar *Perknaster fuscus antarcticus*, *Nat. Prod. Lett.*, 1, 71, 1992.

169. Pathirana, C. and Andersen, R. J., Imbricatine, an unusual benzyltetrahydroisoquinoline alkaloid isolated from the starfish *Dermasterias imbricata*, *J. Am. Chem. Soc.*, 108, 8288, 1986.

170. Burgoyne, D. L., Miao, S., Pathirana, C., Andersen, R. J., Ayer, W. A., Singer, P. P., Kokke, W. C. M. C., and Ross, D. M., The structure and partial synthesis of imbricatine, a benzyltetrahydroisoquinoline alkaloid from the starfish *Dermasterias imbricata*, *Can. J. Chem.*, 69, 20, 1991.

171. Elliott, J. K., Ross, D. M., Pathirana, C., Miao, S., Andersen, R. J., Singer, P., Kokke, W. C. M. C., and Ayer, W. A., Induction of swimming in *Stomphia* (Anthozoa: Actiniaria) by imbricatine, a metabolite of the asteroid *Dermasterias imbricata*, *Biol. Bull.*, 176, 73, 1989.

172. D'Auria, M. V., Paloma, L. G., Minale, L., Riccio, R., and Zampella, A., Isolation and structure characterization of two novel bioactive sulfated polyhydroxysteroids from the Antarctic ophiuroid *Ophioderma longicaudum*, *Nat. Prod. Lett.*, 3, 197, 1993.

173. D'Auria, M. V., Paloma, L. G., Minale, L., Riccio, R., and Zampella, A., On the composition of sulfated polyhyroxysteroids in some ophiuroids and the structure determination of six new constituents, *J. Nat. Prod.*, 58, 189, 1995.

174. Roccatagliata, A. J., Maier, M. S., and Seldes, A. M., New sulfated polyhydroxysteroids from the Antarctic ophiuroid *Astrotoma agassizii*, *J. Nat. Prod.*, 61, 370, 1998.

175. D'Auria, M. V., Riccio, R., Minale, L., La Barre, S., and Pusset, J., Novel marine steroid sulphates from Pacific ophiuroids, *J. Org. Chem.*, 52, 3947, 1987.

176. Fusetani, N., Matsunaga, S., and Konosu, S., Halistanol sulfate; an antimicrobial steroid sulfate from *Halichondria cf. moorei* Bergquist, *Tetrahedron Lett.*, 22, 1985, 1981.

177. De Marino, S., Palagiano, E., Zollo, F., Minale, L., and Iorizzi, M., A novel sulphated steroid with a 7-membered 5-oxalactone B-ring from an Antarctic starfish of the family Asteriidae, *Tetrahedron*, 57, 8625, 1997.

178. Iorizzi, M., De Marino, S., Minale, L., Zollo, F., Le Bert, V., and Roussakis, C., Investigation of the polar steroids from an Antarctic starfish of the family Echinasteridae: isolation of 27 polyhydroxysteroids and steroidal oligoglycosides, *Tetrahedron*, 52, 10997, 1996.

179. Iorizzi, M., Minale, L., Riccio, R., Higa, T., and Tanaka, J., Starfish saponins, part 46. Steroidal glycosides and polyhydroxysteroids from the starfish *Culcita novaguineae*, *J. Nat. Prod.*, 54, 12254, 1991.

180. Adam, G. and Marquardt, V., Brassinosteroids, *Phytochemistry*, 25, 1787, 1986.

181. Vaquez, M. J., Quinoa, E., Riguera, R., San-Martin, A., and Darias, J., Santiagoside, the first asterosaponin from an Antarctic starfish (*Neosmilaster georgianus*), *Tetrahedron*, 48, 6739, 1992.

182. De Marino, S., Minale, L., Zollo, F., Iorizzi, M., Le Bert, V., and Roussakis, C., Starfish saponins: part LIV. Cytotoxic asterosaponins from an Antarctic starfish of the family Echinasteridae, *Gazz. Chim. Ital.*, 126, 667, 1996.

183. Iorizzi, M., de Riccardis, F., Minale, L., and Riccio, R., Starfish saponins, 52. Chemical constituents from the starfish *Echinaster brasiliensis*, *J. Nat. Prod.*, 56, 2149, 1993.

184. Noguchi, Y., Higuchi, R., Marubayashi, N., and Komori, T., Steroid oligoglycosides from starfish *Asterina pectinifera*, 1. Structure of two sapogenins and two oligoglycoside sulfates: pectinioside A and B, *Liebigs Ann. Chem.*, 341, 1987.

185. Iorizzi, M., Minale, L., and Riccio, R., Starfish saponins. part 39. Steroidal oligoglycoside sulphates and polyhydroxysteroids from the starfish *Asterina pectinifera*, *Gazz. Chim. Ital.*, 120, 147, 1990.

186. Ikegami, S., Okano, K., and Muragaki, H., Structure of glycoside B2, a steroidal saponin in the ovary of the starfish *Asterias amurensis*, *Tetrahedron Lett.*, 1769, 1979.

187. Riccio, R., Iorizzi, M., Minale, L., Oshima, Y., and Yasumoto, T., Starfish saponins. part 34. Novel steroidal glycoside sulphates from the starfish *Asterias amurensis*, *J. Chem. Soc. Perkin Trans.*, 1, 1337, 1988.

188. Iorizzi, M., Minale, L., Riccio, R., Debray, M., and Menou, J. L., Starfish saponins; part 23. Steroidal glycosides from the starfish *Halityle regularis*, *J. Nat. Prod.*, 49, 67, 1986.

189. Vazquez, M. J., Quinoa, E., Riguera, R., San-Martin, A., and Darias, J., Antarctic marine metabolites: new polyhydroxylated steroidal glycosides from the starfish *Odontaster validus*, *Liebigs Ann. Chem.*, 1257, 1993.

190. De Marino, S., Iorizzi, M., Palagiano, E., Zollo, F., and Roussakis, C. Starfish saponins 55. Isolation, structure elucidation and biological activity of the steroid oligoglycosides from an Antarctic starfish of the family Asteriidae, *J. Nat. Prod.*, 61, 1319, 1998.

191. Zollo, F., Finamore, E., and Minale, L., Starfish saponins, XXXI. Novel polyhydroxysteroids and steroidal glycosides from the starfish *Sphaerodiscus placenta*, *J. Nat. Prod.*, 50, 794, 1987.

192. Riccio, R., Iorizzi, M., and Minale, L., Starfish saponins XXX. Isolation of sixteen steroidal glycosides and three polyhydroxysteroids from the Mediterranean starfish *Coscinasterias tenuispina*, *Bull. Soc. Chim. Belg.*, 95, 869, 1986.

193. Jayatilake, G. S., Baker, B. J., Thornton, M. P., Leonard, A. C., and Grimwade, J. E., Metabolites from *Pseudomonas aeruginosa* separated from the antarctic sponge *Isodictya setifera*, *J. Nat. Prod.*, 59, 293, 1996.

194. Fenical, W., Chemical studies of marine bacteria: developing a new resource, *Chem. Rev.* 93, 1673, 1993.

195. Guella, G., Dini, F., and Pietra, F., Epoxyfocardin and its putative biogenic precursor, focardin, bioactive, new-skeleton diterpenoids of the marine ciliate *Euplotes focardii* from Antarctica, *Helv. Chim. Acta*, 79, 439, 1996.

196. Goerke, H., Emrich, R., Weber, K., and Duchene, J. C., Concentration and localization of brominated metabolites in the genus *Thelepus* (Polychaeta: Terebellidae), *Comp. Biochem. Physiol.*, 99B, 203, 1991.

197. Loeb, V., Siegel, V., Holm-Hansen, O., Hewitt, R., Fraser, W., Trivelpiece, W., and Trivelpiece, S., Effects of sea-ice extent and krill or salp dominance on the antarctic food web, *Nature*, 387, 897, 1997.

198. Mimura, T., Okabe, M., Satake, M., Nakanishi, T., Inada, A., Fujimoto, Y., Hata, F., Matsumura, Y., and Ikekawa, N., Fatty acids and sterols of the tunicate, *Salpa thompsoni*, from the Antarctic Ocean: chemical composition and hemolytic activity, *Chem. Pharm. Bull.*, 34, 4562, 1986.

199. Schor, L. and Seldes, A. M., Steroids from aquatic organisms — XVII. Sterol composition of the salp *Ihlea racovitzai* from the Antarctic Ocean, *Comp. Biochem. Physiol.*, 92B, 195, 1989.

200. Mezykowski, T. and Ignatowska-Switalska, H., High levels of prostaglandins $PGF_{2\alpha}$ and PGE_2 in antarctic krill *Euphausia superba* Dana, *Meeresforschung*, 29, 64, 1981.

201. Pawlowicz, J. M., Identification and quantification of prostaglandins in antarctic krill (*Euphausia superba* Dana), *Polar Biol.*, 9, 295, 1989.

202. Weinheimer, A. J. and Spraggins, R. L., The occurrence of two new prostaglandin derivatives (15-*epi*-PGA_2 and its acetate methyl ester) in the gorgonian *Plexaura homomalla*. Chemistry of coelenterates XV, *Tetrahedron Lett.*, 59, 5185, 1969.

203. Gerhart, D. J., Prostaglandin A_2: an agent of chemical defense in the Caribbean gorgonian *Plexaura homomalla*, *Mar. Ecol. Prog. Ser.*, 19, 181, 1984.

204. Williams, D. H., Stone, M. J., Hauch, P. R., and Rahman, S. K., Why are secondary metabolites (natural products) biosynthesized?, *J. Nat. Prod.*, 52, 1189, 1989.

205. Williams, D. H. and Maplestone, R. A., Why are secondary metabolites synthesized? Sophistication in the inhibition of cell wall biosynthesis by vancomycin group antibiotics, in *Secondary Metabolites: Their Function and Evolution*, Chadwick, D. J. and Whelan, J., Eds., Ciba Foundation Symposium 171, John Wiley & Sons, New York, 1992, 45.

206. Mauricio, R. and Rausher, M. D., Experimental manipulation of putative selective agents provides evidence for the role of natural enemies in the evolution of plant defense, *Evolution*, 51, 1435, 1997.

207. Agrawal, A. A., Induced responses to herbivory and increased plant performance, *Science* 279, 1201, 1998.

208. Hyman, L. H., *The Invertebrates: Echinodermata*, McGraw-Hill, New York, 1955.

209. Thompson, G. A., Jr. and Lee, P., Studies of the α-glycerol ether lipids occurring in molluscan tissues, *Biochem. Biophys. Acta*, 98, 151, 1965.

210. Gustafson, K. and Andersen, R. J., Chemical studies of British Columbia nudibranchs, *Tetrahedron*, 41, 1101, 1985.

211. Targett, N. M., Bishop, S. S., McConnell, O. J., and Yoder, J. A., Antifouling agents against the benthic marine diatom *Navicula salinicola*: homarine from the gorgonians *Leptogorgia virgulata* and *L. setacea* and analogs, *J. Chem. Ecol.*, 9, 817, 1983.

212. Davis, A. R., Targett, N. M., McConnell, O. J., and Young, C. M., Epibiosis of marine algae and benthic invertebrates: natural products chemistry and other mechanisms inhibiting settlement and overgrowth, in *Bioorganic Marine Chemistry*, Vol. 3, Scheuer, P. J., Ed., Springer-Verlag, Berlin, 1989, 85.

213. Paul, V. J. and Fenical, W., Natural products chemistry and chemical defense in tropical marine algae of the order Chlorophyta, in *Bioorganic Marine Chemistry*, Vol. 1, Scheuer, P. J., Ed., Springer-Verlag, Berlin, 1987, 1.

214. Pawlik, J. R., Chemical ecology of the settlement of benthic marine invertebrates, *Oceanogr. Mar. Biol. Ann. Rev.*, 30, 273, 1992.

215. Bryan, P. J., Rittschof, D., and McClintock, J. B., Bioactivity of echinoderm ethanolic body-wall extracts: an assessment of marine bacterial attachment and macroinvertebrate larval settlement, *J. Exp. Mar. Biol. Ecol.*, 196, 79, 1996.

216. Thornton, M., Characterization of secondary metabolite bioactivity and production in bacteria associated with the antarctic marine sponge *Isodictya setifera*, Master's thesis, Florida Institute of Technology, Melbourne, FL, 1995.

217. Moeller, C. B., Aspects of the chemical ecology of antarctic marine sponges, Master's thesis, University of Alabama at Birmingham, Birmingham, Alabama, 1998.

218. Amsler, C. D., Moeller, C. B., McClintock, J. B., Iken, K. B., and Baker, B. J., Chemical defense against diatom fouling in antarctic marine sponges, *Biofouling*, 16, 29, 2000.

219. Regan, J. D., Carrier, W. L., Gucinski, H., Olla, B. L., Yoshida, H., Fujimura, R. K., and Wicklund, R. I., DNA as a solar dosimeter in the ocean, *Phytochem. Photobiol.*, 56, 35, 1992.

220. Stackowitz, J. J. and Lindquist, N., Chemical defense among hydroids on pelagic *Sargassum*: predator deterrence and absorption of solar UV radiation by secondary metabolites, *Mar. Ecol. Prog. Ser.*, 155, 115, 1997.

221. Karentz, D., McEuen, F. S., Land, M. C., and Dunlap, W. C., Survey of microsporine-like amino acid compounds in Antarctic marine organisms: potential protection from ultraviolet exposure, *Mar. Biol. (Berlin)*, 108, 157, 1991.

222. Karentz, D., Chemical defenses of marine organisms against solar radiation exposure: UV-absorbing mycosporine-like amino acids and scytonemin, in *Marine Chemical Ecology*, McClintock, J. B. and Baker, B. J., Eds., CRC Press, Boca Raton, FL, 2001, 15.

223. McClintock, J. B. and Karentz, D., Mycosporine-like amino acids in 38 species of subtidal marine organisms from McMurdo Sound, Antarctica, *Antarctic Sci.*, 9, 392, 1997.

224. Hoyer, K., Karsten, U., Sawall, T., and Wiencke, C., Photoprotective substances in Antarctic macroalgae and their variation with respect to depth distribution, different tissues and developmental stages, *Mar. Ecol. Prog. Ser.*, 211, 117, 2001.

225. Bandaranayake, W. M., Mycosporines: are they nature's sunscreens?, *Nat. Prod. Rep.*, 15, 159, 1998.

226. Rhoades, D., Evolution of plant chemical defenses against herbivores, in *Herbivores*, Rosenthal, G. A. and Janzen, D. H., Eds., Academic Press, New York, 1979, 4.

227. Gittleman, J. L., Harvey, P. H., and Greenwood, P. J., The evolution of conspicuous coloration: some experiments in bad taste, *Anim. Behav.*, 28, 897, 1980.

228. Guilford, T., The evolution of conspicuous coloration, in *Mimicry and the Evolutionary Process*, Brower, L. P., Ed., University of Chicago Press, Chicago, IL, 1988, 7.

229. Lobban, C. S. and Harrison, P. J., *Seaweed Ecology and Physiology*, Cambridge University Press, Cambridge, UK, 1994.

230. Guilford, T. and Cuthill, J., Evolution of aposematism in marine gastropods, *Evolution*, 45, 449, 1991.

231. Chanas, B. and Pawlik, J. R., Defenses of Caribbean sponges against predatory reef fish. II. Spicules, tissue toughness, and nutritional quality, *Mar. Ecol. Prog. Ser.*, 127, 195, 1995.

232. Furrow, F. B., Tipton, J. D., McClintock, J. B., and Baker, B. J., in preparation.

233. Thompson, J. E., Barrow, K. D., and Faulkner, D. J., Localization of two brominated metabolites, aerothionin and homoaerothionin, in spherulous cells of the marine sponge *Aplysina fistularis* (=*Verongia thiona*), *Acta Zool. (Stockholm)*, 64, 199, 1983.

234. Schupp, P., Eder, C., Paul, V., and Proksch, P., Distribution of secondary metabolites in the sponge *Oceanapia* sp. and its ecological implications, *Mar. Biol.*, 135, 573, 1999.

235. Coley, P. D., Bryant, J. B., and Chapin, F. S., III, Resource availability and plant antiherbivore defense, *Science*, 230, 895, 1985.

236. Bryant, J. P., Chapin, F. S., III, and Klein, D. R., Carbon/nutrient balance of boreal plants in relation to vertebrate herbivory, *Oikos*, 40, 357, 1983.

8 Spatial Patterns in Macroalgal Chemical Defenses

Kathryn L. Van Alstyne, *Megan N. Dethier,*
and David O. Duggins

CONTENTS

I. INTRODUCTION

Secondary metabolites are important for providing marine algae with protection from grazing. There are numerous examples of compounds produced by marine macroalgae that either reduce grazing or have physiological effects on consumers.[1-6] Terpenes, phlorotannins, fatty acids, nitrogenous compounds, and compounds of mixed biogenesis are among the most common secondary metabolites that have been reported for marine macroalgae;[7-13] however, for many of these compounds, no ecological or physiological function has been determined.

Terpenes and phlorotannins are two structural classes of secondary metabolites that are particularly important in mediating chemical interactions between marine plants and herbivores. Terpenes are commonly found in tropical macroalgae such as members of the Dictyotales (Phaeophyta) and

---

* Corresponding author.

the Caulerpales (Chlorophyta) and are deterrent towards a variety of fishes and benthic invertebrates.[1,3,14–18] Phlorotannins are polymers of phloroglucinol that are found in most species of marine brown algae.[19] Isolated phlorotannins are deterrent towards many marine herbivores including the gastropod snails *Littorina littorea*,[20,21] *Littorina obtusata*,[22] *Tegula funebralis*,[23] *Tegula brunnea*,[23] and *Haliotus rufescens*,[24] and the sea urchin *Strongylocentrotus purpuratus*.[23] Phlorotannin-rich methanol extracts deter feeding by tropical Pacific fishes whereas phlorotannin-poor extracts do not.[25] However, not all herbivores are deterred by phlorotannins. Several Australasian herbivores, including the gastropod molluscs *Turbo smaragdus* and *Cookia sulcata* and the urchins *Centrostephanus rodgersii* and *Evechinus chloroticus*,[26] are unaffected by the addition of phlorotannins to diets. The northern Atlantic isopod *Idotea granulosa* is also not deterred by these compounds.[22]

Concentrations of secondary metabolites in marine macroalgae are extremely variable on a wide range of scales. Closely related species can produce vastly different concentrations of secondary metabolites or compounds that are structurally unrelated. Concentrations of secondary metabolites also vary spatially and temporally. Among the best-studied examples of variation in secondary metabolite concentrations are those involving phlorotannins. Variation in phlorotannin concentrations has been reported from virtually every possible spatial scale: between cell layers,[27] among tissues within individuals,[20,27–36] among sites separated by meters to hunderds of kilometers,[33,37,38] and among global regions.[25,26,39–42] Temporal variation has been documented on scales ranging from days to weeks[43–46] to months.[19,41] Patterns of spatial and temporal variation in other types of secondary metabolites are not as well known. However, variation within species[41,47–53] and between populations[18,47,54] has been reported.

An important question challenging researchers in the field of chemical ecology is "Why do plants produce particular concentrations of chemical defenses?" There is clear evidence that the efficacy of defensive compounds is often concentration dependent,[15,24,26,32,41,51,53,55–57] suggesting that variation in secondary metabolite concentrations may affect the outcomes of the ecological interactions they mediate. If an alga's ability to protect itself from herbivores is dependent upon having high concentrations of chemical defenses, why don't all algae produce large quantities of herbivore-deterrent secondary metabolites?

An obvious answer is that the production of chemical defenses incurs a metabolic cost. Resources allocated to chemical defense are not available for other processes such as growth and reproduction.[58] Some of these costs are assumed to include the costs of the raw materials needed to make the compounds, the resources required for producing and storing them, and the costs of preventing autotoxicity.[59] Other costs include those resulting from negative pleiotropy, from specialist herbivores being attracted to defensive chemicals, from natural enemies of herbivores being deterred or poisoned by secondary metabolites, and from a decrease in the plants' resistance to pathogens.[60,61] Such costs are widely assumed, but have rarely been demonstrated for algae.

A second answer to this question lies in the benefits derived from producing large quantities of chemical defenses. Not all algae are constantly subjected to intense grazing pressures. If herbivory has historically been low, there may have been minimal selective pressure to evolve effective grazer deterrents.[26,40,62] Furthermore, algae are made up of a variety of tissues that change during their life history. These tissues may be differentially vulnerable to attacks by grazers. Thus, the costs and benefits of protecting them may also differ, resulting in high degrees of within-plant and temporal variation in secondary metabolite concentrations.

An additional complicating factor in assessing the selective value of secondary metabolites is that not all secondary metabolites are effective against all herbivores.[15,50,56,63–65] Thus, producing large quantities of a defensive compound that is effective against some herbivores may leave an alga vulnerable to attack by others. For example, mesoherbivores, grazers that are small relative to the plants they consume and not highly mobile,[66] have been hypothesized to be unaffected by host plant defenses, whereas larger, more mobile grazers are expected to be more susceptible.[64,67–69] Many of the compounds that are deterrent towards grazers have other functions. These include serving as antibiotics and sunscreens,[70] inhibiting the growth of competitors or fouling organisms,[71,72] and chelating metal ions.[5,19] Selection for increases in secondary metabolites that positively affect one function may produce detrimental or positive results in another.

A final complicating factor is that concentrations of these chemicals can be affected by the organism's environment. Secondary metabolite concentrations in marine macroalgae can be influenced by or correlated with light intensity,[49] particularly in the UV range,[70] grazing pressures,[43–46] nutrient concentrations,[44,45,49,73,74] desiccation,[75] and salinity.[76]

Patterns in the distribution of secondary metabolites across a range of spatial scales can help provide clues to past selective pressures as well as current environmental factors influencing their production. The purpose of this chapter is to seek linkages between known spatial patterns in chemical defense production and hypotheses regarding the evolution of defenses and the role of current environmental factors. This chapter will examine patterns of secondary metabolite production on three spatial scales: (1) among tissues within individuals spanning spatial scales of millimeters to meters, (2) among individuals within species spanning spatial scales of meters to hundreds of kilometers, and (3) among species over large geographical areas spanning thousands of kilometers. Some of the factors believed to be responsible for causing those patterns will also be examined. These include historical patterns of distributions and interactions between algae and herbivores, rates of gene flow among populations, age structure of populations, resource allocation patterns, and the role of environmental factors in altering secondary metabolite concentrations.

II. SPATIAL PATTERNS OF CHEMICAL DEFENSE

A. Patterns within Individuals

Algae are not as well differentiated as terrestrial vascular plants. However, tissues within individual algae can be substantially different in their ecological roles and chemical compositions. Algae can be divided into support and attachment tissues such as holdfasts, stipes, and midribs, reproductive tissues such as receptacles and sporophylls, meristematic tissues, and blade tissues, which are nonreproductive, nonmeristematic tissues that do not function in attachment or support. Within these tissue types are additional morphological divisions such as cortical and medullary tissues and tissues that differ in age.

The tissues most likely to be exposed to grazing by benthic herbivores are holdfasts and stipes. In kelps (Division Phaeophyta, Order Laminariales), stipes and holdfasts frequently, although not always, have higher concentrations of phlorotannins than blade tissues.[27,46] Midrib tissue is more variable; in kelps, midribs have one-half to one-third the phlorotannins of adjacent blade tissues,[77] but in rockweeds (Division Phaeophyta, Order Fucales), concentrations of phlorotannins in midribs are about 10% higher than in the blades.[78] In many tropical systems, fish are the primary grazers, and upright portions of the thallus may be more susceptible to grazing. Uprights of the tropical green alga *Caulerpa* have higher concentrations of caulerpenyne and crude extracts than runners.[51] However, these differences do not affect feeding by fishes. The tropical green alga *Neomeris annulata* produces higher concentrations of brominated sesquiterpenes, compounds that are deterrent towards parrotfish but not towards surgeonfish,[79,80] in the bases of the algae than at the tips.[52]

Tissues in which cell division is taking place or has recently taken place, such as meristems, tissues in juvenile plants, and reproductive tissues, often contain different concentrations of secondary metabolites than nonmeristematic or nonreproductive tissues. In kelps and rockweeds, meristems tend to have higher phlorotannin concentrations than nonmeristematic tissues.[27,28,36] However, in the brown alga *Dictyota ciliolata*, terpene concentrations are lower in meristematic tissues, which are consumed at a higher rate than nonmeristematic tissues by amphipods and sea urchins.[48] Like meristems, newly produced segments of tropical green algae in the genus *Halimeda* appear to compensate for reduced structural defenses by having higher concentrations of potent feeding deterrents. New segments are produced at night[50] when fish, their primary grazers, are inactive. Early in the morning, new segments are unpigmented, lightly calcified, and contain high concentrations of the terpenoid compound halimedatrial, which is highly deterrent towards herbivorous fishes.[18] As the segments age, they become more pigmented, more heavily calcified, and contain lower concentrations of halimedatrial and higher concentrations of halimedatetraacetate, which is less deterrent.[18,50]

Most comparisons of secondary metabolite concentrations between reproductive and vegetative tissues involve phlorotannins in brown algae. Kelps and rockweeds produce distinct reproductive structures, such as receptacles, sori, and sporophylls, which facilitate making such comparisons. Higher phlorotannin concentrations occur in sporophylls of the kelp *Alaria marginata* than in vegetative blade tissues.[31,36] These differences are correlated with lower feeding on sporophylls than blades by the turban snail *Tegula funebralis*. However, these differences in phlorotannin concentrations do not occur in a congener, *A. nana*.[36] Similarly, phlorotannin concentrations are also higher in reproductive tissues of the South African kelps *Ecklonia maxima* and *Macrocystis angustifolia*, but not in *Laminaria pallida*.[27] In contrast to the kelps, phlorotannin concentrations in reproductive tissues of the rockweeds *Fucus vesiculosus*,[20,34] *F. gardneri*,[36] and *Pelvetia compressa*[36] are lower than in vegetative tissues. Similarly, in the red alga *Neorhodomela larix*, concentrations of the bromophenols lanosol and its 1,4-disulfate ester are higher in reproductive portions of the alga than in nearby vegetative tissues.[81]

Most studies of within-plant patterns of secondary metabolite concentrations have been limited to phlorotannin concentrations in rockweeds and kelps. Only a handful have looked at other compounds or groups of seaweeds. There are two likely reasons for this. First, tissues of kelps and rockweeds tend to be better differentiated than tissues from other algal taxa. In kelps and rockweeds, reproductive tissues and primary meristems occur in distinct areas of the alga, whereas in other groups, meristematic and reproductive tissues are often located throughout the thallus.[82,83] Second, phlorotannin concentrations can be estimated using relatively simple extraction procedures and colorometric assays.[19] Quantifying other kinds of secondary metabolites usually requires more complex procedures such as flash chromatography, gas chromatography (GC), and high-pressure liquid chromatography (HPLC). Thus, it cannot yet be assessed whether the general within-thallus patterns seen for phlorotannin concentrations in rockweeds and kelps apply to other taxa and other compounds.

B. Patterns within Species

Environmental conditions change in predictable ways along several gradients in marine ecosystems. One of the sharpest gradients occurs in the intertidal zone. Organisms living in high intertidal communities can experience extreme daily variation in temperature, salinity, desiccation, light, and in the availability of nutrients and gases.[84,85] This variation decreases at lower elevations. Other gradients exist in subtidal communities. Light quality and quantity decrease with depth, while nutrient concentrations generally increase,[86] and the intensity of herbivory sometimes decreases with depth.[87] Lateral gradients in environmental conditions also exist, such as salinity and nutrient concentrations along estuaries, or wave exposure along coastal communities.[85] Any of these gradients may affect concentrations of secondary metabolites either directly, by altering algal physiology, or indirectly by affecting herbivory, which, in turn, can affect the production of chemical defenses (see below). Stress gradients and levels of herbivory may also have interactive effects by synergistically or antagonistically impacting concentrations of secondary metabolites.

Although the rapid changes in environmental conditions along such gradients have numerous effects on algal physiology, surprisingly little is known about their effects on the production of secondary metabolites. Phlorotannin concentrations within a species can differ between the high and low intertidal zone,[45] between intertidal and subtidal individuals,[88] and along a salinity gradient.[76] However, no differences in phlorotannin concentrations of *Fucus gardneri* were seen across a wave exposure gradient off the coast of Oregon, USA.[89]

Differences in secondary metabolite concentrations also can exist among populations over scales of hundreds of kilometers. Phlorotannin concentrations in the Australasian kelp *Ecklonia radiata* are 30% higher in South Australia and 80% higher in New Zealand than in New South Wales.[33] Similar differences among regions occur in *Cystophora moniliformis*, but not in *Hormosira banksii*.[33] In *Lobophora variegata*, phlorotannin concentrations differ six-fold between North Carolina, USA and Aklins Island, Bahamas,[42] but only by 60% within different morphs of the Bahamian *Lobophora*. In *F. vesiculosus*, phlorotannin concentrations differ by an order of magnitude between Delaware, USA and Nova Scotia, Canada.[42]

It is meaningless to seek the processes causing patterns within a species along an environmental or geographic gradient unless the variation at this scale is greater than local, within-population variation. For example, in *A. nodosum*, there is little variation in phlorotannin concentrations at sites located 1000 km apart, but significant variation at scales of meters up to one km.[38] Along the west coast of the United States, phlorotannin concentrations differed among sites for only 25% of the kelps and 75% of the rockweeds examined.[37] In the variable species, there was no north–south geographical cline, and there was often as much variation among the three populations within Washington as between populations from Washington and California.

As seen in patterns among tissues, spatial patterns of secondary metabolite concentrations for groups other than the brown algae and for compounds other than phlorotannins are not well known. Concentrations of the terpenes halimedatrial and halimedatetraacetate in *Halimeda opuntia* differ on reef slopes located several kilometers apart in Pohnpei, FSM and at two sites on Guam, USA, but not in *H. opuntia* from a third reef on Guam or in *H. discoidea* and *H. macroloba*.[18] Concentrations of the terpenoid metabolites dictyol E, dictyodial, and pachydictyol A in the brown alga *Dictyota menstrualis* differ among three sites located within 8 km of one another.[47] These differences change seasonally and from year to year. In the fall of 1991, all three metabolites occurred in higher concentrations in algae from a seagrass bed and mudflat site than from algae collected from a jetty. In the summer of 1992, there were no among-site differences in concentrations of any of the compounds. The following fall, dictyodial occurred in lower concentrations at the mudflat site than the jetty site, and the other two compounds showed no differences in concentrations. Thus, apparent spatial patterns may vanish when a temporal element is considered, making it even harder to search for processes behind these patterns.

At this point, our knowledge of spatial variation in secondary metabolite concentrations across scales of meters to hundreds of kilometers is limited. In some species, there is more variation in chemical defense concentrations within populations than among populations.[37,38] In other species, there can be significant variation in secondary metabolite concentrations among populations.[18,33,37,42,47] However, these differences may be seasonal or vary from year to year.[20,41,47] Further work is needed to better understand the degree of variability that occurs in secondary metabolite concentrations across algal populations and to understand the causes of this variability.

C. GEOGRAPHIC PATTERNS AMONG SPECIES

1. Tropical–Temperate Patterns

The intensity of grazing and predation in tropical marine habitats is thought to be much higher than in temperate habitats.[90–94] Since tropical macroalgae are likely to be at higher risk from herbivores than temperate species, it is assumed they have had stronger selection to evolve chemical and physical defenses. Indeed, most groups of tropical macroalgae appear to be better defended than their temperate counterparts. Tropical green algae produce a diverse group of terpenes, acetogenins, and aromatic compounds[1,7,95] that are uncommon in temperate green algae. Likewise, many tropical red algae produce a variety of secondary metabolites, including terpenes, acetogenins, phenolic compounds, and indoles,[1,7,95] which are not as frequently reported in temperate red algae.

However, the possibility that tropical species seem to produce more secondary metabolites because they have been better studied cannot be ruled out.

The assumption that tropical algae are better defended than their temperate counterparts has not been well tested experimentally. Bolser and Hay[90] tested the hypothesis that tropical algae were less palatable to herbivores than closely related temperate species because of differences in their chemistry. They offered temperate and tropical sea urchins choices of diets that incorporated lyophilized, ground algal tissues embedded in an agar matrix. Each urchin was offered two agar-based diets, one with a tropical alga from the Bahamas and one with a temperate congener from North Carolina. The temperate urchins *Arbacia punctulata* preferentially consumed the diets made from temperate species. In the tropical urchins *Lytechinus variegatus*, diets made from temperate species were significantly preferred in three of nine experiments and diets made from tropical species were significantly preferred in only one. Diets were then made with the lipophilic or water-soluble extracts from species which had been significantly preferred, and urchins were offered pairs of diets with extracts from temperate and tropical species. In 60% of the experiments involving lipophilic extracts, urchin preferences occurred in the same direction as diets containing ground algal tissues. The addition of water-soluble extracts to the diets only affected feeding in 1 of 15 experiments. Thus, the trend was for ground algal tissues of temperate algae to be preferred over ground tissues of tropical algae, with lipophilic secondary metabolites implicated in many of these preferences.

Not all secondary compounds are more abundant in tropical macroalgae. Surprisingly, phlorotannins, which occur in high concentrations in temperate Pacific kelps and rockweeds, are lacking or occur in low concentrations in tropical Pacific brown algae. Phlorotannin concentrations in temperate Pacific kelps typically range from 0.3 to 5.5% dry mass (DM) and in rockweeds from 1.8 to 10.7% DM,[33] whereas *Sargassum* and *Turbinaria* from Tahiti have concentrations of only 0.3 to 1.6% DM.[39] Similarly, tropical Guamanian brown algae that included members of the Dictyotales, Scytosiphonales, and Fucales had phlorotannin concentrations ranging from 0.2 to 1.6% DM.[25] The lack of large quantities of phlorotannins in tropical Pacific algae does not appear to have resulted from a lack of effectiveness of these compounds towards herbivores. In field assays, phlorotannin-rich methanol extracts were deterrent towards herbivorous fishes in Guam, whereas phlorotannin-poor extracts were not, suggesting that phlorotannins could affect grazing by these herbivores.[25]

Another surprising pattern in tropical phlorotannin concentrations is that Caribbean brown algae have much higher concentrations than Pacific species.[42] Concentrations in Caribbean members of the Fucales range from 3.4 to 4.6% DM and in members of the Dictyotales from 1.3 to 14.9% DM. These concentrations are well above those in tropical Pacific brown algae and are comparable to concentrations found in temperate species (see above). The Caribbean data provide evidence that the high temperatures found in tropical waters do not preclude the production of phlorotannins. Unlike Caribbean species, Brazilian brown algae tend to have low phlorotannin levels (from 0.1 to 1.4% DM);[96] four species of *Dictyota*, two species of *Dictyopteris* (Order Dictyotales), and three species of *Sargassum* (Order Fucales) from the Brazilian coast show patterns that are more similar to tropical Pacific species.

Although phlorotannin concentrations in tropical Atlantic macroalgae are high, they do not always affect assimilation efficiencies of herbivores. In the parrotfish *Sparisoma chrysopterum* and *Sparisoma radians* and the brachyuran crab *Mithrax sculptus*, assimilation efficiencies were unaffected by phlorotannins.[97] However, in temperate Pacific fish (*Xiphister mucosus*), assimilation efficiencies were decreased by high molecular weight phlorotannins, but not by lower molecular weight compounds.[98]

2. Northeastern Pacific–Australasian Patterns

A second global-scale pattern in phlorotannin concentrations exists between brown algae in the northeast temperate Pacific and Australasia. Concentrations of phlorotannins in Australasian rockweeds and kelps are considerably higher than concentrations from related species in the northeastern Pacific.[23,26,40] Steinberg[23] reported median concentrations of phlorotannins in brown algae from the northeastern Pacific to be 1.3% DM and those from Australasia to be 6.2% DM. An updated evaluation since Steinberg's[40] review shows a similar pattern (Figure 8.1), with the majority of the northeastern Pacific species having phlorotannin concentrations of less than 2% DM. Only one northeastern Pacific species has phlorotannin concentrations greater than 7% DM, whereas over half the Australasian species have mean concentrations above 5% DM.

Herbivores in Australasia respond differently to phlorotannins than northeastern Pacific herbivores. The latter, including gastropod molluscs, sea urchins, and crustaceans, tend to prefer phlorotannin-poor species such as *Macrocystis*, *Nereocystis*, *Costaria*, *Postelsia*, *Egregia*, and *Laminaria* over phlorotannin-rich species such as *Cystoseira*, *Fucus*, *Agarum*, *Pelvetia*, and *Pelvetiopsis*.[32] Several grazers, including the sea urchin *Strongylocentrotus purpuratus* and the turban snails *Tegula funebralis* and *T. brunnea*, were inhibited from feeding by the addition of phlorotannins to their

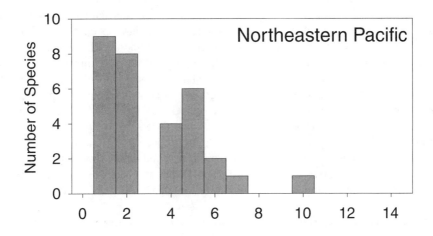

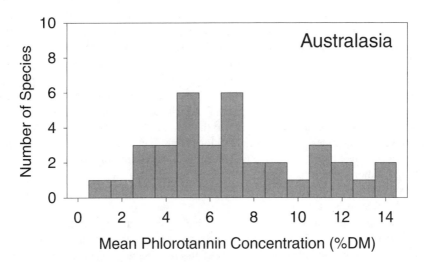

FIGURE 8.1 Histograms of phlorotannin concentrations in northeastern Pacific and Australasian kelps and rockweeds. Data are means of values for each species taken from the literature.[19,24,25,29,31-33,35,37,39,41,42,78,141,150,151]

diets at concentrations of 5% or 13.4%.[26] Feeding by red abalone *Haliotis rufescens* from California was also inhibited by the addition of phlorotannins into an agar diet at concentrations of 2 or 6%.[24]

In contrast, there is little correlation between the phlorotannin content of algae and food preferences of the Australasian sea urchins *Tripneustes gratilla* and *Centrostephanus rodgersii* or the snail *Turbo undulata*.[41] Feeding by *C. rodgersii* and *Evechinus chloroticus* was not deterred by the addition of phlorotannins to diets at 5% DM. Even concentrations of 13.4% DM did not deter *E. chloroticus*.[26] Similarly, the addition of phlorotannins to agar diets did not reduce feeding rates by the gastropod molluscs *Cookia sulcata* or *Turbo smaragdus* at either 5 or 13.4% DM.[26] Growth rates and food conversion efficiencies of Australasian herbivores fed diets of either phlorotannin-rich algae or agar disks supplemented with phlorotannins were similar to growth rates on phloro-tannin-poor diets,[41] providing further evidence that some Australasian herbivores have developed a tolerance for these compounds. The lack of a correlation between food preference and phlorotannin content may also have been partially due to the presence of nonpolar secondary metabolites, such as terpenes, halogenated lactones, and alkylated phenolic compounds, in the Australasian species that could have had greater influence on feeding than the phlorotannins.[41]

The differences between phlorotannin concentrations of Australasian and temperate northeast-ern Pacific brown algae and of herbivore responses to them have been attributed to historical differences in the trophic structures of the two regions.[26,62] Steinberg and coworkers[26] have argued that the northeastern Pacific subtidal kelp forest communities are predominately three-tiered sys-tems, whereas Australasian systems are only two tiered. Kelps, the dominant flora in the northeastern Pacific region, have historically been under low grazing pressures because the populations of the dominant herbivores in the system, sea urchins, have been kept in check by large diving predators such as sea otters.[26] With fewer grazers, kelps have had little selective pressure to produce high concentrations of defensive compounds. The Australasian oceans have never been home to large diving predators, and subtidal habitats have supported large and diverse populations of herbivorous invertebrates, resulting in strong selection for higher concentrations of chemical defenses and more diverse types of defensive compounds in subtidal brown algae.[41,62]

3. A Caveat Regarding Comparisons of Phlorotannin Concentrations across Genera

An inherent flaw in global-scale analyses of phlorotannin concentrations is that they involve comparisons among members of different species, genera, and orders. Work reviewed by Ragan and Glombitza[19] and recent work by Stern et al.[99] demonstrate that comparisons at the generic level should be made very cautiously when phlorotannin assays are done with colorometric techniques that use phloroglucinol as a standard. Stern and colleagues[99] isolated phlorotannins from six species of kelps and rockweeds and derived standards curves from those extracts and from phloroglucinol. The standard curves had slopes that differed by as much as a factor of five and were all less than the slope derived from phloroglucinol (Figure 8.2). Each extract contained a different mixture of phenolic compounds with different reactivities to the Folin reagents. Their results demonstrate that the Folin assays may, at times, underestimate phlorotannin concentrations in marine brown algae. This, along with the results of similar tests with other assays routinely used to measure phenolic concentrations, demonstrates that comparisons across genera of phlorotannin measurements using the Folin, Prussian Blue, and DMBA assays should be interpreted cautiously when a single standard is used for all the assays.

Because the magnitude of the differences between tropical and temperate Pacific phlorotannin concentrations well exceeds differences resulting from problems in standardizing the phlorotannin assays, these patterns are probably not artifacts of the methodology. Mean phlorotannin concen-trations in Australasian brown algae are approximately five times higher than concentrations in northeastern Pacific species, a range that is similar to the most extreme difference in the slopes of the standard curves of phloroglucinol and *Sargassum* found by Stern et al.[99] Because measurement

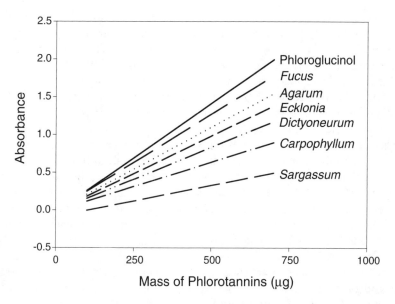

FIGURE 8.2 Relationship between phlorotannin mass (μg phenolic compounds) and absorbance of isolated phlorotannins from six genera of rockweeds (*Carpophyllum maschalocarpum*, *Fucus vesiculosus*, and *Sargassum vestitum*) and kelps (*Agarum cribosum*, *Ecklonia radiata*, and *Dictyoneurum californicum*), and phloroglucinol. Regressions are based on Stern et al.[99]

artifacts for most species are likely to be less extreme than for *Sargassum*, the Australasian–northeastern Pacific pattern is also likely to be real. However, until better standardization procedures are used for these assays, the results of any comparisons across genera should be viewed with caution.

III. EVOLUTIONARY CONSIDERATIONS

A. HISTORY OF THE ALGAL AND HERBIVORE GROUPS AND THEIR INTERACTIONS

It has been hypothesized that over short time scales, spatial differences in grazing intensities can affect the abundances, species composition, and morphological types of algae present in marine communities.[100–103] Variation in grazing intensity can also affect the concentrations and types of defenses encountered by herbivores, as shown by the production of induced and activated defenses in grazed algae.[43,45–47,54,104] Variation among sites on local scales in concentrations of chemical defenses has been shown to correlate with levels of herbivory in a few habitats,[18,43,47] but has not been extensively examined. Over longer time scales, differences in grazing pressure have been suggested to contribute to differences in the concentrations and types of antiherbivore defenses and to the abilities of herbivores to tolerate these defenses.[1–3,26,40,41,62,100,105–107]

Examinations of global-scale patterns of secondary metabolite production frequently look at current interactions between existing algae and herbivores and infer past patterns of selective pressures from them.[26,62,90,91] These studies rarely take into account the evolutionary histories of the algal and herbivore taxa that are common at different sites. Examinations of the evolutionary origins of groups and their dispersal patterns over time may provide valuable insights into the evolution of interactions, patterns of chemical defense concentrations, and selective pressures for the production of chemical defenses.

Differences in the phlorotannin concentrations of northeastern Pacific and Australasian brown algae provide an example of how such an approach may help explain current geographic patterns of chemical defense concentrations. Concentrations of phlorotannins in temperate Pacific rockweeds and kelps appear to be higher in algae from Australasia than from the northeastern Pacific[26,33,40,62]

(Figure 8.1). These differences have been ascribed, in part, to historical differences in herbivore selective pressures between the two hemispheres,[26,62] perhaps driven by the presence of sea otters, a key consumer of herbivores in the northeast Pacific, that are absent in Australasia (see above).

An alternative explanation for the geographic pattern in phlorotannin concentrations was suggested by Steinberg,[40] who states that these differences can be attributed to, in part, the higher diversity[108] of fucoids (typically polyphenolic rich) and the lower diversity[108,109] of laminarians (typically polyphenolic poor) in temperate Australasia than of those in North America. The southern hemisphere is dominated by fucoid species, which are about eight times more diverse than kelps. In the northern hemisphere, kelp species outnumber fucoid species by more than two to one (Table 8.1). This difference in diversity most likely arises from differences in the evolutionary histories of the two orders. The Fucales are thought to have arisen in Australasia sometime during the Mesozoic era,[110] whereas the Laminariales probably radiated from an origin in the North Pacific in the late Cenozoic era.[62] As kelps radiated into Australasia, in all likelihood they encountered a well established and diverse subtidal fucoid assemblage. The paucity of southern hemisphere kelps may, in part, be a reflection of three conditions: (1) the recent nature of the invasion, (2) difficulties in becoming established in a region already dominated by large benthic algae including rockweeds, and (3) difficulties in becoming established in a region with abundant and diverse benthic grazers. This last consideration assumes the Australasian region has had no grazer-consuming predators comparable to sea otters in the northeastern Pacific.

TABLE 8.1
Number of Species of Temperate Pacific Rockweeds and Kelps in the Northern and Southern Hemisphere

Hemisphere	Order Fucales (Rockweeds)	Order Laminariales (Kelp)	Total
Northern[a]	14	37	51
	(27.5%)[c]	(72.5%)	(100%)
Southern[b]	79	10	89
	(88.8%)	(11.2%)	(100%)

[a] Species lists for Northern Hemisphere species were obtained from Abbott and Hollenberg[83] and Gabrielson et al.[147]

[b] Species lists for Southern Hemisphere species were obtained from Womersley[148] and Adams.[149]

[c] Data in parentheses are the percent of species belonging to that grouping in the Northern or Southern Hemisphere out of the total number of kelp and rockweed species reported for that hemisphere. The proportion of rockweeds and kelp are significantly different in the two hemispheres (2×2 contingency table: $X^2 = 35.89$, $P < 0.001$).

Within both the northeastern Pacific and Australasian regions, reported concentrations of phlorotannins are consistently higher in rockweeds than kelps (Figure 8.3), although considerable variability exists across both orders in both regions. Mean phlorotannin concentrations differ significantly between rockweeds and kelps but are less dependent on whether the algae were northeastern Pacific or Australasian species (Table 8.2). These results support the hypothesis that the evolutionary history of the algae may be an important determinant of geographical patterns in chemical defense concentrations. This does not suggest that the history of the algal–herbivore interactions in each region is unimportant, but rather that both factors might play important roles in producing large-scale spatial patterns in chemical defense concentrations.

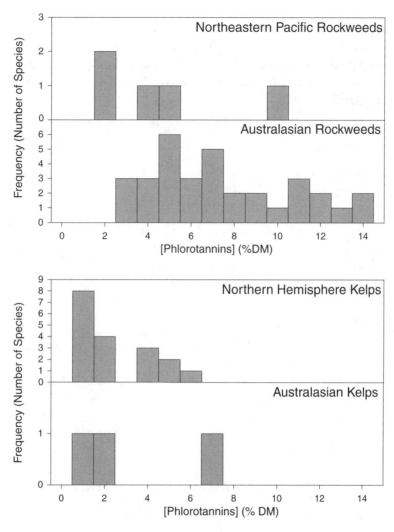

FIGURE 8.3 Frequency distribution of average phlorotannin concentrations of northeastern Pacific and Australasian rockweeds (Order Fucales) and kelps (Order Laminariales). Data are derived from a compilation of phlorotannin concentrations from the literature.[19,24,25,29,31-33,35,37,39,41,42,78,141,150,151]

B. GENE FLOW AND GENETIC DRIFT

The importance of reproductive isolation and genetic drift in determining chemical defense concentrations is unknown but potentially important. The dispersal of algal propagules (gametes and spores) is highly variable across species. Some algae are thought to disperse only a few meters,[111–115] whereas spores of other algae may be found in the water column tens to hundreds of kilometers from potential parent populations.[116–118] In addition, rafting may augment genetic mixing in species whose spores have limited dispersal abilities.[119] Many benthic herbivores, such as urchins, produce larvae that have the potential to travel hundreds of kilometers, whereas others, such as some gastropods, have direct-developing larvae that have much more limited dispersal ranges. Given that many seaweeds and some benthic herbivores have limited dispersal, there is potential for local selection or genetic drift to produce different genetically determined defense concentrations at different sites, contributing to high site-to-site variation. Fine-scale genetic structure has been

TABLE 8.2
Averages (% Dry Mass ± 1 SD) of Reported Phenolic Concentrations across All Sites from which They Were Measured for Northern and Southern Hemisphere Subtidal Rockweeds (Order Fucales) and Kelps (Order Laminariales)

	Order Fucales	Order Laminariales
Northern Hemisphere	4.12 ± 3.53	1.88 ± 1.77
	(N = 5)	(N = 17)
Southern Hemisphere	7.06 ± 3.27	2.81 ± 3.23
	(N = 32)	(N = 3)

ANOVA Table

Source	df	SS	MS	F	P
Order	1	0.050864	0.050864	11.31	0.001
Hemisphere	1	0.015428	0.003439	3.43	0.070
Order*Hemisphere	1	0.003439	0.003439	0.76	0.386
Error	53	0.238359	0.004497		
Total	56				

Note: Data are based on a compilation of phlorotannin values from the literature.[19,24,25,29,31–33,35,37,39,41,42,78,141,150,151] Results from a two-way analysis of variance conducted on the values are given below the species means.

shown for some algae with limited dispersal.[120,121] At this point, too little is known about algal dispersal rates, algal population genetics, and the genetic mechanisms that control concentrations of chemical defenses to determine whether localized selection or genetic drift has an influence on chemical defense concentrations.

IV. RESOURCE ALLOCATION PATTERNS

A. GROWTH–DIFFERENTIATION BALANCE VS. OPTIMAL DEFENSE

Comparisons of defense concentrations between tissues in marine macroalgae have allowed researchers to test predictions derived from resource allocation models. The production of chemical defenses is assumed to incur costs to the plant. Resources used to produce, store, and maintain defenses are assumed to be unavailable for other processes, such as growth and reproduction. Two hypotheses proposed to predict resource allocation to defense in plants[58] are the optimal defense theory (ODT) and the growth-differentiation balance hypothesis (GDBH). The ODT predicts that defenses will be preferentially allocated to tissues with the highest fitness value or tissues at highest risk for attack.[122] Thus, attachment structures, which are valuable and at high risk to benthic grazers, and meristematic and reproductive tissues, which have a high fitness value, should be better defended than nonmeristematic vegetative tissues. The GDBH also predicts trade-offs between growth and defense production.[123,124] According to this hypothesis, actively growing tissues, such as meristems or reproductive tissues, should allocate more resources to growth and fewer resources to defense than nonreproductive vegetative tissues.

Tests of these models using seaweeds have not revealed consistent support for either hypothesis. Higher concentrations of chemical defenses in holdfasts and stipes of kelps,[27,46] in the tips of *Neomeris annulata*,[52] and in the newly produced segments of *Halimeda*,[18,50] tissues at high risk to grazers, support the predictions of the ODT. These predictions are also supported by algae that have higher defense concentrations in reproductive tissues and meristems.[27,28,30,31,36,81] Support for the GDBH comes from algal species in which rapidly growing tissues or reproductive tissues have lower concentrations of defenses than nonmeristematic vegetative tissues. These include the receptacles, but not the meristems, of rockweeds[20,34,36] and the meristems of *Dictyota ciliolata*.[48] Predicting the allocation of chemical defenses in algae is complicated by the fact that tissues are often structurally rather than chemically defended,[101] and the relative costs of these types of defenses are unknown. A more detailed discussion of resource allocation patterns is presented in another chapter in this volume.[125]

B. Ontogenetic Shifts

Little is known about the impacts of age and life-history structure on variability in chemical defense concentrations within populations. Many marine algae produce at least two life-history stages that are morphologically or physiologically dissimilar.[82] This type of alternation of generations is found in all three of the major macroalgal groups: the red algae (Rhodophyta), the green algae (Chlorophyta), and the brown algae (Phaeophyta). Life-history strategies include having stages that are morphologically similar but physiologically different, having microscopic stages that alternate with macroscopic stages, or having encrusting stages that alternate with upright stages.

The production of secondary metabolites is known to differ among life-history stages. For example, kelps alternate between a haploid microscopic gametophyte stage and a diploid macroscopic sporophyte stage. The gametophyte can sometimes be found growing endophytically in filamentous red algae[126] where it may allow the kelp to overwinter and to avoid overgrowth, shading, grazing, or burial in sediments.[127,128] Phlorotannin concentrations of gametophytes grown in culture are low, rarely exceeding 0.5% DM.[77] In the macroscopic sporophytes of these algae, phlorotannin concentrations are as high as 5% DM.

Phlorotannin concentrations can differ in the macroscopic encrusting and upright stages of brown algae. "*Ralfsia californica*" is a brown algal crust that alternates with either *Petalonia* or *Scytosiphon*. On San Juan Island, Washington, USA, *Ralfsia* and upright populations undergo seasonal fluctuations.[129] The crust is most abundant in the spring and least abundant in late summer, while the erect phases are only found in the winter.[129] Phlorotannin concentrations in *Ralfsia* are among the highest measured ($18.3 \pm 3.4\%$ DM, N = 3^{77}) while those in *Petalonia* are much lower ($0.6 \pm 0.3\%$ DM, N = 5^{77}). Herbivorous limpets do not directly consume *Ralfsia*, and their effects on the upright stages are not known.[129] However, this pattern does not occur in all brown algal crusts. In *Analipus japonicus*, phlorotannin concentrations in the crustose stage ($4.2 + 1.1\%$ DM, N = 5^{77}) are similar to those in the upright stage ($5.4 + 2.7\%$ DM, N = 5^{77}). In this alga, however, the crust sprouts directly from the upright stage rather than producing a different life-history stage. Similarly, concentrations of pachydictyol A, dictyol B acetate, and dictyodial do not differ in isomorphic sporophytes and gametophytes of *Dictyota ciliolata*.[130]

Concentrations of defensive metabolites can change during the transition from juveniles to adults of an algal stage. Small sporophytes of the kelps *Alaria marginata*, *Egregia menziesii*, *Fucus gardneri*, *Hedophyllum sessile*, and *Lessoniopsis littoralis* produced higher concentrations of phlorotannins than older or adult stages.[131] Juveniles of the siphonous green alga *Halimeda macroloba* were significantly less calcified and contained higher concentrations of the diterpenoid feeding deterrent halimedatrial than adult tissues;[18] adult *Halimeda* were heavily calcified and contained halimedatetraacetate as their primary defensive compound.[53] Halimedatrial was a more effective feeding deterrent towards herbivorous fishes than halimedatetraacetate, and lipophilic extracts from juveniles were more deterrent than those from adults.[18] In other species, however, defense concentrations were lower in juveniles than adults. Juveniles of the rockweeds *F. vesiculosus* and *F. evenescens*[132] and the kelps *Nereocystis luetkeana* and *Laminaria groenlandica*[46,131] had lower concentrations of phlorotannins than adults. Similarly, concentrations of phlorotannins in *Fucus gardneri* embryos were lower than those in adults.[133]

V. ENVIRONMENTAL CONSIDERATIONS

Changes in environmental factors, including levels of grazing, salinity, nutrients, visible light, UV light and desiccation, can affect concentrations of chemical defenses in marine macroalgae. Work conducted on terrestrial and marine plants suggests that the direct effects of environmental factors on chemical defense concentrations are manifested in three ways:

1. Environmental changes can alter the amounts of materials and energy available for defense production and how those resources are allocated to growth, reproduction, and defense (e.g., the availability of light and nutrients has been linked to changes in resource allocation to defense in many terrestrial plants[123,124,134,135]).
2. Environmental changes can provide cues for algae to alter concentrations of secondary metabolites to concentrations that are more appropriate for the environment at that time and place (e.g., grazing by herbivores can signal plants to produce higher concentrations of defensive compounds[136,137]).
3. Environmental conditions may stress plants, resulting in an overall lower physiological condition. Stressed plants may have lower photosynthetic rates and, thus, fewer resources to allocate to both primary and secondary functions. Consequently, defense concentrations of stressed plants may be lower than concentrations in otherwise healthy individuals.

Environmental conditions may also affect chemical defense concentrations indirectly by changing the intensity of grazing, which can alter defense concentrations in algae with inducible defenses. Evidence for the direct effects of environmental factors on secondary metabolite concentrations are discussed below.

A. ENVIRONMENTAL EFFECTS ON RESOURCE ALLOCATION

Selection should act on plants to configure their resources in a way that maximizes their probability of survival and reproduction.[138] Comparing allocation patterns among species may provide clues about broad evolutionary and environmental constraints, for example, how low-resource environments affect chemical defenses. Patterns within a species may help us understand allocation priorities within a life history and how local processes such as grazing pressures affect these priorities. Understanding allocation patterns requires the measurement of the distribution of resources among functions and the determination of the consequences of alternative allocation strategies.[58] For example, an increase in the allocation to secondary metabolites may decrease growth or reproductive output, but may increase the probability of survival.

The carbon/nutrient balance hypothesis (CNBH) was originally proposed to explain resource allocation patterns in terrestrial vascular plants. It has also been used to examine defense allocation patterns in macroalgae. The CNBH predicts that when carbon is available but nutrients are limited, plants should increase concentrations of carbon-based defenses.[135] When carbon is available but nutrients are abundant, plants should invest resources in growth. If carbon is limiting, such as when shading lowers photosynthetic rates, carbon should be allocated to growth and the production of carbon-based defenses should decrease.

Tests with marine macroalgae suggest that the CNBH tends to be a good predictor of how changes in nutrient and light levels will affect phlorotannin concentrations but not terpene concentrations. Intertidal and shallow subtidal macroalgae are assumed to be limited more often by nitrogen than carbon. Therefore, the CNBH predicts that the production of carbon-based defensive compounds, such as phlorotannins, terpenes, and fatty acids, should be inversely correlated with nitrogen availability. These patterns have been seen in some studies of brown algal phlorotannins. For example, phlorotannin concentrations in *Fucus vesiculosus* at three sites from the northern Baltic Sea were negatively correlated with thallus nitrogen concentrations.[74] Similar patterns were found in *F. vesiculosus* in estuarine sites in Massachusetts, USA.[44] Algae at a low nitrogen site had higher phlorotannin concentrations than algae at a high nitrogen site, although some seasonal variation did occur. When enrichments with an ammonium nitrate and phosphate fertilizer were conducted, tissue nitrogen concentrations increased in algae from both sites, but phlorotannin concentrations decreased only in algae from the low nitrogen site. At both sites, growth rates were higher in fertilized algae, but the response was more pronounced in the low nitrogen site.[44]

In laboratory experiments, the tropical brown alga *Lobophora variegata* fertilized with different concentrations of nitrogen showed patterns consistent with the CNBH. Algae grown in high nitrogen media had higher nitrogen concentrations and lower phlorotannin concentrations than those in low nitrogen media.[73] In another study, *Fucus gardneri* embryos grown in media enriched with nitrogen and phosphorus had lower phlorotannin concentrations than those grown in media with ambient nutrient concentrations.[133] When mixtures of nutrients were added to the media, synergistic effects on growth and phlorotannin concentrations occurred. In the absence of phosphorus enrichment, iron enrichment caused phlorotannins to increase by 2 to 27%. However, when media were enriched with phosphorus and iron together, phlorotannin concentrations decreased by 37 to 64%.[133]

In contrast, field experiments measuring the production of terpenes in algae grown under several nutrient regimes are not in agreement with the predictions of the CNBH. The addition of nitrogen alone, phosphorus alone, and nitrogen and phosphorus together to sites on Guam resulted in no differences in concentrations of the terpene ochtodene or in triglyceride concentrations in the tropical red alga *Portieria hornemannii*.[139] Similarly, terpene concentrations in the brown alga *Dictyota ciliolata* did not change when algae were fertilized with ammonium chloride and a plant food containing nitrogen, phosphorus, and potassium.[49] The addition of a nitrogen–phosphorus–potassium fertilizer also had no effects on growth or phlorotannin concentrations in *Sargassum filipendula*, possibly because nutrients did not limit *Sargassum*'s growth.[49]

Because many subtidal algal species live in a discrete range of depths, with the lower limit set by light limitation, they provide an ideal system for examining the predictions of the CNBH regarding carbon limitation. According to the CNBH, concentrations of carbon-based chemical defense should be lower in algae in light-limited habitats. *Sargassum filipendula* from North Carolina, USA growing in heavily shaded habitats have significantly lower concentrations of phlorotannins than *Sargassum* growing in less-shaded areas.[49] However, experimental manipulations of light levels in the field did not result in differences in phlorotannin concentrations in algae growing in high-light and low-light treatments.

B. ENVIRONMENTAL FACTORS AS CUES

Many intertidal and shallow subtidal habitats are highly variable over space or time. Consequently, the concentrations of chemical defenses that are appropriate at one place or time may be inappropriate at another. A strategy for dealing with this variability is the detection of an environmental cue that can be correlated with an appropriate level of defense for that environment at that time. One of the best examples of environmental cues causing changes in chemical defense concentrations are herbivore-induced defenses in which chemical levels are increased in response to an attack by a consumer.[136,137] Induced defenses require that herbivory be variable and unpredictable, that the inducing cues produce effective increases in defenses, and that the induction is cost-effective and rapid enough to deter further consumption.[140]

Herbivore-induced defenses may explain some spatial variation in macroalgal chemical defense concentrations. Phlorotannin concentrations in *Fucus gardneri* on Tatoosh Island, Washington, USA were 28% higher than concentrations in algae from a site 100 m away.[43] The site where the algae had higher phlorotannin concentrations also had nearly an order of magnitude more herbivorous snails (*Littorina sitkana* and *L. scutulata*) than the site where the algae had lower phlorotannin concentrations. Physical damage to ungrazed algae in a manner that mimicked patterns of snail grazing caused 8 to 25% increases in phlorotannin concentrations over a 2-week period. Damage to *F. gardneri* midribs also results in the production of adventitious branches, which have phloro-tannin concentrations that are 10 to 70% higher than concentrations in blade tissues and are consumed at lower rates by herbivorous snails.[35]

Similarly, the brown alga *Dictyota ciliolata* growing in seagrass and mudflat habitats where the amphipod *Ampithoe longimana* was abundant had higher concentrations of dictyol E, dictyodial, and pachydictyol A than algae growing on a jetty where amphipod grazing was less severe.[47] Algae were collected from the site with low levels of grazing and transplanted to one of the high-grazing sites. Half the transplanted algae were protected from amphipod grazing by caging and half were placed in cages with amphipods. Plants in cages with amphipods had significantly higher concentrations of dictyol E, dictyodial, and sterols and were a lower preference food when offered to amphipods relative to plants from cages without amphipods.

Damage to algae does not always serve as a cue to increase secondary metabolite production. Experimental attempts to induce the production of phlorotannins by damaging the thallus did not increase phlorotannin concentrations in *Ecklonia* or *Sargassum* from Australia.[141] Similar experiments to induce the production of phlorotannins in the northeastern Pacific kelp *Hedophyllum sessile* resulted in decreases in phlorotannin concentrations.[46] Tropical *Halimeda* growing in habitats where grazing pressure from herbivorous fishes is high contain higher quantities of the terpene halimedatrial than algae growing in habitats with lower grazing pressures.[18] However, all attempts to induce the production of halimedatrial by damaging the algae failed when experiments were conducted to determine whether damage could cause a shift in the kinds of secondary metabolites that were present.[142]

The inability of experiments to show an induced defense may occur because the algae do not produce induced defenses or because the wrong cue was used. The specificity of cues that induce secondary metabolite production differs among species. Phlorotannin induction in *Ascophyllum nodosum* occurred when algae were grazed by gastropod snails (*Littorina obtusata*), but not when they were grazed by isopods (*Idotea granulosa*) or when they were mechanically damaged.[22] However, in *Fucus gardneri*,[35] and the kelps *Agarum fimbriatum*, *Pleurophycus gardneri*, *Laminaria complanata*, and *L. groenlandica*,[46] mechanical wounding caused increases in phlorotannin concentrations. Another reason for failing to see an induction may be that the compounds are being measured on the wrong time scale. In *A. fimbriatum*, *P. gardneri*, *L. complanata*, and *L. groenlandica*, wounding increased phlorotannin concentrations by 30 to 90% over 1 to 3 days,[46] but the increased concentrations only persisted for 5 to 7 days before returning to pre-wounding levels.

The induction of increased secondary metabolite concentrations may be cued by environmental factors other than herbivory. In an experimental manipulation of UV light levels and simulated grazing with *Ascophyllum nodosum*, damage to the plants had no effect on phlorotannin concentrations over a 2-week period.[70] However, phlorotannin concentrations increased by approximately 30 to 40% in response to increased UV light levels. Algae exposed to the higher UV light levels were preferentially consumed by the isopod *Idotea granulosa*, despite having higher phlorotannin concentrations. In *Ascophyllum*, the primary function of phlorotannins may be UV light absorption rather than herbivore deterrence.

Changes in water content, nutrient concentrations, and seasons have also been suggested to affect the inducibility of phlorotannins. Simulated grazing on *Fucus vesiculosus* in the low intertidal zone of an oceanic site resulted in increases in phlorotannin concentrations over 3 to 7 days in experiments conducted in June, August, and October, but not in March,[45] when light levels were low. In high intertidal *Fucus*, induction did not occur in June when phlorotannin concentrations were at a seasonal peak. However, induction did occur in August and October when background phlorotannin concentrations were lower. In *Agarum*, induction occurred in the summer but not in the winter, when the basal phlorotannin concentrations were high.[46]

C. ENVIRONMENTAL FACTORS AS STRESSES

Stress is a property of the environment that can be detected only by its effects on organisms; it can be defined as any factor that reduces the production of an organism. Davison and Pearson[143] differentiate limitation stress as the reduction of growth due to an inadequate supply of resources (e.g., light), vs. disruptive stress, which results in damage due to adverse conditions (e.g., desiccation). Environmental stress can affect a plant's resistance to herbivory in a variety of ways. Stressed plants may not be able to compensate for losses to herbivory as well as unstressed plants because their nutrient capital may be poor, resulting in declines in plant fitness.[144]

Differences in secondary metabolite concentrations over environmental gradients suggest that environmental stresses can affect secondary metabolite production. For example, Pedersen[76] demonstrated that phlorotannin concentrations of *Fucus vesiculosus* and *Ascophyllum nodosum* in a Norwegian fjord were lower in algae closer to the freshwater source in the estuary. This suggests that low salinity may constitute a stress and result in low phlorotannin concentrations, although direct tests of the effects of salinity have not been attempted. Prolonged periods of desiccation are stressful to many marine algae. When *Dictyota ciliolata* was exposed for an hour, photosynthetic rates decreased by 47% and respiration rates increased by 40% relative to undesiccated algae.[14] Desiccated algae were consumed at a faster rate by the sea urchin *Arbacia punctulata* and the amphipod *Ampithoe longimana*. The increased grazing rates by the urchins and amphipods were correlated with decreases in the secondary metabolites pachydictyol A, dictyol B acetate, and dictyodial in the desiccated algae. Pachydictyol A and dictyol B acetate both deterred feeding by urchins and amphipods.

Similar tests of the effects of desiccation on herbivore resistance have been performed on the brown alga *Padina gymnospora* and the red alga *Gracilaria tikvahiae*.[75] Sea urchins, *Arbacia punctulata*, significantly preferred to consume *Padina* that had been desiccated for 30 to 90 min over undesiccated algae. When extracts of *Padina* were coated on a preferred food, urchins preferentially consumed algae coated with extracts from desiccated vs. undesiccated algae. This suggests that stressed *Padina* undergo changes in their chemistry that affect urchin feeding; however, it was not determined whether these changes were in the chemical defenses produced by the algae. In contrast, desiccated *Gracilaria* became less susceptible to urchin grazing, which was correlated with a reduction in protein content.[75]

Ultraviolet light reduces secondary metabolite production in some seaweeds. *Dictyota ciliolata* exposed to UV light grew more poorly and suffered higher mortality levels, but were not preferentially consumed by the amphipod *Ampithoe longimama*, relative to algae that had been shielded from UV light.[14] Concentrations of dictyol B acetate and dictyodial were significantly lower in algae exposed to UV light; however, concentrations of pachydictyol A were not different. These results contrast with the effects of UV light on *Ascophyllum nodosum*, in which phlorotannin concentrations increased in response to exposure to UV light.[70] This suggests that the effects of UV radiation on secondary metabolite concentrations in algae depends on the species or compounds being examined. In some cases, UV light can serve as a cue to induce secondary metabolite production. In other cases, it is a physiological stress that decreases secondary metabolite production.

VI. CONCLUSIONS AND DIRECTIONS FOR FUTURE RESEARCH

Our understanding of spatial patterns of chemical defense concentrations and the factors, both historical and current, that produce those patterns is still in its infancy. Spatial patterns in secondary metabolite concentrations are well known only for phlorotannins in the brown algae, and the interpretation of these patterns at the level of genera and higher is confounded by methodological problems. We currently have too little knowledge of spatial patterns in other groups of algae or for other types of metabolites to infer patterns that may apply to larger groups of taxa. From what little is known, it is clear that variation is the rule, not the exception. Documenting patterns in non-phlorotannin compounds and in a more diverse group of algae is a necessary next step to gaining a better understanding of spatial patterns in macroalgal chemical defense concentrations and the factors that drive this variability.

Another area where we need greater understanding is the degree to which genetic and environmental factors affect the production of secondary metabolites. Terrestrial ecologists have made progress in determining the heritability of chemical defense production by conducting controlled breeding experiments and measuring chemical defense levels among siblings whose parentage is known. Such an approach has never been taken with marine macroalgae, in part because algae frequently have complex life histories and can be difficult to culture under controlled conditions. Nonetheless, this approach has the potential to provide valuable insights that would give us a better understanding of the control of defense concentrations, the costs and benefits of producing defenses, and the degree to which algae are able to respond to selective pressures by herbivores and other environmental factors.

The selective pressures that have produced the concentrations of secondary metabolites we see in marine algae today have, in part, resulted from interactions between marine algae and herbivores in the past. To understand how past levels of herbivory resulted in current spatial patterns of secondary metabolites, particularly over large scales, we need to take into account the evolutionary histories of the algae and herbivores in question and look at their distributions and potential interactions over evolutionary time scales. Such an approach has been useful in looking at the evolution of structural defenses in marine invertebrates, such as spine production and shell thickening in gastropod molluscs;[145,146] however, it has not been widely used in examining the evolution of chemical defense concentrations in marine macroalgae. Steinberg and Estes'[26,62] examination of hemispheric-scale patterns in phlorotannin concentrations is a excellent example of the use of this type of approach. However, this approach could be further refined by giving more consideration to the historical distributional patterns of algal orders and how these affected potential interactions with herbivores.

Selective pressures and genetic drift may also cause variation in concentrations of secondary metabolite production. If algal and herbivore dispersal is limited and either the algal or the herbivore population becomes reproductively isolated, as could happen in enclosed bays, then genetically determined traits may potentially become fixed at different levels in different locations, creating spatial variation at a regional level. At this point, relatively little is known about algal gene flow and dispersal distances, and even less is known about the genetics of algal defense production. A better understanding of algal genetics will be needed to determine whether genetic drift and local selection could be responsible for some of the variability in chemical defense concentrations occurring among populations.

Like many traits, concentrations of secondary metabolites in algae are determined by a combination of genetic and environmental factors. Secondary metabolites are known to respond to changes in a variety of environmental conditions, including light,[14,49,70] herbivory,[43–46], nutrients,[44,45,49,73,74] desiccation,[75] and salinity.[76] Most of these studies have looked at the effects of a single environmental stress at once. However, in nature, multiple stresses often occur simultaneously and may have synergistic effects.

Stresses also affect plants in multiple ways. Currently, relatively few studies have taken an integrative approach and simultaneously examined the effects of environmental stresses on resource acquisition, resource allocation, and changes in a plant's ecology. It is also necessary to determine how internal changes in allocation patterns in an alga affect its interactions with the external world. The result of stress-induced changes in secondary metabolite production on food preferences of herbivores is only beginning to be studied. A single secondary metabolite can be a deterrent to one species, have no effect on another, and may enhance feeding by a third. Furthermore, many secondary metabolites have multiple functions, and changes in their abundance can affect an alga's ability to tolerate UV light, prevent microbial infection, or deter settlement of fouling organisms.

Untangling community-level effects when dealing with multiple herbivores and multiple stresses on algae with different physiological responses to those stresses will provide an exciting challenge for future studies of marine algal chemical ecology.

REFERENCES

1. Hay, M.E. and Fenical, W., Marine plant–herbivore interactions: the ecology of chemical defense, *Ann. Rev. Ecol. Syst.*, 19, 111, 1988.
2. Hay, M.E. and Steinberg, P.D., The chemical ecology of plant-herbivore interactions in marine vs. terrestrial communities, in *Herbivores: Their Interactions with Secondary Plant Metabolites: Evolutionary and Ecological Processes*, Vol. 2, Rosenthal, G. and Berenbaum, M., Eds., Academic Press, San Diego, 1992, 371.
3. Paul, V.J., *Ecological Roles of Marine Natural Products*, Cornell University Press, Ithaca, NY, 1992.
4. Hay, M.E., Marine chemical ecology: what's known and what's next?, *J. Exp. Mar. Biol. Ecol.*, 200, 103, 1996.
5. Targett, N.M. and Arnold, T.M., Predicting the effects of brown algal phlorotannins on marine herbivores in tropical and temperate oceans, *J. Phycol.*, 34, 195, 1998.
6. Van Alstyne, K.L. and Paul, V.J., The role of secondary metabolites in marine ecological interactions, *Proc. 6th Int. Coral Reef Symp.*, 1, 175, 1988.
7. Faulkner, D.J., Marine natural products: metabolites of marine algae and herbivorous marine molluscs, *Nat. Prod. Rep.*, 1, 251, 1984.
8. Faulkner, D.J., Marine natural products: metabolites of marine invertebrates, *Nat. Prod. Rep.*, 1, 551, 1984.
9. Faulkner, D.J., Marine natural products, *Nat. Prod. Rep.*, 3, 1, 1986.
10. Faulkner, D.J., Marine natural products, *Nat. Prod. Rep.*, 4, 539, 1987.
11. Faulkner, D.J., Marine natural products, *Nat. Prod. Rep.*, 5, 613, 1988.
12. Faulkner, D.J., Marine natural products, *Nat. Prod. Rep.*, 7, 269, 1990.
13. Faulkner, D.J., Marine natural products, *Nat. Prod. Rep.*, 8, 97, 1991.
14. Cronin, G. and Hay, M.E., Susceptibility to herbivores depends on recent history of both the plant and animal, *Ecology*, 77, 1531, 1996.
15. Hay, M.E., Duffy, J.E., Fenical, W., and Gustafson, K., Chemical defense and the seaweed *Dictyopteris delicatula*: differential effects against reef fishes and amphipods, *Mar. Ecol. Prog. Ser.*, 48, 185, 1988.
16. Paul, V.J. and Hay, M.E., Seaweed susceptibility to herbivory: chemical and morphological correlates, *Mar. Ecol. Prog. Ser.*, 33, 255, 1986.
17. Paul, V.J. and Van Alstyne, K.L., Use of ingested algal diterpenoids by *Elysia halimedae* Macnae (Opisthobranchia: Ascoglossa) as antipredator defenses, *J. Exp. Mar. Biol. Ecol.*, 119, 15, 1988.
18. Paul, V.J. and Van Alstyne, K.L., Chemical defense and chemical variation in some tropical Pacific species of *Halimeda* (Halimediaceae; Chlorophyta), *Coral Reefs*, 6, 263, 1988.
19. Ragan, M.A. and Glombitza, K., Phlorotannins, brown algal polyphenols, in *Prog. Phycol. Res.*, Round, F. E. and Chapman, D. J., Eds., Biopress Limited, Bristol, U.K., 1986, 129.
20. Geiselman, J.A., Ecology of Chemical Defenses of Algae Against the Herbivorous Snail, *Littorina littorea*, in the New England Rocky Intertidal Community, Ph.D. thesis, MIT/ Woods Hole, 1980.
21. Geiselman, J.A. and McConnell, O.J., Polyphenols in brown algae *Fucus vesiculosus* and *Ascophyllum nodosum*: chemical defenses against the marine herbivorous snail, *Littorina littorea*, *J. Chem. Ecol.*, 7, 1115, 1981.
22. Pavia, H. and Toth, G., Inducible chemical resistance to herbivory in the brown seaweed *Ascophyllum nodosum*, *Ecology*, 81, 3213, 2001.
23. Steinberg, P.D., Effects of quantitative and qualitative variation in phenolic compounds on feeding in three species of marine invertebrate herbivores, *J. Exp. Mar. Biol. Ecol.*, 120, 221, 1988.
24. Winter, F.C. and Estes, J.A., Experimental evidence for the effects of polyphenolic compounds from *Dictyoneurum californicum* Ruprecht (Phaeophyta: Laminariales) on feeding rate and growth in the red abalone *Haliotis rufescens* Swainson, *J. Exp. Mar. Biol. Ecol.*, 155, 263, 1992.
25. Van Alstyne, K.L. and Paul, V.J., The biogeography of polyphenolic compounds in marine macroalgae: temperate brown algal defenses deter feeding by tropical herbivorous fishes, *Oecologia*, 84, 158, 1990.

26. Steinberg, P.D., Estes, J.A., and Winter, F.C., Evolutionary consequences of food chain length in kelp forest communities, *Proc. Nat. Acad. Sci.*, 92, 8145, 1995.

27. Tugwell, S. and Branch, G.M., Differential polyphenolic distribution among tissues in the kelps *Ecklonia maxima*, *Laminaria pallida*, and *Macrocystis angustifolia* in relation to plant-defence theory, *J. Exp. Mar. Biol. Ecol.*, 129, 219, 1989.

28. Johnson, C.R. and Mann, K.H., The importance of plant defence abilities to the structure of subtidal seaweed communities: the kelp *Laminaria longicruris* de la Pylaie survives grazing by the snail *Lacuna vincta* (Montagu) at high population densities, *J. Exp. Mar. Biol. Ecol.*, 97, 231, 1986.

29. Pfister, C.A., Costs of reproduction in an intertidal kelp: patterns of allocation and life history consequences, *Ecology*, 73, 1586, 1992.

30. Poore, A.G.B., Selective herbivory by amphipods inhabiting the brown alga *Zonaria angustata*, *Mar. Ecol. Prog. Ser.*, 107, 113, 1994.

31. Steinberg, P.D., Algal chemical defenses against herbivores: allocation of phenolic compounds in the kelp *Alaria marginata*, *Science*, 53, 405, 1984.

32. Steinberg, P.D., Feeding preferences of *Tegula funebralis* and chemical defenses of marine brown algae, *Ecol. Monogr.*, 55, 333, 1985.

33. Steinberg, P.D., Biogeographical variation in brown algal polyphenolics and other secondary metabolites: comparison between temperate Australasia and North America, *Oecologia*, 78, 374, 1989.

34. Tuomi, J., Niemala, P., Siren, S., and Jormalainen, V., Within-plant variation in phenolic content and toughness of the brown alga *Fucus vesiculosus*, *Bot. Mar.*, 32, 505, 1989.

35. Van Alstyne, K.L., Adventitious branching as a herbivore-induced defense and the intertidal brown alga *Fucus distichus*, *Mar. Ecol. Prog. Ser.*, 56, 169, 1989.

36. Van Alstyne, K.L., McCarthy, J.J., III, Hustead, C.L., and Kearns, L.J., Phlorotannin allocation among tissues of northeastern Pacific kelps and rockweeds, *J. Phycol.*, 35, 483, 1999.

37. Van Alstyne, K.L., McCarthy, J.J., III, Hustead, C.L., and Duggins, D.O., Geographic variation in polyphenolic levels of Northeastern Pacific kelps and rockweeds, *Mar. Biol.*, 133, 371, 1999.

38. Pavia, H. and Aberg, P., Spatial variation in polyphenolic content of *Ascophyllum nodosum* (Fucales, Phaeophyta), *Hydrobiologia*, 326/327, 199, 1996.

39. Steinberg, P.D., Chemical defenses and the susceptibility of tropical brown algae to herbivores, *Oecologia*, 69, 628, 1986.

40. Steinberg, P.D., Geographical variation in the interaction between marine herbivores and brown algal secondary metabolites, in *Ecological Roles of Marine Natural Products*, Paul, V.J., Ed., Cornell University Press, Ithaca, NY, 1992, 245.

41. Steinberg, P.D. and van Altena, I.A., Tolerance of marine invertebrate herbivores to brown algal phlorotannins in temperate Australasia, *Ecol. Monogr.*, 62, 189, 1992.

42. Targett, N.M., Coen, L.D., Boettcher, A.A., and Tanner, C.E., Biogeographic comparisons of marine algal phenolics: evidence against a latitudinal trend, *Oecologia*, 89, 464, 1992.

43. Van Alstyne, K.L., Herbivore grazing increases polyphenolic defenses in the intertidal brown alga *Fucus distichus*, *Ecology*, 69, 655, 1988.

44. Yates, J.L. and Peckol, P., Effects of nutrient availability and herbivory on polyphenolics in the seaweed *Fucus vesiculosus*, *Ecology*, 74, 1757, 1993.

45. Peckol, P., Krane, J.M., and Yates, J.L., Interactive effects of inducible defense and resource availability on phlorotannins in the North Atlantic brown alga *Fucus vesiculosus*, *Mar. Ecol. Prog. Ser.*, 138, 209, 1996.

46. Hammerstrom, K., Dethier, M.N., and Duggins, D.O., Rapid phlorotannin induction and relaxation in five Washington kelps, *Mar. Ecol. Prog. Ser.*, 165, 293, 1998.

47. Cronin, G. and Hay, M.E., Induction of seaweed chemical defenses by amphipod grazing, *Ecology*, 77, 2287, 1996.

48. Cronin, G. and Hay, M.E., Within-plant variation in seaweed palatability and chemical defenses: optimal defense theory versus the growth-differentiation balance hypothesis, *Oecologia*, 105, 361, 1996.

49. Cronin, G. and Hay, M.E., Effects of light and nutrient availability on the growth, secondary chemistry, and resistance to herbivory of two brown seaweeds, *Oikos*, 77, 93, 1996.

50. Hay, M.E., Paul, V.J., Lewis, S.M., Gustafson, K., Tucker, J., and Trindell, R.N., Can tropical seaweed reduce herbivory by growing at night? Diel patterns of growth, nitrogen content, herbivory, and chemical versus morphological defenses, *Oecologia*, 75, 233, 1988.

51. Meyer, K.D. and Paul, V.J., Intraplant variation in secondary metabolites concentration in three species of *Caulerpa* (Chlorophyta: Caulerpales) and its effects on herbivorous fishes, *Mar. Ecol. Prog. Ser.*, 82, 349, 1992.

52. Meyer, K.D. and Paul, V.J., Variation in secondary metabolites and aragonite concentrations in the tropical marine seaweed *Neomeris annulata*: effects on herbivory by fishes, *Mar. Biol.*, 122, 537, 1995.

53. Paul, V.J. and Van Alstyne, K.L., Antiherbivore defenses in *Halimeda*, *Proc. 6th Int. Coral Reef Symp.*, 3, 133, 1988.

54. Van Alstyne, K.L., Wolfe, G.V., Freidenburg, T.L., Neill, A., and Hicken, C., Activated defense systems in marine macroalgae: evidence for an ecological role for DMSP cleavage, *Mar. Ecol. Prog. Ser.*, in press, 2001.

55. Harvell, C.D., Fenical, W., and Greene, C.H., Chemical and structural defenses of Caribbean gorgonians (*Pseudopterogorgia* sp.) I. Development of an in situ assay, *Mar. Ecol. Prog. Ser.*, 49, 287, 1988.

56. Hay, M.E., Fenical, W., and Gustafson, K., Chemical defense against diverse coral-reef herbivores, *Ecology*, 68, 1581, 1987.

57. Van Alstyne, K.L., Wylie, C.R., and Paul, V.J., Antipredator defenses in tropical Pacific soft corals (Coelenterata: Alcyonacea) 2. The relative importance of chemical and structural defenses in three species of *Sinularia*, *J. Exp. Mar. Biol. Ecol.*, 178 (1), 17, 1994.

58. Bazzaz, F.A., Chiariello, N.R., Coley, P.D., and Pitelka, L.F., Allocating resources to reproduction and defense, *Bioscience*, 37, 58, 1987.

59. Simms, E.L., Costs of plant resistance to herbivory, in *Plant Resistance to Herbivores and Pathogens*, Fritz, R. and Simms, E., Eds., University of Chicago Press, Chicago, IL, 1992, 392.

60. Agrawal, A.A. and Karban, R., Why induced defenses may be favored over constituative strategies in plants, in *The Ecology and Evolution of Inducible Defenses*, Tollrian, R. and Harvell, C. D., Eds., Princeton University Press, Princeton, NJ, 1999, 45.

61. Berenbaum, M.R. and Zangerl, A.R., Coping with life as a menu option: inducible defenses of the wild person, in *The Ecology and Evolution of Inducible Defenses*, Tollrian, R. and Harvell, C. D., Eds., Princeton University Press, Princeton, NJ, 1999, 10.

62. Estes, J.A. and Steinberg, P.D., Predation, herbivory, and kelp evolution, *Paleobiology*, 14, 19, 1988.

63. Hay, M.E., Duffy, J.E., and Pfister, C.A., Chemical defense against different marine herbivores: are amphipods insect equivalents?, *Ecology*, 68, 1567, 1987.

64. Hay, M.E., Renaud, P.E., and Fenical, W., Large mobile versus small sedentary herbivores and their resistance to seaweed chemical defenses, *Oecologia*, 75, 246, 1988.

65. Paul, V.J., Hay, M.E., Duffy, J.E., Fenical, W., and Gustafson, K., Chemical defense in the seaweed *Ochtodes secundiramea* (Montagne) Howe (Rhodophyta), effects of its monoterpenoid components upon diverse coral-reef herbivores, *J. Exp. Mar. Biol. Ecol.*, 114, 249, 1987.

66. Brawley, S.H. and Fei, X.G., Studies of mesoherbivory in aquaria and in an unbarricaded mariculture farm on the Chinese Coast, *J. Phycol.*, 23, 614, 1987.

67. Hay, M.E., Pawlik, J.R., Duffy, J.E., and Fenical, W., Seaweed–herbivore–predator interactions: host-plant specialization reduces predation on small herbivores, *Oecologia*, 75, 246, 1989.

68. Duffy, J.E. and Hay, M.E., Herbivore resistance to seaweed chemical defense: the roles of mobility and predation risk, *Ecology*, 75, 1304, 1994.

69. Duffy, J.E. and Hay, M.E., Food and shelter as determinants of food choice by an herbivorous marine amphipod, *Ecology*, 72, 1286, 1991.

70. Pavia, H., Cervin, G., Lindgren, A., and Aberg, P., Effects of UV-B radiation and simulated herbivory on phlorotannins in the brown alga *Ascophyllum nodosum*, *Mar. Ecol. Prog. Ser.*, 157, 139, 1997.

71. Schmitt, T.M., Hay, M.E., and Lindquist, N., Constraints on chemically mediated coevolution: multiple functions for seaweed secondary metabolites, *Ecology*, 76, 107, 1995.

72. Wahl, M., Marine epibiosis. I. Fouling and antifouling: some basic aspects, *Mar. Ecol. Prog. Ser.*, 58, 175, 1989.

73. Arnold, T.M., Tanner, C.E., and Hatch, W.I., Phenotypic variation in polyphenolic content of the tropical brown alga *Lobophora variegata* as a function of nitrogen availability, *Mar. Ecol. Prog. Ser.*, 123, 177, 1995.

74. Ilvessalo, H. and Tuomi, J., Nutrient availability and accumulation of phenolic compounds in the brown alga *Fucus vesiculosus*, *Mar. Biol.*, 101, 115, 1987.

75. Renaud, P.E., Hay, M.E., and Schmitt, T.M., Interactions of plant stress and herbivory: intraspecific variation in the susceptibility of a palatable versus an unpalatable seaweed to sea urchin grazing, *Oecologia*, 82, 217, 1990.

76. Pedersen, A., Studies on phenol content and heavy metal uptake in fucoids, *Hydrobiology*, 116/117, 498, 1984.

77. Van Alstyne, K.L., unpublished data.

78. Van Alstyne, K.L., *The Ecology and Evolution of Antiherbivore Defenses in Fucoid Brown Algae*, Ph.D. thesis, University of Washington, 1988.

79. Lumbang, W.A. and Paul, V.J., Chemical defenses of the tropical marine seaweed *Neomeris annulata* Dickie: effects of multiple compounds on feeding by herbivores, *J. Exp. Mar. Biol. Ecol.*, 201, 185, 1996.

80. Pennings, S.C., Puglisi, M.P., Pitlik, T.J., Himaya, A.C., and Paul, V.J., Effects of secondary metabolites and CaCO$_3$ on feeding by surgeonfishes and parrotfishes: within-plant comparisons, *Mar. Ecol. Prog. Ser.*, 134, 49, 1996.

81. Carlson, D.J., Lubchenco, J., Sparrow, M.A., and Trowbridge, C.D., Fine-scale variability of lanosol and its disulfate ester in the temperate red alga *Neorhodomela larix*, *J. Chem. Ecol.*, 15, 1321, 1989.

82. Bold, H.C. and Wynne, M.J., *Introduction to the Algae*, Prentice-Hall, Englewood Cliffs, NJ, 1985.

83. Abbott, I.A. and Hollenberg, G.J., *Marine Algae of California*, Stanford University Press, Stanford 1976.

84. Little, C. and Kitching, J.A., *The Biology of Rocky Shores*, Oxford University Press, Oxford, 1986.

85. Nybakken, J.W., *Marine Biology: An Ecological Approach*, Harper Collins, New York, 1993.

86. Valiela, I., *Marine Ecological Processes*, Springer-Verlag, New York, 1995.

87. Hay, M.E., Spatial patterns of grazing intensity on a Caribbean barrier reef: to herbivory and algal distribution, *Aquat. Bot.*, 11, 97, 1981.

88. Martinez, E.A., Micropopulation differences in phenol content and susceptibility to herbivory in the Chilean kelp *Lessonia nigrescens* (Phaeophyta, Laminariales), *Hydrobiology*, 326/327, 205, 1996.

89. Van Alstyne, K.L., Comparison of three methods for quantifying brown algal polyphenolic compounds, *J. Chem. Ecol.*, 21, 45, 1995.

90. Bolser, R.C. and Hay, M.E., Are tropical plants better defended? Palatability and defenses of temperate vs. tropical seaweeds, *Ecology*, 77, 2269, 1996.

91. Cronin, G., Paul, V.J., Hay, M.E., and Fenical, W., Are tropical herbivores more resistant than temperate herbivores to seaweed chemical defenses? Diterpenoid from *Dictyota acutiloba* as feeding deterrent for tropical versus temperate fishes and urchins, *J. Chem. Ecol.*, 23, 289, 1997.

92. Carpenter, R.C., Partitioning herbivory and its effects on coral reef algal communities, *Ecol. Monogr.*, 56, 345, 1986.

93. Hay, M.E., Fish–seaweed interactions on coral reefs: effects of herbivorous fishes and adaptations of their prey, in *The Ecology of Fishes on Coral Reefs*, Sale, P., Ed., Academic Press, San Diego, CA, 1991, 96.

94. Hatcher, B.G. and Larkum, A.W.D., An experimental analysis of factors controlling the standing crop of the epilithic algal community on a coral reef, *J. Exp. Mar. Biol. Ecol.*, 69, 61, 1983.

95. Fenical, W., Natural products chemistry in the marine environment, *Science*, 25, 923, 1982.

96. Fleury, B.G., Kelecom, A., Perreira, R.C., and Texeira, V.L., Polyphenols, terpenes, and steroids in Brazilian Dictyotales and Fucales (Phaeophyta), *Bot. Mar.*, 37, 457, 1994.

97. Targett, N.M., Boettcher, A.A., Targett, T.E., and Vrolijk, N.H., Tropical marine herbivore assimilation of phenolic-rich plants, *Oecologia*, 103, 170, 1985.

98. Boettcher, A.A. and Targett, N.M., Role of polyphenolic molecular size in reduction of assimilation efficiencies in *Xiphister mucosus*, *Ecology*, 74, 891, 1993.

99. Stern, J.L., Hagerman, A.E., Steinberg, P.D., Winter, F.C., and Estes, J.A., A new assay for quantifying brown algal phlorotannins and comparisons to previous methods, *J. Chem. Ecol.*, 22, 1273, 1996.

100. Vermeij, G.J., *Biogeography and Adaptation*, Harvard University Press, Cambridge, MA, 1978.

101. Steneck, R.S. and Dethier, M.N., A functional group approach to the structure of algal dominated communities, *Oikos*, 69, 476, 1994.

102. Gaines, S.D., Herbivory and between-habitat diversity: the differential effectiveness of defenses in a marine plant, *Ecology*, 66, 473, 1985.

103. Hawkins, S.J. and Hartnoll, R.G., Grazing of intertidal algae by marine invertebrates, *Oceanogr. Mar. Biol. Ann. Rev.*, 21, 195, 1983.

104. Paul, V.J. and Van Alstyne, K.L., Activation of chemical defenses in the tropical marine algae *Halimeda* spp., *J. Exp. Mar. Biol. Ecol.*, 160, 191, 1992.

105. Steneck, R.S., Escalating herbivory and resulting adaptive trends in calcareous algal crusts, *Paleobiology*, 9, 44, 1983.

106. Steneck, R.S., The ecology of coral and algal crusts: convergent patterns and adaptive strategies, *Ann. Rev. Ecol. Syst.*, 17, 273, 1986.

107. Gaines, S.D. and Lubchenco, J., A unified approach to marine plant–herbivore interactions. II. Biogeography, *Ann. Rev. Ecol. Syst.*, 13, 111, 1982.

108. Womersley, H.B.S., A critical survey of the marine benthic algae of Australia. II. Phaeophyta, *Aust. J. Bot.*, 15, 189, 1967.

109. Womersley, H.B.S., Biogeography of Australian marine macroalgae, in *Marine Botany, and Australasian Perspective*, Clayton, N. and King, R., Eds., Longman, Ayreshire, 1981, 292.

110. Clayton, M.N., Evolution of the Phaeophyta with particular reference to the Fucales, *Progr. Phycol. Res.*, 3, 11, 1984.

111. Anderson, E.K. and North, W.J., In situ studies of spore production and dispersal in the giant kelp *Macrocystis*, *Proc. Int. Seaweed Symp.*, 5, 73, 1966.

112. Dayton, P.K., Dispersion, dispersal and persistence of the annual intertidal alga *Postelsia palmeaformis* Ruprecht, *Ecology*, 54, 433, 1973.

113. Paine, R.T., Disaster, catastrophe and local persistence of sea palm *Postelsia palmaeformis*, *Science*, 205, 685, 1979.

114. Pearson, G.A. and Brawley, S.H., Reproductive ecology of *Fucus distichus* (Phaeophyceae): an intertidal alga with successful external fertilization, *Mar. Ecol. Prog. Ser.*, 143, 211, 1996.

115. Kendrick, G.A. and Walker, D.I., Dispersal of propagule of *Sargassum* spp. (Sargassaceae: Phaeophyta): observations of local patterns of dispersal and consequences for recruitment and population structure, *J. Exp. Mar. Biol. Ecol.*, 192, 273, 1995.

116. Reed, D.C., Laur, D.R., and Ebeling, A.W., Variation in algal dispersal and recruitment: the importance of episodic events, *Ecol. Monogr.*, 58, 321, 1988.

117. Fredriksen, S., Sjotun, K., Lein, T.E., and Rueness, J., Spore dispersal in *Laminaria hyperborea* (Laminariales, Phaeophyceae), *Sarsia*, 80, 47, 1995.

118. Zechman, F.W. and Mathieson, A.C., The distribution of seaweed propagules in estuarine, coastal, and offshore waters in New Hampshire, USA, *Bot. Mar.*, 28, 283, 1985.

119. Santelices, B., Patterns of reproduction and recruitment in seaweeds, *Oceanogr. Mar. Biol. Ann. Rev.*, 28, 177, 1990.

120. Coyer, J.A., Olsen, J.L., and Stam, W.T., Genetic variability and spatial separation in the sea palm *Postelsia palmaeformis* (Phaeophyceae) as assessed with M13 fingerprints and RAPDS, *J. Phycol.*, 33, 561, 1997.

121. Williams, S.L. and DiFiori, R.E., Genetic diversity and structure in *Pelvetia fastigiata* (Phaeophyta, Fucales): does a small effective neighborhood size explain fine-scale genetic structure?, *Mar. Biol.*, 126, 371, 1986.

122. Rhoades, D.F., Evolution of plant chemical defense against herbivores, in *Herbivores: Their Interaction with Secondary Plant Metabolites*, Rosenthal, G. A. and Janzen, D. H., Eds., Academic Press, New York, 1979, 1.

123. Tuomi, J., Toward integration of plant defence theories, *Trends Ecol. Evol.*, 7, 365, 1992.

124. Herms, D.A. and Mattson, W.J., The dilemma of plants: to grow or defend, *Quart. Rev. Biol.*, 67, 283, 1992.

125. Cronin, G., Resource Allocation in Seaweeds and Marine Invertebrates: Chemical Defense Patterns in Relation to Defense Theories, in *Marine Chemical Ecology*, McClintock, J.B. and Baker, B.J., Eds., CRC Press, Boca Raton, FL, 2001, chapt. 9.

126. Garbary, D.J., Kim, K.Y., Klinger, T., and Duggins, D., Red algae as hosts for endophytic kelp gametophytes, *Mar. Biol.*, 135, 35, 1999.

127. Lubchenco, J. and Cubit, J., Heteromorphic life histories of certain marine algae as adaptations to variations in herbivory, *Ecology*, 61, 676, 1980.

128. Schiel, D.R. and Foster, M.S., The structure of subtidal algal stands in temperate waters, *Oceanog. Mar. Biol.*, 24, 265, 1986.

129. Dethier, M.N., Heteromorphic algal life histories: the seasonal pattern and response to herbivory of the brown crust, *Ralfsia californica*, *Oecologia*, 49, 333, 1981.

130. Cronin, G. and Hay, M.E., Chemical defenses, protein content, and susceptibility to herbivory of diploid vs. haploid stages of the isomorphic brown alga *Dictyota ciliolata* (Phaeophyta), *Bot. Mar.*, 39, 395, 1996.

131. Van Alstyne, K.L., Ehlig, J.M., and Whitman, S.L., Ontogenetic shifts in phlorotannin production, nutritional quality, and susceptibility to herbivory in marine brown algae, *Mar. Biol.*, in press.

132. Denton, A., Chapman, A.R.O., and Markham, J., Size-specific concentrations of phlorotannins (anti-herbivore compounds) in three species of *Fucus*, *Mar. Ecol. Prog. Ser.*, 65, 103, 1990.

133. Van Alstyne, K.L. and Pelletreau, K.N., The effects of nutrient enrichment on growth and phlorotannin production in *Fucus gardneri* embryos, *Mar. Ecol. Prog. Ser.*, 206, 33, 2001.

134. Coley, P.D., Bryant, J.P., and Chapin, F.S., III, Resource availability and plant antiherbivore defense, *Science*, 230, 895, 1985.

135. Bryant, J.P., Chapin, F.S., III, and Klein, D.R., Carbon/nutrient balance of boreal plants in relation to herbivory, *Oikos*, 40, 357, 1983.

136. Tollrian, R. and Harvell, C.D., *The Ecology and Evolution of Inducible Defenses*, Princeton University Press, Princeton, NJ, 1999.

137. Karban, R. and Baldwin, I.T., *Induced Responses to Herbivory*, University of Chicago Press, Chicago, IL, 1997.

138. Givnish, T.J. and Vermeij, G.J., Sizes and shapes of liane leaves, *Am. Nat.*, 110, 743, 1976.

139. Puglisi, M.P. and Paul, V.J., Intraspecific variation in the red alga *Portieria hornemannii*: monoterpene concentrations are not influenced by nitrogen or phosphorus enrichment, *Mar. Biol.*, 128, 161, 1997.

140. Harvell, C.D. and Tollrian, R., Why inducible defenses?, in *The Ecology and Evolution of Inducible Defenses*, Tollrian, R. and Harvell, C. D., Eds., Princeton University Press, Princeton, NJ, 1999, 3.

141. Steinberg, P.D., Lack of short-term induction of phlorotannins in the Australasian brown algae *Ecklonia radiata* and *Sargassum vestitum*, *Mar. Ecol. Prog. Ser.*, 112, 129, 1994.

142. Paul, V.J. and Van Alstyne, K.L., unpublished data.

143. Davison, I.R. and Pearson, G.A., Review: stress tolerance in intertidal seaweeds, *J. Phycol.*, 32, 197, 1996.

144. Bergelson, J. and Purrington, C.B., Surveying patterns in the cost of resistance in plants, *Amer. Natur.*, 148, 536, 1996.

145. Vermeij, G.J., *Evolution and Escalation: An Ecological History of Life*, Princeton University Press, Princeton, NJ, 1987.

146. Vermeij, G.J., *A Natural History of Shells*, Princeton University Press, Princeton, NJ, 1993.

147. Gabrielson, P.W., Scagel, R.F., and Widdowson, T.B., *Keys to the Benthic Marine Algae of British Columbia, Northern Washington, and Southeast Alaska*, University of British Columbia, Vancouver, Canada, 1987.

148. Womersley, H.B.S., *The Marine Benthic Flora of Southern Australia, Part II*, South Australia Printing Division, Adelaide, Australia, 1987.

149. Adams, N.M., *Seaweeds of New Zealand*, Canterbury University Press, Christchurch, New Zealand, 1994.

150. Irelan, C.D. and Horn, M.H., Effects of macrophyte secondary chemicals on food choice and digestive efficiency of *Cebidichthys violaceus* (Girard), and herbivorous fish of temperate marine waters, *J. Exp. Mar. Biol. Ecol.*, 153, 179, 1991.

151. Steinberg, P.D., Edyvane, K., de Nuys, R., Birdsey, R., and van Altena, I.A., Lack of avoidance of phenolic-rich algae by tropical herbivorous fishes, *Mar. Biol.*, 109, 335, 1991.

9 Resource Allocation in Seaweeds and Marine Invertebrates: Chemical Defense Patterns in Relation to Defense Theories

Greg Cronin

CONTENTS

I. OVERVIEW

To be successful, all organisms must acquire resources. They must also tolerate, avoid, or defend against becoming a resource for consumers, at least until after reproduction. The majority of this volume deals with how organisms use chemicals to defend themselves from consumers, though secondary metabolites affect their interactions with competitors, parasites, commensals, conspecifics, and the abiotic environment.[1–6] This chapter discusses how organisms allocate acquired resources to various life processes, including the production of secondary metabolites.

The finite resources available to organisms must be allocated to several life processes including growth, reproduction, maintenance, defense, and further resource acquisition. It is often assumed that a resource allocated to one process incurs a cost to the remaining processes because the resource is diverted away from them.[7,8] It is also assumed that natural selection acts to optimize the allocation of resources to best suit the life history and environment for a particular organism, of course, within evolutionary and ecological constraints.

While organisms must allocate resources to the above-mentioned processes, resources vary qualitatively and quantitatively among species. For example, autotrophs and heterotrophs have fundamentally different modes of resource acquisition; the former acquires energy from solar radiation or reduced minerals (i.e., abiotic sources), and the latter acquires energy from reduced carbon molecules (i.e., biotic sources). This chapter attempts to address how differences among autotrophs, heterotrophs, and mixitrophs affect resource allocation. Mixitrophic sessile invertebrates share several biological and ecological similarities with autotrophic seaweeds. Both groups are anchored in place, must grow in an open manner to acquire resources (light for both groups and plankton among filter feeding invertebrates), have modular body plans,[9] can regenerate from basal portions following a dormant period,[10] and can regenerate tissue lost to disturbance or partial predation.[11,12] These similar biological traits result in similar ecological strategies; both groups are tolerant of some consumption and are often defended by structural or chemical defenses.[1–6]

This chapter briefly discusses the various processes to which organisms allocate resources and then concentrates on allocation to secondary metabolites, the focus of this volume as a whole. Throughout this chapter, the term predator refers to any consumer of seaweeds or invertebrates. Therefore, herbivores, parasites, and consumers of animal prey are all covered under the general term predators.

II. ALLOCATION OF FINITE RESOURCES

A. THE MODEL

Pie charts are used to represent the total amount of resources (i.e., materials and energy) acquired by an organism, and the size of each pie slice represents the proportion of total resources allocated to a particular process (Figure 9.1). The size of the pie is larger for organisms that allocate a greater proportion of resources to further resource acquisition. The reason chemical defenses are assumed, though rarely determined, to be costly is because the pie slice that represents allocation to defenses subtracts from the amount of resources available to the remaining processes.

The relative success of each allocation pattern is dependent on the biotic and abiotic conditions in which the organism finds itself (Figure 9.2). Allocating resources heavily towards defense may be adaptive under conditions of intense predation pressure, but may be maladaptive under conditions with little predation pressure, as organisms that allocate heavily towards further resource acquisition and growth would likely be superior competitors. Of course, the context in which these interactions take place is far too complicated to address every possible caveat of allocation patterns. For example, in the case just given, if the secondary metabolites produced by the heavily defended species have allelopathic effects on competitors as well as defensive properties against predators, then the defended organism may compete well regardless of the abundance of predators. Under physically stressful conditions, an allocation pattern that favors maintenance may be most adaptive since organisms will be better equipped to maintain life processes under adverse physical conditions.[13]

Important life processes that require energy and material resources include growth, maintenance, reproduction, further resource acquisition, and dealing with natural enemies. It is important to

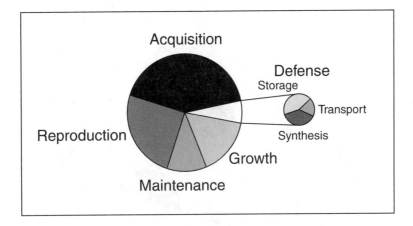

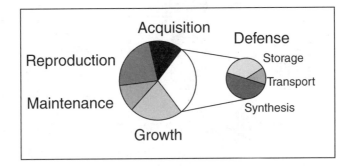

FIGURE 9.1 Conceptual model of allocation of resources among various life processes. The defense pie could be further divided into slices for defense against predator 1, defense against predator 2, defense against parasite 1, secondary metabolite 1, secondary metabolite 2, or secondary metabolite 3, since many organisms produce multiple secondary metabolites that are differentially effective against different natural enemies. The top pie is larger (i.e., has more total resources) than the bottom pie because a greater proportion of resources is allocated to resource acquisition in the former situation.

recognize that these various processes occur simultaneously within organisms and are interrelated in complex ways.

Growth rates vary tremendously among marine organisms. Among seaweeds, filamentous algae have some of the highest growth rates, doubling in size every few days, while some crust-forming coralline algae have very slow growth rates, taking months to double in size.[14] Ecologists have detected correlations among growth form, growth rates, and strategies to deal with herbivores.[15,16] Forms with high surface area to volume ratio have high levels of photosynthesis (i.e., resource acquisition) but are generally susceptible to herbivores. However, rapid growth allows these species to replace tissue rapidly (i.e., they can tolerate consumers). In contrast, seaweeds with a low surface area to volume ratio, a large amount of structural material, or compressed mats or turfs have reduced photosynthesis due to self-shading or limited nutrient uptake, but are typically less susceptible and less tolerant to herbivores.[14,17,18]

Once tissue is grown, it does not stop requiring resources. Material and energy are needed (1) to maintain tissue, whether or not it is actively growing, (2) for reproductive effort (e.g., mate searching, gamete production, and parental care), (3) for the acquisition of additional resources, and (4) for dealing with natural enemies. The allocation of resources to dealing with natural enemies, especially predators, is the focus of this chapter.

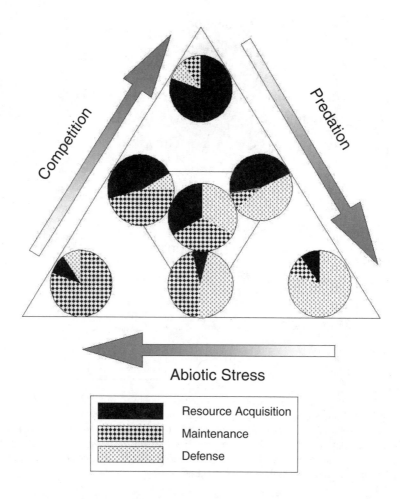

FIGURE 9.2   Different allocation patterns predicted to be adaptive along environmental gradients of abiotic stresses, competition, and predation pressures, which should select for high resource allocation to maintenance, resource acquisition, and defense, respectively. This model predicts allocation patterns for Grime's plant strategies[13] and draws predictions from various chemical defense theories. Allocation to reproduction and growth are not shown for clarity.

B. DEALING WITH NATURAL ENEMIES

All organisms are composed of fixed carbon, biomolecules, and mineral nutrients, and therefore represent energy and nutrient resources for consumers. To be successful (i.e., grow, maintain self, and reproduce), organisms must avoid, tolerate, or defend against natural enemies.[19] Of course, these strategies are not mutually exclusive, and many species use more than one strategy. For example, induction of chemical defenses can be viewed as using an avoidance strategy until being detected by a predator, tolerating a small amount of consumption that serves as the cue to make the switch to the defensive strategy.

Avoiding consumption involves being where and/or when consumers are rare or inactive. Seaweeds and sessile invertebrates can avoid predators spatially, by growing in habitats with low densities of predators, such as reef flats, sand plains, sea grass beds, and mangroves,[20-23] or temporally, by growing when predation pressures are low.[24-26]

Tolerating consumption involves having rates of regeneration that can keep up with losses to consumers. Generally, organisms with a modular body plan such as plants and colonial invertebrates have a greater power of regeneration than organisms with determinant growth such as verte-brates.[11–12] Algal species that tolerate herbivory well are responsible for the high levels of produc-tivity that occur on coral reefs.[14,16] These fast growing filamentous algae have basal portions protected in crevices of the reef (i.e., the basal portions avoid consumption here), but the tops of the filaments are quickly consumed by herbivores shortly after growing above the protective crevice. This consumption is nonlethal, and the alga simply regrows what was lost. This is similar to the strategies of turf grasses: grazers or lawnmowers frequently cut the tops of blades, and the grasses, which have their meristems protected close to the ground or below ground, simply replace the lost portion.[13]

Defending tissue against consumption involves making it less attractive to or shielding it from potential consumers. Defense is a common strategy among sessile or sluggish organisms, especially ones that must grow in an exposed manner to acquire resources (e.g., photosynthetic or filter-feeding organisms). These life history traits place constraints on the ability to avoid consumption, because fleeing or hiding from consumers is difficult or impossible. Four categories of defenses used by marine species are structural, associational, nutritional, and chemical defenses.

1. Structural Defenses

Structures that defend marine organisms from being consumed can act as external shields, sharp spines located externally or internally, or support material that make tissues too hard to easily bite, making them a good first line of defense.[16,27–29] External structures that protect vulnerable internal tissues include the chitonous exoskeleton of crustaceans, calcarious shells of molluscs and barnacles, tests of echinoderms, and the tough tunics of ascidians.[9] Needle-like internal structures such as spicules and sclerites are distributed throughout the soft tissues of some sponges, cnidarians, and ascidians. Calcareous or siliceous spicules of sponges are arranged in specific configurations by a lattice of spongin fibers, creating a tough skeleton that provides support and possible defense for the organism. The effectiveness of these structures as antifeedants is unclear as the sclerites from gorgonians and soft corals deterred fishes,[27,28] but the spicules from several sponges[30] and an ascidian[31] did not deter fishes in bioassays.

Calcium carbonate makes up as much as 90% of the dry mass[14,32] of the hard thallus of seaweeds. This calcium carbonate does not form structures like spicules, but rather precipitates as small spheres that would not likely pierce an herbivore that bit into it. It was long believed that this calcium carbonate served like concrete to make the overall thallus harder to bite (i.e., a structural defense), or perhaps the calcium carbonate could dilute the nutritional value of the seaweed (i.e., a nutritional defense). Recent studies have shown that calcium carbonate reduced feeding by herbivores even when it does not influence the hardness or nutritional value of food. Under these conditions, $CaCO_3$ was likely acting as a chemical defense, perhaps by raising the pH of guts or by increasing the efficacy of secondary metabolites.[33–36]

Defensive structures can represent considerable costs to organisms, but as with chemical defenses (see below), alternative uses of these defenses may help defray their costs. Hard chitonous coverings serve a protective role to arthropods, but these structures also serve as an exoskeleton (likely the main reason for this adaptive covering), which defrays some of the defensive costs. The precipitation of calcium carbonate from seawater is a consequence of photosynthesis altering pH and CO_3^{2-} concentrations,[14,37] hence the cost of producing this defense may be considered low. However, $CaCO_3$ is an opaque powder that could shade chloroplasts, costing the seaweed or mixitrophic invertebrates photosynthetic potential. Calcareous coralline algae are among the slowest growing seaweeds.[16]

2. Associational Defenses

Associational defenses occur when a species gains protection from a natural enemy by associating with a protective host. Mechanisms of protection provided to the defended species can be chemical, structural, camouflage, or aggressive.[5,38–42] The species providing protection can benefit, be unaffected, or be harmed by the association, resulting in the association being termed mutualistic, commensal, or parasitic/predatory, respectively. It is often difficult to place associational defenses into one of these catagories because the impact of the interaction on the defending species is typically not quantified. To add to this difficulty, the category of an interspecific interaction depends on the ecological context. For example, in the presence of light, zooxanthellae are mutualists of many sessile invertebrates,[43] but in darkness, they are commensal or parasitic.

With defensive mutualisms protection from natural enemies is provided by the association with a protective host. Many suspected examples of defensive mutualisms are epibiotic associations, where an unpalatable epibiont benefits from the substrate provided by the basibiont while protecting the basibiont with an unpalatable covering. Such interactions include associations between whelk–bryozoan,[44] kelp–bryozoan,[40] clam–algae,[45] and mussels serving as the basibiont for hydrozoans, sponges, barnacles, and algae.[41,42] Epibionts do not always protect the basibiont, and there are examples of palatable epibionts increasing the susceptibility of the basibiont to predators. This situation was termed "shared doom"[41] as both species in the association are consumed together.

Nonepibiotic defensive mutualisms include clown fish–anemone, amphipod–hydroid,[46] and some seaweed–herbivore interactions.[47–49] For example, the coralline alga *Neogoniolithon strictum* gains protection from fouling organisms (which could interfere with resource acquisition via exploitative competition) from the crab *Mithrax sculptus*, even though *Neogoniolithon* does not directly provide the crab with significant food.[49] *Neogoniolithon* is a very hard, calcified seaweed that is defended structurally, and perhaps chemically, against herbivores, including *Mithrax*, by $CaCO_3$. The crab benefits by consuming epiphytes that it clears from *Neogoniolithon* and by using the hard seaweed as shelter from predators. Branches of the hard seaweed provide a shelter for the crabs: 20% of crabs tethered near the seaweed were consumed within 30 min, while 100% of crabs tethered away from the seaweed were consumed during the same time period.[49]

Predatory associational defenses are the most commonly reported type of associational defense. These involve situations where the protective host is consumed by the defended species (i.e., the host provides both food and habitat to the protected host). Small relatively immobile invertebrates reduce predation by associating with noxious hosts: these associations can be specialized[50–54] or generalized.[33,55–59] Predatory associational defenses are often chemically mediated, including sequestration of host chemical defenses by the defended species. Sequestration is most common among specialists[50–54] but there are examples of larger generalists being able to sequester chemical defenses from their diet.[33,59]

3. Nutritional Defenses

An organism that is of low nutritive value may be protected from consumption because it is not worth eating.[35,60] Optimal foraging theory would dictate that any would-be predator should forage for more valuable prey. This strategy is available to seaweeds and gelatinous animals, but is generally unavailable to most animals, as their tissue is more nutritious.[61] It is difficult to envision how an organism allocates resources to nutritional defenses. However, one can envision costs that may be associated with this strategy; by maintaining itself at a low nutritional value, such an organism may constrain its ability to store nutrients and energy for lean periods or its ability for rapid growth.

4. Chemical Defenses

Organisms defend themselves chemically by using compounds that are distasteful, toxic, or otherwise repulsive to consumers.[1–6] Most defensive compounds are secondary metabolites of unique structures,[62] but can also include more generic compounds such as sulfuric acid[63,64] or $CaCO_3$.[33–36] The production, transport, storage, and maintenance of defensive compounds require the allocation of resources, and, hence, are often assumed to incur a cost to the organism (Figure 9.1).

While this chapter concentrates on the allocation of resources by marine organisms and the related allocation costs of secondary metabolites, it is important to recognize that there are other potential costs of chemical defense that are unrelated to allocation. These nonallocation costs include ecological, autotoxicity, and genetic costs. Ecological costs occur when a secondary metabolite that defends against one consumer makes the prey more susceptible to a second consumer. For example, secondary metabolites often defend against generalist consumers, but can actually stimulate feeding by specialist consumers that have adapted to using a specific noxious host.[56] Autotoxicity, or self-poisoning, is problematic because secondary metabolites can be deterrent by interfering with basic biological processes (thereby making them effective against a large number of consumers), but the compounds can poison the same organism that they are defending if they are not properly handled. Finally, genetic costs occur when genes that confer chemical resistance to natural enemies have negative pleiotrophic effects on other processes (separate from the diversion of resources).[65] Genetic costs of secondary metabolites have not been determined for any marine organism, but there is limited information on ecological, autotoxicity, and allocation costs.

III. ALLOCATION OF RESOURCES TO SECONDARY METABOLITES

For a trait to be selected for, or not selected against, its benefits should increase fitness more than its costs reduce fitness, on average. A growing literature on the evolution of chemical defenses suggests that decreased susceptibility to consumers can be achieved only by diverting materials and energy from other functions.[7,19,65–68] While there are several theoretical reasons to believe that defenses are costly in terms of trade-offs, this common assumption is supported by little direct evidence.[65] However, there is much circumstantial evidence that supports the idea that the production, maintenance, transport, and storage of secondary metabolites have associated costs.

One of the main obstacles for studying the allocation of resources is determining the appropriate currency to measure cost. The different currencies that have been used include (1) the energy stored in chemical bonds,[69] (2) biomass allocated to various tissues or materials responsible for different biological functions,[70] (3) the amount of limiting resource allocated to different processes, (4) the competitive ability of organisms with different allocation patterns, and (5) some measure of fitness trade-off.[71]

Allocation costs of secondary metabolites are discussed below, even though no direct measures of these costs with marine organisms have been made.[69] We tend to equate cost of secondary metabolites with cost of chemical defenses because it is the defensive roles of these compounds that have been most studied. However, it is important to recognize that the benefits of secondary metabolites are not limited to defenses against predators; secondary metabolites can also be used as defenses against fouling organisms,[72–77] microbes,[78–80] competitors,[81] and the damaging effects of UV radiation,[82] as well as aggregation or gamete attractants.[83,84] As a result of serving multiple functions, the total cost of secondary metabolite production could be less than that of these individual functions summed together.

The synthesis of secondary metabolites requires materials in the form of atoms that compose the secondary metabolites, energy to form covalent bonds between the atoms, the enzymes that carry out the formation reactions, the genetic material that codes and synthesizes the enzymes, and the cellular machinery that maintains pH, ionic strength, and redox potentials within a range that allow enzymes to function properly.[85] Some materials such as hydrogen and oxygen are readily available and cheap, and contribute little to the material cost of synthesis. Other materials such as phosphorus and nitrogen are of limited supply in many marine habitats.[86] The use of these rare materials in secondary metabolites is probably costly as they limit many primary metabolic pathways, which helps explain the rarity of secondary metabolites that contain N or P.[2] Fixed carbon can also be limiting under certain conditions (e.g., under shaded conditions for seaweeds). This limitation is not a materials limitation because inorganic carbon (i.e., in the form of HCO_3^-) is abundant in the ocean. Rather, it is an energy limitation, since reduced carbon compounds are the energy currency in living systems.

The principal anabolic pathways for secondary metabolites originate from just a few intermediates of primary metabolic pathways, such as acetyl CoA, shikimic acid, and melvonic acid.[86] Among the important cofactors are ATP, NADPH, and S-adenosylmethionine, which need to be continuously regenerated via primary metabolic pathways of respiration or photosynthesis. The fact that secondary metabolism shares chemical precursors with primary metabolism means that secondary and primary metabolic pathways may compete for substrates and cofactors, strongly suggesting that trade-offs occur at the biochemical level.

Sequestration of chemical defenses synthesized by prey or symbionts is observed in several marine invertebrates. Sequestration is a common phenomenon among molluscs[50,54,59,86–88] but has only rarely been demonstrated among other invertebrate groups.[52,89] By sequestering secondary metabolites, organisms are able to eliminate the costs of synthesis.

There may be additional advantages of sequestration as well, such as providing chemical camouflage.[5,54,90] An interesting example of sequestration that can be interpreted as camouflage is the use of domiciles constructed of a chemically defended seaweed by the amphipod *Pseudamphithoides incurvaria*.[52] The amphipod does not physiologically incorporate its prey defensive compounds into its body, but rather behaviorally sequesters the compounds by remaining inside a bivalved domicile that the amphipod constructs out of the seaweed. The amphipod is palatable to fish once it is removed from the domicile, suggesting that the amphipod gains protection from predators by being camouflaged as seaweed. Such behavioral sequestration not only avoids the cost of synthesis (as does physiological sequestration), but also avoids many costs associated with storage, transport, maintenance, and autotoxicity; costs which physiological sequestration may not avoid. Potential costs associated with the domicile are the allocation of time and effort in constructing the domicile and the burden of carrying it around.

For concentrations of secondary metabolites to be continuously maintained, production must keep up with growth (so increasing biomass does not dilute metabolites) and turnover. Turnover rates are probably species and compound specific.[69,91] Costs associated with maintaining secondary metabolites will depend on turnover rates and the degree at which breakdown products can be recycled by the organism. As with synthesis costs, organisms that sequester defenses may be able to avoid maintenance costs by simply eating another chemically laden meal.

The storage of compounds can differ greatly between seaweeds and invertebrates. Compound storage in seaweeds is not well known, but it likely occurs at the subcellular level, in places such as cytoplasmic vesicles[92] and structures called physodes.[93] Animals can store secondary metabolites within cells as seaweeds do,[94] and they are also able to keep compounds in specialized organs or glands. For example, nudibranchs concentrate secondary metabolites in the dorsal mantle,[50] cerrata,[96] digestive gland,[50,95–97] and in subepidermal structures called mantle dermal formations (MDF).[64,90,98] These structures, whether they are subcellular or storage organs, require materials to construct and energy to maintain.

Related to costs of storage are costs of autotoxicity, or self-poisoning. Besides segregating secondary metabolites away from sensitive areas with storage structures, another way to reduce the cost of autotoxicity is to store defensive metabolites as inactive precursors and convert them to active compounds when they are needed for defense.[99,100] This process is termed activation in order to distinguish it from induction of defenses which refers to the longer term response in which the general defense mechanisms of an organism are increased above constitutive levels (induction is discussed below). Activation of chemical defenses is common among terrestrial plants, but has only been described for *Halimeda* spp.[101] and a sponge.[102] The major compound found in *Halimeda* is halimedatetraacetate, a compound that is not very deterrent against herbivores. Upon damage to the thallus, halimedatetraacetate is rapidly converted to halimedatrial, a much more deterrent compound. This activation of halimedatetraacetate is probably enzymatically mediated, though the enzyme has not been identified. It is also unknown how halimedatetraacetate is compartmentalized, but logic suggests that the compound is stored separately from the enzyme, and damage by an herbivore brings the two components of the reaction together.[101]

Terrestrial ecologists suggested that autotoxicity might constrain the use of chemical defenses in developing organisms when they asked, "Why are embryos so tasty?"[103] It was hypothesized that chemical defenses might be fundamentally incompatible with development because bioactive compounds could have teratogenic effects on the embryo. However, chemical defenses are not completely incompatible with embryo development as many marine invertebrates produce chemically defended eggs and larvae[6,31,50,104–106] as do some beetles and a few other terrestrial organisms.[103] The constraint of producing chemically defended embryos seems to have more to do with adults being capable of producing chemical defenses and provisioning the eggs, and less to do with a physiological incompatibility, since examples of chemically defended eggs and larvae come from species where the adults are known to be chemically defended.[104,105] Questions about the source of secondary metabolites (whether they are autogenic or provisioned by the parent), where compounds are located within larvae, mechanisms that prevent secondary metabolites from poisoning development, or how chemical defenses change ontogenically are only beginning to be addressed in marine organisms.[6,31,50,104–107]

There may also be costs associated with transporting secondary metabolites. The fact that secondary metabolites are compartmentalized or otherwise stored means that they must be actively transported against a concentration gradient into a vesicle, physode, or similar storage structure. To this author's knowledge, the details of these processes have not been determined for marine organisms. However, such intracellular transport of polar metabolites may be similar to the current models of (1) active transport of amino acids, sugars, ions, and other cellular nutrients, (2) receptor mediated endocytosis, or (3) ion trapping.[108] The production of vesicles, carrier proteins, receptors, and ion pumps requires materials and energy, so such intracellular transport likely involves costs. Intercellular transport is advantageous because it allows the movement of defenses to areas that are under attack by natural enemies. However, such capability of efficient transport requires a vascular system, expenditure of energy, and subjects additional cells to risk of autotoxicity.

The prevalence of chemical defenses among marine organisms suggests that the benefits of protection outweigh the various costs listed above. The benefit most often documented by marine ecologists is the decrease in losses to predators.[1–6,109–111] Empirical evidence abounds that secondary metabolites reduce consumption by predators, probably because of the availability of methods to test the antifeedant effects of secondary metabolites against consumers in ecologically realistic manners. It is likely that defensive secondary metabolites serve multiple roles such as allelochemicals,[112,113] antifoulants,[73–76,114] antibiotics,[78,115,116] sex attractants,[82,83,117–119] or settlement cues,[72] but ecologically realistic assays for these functions are in their infancy.[5] As an example, for a secondary metabolite to be beneficial as an antifoulant, it should be located near the surface of the organism so any would-be fouling organism can contact and respond to it. A major drawback for studying such metabolite functions is our lack of knowledge about small-scale storage and deployment of defenses (see above). Such alternative roles of secondary metabolites help defray secondary metabolite costs by receiving benefits from multiple functions.

IV. ALLOCATION MODELS FOR CHEMICAL DEFENSE

Most chemical defense theories were proposed by terrestrial ecologists to explain evolutionary and ecological patterns that were observed in interactions between terrestrial herbivores and vascular plants. There are basic biological differences between the organisms that these theories were proposed for (e.g., largely angiosperms and insects) and the organisms of this volume (e.g., seaweeds, sessile invertebrates, and their noninsect predators).[111] However, for an ecological theory to be a useful predictive tool, it should be applicable to different systems. If observations from an independent system support a model, that model is more robust. Observations from another system that are contrary to the model can provide insights that can allow refinement of current models. An overview of the more prominent chemical defense theories is provided below, and how these models apply to seaweeds and marine invertebrates is addressed.

A. OPTIMAL DEFENSE THEORY

The optimal defense theory (ODT) asserts that organisms allocate defenses in a way that maximizes fitness, and that defenses are costly (in terms of fitness) when enemies are absent. However, in the presence of enemies, defensive secondary metabolites can become beneficial when their costs are outweighed by benefits gained with protection.[7] This theory encompasses both evolutionary and ecological time scales, providing explanations for between-species, within-species, and within-individual variation in defenses.

Several observations suggest that seaweeds allocate more resources to chemical defenses when the probability of attack is high. In tropical latitudes where herbivory is generally more intense than at higher latitudes, a higher concentration and greater diversity of lipophilic secondary metabolites are produced by seaweeds. This pattern is predicted in Figure 9.2. Bolser and Hay[120] found that temperate seaweeds were eaten about twice as much as closely related tropical species, and that secondary chemistry accounted for most of the observed variation in palatability. Within a region, seaweeds from areas of coral reefs where herbivory is intense often produce more potent and higher concentrations of chemical defenses than plants from habitats where herbivory is less intense[26,121] (see Pawlik et al.[23] and Wright et al.[122] for invertebrate examples).

In contrast to patterns for lipophilic secondary metabolites, water-soluble phlorotannins (i.e., polyphenolics produced by brown seaweeds) were initially reported to be more abundant in temperate than tropical Indo-Pacific seaweeds.[123–125] However, this apparent latitudinal pattern has not held when tested in additional locations in the Caribbean.[126,127] The paucity of phlorotannins in some tropical seaweeds was attributed to their ineffectiveness against tropical herbivores,[128–130] although phlorotannins from temperate seaweeds can deter some tropical herbivores.[125] Because geographic patterns of phlorotannins and their impacts on herbivores are unclear, it is difficult to determine how allocation patterns of phlorotannins relate to the ODT.

One prediction of the ODT is that within an organism, costly defenses are allocated to tissues in direct proportion to the vulnerability and the value of the tissue, on a per mass basis, to the fitness of the organism. Most seaweeds and sessile invertebrates can recover from partial predation,[11,12,14,19] so they are able to tolerate some predation and may be able to sacrifice less valuable tissue by allocating fewer defenses to it. Based on tissue value, seaweeds might be expected to preferentially defend meristems that are responsible for the production of new cells, holdfasts that anchor the entire thallus to the substrate, reproductive tissues that are responsible for passing genetic material to the next generation, and young vegetative tissue that represents a greater productive potential than an equivalent amount of older vegetative tissue. Valuable tissues for sessile invertebrates include holdfasts and gonads for reasons similar to those given for seaweeds.

For all organisms, the tissues most exposed to attack by predators are external tissues. Even if a predator seeks internal tissue, it must first penetrate outer tissues to gain access. Additionally, because animals must ingest materials to obtain nutrition, the lining of their gut is made more vulnerable to small predators (i.e., pathogens). Therefore, surfaces that are greatly exposed to consumers should be heavily defended based upon vulnerability.[50,89,95] The Spanish dancer nudibranch is an example of a marine organism that allocates defenses based upon such vulnerability to predation. This nudibranch had higher concentrations of chemical defenses in its dorsal mantle and digestive gland than in its foot, a pattern consistent with the ODT. The digestive gland being combined with the gonad confounds the interpretation, given the fact that the nudibranch provisions its eggs with high concentrations of defensive metabolites.[50] Similarly, the dictyoceratid sponge

Dysidea herbacea, which sequesters brominated biphenyl ethers synthesized by symbiotic bacteria, concentrates these secondary metabolites in their ectosomal area.[89]

Besides the location of the tissue, other characteristics that affect its risk of attack include its nutritive value and its tenderness (i.e., lack of nutritional or physical defenses). In plants, young tissues are usually more nutritious and more delicate than older tissues,[131–133] placing them at higher risk of being attacked, when all other factors are equal. The ODT predicts that these high-risk tissues should be heavily defended with chemicals.

Patterns of within-individual defenses predicted by the ODT are borne out in many terrestrial plants.[66,131,132] Exposed external tissues are usually better defended than internal tissues in roots, stems, leaves, seeds, bulbs, and tubers.[66,131] Because young leaves are often more productive, nutritious, and delicate than older leaves, they are often endowed with higher concentrations of secondary metabolites than older leaves.[66,131,134,135] Even though young foliage is more chemically defended than mature foliage, most herbivore damage occurs to young leaves[135] because insects usually perform better on young nutritious tissue than old tissue.[134] However, most herbivorous insects are feeding specialists that may have adapted to circumvent host allelochemicals.[136]

Although less is known about the within-plant distribution of secondary metabolites in seaweeds, it might be expected that the degree of within-plant variation in seaweeds would be less than in terrestrial plants because seaweeds have fewer differentiated parts than terrestrial plants.[111,133] Terrestrial plants exhibit a greater division of labor among plant parts than seaweeds; roots uptake water and nutrients, leaves capture light energy to nourish themselves and the rest of the plant, and stems and trunks extend the leaves toward the sun, support the weight of leaves, and transport water, nutrient, and photosynthate from sources to sinks. In contrast, the cells or parts of most seaweeds are more independent; they absorb the majority of their own nutrients and make their own photosynthate because most seaweeds lack structures for efficient translocation.[137] However, exceptions occur among siphonous green seaweeds with coenocytic cellular structure and among some kelps with sieve elements, hyphal cells, and sieve tubes that allow photosynthate and other materials to be rapidly (10 to 70 cm h^{-1}) translocated.[138] Most support of seaweed parts comes from seawater and not stiff structural tissue. Although some seaweeds fit this description well, the degree of differentiation varies considerably among species. Kelps provide an example of a high degree of tissue differentiation in seaweeds; they have strong holdfasts that anchor the huge seaweeds, large floats that raise the photosynthetic fronds toward the sunlight, stipes that attach the fronds and floats to the holdfast, and vascular tissue that transports materials from sources to sinks. Similar to terrestrial plants, kelps can display considerable within-plant variation in chemical defenses.[139,140] These large seaweeds allocate more polyphenolics to the thin outer meristoderm than to the inner cortex and medullary tissue.[140] The amphipod *Allorchestes compressa* will consume the inner tissues of *Ecklonia* if the meristoderm is broken, but will not eat the meristoderm.[141] Intercalary meristems as well as holdfasts, stipes, and sporogenous tissue of kelps also have more polyphenolics than most other infertile, nongrowing tissue.[139,140] Within-individual variation in secondary metabolites also occurs in other brown seaweeds,[133,142–144] some green seaweeds,[25,26,145,146] red seaweeds,[92,147,148] bryozoans,[149] sponges,[89] and gorgonians.[150]

1. Inducible Defenses

Vulnerability of prey to attack is positively related to the abundance of predators.[7] Constitutive defenses require resources to synthesize, maintain, and store (plus other potential costs mentioned above) even when consumers are absent and the benefits of protection are not realized. Hence, the ODT predicts that allocation to constitutive defenses should be kept low when predators are absent as a means to reduce the costs of defenses when the benefits of protection are not being realized. However, once predators have been detected and tissues become vulnerable to attack, the allocation to defenses should be increased. The induction of higher levels of defenses is thought to be an adaptation to minimize the cost of defenses until costs can be offset by the benefits of protection.[100,151]

Induction may have advantages in addition to cost savings, such as creating intraspecific variation in defenses and presenting predators with a moving target that may make it more difficult for the evolution of counter adaptations.[152]

Induction of defenses depends on a reliable cue that predicts the probability of future attacks. Attack by a predator is a reliable cue that predators are present and the probability of further attack is high. This cue is useful to seaweeds and sessile invertebrates because these organisms are likely to survive initial attack since they can tolerate partial predation. Using predator attack as a cue to induce defenses is not a good strategy for organisms that suffer high mortality during the initial attack. Organisms less tolerant to predation, such as zooplankton, mobile invertebrates, and vertebrates, often use visual cues to see predators, chemical cues to smell predators, or detect compounds such as alarm pheromones or body contents that conspecifics release when attacked by predators.[153,154]

For seaweeds and colonial invertebrates, the stimulus for induced defenses most likely originates at the point of predator damage, but for other parts of the prey to become less susceptible to further predation, induction of defenses should occur throughout the individual. Therefore, the stimulus should signal the rest of the individual that an attack has occurred. The mobilization of defenses or resources to areas that are under attack would also require an effective translocation system. As a result of these two important roles of translocation in inducible responses, it would seem that non-vascular seaweeds would be highly constrained to responding to a localized attack (i.e., single bite), but perhaps not a diffuse attack.[5] For vascular plants, an easily envisioned mechanism for a systemic response to localized damage is the vascular transport system.

Results of induction experiments are far from consistent. Induction of chemical defenses has been detected in seaweeds that have rudimentary vascular systems such as kelps[155] and rock-weeds.[156,157] Despite the constraints of lacking effective vascular systems, inducible chemical defenses have been demonstrated in the brown alga *Dictyota menstrualis*.[158] This alga had high concentrations of secondary metabolites and a greater number of grazing scars when collected from sites with high densities of herbivorous amphipods, compared to conspecifics collected from a site with few amphipods. When *Dictyota* collected from the site with few amphipods was grazed by artificially increasing the number of amphipods using cages in the field, the alga responded by increasing the concentrations of secondary metabolites and by becoming less palatable to amphipods in laboratory feeding assays.[158]

Other brown seaweeds also display induced resistance to herbivory in response to actual or simulated grazing damage. *Padina gymnospora* became less susceptible to grazing by the sea urchin *Arbacia punctulata* within 0 to 5 days following actual or simulated urchin grazing, but this resistance was lost 9 days after the damage occurred.[159] The role of chemical defenses in the induced resistance of *P. gymnospora* is unknown, but a congener, *Padina jamaicensis*, becomes less susceptible to herbivores following attack by altering its growth form.[160] *P. jamaicensis* exists as a resistant turf growth form when herbivore pressure is great but converts to a susceptible foliose growth form within 4 days of herbivore pressure being reduced.

Herbivore damage can cause the growth of adventitious branches in brown seaweeds.[143,158] ODT predicts that adventitious branches should have elevated levels of defenses given that they grow in response to grazer damage. Consistent with this hypothesis, *Dictyota menstrualis* collected from sites with high densities of amphipods had a high incidence of adventitious branches. Adventitious branches of *Dictyota* had elevated levels of secondary metabolites, although these higher levels of secondary metabolites did not afford the adventitious branches additional protection from amphipods in laboratory feeding assays.[158] Another brown seaweed, *Fucus distichus*, produced adventitious branches at herbivore grazing scars, and these adventitious branches had higher concentrations of phlorotannins relative to apical meristems and were less susceptible to littorine snails.[143]

Many seaweeds apparently do not induce higher levels of secondary metabolites when grazed. Clipping experiments have failed to induce increased defenses in the brown alga *Ascophyllum nodosum*,[81] the green seaweeds *Halimeda*, *Udotea*, and *Caulerpa*,[3,101] and the kelp *Hedophyllum sessile.*[155] Urchin grazing failed to induce defense in the kelp *Ecklonia.*[124] The pattern of better-defended individuals being located in habitats with intense predation pressure could be generated by preferential grazing, local selection, or among-habitat variation in other environmental variables, in addition to inducible defenses.

While there are only a few cases of induction of chemical defenses among seaweeds, the phenomenon is very common among terrestrial plants.[99] This discrepancy in the prevalence of inducible defenses in terrestrial vs. marine plants could be due to the different biologies of vascular terrestrial plants vs. nonvascular seaweeds (e.g., the induction stimulus from localized damage may not be efficiently translocated in seaweeds[133]), complex interactions between grazing damage and environmental conditions,[161] or herbivore taxa studied,[5] but the lesser amount of research on seaweeds relative to terrestrial plants may also explain much of the disparity.

Variation of defenses in invertebrates also suggests that defensive levels and probability of attack are positively related. Inducible structures, morphs, and behaviors have been noted for marine invertebrates,[151,162–169] suggesting allocation costs for these defenses. However, induced resistance in invertebrates has never been demonstrated to be chemically mediated,[162] but inducible physical defenses have been observed. In bivalves, the shell is composed of an outer periostracum covering two to four $CaCO_3$ layers.[9] The periostracum is living tissue that secretes the shell, but it may also protect the outer $CaCO_3$ layer from dissolution and aid in forming a tight seal when the valves are brought together tightly. The $CaCO_3$ crystals are deposited within an organic matrix. Organic material in the matrix and periostracum accounts for 12 to 72% of the shell's dry mass.[170] The costs of producing shells help explain why molluscs in the presence of crushing surf or predators produce thicker shells than conspecifics in calm areas with fewer predators;[169] when crushing surf or predators are present, the costs of producing thick shells are offset by the benefit of protection.

B. Growth–Differentiation Balance Hypothesis

The growth–differentiation balance hypothesis (GDBH) states that resources are allocated between growth processes (e.g., cell division and enlargement) and differentiation processes (e.g., cellular specialization and production of defensive chemicals) and that differentiation tends to occur after growth, although some overlap may occur.[8,171] Treatment of the GDBH has largely dealt with evolutionary and ecological responses among species or whole plants,[8] respectively, along environmental gradients, but the tenets of the GDBH can also be used to predict within-individual allocation of differentiation products of both seaweeds and invertebrates, because cells of each of these groups grow and differentiate.

The GDBH, like the ODT, assumes that a trade-off (i.e., reduced growth rates) is associated with the production of chemical defenses.[8] When considering young, actively growing tissue, the ODT and GDBH predict opposite patterns. The ODT predicts that young, actively growing tissues should be preferentially defended relative to older tissue, all else being equal, because young tissues are more valuable to the fitness of the plant (i.e., meristems make new cells and, thus, contribute much to the future productivity of the plant) and are more vulnerable (i.e., often delicate and nutritious).[66,131,132,135] In contrast, the GDBH predicts that actively growing tissue should be less defended than differentiated tissue because growth processes precede differentiation processes.

There are numerous examples of the concentrations of secondary metabolites being higher in actively growing than in older tissue of terrestrial plants.[66] Although these examples support the ODT, they need not contradict the GDBH. Because terrestrial plants can effectively translocate materials among plant parts,[172] secondary metabolites can be synthesized in differentiated tissues and then be exported to actively growing regions (i.e., site of highest compound concentration need not equal site of synthesis).[8,131] The GDBH may be better tested in organisms with limited ability to translocate secondary metabolites, like some chemically defended seaweeds.

Cronin and Hay[133] used the brown alga *Dictyota ciliolata* to test the GDBH because this alga is defended from sympatric herbivores by diterpenoid secondary metabolites, grows at apical meristems, and does not contain structures associated with translocation.[133,137] The actively growing apices of *Dictyota* contained lower concentrations of secondary metabolites than older parts of the thallus, were more delicate (i.e., had fewer structural defenses), and were highly preferred over older tissues by herbivorous amphipods and urchins. Assays using artificial diets indicated that the higher palatability of apices was chemically mediated and was not due to differences in toughness. Similarly, the actively growing apices of *Zonaria angustata* contained fewer phlorotannin-containing physodes and were more palatable to an amphipod than older regions of the thallus, with the interesting exception that the apical cell maintained a high concentration of physodes, apparently by an uneven sharing of physodes during mitosis; the majority of physodes remained in the cell that was to become the new apical cell, while the other daughter cell began its existence with little chemical protection.[93] In contrast, older tissues which are less involved with growth processes than the apices, contained high concentrations of physodes. Therefore, the apical cells of *Zonaria* did not have to allocate a great proportion of resources to producing physodes (i.e., there was no apparent trade-off between growth and differentiation), but the young, physode-poor daughter cells apparently had to divide, grow, and mature before attaining the level of physodes found in older tissues (i.e., there is a trade-off between growth and differentiation in these cells). The level of dissection of *Dictyota* did not allow comparison of the apical cell with the rest of the apex,[133] but the fact that apical tissue of *Zonaria* (with the exception of the apical cell) had lower concentrations of secondary compounds and/or associated structures suggests that these seaweeds adhere to the GDBH, though *Zonaria* has found a way to optimally defend the apical cell through uneven sharing of physodes. As another example, young apices of *Fucus distichus* were preferred over the older portion by the snail *Littorina sitkana*, and the concentration of phlorotannins in the apices was only half that of the older sections.[143] This example of lower phlorotannins in the meristems of *F. disticus* relative to older sections is contrary to the pattern seen during a broader survey of allocation patterns in rockweeds.[144] These patterns of chemical defenses and susceptibility to herbivores in *Dictyota* and *Zonaria* are contrary to the ODT and the patterns described for terrestrial plants. They are consistent, however, with allocation patterns of the GDBH which predicts that apices, which are actively involved in growth processes, would have low levels of defensive secondary metabolites and would be less tough than older tissue because cell walls have not yet matured.

Other seaweeds show within-thallus pattern of chemical defenses that are different from those seen in *Dictyota* and *Zonaria*, but are more typical of patterns observed in terrestrial plants. As noted above, kelps allocate more secondary metabolites to their meristems and other young tissues than to older tissues. Why should *Dictyota* and *Zonaria* follow the pattern predicted by the GDBH while *Halimeda*,[25-26] *Neomeris*,[146] rockweeds[173] (see Van Alstyne[143]), kelps,[173] and terrestrial plants often do not? One important distinction between the former and latter groups is that the former group does not have structures to effectively translocate materials. All seaweeds probably share some resources among cells, but the ability to translocate materials far distances is most advanced in coenocyctic green seaweeds and some brown seaweeds, including kelps and rockweeds.[137,138] Green algae of the genus *Halimeda* show incredible control over allocation of resources to growth, defense, and resource acquistion. For example, *Halimeda* spp. acquire resources during the day, grow at night,[25] and contain both structural and chemical defenses that are differentially allocated to different portions of the thallus.[25,26,101] The alga initiates these new segments at night while herbivorous fishes are inactive and defends the young, delicate, nutritious (i.e., vulnerable) segments with more potent and higher concentrations of feeding deterrents than older segments.[25-26] These green seaweeds are even capable of translocating organelles; new, non-pigmented segments are produced at night, and the plant waits to translocate chloroplasts into the new segments until daytime when they would be useful for photosynthesis.[25] As the new segments calcify and become less nutritious, the concentrations of chemical defenses diminish (i.e., nutritional and structural defenses partially replace chemical defenses). This tight control over resource allocation is made possible by their coenocytic growth form; multinucleated tubular filaments lacking crosswalls allow rapid translocation of cellular contents.[137]

A distinction needs to be made between allocation of resources and the translocation of previously synthesized components. An increase in the concentration of secondary metabolites as a result of increased synthesis represents a change in allocation to defenses. In contrast, an increase in the concentration of secondary metabolites as a result of translocation does not represent an increased allocation to defenses as far as synthesis is concerned, though there could be increased allocation to transporting costs (see above). Rather, translocation represents a rear-rangement of prior allocations, as an increase in metabolite concentrations at the sink is balanced by a decrease of metabolite concentrations at the source; the amount of metabolite in the whole organism remains unchanged.

Although many terrestrial plants and some seaweeds support the ODT, they are not necessarily in conflict with the GDBH given that the elevated concentrations of secondary metabolites found in meristems or young tissue could have been produced by differentiated tissues and then trans-located to the actively growing regions. *Dictyota* and *Zonaria* may be less capable of translocating secondary metabolites, and patterns seen in these species tend to support the GDBH. A recent survey of within-thallus patterns of phlorotannins in kelps and rockweeds showed that kelps, with well-developed vascular systems compared to other seaweeds, tend to defend tissues as the ODT would dictate, while defenses in rockweeds were less consistent.[173] These observations, taken together, suggest that translocation of defenses may bridge the gap between the ODT and the GDBH. Plants that have limited ability to translocate secondary metabolites may therefore be constrained in their ability to optimally defend their tissues. It would be interesting to quantita-tively compare the ability to translocate secondary metabolites with the occurrence of optimally defended tissues.

Arguments similar to ones used to predict within-individual variation in defenses can also be used to predict ontogenic changes in whole organisms.[103–106,144] GDBH predicts that juveniles, which are actively involved in growth, should be less defended than adults because younger individuals are less differentiated (note the similarity with the tasty embryo hypothesis). Again, this is in contrast to what the ODT might predict, given that a single bite by a consumer could be deadly for a less-than-bite-size juvenile, as opposed to an adult that might be able to tolerate several bites.

The GDBH can also be used to predict among-habitat variation in defenses within a species. Resource-rich habitats that promote rapid growth should be populated by poorly defended individ-uals (Figure 9.2). There is one very important caveat to this aspect of the GDBH: "Because both growth and differentiation are dependent on photosynthate, they are negatively correlated…, and we can speak of a balance of growth and differentiation in which growth is dominant under favorable conditions and differentiation is at a maximum only when conditions *other than supply of photo-synthate are suboptimum for growth*" (Loomis[171]). Hence, among seaweeds, rapid growth due to the high availability of nutrients will result in poorly defended individuals, whereas rapid growth due to high light intensity will result in well-defended individuals. Given this caveat, the GDBH predicts the same patterns as the carbon–nutrient balance hypothesis (CNBH) with respect to light and nutrient availability (Bryant et al.[174]), but the GDBH is more broadly applicable because it predicts levels of defense for any abiotic condition that affects growth rates.[8]

One aspect of the GDBH and CNBH that distinguishes them from other chemical defense theories is that the GDBH and CNBH do not assume that the production of chemical defenses incurs a trade-off with growth with respect to fixed carbon. The GDBH and the CNBH suggest that plants will allocate photosynthate to growth processes as long as growth is light-limited. Only when abiotic conditions limit growth more than photosynthesis (e.g., nutrient or drought/desiccation stress) will excess photosynthate be allocated to the production of differentiation products such as chemical defenses. Thus, defenses do not use photosynthate that could have been used by growth. However, fixed carbon may not be the best currency to measure defensive costs.

C. CARBON–NUTRIENT BALANCE HYPOTHESIS

The CNBH was proposed by Bryant et al.[174] as a framework to explain the response of boreal forest plants to spatial and temporal variation in their environment. This hypothesis focused on intraspecific variation in allocation to defenses that occurs over ecological time scales. Like the GDBH, the CNBH suggests that patterns of intraspecific variation in plant chemical defenses depend on recent ecological conditions. CNBH suggests that plants allocate all photosynthate to growth processes when light is limiting growth, but under conditions where nutrients limit growth more than they limit photosynthesis, excess photosynthate will be allocated to the production of C-based chemical defenses. Therefore, the CNBH covers a subset of the conditions that the GDBH covers when considering C-based defenses. Additionally, the tenets of the CNBH are difficult to apply to within-individual patterns of defense. The CNBH also states that the concentrations of nitrogen-based secondary metabolites (e.g., alkaloids, peptides, and cyanogenic metabolites) should increase under low C/N ratios;[174] however, the logic regarding N-based defenses is less compelling given that nitrogen-based secondary metabolites actually contain much more carbon than nitrogen. Additionally, seaweeds are generally under N-limitation[85] and as a result, rarely incorporate nitrogen into their secondary metabolites.[2] Because this theory is based on allocation of photosynthate, predictions made for seaweeds will be similar for mixitrophic invertebrates, but should not apply to heterotrophic invertebrates.

Although the CNB hypothesis predicts changes in concentrations in all C-based secondary metabolites in relation to the plant C/N status, different classes of these metabolites may respond differently to environmental variation,[175,176] since they are derived from different biosynthetic pathways, represent different proportions of a plant's carbon budget, and may have different turnover rates.[8,176,177] Early successional herbs and woody plants that produce phenolic compounds conform to the CNB hypothesis[178] more predictably than plants that produce other classes of C-based secondary metabolites such as terpenes[176,179] and furanocoumarins.[180]

There are very limited data available about the response of allocation to chemical defenses in seaweeds to altered light and nutrient availability. The concentrations of C-based secondary metabolites in *Dictyota ciliolata* generally did not respond to nutrient additions in a manner consistent with the CNB hypothesis.[181] When nutrient additions increased growth, which the CNB hypothesis predicts will decrease the concentrations of secondary metabolites as photosynthate is shunted to growth processes, the concentration of C-based secondary metabolites increased under low light conditions and remained unchanged under high light conditions. The production of C-based secondary metabolites obviously relies on nutrient-dependent processes, thus the potential exists that nutrient addition can increase the concentrations of C-based secondary metabolites if their production is more limited by a nutrient-dependent process (e.g., enzyme systems or transcription/translation) than substrates.[182] *Dictyota ciliolata* also responded to increased irradiance in a manner inconsistent with the CNBH. In experiments that manipulated light availability, growth of *Dictyota* was always positively correlated with irradiance, and the concentrations of some C-based secondary metabolites decreased when irradiance increased.

High light intensity may have decreased the concentration of secondary metabolites if the plants were stressed by the high irradiance (see the environmental stress theory, Section IV.D). The plants grown at 100% of surface irradiance probably received high doses of ultraviolet radiation as well as photosynthetically active radiation. Ultraviolet radiation can stress *Dictyota ciliolata*, leading to decreased concentrations of secondary metabolites.[183]

Phlorotannins from brown seaweeds seem to change along environmental gradients in a manner that is more consistent with the CNBH. *Sargassum filipendula* allocated resources to C-based secondary metabolites in a manner consistent with the CNBH. *Sargassum* grown at 72% ambient light had twice the concentration of phlorotannins than conspecifics grown just centimeters away under 19% ambient light. However, the increase did not alter *S. filipendula* susceptibility to amphipod grazing.[181] Phlorotannins in *Fucus vesiculosus* were higher at a low nutrient site than a high nutrient site.[157] At the low nutrient site, *F. vesiculosus* generally responded to fertilizer by growing more rapidly and by decreasing concentrations of phlorotannins. At the high nutrient site, the alga responded to fertilization to a lesser degree than it did at the low nutrient site; growth rates generally increased, but phlorotannin concentrations did not change significantly.

In total for seaweeds, it appears that the CNB hypothesis correctly predicted the responses of phlorotannins (three of three species[157,181,184]), but not of terpenes (zero of two species[181,185]). Similarly, terrestrial ecologists have also found that the CNB hypothesis appears to predict the responses of phenolic compounds better than for other C-based secondary metabolites.[176,177]

D. ENVIRONMENTAL STRESS THEORY

Seaweeds and invertebrates live in constantly changing environments in which benign and stressful periods are interspersed; however, we know little about how these periodic stresses affect their interactions with predators. Theories of environmental stress-induced increases in palatability for terrestrial plants have proposed increases in nutritive value[186] or decreases in defensive characters[187] as mechanisms responsible for the increased palatability. The environmental stress theory (EST) is based on the premise that an organism in the stressed state will be less able to acquire resources and will allocate a greater proportion of these reduced resources to maintenance, compared to an unstressed organism. The smaller resources pie and greater size of the maintenance slice (Figures 9.1 and 9.2) result in fewer resources available for defenses. Hence, stressed organisms should be more susceptible to predators than nonstressed ones. A stress can be defined as any environmental condition that reduces the optimal physiological performance of an organism. The discussions of responses to low light or nutrient availability (CNBH and GDBH) pertain to this theory, but note that allocation patterns predicted by the EST are contrary to predictions of the GDBH with respect to nutrients. The GDBH predicts that growth-inhibiting conditions should increase defenses.

The few studies assessing the impacts of abiotic stresses on seaweed–herbivore interactions have produced mixed results. Renaud et al.[159] experimentally assessed the effect of desiccation stress on a normally palatable red seaweed *Gracilaria tikvahiae* and a normally unpalatable brown seaweed *Padina gymnospora*. They found that *Gracilaria* became less palatable and had reduced concentrations of soluble protein following desiccation, while *Padina* became more palatable and less chemically defended following desiccation. They suggested that the impacts of physical stresses on seaweed–herbivore interactions will depend on the initial palatability of the seaweed, and that stresses increase the susceptibility of unpalatable species by compromising their chemical defenses.

The chemically defended brown alga *Dictyota ciliolata* responded to environmental stresses in much the same way that *Padina* responded. Mild desiccation of *Dictyota* reduced photosynthesis by 53% (i.e., resource acquisition), increased respiration by 40% (i.e., possibly indicating increased maintenance costs), reduced the concentrations of defensive secondary metabolites by 7 to 38%, and made plants 2.6 to 3.4 times more susceptible to urchin and amphipod grazing.[183] Additionally, the concentrations of metabolites found in undesiccated *Dictyota* deterred feeding by sea urchins, but concentrations found in the desiccated plants were not deterrent. Desiccated and undesiccated plants did not differ in toughness or nutritive value, indicating that desiccated plants became more palatable because chemical defenses were lost, not because nutritive value was increased. Similarly, near-surface ultraviolet radiation stressed *Dictyota ciliolata*, reduced growth by 84%, increased mortality, decreased the concentrations of secondary metabolites, and tended to increase susceptibility of seaweeds to amphipod grazing.[183] Therefore, in addition to dealing with a physical stress, chemically defended species may be less capable of fending off natural enemies when stressed.

The stress of ultraviolet radiation can also induce increased production of secondary metabolites, as the brown alga *Ascophyllum nodosum* produced 30% more phlorotannins when exposed to additional UV-B radiation under experimental conditions.[81] This induced increase in phlorotannins can be considered adaptive given that the phlorotannins absorb wavelenghts in the UV-B region. However, this increased production of phlorotannnins did not result in a concomitant decrease in susceptibility to an herbivorous isopod.[81]

E.   RESOURCE AVAILABILITY MODEL

The resource availability model (RAM) is a chemical defense theory which suggests that plant species produce levels and types of defenses based on the resource-richness of the environment in which they evolved.[67] In contrast with the CNBH, which explains intraspecific variation in defenses on ecological time scales, the RAM explains interspecifc variation in defenses on evolutionary time scales. RAM suggests that plant species that evolved in low resource environments will be inherently slow growing because of limitations imposed by low resource acquisition, will be long-lived, and will allocate a large proportion of resources to defenses because any tissue lost to an herbivore will be expensive to replace. These slow-growing species should invest in defenses that are immobile; compounds such as polyphenolics that supposedly have little turnover (i.e., low metabolite maintenance cost), allowing the initial investment (i.e., synthesis cost) in the compounds to be recouped over the lifespan of the long-lived plants. Species that evolved in high resource environments will be inherently fast growing (i.e., allocate heavily to resource acquisition and growth), be short-lived or have short-lived foliage, and should allocate minimally to defenses because tissue lost to herbivores can be rapidly replaced by these fast-growing plants. These fast-growing species should be able to tolerate herbivores, whereas slow-growing species should favor the defensive strategy.[19] Fast-growing species, when they do invest in chemical defenses, are predicted to use mobile defenses such as alkaloids and terpenes that are typically effective at low concentrations (i.e., supposedly low synthetic costs) and are easily catabolized, allowing the resources to be easily reallocated to other plant processes.[67]

This model based on interspecific differences in inherent growth rates associated with habitats of different resource availability is difficult to apply to seaweeds because seaweed species with a wide span of potential growth rates can be found within the same habitat.[14,16] To take extreme examples, some of the fastest growing marine plants are filamentous green seaweeds that occur in nutrient-poor tropic waters, and some of the slowest growing marine plants are crustose coralline algae that grow in nutrient-rich temperate waters.[14,16,32] However, if we recall the geographic patterns of secondary metabolites outlined above, we see that so-called immobile phlorotannins tend to be more prominent among temperate seaweeds that grow in nutrient-rich habitats such as upwelling zones and estuaries.[123–125] Seaweeds produce a greater variety and amount of so-called mobile terpenes, acetogenins, and indoles in generally nutrient-poor tropical habitats compared to temperate habitats.[2] These geographic patterns do not conform to RAM.

One pattern that appears to be consistent with RAM is that immobile phlorotannins are often present at much higher concentrations (typically 0 to 18% dry mass) than immobile defenses (0 to 2% dry mass).[111] Recent evidence suggests that turnover of phlorotannins is relatively slow, and that less than 1% of photosynthate from three brown seaweeds is allocated to maintain polyphenolic concentrations of 3 to 5%.[69] Those authors calculated that the phenolic pool would turnover in 120 to 260 days, depending on the species, which is considerably more than the 32 to 110 days calculated in other studies.[69] No comparable data are available for mobile secondary metabolites from seaweeds.

F.   PLANT APPARENCY MODEL

The plant apparency model (PAM) makes predictions about interspecific patterns of chemical defenses that are similar to the predictions of the resource availability model, but evokes differences in the apparency of plants rather than resource availability as the ultimate cause of interspecific differences in plant defenses.[187–189] Apparent plants are large, common, and spatially and/or temporally predictable. These plants are "bound to be found" by a variety of herbivores. As a result, evolution selected for defenses in these plants that were effective against a broad range of herbivores by acting in a generalized manner that would be difficult for herbivores to overcome or counteradapt. Tannins were one such group of chemical defenses that supposedly acted as digestibility reducers by generally binding to proteins in a manner that made the nutritive value of the plant low (i.e., they would tan the tissue, making it indigestible) or made the digestive enzymes of the herbivore inactive. Tannins were termed quantitative defenses because they were believed to act in a dose-dependent manner; quantitative defenses of the PAM are analogous to immobile defenses of the RAM.

Nonapparent plants were fast growing, short-lived, and occurred unpredictably in space and/or time. Thus, many of these plants would escape detection by the majority of herbivores. Because these plants allocated most of their resources to rapid growth and reproduction, they were believed to be defended by relatively low concentrations of toxic compounds that were generally effective against many herbivores but that some specialist herbivores would be able to evolve detoxification mechanisms against (and may even use the qualitative defenses as cues to locate their host). These toxic compounds such as alkaloids and terpenes were termed qualitative defenses because their potency made them effective at low concentrations; qualitative defenses of the PAM are analogous to mobile defenses of the RAM.

Kelps are sessile, many are perennial and extremely large, suggesting that herbivores would find them apparent. These apparent seaweeds produce quantitative-type secondary metabolites (i.e., phlorotannins), but at lower concentrations than are produced by several less-apparent species of brown seaweeds such as intertidal rock-weeds, which are unavailable to many subtidal herbivores.[123,124] Additionally, tropical brown Pacific seaweeds generally produce low concentrations of phlorotannins, even though they are "bound to be found" by the large number of herbivores that occur in these habitats.[123,124] The assumption that phlorotannins act in a general manner as digestibility reducers has been challenged by experimental evidence that shows variable effectiveness of phlorotannins based on molecular size[190] and the inability of phlorotannins to reduce the digestion efficiency of several herbivores.[130] Thus, the PAM does a poor job of explaining the amounts of secondary metabolites produced by seaweeds and the mode of action of the classes of secondary metabolites.

G. Spatial-Variation-in-Consumers Model

The spatial-variation-in-consumers model (SVICM) was proposed by marine ecologists to explain both interspecific and intraspecific patterns of chemical defenses in seaweeds.[111] It states that chemically defended seaweeds will be more evolutionarily persistent than undefended seaweeds in areas subject to significant herbivore impact. Likewise, this model explains intraspecifc variation in chemical defenses, in that well-defended individuals or populations will be more persistent over ecological time scales than undefended seaweeds in areas subject to significant herbivore impact. This model borrows heavily from the optimal defense theory, in that it suggests that seaweeds that occur in areas with high herbivory pressure will be vulnerable to herbivore attack and should thus be heavily defended.

This model says nothing about what type of compounds a species should use to defend itself, such as high concentrations of polyphenolics or low concentrations of terpenes. The authors suggest that since there is little indication that quantitative/immobile and qualitative/mobile defenses differ in cost, effectiveness, or mode of action, there is little need for the model to predict classes of secondary metabolites. This model could be equally applied to invertebrates by replacing the words seaweed and herbivore with invertebrate and predator.

V. FINAL REMARKS

As with many topics in ecology and evolution, allocation patterns are complex due to the fact that organisms are faced simultaneously with multiple variable components of the environment. Additionally, the various processes that receive resources are interrelated in complex and changing ways. An optimal allocation pattern is a moving target, and plasticity in allocation patterns is commonly observed.

It is not surprising that no single unifying model has been developed to encompass the complex aspects of allocation. It can be frustrating that models based on seemingly sound premises can result in opposite predictions for allocation patterns. Sometimes these inconsistencies can be explained by considering the biologies of the organisms that behave as the models predict, such as the case of vasculature in explaining discrepancies between the optimal defense theory and the growth differentiation balance hypothesis. Other discrepancies, such as the prediction of higher (growth–differentiation balance hypothesis) vs. lower (environmental stress theory) allocation to defenses as the result of stress, are more difficult to reconcile, but might depend on whether the stress is nutritive (reduced performance due to decreased resource availability) or physiologically damaging (reduced performance due to cellular damage and increased maintenance costs). This author predicts that the GDBH will best apply to nutritive stresses, while the EST will work better for damaging stresses. Additional studies need to be conducted so that the available models can be refined and hopefully made more accurate.

The model in Figure 9.2 predicts relative allocation patterns using Grime's plant strategy model proposed for terrestrial vegetation. The model incorporates aspects of the ODT, GDBH, PST, RAM, and SVICM and could be useful in predicting both interspecific and intraspecific variation in levels of chemical defenses. Note that disturbance in Grime's model has been replaced with predation, given that the focus of chemical ecology has been on predator–prey interactions. Predation can be replaced with any term (e.g., fouling, disease, or UV radiation) related to the ecological role of the secondary metabolites (e.g., antifoulant, antimicrobial, or sunscreen). Organisms that use secondary metabolites to help them compete (e.g., allelopathics) or handle stress (e.g., phlorotannins ability to screen UV radiation[82] might allocate resources to these defensive secondary metabolites, altering the allocation pie chart. Like the SVICM, the model in Figure 9.2 predicts only levels of defenses, but does not predict types of defenses used.

This model was envisioned from this author's experience with the chemical ecology of the brown seaweeds *Dictyota menstrualis* and *D. ciliolata* at Radio Island Jetty, North Carolina. These seaweeds are apparent paradoxes to chemical defense theories that assume a trade-off between growth and chemical defenses (i.e., GDBH, ODT, and RAM) because they compete very well with non-defended seaweeds, are rapid growers (can double in size in a week), and are well defended chemically.[55,133,158,183] Radio Island Jetty (RIJ) is a temperate habitat with high algal biomass, abundant herbivores, turbid waters, and a 2 m tidal range. *Dictyota* spp. were restricted to subtidal areas of RIJ where they reached great abundance and could apparently cope with the numerous competitors and herbivores. They were not found intertidally where desiccation and UV stress would be intense, and their abundance would decline dramatically following rain storms. Thus, they do not tolerate well the physical stresses of UV radiation, desiccation, or salinity change Cronin and Hay,[113] *D. dichotoma* in Trindell[191]), but they compete and defend quite well under non-stressful situations. It appears that the trade-off of rapid growth and chemical defense in *Dictyota* spp. comes in its inability to tolerate physical stresses. The model assumes that stress-tolerant plants allocate a large proportion of resources to maintaining tissues already grown, that competitive plants allocate a large proportion of resources to further resource acquisition, and that plants faced with predators allocate a large proportion of resources to defenses. The model should be equally applicable to seaweeds and invertebrates. While the model seems fitting for *Dictyota* spp. at RIJ, its utility for other marine organisms remains to be seen.

ACKNOWLEDGMENTS

I would like to thank Cody and Katy Cronin for increasing my understanding of the costs and benefits of reproductive allocation. Niels Lindquist provided resources on allocation patterns of invertebrates. Sujay Kaushal provided useful references. I would also like to thank Jim McClintock and Bill Baker for their encouragement and helpful comments and all the effort they put into this volume. Two anonymous reviewers made many helpful comments and helped focus some of the ideas in the manuscript. Finally, I thank Mark Hay for sparking and nurturing my interest in chemical ecology and for countless conversations about chemical strategies among seaweeds.

REFERENCES

1. Bakus, G.J., Targett, N.M., and Schulte, B., Chemical ecology of marine organisms: an overview, *J. Chem. Ecol.*, 12, 951, 1986.
2. Hay, M.E. and Fenical, W., Marine plant–herbivore interactions: the ecology of chemical defense, *Ann. Rev. Ecol. Syst.*, 19, 111, 1988.
3. Paul, V.J., Seaweed chemical defenses on coral reefs, in *Ecological Roles of Marine Natural Products*, Paul, V.J., Ed., Comstock Publishing, Ithaca, NY, 1992, 24.
4. Pawlik, J.R., Marine invertebrate chemical defenses, *Chem. Rev.*, 93, 1911, 1993.
5. Hay, M.E., Marine chemical ecology: what's known and what's next?, *J. Exp. Mar. Biol. Ecol.*, 200, 103, 1996.
6. McClintock, J.B. and Baker, B.J., A review of the chemical ecology of shallow-water antarctic marine invertebrates, *Amer. Zool.*, 37, 329, 1997.
7. Rhoades, D.F., Evolution of plant chemical defense against herbivores, in *Herbivores: Their Interaction with Secondary Plant Metabolites*, Rosenthal, G.A. and Janzen, D.H., Eds., Academic Press, New York, 1979, 3.
8. Herms, D.A. and Mattson, W.J., The dilemma of plants: to grow or defend, *Quar. Rev. Biol.*, 67, 283, 1992.
9. Barnes, R.D., *Invertebrate Zoology*, 5th ed., Saunders College Publishers, Philadelphia, PA, 1987.
10. Tardent, P., Developmental aspects of regeneration in coelenterates, in *Regeneration in Animals and Related Problems*, Trampusch, H.A.L., Ed., North- Holland Publishing Co., Amsterdam, 1965, 71.
11. MacGinitie, G.E. and MacGinitie, N., *Natural History of Marine Animals*, 2nd ed., McGraw-Hill, New York, 1968.
12. Wahle, C.M., Regeneration of injuries among Jamaican gorgonians: the roles of colony physiology and environment, *Biol. Bull.*,165, 778, 1983.
13. Grime, J.P., *Plant Strategies and Vegetation Process*, John Wiley, New York, 1979.
14. Littler, M.M. and Littler, D.S., The evolution of thallus form and survival strategies in benthic marine macroalgae: field and laboratory tests of a functional form model, *Am. Nat.*, 116, 25, 1980.
15. Steneck, R.S. and Adey, W.H., The role of environment in control of morphology in *Lithophyllum congestum*, a Caribbean algal ridge builder, *Botanica Marina*, 19, 197, 1976.
16. Steneck, R.S. and Dethier, M.N., A functional group approach to the structure of algal-dominated communities, *Oikos*, 69, 476, 1994.
17. Hay, M.E., The functional morphology of turf forming seaweeds: persistence in stressful marine habitats, *Ecology*, 62, 739, 1981.
18. Hay, M.E., Functional geometry of seaweeds: ecological consequences of thallus layering and shape in contrasting light environments, in *On the Economy of Plant Form and Function*, Givnish, T.J., Ed., Cambridge University Press, New York, 1986, 635.
19. Lubchenco, J. and Gaines, S.D., A unified approach to marine plant–herbivore interactions. I. Populations and communities, *Ann. Rev. Ecol. Syst.*, 12, 405, 1981.
20. Hay, M.E., Predictable spatial escapes from herbivory: how do these affect the evolution of herbivore resistance in tropical marine communities?, *Oecologia*, 64, 396, 1984.
21. Russ, G., Distribution and abundance of herbivorous grazing fishes in the central Great Barrier Reef. II. Patterns of zonation of mid-shelf and outer-shelf reefs, *Mar. Ecol. Prog. Ser.*, 20, 35, 1984.

22. Pawlik, J.R., Burch, M.T., and Fenical, W., Patterns of chemical defenses among Caribbean gorgonian corals: a preliminary survey, *J. Exp. Mar. Biol. Ecol.*, 108, 55, 1987.

23. Pawlik, J.R., Chanas, B., Toonen, R.J., and Fenical, W., Defenses of Caribbean sponges against predatory reef fish. I. Chemical deterrency, *Mar. Ecol. Prog. Ser.*, 127, 183, 1995.

24. Lubchenco, J. and Cubit, J., Heteromorphic life histories of certain marine algae as an adaptation to variation in herbivory, *Ecology*, 61, 676, 1980.

25. Hay, M.E., Paul, V.J., Lewis, S.M., Gustafson, K., Tucker, J., and Trindell, R.N., Can tropical seaweeds reduce herbivory by growing at night? Diel patterns of growth, nitrogen content, herbivory, and chemical versus morphological defenses, *Oecologia*, 75, 233, 1988.

26. Paul, V.J. and Van Alstyne, K.L., Chemical defense and chemical variation in some tropical Pacific species of *Halimeda* (Halimedaceae Chlorophyta), *Coral Reefs*, 6, 263, 1988.

27. Van Alstyne, K.L. and Paul, V.J., Chemical and structural defenses in the sea fan *Gorgonia ventilina*: effects against generalist and specialist predators, *Coral Reefs*, 11, 155, 1992.

28. Van Alstyne, K.L., Wylie, C.R., Paul, V.J., and Meyer, K.D., Antipredator defenses in tropical Pacific soft corals (Coelenterata: Alycyonacea). I. Sclerites as defenses against generalist carnivorous fishes, *Biol. Bull.*, 182, 231, 1992.

29. Lodge, D.M., Cronin, G., Van Donk, E., and Froelich, A., Impact of herbivory on plant standing crop: comparisons among biomes, between vascular and non-vascular plants, and among freshwater herbivore taxa, in *The Structuring Role of Submerged Macrophytes in Lakes*, Jeppesen, E., Sondergaard, Ma., Sondergaard, Mo., and Christoffersen, K., Eds., Springer-Verlag, New York, 1998, 149.

30. Chanas, B. and Pawlik, J.R., Defense of Caribbean sponges against predatory reef fish: II. Spicules, tissue toughness, and nutritional quality, *Mar. Ecol. Prog. Ser.*, 127, 195, 1995.

31. Lindquist, N., Hay, M.E., and Fenical, W., Defenses of ascidians and their conspicuous larvae: adult versus larval chemical defenses, *Ecol. Monogr.*, 62, 547, 1992.

32. Steneck, R.S., The ecology of coralline algal crusts: convergent patterns and adaptive strategies, *Ann. Rev. Ecol. Syst.*, 17, 273, 1986.

33. Pennings, S.C. and Paul, V.J., Effect of plant toughness, calcification, and chemistry on herbivory by *Dolabella auricularia*, *Ecology*, 73, 1606, 1992.

34. Pennings, S.C. and Svedberg, J., Does $CaCO_3$ in food deter feeding by sea urchins? *Mar. Ecol. Prog. Ser.*, 101, 163, 1993.

35. Hay, M.E., Kappel, Q.E., and Fenical, W., Synergisms in plant defenses against herbivores: interactions of chemistry, calcification, and plant quality, *Ecology*, 75, 1714, 1994.

36. Pennings, S.C., Puglisi, M.P., Pitlik, T.J., Himaya, A.C., and Paul, V.J., Effects of secondary metabolites and $CaCO_3$ on feeding by surgeonfishes and parrotfishes: within-plant comparisons, *Mar. Ecol. Prog. Ser.*, 134, 49, 1996.

37. Littler, M.M., Calcification and its role among the macroalgae, *Micronesica*, 12, 27, 1976.

38. Hay, M.E., Associational plant defenses and the maintenance of species diversity: turning competitors into accomplices, *Am. Nat.*, 128, 617, 1986.

39. Pfister, C.A. and Hay, M.E., Associational plant refuges: convergent patterns in marine and terrestrial communities result from differing mechanisms, *Oecologia*, 77, 118, 1988.

40. Durante, K.M. and Chia, F., Epiphytism on *Agarum fimbriatum*: can herbivore preferences explain distributions of epiphytic bryozoans?, *Mar. Ecol. Prog. Ser.*, 77, 279, 1991.

41. Wahl, M. and Hay, M.E., Associational resistance and shared doom: effects of epibiosis on herbivory, *Oecologia*, 102, 329, 1995.

42. Wahl, M., Hay, M.E., and Enderlein, P., Effects of epibiosis on consumer–prey interactions, *Hydrobiologia*, 355, 49, 1997.

43. Hill, M.S., Symbiotic zooxanthellae enhance boring and growth rates of the tropical sponge *Anthosigmella varians forma varians*, *Mar. Biol.*, 125, 649, 1996.

44. Barkai, A. and McQuaid, C., Predator–prey role reversal in a marine benthic ecosystem, *Science*, 242, 62, 1988.

45. Vance, R.R., A mutualistic interaction between a sessile marine clam and its epibionts, *Ecology*, 59, 679, 1978.

46. Caine, E.A., First case of caprellid amphipod–hydrozoan mutualism, *J. Crustacean Biol.*, 18, 317, 1998.

47. McQuaid, C.D. and Froneman, P.W., Mutualism between the territorial intertidal limpet *Patella longicosta* and the crustose alga *Ralfsia verrucosa*, *Oecologia*, 96, 128, 1993.
48. Littler, M.M., Littler, D.S., and Taylor, P.R., Selective herbivore increases biomass of its prey: a chiton-coralline reef-building association, *Ecology*, 76, 1666, 1995.
49. Stachowicz, J.J. and Hay, M.E., Facultative mutualism between an herbivorous crab and a coralline alga: advantages of eating noxious seaweeds, *Oecologia*, 105, 377, 1996.
50. Pawlik, J.R., Kernan, M.R., Molinski, T.F., Harper, M.K., and Faulkner, D.J., Defensive chemicals of the Spanish dancer nudibranch *Hexabranchus sanguineus* and its egg ribbons: macrolides derived from a sponge diet, *J. Exp. Mar. Biol. Ecol.*, 119, 99, 1988.
51. Hay, M.E., Duffy, J.E., Paul, V.J., Renaud, P.E., and Fenical, W., Specialist herbivores reduce their susceptibility to predation by feeding on the chemically defended seaweed *Avrainvillea longicaulis*, *Limn. Oceanogr.*, 35, 1734, 1990.
52. Hay M.E., Duffy J.E., and Fenical, W., Host-plant specialization decreases predation on a marine amphipod: an herbivore in plant's clothing, *Ecology*, 71, 733, 1990.
53. Paul, V.J., Lindquist, N., and Fenical, W., Chemical defenses of the tropical ascidian *Atapozoa* sp. and its nudibranch predators *Nembrotha* spp., *Mar. Ecol. Prog. Ser.*, 59, 109, 1990.
54. Cronin, G., Hay, M.E., Fenical, W., and Lindquist, N., Distribution, density, and sequestration of host chemical defenses by the specialist nudibranch *Tritonia hamnerorum* found at high densities on the sea fan *Gorgonia ventalina*, *Mar. Ecol. Prog. Ser.*, 119, 177, 1995.
55. Hay, M.E., Duffy, J.E., Pfister, C.A., and Fenical, W., Chemical defense against different marine herbivores: are amphipods insect equivalents?, *Ecology*, 68, 1567, 1987.
56. Hay, M.E., Renaud, P.E., and Fenical, W., Large mobile versus small sedentary herbivores and their resistance to seaweed chemical defenses, *Oecologia*, 75, 246, 1988.
57. Duffy, J.E. and Hay, M.E., Food and shelter as determinants of food choice by an herbivorous marine amphipod, *Ecology*, 72, 1286, 1991.
58. Duffy, J.E. and Hay, M.E., Herbivore resistance to seaweed chemical defense the roles of mobility and predator risk, *Ecology*, 75, 1304, 1994.
59. Pennings, S.C. and Paul, V.J., Sequestration of dietary secondary metabolites by three species of sea hares: location, specificity and dynamics, *Mar. Biol.*, 117, 535, 1993.
60. Duffy, J.E. and Paul, V.J., Prey nutritional quality and effectiveness of chemical defenses against tropical reef fishes, *Oecologia*, 90, 333, 1992.
61. Mattson, W.J., Herbivory in relation to plant nitrogen, *Ann. Rev. Ecol. Syst.*, 11, 119, 1980.
62. Faulkner, D.J., Marine natural products, *Nat. Prod. Rep.*, 17, 7, 2000.
63. Anderson, R.J. and Velimirov, B., An experimental investigation of the palatability of kelp bed algae to the sea urchin *Parechinus angulosus* Leske, *P.S.Z.N.I.: Mar. Ecol.*, 3, 357, 1982.
64. Karuso, P., Chemical ecology of the nudibranchs, in *Bioorganic Marine Chemistry, Vol. 1*, Scheuer, P.J., Ed., Springer-Verlag, New York, 1987, 31.
65. Rausher, M.D., Genetic analysis of coevolution between plants and their natural enemies, *Trends in Genetics*, 12, 212, 1996.
66. McKey, D., Adaptive patterns in alkaloid physiology, *Am. Nat.*, 108, 305, 1974.
67. Coley, P.D., Bryant, J.P., and Chapin, F.S., III, Resource availability and plant antiherbivore defense, *Science*, 230, 895, 1985.
68. Fagerström, T., Anti-herbivore chemical defense in plants: a note on the concept of cost, *Am. Nat.*, 133, 281, 1989.
69. Arnold, T.M. and Targett, N.M., Quantifying *in situ* rates of phlorotannin synthesis and polymerization in marine brown algae, *J. Chem. Ecol.*, 24, 577, 1998.
70. Russell, M.P., Resource allocation plasticity in sea urchins: rapid, diet induced, phenotypic changes in the green sea urchin, *Strogylocentrotus droebachiensis* (Muller), *J. Exp. Mar. Biol. Ecol.*, 220, 1, 1998.
71. Pavia, H., Toth, G., and Aberg, P., Trade-offs between phlorotannin production and annual growth in natural populations of the brown seaweed *Ascophyllum nodosum*, *J. Ecol.*, 87, 761, 1999.
72. Pawlik, J.P., Chemical ecology of the settlement of benthic marine invertebrates, *Oceanogr. Mar. Biol. Annu. Rev.*, 30, 273, 1992.

73. de Nys, R., Steinberg, P.D., Willemsen, P., Dworjanyn, S.A., Gabelish, C.L., and King, R.J., Broad spectrum effects of secondary metabolites from the red alga *Delisea pulchra* in antifouling assays, *Biofouling*, 8, 259, 1995.

74. Schmitt, T.M., Hay, M.E., and Lindquist, N., Antifouling and herbivore deterrent roles of seaweed secondary metabolites: constraints on chemically-mediated coevolution, *Ecology*, 76, 107, 1995.

75. Schmitt, T.M., Lindquist, N., and Hay, M.E., Seaweed secondary metabolites as antifoulants: effects of *Dictyota* spp. diterpenes on survivorship, settlement, and development of marine invertebrate larvae, *Chemoecology*, 8, 125, 1998.

76. Steinberg, P.D., Schneider, R., and Kjelleberg, S., Chemical defenses of seaweeds against microbial colonization, *Biodegredation*, 8, 211, 1997.

77. Henrikson, A.A. and Pawlik, J.R., Seasonal variation in biofouling of gels containing extracts of marine organisms, *Biofouling*, 12, 245, 1998.

78. Gil-Turnes, M.S., Hay, M.E., and Fenical, W., Symbiotic marine bacteria chemically defend crustacean embryos from a pathogenic fungus, *Science*, 246, 116, 1989.

79. Jensen, P. R., Harvell, C.D., Wirtz, K., and Fenical, W., Antimicrobial activity of extracts of Caribbean gorgonian corals, *Mar. Biol.*, 125, 411, 1996.

80. Newbold, R.W., Jensen, P.R., Fenical, W., and Pawlik, J.R., Antimicrobial activity of Caribbean sponge extracts, *Aquatic Microb. Ecol.*, 19, 279, 1999.

81. de Nys, R., Coll, J.C., and Price, I.R., Chemically mediated interactions between the red alga *Plocamium hamatum* (Rhodophyta) and the octocoral *Sinularia cruciata* (Alcyonacea), *Mar. Biol.*, 108, 315, 1991.

82. Pavia, H., Cervin, G., Lindgren, A., and Aberg, P., Effects of UV-B radiation and simulated herbivory on phlorotannins in the brown alga *Ascophyllum nodosum*, *Mar. Ecol. Prog. Ser.*, 157, 139, 1997.

83. Muller, D.G., Gassmann, G., Boland, W., Marner, F., and Jaenicke, L., *Dictyota dichotoma* (Phaeophyceae): identification of the sperm attractant, *Science*, 212, 1040, 1981.

84. Muller, D.G., Jaenicke, L., Donike, M., and Akintobi, T., Sex attractant in a brown alga: chemical structure, *Science*, 171, 815, 1971.

85. Gershenzon, J., The cost of plant chemical defense against herbivory: a biochemical perspective, in *Insect–Plant Interactions, Vol. 5*, Bernays, E.A., Ed., CRC Press, Boca Raton, FL, 1994, chap. 5.

86. Howarth, R.W., Nutrient limitation of net primary production in marine ecosystems, *Ann. Rev. Ecol. Syst.*, 19, 89, 1988.

87. Kvitek, R.G., DeGange, A.R., and Beitler, M.K., Paralytic shellfish poisoning toxins mediate feeding behavior of sea otters, *Limnol. Oceanogr.*, 36, 393, 1991.

88. Rogers, C.N., Steinberg, P.D., and deNys, R., Factors associated with oligophagy in two species of sea hares (Mollsca: Anaspidea), *J. Exp. Mar. Biol. Ecol.*, 192, 47, 1995.

89. Unson, M.D., Holland, N.D., and Faulkner, D.J., A brominated secondary metabolite synthesized by the cyanobacterial symbiont of a marine sponge and accumulation of the crystalline metabolite in the sponge tissue, *Mar. Biol.*, 119, 1, 1994.

90. Marin, A., Lopez Belluga, M.D., Scognamiglio, G., and Cimino, G., Morphological and chemical camouflage of the Mediterranean nudibranch *Discodoris indecora* on the sponges *Ircinia variabilis* and *Ircinia fasciculata*, *J. Moll. Stud.*, 63, 431, 1997.

91. Gershenzon, J., Murtagh, G.J., and Croteau, R., Absence of rapid terpene turnover in several diverse species of terpene-accumulating plants, *Oecologia*, 96, 583, 1993.

92. Young, D.N., Howard, B.M., and Fenical, W., Subcellular localization of brominated secondary metabolites in the red alga *Laurencia snyderae*, *J. Phycol.*, 16, 182, 1980.

93. Poore, A.G.B., Selective herbivory by amphipods inhabiting the brown alga *Zonaria angustata*, *Mar. Ecol. Prog. Ser.*, 107, 113, 1994.

94. Marin, A., Lopez, M.D., Esteban, M.A., Meseguer, J., Munoz, J., and Fontana, A., Anatomical and ultrastructual studies of chemical defence in the sponge *Dysidea fragilis*, *Mar. Biol.*, 131, 639, 1998.

95. Avila, C., Ballesteros, M., Cimino, G., Crispino, A., Bavagnin, M., and Sodano, G., Biosynthetic origin and anatomical distribution of the main secondary metabolites in the nudibranch mollusc *Doris verrucosa*, *Comp. Biochem. Physiol.*, 97B, 363, 1990.

96. Slattery, M., Avila, C., Starmer, J., and Paul, V.J., A sequestered soft coral diterpene in the aeolid nudibranch *Phyllodesmium guamensis* Avila, Ballesteros, Slattery, Starmer and Paul, *J. Exp. Mar. Biol. Ecol.*, 226, 33, 1998.

97. Pennings, S.C., Interspecific variation in chemical defenses in the sea hares (Opisthobranchia: Anaspidea), *J. Exp. Mar. Biol. Ecol.*,180, 203, 1994.

98. Garcia-Gomez, J.C., Cimino, G., and Medina, A., Studies on the defensive behaviour of *Hypselodoris* species (Gastropoda: Nudibranchia): ultrastructure and chemical analysis of mantle dermal formations (MDFs), *Mar. Biol.*, 106, 245, 1990.

99. Karban, R. and Baldwin, I.T., *Induced Responses to Herbivory*, University of Chicago Press, Chicago, IL, 1997.

100. Tollrian, R. and Harvell, C.D., *The Ecology and Evolution of Inducible Defenses*, Princeton University Press, Princeton, NJ, 1999.

101. Paul, V.J. and Van Alstyne, K.L., Activation of chemical defenses in the tropical green algae *Halimeda* spp., *J. Exp. Mar. Biol. Ecol.*, 160, 191, 1992.

102. Ebel, R., Brenzinger, M., Krunze, A., Gross, H.J., and Proksch P., Wound activation of prototoxins in marine sponge *Aplysina aerophoba*, *J. Chem. Ecol.*, 23, 1451, 1997.

103. Orians, G.H. and Janzen, D.H., Why are embryos so tasty?, *Am. Nat.*, 108, 581, 1974.

104. Lindquist, N. and Hay, M.E., Can small rare prey be chemically defended? The case for marine larvae, *Ecology*, 76, 1347, 1995.

105. Lindquist, N. and Hay, M.E., Palatability and chemical defense of marine invertebrate larvae, *Ecol. Monogr.*, 66, 431, 1996.

106. Lindquist, N., Palatability of invertebrate larvae to corals and sea anemones, *Mar. Biol.*, 126, 745, 1996.

107. Uriz, M.J., Turon, X., Becerro, M.A., and Galera, J., Feeding deterrence in sponges. The role of toxicity, physical defenses, energetic contents, and life-history stage, *J. Exp. Mar. Biol. Ecol.*, 205, 187, 1996.

108. Stein, W.D., *Channels, Carriers, and Pumps: An Introduction to Membrane Transport*, Academic Press, San Diego, CA, 1990.

109. Hay, M.E., The role of seaweed chemical defenses in the evolution of feeding specialization and in the mediation of complex interactions, in *Ecological Roles of Marine Natural Products*, Paul, V.J., Ed., Cornell University Press, Ithaca, NY, 1992, 93.

110. Hay, M.E. and Fenical, W., Chemical mediation of seaweed-herbivore interactions, in *Plant–Animal Interactions in the Marine Benthos*, John, D.M., Hawkins, S.J., and Price, J.H., Eds., Clarendon Press, Oxford, UK, 1992, 319.

111. Hay, M.E. and Steinberg P.D., The chemical ecology of plant–herbivore interactions in marine versus terrestrial communities, in *Herbivores: Their Interactions with Secondary Plant Metabolites: Evolutionary and Ecological Processes*, Vol. 2, Rosenthal, G.A. and Berenbaum, M.R., Eds., Academic Press, New York, 1992, 371.

112. Woodin, S.A., Marinelli, R.L., and Lincoln, D.E., Allelochemical inhibition of recruitment in a sedementary assemblage, *J. Chem. Ecol.*, 19, 517, 1993.

113. Thacker, R.W., Becerro, M.A., Lumbang, W.A., and Paul, V.J., Allelopathic interactions between sponges on a tropical reef, *Ecology*, 79, 1740, 1998.

114. Rittschoff, D., Hooper, I.R., and Costlow, J.D., Barnacle settlement inhibitors from sea pansies, *Renilla reniformis*, *Bull. Mar. Sci.*, 39, 376, 1986.

115. King, G.M., Inhibition of microbial activity in marine sediments by a bromophenol from a hemichordate, *Nature*, 323, 257, 1986.

116. Kim, K., Antimicrobial activity in gorgonian corals (Coelenterata, Octocorallia), *Coral Reefs*, 13, 75, 1994.

117. Coll, J.C., Bowden, B.F., Meehan, G.V., Konig, G.M., Carroll, A.R., Tapiolas, D.M., Alino, P.M., Heaton, A., de Nys, R., Leone P.A., Maida, M., Aceret, T.L., Willis, R.H., Babcock, R.C., Willis, B.L., Florian, Z., Clayton, M.N., and Miller, R.L., Chemical aspects of mass spawning in corals. I. Sperm-attractant molecules in the eggs of the scleractinian coral *Montipora digitata*, *Mar. Biol.*, 118, 177, 1994.

118. Coll, J.C., Leone, P.A., Bowden, B.F., Carroll, A.R., Konig, G.M., Heaton, A., de Nys, R., Maida, M., Alino, P.M., Willis, R.H., Babcock, R.C., Florian, Z., Clayton, M.N., Miller, R.L., and Alderslade, P.N., Chemical aspects of mass spawning in corals. II. (-)-epi-thunbergol, the sperm attractant in the eggs of the soft coral *Lobophytum crassum* (Cnidaria: Octocorallia), *Mar. Biol.*, 123, 137, 1995.

119. Hay, M.E., Piel, J., Boland, W., and Schnitzler, I., Seaweed sex pheromones and their degradation products frequently suppress amphipod feeding but rarely suppress sea urchin feeding, *Chemoecology*, 8, 91, 1998.

120. Bolser, R.C. and Hay, M.E., Are tropical plants better defended? Palatability and defenses of temperate versus tropical seaweeds, *Ecology*, 77, 2269, 1996.

121. Paul, V.J. and Fenical, W., Chemical defense in tropical green algae, order Caulerpales, *Mar. Ecol. Prog. Ser.*, 34, 157, 1986.

122. Wright, J.T., Benkendorff, K., and Davis, A.R., Habitat associated differences in temperate sponge assemblages: the importance of chemical defence, *J. Exp. Mar. Biol. Ecol.*, 213, 199, 1997.

123. Steinberg, P.D., Biogeographical variation in brown algal polyphenolics and other secondary metabolites: comparison between temperate Australasia and North America, *Oecologia*, 78, 373, 1989.

124. Steinberg, P.D., Geographical variation in the interaction between marine herbivores and brown algal secondary metabolites, in *Ecological Roles of Marine Natural Products*, Paul, V.J., Ed., Comstock Publishing, Ithaca, NY, 1992, 51.

125. Van Alstyne, K.L. and Paul V.J., The biogeography of polyphenolic compounds in marine macroalgae: temperate brown algal defenses deter feeding by tropical herbivorous fishes, *Oecologia*, 84,158, 1990.

126. Targett, N.M. and Arnold, T.M., Predicting the effects of brown algal phlorotannins on marine herbivores in tropical and temperate oceans, *J. Phycol.*, 34, 195, 1998.

127. Targett, N.M., Cohen, L.D., Boettcher, A.A., and Tanner, C.E., Biogeographic comparisons of marine algal phenolics: evidence against its latitudinal trend, *Oecologia*, 89, 464, 1992.

128. Steinberg, P. D., Edyvane, K., de Nys, R., Birdsey, R., and Van Altena, I. A., Lack of avoidance of phenolic-rich brown algae by tropical herbivorous fishes, *Mar. Biol.*, 109, 335, 1991.

129. Steinberg, P. D. and Van Altena, I. A., Tolerance of marine invertebrate herbivores to brown algal phlorotannins in temperate Australasia, *Ecol. Monogr.*, 62, 189, 1992.

130. Targett, N.M., Boettcher, A.A., Targett, T.E., and Vrolijk, N.H., Tropical marine herbivore assimilation of phenolic-rich plants, *Oecologia*, 103, 170, 1995.

131. McKey, D., The distribution of secondary compounds within plants, in *Herbivores: Their Interactions with Secondary Plant Metabolites*, Rosenthal, G.A. and Janzen, D.H., Eds., Academic Press, New York, 1979, 55.

132. Denno, R.F. and McClure, M.S., *Variable Plants and Herbivores in Natural and Managed Systems*, Academic Press, New York, 1983.

133. Cronin, G. and Hay, M.E., Within-plant variation in seaweed palatability and chemical defenses: optimal defense theory versus the growth-differentiation balance hypothesis, *Oecologia*, 105, 361, 1996.

134. Raupp, M.J. and Denno, R.F., Leaf age as a predictor of herbivore distribution and abundance, in *Variable Plants and Herbivores in Natural and Managed Systems*, Denno, R.F. and McClure, M.S., Eds., Academic Press, New York, 1983, 91.

135. Coley, P.D. and Aide, T.M., Comparison of herbivory and plant defenses in temperate and tropical broad-leaved forests, in *Plant–Animal Interactions: Evolutionary Ecology in Tropical and Temperate Regions*, Price, P.W., Lewinsohn, T.M., Fernandes, G.W., and Benson, W.W., Eds., John Wiley & Sons, New York, 1991, 25.

136. Brattsten, L.B., Metabolic defenses against plant allelochemicals, in *Herbivores: Their Interactions with Secondary Plant Metabolites, Vol. II: Evolutionary and Ecological Processes*, Rosenthal, G.A. and Berenbaum, M.R., Eds., Academic Press, New York, 1992, 175.

137. Bold, H.C. and Wynne, M.J., *Introduction to the Algae*, 2nd ed., Prentice-Hall, Englewood Cliffs, NJ, 1985.

138. Schmitz, K., Translocation, in *Botanical Monographs, Vol. 17: The Biology of Seaweeds*, Lobban, C.S. and Wynne, M.J., Eds., Blackwell Scientific Publications, Boston, 1981, 534.

139. Steinberg, P.D., Algal chemical defense against herbivores: allocation of phenolic compounds in the kelp *Alaria marginata*, *Science*, 223, 405, 1984.

140. Tugwell, S. and Branch, G.M., Differential polyphenolic distribution among tissues in the kelps *Ecklonia maxima*, *Laminaria pallida*, and *Macrocystis angustifolia* in relation to plant defense theory, *J. Exp. Mar. Biol. Ecol.*, 129, 219, 1989.

141. Robertson, A.I. and Lucas, J.S., Food choice, feeding rates and the turnover of macrophyte biomass by a surf-zone inhabiting amphipod, *J. Exp. Mar. Biol. Ecol.*, 72, 99, 1983.

142. Tuomi, J., Ilvessalo, H., Niemela, P., Siren, S., and Jormalainen, V., Within-plant variation in phenolic content and toughness of the brown alga *Fucus vesiculosus* L., *Botanica Marina*, 32, 505, 1989.

143. Van Alstyne, K.L., Adventitious branching as a herbivore-induced defense in the intertidal brown alga *Fucus disticus*, *Mar. Ecol. Prog. Ser.*, 56, 169, 1989.

144. Van Alstyne, K.L., Ehlig J.M., and Whitman, S.L., Feeding preferences for juvenile and adult algae depend on algal stage and herbivore species, *Mar. Ecol. Prog. Ser.*, 180, 179, 1999.

145. Meyer, K.D. and Paul, V.J., Intraplant variation in secondary metabolite concentration in three species of *Caulerpa* (Chlorophyta: Caulerpales) and its effects on herbivorous fishes, *Mar. Ecol. Prog. Ser.*, 82, 249, 1992.

146. Meyer, K.D. and Paul, V.J., Variation in secondary metabolite and aragonite concentrations in the tropical green seaweed *Neomeris annulata*: effects on herbivory by fishes, *Mar. Biol.*, 122, 537, 1995.

147. Phillips D.W. and Towers, G.H.N., Chemical ecology of red algal bromophenols. I. Temporal, inter-populational, and within-thallus measurements of lanosol levels in *Rhodomela larix* (Turner) C. Agardh., *J. Exp. Mar. Biol. Ecol.*, 58, 285, 1982.

148. Carlson, D.J., Lubchenco, J., Sparrow, M.A., and Trowbridge, C.D., Fine-scale variability of lanosol and its disulfate ester in the temperate red alga *Neorhodomela larix*, *J. Chem. Ecol.*, 15, 1321, 1989.

149. Walls, J.T., Blackman, A.J., and Ritz, D.A., Distribution of amathamide alkaloids within single colonies of the bryozoan *Amathia wilsoni*, *J. Chem. Ecol.*, 17, 1871, 1991.

150. Van Alstyne, K.L., Wylie, C.R., and Paul, V.J., Antipredator defenses in tropical Pacific soft corals (Coelenterata: Alcyonacea). II. The relative importance of chemical and structural defenses in three species of *Sinularia*, *J. Exp. Mar. Biol. Ecol.*, 178, 17, 1994.

151. Harvell, C.D., The ecology and evolution of inducible defenses, *Q. Rev. Biol.*, 65, 323, 1990.

152. Agrawal, A.A. and Karban, R., Why induced defenses may be favored over constitutive strategies in plants, in *The Ecology and Evolution of Inducible Defenses*, Tollrian, R. and Harvell, C.D., Eds., Princeton University Press, Princeton, NJ, 1999, 45.

153. Gilbert, J.J., Kairmone-induced morphological defenses in rotifers, in *The Ecology and Evolution of Inducible Defenses*, Tollrian, R. and Harvell, C.D., Eds., Princeton University Press, Princeton, NJ, 1999, chap. 7, 127.

154. Kuhlmann, H.W., Kusch, J., and Heckmann, K., *Predator-Induced Defenses in Ciliated Protozoa*, Tollrian, R. and Harvell, C.D., Eds., Princeton University Press, Princeton, NJ, 1999, chap. 8, 142.

155. Hammerstrom, K., Dethier, M.N., and Duggins, D.O., Rapid phlorotannin induction and relaxation in five Washington kelps, *Mar. Ecol. Prog. Ser.*, 165, 293, 1998.

156. Van Alstyne, K.L., Herbivore grazing increases polyphenolic defenses in the intertidal brown alga *Fucus distichus*, *Ecology*, 69, 655, 1988.

157. Yates, J.L. and Peckol, P., Effects of nutrient availability and herbivory on polyphenolics in the seaweed *Fucus vesiculosus*, *Ecology*, 74, 1757, 1993.

158. Cronin, G. and Hay, M.E., Induction of seaweed chemical defenses by amphipod grazing, *Ecology*, 77, 2287, 1996.

159. Renaud, P.E., Hay, M.E., and Schmitt, T.M., Interactions of plant stress and herbivory: intraspecific variation in the susceptibility of a palatable versus an unpalatable seaweed to sea urchin grazing, *Oecologia*, 82, 217, 1990.

160. Lewis, S.M., Norris, J.N., and Searles, R.B., The regulation of morphological plasticity in tropical reef algae by herbivory, *Ecology*, 68, 636, 1987.

161. Peckol P., Krane J.M., and Yates, J.L., Interactive effects of inducible defense and resource availability on phlorotannins in the North Atlantic brown alga *Fucus vesiculosus*, *Mar. Ecol. Prog. Ser.*, 138, 209, 1996.

162. Harvell, C.D., Complex biotic environments, coloniality, and heritable variation for inducible defenses, in *The Ecology and Evolution of Inducible Defenses*, Tollrian, R. and Harvell, C.D., Eds., Princeton University Press, Princeton, NJ, 1999, 231.

163. Lively, C.M., Competition, comparative life histories, and maintenance of shell dimorphism in a barnacle, *Ecology*, 67, 858, 1986.

164. Lively, C.M., Developmental strategies in spatially variable environments: barnacle shell dimorphism and strategic models of selection, in *The Ecology and Evolution of Inducible Defenses*, Tollrian, R. and Harvell, C.D., Eds., Princeton University Press, Princeton, NJ, 1999, 245.

165. Appleton, R.D. and Palmer, A.R., Water-bourne stimuli released by predatory crabs and damaged prey induce more predator-resistant shells in a marine gastropod, *Proc. Nat. Acad. Sci. USA*, 85, 4387, 1988.

166. Ayer, D.J. and Grosberg, R.K., Aggression, habituation, and clonal coexistence in the sea anemone *Anthopleura elegantissima*, *Am. Nat.*, 146, 427, 1995.

167. Gruenbaum, D., Hydromechanical mechanisms of colony organization and cost of defense in an encrusting bryozoan, *Membranipora membranacea*, *Limnol. Oceanogr.*, 42, 741, 1997.

168. West, J.M., Plasticity in the sclerites of a gorgonian coral: tests of water motion, light level, and damage cues, *Biol. Bull.*, 192, 279, 1997.

169. Leonard G.H., Bertness M.D., and Yund, P.O., Crab predation, waterbourne cues, and inducible defenses in the blue mussel, *Mytilus edulis*, *Ecology*, 80, 1, 1999.

170. Price, T.J., Thayer G.W., Lacroix, M.W., and Montgomery, G.P., The organic content of shells and soft tissues of selected estuarine gastropods and pelecypods, *Proc. Nat. Shellfish Assoc.*, 65, 26, 1974.

171. Loomis, W.E., Growth and differentiation: an introduction and summary, in *Growth and Differentiation in Plants*, Loomis, W.E., Ed., Iowa State College Press, Ames, IA, 1953, 1.

172. Salisbury, F.B. and Ross, C.W., *Plant Physiology*, 3rd ed., Wadsworth Publishing Co., Belmont, CA, 1985.

173. Van Alstyne K.L., McCarth, J.J., III, Hustead, C.L., and Kearns, L.J., Phlorotannin allocation among tissues of northeastern Pacific kelps and rockweeds, *J. Phycol.*, 35, 483, 1999.

174. Bryant, J.P., Chapin, F.S., III, and Klein, D.R., Carbon/nutrient balance of boreal plants in relation to vertebrate herbivory, *Oikos*, 40, 357, 1983.

175. Waterman, P.G. and Mole, S., Extrinsic factors influencing the production of secondary metabolites in plants, in *Insect–Plant Interactions*, *Vol. 1*, Bernays, E.A., Ed., CRC Press, Boca Raton, FL, 1989, 107.

176. Muzika, R.M., Terpenes and phenolics in response to nitrogen fertilization: a test of the carbon/nutrient balance hypothesis, *Chemoecology*, 4, 3, 1993.

177. Reichardt, P.B., Chapin, F.S., III, Bryant, J.P., Mattes, B.R., and Clausen, T.P., Carbon/nutrient balance as a predictor of plant defense in Alaskan balsam poplar: potential importance of metabolite turnover, *Oecologia*, 88, 401, 1991.

178. Fajer, E.D., Bowers, M.D., and Bazzaz, F.A., The effect of nutrients and enriched CO_2 environments on production of C-based allelochemicals in *Platago*: a test of the carbon/nutrient balance hypothesis, *Am.. Nat.*, 140, 707, 1992.

179. Lincoln, D.E. and Mooney, H.A., Herbivory on *Diplacus aurantiacus* shrubs in sun and shade, *Oecologia*, 64, 173, 1984.

180. Zangerl, A.R. and Berenbaum, M.R., Furanocoumarins in wild parsnip: effects of photosynthetically active radiation, ultraviolet light, and nutrients, *Ecology*, 68, 516, 1987.

181. Cronin, G. and Hay, M.E., Effects of light and nutrient availability on the growth, secondary chemistry, and resistance to herbivory of two brown seaweeds, *Oikos*, 77, 93, 1996.

182. Björkman, C., Larsson, S., and Rolf, G., Effects of nitrogen fertilization on pine needle chemistry and sawfly performance, *Oecologia*, 86, 202, 1991.

183. Cronin, G. and Hay, M.E., Seaweed–herbivore interactions depend on recent history of both the plant and animal, *Ecology*, 77, 1531, 1996.

184. Ilvessalo, H. and Tuomi, J., Nutrient availability and accumulation of phenolic compounds in the brown alga *Fucus vesiculosus*, *Mar. Biol.*, 101,115, 1989.

185. Puglisi, M.P. and Paul, V.J., Intraspecific variation in the red alga *Portieria hornemannii*: monoterpene concentrations are not influenced by nitrogen and phosphorus enrichment, *Mar. Biol.*, 128, 161, 1997.

186. White, T.C.R., The abundance of invertebrate herbivores in relation to the availability of nitrogen in stressed food plants, *Oecologia*, 63, 90, 1984.

187. Rhoades, D. F., Offensive–defensive interactions between insects and plants: their relevance in herbivore population dynamics and ecological theory, *Am. Nat.*, 125, 205, 1985.

188. Feeny, P., Plant apparency and chemical defense, *Recent Adv. Phytochem.*, 10, 1, 1976.
189. Rhoades, D.F. and Cates, R.G., Toward a general theory of plant antiherbivore chemistry, *Recent Adv. Phytochem.*, 10, 168, 1976.
190. Boettcher, A. A. and Targett, N. M., Role of polyphenolic molecular size in reduction of assimilation efficiency of *Xilufaster mucosus*, *Ecology*, 71, 891, 1993.
191. Trindell, R.N., Stress induced intraspecific variation in macrophyte susceptibility to herbivores. Ph.D. thesis, University of North Carolina at Chapel Hill, Chapel Hill, NC, 1991.

10 Chemical Mediation of Surface Colonization

Peter D. Steinberg,* Rocky de Nys,
and Staffan Kjelleberg

CONTENTS

---

* Corresponding author.

I. INTRODUCTION

There is strong evidence for the importance of naturally produced compounds as mediators of ecological interactions between marine benthic consumers and their prey, e.g., in plant–herbivore and predator–prey interactions (reviewed by Hay,[1,2] Hay and Steinberg,[3] McClintock and Baker,[4] Paul,[5] and Pawlik[6]). Dozens of characterized, ecologically relevant feeding deterrents are known from marine benthic organisms (reviewed by Hay[1] and Paul[5]). As a consequence, studies of chemically mediated plant–herbivore or predator–prey interactions have increasingly moved beyond a consideration of simple feeding deterrence to address more complex ecological and evolutionary issues such as induction of defenses,[7,8] specialization of consumers,[3] and geographic variation in defenses.[9–11]

Naturally produced compounds from marine organisms also mediate colonization of surfaces, acting as both positive (inducers) and negative (deterrents) cues for settlement and colonization of animate and inanimate surfaces. However, our understanding of the chemical ecology of colonization of surfaces in marine benthic systems is less advanced than for studies of plant–herbivore and predator–prey interactions. For example, although there is extensive literature on the biology of chemical induction of settlement of invertebrate larvae,[12–14] there is not a single known inducer of invertebrate larval settlement that has been chemically characterized, quantified *in situ,* and demonstrated to be ecologically relevant to the target organism. Characterized, quantified, ecologically relevant examples of deterrent cues — natural antifoulants — are also rare,[15,16] and chemical cues which mediate colonization of surfaces by marine bacteria are largely unknown.

This discrepancy in the amount of direct evidence on the effects of naturally produced compounds in different interactions is probably due to a number of factors intrinsic to these systems. As Pawlik[12] and others have pointed out, invertebrate larvae are often very unpredictable in time and space, making *in situ* ecological studies of many species problematic. In contrast, herbivory and predation are often predictable, pervasive, and intense in marine systems,[2,18–20] and the study organisms often common, macroscopic, and amenable to observations of feeding. The impact of herbivores or predators on their prey in marine benthic systems is well documented,[2,18–21] and strong selection for the evolution of prey chemical defenses is likely to have occurred.[2,4,22] In contrast, demonstrations of the direct[23–26] or indirect[27] impact of colonization by epibiota (fouling) of marine organisms are rare,[28] and the general ecological or evolutionary importance of positive signals (inducers) for colonization is debated, particularly in the context of the effectiveness of chemical cues at different scales in natural flow regimes.[29,30]

Methodologically, both laboratory and field methods for studying the effects of prey chemical defenses on consumers in the laboratory and *in situ* are well established.[31] Similarly, characterizing and then quantifying levels of relevant metabolites in whole marine plants or animals is often relatively straightforward.[32] In contrast, the microscale distribution of chemical cues on or near a surface can crucially determine the efficacy of those cues *in situ*, and methods for the collection and analysis of such samples are not well established. Likewise, for studies of colonization, fewer general, realistic bioassay procedures are available, particularly for *in situ* tests. Many cues for settlement are also water soluble[14,33–35] (see Section III below). Such metabolites are less amenable to traditional techniques of separation chemistry and have thus been harder to characterize and quantify than the small lipophilic secondary metabolites which are the best known examples of chemical mediators of consumer–prey interactions. Finally, for plant–herbivore interactions in particular, there is substantial theoretical literature (mostly derived from terrestrial studies) that has guided empirical studies on the ecology and evolution of these systems.[36]

Perhaps because of these intrinsic difficulties with assessing ecological roles for chemical signals at surfaces, generalities regarding the role of naturally produced metabolites as mediators of surface–based interactions are relatively rare.[34] The generalizations that do exist mostly focus on (1) the physiological mechanisms of action of the metabolites, a focus of much of the research on larval inducers for invertebrates,[37] or (2) contrasting the role of hydrodynamics vs. chemical (and other) settlement cues, and in particular the extent to which hydrodynamic factors modulate the effects of chemical cues.[29] While mechanistic studies of modes of action are valuable in their own right, they do not necessarily speak to the demographic or community effects of natural cues for colonization. Moreover, integrative studies of chemistry and hydrodynamics, while fundamental to our understanding of these systems, are, in all but a few instances, hindered by a lack of sufficiently detailed knowledge of the cues themselves.

This chapter has one main premise and two general goals. The premise is that the chemical ecology of surface-based phenomena such as colonization are different in important ways from the chemical ecology of consumer–prey interactions, requiring different methods, different generalizations, and perhaps different compounds. Our first goal then, is to see whether any generalizations — methodological, chemical, or ecological — might emerge from a broad consideration of colonization phenomena in benthic habitats. This chapter makes no attempt to comprehensively review the field for either positive or negative cues for colonization, as a number of recent reviews are available in each case.[12–14,28,38–42] Rather, it focuses on representative examples from the work of the authors and the broader literature.

Second, in order to try and understand whether there are similarities in these phenomena among macroalgae, invertebrates, and bacteria, this chapter includes examples of chemical mediation of surface colonization for both eukaryotes and prokaryotes. Chemical cues can be fundamental for colonization processes for all three of these groups, but studies of these different taxa often occur largely independently of each other, in disparate literatures. Throughout this chapter, methodological issues relevant to studies of colonization are highlighted, since the field overall is at a stage where progress is particularly reliant on the development of new or improved techniques.

II. DETERRENTS OF COLONIZATION

A. INHIBITION OF FOULING BY SEAWEEDS

The hypothesis that secondary metabolites produced by macroalgae (seaweeds) deter colonization of algal thalli by epibiota (fouling organisms) is at least 50 years old[43] (the chemical claw). Only recently, however, have more rigorous examinations of this hypothesis been done. For a metabolite to be a natural antifoulant, it must be present at the surface of the producing organism, or released, at a concentration which deters ecologically relevant fouling organisms. Our research in this area has focused on the Australian red alga *Delisea pulchra*, which for much of the year is unfouled in the field. *D. pulchra* produces a range of nonpolar halogenated furanones (Figure 10.1), typically between concentrations of 0.5 and 1.5% of the dry weight of the thallus.[44,45] These compounds occur in vesicles in gland cells, which occur at the surface of the plant (as well as in the interior of the thallus), and release furanones to the surface of the thallus.[15] These compounds can be extracted off the surface without lysing cells, enabling surface concentrations of furanones to be quantified.[46] Total levels of furanones on the surface are typically in the range of 100 to 500 ng/cm^2,[15,46] which on average represents between 0.2 and 0.4% of the total amount of furanones in the alga. Naturally occurring concentrations of furanones applied to test surfaces in laboratory or field assays strongly deter ecologically relevant macro- and microbiota.[16,47,48] Surface concentrations of furanones and fouling also vary seasonally and with depth,[48] such that increased fouling on *D. pulchra* on shallow plants in summer corresponds to a significant drop in levels of surface furanones on these plants[48] (relative to plants occurring at greater depths). The experimental and observational data cited above indicates that furanones in *D. pulchra* act as *in situ* natural antifoulants.

For nonpolar secondary metabolites (e.g., metabolites with no known primary function in the organism's physiology) from macroalgae, only two other studies (besides those on *D. pulchra*) are known in which care has been taken to test metabolite concentrations at ecologically realistic levels. The first is the study of Schmitt et al.[17] for terpenoid metabolites from the brown alga *Dictyota menstrualis*. Schmitt et al.[17] found that *D. menstrualis* was typically unfouled in the field and was avoided by larvae of the epiphytic bryozoan *Bugula neretina* in multialgal settlement preference assays in the laboratory. Following these observations, secondary metabolites were "harvested" (as crude extracts) from a known area of the surface of the alga using cotton swabs, and were reapplied to the same areas of bioassay dishes, resulting in significant deterrence of larvae of *B. neretina*. Purified metabolites [pachydictyol A and dictyol E (Figure 10.1)] from the alga also deterred settlement or metamorphosis of larvae. Thus, metabolites appear to be present on the surface of *D. menstrualis* at concentrations high enough to deter an ecologically relevant epiphyte. Caveats to this conclusion are (1) swabbing can damage (lyse) surface cells,[46,48] potentially resulting in the harvesting of nonsurface borne metabolites, and (2) actual surface concentrations of pure metabolites were not determined.

The second example is for terpenoid metabolites from *Laurencia obtusa*.[46] *L. obtusa* in shallow habitats near Sydney is relatively free of fouling when it first appears in the field, becoming increasingly fouled over subsequent weeks. *L. obtusa*, like many species in the genus, produces a variety of sesquiterpenoids which have strong biological activity.[49,50] Two metabolites isolated from *L. obtusa* from the Sydney region, palisadin A and 5 β-hydroxyaplysistatin (Figure 10.1) inhibit settlement of spores of the green alga *Ulva lactuca* and larvae of *B. neretina* (common epiphytes in these habitats) at concentrations of 0.1 and 1 μg/cm^2, respectively.[46,51] However, levels of palisadin A and 5 β-hydroxyaplysistatin on the surface of the plant were less than 5 ng/cm^2 (at most).[46] These levels represent less than 0.1% of the total amount of terpenoids in the alga[46] and were orders of magnitude below levels needed to deter the two species of common epiphytes. The almost complete absence of these metabolites on the surface of the alga is not surprising, in light of the morphology of the plant. Terpenoids in *Laurencia* species are localized in intracellular vesicles known as *corp en cerise*,[52] which — unlike the gland cells in the Bonnemaisoniaceae (including *Delisea pulchra*) — are not known to come to the surface of the thallus. Thus, there is no obvious mechanism for the release of metabolites to the surface of undamaged individuals of *L. obtusa*.

In contrast to the studies described above on nonpolar secondary metabolites, early research on algal secondary metabolites as deterrents of epibiota focused on phlorotannins (Figure 10.1)[53] large water-soluble (polar) metabolites found in brown algae. Phlorotannins would not be efficiently sampled or collected by the dipping or swabbing procedures described above. Phlorotannins in brown algae occur in small vesicles known as physodes, and there are a number of studies demonstrating that phlorotannins are exuded into the water column from brown algae[54,55] (reviewed by Ragan and Glombitza[56]). Several studies have also demonstrated deterrent or toxic effects of phlorotannins against various epibiota or other invertebrates.[56,57]

As for nonpolar metabolites, the effectiveness of phlorotannins as deterrents depends on the relationship between *in situ* concentrations on or near the plant and the concentrations needed to deter fouling organisms. Jennings and Steinberg[54] measured *in situ* exudation of phlorotannins in the sublittoral kelp *Ecklonia radiata*, and found that exudation by undamaged, unstressed plants was much lower than in most previous studies. Based on these exudation rates, they[57] calculated that levels of phlorotannins, either in the boundary layer adjacent to *E. radiata* or in the surrounding water column, were too low to deter settlement of the epiphytic green alga *U. lactuca*. Phlorotannins from a co-occurring alga, *Sargassum vestitum*, also failed to inhibit settlement of *Ulva* at these concentrations. These levels of phlorotannins were much lower than literature values reported to deter other epiphytes (reviewed Jennings and Steinberg[57]), with the exception of peritrichs of the protozoan *Vorticella marine*,[58] and it was concluded that there was no evidence that undamaged *E. radiata* used phlorotannins to inhibit settlement onto their surfaces.

Jennings and Steinberg[57] also pointed out that no other studies of fouling of macroalgae have measured *in situ* rates of exudation of phlorotannins (or any other algal deterrent). Exudation rates of phlorotannins from brown algae in the laboratory, or from stressed thalli, can be substantially elevated,[54,59] and, thus, evidence for the deterrent effects of phlorotannins from other brown algae may be based on unnaturally high concentrations.[60] Possible exceptions to this conclusion are

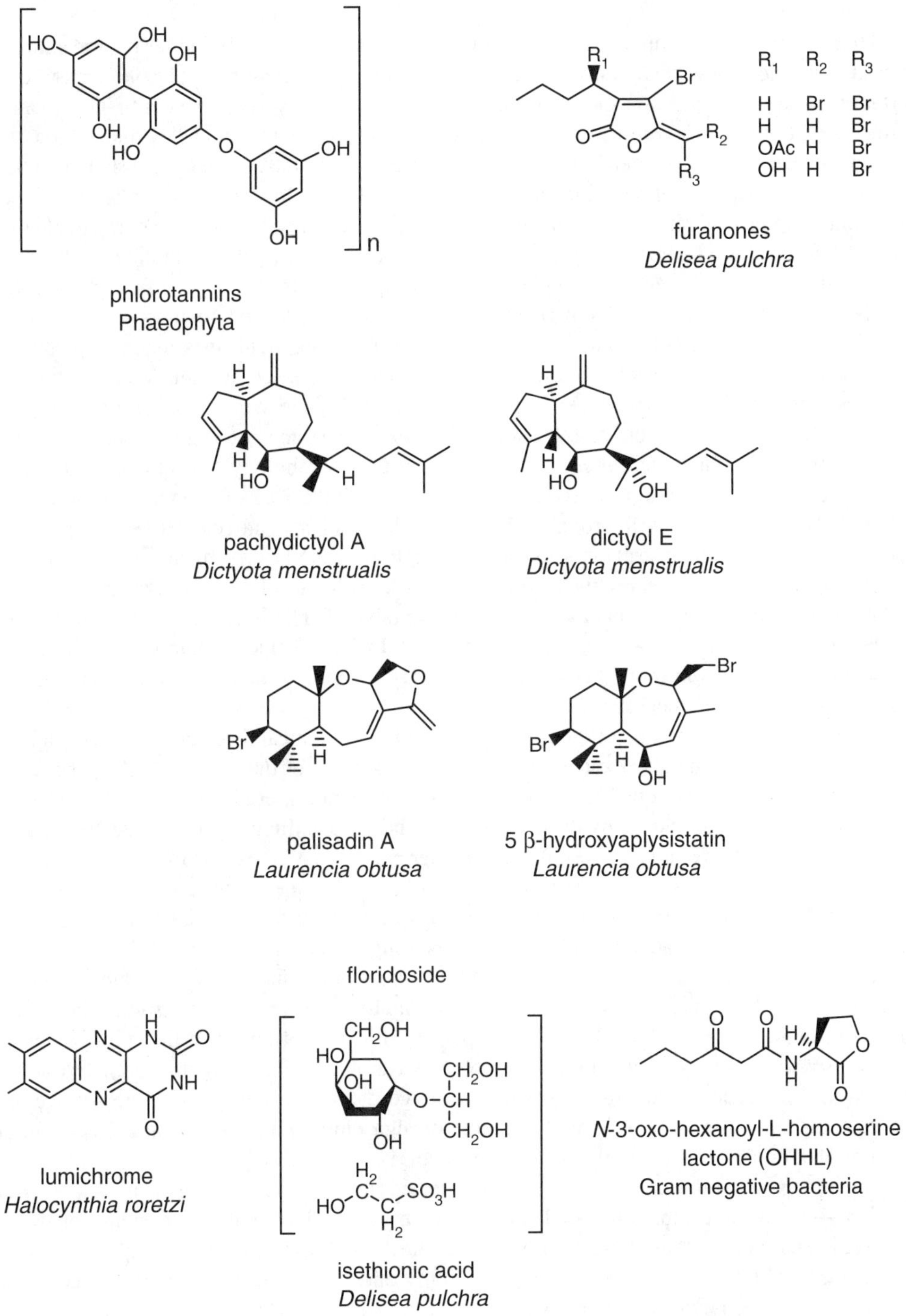

FIGURE 10.1 Structures of selected metabolites investigated for their role as inhibitors or inducers of prokaryote and eukaryote colonization.

intertidal algae, which undergo a burst of exudation following tidal immersion.[55] Such bursts may inhibit settling epibiota for short periods, as may accumulation of phlorotannins in tidepools at low tide.[61]

These contrasting examples for the efficacy of algal secondary metabolites as natural antifoulants raise the question: how commonly do such metabolites act as inhibitors of colonization of algal surfaces? Ideally, this question should be answered on a species-by-species basis first by quantifying surface metabolites in *in situ* or near *in situ* conditions, then performing laboratory bioassays of relevant concentrations of metabolites against relevant epibiota, correlating variation in surface metabolites and fouling of the alga in the field,[48] and finally by field tests of fouling in which metabolites are incorporated into artificial media and released at concentrations comparable to those released by the plant. Unfortunately, development of the suite of methods implied by this protocol is decidedly nontrivial, even for a single species. Formal quantification of surface metabolites has only been done for two species (above), and no published methods are known — with the exception of very short term tests of furanones against bacteria[16] — for field tests using artificial release systems in which metabolites are known to be presented or released at realistic concentrations.

Development of such protocols are necessary for a full understanding of antifouling by algae or other benthic organisms, but simpler alternatives exist that can start to give us an understanding of these processes. One such alternative is to study the morphology and ultrastructure of algae with respect to the production of secondary metabolites (Table 10.1). Given that algae must avoid autotoxicity, metabolites will probably be encapsulated in specialized cells, analogous to the cellular and multicellular structures in which terrestrial secondary plant metabolites are usually found.[62] However, we have relatively little general knowledge of the localization of secondary metabolites for most algae or benthic invertebrates (Table 10.1). Given that in most instances it is not known if the metabolites are able to be presented at the surface of the producing organism, we should be cautious in inferring that antifouling by secondary metabolites from benthic organisms is a general phenomenon.

A second approach to the question of the generality of chemical deterrence of epibiota by algae is to use techniques whereby metabolites can be harvested from the surface of the plant and then reapplied to test surfaces in bioassays. This was used by Schmitt et al.[17] in their study of *Dictyota menstrualis*, and has been used by these authors via the surface dip technique of de Nys et al.[46] (we recommend the latter technique since it has a greater consistency of extraction efficiency and is less damaging to surface cells).[46] In such techniques for harvesting metabolites, the metabolites are not quantified, but as long as metabolites are harvested from a known surface area of an alga, they can be reapplied to the same area of a test surface, resulting in ecologically realistic concentrations.

Using this technique, metabolites from measured areas of the surface of 10 species of chemically rich algae from the Sydney region were harvested (Figure 10.2). These extracts were then reapplied to test surfaces and assayed for their effects against settlement of larvae of the epiphytic bryozoan *Bugula neretina* using standard bioassay protocols (de Nys et al.;[63] Figure 10.2). Extracts were applied to test surfaces (petri dishes) at both average natural concentrations (the surface area of the petri dish was the same as that of the area of alga dipped) and twice natural concentrations (area of alga dipped was twice that of the test surface).

Of the ten species of chemically rich algae tested, only surface extracts from one — *Delisea pulchra* — significantly inhibited settlement of the bryozoan larvae at either natural or two times the average natural concentrations (Figure 10.2). These data are from a single experiment, extracts were collected at only one time and place, and only one species of epibiota was tested (although results for other epiphyte species are similar[64]). However, these algae are both taxonomically and chemically diverse, and the data in Figure 10.2 arguably represent the first broad, ecologically realistic screen for the widespread use of algal nonpolar secondary metabolites as antifoulants. The results of the experiment are not consistent with a ubiquitous role for algal nonpolar metabolites as natural antifoulants.

TABLE 10.1
Localization of Secondary Metabolites in Marine Macroalgae and Invertebrates

Taxa	Metabolite	Localization	*Comes to Surface?	Reference
Algae				
Rhodophyta				
Laurencia spp. (e.g., *L. synderae, L. obtusa*)	Halogenated terpenes	*Corps en cerise*	No	52
				46
Bonnemaisoniaceae (e.g., *Bonnemaisonia nootkana, Delisea pulchra*)	Furanones, other halogenated metabolites	Vesicles within gland cells	Yes	225
				15
Desmarestia firma	H_2SO_4	Cell vacuoles	ND	226
Phaeophyta				
All Phaeophyta	Phlorotannins	Physodes	Yes	56
Invertebrates				
Porifera				
Dysidea herbacea	2-(2',4'-dibromophenyl)-4,6-dibromophenol	Cyanobacterial symbiont-*Oscillatoria spongeliae*	ND	88
	Spirodysin, dihydrodysamide C, didechlorodihydrodysamide C	Cyanobacterial symbiont-*Oscillatoria spongeliae* archaeocytes and choanocytes	No / No	87
Dysidea avara	Avarol	Choanocytes	No	84
Dysidea fragilis	Ent-furodysinin	Vesicles within unspecifed cells	Yes	85
Crambe crambe	Unspecified (measured as toxicity)	Spherulous cells	Yes	86
	Guanidine alkaloids	Spherulous cells	Yes	83
Theonella swinhoei	Theopalauamide	Eubacteria symbiont	No	90
	Swinholide A cyclic peptide (antifungal)	Bacterial symbionts (unicellular heterotroph)	ND	89
		bacterial symbiont (filamentous heterotroph)	ND	
Ascidians				
Ascidia nigra	Vanadium	Vanadocytes	Yes	72

Note: ND = not determined.

* of undamaged cells.

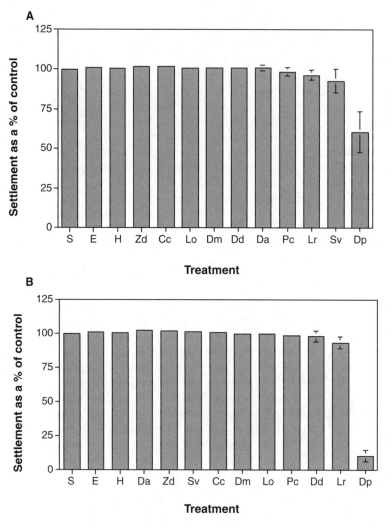

FIGURE 10.2 The effect of surface extracts of ten species of algae from the Sydney area on the settlement and growth of the common fouling bryozoan *Bugula neretina*. The algae used were the red algae *Delisea pulchra* (Greville) Montagne (abbreviated as Dp); *Laurencia obtusa* Lamouroux (Lo); *Laurencia rigidg* Agardh (Lr); *Pterocladia capillacea* (S.G. Gmelin) Bornet (Pc); *Champia compressa* Harvey (Cc); the brown algae *Dictyopteris acrostichoides* (J. Agardh) Boergesen (Da); *Dictyota dichotoma* (Hudson) Lamouroux (Dd); *Dilophus marginatus* J. Agardh (Dm); *Zonaria diesingiana* J. Agardh (Zd); and *Sargassum vestitum* (R. Brown ex Turner) C. Agardh (Sv); All species were collected in the subtidal zone (3 to 5 m depth) at Nielsen Park, Port Jackson (33°51'04"S, 151°16'12"E); Bare Island, Botany Bay (33°59'38"S, 151°14'00"E); or Shark Point (33°55'05"S, 151°16'12"E) New South Wales, Australia. Surface-borne compounds from all species were obtained by extraction in hexane as described by de Nys et al.[46] Pieces from freshly collected individuals with a surface area of 9 or 18 cm^2 were cut and extracted for 20 s in double-distilled hexane (AR grade). After extraction, the algal pieces were removed and the hexane taken to dryness at room temperature. The extracts were re-dissolved in 500 μL of ethanol (99.7% or higher purity) and applied to the surface of treatment dishes (9 cm^2). Extracts were therefore tested at mean natural (A) and twice mean natural (B) concentrations. Seawater (S), and two solvent controls — hexane (H) and ethanol (E) — were also prepared. Larvae of *Bugula neretina* were cultured and prepared for settlement assays as described by de Nys et al.[63] Settlement assays were done by adding 15 larvae to either treatment, solvent control, or untreated dishes, each containing 4 ml of sterilized filtered seawater. Test dishes were incubated for 24 h at 28°C in a 15/9 h light–dark cycle. After this time, the percentage of settlement was determined by counting the number of attached (cont.)

Finally, a strategy quite different from those discussed above for algal chemical defense against microorganisms is the recent suggestion by Weinberger et al.[65,66] that seaweeds use oxidative bursts — specifically the production of hydrogen peroxide (H_2O_2) — to inhibit microbial attack. The mechanism is reliant on bacterial degradation of algal cell wall polysaccharides,[65] and it is perhaps more appropriately considered as a defense against pathogenesis. However, depending on the ecological context[66] of the algal–bacterial interaction, it may also serve to generally inhibit bacterial colonization of an alga's surface. Interestingly, the bioluminescent symbiotic bacterium *Vibrio fisheri* in the light organ of the squid (*Euprymna scolopes*) is also controlled, in part, by oxidative bursts from the squid.[200]

B. INHIBITION OF FOULING BY BENTHIC INVERTEBRATES AND SEAGRASSES

Other benthic organisms such as invertebrates and seagrasses also produce secondary metabolites which deter the settlement of fouling organisms.[28,38,42,68–70] Much of the research in this area has focused on the development of commercial alternatives to current commercial antifouling paints (reviewed by Clare[68]). While some very active metabolites have been discovered, these studies do not generally address the ecological role of these metabolites. For the few studies that have been placed in an ecological context, active deterrents have often been found, but surface concentrations of the metabolites have not been quantified or the compounds have not been tested against ecologically relevant fouling organisms. For example, the ascidian *Eudistoma olivaceum* produces a range of alkaloids including eudistomin G and H, which deter ecologically relevant fouling organisms at low concentrations.[71] However, localization or quantification of these compounds at the surface of the ascidian has yet to be determined. Inorganic vanadium, found in vanadocytes and released to the surface of *Ascidia nigra,* has also been proposed as a mechanism of deterring fouling in ascidians.[69,72] The seagrass *Thalassia testudinum* produces a sulfated flavone glycoside which deters attachment and growth of the marine thraustochytrid protist (fungus) *Schizochytrium aggregatum.*[73] Zosteric acid and other simple phenolic acids from seagrasses deter settlement of barnacles and bacteria.[74] Again, surface concentrations of the seagrass metabolites are not known, although the compound from *T. testudinum* was deterrent at one-fifteenth of whole plant concentrations.[73]

In research that was, in many ways, ahead of its time, Thompson et al.[75] described localization of the brominated secondary metabolites aerothionin and homoaerothionin in the sponge *Aplysina fistularis* in spherulous cells. They further quantified *in situ* release rates of these metabolites,[76,77] although measured release rates may have been at the high range of natural concentrations as they were measured immediately following reimmersion of the sponge.[77] Although the exudates and compounds were not active in antifouling bioassays,[76,78] they did inhibit the feeding response of potential fouling organisms,[76,78] and, thus, may have significant effects on epibiota postcolonization.

A somewhat different example of natural antifouling by invertebrates comes from the work of Woodin, Lincoln, and colleagues[79–81] who have shown that a wide array of infaunal invertebrates in sediment communities produce organohalogens such as bromophenols.[81] In laboratory assays, purified bromophenols at concentrations which occur naturally in the field in beds of the polychaete *Notomastus lobatus*[82] inhibited burrowing activities of recently metamorphosed juveniles of several infaunal species.[80] The taxonomically widespread occurrence of such metabolites and the nature of sediments as a medium capable of accumulating inhibitory metabolites suggest that these compounds may be important in determining spatial distributions of infauna,[80,81] either via inhibition

FIGURE 10.2 (CONTINUED) and unattached larvae. N = six (6) replicates (dishes) were done for all treatments. The results of the assay are expressed as percentage settlement of the seawater (untreated) control. Data are mean ± S.E. Treatments lacking error bars indicate 100% settlement in all replicates. Statistical analysis of the data (separate one-factor analysis of variance ANOVA for each of Figure 10.2A and 10.2B, followed by Tukey's post-hoc comparison among means) showed that only extracts from *D. pulchra* significantly deterred settlement (at both natural and twice natural concentrations).

of settlement or allelopathy (the lines between the two may blur for mobile infauna, particularly juveniles).

As with seaweeds, determining the localization of putative inhibitors and their quantitative distribution is an important step in understanding their ecological roles (Table 10.1). Significant progress has been made in localization of secondary metabolites for sponges in particular (Table 10.1). In some cases, secondary metabolites are in cells that are released to the surface of the producing organism, while in others they remain internal to the organism. In the sponge *Aplysina fistularia*, the secondary metabolites aerothionin and homoaerothionin are contained within spherulous cells adjacent to the exhalent canals.[75] These cells rupture and are the source of the aerothionin and homoaerothionin found in exudates from the sponge.[77] Spherulous cells are also the location of the biologically active secondary metabolites, guanidine alkaloids, in the sponge *Crambe crambe*.[83] At least some of the spherulous cells from *C. crambe* occur outside the sponge exopinacoderm, suggesting release of the compounds and a potential role against fouling organisms.[83,86] In the sponge *Dysidea fragilis*, the furanosesquiterpene, ent-furodysinin, is located in vesicles within cells (cell type not specified). These vesicles appear to open into intercellular spaces, also suggesting release of the compound.[85] No studies of the activity of these furanosesquiterpenes against epibiota have been performed.

In other sponges, secondary metabolites are maintained internally. Avarol, the major active secondary metabolite in *Dysidea avara*, is localized in choanocytes within the sponge.[84] In the related *D. herbacea*, the terpene spirodysin is present within archaeocytes and choanocytes, while another class of secondary metabolite, chlorinated diketopiperazines, occurs within the symbiotic filamentous unicellular cyanobacterium *Oscillatoria spongeliae*.[87] Flowers et al.[87] have also isolated a sample of *O. spongeliae* from *D. herbacea* which do not contain the diketopiperazines. These later examples illustrate the importance of determining the localization of bioactive metabolites for understanding host–symbiont interactions as well as colonization phenomena, and there are now several studies demonstrating the production of secondary metabolites by symbiotic bacteria in host sponges.[88–90] Bacteria may also be associated with the production of secondary metabolites in the bryozoans *Amathia wilsoni* and *Bugula neretina*. In *A. wilsoni*, brominated amathamides occur on the surface of the zooids in association with a rod shaped bacterium. However, the role of these compounds and their origin have yet to be firmly established.[91] Davidson et al.[92] present evidence suggesting that bryostatin from *B. neretina* is produced by a nonculturable bacterial symbiont.

C. INHIBITION OF EPIBIOTA BY BIOFILM BACTERIA

Bacterial biofilms are ubiquitous in the marine environment and play an important role in mediating interactions at surfaces.[93,94] Some invertebrates (e.g., *Hydroides elegans*[95]) require a biofilm for successful settlement and metamorphosis. Settlement in other species can be either facilitated or inhibited by natural biofilms, with the response often specific to the biofilm and the invertebrate investigated. For example, biofilms deter settlement of the bryozoan *Bugula flabellata* but facilitate settlement of the ascidian *Ciona intestinalis*.[96] Similarly, biofilms of differing ages have significantly different effects on settlement, with the effect again dependent on the biofilm and the invertebrate larvae being tested.[97–99] Larvae also respond differentially to biofilms as a function of their metabolic activity, density, and composition.[93]

As with inhibitors produced by eukaryotes, few studies have characterized and quantified inhibitors produced by natural biofilms. Rather, most studies have focused on isolated bacterial strains, or biofilms, cultured in laboratories.[98,100–103] Only one inhibitory metabolite from biofilms has been fully characterized, the natural product ubiquinone-8 from the culture supernatant of the bacteria *Alteromonas* sp.,[103] a bacterium isolated from the sponge *Halichondria okadai*.

Several partially characterized inhibitors have been described from the marine bacterium *Pseudoalteromonas tunicata*, isolated from the tunicate *Ciona intestinalis*. This bacterium produces a diversity of metabolites, each of which specifically inhibits the settlement of invertebrate larvae and algal spores or growth of bacteria, fungi, or diatoms (see Holmström and Kjelleberg[104] for review). Production of the inhibitors is associated with production of a pigment, as appears to be the case for a number of inhibitory strains of bacteria isolated from the surface of marine eukaryotes.[102] Several of the compounds are partially characterized and water-soluble,[101,104,105] in contrast to the hydrophobic ubiquinone-8 and the nonpolar deterrents discussed above. However, treatment of *P. tunicata* cells with periodate to oxidize cell surface polysaccharides enhances the inhibitory effect of the antilarval compound.[101] This suggests that this metabolite is immobilized or associated with the exopolysaccharide coat.

Both *Alteromonas* sp. and *Pseudoalteromonas tunicata* were isolated from eukaryote hosts. This raises the possibility, suggested by several authors,[91,93] that inhibition of fouling by some eukaryotes may be accomplished by specific bacterial biofilms living on their surfaces. In support of this, the known hosts for *P. tunicata* are generally unfouled in the field, but are not known themselves to produce bioactive secondary metabolites. However, the hypothesis that bacteria colonize a host and then provide inhibitory cues to prevent prokaryote or eukaryote fouling is untested in an ecologically realistic context. Experiments using biofilms in the laboratory may be difficult to relate to field conditions[99] given the difficulties both in identifying and measuring cues *in situ* and in identifying and characterizing the biofilm itself.

The fundamental need to characterise and quantify bacterial strains in natural biofilms in order to understand their potential role as producers of settlement signals highlights the need for molecular techniques which can be used to identify and quantify bacteria *in situ*. For example, low abundance of putatively deterrent bacterium on surfaces *in situ* may preclude the production of sufficient quantities of deterrent metabolites. Appropriate techniques for such characterization include denaturing gradient gel electrophoresis (DGGE)[107] and fluorescence *in situ* hybridization (FISH)[108] and would in principle allow for the detection and quantification of all species in the biofilm.

D. Summary of Deterrence of Epibiota

Davis et al.[69] summarized the state of play of natural antifoulants over a decade ago, and some progress has been made since then. We now have a few examples where metabolites at or near the surface of the producing organism have been quantified or extracted and realistic concentrations tested against ecologically meaningful epibiota in the lab or field.[16] Significant methodological challenges remain, but progress has been made in localizing metabolites within or on producing organisms (Table 10.1), quantifying metabolites in a realistic way,[46,57,77] harvesting surface metabolites for bioassays[17] (Figure 10.2), and developing appropriate field tests.[16]

With regard to field tests, one positive development in the search for methods for realistic assessment of natural antifoulants is the use of durable, readily accessible materials for testing metabolites in the field. Examples of such materials are Phytagel[109] and copolymer resins.[110] Hendrikson and Pawlik[109] showed that an extract from the sponge *Aplysilla longispina* incorporated into Phytagel deterred settlement of fouling organisms for up to a month in the field, and Vasishtha et al.[110] used a commercially available resin (VYHH) to measure antifouling activity and release rates of an organic biocide (an isothiazolone). While those are important steps in the right direction, the challenge with such artificial matrices is to achieve a release rate/presentation of metabolites that mimics that of the producing organism. As any manufacturer of antifouling paints can attest, it is quite easy to develop materials that either rapidly release all the incorporated active ingredient or barely release any at all. Different active ingredients (e.g., secondary metabolites) also differ dramatically in their behavior in different media, and, thus, for any long-term field test of a natural antifoulant, it is necessary to calibrate the release rate of a specific metabolite from a specific substance (matrix). This requires the same sort of analytical methods necessary for the quantification of metabolites *in situ* on or around the producing organism.

Are there general patterns in the production of natural antifoulants by marine organisms? Research on seaweeds (reviewed above and by de Nys and Steinberg[28]) and evidence from marine invertebrates[78] suggest that natural antifoulants will primarily be nonpolar secondary metabolites localized within an organism in a fashion that will enable the metabolites to be released to the surface. Polar metabolites such as phlorotannins are likely to be less effective because of the rapid dissolution of such metabolites away from the surface of the producing organism. In contrast, hydrophobic metabolites such as the furanones from *Delisea pulchra* can remain adsorbed to or associated with the organic surfaces of the producing organism, providing a more persistent concentration of deterrent metabolites at the surface. Possible exceptions to this prediction are (1) sponges, which by virtue of their complex system of channels and cavities may be able to concentrate polar metabolites within internal spaces, which may also result in high concentrations of metabolites near their surface, (2) the research of Weinberger et al. [65] on the possible role of hydrogen peroxide as a deterrent of bacterial pathogens, and (3) evidence that at least some bacterially derived deterrents are small, water-soluble metabolites (e.g., *Pseudoalteromonas tunicata*[111]).

III. INDUCTION OF SETTLEMENT OF EUKARYOTE PROPAGULES

Chemical inducers for the settlement or metamorphosis of larvae of marine invertebrates have been comprehensively reviewed during the last decade,[12,13,39] as well as in this volume.[14] Considerably less information is available for algal propagules.[112] Rather than reiterating these reviews, this section focuses its consideration of induction of colonization of surfaces by eukayotes on two questions: (1) what kinds of chemicals are used and why, and, (2) are there habitat or life historical correlates of particular kinds of cues? For the purposes of this chapter, settlement is defined as the sum total of processes that result in the transition from a planktonic existence to a benthic one, which generally incorporates settlement to the bottom per se (a behavioral phenomenon) followed by various morphological/developmental changes such as metamorphosis (animals) or germination (algae). We note, however, that the behavioral components of settlement, vs. metamorphosis or germination, may be affected by different cues.[14]

As with natural antifoulants, there are few (if any) fully documented examples of natural, ecologically relevant inducers of settlement; that is, characterized cues which induce settlement, have been quantified *in situ,* are active at *in situ* concentrations against the relevant target organism, and can be related to patterns in the demography of that organism. However, there are a wealth of examples, particularly for invertebrates, in which various pieces of this puzzle are known, and that information is summarized here. The information that is available points to the widespread use of primary metabolites, particularly water-soluble (polar) peptides and carbohydrates, as natural cues.

A. What Chemicals Are Used for Induction of Invertebrate Settlement?

1. Peptides

Probably the most widespread evidence for the importance of soluble primary metabolites as inducers of invertebrate settlement comes from studies of peptides, amino acids, or other proteinaceous cues. Their importance has been suggested by a number of authors over the years,[33,113–115] and it is now clear that such metabolites affect settlement in a wide taxonomic array of invertebrate larvae. Research on oysters and barnacles are among the more prominent examples of peptides as inducers. Zimmer-Faust and colleagues[115–117] have shown that larvae of the oyster *Crassostrea virginica* respond to a small (approximately 500 to 1000 Da) peptide from oyster shells by dropping rapidly from the water column and starting settlement behavior.[115] The natural peptide is uncharacterized but is mimicked by the tripeptide glycince–glycine–arginine (GGR). The inducer is active in both still water and flow (in a flume).[117,118] The source of the cue — adult oysters or their shells vs. shell-associated biofilms — has not yet been determined.[116,119]

Settlement and metamorphosis of barnacles in response to chemical cues appears more complex than for oysters, with several materials functioning as inducers of settlement (reviewed by Clare and Matsumura[40]). First, barnacles settle and metamorphose in response to a partially characterized proteinaceous cue termed "settlement-inducing protein complex" (SIPC) which is associated with the adult shell. The SIPC is a glycoprotein complex with four subunits, each of which has a similar effect to the intact glycoprotein complex.[120,121] Second, proteinaceous adhesives left by settling cyprids induce settlement of conspecifics in *Semibalanus balanoides* and *Balanus amphitrite*.[122,123] Finally, a smaller, water-soluble peptide settlement cue has also been partially purified from water conditioned by adult barnacles.[124,125] The three cues may be related; Clare and Matsumara[40] have suggested similarities between the adhesive-derived cue and SIPC, and Harrison[126] suggested that the waterborne cue is derived from the SIPC. Interestingly, the same tripeptide — GGR — that induces settlement of oyster larvae also induces metamorphosis in barnacles.[125]

Amino acids and peptides are also implicated in settlement and metamorphosis of abalone (*Haliotis rufescens*). Uncharacterized small peptides (approximately 1000 Da), which are subunits of larger protein moieties associated with the surface of coralline algae, induce settlement of *H. rufescens*.[127,128] The activity of these peptides is mimicked by GABA — γ-amino-butyric acid — in inducing the settlement of abalone larvae.[129] While the signal transduction pathway for GABA and GABA analogues has been a focus of investigation (see Slattery[13] for review), the nature, specific activity, and source of the natural inducer are still not fully understood. In particular, the role of bacteria on the surface of crustose coralline algae in the production of GABA and/or the settlement inducer remains to be resolved. Marine bacteria are able to produce and metabolize GABA,[130,131] and Johnson et al.[106] argue that bacteria associated with the surface of coralline algae may be responsible for the induction of abalone larval settlement (either through production of GABA or other inducers). There also appear to be species-specific responses by abalone to settlement cues, with some abalone species induced to settle by diatom films alone.[132]

One of the most intriguing examples of peptides as cues for settlement and metamorphosis comes from the rotting leaves of the mangrove *Rhizophora mangle*.[133] An inducer of the jellyfish *Cassiopea xamachana* is produced by bacterial decomposition of the mangrove leaves[133] and has been identified as a water-soluble proline- and glycine-rich protein.[134] The inductive compound appears to be a nonspecific by-product of bacterial degradation of proteinaceous plant matter, although the possibility that the inducer is produced by the degradative bacteria themselves cannot be ruled out.

A variety of other amino acid- or peptide-based cues also induce larval settlement and metamorphosis. For example, larvae of the sand dollar *Dendraster excentricus* settle and metamorphose in response to (1) a peptide-based cue from sand associated with adults, and (2) to water conditioned by adults.[135–137] The nudibranch *Adalaria proxima* is induced to settle and metamorphose by a water-soluble-peptide-based cue from its bryozoan host *Electra pilosa*.[138,139] A surface-associated peptide inducer has been proposed for the tube worm *Phragmatopoma lapidosa californica*,[140] but its role as a natural inducer remains to be defined.

2. Carbohydrates

Sugars and their derivatives are increasingly implicated as inducers of settlement. Williamson et al.[35] studied induction of metamorphosis of larvae of the Australian urchin *Holopneustes purpurascens*. This echinoid occurred primarily (in more than 95% of individuals) on two host plants at their study site, the red alga *Delisea pulchra* and the kelp *Ecklonia radiata*. While densities of the urchin on its two host plants did not significantly differ, urchins on *D. pulchra* were significantly smaller than those on *Ecklonia radiata*, and urchins in the smallest size class (e.g., new recruits) were only found on *D. pulchra*.[35] In laboratory bioassays, competent larvae of *H. purpurascens* rapidly metamorphosed (3 to 7 h) when in the presence of pieces of *D. pulchra* or a polar (water-soluble) extract of this alga. No metamorphosis was observed in response to *E. radiata* or its extracts. Bioassay guided fractionation resulted in the characterization of a specific chemical cue from *D. pulchra*, a noncovalently bound complex between the red algal sugar floridoside (Figure 10.1) and the small organic acid, isethionic acid (Figure 10.1), which induced metamorphosis within 2 h at a concentration of 25 µM. Seawater collected *in situ* within centimeters of *D. pulchra* also induced metamorphosis, while seawater collected near *E. radiata* or distant (meters) from other algae did not. Finally, in laboratory assays, other red algae from *H. purpurascen*'s habitat also induced metamorphosis of larvae, although at lower levels than for *D. pulchra*. The larvae did not metamorphose in the presence of brown or green algae, consistent with the fact that floridoside is widespread within the red algae but not found in browns or greens.[141]

Larvae of many species of corals are induced to metamorphose by a carbohydrate-based cue from algal cell walls, in particular those of crustose coralline algae.[142,143] The inducer is apparently conserved across many algal species in the Pacific and Caribbean,[144] and it has been identified as a high molecular weight polysaccharide with an active, water-soluble subunit.[145] The isolated subunit has been adsorbed onto a resin base, which had activity similar to the natural inducer in laboratory and field studies.[145]

Host-plant-derived carbohydrates induce settlement and metamorphosis in Opisthobranch molluscs. The ascoglossan *Alderia modesta* is induced to settle by three carbohydrate cues isolated from the yellow-green alga *Vaucheria longicaulis*, the host of *A. modesta*.[146,147] These include a soluble low molecular weight cue, a soluble high molecular weight cue, and an insoluble surface-associated high molecular weight cue.[146] The nudibranch *Eubranchus doriae* is induced to settle and metamorphose by a soluble carbohydrate isolated from its primary prey, the hydroid *Kirchenpaueria pinnata*.[148] An isolated polysaccharide containing galactosidic residues, as well as purified hexoses and galactosamine, all induced metamorphosis. Only sugars with the hydroxyl groups at carbons 3 and 4 with a *cis*-conformation induced metamorphosis, suggesting a stereospecific response from the larvae.[148] This conformation is the same as that for the sugar component of the floridoside–isethionic acid inducer of *Holopneustes purpurascens*.[35]

Finally, Forward et al.[149] provide further evidence for a role for plant carbohydrates as larval inducers. They showed that humic acids either from a commercial source or extracted from estuarine water (the latter presumably complexes derived from the degradation of plant carbohydrates) enhanced the rate of metamorphosis of blue crab (*Callinectes sapidus*) larvae.[149]

3. Fatty Acids

Arguably, the best known example for induction of invertebrate larvae by fatty acids is the work of Pawlik and others on settlement by *Phragmatopoma lapidus californica* in response to cues associated with sand from adult worm tubes.[12,150,151] Free fatty acids were isolated from the sand surrounding tubes and induced settlement and metamorphosis of the worms.[151–153] However, these findings have been challenged by Jensen and Morse[140,154] and Jensen et al.,[155] who argued that the fatty acids were contaminants, with settlement and metamorphosis actually induced by a polymeric protein containing L-DOPA subunits. The relative role of free fatty acids or L-DOPA in induction of settlement and metamorphosis of *Phragmatopoma* is unresolved to date and warrants further investigation.

Free fatty acids as inducers of settlement and metamorphosis have also been suggested for echinoderms.[156] However, there are, at present, no studies which unequivocally support the role of free fatty acids as settlement inducers for marine invertebrates.

4. Other Metabolites

Lumichrome (Figure 10.1), a water-soluble degradation product of riboflavin,[157] appears to be the first characterized gregarious (produced by conspecifics) cue for induction of metamorphosis of a marine invertebrate. Lumichrome is exuded into aqueous media from cultures of larvae of the ascidian *Halocynthia roretzi*, and was also found in the eggs, gonads, and tunic of the adults following extraction in methanol.[157] Metamorphosis of larvae was induced by purified lumichrome with an EC_{99} value of 100 nM. The effect is species-specific to at least some extent, as lumichrome had no effect on larvae of another ascidian, *Ciona sauvignyi*. As with other chemical inducers, several questions about the activity of lumichrome *in situ* remain. For example, if larvae respond to adult *H. roretzi* in the field, how is lumichrome released from the adults? Moreover, only media from high density larval cultures induced metamorphosis.[157] Thus, although *H. roretzi* produces synchronously developing larvae, unless these larvae remain together, and settle together, it is unclear how larvae alone could be the source of the cue *in situ*.

Several low molecular weight, nonpolar secondary metabolites also induce settlement of invertebrate larvae. These include jacaronone from the red alga *Delesseria sanguinea*, which induces settlement of the scallop *Pecten maximus*,[158] and α-tocopherol epoxide from *Sargassum tortile*, which induces settlement of the hydroid *Coryne uchidai*.[159] However, both jacaronone and the tocopherol epoxide have questionable ecological relevance. Scallop settlement is unrelated to the occurrence or distribution of *D. sanguinea*,[160] and a number of unresolved chemical and biological questions remain regarding the putative inducers of *C. uchidai* (for details, see Pawlik[12]).

There are also a number of examples of chemical cues inducing settlement and metamorphosis, the nature of which have not been determined. An uncharacterized waterborne cue of approximately 500 Da in size from the massive coral *Porites compressa* induces metamorphosis of the nudibranch *Phestilla sibogae* which feeds on these corals.[161,162] The queen conch *Strombus gigas* is induced to settle and metamorphose by red algae, in particular *Laurencia poitei*, the dominant red alga in conch shell nurseries.[163] The cue is a low molecular weight, water-soluble compound (less than 1000 Da) which induces settlement at levels comparable to *L. poitei*. The structure of the cue remains unresolved, although metamorphosis of *S. gigas* is induced by amino acids and peptides.[164] A cue for the metamorphosis of larvae of the mollusc *Haminaea callidegenita* has been partially purified from adult tissue and from the egg masses of this and four other species of opisthobranch molluscs. The cue is a polar nonproteinaceous compound with a molecular weight of less than 1000 Da.[165]

5. Biofilms

Biofilms fundamentally affect settlement for many invertebrate larvae (reviewed by Wieczorek and Todd[94] and Holmström and Kjelleberg[111]), and the prevalence of biofilms on marine benthic surfaces may provide a general habitat signal for settling marine organisms.[106] Settlement and metamorphosis of the tube worm *Hydroides elegans* is (obligately) triggered by bacterial biofilms[95,166] (see Chapter 13 in this volume). Unabia and Hadfield[95] isolated a number of bacterial strains which induced settlement and metamorphosis of the worms. However, this occurred at a lower frequency than for mixed natural biofilms, suggesting the possibility of multiple cues. The related *Hydroides dianthus* also settles in response to biofilms, but this species responds to live *H. dianthus* as well.[167] This polytypic response may provide initial founder and subsequent aggregator communities of *H. dianthus*, thereby supporting the establishment of dense colonies.[167]

Polysaccharides are perhaps the most studied of the partially characterized larval cues from biofilms. Exopolysaccharides from the bacterium *Halomonas* (*Deleya*) *marina* stimulated settlement in the spirobid polychaete *Janua brasiliensis* through surface recognition of a galactose–galactose subunit within the polymer.[168,169] Settlement of *J. brasiliensis* could be blocked by surface coating the biofilm with specific lectins, strongly suggesting that the cue is surface-bound, as opposed to a water-soluble subunit broken off from the polymer. Exopolysaccharides from *Pseudoalteromonas* (*Pseudomonas*) sp. S9 also induce larval settlement, in this case of the ascidian *Ciona intestinalis*. Both recognition of the surface-bound polymer and entrapment of larvae in the exopolymer-enhanced larval settlement and recruitment of *C. intestinalis*.[170] Both of these examples in which larvae respond to surface-bound cues appear to represent exceptions to the more frequent pattern of water-soluble inducers described above.

As discussed earlier for inhibitory biofilms (Section II.C), in general, neither the strains of biofilm bacteria responsible for inducing settlement of invertebrate larvae nor the relevant signals have been characterized. Thus, it is not clear whether biofilm-derived signals differ in general from those produced by eukaryotes. Moreover, responses to monospecific biofilms in the laboratory may not reflect responses of propagules to biofilms in the field. In addition to field biofilms containing a diversity of bacteria, biofilms also contain other microorganisms besides bacteria which may be responsible for induction of settlement. For example, diatom films are commonly used in abalone aquaculture to induce settlement,[132] and fungi appear to be a common component of marine biofilms.[171]

B. INDUCTION OF MACROALGAL SETTLEMENT

While it is generally considered that, "direct evidence for chemical signals associated with settlement is lacking in algae…,"[112] a few examples of such settlement are known. Amsler and Neushul[172] showed that settlement of motile spores of the kelps *Macrocystis pyrifera* and *Pterygophora californica* was enhanced in the presence of inorganic nutrients and the amino acids glycine and aspartate (*M. pyrifera* only). The kelp spores also exhibited positive chemotaxis in response to several simple inorganic and organic nutrients, with iron and high concentrations (1 mM) of ammonium eliciting negative chemotaxis. Nutrients do not elicit chemotaxis or enhanced settlement in spores of *Ectocarpus siliculosus*,[173] however, indicating that such responses are not universal for brown algal spores. Relatively little is known in this regard for other taxa of macroalgae (reviewed by Amsler et al.[173]). Similarly, little is known regarding the role of biofilms in macroalgal settlement. Dillon et al.[174] demonstrated enhanced adhesion of swarmers of the green alga *Enteromorpha* to biofilms, and Thomas and Allsopp[175] showed enhanced attachment and germination of zoospores of *Enteromorpha* in response to several bacterial biofilms.

C. WHY SOLUBLE PRIMARY METABOLITES AS INDUCERS?

The examples above suggest that inducers of settlement differ in some important ways from deterrents. The few ecologically relevant deterrents of settlement that are known are nonpolar secondary metabolites. In contrast to deterrents and some earlier suggestions,[12] that inducers are primarily surface associated, inducers of invertebrate (at least) settlement are generally primary metabolites such as carbohydrates or peptides and are commonly water-soluble.[146] Exceptions to this pattern in regard to the latter criterion (water solubility) appear to be twofold. First, for barnacles[40] and the sea slug *Aldaria*,[146] large, insoluble, surface-associated cues are present in addition to smaller water-soluble ones. Second, cues for coral planula[142] and exopolymers of biofilm bacteria such as *Pseudomonas sp. S9*[170] and *Deleya marina*[168,169] are surface associated. However, some of these cues,[143] when hydrolyzed or otherwise degraded, often reveal active, water-soluble subcomponents, suggesting that there may also be smaller soluble cues present in these examples as well (although this is not supported for the *Deleya marina/Janua brasiliensis* interaction[168,169]). A number of hypotheses may explain the prevalence of water-soluble primary metabolites as inducers.

1. The Signal Should Extend beyond Its Source

Many invertebrate larvae, when competent to settle, and many algal propagules in general, exhibit tropisms (e.g., photonegativity) or other behaviors which cause them to associate with the benthos rather than move more broadly through the water column.[41,176,177] This facilitates the ability of a propagule to detect a surface-associated cue. However, the optimal inducer should be able to be detected as far from the source surface as possible, maximizing the target area for a settling propagule. Water solubility of a cue ensures that it will diffuse readily into the water column, increasing the area in which a settling propagule can encounter the cue. (Note that the distances that a cue diffuses does not need to be far to significantly enhance the area of detection. Diffusion of a cue by 2 cm from a hemispherical source with a radius of 5 cm doubles the surface area of the volume containing the cue.) Nonpolar metabolites are likely to be less effective, because they tend to remain adsorbed to organic surfaces of the producing organisms, and are also likely to rapidly partition to organic aggregates once in the water column. Both phenomena will decrease the volume in which they are available as cues.

In contrast, the optimal strategy for hosts producing deterrents is to maximize the concentration of metabolites at, or as close as possible to, their surface.[57] This is best achieved by producing nonpolar metabolites such as furanones or terpenoids which will adsorb to their surface, or metabolites which are complexed or otherwise immobilized at the surface, as suggested for inhibitory compounds from the bacterium *Pseudoalteromonas tunicata* (reviewed by Holmström and Kjelleberg[111]). The tendency for many nonpolar metabolites to be unique or idiosyncratic to particular taxa of benthic organisms[178] also means that these cues are unlikely to be very widespread in the habitat, lessening the probability of generalist epibiota[179] evolving resistance to common deterrent cues.

These differences between deterrents and inducers highlight differences in selective pressures relevant to the two interactions. For deterrence, selection should be primarily on the host. While consequences of fouling to the host can, at least putatively, be severe, from the colonizer's perspective, there are usually many other sites on which to settle, especially given the generalist nature of many epibiota.[179] In contrast, for induction of settlement, there should be strong selection on colonizers to be able to detect signals and follow concentration gradients as efficiently as possible. In such cases, there may be no selection on the producer of the cue at all; production of the cue may simply be the consequence of natural leakage of metabolites from a producing organism (see below).

A main constraint on the detection of a water-soluble cue *in situ* is the extent to which it is diluted or dispersed by hydrodynamic factors. This topic has been discussed extensively in the literature, and will not be reviewed in depth here (see Chapter 12 in this volume). It has been argued that such cues may not be generally important in the real world because of rapid dilution or dispersion in the field (as Jennings and Steinberg[57] have argued for the deterrent effects of phlorotannins from *Ecklonia radiata*) or because of an inability of weakly swimming larvae to follow a concentration gradient in flow.[180–182] However, such conclusions are generally drawn without much specific knowledge of the cues, although some experiments with isolated cues in flow have been performed.[118,183] With regard to the weakly swimming hypothesis, we note that organisms both larger (larval fish[184]) and smaller (bacteria[185,186]) than invertebrate larvae are capable of strong directional movement in response to cues. Moreover, the observation that settlement patterns of some larvae at a variety of spatial scales do not differ from passive settlement patterns[30,187] is perhaps not surprising, since presumably not all larvae or spores respond strongly (or at all) to chemical signals. It is also not known whether settlement in response to chemical cues is generally concentration dependent, requiring a propagule to follow a diffusion gradient, or a threshold phenomena, in which detection of a cue above a threshold concentration induces settlement. The latter appears to be the case for oyster larvae, which drop out of the water column upon detecting a water-soluble peptide cue.[117,118] Similarly, metamorphosis of the echinoid *Holopneustes purpurascens* in response to the floridoside–isethionic acid (Figure 10.1) complex is only concentration dependent in a narrow range (2.5 to 12.5 µM). At test concentrations above 12.5 µM (125 µM was the highest concentration tested), all larvae rapidly metamorphosed.[35]

A number of authors[29,30] have addressed the potentially contrasting roles of larval cues vs. hydrodynamics by highlighting the need to understand the different spatial scales at which cues are effective. One simple experiment which would address this issue is to collect water samples at varying distances from the putative source of a cue and test their effectiveness at inducing settlement.

2. The Signal Should Be Present (and Accessible to Detection) in as High a Concentration as Possible

A second argument for the prevalence of primary metabolites as inducers is that they are generally the most common kinds of metabolites produced by organisms, and thus, all else being equal, they should be more abundant and therefore easier to detect than less common compounds. Total nonpolar secondary metabolites in many benthic organisms typically represent 1 to 2% or less of the dry weight of the organism.[188] In contrast, primary metabolites, particularly sugars, can constitute significant fractions of the mass of an organism. The red algal sugar floridoside (Figure 10.1), one component of the characterized inducer of metamorphosis for *Holopneustes purpurascens*,[35] can comprise over 30% of the mass of some red algae,[189] and levels of 8 to 10% are common.[141,189] By comparison, furanones, inhibitory compounds in *Delisea pulchra* (from which the floridoside–isethionic acid inducer was first isolated), typically occur at less than 1% by weight, of which 0.2 to 0.4% of the total is on the surface.

Water-soluble primary metabolites should also be more generally accessible to settling propagules than either secondary metabolites or insoluble primary metabolites. Though sequestration of secondary metabolites is much better known in terrestrial[62] than marine organisms (Table 10.1), benthic organisms must also avoid autotoxicity, and their secondary metabolites are also likely to be encapsulated in various ways. As discussed in Section II, sequestered metabolites may not be presented at the surface of undamaged individuals. Insoluble inducers may be contained within cell walls and only released through hydrolysis or other damage to the cell walls.[127,142] In contrast, primary metabolites are often not encapsulated and readily leak or are exuded out of the organism. This is well known for algal sugars and polysaccharides.[190] Peptides and amino acids also leak from algae, biofilms, and invertebrates.[191]

Finally, the strength of a signal from a biological source should also be enhanced the greater the abundance of the producing organism and the greater its tendency to occur in monotypic stands. Not surprisingly, some of the best-known examples of inducers for invertebrates are produced by organisms which occur in dense, nearly monospecific stands, e.g., oysters,[115,116] barnacles,[40] and several species of algae.[35,146]

3. Organisms May Be Pre-Adapted to Use Primary Metabolites as Positive Signals for Colonization

This pre-adaptation of organisms may occur for at least two reasons. First, bacteria, macroalgal spores or gametes and some invertebrate larvae[192] absorb nutrients directly from the environment, including amino acids, peptides, and sugars, and in many cases respond to concentration gradients in these compounds via chemotaxis. Thus, for some organisms, organic nutrients function directly as signals. This is perhaps most evident for bacteria, such as *Myxococcus xantos*, which uses amino acids both as differentiation signals[193] and nutrients. However, echinoderm larvae also use simple waterborne amino acids and sugars as signals for morphological change.[192] Given that nutrients often accumulate at surfaces,[42,194] initial attraction to particular surfaces may, for many organisms, simply be a function of following concentration gradients to high concentrations of nutrients or dissolved organics at a surface. The evolution of recognition of particular individual signals could have then occurred as responses to particular small organic molecules became more specific, even if organisms no longer relied on the absorption of the signal molecules as nutrients.

Second, a significant proportion of the known organic laboratory inducers of invertebrate settlement or metamorphosis are peptides, amino acids, or their derivatives.[37,152] These neurotransmitters and neurotransmitter mimics (e.g., L-DOPA and GABA) may not be presented exogenously to the larvae in the field at active concentrations and, thus, may not be ecologically relevant inducers. Rather, they may act as internal signals in signal transduction pathways. Nonetheless, they provide further evidence for the general importance of peptides and amino acids in the overall process of larval settlement. Existing responses and sensory machinery to internal signals may have pre-adapted organisms to respond to similar metabolites as exogenous signals. Rittschof[114] develops this theme more generally, arguing that peptide cues are widespread, conserved signal molecules which induce a variety of behavioral and ecological responses of marine invertebrates.

4. The Signals Should Be Resistant to Degradation

Though resistance to degradation is not restricted to polar primary metabolites, if such metabolites predominate as signals for other reasons, then it would also be advantageous to be resistant to degradation. Decho et al.[34] have shown that particular peptides — those with a basic amino acid terminus — are more resistant to degradation by marine bacteria than those with acidic termini. As a consequence, particular peptides may be more common as signals than others.[34] Resistance to degradation or uptake by marine bacteria may be a particularly important consideration because marine bacteria typically have a greater uptake affinity for dissolved organics than do marine eukaryotes.[195]

In sum, inducers will be easiest to detect if they are produced in high concentrations by abundant organisms which are persistent features of habitats, leak readily out of organisms, and resist bacterial degradation.

D. ECOLOGICAL CORRELATES OF INDUCERS

1. Habitat-Specific vs. Habitat-General Cues

To date, general classifications of inducers have been primarily based on the sources of the cues, with a variety of authors[12,13] distinguishing between gregarious (conspecifically derived) cues, associative cues derived from species other than the colonizing organism, and cues derived from biofilms. However, the importance of distinguishing between these categories may not always be clear. For example, unless adult organisms or newly settled juveniles have the capability of preferentially attracting related conspecific propagules (thereby enhancing their own extended fitness), there would seem to be few important ecological or evolutionary distinctions between many gregarious and associative cues. That is, a larva attracted to an assemblage of conspecifics or a dense stand of a preferred prey or host plant[35,146] is, in both instances, responding to a signal representing a suitable habitat. The distinction is further blurred by the realization that in many instances it is not known if gregarious cues are, in fact, produced by conspecifics or by a biofilm associated with the organism, in which case it becomes an associative cue (for eukaryote propagules, at any rate).

An alternative approach is to consider whether the cues are produced broadly across the habitat or are specific to particular source organisms. Similarly, one could ask whether the colonizing organisms are habitat generalists or specialists. This approach develops the theme of Johnson et al.,[106] who argue that specificity of cues may be due to the presence of host-specific vs. more generalized biofilms, which are, respectively, the source of habitat-specific or more generalized habitat cues. However, the only two characterized inducers of larval metamorphosis are derived from eukaryotes.[35,157] Thus, the argument of Johnson et al.[106] can be considered more generally, by asking whether cues are broadly distributed across a habitat and whether organisms exhibiting different breadth of settlement preferences use different cues. One possibility is that deterrents will tend to be very specific metabolites associated with specific sources, whereas inducers will be more broadly based (e.g., as suggested for peptides[34,114]). The work of the authors supports this distinction. Negative cues (furanones) from *Delisea pulchra* are unique to the genus, while the components of the inducer from this species — floridoside and isethionic (Figure 10.1) — occur broadly across the red algae. The polysaccharide cues from corallines which induce settlement of planula larvae are also thought to be broadly distributed across reef habitats as well as active against a broad range of species of corals.[144]

2. Life History of the Propagules

The range of life history characteristics among eukaryote propagules is substantial. Algal propagules and invertebrate larvae vary in their sensory capacity, length of life (from hours to months), swimming ability (from nonmotile to active swimmers), ability to absorb nutrients, and many other factors. Algal propagules tend to be short lived relative to larvae and, due to their small size and lack of swimming appendages other than flagella, may be weaker swimmers. Such properties would be expected to affect the ability of propagules to respond to chemical cues. For example, given their inability to respond actively to chemical cues, nonmotile spores (such as those from the red algae) should be more resistant to natural antifoulants than longer-lived propagules which are active swimmers. Given their constraints on active habitat selection, nonmotile propagules are also not expected to rely heavily on positive chemical cues. In a similar vein, Wahl and Mark[179] suggest that sessile epibiota are mostly habitat generalists and facultative with regards to settlement on living vs. nonliving hosts. In one of the few direct comparisons of the response of propagules with different life histories to chemical cues, Krug and Zimmer[147] found minimal differences in the responses of lecithotrophic vs. planktotrophic larvae of the sea slug *Alderia modesta* to chemical cues from their host alga. Given the role that comparisons among propagules with different life histories has played in larval biology, further comparisons of responses to signals by propagules with different life histories would be a fruitful area for further study.

IV. SIGNAL-MEDIATED BACTERIAL COLONIZATION

Effective colonization of a surface by bacteria can occur in a number of different stages and occurs at both the level of the individual cell and in multicellular populations and communities. Pragmatically, the time course of bacterial colonization from initial attachment to complex biofilm formation is also often short in comparison to the time course of settlement and metamorphosis for larvae of many invertebrates. Thus, bacterial colonization is broadly considered here to include directional swimming (e.g., the ability to approach a surface) (see also Chapter 12 in this volume), attachment, various kinds of surface motility, and biofilm formation. Chemical signals can affect all stages of this process for bacteria, although direct evidence is sparse for marine bacteria.

A. BACTERIAL CHEMOTAXIS AND SWIMMING

Standard models for bacterial chemotaxis are based on the behavior of nonmarine enteric bacteria.[196] Chemotactic behavior of nonmarine bacteria consists of discrete steps of short runs interspersed with tumbling, resulting in the random repositioning of the cells, i.e., the classical random walk. As a consequence, the net speed up a chemical gradient via the random-walk response is only a few percent of the swimming speed. The relatively slow speed and mode of chemotaxis displayed by nonmarine enteric bacteria would restrict the ability of marine bacteria to respond to chemical gradients in the sea and hence cast doubt on the importance of chemotaxis for bacteria in turbulent marine environments.

However, Mitchell et al.[185] have recently shown that both heterotrophic marine bacteria from cultures and enriched natural samples of marine bacteria can swim at speeds up to 400 μm s^{-1}. This is at least an order of magnitude faster than enteric bacteria and forces a re-evaluation of the potential role of chemotaxis as a means of approaching nutrient-rich surfaces or patches in the sea. The speed at which marine bacteria move, the pattern at which they travel, and their rapid changes in speed all imply fast reaction times for sensing and responding to nutrients in the environment and the ability to respond to nutrient gradients around micropatches even in turbulent waters.[185,186] Through rapid speed changes and reversals, marine bacteria can detect the gradient edge and maintain themselves in a micropatch, away from turbulent shear that would otherwise remove them from the nutrient patch.

These results have implications for propagules of marine eukaryotes as well as for marine bacteria. The ability of bacteria to rapidly respond to chemical signals as they approach a surface indicates that small organisms can detect and respond in an active way to chemical signals. The chemotactic response of bacteria to nutrients is also consistent with the suggestion above (Section III.C.3) that signal-mediated settlement by propagules of higher organisms could have evolved from chemotactic responses to high localized concentrations of nutrients, such as at surfaces.

B. COLONIZATION: ATTACHMENT, SURFACE MOTILITY, AND BIOFILM FORMATION

Bacteria respond to chemical signals when attaching to and moving about on surfaces and developing biofilms. For example, signal-based regulatory systems such as the AHL (acylated homoserine lactone, Figure 10.1) system are important for colonization of surfaces (including those of higher organisms) for a number of bacteria common in aqueous (though mostly nonmarine) environments.[197] One of the most compelling examples of the role of AHL regulatory systems in bacterial colonization is the formation of multicellular clusters in biofilms of *Pseudomonas aeruginosa*[198] and *Aeromonas hydrophila*.[199] Davies et al.[198] have shown that mutant strains (*lasI-*) of *P. aeruginosa* that lack the ability to produce relevant AHL form flat biofilms in which the bacteria are tightly packed. The addition of exogenous AHL restores the more complex, three-dimensional structure of wild-type biofilms. Similarly, in the best known example of AHL regulation from marine systems, the marine bacterium *Vibrio fischeri* relies on its AHL system for successful colonization of the light organ in squid (reviewed by Visick and Ruby[200]). As with *P. aeruginosa*, mutant strains which lack AHL (*luxI-*) fail to colonize the light organ.[200]

Other specific colonization traits mediated by the AHL signaling system include attachment[201] and surface motility or swarming[202] of *Serratia liquefaciens*. Swarming is apparently common in marine bacteria,[16] although the underlying regulatory control is not known. Bacteria also express extracellular products, via the AHL regulatory system, which facilitate colonization of eukaryotes. For example, exoenzymes accounting for virulence (and thus effective colonization of hosts) by the plant pathogens *Erwinia carotorora* and *E. stewartii* are not produced by mutant strains lacking the AHL system. Similarly, regulation of Ti plasmid conjugation in *Agrobacterium tumefaciens*, and, hence, virulence on host plants, depends on the presence of AHL.[203,204] Colonization by cells in beneficial bacterial–host associations can also rely on AHL-mediated gene regulation. Examples include the nitrogen-fixing symbiotic bacteria *Rhizobium leguminosarum*[203] and *Vibrio fischeri* in the light organ of the squid (above).

Indirect evidence for the importance of AHL regulation for colonization by marine bacteria comes from the observations that furanones from the red alga *Delisea pulchra* specifically interfere with AHL systems[205,206] and inhibit colonization of surfaces (attachment, swarming) by marine bacteria.[16] Involvement of AHL regulation in marine eukaryote–prokaryote interactions is further suggested by recent identification of an AHL-producing bacterium, *Vibrio campbelli*, as a common component of the culturable bacterial fraction from several species of sponges near Sydney.[207]

More generally, there exists a wide range of bacterially produced extracellular signaling molecules.[197] The role of these signals in colonization of surfaces by marine bacteria is largely unknown, although some signals are widespread across many bacteria. These include cyclic dipeptides[208] and the apparently ubiquitous, water-soluble *luxS* encoded, AI2 (AutoInducer 2) signal, first discovered in the marine bacterium *Vibrio harveyi*.[209] One function of the AI2 system may be to facilitate surface colonization by bacteria.[209]

As with eukaryote signals, the presentation and microscale distribution of bacterial signals *in situ* is largely unknown. The data that is available suggest that understanding the distribution of the compounds will be crucial for understanding colonization processes. Charlton et al.[210] measured concentrations of 3-oxo-acylated homoserine lactones (Figure 10.1) in flow cells in which they grew biofilms of *Pseudomonas aeruginosa*. Concentrations of AHL differed strikingly between the biofilm itself and the culture medium exudate, and between bacteria grown separately in bulk liquid cultures (up to one thousand-fold greater concentrations in the biofilm).

V. OTHER CHEMICALLY MEDIATED SURFACE-BASED INTERACTIONS

Many other ecological phenomena besides colonization rely on interactions at surfaces, and many of the issues raised above are also relevant to these phenomena. Allelopathy, the chemical mediation of competition, relies on the production of surface or near-surface cues to deter competitors. Although there are a number of intriguing studies of allelopathy in the literature,[211,212] presentation and surface quantification of the putative allelopathic chemical(s) are largely unknown. Determination of suitable prey by a predator may often occur as a response to an olfactory or gustatory cue present at the surface of the prey. Failure to localize and present a predator deterrent in an appropriate way may, in fact, result in accidental consumption of the prey by a predator, with consequences just as severe as purposeful consumption. The failure of many aplysiid sea hares to localize sequestered algal secondary metabolites at their surface has, in fact, resulted in Pennings and Paul[213] calling into question the utility of sea hare acquired algal defenses. The use of sex pheromones by eggs to attract sperm, well known from both macroalgae and invertebrates,[214,215] is an additional surface-based interaction that is chemically mediated, although, in this instance, the surface in question is the exterior of an egg. Although the inducers in this case are generally nonpolar hydrocarbons, they are effective at concentrations in which they are soluble in water.[214,215]

Finally, an aspect of chemical mediation of surface interactions not covered here in depth is the role that naturally produced chemicals may play as modifiers of the physical characteristics of a surface (e.g., physicochemical effects such as hydrophobicity or surface energy). For example, a number of studies have shown that colonization by some fouling organisms is reduced on artificial surfaces that have surface free energies in a particular range (typically 20 to 30 mN/m[216,217]). Vrolijk et al.,[218] in one of the few instances of an analogous study with marine organisms, showed that the surface free energy of two species of typically unfouled gorgonians corresponded to the energy minima associated with low fouling artificial surfaces. Thus, one chemical strategy for minimizing surface colonization may be to deposit or present at the surface molecules that affect the physical properties of the surface in particular ways. Difficulties in assessing the generality of this proposal in a realistic ecological context for colonization of marine organisms include (1) physical properties of surfaces tend to be rapidly modified by molecular conditioning films[219] or biofilms[220] once immersed, (2) most studies to date have been done on artificial surfaces, often in the context of the development of new antifouling technologies,[221] (3) colonization by propagules of a variety of organisms does not vary predictably with variation in surface free energy,[222,223] (4) measurements of physical characteristics are generally not done when the surface is immersed,[224] which may potentially change hydration and other parameters of the surface relative to the *in situ* condition, and (5) processing of samples (e.g., fixation) prior to analysis may also alter surface characteristics from the *in situ* condition. Vrolijk et al.[218] discuss some of the difficulties associated with measuring and interpreting physicochemical data for living marine surfaces.

VI. OVERVIEW AND CONCLUSIONS

This chapter attempted to understand whether there are common themes in the use of chemical cues to mediate colonization of surfaces by propagules. Few studies have definitively demonstrated that a characterized chemical, at a known concentration, either positively or negatively mediates colonization in the field. Nonetheless, some general patterns appear to emerge (Table 10.2).

Of obvious importance for the success of any cue is that it persists at effective concentrations long enough for the receiving organisms to respond. Where the production of the cue represents a potential cost to the producing organism, there should be selection to achieve this as efficiently as

TABLE 10.2
Comparison between Chemical Deterrents of Colonization vs. Inducers as Proposed in This Chapter

	Deterrents	Inducers
Chemistry	Nonpolar secondary metabolites	Primary metabolites (carbohydrates, peptides) Most water soluble
Taxonomic breadth of production	Narrow	Broad
Quantities produced	Small	Large
Natural selection strongest on	Signal producer	Signal receiver
Importance of hydrodynamic factors for the interaction	Low	High

possible. For deterrents, this argues for deployment of nonpolar metabolites which will adsorb to hydropobic (cell) surfaces of producing organisms with a minimal rate of dissolution into the water column. This may be particularly important for highly pliable organisms such as seaweeds or many benthic invertebrates, for which the establishment of persistent surface boundary layers may be rare.

The ecological context for inducers is different from that of deterrents. First, selection on the receiving organism should be more stringent for positive cues, as the inability to detect a signal may result in a failure to successfully recruit. To maximize the probability of reception by a settling propagule, an inducer should be present in high concentrations and persist in the habitat. These authors suggest that water-soluble primary metabolites, from either benthic eukaryotes or biofilms, best fit these criteria.

The importance of hydrodynamic factors should also differ for inducers vs. deterrents. Hydrodynamic events may greatly reduce the efficacy of inducers,[29] rendering them undetectable by propagules. In contrast, hydrophobic deterrents only need to act right at the point of contact between the settling propagule and the surface of the host. Any propagules that miss the target due to flow or turbulence are irrelevant to the host and its defenses.

Too little is known in a realistic ecological context about the role of chemical signals as mediators of colonization by marine bacteria to generally compare these processes between bacteria and eukaryotes. For at least some (not necessarily marine) bacteria, signals mediate several stages of colonization, including attachment, surface motility, and biofilm formation. In many instances (e.g., between bacterial cells in a biofilm), these signals may be operating over the scale of micrometers, in contrast to the hundreds of micrometers to centimeters that may be more typical of invertebrate larvae.

Finally, new methodologies are likely to lead the way in our exploration of chemically mediated interactions at surfaces. New and improved methods for detection and quantification of signals *in situ* are needed, as are development of *in situ* bioassays and enhanced molecular methods for characterizing bacterial communities.

ACKNOWLEDGMENTS

The authors thank Susan Sennet for providing references on localization of metabolites in sponges, Julie Partridge and Sophia McCloy for their help with preparation of the manuscript, and Jim Mitchell for enlightening us on the topic of chemotaxis in marine bacteria. A special thanks goes to the students and staff of the Centre for Marine Biofouling and Bio-Innovation, whose research provided the seed (and much of the data) for this review.

REFERENCES

1. Hay, M.E., Marine chemical ecology: what's known and what's next?, *J. Exp. Mar. Biol. Ecol.*, 200, 103, 1996.
2. Hay, M.E., The ecology and evolution of seaweed-herbivore interactions on coral reefs, *Proc. 8th Int. Coral Reef Symp.*, 1, 23, 1997.
3. Hay, M.E. and Steinberg, P.D., The chemical ecology of plant–herbivore interactions in marine versus terrestrial communities, in *Herbivores: Their Interaction with Secondary Plant Metabolites: Evolutionary and Ecological Processes*, Vol. 2, Rosenthal G.A. and Berembaum, M., Eds., Academic Press, San Diego, CA, 1992, 372.
4. McClintock, J.B. and Baker, B.J., Palatability and chemical defense of eggs, embryos and larvae of shallow-water Antarctic marine invertebrates, *Mar. Ecol. Prog. Ser.*, 154, 121, 1997.
5. Paul, V.J., Ed., *Ecological Roles of Marine Natural Products*, Comstock Publishing Associates, Ithaca, NY, 1992.
6. Pawlik, J.R., Marine invertebrate chemical defenses, *Chem. Rev.*, 93, 1911, 1993.
7. Cronin, G. and Hay, M.E., Amphipod grazing and induction of seaweed chemical defenses, *Ecology*, 77, 2287, 1996.
8. Van Alstyne, K.L., Herbivore grazing increases polyphenolic defenses in the intertidal brown alga *Fucus distichus*, *Ecology*, 69, 655, 1988.
9. Bolser, R.C. and Hay, M.E., Are tropical plants better defended? Palatability and defenses of temperate versus tropical seaweeds, *Ecology*, 77, 2269, 1996.
10. Steinberg, P.D., Estes, J.A., and Winter, F.C., Evolutionary consequences of food chain length in kelp forest communities, *Proc. Nat. Acad. Sci. USA*, 92, 8145, 1995.
11. Van Alstyne, K.L., Dethier, M.N., Duggins, D., Spatial patterns in macroalgal chemical defenses, in *Marine Chemical Ecology*, McClintock, J.B. and Baker, B.J., Eds., CRC Press, Boca Raton, FL, 2001, chap. 8.
12. Pawlik, J.R., Chemical ecology of the settlement of benthic marine invertebrates, *Oceanogr. Mar. Biol. Annu. Rev.*, 30, 273, 1992.
13. Slattery, M., Chemical cues in marine invertebrate larval settlement, in *Marine Woodboring and Fouling Organisms of the Indian Ocean: A Review*, Naghabushanum, R. and Thompson, J.F., Eds., Oxford and IBH Publishing Co., New Delhi, 1997, 135.
14. Hadfield M.G. and Paul, V.J., Natural chemical cues for settlement and metamorphosis of marine–invertebrate larvae, in *Marine Chemical Ecology*, McClintock, J.B. and Baker, B.J., Eds., CRC Press, Boca Raton, FL, 2001, chap. 13.
15. Dworjanyn, S.A., de Nys, R., and Steinberg, P.D., Localisation and surface quantification of secondary metabolites in the red alga *Delisea pulchra*, *Mar. Biol.*, 133, 727, 1999.
16. Maximilien, R., de Nys, R., Holmström, C., Gram, L., Givskov, M., Crass, K., Kjelleberg, S., and Steinberg, P.D., Chemical mediation of bacterial surface colonization by secondary metabolites from the red alga *Delisea pulchra*, *Aquat. Micro. Ecol.*, 15, 233, 1998.
17. Schmitt, T., Hay, M., and Lindquist, N., Antifouling and herbivore deterrent roles of seaweed secondary metabolites: constraints on chemically-mediated coevolution, *Ecology*, 76, 107, 1995.
18. Gosselin, L.A. and Qian, P.Y., Juvenile mortality in benthic marine invertebrates, *Mar. Ecol. Prog. Ser.*, 146, 265, 1997.
19. John, D.M., Hawkins, S.J., and Price, J.H., Plant–animal interactions in the marine benthos, in *The Systematics Association*, Special Vol. 46, Oxford Scientific Publications, Oxford, UK, 1992.
20. Sih, A., Crowley, P., McPeek, M., Petranka, J., and Strohmeier A., Predation, competition, and prey communities: a review of field experiments, *Annu. Rev. Ecol. Syst.*, 16, 269, 1985.
21. Paine, R.T., Food web complexity and species diversity, *Am. Nat.*, 100, 65, 1966.
22. Cimino, G. and Ghiselin, M., this volume.
23. d'Antonio, C., Epiphytes on the rocky intertidal red alga *Rhodomela larix* (Turner) C Agardh: negative effects on the host and food for herbivores?, *J. Exp. Mar. Biol. Ecol.*, 86, 197, 1985.
24. Dixon, J., Schroeter, S., and Kastandiek, J., Effects of the encrusting bryozoan, *Membranipora membranacea*, on the loss of blades and fronds of the giant kelp *Macrocystis pyrifera*, *J. Phycol.*, 7, 341, 1981.

25. Kain, J.M., The biology of *Laminaria hyperorea*. VII. Reproduction of the sporophyte, *J. Mar. Biol. Assoc. UK*, 55, 567, 1975.

26. Sand-Jensen, K., Effect of epiphytes on eelgrass photosynthesis, *Aq. Bot.*, 3, 55, 1977.

27. Wahl, M. and Hay, M.E., Associational resistance and shared doom: effects of epibiosis on fouling, *Oecologia*, 102, 329, 1995.

28. de Nys, R. and Steinberg, P.D., Role of secondary metabolites from algae and seagrasses in biofouling control, in *Recent Advances in Marine Biotechnology, Vol III.*, Fingerman, M., Nagabhushanam, R., and Thompson, M.F., Eds., Science Publishers Inc., Enfield, NH, 1999, 223.

29. Butman, C.A., Larval settlement of soft-sediment invertebrates: the spatial scales of pattern explained by active habitat selection and the emerging role of hydrodynamic processes, *Oceanogr. Mar. Biol. Ann. Rev.*, 25, 113, 1987.

30. Harvey, M. and Bourget, E., Recruitment of marine invertebrates onto arborescent epibenthic structures: active and passive processes acting at different spatial scales, *Mar. Ecol. Prog. Ser.*, 153, 203, 1997.

31. Hay, M.E., Stachowicz, J.J., Cruz-Rivera, E., Bullard, S., Deal, M.S., and Lindquist, N., Bioassays with marine and freshwater macroorganisms, in *Methods in Chemical Ecology, Vol 2., Bioassay Methods*, Haynes, K.F. and Hillar, J.G., Eds., Chapman & Hall, New York, 1998, 39.

32. Wright, A.E., Isolation of marine natural products, in *Methods in Biotechnology, Vol. 4: Natural Products Isolation*, Cannell, R.J.P., Ed., Humana Press, Inc., Totowa, N.J., 1993, 365.

33. Rittschof, D., Body odors and neutral-basic peptide mimics: a review of responses by marine organisms, *Am. Zool.*, 33, 487, 1993.

34. Decho, A.W., Browne, K.A., and Zimmer-Faust, R.K., Chemical cues: why basic peptides are signal molecules in marine environments, *Limnol. Oceanogr.*, 43, 1410, 1998.

35. Williamson, J.E., de Nys, R., Kumar, N., Carson, D.G., and Steinberg, P.D., Induction of metamorphosis in the sea urchin *Holopneustes purpurascens* by a metabolite complex from the algal host *Delisea pulchra*, *Biol. Bull.*, 198, 332, 2000.

36. Rosenthal, G.A. and Berenbaum, M.R., *Herbivores: Their Interaction with Plant Secondary Metabolites*, Academic Press, New York, 1992.

37. Sarojini, R., Nagabhushanam, R., and Fingerman, M., Induction of larval settlement and metamorphosis by neuroactive compounds in marine invertebrates, in *Recent Advances in Marine Biotechnology, Vol III.*, Fingerman, M., Nagabhushanam, R., and Thompson, M.F., Eds., Science Publishers, Inc., Enfield, NH, 1999, 203.

38. Davis, A.R. and Bremner, J.B., Potential antifouling natural products from ascidians: a review, in *Recent Advances in Marine Biotechnology, Vol III*, Fingerman, M., Nagabhushanam, R., and Thompson, M.F., Eds., Science Publishers, Inc., New Hampshire, 1999, 259.

39. Fusetani, N., Marine natural products influencing larval settlement and metamorphosis of benthic invertebrates, *Curr. Org. Chem.*, 1, 127, 1997.

40. Clare, A.S. and Matsumura, K., Nature and perception of barnacle settlement pheromones, *Biofouling*, 15, 57, 2000.

41. Santelices, B., Patterns of reproduction, dispersal and recruitment in seaweeds, *Oceanogr. Mar. Biol. Annu. Rev.*, 28, 177, 1990.

42. Wahl, M., Marine epibiosis I. Fouling and antifouling: some basic aspects, *Mar. Ecol. Prog. Ser.*, 58, 175, 1989.

43. Walker, F. T. and Smith, M., Seaweed culture, *Nature*, 162, 31, 1948.

44. de Nys, R., Wright A.D., Konig, G.M. and Sticher, O., New halogenated furanones from the marine alga *Delisea pulchra* (cf. *fimbriata*), *Tetrahedron*, 49, 11213, 1993.

45. de Nys, R., Steinberg, P.D., Rogers, C.N., Charlton, T.C., and Duncan, M.W., Quantitative variation of secondary metabolites in the sea hare *Aplysia parvula* and its host plant, *Delisea pulchra*, *Mar. Ecol. Prog. Ser.*, 130, 135, 1996.

46. de Nys, R., Dworjanyn, S.A., and Steinberg, P.D., A new method for determining surface concentrations of marine products on seaweeds, *Mar. Ecol. Progr. Ser.*, 162, 79, 1998.

47. de Nys, R, Steinberg, P.D., Willemsen, P., Dworjanyn, S.A., Gabalish, C.L., and King, R.J., Broad spectrum effects of secondary metabolites from the red alga *Delisea pulchra* in antifouling assays, *Biofouling*, 8, 159, 1995.

48. Dworjanyn, S.A., Chemically mediated antifouling and the cost of producing secondary metabolites in seaweeds, Ph.D. thesis, University of New South Wales, Sydney, Australia, 2000.

49. Hay, M.E., Fenical, W., and Gustafson, K., Chemical defence against diverse coral-reef herbivores, *Ecology*, 68, 1581, 1987.

50. Paul, V.J., Wylie, C.R., and Sanger, H.R., Effects of algal chemical defenses toward different coral-reef herbivorous fishes: a preliminary study, *Proc. 6th Int. Symp. Coral Reef*, 3, 73, 1988.

51. Steinberg, P.D., de Nys, R., and Kjelleberg, S., Chemical inhibition of epibiota by Australian seaweeds, *Biofouling*, 12, 227, 1998.

52. Young, D.N., Howard, B.M., and Fenical, W., Subcellular localization of brominated secondary metabolites in the red algae *Laurencia snydedae*, *Phycologia*, 16, 182, 1980.

53. Conover, J.T. and Sieburth, J., Effects of *Sargassum* distribution on its epibiota and antibacterial activity, *Botanica Marina*, 6, 147, 1964.

54. Jennings, J.G. and Steinberg, P.D., The *in situ* exudation of phlorotannins from the sublittoral kelp *Ecklonia radiata*, *Mar. Biol.*, 121, 349, 1994.

55. Carlson, D.J. and Carlson, M.J., Reassessment of exudation by fucoid macroalgae, *Limnol. Oceanogr.*, 29, 1077, 1984.

56. Ragan, M.A. and Glombitza, K.W., Phlorotannins, brown algal polyphenols, *Prog. Phycol. Res.*, 4, 129, 1986.

57. Jennings, J.G. and Steinberg, P.D., Phlorotannins versus other factors affecting epiphyte abundance on the kelp *Ecklonia radiata*, *Oecologia*, 109, 461, 1997.

58. Langlois, G., Effect of algal exudates on substratum selection by the motile marine telotroch *Vorticella marine*, *J. Protozool.*, 221, 115, 1975.

59. Ragan, M.A. and Jensen, A., Quantitative studies on brown algal phenols. III. Light-mediated exudation of polyphenols from *Ascophyllum nodosum* (L.) Le Jol, *J. Exp. Mar. Biol. Ecol.*, 36, 91, 1979.

60. Lau, S.C.K. and Qian, P.Y., Phlorotannins and related compounds as larval settlement inhibitors of the tube-building polychaete *Hydroides elegans*, *Mar. Ecol. Prog. Ser.*, 159, 219, 1997.

61. Conover, J.T. and Sieburth, J., Effects of tannins excreted from Phaeophyta on planktonic animal survival in tide pools, *Proc. 5th Int. Seaweed Symp.*, 99, 1966.

62. Fahn, A., Secretory tissues in vascular plants, *New Phytol.*, 108, 229, 1988.

63. de Nys, R., Leya, T., Maximilien, R., Asfar, A., Nair, P.S.R., and Steinberg P.D., The need for standardised broad scale bioassay testing: a case study using the red algae *Laurencia rigida*, *Biofouling*, 10, 213, 1996.

64. de Nys, R., Dworjanyn, S., Ison, O., and Steinberg, P., unpublished.

65. Weinberger, F., Friedlander, M., and Hoppe, H.G., Oligoagars elicit a physiological response in *Gracilaria conferta* (Rhodophyta), *J. Phycol.*, 35, 747, 1999.

66. Collén, J., Del Rio, M.J., Garcia-Reina, G., and Pedersen, M., Photosynthetic production of hydrogen peroxide by *Ulva rigida* C. Ag. (Chlorophyta), *Planta*, 196, 225, 1995.

67. Boettcher, A.A. and Targett, N.M., Induction of metamorphosis in queen conch, *Strombus gigas* Linnaeus, larvae by cues associated with red algae from their nursery grounds, *J. Exp. Mar. Biol. Ecol.*, 196, 29, 1996.

68. Clare, A.S., Marine natural product antifoulants: status and potential, *Biofouling*, 9, 211, 1996.

69. Davis, A.R., Targett, N.M., McConell, O.J., and Young, C.M., Epibiosis of marine algae and benthic invertebrates: natural products chemistry and other mechanisms inhibiting settlement and growth, in *Bioorganic Marine Chemistry*, Scheuer, P.J., Ed., Springer-Verlag, Berlin, 1989, 85.

70. Abarzua, S., Kacan, S., and Fuchs P., Status and potential of natural product antifoulants, in *Recent Advances in Marine Biotechnology, Vol. III*, Fingerman, M., Nagabhushanam, R., and Thompson, M.F., Eds., Science Publishers, Inc., Enfield, NH, 1999, 37.

71. Davis, A.R., Alkaloids and ascidian chemical defense: evidence for the ecological role of natural products from *Eudistoma olivaceum*, *Mar. Biol.*, 111, 375, 1991.

72. Stoecker, D., Resistance of a tunicate to fouling, *Biol. Bull.*, 155, 615, 1978.

73. Jensen, P.R., Jenkins, K.M., Porter, D., and Fenical, W., Evidence that a new antibiotic flavone glycoside chemically defends the sea grass *Thalassia testudinum* against zoosporic fungi, *Appl. Environ. Mic.*, 64, 1490, 1998.

74. Todd, J.S., Zimmerman, R.C., Crews, P., and Randall, S.A., The antifouling activity of natural and synthetic phenolic acid sulfate esters, *Phytochemistry*, 34, 401, 1993.

75. Thompson, J.E., Barrow, K.D., and Faulkner, D.J., Localization of two brominated metabolites, aerothionin and homoaerothionin, in spherulous cells of the marine sponge *Aplysina fistularis*, *Acta Zool.*, 64, 199, 1984.

76. Thompson, J.E., Exudation of biologically-active metabolites in the sponge *Aplysina fistularis*, I. Biological evidence, *Mar. Biol.*, 88, 23, 1985.

77. Walker, R.P., Thompson, J.E., and Faulkner, D.J., Exudation of biologically-active metabolites in the sponge *Aplysina fistularis* II. Chemical evidence, *Mar. Biol.*, 88, 27, 1985.

78. Thompson, J.E., Walker, R.P., and Faulkner, D.J., Screening and bioassays for biologically-active substances from forty marine sponges species from San Diego, California, USA, *Mar. Biol.*, 88, 11, 1985.

79. Woodin, S.A., Marinelli, R.L., and Lincoln, D.E., Allelochemical inhibition of recruitment in a sedimentary assemblage, *J. Chem. Ecol.*, 19, 517, 1993.

80. Woodin, S.A., Lindsay, S.M., and Lincoln, D.E., Biogenic bromophenols as negative recruitment cues, *Mar. Ecol. Prog. Ser.*, 157, 303, 1997.

81. Fielman, K.T., Woodin, S.A., Walla, M.D., and Lincoln, D.E., Widespread occurrence of natural halogenated organics among temperate marine infauna, *Mar. Ecol. Prog. Ser.*, 181, 1, 1999.

82. Steward, C.C., Pinckney, J., Piceno, Y., and Lovell, C.R., Bacterial numbers and activity, microalgal biomass and productivity, and meiofaunal distribution in sediments naturally contaminated with biogenic bromophenols, *Mar. Ecol. Prog. Ser.*, 90, 61, 1992.

83. Becerro, M.A., Uri, M.T., and Twan, X., Chemically-mediated interactions in benthic organisms — the chemical ecology of *Crambe crambe* (Porifera Becilosclerida), *Hydrobiologia*, 355, 77, 1997.

84. Uriz, M.J., Turon, X., Galera, J., and Tur, J.M., New light on the cell location of avarol within the sponge *Dysidea avara* (Dendroceratida), *Cell Tissue Res.*, 285, 519, 1996.

85. Marin, A., Lopez, M.D., Esteban, M.A., Meseguer, J., Munoz, J., and Fontana, A., Anatomical and ultrastructural studies of chemical defence in the sponge *Dysidea fragilis*, *Mar. Biol.*, 131, 639, 1998.

86. Uriz, M.J., Becerro, M.A., Tur, J.M., and Turon, X., Location of toxicity within the Mediterranean sponge *Crambe crambe* (Demospongiae:Poecilosclerida), *Mar. Biol.*, 124, 583, 1996.

87. Flowers, A.E., Garson, M.J., Webb, R.I., Dumdei, E.J., and Charan, R.D., Cellular origin of chlorinated diketopiperazines in the dictyoceratid sponge *Dysidea herbacea* (Keller), *Cell Tissue Res*, 292, 597, 1998.

88. Unson, M.D., Holland, N.D., and Faulkner, D.J., A brominated secondary metabolite synthesized by the cyanobacterial symbiont of a marine sponge and accumulation of the crystalline metabolite in the sponge tissue, *Mar. Biol.*, 119, 1, 1994.

89. Bewley, C.A., Holland, N.D., and Faulkner, D.J., Two classes of metabolites from *Theonella swinhoei* are localized in distinct populations of bacterial symbionts, *Experientia*, 52, 716, 1996.

90. Schmidt, E.W., Bewley C.A., and Faulkner, D.J., Theopalauamide, a bicyclic glycopeptide from filamentous bacterial symbionts of the lithistid sponge *Theonella Swinhoei* from Palau and Mozambique, *J. Org. Chem.*, 63, 1254, 1998.

91. Walls, J.T., Blackman, A.J., and Ritz, D.A., Localisation of the amathamide alkaloids in surface bacteria of *Amathia wilsoni* Kirkpatrick, 1888 (Bryozoa: Ctenostomata), *Hydrobiologia* 297, 163, 1995.

92. Davidson, S.K., Allen, S.W., Lim, G.E., Anderson, C., and Haygood, M.G., Evidence that the bacterial symbiont *Candidatus endobugula sertula* plays a role in bryostatin biosynthesis in the bryozoan *Bugula neritina*, Abstract AMS Meeting, 2000.

93. Holmström, C. and Kjelleberg, S., Factors influencing the settlement of macrofoulers, in *Recent Advances in Marine Biotechnology, Vol. III*, Fingerman, M., Nagabhushanam, R., and Thompson, M.F., Eds., Science Publishers, Inc., Enfield, NH, 1999, 173.

94. Wieczorek, S.K. and Todd, C.D., Inhibition and facilitation of settlement of epifaunal marine invertebrate larvae by microbial biofilm cues, *Biofouling*, 12, 81, 1998.

95. Unabia, C.R.C. and Hadfield, M.G., Role of bacteria in larval settlement and metamorphosis of the polychaete *Hydroides elegans*, *Mar. Biol.*, 133, 55, 1999.

96. Wieczorek, S.K. and Todd, C.D., Inhibition and facilitation of bryozoan and ascidian settlement by natural multi-species biofilms: effects of film age and the roles of active and passive larval settlement, *Mar. Biol.*, 128, 463, 1997.

97. Wieczorek, S.K., Clare, A.S., and Todd, C.D., Inhibitory and facilitatory effects of microbial films on settlement of *Balanus amphitrite amphitrite* larvae, *Mar. Ecol. Prog. Ser.*, 119, 221, 1995.

98. Maki, J.S., Rittschof, D., Samuelsson, M.O., Szewzyk, U., Yule, A.B., Kjelleberg, S., Costlow, J.D., and Mitchell, R., Effect of marine bacteria and their exopolymers on the attachment of barnacle cypris larvae, *Bull. Mar. Sci.*, 46, 499, 1990.

99. Keough, M.J. and Raimondi, P.T., Responses of settling invertebrate larvae to bioorganic films: effects of different types of films, *J. Exp. Mar. Biol. Ecol.*, 185, 235, 1995.

100. Maki, J.S., Yule, A.B., Rittschof, D., and Mitchell, R., The effect of bacterial films on the temporary adhesion and permanent fixation of cypris larvai, *Balanus amphitrite* Darwin, *Biofouling*, 8, 121, 1994.

101. Holmström, C., Rittschoff, D., and Kjelleberg, S., Inhibition of settlement by larvae of *Balanus amphitrite* and *Ciona intestinalis* by a surface-colonizing marine bacterium, *Appl. Environ. Microbiol.*, 58, 2111, 1992.

102. Holmström, C., James, S., Egan, S., and Kjelleberg, S., Inhibition of common fouling organisms by marine bacterial isolates with special reference to the role of pigmented bacteria, *Biofouling*, 10, 251, 1996.

103. Kon-ya, K., Shimidzu, N., Otaki, N., Yokoyama, A., Adachi, K., and Miki, W., Inhibitory effect of bacterial ubiquinones on the settling of barnacle, *Balanus amphitrite*, *Experientia*, 51, 153-155, 1995.

104. Holmström, C. and Kjelleberg, S., Marine *Pseudoalteromonas* species are associated with higher organisms and produce biologically active extracellular agents, *FEMS Microbiol. Ecol.*, 30, 285, 1999.

105. James, S., Holmström, C., and Kjelleberg, S., Purification and characterization of a novel antibacterial protein from the marine bacterium D2, *Appl. Environ. Microbiol.*, 62, 2783, 1996.

106. Johnson, C.R., Lewis, R.E., Nichols, D.S., and Degnan, B.M., Bacterial induction of settlement and metamorphosis in marine invertebrates, *Proc. 8th Int. Coral Reef Symp.*, 2, 1219, 1997.

107. Dahllöf, I., Baillie, H., and Kjelleberg, S., *rpoB*-based microbial community analysis avoids limitations inherent to 16S rDNA intraspecies heterogeneity, *Appl. Environ. Microb.*, 66, 3376, 2000.

108. Freidrich, A.B., Merkert, H., Fendert, T., Hacker, J., Proksch, P., and Hentschel, U., Microbial diversity in the marine sponge *Aplysina cavernicola* (formerly *Verongia cavernicola*) analyzed by fluorescence in situ hybridization (FISH), *Mar. Biol.*, 134, 461, 1999.

109. Hendrikson A.A. and Pawlik J.R., A new antifouling method: results from field experiments using extracts of four marine organisms, *J. Exp. Mar. Biol. Ecol.*, 194, 157, 1995.

110. Vasishtha, M., Sundberg, D.C., and Rittschof, D., Evaluation of release rates and control of biofouling using monolithic coatings containing an isothiazolone, *Biofouling*, 9, 1, 1995.

111. Holmström, C. and Kjelleberg, S., Bacterial interactions with marine fouling organisms, *Biofilms: Recent Advances in their Study*, Evans, L.V., Ed., Harwood Academic Publishers, Amsterdam, 101, 2000.

112. Callow, M.E., Callow, J.A., Pickett-Heaps, J.D., and Wetherbee, R., Primary adhesion of *Enteromorpha* (Chlorophyta, ulvales) propagules: quantitative settlement studies and video microscopy, *J. Phycol.*, 33, 938, 1997.

113. Crisp, D.J., Factors influencing the settlement of marine invertebrate larvae, in *Chemoreception in Marine Organisms*, Grant, P.T. and Mackie, A.M., Eds., Academic Press, New York, 1974, 177.

114. Rittschof, D., Peptide-mediated behaviors in marine organisms: Evidence for a common theme, *J. Chem. Ecol.*, 16, 261, 1990.

115. Zimmer-Faust, R.K. and Tamburri, M.N., Chemical identity and ecological implications of a water-borne, larval settlement cue, *Limnol. Oceanogr.*, 39, 1075, 1994.

116. Tamburri, M.N., Zimmer-Faust, R.K., and Tamplin, M.L., Natural sources and properties of chemical inducers mediating settlement of oyster larvae: a re-examination, *Biol. Bull.*, 183, 327, 1992.

117. Turner, E.J., Zimmer-Faust, R.K., Palmer, M.A., Luckenbach, M., and Pentcheff, N.D., Settlement of oyster (*Crassostrea virginica*) larvae: effects of water flow and a water-soluble chemical cue, *Limnol. Oceanogr.*, 39, 1579, 1994.

118. Tamburri, M.N., Finelli, C.M., Wethey, D.S., and Zimmer-Faust, R.K., Chemical induction of larval settlement behavior in flow, *Biol. Bull.*, 191, 367, 1996.

119. Fitt, W.K., Coon, S.L., Walch, M., Weiner, R.M., Colwell, R.R., and Bonar, D.B., Settlement behavior and metamorphosis of oyster larvae (*Crassostrea gigas*) in response to bacterial supernatants, *Mar. Biol.*, 106, 389, 1990.

120. Matsumura, K., Mori, S., Nagano, M., and Fusetani, N., Lentil lectin inhibits adult extract-induced settlement of the barnacle *Balanus amphitrite*, *J. Exp. Zool.*, 280, 213, 1998.

121. Matsumura, K., Nagano, M., and Fusetani, N., Purification of a larval settlement-inducing protein complex (SIPC) of the barnacle *Balanus amphitrite*, *J. Exp. Zool.*, 281, 12, 1998.

122. Yule, A.B. and Walker, G., Settlement of *Balanus balanoides*: the effect of cyprid antennular secretion, *J. Mar. Biol. Assoc.* UK, 65, 707, 1985.

123. Clare, A.S., Freet, R.K., and McClary, M.J., On the antennular secretion of the cyprid of *Balanus amphitrite amphitrite*, and its role as a settlment pheromone, *J. Mar. Biol. Assoc. UK*, 74, 243, 1994.

124. Rittschof, D., Branscomb, E.S., and Costlow, J.D., Settlement of and behaviour in relation to flow and surface in larval barnacles, *Balanus amphitrite* Darwin, *J. Exp. Mar. Biol. Ecol.*, 82, 131, 1984.

125. Tegtmeyer, K. and Rittschof, D., Synthetic peptide analogs to barnacle settlement pheromone, *Peptides*, 9, 1403, 1989.

126. Harrison, P., Barnacle cyprid behaviour, anatomy and neurophysiology, Ph.D. thesis, University of New South Wales, Sydney, Australia, 1998.

127. Morse, A.N.C. and Morse, D.E., Recruitment and metamorphosis of *Haliotis* larvae induced by molecules uniquely available at the surfaces of crustose red algae, *J. Exp. Mar. Biol. Ecol.*, 75, 191, 1984.

128. Morse, A.N.C., Froyd, C.A., and Morse, D.E., Molecules from cyanobacteria and red algae that induce larval settlement and metamorphosis in the mollusc *Haliotis rufescens*, *Mar. Biol.*, 81, 293, 1984.

129. Morse, D.E., Hooker, N., Duncan, H., and Jensen, L., γ- aminobutyric acid, a neurotransmitter, induces planktonic abalone larvae to settle and begin metamorphosis, *Science*, 204, 407, 1979.

130. Mountford, D.O. and Pybus, V., Regulatory influences on the production of gamma-aminobutyric acid by a marine *Pseudomonas*, *Appl. Environ. Microbiol.*, 58, 237, 1992.

131. Mountford, D.O. and Pybus, V., Effect of pH, temperature and salinity on the production of gamma aminobutyric acid (GAMA) from amines by marine bacteria, *FEMS Microbiol. Ecol.*, 101, 237, 1992.

132. Daume S., Brand-Gardner, S., and Woelkerling, W.J., Preferential settlement of abalone larvae: diatom films vs. non-geniculate coralline red algae, *Aquaculture*, 174, 243, 1999.

133. Fleck, J. and Fitt, W.K., Degrading mangrove leaves of *Rhizophora mangle* provide a natural metamorphic cue for the upside down jellyfish *Cassiopea xamachana*, *J. Exp. Mar. Biol. Ecol.*, 234, 83, 1999.

134. Fleck, J., Fitt, W.K., and Hahn, M.G., A proline-rich peptide originating from decomposing mangrove leaves is one natural metamorphic cue of the tropical jellyfish *Cassiopea xamachana*, *Mar. Ecol. Prog. Ser.*, 183, 115, 1999.

135. Highsmith, R.C., Induced settlement and metamorphosis of sand dollar (*Dendraster excentricus*) larvae in predator-free sites: adult sand dollar beds, *Ecology*, 63, 329, 1982.

136. Burke, R.D., Phermonal control of metamorphosis in the pacific sand dollar, *Dendraster excentricus*, *Science*, 225, 440, 1984.

137. Burke, R.D., Pheromones and the gregarious settlement of marine invertebrate larvae, *Bull. Mar. Sci.*, 39, 323, 1986.

138. Lambert, W.J. and Todd, C.D., Evidence for a water-borne cue indicating metamorphosis in the dorid nudibranch mollusc *Adalaria proxima* (Gastropoda: Nudibranchia), *Mar. Biol.*, 120, 264, 1994.

139. Lambert, W.J., Todd, C.D., and Hardege, J.D., Partial characterization and biological activity of a metamorphic inducer of the dorid nudibranch *Adalaria proxima* (Gastropoda: Nudibranchia), *Invertebrate Biol.*, 116, 71, 1997.

140. Jensen, R.A. and Morse, D.E., The bioadhesive of *Phragmatopoma californica* tubes: a silk-like cement containing L-DOPA, *J. Comp. Physiol. B*, 158, 317, 1988.

141. Karsten, U., Barrow, K.D., and King, R.J., Floridoside, 1-iso-floridoside, and d-isofloridoside in the red alga *Porphyra columbina*, *Plant Physiol.*, 103, 485, 1993.

142. Morse, D.E., Hooker, N., Morse, A.N.C., and Jensen, R.A., Control of larval metamorphosis and recruitment in sympatric agariciid corals, *J. Exp. Mar. Biol. Ecol.*, 116, 193, 1988.

143. Morse, D.E. and Morse, A.N.C., Enzymatic characterization of the morphogen recognized by *Agaricia humilis* (scleractinian coral) larvae, *Biol. Bull.*, 181, 104, 1991.

144. Morse, A.N.C., Iwao, K., Baba, M., Shimoike, K., Hayashibara, T., and Omori, M., An ancient chemosensory mechanism brings new life to coral reefs, *Biol. Bull.*, 191, 149, 1996.

145. Morse, D.E., Morse, A.N.C., Raimondi, P.T., and Hooker, N., Morphogen-based chemical flypaper for *Agaricia humilis* coral larvae, *Biol. Bull.*, 186, 172, 1994.

146. Krug, P.J. and Manzi, A.E., Waterborne and surface-associated carbohydrates as settlement cues for larvae of the specialists marine herbivore *Alderia modesta*, *Biol. Bull.*, 197, 94, 1999.

147. Krug, P.J. and Zimmer, R.K., Development dimorphism and expression of chemosensory-mediated behavior: Habitat selection by a specialist marine herbivore, *J. Exp. Biol.*, 203, 1741, 2000.

148. Bahamondes-Rojas, I. and Dherbomez, M., Purification partielle de substances glycoconjuguees capables d'induire la metamorphose des larves competentes de'*Eubranchus doriae* (Trinchese, 1879), mollusque nudibranche, *J. Exp. Mar. Biol. Ecol.*, 144, 17, 1990.

149. Forward, R.B., Jr., Tankersley, R.A., Blondel, D., and Rittschof, D., Metamorphosis of the blue crab *Callinectes sapidus*: effects of humic acids and ammonium, *Mar. Ecol. Prog. Ser.*, 157, 277, 1997.

150. Jensen, R.A. and Morse, D.E., Intraspecific facilitation of larval recruitment: gregarious settlement of the polychaete *Phragmatopoma californica* (Fewkes), *J. Exp. Mar. Biol. Ecol.*, 83, 107, 1984.

151. Pawlik, J.R., Chemical induction of larval settlement and metamorphosis in the reef-building tube worm *Phragmatopoma californica* (Polychaeta: Sabellariidae), *Mar. Biol.*, 91, 59, 1986.

152. Pawlik, J.R., Natural and artificial induction of metamorphosis of *Phragmatopoma lapidosa californica* (Polychaeta: Sabellariidae), with a critical look at the effects of bioactive compounds on marine invertebrate larvae, *Bull. Mar. Sci.*, 46, 512, 1990.

153. Pawlik, J.R. and Faulkner, D.J., Specific free fatty acids induce larval settlement and metamorphosis of the reef-building tube worm *Phragmatopoma californica* (Fewkes), *J. Exp. Mar. Biol. Ecol.*, 102, 301, 1986.

154. Jensen, R.A. and Morse, D.E., Chemically induced metamorphosis of polychaete larvae in both the laboratory and ocean environment, *J. Chem. Ecol.*, 16, 911, 1990.

155. Jensen, R.A., Morse, D.E., Petty, R.L., and Hooker, N., Artificial induction of larval metamorphosis by free fatty acids, *Mar. Ecol. Prog. Ser.*, 67, 55, 1990.

156. Kitamura, H., Kitahara, S., and Koh, H.B., The induction of larval settlement and metamorphosis of two sea urchins, *Pseudocentrotus depressus* and *Anthocidaris crassispina*, by free fatty acids extracted from the coralline red alga *Corallina pilulifera*, *Mar. Biol.*, 115, 387, 1993.

157. Tsukamoto, S., Kato, H., Hirota, H., and Fusetani, N., Lumichrome: a larval metamorphosis-inducing substance in the ascidian *Halocynthia roretzi*, *Eur. J. Biochem.*, 264, 785, 1999.

158. Yvin, J.C., Chevolot, L., Chevolot-Magueur, A.M., and Cochard, J.C., First isolation of jacaranone from an alga, *Delesseria sanguinea*. A metamorphosis inducer of *Pecten* larvae, *J. Nat. Prod.*, 48, 814, 1985.

159. Kato, T., Kumanireng, A.A., Ichinose, I., Kitahara, Y., Kakinuma, Y., Nishihara, M., and Kato, M., Active components of *Sargassum tortile* effecting the settlement of swimming larvae of *Coryne uchidai*, *Experientia*, 31, 433, 1975.

160. Chevolot, L., Cochard, J.C., and Yvin, J.C., Chemical induction of larval metamorphosis of *Pecten maximus* with a note on the nature of naturally occurring triggering substances, *Mar. Ecol. Prog. Ser.*, 74, 83, 1991.

161. Hadfield, M.G. and Scheuer, D., Evidence for a soluble metamorphic inducer in *Phestilla*: ecological, chemical and biological data, *Bull. Mar. Sci.*, 37, 556, 1985.

162. Hadfield, M.G. and Pennington, J.T., Nature of the metamorphic signal and its internal transduction in larvae of the nudibranch *Phestilla sibogae*, *Bull. Mar. Sci.*, 46, 455, 1990.

163. Boettcher, K.J., Ruby, E.G., and McFall-Ngai, M.J., Bioluminescence in the symbiotic squid *Euprymna scolopes* is controlled by a daily biological rhythm, *J. Comp. Physiol.*, 179, 65, 1996.

164. Boettcher, A.A. and Targett, N.M., Role of chemical inducers in larval metamorphis of queen conch, *Strombus gigas*, Linaeus: relationship to other marine invertebrate systems, *Biol. Bull.*, 194, 132, 1998.

165. Gibson, G.D. and Chia F.S., A metamorphic inducer in the opisthobranch *Haminaea callidegenita* — partial purification and biological activity, *Biol. Bull.*, 187, 133, 1994.

166. Beckmann M., Harder T., and Qian, P.Y., Induction of larval attachment and metamorphosis in the serpulid polychaete *Hydroides elegans* by dissolved free amino acids: mode of action in laboratory bioassays, *Mar. Ecol. Prog. Ser*, 190, 167, 1999.

167. Toonen, R.J. and Pawlik, J.R., Settlement of the tube worm *Hydroides dianthus* (Polychaeta: Serpulidae): cues for gregarious settlement, *Mar. Biol.*, 126, 725, 1996.

168. Kirchman, D., Graham, S., Reish, D., and Mitchell, R., Bacteria induce settlement and metamorphosis of *Janua (Dexiospira) brasiliensis* Grube (Polychaeta: Spirorbidae), *J. Exp. Mar. Biol. Ecol.*, 56, 153, 1982.

169. Kirchman, D., Graham, S., Reish, D., and Mitchell, R., Lectins may mediate the settlement and metamorphosis of *Janua (Dexiospira) brasiliensis* Grube (Polychaeta: Spirorbidae), *Mar. Biol. Letters*, 3, 131, 1982.

170. Szewzyk, U., Holmström, C., Wrangstadh, M., Samuelsson, M.O., Maki, J.S., and Kjelleberg, S., Relevance of the exopolysaccharide of marine *Pseudomonas* sp. strain S9 for the attachment of *Ciona intestinalis* larvae, *Mar. Ecol. Prog. Ser.*, 75, 259, 1991.

171. Henschel, J.R. and Cook, P.A., The development of a marine fouling community in relation to the primary film of microorganisms, *Biofouling*, 2, 1, 1990.

172. Amsler, C.D. and Neushul, M., Nutrient stimulation of spore settlement in the kelps *Pterygophora californica* and *Macrocystis pyrifera*, *Mar. Biol.*, 107, 107, 1990.

173. Amsler, C.D., Shelton, K. L., Britton, C.J., Spencer, N.Y., and Greer, S.P., Nutrients do not influence swimming behavior or settlement rates of *Ectocarpus siliculosus* (Phaeophyceae) spores, *J. Phycol.*, 35, 239, 1999.

174. Dillon, P.S., Maki, J.S., and Mitchell, R., Adhesion of *Enteromorpha* swarmers to microbial films, *Microb. Ecol.*, 17, 39, 1989.

175. Thomas, R.W.S.P. and Allsopp, D., The effects of certain periphytic marine bacteria upon the settlement and growth of *Enteromorpha* a fouling alga, *Biodeterioration*, 5, 358, 1983.

176. Forward, R.B., Jr., Swanson, J., Tankersely, R.A., and Welsh, J.M., Endogenous swimming rhythms of blue crab, *Callinectes sapidus*, megalopae: effects of offshore and estuarine cues, *Mar. Biol.*, 127, 621, 1997.

177. Tankersley, R.A., McKelvey, L.M., and Forward, R.B., Jr., Responses of estuarine crab megalopae to pressure, salinity and light: Implication for flood-tide transport, *Mar. Biol.*, 122, 391, 1995.

178. Faulkner, D.J., Marine natural products, *Nat. Prod. Rep.*, 16, 155, 1999.

179. Wahl, M. and Mark, O., The predominantly facultative nature of epibiosis: experimental and observational evidence, *Mar. Ecol. Prog. Ser.*, 187, 59, 1999.

180. Crisp, D.J. and Meadows, P.S., The chemical basis of gregariousness in cirripedes, *Proc. R. Soc. Lond. B*, 156, 500, 1962.

181. Butman, C.A., Larval settlement of soft-sediment invertebrates: some predictions based on analysis of near-bottom velocity profiles, in *Marine Interface Ecohydrodynamics*, Nihoul, J.C.J., Ed., Elsevier, Amsterdam, 487, 1986.

182. Butman, C.A., Sediment-trap experiments on the importance of hydrodynamical processes in distributing settling invertebrate larvae in near-bottom waters, *J. Exp. Mar. Biol. Ecol.*, 134, 37, 1989.

183. Pawlik, J.R., Butman, C.A., and Starczak, V.R., Hydrodynamic facilitation of gregarious settlement of a reef-building tube worm, *Science*, 251, 421, 1991.

184. Leis, J.M. and Carson-Ewart, B.M., In situ swimming and settlement behaviour of larvae of an Indo-Pacific coral-reef fish, the coral trout *Plectropomus leopardus* (Pisces: Serranidae), *Mar. Biol.*, 134, 51, 1999.

185. Mitchell, J.G., Pearson, L., and Dillon, S., Cluster dynamics of marine bacteria in seawater enrichments, *Appl. Environ. Microbiol.*, 62, 3716, 1996.

186. Blackburn, N., Fenchel, T., and Mitchell, J., Microscale nutrient patches in planktonic habitats shown by chemotactic bacteria, *Science*, 282, 2254, 1998.

187. Mullineaux, L.S. and Butman, C.A., Initial contact, exploration and attachment of barnacle (*Balanus amphitrite*) cyprids settling in flow, *Mar. Biol.*, 110, 93, 1991.

188. Hay, M.E. and Fenical, W., Marine plant–herbivore interactions: the ecology of chemical defense, *Annu. Rev. Ecol. Syst.*, 19, 111, 1988.

189. Barrow, K., Karsten, U., King, R.J., and West, J.A., Floridoside in the genus *Laurencia* (Rhodomelaceae: Ceramiales) — a chemosystematic study, *Phycologia*, 343, 279, 1995.

190. Lobban, C.S. and Harrison, P.J., *Seaweed ecology and physiology*, Cambridge University Press, Melbourne, Australia, 1994.

191. Agrawal, S.C. and Sharma, U.K., Chemical and biological properties of culture filtrates of *Westiellopsis prolifica* and *Chaetophora attenuata*, *Israel J. Plant Sci.*, 44, 43, 1996.

192. Shilling, F.M., Morphological and physiological responses of echinoderm larvae to nutritive signals, *Am. Zool.*, 35, 399, 1995.

193. Plamann, L. and Kaplan, H.B., Cell-density sensing during early development in *Myxococcus xanthus*, in *Cell–Cell Signaling Bacteria*, Dunny, G.M. and Winans, S.C., Eds., ASM Press, Washington, D.C., 1999, 67.

194. Baier, R.E., Initial events in microbial film formation, in *Marine Biodetermination: An Interdisciplinary Study*, Costlow, J.D. and Tipper, R.C., Eds., E. and F.N. Spon Ltd., London, UK, 1984, 57.

195. Moriarty, D.J.W. and Bell, R.T., Bacterial growth and starvation in aquatic environments, in *Starvation in Bacteria*, Kjelleberg, S., Ed., Plenum Press, New York, 1993, 25.

196. Berg, H.C. and Brown, D.A., Chemotaxis in *Escherichia coli* analysed by three-dimensional tracking, *Nature*, 239, 500, 1972.

197. Dunny, G.M. and Winans, S.C., Eds., *Cell–Cell Signaling in Bacteria*, ASM Press, Washington, D.C., 1999.

198. Davies, D.G., Parsek, M.R., Pearson, J.P., Iglewski, B.H., Costerton, J.W., and Greenberg, E.P., The involvement of cell-to-cell signals in the development of a bacterial biofilm, *Science*, 280, 295, 1998.

199. Kirke, D.F., Lynch, M.J., Swift, S., Bishop, K., Dodd, C.E.R., Keevil, C.W., Stewart, G.S.A.B., and Williams, P., Quorum sensing in *Aeromonas hydrophila*, Poster, *99th General Meeting* of the *American Society of Microbiology*, Chicago, IL, May 30 to June 3, 1999.

200. Visick, K.L. and Ruby, E.G., The emergent properties of quorum sensing: consequences to bacteria of autoinducer signaling in their natural environment, in *Cell–Cell Signaling in Bacteria*, Dunny, G.M. and Winans, S.C., Eds., ASM Press, Washington, D.C., 1999, 21, 333.

201. Labbatte, M. and Kjelleberg, S., unpublished.

202. Eberl, L., Winson, M.K., Sternberg, C., Stewart, G.S.A.B., Christiansen, G., Chhabra, S.R., Bycroft, B., Williams, P., Molin, S., and Givskov, M., Involvement of N-acyl-L-homoserine lactone autoinducers in control of multicellular behaviour of *Serratia liquefaciens*, *Mol. Microbiol.*, 20, 127, 1996.

203. Pierson, L.S., Wood, D.W. and Beck von Bodman, S., Quorum sensing in plant-associated bacteria, in *Cell–Cell Signaling Bacteria*, Dunny, G.M. and Winans, S.C., Eds., ASM Press, Washington, D.C., 1999, 101.

204. Winans, S.C., Zhu, J., and More, M.I., Cell density-dependent gene expression by *Agrobacterium tumifaciens* during colonization of crown gall tumors, in *Cell–Cell Signaling in Bacteria*, Dunny G.M. and Winans, S.C., Eds., ASM Press, Washington, D.C., 1999, 117.

205. Givskov, M., de Nys, R., Manefield, M., Gram, L., Maximilien, R., Eberl, L., Molin, S., Steinberg, P.D., and Kjelleberg, S., Eukaryotic interference with homoserine lactone mediated prokaryotic signalling, *J. Bacteriol.*, 178, 6618, 1996.

206. Manefield, M., de Nys, R., Kumar, N., Read, R., Givskov, M., Steinberg, P.D., and Kjelleberg, S., Evidence that halogenated furanones from *Delisea pulchra* inhibit acylated homoserine alctone (AHL)-mediated gene expression by displacing the AHL signal from its receptor protein, *Microbiol.*, 145, 283, 1999.

207. Taylor, M., Charlton, T., de Nys, R., Kjelleberg, S., and Steinberg, P., unpublished.

208. Holden, M.T., Ram Chhabra, S., de Nys, R., Stead, P., Bainton, N.J., Hill, P.J., Manefield, M., Kumar, N., Labatte, M., England, D., Rice, S., Givskov, M., Salmond, G.P., Stewart, G.S., Bycroft, B.W., Kjelleberg, S., and Williams, P., Quorum-sensing cross talk: isolation and chemical characterization of cyclic dipeptides from *Pseudomonas aeruginosa* and other gram-negative bacteria, *Mol. Microbiol.*, 33, 1254, 1999.

209. Bassler, B.L., A multichannel two-component signaling relay controls quorum sensing in *Vibrio harveyi*, in *Cell–Cell Signaling in Bacteria*, Dunny, G.M. and Winans, S.C., Eds, ASM Press, Washington, D.C., 1999, 259.

210. Charlton, T.S., de Nys, R., Netting, A., Kumar, N., Hentzer, M., Givskov, M., and Kjelleberg, S., A novel and sensitive method for the quantification of N-3-oxoacyl homoserine lactones using gas chromotography-mass spectrometry: application to a model bacterial biofilm, *Environ. Microbiol.*, 2, 530, 2000.

211. de Nys, R., Coll, J.C. and Price, I.R., Chemically mediated interactions between the red alga *Plocamium hamatum* (Rhodophyta) and the octocoral *Sinularia cruciata* (Alyconacea), *Mar. Biol.*, 108, 315, 1991.

212. Thacker, R.W., Becerro, M.A., Lumbang, W.A., and Paul, V.J., Allelopathic interactions between sponges on a tropical reef, *Ecology*, 79, 1740, 1998.

213. Pennings, S.C. and Paul, V.J., Sequestration of dietary secondary metabolites by three species of sea hares: location, specificity and dynamics, *Mar. Biol.*, 117, 535, 1993.

214. Maier, I. and Muller, D.G., Sexual pheromones in algae, *Biol. Bull. Mar. Biol. Lab.*, 170, 145, 1986.

215. Coll, J.C., Bowden, B.F., Meehan, G.V., Konig, G.M., Carroll, A.R., Tapiolas, D.M., Alino, P.M., Heaton, A., de Nys, R., Leone, P.A., Maida, M., Aceret, T.L., Willis, R.H., Babcock, R.C., Willis, B.L., Florian, Z., Clayton, M.N., and Miller R.L., Chemical aspects of mass spawning in corals. I. Sperm-attractant molecules in the eggs of the scleractinian coran *Montipora digitata*, *J. Chem. Ecol.*, 118, 177, 1993.

216. Baier, R.E., Influence of the initial surface condition of materials on bioadhesion, Proceedings of the Third International Congress on Marine Corrosion and Fouling. National Bureau of Standards, Gaithersburg, MD, 633, 1973.

217. Dexter, S.C., Sullivan, J.D., Williams, J., and Watson, S.W., Influence of substrate wettability on the attachment of marine bacteria to various surfaces, *Appl. Microbiol.*, 30, 298, 1975.

218. Vrolijk, N.H., Targett, N.M., Baier, R.E., and Meyer, A.E., Surface characterization of two gorgonian coral species; implications for a natural antifouling defence, *Biofouling*, 2, 39, 1990.

219. Schneider, R.P., Conditioning-film induced modification of substratum physicochemistry — an analysis by contact angles, *J. Colloid Interface Sci.*, 182, 204, 1996.

220. Maki, J.S., The influence of marine microbes on biofouling, in *Recent Advances in Marine Biotechnology, Vol III*, Fingerman, M., Nagabhushanam, R., and Thompson, M.F., Eds., Science Publishers, Inc., Enfield, NH, 1999, 147.

221. Clarkson, N., The antifouling potential of silicone elastomer polymers, in *Recent Advances in Marine Biotechnology, Vol III*, Fingerman, M., Nagabhushanam, R., and Thompson, M.F., Eds., Science Publishers, Inc., Enfield, NH, 1999, 87.

222. Rittschof, D. and Costlow, J.D., Bryozoan and barnacle settlement in relation to initial surface wettability; a comparison of laboratory and field studies, *Topics in Marine Biology, Proc. 22nd European Marine Biol. Symposium, Instituto de Ciencias del Mar, Barcelona, Spain,* Ros, J.D., Ed., 1989, 411.

223. Becker, K. and Wahl, M., Influence of substratum surface tension on biofouling of artificial substrata in Kiel Bay (Western Baltic): *in situ* studies, *Biofouling*, 4, 275, 1991.

224. Fletcher, M. and Marshall, K.C., Are solid surfaces of ecological significance to aquatic bacteria?, *Adv. Microbial. Ecol.*, 6, 199, 1982.

225. Wolk, C.P., Role of bromine in the formation of the refractile inclusions of the vesicle cells of the *Bonnemaisonales* (Rhodophyta), *Planta*, 78, 371, 1968.

226. Dawes, C.J., Ed., *Marine Botany*, 2nd ed., John Wiley & Sons, New York, 1997.

Section III

Cellular and Physiological Patterns in Marine Chemical Ecology

11 Effects of Secondary Metabolites on Digestion in Marine Herbivores

Nancy M. Targett* and Thomas M. Arnold

CONTENTS

I. INTRODUCTION

Thousands of secondary metabolites have been isolated from marine organisms (e.g., Faulkner[1] and references therein). These compounds are diverse and include chemicals such as phlorotannins, terpenes, sterols, nitrogenous compounds, acetogenins, and compounds of mixed biogenesis. Often, secondary metabolites have been shown to exhibit one or more of a wide range of activities that can be broadly categorized as either attractants,[2–11] deterrents,[12–25] or protectants.[26,27] Much of the work on secondary metabolites has defined organismal responses to compounds through behavioral assays. More recent studies have begun to address the physiological and biochemical mechanisms underlying these responses. Not surprisingly, the mechanistic underpinnings of the behavioral responses are proving to be complex and multivariate.

Given the number and diversity of secondary metabolites and their broad range of pre-ingestive bioactivities, it is of interest to speculate on the effect of these compounds (which can be present in plants and prey at concentrations that constitute approximately 20% of the dry tissue mass) once they are consumed. This chapter reviews what is known about the consumers' post-ingestive response to marine secondary metabolites, focusing on marine herbivores, since

---

* Corresponding author.

most of the work examining the interaction of secondary metabolites and digestion has occurred in plant–herbivore systems.

In reviewing the status of our knowledge of the effects of secondary metabolites on consumer digestion, it is important to recognize that consumer responses are often conditional responses; that is, they are based upon indirect, complex, or synergistic interactions.[25,28–33] Thus, we can expect the relationships between secondary metabolites, diet quality, and consumer digestive physiology to be complex.[33] This means that the activity of secondary metabolites is probably not based solely on structural chemistry. Rather, it is likely a function of multiple factors that include metabolite structure, but also include the digestive physiology of the consumer,[25,34,35] prey nutrient levels,[36,37] and consumer experience.[32]

II. DIGESTIVE PROCESSES

The gut environment determines how efficiently a consumer digests the suite of primary and secondary metabolites present in its prey, for it is here that biochemical interactions are mediated by gut histology, chemistry, and microorganisms.[38,39] Specialization on given food types requires particular sets of morphological and physiological adaptations for ingestion, digestion, and absorption of nutrients.[40] For purposes of this review, this chapter considers marine herbivorous consumers of plants containing secondary metabolites.

A. FISHES

To understand the special constraints of herbivorous fishes, it helps to first understand the carnivore digestive process.[40–42] Carnivorous fishes consume food in which nutrients are generally readily available and of high quality (protein rich). In these fishes, food is usually broken down in the stomach through a combination of muscular contractions of the stomach wall and enzymatic action in an acid medium. Breakdown products are expelled from the stomach through the pyloric sphincter into the intestine in a process called gastric evacuation. Digestion and food absorption are completed in the intestine.[40–42]

In contrast to carnivorous fishes, herbivorous fishes consume food of lesser quality (more structural material, less protein) and must gain access to nutrients stored inside plant cell walls.[33,39,42–45] Unlike the cell walls in vascular plants, the cell walls of marine algae contain a prevalence of sulfated polysaccharides. α-Cellulose, a primary cell wall constituent in higher plants, accounts for only a small portion (1 to 8% of thallus dry weight) of the cell wall in most marine algae.[46] The intracellular matrix components of algal cells are also highly diverse and different from the pectic and hemicellulosic matrix components of vascular plants. In chlorophytes, the matrix is composed of highly branched sulfated heteropolysaccharides, in rhodophytes it is primarily linear sulphated galactans, and in phaeophytes it is alginates.[46] Within each of these taxa, the matrix and cell wall polysaccharides are structurally diverse, and their properties ultimately depend not only on the primary composition of the matrix but also upon its secondary, tertiary, and quaternary structures.[46] It is not surprising, then, that the digestive tracts of herbivorous marine fishes show a range of specializations for handling their algal and seagrass diets.[39,43,44,47–49]

The type of digestive mechanism employed, together with the relative proportions of nutritive and inert material ingested, determines the type of intestinal morphology required for absorbing adequate quantities of nutrients.[33,40] Several distinct herbivore gut types have been identified,[33,39,43,44,50,51] and in 1989, Horn[44] established a broad-based framework to categorize herbivorous fishes according to their adaptations for food processing (Table 11.1). This framework has proven useful when trying to classify the potential effects of secondary metabolites on herbivorous fish consumers, although it is evident that some fish species incorporate characteristics of more than one category.[33,39,40]

TABLE 11.1
Overview of Horn's[44] Framework which Identifies Broad Morphological, Chemical, and Microbial Differences in Marine Herbivore Gut Environments

Gut Type	Gut Characteristics
Type I	• Little or no trituration
	• Acid stomach
	• Neutral to slightly basic intestine
Type II	• Gizzard-like stomach (near neutral pH)
	• Long intestine
Type III	• Pharyngeal mill, no stomach
	• Long intestine (neutral to basic)
Type IV	• Well-defined, acidic stomach
	• Long intestine
	• Hindgut caecum

The primary factors Horn[44] considers in his framework are gut length, gut pH, and microbial activity (Table 11.1). Besides these, other factors are also proving to be part of the complex mixture of variables that are important in herbivorous fish digestion. These other factors include specific endogenous and exogenous digestive enzymes in the gut, redox state, concentration, and activity of surfactants, and concentration and activity of cations. Gut transit time or evacuation rate is also considered an important factor because it influences assimilation efficiency and consumption rates.[39,52,53]

Herbivorous fishes generally, but not always, have longer guts relative to body size than omnivorous or carnivorous species.[39,40,44,45,54] Longer gut length is associated with a larger gut capacity and an increased opportunity for nutrient absorption to accommodate the less digestible foods consumed by herbivores.[39,44] Gut pH ranges from acidic in some herbivores to basic in others. The importance of acid lysis as an effective mechanism for gaining access to algal cell nutrients has been suggested for a number of herbivorous fishes.[39,43,44,48,55,56,57] Lobel[43] first established the variability of gut pH in herbivorous fishes. Since then, several studies have characterized pH and redox parameters along the length of the guts of many herbivorous marine fish species.[25,34,35,57,58]

Many herbivorous marine fishes also contain large populations of microorganisms in their gut. These include acanthurids,[47,59] girellids,[55] odacids,[60] and kyphosids.[50,61,62] The role of these microorganisms in the nutrition of the fishes varies. In some fishes, microbes are responsible for the exogenous production of cellulases and other enzymes that facilitate the breakdown of algal cell wall and matrix material.[33,49,63] The presence of short-chain fatty acids (SCFA) in the guts of several temperate and tropical herbivorous fishes also suggests that microbial symbionts may degrade complex algal carbohydrates and other organic nutrients through fermentation.[47,49,50,60,64,65] Terrestrial vertebrate herbivores typically maintain an extensive microbiota that facilitates the conversion of dietary fiber via fermentation into short-chain fatty acids.[66] Rimmer and Wiebe[50] provided the first confirmation of fermentative digestion in fishes when they isolated SCFA from the posterior guts of the kyphosids, *Kyphosus cornelii* and *Kyphosus sydneyanus*. Other investigators have since confirmed the presence of SCFA in the guts of several other herbivorous fish species.[49,60,64,67] Seeto et al.[67] isolated the microbial consortia from the guts of two herbivorous fishes, *Odax cyanomelas* and *Crinodus lophodon*, and showed that the different patterns of substrate utilization reflected the diets of the host species. Their evidence that the microbes isolated from *Odax cyanomelas* mediated the fermentation of mannitol to short-chain fatty acids such as acetate further supports the fact that these endosymbionts play a role in supplying nutrients to the host. Thus, as noted by Choat and

Clements,[33] it may be no accident that phaeophytes, which contain large amounts of the difficult-to-digest mannitol, tend to be eaten by fish species such as *Kyphosus*, *Odax*, and *Naso* that contain elevated levels of fermentation products in their posterior intestines.

For fishes in general, enzymatic digestion occurs in the stomach (in fishes with stomachs) and the intestine. In the stomach, digestion is primarily through the action of pepsins in an acid medium. These enzymes are endopeptidases which hydrolyse peptide bonds between aromatic and dicarboxylic amino acids.[40,41,42] Proteolytic enzymes other than pepsins have been reported from the stomachs of a number of fish species, and there may also be a variety of nonproteolytic enzymes present. Amylase (a carbohydrase), lipase, and chitinase have been extracted from fish stomachs.[40] However, most of the enzymatic digestion of food consumed by fishes occurs in the intestine.[40] Here, digestive enzymes may be endogenous or exogenous (gut microflora). Endogenous intestinal enzymes include proteases, lipases, and some carbohydrases (e.g., amylase). The carbohydrases are supplemented, particularly in herbivorous fish species, by exogenous production of enzymes such as cellulases.

The secretion of digestive enzymes and fluids in response to the ingestion of a meal is influenced, in part, by the nutrient composition of the meal, suggesting that there is a capacity for enzymatic adaptation to the diet. As a result, there are some generalized interspecific differences in the abilities of fishes to produce and secrete particular types of digestive enzymes.[40] For example, amylase activity tends to be higher in the guts of omnivorous and herbivorous species than in the guts of carnivores.[40] Herbivorous fishes also have relatively high levels of trypsin and chymotrypsin but little or no peptic enzyme activity. Omnivorous species take an intermediate position. Fishes that consume animals with chitinous exoskeletons have high levels of intestinal chitinases, and those that consume planktonic copepods tend to produce relatively large quantities of lipases and esterases.[40] Thus, there is a correspondence between the feeding habits of fish species and production and secretion of digestive enzymes.[40]

B. INVERTEBRATES

Echinoids, gastropods (prosobranchs and opisthobranchs), and amphipods are marine invertebrate groups for which there is a significant body of information regarding their preferences for and consumption of algae rich in secondary metabolites. These invertebrates are known to have varying gut characteristics and to exhibit differences in morphology, pH, enzyme composition, microbial flora, and surfactant characteristics.[68–70] Extracts from the guts of most marine invertebrates display proteinase, amylase, and lipase activity.[71–73]

The gut of regular echinoids is a long, coiled tube that is divided into several regions.[74–77] The stomach is the longest portion and contains five bag-like compartments that make an almost complete circuit of the body. The intestine makes a reverse circuit along the stomach before entering the rectum. Regular echinoids possess a range of enzymes and can digest both α-glucoside and β-glucoside linkages.[72–74,78,79] They are predominantly herbivores although there are numerous reports that some urchins may also act as carnivores. Chymotrypsin appears to function as a general protein-digesting enzyme in echinoderms, including urchins. Levels of protein digestion appear to be the same regardless of whether the species is predominantly carnivorous, omnivorous, or herbivorous.[73] The guts of regular echinoids typically have pH that are slightly acidic to neutral, and they are isosmotic with seawater.[58,80–82] Although urchin guts contain numerous bacteria,[83] studies to date[74,78,84,85] suggest that they do not play a significant role in urchin digestion.

Prosobranch and opisthobranch gastropods[86] have two primary organs involved in digestion: the stomach and the digestive gland.[68,87] In prosobranchs, the stomach is considered the primary site of enzymatic digestion, and it can contain a diverse complement of enzymes.[88,89] The digestive gland functions in the uptake of digestive products and in excretion. Bacteria are usually present in the stomach and in the intestine, but their importance in gastropod digestion is thought to be minimal.[90] In opisthobranchs, the bulk of the midgut is comprised of the digestive gland. Like the prosobranch digestive gland, it is composed of multiple cell types. It is thought to function mainly by producing digestive enzymes but also to be involved with processing secondary metabolites.[91–99] Specialized cells called rhodoplast digestive cells, identified to date only in opisthobranch digestive glands, may be particularly important in acquiring and modifying diet-derived secondary metabolites.[99]

Amphipods, like crustaceans in general, have a digestive system that includes ectodermally derived foregut and hindgut elements lined with cuticle as well as an endodermal midgut.[100] Gut extracts from amphipods have been shown to hydrolyse cellulose, starch, and laminarin.[101–103] Enzyme activity found in amphipods is thought to be primarily the result of endogenous enzymes, although there is also evidence for some exogenous enzyme activity.[103] The ratio of gut enzyme activities has also been shown to change with season.[104]

III. INTERACTION OF SECONDARY METABOLITES AND DIGESTIVE PROCESSES

A. HERBIVORES AND BROWN ALGAL PHLOROTANNINS

1. Phlorotannin Characteristics and Occurrence

Brown algal phlorotannins (Table 11.2) are analogous to terrestrial condensed tannins, which are considered one of the most broadly distributed types of plant natural products.[105–108] Phlorotannins are present at detectable concentrations across almost all brown algal orders. Within the plant, they are stored in vesicles called physodes.[107,109,110] Like many secondary metabolites,[111] phlorotannins have multiple roles, serving as cell wall strengtheners,[112–114] feeding deterrents,[115–120] digestability reducers,[34,35,120–122] and antimicrobial agents.[107,123] They can also function to absorb UV radiation[27] and may act as metal chelators.[107,124]

· Phlorotannins are polymers of phloroglucinol (1,3,5-trihydroxy benzene), and, as a group, span a wide range of molecular sizes (126 Da to 650k Da). They can constitute up to 20% of a brown alga's dry mass.[107] While not as structurally diverse as terrestrial tannins,[107,108] marine phlorotannins can be subdivided into six specific groups: fucols, phlorethols, fucophlorethols, fuhalols, isofuhalols, and eckols.[107] These groups are characterized by differences in the nature of the structural linkages binding the phlorotannin polymers and in the number of hydroxyl groups present (Table 11.2). Specific groups are also often characteristic of specific algal genera, for example, fucols in *Fucus* and eckols in *Ecklonia*.[107] Phlorotannins are thought to polymerize as they age, thus increasing in molecular size. Studies of phlorotannin molecular size distribution within brown algae have shown that most species have phlorotannins dominated by molecular masses greater than 10 kDa.[34,107] *In situ* studies using stable isotopes to examine changes in algal phlorotannins over time support the information obtained from static measurements of molecular size by showing that phlorotannin polymerization takes place rapidly, that is, within a matter of hours of synthesis.[125,126] The nature of the phlorotannin structural linkage, the size range of the phlorotannin polymers within a plant, and their concentrations are all considered important factors in determining the activity of these compounds.

Since the concentration of algal phlorotannins can significantly affect the outcome of the plant–herbivore interactions they mediate,[127] variations in concentration have ecological significance.[25] Phlorotannin concentrations vary over scales ranging from tissues within a single plant to plants from different geographic regions.[35,122,128–130] Phlorotannin concentrations are known to be affected by plant size,[131] age,[127,132] tissue type,[116,133–136] salinity,[107] season,[107,137–139] nutrient levels,[140–145] herbivory,[143,146,147] and light intensity.[126,143–145] As noted by Van Alstyne,[130] the causes and consequences of this high degree of variability are undoubtedly complex, but they need to be considered when making generalizations about the ecological and evolutionary significance of phenolic metabolites.[25,122,148–150]

Table 11.2 Examples of the Six Brown Algal Phlorotannin Structural Types [107]

Category	Description and Occurrence	Example
Fucols	• Phloroglucinol units linked through aryl-aryl bonds • Occur in Fucales and in Ectocarpales	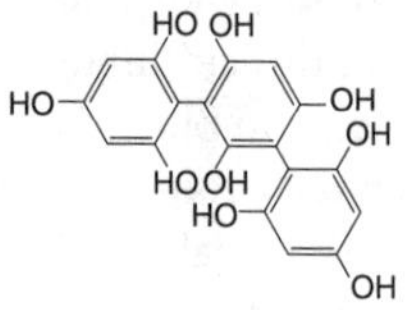
Phlorethols	• Phloroglucinol units linked through diaryl ether bonds • Occur in Fucales and Laminariales	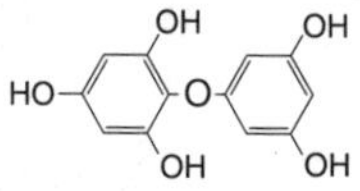
Fucophlorethols	• Dehydrooligomers of phloroglucinol which contain both direct carbon-carbon and diaryl ether bonds • Occur primarily in Fucales, sporadically in Laminariales	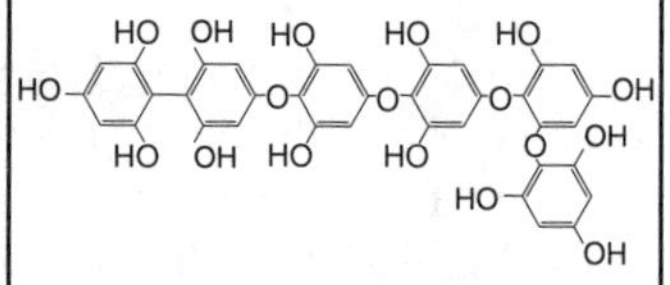
Fuhalols	• Ether-linked phloroglucinol units with an extra hydroxyl group on one unit • Ether bonds are oriented *para* and *ortho* • Occur primarily in the Fucales	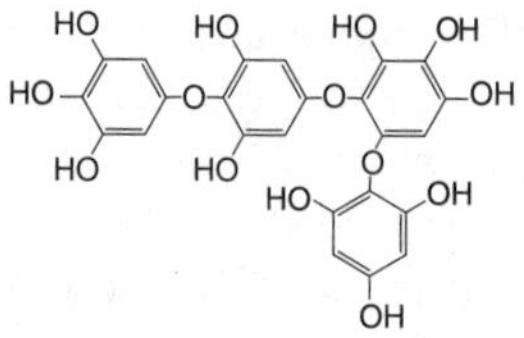
Isofuhalols	• Phloroglucinol units linked through *para*- and *meta*-oriented ether bonds and containing at least one extra hydroxyl group • Limited occurrence in the Laminariales	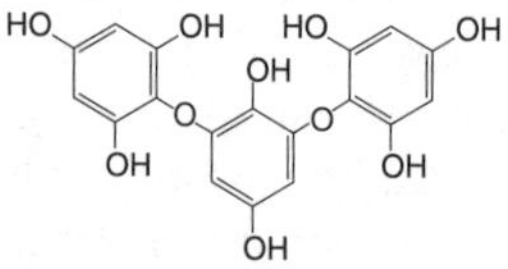
Eckols	• Dehydrooligomerization of three phloroglucinol units, two of which are further cyclized to a dibenzo[1,4]dioxin. • Limited occurrence in the Laminariales	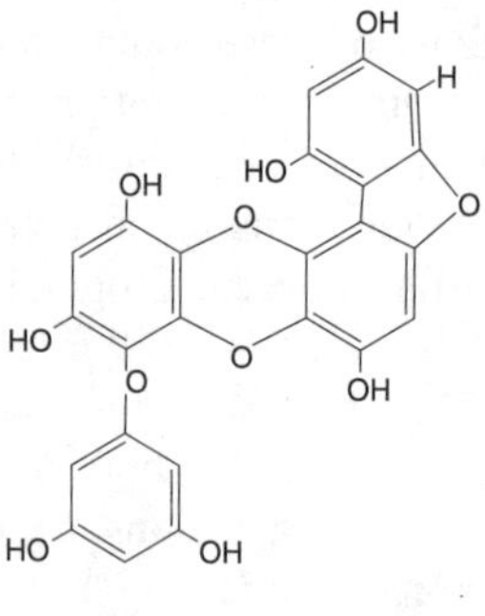

2. Effects of Phlorotannins on Herbivory

There have been conflicting views regarding the effect of phlorotannins on herbivores. Phlorotannin-rich phaeophytes appear in the diets of numerous herbivores and even predominate in the diets of a few species,[33,44] despite evidence that suggests that they are among the most difficult algae to digest. Their difficult digestibility has been attributed to the nature of their storage polysaccharides, laminarin, and mannitol[49] and to the high levels of phlorotannins in some brown algal species.[25,120,122]

Behavioral and physiological assays geared to assessing responses of herbivores to phlorotannin-rich food show mixed results. Many temperate North American marine invertebrate herbivores are consistently deterred by phlorotannins.[115–120,141,149,151] Phlorotannins have also been shown to negatively affect assimilation efficiencies for temperate North American marine fishes[34,121,152] and invertebrates.[120,124] However, many Australian temperate fishes[33,49,50,67] and invertebrates[120,122,124,149,153] are not generally deterred from feeding on phlorotannin-rich algae.

In the tropics, phlorotannin-rich brown algae occur in the Caribbean but not in the Indo-Pacific.[35,128,154] The high phlorotannin levels in many of the tropical western Atlantic brown algae do not seem to deter herbivores like the parrotfish *Sparisoma radians* and *Sparisoma chrysopterum* or the crab *Mithrax sculptus*.[35] For tropical fishes in western Australia, no correlation was found between grazing susceptibility and whole-plant phlorotannin content of different temperate phenolic-rich and tropical phenolic-poor *Sargassum* species.[153] However, studies in Guam, where phlorotannin-rich brown algae are scarce, found that grazing by tropical fishes was generally deterred by temperate plant extracts with high phenolic concentrations, but not by temperate or tropical plant extracts with low phenolic concentrations.[155]

These apparent conflicts can be at least partially resolved by recognizing two factors: (1) phlorotannins are not homogeneous in their structures (i.e., phlorotannin size[34,70,119,120,156] and phlorotannin structural characteristics[124,157]) and thus affect herbivores differently, and (2) herbivore gut environments are not uniform, so that differences, including herbivore nutritional and physiological constraints, may be critical factors driving herbivore responses to phlorotannins.[25,33,35,44,49,120,158–165]

3. The Importance of Herbivore Gut Characteristics in Phlorotannin–Herbivore Interactions

In terrestrial systems, numerous factors have been recognized as essential in determining a herbivore's gut environment, and, hence, its response to tannins. In marine systems, many of these same factors may impact the activity of phlorotannins in the herbivore gut. These include gut morphology,[44] pH,[25] redox potential,[58,124,166] enzyme composition and activity,[33,57] surfactant type and concentration,[70,160,167] cation type and concentration,[70,124,168] proteins or amino acids,[169–175] gut microbial activity,[49,50,67] and nutritional status.[33]

The importance of the herbivore gut environment is easy to reconcile when one considers that phlorotannins, like their terrestrial tannin counterparts, are thought to act primarily by complexing proteins and other macromolecules.[106,108,124,158,159,163,164,176–178] Complexation in tannins can occur through the formation of hydrogen, covalent, or ionic bonds and through hydrophobic interactions.[108,163,179] Hydrogen bonds form between the phenolic hydroxyl of the polyphenolic compound and the free amino group of proteins, or the hydroxyl and carbonyl groups of other polymers.[180] Hydrogen bonding occurs only at acidic to neutral pH. Above a pH of 8.5, autooxidation of the polyphenolic occurs.[163] Under oxidizing conditions and/or in the presence of certain plant enzymes (polyphenol oxidases and peroxidases), polyphenolics undergo oxidation and become available for covalent bonding.[163,166,181–183] The importance of ionic bonds in polyphenolic macromolecule complexation is unknown.[163] Hydrophobic interactions occur between the aromatic central region of polyphenolics and either the aliphatic or aromatic side chains on proteins.[184] This type of interaction is pH-independent.[163] By acting as complexing agents, phlorotannins can bind to plant nutrients (carbohydrates or dietary proteins) and prevent their absorption by herbivores, or they can bind to herbivore digestive enzymes and inactivate their effects.[108,170,185,186]

Using Horn's[44] framework as a basis for differentiating the four teleost herbivore gut types, several studies have examined the *in vivo* effect of phlorotannins or phlorotannin-rich foods on assimilation efficiency and, hence, nutrient availability for fishes. Not surprisingly, the response of herbivores has been varied. Phlorotannins have been shown to negatively affect assimilation efficiency in herbivorous fishes with an acidic stomach.[34,121,152] In the temperate stichaeid *Xiphister mucosus*, the organic assimilation efficiency was reduced to approximately 10% at a concentration of 4% dry mass of high molecular weight phlorotannins, compared to the control which showed an organic assimilation efficiency of approximately 35%.[34] Assimilation efficiencies from *Xiphister* experiments in which phlorotannin-treated and untreated algae were fed to the same fish at the same time indicated that there was no disruption of the gut epithelium.[152,187] This suggests that the mode of phlorotannin action is as a digestibility reducing agent rather than a toxic agent. Assimilation efficiency was also reduced in the girellid, *Girella nigricans*,[152] which has an acid stomach and a gut anatomy that differs little from that of carnivorous species.[49] Irelan and Horn[121] examined the assimilation efficiency of *Fucus* extracts by the temperate stichaeid, *Cebidichthys violaceus*. The phlorotannin-rich treatments did not affect the digestion of carbon but did lower the nitrogen assimilation by about 15% relative to the control. In this type of acidic gut environment, hydrogen bonding between polyphenolics and macromolecules could occur, preventing assimilation of nutrients. However, phlorotannin-rich plant material passing through the acidic stomach of the tropical kyphosid, *Kyphosus incisor,* did not result in a lowered assimilation efficiency relative to the control.[25] Similarly, assimilation efficiencies remained high when the temperate kyphosid, *Hermosilla azurea,* was fed a suite of phaeophytes.[56] It is significant that, in addition to their acidic stomachs, kyphosids have a very specialized hindgut fermentation chamber,[49,50] and it has been speculated that any phlorotannin/macromolecule interaction engendered by the acidic stomach environment is disrupted by the hindgut microbial consortium.[25]

The parrotfish *Sparisoma radians* and *Sparisoma chrysopterum*, which triturate their food, lack a defined stomach, and have a basic gut pH, readily assimilate phlorotannin-rich food at efficiencies ranging from 50 to 70%.[35] In fact, for *Sparisoma chrysopterum*, brown algal food with a higher phenolic content was assimilated more readily than food with a low phenolic content. Although other plant characteristics could be a factor in determining the absolute difference in assimilation efficiencies, the fact remains that phlorotannin-rich foods were assimilated at levels that were comparable to the control in this species.[35] The odacid, *Odax cyanomelas*, which, like *S. radians* and *S. chrysopterum*, has a gut pH of 8 to 9, also readily consumes high phenolic brown algal species with apparent impunity. Like the kyphosid, this species also has a gut microbial consortium.[67] Under basic conditions, hydrogen bonding would not occur. However, ionic and covalent bonds are possible, and the redox state of the gut may be critical in determining the effect of the phlorotannins on digestion.[124,163,164]

Based on information available to date, low assimilation efficiency in fishes fed phlorotannins or phlorotannin-rich food appears to be correlated with an acidic gut, except when the acidic gut co-occurs in fishes with a hindgut fermentation chamber. In the latter case, phlorotannin-rich food appears to be readily assimilated. Basic gut conditions correlate with high assimilation efficiencies in the presence of phlorotannins or phlorotannin-rich food. The actual mechanisms that account for these observations are still to be resolved.

The demonstrated impact of phlorotannins on feeding in marine invertebrates has also been mixed.[115,118–120,122,151,188] A study of the crab *Mithrax sculptus*, with its slightly acidic gut (pH 5.5), had a high assimilation efficiency of phlorotannin-rich food.[35] Likewise, there are no differences in conversion efficiency for the urchin and snail, *Tripneustes gratilla* and *Turbo undulata*, respectively, fed phlorotannin-rich vs. phlorotannin-poor food.[120] In a test of assimilation of phlorotannin-rich and phlorotannin-poor brown algae by the Australian invertebrate herbivores, Stern[124] found that phlorotannins did not affect assimilation in the snails *Turbo torquata* or *Turbo undulatus* or in the urchins *Holopneustes purpurascens* and *Heliocidaris erythrogramma*. However, the snails from California, *Tegula funnebralis* and *Tegula brunnea*, had generally lower assimilation on phlorotannin-rich algae than on phlorotannin-poor algae. Because it was not possible to do reciprocal tests, it was not clear whether the differences in assimilation were the result of differences in herbivore digestive capabilities or due to differences in the digestibility of the algae themselves.[124] However, using an *Ulva* diet that contained phlorotannins in a lipid coating, Stern[124] was able to test specific phlorotannins originating from seaweeds in California and Australia on both California and Australian invertebrate herbivores. He found that, in general, there was a mild reduction in assimilation efficiency for herbivores from both geographic regions, and that selected phlorotannins enhanced the reduction in assimilation efficiency regardless of the herbivore's regional origin.

In a survey of the gut conditions of Californian and Australian invertebrate herbivores, redox conditions in the intestines and digestive glands were found to be significantly more reducing for the Australian herbivores than for the Californian herbivores.[124] As a result, it has been suggested that the Australian invertebrate herbivores cope with high levels of dietary phlorotannins by maintaining them in a reduced state. This complements information available for the hindgut caecum of fishes in the family Kyphosidae, where microbial fermentation processes are thought to occur in a presumably reducing environment.[58]

4. *In Vitro/Ex Vivo* Studies of Phlorotannins

Because of the difficulty of working *in vivo* to assess the mechanisms underlying phlorotannin–herbivore interactions, several studies have worked *in vitro* and/or *ex vivo* using combinations of phlorotannins, proteins, and herbivore digestive fluids. Barwell and co-workers[63] looked at the effects of extracts of several brown algae containing phlorotannins ranging in size from 30 to 100 kDa and found that they inhibited α-amylase, lipase, and trypsin. Tugwell and Branch[70] were the first to show that marine herbivores differ physiologically in their response to ingested phlorotannins. They established the effect of pH and cations on phlorotannin–protein precipitation, correlated the effect with gut conditions of seven invertebrate species (three congeneric isopods, three congeneric limpets, and a rock lobster), and then examined the role of gut surfactants in mitigating the phlorotannin–protein interactions. They found that under acidic conditions, regardless of ion concentration, even low concentrations of phlorotannins caused virtually total precipitation of ribulose-1,5-bisphosphate carboxylate/oxygenase. At neutral pH, cation concentration dramatically impacted phlorotannin–protein precipitation. As the alkalinity of the solution increased further, the overall effect of the phlorotannins was decreased. Cation concentration and type affected phlorotannin–protein precipitation at higher pH, but less so than at neutral pH.

Tugwell and Branch[70] hypothesized that because of the acidic to neutral pH and cation concentration in the guts of marine invertebrates, conditions would favor phlorotannin–protein complexation unless there were other mediating factors. Because of work in terrestrial systems on the effectiveness of surfactants in preventing polyphenol–protein complexation,[167,168] Tugwell and Branch[70] examined the marine invertebrate gut fluids for the presence of surfactants. They found that surfactants were present in all seven of the marine invertebrate species investigated. However, despite the broad presence of surfactants in the invertebrate guts, they found clear herbivore-specific differences in the ability of the surfactants to prevent protein complexation by phlorotannins. Only digestive fluids from two of the isopods, one of which commonly feeds on phlorotannin-rich algae, showed any inhibitory effect on protein precipitation by brown algal phlorotannins. They concluded that phlorotannins are capable of acting as dose-dependent digestibility-reducing substances and that surfactants can be effective in preventing phlorotannin–protein precipitation.

In vitro studies by Stern and co-workers[124,164] have attempted to further define the parameters that control precipitation of phlorotannins by proteins. In general, they found that pH, redox condition, and solution composition influenced phlorotannin–protein interactions. Unlike terrestrial tannins, they found that phlorotannins from marine algae spontaneously oxidized and reacted with protein to form dark-colored complexes that were resistant to dissolution in the absence of a reducing agent. Not surprisingly, Stern and co-workers[164] found that there was considerable variability in terms of the importance of oxidation to interactions of phlorotannins and proteins. They found that the susceptibility of tannins to oxidation could be explained by phlorotannin substituent types and patterns. For example, *in vitro* studies suggested that phlorotannins from the brown alga *Carpophyllum maschalocarpum* were most susceptible to oxidation, while those from *Lobophora variegata* had only a limited tendency to oxidize. Their results suggested that structural differences between phlorotannins may substantially influence the interaction between phlorotannins and proteins.[124,164] Ether-linked phlorotannins with *ortho* substituted hydroxyl groups on one or more rings were thought to be most susceptible to oxidation, and aryl–aryl linked phlorotannins such as the fucols appeared to be least susceptible to oxidation. Interestingly, mannitol, one of the brown algal storage products, is a known antioxidant and may serve to limit the oxidation of phlorotannins under some conditions.[189]

Covalent bonds between oxidized phenolics and proteins were shown to be more likely to form at high pH since phenolics are more readily oxidized at high pH. This was particularly true with basic proteins and less so with acidic proteins.[124,163,164] Because precipitation of phlorotannin–protein complexes was independent of temperature, hydrophobic forces were not considered likely to be important in their formation.[124,164] These studies suggest that, although phlorotannin structure is important, at least part of the variation in sensitivity to dietary tannins is due to variation in herbivore gut chemistry.

In *ex vivo* experiments, Stern[124] also showed that phlorotannins precipitated protein from unaltered gut fluids obtained from both Australian and Californian invertebrate herbivores. Although the results showed no difference in precipitation with gut fluids from different regions, the analysis did indicate that phlorotannins from different seaweeds precipitated at different levels and that this effect depended on the herbivore gut fluid used.

It is difficult to know if the results of the *in vitro* and *ex vivo* studies to date can be accurately extrapolated to *in vivo* systems.[70,124] However, the work clearly demonstrates the subtlety and complexity of the interactions. In addition, it suggests that there is agreement between the phlorotannin–protein complexation that is observed at acid pH *in vitro* and the low assimilation efficiencies of phlorotannin-rich food observed in most acid gut species.[25,34,120,121,124] An exception appears in herbivores where there is the strong likelihood that other factors mitigate or alter the effect of the acid environment on phlorotannin–protein complexes.[25,35,49,56,124]

B. HERBIVORES AND OTHER SECONDARY METABOLITES

Although most work on the effect of secondary metabolites on digestive processes has occurred in the phlorotannin–herbivore system, thousands of other nonphlorotannin metabolites have been identified as putative deterrents.[1] Studies of some of these have shown that the metabolites can have negative post-ingestive consequences.[190,191] And, although little work has been done on these secondary metabolites with respect to consumer digestion processes per se, the complexity of the interactions is apparent from studies that suggest that the effectiveness of the chemical defense is linked to prey nutritional quality,[36,37] calcification,[37,192–195] and recent prey/consumer history.[32]

There have been few studies of the effects of nonphlorotannin secondary metabolites on herbivore assimilation efficiency. Targett and Targett[52] coated extracts containing the feeding deterrents halimedatetraacetate (Structure 11.1) and halimedatrial (Structure 11.2) from *Halimeda incrassata* onto *Thalassia testudinum* blades and found that *H. incrassata* extracts did not alter *S. radians* digestive efficiency or gut evacuation rate. However, they noted that the metabolites were unpalatable and did alter the feeding rate of the fish, although they did not affect the fish's digestive energetics.

Other studies have focused on herbivore growth, which may reflect, at least in part, metabolite impact on digestion. For example, Hay et al.[190] found that the diterpene pachydictyol A

(Structure 11.3) from the brown alga *Dictyota dichotoma* significantly reduced growth of the spottail pinfish, *Diplodus holbrooki*. However, ingestion of this same compound by the sea hare *Aplysia juliana* had no impact on its growth. Ingestion of two other secondary metabolites, malyngamide B (Structure 11.4) from the cyanobacterium *Microcoleus lyngbyaceus* and lufariellolide (Structure 11.5) from the sponge *Hyrtios* spp., did reduce growth of the sea hare *Aplysia juliana* by over 70% relative to the control animals.[191] These variations are clearly related to processes occurring in the digestive system, although the exact mechanisms have not been described. In the above example, the metabolites appear to be differentially sequestered in the sea hare digestive gland, with pachydictyol A (Structure 11.3) and malyngaminde B (Structure 11.4) being sequestered at much higher concentrations in the digestive gland than lufariellolide. The feces of *A. juliana* were also generally depleted of secondary metabolites relative to the diet, again suggesting uptake, alteration, or degradation in the gut.[191]

Working with the ink from *Aplysia californica*, Coelho and co-workers[99] have begun to address the mechanisms by which sea hares acquire and modify pigment from red algal cells. Their study of the ultrastructure of the digestive gland revealed the presence of a previously undescribed cell type, the rhodoplast digestive cell. They speculate that these cells phagocytize red algal chloroplasts and then digest them in digestive vacuoles, modifying the associated pigment. Since the algal diet of *A. californica* provides many of its defensive chemicals,[26] it will be interesting to see if the mechanisms identified for acquisition and modification of red algal cell pigment will also function for the many other secondary metabolites present in the sea hare's diet. Indeed, it has been suggested that sea hares have generic mechanisms for sequestering algal metabolites, rather than mechanisms that are tightly linked to particular compounds.[196] Furthermore, there is evidence that the ingested metabolites can be modified as part of the digestive process.[196,197] Once the metabolites are sequestered, they do not appear to turn over rapidly.[96,97,196] Because they are not located optimally for defense, there has been speculation that it may simply be energetically less expensive for opistho-branchs to store sequestered metabolites rather than detoxify them.[196,198,199]

For sacoglossans, which are invertebrate herbivores that specialize primarily on siphonaceous green algae, there is a more direct link between secondary metabolites and defensive function.[200–204] They are known to sequester functional chloroplasts and secondary metabolites from their algal diets. Of particular interest is evidence that, like sea hares, at least some species can modify dietary-derived metabolites. For example, the ascoglossan *Elysia halimedae*, which feeds preferentially on the green alga *Halimeda macroloba,* modifies a major diterpenoid metabolite, halimedatetraacetate (Structure 11.1), by reducing an aldehyde to a primary alcohol (Structure 11.6). The modified metabolite, which is sequestered and released when the animal is irritated, has been shown to be a fish deterrent.[202] Several other saccoglossans, *Oxynoe olivacea*, *Lobiger serradifalci*, and *Cylin-drobulla fragilis*, which feed primarily on *Caulerpa* species, are able to modify the diet-derived sesquiterpene caulerpenyne (Structure 11.7) into oxytoxin 1 (Structure 11.8), a monoaldehyde, and oxytoxin 2 (Structure 11.9), a 1,4-conjugated dialdehyde.[204,205] These are sequestered in the animals and thought to have a defensive role.[204]

Finally, although not considered a secondary metabolite, it is worth noting the putative role of calcium carbonate as a chemical defense as well as its effect on the digestive processes of herbivores.

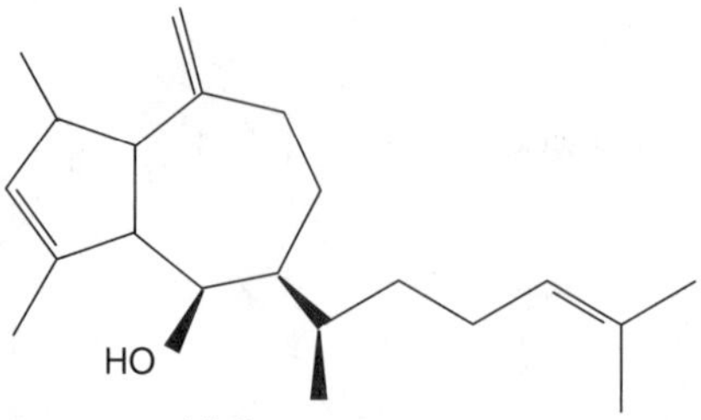

11.3

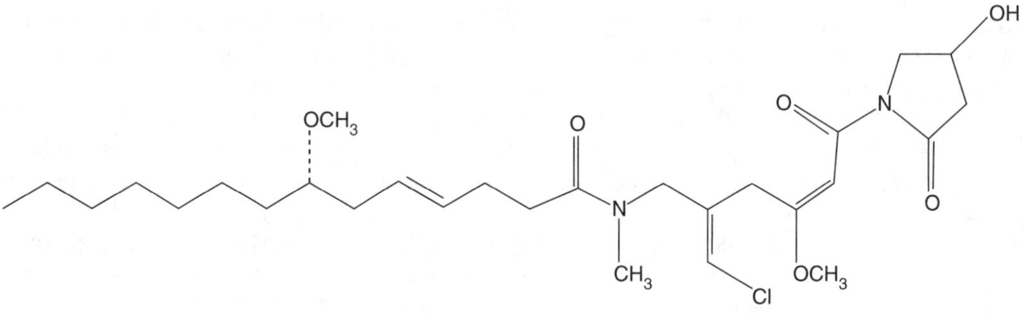

11.4

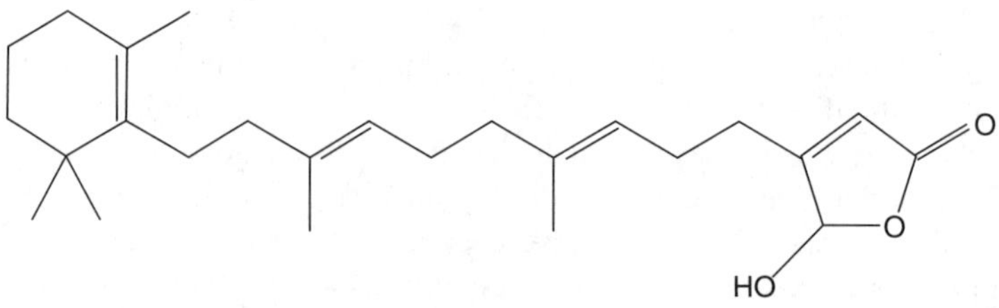

11.5

Calcium carbonate, particularly in marine algae, has long been known to function as a structural defense, limiting consumption of carbonate rich plants by some consumers because of its hardness. It has also been shown to act as a chemical defense by deterring the rate of feeding for consumers with acidic gut environments.[37,192–195,206] In these herbivores, it has been hypothesized that $CaCO_3$ may function in two ways, by neutralizing the low gut pH, thereby affecting nutrient availability, and by causing the release of large quantities of CO_2.

IV. SUMMARY

The effect of plant secondary metabolites on herbivore digestion and nutrition is dependent not only on the characteristics of the metabolites, but also on subtle interactions between the metabolite(s), the gut environment, and plant nutritive qualities. In the marine environment, such interactions are complex. This is due in part to (1) the diversity of marine plants (including vascular and nonvascular species), (2) the diverse array of secondary metabolites that may be encountered by vertebrate and invertebrate grazers, and (3) the diversity of herbivore gut environments.

To date, most studies describing the effect of plant secondary metabolites on herbivore digestion in marine systems have focused upon the macromolecular-complexing activity of algal phlorotannins. While these compounds have demonstrated negative effects on marine herbivore food preference and digestion efficiency in some circumstances, it is clear that they can no longer be

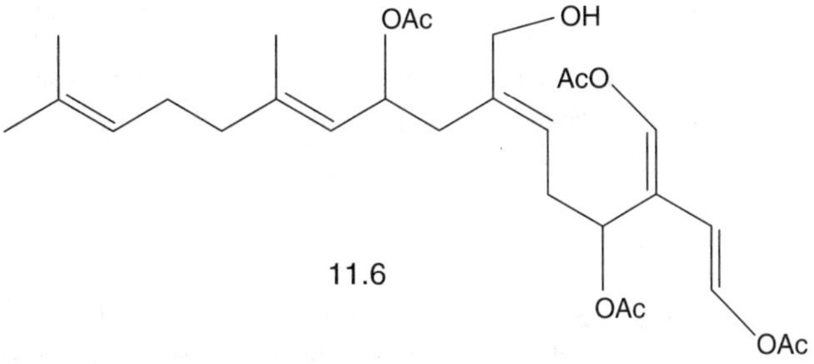

11.6

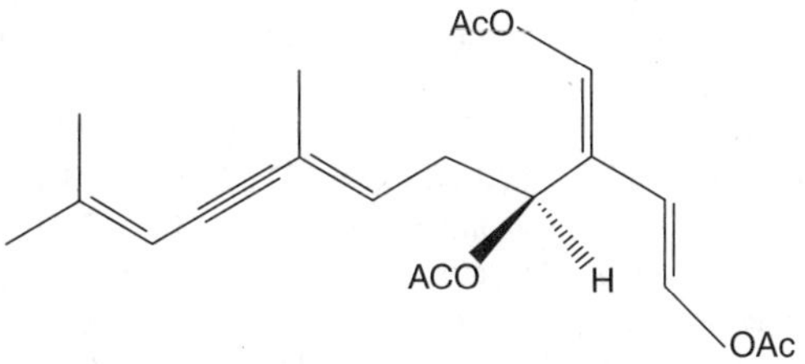

11.7

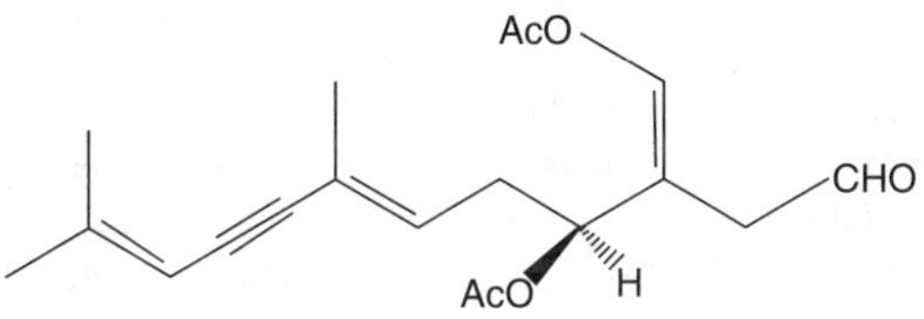

11.8

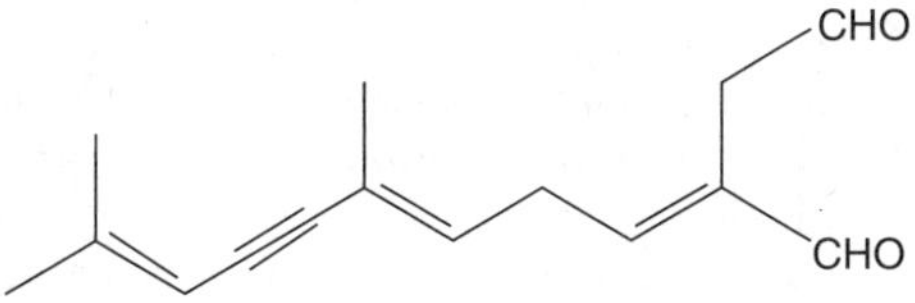

11.9

considered broad-scope plant defenses. Rather, their antidigestive activity can be counteracted in marine herbivores possessing guts with certain chemical (pH, redox condition, cation concentration, surfactants, etc.) and biological (microbes) characteristics. As a result of numerous *in vivo*, *ex vivo*, and *in vitro* studies, a framework linking the bioactivity of dietary phlorotannins with general herbivore gut features has emerged. Additional studies are required to test this framework and to further enhance it by considering other variables such as plant nutritional quality, degree of plant calcification, presence of antioxidants, and the ramifications of multiple defenses.

No such framework yet exists for algal-derived, nonphlorotannin, secondary metabolites. Modifications of diet-derived, nonphlorotannin, secondary metabolites have been identified in some herbivores, and this represents a first step toward understanding how herbivore digestive processes effect metabolite changes.

It is not possible to generalize about consumer digestive responses to secondary metabolites because the activity is inextricably linked to both the structural characteristics of the secondary metabolite and the chemical environment to which the secondary metabolite is exposed in the consumer's gut. It may also be linked to other factors. These complex relationships need to be resolved in order to achieve the goal of accurately predicting the outcome of plant–herbivore interactions.

ACKNOWLEDGMENTS

The authors thank R. Banwarth, K. Ferrari, N. Lopanik, and T. Targett for helpful comments on the manuscript. This work was supported by NSF grant OCE 9618112 to N.M.T.

REFERENCES

1. Faulkner, D.J., Marine natural products, *Nat. Prod. Rep.*, 16, 155, 1999.
2. Pawlik, J.R., Chemical ecology of the settlement of benthic marine invertebrates, in *Oceanogr. Mar. Biol. Ann. Rev.*, 30, 273, 1992.
3. Maier, I., Gamete orientation and induction of gametogenesis by pheromones in algae and plants, *Plant Cell Environ.*, 16, 891, 1993.
4. Maier, I., Brown algal pheromones, *Prog. Phycol. Res.*, 11, 51, 1995.
5. Forward, R.B., Jr., Frankel, D.A.Z., and Rittschof, D., Molting of megalopae from the blue crab *Callinectes sapidus:* effects of offshore and estuarine cues, *Mar. Ecol. Prog. Ser.*, 113, 55, 1994.
6. Boland, W., The chemistry of gamete attraction: chemical structures, biosynthesis, and (a)biotic degradation of algal pheromones, *Proc. Natl. Acad. Sci. U.S.A.*, 92, 37, 1995.
7. Coll, J.C., Bowden, B.F., Meehan, G.V., Konig, G.M., Carroll, A.R., Tapiolas, D.M., Alino, P.M., Heaton, A., and de Nys, R., Chemical aspects of mass spawning in corals. 1. Sperm-attractant molecules in the eggs of the scleractinian coral *Montipora digitata, Mar. Biol.*, 118, 177, 1994.
8. Coll, J.C., Leone, P.A., Bowden, B.F., Carroll, A.R., Koenig, G.M., Heaton, A., de Nys, R., Maida, M., Alino, P.M., Willis, R.H., Babcock, R.C., Florian, Z., Clayton, M.N., Miller, R.L., and Alderslade, P.N., Chemical aspects of mass spawning in corals. 2. (-)-Epi-thunbergol, the sperm attractant in the eggs of the soft coral *Lobophytum crassum* (Cnidaria: Octocorallia), *Mar. Biol.*, 123, 137, 1995.
9. Morse, A.N.C. and Morse, D.E., Flypapers for coral and other planktonic larvae, *Bioscience*, 46, 254, 1996.
10. Clare, A.S., Eicosanoids and egg-hatching synchrony in barnacles: evidence against a dietary precursor to egg-hatching pheromone, *J. Chem. Ecol.*, 23, 2299, 1997.
11. Welch, J.M., Rittschof, D., Bullock, T.M., and Forward, R.B., Jr., Effects of chemical cues on settlement behavior of blue crab *Calllinectes sapidus* postlarvae, *Mar. Ecol. Prog. Ser.*, 154, 143, 1997.
12. Hay, M.E. and Fenical, W., Marine plant–herbivore interactions: the ecology of chemical defense, *Annu. Rev. Ecol. Syst.*, 19, 111, 1988.
13. Hay, M.E. and Fenical, W., Chemical mediation of seaweed–herbivore interactions, in *Plant–Animal Interactions in the Marine Benthos*, John, D.M., Hawkins, S.J., and Price, J.H., Eds., Clarendon Press, Oxford, UK, 1992, 319.
14. Duffy, J.E. and Hay, M.E., Seaweed adaptations to herbivory, *Bioscience*, 40, 368, 1990.
15. Hay, M.E., Fish seaweed interactions of coral reefs: effects of herbivorous fishes and adaptations of their prey, in *The Ecology of Fishes on Coral Reefs*, Sale, P.F., Ed., Academic Press, San Diego, CA, 1991, 96.
16. Hay, M.E., Marine chemical ecology: what's known and what's next?, *J. Exp. Mar. Biol. Ecol.*, 200, 103, 1996.
17. Coll, J.C., The chemistry and chemical ecology of octocorals (Coelenterata, Anthozoa, Octocorallia), *Chem. Rev.*, 92, 613, 1992.
18. Hay, M.E. and Steinberg, P.D., The chemical ecology of plant–herbivore interactions, in *Herbivores their Interaction with Secondary Plant Metabolites: Evolutionary and Ecological Processes*, Vol. 2, 2nd ed., Rosenthal, G.A. and Berenbaum, M., Eds., Academic Press, New York, 1992, 371.
19. Paul, V.J., *Ecological Roles of Marine Natural Products*, Comstock Publishing Associates, Cornell University Press, Ithaca, NY, 1992.
20. Lindquist, N., Hay, M.E., and Fenical, W., Defense of ascidians and their conspicuous larvae: adult vs. larval chemical defenses, *Ecol. Monogr.*, 62, 547, 1992.
21. Proksch, P., Defensive roles for secondary metabolites from marine sponges and sponge-feeding nudibranchs, *Toxicon*, 32, 639, 1994.
22. Pawlik, J.R., Charnas, B., Toonen, R.J., and Fenical, W., Defenses of Caribbean sponges against predatory reef fish. 1. Chemical deterrency, *Mar. Ecol. Prog. Ser.*, 127, 183, 1995.

23. McClintock, J.B. and Baker, B.J., A review of the chemical ecology of Antarctic marine invertebrates, *Am. Zool.*, 37, 329, 1997.

24. Hay, M.E., Piel, J., Boland, W., and Schnitzler, I., Seaweed sex pheromones and their degradation products frequently suppress amphipod feeding but rarely suppress sea urchin feeding, *Chemoecology*, 8, 91, 1998.

25. Targett, N.M. and Arnold, T.M., Predicting the effects of brown algal phlorotannins on marine herbivores in tropical and temperate oceans, *J. Phycol.*, 34, 195, 1998.

26. Nolen, T.G., Johnson, P.M., Kicklighter, C.K., and Capo, T., Ink secretion by the marine snail *Aplysia californica* enhances its ability to escape from a natural predator, *J. Comp. Physiol.*, A176, 239, 1995.

27. Pavia, H., Cervin, G., Lindgren, A., and Åberg, P., Effects of UV-B radiation and simulated herbivory on phlorotannins in the brown alga *Ascophyllum*, *Mar. Ecol. Prog. Ser.*, 157, 139, 1997.

28. Adler, F.R. and Morris, W.F., A general test for interaction modification, *Ecology*, 75, 1552, 1994.

29. Billick, I. and Case, T.J., Higher order interactions in ecological communities: what are they and how can they be detected?, *Ecology*, 75, 1529, 1994.

30. Kareiva, P., Higher order interactions as a foil to reductionist ecology, *Ecology*, 75, 1527, 1994.

31. Wootton, J.T., Putting the pieces together: testing the independence of interactions among organisms, *Ecology*, 75, 1544, 1994.

32. Cronin, G. and Hay, M.E., Susceptibility to herbivores depends on recent history of both the plant and animal, *Ecology*, 77, 1531, 1996.

33. Choat, J.H. and Clements, K.D., Vertebrate herbivores in marine and terrestrial environments: a nutritional ecology perspective, *Annu. Rev. Ecol. Syst.*, 29, 375, 1998.

34. Boettcher, A.A. and Targett, N.M., Role of polyphenolic molecular size in reduction of assimilation efficiency in *Xiphister mucosus*, *Ecology*, 74, 891, 1993.

35. Targett, N.M., Boettcher, A.A., Targett, T.E., and Vrolijk, N.H., Tropical marine herbivore assimilation of phenolic-rich plants, *Oecologia*, 103, 170, 1995.

36. Duffy, J.E. and Paul, V.J., Prey nutritional quality and the effectiveness of chemical defenses against tropical reef fishes, *Oecologia*, 90, 333, 1992.

37. Hay, M.E., Kappel, Q.E., and Fenical, W., Synergisms in plant defenses against herbivores: interactions of chemistry, calcification, and plant quality, *Ecology*, 75, 1714, 1994.

38. Choat, J.H., The biology of herbivorous fishes on coral reefs, in *The Ecology of Fishes on Coral Reefs*, Sale, P.F., Ed., Academic Press, San Diego, CA, 1991, 120.

39. Horn, M.H. and Ojeda, F.P., Herbivory, in *Intertidal Fishes*, Horn, M.H., Martin, K.L.M., and Chotkowski, M.A., Eds., Academic Press, San Diego, CA, 1999, 399.

40. Jobling, M., *Environmental Biology of Fishes*, Chapman and Hall, London, 1995.

41. Bromley, P.J., The role of gastric evacuation experiments in quantifying the feeding rates of predatory fish, *Rev. Fish Biol. Fisher.*, 4, 36, 1994.

42. Gerking, S.D., *Feeding Ecology of Fish*, Academic Press, San Diego, CA, 1994.

43. Lobel, P.S., Trophic biology of herbivorous reef fish: alimentary pH and digestive capabilities, *J. Fish. Biol.*, 19, 365, 1981.

44. Horn, M.H., Biology of marine herbivorous fishes, *Oceanogr. Mar. Biol. Annu. Rev.*, 27, 167, 1989.

45. Horn, M.H. and Messer, K.S., Fish guts as chemical reactors: a model of the alimentary canals of marine herbivorous fishes, *Mar. Biol.*, 113, 527, 1992.

46. Kloareg, B. and Quatrano, R.S., Structure of the cell walls of marine algae and ecophysiological functions of the matrix polysaccharides, *Oceanogr. Mar. Biol. Ann. Rev.*, 26, 259, 1988.

47. Pollak, P.E. and Montgomery, W.L., Giant bacterium (*Epulopiscium fishelsoni*) influences digestive enzyme activity of an herbivorous surgeonfish (*Acanthurus nigrofuscus*), *Comp. Biochem. Physiol.*, 108A, 657, 1994.

48. Ojeda, F.P. and Cáceres, C.W., Digestive mechanisms in *Aplodactylus punctatus* (Valenciennes): a temperate marine herbivorous fish, *Mar. Ecol. Prog. Ser.*, 118, 37, 1995.

49. Clements, K.D. and Choat, J.H., Comparison of herbivory in the closely-related marine fish genera *Girella* and *Kyphosus*, *Mar. Biol.*, 127, 579, 1997.

50. Rimmer, D.W. and Wiebe, W.J., Fermentative microbial digestion in herbivorous fishes, *J. Fish Biol.*, 31, 229, 1987.

51. Clements, K.D., Fermentation and gastrointestinal microorganisms in fishes, in *Gastrointestinal Microbiology*, Vol. 1, Mackie, R.I. and White, B.A., Ed., Chapman and Hall, New York, 1997, 156.

52. Targett, T.E. and Targett, N.M., Energetics of food selection by the herbivorous parrotfish *Sparisoma radians*: roles of assimilation efficiency, gut evacuation rate, and algal secondary metabolites, *Mar. Ecol. Prog. Ser.*, 66, 13, 1990.

53. Horn, M.H., Mailhiot, K.F., Fris, M.B., and McClanahan, L.L., Growth, consumption, assimilation and excretion in the marine herbivorous fish *Cebidichthys violaceus* (Girard) fed natural and high protein diets, *J. Exp. Mar. Biol. Ecol.*, 190, 97, 1995.

54. Gerking, S.D., Assimilation and maintenance ration of an herbivorous fish, *Sarpa salpa*, feeding on a green alga, *Trans. Am. Fish. Soc.*, 113, 378, 1984.

55. Anderson, T.A., Mechanisms of digestion in the marine herbivore, the luderick, *Girella tricuspidata* (Quoy and Gaimard), *J. Fish Biol.*, 39, 535, 1991.

56. Sturm, E.A. and Horn, M.H., Food habits, gut morphology and pH, and assimilation efficiency of the zebraperch *Hermosilla azurea*, an herbivorous kyphosid fish of temperate marine waters, *Mar. Biol.*, 132, 515, 1998.

57. Zemke-White, W.L., Clements, K.D., and Harris, P.J., Acid lysis of macroalgae by marine herbivorous fishes: myth or digestive mechanism?, *J. Exp. Mar. Biol. Ecol.*, 233, 95, 1999.

58. Targett, N.M., unpublished data.

59. Fishelson, L., Montgomery, W.L., and Myrberg, A.A., Jr., A unique symbiosis in the gut of tropical herbivorous surgeonfish (Acanthuridae: Teleostei) from the Red Sea, *Science*, 229, 49, 1985.

60. Clements, K.D., Gleeson, V.P., and Slaytor, M., Short-chain fatty acid metabolism in temperate marine herbivorous fish, *J. Comp. Physiol. B*, 164, 372, 1994.

61. Rimmer, D.W., Changes in the diet and the development of microbial digestion in juvenile buffalo bream, *Kyphosus cornelii*, *Mar. Biol.*, 92, 443, 1986.

62. Clements, K.D. and Choat, J.H., Fermentation in tropical marine herbivorous fishes, *Physiol. Zool.*, 68, 355, 1995.

63. Barwell, C.J., Blunde, G., and Manandhar, P.D., Isolation and characterization of brown algal polyphenols as inhibitors of α-amylase, lipase and trypsin, *J. Appl. Phycol.*, 1, 319, 1989.

64. Kandel, J.S., Horn, M.H., and Van Antwerp, W., Volatile fatty acids in the hindguts of herbivorous fishes from temperate and tropical marine waters, *J. Fish Biol.*, 45, 527, 1994.

65. Stellwag, E.J., Smith, T.D., and Luczkovich, J.J., Characterization and ecology of carboxymethylcellulase-producing anaerobic bacterial communities associated with the intestinal tract of the pinfish, *Lagodon rhomboides*, *Appl. Environ. Micro.*, 61, 813, 1995.

66. Stevens, C.E., *Comparative Physiology of the Vertebrate Digestive System*, Cambridge University Press, Cambridge, UK, 1988.

67. Seeto, G.S., Veivers, P.C., Clements, K.D., and Slaytor, M., Carbohydrate utilisation by microbial symbionts in the marine herbivorous fishes *Odax cyanomelas* and *Crinodous lophodon*, *J. Comp. Physiol. B*, 165, 571, 1996.

68. Purchon R.D., *The Biology of the Mollusca*, Pergamon Press, Oxford, UK, 1977.

69. Dall, W. and Moriarty, D.J.W., Functional aspects of nutrition and digestion, in *The Biology of Crustacea*, Vol. 5, Mantel, L.H., Ed., Academic Press, New York, 1983, 215.

70. Tugwell, S. and Branch, G.M., Effects of herbivore gut surfactants on kelp polyphenol defenses, *Ecology*, 73, 205, 1992.

71. Hammen, C.S., *Marine Invertebrates: Comparative Physiology*, Hammen, C.S., Ed., University Press of New England, Hanover, N.H., 1980.

72. Lawrence, J.M., Digestion, in *Echinoderm Nutrition*, Lawrence, J.M. and Jangoux, M., Eds., A.A. Balkema, Rotterdam, 1982.

73. Klinger, T.S., McClintock, J.B., and Watts, S.A., Activities of digestive enzymes of polar and subtropical echinoderms, *Polar Biol.*, 18, 154, 1997.

74. Anderson, J.M., Aspects of nutritional physiology, in *Physiology of Echinodermata*, Boolootian, R.A., Ed., John Wiley & Sons, New York, 1966, 329.

75. de Ridder, C. and Jangoux, M., Digestive systems: Echinoidea, in *Echinoderm Nutrition*, Jangoux, M. and Lawrence, J.M., Eds., A.A. Balkema, Rotterdam, 1982, 213.

76. de Ridder, C. and Lawrence, J.M., Food and feeding mechanisms: Echinoidea, in *Echinoderm Nutrition*, Jangoux, M. and Lawrence, J.M., Eds., A.A. Balkema, Rotterdam, 1982, 57.

77. Lawrence, J.M., *A Functional Biology of Echinoderms*, Johns Hopkins University Press, Baltimore, MD, 1987.

78. Lawrence, J.M., On the relationships between marine plants and sea urchins, *Oceanogr. Mar. Biol. Annu. Rev.*, 13, 213, 1975.

79. Klinger, T.S., Activities and kinetics of digestive χ- and β-glucosidase and β-galactosidase of five species of echinoids (Echinodermata), *Comp. Biochem. Physiol.*, 78A, 597, 1984.

80. Binyon, J., Salinity tolerance and ionic regulation, in *Physiology of Echinodermata*, Boolootian, R.A., Ed., John Wiley & Sons, New York, 1966, 359.

81. Binyon, J., *Physiology of Echinoderms*, Pergamon Press, Oxford, New York, 1972.

82. Ruppert, E.E. and Barnes, R.D., *Invertebrate Zoology*, Saunders College Publishing Co., Philadelphia, 1994.

83. Unkles, S.E., Bacterial flora of the sea urchin *Echinus esculentus*, *Appl. Env. Microb.*, 34, 347, 1977.

84. Guerinot, M. and Patriquin, D., N_2-fixing vibrios isolated from the gastrointestinal tract of sea urchins, *Can. J. Microbiol.*, 27, 311, 1981.

85. Guerinot, M. and Patriquin, D., The association of N_2-fixing bacteria with sea urchins, *Mar. Biol.*, 62, 197, 1981.

86. Kohn, A.J., Feeding biology of gastropods, in *The Mollusca*, Vol. 5 (Part 2), Wilbur, K.M., Ed., Academic Press, New York, 1983, 1.

87. Fretter, V. and Graham, A., *British Prosobranch Molluscs*, Royal Society of London, London, UK, 1962.

88. Galli, D. and Giese, A., Carbohydrate digestion in a herbivorous snail, *Tegula funebralis, J. Exp. Zool.*, 140, 415, 1959.

89. Gacesa, P., Minireview: enzymic degradation of alginates, *Int. J. Biochem.*, 24, 545, 1992.

90. Owen, G., Digestion, in *Physiology of Mollusca*, Wilbur, K. and Yonge, C., Eds., Academic Press, London, UK, 1966, 53.

91. Eales, N.B., *Aplysia*, Liverpool Marine Biological Committee, *Proc. Trans. Liverpool Biol. Soc. L.M.B.C. Mem.*, 35, 183, 1921.

92. Fretter, V., The structure and function of the alimentary canal of some species of Polyplacophora (Mollusca), *Trans. R. Soc. Edinb.*, 59 (Part I), 119, 1937.

93. Fretter, V., The structure and function of the alimentary canal of some tectibranch molluscs, *Trans. R. Soc. Edinb.*, 59 (Part III), 599, 1939.

94. Graham, A., The structure and function of the alimentary canal of aeolid molluscs, with a discussion on their nematocysts, *Trans. R. Soc. Edinb.*, 59 (Part II), 267, 1938.

95. Howells, H.H., The structure and function of the alimentary canal of *Aplysia punctata, Q. J. Microsc. Sci.*, 83, 357, 1942.

96. Stallard, M.O. and Faulkner, D.J., Chemical constituents of the digestive gland of the sea hare *Aplysia californica*. I. Importance of the diet, *Comp. Biochem. Physiol. B*, 49, 25, 1974.

97. Stallard, M.O. and Faulkner, D.J., Chemical constituents of the digestive gland of *Aplysia californica*. II. Chemical transformation, *Comp. Biochem. Physiol. B*, 49, 37, 1974.

98. Kandel, E.R., *Behavioral Biology of Aplysia*, W.H. Freeman and Co., San Francisco, CA, 1979.

99. Coelho, L., Prince, J., and Nolen, T.G., Processing of defensive pigment in *Aplysia californica*: acquisition, modification and mobilization of the red algal pigment r-phycoerythrin by the digestive gland, *J. Exp. Biol.*, 201, 425, 1998.

100. Shram, F.R., *Crustacea*, Oxford University Press, New York, 1986, 8.

101. Monk, D.C., The digestion of cellulose and other dietary components, and pH of the gut in the amphipod *Gammarus pulex* (L.), *Freshwater Biol.*, 7, 431, 1977.

102. Borowsky, R. and Guarna, M.M., Excess amylase in *Gammarus palustris* (Crustacea: Amphipoda); its release into and possible roles in the environment, *Mar. Biol.*, 101, 529, 1989.

103. Bärlocher, F. and Howatt, S.L., Digestion of carbohydrates and protein by *Gammarus mucronatus* Say (Amphipoda), *J. Exp. Mar. Biol. Ecol.*, 104, 229, 1986.

104. Stuart, V., Head, E.J.H., and Mann, K.H., Seasonal changes in the digestive enzyme levels of the amphipod *Corophium volutator* (Pallas) in relation to diet, *J. Exp. Mar. Biol. Ecol.*, 88, 243, 1985.

105. Rhoades, D.F. and Cates, R.G., Toward a general theory of plant antiherbivore chemistry, in *Recent Advances in Phytochemistry, Volume 10, Biochemical Interaction Between Plants and Insects*, Wallace, J.W. and Mansell, R.L., Eds., Plenum Press, New York, 1976, 168.

106. Swain, T., Tannins and lignins, in *Herbivores: Their Interaction with Secondary Plant Metabolites*, Rosenthal, G.A. and Janzen, D.H., Eds., Academic Press, New York, 1979, 657.

107. Ragan, M.A. and Glombitza, K.W., Phlorotannins, brown algal polyphenols, *Prog. Phycol. Res.*, 4, 130, 1986.

108. Hagerman, A.E. and Butler, L.G., Tannins and lignins, in *Herbivores: Their Interaction with Secondary Plant Metabolites: The Chemical Participants*, Vol. 1, 2nd edition, Rosenthal, G.A. and Berenbaum, M.R., Eds., Academic Press, New York, 1991, 355.

109. Ragan, M.A., Physodes and the phenolic compounds of brown algae. Composition and significance of physodes *in vivo*, *Bot. Mar.*, 19, 145, 1976.

110. Kaur, I. and Vijayaraghavan, M.R., Physode distribution and genesis in *Sargassum vulgare* (*C. Agardh* and *Sargassum johnstonii* Setchell and Gardner), *Aquat. Bot.*, 45, 375, 1992.

111. Schmitt, T.M., Hay, M.E., and Lindquist, N., Constraints on chemically-mediated coevolution: multiple functions for seaweed secondary metabolites, *Ecology*, 76, 107, 1995.

112. Vreeland, V. and Laetsch, W.M., A gelling carbohydrate in algal cell wall formation, in *Organisation and Assembly of Plant and Animal Extracellular Matrix*, Adair, W.S. and Mecham, R.P., Eds., Academic Press, San Diego, CA, 1990, 137.

113. Schoenwaelder, M.E.A. and Clayton, M.N., The secretion of phenolic compounds following fertilization in *Acrocarpia paniculata* (Fucales, Phaeophyta), *Phycologia*, 37, 40, 1998.

114. Schoenwaelder, M.E.A. and Clayton, M.N., Secretion of phenolic substances into the zygote wall and cell plate in embryos of *Hormosira* and *Acrocarpia* (Fucales, Phaeophyceae), *J. Phycol.*, 34, 969, 1998.

115. Geiselman, J.A. and McConnell, O.J., Polyphenols in brown algae *Fucus vesiculosus* and *Ascophyllum nodosum*: chemical defenses against the marine herbivorous snail, *Littorina littorea*, *J. Chem. Ecol.*, 7, 1115, 1981.

116. Steinberg, P.D., Algal chemical defense against herbivores: allocation of phenolic compounds in the kelp *Alaria marginata*, *Science*, 223, 405, 1984.

117. Steinberg, P.D., Phenolic compounds in brown algae: chemical defenses against marine herbivores, Ph.D. thesis, University of California, Santa Cruz, Santa Cruz, California, 1984.

118. Steinberg, P.D., Feeding preferences of *Tegula funebralis* and chemical defenses of marine brown algae, *Ecol. Monogr.*, 5, 333, 1985.

119. Steinberg, P.D., Effects of quantitative and qualitative variation in phenolic compounds on feeding in three species of marine invertebrate herbivores, *J. Exp. Mar. Biol. Ecol.*, 120, 221, 1988.

120. Steinberg, P.D. and Van Altena, I.A., Tolerance of marine invertebrate herbivores to brown algal phlorotannins in temperate Australasia, *Ecol. Monogr.*, 62, 189, 1992.

121. Irelan, C.D. and Horn, M.H., Effects of macrophyte secondary chemicals on food choice and digestive efficiency of *Cebidichthys violaceus* (Girard), an herbivorous fish of temperate marine waters, *J. Exp. Mar. Biol. Ecol.*, 153, 179, 1991.

122. Steinberg, P.D., Geographical variation in the interaction between marine herbivores and brown algal secondary metabolites, in *Marine Chemical Ecology*, Paul, V.J., Ed., Cornell Press, New York, 1992, 51.

123. Scalbert, A., Antimicrobial properties of tannins, *Phytochemistry*, 30, 3875, 1991.

124. Stern, J.L., Brown algal phlorotannins and marine invertebrate herbivore digestion, Ph.D. thesis, University of New South Wales, Sydney, Australia, 1998.

125. Arnold, T.M. and Targett, N.M., Quantifying *in situ* rates of phlorotannin synthesis and polymerization in marine brown algae, *J. Chem. Ecol.*, 24, 577, 1998.

126. Arnold, T.M., *In situ* rates of phlorotannin synthesis, polymerization, and turnover in marine brown algae: an evaluation of terrestrial-derived ecological theories in the marine environment, Ph.D. thesis, University of Delaware, Newark, Delaware, 1998.

127. Van Alstyne, K.L., Ehlig, J.M., and Whitman, S.L., Feeding preferences for juvenile and adult algae depend on algal stage and herbivore species, *Mar. Ecol. Prog. Ser.*, 180, 179, 1999.

128. Targett, N.M., Coen, L.D., Boettcher, A.A., and Tanner, C.E., Biogeographic comparisons of marine algal polyphenolics: evidence against a latitudinal trend, *Oecologia*, 89, 464, 1992.

129. Pavia, H. and Åberg, P., Spatial variation in polyphenolic content of *Ascophyllum nodosum* (Fucales, Phaeophyta), *Hydrobiologia*, 326/327, 199, 1996.

130. Van Alstyne, K.L., McCarthy, J.J., III, Hustead, C.L., and Duggins, D.O., Geographic variation in polyphenolic levels of Northeastern Pacific kelps and rockweeds, *Mar. Biol.*, 133, 371, 1999.

131. Denton, A., Chapman, A.R.O., and Markham, J., Size-specific concentration of phlorotannins (anti-herbivore compounds) in three species of *Fucus*, *Mar. Ecol. Prog. Ser.*, 65, 103, 1990.

132. Pederson, A., Studies on phenol content and heavy metal uptake in fucoids, *Hydrobiologia*, 116/117, 498, 1984.

133. Tugwell, S. and Branch, G.M., Differential polyphenolic distribution among tissues in the kelps *Ecklonia maxima*, *Laminaria pallida* and *Macrocystis angustifolia* in relation to plant-defence theory, *J. Exp. Mar. Biol. Ecol.*, 129, 219, 1989.

134. Tuomi, J., Ilvessalo, H., Miemela, P., Siren, S., and Jormalainen, V., Within-plant variation in phenolic content and toughness of the brown alga *Fucus vesiculosus* L., *Bot. Mar.*, 32, 505, 1989.

135. Pfister, C.A., Costs of reproduction in the intertidal kelp: patterns of allocation and life history consequences, *Ecology*, 73, 1586, 1992.

136. Poore, A.G.B., Selective herbivory by amphipods inhabiting the brown alga *Zonaria angustata*, *Mar. Ecol. Prog. Ser.*, 107, 113, 1994.

137. Ragan, M.A. and Jensen, A., Quantitative studies on brown algal phenols. I. Estimation of absolute polyphenol content of *Ascophyllum nodosum* (L.) Le Jol. and *Fucus vesiculosus* (L.), *J. Exp. Mar. Biol. Ecol.*, 30, 209, 1977.

138. Geiselman, J.A., Ecology of chemical defenses of algae against the herbivorous snail *littorina littorea* in the New England rocky intertidal community, Ph.D. thesis, WHOI, Woods Hole, MA, 1980.

139. Steinberg, P.D., Seasonal variation in the relationship between growth rate and phlorotannin production in the kelp *Ecklonia radiata*, *Oecologia*, 102, 169, 1995.

140. Ilvessalo, H. and Tuomi, J., Nutrient availability and accumulation of phenolic compounds in the brown alga *Fucus vesiculosus*, *Mar. Biol.*, 101, 115, 1989.

141. Yates, J.L. and Peckol, P., Effects of nutrient availability and herbivory on polyphenolics in the seaweed *Fucus vesiculosus*, *Ecology*, 74, 1757, 1993.

142. Arnold, T.M., Tanner, C.E., and Hatch, W.I., Phenotypic variation in polyphenolic content of the tropical brown alga *Lobophora variegata* as a function of nitrogen availability, *Mar. Ecol. Prog. Ser.*, 123, 177, 1995.

143. Peckol, P., Krane, J.M., and Yates, J.L., Interactive effects of inducible defense and resource availability on phlorotannins in the North Atlantic brown alga *Fucus vesiculosus*, *Mar. Ecol. Prog. Ser.*, 138, 209, 1996.

144. Arnold, T.M. and Targett, N.M., unpublished data.

145. Cronin, G. and Hay, M.E., Effects of light and nutrient availability on the growth, secondary chemistry, and resistance to herbivory of two brown seaweeds, *Oikos*, 77, 93, 1996.

146. Van Alstyne, K.L., Grazing increases polyphenolic defenses in the intertidal brown alga *Fucus distichus*, *Ecology*, 69, 655, 1988.

147. Hammerstrom, K., Dethier, M.N., and Duggins, D.O., Rapid phlorotannin induction and relaxation in five Washington kelps, *Mar. Ecol. Prog. Ser.*, 165, 203, 1998.

148. Hay, M.E., The role of seaweed chemical defenses in the evolution of feeding specialization and in the mediation of complex interactions, in *Ecological Roles of Marine Natural Products*, Paul, V.J., Ed., Comstock Publishing Associates, Ithaca, NY, 1992, 93.

149. Steinberg, P.D., Estes, J.A., and Winter, F.C., Evolutionary consequences of food chain length in kelp forest communities, *Proc. Natl. Acad. Sci. USA*, 92, 8145, 1995.

150. Cooper-Driver, G.A. and Bhattacharya, M., Role of phenolics in plant evolution, *Phytochemistry*, 49, 1165, 1998.

151. Steinberg, P.D., Biogeographical variation in brown algal polyphenolics and other secondary metabolites: comparison between temperate Australasia and North America, *Oecologia*, 78, 373, 1989.

152. Boettcher, A.A., The role of polyphenolic molecular size in the reduction of assimilation efficiency in some marine herbivores, Master's thesis, University of Delaware, Newark, Delaware, 1992.

153. Steinberg, P.D., Edyvane, K., de Nys, R., Birdsey, R., and Van Altena, I.A., Lack of avoidance of phenolic-rich algae by tropical herbivorous fishes, *Mar. Biol.*, 109, 335, 1991.

154. Steinberg, P.D. and Paul, V.J., Fish feeding and chemical defenses of tropical brown algae in Western Australia, *Mar. Ecol. Prog. Ser.*, 58, 253, 1990.

155. Van Alstyne, K.L. and Paul, V.J., The biogeography of polyphenolic compounds in marine macroalgae: temperate brown algal defenses deter feeding by tropical herbivorous fishes, *Oecologia*, 84, 158, 1990.

156. Taniguchi, K., Kurata, K., and Suzuki, M., Feeding-deterrent effect of phlorotannins from the brown alga *Ecklonia stolonifera* against the abalone *Haliotis discus hannai*, *Nippon Suisan Gakkaishi*, 57, 2065, 1991.

157. Van Altena, I.A. and Steinberg, P.D., Are differences in the responses between North American and Australasian marine herbivores to phlorotannins due to differences in phlorotannin structure?, *Biochem. Syst. Ecol.*, 20, 493, 1992.

158. Bernays, E.A., Cooper-Driver, G., and Bilgener, M., Herbivores and plant tannins, *Adv. Ecol. Res.*, 19, 263, 1989.

159. Bernays, E.A., Plant tannins and insect herbivores: an appraisal, *Ecol. Entomol.*, 6, 353, 1981.

160. Martin, J.S., Martin, M.M., and Bernays, E.A., Failure of tannic acid to inhibit digestion or reduce digestibility of plant protein in gut fluids of insect herbivores: implications for theories of plant defense, *J. Chem. Ecol.*, 13, 605, 1987.

161. Karowe, D.N., Differential effect of tannic acid on two tree-feeding lepidoptera: implications for theories of plant anti-herbivore chemistry, *Oecologia*, 80, 507, 1989.

162. Clausen, T.P., Provenza, F.D., Burritt, E.A., Reichardt, P.B., and Bryant, J.P., Ecological implications of condensed tannin structure: a case study, *J. Chem. Ecol.*, 16, 2381, 1990.

163. Appel, J., Phenolics in ecological interactions: the importance of oxidation, *J. Chem. Ecol.*, 19, 1521, 1993.

164. Stern, J.L., Hagerman, A.E., Steinberg, P.D., and Mason, P.K., Phlorotannin–protein interactions, *J. Chem. Ecol.*, 22, 1877, 1996.

165. Ayres, M.P., Clausen, T.P., Maclean, S.F., Jr., Redman, A.A.M., and Reichardt, P.B., Diversity of structure and antiherbivore activity in condensed tannins, *Ecology*, 78, 1696, 1997.

166. Appel, H.M. and Martin, M.M., Gut redox conditions in herbivorous lepidopteran larvae, *J. Chem. Ecol.*, 16, 3277, 1990.

167. Martin M.M. and Martin, J.S., Surfactants: their role in preventing the precipitation of proteins by tannins in insect guts, *Oecologia*, 61, 342, 1984.

168. Martin, M.M., Rockholm, D.C., and Martin, J.S., Effects of surfactants, pH, and certain cations on precipitation of proteins by tannins, *J. Chem. Ecol.*, 11, 485, 1985.

169. Austin, P.J., Suchar, L.A., Robbins, C.T., and Hagerman, A.E., Tannin-binding proteins in saliva of deer and their absence in saliva of sheep and cattle, *J. Chem. Ecol.*, 15, 1335, 1989.

170. Mole, S., Butler, L.G., and Iason, G., Defense against dietary tannin in herbivores: a survey for proline rich salivary proteins in mammals, *Biochem. Syst. and Ecol.*, 18, 287, 1990.

171. Hanley, T.A., Robbins, C.T., Hagerman, A.E., and McArthur, C., Predicting digestible protein and digestible dry-matter in tannin-containing forages consumed by ruminants, *Ecology*, 73, 537, 1992.

172. McArthur, C., Sanson, G.D., and Beal, A.M., Salivary proline-rich proteins in mammals: Roles in oral homeostasis and counteracting dietary tannin, *J. Chem. Ecol.*, 21, 663, 1995.

173. Juntheikki, M.R., JulkunenTiitto, R., and Hagerman, A.E., Salivary tannin-binding proteins in root vole (*Microtus oeconomus* Pallas), *Biochem. Syst. Ecol.*, 24, 25, 1996.

174. Konno, K., Hirayama, C., and Shinbo, H., Glycine in digestive juice: a strategy of herbivorous insects against chemical defense of host plants, *J. Insect Physiol.*, 43, 217, 1997.

175. Konno, K., Hirayama, C., and Shinbo, H., Unusually high concentration of free glycine in the midgut content of the silkworm, *Bombyx mori*, and other lepidopteran larvae, *Comp. Biochem. Physiol.*, 115A, 229, 1996.

176. Feeny, P.P., Inhibitory effect of oak leaf tannins on the hydrolysis of proteins by trypsin, *Phytochemistry*, 8, 2119, 1969.

177. Zucker, W.V., Tannins: does structure determine function? An ecological perspective, *Am. Nat.*, 121, 335, 1983.

178. Harborne, J.B., *Introduction to Ecological Biochemistry*, 3rd ed., Academic Press, London, UK, 1988.

179. Hagerman, A.E. and Klucher, K.M., Tannin–protein interactions, in *Plant Flavonoids in Biology and Medicine: Biochemical, Pharmacological, and Structure Activity Relationships*, Middleton, E. and Harborne, J., Eds., Alan R. Liss, Inc., New York, 1996.

180. Haslam, E., Plant polyphenols (syn. vegetable tannins) and chemical defenses — a reappraisal, *J. Chem. Ecol.*, 14, 1789, 1988.

181. Felton, G.W., Donato, K., Del Vecchio, R.J., and Duffey, S.S., Activation of plant foliar oxidases by insect feeding reduces nutritive quality of foliage for noctuid herbivores, *J. Chem. Ecol.*, 15, 2667, 1989.

182. Felton, G.W., Workman, J., and Duffey, S.S., Avoidance of antinutritive plant defense: role of midgut pH in Colorado potato beetle, *J. Chem. Ecol.*, 18, 571, 1992.

183. Felton, G.W., Donato, K.K., Broadway, R.M., and Duffey, S.S., Impact of oxidized plant phenolics on the nutritional quality of dietary protein to anoctuid herbivore, *Spodoptera exigua*, *J. Insect Physiol.*, 38, 277, 1992.

184. Oh, H.I., Hoff, J.E., Armstrong, G.S., and Haff, L.A., Hydrophobic interaction in tannin–protein complexes, *J. Agr. Food Chem.*, 28, 394, 1980.

185. Mole, S. and Waterman, P.G., Tannic acid and proteolytic enzymes: enzyme inhibition or substrate deprivation? *Phytochem.*, 26, 99, 1987.

186. Mole, S., Butler, L.G., Hagerman, A.E., and Waterman, P.G., Ecological tannin assays: a critique, *Oecologia*, 78, 93, 1989.

187. Targett, T., unpublished data.

188. Winter, F.C. and Estes, J.A., Experimental evidence for the effects of polyphenolic compounds from *Dictyoneurum californicum* Ruprecht (Phaeophyta: Laminariales) on feeding rate and growth in the red abalone *Haliotus rufescens* Swainson, *J. Exp. Mar. Biol. Ecol.*, 155, 263, 1992.

189. Hoover, K., Kishida, K.T., DiGiorgio, L.A., Workman, J., Alaniz, S.A., Hammock, B.D., and Duffey, S.S., Inhibition of baculoviral disease by plant-mediated perocidase activity and free radical generation, *J. Chem. Ecol.*, 24, 1949, 1998.

190. Hay, M.E., Duffy, J.E., and Pfister, C.A., Chemical defense against different marine herbivores: are amphipods insect equivalents? *Ecology*, 68, 1567, 1987.

191. Pennings, S.C. and Carefoot, T.H., Post-ingestive consequences of consuming secondary metabolites in sea hares (Gastropoda: Opisthobranchia), *Comp. Biochem. Physiol.*, 111C, 249, 1995.

192. Pennings, S.C. and Paul, V.J., Effect of plant toughness, calcification, and chemistry on herbivory by *Dolabella auricularia*, *Ecology*, 73, 1606, 1992.

193. Schupp, P.J. and Paul, V.J., Calcification and secondary metabolites in tropical seaweeds: variable effects on herbivorous fishes, *Ecology*, 75, 1172, 1994.

194. Meyer, K.D. and Paul, V.J., Variation in secondary metabolite and aragonite concentrations in the tropical green seaweed *Neomeris annulata*: effects on herbivory by fishes, *Mar. Biol.*, 122, 537, 1995.

195. Pennings, S.C., Puglisi, M.P., Pitlik, T.J., Himaya, A.C., and Paul, V.J., Effects of secondary metabolites and CaCO3 on feeding by surgeonfishes and parrotfishes: within-plant comparisons, *Mar. Ecol. Prog. Ser.*, 134, 49, 1996.

196. Pennings, S.C. and Paul, V.J., Sequestration of dietary secondary metabolites by three species of sea hares: Location, specificity and dynamics, *Mar. Biol.*, 117, 535, 1993.

197. Paul, V.J. and Pennings, S.C., Diet-derived chemical defenses in the sea hare *Stylocheilus longicauda* (Quoy et Gaimard 1824), *J. Exp. Mar. Biol. Ecol.*, 151, 227, 1991.

198. Pennings, S.C., Paul, V.J., Dunbar, D.C., Hamann, M.T., Lumbang, W.A., Novack, B., and Jacobs, R.S., Unpalatable compounds in the marine gastropod *Dolabella auricularia*: distribution and effect of diet, *J. Chem. Ecol.*, 25, 735, 1999.

199. de Nys, R., Steinberg, P.D., Rogers, C.N., Charlton, T.S., and Duncan, M.W., Quantitative variation of secondary metabolites in the sea hare *Aplysia parvula* and its host plant, *Delisea pulchra*, *Mar. Ecol. Prog. Ser.*, 130, 135, 1996.

200. Doty, M.S. and Aguilar-Santos, G., Transfer of toxic algal substances in marine food chains, *Pac. Sci.*, 24, 351, 1970.

201. Jensen, K.R., Defensive behavior and toxicity of ascoglossan opisthobranch *Mourgan agermaineae* Marcus, *J. Chem. Ecol.*, 10, 475, 1984.

202. Paul, V.J. and Van Alstyne, K.L., Use of ingested algal diterpenoids by *Elysia halimedae* Macnae (Opisthobranchia:Ascoglossa) as antipredator defenses, *J. Exp. Mar. Biol. Ecol.*, 119, 15, 1988.

203. Jensen, K.R., Behavioural adaptations and diet specificity of sacoglossan opisthobranchs, *Ethol. Ecol. Evol.*, 6, 87, 1994.

204. Gavagnin, M., Marin, A., Castelluccio, F., Villani, G., and Cimino, G., Defensive relationships between *Caulerpa prolifera* and its shelled sacoglossan predators, *J. Exp. Mar. Biol. Ecol.*, 175, 197, 1994.

205. Cimino, G., Crispino, A., Di Marzo, V., Gavagnin, M., and Ros, J.D., Oxytoxins, bioactive molecules produced by the marine opisthobranch mollusc *Oxynoe olivacea* from a diet-derived precursor, *Experientia*, 46, 767, 1990.

206. Hay, M.E., The ecology and evolution of seaweed–herbivore interactions on coral reefs, *Coral Reefs*, 16, S67, 1997.

12 Chemokinesis and Chemotaxis in Marine Bacteria and Algae

Charles D. Amsler* and Katrin B. Iken

CONTENTS

I. INTRODUCTION

Regardless of the size or phylogenetic placement of an organism, the capacity for motility confers a potential ability to exploit spatial heterogeneity in the environment. Motility, however, requires expenditures of organismal resources for the production, maintenance, and utilization of a propulsion mechanism, regardless of whether the organism moves by use of pseudopodia, flagella, fins, legs, or other means. There would appear to be little if any adaptive value to such investments unless an organism has a mechanism to sense environmental gradients in biologically significant resources (or hazards) and move towards more favorable areas. The microscopic organisms and life history stages that are the focus of this chapter often have one or more such mechanisms that allow them to orient their movement. These mechanisms include an ability to sense and respond to light gradients,[1,2] oxygen gradients,[3] and differences in surface topography.[4] They also include an ability to sense and respond to chemical heterogeneity in the environment, as described below.

Strictly defined, chemotaxis refers to directed movement oriented by chemical gradients.[5,6] This allows organisms to move directly towards (positive chemotaxis) or away from (negative chemotaxis) a chemical source. However, chemotaxis is sometimes defined more broadly to include mechanisms that allow organisms to have net movement towards or away from a chemical source by indirect means such as varying their turning frequency and/or swimming speed. These indirect mechanisms are more accurately defined as chemokinesis.[5,6] Klinokinesis refers to the modulation of turning frequency, and orthokinesis refers to modulation of swimming speed. In some cases,

---

* Corresponding author.

most notably with bacteria, such chemokinetic mechanisms are often referred to as chemotaxis for historical reasons. For the purposes of this chapter, all chemokinetic and chemotactic mechanisms by which marine bacteria and algae vary their movement in response to chemicals will be considered. Chemokinesis and chemotaxis will refer to these mechanistically distinct behaviors and will not be used as interchangeable terms.

II. BACTERIA

A. Chemokinetic Mechanisms

To our knowledge, all chemoattractive behaviors in marine bacteria are chemokinetic even though they are most commonly termed chemotactic responses in the literature. Bacteria move through marine environments either by swimming, by swarming, or by gliding motility. Gliding motility involves movements of a cell along a solid or semisolid surface without flagella.[7] Swarming is movement along a solid or semisolid surface by means of flagella.[8] Swimming occurs through more liquid environments, typically requiring the use of flagella, although some marine cyanobacteria swim without flagella or visible surface deformations.[9] The mechanism for this is unknown but may involve propagation of submicroscopic waves along the cell surfaces.[10] Bacterial flagella can be restricted to only one or both cell poles, which is referred to as polar flagellation, or can be inserted at random locations throughout the cell surface, which is referred to as peritrichous flagellation. Some marine species employ both forms of flagellation either simultaneously[11] or under different environmental conditions.[12]

Chemokinetic behavior in bacteria is best understood in peritrichously flagellated enteric bacteria.[13] Peritrichously flagellated bacteria alternate between periods of forward swimming (called runs or smooth swimming) and brief stops when their orientation changes randomly, called tumbles. In the absence of chemokinetic stimuli, the cells tumble every few seconds, which causes them to move randomly about their environment.[13,14] When the concentration of a chemoattractant is increasing (or that of a repellent is decreasing), the cells tumble less frequently. They do continue to tumble and, therefore, to undergo random changes in direction, but because the smooth swimming intervals are longer when attractant concentrations are increasing, the cells' random walk is biased such that net movement is towards the attractant source.[13-16] Although cells with polar flagellation are less commonly studied, they are certainly not uncommon[17] and may well be even more important in marine systems. Cells with polar flagellation make high angle turns, nearly reversing their swimming direction rather than tumbling, but otherwise respond to chemical gradients much like peritrichously flagellated species.[17] The chemokinetic signal transduction pathways of enteric bacteria have been studied in great detail,[13,18] and a good deal of information is also known about such pathways in other bacteria.[17,18]

Some bacteria only make an investment in the cellular machinery necessary for motility in suboptimal environments,[19,20] presumably because there is little adaptive advantage in investing cellular resources in movement when in an optimal environment. However, eventually, the ability of cells in suboptimal environments to utilize their cellular machinery for motility diminishes as conditions become increasingly growth-limiting.[19] Many marine environments, particularly in the plankton, can be severely growth-limiting to bacteria. As an illustration of this, only 10% of cells in a natural assemblage of coastal planktonic bacteria were motile before the addition of carbon and other nutrients, but over 80% had become motile 15 to 30 h after the addition.[21] In combination, these observations suggest that bacterial motility represents a cellular expense that is not necessary in an optimal growth environment and that is not affordable in a severely restrictive one. However, as discussed above, a cell must have the ability to receive and process relevant information about its environment for that expense to be of benefit.

Because of their small body length, bacteria are assumed to have only limited ability to sense spatial chemical gradients, but they are known to integrate chemokinetic stimuli over time through changes in the fractional amount of bound chemoreceptors.[15,22–24] By movement through a spatial gradient, a cell experiences it as a temporal gradient.[16] Dusenbery[25] compared the efficiency of spatial vs. temporal detection of stimuli in free-swimming bacteria using a numerical model and found that spatial detection could be superior to temporal detection for small cells under high concentrations and steep chemical gradients. The problem of a temporal integration of a stimulus is that organisms in a micrometer-scale size range are greatly affected by rotational diffusion due to Brownian motion.[25] A temporal integration does not work efficiently if a bacterium is moved into another random direction by Brownian motion during the time needed to react to the stimulus. However, if small bacteria swim at high speeds, they would be able to detect a directional gradient before being rotated.[26] Cell shape also has a large effect on the ability of a bacterial cell to detect chemical gradients.[27] Cells with rod-like shapes have enhanced capabilities to sense chemical gradients, particularly by temporal detection, and they also have decreased drag while swimming relative to cells of other shapes.[27] These adaptive benefits may help explain why rod-like shapes are more common than other shapes among motile bacteria.[27]

B. Bacterial Chemokinesis in the Plankton: Models and Observations

In marine and other aquatic environments, nutrients are not distributed homogeneously in the plankton but rather exist in patches and gradients. Nutrients including sugars, amino acids, and hydrolyzed macromolecules such as dissolved proteins and polysaccharides are released by phytoplankton into the surrounding water (leaking algae[28–30]). These may serve as nutrients for other microorganisms including heterotrophic bacteria.[31] It is estimated that 10 to 50% of marine primary production is used for heterotrophic bacterial growth.[32] Nutrients released from a cell will diffuse around the cell and, thus, result in an area of higher nutrient concentration compared to the background concentration.[30,33] Nutrient concentration typically decreases with distance from an algal cell.[29,30] Bacteria can only benefit from such nutrient patches if they are able to chemically detect these potential food sources and chemokineticly move and remain there.[23,28] Considering the average abundance of phytoplankton and bacteria in coastal surface waters, bacteria will be within a few hundred μm from a nutrient source,[29] which reflects the distance bacteria will have to cover to encounter a leaking phytoplankton cell.

Over the last 15 years, considerable effort has been expended towards improving our understanding of chemokinesis in marine bacteria through simulation models. Jackson's model[34] simulated the factors influencing marine bacterial chemosensory ability based principally on behavioral and physiological mechanisms known from the enteric bacteria *Escherichia coli* and *Salmonella typhimurium*[14,23] rather than on marine bacteria, since that was the only data base available at the time. The model is based on the simplified assumption that the only physical force active in the interaction between bacteria and leaking algae is diffusion. From an idealized spherical alga, a substrate is released and builds a spherical concentration field around the alga. Parameters considered in the model to influence bacterial chemosensory response are substrate concentration gradient, substrate release rate, algal size, distance of bacteria to alga, binding capacity of bacterial receptors for the substrate, and bacterial swimming speed. Chemokinetic responses in bacteria are especially high at high substrate concentrations or strong gradients. This is directly related to substrate release rates by algae, since higher release rates enhance the substrate concentration surrounding the alga. Bacteria in the model consistently exhibit a higher approach rate towards larger algal cells since those usually have higher specific release rates and build up stronger substrate concentration gradients. Small algal size can be compensated for in the model by an increased release rate, but not infinitely so; there is a minimum size of chemosensitively detectable algae of about 2 μm. The closer a bacterium is to an alga, the stronger the modeled response, and the distance at which an alga is still detectable depends on the size of the substrate cloud around the algal cell. This sensing of a substrate depends directly on the physiological binding capacity of bacterial receptors for that specific substrate. The lower the half-saturation constant (K_D) of a receptor, the higher the sensitivity.

The approach rate of bacteria to algae can be strongly increased by decreasing the K_D. This, however, does not work infinitely since there is a threshold at which the natural concentration of the substrate in the surrounding water (background) will become too high to distinguish. Also, lowering the K_D, will increase the time bacteria need to react to the stimulus. At a very low K_D, this reaction time becomes too large to successfully bias orientation towards the alga cell. The model organism *E. coli* is able to integrate temporal concentration comparisons in a time span of about 1 s, which is a nearly optimal time for reorientation for cells of that size.[23]

Another way to increase the detection of algae in the model is a decrease in bacterial swimming speed.[34] Slower bacteria will stay longer in a substrate cloud and, therefore, increase the available response time and/or detectable algal size. However, again there is a trade-off between increasing sensitivity towards small sources and needing to search a large volume for potential nutrient-rich sources. Decreasing swimming speed increases detection of algae only when bacteria are moved by physical forces (Brownian motion causing rotational diffusion) rather than by chemokinesis. Smaller bacteria would be more easily subjected to rotational diffusion, and Jackson's model[34] suggests that the smallest marine bacteria may not even have any chemokinetic response. This may be compensated by a more efficient nutrient uptake rate from the surrounding water since small bacteria have a high surface-to-volume ratio. This is in accordance with findings of Dusenbery[35] that a minimum size limit exists for useful chemosensory motility. Modeling the effect of size on locomotion in relation to different chemical, light, and temperature stimuli yielded bacterial size limits of around 0.6 μm diameter below which motility does not seem to be beneficial.

In the marine planktonic environment, however, physical factors other than diffusion will also influence distribution of nutrient gradients and chemosensory interactions between bacteria and leaking algae.[30] Turbulence, shear, and sinking of planktonic algae through the water column due to gravitational and other forces cause a distortion in the symmetric distribution of substrate around the alga.[30,36] Mitchell et al.[30] estimated that these physical forces will prevent clustering of bacteria around exudating phytoplankton. In an extended model, Jackson[36] analyzed the chemosensory ability of bacteria to either stay in high substrate concentrations around sinking algae or to attach to algae. During sinking, substrate gradients become steeper upstream and tail off downstream. Water velocity is lowest close to the algal surface because of the dragging force. In order to benefit from algal substrate release, bacteria would need to stay close to these falling algae. This model[36] suggests that latency of the chemokinetic reaction is too long, and net movement in a biased random walk too slow, to keep up with sinking algae. Increasing algal size, leakage rate, and bacterial swimming speed coupled with decreased algal falling velocity, however, could enhance the contact time between bacteria and algae and, thus, enhance bacterial exposure to elevated nutrient concentrations.[36] It has been suggested that significant enhancement of nutrient uptake due to chemokinetic response is only likely in eutrophic waters due to the specific relationships between algal size, algal abundance, nutrient availability, and slow sinking rates,[36,37] and that planktonic interactions in eutrophic and oligotrophic waters may be of fundamentally different nature.[36] Also, in low turbulence zones such as thermoclines, the lifetimes of nutrient gradients around phytoplankton cells can be long enough for bacteria to chemosensitively track and exploit them.[30] Jackson's model,[36] however, indicated that chemosensory reactions are likely to be very important in the context of bacterial attachment to very large falling particles such as marine snow. Independent from physical environmental conditions, attachment depends on the encounter of bacteria with a falling particle. As before, an increase or decrease in parameters such as particle size, algal stickiness, falling rate, bacterial swimming speed, binding capacity for substrate, etc. can increase bacterial attachment, though there is a transition region where the attachment rate cannot be enhanced further.

In contrast to these earlier models, Bowen et al.[38] argue that shear, rather than sinking or Brownian motion, is the more important physical force influencing fluid motion for most natural phytoplankton assemblages in turbulent mixed ocean layers. Nutrient concentrations around leaking algae are distorted by shear forces at irregular time intervals. In modeling bacterial chemotaxis in turbulent waters, Bowen et al.[38] referred to the same parameters as Jackson[34,36] but applied a higher variation in bacterial swimming speed (12 to 80 µm s⁻¹), based on observations of abundant marine bacteria with polar flagellation. Bacterial swimming, however, was still simulated according to the run-and-tumble enteric bacterial pattern. In agreement with the conclusions of Jackson,[36] Bowen et al.[38] found that low water motion, short distance to the leaking phytoplankton, high exudation rates, and a low K_D will considerably enhance bacterial approach rate and exposure time to high substrate concentrations. High bacterial swimming speed strongly increases bacterial approach in turbulent waters. However, even with all factors optimized, the simulation indicated that bacterial clusters will never occupy more than 10% of the substrate cloud volume. Largely independent from strength of shear, individual bacteria are transported into the vicinity of a leaking alga primarily by water motion and random swimming rather than by chemokinetic behavior. Once a bacterium is close to an alga where fluid motion is weakest and concentration gradients are strongest, chemokinesis can significantly increase residence time of bacteria near the alga. However, physical forces prevent a bacterium from remaining in a cluster for more than 1 min.[38] Bacteria transported away from one alga may be transported to the vicinity of another algal cell where chemokinesis can once again be effective. Overall, chemokinesis can enhance exposure time to nutrients by a factor of about 10 over nonchemokinetic behavior, thus giving chemokinetic bacteria a competitive advantage over nonchemokinetic cells, even in turbulent waters.[38] Numerical simulation indicated that bacterial nutrient uptake rates are unlikely to significantly reduce the high substrate concentrations close to algal cells. Likewise, bacterial uptake of inorganic nitrogen does not seem to reduce availability of nutrients for phytoplankton cells.[38]

Most analytical[30] and numerical[34,36,38] models of chemokinesis in marine bacteria to date are based on movement and reaction parameters derived from the well-studied enteric bacteria. These models assume constant movement with changes in speed and direction in the presence of concentration gradients. They also assume mean swimming speeds of 10 to 30 µm per second[23,34,36] or 80 µm per second,[38] but recent analyses of natural assemblages of marine bacteria have yielded much higher bacterial community swimming speeds of approximately 150 µm per second[21,39,40] with individual peak speeds greater than 400 µm per second.[21] This high speed mobility is possible due to lateral and polar flagellation, where the polar flagellum is driven by a sodium-ion motor and lateral flagella by a proton-motive force.[41] Both motors are used simultaneously in marine bacteria, and the sodium-motive force seems to account for about 60% of the swimming speed.[11] Also, marine bacteria are usually smaller (0.2 to 0.6 µm long[26]) than assumed in simulation models (1 to 10 µm).

If small bacteria swim at very high speeds (>100 µm s⁻¹), they are less affected by Brownian motion,[26,39] and chemokinetic reaction time is reduced by an order of magnitude.[21] There may be selection for fast moving cells and short reaction times because nutrient gradients in turbulent ocean waters erode in a matter of tens of seconds,[30] and in order for chemokinesis to function, bacterial reaction time must be equally short or shorter.[21,39,40] High speed, short chemotactic reaction time, and very short turn times at tumbles enable natural assemblages of marine bacteria to cluster around small sources of nutrients.[40] The limiting factor for bacterial residence time close to high nutrient patches seems to be more related to the lifetime of such patches rather than to bacterial chemokinetic mechanisms. However, the efficiency of bacterial clustering around nutrient sources has to be considered within the observations of Mitchell et al.[21] that only about 10% of a natural marine bacteria community was motile when collected. Motility was induced by the presence of nutrients with a lag time of 7 to 10 h and the proportion of motile cells increased to more than 80% of the population, but only after 15 to 30 h. In this sense, on a community level, the significance of bacterial chemokinesis can be limited by the fraction of motile individuals.

Another essential assumption made in almost all previous studies of marine bacterial chemokinesis is the run-and-tumble mode of swimming. The swimming behavior of marine bacteria near air bubbles[40] and in thin layers near sediments,[42] however, suggests that marine bacteria change direction by reversals much more than by tumbles. This back-and-forth strategy was simulated in a numerical model[43] and explained clustering of bacteria around nutrient patches at high shear in a more effective way than the run-and-tumble mode. Furthermore, rotational diffusion actually seems to enhance the efficiency of this back-and-forth swimming mode. The picture evolving is that bacteria are brought into the vicinity of nutrient sources randomly by water flow and then maximize their residence time in the nutrient patch by back-and-forth chemokinetic swimming.[38,43]

Chemokinetic behavior of marine bacteria has primarily been analyzed in isolation. But what role does bacterial chemokinesis play in the microbial food web? Blackburn et al.[44] developed a simulation model consisting of primary producers, nutrients, DOM, bacteria, and flagellate predators in order to evaluate the importance of bacterial chemokinesis in the microbial food web. Release of nutrients was supposed to be related to cell lysis or predation events, i.e., incomplete assimilation or clearance of feeding vacuoles. These nutrient sources around phytoplankton existed long enough before they were physically dispersed that bacteria which clustered around the sources could consume them.[33,40] Simulation predicted that bacterial growth is enhanced by 50% due to chemokinetic clustering in enhanced nutrient patches, emphasizing the adaptive value of chemokinesis in marine bacteria.[44] Results from experimental studies of microbial food webs confirmed the model prediction that swimming speed is the more important factor influencing efficiency of chemosensory behavior and foraging, with speed being directly correlated to nutrient patch size.[45]

C. Chemoattractants of Marine Bacteria

There are a variety of compounds released by phytoplankton that could serve as chemokinetic attractants for planktonic bacteria. Bell and Mitchell[28] showed that a suite of motile (but unidentified) marine bacteria are attracted to filtrates of phytoplankton cultures but only in cultures old enough to contain senescent or lysed cells. They identified a number of amino acids, polyalcohols, and sugars in the filtrates, many of which were chemoattractants for one or more of the bacterial isolates (Table 12.1). Subsequently, amino acids were also shown to be chemoattractants for planktonic isolates of an unidentified psychrophilic vibrio[46] and a *Pseudomonas* sp., which is also attracted to sugars[47] (Table 12.1). Planktonic bacteria can also be chemokineticly repelled by

TABLE 12.1
Chemoattractants in Marine Bacteria

Species	Attractants	Reference
Heterotrophic Bacteria		
Unidentified spp.	ala, arg, lys, met, val, mannitol, sucrose	28
Psychrophilic vibrio	arg	46
Pseudomonas sp.	cys, glu, leu, met, pro, glucose, fructose, mannitol, glactose, ribose, sucrose, lactose, maltose	47
Pseudomonas spp. (pathogenic)	glucose, cellobiose, cellotriose, cellulose	50
Vibrio alginolyticus	acrylate, glycolate, dimethyl sulfide	53
Vibrio anguillarum	glu, gln, gly, his, ile, leu, ser, thr, L-fucose, D-glucose, D-mannose, D-xylose, taurocholic acid, taurochenodeoxycholic acid	55
Vibrio parahaemolyticus	ala, gly, leu, ser	12
Vibrio furnissii	ala, arg, asn, gln, gly, cys, his, ile, leu, lys, met, phe, pro, thr, try, tyr, val, glucose, $(GlcNAc)_n$ with $n = 1–6$, sucrose, trehalose, mannose, mannitol, galactose	57, 58, 59
Alcaligenes strain M3A	dimethylsulfoniopropionate	60
Cyanobacteria		
Synechococcus spp.	NH_4^+, NO_3^-, urea, β-ala, gly	61
Oscillatoria sp.	CO_2, HCO_3^-, O_2	62
Phormidium corallyticum	O_2	63

nontoxic concentrations of copper, lead, tannic acid, benzoic acid, acrylamide, and several organic solvents,[48,49] but the ecological relevance of the behavior is difficult to assess for some of those compounds.

Chemokinesis by pathogenic bacteria can play an important role in host–pathogen relationships. Two *Pseudomonas* spp., one isolated as a pathogen on the marine fungus *Pythium debaryanum* and the other isolated as a pathogen of the marine diatom *Skeletonema costatum*, are attracted to exudates of their hosts.[50] Each bacterial species is also attracted to exudates of the other eukaryote, but to a lesser degree than its own host. A nonpathogenic *Pseudomonas* sp. is not attracted to either host exudate even though it was attracted by nutrient broth at the same level as the pathogenic species.[50] Both pathogens are attracted by cellulose and cellulose degradation products (Table 12.1) with the response much greater in the fungus pathogen. The nonpathogenic *Pseudomonas* sp. is not attracted by cellulose or its derivatives.[50] In another pathogenic relationship, unidentified pseudomonads that attack environmentally stressed corals are chemokineticly attracted to mucus that is produced by the corals as part of their stress response.[51,52]

Vibrio alginolyticus, which can be a pathogen of marine algae, is attracted by the algal extracellular products acrylate and glycolate as well as by dimethyl sulfide[53] (Table 12.1). *V. alginolyticus* is also a common fish pathogen and is chemokineticly attracted to mucus from the skin, gills, and intestines of host fish where it attaches to initiate colonization.[54] Strains of *V. fischeri*, *V. harveyi*, and *V. tubiashii* are also attracted, often more strongly than *V. alginolyticus*, by some or all of the three mucus types but do not adhere.[54] *V. anguillarum* is strongly attracted by these same mucus types as well as mucus from other fish and mammals, and it attaches to fish mucus to initiate host invasion.[54,55] Motility and a functional chemokinetic signal transduction pathway are known to be required for invasion of a host by *V. anguillarum*.[56] Mutants deficient in either motility or chemokinesis have a 500-fold decrease in virulence compared to wild type cells but can cause infections following intraperitoneal injection into a fish.[56] Specific chemoattractants include a variety of amino acids, sugars, and bile acids (Table 12.1). Other amino acids and sugars, as well as other mucus components including a number of lipids, do not elicit significant chemokinetic effects.[55] The marine bacterium and human pathogen *V. parahaemolyticus* is also attracted by several amino acids (Table 12.1) but not by aspartate, acetate, or sugars, and it is repelled by indole.[12]

The chitinivorous marine bacterium *Vibrio furnissii* apparently locates chitin sources initially via chemokinesis to soluble attractants, such as the sugars glucose and trehalose, that would be released by dead animals and fungi[57] as well as amino acids[58] (Table 12.1). Attraction to these sugars is linked to sugar uptake mechanisms and is particularly strong after induction by individual sugars.[57] Chitin is composed of subunits of *N*-acetyl-D-glucosamine (GlcNAc) which are strongly chemo-attractive as monomers or as oligosaccharides of 2 to 6 subunits[58,59] (Table 12.1). Intact chitin is insoluble. It has been hypothesized that new sources of chitin are initially colonized by chitinivorous bacteria that locate it by attraction to soluble cues other than GlcNAc, but that once chitin degradation begins, other bacteria are able to find the source via attraction to GlcNAc and its oligomers released during digestion by the initial colonizers.[57]

Induction of chemokinetic ability also occurs in *Alcaligenes* strain M3A.[60] Dimethyl sulfoniopropionate (DMSP) is an attractant for bacteria that have been induced to produce DMSP lyase. Such cells are attracted to DMSP at ecologically relevant concentrations, but uninduced cells do not respond to DMSP. Since DMSP lyase produces dimethylsulfide, which is important to global climate regulation, Zimmer-Faust et al.[60] suggest that bacterial chemokinesis may play a critical role in global sulfur cycles and climate.

Chemokinetic responses have also been reported in marine cyanobacteria (Table 12.1). Open-ocean isolates of *Synechococcus* spp., which move by swimming without means of flagella, are attracted by ecologically relevant concentrations of ammonium, nitrate, urea, β-alanine, and glycine but not by numerous other amino acids, sugars, or vitamins.[61] A mat-forming *Oscillatoria* sp., which moves by gliding motility, is attracted by carbon dioxide, bicarbonate, and oxygen.[62] *Phormidium corallyticum*, which likewise moves by gliding motility and which is associated with black-band disease of corals, is also attracted by oxygen.[63] Its movement is unaffected by ammonium, phosphate, acetate, glucose, and fructose, but it is strongly repelled by sulfide.[63]

III. ALGAE

A. Responses of Macroalgal Gametes to Sexual Chemoattractants

Gamete chemoattractants (pheromones) are widespread in the brown algae (Class Phaeophyceae). Their structure, function, and biosynthesis have been the subject of a number of comprehensive reviews including those by Müller,[64] Maier,[65,66] and Boland.[67] These chemoattractants have been identified in over 100 brown algal species and include 10 different C_{11} hydrocarbons and one C_8 hydrocarbon[67] (Figure 12.1). A total of over 50 different stereoisomers of these compounds have been identified.[67] All are volatile and, consequently, form steep concentration gradients around female gametes with effective attractive radii that are probably limited to distances ranging from approximately 500 μm in some species[68] to over 1000 μm in others.[69] They can often function as male gamete release stimulants in addition to male gamete attractants,[64,65,67] and some of these compounds or their degradation products may have additional roles in chemical defenses against herbivores.[70]

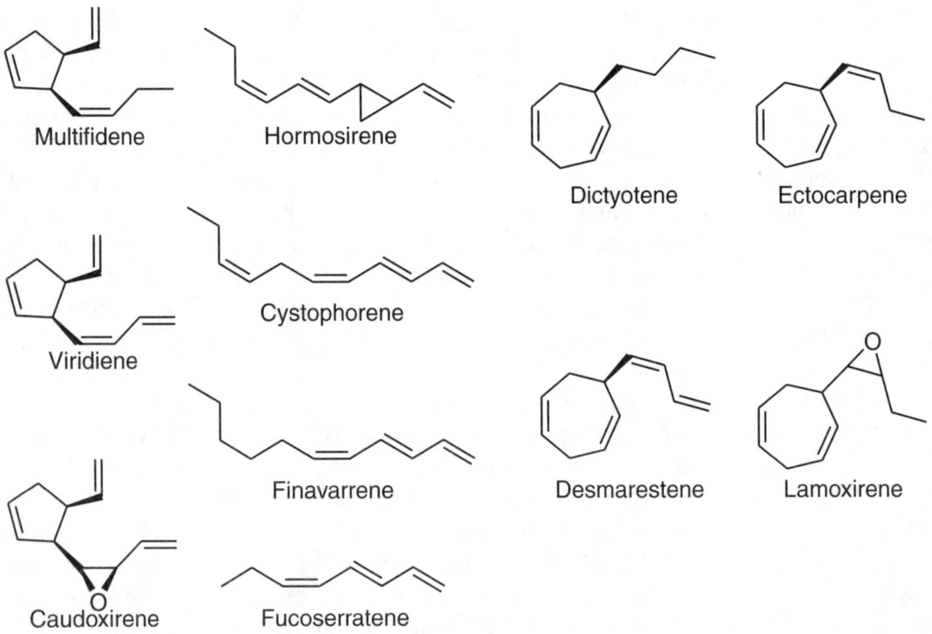

FIGURE 12.1 Pheromone chemoattractants for brown algal male gametes.

All of the known brown algal pheromones are involved in the attraction of male gametes to female gametes. In most cases, the gametes are morphologically anisogamous or oogamous, but in some cases, including in *Ectocarpus siliculosus* (Order Ectocarpales) where brown algal pheromones were first identified, they may be morphologically isogamous but functionally oogamous because of their behavior with respect to pheromones. This differentiation in function between female and male gametes in *E. siliculosus* has been recognized since the work of Berthold and others in the late nineteenth and early twentieth centuries (reviewed by Müller[68] and Fritsch[71]) but was not understood chemically until the pioneering work of Müller and colleagues in the late 1960s.[72,73] The female gametes swim for 30 min or less before settling to the substrate. Upon settlement, female gametes withdraw their flagella and begin to secrete the pheromone ectocarpene[64] (Figure 12.1). This production continues at a rate of 10^5 molecules per second per cell for approximately 7 h.[74]

Male *Ectocarpus siliculosus* gametes respond to ectocarpene secreted by the settled females via a chemokinetic response that has been described as chemothigmoklinokinesis.[69,75] Upon detecting the pheromone, the male gametes exhibit a strong thigmotactic response which causes them to remain in close contact with the substrate (where the female gamete is presumably settled) and to move along it at reduced swimming speeds.[76] Increasing concentrations of pheromone cause the rudder-like posterior flagellum to beat violently at a frequency directly proportional to pheromone concentration. These violent beats cause the gametes to make sharp turns and, therefore, to swim in circles. Because of the concentration effect, the diameter of the circles becomes increasingly smaller as the male gametes approach the settled female, effectively trapping the males in the vicinity of females until contact is made.[76] The posterior flagellum has also been reported to beat rapidly and unilaterally when pheromone concentration is decreasing, which results in the cells reversing direction.[77] This reorientation or phobic response is functionally separate from the chemothigmoklinokinetic attraction response and has been called chemoklinotaxis.[77] In culture, at least, female gametes can become surrounded by swimming male gametes.[71] Initial but firm contact is made by the end of a male gamete's anterior flagella anchoring to the female gamete's plasma membrane. A single male gamete pulls itself in to the female gamete by retracting its anterior flagellum and then fuses with the female. The remaining male gametes release and swim away,[71] presumably because the female gamete has ceased ectocarpene production. Although ectocarpene is the functional pheromone for several species of *Ectocarpus* as well as for members of two other orders of brown algae,[64–67] the initial attachment response is apparently species specific as it does not occur even between reproductively isolated populations of *E. siliculosus*.[78]

Because it relies on modulation of the frequency of swimming direction changes, the chemothigmoklinokinetic response of *Ectocarpus siliculosus* male gametes is superficially similar to the chemoklinokinetic behaviors described above which are used by marine and other bacteria. However, the behaviors differ markedly since *E. siliculosus* gametes increase their turning frequency with increasing concentrations of attractant, while bacteria decrease turning frequency as chemoattractant concentrations increase. This apparent incongruity seems likely to be a result of the very different objectives of the behaviors as well as of the different cell swimming behaviors in the absence of chemoattractants. To be successful, a male gamete must locate the pheromone point source, i.e., the female gamete. The male gametes swim in basically straight paths in the absence of pheromones, and their increased turning frequency in response to increased pheromone concentrations serves to retain them in the vicinity of the settled female gametes until contact is made. Bacterial attraction to carbon or other nutrient sources that are taken-up by the cells does not require contact with an attractant point source. Indeed, on the spatial scale of a bacterium, the source may well not be an isolated point. The signal transduction pathway that controls the behavior of bacterial cells[13,18] ensures that direction changes will occur randomly every few seconds and, consequently, that cells will move about in a random walk in the absence of chemoattractants. By decreasing the frequency of turns in response to increasing chemoattractant concentrations, the random walk is biased such that net movement occurs in the direction of the source.[13]

Even though bacteria and *Ectocarpus siliculosus* male gametes use modulation of direction changes in contrasting ways, the underlying adaptive advantage of chemoklinokinesis over chemotaxis may be the same. Koshland[79] postulated that bacterial chemoklinokinesis is less efficient than the direct approach chemotactic mechanisms utilized by more advanced organisms, but this decreased efficiency is compensated for by the requirement for much less complex sensory and signal-processing mechanisms. Although chemoklinokinesis may not be less efficient than chemotaxis under some conditions,[25] an analogous compromise between efficiency and complexity could favor chemokinetic mechanisms in brown algal gametes, particularly those from the Ectocarpales, which is typically considered the most primitive brown algal order.[80] Male gametes in one of the most advanced brown algal orders, the Laminariales, have a chemotactic response to pheromones, as described below.

The C_8 pheromone fucoserratene (Figure 12.1) is produced by female gametes of *Fucus* spp.[67,69] (Order Fucales). The behavioral response of male gametes has been briefly described based on video recordings of male gametes swimming near an artificial pheromone source[69] and differs from the chemoresponsive mechanisms utilized either by *Ectocarpus siliculosus* male gametes or by bacteria. The *Fucus* male gametes appear to swim randomly except when they are moving away from the source. Gametes that are moving away from the pheromone source execute near-180° turns at a relatively uniform distance from that source. It appears, therefore, that when in a decreasing gradient of pheromone, the gametes make these reversals upon crossing some specific threshold concentration of fucoserratene. This mechanism presumably traps the male gametes in the vicinity of the females until contact is made, a process called chemophobotaxis.[69] A somewhat similar mechanism has been reported for the response of male gametes of the brown alga *Hormosira banksii* (Order Fucales) to the C_{11} pheromone hormosirene (Figure 12.1). In the absence of pheromone, the cells swim in helical patterns and turn only rarely. Increasing concentrations of hormosirene further decrease the frequency of turns, and male gametes make sharp turns when moving away from a pheromone source.[81]

Lamoxirene (Figure 12.1) is known as the male gamete chemotactic attractant in a large number of species in the more advanced families of kelps (brown algae in the Order Laminariales) and also stimulates male gamete release.[82] The specific chemotactic mechanism has been described in detail based on video analysis.[83] In addition to a thigmotactic response that causes the male gametes to remain in contact with the substrate, the gametes exhibit directed movement towards the pheromone source (i.e., true chemotaxis). The specific mechanism is a phobic response. When a male gamete begins to move away from the source, it beats its posterior flagellum such that it makes a greater than 90° turn which helps it reorient towards the general direction of the settled female gamete.[83]

Although a single pheromone is usually (but not always[84]) responsible for the attraction or release of gametes from specific brown algal species, additional brown algal pheromones are often produced by those species. For example, ectocarpene, hormosirene, and dictyotene (Figure 12.1; attractant in *Dictyota* spp., Order Dictyotales) are produced in combination with most of the other brown algal pheromones.[67] While these might only be phylogenetic remnants involved in the biosynthetic pathways of the active pheromones,[67] their potential attractiveness to gametes of competitive (or biofouling) brown algae led Müller[68] to speculate on possible allelopathic roles for them. Though purely conjectural, pheromones produced by gametes could potentially "misguide" male gametes of competing or fouling brown algae away from conspecific female gametes and thereby decrease their reproductive success.[68] Likewise, pheromones such as lamoxirene that stimulate the release of gametes in multiple species could have an allelopathic role if pheromone release by female gametes of one species induced male gametophytes of a competing species to release their gametes before the female gametophytes of the competitor matured.[68] Amsler et al.[4] suggested that such phenomena could explain the competitive dominance of the kelp *Pterygophora californica* over the kelp *Macrocystis pyrifera* in field experiments that investigated the consequences of variable spore settlement on patterns of sporophyte recruitment.[85]

Chemotactic or chemokinetic responses are almost unknown in gametes of other marine macroalgae. Male gametes of the green macroalga *Bryopsis plumosa* are attracted by mature female gametangia and female gametes.[86] The presence of a chemical attractant has been confirmed with cell-free bioassays, but its chemical nature has yet to be elucidated.[86] Similar gamete attraction has been reported in the closely related species *Derbesia tenuissima*,[87] but a chemical basis for this behavior has not been established. Likewise, gamete attraction without known chemical basis has been described in several species of the yellow-green alga *Vaucheria*.[69] Gamete attraction, in some cases mediated by known pheromones, has been reported in a number of freshwater green microalgae and macroalgae.[65]

B. Responses of Macroalgal Spores and Microalgae to Environmental Chemoattractants and Repellents

Gametes are not the only motile reproductive stages produced by brown or other macroalgae. Most marine green and brown macroalgae produce motile spores by meiosis as part of a sexual life history and/or by mitosis either as part of an asexual life history or as an alternative asexual process in addition to a sexual life history. Upon release from the parent alga, these spores may initially become part of the plankton[88] or may be retained near the benthos. However, to be able to succeed as agents of reproduction, they must eventually enter the benthic boundary layer and biofilm community along a substrate and attach in a chemical and physical microenvironment suitable for germination and subsequent growth.

The chemical microenvironment in benthic boundary layers and biofilms are likely to be quite heterogeneous with respect to a number of important chemical compounds including organic and inorganic nutrients.[4] Spores from the brown algae *Macrocystis pyrifera* and *Pterygophora californica* (Order Laminariales) are chemotacticly attracted to or repelled by a number of nutrients[89] (Table 12.2). Spores of one or both species were attracted by biologically significant concentrations of inorganic and organic nitrogen, by organic phosphate, and by five different micronutrients. Perhaps the most intriguing responses were those of *M. pyrifera* spores under varying concentrations of ammonium and ferrous iron. The spores were attracted to concentrations of ammonium that would be stimulatory for growth of the gametophytic generation that the spores germinate into. However, the spores were repelled by ammonium at a higher concentration (1000 µM) that would almost certainly be inhibitory for gametophytic growth.[90,91] Although *Macrocystis* spp. gameto-phytes are able to grow without (conventionally) measurable levels of iron, the presence of iron in

TABLE 12.2
Nonpheromone Chemoattractants in Marine Algae

Species	Attractants	Reference
Brown Macroalgal Spores		
Macrocystis pyrifera	NH_4^+, NO_3^-, asp, gly, Fe^{2+}, BO_3^{3-}, $B_4O_7^{2-}$, Co^{2+}, Mn^{2+}, I^-	89
Pterygophora californica	NH_4^+, NO_3^-, PO_4^{3-}, BO_3^{3-}, $B_4O_7^{2-}$	89
Cryptomonads		
Chroomonas sp.	NO_3^-, ala, asp, glu, met	95
Dinoflagellates		
Symbiodinium microadriaticum	NH_4^+, NO_3^-, urea, ala, arg, gly	96, 97
Gymnodinium fungiforme	ala, arg, asn, asp, cys, gln, glu, gly, hyp, his, ile, leu, met, pro, ser, tau, thr, trp, val, dextrose, trimethylamine-HCl	98
Crypthecodinium cohnii	CO_2	99
Protoperidinium pellucidum	gly	103
Green Microalgae		
Dunaliella tertiolecta	NH_4^+, cys, met, phe, trp, tyr	104, 105
Dunaliella salina	ala, gly, lys	106
Raphidophytes		
Chattonella antiqua	PO_4^{3-}	107
Diatoms		
Amphora coffeaeformis	D-glucose, D-maltose, D-glucoheptose, 3-*O*-methyl-D-glucose	109
Amphora sp.	D-glucose, D-maltose, D-glucoheptose, 3-*O*-methyl-D-glucose	109

a tight concentration range is necessary to allow gametogenesis.[92] *M. pyrifera* spores were attracted to ferrous iron within this stimulatory concentration range (1 μm) but were repelled by a higher concentration (45 μm).[89] Although the spores themselves do not grow and so may have no direct use for nutrients, these data suggest that their chemotactic responses are adaptive by increasing the chances that the spores will settle in chemical microenvironments most advantageous for both the growth and the reproduction of the gametophytic stage they germinate into. This may be particularly important in members of the Laminariales and groups with similar life histories where this obligate gametophytic stage is microscopic and, therefore, must be able to grow to reproductive maturity in the same microenvironment where the spore has settled.

Many of the same nutrients that elicit chemotactic responses in *Macrocystis pyrifera* and *Pterygophora californica* spores also stimulate their settlement.[93] However, specific responses to several of the nutrients differ between the chemotactic and settlement stimulation behaviors, and the settlement stimulation response develops later in the age of the spores.[93] This indicates that these are two mechanistically different behaviors. Both chemotaxis and settlement stimulation responses have been assayed in newly released spores of *Ectocarpus siliculosus* and found to be absent.[94] The *E. siliculosus* spores differ, therefore, from those of kelps in not being chemotactic. Amsler et al.[94] suggested that this could be related to microenvironmental differences resulting from the size of the plants the spores germinate into and/or to differences in the cellular mechanisms available to the spores.

Chemotactic or chemokinetic attraction to nutrients has also been observed in marine microalgae (Table 12.2). In most cases, the specific behavioral pattern (kinesis vs. taxis) has not been described, and so these behaviors are referred to below as chemoattraction. The behavior has been well described in the cryptomonad *Chroomonas* sp., which swims toward nitrogen sources utilizing a behavioral pattern very similar to bacterial chemoklinokinesis.[95] Cells cultured with glycine as their sole nitrogen source are significantly attracted by several different amino acids (Table 12.2) but not by nitrate, while cells cultured with nitrate are significantly attracted by nitrate but not amino acids,[95] suggesting that perhaps different chemoreceptors or other signal pathway molecules are induced under different nutrient regimes. The symbiotic dinoflagellate *Symbiodinium microadriaticum* also shows a variable chemoattraction response to nitrogen. Cells cultured under nitrogen limitation are attracted by nitrate, ammonium, urea, and several amino acids (Table 12.2), while cells cultured in nitrogen-replete conditions are not attracted by nitrogen.[96] Phosphate, sulfate, sugars, some other amino acids, trace metals, and vitamins do not elicit significant responses under any conditions tested.[96] It is likely that *S. microadriaticum* utilizes ammonium as a cue in finding suitable symbiotic hosts within an effective range of about 1 cm.[96,97]

Chemoattraction to a wide variety of amino acids and to dextrose (Table 12.2) has also been reported as a prey-finding mechanism in the heterotrophic dinoflagellate *Gymnodinium fungiforme*.[98] Another heterotrophic dinoflagellate, *Crypthecodinium cohnii*, is stimulated to attach to agar by some amino acids and sugars but apparently not via chemotactic or chemokinetic attraction,[99] even though swimming cells are repelled by CO_2.[100] Prey-finding behaviors have been described in numerous predatory dinoflagellates,[101,102] and although chemoattractants have usually not been identified, many of the approach patterns are consistent with chemokinetic mechanisms. The approach behavior of *Protoperidinium pellucidum* has been studied quantitatively and shown to consist of both chemoklinokinetic and chemoorthokinetic components.[103] The cells respond both to a filtrate of prey species culture media and to glycine by increasing their swimming speed and decreasing their turning frequency.[103] A very different approach has been described in several other *Protoperidinium* spp. and species from two other genera which slow down and increase their turning frequency around prey species.[101]

The green microalga *Dunaliella tertiolecta* is attracted to ammonium and to several amino acids[104,105] (Table 12.2). Calcium is required for full development of the response,[104] and neither nitrate, urea, other amino acids, phosphate, vitamins, nor sugars elicit significant responses.[105] Chemoattraction in a second species, *D. salina*, has been reported to several amino acids

(Table 12.2) and small oligopeptides thereof.[106] Chemoattraction to phosphate has been reported in the raphidophyte *Chattonella antiqua*.[107] The diatom *Amphora coffeaeformis* and a second, unidentified species of *Amphora* show positive chemotactic responses to D-glucose and similar sugars (Table 12.2) which can be taken up to support heterotrophic or mixotrophic growth.[108,109] D-mannose, which is toxic, and L-glucose elicit negative chemotactic behaviors.[109,110] Pretreatment of cells with glucose eliminates the cells' chemotactic response to the glucose.[109]

IV. CONCLUDING REMARKS

Because of their size and swimming speeds, marine bacteria, microalgae, and the microscopic stages of marine macroalgae are similar not only in being primarily unicellular organisms but also in residing within a somewhat counterintuitive physical world dominated by viscous forces.[4,111] Often they are also similar, as described throughout this chapter, in using flagella and an ability for chemokinesis or chemotaxis to exploit chemical heterogeneity in their microenvironments. Heterotrophic cells are able to locate carbon or other nutrient sources whether they are dissolved, suspended, or, for predators or pathogens, prey or host organisms. Autotrophic cells are able to locate inorganic and organic sources of nitrogen, phosphorous, and other nutrients necessary for growth and reproduction. Male gametes are able to locate female gametes by responding to pheromones.

Although bacteria and algae are phylogenetically quite distinct, their chemokinetic and chemotactic behaviors have a number of similarities. The authors believe that comparisons and contrasts of these behaviors within and between the groups, as we have begun to make here, can provide insights into the ecological and ecophysiological relevance of the responses. It is critical, however, for there to be more information on these behaviors, specifically in marine bacteria and algae. This is particularly well illustrated by the shortcomings of using behavioral data from enteric bacteria in models of planktonic marine bacteria, as discussed above (Section II.B). Computer-assisted analysis of behavioral patterns is a relatively new tool that has been applied to bacteria, microalgae, and macroalgal gametes and spores.[19,60,94,95,103,112–120] This quantitative approach should continue to facilitate detailed descriptive and experimental studies of cell behavior in the marine environment. In combination with advanced mathematical modeling, modern cell physiological techniques, and modern tools for microbial ecology, these studies should greatly enhance our understanding of the roles of bacterial and algal chemokinesis and chemotaxis in marine chemical ecology.

ACKNOWLEDGMENTS

We are grateful to Dr. B.J. Baker for preparing Figure 12.1 and to Dr. D.C. Reed and an anonymous reviewer for their constructive comments on the manuscript. This work was supported in part by grants OPP-9814538 from the National Science Foundation and R/MT-40 from the Mississippi–Alabama Sea Grant Consortium.

REFERENCES

1. Hegemann, P., Vision in microalgae, *Planta*, 203, 265, 1997.
2. Spudich, J. L., Variations on a molecular switch: transport and sensory signaling by archaeal rhodopsins, *Mol. Microbiol.*, 28, 1051, 1998.
3. Taylor, B. L., Zhulin, I. B., and Johnson, M. E., Aerotaxis and other energy-sensing behavior in bacteria, *Annu. Rev. Microbiol.*, 53, 103, 1999.
4. Amsler, C. D., Reed, D. C., and Neushul, M., The microclimate inhabited by macroalgal propagules, *Br. Phycol. J.*, 27, 253, 1992.
5. Fraenkel, G. S. and Gunn, D. L., *The Orientation of Animals, Kineses, Taxes and Compass Reactions*, Dover Publications, New York, 1961.

6. Dusenbery, D. B., *Sensory Ecology*, W.H. Freeman and Company, New York, 1992.

7. Spormann, A. M., Gliding motility in bacteria: insights from studies of *Myxococcus xanthus*, *Microbiol. Mol. Biol. Rev.*, 63, 621, 1999.

8. Belas, R., *Proteus mirabilis* and other swarming bacteria, in *Bacteria as Multicellular Organisms*, Shapiro, J. and Dworkin, M., Eds., Oxford University Press, New York, 1997, 183.

9. Waterbury, J. B., Willey, J. M., Franks, D. G., Valois, F. W., and Watson, S. W., A cyanobacterium capable of swimming motility, *Science*, 230, 74, 1985.

10. Pitta, T. P., Sherwood, E. E., Kobel, A. M., and Berg, H. C., Calcium is required for swimming by the nonflagellated cyanobacterium *Synechococcus* strain WH8113, *J. Bacteriol.*, 179, 2524, 1997.

11. Mitchell, J. G. and Barbara, G. M., High speed marine bacteria use sodium-ion and proton driven motors, *Aquat. Microb. Ecol.*, 18, 227, 1999.

12. Sar, N., McCarter, L., Simon, M., and Silverman, M., Chemotactic control of the two flagellar systems of *Vibrio parahaemolyticus*, *J. Bacteriol.*, 172, 334, 1990.

13. Amsler, C. D. and Matsumura, P., Chemotactic signal transduction in *Escherichia coli* and *Salmonella typhimurium*, in *Two-Component Signal Transduction*, Hoch, J.A. and Silhavy, T.J., Eds., ASM Press, 1995, 89.

14. Berg, H. C. and Brown, D. A., Chemotaxis in *Escherichia coli* analyzed by three-dimensional tracking, *Nature*, 239, 500, 1972.

15. Brown, D. A. and Berg, H. C., Temporal stimulation of chemotaxis in *Escherichia coli*, *Proc. Natl. Acad. Sci. USA*, 71, 1388, 1974.

16. Macnab, R. and Koshland, D. E., Jr., The gradient-sensing mechanism in bacterial chemotaxis, *Proc. Natl. Acad. Sci. USA*, 69, 2509, 1972.

17. Armitage, J. P. and Lackie, J. M., *Biology of the Chemotactic Response*, Society for General Microbiology Symposium 46, Cambridge University Press, Cambridge, 1990.

18. Manson, M. D., Armitage, J. P., Hoch, J. A., and Macnab, R. M., Bacterial locomotion and signal transduction, *J. Bacteriol.*, 180, 1009, 1998.

19. Amsler, C. D., Cho, M., and Matsumura, P., Multiple factors underlying the maximum motility of *Escherichia coli* as cultures enter post-exponential growth, *J. Bacteriol.*, 175, 6238, 1993.

20. Staropoli, J. F. and Alon, U., Computerized analysis of chemotaxis at different stages of bacterial growth, *Biophys. J.*, 78, 513, 2000.

21. Mitchell, J. G., Pearson, L., Bonazinga, A., Dillon, S., Khouri, H., and Paxinos, R., Long lag times and high velocities in the motility of natural assemblages of marine bacteria, *Appl. Environ. Microbiol.*, 61, 877, 1995.

22. Spudich, J.L. and Koshland, D.E., Jr., Quantitation of the sensory response in bacterial chemotaxis, *Proc. Nat. Acad. Sci. USA*, 72, 710, 1975.

23. Berg, H. C. and Purcell, E. M., Physics of chemoreception, *Biophys. J.*, 20, 193, 1977.

24. Segall, J. E., Block, S. M., and Berg, H. C., Temporal comparisons in bacterial chemotaxis, *Proc. Natl. Acad. Sci. USA*, 83, 8987, 1986.

25. Dusenbery, D. B., Spatial sensing of stimulus gradients can be superior to temporal sensing for free-swimming bacteria, *Biophys. J.*, 74, 2272, 1998.

26. Mitchell, J. G., The influence of cell size on marine bacterial motility and energetics, *Microb. Ecol.*, 22, 227, 1991.

27. Dusenbery, D. B., Fitness landscapes for effects of shape on chemotaxis and other behaviors of bacteria, *J. Bacteriol.*, 180, 5978, 1998.

28. Bell, W. and Mitchell, R., Chemotactic and growth responses of marine bacteria to algal extracellular products, *Biol. Bull.*, 143, 265, 1972.

29. Azam, F. and Ammerman, J. W., Cycling of organic matter by bacterioplankton in pelagic marine ecosystems: microenvironmental considerations, in *Flows of Energy and Material in Marine Ecosystems*, Fasham, M. J. R., Ed., NATO Conf. Ser. 4, Mar. Sci. V. 13, Plenum Publishing Corp., New York, 1984, 345.

30. Mitchell, J. G., Okubo, A., and Fuhrmann, J. A., Microzones surrounding phytoplankton form the basis for a stratified marine microbial ecosystem, *Nature*, 316, 58, 1985.

31. Baines, S. B. and Pace, M. L., The production of dissolved organic matter by phytoplankton and its importance to bacteria: patterns across marine and freshwater systems, *Limnol. Oceanogr.*, 17, 1817, 1991.

32. Fuhrman, J. A. and Azam, F., Thymidine incorporation as a measure of heterotrophic bacterioplankton production in marine surface waters: evaluation and field results, *Mar. Biol.*, 66, 109, 1982.

33. Blackburn, N., Fenchel, T., and Mitchell, J., Microscale nutrient patches in planktonic habitats shown by chemotactic bacteria, *Science*, 282, 2254, 1998.

34. Jackson, G. A., Simulating chemosensory responses of marine microorganisms, *Limnol. Oceanogr.*, 32, 1253, 1987.

35. Dusenbery, D. B., Minimum size limit for useful locomotion by free-swimming microbes, *Proc. Natl. Acad. Sci. USA*, 94, 10949, 1997.

36. Jackson, G. A., Simulation of bacterial attraction and adhesion to falling particles in an aquatic environment, *Limnol. Oceanogr.*, 34, 514, 1989.

37. Bienfang, P. K. and Harrison, P. J., Sinking-rate response of natural assemblages of temperate and subtropical phytoplankton to nutrient depletion, *Mar. Biol.*, 83, 293, 1984.

38. Bowen, J. D., Stolzenbach, K. D., and Chisholm, S. W., Simulating bacterial clustering around phytoplankton cells in a turbulent ocean, *Limnol. Oceanogr.*, 38, 36, 1993.

39. Mitchell, J. G., Pearson, L., Dillon, S., and Kantalis, K., Natural assemblages of marine bacteria exhibiting high-speed motility and large accelerations, *Appl. Environ. Microbiol.*, 61, 4436, 1995.

40. Mitchell, J. G., Pearson, L., and Dillon, S., Clustering of marine bacteria in seawater enrichments, *Appl. Environ. Microbiol.*, 62, 3716, 1996.

41. Atsumi, T., McCarter, L., and Imae, Y., Polar and lateral flagellar motors of marine *Vibrio* are driven by different ion-motive forces, *Nature*, 355, 182, 1992.

42. Barbara, G. M. and Mitchell, J. G., Formation of 30- to 40-micrometer-thick laminations by high-speed marine bacteria in microbial mats, *Appl. Environ. Microbiol.*, 62, 3985, 1996.

43. Luchsinger, R. H., Bergersen, B., and Mitchell, J. G., Bacterial swimming strategies and turbulence, *Biophys. J.*, 77, 2377, 1999.

44. Blackburn, N., Azam, F, and Hagström, Å., Spatially explicit simulations of a microbial food web, *Limnol. Oceanogr.*, 42, 613, 1997.

45. Blackburn, N. and Fenchel, T., Influence of bacteria, diffusion and shear on micro-scale nutrient patches, and implications for bacterial chemotaxis, *Mar. Ecol. Prog. Ser.*, 189, 1, 1999.

46. Geesey, G. G. and Morita, R. Y., Capture of arginine at low concentration by a marine psychrophilic bacterium, *Appl. Environ. Microbiol.*, 38, 1092, 1979.

47. Chet, I. and Mitchell, R., The relationship between chemical structure of attractants and chemotaxis by a marine bacterium, *Can. J. Microbiol.*, 22, 1206, 1976.

48. Young, L. Y. and Mitchell, R., Negative chemotaxis of marine bacteria to toxic chemicals, *Appl. Microbiol.*, 25, 972, 1973.

49. Chet, I., Asketh, P., and Mitchell, R., Repulsion of bacteria from marine surfaces, *Appl. Microbiol.*, 30, 1043, 1975.

50. Chet, I., Fogel, S., and Mitchell, R., Chemical detection of microbial prey by bacterial predators, *J. Bacteriol.*, 106, 863, 1971.

51. Mitchell, R. and Chet, I., Bacterial attack of corals in polluted seawater, *Microb. Ecol.*, 2, 227, 1975.

52. Chet, I. and Mitchell, R., Ecological aspects of microbial chemotactic behavior, *Annu. Rev. Microbiol.*, 30, 221, 1976.

53. Sjoblad, R. D. and Mitchell, J., Chemotactic responses of *Vibrio alginolyticus* to algal extracellular products, *Can. J. Microbiol.*, 25, 964, 1979.

54. Bordas, M. A., Balebona, C., Rodriguez-Maroto, J. M., Borrego, J. J., and Moriñigo, M. A., Chemotaxis of *Vibrio* strains towards mucus surfaces of guilt-head sea bream (*Sparus aurata* L.), *Appl. Environ. Microbiol.*, 64, 1573, 1998.

55. O'Toole, R., Lundberg, S., Fredriksson, S. Å., Jansson, A., Nilsson, B., and Wolf-Watz, H., The chemotactic response of *Vibrio anguillarum* to fish intestinal mucus is mediated by a combination of multiple mucus components, *J. Bacteriol.*, 181, 4308, 1999.

56. O'Toole, R., Milton, D. L., and Wolf-Watz, H., Chemotactic motility is required for invasion of the host by the fish pathogen *Vibrio anguillarum*, *Mol. Microbiol.*, 19, 625, 1996.

57. Yu, C., Bassler, B. L., and Roseman, S., Chemotaxis of the marine bacterium *Vibrio furnissii* to sugars. A potential mechanism for initiating the chitin catabolic cascade, *J. Biol. Chem.*, 268, 9405, 1993.

58. Bassler, B. L., Gibbons, P. J., Yu, C., and Roseman, S., Chitin utilization by marine bacteria. Chemotaxis to chitin oligosaccharides by *Vibrio furnissii*, *J. Biol. Chem.*, 266, 24268, 1991.

59. Bassler, B. L., Gibbons, P. J., and Roseman, S., Chemotaxis to chitin oligosaccharides by *Vibrio furnissii*, a chitinivorous marine bacterium, *Biochem. Biophys. Res. Com.*, 161, 1172, 1989.

60. Zimmer-Faust, R. K., de Souza, M. P., and Yoch, D. C., Bacterial chemotaxis and its potential role in marine dimethylsulfide production and biogeochemical sulfur cycling, *Limnol. Oceanogr.*, 41, 1330, 1996.

61. Willey, J. M. and Waterbury, J. B., Chemotaxis toward nitrogenous compounds by swimming strains of marine *Synechococcus* spp., *Appl. Environ. Microbiol.*, 55, 1888, 1989.

62. Malin, G. and Walsby, A. E., Chemotaxis of a cyanobacterium on concentration gradients of carbon dioxide, bicarbonate and oxygen, *J. Gen. Microbiol.*, 131, 2643, 1985.

63. Richardson, L. L, Horizontal and vertical migration patterns of *Pormidium corallyticum* and *Beggiatoa* spp. associated with black-band disease of corals, *Microb. Ecol.*, 32, 323, 1996.

64. Müller, D. G., The role of pheromones in sexual reproduction of brown algae, in *Algae as Experimental Systems*, Coleman, A. W., Goff, L. J., and Stein-Taylor, J. R., Eds., Alan R. Liss, Inc., New York, 1989, 201.

65. Maier, I., Gamete orientation and induction of gametogenesis by pheromones in algae and plants, *Plant Cell Environ.*, 16, 891, 1993.

66. Maier, I., Brown algal pheromones, *Prog. Phycol. Res.*, 11, 51, 1995.

67. Boland, W., The chemistry of gamete attraction: chemical structures, biosynthesis, and (a)biotic degradation of algal pheromones, *Proc. Natl. Acad. Sci. USA*, 92, 37, 1995.

68. Müller, D. G., Sexuality and sexual attraction, in *The Biology of Seaweeds*, Lobban, C. S. and Wynne, M. J., Eds., Blackwell Scientific Publications, Oxford, UK, 1981, 661.

69. Maier, I. and Müller, D. G., Sexual pheromones in algae, *Biol. Bull.*, 170, 145, 1986.

70. Hay, M. E., Piel, J., Boland, W., and Schnitzler, I., Seaweed sex pheromones and their degradation products frequently suppress amphipod feeding but rarely suppress sea urchin feeding, *Chemoecology*, 8, 91, 1998.

71. Fritsch, F. E., *The Structure and Reproduction of the Algae, Volume II*, Cambridge University Press, Cambridge, 1945.

72. Müller, D. G., Versuche zur Charakterisierung eines Sexuallockstoffes bei der Braunalge *Ectocarpus siliculosus*. I. Methoden, Isolierung, und gaschromatographischer Nachweis, *Planta*, 81, 160, 1968.

73. Müller, D. G., Jaenicke, L., Donike, M., and Akintobi, T., Sex attractant in a brown alga: chemical structure, *Science*, 171, 815, 1971.

74. Müller, D. G. and Schmid, C. E., Qualitative and quantitative determination of pheromone secretion in female gametes of *Ectocarpus siliculosus* (Phaeophyceae), *Biol. Chem. Hoppe-Seyler*, 369, 647, 1988.

75. Müller, D. G., Locomotive responses of male gametes to the species-specific sex attractant of *Ectocarpus siliculosus* (Phaeophyta), *Arch. Protistenk.*, 120, 371, 1978.

76. Geller, A. and Müller, D. G., Analysis of the flagellar beat pattern of male *Ectocarpus siliculosus* gametes (Phaeophyta) in relation to chemotactic stimulation by female cells, *J. Exp. Biol.*, 92, 53, 1981.

77. Maier, I. and Calenberg, M., Effect of extracellular Ca^{2+} and Ca^{2+} antagonists on the movement and chemoorientation of male gametes of *Ectocarpus siliculosus* (Phaeophyceae), *Bot. Acta*, 107, 451, 1994.

78. Müller, D. G., Sexual isolation between a European and American population of *Ectocarpus siliculosus* (Phaeophyta), *J. Phycol.*, 12, 252, 1976.

79. Koshland, D. E., *Bacterial Chemotaxis as a Model Behavioral System*, Distinguished Lecture Series of the Society of General Physiologists, Vol. 2, Raven Press, New York, 1980.

80. Graham, L. E. and Wilcox, L. W., *Algae*, Prentice-Hall, Englewood Cliffs, NJ, 2000.

81. Maier, I., Wenden, A., and Clayton, M. N., The movement of *Hormosira banksii* (Fucales, Phaeophyta) spermatozoids in response to sexual pheromone, *J. Exp. Bot.*, 43, 1651, 1992.

82. Müller, D. G., Maier, I., and Gassman, G., Survey on sexual pheromone specificity in Laminariales, *Phycologia*, 24, 475, 1985.

83. Maier, I. and Müller, D. G., Chemotaxis in *Laminaria digitata* (Phaeophyceae). I. Analysis of spermatozoid movement, *J. Exp. Bot.*, 41, 869, 1990.

84. Müller, D. G., Kawai, H., Stache, B., Fölster, E., and Boland, W., Sexual pheromones and gamete chemotaxis in *Analipus japonicus* (Phaeophyceae), *Experientia*, 46, 534, 1990.

85. Reed, D.C., The effects of variable settlement and early competition on patterns of kelp recruitment, *Ecology*, 71, 776, 1990.
86. Togashi, T., Motomura, T., and Ichimura, T., Gamete dimorphism in *Bryopsis plumosa*. Phototaxis, gamete motility and pheromonal attraction, *Bot. Mar.*, 41, 257, 1998.
87. Ziegler, J. R. and Kingsbury, J. M., Cultural studies on the marine green alga *Halicystis parvula–Derbesia tenuissima*. I. Normal and abnormal sexual and asexual reproduction, *Phycologia*, 4, 105, 1964.
88. Amsler, C. D. and Searles, R. B., Vertical distribution of seaweed spores in a water column offshore of North Carolina, *J. Phycol.*, 16, 617, 1980.
89. Amsler, C. D. and Neushul, M., Chemotactic effects of nutrients on spores of the kelps *Macrocystis pyrifera* and *Pterygophora californica*, *Mar. Biol.*, 102, 557, 1989.
90. Haines, K. C. and Wheeler, P. A., Ammonium and nitrate uptake by the marine macrophytes *Hypnea musciformis* (Rhodophyta) and *Macrocystis pyrifera* (Phaeophyta), *J. Phycol.*, 14, 319, 1978.
91. DeBoer, J. A., Nutrients, in *The Biology of Seaweeds*, Lobban, C. S. and Wynne, M. J., Eds., University California Press, Berkeley, 1981, 356.
92. Steckol, M. S. and Neushul, M., unpublished data, 1988.
93. Amsler, C. D. and Neushul, M., Nutrient stimulation of spore settlement in the kelps *Pterygophora californica* and *Macrocystis pyrifera*, *Mar. Biol.*, 107, 297, 1990.
94. Amsler, C. D., Shelton, K. L., Britton, C. J., Spencer, N. Y., and Greer, S. P., Nutrients do not influence swimming behavior or settlement rates of *Ectocarpus siliculosus* (Phaeophyceae) spores, *J. Phycol.*, 35, 239, 1999.
95. Lee, E. S., Lewitus, A. J., and Zimmer, R .K., Chemoreception in a marine cryptophyte: behavioral plasticity in response to amino acids and nitrate, *Limnol. Oceanogr.*, 44, 1571, 1999.
96. Fitt, W. K., Chemosensory responses of the symbiotic dinoflagellate *Symbiodinium microadriaticum* (Dinophyceae), *J. Phycol.*, 21, 62, 1985.
97. Fitt, W. K., The role of chemosensory behavior of *Symbiodinium microadriaticum*, intermediate hosts, and host behavior in the infection of coelenterates and molluscs with zooxanthellae, *Mar. Biol.*, 81, 9, 1984.
98. Spero, H. J., Chemosensory capabilities in the phagotrophic dinoflagellate *Gynmnodinium fungiforme*, *J. Phycol.*, 21, 181, 1985.
99. Hauser, D. C. R., Levandowsky, M., Hunter, S. H., Chunosoff, L., and Hollwitz, J. S., Chemosensory responses by the heterotrophic marine dinoflagellate *Crypthecodinium cohnii*, *Microb. Ecol.*, 1, 246, 1975.
100. Hauser, D. C. R., Petrylak, D., Singer, G., and Levandowsky, M., Calcium-dependent sensory-motor response of a marine dinoflagellate to CO_2, *Nature*, 273, 230, 1978.
101. Jacobson, D. M. and Anderson, D. M., Thecate heterotrophic dinoflagellates: feeding behavior and mechanisms, *J. Phycol.*, 22, 249, 1986.
102. Hansen, P. J. and Calado, A. J., Phagotrophic mechanisms and prey selection in free-living dinoflagellates, *J. Eukaryot. Microbiol.*, 46, 382, 1999.
103. Buskey, E. J., Behavioral components of feeding selectivity of the heterotrophic dinoflagellate *Protoperidinium pellucidum*, *Mar. Ecol. Prog. Ser.*, 153, 77, 1997.
104. Sjoblad, R. D., Chet, I., and Mitchell, R., Quantative assay for algal chemotaxis, *Appl. Environ. Microbiol.*, 36, 847, 1978.
105. Sjoblad, R. D., Chet, I., and Mitchell, R., Chemoreception in the green alga *Dunaliella tertiolecta*, *Curr. Microbiol.*, 1, 305, 1978.
106. Kohidai, L., Kovacs, P., and Csaba, G., Chemotaxis of the unicellular green alga *Dunaliella salina* and the ciliated *Tetrahymena pyriformis* B effects of glycine, lysine, and alanine, and their oligopeptides, *Biosci. Reports*, 16, 467, 1996.
107. Ikegami, S., Imai, I., Kato, J., and Ohtake, H., Chemotaxis toward inorganic-phosphate in the red tide alga *Chattonella antiqua*, *J. Plankton Res.*, 17, 1587, 1995.
108. Cooksey, B. and Cooksey, K. E., Adhesion of fouling diatoms to surfaces: some biochemistry, in *Algal Biofouling*, Evans, L.V. and Hoagland, K.D., Eds., Elsevier, Amsterdam, 1986, 41.
109. Cooksey, B. and Cooksey, K. E., Chemical signal-response in diatom of the genus *Amphora*, *J. Cell Sci.*, 91, 523, 1988.
110. Wigglesworth-Cooksey, B. and Cooksey, K. E., Can diatoms sense surfaces? State of our knowledge, *Biofouling*, 5, 227, 1992.

111. Vogel, S., *Life in Moving Fluids: The Physical Biology of Flow*, 2nd ed., Princeton University Press, Princeton, NJ, 1994.
112. Kondo, T., Kubota, M., Aono, Y., and Watanabe, M., A computerized video system to automatically analyze movements of individual cells and its application to the study of circadian rhythms in phototaxis and motility in *Chlamydomonas reinhardtii*, *Protoplasma*, Suppl. 1, 185, 1988.
113. Marwan, W. and Oesterhelt, D., Quantitation of photochromism of sensory rhodopsin-I by computerized tracking of *Halobacterium halobium* cells, *J. Mol. Biol.*, 215, 277, 1990.
114. Kawai H., Müller, D. G., Fölster, E., and Häder, D. P., Phototactic responses in the gametes of the brown alga *Ectocarpus siliculosus*, *Planta*, 182, 292, 1990.
115. Kawai, H., Kubota, M., Kondo, T., and Watanabe, M., Action spectra for phototaxis in zoospores of the brown alga *Pseudocorda gracilis*, *Protoplasma*, 161, 17, 1991.
116. Khan, S., Castellano, F., Spudich, J. L., McCrary, J. A., Goody, R. S., Reid, G. P., and Trentham, D. R., Excitatory signaling in bacteria probed by caged chemoeffectors, *Biophys. J.*, 65, 2368, 1993.
117. Zacks, D. N., Derguini, F., Nakanishi, K., and Spudich, J. L., Comparative study of phototactic and photophobic receptor chromophore properties in *Chlamydomonas reinhardtii*, *Biophy. J.*, 65, 508, 1993.
118. Lopez-de-Victoria, G., Zimmer-Faust, R. K., and Lovell, C., Computer-assisted video motion analysis: a powerful technique for investigating motility and chemotaxis, *J. Microb. Meth.*, 23, 329, 1995.
119. Amsler, C. D., Use of computer-assisted motion analysis for quantitative measurements of swimming behavior in peritrichously flagellated bacteria, *Anal. Biochem.*, 235, 20, 1996.
120. Iken, K. B., Amsler, C. D., Greer, S. P., and McClintock, J. B., Quantitative and qualitative studies of the swimming behavior of *Hincksia irregularis* (Phaeophyceae) spores: ecological implications and parameters for quantitative swimming assays, *Phycol.*, 40, 2001.

13 Natural Chemical Cues for Settlement and Metamorphosis of Marine-Invertebrate Larvae

Michael G. Hadfield* and Valerie J. Paul

CONTENTS

---

* Corresponding author.

I. INTRODUCTION

Most benthic marine invertebrates produce planktonic larvae that remain in their larval phases for minutes to months, sometimes dispersing over great distances before settling on suitable substrata. Abundant literature attests that larvae of many species are influenced by specific chemical and physical cues to settle and metamorphose in appropriate sites for juvenile growth and eventual reproduction.[1-4] Understanding the environmental factors that influence settlement and metamorphosis of marine invertebrate larvae is important to critical understanding in such disparate fields as developmental biology, marine benthic community ecology, aquaculture, and biofouling.

Considerable experimental evidence suggests that chemical cues are very important in substrate selection by larvae. In nature, chemical cues may interact with physical or hydrodynamic factors to induce larval settlement.[5-7] Despite the evidence that chemical cues are extremely important for settling larvae, the complete chemical identity of the natural inducer molecules is known in very few cases.[3,8-11] More commonly, partial chemical characterization has provided clues to the chemical identity of the natural inducers. These partially purified inducers are useful for studying the biology of larval settlement and metamorphosis.[9,12-18]

Chemical settlement cues originate from a variety of sources in the marine environment. Such cues may be waterborne or adsorbed to surfaces and associated with conspecific individuals, specific prey, or the surfaces of substrata. Some marine invertebrates settle preferentially on or among individuals of their own species, resulting in gregarious or aggregative settlement (see examples in Pawlik[3]). In gregarious settlement, conspecific adults may be the source of specific chemical cues.[19-21] A recently detailed example of gregarious settlement is that of oyster larvae responding to waterborne cues from adult oysters (with intact biofilms).[18] Associative settlement[1] involves the settlement of invertebrate larvae on another specific organism, usually either a food, host, symbiont, or requisite substratum. Chemical cues are probably involved whenever invertebrates have restricted host ranges (such as specialist feeders, but see Hubbard[22]). A well-studied example of a chemical inducer coming from a prey organism is the waterborne cue from *Porites* corals that induces metamorphosis in larvae of the coralivorous nudibranch *Phestilla sibogae*.[12,13]

Another source of cues for larval settlement are the films of microorganisms (biofilms) found on most underwater surfaces. Microbial films have long been recognized as necessary for the settlement of some invertebrate larvae,[23] and the settlement of many, but not all, groups of invertebrate larvae are facilitated by the presence of microbial films.[3,24] In some cases, specific types of bacteria present in biofilms may be responsible for facilitating settlement,[24-26] and chemicals bound to or released from bacteria may function as settlement inducers.

There are many reasons why so few natural chemical inducers have been fully characterized. Larvae may respond to specific cues that are present only at very low concentrations; therefore, obtaining adequate amounts of these inducers for purification and structural work may be problematic. The problem of adequate material may be particularly troublesome for the waterborne inducers and has hampered the study of the water-soluble cue from *Porites* corals that induces settlement in *Phestilla sibogae*. A second problem is the low number of collaborations between natural products chemists and larval biologists in the study of natural chemical inducers. However, even a good collaboration between chemists and biologists cannot resolve the difficulties inherent in research on natural inducers that are waterborne molecules, other macromolecules that are not soluble in organic solvents, or membrane-bound inducers. Despite the problems inherent in chemical studies of natural inducers, recent progress has been made in characterizing a number of inducer molecules and demonstrating their biological importance.[10,11,17,18,27,28]

It is not the intention of this review to reconsider all of the abundant and rapidly increasing literature on chemical interactions in larval settlement of marine invertebrate larvae. Excellent summaries and analyses of the earlier literature are available.[1,29] Indeed, Crisp's 1974 review[1] provided both the framework and the terminology for subsequent work in this area over the last 25 years. Several symposia have resulted in published volumes whose contributions summarized the most recent data at the time on the settlement biology of many marine invertebrate groups.[30-34] In addition, later reviews[9,35-37] have added new viewpoints or stressed subsets of the topic. The last comprehensive review of settlement of benthic marine invertebrates was that of Pawlik.[3] This chapter attempts principally to summarize information that has been published subsequent to Pawlik's review, returning to older literature only where necessary to provide a context for discussions of later work or to make comparisons that may lead to useful generalizations.

This review focuses on the biological and chemical aspects of marine invertebrate settlement. With this limitation, an abundance of recent data on physical factors, especially from flume studies of the effects of flow on larval settlement, is not convered in depth. This chapter also does not include the burgeoning literature on artificial inducers for metamorphosis; while of great importance to understanding developmental aspects and internal signaling systems in metamorphosis, these artificial inducers do not, in general, contribute to understanding the chemical ecology of invertebrate recruitment.

II. SETTLEMENT VS. METAMORPHOSIS: THE SAME OR DIFFERENT INDUCERS?

While sometimes used as synonyms, the terms settlement and metamorphosis are usually considered separate phenomena. Settlement is defined as the behavioral act performed when a pelagic larva leaves the plankton, descends to the benthos, and moves upon a substratum with or without attaching to it. Many authors have stressed that settlement is reversible; that is, a larva can swim up from the benthos to settle again in a new location. By contrast, metamorphosis is a definitive morphogenetic event. For all larvae, metamorphosis includes loss of larva-specific organs — typically those used in planktonic swimming — and emergence of juvenile/adult-specific structures. For sessile invertebrates, a firm and often permanent attachment may represent the first stage of metamorphosis. For example, the settling cyprid larva of a barnacle moves from swimming in the water column to the proximity of a firm surface and then "walks" across the surface on the tips of its antennules[38] (for details of settlement behavior, see Walters et al.[39]). The antennules are rich in receptors that are sensitive to physical and especially chemical characters of the benthos (see Clare[40] for scanning electron micrographs). If a cyprid does not encounter cues on the surface that stimulate attachment and metamorphosis, it may resume swimming and eventually settle in a new location, a process that can be repeated numerous times. At some point, the cyprid attaches firmly and irrevocably by secreting a tough cement from the tips of its antennules. Thus anchored, the ostracod-like larva metamorphoses, in one molt, into a small stalked or balanomorph barnacle which remains sessile for life. In contrast to these distinctions, Rodriguez et al.[37] defined settlement as a process with two phases, the first a "behavioral searching phase," and the second "a phase of permanent residence or attachment to the substratum, which triggers metamorphosis and in which morphogenetic events take place." In the settlement definition of those authors, the process begins "with the onset of a behavioral search for suitable substratum," (p. 194), adding a teleological implication which seems unnecessary. Clare,[41] reviewing the subject for barnacle larvae, defined settlement as "attachment and metamorphosis." The term metamorphic induction, used in most modern discussions of settlement and metamorphosis of marine invertebrates, emphasizes the importance of mechanistic developmental processes rather than the implication that small larvae with simple nervous systems are exercising complex substratum choices.

Having distinguished two distinct processes critical for larval recruitment to benthic marine habitats, it can be asked: does the same cue induce both settlement and metamorphosis, or are different cues involved? While the answer to this question remains unknown — indeed, unexplored — for most marine invertebrate species a single cue is known, while for others, separate cues are indicated. In fact, for the greater number of species that have been intensively investigated, larvae are brought into proximity with the benthos, not by responses to chemical stimuli, but by developmental changes in tropistic behavior.[1,42] Ontogenetic onset or alteration of tropistic behavior drives larvae toward the bottom or shore or up estuaries. Negative phototaxis or positive geotaxis or both, may be responsible, and megalopae of the blue crab *Callinectes sapidus* apparently combine negative photokinesis with sensitivity to tidal currents (rheotaxis) to effect movement into estuaries where they settle and metamorphose.[43,44] In most of these cases, settlement, as defined above, occurs after larvae are already close to appropriate settlement substrata and thus able to detect surface-adsorbed or diffusing soluble settlement inducers.

Most abundant are examples where a single inducer is known to affect the entire recruitment process, including settlement and metamorphosis. Examples, discussed more thoroughly in other parts of this chapter, include sand dollars[20] and other echinoids,[45] ascidians,[46] prosobranchs,[47] and perhaps some nudibranchs.[48] Abalone larvae appear to respond only to cues associated with the surfaces of coralline algae,[49] although their sensitivity to a soluble artificial mimic of the surface-complexed inducer (the amino acid GABA) is significantly increased by the presence of lysine in seawater, a phenomenon thought to have significance to settlement of abalone larvae in the field.[50] It must be cautioned that in nearly all cases where only substrate-conditioned water, crude extracts, or partially purified compounds are reported to induce both settlement and metamorphosis, two or more chemicals might be present and influencing the two recruitment processes independently.

Separate cues for settlement and metamorphosis are reported for a disparate group of invertebrates. Larvae of the barnacle-eating nudibranch *Onchidoris bilamellata* (frequently referred to as *O. fusca* in the literature) settle from the water column in response to a water-soluble metabolite released by living barnacles, which are the prey of the adult nudibranchs.[51] However, the larvae metamorphose only if they make surface contact with the shell of a barnacle, leading the authors to conclude that the metamorphic inducer is a chemical associated with the surfaces of barnacles. The settlement process is reversible and repeatable. Curiously, Chia and Koss[51] reported that larvae of *O. bilamellata* will metamorphose on contact with shells of dead barnacles only if they are in seawater conditioned by live barnacles. This observation seems to suggest that a single soluble metabolite induces behavioral settlement, but that metamorphosis requires the inductive action of both that same factor plus one or more others, associated with barnacle shells.

Two-stimulus systems for settlement and metamorphosis may also be required for recruitment of larvae of at least some ascidian, oyster, and barnacle species. Wieczorek and Todd[52] reported that settlement and metamorphosis of the solitary ascidian *Ciona intestinalis* "can be temporally separate and are possibly induced by different environmental cues." Those authors note, however, that "the exact nature of the respective cue(s)....need(s) yet to be further characterized before any conclusions can be reached."

Settlement of barnacles, especially the now cosmopolitan species *Balanus amphitrite*, has received extensive attention. Rittschof[53] pioneered the investigation of the role of specific soluble peptides released as metabolites into seawater by adult barnacles as larval-settlement stimuli for invertebrate larvae, especially those of barnacles. Cyprid larvae of *B. amphitrite* are powerfully stimulated to settle to surfaces and walk upon them by the presence of very low concentrations of dissolved peptides with a C-terminal arginine or lysine residue.[53,54] However, there are reports of attachment and metamorphosis of cyprids of *B. amphitrite*, in the absence of a soluble peptide cue, in response to surface-bound inducers associated with biofilms, adult barnacle shells, and the footprints left when other cyprids walk across the surface (reviewed by Clare and Matsumura[55]). Whether barnacle settlement and metamorphosis in the field are achieved by one or more cues, and whether a single generalization applies to all acorn-barnacle species, awaits further investigation.

As was earlier shown for barnacles, larvae of the oyster *Crassostrea virginica* are strongly induced to settle by small peptides with a C-terminal arginine residue released into seawater both by living adult *C. virginica* and biofilm bacteria.[7,18,56] That the same peptides, adsorbed to exopolymers of marine biofilms, act as contact-mediated attachment and metamorphosis stimuli was proposed by Zimmer-Faust et al.[57] (see also Decho et al.[58]). This would imply that different receptor systems operate for detecting the dissolved cue (i.e., olfaction) and for sensing the adsorbed cue (i.e., contact chemoreception). The chemical nature of oyster-settlement cues is discussed in greater detail in Section III.C, below.

It is difficult to compare directly recruitment processes of decapod crustacea with those of other invertebrate larvae because pre- and postmetamorphic stages in crabs and shrimp are not equivalent to those in nonarthropod invertebrates (nor in cirripedes), and metamorphosis may, in fact, precede settlement. Although megalopa-stage blue crabs are typically referred to as postlarvae, their molt to a first crab bottom stage is termed metamorphosis (e.g., Forward et al.[59]) (For other animals, metamorphosis is, by definition, a phenomenon that occurs only in larvae and transforms them into postlarvae, usually referred to as juveniles. See, for example, Giese and Pearse.[60]) In the blue crab *Callinectes sapidus*, molting from the last zoeal stage to the megalopa (postlarva) occurs in offshore waters. Behavioral responses of megalopae cause them to be carried to inshore waters, and from there into estuaries.[43,44,61–63] In estuarine waters, megalopae are specifically attracted to seagrasses where they typically settle and molt into the benthic first crab stage (metamorphosis).[64] This molt is hastened by the reduced salinity of estuarine waters, substances associated with sea grasses, and humic substances transported down rivers and into estuarine waters.[59,65,66]

With the wealth of studies of settlement and metamorphosis of marine invertebrates currently underway, it is very likely that additional examples of larvae relying on two or more cues for settlement and metamorphosis will be described.

III. GREGARIOUS SETTLEMENT

A. BIOLOGICAL RELATIONSHIPS

Gregarious settlement — that is, larvae choosing to settle in response to the presence of adults, juveniles, or recent recruits of the same species — has been reported for many phyla, especially for tube-dwelling polychaete worms, oysters, and barnacles. Studies of the polychaetes *Phragmatopoma lapidosa californica*, *Sabellaria alveolata*, *Hydroides dianthus*, and *H. ezoensis* have found adult-associated compounds to be responsible for larval settlement into adult aggregations (described in greater detail below). Barnacle cyprid larvae have long been known to settle specifically among adults of their own species.[67] Studies of gregarious settlement of oysters date at least from the 1940s,[68] and recent studies, cited below, have significantly expanded our understanding of the chemical nature of these processes. Note that all of the groups cited above are sessile after metamorphosis, and in some species for each group, factors arising from adult organisms have been invoked as larval-settlement stimuli. However, other examples, such as the sand dollar *Dendraster excentricus*, demonstrate that gregarious settlement can occur into populations of free-living animals that live in dense aggregations.[20]

Because marine invertebrates demonstrating chemically specific gregarious settlement are among the best studied examples for purified or partially purified inducers, they are discussed at greater length below. However, it is dangerous to assume that all aggregations of sessile or attached marine invertebrates arose through intraspecific chemical cues for settlement and metamorphosis. Asexual propagation can produce such masses in anthozoans, polychaetes (e.g., *Salmacina* spp.),

and, of course, encrusting bryozoans and compound ascidians. Dense aggregations of other polychaetes (e.g., *Hydroides elegans*, see below under discussion of biofilms), sea urchins, and other
mobile species may arise through a settlement response to an abundant benthic cue (associative
settlement, see below) or postsettlement migration (e.g., crustaceans[69]). In fact, based on extensive
field experiments on settlement onto manipulated panels attached to a pier, Keough[70] concluded
that early recruits to the substratum had little effect on the density of later conspecific settlers for
the fouling community at the site.

B. FULLY CHARACTERIZED CHEMICAL CUES

Recent studies of the chemistry of compounds inducing settlement of ascidian larvae by Fusetani
and coworkers have identified the structures of several different metabolites including urochordamine A (Figure 13.1),[27] the narains,[71] and anthosamines A and B.[72] Urochordamines A and B
were isolated from the tunic of the ascidian *Ciona savignyi* and were later also found in the tunic
of the ascidian *Botrylloides* sp. Only urochordamine A (Figure 13.1) was active in inducing settlement and metamorphosis in laboratory assays, and it could conceivably function in gregarious
settlement in nature.[27] The narains and anthosamines were isolated from sponges, and it is not clear
whether the ascidian larvae settle on these sponges naturally.[71,72] For all these studies, more
biological and ecological data are needed before any conclusions can be drawn about the natural
functions of these compounds. It now appears that at least some of these compounds, such as
urochordamine A, may simply mimic the natural inducers.[10,73]

Many ascidians live gregariously, and for some of these species conspecific chemical cues may
play an important role in gregarious settlement of the larvae. The extracts of conspecific adults,
larvae, or their conditioned seawater have been shown to contain metamorphosis inducers which
have never been characterized.[74,75] Recently, Fusetani and co-workers elucidated the structure of
the metamorphosis inducer for the solitary ascidian *Halocynthia roretzi*, which was isolated from
seawater conditioned by ascidian larvae.[10] The compound isolated from the medium was identical
to lumichrome, a compound known to be a degradation product of riboflavin (Figure 13.1). The
origin of lumichrome in *H. roretzi* is not known at present.

The conditioned medium containing larvae of *Halocynthia roretzi* was chromatographed by
reverse-phase chromatography, Sephadex LH-20 chromatography, and, finally, reverse-phase HPLC
to yield the active compound, which was then characterized by ultraviolet (UV) and proton and
carbon nuclear magnetic resonance (NMR) spectroscopy.[10] Standard bioassay-guided fractionation
was used; the chromatographic fractions were assayed for their ability to induce metamorphosis at
each stage of the isolation process. The active compound could also be isolated from *H. roretzi*
larvae and from the tunic dissected from adults. Lumichrome was also present in gonads and
unfertilized eggs of adults, but not in hemolymph, muscles, or hepatopancreas. Whole larvae were
examined by epifluorescence microscopy to determine where lumichrome was localized. The larvae
emitted fluorescence indicative of lumichrome from the basal region of the adhesive organ and the
posterior part of the trunk. Lumichrome is commercially available, and, therefore, the compound
isolated from the ascidian could be compared with authentic material. Its ability to induce larval
metamorphosis seems to be species specific.

C. PARTIALLY CHARACTERIZED CHEMICAL CUES

Marine polychaete worms of the family Sabellariidae live in tubes of cemented sand grains.
Gregarious sabellariid species often form extensive colonies of these sand tubes by recruiting larvae
from the plankton. Evidence that a chemical cue in the tube cement triggers larval settlement was
first reported by Wilson[76,77] who worked on species from British waters. Larvae of *Phragmatopoma
lapidosa californica,* a sand tube-building worm from California, settle and metamorphose preferentially on anterior portions of sand tubes of adult worms,[78] again supporting the presence of a

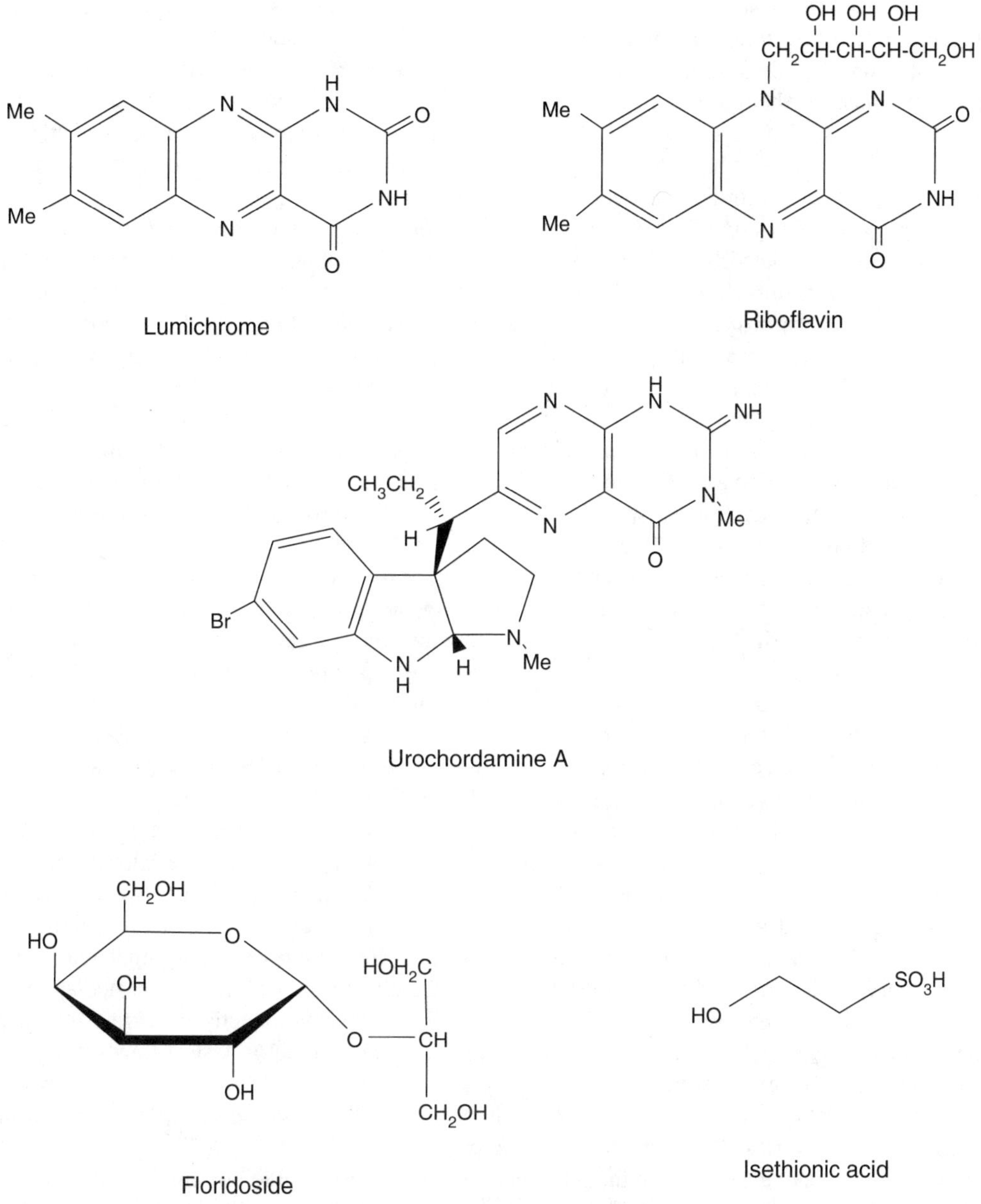

FIGURE 13.1   Structures of metamorphic inducers and related compounds.

chemical cue. Glass beads with which conspecifics had previously built tubes were also highly inductive (over 80% metamorphosis), and the stimulus for metamorphosis was inactivated by boiling.[78] The chemical nature of the settlement cue from adult tubes that stimulates settlement and metamorphosis in this species has been controversial.

Pawlik[79,80] and Pawlik and Faulkner[81,82] showed that certain free fatty acids (FFA) that are present in the tube sand of *Phragmatopoma lapidosa californica* can induce larval settlement and metamorphosis (reviewed by Pawlik[3]). Palmitoleic, linoleic, arachidonic, and eicosapentaenoic acids were found in the original inductive fraction. The individual FFA induced metamorphosis over a range of concentrations, although abnormal metamorphosis was observed for some FFA at high concentrations (above 300 μg FFA/g sand). Individual FFA showed relatively low levels of activity

(5 to 15% metamorphosis) at the concentrations at which they were isolated (10 to 50 µg/g sand). In further tests of 37 different commercially available FFA, it was found that *cis* stereochemistry of double bonds, chain length, and the presence of a free carboxyl group were all important for larval settlement.[81] About a dozen different FFA induced some level of metamorphosis (5 to 25%) at concentrations of 100 µg FFA/g sand. No single FFA could completely explain the settlement behavior in response to tube sand. Again, high levels of abnormal metamorphosis were observed for many FFA at concentrations above 300 µg/g sand.

Larvae of the gregarious sabellariid subspecies from the Atlantic, *Phragmatopoma lapidosa lapidosa*, were stimulated to settle by the same group of FFA that induced settlement in *Phragmatopoma l. californica*.[83] Again, the stereochemistry of double bonds, chain length, and the presence of a free carboxyl group were all important for larval settlement and metamorphosis.

Sabellaria alveolata, a reef-building tube worm from European waters, also settled in response to conspecific tube sand. However, *S. alveolata* did not metamorphose normally in response to any FFA, and an inducer from the tube sand of these animals was not isolated or identified.[84] These larvae were much less discriminating in their choice of substrata than were those of *Phragmatopoma lapidosa californica*. Larvae of the nongregarious species *Sabellaria floridensis* did not metamorphose to any greater extent on conspecific tube sand than on control sand. Metamorphosis was also not enhanced upon exposure to FFA.[83]

To examine the interactions between hydrodynamic processes and chemical cues, Pawlik and co-workers studied the responses of larval *Phragmatopoma lapidosa californica* to inductive substrates in flume experiments.[5,6] Five different treatments were presented including two known to be inductive (conspecific tube sand and sand treated with palmitoleic acid) and three that were noninductive in previous experiments (sand held in aquaria with worms but not used to build tubes, sand treated with palmitic acid, and untreated control sand). Larvae preferentially metamorphosed on the two inductive substrates in both fast and slow flow regimes. The number of larvae metamorphosing in inductive treatments was much higher in fast than slow flow, because significantly more larvae reached the treatment array in fast flow.[5] In another study, settlement experiments were carried out over a range of turbulent flume flows (5 to 35 cm s^{-1}). Larvae were allowed one pass over a sediment array with two treatments: inductive tube sand and noninductive sand. Metamorphosed juveniles and total animals (larvae + juveniles) in the array were highest at intermediate flows (15 to 25 cm s^{-1}).[6] In both series of experiments, the authors concluded that interactions between larval behavior and flow regimes determined patterns of settlement. At intermediate flows, larvae tumbled along the bottom where they more readily contacted inductive substrate. At slower flows, larvae actively left the bottom and stayed in the water column; at fastest flow rates, larvae were swept away from the bottom.

Jensen and Morse investigated settlement inducers for *Phragmatopoma lapidosa californica* concurrently and concluded (1) that a polymeric protein containing L-β-3,4-dihydroxy phenylalanine (L-DOPA) residues, present in the tube sand cement, is responsible for stimulating settlement in these polychaetes, and (2) that FFA in Pawlik's preparations were contaminants that induce settlement nonspecifically.[85–87] They demonstrated that clean preparations of the natural inducer (made by providing adults with clean glass beads with which they built tubes) do not contain FFA, that all of the carbon in the inducer can be accounted for as protein, that the natural inducer is a labile substance that can be inactivated by extraction, and that the kinetics of natural metamorphosis in response to tube sand is identical to that in response to tube glue, but not to FFA.[87] They suggested that the FFA induce metamorphosis by operating physiologically downstream or parallel to the natural inducer, but that they are not the natural inducers of metamorphosis.

Jensen and Morse[86] also tested the capacity of 2,6-di *tert*-butyl-4-methylphenol (DBMP), which they considered a possible aromatic analog of cross-linked L-DOPA, to induce settlement and metamorphosis of *Phragmatopoma lapidosa californica* in the laboratory and field. Plexiglass plates were coated with underwater epoxy and then a solution of DBMP in diethyl ether prior to deployment in the ocean. Significantly higher numbers of *P. l. californica* metamorphosed on surfaces with DBMP in the laboratory and the ocean than on control surfaces without DBMP. This is one of the few studies to demonstrate that a defined organic molecule can induce settlement and metamorphosis in the ocean.

Thus, questions remain about the natural role of FFA as settlement inducers in *Phragmatopoma lapidosa californica,* and it is very likely that the FFA are another type of artificial inducer, such as cations, GABA, and organic solvents.[87,88] Additional research is warranted to resolve the roles of FFA and polymeric proteins containing L-DOPA residues in inducing settlement and metamorphosis of *P. l. californica* and to determine whether either or both types of molecules are important in nature.

Several other researchers have studied chemical cues influencing gregarious settlement and metamorphosis of polychaete larvae. An unidentified waterborne cue was responsible for gregarious settlement of *Hydroides dianthus.*[89] The inducer was not associated with the tubes, but rather with the bodies of adult worms. Extractions of adult worms with organic solvents removed the inductive capacity of the tissue. The activity was found in both polar (methanol) and nonpolar (dichloromethane) fractions of an extraction series.[89]

Conspecific adult tubes contained a chemical cue involved in larval metamorphosis of *Hydroides ezoensis,* a common serpulid polychaete in Japan.[90,91] Methanol extracts of adult tube clumps were fractionated by Sephadex LH-20 and then silica column chromatography; the separations were followed by larval bioassays. Ultimately, a new monoacyl glycerol was isolated by HPLC from the active fraction.[91] The chemical structure was determined to be 1-(4Z, 7Z, 10Z, 13Z, 16Z, 19Z)-dochosa hexanoyl-X-glycerol by spectroscopic analysis and comparison with an enzymatically prepared authentic sample. The stereochemistry of the glycerol moiety is not known. A semisynthetic sample induced more than 50% metamorphosis at 3×10^{-6} M. The authors suggested that acyl glycerols may possibly act as second messengers and not as the primary cue of metamorphosis.[91]

Sand dollars, *Dendraster excentricus*, exhibit gregarious settlement, and larvae undergo settlement and metamorphosis in the presence of adult-associated sand.[92] Adults produce a chemical cue that remains associated with the sand and is stable for at least 7 weeks. Thus, settlement occurs within or near existing sand dollar beds. Highsmith[92] suggested that the cue was a small peptide. This was confirmed by Burke,[20,21] who isolated an active peptide of 980 Da from extracts of sands from sand dollar beds.

Gregarious settlement has long been recognized for barnacle larvae.[67,93] A factor identified as arthropodin, a glycoprotein component of arthropod cuticles, induced settlement in cyprid larvae of *Balanus balanoides* when they contacted factor-treated surfaces.[19,94,95] Further purification and characterization of the settlement-inducing factor from adult shells led to the conclusion that it was a series of closely related acidic proteins, with an amino acid composition similar to that of actin.[96,97] Yule and Walker[98] reported that cyprids of *Balanus balanoides* were stimulated to attach by the antennular secretions left as footprints (actually, antennule prints) when earlier settling cyprids explored a surface. Similarly, attachment of cyprid larvae of *B. amphitrite* was significantly enhanced by the presence of antennular secretion left behind on a surface by other cyprids,[99] but in neither case has the chemical nature of the adsorbed substance been elucidated. Pawlik[3] provided a good summary of the extensive research on the chemical cues for gregarious settlement in barnacle larvae up through 1990.

The settlement-inducing protein complex from adult shells of the barnacle *Balanus amphitrite* has been further purified and characterized.[100] A settlement-inducing glycoprotein complex with high molecular weight was purified from homogenates of whole adult barnacles by ammonium sulfate precipitation, ion-exchange chromatography, gel filtration, and lectin-affinity chromatography on lentil lectin-Sepharose. A new bioassay with nitrocellulose membranes was used to monitor cyprid settlement in response to adsorbed proteins.[100,101] SDS-polyacrylamide gel electrophoresis (SDS-PAGE) of the purified protein complex showed three major bands of 76, 88, and 98 KDa and a minor band of 32 KDa. The authors concluded that the active protein is a high molecular weight complex composed of four subunits.[100] Each individual major subunit, when isolated by

SDS-PAGE, had comparable settlement-inducing activity to the intact glycoprotein. Lentil lectin, a lectin that inhibits adult extract-induced settlement,[101] could bind to each of the three major subunits. The authors suggest that specific sugar chains on each subunit of the protein play an important role in the settlement of *B. amphitrite*.[100]

For *Balanus amphitrite*, a waterborne settlement inducer that is more active than the surface-associated settlement factors has been extracted from seawater conditioned with adult barnacles.[53] The cyprid settlement responses could be stimulated by peptides bearing C-terminal arginine or lysine residues.[54,102] Interestingly, the same barnacle settlement-inducing peptides attract a variety of barnacle predators such as oyster drills.[53,103]

Recognition of the importance of chemical cues in barnacle larval settlement has been important for studies of how physical factors such as salinity[104,105] and water flow interact with chemical cues to mediate settlement behavior.[106,107] In a recent study, Wright and Boxshall[107] showed that larvae of two species of barnacles, *Elminius modestus* and *E. covertus*, responded positively to soluble conspecific chemical cues for settlement; however, the physical presence of barnacles also influenced settlement patterns in these species. An excellent review of the chemical basis for barnacle settlement has recently been published.[55]

Because of the commercial importance of oysters, many studies have examined the settlement behavior of oyster larvae. Earlier studies demonstrated that substances in the outer layers of the shells of *Crassostrea virginica*,[108] aqueous extracts of whole oysters,[109] and water held between the valves of living adults[110] could induce settlement of oyster larvae. Proteinaceous compounds have been implicated as the natural inducers.[111,112] It was suggested that the settlement inducers for *Crassostrea gigas* originate from microbial biofilms present in oyster beds.[113,114] However, Tamburri et al.[56] found that oyster larvae responded similarly to waterborne substances released by adult oysters or biofilms. Coon et al.[115] proposed that ammonia was important in the initial settlement behavior of oyster larvae,[115] and that enough ammonia is present in oyster beds to presumably trigger this behavior in nature.[116] Bonar et al.[117] worked with supernatants from biofilm bacteria and proposed that the ammonia is bacterially produced. However, other researchers have suggested that ammonia is simply a poison, and responses to it may not reflect natural settlement activity.[18]

Zimmer-Faust and Tamburri[18] investigated waterborne settlement inducers released by adult *Crassostrea virginica* to try to understand better the chemical identity of the natural settlement inducer. They prepared oyster "bath water" by soaking adult oysters with intact biofilm in artificial seawater for 2 h. Molecular weight fractionations of the oyster bath water indicated the presence of waterborne inducers between 500 and 1000 Da. The inducers were degraded by some proteases but not by carbohydrases or by lipase, indicating their peptide nature. Proteases that cleave basic amino acids (lysine and arginine) from the peptide C-terminal ends destroyed the activity of the inductive fraction. The enzyme arginase also destroyed the activity of the 500 to 1000 Da fraction, indicating that arginine was at the C-terminal of the active peptides. A tripeptide having arginine at the C-terminal, glycyl-glycyl-L-arginine (GGR), induced settlement at a concentration as low as 10^{-10} M. Dose-response curves for GGR and the active 500 to 1000 Da fraction of oyster water were essentially identical. Tests of 21 different amino acids identified only lysine and arginine as settlement cues; however, peptides with arginine at the C-terminal were more active than the amino acids. Thus, the authors proposed that low molecular weight peptides with arginine at the C-terminal were the water-soluble cues that induce oyster settlement.[18] Because the authors used the settlement process (attachment of the larval foot to the substrate) and not metamorphosis for their bioassays, it is not clear if these same peptides induce metamorphosis of oyster larvae. During the experiments, larvae were induced to metamorphose with an artificial inducer, epinephrine, to ensure that they were competent.

Flume experiments on settlement of oyster veligers were conducted with GGR at 10^{-7} M concentrations released by peristaltic pumps into small target wells lined with crushed oyster shells.[7] The crushed shells acted as a diffuser, allowing the test solution to slowly seep upward. Settlement of the oyster larvae (*Crassostrea virginica*) on peptide-containing wells and adjacent control wells (seawater alone) was compared for several runs with different flow speeds of the flume. The wells were arranged in a Latin-square design so that there were both upstream and downstream peptide and control wells. Settlement was enhanced in the upstream wells releasing GGR compared to control wells without cue, but not in the downstream wells at the two flow speeds that were tested. These experiments demonstrated that oyster larvae settle in response to a waterborne chemical cue in both still and flowing water. The researchers proposed that larvae swim or sink rapidly upon contact with the chemical cue, which brings them quickly into proximity with benthic surfaces for subsequent settlement and metamorphosis.[7] These predictions were confirmed in a study that used computer-assisted video motion analysis to quantify the movements of individual oysters in response to chemical cues in flow. Waterborne chemical cues, either oyster bath water or GGR, but not plain scawater, caused larvae to move downward in the water column and swim in slow curved paths before attaching to the bottom.[57,118]

IV. ASSOCIATIVE SETTLEMENT

A. Biological Relationships

Associative settlement is the term used by Crisp[1] to describe settlement of invertebrate larvae specifically upon other species, including both plants and animals. Often, associative settlement results in small predators recruiting to colonial prey (e.g., opisthobranchs that feed on only one or a few species of hydrozoans, corals, bryozoans, barnacles, or algae; see Hadfield and Switzer-Dunlap[119]). Crisp's review[1] provided comprehensive tables for such recruitment relationships recorded up to that time. Pawlik[3] divided this category into several subcategories, including parasitic relationships, nonparasitic relationships, and herbivorous/predatory relationships, and summarized data then available. Since 1992, biologists have continued to work on earlier recognized models and list new associations across the marine-invertebrate phyletic spectrum. Recent papers on this topic, discussed in greater depth below, deal with cnidarians (planulae of corals and the jellyfish *Cassiopea xamachana*), molluscs (a chiton and several types of gastropods), tube-dwelling polychaetes, sea urchins, and Crustacea (blue crab and a prawn). Pawlik[3] summarized data on settlement of shipworms, to which we can add the recent observations of Gara et al.,[120] who reported that veligers of *Bankia setacea* attack Douglas fir logs at sites where the bark has peeled back, possibly indicating a chemical cue from freshly exposed wood. In addition, wood already attacked by *B. setacea* is more heavily invaded by new veligers than unattacked wood, suggesting that gregarious factors may also be present. The special case of parasitic infections is not reviewed in this chapter.

B. Fully Characterized Chemical Cues

There are several fully identified compounds that induce larval settlement behavior; however, the ecological relevance of many of these chemical inducers remains unclear.[3] For example, the metabolite jacaranone from the red alga *Delesseria sanguinea* induces settlement in *Pecten maximus*; however, *P. maximus* is not known to settle with any specificity on *D. sanguinea*.[121,122] Also, several δ-tocopherol epoxides were identified as being responsible for settlement of the hydroid *Coryne uchidai* on the brown alga *Sargassum tortile*;[123] however, these metabolites are lipophilic, whereas a previous study had suggested that water-soluble molecules were responsible for settlement.[124]

A recent study has identified the substances that induce metamorphosis in the sea urchin *Holopneustes purpurascens*.[11] Although adult urchins live in the canopy of several different subtidal algae, including the red alga *Delisea pulchra* and the kelp *Ecklonia radiata*, recent recruits were found predominantly on *D. pulchra*. Competent larvae metamorphosed in the presence of *D. pulchra*, seawater surrounding this alga, and polar extracts of *D. pulchra*. The cue for metamorphosis was isolated by reverse-phase HPLC from an aqueous partition of the MeOH extract of *D. pulchra*. It was characterized primarily by NMR and found to be a water-soluble complex of the sugar floridoside and isethionic acid (Figure 13.1) in a 1:1 molar ratio. The rate of metamorphosis of larvae exposed to the floridoside–isethionic acid complex was concentration dependent with larvae exposed to 25 µM metamorphosing significantly faster than those exposed to 2.5 µM. The floridoside–isethionic acid complex was important; neither compound in isolation induced metamorphosis. Although *Holopneustes purpurascens* responded most strongly to *Delisea pulchra*, it also metamorphosed in response to several other species of red (but not brown or green) algae from its habitat.

C. PARTIALLY CHARACTERIZED CHEMICAL CUES

1. Inducers from Algae and Higher Plants

Abalone larvae settle preferentially on crustose coralline red algae of the genera *Lithothamnion*, *Lithophyllum*, and *Hildenbrandia*,[49,125–127] although mucus from juvenile conspecifics has also been shown to induce larval settlement.[128] Water-soluble extracts of these coralline algae also induce settlement. The settlement inducing activity was found in a macromolecular fraction isolated by gel-filtration chromatography from several species of red algae and cyanobacteria, but was only detectable on the surfaces of crustose coralline algae. The inducers appear to be associated with the phycobiliproteins of red algae and cyanobacteria. The inducers have been partially purified and characterized. They are small peptides, 640 to 1250 Da, with unusual composition, complexed to proteins at the surfaces of coralline algae.[49,125,129,130]

Chemicals on the surfaces of crustose coralline algae are also important cues for the settlement of coral larvae, although different species of corals display different degrees of specificity in their requirements for crustose coralline algae to induce metamorphosis.[17,28,131] An insoluble, cell-wall polysaccharide (a type of sulfated lipoglycosaminoglycan containing multiple *N*-acetyl lactosamine sulfate residues) is the type of compound that induces the settlement of many species of coral larvae including *Agaricia* spp. in the Caribbean and *Acropora* spp. in the Pacific.[15,28] The soluble but unstable fragment of the inducing molecule can be liberated by hydrolysis of the cell wall associated polysaccharides with specific enzymes,[15] while a more stable fragment can be liberated by decalcification of the coralline algal cell walls.[17] It appears that many different corals require the same type of algal cue for the induction of settlement and metamorphosis. The algal species (*Hydrolithon reinboldii* and *Peysonnelia*) that induce metamorphosis of acroporid corals in the Pacific have congeners in the Caribbean (*H. boergesenii*) that induce metamorphosis in the agariciid corals. These algae all contain polymeric morphogens that are apparently of the same class of polysaccharide compounds.[132]

In further studies of the inducer from crustose coralline algae, an active polysaccharide could be prepared by decalcification of the coralline algal cell walls with the chelators EGTA or EDTA. The solubilized morphogen was then purified by hydrophobic interaction chromatography and anion exchange (DEAE) chromatography guided by larval bioassays. This active polysaccharide fragment, with an apparent molecular weight in the range of 6000 to 8000 Da, contained both hydrophobic and anionic domains.[17] Because of the presence of both hydrophilic and hydrophobic portions, the polysaccharide could be readily absorbed onto hydrophobic resins. The inducer retains its activity after absorption, and the activity is identical in the laboratory and ocean.[17] The resin-bound inducer can be attached to surfaces with silicone adhesive to create an artificial recruitment surface (chemical flypaper) that should have many applications for field studies of chemical cues in larval settlement.[28]

Recent studies with larvae of *Acropora millepora,* a common Indo-Pacific coral species, and coral larvae collected from natural slicks after mass spawning events also examined the role that coralline algae play in inducing settlement and metamorphosis of acroporid larvae.[133] Four species of crustose coralline algae, one non-coralline crustose alga and two branching coralline algae, as well as the skeleton of the massive coral *Goniastrea retiformis* induced metamorphosis. Chemical extracts were prepared from the crustose red alga *Peyssonnelia* sp. and the skeleton of *G. retiformis* in two ways: (1) by closely following the methods of Morse et al.,[17] and (2) by methanol extraction. Extracts from both the crustose red alga and the coral skeleton were highly active inducers with up to 80% of larvae metamorphosing in 24-h assays. Researchers concluded that several different crustose algal species can induce metamorphosis of coral larvae, that some inducers may not be strongly associated with calcified algal cell walls, and that inducer sources in reef habitats may be more diverse than previously recognized.[133]

Many different marine invertebrates settle preferentially on species of red algae including crustose corallines (reviewed by A. Morse[130]). Crustose coralline algae induce settlement and metamorphosis in various marine invertebrates including the green sea urchin *Strongylocentrotus droebachiensis*,[45] the chiton *Tonicella lineata*,[134] and the serpulid worm *Spirorbis rupestris*.[135] Two commercially important sea urchins in Japan, *Pseudocentrotus depressus* and *Anthocidaris crassispina*, have been reported to settle and metamorphose in response to free fatty acids, especially eicosapentaenoic acid, extracted from the coralline red alga *Corallina pilulifera*.[136]

Other red algae have been suggested as a source of inducers of settlement and metamorphosis for the sea hare, *Aplysia californica*[137] (but see Pawlik[138]), and the queen conch *Strombus gigas*.[139] Various detrital substrates as well as the red alga *Laurencia poitei* and aqueous extracts of the alga were potent inducers of settlement and metamorphosis for *S. gigas*. The cue from *L. poitei* was water soluble, of low molecular size (<1 kDa), and stable to boiling.[139] Seventeen amino acids were detected in hydrolyzed extracts of *L. poitei*, among which isoleucine and valine produced significant levels of metamorphosis in larval *S. gigas*, although never as high as those produced by extracts of the alga.[140]

Other marine plants induce the settlement and metamorphosis of marine invertebrate larvae. Postlarvae of the blue crab *Callinectes sapidus* preferentially settled in the field on collectors containing odors of the seagrasses *Zostera marina* and *Halodule wrightii*.[64] Large postlarvae and juveniles of the tiger prawn *Penaeus semisulcatus* showed a preference for settling on the seagrass *Zostera capricorni* over bare sand and artificial seagrass, especially during the day.[141] The authors suggested that a chemical cue released from *Zostera capricorni* might explain the greater numbers of prawns in natural rather than artificial seagrass in their experiments.

Planula larvae of the jellyfish *Cassiopea xamachana* settle and metamorphose on decomposing mangrove leaves of *Rhizophora mangle* that lie submerged in shallow water mangrove ecosystems. Marine bacteria were found to be involved in the production of water-soluble peptides during the decomposition of the mangrove leaves.[142] The homogenate of decaying mangrove leaves was subject to gel filtration chromatography and reversed-phase HPLC yielding several fractions that induced high rates of metamorphosis of *C. xamachana*.[143] The most active fraction was further characterized and found to contain a proline-rich peptide with a molecular weight of approximately 5.8 KDa, probably a byproduct of nonspecific bacterial degradation of proline-rich cell wall proteins of plants.[143]

Larvae of the specialist sacoglossan *Alderia modesta* settle in response to a chemical cue from their algal food, the yellow-green alga *Vaucheria longicaulis*. Water conditioned by *V. longicaulis*, frozen or homogenized algal tissue, and boiling water extracts of the alga as well as residual unextracted material all induced high rates of metamorphosis in *A. modesta*.[144] The boiling water (aqueous) extract was fractionated by ethanol precipitation into a supernatant and precipitate; both induced metamorphosis. Further fractionation by gel filtration chromatography of the supernatant and precipitate indicated that the supernatant contained a low molecular weight (<2000 Da) carbohydrate fraction that was inductive, and the precipitate contained a high molecular weight (>100,000 Da) carbohydrate-containing fraction that was inductive. The aqueous extract was treated with proteinase K to determine if proteinaceous components were important for bioactivity; proteinase K treatment did not affect the activity of the extract. In contrast, treatment with periodate, an agent which cleaves sugar residues, significantly decreased the activity of the aqueous extract. The authors propose that the larvae of *A. modesta* metamorphose in response to high and low molecular weight carbohydrates. Water-soluble, low molecular weight carbohydrates and insoluble, surface-associated carbohydrates from the host alga both seem to induce metamorphosis in this specialist herbivore.[144] Further structural information will be necessary to determine the chemical nature of the carbohydrate cues and whether the soluble and insoluble inducers are related types of polysaccharides.

2. Inducers from Invertebrate Animals

The coralivorous nudibranch *Phestilla sibogae* settles preferentially on *Porites* corals, which are its preferred food.[12,13,145] A waterborne molecule induces metamorphosis in *P. sibogae*; it is released naturally from *Porites compressa*, a preferred food item in Hawaii, and other *Porites* corals.[12,13,146] This small, polar molecule (molecular weight 300 to 500 Da) has defied attempts at full structural characterization for over two decades, primarily because it is difficult to isolate. It is a very minor component of a complex mixture of metabolites released by the coral into seawater; therefore, obtaining enough material for structural elucidation has been difficult. Although attempts to fully characterize this molecule have not been successful, seawater containing the inducer has been a useful tool for studying developmental processes that occur during settlement and metamorphosis in *Phestilla sibogae*.[12,126,147–150]

Other opisthobranch molluscs settle and metamorphose in response to prey items. The dorid nudibranch *Adalaria proxima* feeds on the cheilostome bryozoan *Electra pilosa*. Soluble cues secreted by live colonies of the bryozoan, but not dead, frozen, or sonicated *E. pilosa,* induced metamorphosis.[151] Ultrafiltration showed that the active inducer is less than 1000 Da in size.[48] Trypsin eliminated the inductive cue; Lambert et al.[48] suggested that the active inducer is a small, arginine- or lysine-containing peptide.

Larvae of the cephalaspidean mollusc *Haminaea callidegenita* were induced to metamorphose by a small (<1000 Da), polar, nonproteinaceous compound found in the gelatinous matrix of the egg masses.[16] The inducer was purified from the gelatinous egg masses by extraction with methanol and bioassay-guided fractionation by reversed-phase HPLC. A similar inductive compound could also be isolated from other tissues of adult animals and the gelatinous egg masses of four other opisthobranchs. The presence of the inducer within egg masses caused an unusual developmental pattern in *H. callidegenita*; both swimming, lecithotrophic veliger larvae and crawling juveniles were produced.[16]

V. SETTLEMENT IN RESPONSE TO BIOFILMS

A. GENERAL CHARACTERISTICS OF BIOFILMS

It has long been known that dense aggregations of microorganisms begin to accumulate within a few minutes to a few hours when clean surfaces are placed in the sea.[23] These films usually originate as adsorbed organic molecules, which may serve as a stimulus for the attachment of bacteria, fungi, and single-celled algae.[152,153] Most research has focused on the role of biofilm bacteria in enhancing the acceptability of surfaces for settlement of algal spores and invertebrate larvae, and extensive literature on the topic now exists. Some communities of organisms, most notably the fouling community of mostly sessile invertebrates that coat piers, pilings, and ships' hulls in bays and harbors, may be almost completely established by the attraction of their larval stages to biofilmed surfaces. Johnson et al.[24] pose hypotheses that induction of invertebrate metamorphosis by bacteria may, in fact, be far more widespread than suspected. According to their hypothesis, generalist species may respond to a series of biofilm bacterial species which are common to many substrata, while specialist settlers (e.g., those settling in response to conspecific individuals, prey or specific algae for living substrates) settle in response to bacteria that are specifically adapted to those surfaces, rather than to metabolites of living substrata themselves.

Most of the research reviewed below consists of controlled laboratory experiments. However, in a series of excellent papers, Todd and Keough,[154] Keough and Raimondi,[155] and Keough[70] report on experiments conducted largely in the field in Victoria, Australia, where panels were allowed to develop marine biofilms in seawater tanks and were then placed in the field to measure settlement on the panels as a function of biofilm age. In nearly all cases, recruitment of sessile invertebrates onto the panels positively correlated with age of the biofilm when films had been allowed to develop for 1, 3, or 6 days prior to placing the panels in the field.[155] Although some taxa appeared to show little or no relationship to film-age in their settlement density, the first observation of the panels, three days after their emplacement in the field, would have allowed for 1 to 3 days of biofilm development and at least 1 to 2 days of larval settlement prior to their examination. About the only way to actually determine the results of biofilm age on settlement in the field would seem to necessitate some mechanism for almost continuous recording of settlement from the moment the panels are put in place. Perhaps high-magnification video recording will offer such opportunities.

Rather than attempt an exhausting review of this literature, this chapter focuses mainly on papers published after 1990 that offer new insights into the role(s) of biofilms on hard surfaces in inducing settlement of larvae of marine invertebrate animals. See Pawlik[3] for a good summary of earlier literature on the role of biofilms in the recruitment of invertebrate larvae. It is probably valid to say that most, if not all, members of the common fouling communities that have become widely distributed around the world due to their transport on ships are initiated when their larvae respond to bacterial films on submerged surfaces.[29]

B. Biofilm-Specific Settlement and Metamorphosis

1. Porifera

Recruitment of sponge larvae to experimental substrata requires the prior accumulation of a complex biofilm.[156–158] Studies on the inductive effects of potassium ion on sponge larvae revealed that larvae of *Aplysilla* sp. would not settle on clean substrata, but settled selectively — albeit in percentages rarely exceeding 20% — on pieces of membrane peeled from the inner surfaces of hens' eggs that had been allowed to accumulate a biofilm in flowing seawater.[159]

2. Cnidaria

Two organisms, the hydrozoan *Hydractinia echinata* and the scyphozoan *Cassiopea andromeda* (and also the congener *C. xamachana*), provide most of the available analytical data on the roles of marine biofilms in larval settlement in cnidarians. *Hydractinia echinata* forms dense colonies on the outer surfaces of gastropod shells that are inhabited by hermit crabs. The species has become a major model organism for studies on metamorphic induction, metamorphic-signal transduction, and morphogenesis in cnidarians. Research begun by Müller[160–162] on the role of surface bacteria in metamorphic induction of *H. echinata* has led to a large literature recently reviewed by Leitz.[88,163] Leitz and Wagner[164] isolated the bacterium *Alteromonas espejiana* from the surface of hermit crab shells and found that it would induce attachment and metamorphosis of planulae of *H. echinata*. However, other bacterial species, including additional *Alteromonas spp.* and *Oceanospirillum* spp., have also been found to have this inductive capacity (cited in Leitz[88]). Leitz[88] points out that there is, as yet, no solid information on the factors that bring larvae of *H. echinata* to the surfaces of hermit crab inhabited shells, although factors such as shell vibrations have been proposed by Cazaux,[165,166] but doubted by Müller.[161] Ultimate attachment of planulae to the surface is made by the firing of specific nematocysts, which apparently anchor the larva in the biofilm–bacterial matrix until more secure adhesion is established.[167,168] The bacterial substance responsible for metamorphic induction has not been fully characterized. An active lipophilic fraction is thought to be a component of the bacterial cell wall.[88]

The scyphozoans *Cassiopea andromeda* and *C. xamachana* produce both sexual planulae, which develop from fertilized eggs, and asexual buds, liberated from the polyp-stage scyphistomae, which swim, behave, and settle in much the same manner as the true planulae.[169,170] Settlement of these planulae and buds is dependent on the presence of biofilm bacteria, one of which, *Vibrio alginolyticus*, was shown to produce peptides that induced metamorphosis in buds.[169] In the field, planulae of *Cassiopea xamachana* may prefer to settle upon decaying leaves of the red mangrove.[170,171] Laboratory tests revealed that larvae rarely settle on fresh mangrove leaves and that antibiotics inhibited the release of metamorphosis-inducing compounds.[170] Rather than suspecting that bacteria associated with decaying mangrove leaves are the source of a metamorphic inducer, Hofmann et al.[170] and Fleck and Fitt[142] propose that they are necessary for the decay processes that release compound(s) present in the mangrove leaves, which stimulate settlement of planulae of *C. xamachana*.

Settlement of the planulae of scleractinian corals has been the focus of intensive research, much of it summarized in the above sections on chemical characterization of inducer substances. The investigations of Morse and Morse,[15] Morse et al.,[17] and Morse and Morse[28] identify polysaccharides produced at the surfaces of coralline algae as the active agents in inducing settlement of several coral species (see Section IV.C). However, Johnson et al.[172] noted that unique bacteria occur on the surfaces of crustose coralline algae, and these bacteria could serve as the sources of inducers for settlement of corals and other invertebrates. Golbuu and Richmond (manuscript in preparation; provided by R. Richmond, University of Guam) explored settlement preferences of planulae of four western Pacific coral species and found them divided. Planulae of *Goniastrea retiformis* and *Acropora wardii* settled preferentially on the crustose coralline alga *Hydrolithon reinboldii*. Larvae of *Leptoria phrygia* selected biofilms, without distinguishing whether those films coated carbonate reef rock or coralline algal crusts, and larvae of *Stylaraea punctata* showed higher settlement on microbial films on coral rubble.

3. Polychaeta

The importance of marine biofilms for settlement of polychaete larvae has long been recognized. The spirorbid *Janua brasiliensis* releases larvae that show strong preferences for settlement on the green alga *Ulva lobata*. However, experimental evidence demonstrated that microbial films growing specifically on the surfaces of *U. lobata* contain the inductive cue for larval settlement of *J. brasiliensis*.[173]

As a very common component of the fouling community in tropical and subtropical bays and harbors around the world, the microserpulid polychaete *Hydroides elegans* has received intensive study in the last decade. A congener, *H. dianthus*, known from the temperate east coast of North America, has also been studied in several laboratories. *Hydroides elegans* can rapidly form dense, nearly monospecific layers on ships and pilings in locations such as Pearl Harbor, Hawaii.[25] Investigations in Hawaii by the Hadfield group have revealed the following information relative to recruitment in this species: (1) the rate and density of settlement and metamorphosis of its larvae are strictly regulated by the density of biofilm bacteria, and small rod-shaped bacteria are the types most closely correlated with density of settlement of the polychaete larvae,[25] and (2) in both laboratory and field experiments larvae of *H. elegans* were not found to be specifically attracted to settle on or beside living members of their own species.[174] In static laboratory experiments, larval settlement was random across surfaces despite the presence of living *H. elegans,* empty tubes of *H. elegans*, tube mimics, or living juveniles of *H. elegans*, while in the field, there was a tendency for settlement to be somewhat denser, but equal, adjacent to tubes containing living adults, empty tubes of *H. elegans*, or teflon tubes of approximately the same diameter as adult worm tubes.[174] The authors interpreted the latter distribution of settlers to hydrodynamic factors, since there was no tendency of larvae to settle in greater numbers near tubes inhabited by living adult worms than to models. The research in Hawaii also revealed that single strains of bacteria, isolated from marine biofilms and coated onto various sterile substrata, induce settlement and metamorphosis in larvae of *H. elegans*.[26] The latter authors report that living bacteria were required to induce settlement and that single-species films were never as effective as complex, multispecies biofilms (e.g., 60 vs. 80% settlement). The response of competent larvae of *H. elegans* to living biofilms is very rapid; within 15 min they have attached and formed a primary tube, by 1 h they have completed metamorphosis and secreted a definitive calcareous tube, and by 11 h they have developed branchial plumes and are feeding.[175] Thus, assays for metamorphic induction by natural inducers should be very brief for larvae of *H. elegans*; after 24 h, the potential for growth of bacterial films in the test dishes is great.

Researchers in Hong Kong have published a series of papers dealing with settlement of larvae of *Hydroides elegans* from that region. Starting from the assumption that dense aggregations of this polychaete could only arise from specific adult-associated cues, these researchers have tested the metamorphosis-inducing capacities of homogenates of adult worms and achieved up to 82% larval settlement after 4 days.[176] Subsequently, Harder and Qian[177] investigated the composition of inductive fractions of homogenized adult worms and found them to be a mixture of amino acids. Using 48-h bioassays, Harder and Qian observed similar rates of metamorphosis in response to adult extracts, purified extracts that consisted of mixtures of free amino acids, and artificial mixtures of the same amino acids. Including antibiotics in the assays led to significant reduction of the metamorphic response in all comparisons except the purified fraction of adult homogenates. Harder and Qian also tested the inductive capacity of water samples taken within 5 mm of dense adult colonies of *H. elegans* in the field. They were unable to detect free amino acids in these samples, and the samples induced negligible metamorphosis in the laboratory. Given the lengthy bioassays required to achieve significant levels of metamorphosis (2 days), the reduced levels of settlement typical of the assays (30 to 60% vs. 80 to 100% typically seen with this species when exposed to natural biofilms), the reduction of inductive activity in most assays when antibiotics were included, and the absence of detectable levels of free amino acids and inductive activity in seawater taken directly from dense clusters of adult *H. elegans*, it is difficult to ascribe natural significance to the data provided by Harder and Qian.[177] In a recent study, Beckman et al.[178] demonstrated that the major effect of dissolved free amino acids in their assay dishes was stimulation of growth of antibiotic-resistant bacteria, and concluded that metamorphic induction was due to bacterial films rather than larval perception of the free fatty acids.

An interesting case of polytypic cue dependence is reported for *Hydroides dianthus*.[179] In sibling cohorts of this species, 10 to 20% (varying from 0 to 50% per sibling group) settle in response to common biofilms, while the remainder depend on cues from living post-settlement *H. dianthus*.[89] Toonen and Pawlik[99] propose that this duality provides, in each cohort, a group of "founders" or colonizing larvae that settle onto surfaces characterized by biofilms. Once established, the founders serve as the stimulus for recruitment of the larger proportion of larvae ("aggregators"), leading to the buildup of dense aggregations of *H. dianthus* on submerged surfaces. Studies of *Hydroides ezoensis*, a Japanese species, began with the assumption that the dense clumps of the species found in the field must arise through metamorphic induction by products from adult worms.[90,91] Compounds were obtained from worm homogenates that induced metamorphosis, but tests of simple biofilms were never carried out, and results were determined after 48-h assays, allowing more than sufficient time for bacterial films to develop in the assay chambers. It thus remains to be determined if, in the field, larvae of *H. ezoensis* are strictly induced to settle by adult worm metabolites.

4. Mollusca

Larvae of the scallop *Placopecten magellanicus* actively selected monofilament substrates with dense coatings of bacteria, microalgae, and detritus.[180] However, there was also significant settlement on noncoated monofilaments in these experiments, suggesting that "preference" does not imply a necessary inductive process by biofilm organisms for the scallop larvae to settle and metamorphose. On the other hand, these experiments were conducted by placing the monofilament collectors in the field and leaving them undisturbed for 2 weeks to attract naturally settling larvae. In 2 weeks time, a rich biofilm could have collected on the control filaments. Satuito et al.[181] found settlement of the mussel *Mytilus galloprovincialis* to be strongly dependent on the presence of bacterial films.

Grazing marine gastropods that inhabit rocky substrata, including trochaceans, limpets, and keyhole limpets, are easily induced to metamorphose by small pieces of rocky substratum naturally coated with complex films of bacteria, microalgae, and probably fungi.[182] It is likely that such induction is widespread in these groups of snails.

5. Echinodermata

Bacterial films have been demonstrated to induce settlement in competent larvae of the sea urchin *Lytechinus pictus*.[183,184] Pearse and Cameron[185] summarize information indicating that biofilm bacteria are instrumental in inducing settlement and metamorphosis in several additional echinoid species. Johnson et al.[172] and Johnson and Sutton[186] determined that specific settlement of larvae of the starfish *Acanthaster planci* on crustose coralline algae was dependent on bacteria present on the algal surfaces. The bacteria are apparently specifically associated with the crustose coralline algae. In an aquaculture setting, competent larvae of the holothurian *Stichopus japonicus* are induced to settle on biofilms enriched with adherent diatoms of densities greater than 200,000 cells per cm.[187]

6. Bryozoa

The importance of biofilms for larval settlement of bryozoans has been well demonstrated.[188,189] Brancato and Woollacott[189] found that larvae of three species of the common bryozoan genus *Bugula*, *B. simplex*, *B. stolonifera*, and *B. turrita*, all settled preferentially on biofilmed substrata when given a choice. Of the three species, only larvae of *B. stolonifera* would settle on unfilmed substrata. Contrasting with these results, experimental results first reported by Crisp and Ryland[156] and recently confirmed by Wieczorek and Todd[52] demonstrate that settlement of *Bugula flabellata* is significantly reduced, but not eliminated, by the presence of biofilms. Differences in the ages of the biofilms, ranging between 1 and 12 days, had no significant effect on the level of settlement of *B. flabellata* upon them. The membrane from the inside of hens' egg shells, peeled away, washed, and coated with a biofilm by submergence in running seawater for several days, has become a standard for experimental work on bryozoan metamorphosis in several laboratories (e.g., Reed and Woollacott[190] and Wendt and Woollacott[191]).

7. Brachiopoda

Larvae of at least some species of articulate brachiopods will settle only on glass surfaces or adult shells bearing biofilms.[192] Long[192] summarized observations on brachiopod settlement as follows: adult shells collected from the field frequently bear attached postlarval stages; larvae of *Frenulina* sp. settled preferentially on the coralline alga *Lithothamnion* sp.; and larvae of *Pumulis* sp. chose weathered rock surfaces over newly exposed ones. Consistent with many of these observations are those of Pennington et al.,[193] who found preferential settlement of larval *Laqueus californianus* on shells of living conspecific adults.

8. Phoronida

Herrmann[194–196] was able to stimulate metamorphosis in *Phoronis muelleri* and *P. psammophila* with dense suspensions of bacteria. It is impossible to know if these results are indicative of a natural response to biofilm bacteria in the field.

9. Ascidiacea

Tadpole larvae of *Ciona intestinalis* settle in response to marine bacteria and their extracellular products.[197] Furthermore, larvae of this species settle on complex biofilms in frequencies that correlate with the age of the biofilm and, presumably, with its density.[52] Szewzyk et al.[197] suggested that recruitment of larval *C. intestinalis* might result from their being trapped in exopolysaccharides of biofilm bacteria, rather than through inductive-ligand binding to larval receptors. Studies of recruitment of *C. intestinalis* in different parts of the world do not yield the same results. Keough and Raimondi,[155] conducting field experiments with filmed vs. unfilmed panels in southern Australia, found no positive response to biofilms.

10. Crustacea

The role of biofilm bacteria in barnacle settlement has been a topic of considerable investigation and discussion (see Sections II and III.C above). Despite numerous studies finding positive associations between biofilms and the settlement of barnacle cyprid larvae, Roberts et al.,[198] using short-term field tests of treated surfaces, found, "Although bacterial films can have dramatic effects on settlement of barnacles and bryozoans...they are not essential for colonization by barnacles, bryozoans or hydroids."

C. Partially Characterized Cues

Definitive chemical analyses of the nature of compounds produced by biofilm bacteria that induce settlement and metamorphosis of invertebrates are generally lacking. Peptides, supposedly arising from biofilm bacteria and having the capacity to induce settlement of oyster larvae, were described by Zimmer-Faust and Tamburri[18] (discussed above in Section III.C). Studies of larval settlement in the spirorbid polychaete *Janua brasiliensis* have indicated that its alga-specific settlement is, in fact, mediated by carbohydrate moieties of the exopolymers of biofilm bacteria.[173]

VI. SETTLEMENT IN SOFT SEDIMENTS

While not defined as biofilms in the usual sense, much of the organic enrichment of marine sediments is undoubtedly due to bacteria. The role of organic constituents of sediments in specific recognition of appropriate habitats for settlement by larvae of polychaetes, bivalves, and snails was among the first to be recognized (e.g., Wilson,[199] Gray,[200] Calabrese,[201] and Scheltema[202]). Settlement by infaunal species, including many bivalves (e.g., Grassle et al.[203]) and polychaete worms (e.g., Wilson[199]), into soft substrates is dependent on such organic content (e.g., Snelgrove et al.[204]). Cohen and Pechenik[205] concluded from settlement experiments with larvae of *Capitella* sp. I that "the cue for metamorphosis is probably organic and bound to fine particles" allowing for the strong possibility that living bacteria are important. Also studying larvae of *Capitella* sp. I, Biggers and Laufer[206,207] found that metamorphosis was stimulated by insect juvenile hormones and related compounds and that chemicals with juvenile hormone activity could be detected in marine sediments that stimulated settlement. Larvae of at least some sipunculan species are also stimulated to settle and metamorphose in response to specific organic components of soft substrata.[208]

A particularly good summary of the settlement habits of soft-sediment species, contrasting physical and biological/chemical processes and critiquing methods and some prior conclusions, was provided by Butman.[2] Gray's[209] summary of factors influencing recruitment into soft sediments is still useful. More recent investigations testing substrate selection in flumes (e.g., Butman[210] and Snelgrove et al.[211]) round out current understanding of the relative roles of larval substrate selection and physical factors in determining recruitment of invertebrates to soft-sediment communities. The absence of significant new work on the identities, biological or biochemical, of settlement inducers in soft substrata is notable.

VII. NEGATIVE CUES TO SETTLEMENT

When larvae fail to settle in a particular test situation, is it due to the absence of positive cues, or are there other factors that prohibit settlement, even when positive cues may be present? This is a difficult question to answer because of the possibility that a negative cue may, in fact, be a neutral substance that simply makes the positive cue undetectable by a larva, rather than a substance that causes a larva to reject a potential settlement site. Much effort to demonstrate negative cues to settlement has come from the search for antifouling natural products from marine organisms (reviewed by Pawlik;[3] see also Chapters 10 and 17 in this volume). However, there is little reason to believe that extracted compounds which are toxic or aversive to invertebrate larvae have any such function in nature. The best evidence to date for chemical deterrence of larval settlement comes from the work of Woodin and co-workers.[212–214] Focusing on halogenated compounds released by many marine polychaete and enteropneust worms,[212] they have shown inhibition of settlement of larvae of other species in the presence of the compounds.[213,214]

Twelve common Hawaiian marine algae were tested by Walters et al.[215] to determine if they released compounds that would inhibit settlement of the polychaete *Hydroides elegans* or the bryozoan *Bugula neritina*. Eight of the algal species released nontoxic substances that inhibited the settlement of one or the other of the two invertebrate species. However, none of the algae produced substances that inhibited settlement by both invertebrate species. Walters et al.[215] concluded that chemical substances provide at least one cause for the general absence of surface fouling of Hawaiian shallow-water algae. Although Lau and Qian[216] reported that phlorotannins were instrumental in inhibiting settlement of the polychaete *Hydroides elegans* on surfaces of brown algae, Jennings and Steinberg[217] found no relationship between phlorotannins and epiphytism on the kelp *Ecklonia radiata*.

Megalopae of the blue crab *Callinectes sapidus* recruit into suspended cages containing the seagrasses *Zostera marina* and *Halodule wrightii* in the field.[64] However, when one of a series of decapod predators was included in the cage, there was a 40% decline in the number of megalopae recruiting to it, demonstrating the ability of these postlarvae to detect odors from the predatory crustaceans and avoid them at the time of settlement.

Biofilms, sources of powerful settlement stimulation for many benthic invertebrates, have been found to repel settlement of some bryozoans[52] and barnacles,[218] although in the latter case, older biofilms were found to stimulate settlement.[219] Biofilm bacteria were reported by Maki et al.[218,220] to inhibit settlement of the barnacle *Balanus amphitrite*, a view subsequently contradicted by Wieczorek et al.,[219] who achieved strong induction of cyprid attachment and metamorphosis (>60%) on biofilms that had accumulated on test substrata in running seawater for more than 18 days. Wieczorek et al.,[219] suggest that testing the settlement of larvae of *B. amphitrite* with young biofilms yielded the results reported by Maki et al.[218,220] If true, these observations suggest interesting future studies of the biological characteristics of biofilms that change them so dramatically, in only a few days, from providing negative cues to settlement to just the opposite.

VIII. CONCLUSIONS

A. TYPES OF COMPOUNDS THAT INDUCE SETTLEMENT AND METAMORPHOSIS

1. Fatty Acids

Fatty acids have been implicated in the settlement and metamorphosis of the gregarious tube worm *Phragmatopoma lapidosa californica*, which settles in response to cues deposited in the tube sand cement, and two sea urchins, *Pseudocentrotus depressus* and *Anthocidaris crassispina*, which metamorphose in response to eicosapentaenoic acid extracted from the coralline red alga *Corallina pilulifera*. However, Jensen et al.[87] provided good evidence that fatty acids were contaminants that induce settlement nonspecifically in *P. lapidosa californica.* Fatty acids have not been reported as metamorphic inducers for other sea urchin larvae; instead, water-soluble inducers have been described.[45] It is possible that fatty acids are not natural inducers for these larvae, but may act physiologically downstream in a manner similar to many types of artificial inducers. For example, arachidonic acid and eicosanoids are generated after induction of metamorphosis in *Hydractinia echinata* larvae.[88] Further studies will be necessary to clarify the roles of fatty acids in larval settlement.

2.   Peptides and Proteins

Small peptides have received much recent attention as chemical inducers. Larvae of *Balanus amphitrite*,[54] *Crassostrea virginica*,[18] *Dendraster excentricus*,[20] and *Adalaria proxima*[48] all respond to waterborne peptides of more than 1000 Da molecular weight. Peptides with arginine and lysine carboxyl terminals may be very important cues for certain species of larvae such as barnacles and oysters.[103]

Small peptides associated with proteins at the surface of crustose coralline algae induce the settlement of abalone larvae.[49] A proline-rich peptide formed during bacterial decomposition of mangrove leaves triggers metamorphosis in *Cassiopea xamachana*.[142] Larger proteins are also important for settlement of barnacles[100] and *Phragmatopoma lapidosa californica*.[85,87] The sugar portions of the glycoprotein that induces barnacle settlement are considered important components of the active inducer.[100,101]

3.   Carbohydrates

Several studies now point to the importance of carbohydrates as settlement cues. Many species of coral larvae that settle on crustose coralline algae metamorphose in response to an algal cell wall polysaccharide.[17,132] The sacoglossan *Alderia modesta* metamorphoses in response to carbohydrates of different size classes from its algal food *Vaucheria longicaulis*.[144] The sea urchin *Holopneustes purpurascens* metamorphoses in response to a complex of the sugar floridoside and isethionic acid from the red alga *Delisea pulchra*.[11] In this case, the combination of compounds was important because neither induced metamorphosis alone.

B.   The State of the Art

Exciting new information on settlement and metamorphosis of invertebrate larvae continues to accumulate at a great rate. Not only are more species included in this effort, but also members of groups that have eluded such study in the past, as well as inhabitants of difficult-to-study habitats such as deep benthic environments. How larvae disperse from dying deep-sea vents and recruit to new ones has been a puzzle that has recently become almost tractable. Invertebrate larvae have been studied from nearly all possible types of shallow habitats, including protected and exposed rocky shores, sandy and muddy beaches, coral reefs, seagrass beds, and the subtidal bottoms of bays. Generalizations are beginning to emerge, for example, the near universality of biofilm stimulation of settlement of fouling organisms, the responses of infauna to organic products of sediments, and the necessity of adult prey to settlement of larvae of species with great prey specificity. The latter is especially true for larvae of animals that prey on sessile, colonial animals and algae.

The cellular mechanisms underlying the abilities of larvae to detect and respond to both adsorbed and soluble settlement cues are yielding to intense study. It has been shown that larvae of many types have the ability to utilize dissolved cues to settlement, which was thought impossible only 20 to 25 years ago. Understanding the details of larval receptor systems will require the use of most of the modern techniques for studying subcellular biology, but critical breakthroughs have already occurred in this area and more are expected in the next decade.

Despite more than 60 years of effort, very few environmental inducers of settlement and metamorphosis of marine-invertebrate larvae have been unequivocally identified (e.g., Williamson et al.[11]). For example, we know that the inducer for larvae of the sand dollar *Dendraster excentricus* is an approximately 1000 Da peptide,[20] but we know nothing of the amino acid composition or structure of this peptide that provides its specificity. Although intensely studied for nearly 30 years, the metamorphic inducers for the nudibranch *Phestilla sibogae*, a soluble metabolite released by its coral prey,[13] and the prosobranch *Haliotis rufescens*, a surface-bound part of an algal protein,[49] remain elusive. Other examples abound. The problems in carrying out these purifications and characterizations are many, but the predominant ones are (1) the molecules are often polar, and (2) the compounds, even when highly inductive, are present in very low concentrations. Additionally, the low molecular weights of many inducers make their isolations more difficult. Many of these compounds are as potent as insect pheromones and perhaps more potent than many hormones. Their water-soluble nature makes them difficult to separate from the myriad water-soluble compounds present in most organisms. The inducers may also be quickly degraded by bacteria or physical processes once free of cells. Completing these identifications may require the development of new methodologies for both their purification and concentration. Yet the endeavor is worth the effort, both for greater understanding of the developmental and ecological processes in larval settlement and metamorphosis, and because unique compounds with fascinating applications may be revealed.

ACKNOWLEDGMENTS

Research on larval settlement in MGH's laboratory is currently supported by NSF grant OCE-9907545. VJP received support from NSF grant HRD-9626896 in the Visiting Professorships for Women Program for her research on the chemistry of larval settlement cues. Both authors gratefully acknowledge the assistance and moral support of the students and postdoctoral fellows in their laboratories while their mentors write book chapters.

REFERENCES

1. Crisp, D. J., Factors influencing the settlement of marine invertebrate larvae, in *Chemoreception in Marine Organisms*, Grant, P. T. and Mackie, A. M., Eds., Academic Press, New York, 1974, 177.
2. Butman, C. A., Larval settlement of soft-sediment invertebrates: the spatial scales of pattern explained by active habitat selection and the emerging role of hydrodynamic processes, *Oceanogr. Mar. Biology Annu. Rev.*, 25, 113, 1987.
3. Pawlik, J. R., Chemical ecology of the settlement of benthic marine invertebrates, in *Oceanogr. Mar. Biol. Annu. Rev.*, 30, 273, 1992.
4. McEdwards, L. D., *Ecology of Marine Invertebrate Larvae*, CRC Press, Boca Raton, FL, 1995.
5. Pawlik, J. R., Butman, C. A., and Starczak, V. R., Hydrodynamic facilitation of gregarious settlement of a reef-building tube worm, *Science*, 251, 421, 1991.
6. Pawlik, J. R. and Butman, C. A., Settlement of a marine tube worm as a function of current velocity: interacting effects of hydrodynamics and behavior, *Limnol. Oceanogr.*, 38, 1730, 1993.
7. Turner, E. J., Zimmer-Faust, R. K., Palmer, M. A., Luckenbach, M., and Pentcheff, N. D., Settlement of oyster (*Crassostrea virginica*) larvae: effects of water flow and a water-soluble chemical cue, *Limnol. Oceanog.*, 39, 1579, 1994.
8. Slattery, M., Chemical cues in marine invertebrate larval settlement, in *Marine Woodboring and Fouling Organisms of the Indian Ocean: A Review*, Naghabushanum, R. and Thompson, J. F., Eds., Oxford and IBH Publishing Co., New Delhi, 1997, 135.
9. Hadfield, M. G., The D P Wilson lecture. Research on settlement and metamorphosis of marine invertebrate larvae: past, present and future, *Biofouling*, 12, 9, 1998.

10. Tsukamoto, S., Kato, H., Hirota, H., and Fusetani, N., Lumichrome: a larval metamorphosis-inducing substance in the ascidian *Halocynthia roretzi*, *Euro. J. Biochem.*, 264, 785, 1999.

11. Williamson, J. E., de Nys, R., Kumar, N., Carson, D. G., and Steinberg, P. D., Induction of metamorphosis of a marine invertebrate by the metabolites floridoside and isethionic acid from its algal host, *Biol. Bull.*, 198, 332, 2000.

12. Hadfield, M. G. and Scheuer, D., Evidence for a soluble metamorphic inducer in *Phestilla*: ecological, chemical and biological data, *Bull. Mar. Sci.*, 37, 556, 1985.

13. Hadfield, M. G. and Pennington, J. T., Nature of the metamorphic signal and its internal transduction in larvae of the nudibranch *Phestilla sibogae*, *Bull. Mar. Sci.*, 46, 455, 1990.

14. Bahamondes-Rojas, I. and Dherbomez, M., Purification partielle des substances glycoconjuguees capable d'induire la metamorphose des larves competentes d'*Eubranchus doriae* (Trinchese, 1879), mollusque nudibranche, *J. Exp. Mar. Biol. Ecol.*, 144, 17, 1990.

15. Morse, D. E. and Morse, A. N. C., Enzymatic characterization of the morphogen recognized by *Agaricia humilis* (scleractinian coral) larvae, *Biol. Bull.*, 181, 104, 1991.

16. Gibson, G. D. and Chia, F. S., A metamorphic inducer in the opisthobranch *Haminaea callidegenita*: partial purification and biological activity, *Biol. Bull.*, 187, 133, 1994.

17. Morse, D. E., Morse, A. N. C., Raimondi, P. T., and Hooker, N., Morphogen-based chemical flypaper for *Agaricia humilis* coral larvae, *Biol. Bull.*, 186, 172, 1994.

18. Zimmer-Faust, R. K. and Tamburri, M. N., Chemical identity and ecological implications of a waterborne, larval settlement cue, *Limnol. Oceanogr.*, 39, 1075, 1994.

19. Crisp, D. J. and Meadows, P. S., The chemical basis of gregariousness in cirripedes, *Proc. R. Soc. Lond. B*, 156, 500, 1962.

20. Burke, R. D., Phermonal control of metamorphosis in the pacific sand dollar, *Dendraster excentricus*, *Science*, 225, 440, 1984.

21. Burke, R. D., Pheromones and the gregarious settlement of marine invertebrate larvae, *Bull. Mar. Sci.*, 39, 323, 1986.

22. Hubbard, E. J. A., Larval growth and the induction of metamorphosis of a tropical sponge-eating nudibranch, *J. Moll. Stud.*, 54, 259, 1988.

23. Zobell, C. E. and Allen, E. C., The significance of marine bacteria in the fouling of submerged surfaces, *J. Bacteriol.*, 29, 239, 1935.

24. Johnson, C. R., Lewis, T. E., Nichols, D. S., and Degnan, B. M., Bacterial induction of settlement and metamorphosis in marine invertebrates, in *Proceedings of the 8th International Coral Reef Symposium*, Vol. 2, 1997, 1219.

25. Hadfield, M. G., Unabia, C. C., Smith, C. M., and Michael, T. M., Settlement preferences of the ubiquitous fouler *Hydroides elegans*, in *Recent Developments in Biofouling Control*, Thompson, M. F., Nagabhushanam, R., Sarojini, R., and Fingerman, M., Eds., Oxford and IBH Publishing Co., New Delhi, 1994, 65.

26. Unabia, C. R. C. and Hadfield, M. G., Role of bacteria in larval settlement and metamorphosis of the polychaete *Hydroides elegans*, *Mar. Biol.*, 133, 55, 1999.

27. Tsukamoto, S., Hirota, H., Kato, H., and Fusetani, N., Urochordamines A and B: larval settlement/metamorphosis-promoting, pteridine-containing physostigmine alkaloids from the tunicate *Ciona savignyi*, *Tetrahedron Lett.*, 34, 4819, 1993.

28. Morse, A. N. C. and Morse, D. E., Flypapers for coral and other planktonic larvae, *Bioscience*, 46, 254, 1996.

29. Scheltema, R. S., Biological interactions determining larval settlement of marine invertebrates, *Thalassia Jugoslavica*, 10, 263, 1974.

30. Chia, F. S. and Rice, M. E., *Settlement and Metamorphosis of Marine Invertebrate Larvae*, Elsevier, New York, 1978, 290.

31. *Bull. Mar. Sci.*, 37, 1985.

32. *Bull. Mar. Sci.*, 39, 1986.

33. *Bull. Mar. Sci.*, 46, 1990.

34. *Biofouling*, 12, 1998.

35. Crisp, D. J. and Meadows, P. S., Factors controlling cold tolerance and breeding in *Balanus balanoides*, *J. Mar. Biol.*, 64, 125, 1984.

36. Young, C. M., Larval ecology of marine invertebrates: a sesquicentennial history, *Ophelia*, 32, 1, 1990.

37. Rodriguez, S. R., Ojeda, F. P., and Inestrosa, N. C., Settlement of benthic marine invertebrates, *Mar. Ecol. Prog. Ser.*, 97, 193, 1993.

38. Knight-Jones, E. W. and Crisp, D. J., Gregariousness in barnacles in relation to the fouling of ships and to anti-fouling research, *Nature*, 171, 1109, 1953.

39. Walters, L. J., Miron, G., and Bourget, E., Endoscopic observations of invertebrate larval substratum exploration and settlement, *Mar. Ecol. Prog. Ser.*, 182, 95, 1999.

40. Clare, A. S., Chemical signals in barnacles: old problems, new approaches, in *New Frontiers in Barnacle Evolution*, Schram, F. R. and Hoeg, J. T., Eds., A. A. Balkema Publishers, Rotterdam, 1995, 49.

41. Clare, A. S., Marine natural product antifoulants: status and potiental, *Biofouling*, 9, 211, 1996.

42. Thorson, G., Reproduction and larval development of Danish marine bottom invertebrates, with special reference to the planktonic larvae in the sound (Oresund), *Medd. Dansk Fisk. Havunders.*, 4, 1, 1946.

43. Tankersley, R. A. and Forward, R. B., Jr., Endogenous swimming rhythms in estuarine crab megalopae: implications for flood-tide transport, *Mar. Biol.*, 118, 415, 1994.

44. Tankersley, R. A., McKelvey, L. M., and Forward, R. B., Jr., Responses of estuarine crab megalopae to pressure, salinity and light: implications for flood-tide transport, *Mar. Biol.*, 122, 391, 1995.

45. Pearce, C. M. and Scheibling, R. E., Induction of metamorphosis of larvae of the green sea-urchin, *Strongylocentrotus droebachiensis*, by coralline red algae, *Biol. Bull.*, 179, 304, 1990.

46. Young, C. M. and Braithwaite, L. F., Orientation and current-induced flow in the stalked ascidian *Styela montereyensis*, *Biol. Bull.*, 159, 428, 1980.

47. McGee, B. L. and Targett, N. M., Larval habitat selection in Crepidula (L.) and its effect on adult distribution patterns, *J. Exp. Mar. Biol. Ecol.*, 131, 195, 1989.

48. Lambert, W. J., Todd, C. D., and Hardege, J. D., Partial characterization and biological activity of a metamorphic inducer of the dorid nudibranch *Adalaria proxima* (Gastropoda: Nudibranchia), *Invertebrate Biol.*, 116, 71, 1997.

49. Morse, A. N. C. and Morse, D. E., Recruitment and metamorphosis of *Haliotis* larvae induced by molecules uniquely available at the surfaces of crustose red algae, *J. Exp. Mar. Biol. Ecol.*, 75, 191, 1984.

50. Morse, D. E., Recent progress in larval settlement and metamorphosis: closing the gaps between molecular biology and ecology, *Bull. Mar. Sci.*, 46, 465, 1990.

51. Chia, F. S. and Koss, R., Induction of settlement and metamorphosis of the veliger larvae of the nudibranch, *Onchidoris bilamellata*, *Int. J. Invertebrate. Reprod. Dev.*, 14, 53, 1988.

52. Wieczorek, S. K. and Todd, C. D., Inhibition and facilitation of bryozoan and ascidian settlement by natural multi-species biofilms: effects of film age and the roles of active and passive larval settlement, *Mar. Biol.*, 128, 463, 1997.

53. Rittschof, D., Oyster drills and the frontiers of chemical ecology: unsettling ideas, *Am. Malac. Bull.*, special ed., 1, 111, 1985.

54. Tegtmeyer, K. and Rittschof, D., Synthetic peptide analogs to barnacle settlement pheromone, *Peptides*, 9, 1403, 1989.

55. Clare, A. S. and Matsumura, K., Nature and perception of barnacle settlement pheromones, *Biofouling*, 15, 57, 2000.

56. Tamburri, M. N., Zimmer-Faust, R. K., and Tamplin, M. L., Natural sources and properties of chemical inducers mediating settlement of oyster larvae: a re-examination, *Biol. Bull.*, 183, 327, 1992.

57. Zimmer-Faust, R. K., Tamburri, M. N., and Decho, A. W., Chemosensory ecology of oyster larvae: benthic-pelagic coupling, in *Zooplankton: Sensory Ecology and Physiology*, Lenz, P. H., Hartline, D. K., Purcell, J. E., and MacMillan, D. L., Eds., Gordon and Breach, Amsterdam, 1996, 37.

58. Decho, A. W., Browne, K. A., and Zimmer-Faust, R. K., Chemical cues: why basic peptides are signal molecules in marine environments, *Limnol. Oceanog.*, 43, 1410, 1998.

59. Forward, R. B., Jr., Frankel, D. A. Z., and Rittschof, D., Molting of megalopae from the blue crab *Callinectes sapidus*: effects of offshore and estuarine cues, *Mar. Ecol. Prog. Ser.*, 113, 55, 1994.

60. Giese, A. C. and Pearse, J. S., Introduction: general principles, in *Reproduction of Marine Invertebrates, Vol. 1*, Giese, A. C. and Pearse, J. S., Eds., Academic Press, New York, 1974, 1.

61. De Vries, M. C., Tankersley, T. A., Forward, R. B., Jr., Kirby-Smith, W. W., and Luettich, R. A., Abundances of crab megalopae are associated with estuarine tidal hydrologic variables, *Mar. Biol.*, 118, 403, 1994.

62. Olmi, E. J., III, Vertical migration of blue crab *Callinectes sapidus* megalopae: implications for transport in estuaries, *Mar. Ecol. Prog. Ser.*, 113, 39, 1994.
63. Wolcott, D. L. and De Vries, M. C., Offshore megalopae of *Callinectes sapidus*: depth of collection, molt stage and response to estuarine cues, *Mar. Ecol. Prog. Ser.*, 109, 157, 1994.
64. Welch, J. M., Rittschof, D., Bullock, T., and Forward, R. B., Jr., Effects of chemical cues on settlement behavior of blue crab *Callinectes sapidus* postlarvae, *Mar. Ecol. Prog. Ser.*, 154, 143, 1997.
65. Forward, R. B., Jr., DeVries, M. C., Rittschof, D., Frankel, D. A. Z., Bischoff, J. P., Fisher, C. M., and Welch, J. M., Effects of environmental cues on metamorphosis of the blue crab *Callinectes sapidus*, *Mar. Ecol. Prog. Ser.*, 131, 165, 1996.
66. Forward, R. B., Jr., Tankersley, R. A., Blondel, D., and Rittschof, D., Metamorphosis of the blue crab *Callinectes sapidus*: effects of humic acids and ammonium, *Mar. Ecol. Prog. Ser.*, 157, 277, 1997.
67. Knight-Jones, E. W., Laboratory experiments on gregariousness during settling in *Balanus balanoides* and other barnacles, *J. Exp. Biol.*, 30, 584, 1953.
68. Cole, H. A. and Knight-Jones, E. W., The setting behaviour of larvae of the European oyster *Ostrea edulis* L. and its influence on methods of cultivation and spat collection, *Fish. Invest. Lond. Ser. II*, 17, 1, 1949.
69. Lee, W. L., Color change and the ecology of the marine isopod *Idothea (Pentidotea) montereyensis* Maloney, 1933, *Ecology*, 47, 930, 1966.
70. Keough, M. J., Responses of settling invertebrate larvae to the presence of established recruits, *J. Exp. Mar. Biol. Ecol.*, 231, 1, 1998.
71. Tsukamoto, S., Kato, H., Hirota, H., and Fusetani, N., *N,N*-Dimethylguanidium styryl sulfates, metamorphosis inducers of ascidian larvae from a marine sponge *Jaspis* sp., *Tetrahedron Lett.*, 35, 5873, 1994.
72. Tsukamoto, S., Kato, H., Hirota, H., and Fusetani, N., Pipecolate derivatives, anthosamines A and B, inducers of larval metamorphosis in ascidians, from a marine sponge *Anthosigmella* aff. *raromicrosclera*, *Tetrahedron*, 51, 6687, 1995.
73. Fusetani, N., Marine natural products influencing larval settlement and metamorphosis of benthic invertebrates, *Curr. Org. Chem.*, 1, 127, 1997.
74. Svane, I., Havenhand, J. N., and Jorgensen, A. J., Effects of tissue extract of adults on metamorphosis in *Ascidia mentula* (O.F. Muller) and *Ascidiella scabra* (O.F. Muller), *J. Exp. Mar. Biol. Ecol.*, 110, 171, 1987.
75. Svane, I. and Young, C. M., The ecology and behaviour of ascidian larvae, *Oceanog. Mar. Biol. Annu. Rev.*, 27, 45, 1989.
76. Wilson, D. P., The settlement behavior of the larvae of *Sabellaria alveolata* (L.), *J. Mar. Biol. Assoc. U.K.*, 48, 387, 1968.
77. Wilson, D. P., *Sabellaria* colonies at Duckpool, North Cornwall, 1971-1972, with a note for May 1973, *J. Mar. Biol. Assoc. U.K.*, 54, 393, 1974.
78. Jensen, R. A. and Morse, D. E., Intraspecific facilitation of larval recruitment: gregarious settlement of the polychaete *Phragmatopoma californica* (Fewkes), *J. Exp. Mar. Biol. Ecol.*, 83, 107, 1984.
79. Pawlik, J. R., Chemical induction of larval settlement and metamorphosis in the reef-building tube worm *Phragmatopoma californica* (Polychaeta: Sabellariidae), *Mar. Biol.*, 91, 59, 1986.
80. Pawlik, J. R., Natural and artificial induction of metamorphosis of *Phragmatopoma lapidosa californica* (Polychaeta: Sabellariidae), with a critical look at the effects of bioactive compounds on marine invertebrate larvae, *Bull. Mar. Sci.*, 46, 512, 1990.
81. Pawlik, J. R. and Faulkner, D. J., Specific free fatty acids induce larval settlement and metamorphosis of the reef-building tube worm *Phragmatopoma californica* (Fewkes), *J. Exp. Mar. Biol. Ecol.*, 102, 301, 1986.
82. Pawlik, J. R. and Faulkner, D. J., The gregarious settlement of sabellariid polychaetes: new perspectives on chemical cues, in *Marine biodeterioration*, Thompson, M. F., Sarojini, R. and Nagabhushanam, R., Eds., Oxford & IBH Publishing Co., New Delhi, 1988, 475.
83. Pawlik, J. R., Larval settlement and metamorphosis of sabellariid polychaetes, with special reference to *Phragmatopoma lapidosa*, a reef-building species, and *Sabellaria floridensis*, a non-gregarious species, *Bull. Mar. Sci.*, 43, 41, 1988.
84. Pawlik, J. R., Larval settlement and metamorphosis of two gregarious sabellariid polychaetes: *Sabellaria alveolata* compared with *Phragmatopoma californica*, *J. Mar. Biol. Assoc. U.K.*, 68, 101, 1988.

85. Jensen, R. A. and Morse, D. E., The bioadhesive of *Phragmatopoma californica* tubes: a silk-like cement containing L-DOPA, *J. Comp. Physiol. B.*, 158, 317, 1988.

86. Jensen, R. A. and Morse, D. E., Chemically induced metamorphosis of polychaete larvae in both the laboratory and ocean environment, *J. Chem. Ecol.*, 16, 911, 1990.

87. Jensen, R. A., Morse, D. E., Petty, R. L., and Hooker, N., Artificial induction of larval metamorphosis by free fatty acids, *Mar. Ecol. Prog. Ser.*, 67, 55, 1990.

88. Leitz, T., Induction of settlement and metamorphosis of Cnidarian larvae: signals and signal transduction, *Invertebrate Reprod. Dev.*, 109, 1997.

89. Toonen, R. J. and Pawlik, J. R., Settlement of the tube worm *Hydroides dianthus* (Polychaeta: Serpulidae): cues for gregarious settlement, *Mar. Biol.*, 126, 725, 1996.

90. Okamoto, K., Watanabe, A., Sakata, K., and Watanabe, N., Chemical signals involved in larval metamorphosis in *Hydroides ezoensis* (Serpulidae; Polychaeta). Part I: Induction of larval metamorphosis by extract of adult tube clumps, *J. Mar. Biotech.*, 6, 7, 1998.

91. Watanabe, N., Watanabe, S., Ide, J., Watanabe, Y., Sakata, K., and Okamoto, K., Chemical signals involved in larval metamorphosis in *Hydroides ezoensis* (Serpulidae; Polychaeta). Part II: Isolation and identification of a new monoacyl glycerol from adult tube clumps as a metamorphosis-inducing substance, *J. Mar. Biotech.*, 6, 11, 1998.

92. Highsmith, R. C., Induced settlement and metamorphosis of sand dollar (*Dendraster excentricus*) larvae in predator-free sites: adult sand dollar beds, *Ecology*, 63, 329, 1982.

93. Knight-Jones, E. W. and Stevenson, J. P., Gregariousness during settlement in the barnacle *Elminius modestus* Darwin, *J. Mar. Biol. Assoc. U.K.*, 29, 281, 1950.

94. Crisp, D. J. and Meadows, P. S., Adsorbed layers: the stimulus to settlement in barnacles, *Proc. R. Soc. Lond. B*, 158, 364, 1963.

95. Larman, V. N. and Gabbott, P. A., Settlement of cyprid larvae of *balaunus Balanoides* and *Elminus modestus* induced by extracts of adult barnacles and other marine animals, *J. Mar. Biol. Assoc. U.K.*, 55, 183, 1975.

96. Larman, V. N., Gabbott, P. A., and East, J., Physico-chemical properties of the settlement factor proteins from the barnacle *Balanus balanoides*, *Comp. Biochem. Physiol. B.*, 72, 329, 1982.

97. Larman, V. N., Protein extracts from some marine animals which promote barnacle settlement: possible relationship between a protein component of arthropod cuticle and actin, *Comp. Biochem. Physiol. B.*, 77, 73, 1984.

98. Yule, A. B. and Walker, G., Settlement of *Balanus balaniodes*: the effect of cyprid antennular secretion, *J. Mar. Biol. Assoc. U. K.*, 65, 707, 1985.

99. Clare, A. S., Freet, R. C., and McClary, M., Jr., On the antennular secretion of the cyprid of *Balanus amphitrite amphitrite*, and its role as a settlement pheromone, *J. Mar. Biol. Assoc. U.K.*, 74, 243, 1994.

100. Matsumura, K., Nagano, M., and Fusetani, N., Purification of a larval settlement-inducing protein complex (SIPC) of the barnacle, *Balanus amphitrite*, *J. Exp. Zool.*, 281, 12, 1998.

101. Matsumura, K., Mori, S., Nagano, M., and Fusetani, N., Lentil lectin inhibits adult extract-induced settlement of the barnacle, *Balanus amphitrite*, *J. Exp. Zool.*, 280, 213, 1998.

102. Rittschof, D., Forward, R. B., Jr., Cannon, G., Welch, J. M., McClary, M., Holm, E. R., Clare, A. S., Conova, S., McKelvey, L. M., Bryan, P., and Van Dover, C. L., Cues and context: larval responses to physical and chemical cues, *Biofouling*, 12, 31, 1998.

103. Rittschof, D., Body odors and neutral-basic peptide mimics: a review of responses by marine organisms, *Am. Zool.*, 33, 487, 1993.

104. Dineen, J. F., Jr. and Hines, A. H., Larval settlement of the polyhaline barnacle *Balanus eburneus* (Gould): cue interactions and comparisons with two estuarine congeners, *J. Exp. Mar. Biol. Ecol.*, 179, 223, 1994.

105. Dineen, J. F., Jr. and Hines, A. H., Effects of salinity and adult extracts on settlement of the oligohaline barnacle *Balanus subalbidus*, *Mar. Biol.*, 119, 423, 1994.

106. Rittschof, D., Branscomb, E. S., and Costlow, J. D., Settlement and behavior in relation to flow and surface in larval barnacles, *Balanus amphitrite* Darwin, *J. Exp. Mar. Biol. Ecol.*, 82, 131, 1984.

107. Wright, J. R. and Boxshall, A. J., The influence of small-scale flow and chemical cues on the settlement of two congeneric barnacle species, *Mar. Ecol. Prog. Ser.*, 183, 179, 1999.

108. Crisp, D. J., Chemical factors inducing settlement in *Crassostrea virginica* (Gmelin), *J. Anim. Ecol.*, 36, 329, 1967.

109. Keck, R., Maurer, D., Kauer, J. C., and Shepperd, W. A., Chemical stimulants affecting larval settlement in the American oyster, *Proc. Natl. Shellfish Assoc.*, 61, 24, 1971.

110. Hidu, J., Gregarious setting in the American oyster *Crassostrea virginica* Gmelin, *Chesapeake Sci.*, 10, 85, 1969.

111. Bayne, B. L., The gregarious behaviour of the larvae of *Ostrea edulis* L. at settlement, *J. Mar. Biol. Assoc. U.K.*, 49, 327, 1969.

112. Veitch, F. P. and Hidu, J., Gregarious settling in the American oyster *Crassostrea virginica* Gmelin. I. properties of a partially purified settling factor, *Chesapeake Sci.*, 12, 173, 1971.

113. Fitt, W. K., Labare, M. P., Fuqua, W. C., Walch, M., Coon, S. L., Bonar, D. B., Colwell, R. R., and Weiner, R. M., Factors influencing bacterial production of inducers of settlement behavior of larvae of the oyster *Crassostrea gigas*, *Microb. Ecol*, 17, 287, 1989.

114. Fitt, W. K., Coon, S. L., Walch, M., Weiner, R. M., Colwell, R. R., and Bonar, D. B., Settlement behavior and metamorphosis of oyster larvae (*Crassostrea gigas*) in response to bacterial supernatants, *Mar. Biol.*, 106, 389, 1990.

115. Coon, S. L., Walch, M., Fitt, W. K., Weiner, R. M., and Bonar, D. B., Ammonia induces settlement behavior in oyster larvae, *Biol. Bull.*, 179, 297, 1990.

116. Fitt, W. K. and Coon, S. L., Evidence for ammonia as a natural cue for recruitment of oyster larvae to oyster beds in a Georgia salt marsh, *Biol. Bull.*, 182, 401, 1992.

117. Bonar, D. B., Coon, S. L., Walch, M., Weiner, R. M., and Fitt, W., Control of oyster settlement and metamorphosis by endogenous and exogenous chemical cues, *Bull. Mar. Sci.*, 46, 484, 1990.

118. Tamburri, M. N., Finelli, C. M., Wethey, D. S., and Zimmer-Faust, R. K., Chemical induction of larval settlement behavior in flow, *Biol. Bull.*, 191, 367, 1996.

119. Hadfield, M. G. and Switzer-Dunlap, M., Opisthobranchs, *The Mollusca*, 7, 209, 1984.

120. Gara, R. I., Greulich, F. E., and Ripley, K. L., Shipworm (*Bankia setacae*) host selection habits at the Port of Everett, Washington, *Estuaries*, 20, 441, 1997.

121. Chevolot, L., Cochard, J. C., and Yvin, J. C., Chemical induction of larval metamorphosis of *Pecten maximus* with a note on the nature of naturally occurring triggering substances, *Mar. Ecol. Prog. Ser.*, 74, 83, 1991.

122. Nicolas, L., Robert, R., and Chevolot, L., Comparative effects of inducers on metamorphosis of the Japanese oyster *Crassostrea gigas* and the great scallop *Pecten maximus*, *Biofouling*, 12, 189, 1998.

123. Kato, T., Kumanireng, A. A., Ichinose, I., Kitahara, Y., Kakinuma, Y., Nishihara, M., and Kato, M., Active components of *Sargassum tortile* affecting the settlement of swimming larvae of *Coryne uchidai*, *Experientia*, 31, 433, 1975.

124. Nishihira, M., Brief experiments on the effect of algal extracts in promoting the settlement of the larvae of *Coryne uchidai* Stechow (Hydrozoa), *Bull. Mar. Biol. Sta. Asamushi*, 23, 91, 1968.

125. Morse, D. E., Tegner, M., Duncan, H., Hooker, N., Trevelyan, G., and Cameron, A., Induction of settling and metamorphosis of planktonic molluscan (*Haliotis*) larvae. III: Signalling by metabolites of intact algae is dependent on contact, in *Chemical Signals in Vertebrate and Aquatic Animals*, Müller-Schwarze, D. and Silverstein, R. M., Eds., Plenum Press, New York, 1980, 67.

126. Miller, S. E. and Hadfield, M. G., Developmental arrest during larval life and life-span extension in a marine mollusc, *Science*, 248, 356, 1990.

127. Shepherd, S. A. and Turner, J. A., Studies on Southern Australian abalone (genus *Haliotis*). VI. Habitat preference, abundance and predators on juveniles, *J. Exp. Mar. Biol. Ecol.*, 93, 285, 1985.

128. Slattery, M., Larval settlement and juvenile survival in the red abalone (*Haliotis rufescens*): an examination of inductive cues and substrate selection, *Aquaculture*, 102, 143, 1992.

129. Morse, A. N. C., Froyd, C. A., and Morse, D. E., Molecules from cyanobacteria and red algae that induce larval settlement and metamorphosis in the mollusc *Haliotis rufescens*, *Mar. Biol.*, 81, 293, 1984.

130. Morse, A. N. C., Recruitment of marine invertebrate larvae, in *Plant–Animal Interactions in the Marine Benthos*, John, D. M., Hawkins, S. J., and Price, J. H., Eds., Clarendon Press, Oxford, UK, 1992, 383.

131. Morse, D. E., Hooker, N., Morse, A. N. C., and Jensen, R. A., Control of larval metamorphosis and recruitment in sympatric agariciid corals, *J. Exp. Mar. Biol. Ecol.*, 116, 193, 1988.

132. Morse, A. N. C., Iwao, K., Baba, M., Shimoike, K., Hayashibara, T., and Omori, M., An ancient chemosensory mechanism brings new life to coral reefs, *Biol. Bull.*, 191, 149, 1996.

133. Heyward, A. J. and Negri, A. P., Natural inducers for coral larval metamorphosis, *Coral Reefs*, 18, 273, 1999.

134. Barnes, J. R. and Gonor, J. J., The larval settling response of the lined chiton *Tonicella lineata*, *Mar. Biol.*, 201, 259, 1973.

135. Gee, J. M., Chemical stimulation of settlement of larvae of *Spirorbis rupestris* (Serpulidae), *Anim. Behav.*, 13, 181, 1965.

136. Kitamura, H., Kitahara, S., and Koh, H. B., The induction of larval settlement and metamorphosis of two sea urchins, *Pseudocentrotus depressus* and *Anthocidaris crassispina*, by free fatty acids extracted from the coralline red alga *Corallina pilulifera*, *Mar. Biol.*, 115, 387, 1993.

137. Nadeau, L., Paige, J. A., Starczak, V., Capo, T., Lafler, J., and Bidwell, J. P., Metamorphic competence in *Aplysia californica* Cooper, *J. Exp. Mar. Biol. Ecol.*, 131, 171, 1989.

138. Pawlik, J. R., Larvae of the sea hare *Aplysia californica* settle and metamorphose on an assortment of macroalgal species, *Mar. Ecol. Prog. Ser.*, 51, 195, 1989.

139. Boettcher, A. A. and Targett, N. M., Induction of metamorphosis in queen conch, *Strombus gigas* L., larvae by cues associated with red algae from their nursery grounds, *J. Exp. Mar. Biol. Ecol.*, 196, 29, 1996.

140. Boettcher, A. A. and Targett, N. M., Role of chemical inducers in larval metamorphosis of Queen Conch, *Strombus gigas* Linneaus: relationship to other marine invertebrate systems, *Biol. Bull.*, 194, 132, 1998.

141. Liu, H. and Loneragan, N. R., Size and time of day affect the response of postlarvae and early juvenile grooved tiger prawns *Penaeus semisulcatus* De Haan (Decapoda: Penaeidae) to natural and artificial seagrass in the laboratory, *J. Exp. Mar. Biol. Ecol.*, 211, 263, 1997.

142. Fleck, J. and Fitt, W. K., Degrading mangrove leaves of *Rhizophora mangle* provide a natural metamorphic cue for the upside down jellyfish *Cassiopea xamachana*, *J. Exp. Mar. Biol. Ecol.*, 234, 83, 1999.

143. Fleck, J., Fitt, W. K., and Hahn, M. G., A proline-rich peptide originating from decomposing mangrove leaves is one natural metamorphic cue of the tropical jellyfish *Cassiopea xamachana*, *Mar. Ecol. Prog. Ser.*, 183, 115, 1999.

144. Krug, P. J. and Manzi, A. E., Waterbourne and surface-associated carbohydrates as settlement cues for larvae of the specialist marine herbivore *Alderia modesta*, *Biol. Bull.*, 197, 94, 1999.

145. Harris, L. G., Studies on the life history of two coral-eating nudibranchs of the genus *Phestilla*, *Biol. Bull.*, 149, 539, 1975.

146. Hadfield, M. G., Chemical interactions of larval settling of a marine gastropod, in *Marine Natural Products Chemistry*, Faulkner, D. J. and Fenical, W., Eds., Plenum Press, New York, 1977, 403.

147. Hadfield, M. G., Settlement requirements of molluscan larvae: new data on chemical and genetic roles, *Aquaculture*, 39, 293, 1984.

148. Hirata, K. Y. and Hadfield, M. G., The role of choline in metamorphic induction of *Phestilla* (Gastropoda, Nudibranchia), *Comp. Biochem. Physiol.*, 84C, 15, 1986.

149. Pires, A. and Hadfield, M. G.. Responses of isolated vela of nudibranch larvae to inducers of metamorphosis, *J. Exp. Zool.*, 266, 234, 1993.

150. Hadfield, M. G., Meleshkevitch, E. A., and Boudko, D. Y., The apical sensory organ of a gastropod veliger is receptor for settlement cues, *Biol. Bull.*, 198, 67, 2000.

151. Lambert, W. J. and Todd, C. D., Evidence for a water-borne cue indicating metamorphosis in the dorid nudibranch mollusc *Adalaria proxima* (Gastropoda: Nudibranchia), *Mar. Biol.*, 120, 264, 1994.

152. Baier, R. E., Surface properties influencing biological adhesion, in *Adhesion in Biological Systems*, Manley, R. S., Ed., Academic Press, New York, 1970, 15.

153. Little, B. J., Succession in microfouling, in *Marine Biodeterioration: An Interdisciplinary Study*, Costlow, J. D. and Tipper, R. C., Eds., Naval Institute Press, Annapolis, MD, 1984, 63.

154. Todd, C. D. and Keough, M. J., Larval settlement in hard substratum epifaunal assemblages: a manipulative field study of the effects of substratum filming and the presence of incumbents, *J. Exp. Mar. Biol. Ecol.*, 181, 159, 1994.

155. Keough, M. J. and Raimondi, P. T., Responses of settling invertebrate larvae to bioorganic films: effects of different types of films, *J. Exp. Mar. Biol. Ecol.*, 185, 235, 1995.

156. Crisp, D. J. and Ryland, J. S., Influence of filming and of surface texture on the settlement of marine organisms, *Nature*, 185, 119, 1960.

157. Bergquist, P. R., *Sponges*, University of California Press, Berkeley, CA, 1978.

158. Zea, S., Estimation of desmosponge (Porifera, Desmospongiae) larval settlement rates from short-term recruitment rates: preliminary experiments, *Helgol. Wissen. Meeresunters.*, 46, 293, 1992.

159. Woollacott, R. M. and Hadfield, M. G., Induction of metamorphosis in a larvae of a sponge, *Invert. Biol.*, 115, 257, 1996.

160. Müller, W. A., Auslosung der Metamorphose durch Bakterien bei den Larven von *Hydractinia echinata*, *Zool. Jb. Anat.*, 86, 84, 1969.

161. Müller, W. A., Induction of metamorphosis by bacteria and ions in the planulae of *Hydractinia echinata*; an approach to the mode of action, *Publ. Seto Mar. Biol. Lab.*, 20, 195, 1973.

162. Müller, W. A., Metamorphoseinduktion bei Planulalarven. I. Der bakterielle Induktor, *Roux's Arch. Dev. Biol.*, 173, 107, 1973.

163. Leitz, T., Induction of metamorphosis of the marine hydrozoan *Hydractinia echimata* Fleming, 1828, *Biofouling*, 12, 173, 1998.

164. Leitz, T. and Wagner, T., The marine bacterium *Alteromonas espejiana* induces metamorphosis of the hydroid *Hydractinia echinata*, *Mar. Biol.*, 115, 173, 1993.

165. Cazaux, C., Facteurs de la morphogenese chez un Hydractine polymorphe *Hydractinia echinata* Fleming, *C. R. Hebd. Sean.c Acad. Sci. (Paris)*, 247, 2195, 1958.

166. Cazaux, C., Signification et origine de l'assocation entre *Hydractinia* et Pagure. Role des tropismes larvaires dans le developpement de l'Hydraire, *Bull. Sta. Biol. Arcachon*, 13, 1, 1961.

167. Muller, W. A., Wieker, F., and Eiben, R., Larval adhesion, releasing stimuli and metamorphosis, in *Coelenterate Ecology and Behavior*, Mackie, G. O., Ed., Plenum Press, New York, 1976, 339.

168. Weis, V. M. and Buss, L. W., Ultrastructure of metamorphosis in *Hydractinia echinata*, *Postilla*, 199, 1, 1986.

169. Hofmann, D. K. and Brand, U., Induction of metamorphosis in the symbiotic Scyphozoan *Cassiopea andromeda*: role of marine bacteria and of biochemicals, *Symbiosis*, 4, 99, 1987.

170. Hofmann, D. K., Fitt, W. K., and Fleck, J., Checkpoints in the life-cycle of *Cassiopea* spp.: control of metagenesis and metamorphosis in a tropical jellyfish, *Int. J. Dev. Biol.*, 40, 331, 1996.

171. Fitt, W. K., Natural metamorphic cues of larvae of a tropical jellyfish, *Am. Zool.*, 31, 106A (Abstract), 1991.

172. Johnson, C. R., Muir, D. G., and Reysenbach, A. L., Characteristic bacteria associated with surfaces of coralline algae: a hypothesis for bacterial induction of marine invertebrate larvae, *Mar. Ecol. Prog. Ser.*, 74, 281, 1991.

173. Kirchman, D., Graham, S., Reish, D., and Mitchell, R., Bacteria induce settlement and metamorphosis of *Janua (Dexiospira) brasiliensis* Grube (Polychaeta: Spirorbidae), *J. Exp. Mar. Biol. Ecol.*, 56, 153, 1982.

174. Walters, L. J., Hadfield, M. G., and del Carmen, K. A., The importance of larval choice and hydrodynamics in creating aggregations of *Hydroides elegans* (Polychaeta: Serpulidae), *Invert. Biol.*, 116, 102, 1997.

175. Carpizo-Ituarte, E. and Hadfield, M. G., Stimulation of metamorphosis in the polychaete *Hydroides elegans* Haswell (Serpulidae), *Biol. Bull.*, 194, 14, 1998.

176. Bryan, P. J., Qian, P. Y., Kreider, J. L., and Chia, F. S., Induction of larval settlement and metamorphosis by pharmacological and conspecific association compounds in the serpulid polychaete *Hydroides elegans*, *Mar. Ecol. Prog. Ser.*, 146, 81, 1997.

177. Harder, T. and Qian, P. Y., Induction of larval attachment and metamorphosis in the serpulid polychate *Hydroides elegans* by dissolved free amino acids: isolation and identification, *Mar. Ecol. Prog. Ser.*, 179, 259, 1999.

178. Beckman, M., Harder, T., and Qian, P. Y., Induction of larval attachment and metamorphosis in the serpulid polychaete *Hydroides elegans* by dissolved free amino acids: mode of action in laboratory bioassays, *Mar. Ecol. Prog. Ser.*, 190, 167, 1999.

179. Toonen, R. J. and Pawlik, J. R., Foundations of gregariousness, *Nature*, 370, 511, 1994.

180. Parsons, G. J., Dadswell, M. J., and Roff, J. C., Influence of biofilm on settlement of sea scallop, *Placopecten magellanicus* (Gmelin, 1791), in Passamaquoddy Bay, New Brunswick, Canada, *J. Shellfish Res.*, 12, 279, 1993.

181. Satuito, C. G., Shimizu, K., and Fusetani, N., Studies on the factors influencing larval settlement in *Balanus amphitrite* and *Mytilus galloprovincialis*, *Hydrobiologia*, 358, 275, 1997.

182. Hadfield, M. G. and Strathmann, M. F., Variability, flexibility and plasticity in life histories of marine invertebrates, *Oceanologica Acta*, 19, 323, 1996.

183. Cameron, A. R. and Hinegardner, R. T., Initiation of metamorphosis in laboratory cultured sea urchins, *Biol. Bull.*, 146, 335, 1974.

184. Chen, C. P. and Run, J. Q., Larval growth and bacteria-induced metamorphosis of *Arachnoides placenta* (L.) (Echinodermata: Echinoidea), in *Reproduction, Genetics and Distributions of Marine Organisms*, Ryland, J. S. and Tyler, P. A., Eds., Olsen & Olsen, Fredensborg, 1989, 55.

185. Pearse, J. S. and Cameron, R. A., Echinodermata: Echinoidea, in *Reproduction of Marine Invertebrates*, Giese, A. C., Pearse, J. S., and Pearse, V. B., Eds., Boxwood Press, Pacific Grove, 1991, 513.

186. Johnson, C. R. and Sutton, D. C., Bacteria on the surface of crustose coralline algae induce metamorphosis of the crown-of-thorns starfish *Acanthaster planci*, *Mar. Biol.*, 120, 305, 1994.

187. Ito, S. and Kitamura, H., Induction of larval metamorphosis in the sea cucumber *Stichopus japonicus* by periphitic diatoms, *Hydrobiologia*, 358, 281, 1997.

188. Mihm, J. W., Banta, W. C., and Loeb, G. I., Effects of adsorbed organic and primary fouling on bryozoan settlement, *J. Exp. Mar. Biol. Ecol.*, 54, 167, 1981.

189. Brancato, M. S. and Woollacott, R. M., Effect of microbial films on settlement of bryozoan larvae *Bugula simplex*, *B. stolonifera*, and *B. turrita*, *Mar. Biol.*, 71, 51, 1982.

190. Reed, C. G. and Woollacott, R. M., Mechanisms of rapid morphogenetic movements in the metamorphosis of the bryozoan "*Bugula neritina*" (Cheilostomata, Cellularioidea), *J. Morph.*, 172, 335, 1982.

191. Wendt, D. E. and Woollacott, R. M., Induction of larval settlement by KCl in three species of *Bugula* (Bryozoa), *Invertebrate Biol.*, 114, 345, 1995.

192. Long, J. A. and Stricker, S. A., Brachiopoda, in *Reproduction of Marine Invertebrates*, Giese, A. C., Pearse, J. S., and Pearse, V. B., Eds., Boxwood Press, Pacific Grove, 1991, 47.

193. Pennington, J. T., Tamburri, M. N., and Barry, J. P., Development, temperature tolerance, and settlement preference of embryos and larvae of the articulate brachiopod *Laqueus californianus*, *Biol. Bull.*, 196, 245, 1999.

194. Herrmann, K., Einfluss von Bakterien auf die Metamorphose-Auslosung und deren Verlauf bei *Actinotrocha branchiata* (*Phoronis muelleri*), *Verh. Dtsch. Zool. Ges.*, 1974, 112, 1975.

195. Herrmann, K., Untersuchungen uber Morphologie, Physiologie, und Okologie der Metamorphose von *Phoronis mulleri* (Phoronida), *Zool. Jahrb. Anat.*, 95, 354, 1976.

196. Herrmann, K., Larvalentwicklung und Metamorphose von *Phoronis psammophila* (Phoronida, Tentaculata), *Helgol. Wiss. Meeresunters.*, 32, 550, 1979.

197. Szewzyk, U., Holmstom, C., Wrangstadh, M., Samuelsson, M. O., Maki, J. S., and Kjelleberg, S., Relevance of the exopolysaccharide of marine *Pseudomonas* sp. strain S9 for the attachment of *Ciona intestinalis* larvae, *Mar. Ecol. Prog. Ser.*, 75, 259, 1991.

198. Roberts, D., Rittschof, D., Holm, E., and Schmidt, A. R., Factors influencing initial larval settlement: temporal, spatial and surface molecular components, *J. Exp. Mar. Biol. Ecol.*, 150, 203, 1991.

199. Wilson, D. P., The influence of the substratum on the metamorphosis of *Notomastus* larvae, *J. Mar. Biol. Ass. U.K.*, 22, 227, 1937.

200. Gray, J. S., The attractive factor of intertidal sands to *Protodrilus symbioticus*, *J. Mar. Biol. Assoc. U.K.*, 46, 627, 1966.

201. Calabrese, A., The pH tolerance of embryos and larvae of the coot clam, *Mulinia lateralis* (Say), *Veliger*, 13, 122, 1970.

202. Scheltema, R. S., Metamorphosis of the veliger larvae of *Nassarius obsoletus* (Gastropoda) in response to bottom sediment, *Biol. Bull.*, 120, 92, 1961.

203. Grassle, J. P., Snelgrove, P. V. R., and Butman, C. A., Larval habitat choice in still water and flume flows by the opportunistic bivalve *Mulinia lateralis*, *Netherlands J. Sea Res.*, 33, 1992.

204. Snelgrove, P. V. R., Grassle, J. P., Grassle, J. F., Petrecca, R. F., and Ma, A., In situ habitat selection by settling larvae of marine soft-sediment invertebrates, *Limnol. Oceanog.*, 44, 1341, 1999.

205. Cohen, R. A. and Pechenik, J. A., Relationship between sediment organic content, metamorphosis, and postlarval performance in the deposit-feeding polychaete *Capitella* sp. 1, *J. Exp. Mar. Biol. Ecol.*, 240, 1, 1999.

206. Biggers, W. J. and Laufer, H., Chemical induction of settlement and metamorphosis of *Capitella capitata* Sp. 1 (Polychaeta) larvae by juvenile hormone-active compounds, *Invertebrate Reprod. Dev.*, 22, 39, 1992.

207. Biggers, W. J. and Laufer, H., Detection of juvenile hormone-active compounds by larvae of the marine annelid *Capitella* sp. 1, *Arch. Insect Biochem. Physiol.*, 32, 475, 1996.

208. Rice, M. E., Factors influencing larval metamorphosis in *Golfinga misakiana* (Sipuncula), *Bull. Mar. Sci.*, 39, 362, 1986.

209. Gray, J. S., Animal–sediment relationships, *Oceanogr. Mar. Biol. Ann. Rev.*, 12, 223, 1974.

210. Butman, C. A., Sediment-trap experiments on the importance of hydrodynamical processes in distributing settling invertebrate larvae in near-bottom waters, *J. Exp. Mar. Biol. Ecol.*, 134, 37, 1989.

211. Snelgrove, P. V. R., Butman, C. A., and Grassle, J. P., Hydrodynamic enhancement of larval settlement in the bivalve *Mulinia lateralis* (Say) and the polychaete *Capitella* sp. 1 in microdepositional environments, *J. Exp. Mar. Biol. Ecol.*, 168, 71, 1993.

212. Fielman, K. T., Woodin, S. A., Walla, M. D., and Lincoln, D. E., Widespread occurrence of natural halogenated organics among temperate marine infauna, *Mar. Ecol. Prog. Ser.*, 181, 1, 1999.

213. Woodin, S. A., Marinelli, R. L., and Lincoln, D. E., Allelochemical inhibition of recruitment in a sedimentary assemblage, *J. Chem. Ecol.*, 19, 517, 1993.

214. Woodin, S. A., Lindsay, S. M., and Lincoln, D. E., Biogenic bromophenols as negative recruitment cues, *Mar. Ecol. Prog. Ser.*, 157, 303, 1997.

215. Walters, L. J., Hadfield, M. G., and Smith, C. M., Waterbourne chemical compounds in tropical macroalgae: positive and negative cues for larval settlement, *Mar. Biol.*, 126, 283, 1996.

216. Lau, S. C. K. and Qian, P. Y., Phlorotannins and related compounds as larval settlement inhibitors of the tube-building polychaete *Hydroides elegans*, *Mar. Ecol. Prog. Ser.*, 159, 219, 1997.

217. Jennings, J. G., and Steinberg, P. D., Phlorotannins vs. other factors affecting epiphyte abundance on the kelp *Ecklonia radiata*, *Oecologia*, 109, 461, 1997.

218. Maki, J. S., Rittschof, D., Costlow, J. D., and Mitchell, R., Inhibition of attachment of larval barnacles, *Balanus amphitrite*, by bacterial surface films, *Mar. Biol.*, 97, 199, 1988.

219. Wieczorek, S. K., Clare, A. S., and Todd, C. D., Inhibitory and facilitatory effects of microbial films on settlement of *Balanus amphitrite amphitrite* larvae, *Mar. Ecol. Prog. Ser.*, 119, 221, 1995.

220. Maki, J. S., Rittschof, D., Samuelsson, M. O., Szewzyk, U., Yule, A. B., Kjelleberg, S., Costlow, J. D., and Mitchell, R., Effect of marine bacteria and their exopolymers on the attachment of barnacle cypris larvae, *Bull. Mar. Sci.*, 46, 499, 1990.

14 Contributions of Marine Chemical Ecology to Chemosensory Neurobiology

Henry G. Trapido-Rosenthal

CONTENTS

I. INTRODUCTION

The concept of specific receptors for bioactive chemical substances originated around the turn of the last century, as a consequence of Langley's studies on the actions of plant alkaloids on animal tissues. In analyzing the results of his studies of the effects of nicotine and curare on the contraction of vertebrate skeletal muscle, he maintained that those substances must be interacting, not directly with the contractile machinery of the tissue, but rather with "receptive substance" of the muscle.[1] Langley came to the general conclusion that for bioactive molecules to effect specific actions, they must interact with specific entities; these have come to be known as receptors. The receptor concept was used to explain the effects of substances such as hormones, neurotransmitters, drugs, and poisons on cells within multicellular organisms.[2] The concept of receptor-mediated responses to waterborne environmental chemical signals was postulated by Haldane[3] in 1954, based on these developing concepts of receptor-mediated communication between the component cells of metazoan organisms. In the subsequent 20 years, it was not uncommon for chemical ecologists to hypothesize that observed chemically stimulated behaviors were mediated by specific chemoreceptors. In a

1974 article on chemoreception in the marine environment, Laverack[4] reviewed studies of organismal and cellular responses to environmental chemical signals and concluded that mediation by chemoreceptors was the most parsimonious way of accounting for these responses. However, experimental demonstration of the existence of such receptors had not yet been achieved, and Laverack used the existing knowledge of neurotransmitter receptors as a heuristic device to demonstrate to his readers how signal detection and transduction might operate in chemoreceptor cells. Only relatively recently have electrophysiological and biochemical characterizations of chemoreceptors for environmental chemical signals been accomplished, and the molecular characterization of such receptors is just beginning. This chapter reviews the past five decades of work devoted to the study of chemoreceptors in aquatic organisms. Since, during this period of time, various aspects of this subject have been subjected to review by other writers, a temporal bias towards more recent work will be detected. It is the hope of this author that this bias will be overcome by directing the reader to the important reviews of work in marine chemoreception that precede this one.

II. THE NATURE OF CHEMICAL SIGNALS IN THE MARINE ENVIRONMENT

The chemical signals encountered by organisms in marine and other aquatic environments can be conceptually distributed among four categories. They can be chemically characterized as being either primary metabolites (roughly defined as substances used in the basic metabolic processes of organisms) or secondary metabolites (substances constructed by the condensation of primary metabolites into more complex structures, and which can be used as chemical signals that regulate both intracellular and intercellular processes), with the distinction between these two classifications being somewhat arbitrary and dependent on the interests and definitions of the classifier.[5,6] Substances can be more positively characterized according to the way in which they are presented, or made accessible, to a detecting organism. An organism can detect the signal molecule either in the three-dimensional space of solution or on the two-dimensional space of a solid surface.

A comprehensive review of the nature of chemical signals that are encountered by organisms in aquatic environments is presented by Carr.[7] In his review, Carr points out that many of the substances that we know to be important chemical signals in the marine environment are, in fact, also potent neuroactive agents, and their neuroactive properties are initiated by specific interactions with receptors. Thus, our understanding of the chemical nature of many of the substances that serve as signal molecules in the marine environment and occupants of cell-surface receptors in the internal environment of metazoans has led to the creation and testing of hypotheses concerning the receptor-mediated nature of cellular and organismal responses to environmentally important chemical signals encountered in marine and aquatic environments. Nevertheless, there has been some controversy on the subject of whether or not substances that are indeed neuroactive in the context of a multicellular organism's central nervous system are in fact likely to be chemical signals in certain different contexts of an organism's external environment;[8,9] these controversies have likewise contributed to the scientific investigation of the molecular mechanisms underlying the detection of environmental chemical signals.

Among the substances that serve as chemical signals in both internal and external aqueous environments are nucleotides (such as AMP, ADP, and ATP), amino acids (such as glycine, glutamate, arginine, and taurine), and peptides (of which an astronomically large variety can exist due to the vast combinatorial possibilities that just the standard 20 protein-forming amino acids afford — 20^n, with the exponent n indicating the chain length of a peptide). In both internal and external environments, nucleotides and amino acids are typically presented to a receptor in solution while peptides can be presented either in the three-dimensional context of solution, or the two-dimensional context of solid-phase attachment to a surface. Depending on the particular identity and sequence of component amino acid residues, individual peptide molecules can also have higher signal-to-noise ratios[10] and longer lifetimes as signal molecules than the smaller amino acids or nucleotides, due to slower biotic (enzymatic degradation and uptake) and abiotic (removal from solution by adsorption to colloids) clearance mechanisms; such increased residence time in the environment can be a distinct advantage in a chemical signaling system.[11]

III. CHEMORECEPTION IN BACTERIA

Chemically mediated behavior in bacteria was first noted by Engelmann in 1881 and Pfeffer in 1884 (see Paoni et al.[12]). At the time that he wrote his 1969 article entitled "Chemoreceptors in Bacteria," Adler[13] maintained that answers to the questions of how bacteria detect and respond to chemical signals was still almost entirely unknown. In the decades since then, a great deal of progress has been made in determining the mechanisms by which bacterial responses to environmental chemicals are initiated and executed. Although much of this progress has been made using the "lab rat" bacteria *Escherichia coli* and *Salmonella typhimurium*, the results are thought to be broadly applicable to most bacterial species.[12]

Chemotaxis is perhaps the best studied bacterial responses to environmental chemical signals (see Chapter 12, this volume). Bacteria typically move up concentration gradients of nutrient molecules and down concentration gradients of noxious compounds. Berg and Brown[14] showed that directionality is conferred by alteration of two behaviors, one a straight-ahead swimming behavior and the other a direction-changing tumbling behavior. When moving in a desired direction (up a concentration gradient of a nutrient, for example), swimming is rarely interrupted by tumbling episodes. When moving in a direction interpreted as undesirable, tumbling becomes more frequent; after each tumble, swimming begins anew, and, since tumbling results in a random reorientation of the bacterium, the chances are good that the new direction will be away from the source of the noxious chemical.

Using behavioral and genetic assays, Adler[13] concluded that detection of chemical signals was by means of receptors that recognized the chemical structures of these signals. This work was then expanded by Koshland and others[15-17] to directly measure the interaction of various nutrient sugars and amino acids with their respective receptors.

The molecular mechanisms by which enteric bacteria respond to occupation of chemoreceptors have been worked out in substantial detail by combining the techniques of classical genetics, molecular genetics, and biochemistry. Upon occupation of a receptor by an appropriate ligand, conformational changes in the structure of the receptor transmit information to the cell's interior by altering the activities of enzymes that affect the methylation and phosphorylation states of proteins involved in the signal detection and signal transduction processes.[18] Of particular importance among these chemotaxis (Che) proteins is Che-Y, the phosphorylation state of which governs the direction of rotation of the bacterial flagellum and, thus, determines whether the cell is swimming or tumbling. Other Che proteins are involved with adaptation, directly affecting the ability of the receptors to interact with their ligands.[19] Marine bacteria can respond more rapidly than enteric bacteria to environmentally encountered chemical signals,[20] suggesting that signal detection and transduction mechanisms that are both qualitatively and quantitatively different than those characterized in *E. coli* await elucidation.

Bacteria also use chemical signals to communicate with each other. The observation that the marine bacteria *Vibrio fischeri* were brightly luminescent at high population densities but dim when densities were low led to the identification of bacterial metabolites that have become termed quorum-sensing factors. The quorum-sensing factors of marine *Vibrio* sp. that regulate luminescence are acylated homoserine lactones that are synthesized by the bacteria and released into the surrounding medium. Work by Bassler and colleagues[21] has shown that the quorum-sensing factor is detected by a transmembrane receptor-transducer molecule that has both kinase and phosphatase activities. When unoccupied, the receptor's kinase activities bring about both autophosphorylation and the phosphorylation of a series of response regulator proteins; when phosphorylated, these proteins repress the transcription of the operon that codes for the proteins involved in light production (luciferase and the enzymes that synthesize the substrate from which luciferase generates light). When the receptor is occupied, it becomes a phosphatase; phosphate groups are removed from the response regulation enzymes, which lose their repressor functions, permitting the transcription of the genes in the light-generating operon, with the ultimate result of bacterial luminescence.

A growing number of bacterial quorum-sensing factors are now being discovered. These include not only a number of variations on the homoserine lactone theme, but also a variety of peptides, as well as specific cocktails of amino acids.[22] They appear to be used to measure population densities of other, perhaps competing, bacteria, as well as of conspecifics, and many of them clearly function as regulators of transcriptional activity. The development, organization, and functional maintenance of bacterial biofilms appears to be mediated, in large part, by the generation, release, and detection of quorum-sensing factors.[23] Thus, the ability to detect and respond to these environmentally encountered chemical signals clearly has not only tremendous adaptive value for bacteria, but will be of fundamental importance to our understanding of the many biofilm-based communities that are important components of marine ecosystems.

IV. CHEMORECEPTION IN EUKARYOTIC MICROORGANISMS

In eukaryotic microorganisms, as in bacteria, detection and evaluation of environmental chemical signals, as well as responses to those signals, are all accomplished by the same cell. This enables the tight coupling of behavioral data with biochemical, physiological, and molecular investigations into the cellular and molecular mechanisms involved in chemoreception. A number of model systems, including the single-celled gametes of various multicellular organisms, slime molds, yeast, and paramecia, have been used in such studies. The latter well demonstrates the research value of eukaryotic microorganisms for chemosensory research and will be touched on here.

Paramecium tetraurelia is a diploid eukaryotic unicellular organism that alters its swimming behavior when it encounters certain environmental chemicals. Like bacteria, paramecia move towards attractants and away from irritants by altering the ratios of turning behavior to swimming behavior. However, many of the molecular mechanisms by which these behavioral changes are brought about are significantly different.

By reducing the amount of turning, paramecia move towards a number of compounds such as lactate, acetate, folate, cyclic AMP (cAMP), or the excreted bacterial metabolite biotin; these substances can be considered either of direct nutritional value or of informational value, as indicators of the presence of nutritional resources. By increasing the amount of turning, they move away from irritants such as quinidine-HCl. By combining series of electrophysiological, biochemical, and molecular experiments, Van Houten and colleagues[24–28] have made a great deal of progress in elucidating the molecular mechanisms that underlie the cellular (and in this case organismal) response to environmental chemical signals.

The responses are typically initiated by the specific interaction of the environmentally encountered chemical with receptors that are deployed on the cell surface. Radiolabeled biotin, for example, interacts with a structurally selective receptor with an estimated affinity (as represented by the K_D) of 400 μM; this compares with a behavioral EC_{50} for this substance of 300 μM. Compounds that are structurally similar to biotin can compete for the binding of the radiolabeled molecule, whereas compounds that are structurally different cannot.

Upon occupation of biotin receptors, the cell membrane becomes hyperpolarized. This hyperpolarization causes an increase in the posteriorly directed beating of the cell's propulsive cilia, and the cells move smoothly up the concentration gradient. Importantly, the hyperpolarization also decreases the likelihood of membrane depolarization, and if mild, slows the ciliary beat frequency and slows the cells, while if large, brings about a calcium action potential that reverses the ciliary beat and causes the cells to turn sharply. The linkage of receptor occupation with membrane potential and ciliary motion appears to be, in part, an ATP-powered Ca^{++} pump that resides in the plasma membrane; when active, it serves to extrude Ca^{++} from the cells, thus maintaining a low intracellular concentration (in the vicinity of 10^{-8} M) of this cation. During action potentials, Ca^{++} floods down its electrochemical gradient into the cells, resulting in a transitory increase in concentration to as high as 10^{-6} M in the motile cilia; at this concentration, the interaction of Ca^{++} with the cilia's axonemal machinery results in a directional change in ciliary beat.

An example of molecularly mediated social interaction is provided by the ciliate *Euplotes raikovi*. In this organism, mating is coordinated by a family of water-soluble peptide pheromones of modular construction, with highly conserved residues and regions (that, importantly, either consist of or include six cysteine residues that provide these peptides with three intramolecular disulfide bonds) mixed with variable regions that are presumed to provide a given pheromone its functional specificity.[29] Although receptors for these pheromones remain to be elucidated, biochemical, molecular, and behavioral data are consistent with the hypothesis that the cellular and organismal actions of these molecules are initiated by means of interaction with specific receptor molecules.

V. CHEMORECEPTION IN MULTICELLULAR ORGANISMS

The organismal division of labor that resulted from the development of multicellularity brought about behavioral repertoires that, by the standards of single-celled life forms, can be considered complex. The study of the organismal, cellular, and molecular ways in which environmentally encountered chemical signals influence behaviors associated with feeding, development, and social interactions has made important contributions to our understanding of chemoreception.

A. Feeding

1. Behavioral Observations and Studies

Chemically initiated feeding behavior has long been observed and studied in a large variety of marine organisms. A well-studied example in an evolutionarily ancient metazoan that links this behavior to chemoreceptors has been the study of the responses of cnidarians to particular chemicals. Loomis[30] reported that reduced glutathione (GSH) initiated feeding behavior in *Hydra littoralis*. Subsequently, it was shown that representatives of every class of cnidarian exhibit a feeding response when exposed to one of a few small compounds,[31] with GSH, the amino acid proline, a variety of other amino acids, and the quaternary ammonium compound betaine being the most typical initiators of the behavior.[32] The apparent specificity led to the conclusion that the observed behaviors were likely to be receptor mediated.[33]

Similar observations have been made, and similar conclusions drawn, with members of many other metazoan phyla including annelids, molluscs, and echinoderms (reviewed by Lenhoff and Lindstedt[34]), arthropods (reviewed by Ache[35]), and vertebrates (reviewed by Sorensen and Caprio[36]).

In addition to feeding attractants, behavioral observations have made it clear that many organisms are deterred from eating certain other plants and animals, and this deterrence is often chemical in nature. Whereas feeding attractants are often small molecules such as nucleotides, sugars, and amino acids that are components of important metabolic pathways and can be considered primary metabolites, feeding deterrents are often somewhat larger, more complex molecules that play no obvious role in basic metabolic pathways and, as mentioned earlier, are termed secondary metabolites.[37] Some feeding deterrents function by interacting with the consuming organism's peripheral chemosensory systems, and chemoreceptors are implicated in subsequent behavioral responses, but many function in a different manner entirely, by affecting one aspect or another of the physiology of the consuming organism (see also Chapter 11, this volume).[38]

2. Physiological, Biochemical, and Molecular Studies

Building upon the behavioral observations of glutathione-initiated feeding in *Hydra*, Rushforth and colleagues[39,40] demonstrated the effects of this chemical on the electrophysiological activity of the animal. Subsequently, a number of studies have been performed to biochemically evaluate the nature of the interaction of this feeding stimulant with this organism. Bellis et al.[41] demonstrated a reversible interaction of glutathione with *Hydra* plasma membrane preparations that was characterized by a K_D of 3.4 μM, a value in close agreement with the EC_{50} of glutathione-induced feeding behavior.

Although electrophysiological and biochemical studies of chemoreception in other aquatic invertebrates demonstrated similar support for chemoreceptor-mediated detection of environmental chemical signals associated with feeding (e.g., Croll[42]), the bulk of this sort of data has been collected from crustaceans (the lobsters *Panulirus argus*, *P. californicus*, and *Homarus americanus*, the crayfish *Austropotamobius torrentium*, and various crabs, such as *Callinectes sapidus*). The relatively large size and accessibility of the chemosensory organs of these animals have led to their use as model systems to study the cellular electrophysiology of chemoreception.[43–46] The same attributes make them attractive organisms for biochemical and molecular studies.[47–49]

Anatomically, the chemosensory cells of these animals share a unifying set of characteristics: they are bipolar neurons with ciliated dendrites closely apposed to the environment and axons that project into the central nervous system from a peripherally located cell body. This is a cellular bodyplan that is characteristic of chemosensory cells from a broad range of metazoan phyla, so much that has been learned by the study of crustacean chemosensory neurophysiology has been of heuristic value to the understanding of chemoreception in other organisms.

Knowing that crustacea respond to the amino acids, nucleotides, and other compounds present in the food odors that stimulate feeding behavior in these animals, a number of researchers began studying the electrophysiological responses of crustacean chemosensory cells to these chemicals. Ache and colleagues[43,50–53] demonstrated that chemosensory cells responded to various amino acids and nucleotides. The structural specificities exhibited by receptor cells for stimulatory compounds were consistent with the hypothesis that the compounds were interacting with cell-surface receptors. In many cases, the structure–activity relationships were strikingly similar to the structure–activity relationships that had been described for internal receptors for these compounds. These similarities led to a restating of the Haldane hypothesis that there is an important evolutionary link between chemoreceptors that monitor the chemical composition of the external environment and those that monitor the chemical composition of the internal (but extracellular) environment of metazoans.[54]

Derby and colleagues designed studies to characterize the interaction of amino acid and nucleotides with putative lobster olfactory receptors for these substances. They prepared plasma membrane fractions from the chemosensory dendrite-rich sensilla of the spiny lobster, and demonstrated specific, saturable, and reversible binding of the sulfonic amino acid taurine and the adenine nucleotide AMP to this material;[55,56] ultrastructural localization of binding sites on the dendritic membrane for AMP were also demonstrated.[57] In subsequent studies combining electrophysiological and biochemical experiments, multiple receptor types for L-glutamic acid were characterized.[58] Separate binding sites for L-alanine and D-alanine were characterized by Michel et al.[59] The interactions of mixtures of amino acid and nucleotides with receptors for individual amino acids have also been characterized and shown to bear close relationships to the inhibitory effects of mixtures upon electrophysiological and behavioral responses to individual amino acid and nucleotide odorants.[60,61]

Ache and coworkers demonstrated that both cyclic nucleotides and inositol phosphates mediate the transduction of environmental chemical signals by the olfactory neurons of *P. argus*.[62–65] Both biochemical and molecular biological techniques have shown that the receptor cells contain various G-protein subunits that would be necessary for signal detection by G-protein-associated chemoreceptors.[48,49,66–69] In combination with electrophysiological studies, these techniques have demonstrated the existence of an ensemble of ion channels in these cells, the opening and closing of which are mediated by the second messengers that are generated by the occupancy of receptors by odorants. It is interesting to note that individual receptor cells can be depolarized by one amino acid and hyperpolarized by another.[70] These results indicate that both cAMP- and IP$_3$-gated channels are present in a single neuron. More importantly, these results are consistent with the hypothesis that in this marine invertebrate, single olfactory neurons can express more than one receptor. This is in apparent contrast with the situation in vertebrates, where it appears that an individual receptor cell deploys only a single chemoreceptor.[70–72]

The demonstration of G-protein-mediated signal transduction of amino acid signals suggests that the chemoreceptors of the lobster olfactory organ for these substances are of the seven transmembrane-segment, G-protein-coupled (GPC) variety. Although the lobster olfactory organ contains mRNA transcripts, the sequences of which bear reasonable homology to GPC receptors from other organisms that are presumed to be chemosensory, the functional demonstration that these transcripts code for chemosensory receptors in the lobster has not yet been achieved.

Electrophysiological studies of the smell and taste systems of fish have likewise demonstrated chemoreceptor cells that are responsive, with varying degrees of specificity, to the amino acids known to elicit feeding behavior.[73–75] In addition, a number of fish have receptor cells that respond to bile acids, amphipathic steroid compounds that are used as digestive detergents and that can be released into the environment in substantial quantities. Responses can exhibit both exquisite specificity for the structure of a bile acid, and extreme sensitivity, as best exemplified by the sea lamprey.[76,77]

Membrane preparations of fish olfactory and gustatory organs have been used to test the hypothesis that receptors for odorants and tastants are resident in these membranes. Scientists at the Monell Chemical Senses Center have published an extensive series of papers on the biochemical characterization of trout and catfish receptors for amino acids. Krueger and Cagan[78] demonstrated a structurally specific, reversible interaction of the amino acid L-alanine with plasma membrane fractions of catfish taste epithelium, and Brand, Bryant, Kalinoski and colleagues comprehensively characterized a catfish taste receptor for L-arginine.[79–82] These scientists further reported the presence of G-proteins in catfish olfactory cilia,[83,84] with cyclic AMP (cAMP) being implicated as a second messenger involved in the transduction of amino acids binding to olfactory receptors.[85] Lo et al.[86] have shown that bile acids bring about increases in intracellular second messengers in the olfactory system of salmon, and they hypothesized that this second-messenger generation is, at least in part, receptor mediated. Recently, the cloning and functional expression of a goldfish odorant receptor that specifically interacts with basic amino acids has been achieved; analysis of the sequence of nucleotides that codes for this receptor demonstrates that it is a member of the G-protein-coupled family of receptors.[87]

B. Larval Development

1. Behavioral Observations and Studies

For many marine organisms, a larval period is an evolutionarily important component of the life cycle. In many case, the developmental transition from the larval stage to the juvenile stage is initiated by an appropriate environmental signal. Upon detection of this signal, appropriate internal developmental processes will be triggered or released; if the signal is not detected, larvae remain in a state of developmental arrest.[88] Behavioral observations have indicated that, although phenomena such as light, substrate surface texture, and hydrostatic pressure can be the metamorphosis-inducing trigger for selected species,[89] more frequently a trigger of a chemical nature has been implicated.[88–94]

In some cases, the nature of the chemical signal is known as well. Larvae of the tube worm *Phragmatopoma californica* undergo metamorphosis in response to a proteinaceous substance present in the tubes built by conspecific adults.[95] Larvae of the sand dollar *Dendraster excentricus* respond metamorphically to material from adults of this species (a peptide of about 1000 Da); this material can be found in the sand beds occupied by adults.[92] Larvae of the opisthobranch mollusc *Phestilla sibogae* are induced to metamorphose by a low molecular weight (300 to 500 Da) water-soluble material that is released from the coral *Porites compressa*; this coral is fed upon by the adult form of this nudibranch.[91,96,97] Recently metamorphosed juvenile red abalone (*Haliotis rufescens*) are typically found on rocks encrusted with the red alga *Lithothamnium californicum*;[98] a peptide with a molecular weight of about 1000 Da that is present on the surface of this alga appears to be the molecule that induces larval abalone to metamorphose to the juvenile state.[93,99] Red algae also provide the chemical cues that induce larvae of the conch *Strombus gigas* to undergo metamorphosis,[100] and larvae of the sea urchin *Holopneusteus purpurascens* are induced to metamorphose by a water-soluble complex of the red algal metabolites floridiside and isethionic acid.[101] Larvae of agariciid corals are induced to metamorphose by sulfated polysaccharides found at the surfaces of tropical species of corraline red algae.[102] Oyster larvae can be induced to settle by small, soluble peptides containing C-terminal arginine residues;[103] both adult conspecifics and microbial biofilms found in association with the adults could serve as the source of these inducing peptides.[104] Variations in peptide amino acid composition leads to alterations in the efficacy of a molecule as an inducer; the resulting structure–activity relationships strongly suggest interaction with specific chemoreceptors.[105] The ability of specific exogenous compounds to initiate and modulate larval metamorphosis has led many students of this developmental phenomenon to implicate larval chemoreceptor molecules, deployed at the environment-facing surfaces of chemosensory cells, as key components that serve as an interface between the larval nervous system and the marine environment.

2. Physiological, Biochemical, and Molecular Studies

The small size and challenging anatomy of molluscan larvae have made electrophysiological studies of chemically induced settlement and metamorphosis considerably more difficult than similar studies of the effects of feeding stimulants on the olfactory neurons of adult crustaceans. Compounds that induce the larvae of the abalone *Haliotis rufescens* to settle and metamorphose affect the firing of the motile ciliated velar cells that, in aggregate, comprise the swimming organ of the veliger.[106] However, this phenomenon appears to be mediated by the larval nervous system rather than by the inducing molecule itself — the velar cells are not themselves sensing metamorphosis-inducing chemicals in the environment. Arkett et al.[107] showed that cells on the propodium of the larval nudibranch *Onchidoris bilamellata* were depolarized by exposure to barnacle-derived compounds that induce these larvae to settle and undergo metamorphosis. Another approach to electrophysiological studies of chemically induced larval settlement and metamorphosis has been to focus on neurons one or more synapses away from the actual chemosensory neurons; Leise and Hadfield[108] demonstrated that cells in the central ganglia of *Ilyanassa obsoleta* larvae alter their firing patterns in response to compounds that induce the metamorphosis of these larvae. The results of these studies infer, as did behavioral assays, the existence of chemosensory cells with receptors for inducing molecules.

In a series of imaginative experiments combining electrophysiological principals with behavioral observations, Yool (née Baloun) and colleagues subjected competent larvae from a number of marine genera to treatment with artificial seawaters containing ionic additions, substitutions, or deletions designed to either bring about or prevent the depolarization of neurons.[109,110] The results of these experiments, as exemplified by the finding that a brief period of larval exposure to elevated concentrations of K^+ would induce metamorphosis, were consistent with the hypothesis that the depolarization of larval neurons, perhaps but not necessarily chemosensory neurons, was a necessary step between the encountering of a metamorphosis-inducing environmental cue and the subsequent behavioral and developmental metamorphic events.

There are few reports of direct biochemical characterization of larval chemoreceptors. Following the finding that abalone larvae could be induced to metamorphose by γ-aminobutyric acid (GABA) as well as by peptidic materials extracted from the red alga *Lithothamnium californicum*,[111,112] Trapido-Rosenthal and Morse[113] used a radiolabeled GABA analog, β-chlorophenyl GABA (baclofen), to characterize the interactions of settlement-inducing compounds with larval abalone. They demonstrated the existence of reversible binding of this metamorphosis-inducing compound to larvae, an interaction that is characterized by a K_D (the concentration of inducer at which half of the binding sites are occupied) on the order of 1 μM and a B_{max} (the total number of binding sites) of 15 fmoles/larva. It was further shown that this binding could be competed for by preparations of morphogenic peptides isolated from *L. californicum*.[114] These binding sites disappear from larvae at the time of metamorphosis, when the larvae shed both their velar cilia and the cilia of their apical tuft, to which chemosensory functions have been attributed on anatomical grounds.[115–117] In further experiments, Baxter and Morse[118,119] isolated velar and apical tuft cilia from competent abalone larvae and demonstrated that the cilia in these preparations specifically and reversibly bound the diamino acid lysine, a substance which itself is nonmorphogenic but which dramatically modulates the effectiveness of both algal and amino acid morphogens.[120]

The transduction of the chemical signals that initiate and regulate metamorphosis has been investigated using imaginative combinations of a variety of techniques. The above-mentioned demonstrations by Yool and colleagues[109,110] (that conditions that bring about neuronal depolarization induce abalone larvae to undergo metamorphosis) clearly corroborate the hypothesis that transmembrane ion fluxes are obligatory components of metamorphic responses. Results from experiments with tetraethylammonium, a membrane-impermeant blocker of chloride ion channels, in which the presence of this compound in seawater prevents larvae from responding to metamorphic signals, suggest that depolarization of cells exposed to the environment are necessary for the initiation of metamorphosis. However, Hadfield and colleagues[121] have demonstrated that the selective ablation of putative environment-contacting chemosensory cells does not prevent *Phestilla* larvae from undergoing metamorphosis when subjected to depolarizing concentrations of potassium or cesium ions; these results make it clear that depolarizations of cells one or more synapses downstream from the chemoreceptor cells are also required for metamorphosis.

The modulatory effect of lysine on the induction of abalone metamorphosis by GABA or by appropriate algal peptides is mediated by receptors located on larval cilia.[118–120] When these receptors are occupied by an appropriate ligand, signal transduction is brought about by the interaction of the receptor–ligand complex with G-proteins; this interaction in turn activates a second messenger cascade involving phospholipase C and protein kinase C (PKC).[122] The ways in which the proteins phosphorylated by PKC enhance responses to metamorphic signals remain unknown.

C. Social Interactions

1. Behavioral Observations and Studies

As important to an organism as eating and developing is staying alive. Detection of chemicals emanating from potential predators, or from the dead or damaged prey of these predators, can lead to a behavioral response that removes an animal from the predator's environment. Some of the chemicals that induce escape responses are identical to compounds that, in a different context, serve as feeding deterrents. Thus, the starfish saponins that are feeding deterrents to animals that prey upon starfish warn molluscs that would be preyed upon by the starfish that they are in a dangerous environment. In other cases, an organism that is molested by a predator will release a compound that will, if detected by its conspecifics, induce an escape or avoidance behavior. An example of this is the release of navenones into the slime trail produced by an aggravated specimen of the nudibranch *Navanax inermis*; conspecifics detecting this signal will turn off of this trail.[123] Another example is the release of anthopleurines into the water by a damaged specimen of the anemone *Anthopleura elegantissima*; detection of this compound by nearby conspecifics will induce them to contract into a conformation less vulnerable to predatory damage.[124] Crustaceans can recognize chemical cues emanating from predators, at which point they will engage in avoidance behavior.[125] In addition, they can deliver, via their urine, chemical signals that can indicate to nearby conspecifics the presence of a nearby stressor such as a predator.[126,127] Fish likewise recognize a variety of alarm substances, some of which are released by members of their own species and others that may emanate from other organisms, and signify a predatorial presence.[128–130]

Crustaceans are known to use pheromones in behavioral contexts other than avoidance, including reproduction and social interactions. For example, at least one spiny lobster, the California spiny lobster *Panulirus interruptus*, has been shown to be attracted to the odor of conspecifics.[131] This pheromonal phenomenon is taken advantage of by workers in the lobster fishery, who use live lobsters as bait in their pots.[132] However, relatively little research has been performed on the topic of aggregation pheromones, including the nature of the signal and its sensory reception. Pheromonal chemical signals are also involved in the establishment and maintenance of social hierarchies in crustaceans.[133,134]

Using a comprehensive series of behavioral, biochemical, and molecular biological experiments, Painter and colleagues[135] have identified and chemically characterized an aggregation pheromone, which they named attractin, from the opisthobranch mollusc *Aplysia californica*. This molecule, a glycosylated 58-residue peptide, is produced by the albumin gland and released into the environment with the material that this gland adds to the animal's egg cordons. There is a striking structural similarity between *Aplysia* attractin and the peptide pheromones of the ciliate protozoan *Euplotes*.[29] Future research may reveal that molecules such as these, which have the possibilities of mixing highly conserved domains with variable domains, may well be used as pheromones by a number of marine organisms.

Fish provide numerous examples of other chemically mediated social behaviors. A dramatic example is the ability of many fish to "home" or return to a particular geographic location, most typically the site of their nativity. The chemical signals used in homing behavior have not been comprehensively identified but are thought to include both molecules of plant origin that are characteristic of the natal site as well as odorants, including bile acids, that derive from conspecific fish.[136,137]

Reproductive behavior in fish is a phenomenon that is synchronized by chemical means.[138–140] A fraction of the steroids that are involved in the internal development of oocytes are released by females into the environment, where they are encountered by males — detection of this steroid induces internal hormonal changes in males that bring about enhanced sperm production. At a later time in the reproductive cycle, prostaglandins in the female that are associated with the follicular rupture of mature egg cells are released. Upon detection of the appropriate prostaglandin, males begin mating behaviors that culminate in the release of gametes by both sexes.

2. Physiological, Biochemical, and Molecular Studies

The goldfish has been established as a model system for the study of chemically mediated reproductive phenomena in aquatic vertebrates. Sorensen and colleagues[139,141–145] have performed extensive studies of the electrophysiological responses of the olfactory systems of males to the pheromonal steroids (preovulatory signals that prime males for subsequent sexual activity) and prostaglandins (released into the environment after ovulation, the function of which remains to be completely elucidated). Their results have shown that goldfish have receptor cells for steroids that are highly specific and sensitive, with minute changes in molecular structure resulting in one hundred-fold increases in nanomolar threshold concentrations.[139,145] The results of this work have been consistent with receptor mediation of the behavioral responses to these environmentally encountered chemical signals.

Likewise, in goldfish, Rosenblum et al.[146] have characterized the interaction of the steroidal pheromone 17α,β-dihydroxy-4-pregnen-3-one to a plasma membrane isolated from the animal's olfactory epithelium. In an attempt to elucidate the molecular basis of pheromone recognition, Cao et al.[147] have cloned and localized two multigene receptor families from mRNA isolated from goldfish olfactory epithelium. Analyses of the structures of the proteins coded for by these RNA indicates that the members of one of the families (the GFB family) are homologous to the putative pheromone receptors found in mammals. This, together with the fact that the members of the two families are expressed in different areas of the olfactory epithelium, has led the authors to postulate that the GFB receptors are responsible for interacting with pheromones.

VI. CONCLUSIONS

At this point in time, only a small number of chemoreceptor–odorant pairs have been experimentally characterized at behavioral, physiological, and molecular levels. The majority of the odorants in these pairs have been primary metabolites such as amino acids and nucleotides, that, when encountered in the context of a metazoan nervous system, are also neuroactive molecules. Likewise, the chemoreceptors for these identified odorants have characteristics that suggest that they are qualitatively similar to the receptors for neurotransmitters and other molecules that have a role to play in cell–cell communication in multicellular organisms. Nevertheless, many environmentally encountered molecules that function as odorants at the behavioral or physiological levels remain to be identified chemically, and genes coding for a large number of chemoreceptors for as yet unidentified odorants certainly exist in the genomes of marine organisms. As research in molecular aspects of chemical ecology progresses, an increase in the number of fully characterized odorant-chemoreceptor pairs can be expected. Among this number will almost certainly be some odorants with well-known chemical structures, others with novel chemical structures, and the chemoreceptors that have been developed to recognize them as compounds of ecological significance.

ACKNOWLEDGMENTS

I gratefully acknowledge the intellectual contributions of Drs. W. E. S. Carr, C. D. Derby, R. A. Gleeson, M. G. Hadfield, D. E. Morse, and J. Van Houten, made during the decades-long conversation that we have been engaged in on the subject of chemoreception in aquatic environments. This is contribution number 1581 from the Bermuda Biological Station for Research, Inc.

REFERENCES

1. Langley, J. N., On nerve endings and on special excitable substances in cells, *Proc. Royal Soc. Lond. B*, 78, 170, 1906.
2. Levitzki, A., *Receptors, A Quantitative Approach*, Benjamin/Cummings, Menlo Park, CA, 1984.
3. Haldane, J. B. S., La signalisation animale, *Année Biologique*, 58, 89, 1954.
4. Laverack, M. S., The structure and function of chemoreceptor cells, in *Chemoreception in the Marine Environment*, Grant, P. T. and Mackie, A. M., Eds., Academic Press, London, UK, 1974, 1.
5. Davies, J., What are antibiotics? Archaic functions for modern activities, *Mol. Microbiol.*, 4, 1227, 1990.
6. Betina, V., *Bioactive Secondary Metabolites of Microorganisms. Progress in Industrial Microbiology*, Elsevier, Amsterdam, 1994.
7. Carr, W. E. S., The molecular nature of chemical stimuli in the aquatic environment, in *Sensory Biology of Aquatic Animals*, Atema, J., Fay, R. R., Popper, A. N., and Tavolga, W.N., Eds., Springer-Verlag, New York, 1988, 3.
8. Pawlik, J. R., Natural and artificial induction of metamorphosis of *Phragmatopoma lapidosa californica* (Polychaeta: Sabellariidae), with a critical look at the effects of bioactive compounds on marine invertebrate larvae, *Bull. Mar. Sci.*, 46, 512, 1990.

9. Morse, D. E., Recent progress in larval settlement and metamorphosis: closing the gaps between molecular biology and ecology, *Bull. Mar. Sci.*, 46, 465, 1990.

10. Rittschof, D. and Bonaventura, J., Macromolecular cues in marine systems, *J. Chem. Ecol.*, 12, 1013, 1986.

11. Decho, A.W., Brown, K. A., and Zimmer-Faust, R. K., Chemical cues: why basic peptides are signal molecules in marine environments, *Limnol. Oceanogr.*, 43, 1410, 1998.

12. Paoni, N. F., Maderis, A. M., and Koshland, D. E., Chemical sensing by bacteria, *Biochemistry of Taste and Olfaction*, Cagan, R. H. and Kare, M. R., Academic Press, New York, 1981, 459.

13. Adler, J., Chemoreceptors in bacteria, *Science*, 166, 1588, 1969.

14. Berg, H. C. and Brown, D. A., Chemotaxis in *Escherichia coli* analyzed by three-dimensional tracking, *Nature,* 239, 500, 1972.

15. Aksamit, R. and Koshland, D. E., Identification of the ribose binding protein as the receptor for ribose chemotaxis in *Salmonella typhimurium*, *Biochemistry*, 13, 4473, 1974.

16. Clarke, S. and Koshland, D. E., Membrane receptors for aspartate and serine in bacterial chemotaxis, *J. Biol. Chem.* 254, 9695, 1979.

17. Hazelbauer, G. L., The maltose chemoreceptor of *Escherichia coli*, *J. Bacteriol.,* 122, 206, 1975.

18. Otteman, K. M. and Koshland, D. E., Converting a transmembrane receptor to a soluble receptor: recognition domain to effector domain signaling after excision of the transmembrane domain, *Proc. Natl. Acad. Sci. USA*, 94, 11201, 1997.

19. Yi, T. M., Huang, Y., Simon, M. I., and Doyle, J., Robust perfect adaptation in bacterial chemotaxis through integral feedback control, *Proc. Natl. Acad. Sci. USA*, 97, 4649, 2000.

20. Mitchell, J. G., Pearson, L., and Dillon, S., Clustering of marine bacteria in seawater enrichments, *Appl. Environ. Microbiol.*, 62, 3716, 1996.

21. Surette, M. G., Miller, M. B., and Bassler, B. L., Quorum sensing in *Escherichia coli, Salmonella typhimurium* and *Vibrio harveyi*: a new family of genes responsible for autoinducer production, *Proc. Natl. Acad. Sci. USA*, 96, 1639, 1999.

22. Kuspa, A., Plamann, L., and Kaiser, D., Identification of heat-stable A-factor from *Myxococcus xanthus*, *J. Bacteriol.*, 174, 3319, 1992.

23. Parsek, M. R. and Greenberg, E. P., Acyl-homoserine lactone quorum sensing in Gram-negative bacteria: a signaling mechanism involved in associations with higher organisms, *Proc. Natl. Acad. Sci. USA*, 97, 8789, 2000.

24. Van Houten, J. L., Two mechanisms of chemotaxis in *Paramecium*, *J. Com. Physiol. A*, 127, 167, 1978.

25. Preston, R. R. and Van Houten, J. L., Localization of the chemoreceptive properties of the surface membrane of *Paramecium tetraurelia*, *J. Comp. Physiol. A*, 160, 537, 1997.

26. Yang, W. Q., Braun, C., Plattner, H., Purvee, J., and Van Houten, J. L., Cyclic nucleotides in glutamate chemosensory signal transduction of *Paramecium*, *J. Cell Sci.*, 110, 2567, 1997.

27. Bell, W. E., Karstens, W., Sun, Y., and Van Houten, J. L., Biotin chemoresponse in *Paramecium*, *J. Comp. Physiol. A*, 183, 361, 1998.

28. Van Houten, J. L., Chemosensory transduction in *Paramecium*, *Eur. J. Protistol.*, 34, 301, 1998.

29. Raffioni, S., Miceli, C., Vallesi, A., Chowdhury, S. K., Chait, B. T., Luporini, P., and Bradshaw, R. A., The primary structure of *E. raikovi* pheromones: a comparison of five sequences of pheromones of cells with variable mating interactions, *Proc. Natl. Acad. Sci. USA*, 89, 2071, 1992.

30. Loomis, W. F., Glutathione control of the specific feeding reactions of *Hydra*, *Ann. NY Acad. Sci.*, 62, 209, 1955.

31. Lenhoff, H. M., Heagy, W., and Daner, J., A view of the evolution of chemoreceptors based on research with cnidarians, in *Coelenterate Ecology and Behavior*, Mackie, G. O., Ed., Plenum Press, New York, 1976, 571.

32. Schick, J. M., *A Functional Biology of Sea Anemones*, Chapman and Hall, London, UK, 1991.

33. Lenhoff, H. M. and Heagy, W., Aquatic invertebrates: model systems for the study of receptor activation and evolution of receptor proteins, *Ann. Rev. Pharmacol. Toxicol.*, 17, 243, 1977.

34. Lenhoff, H. M. and Lindstedt, K. J., Chemoreception in aquatic invertebrates with special emphasis on the feeding behavior of coelenterates, in *Chemoreception in the Marine Environment*, Grant, P. T. and Mackie, A. M., Eds., Academic Press, London, UK, 1974, 143.

35. Ache, B. W., Chemoreception and thermoreception, in *The Biology of Crustacea, Vol. 3, Neurobiology: Structure and Function*, Atwood, H. L. and Sandeman, D. C., Eds., Academic Press, New York, 1982, 369.
36. Sorensen, P. W. and Caprio, J., Chemoreception, in *The Physiology of Fishes,* Evan, D. H., Ed., CRC Press, Boca Raton, FL, 1998, 375.
37. Paul, V. J., Chemical defenses of benthic marine invertebrates, in *Ecological Roles of Marine Natural Products*, Paul, V. J., Ed., Cornell University Press, Ithaca, NY, 1992, 164.
38. Hay, M. E., Marine chemical ecology: what's known and what's next?, *J. Exp. Mar. Biol. Ecol.*, 200, 103, 1996.
39. Rushforth, N. B. and Burke, D. S., Behavioral and electrophysiological studies of *Hydra*. II. Pacemaker activity of isolated tentacles, *Biol. Bull.*, 140, 502, 1971.
40. Rushforth, N. B. and Hofman, F., Behavioral and electrophysiological studies of *Hydra* III. Components of feeding behavior, *Biol. Bull.*, 142, 110, 1972.
41. Bellis, S. L., Kass-Simon, G., and Rhoads, D. E., Partial characterization and detergent solubilization of the putative glutathione chemoreceptor from *Hydra*, *Biochemistry*, 31, 9838, 1992.
42. Croll, R. P., Gastropod chemoreception, *Biol. Rev.*, 58, 293, 1983.
43. Johnson, B. R. and Ache, B. W., Antennular chemosensitivity in the spiny lobster, *Panulirus argus*: amino acids as feeding stimuli, *Mar. Behav. Physiol.*, 5, 145, 1978.
44. Derby, C. D. and Atema, J., Chemoreceptor cells in aquatic invertebrates: peripheral filtering mechanisms in decapod crustaceans, *Sensory Biology of Aquatic Animals*, Atema, J., Fay, R. R., Popper, A. N. and Tavolga, W. N., Eds., Springer-Verlag, New York, 1988, 365.
45. Derby, C. D., Single unit electrophysiological recordings from crustacean chemoreceptor neurons, in *Experimental Biology of Taste and Olfaction: Current Techniques and Protocols*, Spielman, A. I. and Brand, J. G., Eds., CRC Press, Boca Raton, FL, 1995, 241.
46. Grünert, U. and Ache, B. W., Ultrastucture of the aesthetasc (olfactory) sensilla of the spiny lobster, *Panulirus argus*, *Cell Tiss. Res.*, 251, 95, 1988.
47. Trapido-Rosenthal, H. G. The use of the spiny lobster as a tool for the study of chemosensory biochemistry, in *Experimental Biology of Taste and Olfaction: Current Techniques and Protocols*, Spielman, A. I. and Brand, J. G., Eds., CRC Press, Boca Raton, FL, 1995, 169.
48. McClintock, T. S., Byrnes, A. P., and Lerner, M. R., Molecular cloning of a G-protein α_i subunit from the lobster olfactory organ, *Mol. Brain. Res.*, 14, 273, 1992.
49. Xu, F. and McClintock, T. S., A lobster phospholipase C-β that associates with G-proteins in response to odorants, *J. Neurosci.*, 19, 4881, 1999.
50. Fuzessery, Z. M., Carr, W. E. S., and Ache, B. W., Antennular chemosensitivity in the spiny lobster, *Panulirus argus*: studies of taurine sensitive receptors, *Biol. Bull.*, 154, 226, 1978.
51. Derby, C. D. and Ache, B. W., Quality coding of a complex odorant in an invertebrate, *J. Neurophysiol.*, 51, 906, 1984.
52. Derby, C. D., Carr, W. E. S., and Ache, B. W., Purinergic olfactory receptor cells of crustaceans: response characteristics and similarities to internal purinergic cells of vertebrates, *J. Comp. Physiol. A*, 155, 341, 1984.
53. Gleeson, R. A. and Ache, B. W., Amino acid suppression of taurine-sensitive chemosensory neurons, *Brain Res.*, 335, 99, 1985.
54. Carr, W. E. S., Gleeson, R. A., and Trapido-Rosenthal, H. G., Olfactory receptors of crustaceans with similarities to internal receptors for neuroactive substances, in *Biomedical Importance of Marine Organisms*, Fautin, D. G., Ed., California Academy of Sciences, San Francisco, CA, 1988, 115.
55. Olson, K. S., Trapido-Rosenthal, H. G., and Derby, C. D., Biochemical characterization of independent olfactory receptor sites for 5′AMP and taurine in the spiny lobster, *Brain Res.*, 583, 262, 1992.
56. Olson, K. S. and Derby, C. D., Inhibition of taurine and 5′AMP olfactory receptor sites of the spiny lobster *Panulirus argus* by odorant compounds and mixtures, *J. Comp. Physiol. A*, 176, 527, 1995.
57. Blaustein, D. N., Simmons, R. B., Burgess, M. F., Derby, C. D., Nishikawa, M., and Olson, K. S., Ultrastructural localization of 5′AMP odorant receptor sites on the dendrites of olfactory receptor neurons of the spiny lobster, *J. Neurosci.*, 7, 13, 2821.
58. Burgess, M. F. and Derby, C. D., Two novel types of L-glutamate receptors with affinities for NMDA and L-cysteine in the olfactory organ of the Caribbean spiny lobster *Panulirus argus*, *Brain Res.*, 771, 292, 1997.

59. Michel, W. C., Trapido-Rosenthal, H. G., Chao, E. T., and Wachowiak, M., Stereoselective detection of amino acids by lobster olfactory receptor neurons, *J. Comp. Physiol. A*, 171, 705, 1993.

60. Gentilcore, L. R. and Derby, C. D., Complex binding interactions between multicomponent mixtures and odorant receptors in the olfactory organ of the Caribbean spiny lobster *Panulirus argus*, *Chem. Senses*, 23, 269, 1998.

61. Cromarty, S. I. and Derby, C. D., Multiple excitatory receptor types on individual olfactory neurons: implications for coding of mixtures in the spiny lobster, *J. Comp. Physiol. A*, 180, 481, 1997.

62. Fadool, D. A. and Ache, B. W., Plasma membrane inositol 1,4,5-trisphosphate-activated channels mediate signal transduction in lobster olfactory receptor neurons, *Neuron*, 9, 907, 1992.

63. Boekhoff, I., Michel, W. C., Breer, H., and Ache, B. W., Single odors differentially stimulate dual second messenger pathways in lobster olfactory cells, *J. Neurosci.*, 14, 3304, 1994.

64. Hatt, H. and Ache, B. W., Cyclic nucleotide- and inositol phosphate-gated ion channels in lobster olfactory receptor neurons, *Proc. Natl. Acad. Sci. USA*, 91, 6264, 1994.

65. Ache, B. W. and Zhainazarov, A., Dual second-messenger pathways in olfactory transduction, *Curr. Op. Neurobiol.*, 5, 461, 1995.

66. Fadool, D. A., Estey, S. J., and Ache, B. W., Evidence that a G_q-protein mediates excitatory odor transduction in lobster olfactory receptor neurons, *Chem. Senses*, 20, 489, 1995.

67. McClintock, T. A., Xu, F., Quintero, J., Gress, A. M., and Landers, T. M., Molecular cloning of a lobster $G_{\alpha q}$ protein expressed in neurons of olfactory organ and brain, *J. Neurochem.*, 68, 2248, 1997.

68. Xu, F., Hollins, B., Gress, A. M., Landers, T. M., and McClintock, T. S., Molecular cloning and characterizations of a lobster $G_{\alpha o}$ protein expressed in neurons of olfactory organ and brain, *J. Neurochem.*, 69, 1793, 1997.

69. Xu, F., Hollins, B., Landers, T. M., and McClintock, T. S., Molecular cloning of a lobster G_β subunit enriched in neurites of olfactory receptor neurons and brain interneurons, *J. Neurobiol.*, 36, 525, 1998.

70. Michel, W. C., McClintock, T. S., and Ache, B. W., Inhibition of lobster olfactory receptor cells by an odor-activated potassium conductance, *J. Neurophysiol.*, 65, 446, 1991.

71. Malnic, B., Hieono, J., Sato, T., and Buck, L. B., Combinatorial receptor codes for odors, *Cell*, 96, 713, 1999.

72. Touhara, K., Sengoku, S., Inaki, K., Tsuboi, A., Hirono, J., Sato, T., Sakano, H., and Haga, T., Functional identification and reconstitution of an odorant receptor in single olfactory neurons, *Proc. Natl. Acad. Sci. USA*, 96, 4040, 1999.

73. Caprio, J., Olfaction and taste in the channel catfish: an electrophysiological study of the responses to amino acids and derivatives, *J. Comp. Physiol. A*, 123, 357, 1978.

74. Michel, W. C. and Caprio, J., Responses of single facial taste fibers in the sea catfish, *Arius felis*, to amino acids, *J. Neurophysiol.*, 66, 247, 1991.

75. Valenticic, T., Wegert, S., and Caprio, J., Learned olfactory discrimination versus innate taste responses to amino acids in channel catfish (*Ictalurus punctatus*), *Physiol. Behav.*, 55, 865, 1994.

76. Li, W., Sorensen, P. W., and Gallaher, D., The olfactory system of migratory adult sea lamprey (*Petromyzon marinus*) is specifically and acutely sensitive to unique bile acids released by conspecific larvae, *J. Gen. Physiol.*, 105, 569, 1995.

77. Li., W. and Sorensen, P. W., Highly independent olfactory receptor sites for conspecific bile acids in the sea lamprey, *Petromyzon marinus*, *J. Comp. Physiol. A*, 180, 429, 1997.

78. Krueger, J. M. and Cagan, R. H., Biochemical studies of taste sensation. IV. Binding of L-[³H]alanine to a sedimentable fraction from catfish barbel epithelium, *J. Biol. Chem.*, 251, 88, 1976.

79. Brand, J. G., Bryant, B. P., Cagan, R. H., and Kalinoski, D. L., Biochemical studies of taste sensation. XIII. Enantiomeric specificity of alanine taste receptor sites in catfish, *Icatalurus punctatus*, *Brain Res.*, 416, 119, 1987.

80. Bryant, B. P., Brand, J. G., Kalinsoki, D. L., Bruch R. C., and Cagan, R. H., Use of monoclonal antibodies to characterize amino acid taste receptors in catfish, in *Olfaction and Taste IX*, Roper, S. D. and Atema, J., Eds., The New York Academy of Sciences, New York, 1987, 208.

81. Kalinoski, D. L., Bryant, B. P., Shaulsky, G., Brand, J. G., and Harpaz, S., Specific L-arginine taste receptor sites in the catfish, *Ictalurus punctatus*: biochemical and neurophysiological characterization, *Brain Res.*, 488, 163, 1989.

82. Kalinoski, D. L., Hohnson, L. C., Bryant, B. P., and Brand, J. G., Selective interactions of lectins with amino acid taste receptor sites in the channel catfish, *Chem. Senses*, 17, 381, 1992.

83. Huque, T. and Bruch, R. C., Odorant- and guanine nucleotide-stimulated phosphoinositide turnover in olfactory cilia, *Biochem. Biophys. Res. Commun.*, 137, 36, 1986.
84. Bruch, R. C. and Kalinoski, D. L., Interaction of GTP-binding regulatory proteins with chemosensory receptors, *J. Biol. Chem.*, 262, 2401, 1987.
85. Bruch, R. C. and Teeter, J. H., Second-messenger signaling mechanisms in olfaction, in *Receptor Events and Transduction in Taste and Olfaction*, Brand, J. G., Teeter, J. H., Cagan, R., and Kare, M. R., Eds., Marcel Dekker, New York, 1988, 283.
86. Lo, Y. H., Bellis, S. L., Cheng, L. J., Pang, J., Bradley, T. M., and Rhoads, D. E., Signal transduction for taurocholic acid in the olfactory system of Atlantic salmon, *Chem. Senses*, 19, 371, 1994.
87. Speca, D. J., Lin, D. M., Sorensen, P. W., Isacoff, E. Y., Ngai, J., and Dittman, A. H., Functional identification of a goldfish odorant receptor, *Neuron*, 23, 487, 1999.
88. Miller, S. E. and Hadfield, M. G., Developmental arrest during larval life and life-span extension in a marine mollusc, *Science*, 248, 356, 1990.
89. Crisp, D. J., Factors influencing the settlement of marine invertebrate larvae, in *Chemoreception in Marine Organisms*, Grant, P. T. and Mackie, A. M., Eds., Academic Press, New York, 1974, 177.
90. Chia, F. S., Perspectives: settlement and metamorphosis of marine invertebrate larvae, in *Settlement and Metamorphosis of Marine Invertebrate Larvae*, Chia, F. S. and Rice, M. E., Eds., Elsevier, New York, 1978, 283.
91. Hadfield, M. G., Metamorphosis in marine molluscan larvae: an analysis of stimulus and response, in *Settlement and Metamorphosis of Marine Invertebrate Larvae*, Chia, F. S. and Rice, M. E., Eds., Elsevier, New York, 1978, 165.
92. Burke, R. D., Pheromonal control of metamorphosis in the Pacific sand dollar, *Dendraster excentricus*, *Science*, 225, 442, 1984.
93. Morse, A. N. C. and Morse, D. E., Recruitment and metamorphosis of *Haliotis* larvae induced by molecules uniquely available at the surfaces of crustose red algae, *J. Exp. Mar. Biol. Ecol.*, 75, 191, 1984.
94. Pawlik, J. R., Chemical ecology of the settlement of benthic marine invertebrates, *Oceanogr. Mar. Biol. Annu. Rev.*, 30, 273, 1992.
95. Jensen, R. A. and Morse, D. E., Chemically induced metamorphosis of polychaete larvae in both the laboratory and ocean environment, *J. Chem. Ecol.*, 16, 911, 1990.
96. Hadfield, M. G., Chemical interactions in larval settling of a marine gastropod, in *Marine Natural Products Chemistry*, Faulkner, D. J. and Fenical, W. H., Eds., Plenum Press, New York, 1977, 403.
97. Hadfield, M. G., Settlement requirements of molluscan larvae: new data on chemical and genetic roles, *Aquaculture*, 39, 283, 1984.
98. Morse, D. E., Tegner, M., Duncan, H., Hooker, N., Trevelyan, G., and Cameron, A., Induction of settling and metamorphosis of planktonic molluscan (*Haliotis*) larvae. III: Signaling by metabolites of intact algae is dependent on contact, in *Chem. Signals*, Müller-Schwarze, D. and Silverstein, R. M., Eds., Plenum Press, New York, 1980, 67.
99. Morse, A. N. C., Froyd, C. A., and Morse, D. E., Molecules from cyanobacteria and red algae that induce larval settlement and metamorphosis in the mollusc *Haliotis rufescens*, *Mar. Biol.*, 81, 293, 1984.
100. Boettcher, A. A. and Targett, N. M., Role of chemical inducers in larval metamorphosis of queen conch, *Strombus gigas* Linnaeus: relationship to other marine invertebrate systems, *Biol. Bull.*, 194, 132, 1998.
101. Williamson, J. E., De Nys, R., Kumar, N., Carson, D. G., and Steinberg, P. D., Induction of meta-morphosis in the sea urchin *Holopneustes purpurascens* by a metabolite complex from the algal host *Delisea pulchra*, *Biol. Bull.*, 198, 332, 2000.
102. Morse, D. E. and Morse, A. N. C., Enzymatic characterization of the morphogen recognized by *Agaricia humilis* (scleractinian coral) larvae, *Biol. Bull.*, 104, 1991.
103. Zimmer-Faust, R. K. and Tamburri, M. N., Chemical identity and ecological implications of a waterborne, larval settlement cue, *Limnol. Oceanogr.*, 39, 1075, 1994.
104. Tamburri, M. N., Zimmer-Faust, R. K., and Tamplin, M. L., Natural sources and properties of chemical inducers mediating settlement of oyster larvae: a re-examination, *Biol. Bull.*, 183, 327, 1992.

105. Browne, K. A., Tamburri, M. N., and Zimmer-Faust, R. K., Modelling quantitative structure-activity relationships between animal behaviour and environmental signal molecules, *J. Exp. Biol.*, 201, 245, 1998.

106. Barlow, L. A., Electrophysiological and behavioral responses of larvae of the red abalone (*Haliotis rufescens*) to settlement-inducing substances, *Bull. Mar. Sci.*, 46, 537, 1990.

107. Arkett, S. A., Chia, F. S., Goldberg, J. I., and Koss, R., Identified settlement receptor cells in a nudibranch veliger respond to specific cue, *Biol. Bull.*, 176, 155, 1989.

108. Leise, E. M. and Hadfield, M. G., An inducer of molluscan metamorphosis transforms activity patterns in a larval nervous system, *Biol. Bull.*, 199, 241, 2000.

109. Baloun, A. J. and Morse, D. E., Ionic control of settlement and metamorphosis in larval *Haliotis rufescens* (Gastropoda), *Biol. Bull.*, 167, 124, 1984.

110. Yool, A. J., Grau, S. M., Hadfield, M. G., Jensen, R. A., Markell, C. A., and Morse, D. E., Excess potassium induces larval metamorphosis in four marine invertebrate species, *Biol. Bull.*, 170, 255, 1986.

111. Morse, D. E., Hooker, N., Duncan, H., and Jensen, L., γ-Aminobutyric acid, a neurotransmitter, induces planktonic abalone larvae to settle and begin metamorphosis, *Science*, 204, 407, 1979.

112. Morse, D. E., Hooker, N., and Duncan, H., GABA induces metamorphosis in *Haliotis*, V: stereochemical specificity, *Brain Res. Bull.*, Suppl. 2, 5, 381, 1980.

113. Trapido-Rosenthal, H. G. and Morse, D. E., Availability of chemoreceptors is down-regulated by habituation of larvae to a morphogenetic signal, *Proc. Natl. Acad. Sci. USA*, 83, 7658, 1986.

114. Trapido-Rosenthal, H. G., Initial characterization of receptors for molecules that induce the settlement and metamorphosis of *Haliotis rufescens* larvae. Ph.D. thesis, University of California, Santa Barbara, CA, 1985.

115. Bonar, D. B., Ultrastructure of a cephalic sensory organ in larvae of the gastropod *Phestilla sibogae* (Aeolidacea, Nudibranchia), *Tissue Cell*, 10, 153, 1978.

116. Kempf, S. C., Page, L. R., and Pires, A., Development of serotonin-like immunoreactivity in the embryos and larvae of nudibranch mollusks with emphasis on the structure and possible function of the apical sensory organ, *J. Comp. Neurol.*, 386, 507, 1997.

117. Marois, R. and Carew, T. J., Fine structure of the apical ganglion and its serotonergic cells in the larva of *Aplysia californica*, *Biol. Bull.*, 192, 388, 1997.

118. Baxter, G. and Morse, D. E., G protein and diacylglycerol regulate metamorphosis of planktonic molluscan larvae, *Proc. Natl. Acad. Sci. USA*, 84, 1867, 1987.

119. Baxter, G. and Morse, D. E., Cilia from abalone larvae contain a receptor-dependent G protein transduction system similar to that in mammals, *Biol. Bull.*, 183, 147, 1992.

120. Trapido-Rosenthal, H. G. and Morse, D. E., L-α,ω-diamino acids facilitate GABA induction of larval metamorphosis in a gastropod mollusc (*Haliotis rufescens*), *J. Comp. Physiol. B*, 155, 403, 1985.

121. Hadfield, M. G., Meleshkevitch, E. A., and Boudko, D. Y., The apical sensory organ of a gastropod veliger is a receptor for settlement cues, *Biol. Bull.*, 198, 67, 2000.

122. Wodicka, L. M. and Morse, D. E., cDNA sequences reveal mRNAs for two G_a signal transducing proteins from larval cilia, *Biol. Bull.*, 180, 1991.

123. Sleeper, H. L., Paul, B. J., and Fenical, W., Alarm pheromones from the marine opisthobranch *Navanax inermis*, *J. Chem. Ecol.*, 6, 57, 1980.

124. Howe, N. R. and Sheikh, Y. M., Anthopleurine: a sea anemone alarm pheromone, *Science*, 189, 386, 1975.

125. Petranka, J. W., Kats, L. B., and Sih, A., Predator–prey interactions among fish and larval amphibians: use of chemical cues to detect predatory fish, *Anim. Behav.*, 35, 420, 1987.

126. Breithaupt, T., Lindstrom D. P., and Atema, J., Urine release in freely moving catheterized lobsters (*Homarus americanus*) with reference to feeding and social activities, *J. Exp. Biol.*, 202, 837, 1999.

127. Zulandt Schneider, R. A. and Moore, P. A., Urine as a source of conspecific disturbance signals in the crayfish *Procambrus clarkii*, *J. Exp. Biol.*, 203, 765, 2000.

128. Smith, R. J. F., Alarm substances in fish, *Rev. Fish. Biol.*, 2, 33, 1992.

129. Mathias, A. and Smith, R. J. F., Chemical labeling of northern pike (*Esox lucias*) by the alarm pheromone of fathead minnows (*Pimpromelas sp.*), *J. Chem. Ecol.*, 19, 1967, 1993.

130. Brown, G. E. and Smith, R. J. F., Conspecific skin extracts elicit antipredator responses in juvenile rainbow trout (*Onchorrhynchys mikiss*), *Can. J. Zool.*, 75, 1916, 1997.

131. Zimmer-Faust, R. F., Tyre, J. E., and Case, J. F., Chemical attraction causing aggregation in the spiny lobster, *Panulirus interruptus* (Randall), and its probable ecological significance, *Biol. Bull.*, 169, 106, 1985.

132. Hunt, J. H., Lyons, W. G., and Kennedy, F. S., Effects of exposure and confinement on spiny lobsters, *Panulirus argus*, used as attractants in the Florida trap fishery, *Fish. Bull.*, 84, 69, 1985.

133. Karanavich, C. and Atema, J., Individual recognition and memory in lobster dominance, *Anim. Behav.*, 56, 1553, 1998.

134. Zulandt Schneider, R. A., Schneider, R. W. S., and Moore, P. A., Recognition of dominance status by chemoreception in the red swamp crayfish, *Procambrus clarkii*, *J. Chem. Ecol.*, 25, 781, 1999.

135. Painter, S. D., Clough, B., Gardern, R. W., Sweedler, J. V., and Nagle, G. T., Characterization of *Aplysia* attractin, the first water-borned peptide pheromone in invertebrates, *Biol. Bull.*, 194, 120, 1998.

136. Nevitt, G. A., Dittman, A., Quinn, T., and Moody, W., Evidence for a peripheral olfactory memory in imprinted salmon, *Proc. Nat. Acad. Sci. USA*, 91, 4288, 1994.

137. Dittman, A., Quinn, T., and Nevitt, G., Timing of imprinting to natural and artificial odors by coho salmon (*Oncorhynchus kisutch*), *Can. J. Fish. Aquat. Sci.*, 53, 434, 1996.

138. Sorensen, P. W., Hara, T. J., Stacey, N. E., and Goetz, F. W., F prostaglandins function as potent olfactory stimulants that comprise the postovulatory female sex pheromone in goldfish, *Biol. Reprod.*, 39, 1039, 1988.

139. Sorensen, P. W., Hara, T. J., Stacey, N. E., and Dulka, J. G., Extreme olfactory specificity of male goldfish to the preovulatory pheromone 17α,20β-dihydroxy-4-pregnen-3-one, *J. Comp. Physiol. A*, 166, 373, 1990.

140. Bjerselius, R. and Olsén, K. H., A study of the olfactory sensitivity of crucian carp (*Carassius carassius*) and goldfish (*Carassius auratus*) to 17α,20β-dihydroxy-4-pregnen-3-one and prostaglandin $F_{2\alpha}$, *Chem. Senses*, 18, 427, 1993.

141. Sorensen, P. W., Hormones, pheromones, and chemoreception, in *Fish Chemoreception*, Hara, T. J., Ed., Chapman and Hall, London, UK, 1992, 199.

142. Sorensen, P. W. and Goetz, F. W., Pheromonal function of prostaglandin metabolites in teleost fish, *J. Lipid Mediators*, 6, 386, 1993.

143. Sorensen, P. W. and Scott, A. P., The evolution of hormonal sex pheromones in teleost fish. Poor correlation between the pattern of steroid release by goldfish and olfactory sensitivity suggests that these cues evolved as a result of chemical spying rather than signal specialization, *Acta Scand. Physiol.*, 152, 191, 1994.

144. Sorensen, P. W., Hara, T. J., and Stacey, N. E., Sex pheromones selectively stimulate the medial olfactory tracts of male goldfish, *Brain Res.*, 558, 343, 1991.

145. Sorensen, P. W., Scott, A. P., Stacey, N. E., and Bowdin, L., Sulfated 17α,20β-dihydroxy-4-pregnen-3-one functions as a potent and specific olfactory stimulant with pheromonal actions in the goldfish, *Gen. Comp. Endocrinol.*, 100, 128, 1995.

146. Rosenblum, P. M., Sorensen, P. W., Stacey, N.E., and Peter, R. E., Binding of the steroidal pheromone 17α, 20β-dihydroxy-4-pregnen-3-one to goldfish (*Carassius auratus*) olfactory epithelium membrane preparations, *Chem. Senses*, 16, 143, 1991.

147. Cao, Y., Oh, B. C., and Stryer, L., Cloning and localization of two multigene receptor families in goldfish olfactory epithelium, *Proc. Nat. Acad. Sci. USA*, 95, 11987, 1998.

15 Chemical Defenses of Marine Organisms Against Solar Radiation Exposure: UV-Absorbing Mycosporine-Like Amino Acids and Scytonemin

Deneb Karentz

CONTENTS

I. INTRODUCTION

Ultraviolet radiation (UV, 100 to 400 nm) comprises the shortest wavelengths of the solar spectrum that reach the Earth's surface (~290 to 400 nm). Although UV spans a very small range within the solar spectral band, it can elicit a wide variety of biological responses and it is impossible for most organisms to avoid UV exposure. UVB wavelengths (280 to 320 nm) are injurious to cells (e.g., cause direct molecular damage to DNA and proteins), while longer UVA wavelengths (320 to 400 nm) can be both harmful (e.g., initiate photo-oxidative damage) and beneficial (e.g., required for vitamin D synthesis and DNA repair).[1–3]

It is generally accepted that incident UV wavelengths and intensities were much more hazardous under ancient Earth atmospheres than they are today.[4,5] As a result, nearly all organisms have some capability for protection against UV exposure and for repair of UV-induced damage; many of these biological defenses are common across very diverse taxa.[2,6,7] Secondary metabolites that absorb UV radiation and provide protection from UV damage occur in most phylogenetic groups.[8] For example, melanins in bacteria, fungi, and animals and flavanoids in plants significantly reduce the potential damage caused by direct exposure to UV.[9–11] In aquatic organisms, mycosporine-like amino acids (MAAs) and scytonemin are assumed to serve a complementary or analogous sunscreen function.[6,12–15]

A. ULTRAVIOLET RADIATION AND THE SOLAR SPECTRUM

Within the electromagnetic spectrum, UV is the wavelength band between X-rays (1 to 100 nm) and visible light (400 to 700 nm). UV wavelengths are further subdivided into four categories based on physical properties and the biological consequences of exposure.[3]

- Vacuum UV (100 to 200 nm): Air and water absorb this portion of the UV spectrum; therefore, these wavelengths do not penetrate past the upper reaches of the Earth's atmosphere.
- UVC (far UV, 200 to 280 nm): UVC wavelengths are the most damaging to organisms because they are most efficiently absorbed by nucleic acids and proteins. Fortunately, as sunlight is attenuated through the atmosphere, the entire UVC component is absorbed and these wavelengths do not penetrate past the stratosphere. Although UVC was probably a biologically important component of incident solar radiation during earlier geologic eras when life on Earth first originated and began to evolve (without benefit of an ozone layer), UVC is not ecologically relevant in present day environments. However, much of what is known about UV photobiology is the result of research investigating the response of cells to UVC exposure from artificial sources (e.g., germicidal lamps emitting 254 nm radiation).
- UVB (middle UV, 280 to 320 nm): UVB is extremely harmful to organisms and the primary cause of erythema (sunburning of human skin). A large proportion of the UVB below 300 nm is absorbed by stratospheric ozone, but the small amount that does reach the Earth's surface is sufficient to cause significant damage and can be lethal. The primary consequence of ozone depletion is an increase in the bandwidth and intensity of shorter wavelengths of incident UVB.

- UVA (near UV, 320 to 400 nm): These longest wavelengths of UV can be both harmful and beneficial. UVA is implicated in many photosensitive reactions and can compound the damage caused by UVB. Despite the negative impact of UVA on cells, these wavelengths trigger an array of fundamental responses in organisms through the action of cryptochromes.[16] UVA wavelengths are not significantly affected by stratospheric ozone.

B. ULTRAVIOLET RADIATION IN MARINE ENVIRONMENTS

After atmospheric attenuation of UV, marine organisms have an additional environmental UV filter of the overlying water column. Although intertidal species have the greatest risk of exposure and experience the highest doses of UV, planktonic and subtidal benthic organisms are also subject to harmful levels of UV in surface waters and at shallow depths. Pure water is transparent to UV wavelengths, but dissolved substances and particulate matter present in natural waters cause significant absorption, reflection, and diffusion of UV within the water column. This results in variable absorption of wavelengths from the incident UV spectrum and wide variation in attenuation coefficients between different water masses (Figure 15.1).[17–20] Even in the clearest waters, UVB is usually attenuated within the upper 10 m of the water column, although UVB wavelengths have been detected up to 60 m in the very transparent waters of the Southern Ocean during springtime ozone depletion.[18,20–23] UVA wavelengths generally penetrate to depths of approximately 50 m or more.

Characterization of the underwater UV regime is not extensive, although documentation of UV penetration to ecologically significant depths has been known since at least the 1950s.[24] The importance of daily and seasonal variations in UV or the role of vertical mixing in the temporal variability of exposure have not yet been comprehensively studied or evaluated but are the focus of current research efforts.

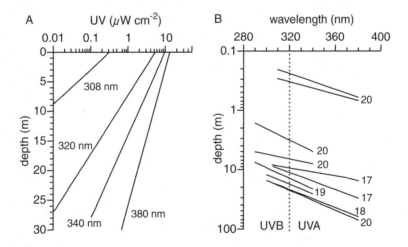

FIGURE 15.1 A. In-water attenuation of four indicated UV wavelengths in coastal Antarctic waters on 16 Oct. 1996 measured with a PUV-500 spectroradiometer (Biospherical Instruments, San Diego, CA). B. Spectral depth range of attenuation to 10% of incident UV levels in various water masses. Number labels on lines refer to reference citations. Modified from Booth, C.R. and Morrow, J.E., Impacts of solar UVR on aquatic microorganisms: the penetration of UV into natural waters, *Photochem. Photobiol.*, 65:254–257, 1997, Figure 3. With permission from the Amercian Society for Photobiology, Augusta, GA.

C. Biological Consequences of Ultraviolet Exposure

1. Absorption of UV by Organic Molecules

UV (especially UVB) is absorbed by two major groups of organic molecules — nucleic acids and proteins. Absorption of UV photons causes these molecules to undergo conformational changes that subsequently interfere with or destroy their ability to participate in vital metabolic functions. The formation of UV photoproducts in DNA can significantly compromise the accuracy of transcription and replication; therefore, UV damage to DNA molecules by UV exposure can result in debilitating, mutagenic, and lethal effects.[3] Damage to proteins is also problematic as these molecules function as enzymes, hormones, and structural components of cells. Generally, protein damage is not considered as important as damage to DNA.[25] Protein molecules are present in numerous copies and are readily degraded and resynthesized; however, prevention of UV-induced damage to defray the energetic costs of repair and replacement of molecules is certainly advantageous.

2. Photo-Oxidative Stress

In addition to the direct absorption as a biological hazard, UV can have additional indirect effects on organisms.[26,27] A number of UV photochemical reactions occur in solutions, both within cells and in the external aquatic environment. In the presence of UV, water itself is hydrolyzed, producing hydroxyl ions. Related reactions involving dissolved substances and mediated by UV lead to the formation of peroxides, super oxide, and other radicals. These reactive products are toxic by causing oxidative damage to biological molecules.[28–31]

D. Biological Defenses against Ultraviolet Radiation

In response to the ubiquitous presence of UV, most organisms have developed a variety of defenses to tolerate exposure.[2] These include adaptations to minimize UV exposure and to repair UV-induced damage when protective measures are not adequate.[2,6] Characteristics that are useful for evading UV damage include avoidance, screening of UV, and antioxidant activity.[6] The focus of this chapter is on specific chemical defenses used by marine organisms to reduce the amount of UV radiation that reaches vital molecular targets, but a very brief overview of other UV defense strategies is presented below.

1. Avoidance

Physically moving away from UV radiation is one of the most effective means of minimizing exposure. This can be accomplished in a number of ways and does not necessarily require moving great distances. However, it does require that an organism have the ability to detect the presence of UV wavelengths (directly or indirectly) and that the organism is capable of movement. UV can alter the behavior (motility and photo-orientation) of unicellular organisms and metazoans.[32–34] Generally, organisms tend to avoid high light intensities, and many of the photoreceptors responsible for light responses detect UVA and visible, not the more harmful, UVB radiation.[16] In terrestrial plants, there are many examples of blue light receptors and numerous genes that are regulated by blue light exposure.[35] In some cases (e.g., chalcone synthesis), the roles of various wavelengths from the UV and visible spectra can be clearly distinguished on both a genetic and biochemical basis.

While receptors for visible light are very common in marine species, less is known about the detection of UV. In aquatic invertebrates, UV photoreception is primarily in the UVA wavelengths, although variable UVB phototropic responses have been observed.[36–38] Visual UV photosensitivity is a specific characteristic in some animal species, including arthropods, reptiles, fish, birds, and mammals.[39-41] It is presumed that the potential for ocular damage caused by UV is offset by the benefits of UV vision (e.g., increased visual acuity for capture of prey).[41,42]

Obligate phototrophs have a dilemma with regard to solar exposure. UVA and/or visible wavelengths (also referred to as photosynthetically available radiation, PAR, 350 to 700 nm[43]) are required for photosynthesis, but exposure to UV can be harmful. Benthic autotrophs (e.g., algal macrophytes and seagrasses) cannot relocate once spores or seeds have germinated. Phytoplanktonic species have more options for avoiding UV through passive transport by vertical mixing or active vertical migration behavior.

Some biological processes are on a diurnal cycle and may be entrained to a circadian rhythm. When events are scheduled at night, this provides an opportunity for UV avoidance. For example, dinoflagellates tend to undergo mitosis and cytokinesis during the dark, thereby avoiding UV exposure during a vulnerable period of the cell cycle. A common temporal avoidance strategy in invertebrates is spawning after sunset.[44] In the ascidian *Corella inflata*, not only are embryos shielded within the adult body cavity during development, but release and settlement of competent larvae are nocturnal events.[45]

Usually, intensities of UV and visible light co-vary so that an avoidance response to one includes avoiding the other. However, under ozone depletion, incident UVB intensities increase while UVA/visible light fields remain unchanged. Therefore, if organisms are relying on visible or UVA wavelengths to provide accurate proportional cues for changes in UVB, ozone depletion can exert an undue and unexpected selective pressure on populations and communities.

2. Sunscreening

The outer covering of organisms can provide substantial protection against UV exposure. At the cellular level, cell walls and membranes offer protection by blocking or attenuating incident UV before it reaches organelles and other intracellular components. For example, it is estimated that the outer silicate wall of diatoms can absorb up to 30% of incident UVB radiation, affording a significant primary UV defense for the cell.[46] For metazoans, cuticles, carapaces, shells, scales, feathers, and fur all provide an effective optical barrier between incident UV and internal tissues. These external surfaces can absorb more than 95% of incident UVB, providing an effective radiation shield for internal cells and tissues.[47,48] The environment can also provide passive shading from UV. In benthic cyanobacterial communities, the extracellular presence of ferric chloride in sediments functions as an adequate UV filter for cells.[49]

Across diverse taxonomic groups of marine organisms there are several classes of compounds that absorb UV and act as putative sunscreens. These include scytonemin (Figure 15.9), an extracellular cyanobacterial sheath pigment, and the mycosporine-like amino acids (MAAs, Figures 15.3–15.6) that are usually located intracellularly in cyanobacteria, algae, invertebrates, and fish. These compounds are the major focus of this chapter and will be discussed in detail below.

Many marine species also possess the tyrosinase-mediated pathway to synthesize the UV-absorbing pigment melanin. Melanin occurs in a wide range of taxa including bacteria, fungi, invertebrates, and chordates. While much is known about the role of melanin in the UV protection of mammalian skin, very little research has been conducted to examine the efficiency of melanin as a UV-protective mechanism in aquatic taxa.[9] It is known that melanin levels in juvenile hammerhead sharks, *Sphyrna lewini*, are directly correlated to solar UV exposure; in the freshwater crustacean *Daphnia pulex*, melanin concentrations are genetically determined within populations and are correlated to UV sensitivity.[50,51] The few studies that have been undertaken suggest that melanin has an important role in UV protection in aquatic environments.

Several other UV-absorbing compounds have been implicated in the protection of aquatic organisms from UV exposure. These compounds do not seem to be as common as MAAs (Figures 15.3–15.6) or scytonemin (Figure 15.9), possibly because less effort has been made to study them. One such molecule is biopterin glucoside (BG), a UVA-absorbing compound isolated from cyanobacteria and related to pteridine pigments.[52] The synthesis of BG in *Oscillatoria* sp. is induced by exposure to UVA wavelengths, and the presence of BG has been shown to confer increased UVA resistance to cells.[53,54] However, in intertidal cyanobacterial mat communities, pterin concentrations remain unchanged across seasons, suggesting a minor role in UV protection.[55] Further studies suggest that pterins may have a regulatory role in UV protection. In *Chlorogloeopsis* sp., pterins appear to function as UVB receptors and may be involved with signaling the induction of MAA synthesis.[56] Further research on these compounds is required to better understand their contribution to cell survival under solar stress.

Other compounds such as phlorotannins, sporopollenin, coumarins, tridentatols, polyphenolics, and several as yet unidentified substances (e.g., P380) have also been implicated as UV protectants that can increase UV tolerance.[57–63] With the rapidly accelerating rate of research in the area of aquatic UV photobiology, it is highly likely that additional new types of UV-screening compounds will continue to be discovered. Many of these secondary metabolites probably have multiple protective functions. For example, tridentatols serve as allelopathic agents, antioxidants, and sunscreens.[57,58,64]

3. Antioxidants

Cells have substantial chemical defenses against the UV photoproducts produced in seawater and intracellular fluids. Many organisms have antioxidants (e.g., carotenoids, ascorbate, tocopherols, anthocyanins, and tridentatols) that quench photo-oxidative reactions.[64–67] Cells also have enzymes (e.g., catalase and superoxide dismutase) that can counteract the oxidative nature of peroxides and other radicals.[26] Some compounds, such as the UV-absorbing pigment melanin, can act as both optical filter and antioxidant.[68] The MAA mycosporine-glycine (Figure 15.3) functions in a similar dual capacity.[69] The role of UV-mediated reactions in seawater relative to biological effects is an important current area of study.

II. UV-ABSORBING COMPOUNDS IN MARINE ORGANISMS

A. MYCOSPORINE-LIKE AMINO ACIDS (MAAs)

In the late 1960s, studies of water extracts from marine cyanobacteria and cnidarians detected unknown UV-absorbing compounds that were initially named for their maximum wavelengths of absorbance, e.g., substance-320 (S-320) exhibited maximum absorbance at 320 nm.[70,71] Relatively high concentrations of S-320 and related compounds were observed in a variety of marine organisms. It was speculated that these compounds either have a solar protective function or are precursors to common pigments. During the subsequent decade, the widespread and taxonomically diverse distribution of S-320 and similar UV-absorbing compounds in marine organisms was confirmed.[72–78] Additional investigations strengthened the notion that these compounds serve as sunscreens.[73,79–81] With the subsequent elucidation of molecular structures (Figures 15.3–15.6), the S compounds were identified as mycosporine-like amino acids (MAAs).[80,82–87]

1. MAA Structure

MAAs found in aquatic organisms are closely related to fungal mycosporines that were first isolated from sporulating mycelia.[88–91] (See Bandaranayake[12] for a detailed comparison of MAA and mycosporine structure.) MAAs are colorless water-soluble compounds with absorption maxima (309 to 360 nm) within the UVB and UVA (Table 15.1).[92] They are derivatives of aminocyclohexenone or aminocyclohexenimine rings (Figure 15.2).[93] Nineteen known MAA compounds result from *N*-substitutions of different amino acid moieties to the cyclohexenone or cyclohexenimine chromophore. There are only two aminocyclohexenone-derived MAAs from marine organisms:

mycosporine-glycine and mycosporine-taurine (Figure 15.3).[12] These are most similar in structure to the fungal mycosporines and the only two known MAAs with maximum absorbance in the UVB. The remaining MAAs are iminomycosporines and absorb maximally at UVA wavelengths. The majority of iminomycosporines contain glycine (12 MAAs, including shinorine, porphyra-334, palythine, asterina-330, and palythinol), and the other five iminomycosporines have serine or threonine substitutions (Figures 15.4 and 15.5).[94,95] Two MAAs isolated from reef-building corals are sulfate esters (palythine:threonine-sulfate and palythine:serine-sulfate).[96] It is most likely that more types of MAAs and related compounds occur in marine organisms. With increasing interest in these molecules and further investigation, identification of additional forms is expected.

Related UV-absorbing compounds found in marine organisms are gadusol (1,4,5-trihydroxy-5-hydroxymethyl-2-methoxycyclohex-1-en-3-one) and deoxy-gadusol (Figure 15.6).[97–99] Gadusol is a colorless oil first observed in the eggs of fish and sea urchins.[97–99] It is a derivative of cyclohexane and is structurally similar to both the fungal mycosporines and MAAs. An isomer of gadusol (spinulosin quinol-hydrate) is synthesized in fungi and arises from an acetate precursor.[97] Other fungal mycosporines are synthesized via products of the shikimate pathway, and it is speculated that the shikimate pathway is also a plausible synthetic route for gadusol.[97]

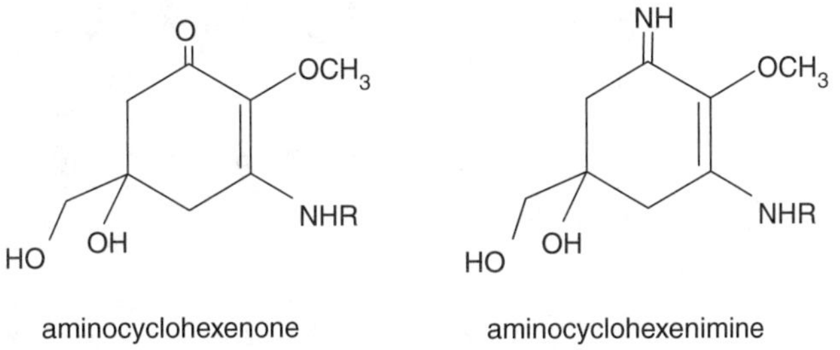

FIGURE 15.2 Aminocyclohexenone and aminocyclohexenimine ring structures.

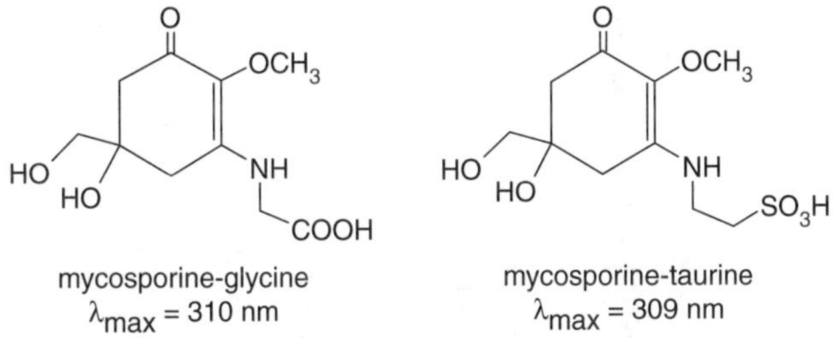

FIGURE 15.3 Molecular structures and wavelengths of maximum absorbance (λ_{max}) for two aminomycosporines from marine organisms.

2. MAA Synthesis

Metabolic pathways for MAA synthesis, conversion, and degradation have not yet been elucidated. Based on structural affinity with gadusol and fungal mycosporines, MAAs are most likely synthesized through the shikimate pathway.[74,97] This is confirmed for the production of MAAs by the zooxanthellae of the coral *Stylophora pistillata* and is probably true for other organisms as well.[100] The shikimate pathway is present in a variety of taxa including bacteria, fungi, algae, and plants. The products of this pathway are involved in the synthesis of the aromatic amino acids tyrosine, phenylalanine, and tryptophan. There is no complementary or analogous pathway for the synthesis of aromatic compounds in animals.[101] Organisms that do not have the shikimate pathway have an obligate requirement for ingestion of the aromatic amino acids phenylalanine and tryptophan

TABLE 15.1
Chemical and Optical Characteristics of Deoxy-Gadusol, Gadusol, and 19 MAAs that Have Been Identified in Marine Organisms

Compound	Abbreviation	Formula	mw	λ_{max}	ε	Species
Deoxy-gadusol	DG	$C_8H_{12}O_5$	188	268[a] 294[b]		Auxis thazard[76]
Gadusol	GD	$C_8H_{12}O_6$	204	269[a] 296[b]	12400[97] 21800[76]	Gadus morhua[97] Auxis thazard[76]
Palythine	PI	$C_{10}H_{16}N_2O_5$	244	320	36200[204] 29400 @ 310 nm[153]	Palythoa tuberculosa[204] Chondrus yendoi[84]
Mycosporine-glycine	MG	$C_{10}H_{15}NO_6$	245	310	28100 @ 310 nm[153] 22400 @ 320 nm[153]	Palythoa tuberculosa[83]
Palythine-serine	PS	$C_{11}H_{18}N_2O_6$	274	320	10500[95]	Pocillopora eydouxi[95]
Usujirene	US	$C_{13}H_{20}N_2O_5$	284	357		Palmaria palmata[205]
Palythene	PE	$C_{13}H_{20}N_2O5$	284	360	50000[75,85]	Palythoa tuberculosa[75,85]
Mycosporine-methylamine:serine (N-methylpalythine:serine)	MS	$C_{12}H_{20}N_2O_6$	288	325	16600[95]	Pocillopora eydouxi[95]
Asterina-330	AS	$C_{12}H_{20}N_2O_6$	288	330	23030 @ 310 nm[160] 37260 @ 320 nm[160]	Asterina pectinifera[206]
Mycosporine-taurine	MT	$C_{10}H_{17}NSO_7$	295	309	-----[c]	Anthopleura elegantissima[120]
Mycosporine-methylamine:threonine (N-methylmycosporine:threonine)	MM	$C_{13}H_{22}N_2O_6$	302	330	3300[94] 2800 @ 320 nm[94] 1900 @ 313 nm[94]	Pocillopora damicornis[94] Stylophora pistillata[94]
Mycosporine-2-glycine	M2	$C_{12}H_{18}N_2O_7$	302	331	-----[d]	Anthopleura elegantissima[102,120]

Compound	Abbr.	λ	λ	Formula	ε	Species
Palythinol	PL	302	332	$C_{13}H_{22}N_2O_6$	43500[75,85]	*Palythoa tuberculosa*[75,85]
Z-palythenic acid	PA	328	337	$C_{14}H_{20}N_2O_7$	29200[92]	*Halocynthia roretzi*[92]
E-palythenic acid		328	337	$C_{14}H_{20}N_2O_7$	29200[206]	*Halocynthia roretzi*[206]
Shinorine (mytilin A)	SH	332	334	$C_{13}H_{20}N_2O_8$	44668[87]	*Chondrus yendoi*[87]
						Mytilus galloprovincialis[74]
Mycosporine-glycine:valine	MV	344	335	$C_{15}H_{24}N_2O_7$		*Euphausia superba*[107]
Porphyra-334 (mytilin B)	PR	346	334	$C_{14}H_{22}N_2O_8$	42300[86]	*Porphyra tenera*[86]
						Mytilus galloprovincialis[74]
Palythine-serine-sulfate	SS	354	321	$C_{11}H_{18}N_2O_9S$		*Stylophora pistillata*[96]
Palythine-threonine-sulfate	TS	368	321	$C_{12}H_{20}N_2O_9S$		*Stylophora pistillata*[96]
Mycosporine-glutamic acid:glycine	GG	374	330	$C_{15}H_{22}N_2O_9$	43900[122]	*Dysidea herbacea*[122]

Note: Abbreviations used in subsequent tables are shown (Abb.) along with chemical formulae, molecular weights (mw), wavelengths of maximum absorbance (λ_{max}, in nm), extinction coefficients (ε) at λ_{max} or other indicated wavelengths, and species from which original identifications were made. (Compounds are listed in order of molecular weight.)

[a] at pH < 2.
[b] at pH > 7.
[c] value for mycosporine-glycine methyl ester (ε = 28000) can be substituted.[120]
[d] value for shinorine (ε = 44668) can be substituted.[120]

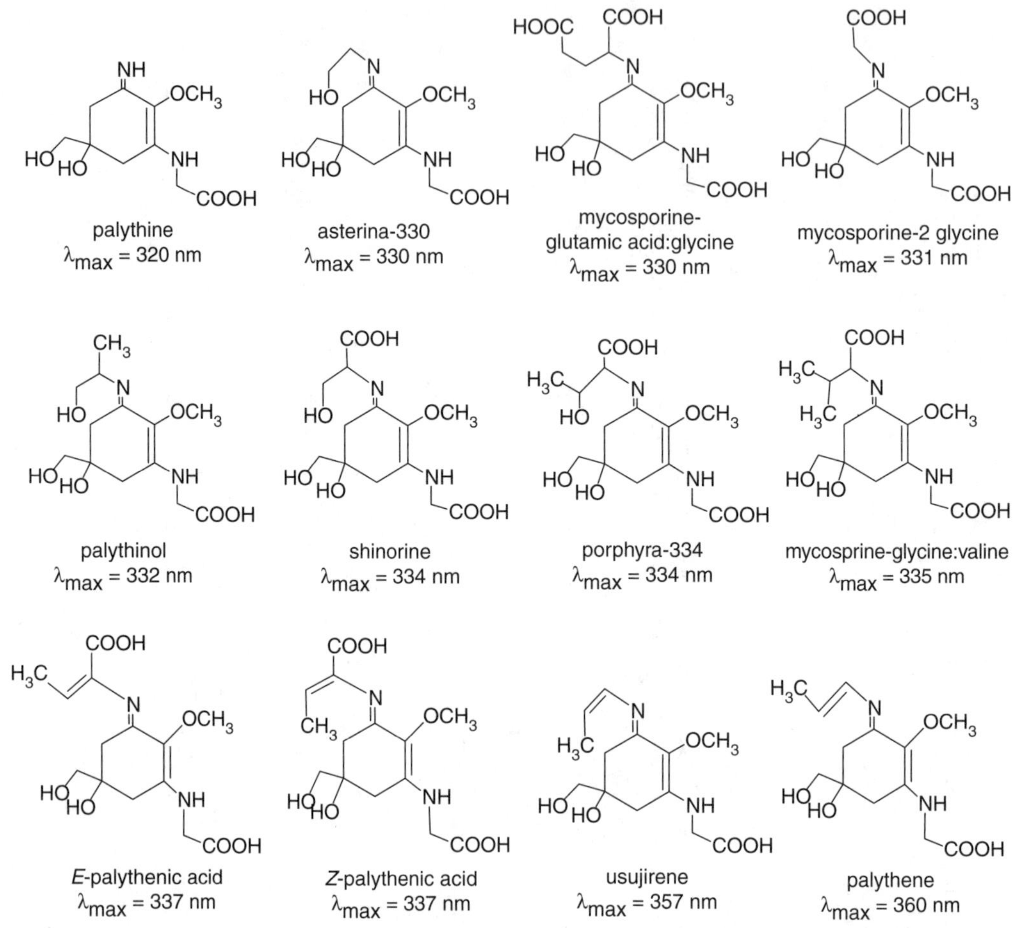

FIGURE 15.4 Molecular structures and wavelengths of maximum absorbance (λ_{max}) for 12 glycine-containing iminomycosporines from marine organisms.

(phenylalanine can be converted into tyrosine by animal metabolism). The dietary requirement for aromatic amino acids probably precludes *de novo* synthesis of gadusol or MAAs in invertebrates and chordates.[101] It has been proposed that ingested MAAs could serve as precursors for conversion to gadusol and can be interconverted into different MAAs by animal metabolism or enteric bacteria.[14,101–103]

If MAAs are synthesized via the shikimate pathway, then in the marine environment they would be produced by bacteria and primary producers (algae) and transferred by ingestion/assimilation to consumer organisms. It has been demonstrated that diet can regulate MAA content in invertebrates and fish.[104–106] The first direct evidence of this was obtained from controlled feeding experiments with the temperate sea urchin *Strongylocentrotus droebachiensis*.[104] Furthermore, the assimilation of MAAs from food sources can be very efficient. In medaka fish, the MAA concentrations in ocular tissues increase by nearly 800% over a 5-month period by providing fish with a MAA-enriched diet.[106]

MAA acquisition by ingestion/assimilation and not *de novo* synthesis in consumer organisms is further supported by observed gradients of MAAs along the digestive tract of sea urchins. In *Strongylocentrotus droebachiensis*, the posterior portion of the gut has over three times the concentration of MAAs as the anterior portion of the digestive tract, indicating sequential absorption of ingested material.[104] Similar observations have been made in holothuroid species.[102]

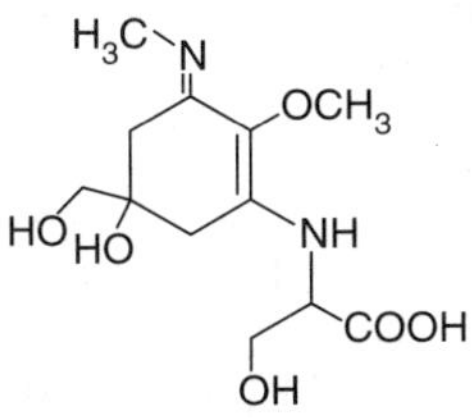

mycosporine-methylamine:serine
λ_{max} = 325 nm

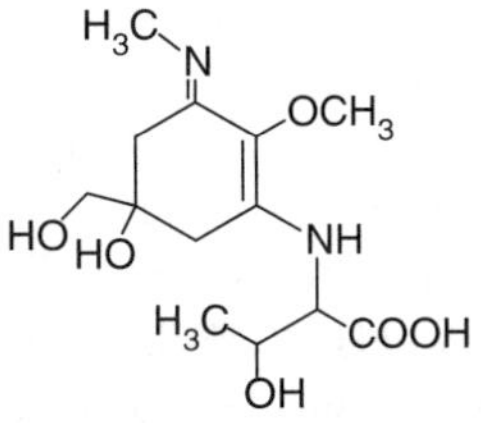

mycosporine-methylamine:threonine
λ_{max} = 330 nm

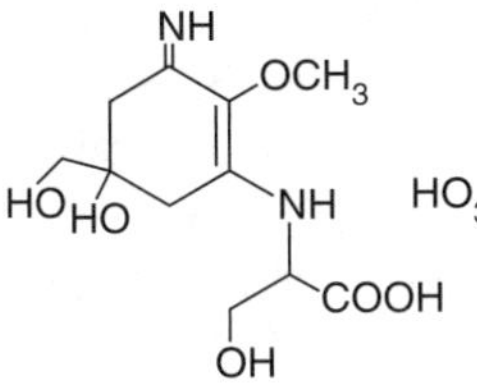

palythine-serine
λ_{max} = 320 nm

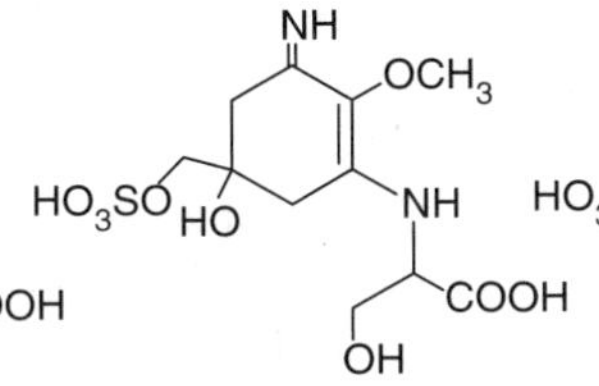

palythine-serine suflate
λ_{max} = 321 nm

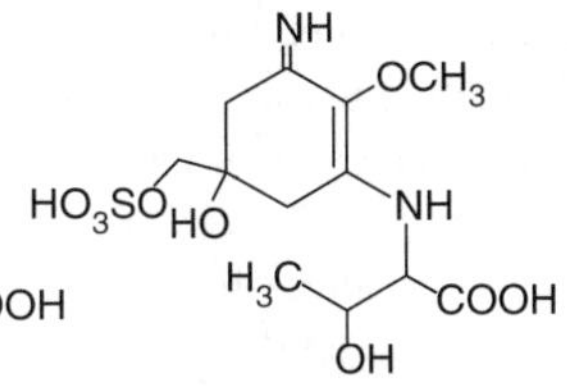

palythine-threonine sulfate
λ_{max} = 321 nm

FIGURE 15.5 Molecular structures and wavelengths of maximum absorbance (λ_{max}) for five nonglycine containing iminomycosporines from marine organisms.

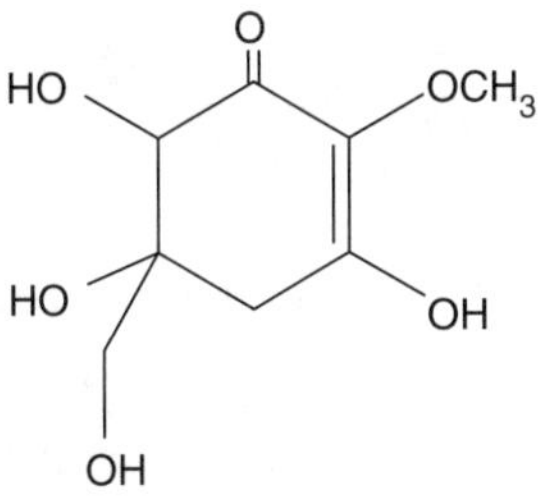

gadusol
λ_{max} = 269 nm (pH<2)
λ_{max} = 296 nm (pH>7)

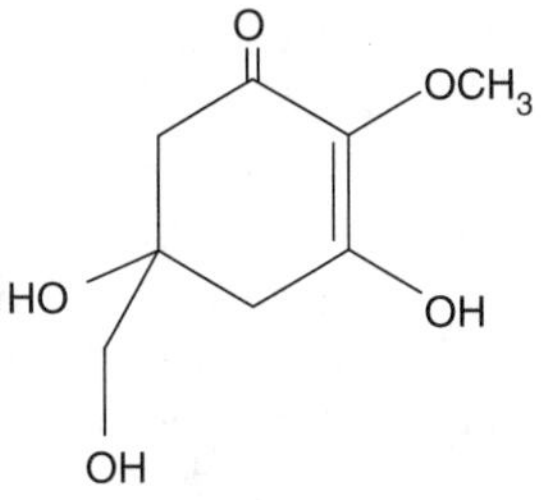

deoxy-gadusol
λ_{max} = 268 nm (pH<2)
λ_{max} = 294 nm (pH>7)

FIGURE 15.6 Molecular structures of gadusol and deoxy-gadusol with wavelengths of maximum absorbance (λ_{max}).

Degradation by invertebrate digestive enzymes cannot be excluded as an alternative explanation for the observed intestinal gradients, although this is unlikely since MAAs are not susceptible to chemical digestion by mammalian digestive enzymes (virtually all of the shinorine eaten by mice appears in their feces).[106]

3. Phylogenetic Patterns of MAA Occurrence

Nineteen MAA compounds have been reported from 382 species of marine organisms collected from tropical, temperate and polar latitudes (Tables 15.1, 15.2, and 15.3). Approximately 75% of the data available on MAA presence/absence and quantification in individual taxa come from only one or several specimens or cultures. While a complete evaluation of intraspecific variation, geographical distribution or phylogenetic trends is not possible, a number of generalizations about MAA occurrence are evident.

The majority of marine cyanobacteria, algae, invertebrates, ascidians, and fish probably contain MAAs, as 87% of marine taxa examined have detectable levels (Table 15.2). Palythine (see Figure 15.4) is the most common MAA found in marine organisms (61% of all species analyzed), followed by shinorine (57%), mycosporine-glycine (52%), porphyra-334 (48%), asterina-330 (35%), palythinol (29%), and palythene (23%) (Figure 15.7). Chordata (68 species examined, no birds or mammals included), Cnidaria (59 species), and Rhodophyta (49 species) have been more extensively studied than other groups. One hundred percent of cnidarians, 96% of chordates, and 82% of rhodophytes analyzed contain MAAs (Table 15.2). MAAs have not been observed in the Chaetognatha, Ctenophora, or Protozoa, but these phyla are, thus far, only represented by one or two specimens from one geographic location (Antarctica).[107] In all other phyla and divisions, 57 to 100% of species examined contain MAAs.

Most species have a complement of at least several MAA compounds (Figure 15.8A). Fifty-eight percent of all species examined contain at least three MAAs; 36% of algal species, 74% of invertebrate taxa, and 62% of chordates have a suite of three or more MAAs. The maximum reported number of MAAs in a single individual is ten in the coral *Stylophora pistillata*.[100] Two species have been reported with eight MAAs in individual specimens: the Antarctic krill, *Euphausia superba*, and the Antarctic amphipod, *Pontogeneia* sp.[107] Since wavelengths of maximum absorbance of MAAs span a 55-nm range across the UVB and UVA, the presence of multiple MAAs is assumed to provide a broader band and more effective optical filter than the presence of one MAA alone (Figure 15.8B).[108,109]

In comparing algal divisions (total of 143 species examined, including cyanobacteria), invertebrate phyla (171 species), and the chordates (68 species), there are distinct differences in the occurrence of MAAs (Table 15.2). Palythine, shinorine, porphyra-334, and mycosporine-glycine are the four most abundant MAAs within each group, but their rankings are not consistent. Shinorine is the most common MAA in algae (found in 59% of species), palythine is the most frequent MAA in invertebrates (occurs in 80% of species), and asterina-330 is the predominant MAA in chordates (present in 81% of species) (Figure 15.7).

There are sufficient data on MAA distribution from three algal divisions (Chlorophyta, green algae; Phaeophyta, brown algae; and Rhodophyta, red algae) to allow for a phylogenetic comparison of MAA content. The Rhodophyta have the highest levels of MAAs, particularly in shallow benthic environments.[107,110,111] Nearly the same proportion of rhodophyte species posses MAAs (82%) as found in the Chlorophyta (81%), but concentrations of MAAs in red algae are an order of magnitude larger (<1300 nmol mg^{-1} protein) than maximum concentrations found in chlorophytes (<240 nmol mg^{-1} protein). Only one examined chlorophyte species, *Halimeda opuntia*, has a high concentration of MAAs (885 nmol MAAs mg^{-1} protein).[112] Many chlorophtes have other types of UV-absorbing compounds (e.g., sporopollenin and coumarins).[60,113,114] In the macrophyte *Dasycladus vermicularis*, colored substances are excreted during UV stress and release is correlated with light intensity.[63] The exudate does not contain known MAAs, and the colored substances are most likely coumarins.

A smaller percentage of the Phaeophyta (59% of taxa) contain MAAs and concentrations in thalli are relatively low (<300 nmol mg^{-1} protein). The Phaeophyta also excrete colored compounds with UV-absorbing properties.[62] Some of these exudates are polyphenolic substances that are usually associated with alleopathy. However, in *Ascophyllum nodosum*, thallus concentrations are regulated by UV exposure and small herbivores are not deterred, rather they feed preferentially on irradiated algae.[59] Since the Chlorophyta and Phaeophyta successfully inhabit intertidal and shallow subtidal areas, they have apparently evolved very efficient protective and repair mechanisms for dealing with UV exposure, but MAAs are probably not the key to their fitness in high light environments.

A general survey of UV-absorbing characteristics in 152 species (206 strains) of microalgae shows that all taxa absorb within the UVA (between 320 and 340 nm, with most maxima near 337 nm), but only 11% absorb in the UVB.[115] While there is a large range of variation among strains, the highest ratios of UV attenuation to chlorophyll *a* concentration (UV:chl) are in bloom-forming flagellate taxa. In this study, MAAs were analyzed in only five species, and a variety of MAA combinations (0–4 of five identified MAAs) occur. From these results, the authors conclude that UV:chl values greater than 1 are indicative of the presence of MAAs in unicellular algae.

There are varied patterns in MAA occurrence within phylogenetically related invertebrates and fish. In a wide geographic study (including temperate and tropical regions), similar complements of UV-absorbing compounds occur in the eye lenses of fish at the family level. Concentration is not a taxonomic feature, but is related to the radiation regime of the habitat.[116] On a smaller geographic scale, little taxonomic diversity is evident in ocular tissues of 52 species (19 families) of fishes and sharks from the Great Barrier Reef.[117] These taxa generally contain the same four MAAs: palythine, asterina-330, palythinol, and palythene. Sea cucumbers (12 species examined) from the Great Barrier Reef have three MAAs in common with local fishes (no palythene), and an additional four MAAs (shinorine, porphyra-334, mycosporine-2 glycine, and mycosporine-glycine) are consistently present in various tissues.[102] Corals (23 species) from the South Pacific (French Polynesia) typically contain palythine, shinorine, porphyra-334, mycosporine-glycine, and palythinol.[118] Four species of the temperate sea anemone *Anthopleura* have been found to have the same complement of MAAs (shinorine, porphyra-334, mycosporine-2 glycine, and mycosporine-taurine), although proportions of each compound vary among taxa.[119–121] Mycosporine-taurine has only been reported from these four species of *Anthopleura* and may be unique to this genus. A more detailed assessment of phylogenetic patterns of MAA occurrence will require more extensive data sets.

4. Geographic Distribution of MAAs

Several surveys have been undertaken to evaluate possible habitat or behavioral patterns in MAA distribution and content, but only a few general trends in MAA occurrence are evident.[107,110,112,117] Analyses of 382 species reveal that 95% of tropical species, 80% of temperate species and 82% of polar species have detectable levels of MAAs (Table 15.3). Palythine, shinorine, porphyra-334, mycosporine-glycine, asterina-330, palythinol, palythene, and mycosporine-2-glycine have been found at all latitudes. Other MAAs are less frequently reported and have more limited distributions; however, this may be a function of insufficient data and not a true representation of MAA occurrence.

Tropical marine organisms are not only more likely to have MAAs, but also to have the highest MAA concentrations.[110] Among 25 of the highest reported MAA values, all but one sample (the temperate red macrophyte, *Cystoclonium purpureum*[112]) are from tropical areas (Table 15.4). While high values for MAA concentrations are in keeping with a photoprotective function given that tropical environments experience the highest radiation exposures, not all tropical species have high MAA content.[103,117,122]

Forty-eight invertebrate species examined from the Antarctic Peninsula most often have various combinations of mycosporine-glycine, shinorine, porphyra-334, palythine and mycosporine-glycine:valine.[107] Thirty-eight species of invertebrates from farther south (McMurdo Sound) have a similar (but less concentrated) suite of MAAs with palythine the most common and, at times, the only MAA present.[123] While many of the species are found at both locations, there are distinct differences between the benthic habitats of the Antarctic Peninsula and McMurdo Sound. In the Sound, benthic organisms are restricted to depths deeper than 20 m because of more intense seasonal ice scour, and this results in lower UV exposure for these organisms. Coastal ice cover also causes lower PAR intensities, resulting in lower macroalgal biomass, so most of the local invertebrates are not herbivores. Since MAAs must be ingested by animals, this would further contribute to lower MAA contents. Antarctic diatoms (26 species examined) tend to have only shinorine and porphyra-334 with less frequent occurrence of mycosporine-glycine.[107,124–127] Antarctic benthic macroalgae (eight species studied) have a larger complement of MAAs (mycosporine-glycine, shinorine, porphyra-334, palythine, asterina-330, palythinol, and palythene), and these occur in much higher concentrations than observed in the local microalgae.[107]

TABLE 15.2
Summary of the Phylogenetic Distribution of MAAs Reported in the Literature (Data Compiled from over 500 Samples of 382 Marine Taxa)

Phylum/Division	#T		PI	SH	PR	MG	AS	PL	PE	MV	M2	US	MM	PA	MT	PS	MS	SS	GG	TS	Σ
Prokaryotes																					
Prochlorophyta[152]	1	#	0	1	0	0	0	0	0	0	0	0	0	0	0	0	0	0	0	0	1
		%	0	100	0	0	0	0	0	0	0	0	0	0	0	0	0	0	0	0	100
Cyanophyta[112,183,207]	4	#	3	4	3	3	2	1	0	0	0	0	0	0	0	0	0	0	0	0	4
		%	75	100	75	75	50	25	0	0	0	0	0	0	0	0	0	0	0	0	100
Total	5	#	3	5	3	3	2	1	0	0	0	0	0	0	0	0	0	0	0	0	5
		%	60	100	60	60	40	20	0	0	0	0	0	0	0	0	0	0	0	0	100
Algae																					
Bacillariophyta[46,115,124-126,171]	28	#	1	19	22	2	0	0	1	1	1	0	0	0	0	0	0	0	0	0	23
		%	4	68	79	7	0	0	4	4	4	0	0	0	0	0	0	0	0	0	82
Chlorophyta[77,105,107,110,112,171,195]	21	#	7	11	11	11	3	6	0	0	0	0	0	0	0	0	0	0	0	0	17
		%	33	52	52	52	14	29	0	0	0	0	0	0	0	0	0	0	0	0	81
Dinophyta[31,115,119,129,138,153,171,172,195]	9	#	3	5	6	5	2	1	3	1	0	1	0	1	0	0	0	0	0	0	8
		%	33	56	67	56	22	11	33	11	0	11	0	11	0	0	0	0	0	0	89
Haptophyta[171]	2	#	0	0	0	2	0	0	0	0	0	0	0	0	0	0	0	0	0	0	2
		%	0	0	0	100	0	0	0	0	0	0	0	0	0	0	0	0	0	0	100
Phaeophyta[77,107,110,112,195]	28	#	3	8	9	10	1	3	0	0	0	0	0	0	0	0	0	0	0	0	16
		%	11	29	32	36	4	11	0	0	0	0	0	0	0	0	0	0	0	0	57
Raphidophyta[115]	2	#	1	1	1	1	2	0	0	0	0	0	0	0	0	0	0	0	0	0	2
		%	50	50	50	50	100	0	0	0	0	0	0	0	0	0	0	0	0	0	100
Rhodophyta[69,77,86,87,104,105,107,110-112,123,141,142,167,195,205]	48	#	24	37	28	19	19	16	9	0	2	1	0	0	1	0	0	0	0	0	39
		%	50	77	58	40	40	33	19	0	4	2	0	0	2	0	0	0	0	0	81
Total	138	#	39	81	77	50	27	26	13	2	3	2	0	1	1	0	0	0	0	0	107
		%	28	59	56	36	20	19	9	1	2	1	0	1	1	0	0	0	0	0	78
Invertebrates																					
Annelida[107]	6	#	6	6	6	6	1	1	3	3	0	0	0	0	0	0	0	0	0	0	6
		%	100	100	100	100	17	17	50	50	0	0	0	0	0	0	0	0	0	0	100

Taxon		#T														Σ
Arthropoda[77,93,107,123]	#	16	15	15	14	13	4	2	11	12	0	0	0	1	0	16
	%		94	94	88	81	25	13	69	75	0	0	0	6	0	100
Bryozoa[107,123]	#	3	2	2	2	1	0	0	1	1	0	0	0	0	0	3
	%		67	67	67	33	0	0	33	33	0	0	0	0	0	100
Chaetognatha[107]	#	1	0	0	0	0	0	0	0	0	0	0	0	0	0	0
	%		0	0	0	0	0	0	0	0	0	0	0	0	0	0
Cnidaria[29,69,77,94,96,100,107,108,112,118-121,123,150,155,158,160,163,176]	#	54	43	41	38	50	16	29	9	2	15	9	5	4	3	54
	%		80	76	70	93	30	54	17	4	28	17	9	7	6	100
Ctenophora[107]	#	2	0	0	0	0	0	0	0	0	0	0	0	0	0	0
	%		0	0	0	0	0	0	0	0	0	0	0	0	0	0
Echinodermata[77,102-104,107,112,123,137,145,146,206,208]	#	38	27	23	22	24	19	16	1	3	17	0	0	0	0	27
	%		71	61	58	63	50	42	3	8	45	0	0	0	0	71
Mollusca[48,74,77,105,107,123,144,153,203]	#	25	19	19	16	16	24	6	3	3	4	1	1	0	0	23
	%		76	76	64	64	64	24	12	12	16	4	4	0	0	92
Nemertinea[107,123]	#	3	3	3	3	3	0	0	0	0	0	0	0	0	0	3
	%		100	100	100	100	0	0	0	0	0	0	0	0	0	100
Platyhelminthes[107]	#	2	2	2	2	2	0	0	0	0	0	0	0	0	0	2
	%		100	100	100	100	0	0	0	0	0	0	0	0	0	100
Porifera[77,107,122,123]	#	20	14	14	12	8	2	3	2	6	15	1	5	1	0	20
	%		70	70	60	40	10	15	10	30	5	0	0	0	0	100
Protozoa[107]	#	1	0	0	0	0	0	0	0	0	0	0	0	0	0	0
	%		0	0	0	0	0	0	0	0	0	0	0	0	0	0
Total	#	171	137	125	115	123	48	54	30	36	33	10	5	3	3	154
	%		80	73	67	72	28	32	18	21	19	6	3	2	2	90

Vertebrates

Taxon		#T														Σ
Chordata[5,69,77,107,116,117,123,149,152]	#	68	53	8	4	7	55	29	43	2	0	0	0	0	0	65
	%		78	12	6	10	81	43	63	3	0	0	0	0	0	96
Total for all taxa	#	382	232	219	199	183	132	110	86	40	36	36	33	12	4	331
	%		61	57	52	48	35	29	23	10	9	21	19	3	1	87

Note: Total number of taxa examined for each group (#T), number (#) and percent (%) of species containing a specific MAA (see Table 15.1 for key to MAA acronyms), and number and percentage of species that contain any MAAs (Σ) are indicated. Where multiple samples of a species are available and show variable presence of specific MAAs, observations were pooled to represent the maximum total number of MAAs that can occur in a given taxon.

TABLE 15.3
Regional Distribution of MAAs

Region	#T		PI	SH	PR	MG	AS	PL	PE	MV	M2	US	MM	PA	MT	PS	MS	SS	GG	TS	Σ
Polar[46,77,107,110-112,123-126,142,144]	138	#	80	89	91	61	20	14	29	39	1	0	0	1	0	0	0	0	0	0	113
		%	58	64	66	44	14	10	21	28	1	0	0	1	0	0	0	0	0	0	82
Temperate[31,45,48,74,77,87,104,105,110,112,115,116,119-121,129,138,141,145,149,167,171,172,195,203,205-207]	97	#	37	59	43	42	22	22	10	1	6	2	0	4	4	0	0	0	0	0	78
		%	38	60	44	43	22	22	10	1	6	1	0	4	4	0	0	0	0	0	80
Tropical[29,69,77,86,93,94,96,100,102,103,105,108,112,116-119,122,143,146,150,152,153,155,158,160,163,164,176,183,208]	147	#	115	71	65	80	90	74	47	0	29	10	5	0	0	2	3	1	1	1	140
		%	78	48	44	54	61	50	32	0	20	7	3	0	0	1	2	1	1	1	95
All taxa	382	#	232	219	199	183	132	110	86	40	36	12	5	5	4	2	3	1	1	1	331
		%	61	57	52	48	35	29	23	10	9	3	3	5	4	2	3	1	0	0	87

Note: Total number of species examined (#T), number (#) and percent (%) and percent of species containing a specific MAA (see Table 15.1 for key to MAA acronyms), and number and percent of species containing any MAAs (Σ) are shown.

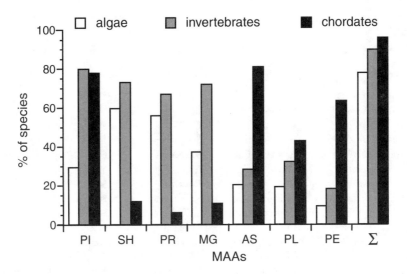

FIGURE 15.7 Phylogenetic distribution of MAAs by percent occurrence in individual species within algal (this category includes photosynthetic prokaryotic taxa indicated in Table 15.2), invertebrate, and chordate groups (see Table 15.2 for reference citations and groups included in these categories; Σ = percent values for all taxa and all MAAs combined).

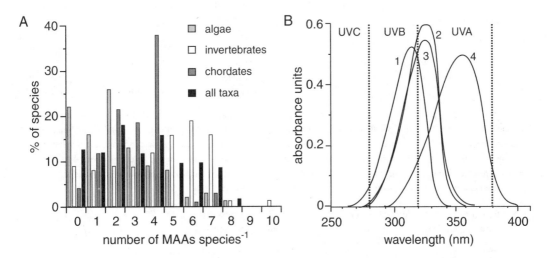

FIGURE 15.8 A. Number of MAAs occurring in individual species within algal, invertebrate, and chordate groups (see Table 15.2 for reference citations and groups included in these categories). B. Broad spectrum UV screening afforded by the presence of multiple MAA compounds (1 = palythine, 2 = asterina-330, 3 = palythinol, 4 = palythene). From Dunlap, W. C., Williams, D. M., Chalker, B., and Banaszak, A., Biochemical photoadaptation in vision: UV-absorbing pigments in fish eye tissues, *Comparative Biochemistry and Physiology B*, Elsevier Science, 93B(3), 1989, 604, Figure 4, reprinted with permission from Elsevier Science, Oxford, UK.

An MAA with a possible restricted geographic occurrence is mycosporine-glycine:valine. This MAA was first described from a survey of Antarctic species and was initially observed only in consumer organisms (invertebrates and fish), not any of the sampled primary producers.[107] Subsequent studies in the Antarctic have identified this MAA in the diatom *Fragilariopsis cylindrus*, the haptophyte *Phaeocystis* sp., and mixed assemblages of Southern Ocean phytoplankton.[124,126,128] There have been a few tentative identifications of mycosporine-glycine:valine in organisms from

TABLE 15.4
Twenty-Five Maxima Reported for MAA Concentrations in Marine Species

	Species	Phylum/Division	Common Name	[MAA]	Notes	Value
1	*Diphora strigosa*[112]	Cnidaria	Coral	8852	Whole	Max
2	*Montastrea cavernosa*[112]	Cnidaria	Coral	3743	Whole	Max
3	*Pocillopora damicornis*[143]	Cnidaria	Coral	2250	Whole[a]	Max
4	*Pearsonothuria graeffe*[102]	Echinodermata	Sea cucumber	2100	Epidermis	Single
5	*Holothuria nobilis*[102]	Echinodermata	Sea cucumber	1600	Respiratory tree	Single
6	*Cassiopeia frondosa*[112]	Cnidaria	Jellyfish	1520	Whole	Single
7	*Hypnea cervicornis*[112]	Rhodophyta	Red alga	1306	Whole	Single
8	*Actinopyga lecanora*[102]	Echinodermata	Sea cucumber	1300	Epidermis	Single
9	*Holothuria nobilis*[102]	Echinodermata	Sea cucumber	1250	Epidermis	Single
10	*Montastrea annularis*[112]	Cnidaria	Coral	1179	Whole	Max
11	*Actinopyga echinites*[102]	Echinodermata	Sea cucumber	1000	Epidermis	Single
12	*Halimeda opuntia*[112]	Chlorophyta	Green alga	885	Whole	Single
13	*Bohadschia argus*[102]	Echinodermata	Sea cucumber	850	Cuverian tubules	Single
14	*Agarica tenufolia*[112]	Cnidaria	Coral	617	Whole	Max
15	*Porites porites*[112]	Cnidaria	Coral	601	Whole	Max
16	*Bohadschia argus*[102]	Echinodermata	Sea cucumber	550	Epidermis	Single
17	*Holothuria edulis*[102]	Echinodermata	Sea cucumber	550	Epidermis	Single
18	*Porites asteroides*[160]	Cnidaria	Coral	525	Whole[a]	Mean
19	*Synapta maculata*[102]	Echinodermata	Sea cucumber	520	Epidermis	Single
20	*Synapta maculata*[102]	Echinodermata	Sea cucumber	440	Tentacles	Single
21	*Cassiopea xamachama*[112]	Cnidaria	Sea anemone	395	Whole	Single
22	*Aipastia pallida*[112]	Cnidaria	Sea anemone	354	Whole	Single
23	*Cystoclonium pupureum*[112]	Rhodophyta	Red alga	333	Whole[b]	Single
24	*Trichodesmium* sp.[112]	Cyanophyta	Cyanobacterium	307	Whole	Single
25	*Clavularia* sp.[29]	Cnidaria	Octocoral	305	Whole[a]	Mean

Note: Species name, phylum or division, common name, and concentration of total MAAs ([MAA], in nmol mg^{-1} protein) are indicated. "Notes" describe whether samples are from a whole organism or specific tissues, and "value" indicates if data are from a single measurement, a maximum value from several samples (max), or a maximum mean value (mean) reported. See text Section II.A.6 for an explanation of how rankings were determined. The comparison of values from different research laboratories may be somewhat problematic as there are currently no commercial standards available for MAAs and calibration of instruments is achieved by a variety of means.

[a] Not from a natural collection, value from experimental radiation exposure.
[b] The only temperate species in this table, all others are from tropical latitudes.

outside the Antarctic; these include the temperate dinoflagellate *Lingulodinium* (*Gonyaulax*) *polyedra* and the tropical corals *Porites compressa* and *Pocillopora* spp.[112,129] Further investigations will be necessary to establish if this MAA may have a more widespread geographic distribution than originally assumed.

5. MAAs in Freshwater Taxa

MAAs have not been studied as extensively in freshwater as in marine species, but these compounds have been reported in freshwater cyanobacteria, microalgae, invertebrates, and chordates (fish).[106,116,130–133] There may be several MAAs unique to freshwater organisms. Aqueous extracts from the cyanobacterium *Nostoc commune* contain a mixture of UV-absorbing compounds with two distinct chromophores that have maximum absorbance at 312 and 335 nm.[134,135] These compounds are comprised of the usual mycosporine cyclohexenone ring structure; however, it is conjugated to both amino acids and oligosaccharides. MAAs not found in marine species also occur in the terrestrial chlorophyte *Prasiola crispa*, but these MAAs have not yet been characterized.[136] Since limited data are available, the actual extent of MAA habitat specificity (if any) relative to geographic region or marine/freshwater environments is not known.

6. Concentration of MAAs in Cells and Tissues

Absolute concentrations of MAAs have been reported in a variety of units, most commonly as µg MAAs g^{-1} dry weight or nmol MAAs mg^{-1} protein. Unfortunately, reliable comparisons of concentrations standardized by dry weight to those expressed on a mg^{-1} protein basis are not possible. Assuming a mean MAA molecular weight of 280 and making a broad estimation of the mean protein content of organisms at 50% of dry weight, more than 350 published concentrations of total MAA content from individual tissues and whole organisms were standardized to µg MAAs g^{-1} dry weight. Based on these calculations more than 50 of the highest values were from measurements originally made in nmol mg^{-1} protein. The samples with the 25 highest concentrations of MAAs are listed in Table 15.4.

The highest reported MAA concentration is 8852 nmol mg^{-1} protein in the tropical coral *Diphora strigosa*.[112] However, this equates to a remarkably high concentration of 2.5 mg MAAs mg^{-1} protein and bears further investigation. Within the 25 highest reported values for MAAs, all but one is from a tropical organism, and 21 of the 25 values are from invertebrates. The remaining four species are primary producers (two Rhodophyta, one Chlorophyta, and one Cyanophyta species). The observation of highest concentrations in consumer organisms further supports the notion that MAAs are bioaccumulated, as this fits the pattern expected for biological magnification through sequential trophic levels. However, the assimilation of MAAs is apparently restricted to invertebrates, lower chordates, and fish (including sharks). Based on a study of mice, mammals do not have the metabolic pathways necessary to assimilate MAAs from their diet.[106] (There have been no studies of birds.)

The concentration of MAAs among individual organisms or strains of the same species can be extremely variable, even if individuals are collected at the same location and depth.[118,137,138] Moreover, MAA concentrations can vary across a considerable range within the tissues of a single individual. In rhodophytes, meristematic apices are not fully pigmented (they are green rather than red) and can have as much as eight times the MAA concentration of mature red-pigmented portions of the thallus.[111,139–142] A similar pattern of MAA distribution occurs in the tropical coral *Pocillopora damicornis*, where branch tips can also have an eight-fold higher concentration of MAAs than branch portions closer to the central part of the colony.[143] (The MAA gradient observed along coral branches is evident for palythine and palythinol, but not obvious for mycosporine-glycine.)

In more complex metazoans, MAA concentrations can vary by orders of magnitude between various tissues. For example, within one individual of the Antarctic limpet *Nacella concinna*, the ovary has 5778 µg MAAs g^{-1} dry weight, the digestive tract has 525 µg MAAs g^{-1} dry weight, and the remaining body tissues (minus shell) have 69 µg MAAs g^{-1} dry weight.[144] A number of invertebrate species show a similar pattern of MAA allocation with maximum amounts of MAAs concentrated in the female gonads.[104,107,137,144–146] In cod, gadusol is primarily concentrated in mature ovaries, with barely detectable levels in liver and muscle tissue and no evidence in testes.[98] Little is known or understood about how the selective partitioning of MAAs into specific cells and tissues is achieved.

Testes and sperm have very low or nondetectable levels of MAAs, even though sperm are expected to be more vulnerable to UV damage because of their small size. The high MAA content in ovaries translates into high MAA content for eggs that is passed on to the developing embryo. In the temperate sea urchin *Strongylocentrotus droebachiensis*, eggs have over twice the MAA concentration as ovary tissue.[145] In tropical corals, the suite of MAAs in eggs is different from adult tissue, and total MAA concentrations in eggs can be nearly seven times higher than in the adult coral.[147] In the Antarctic limpet, *Nacella concinna*, eggs and ovary tissue have the same MAA complement in equally high concentrations.[148]

There is a significant ecological advantage to packaging MAAs into eggs, and consequently into developing embryos, since these early developmental stages are often exposed to higher UV doses than benthic adults, particularly in species with planktonic development. In the tunicate *Ascidia ceratodes*, MAAs are distinctly partitioned between the egg and the surrounding cells that form the follicle and test.[149] Eggs contain predominantly mycosporine-glycine and palythine, while follicle and test cells contain shinorine and palythine. Removal of the external cell layers results in a 60% delay in cell division under experimental UV exposures. Several other studies have demonstrated that successful development under UV exposure is directly correlated to MAA concentrations in embryos and larvae (see also Section II.A.9).[105,145,150,151]

Some adult invertebrates in tropical areas have relatively high MAA concentrations in external surfaces. Holothuroids preferentially accumulate MAAs in the epidermis, giant clams have highest concentrations of MAAs in the outermost layers of siphonal mantle tissue (more than four times the concentrations in subsurface mantle layers), sea hares have high MAA levels in skin, and ascidian tunics have higher MAA concentrations in the surface cells than in basal dermal layers.[102,152–154] In corals, the upper exposed surface of colonies can have five-fold higher UV absorption than the less irradiated vertical faces.[155] These topical distribution patterns reinforce the premise that MAAs have a photoprotective function.

Discrepancies between the types of MAAs found in algae (or other food sources such as sediments for deposit feeders) and animals from the same region have been observed in a number of studies.[104–107,112,123,130,154,156] Even when the MAAs in primary producers and consumers are the same, the relative concentrations often are not. At present, the commonly accepted explanation for such differences is that intestinal microflora or animal enzymes are facilitating interconversions among various MAAs.[14,101–103,156]

There are distinct patterns of MAA distribution between algae and hosts in marine symbiotic relationships, with higher MAA concentrations in the host tissue than in the resident algal cells. Endosymbionts of the anemone *Anthopleura elegantissima* contain approximately 12% of the MAAs present in the host tissue.[120] (Note: isolated zooxanthellae in this study may have been contaminated with MAAs from host tissue, as a previous study detected no MAAs in isolated or cultured endosymbionts from *Anthopleura*.[119,157]) In the coral *Acropora microphthalma*, animal tissues contain more than 90% of the MAA complement within the host–algal association.[158] In this species, the MAA content of endosymbiotic algae is constant with depth, while MAA concentrations in host tissues exhibit a vertical gradient with highest concentrations in shallow specimens. Cells of *Prochloron* sp. that are endosymbiotic with the ascidian *Lissoclinum patella* have only half the MAA concentration of the host.[152] Zooxanthellae (*Symbiodinium* sp.) of the giant clam *Tridacna crocea* contain no detectable MAAs, while the host tissue has appreciable amounts of four MAAs.[153]

It is presumed that MAAs are synthesized by zooxanthellae and released into the host tissue environment where they are available for uptake and/or assimilation into host cells. The endosymbionts then rely on the host for UV protection rather than their own possession of screening compounds.[152,153,158] It is of interest that endosymbionts are not required for MAAs to be present in animal tissues. MAAs occur in both symbiotic and aposymbiotic species of clams and in symbiotic and aposymbiotic anemones of the same species.[120,121,153]

7. Distribution of MAAs Relative to Radiation Exposure (Depth and Season)

In a number of species, the highest concentrations of MAAs occur in individuals inhabiting shallow waters where solar exposures are most intense or they occur during summer when exposure to solar radiation is maximal relative to day length and sun angle.[108,111,137,141,150,158–162]

These observations have prompted speculation about the role of radiation quality and intensity in regulating MAA content.

In hermatypic corals (*Acropora* spp.) from the Great Barrier Reef, concentrations of palythine are 16 times higher at the surface than at 20 m depths.[108] The co-occurring MAA, mycosporine-glycine, is over twice as concentrated in *Acropora* tissues from individuals at 1 m than in corals living at 20 m. *A. microphthalma* colonies exhibit a sharp gradient of MAA concentrations with mycosporine-glycine concentrations eight times higher in corals at 2 m compared to 30 m depths.[158] Similarly, in the coral *Pocillopora damicornis*, colonies at 1 m depths have five times the MAA content of corals at 3 m.[143] Vertical gradients are also present in the Caribbean coral *Porites asteroides*, where colonies at 1 m depths have more than double the MAA content of colonies sampled at 6 m.[160] In this species, variations in MAA content are accompanied by phenotypic variation in colony color. Shallow MAA-rich corals are green, and deeper corals with lower MAA content are brown. Urchin species from the Caribbean and benthic macroalgae from several latitudes also exhibit a negative correlation between MAA content and depth.[110,111,146] The rhodophyte *Polysiphonia arctica*, collected from a depth of 1 m, has five times the MAA concentration as plants collected from 7 m.[110] These observations of vertical gradients of MAA concentration strongly suggest that both spectral quality and radiation intensity are involved in the determination of MAA content.

The Antarctic sea urchin *Sterechinus neumayeri* exhibits a concentration gradient of MAAs with highest levels in intertidal animals and lowest concentrations in urchins collected at 24 m.[137] Even though there is a large degree of variability in the MAA concentration of individual animals at the same depth, the vertical gradient is evident. However, the MAA content of this species does not respond to the pronounced seasonal changes in solar exposure that are characteristic of this region. The onset of austral spring results in rapidly increasing day lengths and high levels of ozone depletion (>50%). Both of these phenomena contribute to increases in incident radiation, especially UVB, with no effect on MAA concentrations in *S. neumayeri*. In contrast, the MAA content of some tropical and temperate species is closely linked to seasonal changes in ambient radiation (e.g., MAA concentrations in mucus from the tropical coral *Fungia repanda* and in eggs of the temperate ascidian *Ascidia ceratodes* fluctuate with seasonal changes in sunlight).[111,149,161–163]

In most studies, MAA concentrations *in vivo* remain stable under UV, heat stress, and darkness.[125,145,164] Phytoplankton cells maintained in dark culture conditions for a period of 2 months show no change in cellular MAA concentration.[125] When moved from the dark into lighted conditions that support photosynthesis and growth, existing MAA pools are partitioned equally between the newly divided cells of each successive generation, resulting in a sequential decrease in cellular MAA content. In temperate sea urchins with both field- and laboratory-manipulated levels of MAAs, concentrations remain unchanged for a period of at least 9 months.[104] As discussed in the following section, UV exposure can cause increases in MAAs, but only in cyanobacteria have MAA concentrations been reported to decrease under UV exposure.[165]

8. Regulation of MAA Concentrations

Little is known or understood about the regulation of MAA concentrations within organisms. Is MAA synthesis constitutive or induced? Is synthesis stimulated by exposure to solar radiation? Is radiation intensity a controlling factor? Do particular wavelengths within the solar spectrum stimulate MAA production or accumulation? Does UV-induced damage, rather than the perception of UV exposure, trigger the protective response?

A number of studies have addressed questions related to how radiation spectral quality and intensity regulate MAA composition and concentration (Table 15.5). Unfortunately, observations vary in duration and there is no standardization of radiation conditions, so combined results are inconclusive. There is a wide range of variation in the responses within and between species to the questions of spectral quality and intensity of incident radiation. Moreover, individual MAAs are differentially affected by the radiation regime, and these responses are not consistent across species. UVB, UVA, and visible light, synergistically and individually, have all been shown to cause changes in MAA concentrations in aquatic organisms. The time course of monitored changes in MAA content is also variable, spanning hours, days, weeks, or months.[29,56,100,111,113,126,141,160,166,167]

In 13 strains of cyanobacteria (from mostly freshwater and terrestrial habitats) grown under artificial UV (λ_{max} = 320 nm), MAA concentrations increase under UV exposure relative to controls under PAR alone (Table 15.5).[132] A similar response occurs for MAAs in the sheath matrix of *Nostoc commune*.[135,168] Additionally, the increased synthesis of extracellular MAAs is accompanied by an increased production of glycan that thickens the cellular sheath.[168] In the cyanobacterium *Chlorogloeopsis* sp., UV irradiation is not an obligate requirement for MAA synthesis.[56] MAA synthesis in this species is not constitutive but is induced by UVB or increased osmotic stress (see also Section II.A.10).

TABLE 15.5
Summary of Studies Investigating the Effects of Artificial or Ambient Radiation on MAA Concentrations

Species	UVB	UVA	UVB + UVA	VIS	Radiation Source
Cyanophyta					
Chlorogloeopsis sp.[56,191]	+			0	Artificial
Nostoc commune[168]	+	0			Artificial
Nostoc commune[135]	+	+		0	Artificial
13 strains[132]			+		Artificial
Chlorophyta					
Ankistrodesmus spiralis[131]	0				Artificial
Chlorella sorokiniana[131]	0				Artificial
Dunaliella tertiolecta[171]	+	0		0	Artificial
Ennallax coelastroides[131]	0				Artificial
Pseudococcomyxa sp.[131,a]	+				Artificial
Pyramimonas parkae[171]	+	+		+	Artificial
Scendesmus sp.[131,b]	+				Artificial
Scotiella chlorelloidea[131]	+				Artificial
Haptophyta					
Isochrysis sp.[171]	+	+		0	Artificial
Pavlova gyrans[171]	0	0		+	Artificial
Phaeocystis antarctica[125,c]		+		0	Artificial
Bacillariophyta					
Corethron cryophilum[126]			+	+	Ambient
Fragilariopsis cylindrus[126]			+	0	Ambient
Porosira pseudodenticulata[125,c]		+		+	Artificial
Pseudonitzschia sp.[126]			+	0	Ambient
Thalassiosira antarctica[125,c]		+		+	Artificial
Thalasssiosira tumida[125,c]		+		+	Artificial
Thalassiosira weissflogii[171]	0	+		+	Artificial
Thalassiosira sp.[126]			+	+	Ambient
Dinophyta					
Alexandrium excavatum[109,c]		++		+	Artificial
Amphidinium caterae[171]	0	+		++	Artificial

TABLE 15.5 (CONTINUED)
Summary of Studies Investigating the Effects of Artificial or Ambient Radiation on MAA Concentrations

Species	UVB	UVA	UVB + UVA	VIS	Radiation Source
Dinophyta (continued)					
Gymnodinium sanguineum[172]				+	Artificial
Heterocapsa triquetra[170]	++			+	Artificial
Lingulodinium polyedra[129]			+	-	Artificial
Prorcentrum mican[166]				+	Artificial
Prorcentrum micans[31]			+		Artificial
Phytoplankton assemblage[173,d]			++	+	Ambient
Phytoplankton assemblage[127]			++	+	Ambient
Rhodophyta					
Chondrus crispus[167]		+	+		Ambient
Chondrus crispus[141]	0	+	+	+	Ambient
Devaleraea ramentacea[111,e]	+	0		0	Ambient
Euchema striatum[140]			+		
Gracilaria cornea[164]			0	0	Artificial
Palmaria palmata[142,f]	+	+		+	Ambient
Gracilaria chilensis[139]	+				Artificial
Porphyra yezoensis[81]		+			Artificial
Cnidaria					
Anthopleura elegantissima[119] (aposymbiotic)			+	0	Artificial
Anthopleura elegantissima[119] (symbiotic)			0	0	Artificial
Clavularia sp.[29,g]			+	+	Ambient
Montipora verrucosa[209]			+	+	Ambient
Phyllodiscus semoni[29,h]			+		
Porcillopora damicornis[143,i]			+	+	Ambient
Porites asteroides[160,j]			+	0	Ambient
Stylophora pistillata[100]		0	+	0	Artificial

Note: Key for abbreviations and symbols: UVB: 280–320 nm, UVA: 320–400 nm, UVB+UVA: 280–400 nm, VIS: 400–750 nm, +: increase in MAA concentration observed after exposure to experimental wavelengths, ++: relatively larger increase in MAA concentration if increases observed under both visible and UV, 0: no change in MAA concentration observed, -: decrease in MAA concentration observed, blank: no observations reported for these radiation conditions.

[a] Observed changes in palythine, not mycosporine-glycine.

[b] Observed changes in shinorine, not other MAAs.

[c] Visible light response from blue wavelengths (400–500 nm).

[d] Observed changes in shinorine and porphyra-334, not mycosporine-glycine:valine or palythine.

[e] Observed changes in mycosporine-glycine and palythine, not shinorine, porphyra-334 or palythinol.

[f] Individual MAAs have waveband-specific responses.

[g] Observed changes in palythine and palythene, not asterina-330.

[h] Observed changes in palythine, not mycosporine-glycine.

[i] Observed changes in palythene and palythinol, not mycosporine-glycine.

[j] Observed changes in asterina-330 and shinorine, not palythine.

In cultures of the temperate dinoflagellate *Alexandrium excavatum* grown under artificial UV, the concentration of MAAs is most greatly influenced by UVA wavelengths but can also be affected by visible light alone, specifically wavelengths between 400 and 500 nm.[109,138,166] Shifts from low to high light intensity result in significant increases in MAA concentration and changes in the relative proportions of individual MAAs that can be detected after several hours. MAA synthesis does not occur in *A. excavatum* if photosynthesis is disrupted. This may be related to the role of chloroplasts as the location of enzymes involved in the initial steps of the shikimate pathway.[101,169]

MAA synthesis in the dinoflagellate *Heterocapsa triquetra* is slightly increased by visible light, but is more strongly stimulated by exposure to UVB radiation.[170] MAA synthesis in *H. triquetra* is apparently regulated by or restricted to specific phases of the cell cycle. Similar to the schedule of synthesis for many other biological molecules, increases in MAA content of *H. triquetra* cells occur only during interphase and not while cells are undergoing mitosis and cytokinesis.

Observations of MAA synthesis relative to radiation exposure have been reported for several other dinoflagellate species (Table 15.5). In all but one study, increases in visible light intensities caused increases in the MAA concentration of cells.[166,171,172] The exception was the dinoflagellate, *Lingulodinium (Gonyaulax) polyedra*, where exposure to visible light plus UV stimulated a 70% increase in the UV absorption of cells, but exposure to visible light alone resulted in an estimated 30% decline in cellular UV absorption.[129]

In Antarctic diatoms, there are distinct differences in the level of MAA content and the action spectrum of MAA synthesis between centric (radially symmetric cells with high MAA concentrations) and pennate (bilaterally symmetric cells with low MAA concentrations) species.[126] In experiments conducted under both total and partitioned (minus UVB) ambient sunlight, increases in the MAA content of centric diatoms were initiated within 2 days and continued for at least a week to the end of the observational period. As noted for *Alexandrium excavatum*, visible light alone can stimulate MAA synthesis in centric diatom species. In contrast, increases in the much lower concentrations of MAAs in pennate diatoms (*Fragilariopsis cylindrus* and *Pseudonitzschia* sp.) are evoked by UV wavelengths (<400 nm, UVB + UVA) and not by exposure to visible light alone. In laboratory experiments with artificial lamps, MAAs in centric Antarctic diatoms are most strongly induced by UVA and shorter wavelength visible light (350 to 470 nm).[125] The discrepancy between these observations and those from ambient light might be related to species-specific variations or spectral differences between ambient and artificial radiation exposures.

Increased synthesis of MAAs by exposure to high intensity artificial visible light also occurs in the Antarctic diatom *Thalassiosira weissflogii*, the prasinophyte *Pyramimonas parkae*, and most markedly in the dinoflagellate *Amphidinium carterae* (six-fold increase over control).[171] However, high visible light exposure does not affect the MAA content of two other unicellular algae, *Dunaliella tertiolecta* (Chlorophyta) and *Isochrysis* sp. (Haptophyta). Supplemental exposures with UVA and UVB in combination and alone result in a variety of species-specific responses.

Various MAA compounds have different responses to partitioned radiation exposures. In whole water samples of Antarctic phytoplankton monitored over a 2-week period, shinorine and porphyra-334 increased in concentration with exposure to either total sunlight or visible light alone.[173] In contrast, mycosporine-glycine and palythine concentrations increased only under total sunlight treatments and not when exposure was limited to the visible band. MAA-specific increases also occur in unicellular freshwater Chlorophyta species.[131]

For benthic algae, transplantation experiments have been employed to alter the intensity and spectral quality of ambient solar illumination in order to observe changes in MAA concentrations. Increasing exposure to UVB wavelengths by relocating thalli of the rhodophyte *Devaleraea ramentacea* from 2 m depths to surface radiation intensities results in an increase in total MAA concentration.[111] Observed increases are attributable to changes in concentrations of mycosporine-glycine and palythine; the tissue content of shinorine, porphyra-334, and palythinol remains unchanged. This is in contradiction to other results reported in the same study and results from Antarctic phytoplankton described above.[173,111]

Review of results from radiation exposure studies indicates great variability in species-specific and MAA-specific responses and precludes any general conclusions as to how radiation intensity or spectral quality might regulate MAA synthesis in algae. In higher plants, UVB wavelengths induce the expression of specific genes that code for enzymes involved in the synthesis of UV-protective compounds.[174] It is possible that similar wavelength-specific activation of genes is involved in the regulation of MAA synthesis, but there has been very little research in this area.

In invertebrates and chordates, the issue of regulation of MAA content is more complicated since both the MAA content of available food and the kinetics of MAA translocation between ingested material and animal tissues are involved. Given the contrasting results of investigations on algae, radiation spectral quality and intensity may not be the most appropriate variables to study. In mammalian cells it has been shown that DNA damage alone (without radiation exposure) can trigger the production of melanin to darken the skin.[9] Similar linkages may be involved in the stimulation or induction of MAA synthesis.

An interesting relationship has been observed between MAA concentration and water flow.[143] When exposed to higher water velocities but the same radiation environment, corals accumulate larger amounts of MAAs. This positive correlation has been attributed to the stimulatory effect of increased water velocity on photosynthesis that, in turn, supplies the necessary substrates for MAA synthesis. If this is true, then the observed vertical gradient of MAAs may also be a function of photosynthetic rate relative to the attenuation of PAR and not only a direct signaling of MAA synthesis by radiation intensity. Alternatively, there could be a mechanical stimulatory mechanism related to a damage-induction response for MAA synthesis in the zooxanthellae.

9.   Effectiveness of MAAs for UV Protection

The location and concentration of MAAs within cells and various tissues are prime determinants of effective UV screening. If UV absorption equates to a protective function, then sunscreening compounds would provide maximum benefit if located in external surfaces of cells and organisms.[175] In unicellular taxa, MAA distribution and localization are vital factors for cell survival. Since MAAs are water soluble, it is not likely that they would be an integral part of cell membranes, but it is plausible that they are dissolved in the cytoplasm.

Cell fractionation studies of five strains of cyanobacteria indicate that MAAs are located primarily (>90%) within the cytoplasm and not the cell sheaths, walls, or membranes.[132] Extracellular placement of MAAs does occur in some cyanobacterial species that posses cellular sheath layers.[134,135] Extracellular MAAs are covalently bonded to oligosaccharide molecules embedded in the cyanobacterial sheath matrix and provide substantial protection to prevent photobleaching of chlorophyll within the cell. Intracellular or extracellular distributions of MAAs in eukaryotic cells have not been investigated. Based on the high MAA concentrations of *Phaeocystis antarctica* colonies, it has been suggested that MAAs are associated with the extracellular mucopolysaccharide matrix of the colony.[125] This may be a more common phenomenon than currently recognized, and future research efforts will be necessary to further document extracellular occurrence of MAAs in cyanobacteria and algae.

Extracellular occurrence of MAAs is common in corals where these UV-absorbing compounds are found in the external mucus layer of the colony.[163,176] MAA concentrations in mucus layers are closely matched to or occasionally less than the MAA concentrations in the coral tissues, indicating that MAAs are probably passively and nonselectively excreted.[176] Consideration of the MAA content and optical path length of the mucus layer (~1 mm) gives an estimated absorbance efficiency of approximately 7% of incident UV. Such low attenuation of UV radiation probably does not provide substantial protection for the living polyps, and it may be the antioxidant properties of mycosporine-glycine that are more important in the mucus layer (see also Section II.A.10).

In addition to location, the concentration of MAAs is critical for screening efficiency. Correlation of concentration to UV sensitivity has been established for many biological UV-absorbing compounds (e.g., melanin, flavanoids, anthocyanin, and mycosporines).[9–11,51,177–179] Concentration-dependent protection has also been demonstrated with artificial encapsulation or shielding of cells in UV protective layers.[49,180,181] A theoretical optical model of MAA screening potential in unicells has been proposed based on the assumption of homogeneous MAA distribution within cells.[182] The model predicts that for nanoplankton-size cells (1–10 μm equivalent spherical diameter), the potential benefits of MAA screening may not warrant the energetic costs of MAA synthesis, maintenance, and storage within the cell. According to the model parameters, only cells larger than 200 μm would benefit from having MAAs. A study of MAA efficiency in the dinoflagellate *Gymnodinium sanguineum* has demonstrated that intracellular MAAs can provide more effective UV screening than predicted.[172] Also, in the tropical planktonic cyanobacteria *Trichodesmium* spp., MAAs can provide approximately 10 times greater efficiency in screening of UVA wavelengths than photosynthetic pigments can absorb PAR.[183] Therefore, packaging of MAAs into even very small cells may be of considerable benefit.

Several studies have demonstrated the ability of MAAs to prevent UV-induced photoinhibition of photosynthesis and thereby enhance survival under UV exposure. Photosynthetic rate is positively correlated to MAA concentration in natural phytoplankton assemblages, the dinoflagellate *Gymnodinium sanguineum*, zooxanthellae of the anemone *Phyllodiscus semonii*, and the rhodophyte *Devaleraea ramentacea*.[29,111,126,128,172,173] The UV photoinhibition of algae *in hospite* is also proportional to host MAA concentrations, and isolated algal cells with only their own (or no) MAAs for protection are much more vulnerable to UV exposure than when shielded by MAA-containing host tissues.[152,153,158] Other studies with Antarctic phytoplankton and freshwater microalgae show no effect of MAA concentration on UV photoinhibition of photosynthesis.[127,131] Therefore, MAAs may not always provide sufficient UV screening to protect the photosynthetic apparatus of cells.

UV-irradiation of embryos often results in delay of developmental events (e.g., cleavage and hatching) in invertebrate embryos and larvae.[105,184,185] The delay in irradiated sea urchin embryos is proportional to MAA concentration.[145,151] Planula larvae of the coral *Agarica agaricites* collected from adults inhabiting a depth of 24 m contain 33% of the MAA concentration of larvae collected from adults at 3 m, and larvae with higher MAA content have higher survival rates.[150] As mentioned earlier, MAAs in egg follicle and test layers of the ascidian *Ascidia ceratodes* provide considerable UV protection for zygotes. The absence of these layers can result in a 60% delay in cell division.[149] From these studies, it is apparent that MAAs have an important role in the protection of early developmental stages of invertebrates and ascidians, and probably also fish.

10. Other Functions of MAAs in Marine Organisms

While UV protection has been assumed to be the primary function of MAAs in marine organisms, these compounds can be involved in other metabolic roles. At least one MAA is also an effective antioxidant.[69] Mycosporine-glycine reacts with and eliminates peroxyl radicals *in vitro*, providing protection from oxidative damage caused by products of UV photochemical reactions in intra- and extracellular environments. An antioxidant role of mycosporine-glycine may be of special significance in symbiotic relationships where the generation of oxygen by endosymbionts exacerbates the oxidative stress on host tissues.[69] It is not yet known if the other marine aminomycosporine, mycosporine-taurine, also has antioxidative properties.

While MAAs are most often associated with UV protection, there are some observations that do not support this role. Cyanobacterial mats from meltwater pools of the McMurdo Ice Shelf (Antarctica) are exposed to the highest levels of UV in this region. Observations of the UV responses of two species isolated from these communities has shown that the less UV-tolerant taxa (*Oscillatoria priestleyi*) contains UV-absorbing compounds (probably MAAs and scytonemin), while the more resistant species (*Phormidium murrayi*) does not show any evidence of UV screening.[186] Moreover, the concentration of UV-absorbing compounds in *O. priestleyi* does not change during the course of controlled exposure experiments, suggesting that MAA synthesis may be a constitutive process and not an inducible one. The authors conclude that *P. murrayi* must have alternative UV protective mechanisms to ensure survival.

In fungi, mycosporines are involved in both the regulation of fungal spore development and spore release.[187,188] In nature, radiation exposure is required to induce mycosporine synthesis and initiate sporulation, but under laboratory conditions, addition of mycosporines to dark-grown cultures will mimic the radiation-induced response. Therefore, mycosporines have a physiological as well as protective function in the process of fungal spore development. In the tropical sponge *Dysidea herbacea*, temporal shifts in MAA composition and concentration more closely correlate to the reproductive cycle than to the ambient radiation regime.[122,189] Based on this observation, the authors suggest that MAAs might be directly involved in the induction of spawning, analogous to the role of mycosporines in the sporulation of fungi. Direct evidence to test this hypothesis is not yet available.

In *Artemia* sp., MAA concentrations decline rapidly during the first 8 h of embryological development.[93] This decline is followed by an increase in gadusol concentration within the developing cysts. It has been suggested that the enzymatic degradation of MAAs may provide the embryo with an early source of amino acids before yolk storage compounds become available for assimilation.[93] It is also possible that the resulting cyclohexenone structure of the MAAs provides the basis for gadusol synthesis.[101] In contrast, embryos of the sea urchin *Strongylocentrotus droebachiensis* maintain stable MAA concentrations until the pluteus stage when MAA content declines by approximately 30 to 40%.[145] The loss of MAAs during later development is considered to be primarily due to leakage from cells and not enzymatic degradation, since embryos and larvae with very low MAA contents do not exhibit any changes in concentration over time.

MAAs can have an inhibitory effect on macroalgae. MAAs added to culture medium retard the growth of *Porphyra yezoensis*.[79] The response is additive: the higher the concentration of MAAs, the greater the inhibition of growth. The decline in growth is attributed in part to MAA interference with chloroplast metabolism. Whether this is a direct inhibitory effect or interference with feedback regulatory mechanisms is not yet known.

MAAs (and scytonemin) have been implicated in the process of osmotic regulation in cyanobacteria. In the freshwater species *Nostoc commune*, a water stress protein is secreted from cells in association with MAA compounds.[134,190] In *Chlorogloeopsis* sp., MAA synthesis can be induced in the dark when salt is added to the culture medium.[191] Euryhaline species from a saltwater lake have been observed releasing MAAs when the external medium is diluted with freshwater.[192] The concentrations of MAAs released is correlated to the level of osmotic stress. The response is a unidirectional process, as MAAs are not re-absorbed if salinity is increased. Excretion of MAAs has also been reported for the dinoflagellate *Lingulodinium (Gonyaulax) polyedra*.[129] In this case, a direct metabolic function is not attributed to the excretion process. The authors suggest that the release of MAAs increases the attenuation of UV wavelengths within the water column, extending the screening capacity of MAAs to an extracellular filter of the surrounding seawater.

The possible role of MAAs in the biogeochemical cycling of carbon in marine systems has received little attention. Cells that excrete MAAs enrich the concentration of dissolved organic matter (DOM) in seawater. Observations of differential UV absorbance characteristics in contaminated and axenic cultures of the colonial stage of the haptophyte *Phaeocystis pouchetii* suggest that MAAs released by cells are utilized as a carbon and nitrogen source by bacteria.[193] In the dinoflagellate *Lingulodinium (Gonyaulax) polyedra*, excreted MAAs undergo photo-oxidation and may further contribute to bacterial growth.[129] There is little information on the distribution of MAAs in the oceans, their contribution to DOM, or the fate of MAAs once they are excreted by live cells or released by cell death.

The full range of MAA functions is not yet known. These compounds may be involved in a variety of metabolic processes in addition to their most commonly assumed role as sunscreens. It is possible that UV absorption may prove to be an ancillary and not primary function of these compounds, at least in some organisms.

B. SCYTONEMIN

1. Scytonemin Structure and Localization

The UV-absorbing property of cyanobacterial sheaths was first noted in the mid-1800s, and the term scytonemin was adopted at that time to describe the UV-absorbing pigment.[194,195] Scytonemin is a dimer derived from the aromatic amino acid tryptophan and a series of phenylpropanoid rings (Figure 15.9A).[196] Scytonemin ($C_{36}H_{22}N_2O_4$) is larger than MAAs with a molecular weight of 546. Unlike MAAs that are colorless and technically not pigments, scytonemin has a yellow-brown color. The pigment becomes green in the oxidized state and red when the molecule is reduced.[194] Also in contrast to MAAs, scytonemin is not water soluble but dissolves readily in lipids. Scytonemin has a bimodal absorption spectrum with peaks at 252 and 386 nm when dissolved in tetrahydrofuran (Figure 15.9B).[196] The *in vivo* absorption maximum is 370 nm.[194] Scytonemin absorbs radiation very effectively across UVB, UVA, and the shorter wavelengths of visible light. The pigment is transparent to a large portion of the PAR wavelengths that are involved in the energy transfer reactions of photosynthesis.

Most MAAs are intracellular, although MAAs can be translocated to external surfaces of cells and organisms because MAAs are found in significant concentrations in the extracellular sheath matrix of some cyanobacteria and in coral mucus.[134,163,176] Scytonemin is not found intracellularly. It is located extracellularly in association with the glycan sheath but is not a structural component of the sheath layer.[194,197] Scytonemin concentrations are not uniform within the sheath layer, and variations are related to the age of the sheath. In *Scytonema myochrous*, higher concentrations of scytonemin occur near the base of filaments where the sheath is older and thicker, while at newly formed apical cells, the sheath is usually absent or devoid of pigment.[198]

2. Scytonemin Distribution

Scytonemin occurs in all cyanobacterial groups represented by taxa that posses extracellular sheaths.[132,194] This includes marine, freshwater, and terrestrial species, although less than 50 species of cyanobacteria have been examined. Scytonemin seems to be restricted to sheathed species that are exposed to very high radiation intensities (e.g., intertidal areas, shallow ponds, and exposed thermal springs), but is not common in planktonic species.[194] In 22 strains (20 species) of predominantly freshwater and terrestrial taxa surveyed, MAAs and scytonemin tend to co-occur.[132] Thirteen of the 22 strains contain MAAs, and 12 of the MAA-containing strains also have scytonemin. Only one species has scytonemin and no MAAs, and the eight remaining taxa give no indication of having any UV-absorbing compounds. MAAs and scytonemin are also present in cyanobacterial lichens.[133] In eight lichen species examined, all contain MAAs and seven also have scytonemin.

There is no obvious correlation between MAA and scytonemin concentrations in free-living or symbiotic cyanobacteria. This lack of covariance suggests that although both compounds have UV-absorbing properties, they may respond to different stimuli. It has been suggested that scytonemin and MAAs have different UV photoreceptors.[168] It is also possible that primary or secondary functions may differ. In addition to UV absorption, MAAs may have physiological functions related to antioxidant activity, regulation of reproductive processes, or maintenance of osmotic balance. Scytonemin is more likely to provide only radiation protection, and, therefore, there may be little reason to expect concentrations of these compounds to be correlated.

3. Regulation of Scytonemin Concentration

Field and laboratory studies have not yet resolved the question of environmental regulation of scytonemin content. Either UV or high visible light exposure can affect scytonemin concentrations with both positive and negative correlations. Scytonemin concentrations monitored for a 20-month period in an intertidal cyanobacterial mat community have ten-fold temporal fluctuations that

A

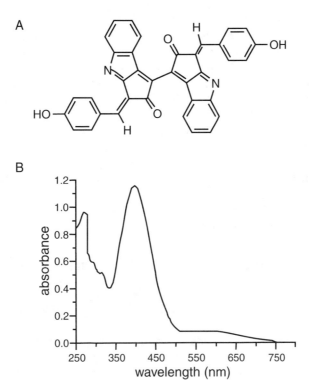

B

FIGURE 15.9 A. Molecular structure of scytonemin. B. Absorbance spectrum of scytonemin. From Proteau, P. J., Gerwick, W. H., Garcia-Pichel, F., and Castenholz, R. W., The structure of scytonemin, an ultraviolet sunscreen pigment from the sheaths of cyanobacteria, *Experientia*, 49, 826 (Figure 2), 828 (Figure 6), 1993, reprinted with permission from Springer-Verlag, New York.

correlate to seasonal changes in incident solar radiation.[55] Concurrent measurements of pterins and MAAs show no response to seasonal gradients of radiation exposure. Contradicting results are reported for *Scytonemin myochrous* and *Rivularia* sp.[198] In natural communities, scytonemin content of *Scytonema* inversely correlates to UV exposure, while in *Rivularia*, scytonemin exhibits the expected pattern of increase with increased UV exposure.

Generally, laboratory cultures grown without UV tend to have very low scytonemin content, and cultures exposed to ambient sunlight or high intensities of artificial radiation show significant increases in scytonemin within a period of several days.[7] High radiation intensities stimulate scytonemin synthesis with or without the presence of UV, but UVA wavelengths have the strongest influence on scytonemin content.[194,199] Additional evidence that UVA wavelengths induce scytonemin synthesis comes from observations of *Calothrix* mats.[200] When mats are exposed to high intensity radiation (with or without UV), scytonemin content increases; however, under low intensity radiation, synthesis of scytonemin requires the presence of low levels of UVA. Stimulation of scytonemin in cultures of *Nostoc commune* also depends primarily on UVA (375 nm) exposure.[168]

4. Effectiveness of Scytonemin for Ultraviolet Protection

Scytonemin-containing cyanobacterial sheaths are estimated to attenuate more than 88% of incident UVA radiation (at 370 nm).[194] Several studies have demonstrated the ability of scytonemin to prevent UV-induced damage in cells, and these results have generally demonstrated that the presence of scytonemin does increase the UV tolerance of cyanobacteria. Dose response experiments with strains of the terrestrial/freshwater cyanobacterium *Tolypothrix*, with and without scytonemin in the filament sheaths, show that a nonpigmented sheath results in greater UV sensitivity of cells.[165] This study also demonstrates the photostability of scytonemin since exposure to UV radiation causes no change in the absorption of scytonemin in *Tolypothrix*, even though absorption by MAAs, carotenoids, and chlorophyll is significantly diminished after exposure. Comparison of cultures with different concentrations of scytonemin have also indicated that scytonemin can be a significant factor in reducing the level of UV-induced inhibition of photosynthesis in cyanobacteria.[7,199] Experiments with artificially produced sheathless cells of *Chlorogloeopsis* sp. further verify that higher concentrations of scytonemin enhance the ability of cells to photosynthesize and grow under UVA exposure.[199]

There is at least one report where the presence of scytonemin does not provide enhanced protection for UVB exposure. In *Nostoc commune*, the presence of extracellular MAAs provides substantial protection against the photobleaching of chlorophyll by UVB, but in strains that contain scytonemin and no MAAs, UVB exposure causes complete destruction of cellular chlorophyll.[168] It is most probable that MAAs and scytonemin serve as complementary UV protectants and do not have sufficient overlap in their UV screening ranges to act as interchangeable filters.

III. EVOLUTIONARY ASPECTS OF ULTRAVIOLET PROTECTION

Under earlier Earth atmospheres, cells were subjected to much higher intensities and a broader spectral band of ambient UV (including UVC), and biological defenses against UV would have been a primary prerequisite for the early evolution of life on this planet.[4,5,201] The widespread geographic and taxonomic occurrence of MAAs is commonly accepted as evidence that UV protection has been an obligate requirement for the existence of life on Earth throughout geologic time.[6,7,13,15,78,107,118,194,196,202] The ability of scytonemin to absorb wavelengths in the UVC range further substantiates this line of reasoning.[7]

The release of oxygen via photosynthesis is, in large part, responsible for the long-term gradual changes that have occurred in the chemical composition of the Earth's atmosphere. The formation of the ozone layer in the stratosphere has eliminated all of the UVC and most of the UVB from incident sunlight, but the small fraction of UVB reaching the Earth's surface is still a biological hazard. Consequently, species have retained effective means of surviving under the inevitable exposure to the harmful components of solar radiation.

Today, there are several UV tolerance strategies that are almost universal among organisms. These include metabolic pathways for DNA repair and the synthesis of compounds that absorb UV. However, even with these means to tolerate UV, species have threshold limits of exposure beyond which screening is insufficient and damage incurred is greater that can be repaired without debilitating or lethal effects. These thresholds vary at intra- and interspecific levels and establish the limitations by which natural selection acts on populations and restructures communities. The ubiquitous occurrence of UV-absorbing compounds (e.g., melanin, flavanoids, and MAAs) in all the kingdoms supports the notion that biochemical screening of UV is a very effective defense against solar radiation. There are also numerous studies that demonstrate increased tolerance to UV with increases in the concentrations of UV-absorbing secondary metabolites. An important point for consideration is that while UV-absorbing compounds are widespread, they do not provide complete protection from UV exposure.[7,127,132,141]

Although the large majority of compounds that have been identified as UV protectants are not necessarily structurally related in their final form, they do originate from similar precursor molecules that are linked to products of the shikimate pathway. This further corroborates a common ancestral origin. The synthesis of many UV-absorbing compounds occurs through the degradation of the aromatic amino acids tyrosine and tryptophan. This is true for melanins, flavanoids, pteridines, MAAs, mycosporines, and scytonemin, as well as less prevalent UV screening molecules. For example, the UV-absorbing compounds in the lenses of vertebrate eyes (including fish) are tryptophan derivatives (e.g., kynurenine and 3-hydroxy kynurenine).[116,203]

MAAs and scytonemin are important groups of secondary metabolites in aquatic organisms, and there is still much to learn about the physiological characteristics and ecological aspects of these compounds. It is expected that continued research will lead to the discovery of new MAAs and other sunscreening compounds, further characterization of biological and environmental factors that regulate MAA and scytonemin synthesis and assimilation, identification of mechanisms for translocation and interconversion by consumer organisms, elucidation of additional physiological functions, and determination of the role of these compounds in biogeochemical cycling.

ACKNOWLEDGMENTS

I would like to extend sincere thanks to Ulf Karsten, Malcom Shick, and Walter Dunlap for their comments and suggestions on improving this chapter and to the editors, Jim McClintock and Bill Baker, for organizing this volume.

REFERENCES

1. Webb, A. R., Vitamin D synthesis under changing UV spectra, in *Environmental UV Photobiology*, Young, A. R., Björn, L. O., Moan, J., and Nultsch, W., Eds., Plenum Press, New York, 1993, 185.
2. Mitchell, D. L. and Karentz, D., The induction and repair of DNA photodamage in the environment, in *Environmental UV Photobiology*, Young, A. R., Björn, L. O., Moan, J., and Nultsch, W., Eds., Plenum Press, New York, 1993, 345.
3. Friedberg, E. C., Walker, G. C., and Siede, W., *DNA Repair and Mutagenesis*, ASM Press, Washington, D.C., 1995.
4. Cleaves, H. J. and Miller, S. L., Oceanic protection of prebiotic organic compounds from UV radiation, *Proc. Nat. Acad. Sci. USA*, 95, 7260, 1998.
5. Kasting, J. F., Earth's early atmosphere, *Science*, 259, 920, 1993.
6. Karentz, D., Ultraviolet tolerance mechanisms in Antarctic marine organisms, in *Ultraviolet Radiation and Biological Research in Antarctica*, Weiler, C. S. and Penhale, P. A., Eds., American Geophysical Union, Washington, D.C., 1994, 93.
7. Dillion, J. G. and Castenholz, R. W., Scytonemin, a cyanobacterial sheath pigment, protects against UVC radiation: implications for early photosynthetic life, *J. Phycol.*, 35, 673, 1999.
8. Cockell, C. S., Biological effects of high ultraviolet radiation on early earth — a theoretical evaluation, *J. Theoret. Biol.*, 193, 717, 1998.
9. Gilchrest, B. A., Park, H. Y., Eller, M. S., and Yaar, M., Mechanisms of ultraviolet light-induced pigmentation, *Photochem. Photobiol.*, 63, 1, 1996.
10. Kootstra, A., Protection from UV-B-induced DNA damage by flavonoids, *Plant Mol. Biol.*, 26, 771, 1994.
11. Stapleton, A. E. and Walbot, V., Flavonoids can protect maize DNA from the induction of ultraviolet radiation damage, *Plant Physiol.*, 105, 881, 1994.
12. Bandaranayake, W. M., Mycosporines: are they nature's sunscreens?, *Nat. Prod. Rep.*, 15, 159, 1998.
13. Cockell, C. S. and Knowland, J., Ultraviolet radiation screening compounds, *Biol. Rev. Camb. Philos. Soc.*, 74, 311, 1999.
14. Dunlap, W. C. and Shick, J. M., Ultraviolet radiation-absorbing mycosporine-like amino acids in coral reef organisms: a biochemical and environmental perspective, *J. Phycol.*, 34, 418, 1998.
15. Garcia-Pichel, F., Solar ultraviolet and the evolutionary history of cyanobacteria, *Orig. Life Evol. Biosph.*, 28, 321, 1998.
16. Cashmore, A. R., Jarillo, J. A., Wu, Y. J., and Liu, D., Cryptochromes: blue light receptors for plants and animals, *Science*, 284, 760, 1999.
17. Helbling, E. W., Villafañe, V., and Holm-Hansen, O., Effects of ultraviolet radiation on Antarctic marine phytoplankton photosynthesis with particular attention to the influence of mixing, in *Ultraviolet Radiation and Biological Research in Antarctica*, Weiler, C. S. and Penhale, P. A., Eds., American Geophysical Union, Washington, D.C., 1994, 207.

18. Booth, C. R. and Morrow, J. E., Impacts of solar UVR on aquatic microorganisms: the penetration of UV into natural waters, *Photochem., Photobiol.*, 65, 254, 1997.

19. Piazena, H. and Häder, D. P., Penetration of solar UV irradiation in coastal lagoons of the southern Baltic Sea and its effect on phytoplankton communities, *Photochem. Photobiol.*, 60, 463, 1994.

20. Smith, R. C. and Baker, K. S., Penetration of UV-B and biologically effective dose-rates in natural waters, *Photochem. Photobiol.*, 29, 311, 1979.

21. Karentz, D. and Lutze, L. H., Evaluation of biologically harmful ultraviolet radiation in Antarctica with a biological dosimeter designed for aquatic environments, *Limnol. Oceanogr.*, 35, 549, 1990.

22. Baker, K. S., Smith, R. C., and Green, A. E. S., Middle ultraviolet radiation reaching the ocean surface, *Photochem. Photobiol.*, 32, 367, 1980.

23. Smith, R. C., Prézelin, B. B., Baker, K. S., Bidigare, R. R., Boucher, N. P., Coley, T., Karentz, D., MacIntyre, S., Matlick, H. A., Menzies, D., Ondrusek, M., Wan, Z., and Waters, K. J., Ozone depletion: ultraviolet radiation and phytoplankton biology in Antarctic waters, *Science*, 255, 952, 1992.

24. Jerlov, N. G., Ultraviolet radiation in the sea, *Nature*, 166, 111, 1950.

25. Harm, W., *Biological Effects of Ultraviolet Radiation*, Cambridge University Press, Cambridge, UK, 1980.

26. Sies, H., *Oxidative Stress: Oxidants and Antioxidants*, Academic Press, San Diego, CA, 1991.

27. Mopper, K. and Zhou, X., Hydroxyl radical photoproduction in the sea and its potential impact on marine processes, *Science*, 250, 661, 1990.

28. Dykens, J. A., Shick, J. M., Bendit, G., Buettner, G., and Winston, G. W., Oxygen radical production in the sea anemone *Anthopleura elegantissima* and its endosymbiotic algae, *J. Exp. Biol.*, 168, 219, 1992.

29. Shick, J. M., Lesser, M. P., and Stochaj, W. R., Ultraviolet radiation and photooxidative stress in zooxanthellate Anthozoa: the sea anemone *Phyllodiscus semoni* and the octocoral *Clavularia* sp., *Symbiosis*, 10, 145, 1991.

30. Lesser, M. P. and Stochaj, W. R., Photoadaptation and protection against active forms of oxygen in the symbiotic procaryote *Prochloron* sp. and its ascidian host, *Appl. Environ. Microbiol.*, 56, 1530, 1990.

31. Lesser, M. P., Acclimation of phytoplankton to UV-B radiation — oxidative stress and photoinhibition of photosynthesis are not prevented by UV-absorbing compounds in the dinoflagellate *Prorocentrum micans*, *Mar. Ecol. Prog. Ser.*, 132, 287, 1996.

32. Häder, D. P., The effect of enhanced solar UV-B radiation on motile organisms, in *Stratospheric Ozone Reduction, Solar Ultraviolet Radiation and Plant Life*, Worrest, R. C. and Caldwell, M. M., Eds., Springer-Verlag, Berlin, 1986, 223.

33. Häder, D. P., Effects of enhanced solar ultraviolet radiation on aquatic ecosystems, in *UV-B Radiation and Ozone Depletion Effects on Humans, Animals, Plants, Microorganisms, and Materials*, Tevini, M., Ed. Lewis Publishers, Boca Raton, FL, 1993, 155.

34. Sharp, D. T. and Gray, I. E., Studies on the factors affecting the local distribution of two sea urchins, *Arbacia punctulata* and *Lytechinus variegatus*, *Ecology*, 43, 309, 1962.

35. Fankhauser, C. and Chory, J., Light control of plant development, *Annu. Rev. Cell Dev. Biol.*, 13, 203, 1997.

36. Barcelo, J. A. and Calkins, J., Positioning of aquatic microorganisms in response to visible light and simulated solar UV-B irradiation, *Photochem. Photobiol.*, 29, 75, 1979.

37. Barcelo, J. A. and Calkins, J., The kinetics of avoidance of simulated solar UV radiation by two arthropods, *Biophys. J.*, 32, 921, 1980.

38. Shashar, N., UV vision by marine animals: mainly questions, in *Ultraviolet Radiation and Coral Reefs*, Gulko, D. and Jokiel, P. L., Eds., University of Hawaii, School of Ocean and Earth Science and Technology, Hawaii Institute of Marine Biology, Technical Report 41, Sea Grant Publication UNIHI-SEAGRANT-CR-95-03, 1995, 201.

39. Jacobs, G. H., Ultraviolet vision in vertebrates, *Am. Zool.*, 32, 544, 1992.

40. Kunz, Y. W., Tracts of putative ultraviolet receptors in the retina of the two-year-old brown trout (*Salmo trutta*) and the Atlantic salmon (*Salmo salar*), *Experientia*, 43, 1202, 1987.

41. Browman, H. I., Novales-Flamarique, I., and Hawryshyn, C. W., Ultraviolet photoreception contributes to prey search behaviour in two species of zooplanktivorous fishes, *J. Exp. Biol.*, 186, 187, 1994.

42. Chou, B. R. and Hawryshyn, C. W., Spectral transmittance of the ocular media of the bluegill *(Lepomis macrochirus)*, *Can. J. Zool.*, 65, 1214, 1987.

43. Falkowski, P. G. and Raven, J. A., *Aquatic Photosynthesis*, Blackwell Science, Malden, MA, 1997.

44. Gulko, D. and Jokiel, P. L., Eds., *Ultraviolet Radiation and Coral Reefs*, University of Hawaii, School of Ocean and Earth Science and Technology, Hawaii Institute of Marine Biology, Technical Report 41, Sea Grant Publication UNIHI-SEAGRANT-CR-95-03, 1995.

45. Bingham, B. L. and Reyns, N., Ultraviolet radiation and distribution of the solitary ascidian *Corella inflata* (Huntsman), *Biol. Bull.*, 196, 94, 1999.

46. Davidson, A. T., Bramich, D., Marchant, H. J., and Mcminn, A., Effects of UV-B irradiation on growth and survival of Antarctic marine diatoms, *Mar. Biol.*, 119, 507, 1994.

47. Karentz, D., Prevention of ultraviolet radiation damage in Antarctic marine invertebrates, in *Stratospheric Ozone Depletion/UVB Radiation in the Biosphere: Proceedings of NATO Advanced Research Workshop*, Biggs, R. H. and Joyner, M., Eds., Springer-Verlag, Berlin, 1994, 175.

48. Rawlings, T. A., Shields against ultraviolet radiation: an additional protective role for the egg capsules of benthic marine gastropods, *Mar. Ecol. Prog. Ser.*, 136, 81, 1996.

49. Kumar, A., Tyagi, M. B., Srinivas, G., Singh, N., Kumar, H. D., Sinha, R. P., and Häder, D. P., UVB shielding role of $FeCl_3$ and certain cyanobacterial pigments, *Photochem. Photobiol.*, 64, 321, 1996.

50. Lowe, C. and Goodman-Lowe, G., Suntanning in hammerhead sharks, *Nature*, 383, 677, 1996.

51. Hessen, D. O., Competitive trade-off strategies in Arctic *Daphnia* linked to melanism and UV-B stress, *Polar Biol.*, 16, 573, 1996.

52. Matsunaga, T., Burgess, J. G., Yamada, N., Komatsu, K., Yoshida, S., and Wachi, Y., An ultraviolet (UV-A) absorbing biopterin glucoside from the marine planktonic cyanobacterium *Oscillatoria* sp., *Appl. Microbiol. Biotechnol.*, 39, 250, 1993.

53. Yamazawa, A., Takeyama, H., Takeda, D., and Matsunaga, T., UV-A-induced expression of GroEL in the UV-A-resistant marine cyanobacterium *Oscillatoria* sp. NKBG 091600, *Microbiol.*, 145, 949, 1999.

54. Wachi, Y., Burgess, J. G., Iwamoto, K., Yamada, N., Nakamura, N., and Matsunaga, T., Effect of ultraviolet-A (UV-A) light on growth, photosynthetic activity and production of biopterin glucoside by the marine UV-A resistant cyanobacterium *Oscillatoria* sp., *Biochim. Biophys. Acta*, 1244, 165, 1995.

55. Karsten, U., Maier, J., and Garcia-Pichel, F., Seasonality in UV-absorbing compounds of cyanobacterial mat communities from an intertidal mangrove flat, *Aquat. Microb. Ecol.*, 16, 37, 1998.

56. Portwich, A. and Garcia-Pichel, F., Ultraviolet and osmotic stresses induce and regulate the synthesis of mycosporines in the cyanobacterium *Chlorogloeopsis*, *Arch. Microbiol.*, 172, 187, 1999.

57. Lindquist, N., Lobkovsky, E., and Clardy, J., Tridentatols A–C, novel natural products of the marine hydroid *Tridentata marginata*, *Tetrahedron Lett.*, 37, 9131, 1996.

58. Stachowicz, J. J. and Lindquist, N., Chemical defense among hydroids on pelagic *Sargassum*: predator deterrence and absorption of solar UV radiation by secondary metabolites, *Mar. Ecol. Prog. Ser.*, 155, 115, 1997.

59. Pavia, H., Cervin, G., Lindgren, A., and Aaberg, P., Effects of UV-B radiation and simulated herbivory on phlorotannins in the brown alga *Ascophyllum nodosum*, *Mar. Ecol. Prog. Ser.*, 157, 139, 1997.

60. Xiong, F., Komenda, J., Kopecky, J., and Nedbal, L., Strategies of ultraviolet-B protection in microscopic algae, *Physiol. Plant.*, 100, 378, 1997.

61. Llewellyn, C. A. and Mantoura, R. F. C., A UV absorbing compound in HPLC pigment chromatograms obtained from Icelandic Basin phytoplankton, *Mar. Ecol. Prog. Ser.*, 158, 283, 1997.

62. Craigie, J. S. and McLachlan, J., Excretion of colored ultraviolet-absorbing substances by marine algae, *Can. J. Bot.*, 42, 23, 1964.

63. Menzel, D., Kazlauskas, R., and Reichelt, J., Coumarins in the siphonalean green algal family Dasycladaceae Kuetzing (Chlorophyceae), *Bot. Mar.*, 26, 23, 1983.

64. Johnson, M. K., Alexander, K. E., Lindquist, N., and Loo, G., Potent antioxidant activity of a dithiocarbamate-related compound from a marine hydroid, *Biochem. Pharmacol.*, 58, 1313, 1999.

65. Krinsky, N. I., Carotenoid protection against oxidation, *Pure Appl. Chem.*, 51, 649, 1979.

66. Chalker-Scott, L., Environmental significance of anthocyanins in plant stress responses, *Photochem. Photobiol.*, 70, 1, 1999.

67. Mathews-Roth, M. M., Carotenoids and photoprotection, *Photochem. Photobiol.*, 65S, 148S, 1997.

68. Yohn, J. J., Lyons, M. B., and Norris, D. A., Cultured human melanocytes from black and white donors have different sunlight and ultraviolet A radiation sensitivities, *J. Invest. Dermatol.*, 99, 454, 1992.

69. Dunlap, W. C. and Yamamoto, Y., Small-molecule antioxidants in marine organisms: antioxidant activity of mycosporine-glycine, *Comp. Biochem. Physiol.*, 112B, 105, 1995.

70. Shibata, K., Pigments and a UV-absorbing substance in corals and a blue-green alga living in the Great Barrier Reef, *Plant Cell Physiol.*, 10, 325, 1969.

71. Price, J.H. and Forrest, H. S., 310 μm absorbance in *Physalia physalis* - distribution of the absorbance and isolation of a 310 μm absorbing compound, *Comp. Biochem. Physiol.*, 30, 879, 1969.

72. Iwamoto, K. and Aruga, Y., Distribution of the UV-absorbing substance in algae with reference to the peculiarity of *Prasiola japonica* Yatabe, *J. Tokyo Univ. Fish*, 60, 43, 1973.

73. Sivalingam, P. M., Ikawa, T., Yokohama, Y., and Nisizawa, K., Distribution of a 334 UV-absorbing substance in algae, with special regard of its possible physiological roles, *Bot. Mar.*, 17, 23, 1974.

74. Chioccara, F., Misuraca, G., Novellino, E., and Prota, G., Occurrence of two new mycosporine-like amino acids, mytilins A and B in the edible mussel, *Mytilis galloprovincialis*, *Tetrahedron Lett.*, 34, 3181, 1979.

75. Hirata, Y., Uemura, D., Ueda, K., and Takano, S., Several compounds from *Palythoa tuberculosa* (Colenterata), *Pure Appl. Chem.*, 51, 1875, 1979.

76. Chioccara, F., Della Gala, A., De Rosa, M., Novellino, E., and Prota, G., Mycosporine amino acids and related compounds from the eggs of fishes, *Bull. Soc. Chim. Belg.*, 89, 1101, 1980.

77. Nakamura, H., Kobayashi, J., and Hirata, Y., Separation of mycosporine-like amino acids in marine organisms using reverse phase high-performance liquid chromatography, *J. Chromatogr.*, 250, 113, 1982.

78. Yentsch, C. S. and Yentsch, C. M., The attenuation of light by marine phytoplankton with specific reference to the absorption of near-UV radiation, in *The Role of Solar Ultraviolet Radiation in Marine Ecosystems*, Calkins, J., Ed., Plenum Press, New York, 1982, 691.

79. Sivalingam, P. M., Ikawa, T., and Nisizawa, K., Possible physiological roles of a substance showing characteristic UV-absorbing patterns in some marine algae, *Plant Cell Physiol.*, 15, 583, 1974.

80. Sivalingam, P. M., Ikawa, T., and Nisizawa, K., Isolation and physico-chemical properties of a substance 334 from the red alga, *Porphyra yezoensis* Ueda, *Bot. Mar.*, 19, 1, 1976.

81. Sivalingam, P. M., Ikawa, T., and Nisizawa, K., Physiological roles of a substance 334 in algae, *Bot. Mar.*, 19, 9, 1976.

82. Yoshida, T. and Sivalingham, P. M., Isolation and characterization of the 337 mμ UV-absorbing substance in red alga, *Porphyra yezoensis* Ueda, *Plant Cell Physiol.*, 11, 427, 1970.

83. Ito, S. and Hirata, S., Isolation and structure of a mycosporine from the zoanthid *Palythoa tuberculosa*, *Tetrahedron Lett.*, 1977, 2429, 1977.

84. Tsujino, I., Yabe, K., Sekikawa, I., and Hamanaka, N., Isolation and structure of a mycosporine from the red alga *Chondrus yendoi*, *Tetrahedron Lett.*, 16, 1401, 1978.

85. Takano, S., Uemura, D., and Hirata, Y., Isolation and structure of two new amino acids, palythinol and palythene, from the zoanthid *Palythoa tuberculosa*, *Tetrahedron Lett.*, 49, 4909, 1978.

86. Takano, S., Nakanishi, A., Uemura, D., and Hirata, Y., Isolation and structure of a 334 nm UV-absorbing substance, porphyra-334 from the red alga *Porphyra tenera* Kjellman, *Chem. Lett.*, 419, 1979.

87. Tsujino, I., Yabe, K., and Sekikawa, I., Isolation and structure of a new amino acid, shinorine, from the red alga *Chondrus yendoi* Yamada et Mikami, *Bot. Mar.*, 23, 65, 1980.

88. Leach, M., Ultraviolet-absorbing substances associated with light-induced sporulation in fungi, *Can. J. Bot.*, 43, 185, 1965.

89. Trione, E. J., Leach, C. M., and Mutch, J. T., Sporogenic substances isolated from fungi, *Nature*, 212, 163, 1966.

90. Favre-Bonvin, J., Arpin, N., and Brevard, C., Structure de la mycosporine (P 310), *Can. J. Chem.*, 54, 1105, 1976.

91. White, J. D., Cammack, J. H., and Sakuma, K., The synthesis and absolute configuration of mycosporins — a novel application of the staudinger reaction, *J. Am. Chem. Soc.*, 111, 8970, 1989.

92. Kobayashi, J., Nakamura, H., and Hirata, Y., Isolation and structure of a UV-absorbing substance 337 from the ascidian *Halocynthia roretzi*, *Tetrahedron Lett.*, 22, 3001, 1981.

93. Grant, P. T., Middleton, C., Plack, P. A., and Thomson, R. H., The isolation of four aminocyclohex-enimines (mycosporines) and a structurally related derivative of cyclohexane-1:3-dione (gadusol) from the brine shrimp, *Artemia*, *Comp. Biochem. Physiol.*, 80B, 755, 1985.

94. Wu Won, J. J., Rideout, J. A., and Chalker, B. E., Isolation and structure of a novel mycosporine-like amino acid from the reef-building corals *Pocillopora damicornis* and *Stylophora pistillata*, *Tetrahedron Lett.*, 36, 5255, 1995.

95. Teai, T. T., Raharivelomanana, P., Bianchini, J. P., Faure, R., Martin, P. M. V. and Cambon, A., Structure de deux nouvelles iminomycosporines isolées de *Pocillopora eydouxi*, *Tetrahedron Lett.*, 38, 5799, 1997.

96. Wu Won, J. J., Chalker, B. E., and Rideout, J. A., Two new UV-absorbing compounds from *Stylophora pistillata*: sulfate esters of mycosporine-like amino acids, *Tetrahedron Lett.*, 38, 2525, 1997.

97. Grant, P. T., Plack, P. A., and Thomson, R. H., Gadusol, a metabolite from fish eggs, *Tetrahedron Lett.*, 21, 4043, 1980.

98. Plack, P. A., Fraser, N. W., Grant, P. T., Middleton, C., Mitchell, A. I., and Thomson, R. H., Gadusol, an enolic derivative of cyclohexane-1,3-dione present in the roes of cod and other marine fish, *Biochem. J.*, 199, 741, 1981.

99. Chioccara, F., Zeuli, L., and Novellino, E., Occurrence of mycosporine related compounds in sea urchin eggs, *Comp. Biochem. Physiol.*, 85B, 459, 1986.

100. Shick, J. M., Romaine-Lioud, S., Ferrier-Pagès, C., and Gattuso, J. P., Ultraviolet-B radiation stimulates shikimate pathway-dependent accumulation of mycosporine-like amino acids in the coral *Stylophora pistillata* despite decreases in its population of symbiotic dinoflagellates, *Limnol. Oceanogr.*, 44, 1667, 1999.

101. Bentley, R., The shikimate pathway — a metbolic tree with many branches, *Crit. Rev. Biochem. Molec. Biol.*, 25, 307, 1990.

102. Shick, J. M., Dunlap, W. C., Chalker, B. E., Banaszak, A. T., and Rosenzweig, T. K., Survey of ultraviolet radiation-absorbing mycosporine-like amino acids in organs of coral reef holothuroids, *Mar. Ecol. Prog. Ser.*, 90, 139, 1992.

103. Bandaranayake, W. M. and Des Rocher, A., Role of secondary metabolites and pigments in the epidermal tissues, ripe ovaries, viscera, gut contents and diet of the sea cucumber *Holothuria atra*, *Mar. Biol.*, 133, 163, 1999.

104. Carroll, A. K. and Shick, J. M., Dietary accumulation of UV-absorbing mycosporine-like amino acids (MAAs) by the green sea urchin (*Strongylocentrotus droebachiensis*), *Mar. Biol.*, 124, 561, 1996.

105. Carefoot, T. H., Harris, M., Taylor, B. E., Donovan, D., and Karentz, D., Mycosporine-like amino acids: possible UV protection in eggs of the sea hare *Aplysia dactylomela*, *Mar. Biol.*, 130, 389, 1998.

106. Mason, D. S., Schafer, F., Shick, J. M., and Dunlap, W. C., Ultraviolet radiation-absorbing mycosporine-like amino acids (MAAs) are acquired from their diet by medaka fish (*Oryzias latipes*) but not by SKH-1 hairless mice, *Comp. Biochem. Physiol.*, 120A, 587, 1998.

107. Karentz, D., McEuen, F. S., Land, K. M., and Dunlap, W. C., Survey of mycosporine-like amino acid compounds in Antarctic marine organisms: potential protection from ultraviolet exposure, *Mar. Biol.*, 108, 157, 1991.

108. Dunlap, W. C., Chalker, B. E., and Oliver, J. K., Bathymetric adaptations of reef-building corals at Davies Reef, Great Barrier Reef, Australia. III. UV-B absorbing compounds, *J. Exp. Mar. Biol. Ecol.*, 104, 239, 1986.

109. Carreto, J. I., Lutz, V. A., De Marco, S. G., and Carignan, M. O., Fluence and wavelength dependence of mycosporine amino acid synthesis in the dinoflagellate *Alexandrium excavatum*, in *Toxic Marine Phytoplankton*, Granéli, E., Sundström, B., Edler, L., and Anderson, D. M., Eds., Elsevier, New York, 1990, 275.

110. Karsten, U., Sawall, T., Hanelt, D., Bischof, K., Figueroa, F., Flores-Moya, A., and Wiencke, C., An inventory of UV-absorbing mycosporine-like amino acids in macroalgae from polar to warm-temperate regions, *Bot. Mar.*, 41, 443, 1998.

111. Karsten, U., Bischof, K., Hanelt, D., Tüg, H., and Wiencke, C., The effect of ultraviolet radiation on photosynthesis and ultraviolet-absorbing substances in the endemic Arctic macroalga *Devaleraea ramentacea* (Rhodophyta), *Physiol. Plant.*, 105, 58, 1999.

112. Banaszak, A. T., Lesser, M. P., Kuffner, I. B., and Ondrusek, M., Relationship between ultraviolet (UV) radiation and mycosporine-like amino acids (MAAs) in marine organisms, *Bull. Mar. Sci.*, 63, 617, 1998.

113. Goméz, I., Pérez-Rodriguez, E., Viñegla, B., Figueroa, F. L., and Karsten, U., Effects of solar radiation on photosynthesis, UV-absorbing compounds and enzyme activities of the green alga *Dasycladus vermicularis* from southern Spain, *J. Photochem. Photobiol. B: Biol.*, 47, 46, 1998.

114. Pérez-Rodriguez, E., Goméz, I., Karsten, U., and Figueroa, F. L., Effects of UV radiation on photosynthesis and excretion of UV-absorbing compounds of *Dasycladus vermicularis* (Dasycladales, Chlorophyta) from southern Spain, *Phycologia*, 37, 379, 1998.

115. Jeffrey, S. W., MacTavish, H. S., Dunlap, W. C., Vesk, M., and Groenewoud, K., Occurrence of UVA- and UVB-absorbing compounds in 152 species (206 strains) of marine microalgae, *Mar. Ecol. Prog. Ser.*, 189, 35, 1999.

116. Thorpe, A., Douglas, R. H., and Truscott, R. J. W., Spectral transmission and short-wave absorbing pigments in the fish lens — I. Phylognetic distribution and identity, *Vision Res.*, 33, 289, 1993.

117. Dunlap, W. C., Williams, D. M., Chalker, B., and Banaszak, A., Biochemical photoadaptation in vision: UV-absorbing pigments in fish eye tissues, *Comp. Biochem. Physiol.*, 93B, 601, 1989.

118. Teai, T., Drollet, J. H., Bianchini, J. P., Cambon, A., and Martin, P. M. V., Widespread occurence of mycosporine-like amino acid compounds in scleractinians from French Polynesia, *Coral Reefs*, 16, 169, 1997.

119. Banaszak, A. T. and Trench, R. K., Effects of ultraviolet (UV) radiation on marine microalgal-invertebrate symbioses. II. The synthesis of mycosporine-like amino acids in response to exposure to UV in *Anthopleura elegantissima* and *Cassiopeia xamachana*, *J. Exp. Mar. Biol. Ecol.*, 194, 233, 1995.

120. Stochaj, W. R., Dunlap, W. C., and Shick, J. M., Two new UV-absorbing mycosporine-like amino acids from the sea anemone *Anthopleura elegantissima* and the effects of zooxanthellae and spectral irradiance on chemical composition and content, *Mar. Biol.*, 118, 149, 1994.

121. Shick, J. M., Dunlap, W. C., and Carroll, A. K., Mycosporine-like amino acids and small-molecule antioxidants in *Anthopleura* spp.: taxonomic and environmental patterns, *Am. Zool.*, 36, 72A, 1996.

122. Bandaranayake, W. M., Bourne, D. J., and Sim, R. G., Chemical composition during maturing and spawning of the sponge *Dysidea herbacea* (Porifera: Demospongiae), *Comp. Biochem. Physiol.*, 118B, 851, 1997.

123. McClintock, J. B. and Karentz, D., Mycosporine-like amino acids in 38 species of subtidal marine organisms from McMurdo Sound, Antarctica, *Antarctic Sci.*, 9, 392, 1997.

124. Bidigare, R. R., Iriarte, J. L., Kang, S. H., Ondrusek, M. E., Karentz, D., and Fryxell, G. A., Phytoplankton: quantitative and qualitative assessments, in *Foundations for Ecosystem Research in the Western Antarctic Peninsula Region*, Ross, R., Hofmann, E., and Quetin, L., Eds., American Geophysical Union, Washington, D.C., 1996, 173.

125. Riegger, L. and Robinson, D., Photoinduction of UV-absorbing compounds in Antarctic diatoms and *Phaeocystis antarctica*, *Mar. Ecol. Prog. Ser.*, 160, 13, 1997.

126. Helbling, E. W., Chalker, B. E., Dunlap, W. C., Holm-Hansen, O., and Villafañe, V. E., Photoacclimation of antarctic marine diatoms to solar ultraviolet radiation, *J. Exp. Mar. Biol. Ecol.*, 204, 85, 1996.

127. Lesser, M. P., Neale, P. J., and Cullen, J. J., Acclimation of antarctic phytoplankton to ultraviolet radiation: ultraviolet-absorbing compounds and carbon fixation, *Mol. Mar. Biol. Biotech.*, 5, 314, 1996.

128. Dunlap, W. C., Rae, G. A., Helbling, E. W., Villafañe, V. E., and Holm-Hansen, H., Ultraviolet-absorbing compounds in natural assemblages of antarctic phytoplankton, *Antarctic J. U.S.*, 30, 323, 1995.

129. Vernet, M. and Whitehead, K., Release of ultraviolet-absorbing compounds by the red-tide dinoflagellate *Lingulodinium polyedra*, *Mar. Biol.*, 127, 35, 1996.

130. Sommaruga, R. and Garcia-Pichel, F., UV-absorbing mycosporine-like compounds in planktonic and benthic organisms from a high-mountain lake, *Arch. Hydrobiol.*, 144, 255, 1999.

131. Xiong, F., Kopecky, J., and Nedbal, L., The occurrence of UV-B absorbing mycosporine-like amino acids in freshwater and terrestrial microalgae (Chlorophyta), *Aquat. Bot.*, 63, 37, 1999.

132. Garcia-Pichel, F. and Castenholz, R. W., Occurrence of UV-absorbing, mycosporine-like compounds among cyanobacterial isolates and an estimate of their screening capacity, *Appl. Environ. Microbiol.*, 59, 163, 1993.

133. Büdel, B., Karsten, U., and Garcia-Pichel, F., Ultraviolet-absorbing scytonemin and mycosporine-like amino acid derivatives in exposed, rock-inhabiting cyanobacterial lichens, *Oecologia*, 112, 165, 1997.

134. Böhm, G. A., Pfleiderer, W., Böger, P., and Scherer, S., Structure of a novel oligosaccharide-mycosporine-amino acid ultraviolet A-B sunscreen pigment from the terrestrial cyanobacterium *Nostoc commune*, *J. Biol. Chem.*, 270, 8536, 1995.

135. Scherer, S., Chen, T. W., and Böger, P., A new UV-A/UV-B protecting pigment in the terrestrial cyanobacterium *Nostoc commune*, *Plant Physiol.*, 88, 1055, 1988.

136. Dunlap, W. C. and Karentz, D., unpublished.

137. Karentz, D., Dunlap, W. C., and Bosch, I., Temporal and spatial occurrence of UV-absorbing mycosporine-like amino acids in tissues of the Antarctic sea urchin *Sterechinus neumayeri* during springtime ozone depletion, *Mar. Biol.*, 129, 343, 1997.

138. Carreto, J. I., Carignan, M. O., Daleo, G., and De Marco, S. G., Occurrence of mycosporine-like amino acids in the red tide dinoflagellate *Alexandrium excavatum*. UV-photoprotective compounds?, *J. Plankton Res.*, 12, 909, 1990.

139. Molina, X. and Montecino, V., Acclimation to UV irradiance in *Gracilaria chilensis* Bird, McLachlan and Oliveira (Gigratinales, Rhodophyta), *Hydrobiologia*, 326/327, 415, 1996.

140. Wood, W. F., Photoadaptive responses of the tropical red alga *Euchema striatum* Schmitz (Gigartinales) to ultra-violet radiation, *Aquatic Bot.*, 33, 41, 1989.

141. Franklin, L. A., Yakovleva, I., Karsten, U., and Lüning, K., Synthesis of mycosporine-like amino acids in *Chondrus crispus* (Florideophyceae) and the consequences for sensitivity to ultraviolet B radiation, *J. Phycol.*, 35, 682, 1999.

142. Karsten, U. and Wiencke, C., Factors controlling the formation of UV-absorbing mycosporine-like amino acids in the marine red alga *Palmaria palmata* from Spitsbergen (Norway), *J. Plant Physiol.*, 155, 407, 1999.

143. Jokiel, P. L., Lesser, M. P., and Ondrusek, M. E., UV-absorbing compounds in the coral *Pocillopora damicornis*: interactive effects of UV radiation, photosynthetically active radiation, and water flow, *Limnol. Oceanogr.*, 42, 1468, 1997.

144. Karentz, D., Dunlap, W. C., and Bosch, I., Distribution of UV-absorbing compounds in the Antarctic limpet, *Nacella concinna*, *Antarctic J. U.S.*, 27, 121, 1992.

145. Adams, N. L. and Shick, J. M., Mycosporine-like amino acids provide protection against ultraviolet radiation in eggs of the green sea urchin *Strongylocentrotus droebachiensis*, *Photochem. Photobiol.*, 64, 149, 1996.

146. Bosch, I., Janes, P., Schack, R., Steves, B., and Karentz, D., Survey of UV-absorbing compounds in sub-tropical sea urchins from Florida and the Bahamas, *Am. Zool.*, 34, 102A, 1994.

147. Krupp, D. A. and Blanck, J., Preliminary report on the occurrence of mycosporine-like amino acids in the eggs of the Hawaiian scleractinian corals *Montipora verrucosa* and *Fungia scutaria*, in *Ultraviolet Radiation and Coral Reefs*, Gulko, D. and Jokiel, P. L., Eds., University of Hawaii, School of Ocean and Earth Science and Technology, Hawaii Institute of Marine Biology, Technical Report 41, Sea Grant Publication UNIHI-SEAGRANT-CR-95-03, 1995, 129.

148. Karentz, D., Wardle, G. S., and Bosch, I., unpublished.

149. Epel, D., Hemela, K., Shick, M., and Patton, C., Development in the floating world: defenses of eggs and embryos against damage from UV radiation, *Am. Zool.*, 39, 271, 1999.

150. Gleason, D. and Wellington, G. M., Variation in UVB sensitivity of planula larvae of the coral *Agaricia agaricites* along a depth gradient, *Mar. Biol.*, 123, 693, 1995.

151. Karentz, D. and Bosch, I., unpublished.

152. Dionisio-Sese, M. L., Ishikura, M., Miyachi, S., and Maruyama, T., UV-absorbing substances in the tunic of a colonial ascidian protect its symbiont, *Prochloron* sp., from damage by UV-B radiation, *Mar. Biol.*, 128, 455, 1997.

153. Ishikura, M., Kato, C., and Maruyama, T., UV-absorbing substances in zooxanthellate and azooxanthellate clams, *Mar. Biol.*, 128, 649, 1997.

154. Carefoot, T. H., Karentz, D., Pennings, S. C., and Young, C. L., Distribution of mycosporine-like amino acids in the sea hare *Aplysia dactylomela*: effect of diet on amounts and types sequestered over time in tissues and spawn, *Comp. Biochem. Physiol.*, 126, 91, 2000.

155. Muszynski, F. Z., Bruckner, A., Armstrong, R. A., Morell, J. M., and Corredor, J. E., Within-colony variations of UV absorption in a reef building coral, *Bull. Mar. Sci.*, 63, 589, 1998.

156. Dunlap, W. C., Banaszak, A., Roseweig, T. T., and Shick, J. M., Ultraviolet light-absorbing compounds in coral reef holothurians: organ distribution and possible sources, in *Proceedings of the 7th International Echinoderm Conference*, Yanagisawa, T., Yasumasu, I., Ogura, C., Suzuki, N., and Motokawa, T., Eds., Balkema and Rotterdam, Atami, Japan, 1991, 560.

157. Shick, J. M., personal communication, 1999.

158. Shick, J. M., Lesser, M. P., Dunlap, W. C., Stochaj, W. R., Chalker, B. E., and Wu Won, J., Depth-dependent responses to solar ultraviolet radiation and oxidative stress in the zooxanthellate coral *Acropora microphthalma*, *Mar. Biol.*, 122, 41, 1995.

159. Dunlap, W. C., Chalker, B. E., and Bandaranayake, W. M., Ultraviolet light absorbing agents derived from tropical marine organisms of the Great Barrier Reef, Australia, in *6th International Coral Reef Symposium*, Choat, J. H., Barnes, D., Borowitzka, M. A., Coll, J. C., Davies, P. J., Flood, P., Hatcher, B. G., Hopley, D., Hutchings, P. A., Kinsey, D., Orme, G. R., Pichon, M., Sale, P. F., Sammarco, P., Wallace, C. C., Wilkinson, C., Wolanski, E., and Bellwood, O., Eds., 6th International Coral Reef Symposium Executive Committee, Australia, 1988, 89.

160. Gleason, D. F., Differential effects of ultraviolet radiation on green and brown morphs of the Caribbean coral *Porites astreoides*, *Limnol. Oceanogr.*, 38, 1452, 1993.

161. Wood, W. F., Effect of solar ultra-violet radiation on the kelp *Ecklonia radiata*, *Mar. Biol.*, 96, 143, 1987.

162. Post, A. and Larkum, A. W. D., UV-absorbing pigments, photosynthesis and UV exposure in Antarctica: comparison of terrestrial and marine algae, *Aquat. Bot.*, 45, 231, 1993.

163. Drollet, J. H., Teai, T., Faucon, M., and Martin, P. M. V., Field study of compensatory changes in UV-absorbing compounds in the mucus of the solitary coral *Fungia repanda* (Scleractinia: Fungiidae) in relation to solar UV radiation, sea-water temperature, and other coincident physico-chemical parameters, *Mar. Freshwater Res.*, 48, 329, 1997.

164. Sinha, R. P., Klisch, M., and Häder, D.-P., Mycosporine-like amino acids in the marine red alga *Gracilaria cornea* — effects of UV and heat, *Environ. Exp. Bot.*, 43, 33, 2000.

165. Adhikary, S. P. and Sahu, J. K., UV protecting pigment of the terrestrial cyanobacterium *Tolypothrix byssoidea*, *J. Plant Physiol.*, 153, 770, 1998.

166. Carreto, J. I., De Marco, S. G., and Lutz, V. A., UV-absorbing pigments in the dinoflagellates *Alexandrium excavatum* and *Prorocentrum micans*. Effects of light intensity, in *Red Tides Biology, Environmental Science, and Toxicology*, Okaichi, T., Anderson, D. M., and Nemoto, T., Eds., Elsevier, New York, 1989, 333.

167. Karsten, U., Franklin, L. A., Lüning, K., and Wiencke, C., Natural ultraviolet radiation and photosynthetically active radiation induce formation of mycosporine-like amino acids in the marine macroalga *Chondrus crispus* (Rhodophyta), *Planta*, 205, 257, 1998.

168. Ehling-Schulz, M., Bilger, W., and Scherer, S., UV-B-induced synthesis of photoprotective pigments and extracellular polysaccharides in the terrestrial cyanobacterium *Nostoc commune*, *J. Bacteriol.*, 179, 1940, 1997.

169. Dennis, D. T. and Turpin, D. H., *Plant Physiology, Biochemistry and Molecular Biology*, Longman Scientific and Technical with John Wiley & Sons, Inc., New York, 1990.

170. Wängberg, S. Å., Persson, A., and Karlson, B., Effects of UV-B radiation on synthesis of mycosporine-like amino acid and growth in *Heterocapsa triquetra* (Dinophyceae), *J. Photochem. Photobiol. B: Biol.*, 37B, 141, 1997.

171. Hannach, G. and Sigleo, A. C., Photoinduction of UV-absorbing compounds in six species of marine phytoplankton, *Mar. Ecol. Prog. Ser.*, 174, 207, 1998.

172. Neale, P. J., Banaszak, A. T., and Jarriel, C. R., Ultraviolet sunscreens in *Gymnodinium sanguineum* (Dinophyceae): mycosporine-like amino acids protect against inhibition of photosynthesis, *J. Phycol.*, 34, 928, 1998.

173. Villafañe, V. E., Helbling, E. W., Holm-Hansen, O., and Chalker, B. E., Acclimitization of Antarctic natural phytoplankton assemblages when exposed to solar ultraviolet radiation, *J. Plankton Res.*, 17, 2295, 1995.

174. Ouwerkerk, P. B. F., Hallard, D., Verpoorte, R., and Memelink, J., Identification of UV-B light-responsive regions in the promoter of the tryptophan decarboxylase gene from *Catharanthus roseus*, *Plant Mol. Biol.*, 41, 491, 1999.

175. Cheng, L., Douek, M., and Goring, D. A. I., UV absorption by gerrid cuticles, *Limnol. Oceanogr.*, 23, 554, 1978.

176. Teai, T., Drollet, J. H., Bianchini, J. P., Cambon, A., and Martin, P. M. V., Occurrence of ultraviolet radiation-absorbing mycosporine-like amino acids in coral mucus and whole corals of French Polynesia, *Mar. Freshwater Res.*, 49, 127, 1998.

177. Middleton, E. M. and Teramura, A. H., The role of flavonol glycosides and carotenoids in protecting soybean from ultraviolet-B damage, *Plant Physiol.*, 103, 741, 1993.

178. Takahashi, A., Takeda, K., and Ohnishi, T., Light-induced anthocyanin reduces the extent of damage to DNA in UV-irradiated *Centaurea cyanus* cells in culture, *Plant Cell Physiol.*, 32, 541, 1991.

179. Brook, P. J., Protective function of an ultraviolet absorbing compound associated with conidia of *Glomerella cingulata*, *NZ J. Bot.*, 19, 299, 1981.

180. Dunkle, R. L. and Shasha, B. S., Response of starch-encapsulated *Bacillus thuringiensis* containing ultraviolet screens to sunlight, *Environ. Entomol.*, 18, 1035, 1989.

181. Ignoffo, C. M., Shasha, B. S., and Shapiro, M., Sunlight ultraviolet protection of the *Heliothis* nuclear polyhedrosis virus through starch-encapsulation technology, *J. Invertebrate Pathol.*, 57, 134, 1991.

182. Garcia-Pichel, F., A model for internal self-shading in planktonic organisms and its implications for the usefulness of ultraviolet sunscreens, *Limnol. Oceanogr.*, 39, 1704, 1994.

183. Subramaniam, A., Carpenter, E. J., Karentz, D., and Falkowski, P. G., Bio-optical properties of the marine diazotrophic cyanobacteria *Trichodesmium* spp. I. Absorption and photosynthetic spectra, *Limnol. Oceanogr.*, 44, 608, 1999.

184. Amemiya, S., Yonemura, S., Kinosita, S., and Shiroya, T., Biphasic stage sensitivity to UV suppression of grastrulation in sea urchin embryos, *Cell Differ.*, 18, 45, 1986.

185. Rustad, R., Irradiation, chemical treatment and the cleavage-cycle and cleavage pattern alteration, in *The Sea Urchin Embryo: Biochemistry and Morphogenesis*, Czihak, G., Ed., Springer-Verlag, Berlin, 1975, 345.

186. Quesada, A. and Vincent, W. F., Strategies of adaptation by Antarctic cyanobacteria to ultraviolet radiation, *Eur. J. Phycol.*, 32, 335, 1997.

187. Dehorter, B., Induction des périthèces de *Nectria galligena* Bres. par un photocomposé mycélien absorbant à 310 nm, *Can. J. Bot.*, 54, 600, 1976.

188. Dehorter, B. and Bernillon, J., Photoinduction des périthèces de *Nectria galligena*: production et activité photomorphogène des mycosporines, *Can. J. Bot.*, 61, 1435, 1983.

189. Bandaranayake, W. M., Bemis, J. E., and Bourne, D. J., Ultraviolet absorbing pigments from the marine sponge *Dysidea herbacea*: isolation and structure of a new mycosporine, *Comp. Biochem. Physiol.*, 115C, 281, 1996.

190. Hill, D. R., Hladun, S. L., Scherer, S., and Potts, M., Water stress proteins of *Nostoc commune* (Cyanobacteria) are secreted with UV-A/B-absorbing pigments and associate with 1,4-β-D-xylanxylanohydrolase activity, *J. Biol. Chem.*, 269, 7726, 1994.

191. Portwich, A. and Garcia-Pichel, F., A novel prokaryotic UVB photoreceptor in the cyanobacterium *Chlorogloeopis* PCC 6912, *Photochem. Photobiol.*, 71, 493, 2000.

192. Oren, A., Mycosporine-like amino acids as osmotic solutes in a community of halophilic cyanobacteria, *Geomicrobiol. J.*, 14, 231, 1997.

193. Marchant, H. J., Davidson, A. T., and Kelly, G. J., UV-B protecting compounds in the marine alga *Phaeocystis pouchetii* from Antarctica, *Mar. Biol.*, 109, 391, 1991.

194. Garcia-Pichel, F. and Castenholz, R. W., Characterization and biological implications of scytonemin, a cyanobacterial sheath pigment, *J. Phycol.*, 27, 395, 1991.

195. Sinha, R. P., Klisch, M., Gröniger, A., and Häder, D. P., Ultraviolet-absorbing/screening substances in cyanobacteria, phytoplankton and macroalgae, *J. Photochem. Photobiol. B: Biol.*, 47B, 83, 1998.

196. Proteau, P. J., Gerwick, W. H., Garcia-Pichel, F., and Castenholz, R. W., The structure of scytonemin, an ultraviolet sunscreen pigment from the sheaths of cyanobacteria, *Experientia*, 49, 825, 1993.

197. Robbins, R. A., Bauld, J., and Chapman, D. J., Chemistry of the sheath of the cyanobacterium *Lyngbya aestuarii* Lieb., *Cryptog. Algol.*, 19, 169, 1998.

198. Pentecost, A., Field relationships between scytonemin density, growth and irradiance in cyanobacteria occurring in low illumination regimes, *Microb. Ecol.*, 26, 101, 1993.

199. Garcia-Pichel, F., Sherry, N. D., and Castenholz, R. W., Evidence for an ultraviolet sunscreen role of the extracellular pigment scytonemin in the terrestrial cyanobacterium *Chlorogloeopsis* sp., *Photochem. Photobiol.*, 56, 17, 1992.

200. Brenowitz, S. and Castenholz, R. W., Long-term effects of UV and visible irradiance on natural populations of a scytonemin-containing cyanobacterium (*Calothrix* sp.), *FEMS Microbiol. Ecol.*, 24, 343, 1997.

201. Rozema, J., Gieskes, W. W. C., van de Geijn, S. C., Nolan, C., and de Boois, H., Eds., *UV-B and Biosphere*, Kluwer Academic Publishers, Dordrecht, 1997.

202. Karentz, D., Evolution and ultraviolet light tolerance in algae, *J. Phycol.*, 35, 629, 1999.

203. Shashar, N., Harosi, F. I., Banaszak, A. T., and Hanlon, R. T., UV radiation blocking compounds in the eye of the cuttlefish *Sepia officinalis*, *Biol. Bull.*, 195, 187, 1998.

204. Takano, S., Uemura, D., and Hirata, Y., Isolation and structure of a new amino acid, palythine, from the zoanthid *Palythoa tuberculosa*, *Tetrahedron Lett.*, 49, 2299, 1978.

205. Sekikawa, L., Kubota, C., Hiraoki, T., and Tsujino, I., Isolation and structure of a 357 nm UV-absorbing substance, usujirene, from the red alga, *Palmaria palmata* (L) O. Kintzie, *Jpn. J. Phycol.*, 34, 185, 1986.

206. Nakamura, H., Kobayashi, J., and Hirata, Y., Isolation and structure of a 330 nm UV-absorbing substance, asterina-330 from the starfish *Asterina pectinifera*, *Chem. Lett.*, 28, 1413, 1981.

207. Karsten, U. and Garcia-Pichel, F., Carotenoids and mycosporine-like amino acid compounds in members of the genus *Microcoleus* (cyanobacteria): a chemosystematic study, *Syst. Appl. Microbiol.*, 19, 285, 1996.

208. Bosch, I. and Karentz, D., unpublished.

209. Scelfo, G., Relationship between solar radiation and pigmentation of the coral *Monotipora verrucosa* and its zooxanthellae, in *Coral Reef Population Biology*, Jokiel, P. L., Richmond, R. H., and Rogers, R. A., Eds., Hawaii Institute Marine Biology Technical Report #37, 1986, 440.

Section IV

Applied Marine Chemical Ecology

16 Marine Chemical Ecology: Applications in Marine Biomedical Prospecting

Susan H. Sennett

CONTENTS

I. INTRODUCTION

Natural products have long been a source of bioactive metabolites utilized for the treatment of human disease. In spite of advancing technologies and the development of compound sources such as combinatorial chemistry, genomics, and screening programs utilizing high throughput technology, natural products remain a valuable source of bioactive compounds with therapeutic potential.[1–4] It has been reported that 57% of the 150 most prescribed drugs have their origins in natural products.[5] Furthermore, it has been estimated that more than 60% of drugs approved for use as anticancer and antiinfective agents are derived from natural sources including semisynthetic derivatives, synthetic analogs, as well as the original natural products.[6] A major advantage in using natural products as a source for drug discovery is the structural diversity of these compounds when compared to other sources in use.[7]

The marine environment is a vast resource for the discovery of structurally unique bioactive secondary metabolites, some belonging to totally novel chemical classes.[8] Sessile benthic organisms including the Porifera, Cnidaria, Bryozoa, and Tunicata as well as marine algae have developed an arsenal of compounds which have been demonstrated to confer a competitive advantage in ecosystems characterized by extreme resource limitations. Interactions of these organisms at the genetic, species, and ecosystem levels provide the basis for the production of these compounds. Hay and Fenical[9] provide detailed examples of chemically mediated effects on ecosystems, species diversity, as well as intraspecific diversity in production of marine natural products, and how these interactions affect the biological and chemical diversity of the oceans.

To date, there are more than 10,000 publications relating to marine natural products and more are being added to the list at an increasing rate.[2] However, it is estimated that discoveries to date only reflect a small percentage of the biological and chemical diversity of the marine environment, including macro- and microorganisms. Advances in technology for collection of marine organisms, isolation and cultivation of marine microorganisms, and isolation and structure elucidation of marine natural products will facilitate the discovery of numerous other compounds.

II. TARGETS FOR MARINE NATURAL PRODUCTS DRUG DISCOVERY

There is a growing urgency to identify novel drug leads in disease areas for which treatment or cure has been elusive. A recent review of the use of natural products (predominantly plant and microbial) in pharmaceutical screens summarized the results of drug discovery efforts in a number of these critical areas, including infectious (e.g., drug resistance), neurological (e.g., Parkinsons, Alzheimer's), cardiovascular, immunological (e.g., transplant rejection), anti-inflammatory (e.g., arthritis), antiviral (particularly HIV), and oncological (particularly solid tumors) diseases.[3] This review provides compelling evidence that supports the use of natural products in pharmaceutical screening programs. To date, active marine natural products have been identified in a number of these disease targets, including anti-infective, immunomodulatory, anti-inflammatory, antiviral, and anticancer areas. Recent reviews provide examples of bioactive compounds correlated with taxonomic source and therapeutic group.[10,11]

Traditionally, targets have been cell-based, *in vitro* assays used to detect antiinfective and cytotoxic activities. Increasingly, assays for specific molecular targets in the above disease areas are being developed for use in high throughput screening. Mechanistically, marine derived compounds are known to interact with a variety of molecular targets (Table 16.1).

TABLE 16.1
Molecular Targets for Marine Natural Products

Target	Compound	Source	Therapeutic Area
Actin	Jasplakinolide (Structure 16.1)	Sponge, *Jaspis* sp.	Anticancer[12]
	Latrunculin A (Structure 16.2)	Sponge, *Latrunculia* sp.	Anticancer[13]
Tubulin	Discodermolide (Structure 16.3)	Sponge, *Discodermia* sp.	Anticancer[14]
	Curacin (Structure 16.4)	Cyanophyte, *Lyngbya majuscula*	Anticancer[15]
Phospholipase A$_2$	Manoalide (Structure 16.5)	Sponge, *Luffariella variabilis*	Anti-inflammatory[16]
Protein phoshatases	Okadaic acid (Structure 16.6)	Dinoflagellate, *Prorocentrum lima*	Anticancer[17]
	Dysidiolide(cdc25) (Structure 16.7)	Sponge, *Dysidea etheria*	Anticancer[18]
	Discorhabdin P (Structure 16.8) (calcineurin)	Sponge, *Batzella* sp.	Heart disease[19]
Protein kinase C	Bryostatin 1 (Structure 16.9)	Bryozoan, *Bugula neritina*	Anticancer[20]
Ion channels	Saxitoxin (Structure 16.10)	Dinoflagellate, *Alexandrium* spp.	Pain[21]
Nicotinic acetylcholine receptor	Conus toxins	Cone snails, *Conus* sp.	Pain[22]
Topoisomerase II	Makaluvamines (Structure 16.11)	Sponge, *Zyzzya* sp.	Anticancer[23,24]

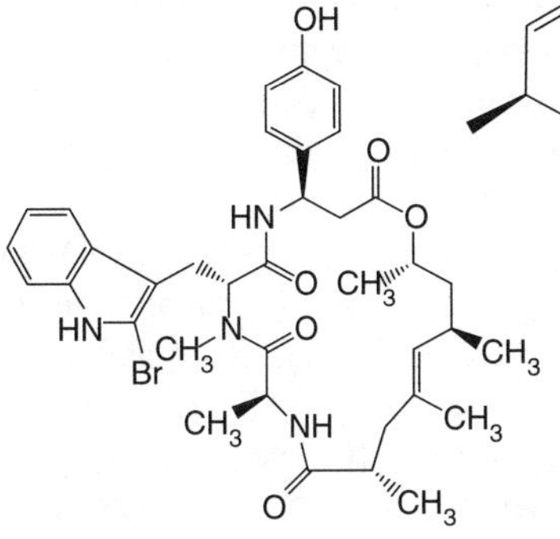

16.1

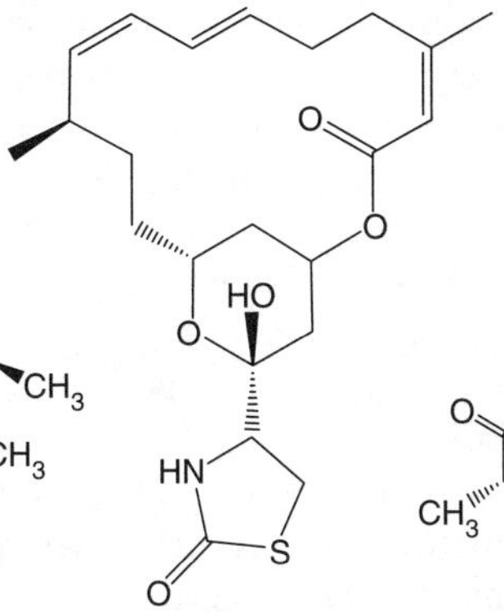

16.2

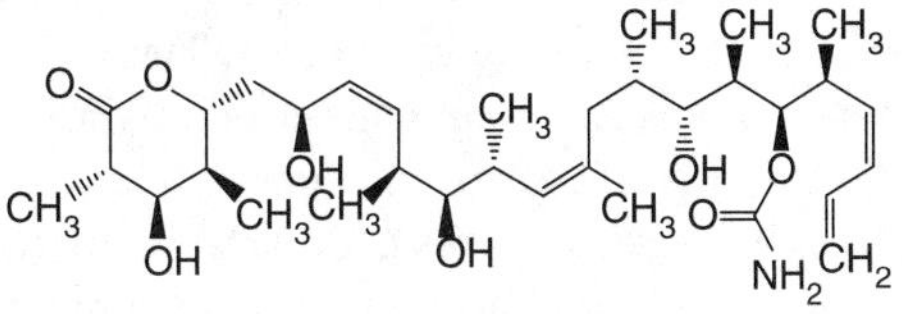

16.3

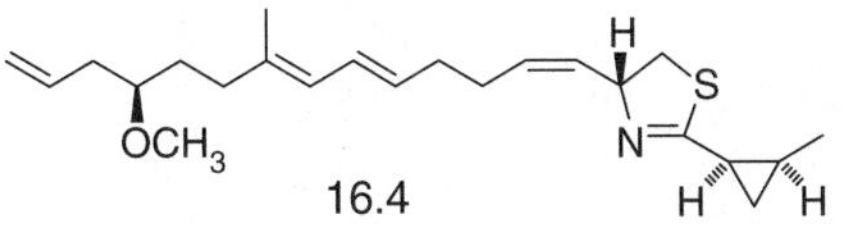

16.4

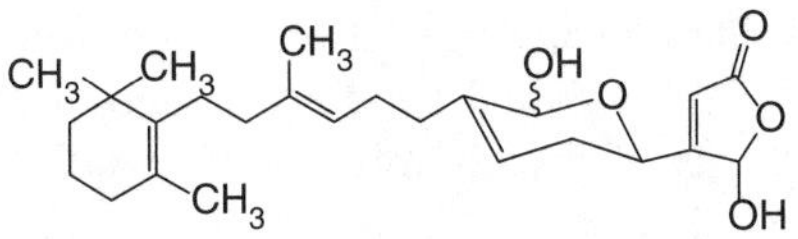

16.5

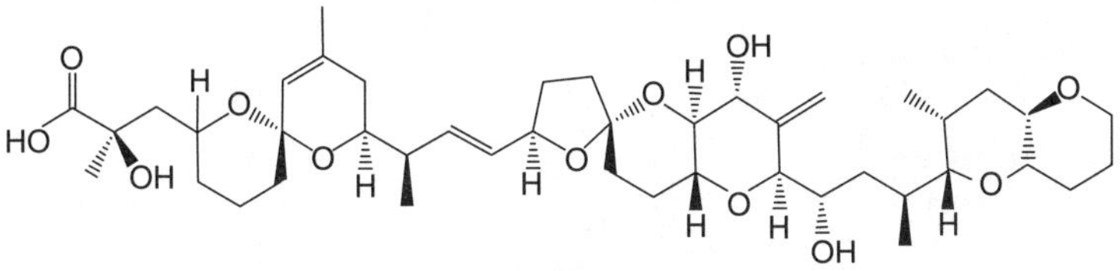

16.6

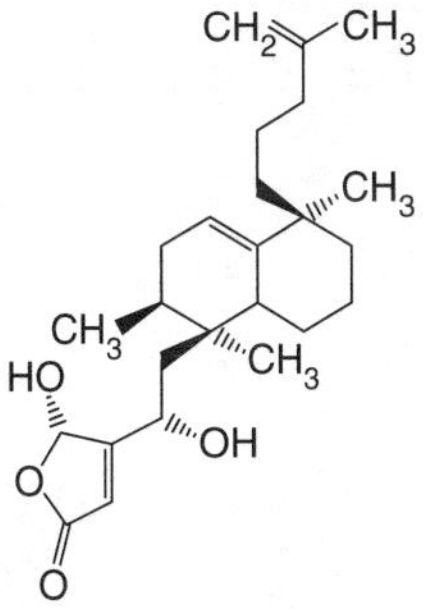

16.7

16.8

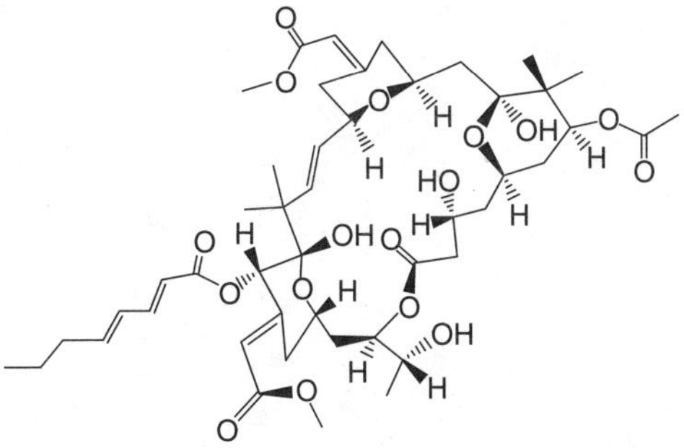

16.9

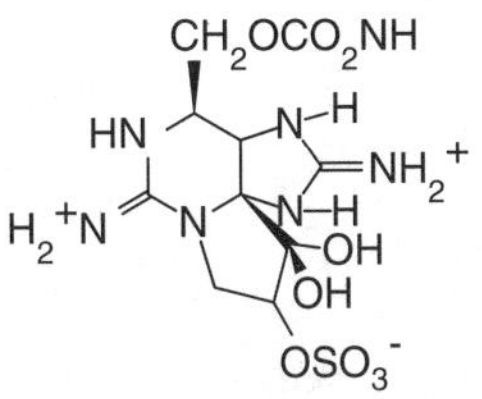

16.10

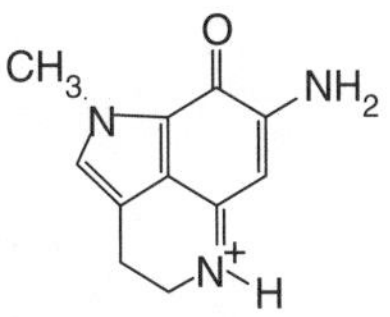

16.11

Several other molecular targets of interest for natural products screening include: the dopamine receptor for Parkinson's disease and acetylcholinesterase for Alzheimer's disease. In addition, inhibitions of protein and peptidoglycan syntheses are targets for infectious diseases. As we gain a greater understanding of the molecular basis of these disease areas, additional targets will become available.

A number of compounds listed in Table 16.1 have not proved suitable as drug candidates but are being used as molecular probes that may facilitate the identification of new targets and the discovery of new drug candidates. These compounds include manoalide, okadaic acid, and neurotoxins such as saxitoxin and tetrodotoxin.[25]

Currently, there are four marine-derived compounds in clinical trials for anticancer use: bryostatin 1 (Structure 16.9), dolastatin 10 (Structure 16.12), ecteinascidin 743 (Structure 16.13) (ET743), and aplidine (dehydrodideminin B) (Structure 16.14). Bryostatin 1, isolated from the bryozoan *Bugula neritina*,[26] is in NCI sponsored Phase II trials against melanoma, non-Hodgkins lymphoma, and renal cancer. Dolastatin 10 isolated from the sea hare *Dolabella auricularia*[27] is in Phase I trials against breast and liver cancers, solid tumors, and leukemia, also sponsored by the NCI. ET743 and aplidine, isolated from the tunicates *Ecteinascidia turbinata*[28,29] and *Aplidium albicans*,[30] respectively, are anticancer agents[31,32] undergoing evaluation conducted by PharmaMar (based in Spain).

Although these compounds are all anticancer agents, they act through different mechanisms. Bryostatin 1, a cyclic macrolide, inhibits protein kinase C tumor promotion while aplidine is a protein synthesis inhibitor. Dolastatin 10, a linear peptide, and ET743, a tetrahydroisoquinoline

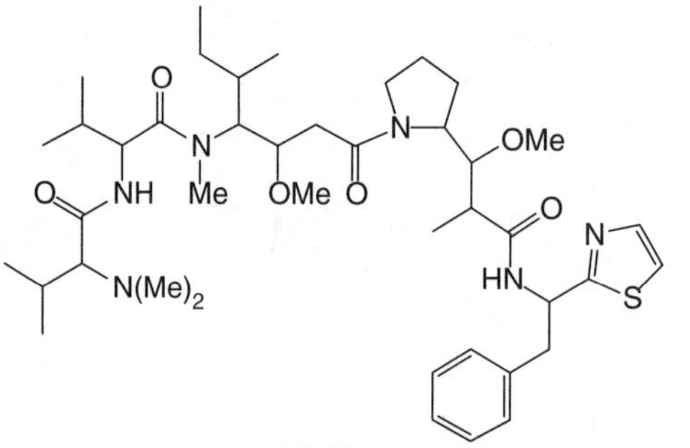

16.12

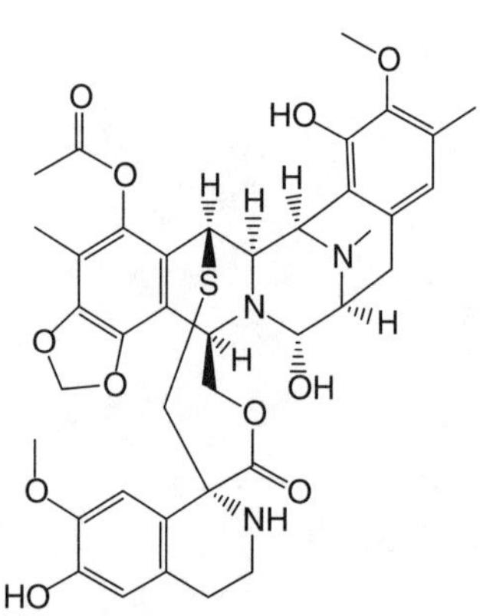

16.13

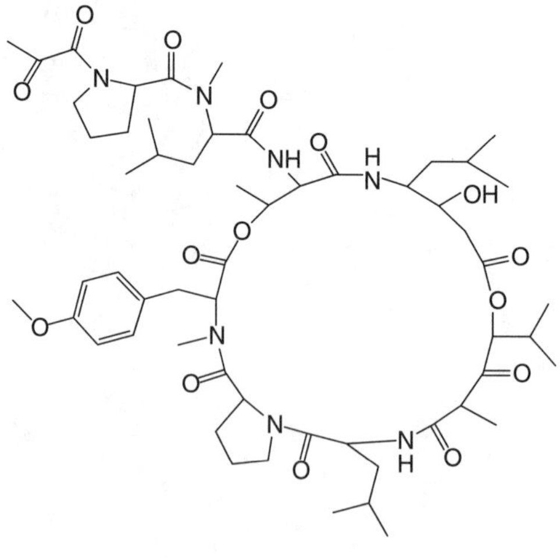

16.14

alkaloid, are both antimitotic agents. However, dolastatin 10 acts at the GTP binding site, while ET743 acts through the disorganization of the microtubule network. It is interesting to note the phyletic, structural, and mechanistic diversity among these compounds.[2]

In addition to the compounds currently in clinical trials, there are a number of compounds in advanced pre-clinical trials. Discodermolide (Structure 16.3) is a polyhydroxylated lactone isolated from the deep-water sponge *Discodermia dissoluta*.[33] An antimitotic compound which stabilizes microtubules,[14] discodermolide has been licensed to Novartis Pharma AG for development as an anticancer drug. Halichondrin B (Structure 16.15), isolated from the deep-water sponge *Lissoden-doryx*,[34] is in pre-clinical trials at the NCI for the treatment of melanoma and leukemia. Kahalalide F (Structure 16.16), isolated from the mollusc *Elysia rubefescens*,[35] is being evaluated as an anticancer agent against colon and prostate cancer. The pseudopterosins (e.g., Structure 16.17), anti-inflammatory agents from the Caribbean sea whip *Pseudopterogorgia elisabethae*,[36] have been licensed by Nereus Pharmaceuticals (La Jolla, CA) for evaluation of their activity in pharmaceutical assays.[37] The pseudopterosins are currently on the market as a component of Estee Lauder's Resilience® line of skin care products.

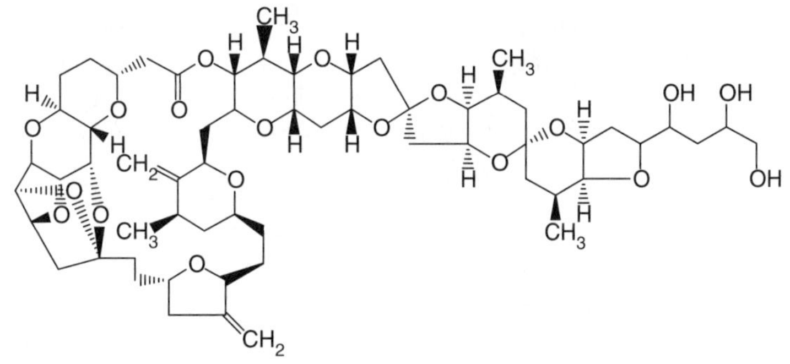

16.15

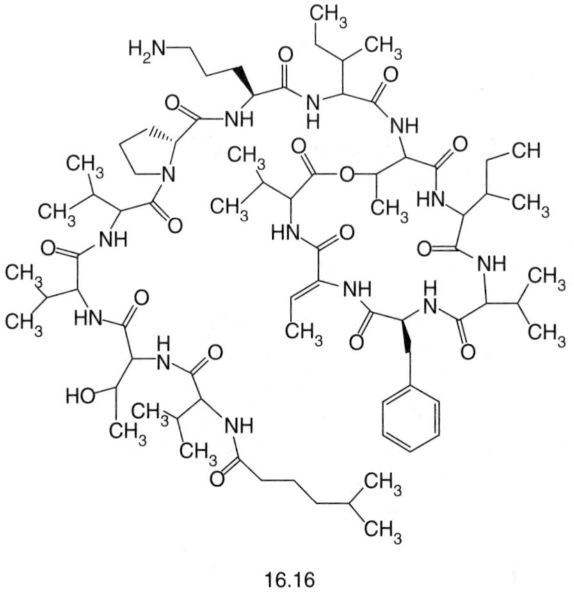

16.16

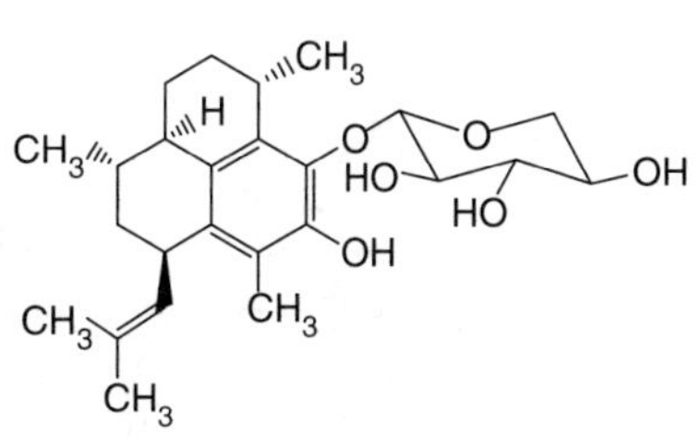

16.17

In several cases, series of structurally related analogs such as the didemnins have been discovered from a single source, the compounds within the series demonstrating varying levels of activity in pharmaceutical assays. Evaluation of such a series of analogs provides an opportunity for examination of structure–activity relationships and selection of the most active structure.[11] Other series of analogs have been evaluated for the bryostatins, halichondrins, dolastatins, and cryptophycins. Interestingly, the sources for these groups of compounds are taxonomically distinct. Ecologically, it may be hypothesized that these variants are produced to be effective against specific predators or fouling agents rather than producing one potent, generally active compound. Alternatively, a group of such closely related compounds may be used to target a specific group of isozymes.

Marine natural products have also been shown to have multiple activities in pharmaceutical assays. For example, dercitin (Structure 16.18), a heterocyclic acridine alkaloid isolated from the deep-water sponge *Dercitus*, has *in vitro* antiviral, antitumor, and immunomodulatory activity.[38,39] Topsentin (Structure 16.19), a bis(indole)-alkaloid isolated from the deep water sponge *Spongosorites*, is a potent anti-inflammatory agent[40] in addition to having antitumor and antiviral activity.[41,42] The antitumor compound discodermolide, produced by the deep-water sponge *Discodermia,* was originally isolated as an immunosuppressive agent[30] before its antitumor activity was discovered. Multiple activities might suggest interaction with multiple receptors or targets. It may be possible that these nonspecific compounds might have multiple ecological targets as well.

Marine-derived secondary metabolites have clearly had an impact on the discovery of novel agents with pharmaceutical potential. Continued prospecting in the marine environment will surely result in the identification of additional drug leads.

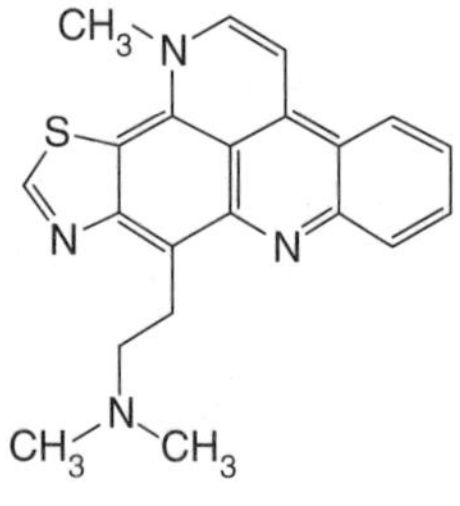

16.18 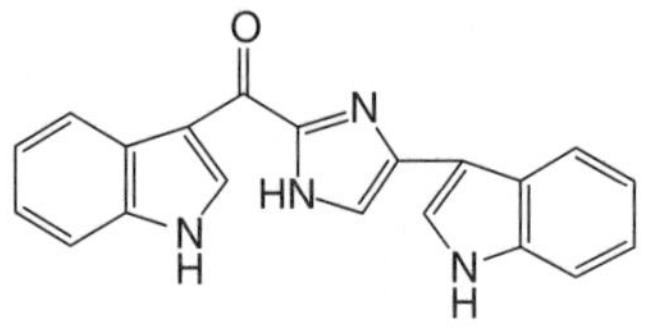   16.19

III. MARINE NATURAL PRODUCTS AS A SOURCE FOR DRUG DISCOVERY

A. HISTORICAL ASPECTS

The use of marine natural products in drug discovery began serendipitously in the 1950s with the isolation of the arabinosyl nucleosides spongothymidine and spongouridine from the sponge *Crytpotethya crypta*.[43,44] These compounds were used as models for the production of Ara A and Ara C, antiviral and anticancer drugs that are currently the only marine-derived compounds that have been marketed as pharmaceuticals. During the 1960s, the driving force for marine natural products research was the structural novelty and diversity of marine-derived compounds.[8,10] Interest in these unique compounds was heightened by the realization that many had biomedical potential, and so began a trend in chemical prospecting that continues to accelerate.[45]

B. ESTABLISHING TRENDS IN INCIDENCE OF ACTIVITY

As there was no ethnomedicinal history to guide the selection of organisms to probe for bioactive natural products, two basic approaches were taken. Early discovery programs sought to establish trends in incidence of activity to use as guides for subsequent collections.[46–48] This was accomplished by screening crude extracts of large numbers of organisms in a variety of bioassays. Resultant trends were observed in the following areas: taxonomy, geography, bioactivity, and structural class. Although the bioassays used in these studies ranged from pharmaceutical screens in human disease areas, such as anticancer, antiviral, and anti-infective, to more ecologically relevant assays including inhibition of settlement of larvae and behavioral modification of invertebrate adults,[46] there were several trends in common. However, there were some discrepancies in the trends reported (e.g., no latitudinal trend[49,50]). In general, trends suggest that the tropics, especially the Indo-Pacific, are the greatest source of bioactive natural products.[51] However, temperate,[52] Antarctic,[53] and deep-water habitats[33,38,41] have also proven to be good sources. Although the logistics of obtaining organisms from these environments may be more difficult, the extreme conditions would suggest the potential for even greater diversity of bioactive metabolites. The tropics, in general, are characterized by high diversity but low abundance of individuals with bioactive metabolites. Conversely, polar regions support lower diversity but higher abundance of source organisms.[54] Greater biomass of source organisms is an important consideration as supply for biological evaluation becomes an issue for compounds identified as viable drug candidates. This supply issue is discussed in more detail later.

Taxonomically, data suggest that the Porifera are the most prolific in the production of bioactive metabolites.[55] Other groups rich in bioactive compounds include the cnidarians, bryozoans, tunicates, and algae. It is interesting to note that in spite of the number and diversity of compounds produced by sponges, many sponge metabolites have not been suitable drug leads because of their extreme toxicity. It may be possible that the primitive nature of these sessile, benthic organisms requires production of extremely potent defensive compounds such as the cytotoxic mycalamides isolated from *Mycale* sp.[56] and the anti-inflammatory manoalide isolated from *Luffariella variabilis*,[57] both of which have proven to be too toxic to be clinically useful to date. Another point of interest with regard to the Porifera is that discodermolide and halichondrin B, currently in advanced preclinical trials, have been isolated from deep-water organisms. It is possible that differences in resource limitations and predation pressure account for differences in the level of activity.

On a finer scale, one study found correlations between antiviral and cytotoxic activities and growth forms in the Porifera and concluded that encrusting individuals had higher antiviral activity, perhaps enhancing their ability to rapidly colonize substrates, overgrow other individuals, and prevent being overgrown by other encrusting organisms. The solitary, erect forms, which were more often grazed upon than the encrusting forms, had greater cytotoxic activity, suggesting a possible antipredatory role.[47] Uriz et al.[48] identified chemical strategies among groups of organisms in the Mediterranean using nonmarine assays. Correlations were observed between antitumor and cytotoxic activities and between antibacterial and antifungal activities. It was suggested that the antimicrobial activities might be correlated with different levels of fouling.

More recently, Reed et al.[58] used a database linking biological descriptors (e.g., taxonomy, morphology, and observed associations), physical data (e.g., location, depth, and salinity), and bioactivity data to examine trends for a worldwide collection of over 25,000 marine macroinvertebrates and algae. Bioassays used in this study included whole-cell and receptor-based assays, which are generally more specific and have lower hit rates than the whole-cell assays. The greatest incidence of activity was found in the Porifera, as was observed in previous studies. Comparison of bioactivity with depth of collection suggests that, although different for each type of assay, activity is observed throughout the depth range of this study (0–3000 ft). Latitudinal trends were observed within assays as well as within some of the phyla. These data indicate that the negative correlation between bioactivity and latitude is not without exception, and other factors such as *in situ* temperature and depth should be considered. Further analysis of this large data set may be used to guide further collections.

As demonstrated, random screening can provide insight into a variety of trends in bioactivity. However, these investigations are often restricted to a particular geographic region, taxonomic group, or screening target, limiting the ability to apply findings to benthic communities in general. It has been suggested that finding unique structural classes is enhanced by screening organisms from many taxa when compared to screening samples within a single taxon.[59]

C. BIOACTIVITY AND ECOLOGICAL INTERACTIONS

A second, perhaps more rational, approach to marine chemical prospecting is based upon observation of ecological interactions *in situ*. Examples of these chemical ecological interactions and corresponding bioactivity are the subject of reviews by Bakus et al.,[60] Paul,[51] and Hay,[61] as well as in other chapters in this volume. Trends in the incidence of bioactivity reported in these studies are similar to those established by random screening. The most studied interactions have been with predators/herbivores as these are believed to exert the greatest pressure driving the evolution of chemical diversity, especially in tropical systems.[9] More recently, the focus has shifted to receptor-based activities including antifouling,[62,63] gamete settlement,[64,65] metamorphic cues,[66] and allelo-pathic interactions.[67-69] Considering the variety of stimuli, it is not surprising that some metabolites have more than one function.[67,70,71] These compounds may act independently or have additive or synergistic effects with other metabolites produced by the same organism. For example, Thacker et al.[67] reported that the crude extract of a *Dysidea* sp. was more deterrent against predators and caused a greater amount of necrosis than the pure compound, 7-deacetoxyolepupuane. Would compounds with multiple ecological roles and, perhaps, multiple effects on one or several receptors be appropriate as drug leads? Is it possible that, mechanistically, these compounds act in a non-specific manner in defending against predators, fouling organisms, and competitors for substratum? Perhaps nonspecificity would allow interaction with several molecular receptors that, synergistically, would provide the desired organismal response. Would the activity be nonspecific in mammalian systems as well? Although several of the compounds listed in Table 16.1 have multiple pharmaco-logical roles, ecological roles and specificity of these compounds are not known.

Marine macroinvertebrates may produce a number of different compounds which may or may not have the same biosynthetic origins. These compounds may be products of the invertebrate host, microbial associates, or a combination of the two.[72,73] The true role of the compound may differ depending on the origin. For example, production of an antibiotic by a microbial associate might suggest a role in competition against nonsymbiotic strains. However, if the host invertebrate is the source, the compound may aid in feeding efficiency by clumping bacteria, as was suggested by Thompson et al.[74] as a role for aerothionin in *Aplysina fistularis*. Regardless of the biogenetic origin of these compounds, the intact association provides benefits.

Another consideration in targeting potential sources of bioactive compounds is that a variety of organisms at higher trophic levels may not synthesize their own defenses but may acquire them instead from dietary sources. Examples include molluscs ingesting sponge metabolites and storing them in gut or mantle tissues or transferring the compound to larvae.[75] Compounds may be sequestered in the form in which they are ingested,[76] or they may be modified by the consumer, altering the activity.[77] Comparison of the original natural product with modified analogs may provide insight into structure–activity relationships and suggest the source with the greatest potential for the desired bioactivity. The route of biotransformation might also provide insight for further modification after isolation.

Finally, recent reviews indicate that marine microorganisms are potentially a greater source of bioactive compounds than marine macroorganisms.[78-81] Marine microorganisms can be found in seawater or sediment, associated with macroorganisms either on the surface or symbiotically, and in extreme environments. Extremophiles, in particular, may have the greatest capacity for the production of unique bioactive metabolites.[78]

IV. CONSIDERATIONS FOR DEVELOPMENT

When a compound from a marine source is identified as a drug lead or candidate, there are several issues to consider related to supply of the compound: yield of the compound, abundance of the source organism, ease and cost of recollection, and, finally, the issue of sovereign rights over biological resources. The UN Convention on Biological Diversity requires that member countries be committed to the conservation and sustainable use of natural resources and the fair and equitable sharing of revenues resulting from commercialization of products based on natural resources.[82] These requirements are having a significant impact on where collections can be made and who will benefit from subsequent discoveries.

V. ADDRESSING THE SUPPLY ISSUE

There are many reports of intraspecific variation in the quantitative and qualitative production of secondary metabolites. These variations can occur within a single member in a population,[83] among different members of a population,[84] or among different populations of a given species.[85,86] Differences may be due to differences in levels of predation, habitat,[87] ontogeny,[88] geographic location,[89,90] and depth,[91] though it is not known whether the differences are genetically or environmentally mediated.

As a result of this sometimes ephemeral production, or simply the naturally low yield of some compounds, adequate supply for preclinical and clinical evaluation is a critical issue.[92] It has been demonstrated that obtaining sufficient quantities by bulk collection from natural populations is not economically or ecologically feasible.[93] The long-term impact of massive collections on the survival of wild populations is unknown, but it is obvious that if marine-derived compounds prove to be clinically successful, demand will exceed what the natural populations can provide.

Also at issue is the preservation of marine biodiversity that relates, in turn, to chemical diversity. As stated previously, discoveries to date reflect a small percentage of the resources available. Measures must be taken to ensure sustainable use of these resources. Alternative renewable sources need to be identified to supply pharmacological evaluation. Due to the critical nature of this aspect of marine natural products drug discovery, several alternative approaches are discussed in detail.

There are several approaches that can be taken to supply material for pharmaceutical evaluation. Two of these, invertebrate cell culture and fermentation of associated microorganisms, begin with determining the biogenetic origin of the compound. There are numerous reports of metabolite localization, primarily in sponges, in which production of a bioactive metabolite has been inferred or demonstrated to be localized either in a host invertebrate cell or in microbial associates which include cyanobacteria and heterotrophic bacteria and fungi.[72,73]

A. MICROBIAL LOCALIZATION BY INFERENCE

There are a number of examples in which the production of a metabolite isolated from a marine invertebrate has been attributed to an associated microorganism.[94] Some of these reports are based on circumstantial evidence such as similarity of the metabolite to a compound produced by a microorganism. For example, a bacterium of the genus *Alteromonas* isolated from the sponge *Halichondria okadai* produced the macrocyclic lactam alteramide A (Structure 16.20) in culture.[95] This compound, which was not found in extracts of the host sponge, appears to be biogenetically related to ikarugamycin (Structure 16.21), a peptide-like antibiotic isolated from a terrestrial *Streptomyces* sp.[96] Discodermide (Structure 16.22), isolated from the marine sponge *Discodermia dissoluta*,[97] also appears to be structurally related to these compounds. Based on this structural similarity, it has been suggested that discodermide is a microbial metabolite.

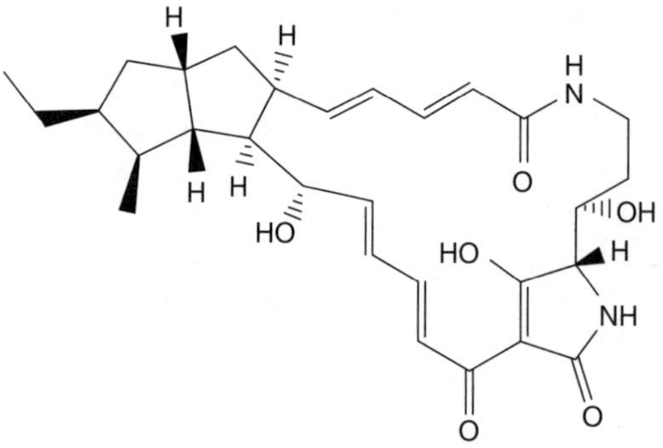

16.20

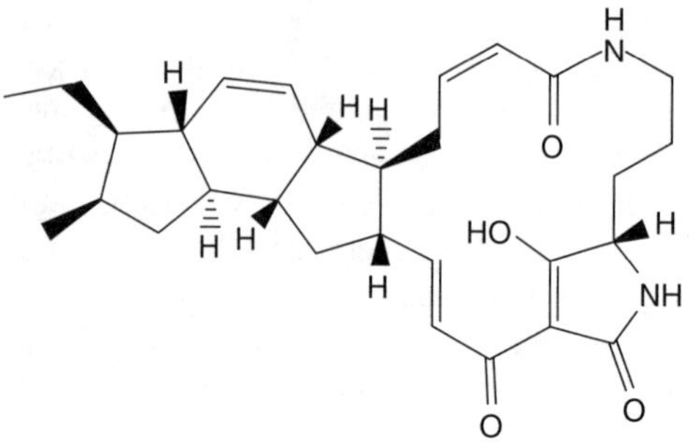

16.21

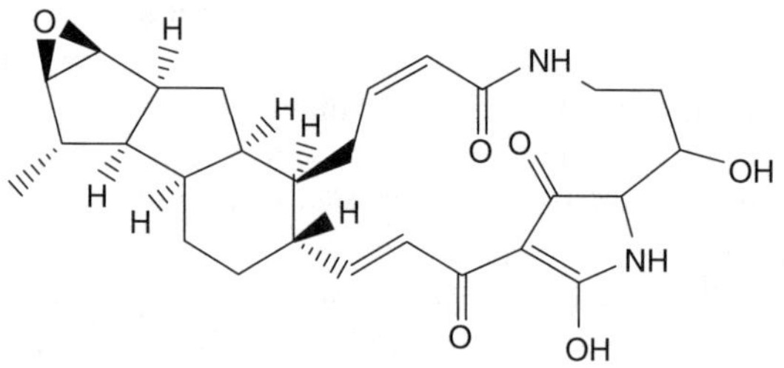

16.22

The occurrence of structurally related metabolites in unrelated taxa has also been cited as evidence for production of metabolites by microbial associates. A number of fused polycyclic aromatic alkaloids have been isolated, including dercitin from the sponge *Dercitus* sp.,[38] the cystodytins from the tunicate *Cystodytes* sp.,[98] and calliactine from the sea anemone *Calliactis* sp.,[99] among others. It has been suggested that these might originate from associated microbes, but a microbe common to these organisms has not yet been identified. An alternative explanation for the occurrence of similar compounds in diverse taxa might be that of convergent evolution depending on the structure class and complexity of biosynthetic starting materials.

In other cases, associated microorganisms have been isolated from their hosts and demonstrated to produce the same metabolites originally isolated from the intact association.[72,100–102] One of the first studies involved a bright-orange-pigmented Gram-positive bacterium of the genus *Micrococcus* which was isolated from the sponge *Tedania ignis*. The bacterium was successfully cultured, and three diketopiperazines were isolated from the culture media.[100] The compounds were identical to those previously isolated from the intact association,[103] suggesting that the compounds were of microbial origin. It has since been suggested that the diketopiperazines were degradation products of peptides in both the sponge and the bacteria, or were perhaps produced by the bacteria and merely accumulated by the sponge.[104] In addition, it has been reported that greater than 90% of all Gram-negative bacteria produce these diketopiperazines in nutrient-rich conditions.[78] There are other similar examples suggesting a microbial origin for secondary metabolites,[101,102] but very few have been clearly demonstrated.

B. DIRECT EVIDENCE FOR METABOLITE LOCALIZATION IN HOST CELLS

Direct evidence for localization in sponge cells or associated microorganisms has been obtained using a variety of separation and detection techniques, including density gradient fractionation, differential centrifugation, and auto- and induced-fluorescence combined with flow cytometric sorting as well as confocal or electron microscopy. The following examples detail the methods used and the site of production of the target compounds.

Cellular localization of a secondary metabolite in a sponge was first reported by Thompson et al.[74] Following separation and enrichment of the different sponge cell types using Ficoll density gradient centrifugation, two brominated compounds with antibiotic properties, aerothionin and homoaerothionin, were detected by energy dispersive x-ray microanalysis in the spherulous cells of the sponge *Aplysina fistularis*.

Density gradient fractionation followed by transmission electron microscopy and chemical methods were used to determine that the antimitotic agent avarol was produced in the spherulous cells of the sponge *Dysidea avara*.[105] It was suggested that this compound may be involved in the regulation of bacterial flora. Uriz et al.[106] re-examined the cellular localization of avarol in *Dysidea avara*. Cytological characterization of cell types present prior to fractionation was carried out on fresh and fixed sponge material using light and electron microscopy. Subsequent fractionation and chromatographic analysis revealed that avarol was present in choanocytes and that no true spherulous cells were present in this species. A difference in location of the compound in the intact sponge may suggest an alternative role for avarol *in situ*. Uriz et al.[107] also used these methods to localize the toxic crambines and crambescidines in the spherulous cells of the Mediterranean sponge *Crambe crambe*. Spherulous cells are found in the ectosomal layer of the sponge prompting the suggestion that the compounds play a defensive role in the sponge.

Density gradient fractionation followed by chromatographic analysis were also used to localize the antibiotic compound 5-hydroxy tryptophan (5HTP) in the sponge *Hymeniacidon heliophila*.[108] Enriched fractions of several sponge cell types were obtained using a discontinuous Percoll gradient, and the cell types in each fraction were identified using transmission electron microscopy. HPLC analysis of methanol extracts of the enriched cell fractions and co-injection of extracts with the pure compound revealed that 5HTP was localized in the archaeocytes of *H. heliophila*.

The biosynthesis of the halichondrins, a family of compounds isolated from *Lissodendoryx* sp., was found to be associated with Percoll-enriched fractions of choanocytes and spherulous cells.[109] The discorhabdins, isolated from *Latrunculia brevis*, were also observed to be associated with sponge cells using similar techniques.

Fluorescence may by induced using labeled mono- or polyclonal antibodies raised against the compound of interest. This technique has been used successfully to detect the presence of okadaic acid in cultures of *Prorocentrum lima*, and, further, to estimate quantities of the compound in individual cells.[110] Immunofluoresence in combination with thin layer chromatography and ELISA techniques have also been used to detect multiple haptens in mycotoxin families.[111,112]

Ilan[113] reported the localization of latrunculin B, isolated from *Negombata magnifica*, using antibodies raised against the compound. Transmission electron microscopy of immunohistochemical and immunogold labeled cells revealed that the compound was concentrated in archaeocytes and choanocytes in the ectosome. Presence of latrunculin B in the outer layer of the sponge is suggested to correlate with the defensive role of this toxic compound.

1. *In Vitro* Production of Bioactive Compounds

All of the examples cited above reported the production of the target compound in invertebrate cells. Isolation and culture of the source cells could provide a renewable source of bioactive compounds. A review of the last decade of research in invertebrate cell culture summarizes the successes and difficulties encountered in this developing field.[114] To date, only primary cultures have been established for a limited number of species in six phyla including the Cnidaria, Crustacea, Echinodermata, Mollusca, Porifera, and Urochordata. However, no continuous cell lines have been established.

These studies have revealed problems unique to each group as well as difficulties common among the different phyla. There are several obstacles specific to the Porifera, which may negatively impact the feasibility of *in vitro* production of bioactive compounds by sponges. Perhaps the greatest obstacle is the lack of any area of a sponge from which axenic primary cultures can be initiated.

In addition, sponges harbor symbiotic intracellular microorganisms which may or may not be obligate symbionts and required for production of the desired metabolites. Maintenance of sponge cell cultures may require the use of antibiotics; however, use of antibiotics may be harmful to the sponge cells[115] and may alter the biosynthesis of the desired metabolite.

Although particularly problematic in the Porifera, the problem of contamination, mainly by unicellular eukaryotic organisms, is an area of focus. This obstacle necessitates the validation of cell type in the culture system. Other areas in which new approaches are being sought include the selection of cell sources for the initiation of cultures and cryopreservation protocols for the maintenance of source cells for long-term studies.

In spite of the difficulties reported, model primary culture systems have been established for shallow-water sponges that produce bioactive secondary metabolites. It has been demonstrated that cultured cells can be stimulated to divide by mitogens and retain the ability to synthesize bioactive compounds.[115,116] These successes suggest that, with a greater understanding of requirements for growth and compound production, invertebrate cell culture may become a viable source for bioactive marine natural products.

C. Direct Evidence for Microbial Localization

Isolation and culture of putative symbiotic microorganisms to obtain the bioactive metabolites isolated from the invertebrate–microbe assemblages has been particularly problematic. It has been estimated that only 1% of marine microbes can be isolated and fermented using techniques modified to approximate marine conditions. However, to address this problem, several groups have used differential fractionation or flow cytometric techniques to separate populations of extracellularly associated microorganisms from invertebrate host cells. Chromatographic and spectroscopic analyses of the enriched cell fractions have been used to determine which cell type is producing the compounds. The assumption is that the metabolites are produced in the cell types in which they are detected.

The symbiotic association between *Dysidea herbacea* and the cyanobacterium *Oscillatoria spongeliae* is well documented.[117] In addition, the secondary chemistry of this association is well known.[118–120] Polychlorinated amino acid derivatives and sesquiterpenes have been reported from *Dysidea herbacea*. Sesquiterpenes are known throughout the genus *Dysidea*, while the polychlorinated amino acid derivatives appear to be related to compounds isolated from free-living cyanobacteria. To investigate the cellular localization of these compounds, sponge and microbial cells were dissociated, fixed in formalin or glutaraldehyde, and separated by flow cytometry using the fluorescent photosynthetic pigments in the cyanobacteria. Cells were then extracted and analyzed by ^{1}H NMR spectroscopy, revealing that the polychlorinated compounds were found exclusively in the cyanobacterial cells while the sesquiterpenoid metabolites were limited to the sponge cells.[72] This study was the first demonstration that secondary metabolites attributed to a sponge were localized in prokaryotic symbiont cells.

Bewley et al.[121] used a similar strategy to localize the source of bioactive compounds in the lithistid sponge *Theonella swinhoei*. Gross separation of the endosomal and ectosomal tissues followed by differential centrifugation resulted in a fraction enriched in sponge cells and three microbial fractions — heterotrophic unicellular and filamentous bacteria and unicellular cyanobacteria. Chemical analysis of the fractions revealed that the bioactive metabolites swinholide A (Structure 16.23) and theopaluamide (Structure 16.24), previously identified from the sponge, were actually localized in two heterogeneous fractions of microorganisms. Although the techniques were not suitable for subsequent culture of the microorganisms, they may form the basis for development of methods to selectively isolate and culture associated microorganisms for bulk production of bioactive metabolites.

Invertebrate cell culture and fermentation of associated microbes for the production of secondary metabolites appear to be promising. These technologies are attractive as they may

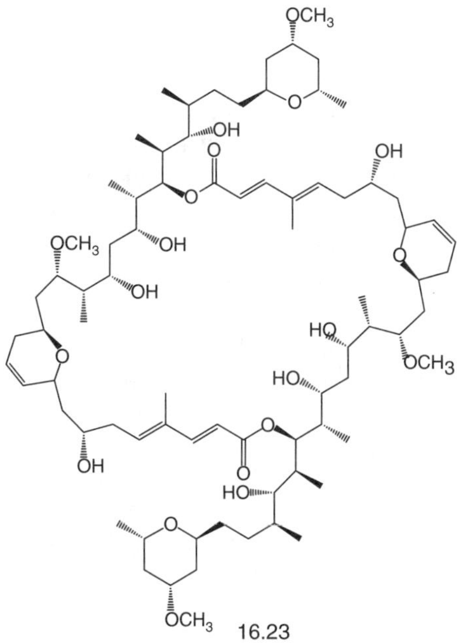

16.23

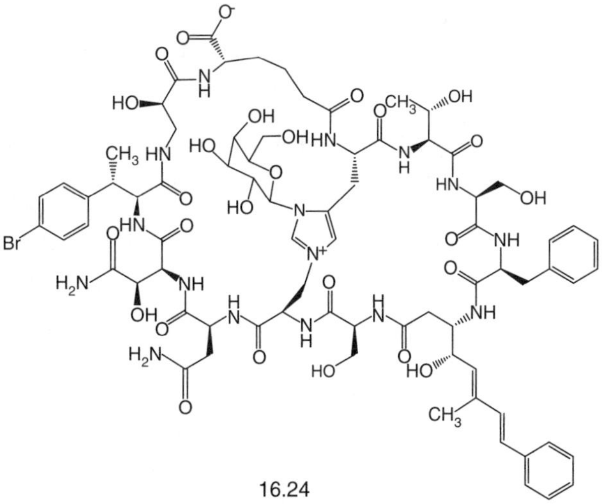

16.24

allow for manipulation of growth and metabolite production at the molecular level. However, these cannot yet be considered viable options for supply. There are several alternatives that are currently being used, including chemical synthesis, controlled harvest, and land-based and in-the-sea aquaculture.

D. ESTABLISHED METHODS FOR ALTERNATIVE SUPPLY

Chemical synthesis may be the first consideration in addressing the supply issue. However, the complexity of many marine natural products often eliminates this as an alternative. Even if synthesis is possible, a distinction must be made between academic and industrial syntheses, the latter necessarily being straightforward and low cost.[2] Medicinal chemistry techniques and structure–activity relationship studies might identify analogs or the pharmacophore responsible for the activity.

These structures may be more amenable to providing an alternative to synthesis of the original natural product.

Controlled harvest may be a viable option for fast-growing species. However, the growth rates of many of the organisms currently of interest are not known. In addition, this method may be hampered by changes in environmental variables which may negatively impact growth of the organism and production of the target compound. For example, early clinical evaluations of bryostatin 1 were made possible by bulk harvest of Pacific populations of *Bugula neritina*.[122] Low yield and patchy distribution of production among different populations required large-scale collections. Unfortunately, target populations had already been stressed by El Niño conditions which exacerbated the negative impact of the collections.[93]

In-the-sea and land-based aquaculture can be cost-effective production methods which afford increased control over environmental conditions. CalBioMarine Technologies (Carlsbad, CA) has developed both land- and sea-based systems for the culture of the shallow water bryozoan, *Bugula neritina*, and the ascidian *Ecteinascidia turbinata*, the sources of compounds bryostatin 1 and ET743, respectively.[123] The life histories of these organisms have been studied, but the factors controlling the synthesis of the target compounds have not been fully characterized. Recent studies have identified two chemotypes of *Bugula*, with type-specific microbial associations. Chemotype O contains bryostatin 1 and related compounds with an octa-2,4-dienoate substituent. Chemotype M compounds lacks bryostatins with the octa-2,4-dienoate moiety. A better understanding of the conditions related to the differences in production may lead to optimization and increased production of bryostatin 1.[124]

Lissodendoryx is a deep-water sponge occurring at depths from 85 to 105 m and is the source of the halichondrins. Although this is a deep-water species, Battershill et al.[125] have been successful in culturing cuttings of this sponge on lantern arrays at shallower depths while maintaining production of the halichondrins. This model suggests that culture of some deep-water organisms may be feasible, though species from greater depths may require consideration of low light, high pressure, and low temperature. Battershill et al.[125] have also been successful in in-the-sea aquaculture of sponges of the genera *Mycale* and *Latrunculia* for the production of bioactive metabolites.[125]

Although in-the-sea culture may be cost-effective, closed systems would allow greater control over environmental parameters and manipulation of those parameters to increase production of the compounds of interest. Osinga et al.[126] have demonstrated the feasibility of culturing explants of shallow water sponges in a closed-system bioreactor.

The examples cited above illustrate that there are currently several alternatives to bulk harvest for the supply of bioactive marine natural products and several more in development. However, as the ability to access organisms in more extreme environments improves, conditions required for growth of these extremophiles will be more complex and difficult to reproduce.

VI. A MULTIDISCIPLINARY APPROACH TO MARINE NATURAL PRODUCTS DRUG DISCOVERY

The marine environment has been and will continue to be a rich source of diverse and structurally unique bioactive metabolites for drug discovery. Effective utilization of this resource will require advances in technologies for collection of organisms from more extreme environments as well as more timely isolation and structure elucidation of potential drug candidates from crude extracts. The current time frame for isolation of bioactive pure compounds from crude extracts is approximately that of use of a target in high throughput screening,[3] generally around 6 months for major pharmaceutical companies. Biotechnological advances for identification and development of alternative supplies will also be necessary for the conservation and sustainable use of marine natural products.[123]

Pharmacologists have acquired a tremendous amount of information regarding the activity of marine natural products in mammalian systems. However, ecological roles have not been determined for any of the compounds currently in advanced pre-clinical or clinical development. Considering the evolutionary conservation of signaling pathways, can we use what is known about molecular targets in mammalian systems to better understand the ecological roles of these compounds in the marine environment? An increased understanding of ecological roles will allow for better selection of potential source organisms as well as the development of more relevant screens.

Exploration of the molecular basis of human disease will continue to provide additional targets for drug discovery, while a better understanding of marine ecological interactions at the molecular level can be used to guide the selection of novel sources and novel screens. This multidisciplinary approach to discovery of marine-derived drugs will likely result in the continued discovery of unique bioactive chemical entities and new ways to address the treatment of human disease.

REFERENCES

1. Cragg, G. M. and Newman, D. J., Discovery and development of antineoplastic agents from natural sources, *Cancer Invest.*, 17, 153, 1999.
2. Munro, M. H. G., Blunt, J. W., Dumdei, E. J., Hickford, S. J. H., Lill, R. E., Li, S., Battershill, C. N., and Duckworth, A. R., The disocvery and development of marine compounds with pharmaceutical potential, *J. Biotech.*, 70, 15, 1999.
3. Shu, Y. Z., Recent natural products based drug development: a pharmaceutical industry perspective, *J. Nat. Prod.*, 61, 1053, 1998.
4. Zurer, P., When it comes to diversity, nature comes out ahead, *Chem. Eng. News*, 17, 13, 1999.
5. Wallace, R. W., Drugs from the sea: harvesting the results of aeons of chemical evolution, *Mol. Med. Today*, 3, 277, 1997.
6. Cragg, G. M., Newman, D. J., and Snader, K. M., Natural products in drug discovery and development, *J. Nat. Prod.*, 60, 52, 1997.
7. Harvey, A. L., Medicines from nature: are natural products still relevant to drug discovery?, *TIPS*, 20, 196, 1999.
8. Carte, B. K., Marine natural products as a source of novel pharmacological agents, *Current Opinion. Biotech.*, 4, 275, 1993.
9. Hay, M. E. and Fenical, W., Chemical ecology and marine biodiversity: insights and products from the sea, *Oceanogr.*, 9, 10, 1996.
10. Deitzman, G. R., The marine environment as a discovery resource, in *High Throughput Screening: The Discovery of Bioactive Substances*, Devlin, J. P., Ed., Marcel Dekker, New York, 1997, 99.
11. Ireland, C. M., Copp, B. R., Foster, M. P., McDonald, L. A., Radisky, D. C., and Swersy, J. C., Biomedical potential of marine natural products, in *Marine Biotechnology: Pharmaceutical and Bioactive Natural Products*, Attaway, D. H. and Zaborsky, O. R., Eds., Plenum Press, New York, 1993, 1.
12. Bubb, M. R., Senderowicz, M. J., Sausville, E. A., Duncan, K. L. K., and Korn, E. D., Jasplakinolide, a cytotoxic natural product, induces actin polymerization and competitively inhibits the binding of phalloidin to F-actin, *J. Biol. Chem.*, 269, 14869, 1994.
13. Spector, I. and Shochet, N. R., Latrunculins: novel marine toxins that disrupt microfilament organization in cultured cells, *Science*, 219, 493, 1983.
14. ter Haarr, E., Kowalski, R., Hamel, E., Lin, C. M., Longley, R. E., Gunasekera, S. P., and Rosenkranz, H. S., Discodermolide, a cytotoxic agent that stabilized microtubules more potently than taxol, *Biochemistry*, 35, 243, 1996.
15. Blokhin, A .V., Yoo, H., Geralds, R. S., Nagle, D. G., Gerwick, W. H., and Hamel, E., Characterization of the interaction of the marine cyanobacterial natural product Curacin A with the colchicine site of tubulin and initial structure-activity studies with analogues, *Mol. Pharmcol.*, 48, 523, 1995.
16. Potts, B. C. M., Faulkner, D. J., de Carralho, M. S., and Jacobs, R. S., Chemical mechanism of inactivation of bee venom phospholipase A_2 by the marine natural products manoalide, luffariellolide, and scalaradial, *J. Am. Chem. Soc.*, 114, 5093, 1992.

17. Bialojan, C. and Takai, A., Inhibitory effect of a marine-sponge toxin, okadaic acid, on protein phosphatases, *Biochem. J.*, 256, 283–290, 1988.

18. Gunasekera, S. P., McCarthy, P. J., Kelly-Borges, M., Lobkovsky, E., and Clardy, J., Dysidiolide: a novel protein phosphatase inhibitor from the Caribbean sponge *Dysidea etheria* de Laubenfels, *J. Am. Chem. Soc.*, 118, 8759, 1996.

19. Gunasekera, S. P., McCarthy, P. J., Longley, R. E., Pomponi, S. A., Wright, A. E., Lobkovsky, E., and Clardy, J., Discorhabdin P, a new enzyme inhibitor from a deep-water Caribbean sponge of the genus *Batzella*, *J. Nat. Prod*, 62, 173, 1999.

20. Hennings, H., Blumberg, P. M., Pettit, G. R., Herald, C. L., Shores, R., and Yuspa, S. H., Bryostatin 1, an activator of protein kinase C, inhibits tumor promotion by phorbol esters in SENCAR mouse skin, *Carcinogenesis*, 8, 1343, 1987.

21. Kao, C. Y., Pharmacology of tetrodotoxin and saxitoxin, *Fed. Proc.*, 31, 1117, 1972.

22. Leutje, C. W., Wada, K., Rogers, W., Abramson, S.N., Tsuji, K., Heinemann, S., and Patrick, J., Neurotoxins distinguish between different neuronal nicotinic acetylcholine receptor subunit combinations, *J. Neurochem.*, 55, 632, 1990.

23. Matsumoto, S. S., Haughey, H. M., Schmehl, D. M., Venables, D. A., Ireland, C. M., Holden, J. A., and Barrows, L. R., Makaluvamines vary in ability to induce dose-dependent DNA cleavage via topoisomerase II interaction, *Anti-Cancer Drugs*, 10, 39, 1999.

24. Barrows, L. R., Radisky, D. C., Copp, B. R., Swaffar, D. S., Kramer, R. A., Warters, R. L., and Ireland, C. M., Makaluvamines, marine natural products, are active anti-cancer agents and DNA topoisomerase II inhibitors, *Anti-Cancer Drug Design*, 8, 333, 1993.

25. McConnell, O. J., Longley, R. E., and Koehn, F. E., The discovery of marine natural products with therapeutic potential, in *The Discovery of Natural Products with Therapeutic Potential*, Gullo, V. P., Ed., Butterworth–Heinemann, Boston, MA, 1994, 109.

26. Pettit, G. R., Herald, C. L., Doubek, D. L., and Herald, D. L., Isolation and structure of bryostatin 1, *J. Am. Chem. Soc.*, 104, 6846, 1982.

27. Poncet, J., The dolastatins, a family of promising antineoplastic agents, *Curr. Pharm. Des.*, 5, 139, 1999.

28. Rinehart, K. L., Holt, T. G., Fregeau, N. L., Stroh, J. G., Keifer, P. A., Sun, F., Li, L. H., and Martin, D. G., Ecteinascidins 729, 743, 745, 759B, and 770: potent antitumor agents from the Caribbean tunicate, *Ecteinascidia turbinata*, *J. Org. Chem.*, 55, 4512, 1990.

29. Wright, A. E., Forleo, D. A., Gunawardana, G. P., Gunasekera, S. P., Koehn, F. E., and McConnell, O. J., Antitumor tetrahydroisoquinoline alkaloids from the colonial ascidian *Ecteinascidia turbinata*, *J. Org. Chem.*, 55, 4508, 1990.

30. Rinehart, K. L. and Lithgow-Bertelloni, A. M., PCT International Patent WO 9104985 A1, 1991.

31. Valoti, G., Nicoleetti, M. I., Faircloth, G., Jimeno, J., and Giavazzi, R., Antitumor effect of ecteinascidin-743 (ET743) on human ovarian carcinoma xenografts, in *Proceedings of the Annual Meet.ing of the American Association of Cancer Research*, Mario Negri Institute for Pharmacological Research, Bergamo, Italy, 1997.

32. Depenbrock, H., Peter, R., Faircloth, G. T., Manzanares, I., Jimeno, J., and Hanauske, A. R., In vitro activity of aplidine, a new marine-derived anticancer compound, on freshly explanted clonogenic human tumor cells and hematopoietic precursor cells, *Br. J. Cancer*, 78, 739, 1998.

33. Gunasekera, S. P., Gunasekera, M., Longley, R. E., and Schulte, G. K., Discodermolide: a new bioactive polyhydroxylated lactone from the marine sponge, *Discodermia dissoluta*, *J. Org. Chem.*, 55, 4912, 1990.

34. Litaudon, M., Hickford, S. J. H., and Lill, R. E., Antitumor polyether macrolides: new and hemisynthetic halichondrins from the New Zealand deep-water sponge *Lissodendoryx* sp., *J. Org. Chem.*, 62, 1868, 1997.

35. Hamman, M. T., Otto, C. S., Scheuer, P. J., and Dunbar, D. C., Kahalalides: bioactive peptides from a marine mollusc *Elysia rufescens* and its algal diet *Bryopsis* sp., *J. Org. Chem.*, 61, 6594, 1996.

36. Mayer, A. M. S., Jacobson, P. B., Fenical, W., Jacobs, R. S., and Glaser, K. S., Pharmacological characterization of the pseudopterosins: novel anti-inflammatory natural products isolated from the Caribbean soft coral, *Pseudopterogorgia elisabethae*, *Pharmacol. Lett.*, 62, 401, 1998.

37. Mestel, R., Drugs from the sea, *Discovery*, 20, 70, 1999.

38. Gunawardana, G. P., Kohmoto, S., Gunasekera, S. P., McConnell, O. J., and Koehn, F. E., Dercitin, a new biologically active acridine alkaloid from a deep water marine sponge, *Dercitus* sp., *J. Am. Chem. Soc.*, 110, 4856, 1988.

39. Gunawardana, G. P., Koehn, F. E., Lee, A. Y., Clardy, J., He, H., and Faulkner, D .J., Pyridoacridine alkaloids from deep-water sponges of the family Pachastrellidae: structure revision of dercitin and related compounds and correlation with kuanoniamines, *J. Org. Chem.*, 57, 1523, 1992.

40. McConnell, O. J., Saucy, G., Jacobs, R. S., and Gunasekera, S. P., Use of bis-heterocyclic compounds and pharmaceutical compositions containing same, U. S. Patent 5,496,950, 1996.

41. Tsujii, S., Rinehart, K. L., Gunasekera, S. P., Kashman, Y., Cross, S. S., Lui, M. S., Pomponi, S. A., and Diaz, M. C., Topsentin, bromotopsentin, and dihydrodeoxybromotopsentin: antiviral and antitumor bis(indolyl)imidazoles from the Caribbean deep-sea sponges of the family Halichondriidae. Structural and synthetic studies, *J. Org. Chem.*, 53, 5446, 1988.

42. Burres, N. S., Barber, D. A., Gunasekera, S. P., Shen, L. L., and Clement, J. J., Antitumor activity and biochemical effects of topsentin, *Biochem. Pharm.*, 42, 745, 1991.

43. Bergmann, W. and Burke, D. C., Contributions to the study of marine products XXXIX. The nucleosides of sponges III. Spongothymidine and spongouridine, *J. Org. Chem.*, 20, 1501, 1955.

44. Bergmann, W. and Feeney, R. J., Contributions to the study of marine products XXXII. The nucleosides of sponges. I., *J. Org. Chem.*, 16, 981, 1951.

45. Webber, H. H. and Ruggieri, G. D., in *Proceedings, Food-Drugs from the Sea*, Marine Technology Society, 1974.

46. Thompson, J. E., Walker, R. P., and Faulkner, D. J., Screening and bioassays for biologically-active substances from forty marine sponges from San Diego, California, USA. *Mar. Biol.*, 88, 11, 1985.

47. Munro, M. H. G, Blunt, J. W., Barnes, G., Battershill, C. N., Lake, R. S., and Perry, N.B., Biological activity in New Zealand marine organisms, *Pure Appl. Chem.*, 61, 529, 1989.

48. Uriz, M. J., Martin, D., Turon, X., Ballesteros, E., Hughes, R., and Acebal, C., An approach to the ecological significance of chemically mediated bioactivity in Mediterranean benthic communities, *Mar. Ecol. Prog. Ser.*, 70, 75, 1991.

49. Bergquist, P. R. and Bedford, J. J., The incidence of antibacterial activity in marine Demospongiae; systematic and geographic considerations, *Mar. Biol.*, 46, 215, 1978.

50. Green, G., Ecology of toxicity in marine sponges, *Mar. Biol.*, 40, 207, 1977.

51. Paul, V. J., Chemical defenses of benthic marine invertebrates, in *Ecological Roles of Marine Natural Products*, Paul, V. J., Ed., Comstock Press, Ithaca, NY, 1992, 164.

52. Steinberg, P. D., Geographical variation in the interaction between marine herbivores and brown algal secondary metabolites, in *Ecological Roles of Marine Natural Products*, Paul, V. J., Ed., Comstock Press, Ithaca, NY, 1992, 51.

53. McClintock, J. B., and Baker, B. J., Chemical ecology in Antarctic seas, *Am. Sci.*, 86, 254, 1998.

54. Rex, M. A., Stuart, C. T., Hessler, R. R., Allen, J. A., Sanders, H. L., and Wilson, G. D. F., Global-scale latitudinal patterns of species diversity in the deep sea benthos, *Nature*, 365, 636, 1993.

55. Ireland, C. M., Roll, D. M., Molinski, T. F., McKee, T. C., Zabriskie, T. M., and Swersy, J. C., Uniqueness of the marine chemical environment: categories of marine natural products from invertebrates, in *Biomedical Importance of Marine Organisms* (Memoirs of the California Academy of Sciences Number 13), Fautin, D. G., Ed., California Academy of Sciences, San Francisco, CA, 1988, 41.

56. Perry, N. B., Blunt, J. W., Munro, M. H. G., and Pannell, L. K., Mycalamide A, an antiviral compound from the New Zealand sponge of the genus *Mycale*, *J. Am. Chem. Soc.*, 110, 4850, 1988.

57. De Silva, E. D., and Scheuer, P. J., Manoalide, an antibiotic sesterterpenoid from the marine sponge *Luffariella variabilis* (Polejaeff), *Tetrahedron Lett.*, 21, 1611, 1980.

58. Reed, J. K., Sennett, S. H., McCarthy, P. J., Pitts, T. A., Wright, A. E., and Pomponi, S. A., Bioactivity of marine organisms: relationships with taxonomy, geography and depth, in *Proceedings of the American Academy of Underwater Sciences*, Hartwick, B., Banister, E., and Morariu, G., Eds., American Academy of Underwater Sciences, Nahant, MA, 1998, 50.

59. Devlin, J. P., Chemical diversity and genetic equity: synthetic and naturally derived compounds, in *High Throughput Screening*, Devlin, J. P., Ed., Marcel Dekker, New York, 1997, 3.

60. Bakus, G. J., Targett, N. M., and Schulte, B., Chemical ecology of marine organisms: an overview, *J. Chem. Ecol.*, 12, 951, 1986.

61. Hay, M. E., Marine chemical ecology: what's known and what's next?, *J. Exp. Mar. Bio. Ecol.*, 200, 103, 1996.

62. Henrikson, A. A. and Pawlik, J. R., A new antifouling assay method: results from field experiments using extracts of four marine organisms, *J. Exp. Mar. Biol. Ecol.*, 194, 157, 1995.

63. Sears, M. A., Gerhart, D. J., and Rittschof, D., Antifouling agents from marine sponge *Lissodendoryx isodictyalis* Carter, *J. Chem. Ecol.*, 16, 791, 1990.

64. Pawlik, J. R., Induction of marine invertebrate larval settlement: evidence for chemical cues, in *Ecological Roles of Marine Natural Products*, Paul, V. J., Ed., Comstock Publishing Associates, Ithaca, NY, 1992, 189.

65. Pawlik, J. R., Chemical ecology of the settlement of benthic marine invertebrates, *Oceanogr. Mar. Biol. Annu. Rev.*, 30, 273, 1992.

66. Morse, D. E., Recent progress in larval settlement and metamorphosis: closing the gaps between molecular biology and ecology, *Bull. Mar. Sci.*, 46, 465, 1990.

67. Thacker, R. W., Becero, M. A., Lumbang, W. A., and Paul, V. J., Allelopathic interactions between sponges on a tropical reef, *Ecology*, 79, 1740, 1998.

68. Porter, J. W. and Targett, N. M., Allelochemical interactions between sponges and corals, *Biol. Bull.*, 175, 230, 1988.

69. Bingham, B. L. and Young, C. M., Influence of sponges on invertebrate recruitment: a field test of allelopathy, *Mar. Biol.*, 109, 19, 1991.

70. Sammarco, P. W. and Coll, J. C., Chemical adaptations in the Octocorallia: evolutionary considerations, *Mar. Ecol. Prog. Ser.*, 88, 93, 1992.

71. Schmitt, T .M., Hay, M. E., and Lindquist, N., Constraints on chemically mediated coevolution: multiple functions for seaweed secondary metabolites, *Ecology*, 76, 107, 1995.

72. Unson, M. D. and Faulkner, D. J., Cyanobacterial symbiont biosynthesis of chlorinated metabolites from *Dysidea herbacea* (Porifera), *Experientia*, 49, 349, 1993.

73. Bewley, C. A., Holland, N. D., and Faulkner, D. J., Two classes of metabolites from *Theonella swinhoei* are localized in distinct populations of bacterial symbionts, *Experientia*, 52, 716, 1996.

74. Thompson, J. E., Barrow, K. D., and Faulkner, D. J., Localization of two brominated metabolites, aerothionin and homoaerothionin, in spherulous cells of the marine sponge *Aplysina fistularis*, *Acta Zool.*, 64, 199, 1983.

75. Pawlik, J. R., Kernan, M. R., Molinski, T. F, Harper, M. K., and Faulkner, D. J., Defensive chemicals of the Spanish dancer nudibranch, *Hexabranchus sanguineus*, and its egg ribbons: macrolides derived from a sponge diet, *J. Exp. Mar. Bio. Ecol.*, 19, 99, 1988.

76. Hay, M. E., Duffy, J. E., Paul, V. J., Renaud, P. E., and Fenical, W., Specialist herbivores reduce their susceptibility to predation by feeding on the chemically defended seaweed *Avrainvillea longicaulis*, *Limnol. Oceanogr.*, 35, 1734, 1990.

77. Pennings, S. C., Weiss, A. M., and Paul, V. J., Secondary metabolites of the cyanobacterium *Microcoleus lyngbyaceus* and the sea hare *Stylocheilus longicauda*: palatability and toxicity, *Mar. Biol.*, 126, 735, 1996.

78. Fenical W., Chemical studies on marine bacteria: developing a new resource, *Chem. Rev.*, 93, 1673, 1993.

79. Jensen, P. J. and Fenical, W., Strategies for the discovery of secondary metabolites from marine bacteria: ecological perspectives, *Annu. Rev. Microbiol.*, 48, 559, 1994.

80. Sponga, F., Cavaletti, L., Lazzarini, A., Borghi, A., Ciciliate, D., Losi, D., and Marinelli, F., Biodiversity and potentials of marine-derived microorganisms, *J. Biotech.*, 70, 65, 1999.

81. Davidson, B. S., New dimensions in natural products research: cultured marine microorganisms, *Current Opinion. Biotech.*, 6, 284, 1995.

82. Gollin, M. A., New rules for natural products research, *Nat. Biotech.*, 17, 921, 1999.

83. Paul, V. J. and Van Alstyne, K. L., Chemical defense and chemical variation in some tropical Pacific species of *Halimeda* (Halimedaceae, Chlorophyta), *Coral Reefs*, 6, 263, 1988.

84. Meyer, K. D. and Paul, V. J., Intraplant variation in secondary metabolite concentration in three species of *Caulerpa* (Chlorophyta: Caulerpales) and its effects on herbivorous fishes, *Mar. Ecol. Prog. Ser.*, 82, 249, 1992.

85. Bolser, R. C. and Hay, M. E., Are tropical plants better defended? Palatability and defenses of temperate vs. tropical seaweeds, *Ecology*, 77, 2269, 1996.

86. Pettit, G. R., The bryostatins, *Prog. Chem. Org. Nat. Prod.*, 57, 153, 1991.

87. Swearingen, D. C., III and Pawlik, J. R., Variability in the chemical defense of the sponge *Chondrilla nucula* against predatory reef fishes, *Mar. Biol.*, 131, 619, 1998.

88. Lindquist, H., Hay, M. E., and Fenical, W., Defense of ascidians and their conspicuous larvae: adult vs. larval chemical defenses, *Ecol. Mono.*, 62, 547, 1992.

89. Harvell, C. D., Fenical, W., Oussis, V., Ruesink, J. L., Griggs, C. G., and Breene, C. H., Local and geographic variation in the defensive chemistry of a West Indian gorgonian coral (*Briareum asbestinum*), *Mar. Ecol. Prog. Ser.*, 93, 165, 1993.

90. Targett, N. M., Coen, L. D., Boettcher, A. A., and Tanner, C. E., Biogeographic comparisons of marine algal polyphenolics: evidence against a latiudinal trend, *Oecologia*, 89, 464, 1992.

91. Sennett, S. H., Pomponi, S. A., and Wright, A. E., Diterpene metabolites from two chemotypes of the marine sponge *Myrmekioderma styx*, *J. Nat. Prod.*, 55, 1421, 1992.

92. Cragg G. M., Schepartz, S. A., Suffness, M., and Grever, M.R., The Taxol supply crisis. New NCI policies for handling the large scale production of novel natural product anticancer and anti-HIV agents, *J. Nat. Prod.*, 56, 1657, 1993.

93. Olson, S. G., Curing cancer through aquaculture, *Sea Technol.*, August 1996, 89.

94. Kobayashi, J. and Ishibashi, M., Bioactive metabolites of symbiotic marine microorganisms, *Chem. Rev.*, 93, 1753, 1993.

95. Shigemori, H., Bae, M. A., Yazawa, K, Sasaki, T., and Kobayashi, J., Alteramide A, a new tetracyclic alkaloid from a bacterium *Alteromonas* sp. associated with the marine sponge *Halichondria okadai*, *J. Org. Chem.*, 57, 4317, 1992.

96. Ito, S. and Hirata, Y., The structure of ikarugamycin, an acyltetramic antibiotic possessing a unique as-hydrindacane skeleton, *Bull. Chem. Soc. Jpn.*, 50, 1813, 1977.

97. Gunasekera, S. P., Gunasekera, M., and McCarthy, P. J., Discodermide: a new bioactive macrocyclic lactam from the marine sponge *Discodermia dissoluta*, *J. Org. Chem.*, 56, 4830, 1991.

98. Kobayashi, J., Cheng, J. F., Walchli, M. R., Nakamura, H., Hirata, Y., Saski, T., and Ohizumi, Y., Cysodytins A, B, and C, novel tetracyclic aromatic alkaloids with potent antineoplastic activity from the Okinawan tunicate *Cystodytes dellechiajei*, *J. Nat. Prod.*, 53, 1800, 1988.

99. Cimino, G., Crispino, S., De Rosa, S., De Stafano, S., Gavagnin, M., and Sodano, G., Studies on the structure of calliactine, the zoochrome of the sea anemone, *Calliactis parasitica*, *J. Org. Chem.*, 56, 804, 1987.

100. Stierle, A. A., Cardellina, J. H., II, and Singleton, F. L., A marine Micrococcus produces metabolites ascribed to the sponge *Tedania ignis*, *Experientia*, 44, 1021, 1988.

101. Elyakov, G. B., Kuznetsova, T., Mikahilov, V. V., Maltsev, I. I., Voinov, V. G., and Fedoreyev, Brominated diphenyl ethers from a marine bacterium associated with the sponge *Dysidea* sp., *Experientia*, 47, 632, 1991.

102. Miki, W., Otaki, N., Yokoyama, A., and Kusumi, T., Possible origin of zeaxanthin in the marine sponge *Reniera japonica*, *Experientia*, 52, 93, 1996.

103. Schmitz, F. J., Vanderah, D. J., Hollenbeck, K. H., Enwall, C. E., Gopichand, Y., Sengupta, P. K., Hossam, M. B., and van der Holm, D., Metabolites from the marine sponge *Tedania ignis*. A new atisanediol and several known diketopiperazines, *J. Org. Chem.*, 48, 3941, 1983.

104. Faulkner, D. J., He, H., Unson, M. D., Bewley, C. A., and Garson, M. J., New metabolites from marine sponges: are symbionts important? *Gazzetta Chimica Italiana*, 123, 301, 1993.

105. Muller, W. E. G., Diehl-Seifert, B., Sobel, C., Bechtold, A., Kljajic, Z., and Dorn, A., Sponge secondary metabolites: biochemical and ultrastructural localization of the antimitotic agent avarol in *Dysidea avara*, *J. Histochem. Cytochem.*, 34, 1687, 1986.

106. Uriz, M. J., Turon, X., Galera, J., and Tur, J. M., New light on the cell location of avarol within the sponge *Dysidea avara* (Dendroceratida), *Cell Tissue Res.*, 285, 519, 1996.

107. Uriz, M. J., Becerro, M. A., Tur, J. M., and Turon, X., Location of toxicity within the Mediterranean sponge *Crambe crambe* (Demospongiae: Poecilosclerida), *Mar. Biol.*, 124, 583, 1996.

108. Sennett, S. H., Wright, A. E., Pomponi, S. A., Armstrong, J. E., Willoughby, R., and Bingham, B. L., Cellular localization and ecological role of a secondary metabolite from the sponge *Hymeniacidon heliophila*, in *Proceedings of the International Society of Chem. Ecol. Ann. Mtg*, Abstract, Quebec, 1990.

109. Battershill, C. N., Pomponi, S. A., Willoughby, R., Northcote, L., McClean M. L., Munro, M. H. G., Blunt, J. W., and Garson, M. J., Sponge cells or symbionts? The site of biosynthesis of biologically active metabolites in two species of sponges and the feasibility of cell culture as a production option, in *9th International Symposium on Marine Natural Products, 1998*, Symposium Proc., Tonnville, Australia.

110. Costas, E., San Andres, M. I., Gonzalez-Gil, S., Aguilera, A., and Lopez-Rodas, V., A procedure to estimate okadaic acid in whole dinoflagellate cells using immunological techniques, *J. Appl. Phycol.*, 7, 407, 1995.

111. Miles, C. O., Wilkins, A. L., Garthwaite, I., Ede, R. M., and Munday-Finch, S. C., Immunochemical techniques in natural products chemistry: isolation and structure determination of a novel indole-diterpenoid aided by TLC-ELISAgram, *J. Org. Chem.*, 60, 6067, 1995.

112. Pestka, J. J., High performance thin layer chromatography ELISAGRAM, *J. Immunol. Methods*, 136, 177, 1991.

113. Ilan, M., *Negombata magnifica* — a magnificent pet, in *Origin and Outlook: 5th International Sponge Symposium 1998*, Book of Abstracts, Queensland Museum, Brisbane, Australia, 1998, 29.

114. Rinkevich, B., Cell cultures from marine invertebrates: obstacles, new approaches and recent improvements, *J. Biotech.*, 70, 133, 1999.

115. Pomponi, S. A. and Willoughby, R., Sponge cell culture for production of bioactive metabolites, in *Sponges in Time and Space*, van Soest, R. W. M., van Kempen, T. M. G., and Braekman, J.C., Eds., 4th Intl. Porifera Congress, Amsterdam, Rotterdam, Netherlands, 1994, 395.

116. Pomponi, S. A., Willoughby, R., Kaighn, M. E., and Wright, A. E., Development of techniques for in vitro production of bioactive natural products from marine sponges, in *Invertebrate Cell Culture: Novel Directions and Biotechnology Applications*, Maramorosch, K. and Mitsuhashi, J., Eds., Academic Press, New York, 1997.

117. Berthold, R. J., Borowitzka, M. A., and Mackay, M.A., The ultrastructure of *Oscillatoria spongeliae*, the blue-green algal endosymbiont of the sponge *Dysidea herbacea*, *Phycologia*, 21, 327, 1982.

118. Faulkner, D. J., Marine natural products: metabolites of marine invertebrates, *Nat. Prod. Rep.*, 1, 551, 1984.

119. Faulkner, D. J., Marine natural products, *Nat. Prod. Rep.*, 5, 613, 1988.

120. Faulkner, D. J., Marine natural products, *Nat. Prod. Rep.*, 9, 323, 1992.

121. Bewley, C. A. and Faulkner, D. J., Lithistid sponges: star performers or hosts to the stars, *Angew. Chem. Int. Ed.*, 37, 2162, 1998.

122. Schaufelberger, D. E., Koleck, M. P., Beutler, J. A., Vatakis, A. M., Alvarado, A. B., Andrews, P., Marzo, L. V., Muschik, G. M., Roach, J., Ross, J. T., Lebherz, W. B., Reeves, M. P., Eberwein, R. B., Rodgers, L. L., Testerman, R. P., Snader, K. M., and Forenza, S., The large-scale isolation of bryostatin 1 from *Bugula neritina* following good manufacturing practices, *J. Nat. Prod.*, 54, 1265, 1991.

123. Pomponi, S. A., The bioprocess-technological potential of the sea, *J. Biotech.*, 70, 5, 1999.

124. Davidson, S. K., and Haygood, M. G., Identification of sibling species of the bryozoan *Bugula neritina* that produce different anticancer bryostatins and harbor distinct strains of the bacterial symbiont *Candidatus endobugula sertula*, *Biol. Bull.*, 196, 273, 1999.

125. Battershill, C. N., Page, M. J., Duckworth, A. R., Miller, K. A., Bergquist P. R., Blunt, J. W., Munro M. H. G., Northcote, P. T., Newman D., and Pomponi, S. A., Discovery and sustainable supply of marine natural products as drugs, industrial compounds and agrochemicals: chemical ecology, genetics, aquaculture and cell culture, in *Origin and Outlook: 5th International Sponge Symposium 1998*, Book of Abstracts, Queensland Museum, Brisbane, Australia, 1998, 16.

126. Osinga, R., de Beukelaer, P. B., Meijer, E. M., Tramper, J., and Wijffels, R. H., Growth of the sponge *Pseuberites* (aff.) *andrewsi* in a closed system, *J. Biotech.*, 70, 155, 1999.

17 Natural Product Antifoulants and Coatings Development

Dan Rittschof

CONTENTS

I. OVERVIEW

The intent of this section is to provide an uncluttered overview and rapid reference to specific issues of interest. Every effort has been made to reference the rest of the text to enable entry into an extensive and detailed literature. My apologies to any and all researchers whose work has been unintentionally slighted.

Fouling, the colonization of surfaces by abiotic and biotic substances and organisms, has molecular, microbial, and organismal levels of organization. All mechanisms of fouling involve molecular bonding interactions and molecules that act as inorganic and biological adhesives. Existing commercial technology intended to minimize or control impacts of fouling includes antifouling and foul-release approaches.

Antifouling coatings technology is based upon mechanisms in which broad-spectrum biocides, usually toxic metal ions, kill organisms that settle on coatings. Many toxic coatings are designed

0-8493-9064-8/01/$0.00+$1.50
© 2001 by CRC Press LLC

to slowly hydrolize so that surface erosion continuously presents toxic additives as well as polishes the surface to reduce drag forces. Foul-release coatings are usually based upon dimethyl silicone polymer technology. These coatings foul, but are designed to be cleaned easily. Foul-release coatings are usually catalyzed with organotin catalysts and are toxic until the catalyst leaches from the coating. These coatings often contain additives, such as oils and surfactants, that function as biocides. Antifouling coatings often have foul-release coating properties, and foul-release coatings have antifouling coating properties. Environmentally unacceptable consequences of the use of toxic organotins and heavy metals have prompted research on natural antifoulants. Natural antifoulants may result in impacts and issues of similar importance to the environment as those resulting from toxic metals.

Laboratories worldwide now use bioassays with target fouling organisms to direct purification, identification, and development of antifoulant compounds. Most living organisms employ a variety of antifouling strategies, many of which are chemically based. Reports of natural compounds with antifoulant activity span over 40 years. Chemically, natural antifoulant compounds represent most classes of organic compounds. Although specific mechanisms of action are rarely reported, general mechanisms of action include toxins, anesthetics, surface-active agents, attachment and/or meta-morphosis inhibitors, and repellents.

One concept addressed repeatedly by academic researchers is the use of compounds found in living organisms as active agents in antifouling coatings. Although natural antifoulants are common, development of functional coatings based upon natural products is a technological, financial, temporal, and regulatory nightmare. This is, in part, due to the biological impacts and chemical nature of the compounds. Other than broad-spectrum highly toxic compounds, individual additives are usually narrow spectrum in effectiveness. Prevention of fouling by one kind of organism is routinely supplanted by fouling of another type. Other challenges related to the development of viable, environmentally benign coatings that contain natural products relate to the chemical nature of compounds. These challenges include developing polymer systems compatible with the additives as well as with anticorrosive undercoats and appropriate mechanical and application properties.

Development of commercial coatings using natural products is blocked by cost, the time horizon to meet government regulations, and meeting performance standards which are judged by compar-ison to coatings with unacceptable environmental impacts. For example, one new organic biocide was registered for use as an antifoulant in the United States in the last decade. In addition to taking 10 years, registration of this biocide cost millions of dollars. When one combines these time and monetary factors with the practice of holding new coatings to performance standards of coatings with environmentally unacceptable impacts, it is clear why there is little progress. Even if these political and economic constraints are addressed, it is unlikely, due to the diversity of fouling mechanisms, that nontoxic natural products will become the hoped-for broad-spectrum solutions. However, the potential is high for the development of environmentally benign solutions to fouling which combine natural products with degradable organic biocides.

II. FOULING AND ANTIFOULING

A. THE SCOPE OF FOULING

Fouling encompasses processes that range from purely physicochemical and electrochemical to those of complex biology. At the molecular level, fouling consists of all standard bonding interac-tions and plating phenomena between molecules in solution and on a surface.[1,2,3] In natural waters, there are large numbers of biologically derived surfactants that bind to any surface. Surface-active molecules also partition at the air/water interface to form a fouling layer.[4] All surfaces submerged in the ocean are immediately subjected to molecular fouling.

The physics of how moving water interacts with surfaces both explains why it is advantageous for organisms to attach to surfaces and provides insights into fouling, control limits, and antifouling technology. Because water is viscous, considerable energy must be expended to move it. The combination of friction and high viscosity results in a layer (several centimeters to less than a millimeter) of water near a surface that is usually not moving and where diffusional processes dominate. This low energy environment provides a refuge from flow for weakly swimming planktonic organisms during the settlement phase.

Planktonic organisms have adaptations that enable them to move out of water masses in which nutrients are depleted. Many plankton have the ability to migrate vertically and change water masses.[5,6] A large number of microorganisms (bacteria and diatoms) stop replicating and produce sticky exopolymers when nutrient levels drop in the water mass in which they are traveling. Propagules and micoorganisms are routinely small enough to fit in the boundary layer over most surfaces.[7] Even those with very poor bioadhesives are not exposed to forces that can dislodge them. This passive attachment has no behavioral component and occurs continuously in natural waters.[8] A final wrinkle in the mechanisms of macrofouling attachment is that molecular, micro-, and macrofouling propagules routinely aggregate in the water column in the absence of surfaces, forming long sticky strands that are carried to surfaces by flow and gravity.[9] If the new condition is high in nutrients, the phenotype of the microbe changes from sticky polymer/exopolymer production back to growth and replication.[10,11]

Taking physics and energetics into account, attachment of living organisms to surfaces is advantageous because organisms can use environmentally generated water movement (energetically free to the organism) for feeding and waste removal. Delivery of toxic compounds from a surface is diffusionally driven and compromised by the decrease in toxin concentration proportional with the square of the distance from the surface. Even in thick boundary layers such as those found in very still conditions, convective flow further dilutes and carries toxins away. A consequence of these processes is that an organism attached to the surface, but that breathes and feeds in the water either above the boundary layer or at some distance from the diffusional source of toxin, is minimally exposed to toxins. This is part of the reason that organisms such as barnacles, with calcareous base plates that act as diffusion barriers, are such tenacious foulers and among the first to appear on a failing coating. Barnacles that survive the first few days on a surface grow sufficiently to feed and breathe in water with ineffective toxin levels.

The spectrum of cues for attachment and attachment mechanisms that are used by macrofoulers is daunting,[8] as is the tenacity of many of their bioadhesives.[12–14] Attachment mechanisms can be characterized as a continuum representing successional fouling at one end of the spectrum and probability-driven fouling at the other extreme. In successional fouling, settlement of a macrofouling organism is dependent upon prior microfouling. Successional fouling has been clearly demonstrated for hydroids specialized to live on macroalgae and shells of hermit crabs,[15,16] as well as Hawaiian populations of the tube-building polychaete worm *Hydroides elegans*.[17] In contrast, with probability-driven fouling, organisms settle based upon the probability of their contact with surfaces and the probability of their settlement. Probability of contact depends upon the numbers of propagules available. Probability of settlement depends upon the physiological state of the organism and physical and physicochemical properties of the surface. Organisms such as barnacles, e.g., *Balanus amphitrite*, settle best on high surface energy unfouled surfaces.[18,19] In contrast to barnacles, there are organisms, such as abalone,[20] that require location of another species for settlement, and others, such as many hydroids (*Tubularia* spp. and *Eudendrum* spp.), that settle passively on all surfaces.[21]

Because fouled ships have visited various ports for years, many of the most common fouling organisms are now found in ports throughout the world.[22] Many of these cosmopolitan species — such as *Enteromorpha* spp.,[23,24] calcareous tube worms such as *Hydroides elegans*,[17] bryozoans such as *Bugula neritina*,[25] hydroids such as *Tubularia crocea*,[21] and barnacles such as *Balanus amphitrite*[18] — are reproductive much of the year. These common foulers disperse as short-lived microscopic larvae, spores, or propagules that settle, metamorphose, and complete their life cycle in less than a month. Short propagule duration and rapid generation times may, in part, explain their successful colonization of the world's harbors. These organisms are the weeds of the sea, thriving out of their original ecological context. It is likely that new environmentally sound anti-fouling technology will be based upon an understanding of the physics, chemistry, and biology controlling colonization of surfaces by these organisms.

A major factor controlling initial colonization is the nature of the surface. All surfaces have a physicochemical property called surface energy.[21] Surface energy is described by the way in which solvent molecules interact with a surface. It is surface energy, for example, that is responsible for water beading on some surfaces and spreading on others. Some surfaces, those with surface energies similar to dimethyl silane, have little ability to interact with biological adhesives and, thus, form poor adhesive bonds.[2] These surfaces are relatively easy to clean. The responses of fouling organisms to surface energy have been studied extensively,[4,7,9,21,26–31] and two generalizations can be made: (1) all surface energies foul, and (2) all possible patterns of colonization with respect to surface energy occur. Some organisms colonize surfaces with a range of surface energies. Some only colonize surfaces with a specific surface energy, and some colonize surfaces independent of surface energy.[21] Thus, if the local fauna is known, one can predict the kinds of fouling a surface with known surface energy will initially experience. However, the relationship gets rapidly muddled as time passes. Over time, surface energies on all submerged surfaces converge.[4,28] At the developing community level, other factors such as predation, competition between fouling organisms, and temporal larval availability further complicate the picture.[30]

B. ANTIFOULING

Fouling is combatted by a variety of cleaning techniques and by killing propagules. Killing is most effective either immediately before or immediately after propagules attach. By far, the most common antifouling techniques are based upon broad-spectrum biocides that kill settling organisms. Common biocides include strong oxidants and metals such as copper, zinc, and tin. There are two basic mechanisms of toxic metal action. One is death by metal ion overload. Free metal ions are essential cofactors and usually in short supply. As a result, organisms have efficient active transport mechanisms for obtaining metal ions, but mechanisms for shutting off their metal ion pumps have not evolved. In the presence of metal-based antifouling technology, organisms overload themselves with metals. The metals, in turn, disrupt their normal enzymatic and metabolic functions, and the organisms die.[32–34] The other mechanism is death by uncoupling of oxidative phosphorylation and electron transport. Organometal compounds such as tributyltin (TBT) are lipophylic. Lipophylic molecules partition into membranes and disrupt essential membrane functions, such as the electron transport processes required for generation of cellular energy through oxidative phosphorylation.[35] A fascinating sublethal effect of tributyltin is imposex, the development of male secondary and, in extreme cases, primary characteristics by females. Imposex leads to sterility in many molluscs or lack of reproduction because the population becomes all male.[36–40]

C. ANTIFOULING AND THE ENVIRONMENT

Toxic metals have long-term impacts on freshwater and marine environments. This is because metals are biologically conserved and recycled. Two major biological processes result in buildup of toxic metals in the environment: (1) continuous conservation by organisms of free ions such as copper and zinc,[32] and (2) "reorganification" of metals like tin.[35] Both processes result in conservation and buildup of toxic compounds until nontarget species are impacted.

Although antifouling coatings containing TBT are very effective,[41] the negative impact of TBT and related metabolites on mariculture and wild-caught shellfish has resulted in regulations and bans on its use in countries that have established environmental policies.[42] The conclusion reached by regulators in most developed countries is that TBT released from antifouling coatings damages environmental health, impacting fisheries and aquaculture. Nonspecific effects are detrimental to quality of life and may threaten human health. The result is a pending worldwide ban to be in place in 2003 and raging debate by members of the International Maritime Organization (IMO) on use of TBT antifouling coatings.

Copper, though currently receiving less attention than organotin, will probably face similar restrictive legislation in the future. At this point, one developed country, Sweden, has banned all toxic metal antifouling coatings in its territorial waters,[43] and other countries such as the United States regulate application, removal, and waste disposal of toxic metal coatings.

D. Experimental Approaches to Nontoxic Antifouling

Social and governmental responses to negative environmental impacts of toxic coatings have resulted in pressure to find alternatives. This pressure, combined with basic scientific curiosity about natural chemical control of fouling and the capability of laboratory rearing of fouling organisms for testing, led academic researchers to the study of natural antifouling mechanisms. One common hypothesis supported repeatedly is that many organisms continuously produce antifoulants, which is how they remain unfouled.[44–48] In some instances, the antifoulant produced is mucus.[49] In other instances, extractable organic compounds are known[50] or are novel secondary metabolites.[46,51]

Initial searches for natural antifoulants involved biologists and natural products chemists.[52] Source organisms, such as tropical sponges and octocorals, were chosen because they do not foul when alive and because they are rich sources of novel secondary metabolites.[53–55] Similarly, organisms that were relatively easily cultured (diatoms, nudibranch larvae, and barnacles) were used for bioassays. Although the surfaces of living intact sponges and corals did not foul, molecules were extracted with organic solvents from the whole colony.[44,45] Only a few researchers[49,56–63] have concerned themselves with the activity found in the water surrounding, or on the surface, of intact organisms. Thus, there are really two broad classes of natural antifouling compounds: (1) compounds extracted from organisms that have antifouling activity which may never reach the surface of the organism, and (2) compounds found in the water-bathing organisms that have antifouling activity and are likely to serve a role in deterring growth of epibiota.[55] Many biological scholars consider that this second, more restrictive functional ecology approach is more likely to result in discovery of commercially viable compounds.[49,55,64] However, the relative merits of these approaches are unresolved, and the vast majority of compounds reported to possess antifouling activity are from organic extracts of whole colonies or organisms.

Although the quest for natural antifouling compounds began about 50 years ago, citations in literature reviews are rare until the early and mid-1980s.[54,65] However, substantial funding for the natural product antifoulants research became available in the United States in the late 1970s and early 1980s. The Duke Marine Laboratory research team was the first to use mass-reared barnacles, *Balanus amphitrite*, in bioassays[45] and to direct isolation of antifoulant compounds.[18] *Balanus amphitrite* is a cosmopolitan hard-fouling organism introduced by shipping around the world. Barnacles and other hard foulers, such as oysters and tube worms, are logical targets because their tests are glued permanently to surfaces and, once attached, act as platforms for other fouling organisms.

Since the late 1980s and early 1990s, funding for antifouling studies using natural products has been available around the world. Major contributions have been made, especially by Japanese and Australian research groups. The first comprehensive review of the topic is by Davis et al.,[55] which compiles and synthesizes information from over 200 publications. Clare[46] adds substantial and comprehensive new information. Table 17.1 provides additional references from 1997 to the present.

Since the development of bioassays, there has been little change in the approach to discovering natural product antifoulants. The first step is showing that an organic extract of some kind of organism prevents settlement of a target species.[66] Then, depending upon the potency, stability, and maturity of chemical techniques for determining that particular class of natural product and the availability of the techniques to the investigators, there are varying levels of success in determining the structure of the active molecule. Initially, steroids,[67] terpenes,[44,51] phenolics,[68] bromonated

TABLE 17.1
Some Reports of Natural Product Antifoulants from 1995–Present

Extract or Parent Compound (Number of New Structures)	Source	Bioassay	Potency (EC$_{50}$)	Reference
Diterpene (3)	Sponge	Barnacle Settlement	0.45-1.1 µg/ml	147
Ca(OH)$_2$	Concrete	Oysters	—	148
Nitrogen Hetrocycle	Sponge	Barnacle Settlement	25 µg/cm^2	121
		Bryozoan Settlement	1.1 µg/cm^2	
		Byssal Thread Attach.	1.2 4.4 µg/cm^2	
Terpenoids (32)	Sponge Nudibranch	Barnacle Settlement	0.08–>50 µg/ml	142
Terpenoids (14)	Sponge	Barnacle Settlement	<0.5–0.05 µg/ml	149
Film (3)	Bacteria	Barnacle Settlement	—	75
		Bryozoan	—	
Diterpene (9)	Sponge	Barnacle Settlement	0.08–4.6 µg/ml	150
Sesquiterpene (3)	Nudibranch	Barnacle Settlement	0.13–0.14 µg/ml	150
Spermidine (10)	Sponge	Barnacle Settlement	0.1–15 µg/ml	151
Cyanoformamide (2)	Sponge	Barnacle Settlement	4.3–5 µg/ml	152
		Cytotoxicity	2.1–3.4 µg/ml	
BromoTyrosine (8)	Sponge	Barnacle Settlement (7)	0.1–8.0 µg/ml	77
		Ascidian Met.	1.2–25 µg/ml	
		Promote (2)	2.1–3.4 µg/ml	
		Cytoxicity (2)		
Oroidin (2) (1)	Sponge	Barnacle Settlement	15–19 µg/ml	153
		Ascidian (Promoter)	2.5 µg/ml	
Steroid (2)	Sponge	Antimicrobial	10 µg/disc	154,155
Aqueous Extract Film	Bacteria	Bryozoan Settlement	—	97
Phlortannins (3)	Marine Algae	Polychaete Worms	0.5–5 µg/ml	156
		Bacteria	—	
Aqueous Extract (Homarine)	Soft Coral	Antimicrobial	1.26 µg/ml	60
Polyacetylene (7)	Sponge	Barnacle Settlement	0.24–4.5 µg/ml	78
Terpenes (6) and Steroids (2)	Sponge	Barnacle Settlement	0.24–4.0 µg/ml	154
Bromophenol (3)	Polychaete	Bivalves and Polchaete Settlement Field	5–240 ng/g	59
Lentil Lectin	Plants	Barnacle Settlement Inhibition	50–100 µg/ml	157
Protein Complex (SIPC)	Barnacle	Larval Settlement Inducer	10–100 µg/ml	106
Spermidine (10)	Sponge/Synthetic	Antimicrobial	8–256 µg/ml	115
Diterpene (4)	Algae	Bryozoans, Hydroid	>1 µg/ml	61
Steroid (4)	Sponge	Anti-barnacle Settlement	5 µg/ml	113
Terpenoids (14)	Sponge	Barnacle Settlement	50 mg/ml	113
Cyanoformamide (2)	Sponge	Barnacle Settlement	5 µg/ml	158
Calmodulin Inhibitors	Synthetic	Barnacle Settlement Inhibition	1–10 µM	159
Neurotransmitter Blockers (3)	Synthetic	Barnacle Settlement	30–300 µg/ml	111
Terpene Analogues (19)	Synthetic	Barnacle Settlement	1 ng–5 µg/ml	84
Neuroactive Compounds	Synthetic	Mussel Metamorphosis Induction	100 µM	160
Steroid (4)	Octocoral	Anti-barnacle Settlement	2 µg/ml	161
Steroid (4)	Octocoral	Lethal to Barnacle Larvae	100 µg/ml	162
Lumichrome	Ascidian	Ascidian Metamorphosis Inducer	100 nM	112

TABLE 17.1 (CONTINUED)
Some Reports of Natural Product Antifoulants from 1995–Present

Extract or Parent Compound (Number of New Structures)	Source	Bioassay	Potency (EC_{50})	Reference
Neurotransmitter Blockers (6)	Synthetic	Barnacle Settlement	0.1–100 μM	163
Homoserine Lactone	Algae	Bacterial Movement	—	127
Aqueous Extracts	Synthetic	Barnacle Settlement	—	19
Cardenolide (5)	Insects/Plants	Barnacle Settlement	10–50,000 μg/ml	48
Biogenic Amine	Synthetic	Bryozoan Metamorphosis Inhibitor	100 μM	164

hydrocarbons,[20,59] brominated tyrosine derivatives,[69] and saponins[70] were reported to act as antifoulants at some level.

Since the early 1990s, there has been a dramatic increase in research looking for potential antifoulants.[46] There are, or were until very recently, very active research groups in Australia (Kjelleberg, Steinberg, and associates), the Netherlands (TNO), Japan (Fusetani Biofouling Project), and the United Kingdom (Callow, Clare, and others). Additionally, there are smaller efforts in the United States, Singapore, India, New Caledonia, and Hong Kong. Major productive efforts, especially by the Fusetani research group in Japan and the European and Australian research communities employing bioassays-directed purifications, have resulted in identification of many new antifoulant compounds from a variety of marine invertebrates. In a review, Clare[46] reported over 50 core natural product structures with antifouling activity. Since 1995, the research community has reported the identity of over 100 additional natural products with antifouling activity (Table 17.1). Reports of many more compounds are delayed because of issues related to the process of patent protection, required for any subsequent commercialization.

Inspection of Table 17.1 is useful in understanding the strengths and weaknesses of what has become a worldwide quest. Of the more than one hundred compounds shown in Table 17.1, the vast majority, about four-fifths, are terpenoid compounds, their relatives, or analogues. Most of these compounds are new to science since 1995. In over half of the studies, the bioassay used is for prevention of settlement of an easily culturable barnacle,[71] a major fouling species common to temperate and tropical harbors around the world, *Balanus amphitrite*.[72] Although *B. amphitrite* is a dominant fouling barnacle and an excellent test organism, it is known that it cannot be used as a representative of all fouling organisms or even, for that matter, of all balanoid barnacles.[73–75] Several authors have observed that potent antibarnacle settlement has no effect on other fouling species[76] or actually stimulates settlement.[77,78]

From the perspective of potency for theoretical use in an antifouling coating, the vast majority of compounds in Table 17.1 exceed the potency criterion for future study that was followed in the U.S. Navy program, that of being active at less than 25 μg/mL in static bioassays. However, the notion of harvesting organisms and purifying commercial amounts of natural products for virtually any commercial use is probably untenable[79] and extremely remote for natural product antifoulants.[46,80–82] One potential solution is the use of synthetic analogues developed from structure-function studies.[44,80–84] The idea here is to find commercially practical alternatives. In general, it is unlikely that natural products will be in sufficient quantity that they can be harvested, especially from the natural sources that are often exotic rare corals, sponges, and other invertebrates. Most potent natural product compounds are often too structurally complex to be commercially synthesized. Alternative compounds must have a potency that makes them practical and must be amenable to cost effective synthesis.

III. NATURAL PRODUCTS ANTIFOULANTS

A. Biological Targets and a Brief History of Natural Product Antifouling Studies

As the natural products antifouling field matures, the nature and diversity of organisms used in fouling assays is increasing, as is our understanding of the complexity of the biological processes involved. Initially, researchers assumed that fouling was a successional process and that microbial fouling was a requirement for macrofouling. This concept originated mainly from two sources: (1) a Meadows and Williams paper in *Nature*,[85] and (2) a paper by Corpe.[86] The *Nature* paper[85] specifically addressed settlement by polychaetes, and the assumption of successional fouling appears correct for many polychaetes.[17,85,87,88] Corpe[86] never suggested that microfouling was requisite for macrofouling, but his work was interpreted that way. Thus, initially it was postulated, that if microbial fouling could be controlled, then macrofouling would also be controlled. A statement to this effect can be found in virtually every report on control of microfouling far into the 1990s (see for example, Readman[89]). There is also ample evidence that bacterial films can stimulate settlement of bryozoans on some surfaces[90] and settlement of bivalves like oysters[91,92] on others, and that bacterial films may have a variety of effects, ranging from nothing to inhibiting barnacle settlement.[19,75,93–97] However, temporal aspects on the level of weeks may result in changes in the effects of films on settlement. Perhaps the best way to view this phenomenon is that bacterial films have a major impact on most fouling organisms,[98,99] but preventing microfouling will not prevent macrofouling. Other chapters in this volume treat this topic in detail.

Many macrofoulers readily settle on surfaces whether or not they are filmed with bacteria, and chemical surface characteristics mediate the settlement of many common macrofoulers.[9,21,28,29] Finally, at all levels of fouling, there are some organisms that settle like dust on all surfaces.[8,21,55]

The major early exception to using microfoulers instead of macrofouling organisms in fouling studies was the use of settlement stage barnacles. Barnacles were targeted initially because they were recognized as central to the failing of antifouling coatings.[71,72,100] Barnacles are dominant members of fouling communities around the world and are easily recognized during visual inspection and cleaning of fouled hulls and surfaces. Barnacles are calcareous (hard foulers) and have a relatively robust adhesive.

Since cyprid glue contains barnacle settlement pheromones,[101,102] even if the first wave of larvae dies, the glue and tests left behind stimulate settlement of other barnacle larvae while at the same time acting as a barrier to toxin diffusion. If the relatively vulnerable cyprid survives long enough to settle and metamorphose, the newly recruited "pin head" barnacle has its own larval glue and uncalcified base plate to act as barriers to diffusion from below. The top of the pin head barnacle is elevated approximately 500 μm over the surface, just high enough to enable the barnacle to ventilate in water with reduced levels of toxins.[7,103] The base plate and a growth form, which results in breathing and feeding in diffusionally limited toxicity associated with the boundary layer of the hull surface, often results in barnacles being among the first to settle and survive on toxic surfaces. In addition to cyprids,[101,104] settled barnacles also produce settlement pheromones[102,105,106] thereby furthering recruitment.

The result is that barnacles are often the first macrofoulers to colonize a protected surface.[100] Once growth of barnacles on toxic metal coatings is initiated, the surface rapidly fails because barnacles act as nontoxic platforms for other organisms. As a result of these kinds of field evidence, the U.S. Navy supported basic barnacle research for almost four decades, and barnacles were the only macrofoulers included in initial studies of natural product antifoulants supported by the U.S. Navy. Most of this work was conducted by the Costlow group at the Duke University Marine Laboratory.[47]

At Duke in the early 1980s, mass cultured barnacle larvae were used as the target organism in bioassay directed studies of natural inhibitors of fouling. Approximately 70% of the marine organisms in the temperate to subtropical estuarine communities around the Marine Laboratory were found to contain compounds with readily detectable antibarnacle settlement activity. Many compounds are extremely potent in the settlement assay, even though they have low toxicity.[94,107,108]

The prevalence of compounds that interfere with settlement and metamorphosis of complex organisms like barnacles is a logical consequence of the complexity of the biochemical pathways controlling metamorphosis. There is evidence that these pathways begin with chemoreceptors coupled to neuronal, hormonal, and metabolic control processes through classic amplification cascades that include second messengers[20,22,88,102] that are stimulated by changes in ion permeability.[109]

B. Mechanism of Action of Natural Product Antifoulants

The mechanisms of action of most compounds is not obvious from their structures. Although certain major classes of molecules, such as steroids, are recognized to have known functions, molecules are usually multifunctional. Examination of the structures presented illustrate this point. Compound 17.1 is a sodium/potassium ATPase inhibitor and a potent natural product toxin.[110] Compound 17.2 is a component of the fragrance of peaches and apricots and somehow anesthetizes barnacle larvae.[84] Compound 17.3 is reported to be a repellant of polychaete larvae. Compounds 17.4–17.7 have unreported mechanisms of action. One might recognize that Compound 17.4 has many of the characteristics of sesquiterpene antifoulants, which act via an unknown mechanism, and hazard that Compounds 17.5–17.9 might function through interaction with a surface or act like detergents.

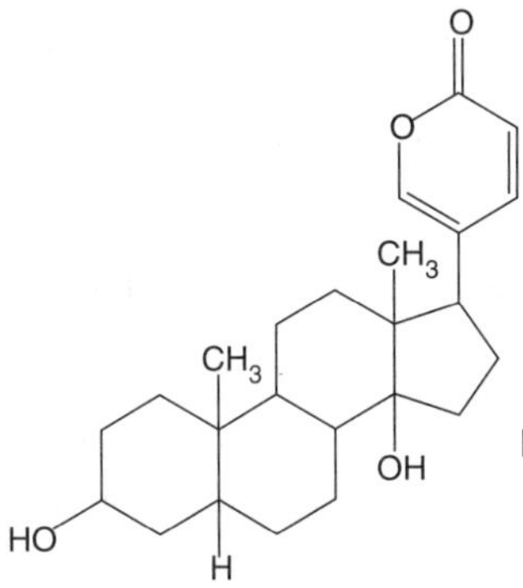

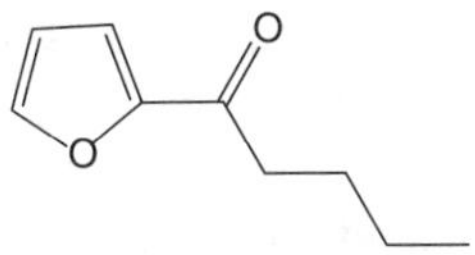

Bufalin 10pg/mL (17.1)

Pentyl 2-Furyl Ketone 1ng/mL (17.2)

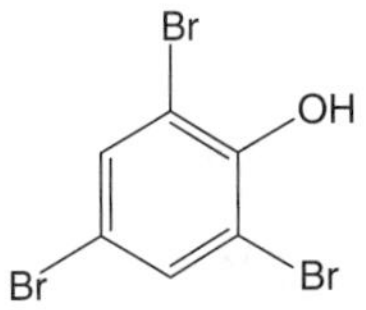

Tribromophenol 3 ng/g (17.3)

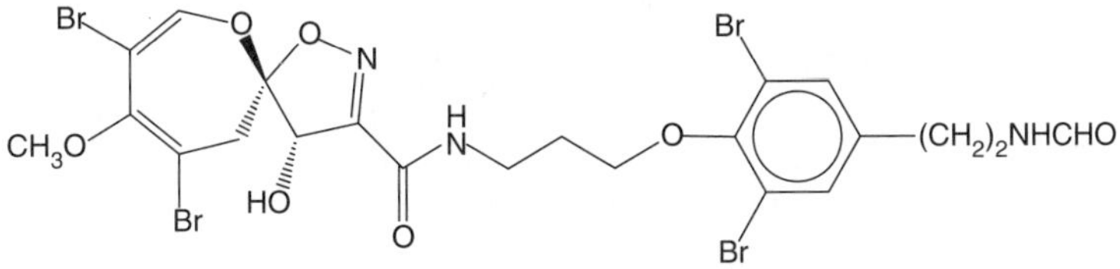

Ceratinamide A 100 ng/mL (17.4)

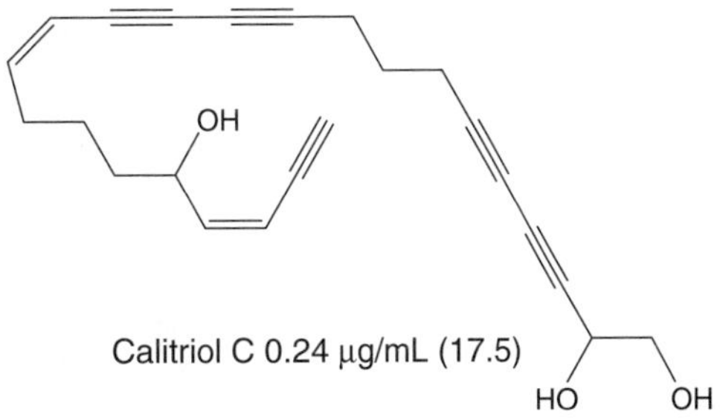

Calitriol C 0.24 µg/mL (17.5)

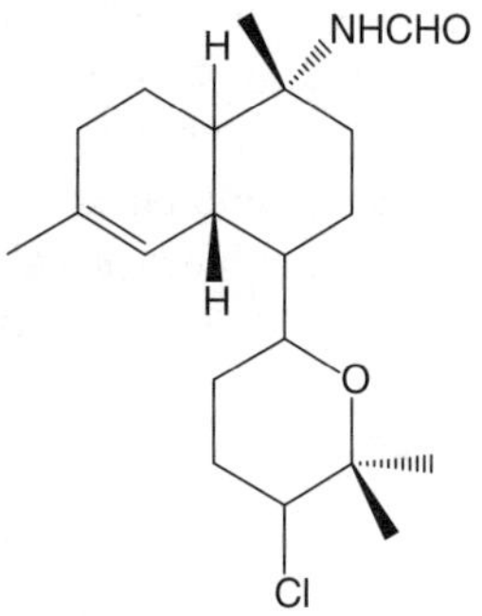

Kalihinene X 0.5 µg/mL (17.6)

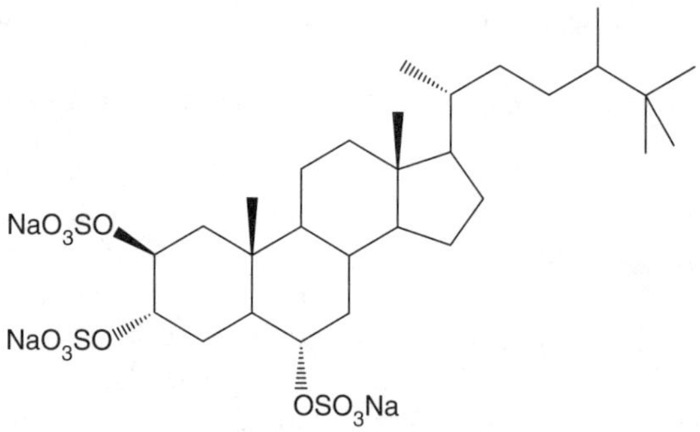

Halistanol Sulfate 10 µg/disk (17.7)

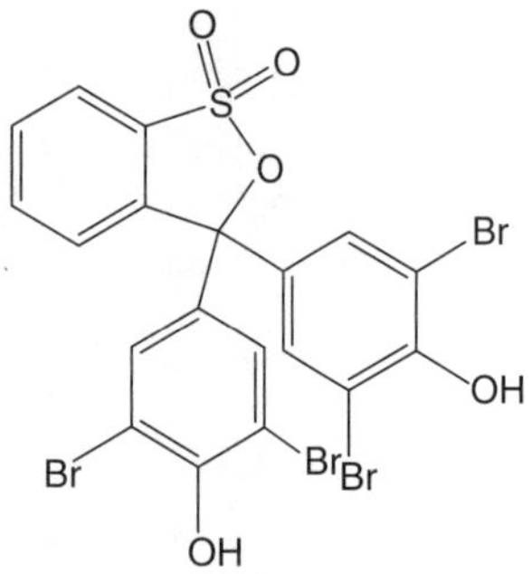

Bromophenol (17.8)

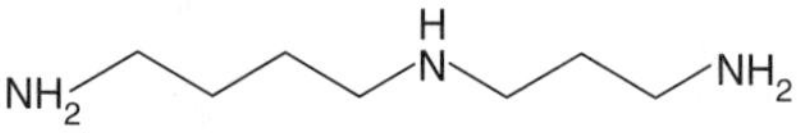

Spermadine-antimicrobial (17.9)

As molecules increase in complexity, the number of possibilities for mechanism of action also increase. For example, Structure 17.10 inhibits a specific kinase and specific cellular second messenger.[111] In contrast, Structure 17.11 is specifically related to vitamin B2,[112] while Structure 17.12, a steroid peroxide,[113] is, in the imagination of this author, a molecule likely to impact steroid receptors and enzymes like cytochrome p450s involved in steroid metabolism. However, it is likely that the multiple functions observed for many complex natural products are a result of their interaction with multiple pathways and mechanisms. The environmental fates and

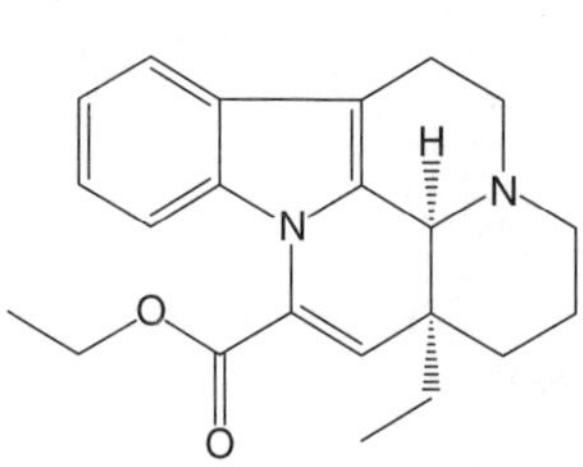

Vinocetine (17.10)

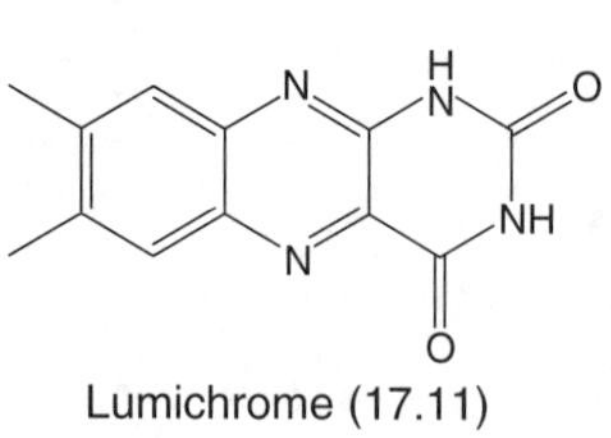

Lumichrome (17.11)

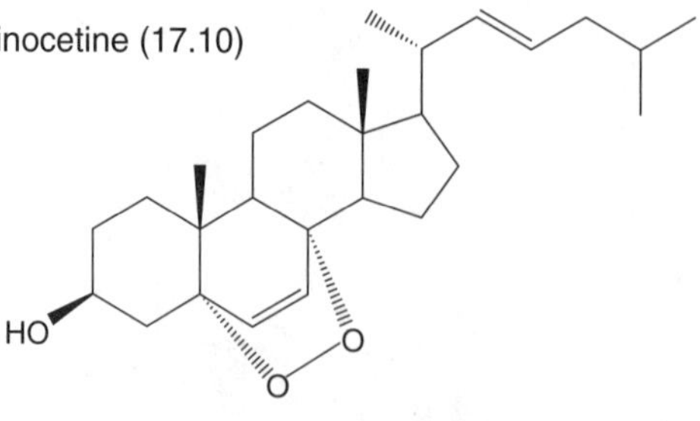

Sterol Peroxide (17.12)

effects of these molecules have not been studied. Anyone seriously considering commercial applications of such molecules should simultaneously consider environmental consequences.

Potentially as confounding and just as dangerous a pitfall is the assumption that a known pathway is the mechanism of action for a known molecule. Structures 17.13–17.17 all function in vertebrate and invertebrate biogenic amine-mediated neurotransmission. These molecules also alter settlement and metamorphosis of larvae of fouling organisms. Although these molecules may or may not have commercial potential as antifouling agents, they do point to the possibility that pharmaceuticals and natural products that specifically alter biogenic amine-mediated pathways may be candidates for antifouling technology. One need only consider that drugs used to alter moods in humans impact biogenic amine pathways to grasp the magnitude of the issues that must be considered for developing organic alternatives to the use of toxic metals. On one hand, the metals have clear environmental impact, but on the other hand, their impact on food supplies and human health is understood. This is not the case for organic additives.

Historically, funding sources that have supported natural product antifoulant studies have not funded inquiry into detailed mechanisms of action of a particular candidate compound. The logic has been that knowledge of the mechanism is not important for developing a product. This ignores many significant issues, such as improvement of performance by quantitative structure–function studies,[80–82,114,115] prediction of effects on nontarget species, chronic long-term studies, and fates and effects of bioactive additives. A few examples resulting from similar approaches to biocides include dichloro-diphenyl-trichloroethane (DDT), mercury, lead arsenic, cadmium, and TBT. A

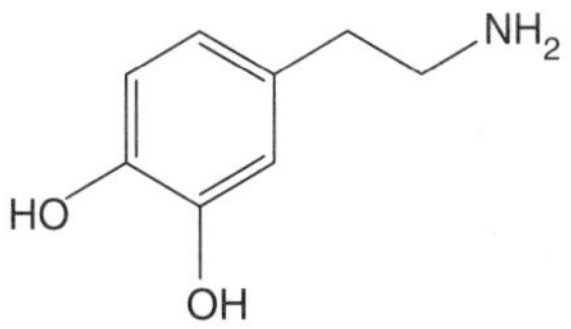

DOPA (17.13)

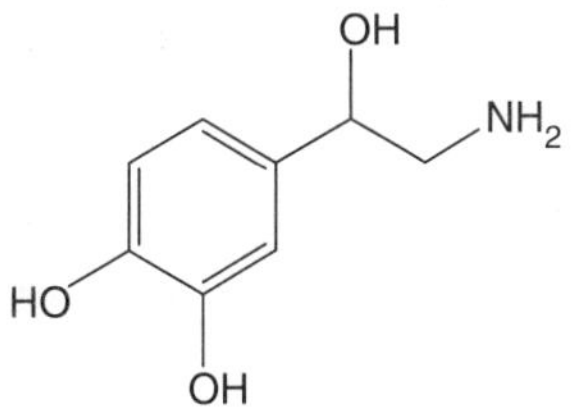

Norepinephrine (17.14)

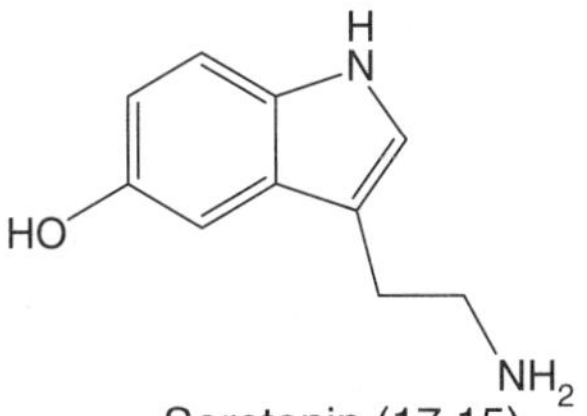

Serotonin (17.15)

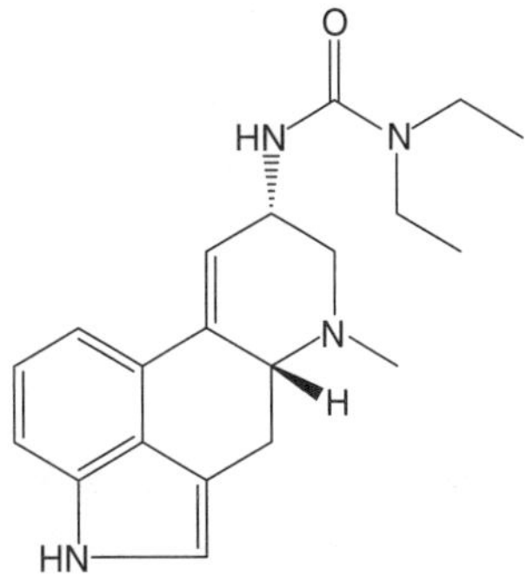

Lisuride (17.16)

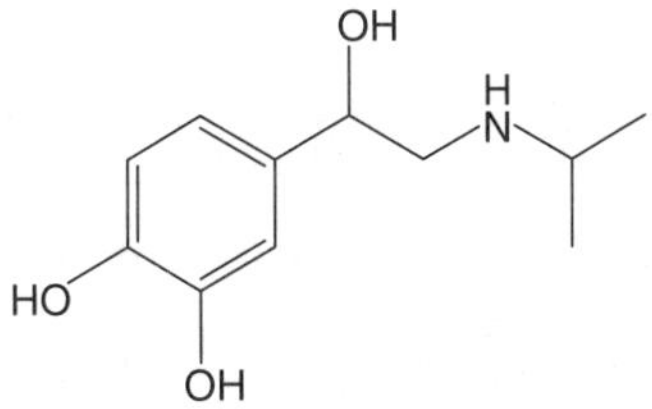

Isoproteranol (17.17)

consequence of this approach is that major environmental damage and monetary commitment are involved before problems are discovered and understood. Many researchers hope that a change in approach is forthcoming.

Those intimately familiar with bioassays have pushed these assays to their limits with respect to determining mechanisms of action at the whole animal level. Three obvious mechanisms are easy to discern. Two mechanisms, internal to the larvae, are toxicity and anesthesia.[84,107] The third mechanism is external to the larva and involves preventing settlement by molecules adsorbing to the surface and changing its characteristics.[29]

Adsorption onto a surface may also impact settling organisms by altering delivery of a bioactive compound.[47,84] For example, if toxicity assays are performed using both presettlement swimming barnacle larvae with hydrophyllic surfaces (nauplii) and the settlement-stage barnacle larvae with a lipophylic surface, the results depend on the chemical surface properties of the test container. Assays in glass show that both kinds of larvae are susceptible to similar levels of toxin. However, assays in polystyrene containers suggest that nauplii are more resistant to TBT. This result is obtained because TBT is lipophylic and adsorbs from solution onto plastic. The concentration of TBT is reduced in solution, where the nauplii are exposed and concentrated on the polystyrene container surface, where it partitions from the surface into the cyprid larvae through its lipophylic surface.

Since little or no attention is paid to the detailed mechanisms of action of existing additives, many commercial coatings, even those that are advertised as nontoxic, contain potent nonspecific biocides.[47] Performance of coatings is enhanced by adding oils[116] and surface-active agents. The author's research group at Duke was surprised to discover, for example, that silicone detergents were considered nontoxic. This is surprising insofar as a variety of surfactants and detergents are routinely used as insecticides and bacteriocides. When tests were conducted with silicone detergents, these detergents were found more toxic to larvae than copper on a weight basis (Rittschof, Bonaventura, Gerhart, and Clare, unpublished data). Also surprising was the revelation that many dimethyl silicone coatings are polymerized using organotin catalysts in sufficient concentration ($\approx 250\ \mu g/cm^2$ in experimental coatings tested) to render the coating initially toxic.[47] Dimethylsilane coatings polymerized by dibutyltin have been declared exempt from the Federal Insecticide Fungacide and Rodentacide Act (FIRFA) by the EPA Office of Pesticides.

In summary, natural products have a few known general mechanisms of action. These include toxins, inhibitors of growth, surface-energy modifiers, nervous pathway interference (anesthetics, including nitric oxide synthase inhibitors and neurotransmitter blockers), inhibitors of attachment, inhibitors of metamorphosis, and repellants.[46,55,117,118] It is likely that many compounds have multiple mechanisms of action.

C. Toxic Mechanisms

Should the final outcome of antifouling research be that toxins are necessary components of antifouling compounds, some biological compounds are among the most toxic known. For example, Donald J. Gerhart, then at Duke University Marine Laboratory, considered the antisettlement activity and toxicity of a set of natural toxins including a potent sodium/potassium ATPase inhibitor Bufalin.[48] Bufalin, the most potent natural product tested, was over 100 times more toxic than TBT and over 6000 times more potent than TBT with respect to antisettlement activity. The absence of a strict correlation between toxicity and antisettlement activity can be partially attributed to the relative solubility of the compounds in seawater and differences in the uptake kinetics between different life stages. However, the major difference can be attributed to the relationship between the speed at which the toxin kills and the speed at which metamorphosis occurs. For slow-acting toxins such as metals, settlement, and metamorphosis often precede death.[107]

D. NONTOXIC MECHANISMS

The antifouling research and coatings community considers agents and processes without acute toxicity as nontoxic. In fact, all compounds that prevent settlement or metamorphosis increase the probability of death of the target organism because the organism must spend more time in the plankton. An increase in larval time in the water column increases larval mortality. Interruption of attachment and metamorphosis can lead directly to death. Nontoxic mechanisms include disruption of attachment and metamorphosis,[18] reversible anesthesia[84,119] and repulsion,[63,80–82] either by modification of surface energy or use of a kairomone,[120] and bacteriostatic activity.[44,94] Complex natural products often display multiple activities.[121] For example, juncellins are bacteriostatic as well as inhibiting attachment and metamorphosis of macroinvertebrate larvae.[94]

E. QUANTITATIVE STRUCTURE–FUNCTION STUDIES

Clare[119] succinctly stated a conclusion that most researchers interested in the development of commercially viable derivatives of natural products eventually reach when summarizing the present state of knowledge: "The isolation of NPA (natural product antifoulants) through screening programs employing various bioassays is likely to continue into the forseeable future, but what is now needed is a more rational approach to these endeavours." This author's entire research group reached this conclusion in the mid-1980s when we realized that our interest in developing antifouling coatings using compounds based upon natural products was restricted to the compounds compatible with the coating matrix. Our solution to these practical issues was to use bioassays to perform quantitative structure–function studies on compounds that were structurally simple enough to be commercially synthesized.[80–82,122] Because these studies were part of a slow patenting process, the results were slow to reach the public domain. For example, although six patents were initiated and granted on this set of concepts by 1994, the first published reports did not appear until 1995 and then 1999.[84,95] Occasionally, researchers report basic structure–function studies in a more timely fashion.[115,123] Conflicts between patent protection and timely reporting need to be resolved if commercial potential of natural products is to be realized.

Examples of commercially practical compounds are lactones and furans.[62,64,80–82,114,122,124] These compounds are examples of multiple antifouling activity in relatively simple and commercially practical compounds. The lactones prevent barnacle settlement possibly through an anesthetic action (again suggesting interaction with transduction pathways).[84] When working with them, we were struck by the fact that many of the compounds have fruity odors. The fact that we can smell them shows that they are also biologically active in vertebrates. The basic reasoning that led us to study lactones and furans in the mid 1980s is described below. The idea of these molecules functioning in biochemical pathways that are common to virtually all organisms is supported by research on lactones that shows that they are involved in bacterial motility.[125–127] While an exciting possibility, the concept of interfering with common biochemical pathways also presents the potential for unknown impact on nontarget species such as food species and man (see also the section on biogenic amines). Historically, nontarget impacts of natural products have not been addressed. Priority is focused instead on detailed studies that provide information for patenting and assessing direct commercial potential.

F. PROOF-OF-PRINCIPLE FOR NATURAL PRODUCT ANTIFOULANTS

At present, compounds with promise must move through a torturous and daunting process if they are to be commercialized. The process begins with proof-of-principle testing for patenting. A standard predicament is that the researcher with the natural product is not familiar with polymer coatings and associated technology. Thus, proof-of-principle is generally done in short-term studies using either no polymer matrix at all[128] or matrices like abscisic acid,[129] phytogel[130,131] or, in the more sophisticated cases, derivatives of commercial polymers like vinyl,[132] resin rosin systems,[80–82,122] or ablative antifouling coatings without toxic metal additions.[133] Most recently, compounds have been directly impregnated into plastic matrices with practical potential. Intrinsic for commercialization is the patent protection of compounds.[134]

IV. IRONIES AND PITFALLS

A. THE IRONY OF SCHOLARLY WORK

A consequence of looking for novel secondary metabolites with bioassay driven approaches is that novel natural compounds are generally structurally complex and are often bound only in very small amounts. The first three groups of compounds that we identified were extremely potent at preventing barnacle settlement. However, they occurred in mg/kilogram dry weight of starting material amounts, and each had approximately seven chiral centers.[51,94] Compounds with this abundance and complexity have little practical commercial potential.

Considering the biological activity of rare and complex natural products, there is a convoluted path of reasoning that provides a way out of the dilemma. First, one assumes that since the nontoxic compounds do not kill larvae, even at saturating levels in assays, they are working to inhibit a particular pathway, possibly even by working through a receptor. Next, one assumes that if the above assumption is correct, then it is likely that only a part of the molecule is required for activity. The question then becomes, what parts of these complex molecules might be active? Don Gerhart came up with a testable postulate that worked. Gerhart reasoned, based upon the evolutionary relationships between arthropods, that insect pheromones found imbedded within the more complex natural products might be effective as antifoulants against barnacles. Accordingly, he looked at the structures of the natural products and identified a series of lactone and furan molecules that were biologically active in insects. Barnacle settlement tests with three series of compounds chosen in this way showed them to be very potent settlement inhibitors.[80–82,84,122] Interestingly, structural relatives of one of these series have also been shown to be involved in bacterial motility pathways.[127] However, one consequence of finding smaller, less complex molecules with antibarnacle activity was that the multiple biological activities observed in the complex natural products[94] were reduced or lost when simple analogues were found.[124]

Once simple analogues are discovered, the process of performing bioassay directed quantitative structure–function studies is rapid,[62,84] even if the specific pathway or mechanism is not known. The impression that this author has from these studies is that changes in potency due to structural changes are related to efficiency of delivery. The potential of this area should be high as the overall knowledge of how natural products function improves, especially if biology, pharmacology, and antifouling research are combined. For example, Ware[135] observed, in a doctoral thesis, that larvae moved near the surface of a sponge were reversibly anesthetized. A subsequent study on natural products from the same sponge genus in the Indian Ocean revealed a variety of natural products that were biologically active.[136] A later pharmacology study using rat brain nitric oxide (NO) synthase showed that these marine natural products inhibit NO synthase.[137] Since it has been known for over 10 years that changes in NO levels reversibly anesthetize barnacle larvae, it is becoming accepted that NO is a universal signal pathway in organisms. Thus, it is a logical extension to speculate that this pathway may be involved in reported anesthetic effects of natural products on larvae.

B. ADDITIONAL TECHNICAL PITFALLS

Once effective additives are identified, researchers encounter additional pitfalls on the way to commercialization. Especially with the use of simple organic molecules, but to some extent with any additive, additional technical problems of compatibility may arise. Compatibility is an issue for the compounds with antifouling polymer coating technology as well as other additives in the coating.[124] These effects span the gamut of physical and chemical properties of the coating. For example, organic compounds containing rings and double or triple bonds may interact with polymer substituents. Additives may interfere with, or alter, polymerization or final properties of the coating, and they may also lose biological activity due to chemical modification during the polymerization and curing process. Effects of interactions between additives are equally troublesome. Because there is usually little or no information on the detailed mechanism of action, the active structure, or chemical reactivity of the additive, it is difficult to predict the outcome of mixing additives into polymer coatings. As a result, effects of mixtures of additives do not follow simple rules. For example, effects of mixtures of two very effective antibarnacle settlement compounds are dependent on the ratio of the mixture of the two compounds. Some ratios of mixtures of individually active compounds are more effective than either compound alone, while other ratios are ineffective.[124] The most common solution to address both issues simultaneously is to encapsulate additives in such a way that they do not interfere with polymerization and can be mixed into and delivered predictably from a coating.[133] Often, such technical details are not addressed in the proof-of-principle stage and are targeted for the technology transfer stage.

V. WORK WITH THE COMMERCIAL SECTOR

A. TECHNOLOGY TRANSFER

Once an additive is patented, the next logical step is incorporation into a potentially viable commercial coating and the standard ASTM and approved industry methods of demonstrating effectiveness. Usually, it is at this point that questions of fates and effects of additives are posed. In most cases, the first issue addressed is that of insertion of the additives into a coating system and documentation of effectiveness. Most academic researchers never attempt this step because it requires expertise in coating polymer chemistry that they do not have, and most agencies that fund academics do not fund this kind of work. Those agencies that might fund the work often have a daunting governmental bureaucracy and associated regulations that make them unattractive to the business partner of the academic with the commercial polymer expertise. A more effective approach might be a collaboration between academics and industry, funded in a hands-off way by government. The U.S. Office of Naval Research (ONR) attempted to conduct such a program.

This ONR programmatic effort was made in the late 1980s and 1990s.[138] ONR funded a formal two-phase program that can be considered a case study of a starting point for future technology-transfer programs. Phase one was intended as a rapid empirical development phase. Phase two was intended to take experimental coatings to the point of commercialization. In phase one, rapid tests using miniaturized panels were performed over a 1-month interval to conserve materials and time.[107,132,139] In phase two, important issues such as fates and effects and governmental regulations were addressed, as was full-scale coatings tests on ships.

Phased miniature short-term tests were developed to save time in development of testable coatings. Experimental coatings failed for the variety of technical reasons mentioned above related to attempting to deliver organic additives in polymer coatings.[139] The short-term tests also quickly indicated any failure of the active ingredient because they were monitored weekly.[139] A variety of other very valuable tests, procedures, and issues were also developed and/or identified (see *Biofouling*, vol. 6, number 2 for a special issue devoted to the ONR program). In phase two, the best candidate coatings from phase one advanced to the standard panel dynamic bilge, keel, and other tests used by the industry to demonstrate that coatings were effective. Before any coatings finished phase two testing, the concept of natural product antifoulants had been dropped in favor of foul-release coatings containing additives such as silicone oils and detergents. (Progress and issues for this and related topics can be found in a new special issue of *Biofouling*, vol. 15, 2000, that is a compilation of reports from the 10th International Congress on Marine Corrosion and Fouling in Melbourne, Australia, 1999.) It is clear from the Melbourne meeting that there is an elevated level of sophistication as well as politicization of the issues. The most important of these issues are synthesized and discussed below.

B. COMMERCIALIZATION

Part of the reason for the lack of natural product-based antifouling alternatives is the necessity for expertise that historically is only found in an academic–industry research collaboration. The cross purposes of academia and business make it difficult to collaborate. Both academia and industry usually frown upon such research collaborations. The coatings industry usually does its research in-house and shies away from joint research and development programs. At the 10th International Congress, there was one report on isolation of natural products for prevention of larval settlement from an industrial source.[140] This report was from scientists trained in academia and hired by a coatings company to do in-house studies of natural products.

In addition to what can be considered cultural and philosophical differences between academia and industry, there are also very practical impediments. Business issues related to control of products and technology as well as intellectual property rights often conflict with requirements for receiving government funding. Our U.S. industry collaborators in the 1980s routinely turned down opportunities for government funding because of loss of control of intellectual property issues. Academic and industry researchers in other countries appear to be successfully collaborating on antifouling technology. These interactions are fostered in countries such as England,[141] Australia, Japan,[142] and the Netherlands.

Because of the cost of potential failures, the antifouling coatings industry and coatings users are very conservative. Coatings are tested exhaustively prior to commercialization (*Biofouling*, vol. 15, 1992, special issue). A major flaw in the process is the practice of evaluation by comparing experimental coatings to toxic metal coatings. Using the performance of toxic coatings with environmentally unacceptable consequences to judge the performance of experimental coatings clearly stifles the development of alternative technologies. The field would advance more rapidly if the industry were forced to find acceptable alternatives. For example, if TBT and copper-based antifouling paints were banned, the demands from all sections on governments and the coatings industry for new solutions could result in reform that cleared the way for productive collaborative interactions and progress.

At this time, the single largest stumbling block to the development of natural product antifouling coatings in the United States is the process of registration of the active ingredients. Registration is designed to prevent unacceptable alternatives and to avoid consequences and costly litigation due to incomplete or improper testing. The registration process is so time-consuming and costly that, if all other problems were solved, no natural product antifoulants would be commercialized. A documented example of the problem is the time and expenditure made by Rohm and Haas to register its organic biocide SEA-NINE® 211. SEA-NINE® 211 is an isothiazolone biocide that is broad-spectrum but that degrades.[143] It took approximately 10 years and over 10 million dollars to register this biocide in the United States.

Even if a compound is successfully registered, it is only when the product is commercialized and used for several years that environmental issues begin to emerge. The most recent example is Irgarol 1051, an herbicidal triazine additive used in copper-based antifouling paints in Europe[89,144,145] and Japan.[146] Irgarol ihhibits algal growth at approxiamtely 50 ng per liter. Levels substantially above 50 ng per liter are now reported from a variety of European locations and in the Seto Inland Sea of Japan. No one knows what the impact of this potent nonspecific herbicide will be on the affected ecosystems.[146]

Thus, among the practical challenges to be met for successful introduction of antifouling solutions with reduced environmental impact is to find ways to promote industry–academic interactions, accelerate testing, and reduce time and expense of registration and environmental fates and effects assessment. Unless these issues are resolved, there will be no substantial progress with novel approaches.

A study of the consequences of the result of banning DDT might be a good way to begin thinking about the actual consequences of banning TBT. Although there is substantial evidence of many of the short-term effects of TBT on marine organisms, it is clear that the scope of the problem extends further than acute losses to aquaculture and the shellfish industry. Because these larger issues have not been addressed, the consequences of introducing measurable levels of organotins to the world's oceans and food supplies cannot be fully understood. We do know that TBT has chronic effects in all organisms tested because of its interaction with cytochrome P_{450} enzymes central to steroid metabolism.[32] Perhaps in the future, the TBT industry and countries around the world will face TBT environmental and human-health-related issues roughly comparable to those faced by the U.S. tobacco industry with respect to worldwide health problems associated with chronic smokers.

ACKNOWLEDGMENTS

Thanks to Alva R. Schmidt, Eric R. Holm, Anthony S. Clare, Donald J. Gerhart, Joseph Bonaventrua, John D. Costlow, Jr., Celia Bonaventura, Cazlyn G. Bookhout, Irving R Hooper, James S. Maki, Dennis J. Crisp, John P. Sutherland, Richard B. Forward, Jr., Michael G. Hadfield, Nancy M. Targett, Patrick J. Bryan, E. Sanford Branscomb, Sister Avelin Mary, Bernard Zahuranec, Robert E. Baier, Ann Meyer, Helen E. Nearing, Gail W. Cannon, Kelly Eisenman, Dierdre Roberts, David M. Talbert, Dianne R. Gagnon, and Lilian I. Lorenzsonn-Willis for assistance in every aspect of this effort.

REFERENCES

1. Little, B.J., Wagner, P., Maki, J.S., Walch, M., and Mitchell, R., Factors influencing the adhesion of microorganisms to surfaces, *J. Adhesion*, 20, 187, 1986.
2. Brady, R.F., Jr. and Singer, I.L., Mechanical factors favoring release from fouling release coatings, *Biofouling*, 15, 73, 2000.
3. Franklin, M., White, D.C., Little, B., Ray, R., and Pope, R., The role of bacteria in pit propagation of carbon steel, *Biofouling*, 15, 13, 2000.
4. Meyer, A.E., Baier, R.M., and King, R.W., Initial fouling of nontoxic coatings in fresh brackish and sea water, *Can. J. Chem. Eng.*, 66, 5562, 1987.
5. Forward, R.B., Jr. and Cronin, T.W., Crustacean larval vertical migration: a perspective, in *Signposts in the Sea*, Herrnkind, W.F. and Thisle, A.B., Eds., Florida State University, Tallahassee, FL, 1987, 29.
6. Forward, R.B., Jr., Diel vertical migration: zooplankton photobiology and behavior, *Oceanogr. Mar. Biol. Annu. Rev.*, 26, 361, 1988.
7. Crisp, D.J., Overview of research on marine invertebrate larvae, 1940-1980, in *Marine Biodeterioration: an Interdisciplinary Study*, Costlow, J.D. and Tipper, R.C., Eds., Naval Institute Press, Annapolis, MD, 1984, 103.
8. Clare, A.S., Rittschof, D., Gerhart, D.J, and Maki, J.S., Molecular approaches to nontoxic antifouling, *J. Invert. Reprod. Dev.*, 22, 67, 1992.
9. Rittschof, D. and Costlow, J.D., Bryozoan and barnacle settlement in relation to initial surface wettability: a comparison of laboratory and field studies, in *Topics in Marine Biology*, *Proceedings of the 22nd European Marine Biology Symposium*, Ros, J.D., Ed., Instituto de Ciencias del Mar, Barcelona, Spain, 1989, 411.
10. Characklis, W.G., Trulear, M.G., Bryers, J.D., and Zelver, N., Dynamics of biofilm processes: methods, *Wat. Res.*, 16, 1207, 1982.
11. Characklis, W.G. and Cooksey, K.E., Biofilms and microbial fouling, *Adv. Appl. Microbiol.*, 29, 93, 1983.

12. Waite, J.H., Adhesion in bysally attached bivalves, *Biol. Rev.*, 58, 209, 1983.

13. Waite, J.H., Jensen, R.A., and Morse, D.E., Cement precursor proteins of the reef-building polychaete *Phragmatopoma californica* (Fewkes), *Biochem.*, 31, 5733, 1992.

14. Yamamoto, H., Marine adhesive proteins and some biotechnological applications, *Biotechnol. Eng. Rev.*, 13, 133, 1995.

15. Spindler, K.D. and Muller, W.A. Introduction of metamorphosis by bacteria and by Australian seaweeds, *Biofouling*, 169, 271, 1972.

16. Kato, T., Kumanireng, A.S., Ichninose, I., Kitahara, Y., Kakinuma, Y., Nishihara, M., and Kato, M., Active components of *Sargassum tortile* effecting the settlement of swimming larvae of *Coryne uchidai*, *Experientia*, 31, 433, 1975.

17. Hadfield, M.G., Unabia, C.C., Smith, C.M., and Michael, T.M., Settlement preferences of the ubiquitous fouler *Hydroides elegans*, in *Recent Developments in Biofouling Control*, Thompson, M.F., Nagabushanam, R., Sarojini, R., Fingerman, M., Eds., Oxford and IBH Publ. Co., New Delhi, 1994, 65.

18. Rittschof, D., Branscomb, E.S., and Costlow, J.D., Settlement and behavior in relation to flow and surface in larval barnacles, *Balanus amphitrite* Darwin, *J. Exp. Mar. Biol. Ecol.*, 82, 131, 1984.

19. Maki, J.S., Rittschof, D., Costlow, J.D., and Mitchell, R., Inhibition of attachment of larval barnacles, *Balanus amphitrite*, by bacterial surface films, *Mar. Biol.*, 97, 199, 1988.

20. Morse, A.N.C., Froyd, C.A., and Morse, D.E., Molecules from cyanobacteria and red algae that induce larval settlement and metamorphosis in the mollusc *Haliotis rufescens*, *Mar. Biol.*, 81, 293, 1984.

21. Roberts, D., Rittschof, D., Holm, E., and Schmidt, A.R., Factors influencing larval settlement: temporal, spatial and molecular components of initial colonization, *J. Exp. Mar. Biol. Ecol.*, 150, 203, 1991.

22. Hadfield, M.G., Carpizo-Ituarte, E., Holm, E., Nedved, B., and Unabia, C., Macrofouling processes: a developmental and evolutionary perspective, Abstract 10th International Congress on Marine Corrosion and Fouling, 1999.

23. Callow, M.E., Callow, J.A., Pickett-Heaps, J.C., and Wetherbee, R., Primary adhesion of *Enteromorpha* (Chlorophyta, Ulvales) propagules: quantitative settlement studies and video microscopy, *J. Phycol.*, 33, 938, 1997.

24. Callow, M.E. and Callow, A.J., Substratum location and zoospore behaviour in the fouling alga *Enteromorpha*, *Biofouling*, 15, 49, 2000.

25. Woollacott, R.M., Environmental factors in bryozoan settlement, in *Marine Biodeterioration: an Interdisciplinary Study*, Costlow, J.D. and Tipper, R.C., Eds., Naval Institute Press, Annapolis, MD, 1984, 149.

26. Baier, R.E., Initial events in microbial film formation, in *Marine Biodeterioration: an Interdisciplinary Study*, Costlow, J.D. and Tipper, R.C., Eds., Naval Institute Press, Annapolis, MD, 1984, 52.

27. Mihm, J.W., Banta, W.C., and Loeb, G.I., Effects of adsorbed organic and primary fouling films on bryozoan settlement, *J. Exp. Mar. Biol. Ecol.*, 54, 167, 1981.

28. Rittschof, D. and Costlow, J.D., Surface determination of macroinvertebrate larval settlement, in *Proceedings of the 21st European Marine Biology Symposium*, Styczynska-Jurewicz, E., Ed., Gdansk Polish Academy of Sciences, Institute of Oceanology, Gdansk, Poland, 1989, 155.

29. Gerhart, D.J., Rittschof, D., Hooper, I.R., Eisenman, K., Meyer, A.E., Baier, R.E., and Young, C., Rapid and inexpensive quantification of the combined polar components of surface wettability: application to biofouling, *Biofouling*, 5, 251, 1992.

30. Holm, E.R., Cannon, G., Roberts, D., Schmidt, A.R., Sutherland, J.P., and Rittschof, D., The influence of initial surface chemistry on the development of the fouling community at Beaufort, North Carolina, *J. Exp. Mar. Biol. Ecol.*, 215, 189, 1997.

31. Holm, E.R., Nedved, B.T., Phillips, N., DeAngelis, K.L., Hadfield, M.G., and Smith, C.M., Temporal and spatial variation in the fouling of silicone coatings in Pearl Harbor, *Biofouling*, 15, 95, 2000.

32. Goyer, R.A., Toxic effects of metals, in *Casarett and Doull's Toxicology: The Basic Science of Poisons*, 4th ed., Amdur, M.O., Doull, J.D., and Klaassen, C.D., Eds., Pergamon Press, New York, 1991, 623.

33. Stadtman, E.R. and Oliver, C.N., Metal-catalyzed oxidation of proteins, *J. Biol. Chem.*, 266, 2005, 1991.

34. Cheney, M.A. and Criddle, R.S., Heavy metal effects on the metabolic activity of *Elliptio complanata*: a calorimetric method, *J. Environ. Qual.*, 25, 235, 1996.

35. Boyer, I.J., Toxicity of dibutyltin, tributyltin and other organotin compounds to humans and experimental animals, *Toxicology*, 55, 253, 1989.

36. Smith, B.S., Sexuality in the American mud snail, *Nassarius obsoletus* Say, *Proc. Malac. Soc. Lond.*, 39, 378, 1971.

37. Jenner, M.G., Pseudohermaphroditism in *Ilyanassa obsoleta* (Mollusca: Neogastropoda), *Science*, 205, 1407, 1979.

38. Curtis, L.A. and Barse, A., Sexual anomalies in estuarine snail *Ilyanassa obsoleta*: imposex in females and associated phenomena in males, *Oecologia*, 84, 371, 1990.

39. Curtis, L.A., A decade-long perspective on the bioindicator of pollution: imposex in *Ilyanassa obsoleta* on Cape Henlopen, Delaware Bay, *Mar. Environ. Res.*, 30, 291, 1994.

40. Schulte-Oehlmann, U., Oehlmann, J., Fioroni, P., and Bauer, B., Imposex and reproductive failure in *Hydrobia ulvae* (Gastropoda: Prosobranchia), *Mar. Biol.*, 28, 7, 1997.

41. de la Court, F.H., Erosion and hydrodynamics of biofouling coatings, in *Marine Biodeterioration: an Interdisciplinary Study*, in Costlow J.D. and Tipper, R.C., Eds., U.S. Naval Institute Press, Annapolis, Maryland, 1984, 230.

42. Brancato, M.S., Toll, J., DeForest, D., and Tear, L., Aquatic ecological risks posed by tributyltin in United States surface waters: pre-1989 to 1996 data, *Environ. Toxicol. Chem.*, 18, 567, 1999.

43. Rosenthal, H., Gollasch, S., Laing, I., Leppakoski, E., Macdonald, E., Minchin, D., Nauke, M., Olenin, S., Utting, S., Voigt, M., and Wallentinus, I., Exotics across the ocean: testing monitoring systems for risk assessment of harmful introductions by ships to European waters, *Biofouling*, Abstract 10th International Congress on Marine Corrosion and Fouling, Melbourne, Australia, 1999.

44. Targett, N.M., Bishop, S.S., McConnell, O.J., and Yoder, J.A., Antifouling agents against the benthic marine diatom *Navicula salinicola:* Homarine from the gorgonian *Leptogorgia virgulata* and *L. setacea* and analogs, *J. Chem. Ecol.*, 9, 817, 1983.

45. Standing, J., Hooper, I.R., and Costlow, J.D., Inhibition and induction of barnacle settlement by natural products present in octocorals, *J. Chem. Ecol.*, 10, 823, 1984.

46. Clare, A.S., Marine natural product antifoulants: status and potential, *Biofouling*, 9, 211, 1996.

47. Rittschof, D. and Holm, E.R., Antifouling and foul-release: a primer, in *Recent Advances in Marine Biotechnology, Vol. I. Endocrinology and Reproduction*, Fingerman, M., Nagabhushanam, R., and Thompson, M.F., Eds., Oxford and IBH Publishing, New Delhi, India, 1997, 497.

48. Rittschof, D., Natural product antifoulants: one perspective on the challenges related to coatings development, *Biofouling*, 15, 119, 2000.

49. Vrolijk, N.H., Targett, N.M., Baier, R.E., and Meyer, A.E., Surface characterisation of two gorgonian coral species: implications for natural antifouling defense, *Biofouling*, 2, 39, 1990.

50. Gerhart, D.J., Rittschof, D., and Mayo, S.W., Chemical ecology and the search for marine antifoulants: studies of a predator–prey symbiosis, *J. Chem. Ecol.*, 14, 1903, 1988.

51. Keifer, P.A., Rinehart, K.L., Jr., and Hooper, I.R., Renillafoulins, antifouling diterpenes from the sea pansy *Renilla reniformis* (Octocorallia), *J. Org. Chem.*, 51, 4450, 1986.

52. Hadfield, M.G. and Ciereszko, L.S., Action of cembranolides derived from octocorals on larvae of the Nudibranch *Phestilla sibogae*, in *Drugs and Food from the Sea*, Kaul, P. K. and Sindermann, C. J., Eds., University of Oklahoma Press, Norman, OK, 1978, 145.

53. Bakus, G.J. and Green, G., Toxicity in sponges and holothurians: a geographic pattern, *Science*, 185, 951, 1974.

54. Bakus, G.J., Targett, N.M., and Schulte, B., Chemical ecology of marine organisms: an overview, *J. Chem. Ecol.*, 12, 951, 1986.

55. Davis, A.R., Targett, N.M., McConnell, O.J., and Young, C.M., Epibiosis of marine algae and benthic invertebrates: natural products chemistry and the mechanisms inhibiting settlement and overgrowth, in *Marine Bioorganic Chemistry*, Vol. 3., Scheuer, P.J., Ed., Springer-Verlag, Berlin, 1989, 85.

56. Coll, J.C., Bowden, B.J., Tapiolas, D.M., and Dunlap, W.C., *In situ* isolation of allelochemicals released from soft corals (Coelenterata: Octocorallia): a totally submersible sampling apparatus, *J. Exp. Mar. Biol. Ecol.*, 17, 69, 1982.

57. Targett, N.M., Allelochemistry in marine organisms: chemical fouling and antifouling strategies, in *Marine Biodeterioration*, Thompson, M., Sarojini, R., and Nagabhushanam, R., Eds., Oxford & IBH Publishing, New Delhi, India, 1988, 609.

58. Woodin, S.A., Marinelli, R.L., and Lincoln, D.E., Allelochemical inhibition of recruitment in a sedimentary assemblage, *J. Chem. Ecol.*, 19, 517, 1993.

59. Woodin, S.A., Lindsay, S.M., and Lincoln, D.E., Biogenic bromophenols as negative recruitment cues, *Mar. Ecol. Prog. Ser.*, 157, 303, 1997.

60. Slattery, M., Hamann, M.T., McClintock, J.B., Perry, T.L., Puglisi, M.P., and Yoshida, W.Y., Ecological roles for water-borne metabolites from Antarctic soft corals, *Mar. Ecol. Prog. Ser.*, 161, 133, 1997.

61. Schmitt, T.M., Lindquist, N., and Hay, M.E., Seaweed secondary metagolites as antifoulants: effects of *Dictyota* spp. diterpenes on survivorship, settlement, and development of marine invertebrate larvae, *Chemoecology*, 8, 125, 1998.

62. Steinberg, P.D., de Nys, R., and Kjelleberg, S., Chemical inhibition of epibiota by Australian seaweeds, *Biofouling*, 12, 227, 1998.

63. Walters, L.J., Hadfield, M.G., and del Carmen, K.A., The importance of larval choice and hydrodynamics in creating aggregations of *Hydroides elegans* (Polychaeta: Serpulidae), *Invertebrate Biol.*, 116, 102, 1997.

64. de Nys R., Steinberg, P.D., Willemsen, P., Dworjanyn, S.A., Gabelish, C.B., and King, R.J., Broad spectrum effects of secondary metabolites from the red alga *Delisea pulchra* in antifouling assays, *Biofouling*, 8, 259, 1995.

65. Colwell, R.R., Biotechnology in the marine sciences, *Science*, 222, 19, 1983.

66. Bryan, P.J., Rittschof, D., and McClintock, B.J., Bioactivity of echinoderm ethanolic body-wall extracts: an assessment of marine bacterial attachment and macroinvertebrate larval settlement, *J. Exp. Mar. Biol. Ecol.*, 196, 79, 1996.

67. Nakatsu, T., Walker, R.P., Thompson, J.E., and Faulkner, D.J., Biologically active sterol sulfates from the marine sponge *Toxadocia zumi*, *Experientia*, 39, 759, 1983.

68. Sieburth, J.M. and Conover, J.T., *Sargassum* tannins, an antibiotic which retards fouling, *Nature*, 208, 52, 1965.

69. Walker, R.P., Thompson, J.E., and Faulkner, D.J., Exudation of biologically active metabolites in a sponge (*Aplysina fistularis*). II. Chemical evidence, *Mar. Biol.*, 88, 27, 1985.

70. Russel, F.E., Marine toxins and venomous and poisonous marine plants and animals (invertebrates), *Adv. Mar. Biol.*, 21, 59, 1984.

71. Bookhout, C.G. and Costlow, J.D., Jr., Feeding, molting and growth in barnacles, in *Marine Boring and Fouling Organisms*, Ray, D. L., Ed., University of Washington Press, Seattle, WA, 1959, 212.

72. Henry, D.P., The distribution of the *Amphitrite* series of *Balanus* in North American waters, in *Marine Boring and Fouling Organisms*, Ray, D.L., Ed., University of Washington Press, Seattle, WA, 1959, 190.

73. Dineen, J.F., Jr. and Hines, A.H., Interactive effects of salinity and adult extract upon settlement of the estuarine barnacle *Balanus improvisus* (Darwin, 1854), *J. Exp. Mar. Biol. Ecol.*, 56, 239, 1992.

74. O'Connor, N.J. and Richardson, D.L., Comparative attachment of barnacle cyprids (*Balanus amphitrite* Darwin, 1854; *B. improvisus* Darwin, 1854; and *B. eburneus* Gould, 1841) to polystyrene and glass substrata, *J. Exp. Mar. Biol. Ecol.*, 183, 213, 1994.

75. O'Connor, N.J. and Richardson, D.L., Effects of bacterial films on attachment of barnacle (*Balanus improvisus* Darwin) larvae: laboratory and field studies, *J. Exp. Mar. Biol. Ecol.*, 206, 69, 1996.

76. Rittschof, D., Hooper, I. R., and Costlow, J. D., Settlement inhibition of marine invertebrate larvae: comparison of sensitivities of bryozoan and barnacle larvae, in *Marine Biodeterioration*, Thompson, M., Sarojini, R., and Nagabhushanam, R., Eds., Oxford & IBW Publishing Co., Bombay, 1988, 599.

77. Tsukamoto, S., Kato, H., Hirota, H., and Fusetani, N., Ceratinamides A and B: new antifouling dibromotyrosine derivatives from the marine sponge *Pseudoceratina purpurea*, *Tetrahedron*, 52, 8181, 1996.

78. Tsukamoto, S., Kato, H., Hirota, H., and Fusetani, N., Seven new polyacetylene derivatives, showing both potent metamorphosis-inducing activity in ascidian larvae and antifouling activity against barnacle larvae, from the marine sponge *Callyspongia truncata*, *J. Nat. Prod.*, 60, 126, 1997.

79. Faulkner, D.J., Biomedical uses for natural marine chemicals, *Oceanus*, 29, Spring 1992.

80. Gerhart, D.J., Rittschof, D., and Hooper, I.R., Antifouling Coating Composition Comprising Lactone Compounds, Method for Protecting Aquatic Structures, and Articles Protected Against Fouling Organisms, U.S. Patent No. 5,248,221, 1993.

81. Gerhart, D.J., Rittschof, D., and Bonaventura, J., Antifouling Coating Comprising Steroidal Compounds and Method for Using Same, U.S. Patent No. 5,252,630, 1993.

82. Gerhart, D.J., Rittschof, D., Hooper, I.R., and Clare, A.S., Antifouling Coating Composition Comprising Furan Compounds, Method for Protecting Aquatic Structures, and Articles Protected Against Fouling Organisms, U.S. Patent No. 5,259,701, 1993.

83. Kon-ya, K., Shimidzu, N., Miki, W., and Endo, M., Indole derivatives as potent inhibitors of larval settlement by the barnalce, *Balanus amphitrite*, *Biosci. Biotech. Biochem.*, 58, 2178, 1994.

84. Clare, A.S., Rittschof, D., Gerhart, D.J., Hooper, I.R., and Bonaventura, J., Antisettlement and narcotic action of analogues of diterpene marine natural product antifoulants from octocorals, *Mar. Biotechnol.*, in press.

85. Meadows, P.S. and Williams, G.B., Settlement of *Spirorbis borealis* Daudin larvae on surfaces bearing films of micro-organisms, *Nature*, 198, 610, 1963.

86. Corpe, W.A., Attachment of marine bacteria to solid surfaces, in *Adhesion in Biological Systems*, Manly, R.S., Ed., Academic Press, New York, 1970, 73.

87. Kirchman, D. and Mitchell, R., Possible role of lectins in the settlement and metamorphosis of marine invertebrate larvae on surfaces coated with bacteria, *Bacteriol. Mar.*, 331, 173, 1982.

88. Hadfield, M.G., Research on settlement and metamorphosis of marine invertebrate larvae: past, present and future, *Biofouling*, 12, 9, 1998.

89. Readman, J.W., Antifouling herbicides — a threat to the marine environment?, *Mar. Pollut. Bull.*, 32, 320, 1996.

90. Brancato, M.S. and Woollacott, R.M., Effect of microbial films on settlement of bryozoan larvae (*Bugula simplex, B. stolonifera* and *B. turrita*), *Mar. Biol.*, 71, 51, 1982.

91. Weiner, R.M., Segall, A.M., and Colwell, R.R., Characterization of a marine bacterium associated with *Crassostrea virginica* (the Eastern Oyster), *Appl. Env. Microbiol.*, 49, 83, 1985.

92. Bonar, D.B., Coon, S.L., Walch, M., Weiner, R.M., and Fitt, W., Control of oyster settlement and metamorphosis by endogenous and exogenous chemical cues, *Bull. Mar. Sci.*, 46, 484, 1990.

93. Vitalina Mary, S., Avelin Mary, S., Rittschof, D., Sarojini, R., and Nagabhushanam, R., Compounds from octocorals that inhibit barnacle settlement: isolation and biological potency, in *Bioactive Compounds from Marine Organisms*, Thompson, M.F., Sarojini, R., and Nagabhushanam, R., Eds., Oxford and IBH Publishing Co., New Delhi, India, 1991, 331.

94. Avelin Mary, S., Vitalina Mary, S., Rittschof D., and Nagabhushanam R., Bacterial–barnacle interaction: potential of using juncellins and antibiotics to alter structure of bacterial communities, *J. Chem. Ecol.*, 19, 2155, 1993.

95. Clare, A.S., Rittschof, D., Price, R.R., and Gerhart, D.J., Khellin, a natural product analogue with antifouling activity: laboratory and field studies, in *Biodeterioration and Biodegradation*, Bousher, A. and Edyvean, R.G.J., Eds., Rugby, Institution of Chemical Engineers, 1995, 573.

96. Wieczorek, S.K., Clare, A.S., and Todd, C.D., Inhibitory and facilitatory effects on microbial films on settlement of *Balanus amphitrite* larvae, *Mar. Ecol. Prog. Ser.*, 119, 221, 1995.

97. Bryan, P.J., Rittschof, D., and Qian, P.Y., Settlement inhibition of bryozoan larvae by bacterial films and aqueous leachates, *Bull. Mar. Sci.*, 61, 849, 1997.

98. Holmström, C. and Kjelleberg, S., The effects of external biological factors of settlement of marine invertebrate and new antifouling technology, *Biofouling*, 8, 147, 1994.

99. Wieczorek, S.K. and Todd, C.D., Inhibition and facilitation of settlement of epifaunal marine invertebrate larvae by microbial biofilm cues, *Biofouling*, 12, 81, 1997.

100. Fisher, E.C., Castelli, V.J., Rodgers, S.D., and Bleile, H.R., Technology for control of marine biofouling — a review, in *Marine Biodeterioration: an Interdisciplinary Study*, Costlow, J.D. and Tipper, R.C., Eds., Naval Institute Press, Annapolis, MD, 1984, 261.

101. Yule, A.B. and Walker, G., Settlement of *Balanus balanoides*: the effect of cyprid antennular secretion, *J. Mar. Biol. Assoc. UK*, 65, 707, 1985.

102. Clare, A.S. and Matsumara, K., Nature and perception of barnacle settlement pheromones, *Biofouling*, 15, 57, 2000.

103. Lindner, E., The attachment of macrofouling invertebrates, in *Marine Biodeterioration: an Interdisciplinary Study*, Costlow, J.D. and Tipper, R.C., Eds., Naval Institute Press, Annapolis, MD, 1984, 183.

104. Clare, A.S., Freet, R.K., and McClary, M., On the antennular secretion of the cyprid *Balanus amphritite amphritite*, and its role as a settlement pheromone, *J. Mar. Biol. Assoc. UK*, 74, 243, 1994.

105. Rittschof, D., Oyster drills and the frontiers of chemical ecology: unsettling ideas, *Am. Malac. Bull.*, special ed., 1, 111, 1985.

106. Matsumara, K., Nagano, M., and Fusetani, N., Purification of a larval settlement-inducing protein complex (SIPC) of the barnacle, *Balanus amphitrite*, *J. Exp. Zool.*, 281,12, 1998.

107. Rittschof, D., Clare, A.S., Gerhart, D.J., Avelin Mary, S., and Bonaventura, J., Barnacle in vitro assays for biologically active substances: toxicity and settlement assays using mass cultured *Balanus amphitrite amphitrite* Darwin, *Biofouling*, 6, 115, 1992.

108. Rittschof, D., Hooper, I.R., and Costlow, J.D., Barnacle settlement inhibitors from sea pansies (*Renilla reniformis*),. *Bull. Mar. Sci.*, 39, 376, 1986.

109. Holm, E.R., Nedved, B.T., Carpizo-Ituarte, E., and Hadfield, M.G., Metamorphic-signal transduction in *Hydroides elegans* (Polychaeta: Serpulidae) is not mediated by a G protein, *Biol. Bull.*, 195, 21, 1998.

110. Jing, Y., Watabe, M., Hashimoto, S., Nakajo, S., and Nakaya, K., Cell cycle arrest and protein kinase modulating effect of bufalin on human leukemia ML1 cells, *Anticancer Res.*, 14, 1193, 1994.

111. Yamamoto, H., Satuito, C.G., Yamazaki, M., Natoyama, K., Tachibana, A., and Fusetani, N., Neurotransmitter blockers for antifoulants against planktonic larvae of the barnacle *Balanus amphitrite* and the mussel *Mytilus galloprovincialis*, *Bioufouling*, 13, 69, 1998.

112. Tsukamoto, S., Kato, H., Hirota, H., and Fusetani, N., Lumichrome — a larval metamorphosis-inducing substance in the ascidian *Halocynthia roretzi*, *Eur. J. Biochem.*, 264, 785, 1999.

113. Tomono, Y., Hirota, H., and Fusetani, N., Antifouling compounds against barnacle (*Balanus amphitrite*) larvae from the marine sponge *Acanthella cavernosa*, in *Sponge Sciences — Multidisciplinary Perspectives*, Watanabe, Y. and Fusetani, N., Eds., Springer-Verlag, Tokyo, 1998, 413.

114. Targett, N.M. and Sojac, W.R., Natural antifoulants and their analogues: applying nature's defense strategies to problems of biofouling control, in *Recent Developments in Biofouling Control*, Thomson, M.F., Nagabushanam, R., Sarojini, R., and Fingerman, M., Eds., A.A. Balkema, Rotterdam, 1994, 222.

115. Ponasik, J.A., Conova, S., Kinghorn, D., Kinney, W.A., Rittschof, D., and Ganem, B., Pseudoceratidine, a marine natural product with antifouling activity: synthetic and biological studies, *Tetrahedron*, 54, 6977, 1998.

116. Truby, K.I., Darkangelo Wood, C., Stein, J., Cella, J., Carpenter, J., Kavenaugh, C., Swain, G., Wiebe, D., Lapota, D., Meyer, A., Holm, E., Wendt, D., Smith, C., and Montemaranao, J., Evaluation of the performance enhancement of silicone biofouling release coatings by oil deposition, 10th International Congress on Marine Corrosion and Fouling, Melbourne, Australia, 1999.

117. Walters, L.J., Hadfield, M.G., and Smith, C.M., Waterbourne chemical compounds in tropical macro algae. Positive and negative cues for larval settlement, *Mar. Biol.*, 126, 383, 1996.

118. Rittschof, D., Forward, R.B., Jr., Cannon, G., Welch, J.M., McClary, M.M., Jr., Holm, E.R., Clare, A.S., Conova, S., McKelvey, L.M., Bryan, P., and Van Dover, C.L., Cues and context: larval responses to physical and chemical cues, *Biofouling*, 12, 31, 1998.

119. Clare, A.S., Towards nontoxic antifouling, *J. Mar. Biochenol.*, 6, 3, 1998.

120. McKelvey, L.M. and Forward, R.B., Jr., Activation of brine shrimp photoresponses in diel vertical migration by chemical cues from visual and nonvisual predators, *J. Plankton Res.*, 17, 2191, 1995.

121. Butler, A.J., van Altena, I.A., and Dunne, S.J., Antifouling activity of lyso-platelet-activating factor extracted from Australian sponge *Crella incrustans*, *J. Chem. Ecol.*, 22, 2041, 1996.

122. Gerhart, D.J., Rittschof, D., and Bonaventura, J., Antifouling Coating and Method for Using Same, U.S. Patent 5,314,932, 1994.

123. Ponasik, J.A., Kassab, D.J., and Ganem, B., Synthesis of the antifouling polyamine pseudoceratidine and its analogs: factors influencing biocidal activity, *Tetrahedron Lett.*, 37, 6041, 1996.

124. Rittschof, D., Sasikumar, N., Murlless, D., Clare, A.S., Gerhart, D.J., and Bonaventura, J., Mixture interactions of lactones, furans, and a commercial biocide: toxicity and antibarnacle settlement activity, in *Recent Developments in Biofouling Control*, Thompson, M.F., Nagabhushanam, R., Sarojini, R., and Fingerman, M., Eds., Oxford and IBH Publishing, New Delhi, India, 1994, 269.

125. Stretton, S., Danon, S.J., Kjelleberg, S., and Goodman, A.E., Changes in cell morphology and motility in the marine *Vibrio* sp. strain S14 during conditions of starvation and recovery, *FEMS Microbiol. Lett.*, 146, 23, 1997.

126. Givskov, M., Ostling, J., Eberl, L., Lindum, P.W., Christensen, A.B., Christiansen, G., Molin, S., and Kjelleberg, S., Two separate regulatory systems participate in control of swarming motility of *Serratia liquefaciens* MG1, *J. Bacteriol.*, 180, 742, 1998.

127. Kjelleberg, S., Signal mediated bacterial colonization, Abstract 10th International Symposium on Marine Corrosion and Fouling, Melbourne, Australia, 1999.

128. Costlow, J.D., Hooper, I.R., and Rittschof, D., Anti-Fouling Compound and Method of Use, U.S. Patent No. 4,788,302, 1988.

129. Bakus, G.J., Wright, M., Khan, A.K., Ormsby, B., Gulko, D.A., Licuanan, C.E., Oritz, A., Chan, D., Lorenzana, D., and Huxley, M.P., Experiments seeking marine natural antifouling compounds, in *Recent Developments in Biofouling Control*, Thompson M. F., Nagabhushanam R., Sarojini, R., and Fingerman, M., Eds., Oxford and IBH Publishing Co., New Delhi, India, 1994, 373.

130. Pawlik, J.R., Chemical ecology of the settlement of benthic marine invertebrates, *Oceanogr. Mar. Biol. Annu. Rev.*, 30, 273, 1992.

131. Bryan, P.J., Kreider, J.L., and Qian, P.Y., Settlement of the serpulid polychaete *Hydroides elegans* (Haswell) on the arborescent bryozoan *Bugula neritina* (L.): evidence of a chemically mediated relationship, *J. Exp. Mar. Biol. Ecol.*, 220, 171, 1998.

132. Vasishtha, N., Sundberg, D.C., and Rittschof, D., Evaluation of release rates and control of biofouling using monolithic coatings containing an isothiazolone, *Biofouling*, 9, 1, 1995.

133. Price, R.R., Patchen, M., Rittschof, D., Clare, A.S., and Bonaventura, J., Performance enhancement of natural antifouling compounds and their analogs through microencapsulation and controlled release, *Biofouling*, 6, 207, 1992.

134. Hodson, S., Christov, V., de Nys, P.C., Steinberg, P., and Christie, G.B., Antifouling polymers, PCT/AU98/00509 (WO99/01514) — patent application, 1998.

135. Ware, G. N., The patterns and mechanisms of antifouling in some temporate marine sponges, Ph.D. thesis, Duke University, Beaufort, NC, 1984.

136. Venkateswarul, Y., Reddy, V.R.M., and Venkateswara Rao, J., Bis-1-oxaquinolidizines from the sponge *Haliclona exigua*, *J. Nat. Prod.*, 57, 1283, 1994.

137. Venkateswara Rao, J., Desaiah, D., Vig, P.J.S., and Venkateswarul, Y., Marine biomolecules inhibit rat brain nitric oxide synthase activity, *Toxicology*, 129, 103, 1998.

138. Alberte, R.S., Snyder, S., and Zahuranec, B., Biofouling research needs for the United States Navy: program history and goals, *Biofouling*, 6, 91, 1992.

139. Rittschof, D., Clare, A.S., Gerhart, D.J., Bonaventura, J., Smith, C., and Hadfield, M., Rapid field assessment of antifouling and foul-release coatings, *Biofouling*, 6, 181, 1992.

140. Christoffersen, M.W., Isolation of natural compounds from marine organisms, Abstract 10th International Symposium on Marine Corrosion and Fouling, Melbourne, Australia, 1999.

141. Callow, M.E. and Willingham, G.L., Degradation of anitfouling biocides, *Biofouling*, 10, 239, 1996.

142. Fusetani, N., Hiroto, H., Okino, T., Tomono, Y., and Yoshumura, E., Antifouling activity of isocyanoterpenoids and related compounds isolated from a marine sponge and nudibranchs, *J. Nat. Toxins*, 5, 249, 1996.

143. Willingham, G.L. and Jacobson, A.H., Designing an environmentally safe marine antifoulant, *ACS Symp. Ser.*, 640, 224, 1996.

144. Gough, M.A., Fothergill, J., and Hendrie, J.D., A survey of southern England coastal waters for the s-triazine antifouling compound Irgarol 1051, *Mar. Pollut. Bull.*, 28, 613, 1994.

145. Tolosa, I.J., Readman, W., Blaevoet, A., Ghilini, S., Bartocci, J., and Horvat, M., Contamination of Mediterranean (Cote d' Azur) coastal waters by organotins and Irgarol 1051 used in antifouling paints, *Mar. Pollut. Bull.*, 32, 335, 1996.

146. Liu, D., Pacepavicius, G.J., Maguire, R.J., Lau, Y.L., Okamura, H., and Aoyama, I., Survey for the occurrence of the new antifouling compound Irgarol 1051 in the aquatic environment, *Wat. Res.*, 33, 2833, 1999.

147. Okino, T., Yoshimura, E., Hirota, H., and Fusetani, N., Antifouling kalihinenes from the marine sponge *Acanthella cavernose*, *Tetrahedron Lett.* 36, 8637, 1995.

148. Anderson, M.J., A chemical cue induces settlement of Sydney Rock oysers, *Crassostrea commercialis* in the laboratory and in the field, *Biol. Bull.*, 190, 350, 1996.

149. Hirota, H., Tomono, Y., and Fusetani, N., Terpenoids with antifouling activity against barnacle larvae from the marine sponge *Acanthella cavernosa*, *Tetrahedron*, 52, 2359, 1996.

150. Okino, T., Yoshimura, E., Hirota, H., and Fusetani, N., New antifouling kalihipyrans from the marine sponge *Acanthella cavernosa*, *J. Nat. Prod.*, 59, 1081, 1996.

151. Tsukamoto, S., Kato, H., Hirota, H., and Fusetani, N., Pseudoceratidine: a new antifouling spermidine derivative from the marine sponge *Pseudoceratina purpurea*, *Tetrahedron Lett.*, 37, 1439, 1996.

152. Tsukamoto, S., Kato, H., Hirota, H., and Fusetani, N., Ceratinamine: an unprecedented antifouling cyanoformamide from the marine sponge *Pseudoceratina purpurea*, *J. Org. Chem.*, 61, 2936, 1996.

153. Tsukamoto, S., Kato, H., Hirota, H., and Fusetani, N., Mauritiamine, a new antifouling oroidin dimer from the marine sponge *Agelas mauritiana*, *J. Nat. Prod.*, 59, 501, 1996.

154. Tsukamoto, S., Kato, H., Hirota, H., and Fusetani, N., Antifouling terpenes and steroids against barnacle larvae from marine sponges, *Biofouling*, 11, 283, 1997.

155. Tsukamoto, S., Kato, H., Hirota, H., and Fusetani, N., Isolation of an unusual 2-aminoimidazolium salt of steroid trisulfate from a marine sponge *Topsentia* sp., *Fish. Sci.*, 63, 310, 1997.

156. Lau, S.C.K. and Qian, P.Y., Phlorotannins and related compounds as larval settlement inhibitors of the tube-building polychaete *Hydroides elegans*, *Mar. Ecol. Prog. Ser.*, 159, 219, 1997.

157. Matsumara, K., Mori, S., Nagano, M., and Fusetani, N. Lentil lectin inhibits adult extract-induced settlement of the barnacle, *Balanus amphitrite*, *J. Exp. Zool.*, 280, 213, 1998.

158. Tsukamoto, S., Kato, H., Hirota, H., and Fusetani, N., Antifouling and metamorphosis-promoting compounds from the marine sponges *Pseudoceratina purpurea* and *Agelas mauritiana*, in *Sponge Sciences — Multidisciplinary Perspectives*, Watanabe, Y. and Fusetani., N., Eds., Springer-Verlag, Tokyo, 1998, 399.

159. Yamomoto, H., Tachibana, A., Saikawa, W., Nagano, M., Matsumara, K., and Fusetani, N., Effects of calmodulin inhibitors on cyprid larvae of the barnacle, *Balanus amphitrite*, *J. Exp. Zool.*, 280, 8, 1998.

160. Satuito, C.G., Natoyama, K., Yamazaki, M., Shimizu, K., and Fusetani, N., Induction of metamorphosis in the pediveliger larvae of the mussel *Mytilus galloprovincialis* by neuroactive compounds, *Fish Sci.*, 65, 384, 1999.

161. Tomono, Y., Hirota, H., and Fusetani, N., Isogosterones A–D, antifouling 13,17-secosteroids from an octocoral *Dendronephthya* sp., *J. Org. Chem.*, 64, 2272, 1999.

162. Tomono, Y., Hirota, H., Imahara, Y., and Fusetani, N., Four new steroids from octocorals, *J. Nat. Prod.*, 62, 1538, 1999.

163. Yamamoto, H., Shimizu, K., Tachibana, A., and Fusetani, N., Roles of dopamine and serotonin in larval attachment of the barnacle *Balanus amphitrite*, *J. Exp. Zool.*, 284, 746, 1999.

164. Shimizu, K., Hunter, E., and Fusetani, N., Localisation of biogenic amines in larvae of *Bugula neritina* (Bryozoa: Cheilostomatida) and their effects on settlement, *Mar. Biol.*, 136, 1, 2000.

18 Metabolites of Free-Living, Commensal, and Symbiotic Benthic Marine Microorganisms

Valerie S. Bernan

CONTENTS

I. INTRODUCTION

This chapter describes the natural product chemistry that has been identified or associated with marine microorganisms from the benthos and focuses on the marine eubacteria. Most of the compounds described in this chapter resulted from a detailed search of the literature for microbially derived natural products. This approach is biased in many respects, since it only describes bioactive compounds and small molecules. However, due to the growing demand for new therapeutic agents, the discovery of new chemical entities is being driven by these efforts.[1] While comparatively little research has been directed toward the study of natural products from marine microorganisms, the results to date have been encouraging.[2] Data from these investigations demonstrate complex chemical interactions between marine bacteria and their hosts,[2] including systems of signaling and territorial marking. It is estimated that less than 1% of potentially useful chemicals have been discovered from the marine environment, with microbial products representing 1% of that total number.[1] Compounds isolated from marine microorganisms have demonstrated antibiotic, anticancer, anti-inflammatory, and other pharmacological activites.[2,3,4] Hopefully a greater understanding of microbial metabolic diversity, coupled with ecological information, will yield a greater understanding of the complexity of marine environments. This chapter reviews metabolites of microorganisms that occur in sediments, as well as those in commensal associations and symbiotic relationships.

II. MARINE SEDIMENTS

Marine sediments are composed of organic debris resulting from ongoing, seasonal, or catastrophic die-offs of macrobiotic and microbial populations.[2] This sedimenting debris provides energy for the benthic organisms on the sea floor. Organic mineral aggregates that are rich in carbohydrates are produced by deposit-feeding animals such as bivalves, which rework the sediments as they feed. This extensive reworking of the sediments makes them unstable so that the organic aggregates are resuspended by the tidal flow to produce turbid estuarine waters which are important in recycling. Consumption of resuspended organic debris and phytoplankton by bivalve molluscs and other benthic invertebrates produces rich deposits of feces and pseudofeces that coat mineral fragments. These mineral fragments yield additional organic mineral aggregates, which in turn are heavily colonized by microorganisms. Other sources of organic matter include those derived from cordgrass and from beds of seagrasses, such as eelgrass and turtle grass as well as mangroves. As these plants decompose, they are reworked into smaller fragments by deposit feeders, which can support increased populations of microbiota.

Within interstitial habitats of sandy beaches, particles are trapped in the upper 5-cm surface layer and give rise to a bacterial–protozoan community.[5] Below this level, a bacterial flora attached to sand grains removes some of the dissolved organic carbon while supporting a meiofauna community comprised of nematods and copepods. The biotic community of these intertidal sandflats is supplemented by the production of organic matter via benthic diatoms which migrate vertically with the tides.

Since terrestrial actinomycetes have been such prolific producers of bioactive molecules, it was natural to investigate marine species of the same group. Therefore, it was no surprise that actinomycetes isolated from the marine sediments have proven to be one of the most prolific sources of bioactive secondary compounds.[6,7] Their distribution in sediments varies depending on the depth from which the sample was collected. In several studies, *Streptomyces* predominated in near-shore marine sediments, but decreased dramatically past the sublittoral zone. In contrast, actinoplanetes are found in greater numbers as the distance from the shoreline increases.

A. SHALLOW MARINE SEDIMENTS

The study of the metabolites produced by marine actinomycetes was pioneered by researchers at the Institute of Microbial Chemistry in Tokyo in the early 1970s. One of the first compounds described by Okami was a benzanthraquinone antibiotic isolated from the actinomycete *Chainia purpurgensa* SS-228 collected in mud samples from Sagami Bay, Japan.[8] This antibiotic selectively inhibited Gram-positive bacteria and was active against Ehrlich ascites tumor cells in mice. It also produced a hypotensive effect in mice, probably due to its inhibition of dopamine hydroxylase in the pathway of epinephrine biosynthesis. Most interestingly, this bacterium only produced the antibiotic when it was fermented in diluted yeast extract containing "Kobu Cha" (the brown seaweed *Laminaria*) and with the addition of 3% NaCl. This observation demonstrated that marine micro-organisms have nutritional requirements corresponding to nutrients in their natural habitats. This study was also one of the first to report that bacteria from the order Actinomycetales could be isolated from the marine environment. Actinomycetes were originally thought to have entered the marine environment via rivers or runoff or to have existed as spores of terrestrial species. However, work by Moran et al.,[9] using a 16S rRNA genus-specific probe, demonstrated that *Streptomyces* occurred as indigenous populations, and that populations increased in relative abundance in response to the availability of certain nutrients. Greater abundances of culturable streptomycetes found in coastal environments vs. deepwater marine systems may be attributed to higher amounts of organic detritus, much of which is derived from vascular plants concentrated in shallow marine systems.

Unlike the terrestrial actinomycetes, marine actinomycetes have been shown to produce macrolides only rarely. One example of this class of compounds is the aplasmomycins A–C

(Structure 18.1a–c). These compounds were isolated from the fermentation of *Streptomyces griseus* SS-20 collected from shallow mud in Sagami Bay, Japan. The antibiotic was only produced under low-nutrient media containing "Kobu Cha" and 3% NaCl. The aplasmomycins are inhibitory against Gram-positive bacteria and the bacteria *Corynebacterium smegmatis*. More importantly, aplasmomycin is an effective antimalarial agent in *Plasmodium berghei*-infected mice. X-ray crystallography revealed that the active compound contained a symmetric ring in which boron was coordinated in the center with a crown ether-like structure.[10] Fenical's group[11] from Scripps Institute of Oceanography also isolated a new member of a rare class of macrolides, maduralide (Structure 18.2). Maduralide was isolated from the fermentation of a Maduramycete isolated from shallow sediments from Bodega Bay, California. This compound is a member of a rare 24-membered ring lactone group represented by rectilavendomycin.

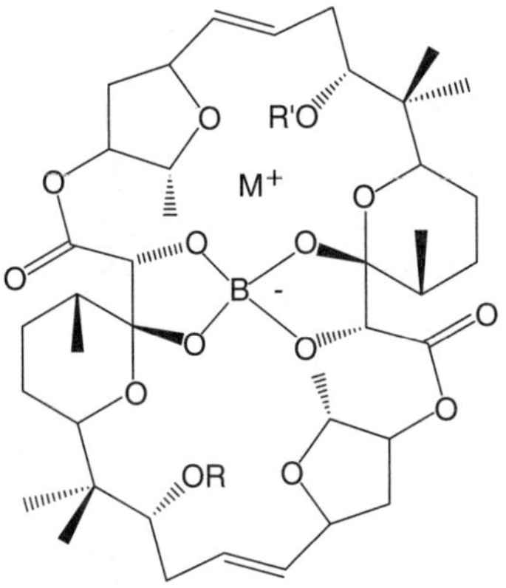

18.1a: Aplasmomycin A R = R' = H
18.1b: Aplasmomycin B R = H, R' = Ac
18.1c: Aplasmomycin C R = R' = Ac

Among the alkaloids, the most unusual example is an acaricidal (lethal to arachnids) monoterpene derivative, altemicidin (Structure 18.3). This novel alkaloid was purified from a marine strain of *Streptomyces sioyaensis* SA-1758 isolated from marine sediments collected from the northern part of Japan. It yielded potent antitumor activity *in vitro* against L1210 murine leukemia and IMC carcinoma cell lines, but was toxic *in vivo* in mice. Altemicidin is a novel sulfur- and nitrogen-containing microbial metabolite with a monoterpene carbon skeleton.[12]

A sediment-derived *Streptomyces* sp. was isolated from Laguna de Terminos from the Gulf of Mexico.[13] When the culture was fermented in 50% seawater, a new anti-algal anthranilamide derivative was isolated that was active against several algae including *Chlorella* spp. and *Scenedesmus subspicatus*. This new compound was shown to be an *N*-methyl anthranilamide derivative of phenylpropionic acid (Structure 18.4). A series of analogs was synthesized and their anti-algae properties examined. These methyl ester analogs were found to be more active against algae than the free acids in both agar diffusion and liquid test systems. The compounds did not exhibit any antimicrobial activity against *Staphylococcus aureus*, *Escherichia coli*, or *Mucor miehei* at concentrations up to 200 µg/mL.

Another *Streptomyces* sp. obtained from black anaerobic intertidal sediment collected near Christchurch, New Zealand was fermented in a saline medium and found to produce modest antibacterial activity against *Bacillus subtilis*.[14] The isolation of the active compound revealed it to be an actinoflavoside, a molecule of an unprecedented structure class (Structure 18.5). Actinoflavoside resembles the common plant-derived flavonoid glycosides, but this compound contains an

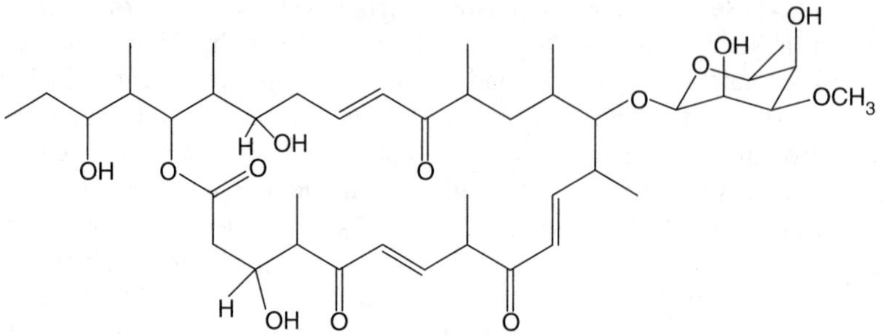

18.2: Maduralide

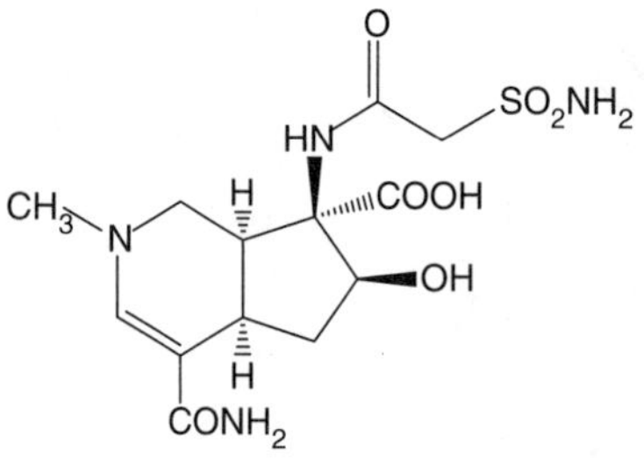

18.3: Altemicidin

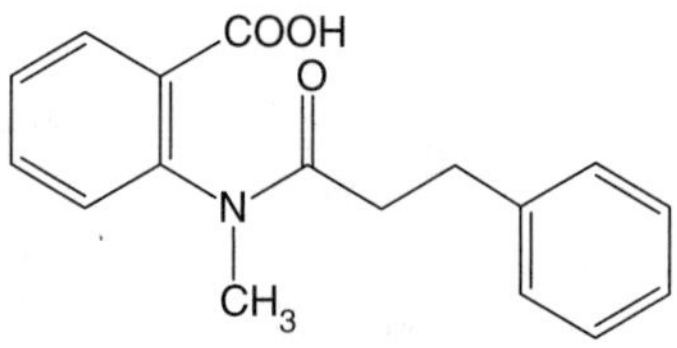

18.4: Anthranilamides

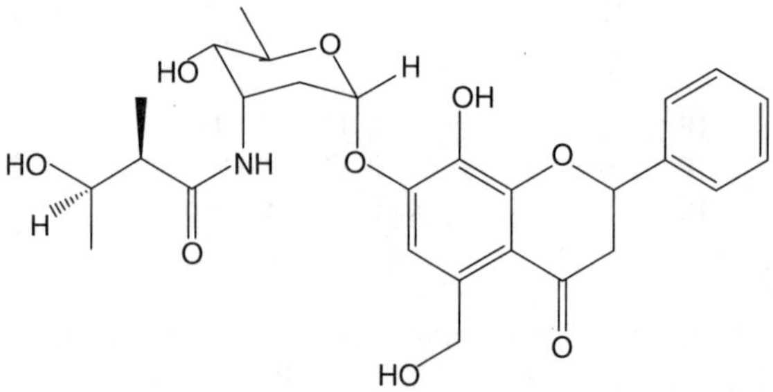

18.5: Actinoflavoside

additional alkylation at C-5. Due to the general conclusion that prokaryotes do not produce this class of compounds, its origin via the flavinoid biosythetic pathway seems questionable.

A marine actinomycete, *Streptomyces virdostaticus* spp. "*litoralis*" was isolated from an intertidal sample collected in Key West, Florida. Fermentation samples exhibited both antibacterial and DNA-damaging activites. Although activity was observed in a tap-water based medium, the addition of 2% NaCl increased the biomass by 33% and increased activity four-fold. Purification of the active materials revealed four related bioxalomycins (Structure 18.6a and b). The β species were found to be the quinone forms of the corresponding α components. The β components are distinguished from the antibiotic naphthyridinomycin by the presence of a second oxazolidine ring in a region of the molecule analogous to quinocarin. Bioxalomycin α2 was the most potent antibiotic of the group, showing MIC values between 0.002–0.25 µg/mL, and also exhibited excellent *in vitro* activity against neoplastic cell lines. This compound was active *in vivo* in a mouse P388 leukemia model demonstrating an 80% increase in life survival (ILS). Mechanistic studies have shown that following metabolic reduction of the quinone, bioxalomycin α2 cross-links DNA through alkylation of guanine residues in the minor groove of DNA.[15]

A new series of phenazines has been isolated from a strain of marine-derived *Streptomyces* by Fenical's group.[16] A study from the shallow sediments of Bodega Bay, California resulted in the

18.6a: Bioxalomycin α1, R = H
18.6b: Bioxalomycin α2, R = CH$_3$

isolation of an unknown *Streptomyces* sp. that was found to produce compounds with broad antibacterial activity. Subsequent fermentation of this isolate in a saltwater based medium produced four new alkaloid esters of the phenazine class which contained the sugar L-quinovose at the 2′ and 3′ positions (Structures 18.7a and b). These compounds were found to exhibit antibacterial activity against Gram-negative and Gram-positive bacteria. Another set of phenazine derivatives was isolated from the fermentation of a *Streptomyces* sp. isolated from a sediment sample collected in Laguna de Terminos in the Gulf of Mexico.[17] Cultivation and production occurred in enriched 50% seawater medium and produced the novel 5,10-dihydrophencomycin methyl ester and the known microbial metabolites (2-hydroxyphenyl)acetamide, meanquinone MK9, and phencomycin. The new 5,10-dihydrophencomycin methyl ester (Structure 18.8) exhibited less antimicrobial activity than phencomycin, and the dimerization of identical *m*-C$_7$N units may explain its origin.

Four new α-pyrone-containing metabolites, wailupemycins A–C and 3-epi-5-deoxyenterocin (Structures 18.9 a–e) were isolated together with the known compounds enterocin and 5-deoxyenterocin from the fermentation broth of a new *Streptomyces* sp.[18] The strain was isolated from the shallow marine sediments collected at Wailupe beach park on the southeast shore of Oahu, Hawaii. The α-pyrone moeity is commonly observed in many antibiotics and toxins. Interestingly, enterocin and 5-deoxyenterocin, along with the 5-behenate and 5-arachidate esters of enterocin, has been isolated from a marine ascidian of the genus *Didemnum*. The occurrence of the same compounds in both prokaryotes and chordates raises the question of whether symbiotic or associated microorganisms are responsible for the production of the metabolites isolated from some marine invertebrates.

In a study of estuarine microorganisms isolated from Torrey Pines, La Jolla, California, Fenical et al.[19] isolated marinone (Structure 18.10) and its debromo analogue debromomarinone from the fermentation broth of an unidentified actinomycete. Both compounds possess a common naphthoquinone with rare sesquiterpenoid structural components. In addition, marinone possesses a bromine substituent in the dihydroxybenzene ring, a position typical for bromination in marine metabolites. These new molecules are among a rare group of bacterial metabolites produced via mixed biosynthesis involving both acetate and terpene pathways. Marinone and debromomarinone exhibit significant *in vitro* antibacterial activity against Gram-positive bacteria. More recently, the same group isolated a different unidentified actinomycete, from a sediment sample collected at 1m depth in Batiquitos Lagoon, California, that also produced marinone in addition to several cytotoxic metabolites related to marinone. One compound, neomarinone (Structure 18.11), is a novel metabolite possesssing a new sesquiterpene- and polyketide-derived carbon skeleton. The other two derivatives are isomarinone and methoxydebromomarinone. All three compounds are also derived from a mixed biosynthetic pathway involving polyketide and terpene pathways. Connection of the sesquiterpenoid side-chain to the naphthoquinone core occurs on the nonquinone side in neomarinone. The origin of the sesquiterpenoid side-chain appears complex; it is possibly derived from a cation-induced methyl migration as observed in the trichothecenes. All three compounds exhibited moderate *in vitro*

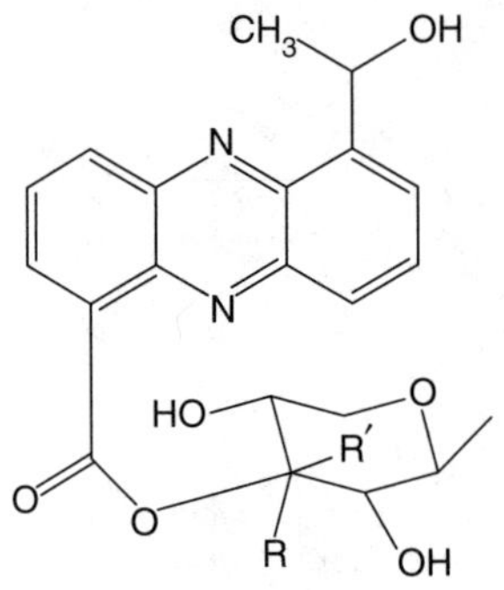

18.7a: R=OH, R´=H

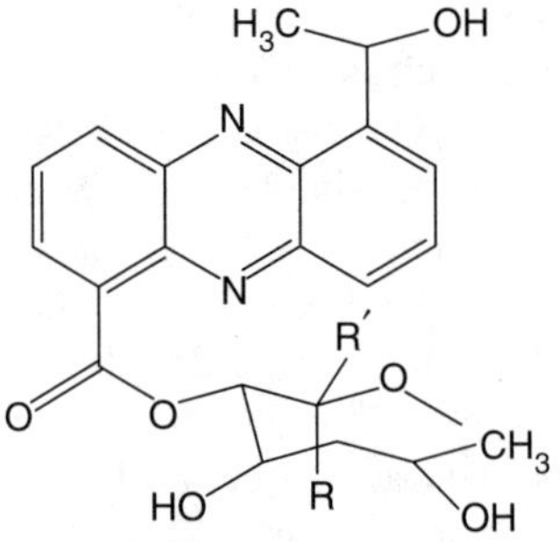

18.7b: R=H, R´=OH

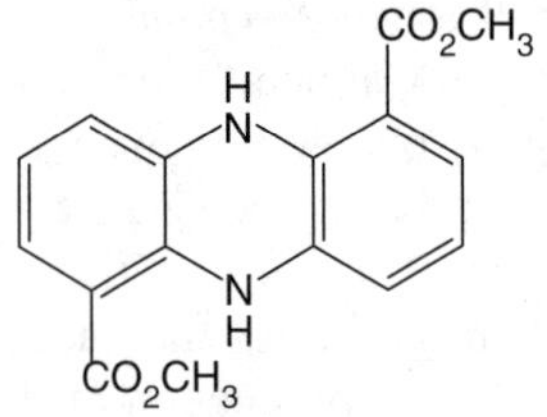

18.8: 5,10-dihydrophencomycin methyl ester

cytotoxicity in the National Cancer Institute's 60 cancer cell line panel with a mean IC_{50} value of 10 μM.[20]

As part of the continuing interest in isolating secondary metabolites from marine estuaries, a Streptomycete was isolated from a sediment sample collected in Mission Bay, California. When the culture was fermented under saline culture conditions, it was found to produce a family of novel cyclic heptapeptides, cyclomarines A, B, and C. While the major metabolite, cyclomarine A (Structure 18.12), is cytotoxic *in vitro* toward cancer cells, it is more interesting for its significant *in vitro* and *in vivo* anti-inflammatory properties. The compound displays significant topical anti-inflammatory activity in the phorbol ester-induced mouse edema assay, showing 92% inhibition at the standard test dose. Cyclomarine A contains three common and four unusual amino acids. The four unusual amino acids are *N*-methylhydroxyleucine, β-methoxyphenylalanine, 2-amino-3,5-dimethylhex-4-enoic acid, and *N*-(1,1-dimethyl-2,3-epoxypropyl)-β-hydroxytryptophan. The latter two amino acids have not been previously described, although similar *N*-prenyltryptophan amino acids have been observed in the ilamycins. The amino acid β-methoxyphenylalanine is a well known synthetic building block, but is a rare constituent of natural products, found in only

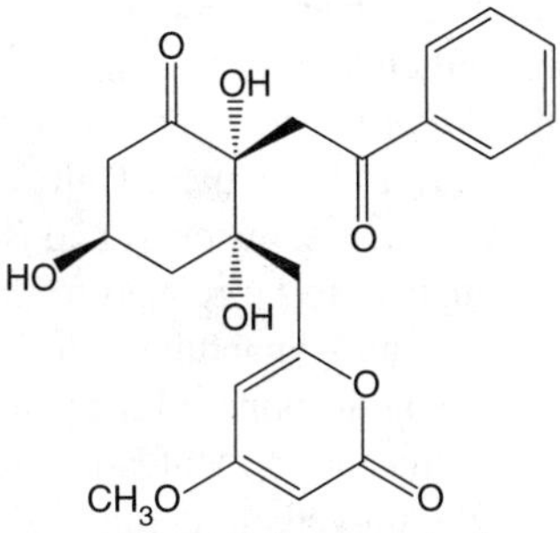

18.9a: Wailupemycin A

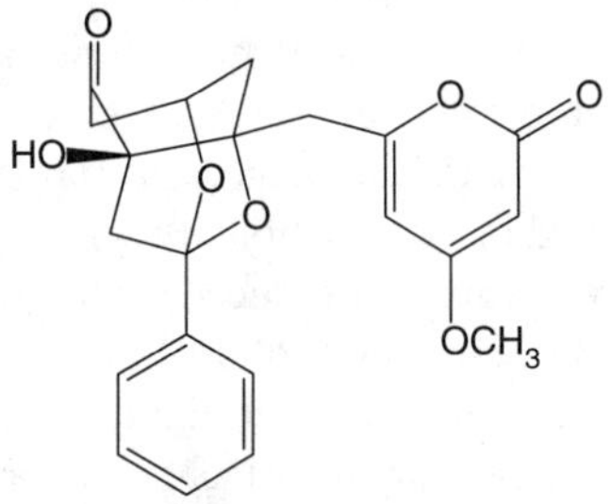

18.9b: Wailupemycin B

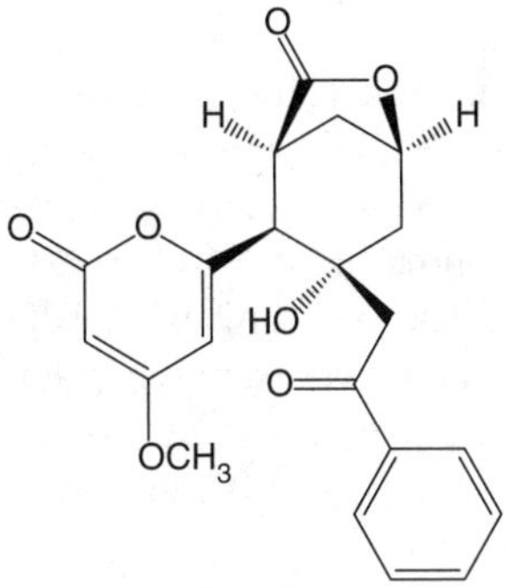

18.9c: Wailupemycin C

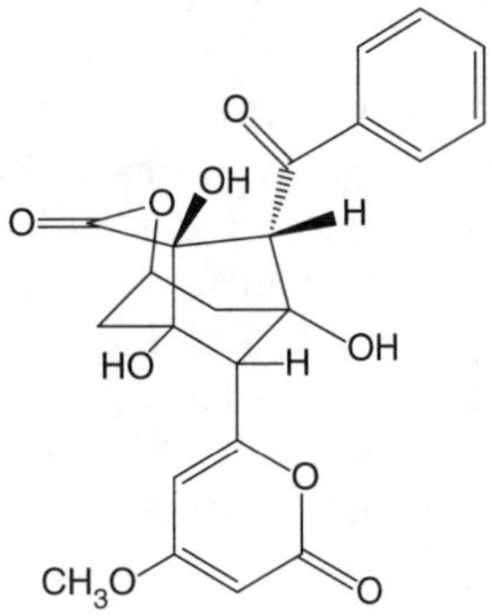

18.9d: Wailupemycin D

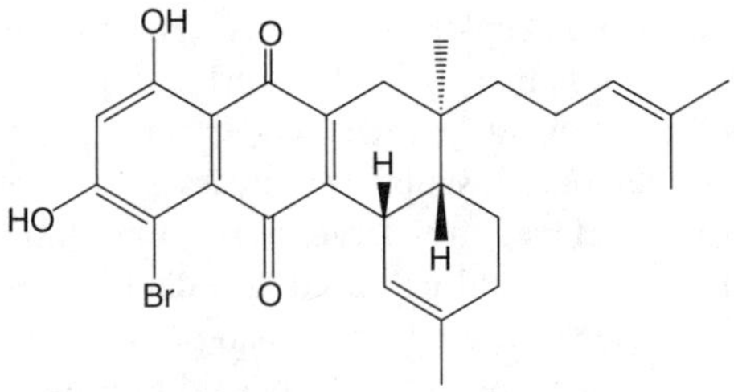

18.10: Marinone

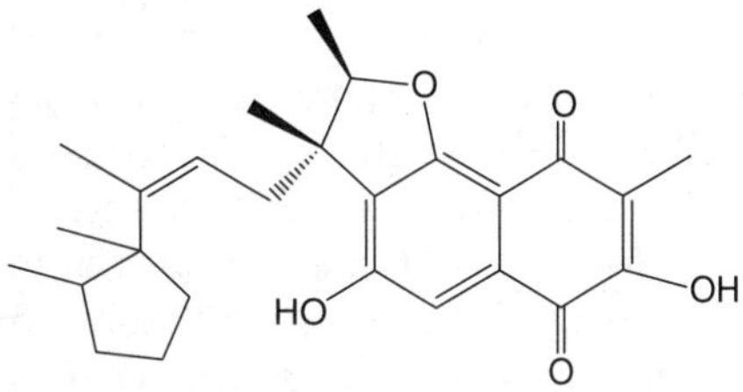

18.11: Neomarinone

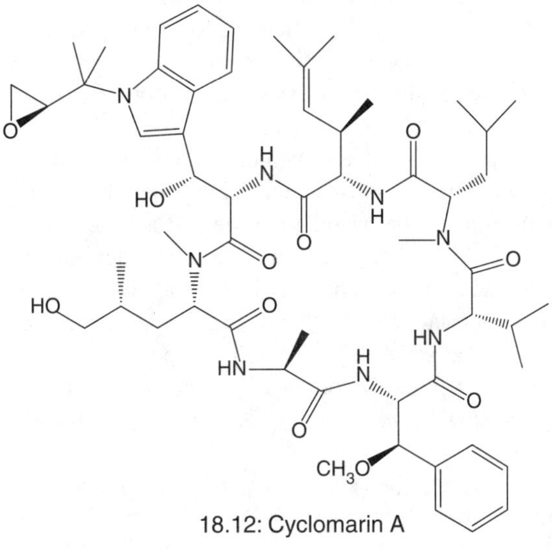

18.12: Cyclomarin A

the discokiolides, which are cyclic depsipeptides from the marine sponge *Discodermia kiiensis*. These amino acids may be products of unusual biosynthetic pathways, and it will be interesting to elucidate their biosynthesis.[21] Another estuarine *Streptomyces* was isolated from the sandy sediment collected near San Diego, California in the San Luis Estuary. Cultivation of the *Streptomyces* sp. resulted in the isolation of two new aromatic tetraols, luisols A and B (Structure 18.13a and b). Luisol A, formally a reduced hydroquinone, appears to be related to the quinones of the granaticin class. The structure of luisol B contains the rare epoxynaphtho[2,3c]furan which is only found in one other natural product, the fungal metabolite anthrinone.[22] Lastly, a *Streptomycete* sp. was isolated from an estuary near Doheny Beach, California and fermented in a salt-based medium. The culture broth was found to contain a new pentacyclic polyether, arenic acid (Structure 18.14), which is related to two known polyether antibiotics, K41-A and oxolonomycin. The structure of arenaric acid was established by spectroscopic methods involving comprehensive two-dimensional NMR measurements.[23]

Although marine actinomycetes are the most prolific source for bioactive metabolites from shallow sediments, marine *Bacilli* spp. have also been isolated, and unusual secondary metabolites have been reported. For example, a *Bacillus* sp. was isolated from marine sediment and cultured in an enriched seawater medium to yield a new isocoumarin, PM-94128 (Structure 18.15).[24] Comparison of IR and UV data with known substances and one-dimensional ^1H and ^{13}C NMR data and a ^1H–^1H COSY suggested a dihydroisocoumarine derivative with a prenyl group substituted with a ketide extended leucine. PM-94128 was quite potent with cytoxicity against several tumor cell lines (IC_{50} 0.05 μM to P388, A-549, HT-29, and MEL-28) and may act by inhibiting protein synthesis.

B. DEEP-SEA MARINE SEDIMENTS

Many interesting metabolites have also been isolated from deep-sea sediments.[2] The energy input to the sea floor below a depth of 2 km is thought to be less than 10% of the primary productivity in the euphotic zone. At these depths, food consists of a slow rain of fecal pellets and zooplankton along with the carcasses of larger organisms from the nekton. Most of these carcasses are consumed quickly by fish who scatter the remains in the form of feces over large areas to be utilized by benthic microorganisms. A bacterial isolate from deep-sea mud collected at a depth of 3300 m off the Aomori coast of the Japan Sea required a seawater medium to grow and produce bioactive substances. Even though the strain was isolated at 700 atm pressure, it appeared to grow well at surface pressure and temperature. The strain was identified as *Alteromonas haloplanktis* and was found to produce a new bioactive siderophore metabolite called bisucaberin (Structure 18.16).[25] This compound had little cytotoxicity but, when added to a mixed macrophage-tumor cell culture, induced macrophage-mediated cytolysis of tumor cells. Its dimeric structure contains two hydroxamates and two amide functionalitites and is similar to other siderophores such as nocardimin and desferioxamine.

While screening for antitumor effects, Fenical et al.[26] isolated a deep-sea bacterium from a sediment sample obtained from a 1000-m depth along the California coast. Fermentation of the slow-growing bacterium in a salt-based medium yielded a series of novel cytotoxic and antiviral macrolides, the macrolactins A–F. This bacterium was an unidentified Gram-positive organism that produced six macrolides and two open-chain hydroxy acids when fermented in the presence of salt at atmospheric pressure. Macrolactin A (Structure 18.17) was the predominate compound produced, showing moderate antibacterial activity, yet it was quite potent against B16-F10 murine melanoma *in vitro* with an IC_{50} of 3.5 μg/ml. Of potentially greater significance, macrolactin A inhibited several viruses, including *Herpes simplex* ($IC_{50} = 5.0$ μg/ml) and HIV, the human immunodeficiency virus ($IC_{50} = 10.0$ μg/mL).

Two new caprolactams were isolated by Davidson and Schumacher[27] from an unidentified Gram-positive bacterium cultured from deep-sea sediments. The caprolactams A and B (Structure 18.18a

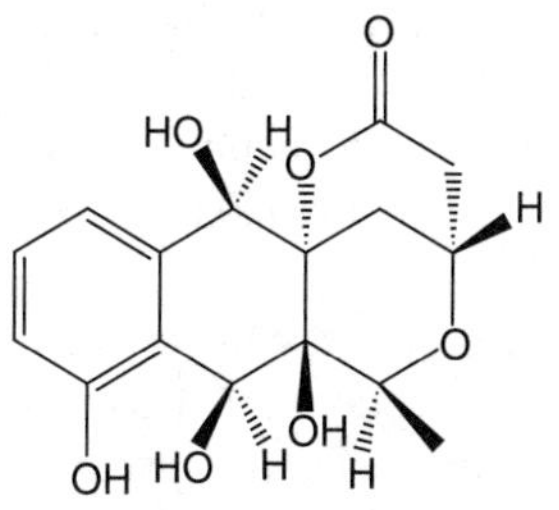

18.13a: Luisol A 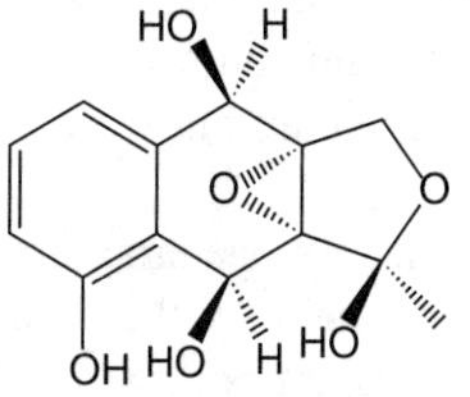   18.13b: Luisol B

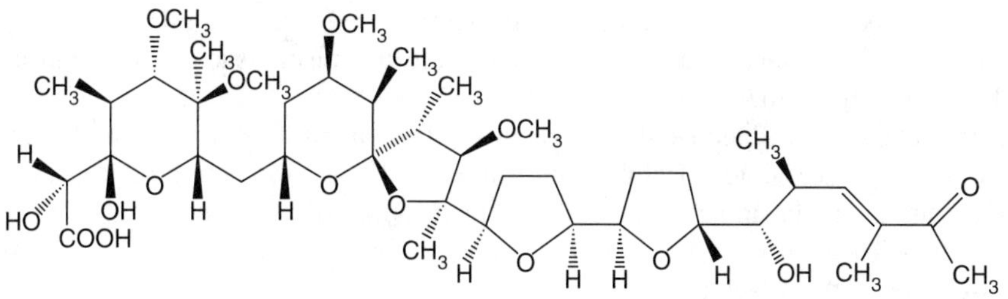

18.14: Arenic acid

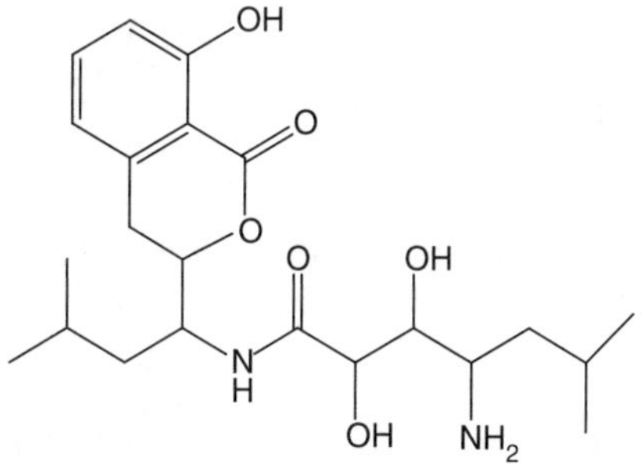

18.15: PM-94128

and b) were obtained as an inseparable mixture and were composed of cyclic-L-lysine linked to 7-methyloctanoic acid and 6-methyloctanoic acid, respectively. Natural products containing a cyclized lysine are uncommon and have only been reported in several sponges and fungi. In fact, these structures are quite similar to the bengamides, which are sponge-derived caprolactams with oxidized acyl side-chains. Thus, one may speculate that the bengamides isolated from sponges may truly be produced by a symbiotic microorganism. The compounds are mildly cytotoxic toward human epidermoid carcinoma and colorectal adenocarcinoma cells with MIC values of 10 and 5 μg/mL, respectively, and exhibit antiviral activity toward *Herpes simplex* type II virus at a concentration of 100 μg/mL. The same group also isolated a new pluramycin metabolite, γ-indomycinone (Structure 18.19), from a *Streptomyces* species isolated from a deep-sea sediment core sample.[28] γ-Indomycinone is composed of an anthraquinone-γ-pyrone nucleus with a 1-hydroxy-1-methylpropyl side-chain. The compound exhibits only mild cytotoxicity against a human colon cancer cell line HCT-116 but shows a differential cytotoxicity against the Chinese hamster ovary cell lines UV20 (deficient in DNA excision repair) and BR1 (proficient in DNA repair). These results suggests that γ-indomycinone may act by forming a bulky DNA adduct.

Another Streptomycete species was isolated from deep-sea sediments collected at 1500 m in the sea surrounding Tokyo, Japan. When this culture was fermented in the presence of seawater, growth was enhanced as was the production of two β-glucosidase inhibitors which have the potential to be developed as antimetastasis or anti-HIV drugs. The two inhibitors were characterized as D-glucono-1,5-lactam and D-mannono-1,5-lactam (Structure 18.20). The two lactams had never been isolated from nature but were found in marine organisms, suggesting that they are of microbial origin.[29]

Simidu et al.[30] isolated 49 bacterial strains from deep-sea core sediments collected at a depth of 4000 m and examined them for the production of tetrodotoxin. This study indicates that tetrodotoxin-producing bacteria are not restricted to certain taxonomic groups. A variety of groups of bacteria, including *Bacillus, Micrococcus, Acinetobacter, Aeromonas, Alteromonas, Moraxella, Vibrio*, and one unidentified bacterium, all produce tetrotoxin. Although the strains are limited in number, the tetrodotoxin-producing bacteria are quite widespread among various bacterial groups in marine sediments. It has been postulated that the tetrodotoxins are synthesized solely by bacteria in sediments and subsequently accumulated by benthic organisms, such as fish and crabs, that acquire them through the food web.

Another cyclic peptide, halobacillin, was isolated from a marine-derived *Bacillus* sp. isolated from a deep-sea sediment core.[31] Halobacillin (Structure 18.21) was only produced in sea-water-based media and is similar in structure to the surfactins and iturins. Interestingly, the compound exhibits cytotoxicity against the human colon tumor cell line HCT-116 but lacks the antibacterial activity associated with surfactin.

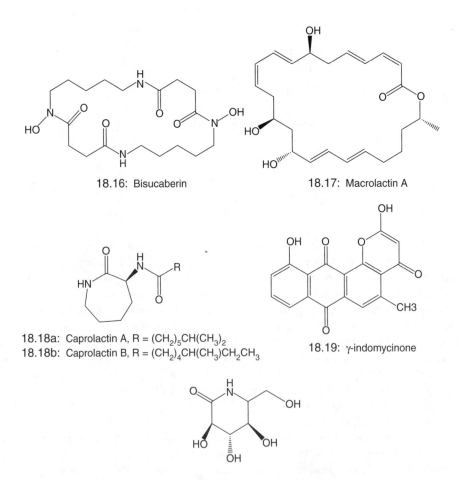

18.16: Bisucaberin

18.17: Macrolactin A

18.18a: Caprolactin A, R = (CH₂)₅CH(CH₃)₂
18.18b: Caprolactin B, R = (CH₂)₄CH(CH₃)CH₂CH₃

18.19: γ-indomycinone

18.20: D-Glucono-1,5-lactam

III. COMMENSAL MARINE MICROORGANISMS

Commensal marine bacteria inhabit surfaces, tissues, and internal spaces of other organisms and plants, but, often, the exact associations are transient and the interactions are not well understood. Since bacteria occur in seawater at concentrations of one million cells per milliliter, marine plants and animals are constantly exposed to extremely high concentrations of bacteria. Many of these bacteria are motile, chemotactic, and/or pathogenic, and readily colonize a variety of surfaces. Bacteria are the first detectable microorganisms to colonize a surface, which they do by two processes. The first process is an instantaneous but reversible adsorption of the bacteria onto a surface from which the bacteria can easily be removed.[32] The second process is an irreversible adsorption, in which different populations develop with time. These bacterial films often play a vital role in the growth and development of macroscopic marine plants. A number of seaweeds have distinctive epibacterial floras that may supply growth factors and contribute to the destruction of algal autoinhibitory substances. Animal surfaces also appear to selectively enhance or inhibit microbial colonization. Documentation of these coexistances is increasing, but evidence supporting true symbiosis rather than a nonobligate association is still lacking. Nonetheless, these associations are significant in the quest for new bioactive natural products because of the potential cooperative role in the production of novel metabolites. These surfaces are richer in nutrients than seawater and most sediments, thus providing a unique niche for the isolation of many diverse bacteria, as described below.[32]

Umezawa and co-workers screened fermentation broths containing various bacterial species, isolated from the surface of seaweeds, for the production of polysaccharides.[33] One genus, *Flavobacterium uglignosum*, was found to produce marinactan, which is capable of suppressing sarcoma-180 tumors in mice. At daily doses of 10 to 50 mg/Kg in mice, marinactan inhibited 75–95% of the growth of these tumors. Marinactan is a neutral heteroglycan consisting of fucose, mannose, and glucose. In an antibacterial screening program by Sano et al.,[34] a *Pseudoalteromonas* sp. was found to produce the novel antibiotic korormicin (Structure 18.22). The organism was isolated from the tropical green alga *Halimeda* sp., collected in Palau and required a seawater-based medium for growth and survival. Korormicin is a combination of an oxidized fatty acid and an unusual lactonized amino acid. This unusual amino acid is without biosynthetic precedent and poses an interesting question as to its origin. Interestingly, korormicin was harmless to terrestrial bacteria and Gram-positive marine bacteria, but was active against 11 marine Gram-negative bacteria. During the course of an anticancer screening program, a new marine bacterium, *Pelagiobacter variabilis*, was isolated from the blades of the tropical brown alga *Pocockiella variegata*.[35] This Gram-negative, pleomorphic, halophillic bacterium represents a new genus of marine eubacteria. Fermentation extracts of this culture were found to contain the known compound, griseolutic acid, and three new phenazine antibiotics, the pelagiomicins A–C. Pelagiomicin A (Structure 18.23) demonstrated strong antibacterial activity to several Gram-negative and Gram-positive bacteria and inhibited several cancer cell lines *in vitro* [ID$_{50}$ values between 0.04 to 0.2 μg/mL].

Interest in finding new antibiotics against multidrug resistant *Mycobacterium tuberculosis* and *M. avium-intracellulare* led Andersen et al.[36] to isolate one *Pseudomonas* sp. from the surface of an unidentified leafy red alga collected in Masset Inlet, British Columbia, and another *Pseudomonas* sp. from an unidentified tube worm collected near Moira Island, British Columbia. When cultured on solid agar in the presence of salt, the two strains produced the novel cyclic depsipeptides, massetolides A–H, and the known compound viscosin. Massetolide A (Structure 18.24) was two to four times more potent than viscosin in its *in vitro* inhibition of *M. tuberculosis* (MIC = 5–10 μg/mL) and *M. avium-intracellulare* (MIC = 2.5–5.0 μg/mL), and a single intraperitoneal injection of 10 mg/Kg of massetolide A was found to be nontoxic to mice. No activity was observed for either compound against a panel of other human pathogens. A number of unnatural and intriguing massetolides were produced by feeding with various leucine analogs (L-butyrine, L-norvaline, L-cyclopropylalanine, etc.) but the supply was not sufficient for testing. The same group also isolated

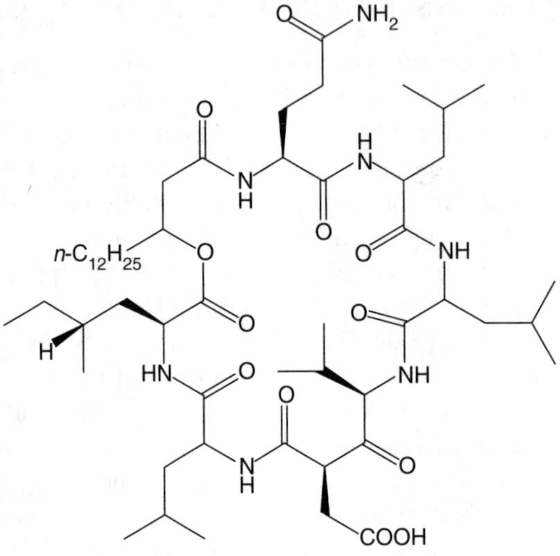

18.21: Halobacillin

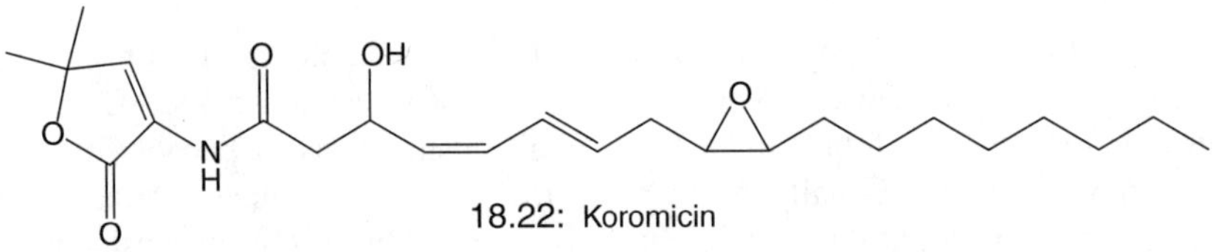

18.22: Koromicin

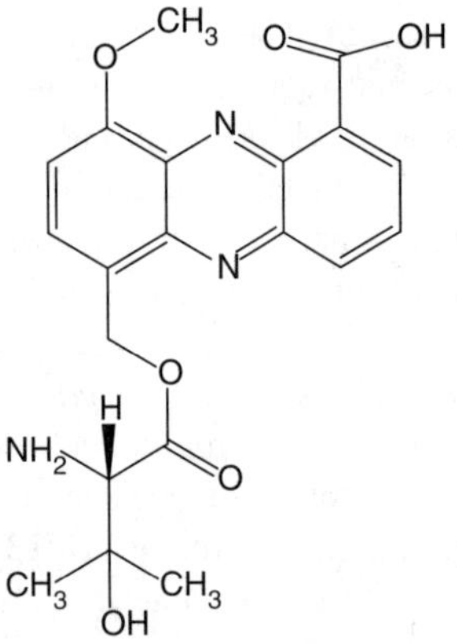

18.23: Pelagiomicin A

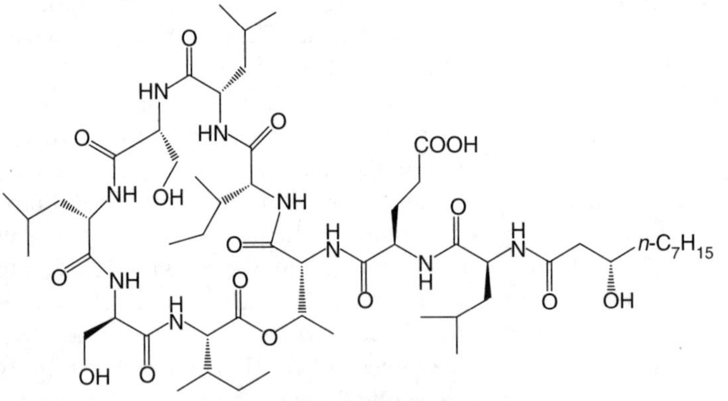

18.24: Massetolide A

a family of novel cyclic decapeptide antibiotics, loloatin A–D (Structure 18.25a–d) from a *Bacillus* sp. isolated from a subtidal tube worm collected off Loloata Island, Papua, New Guinea.[37] The *Bacillus* sp. was cultured on solid agar containing salt, and the extract from the cells was found to produce related antibiotics that were active against several strains of antibiotic resistant bacteria. Loloatins A–C inhibited the growth of methicillin-resistant *S. aureus*, vancomycin-resistant *Enterococcus* sp., and penicillin-resistant *Streptococcus pneumoniae* with MIC of 0.5 to 4 μg/mL. Interestingly, only loloatin C exhibited antibacterial activity against the Gram-negative bacterium *Escherichia coli* (MIC 1 μg/mL), and loloatin D was four times less active than A–C against Gram-positive bacteria. These results demonstrate that subtle changes in cyclic decapeptide structure can have a significant impact on antimicrobial activity. Even though the loloatins share structural

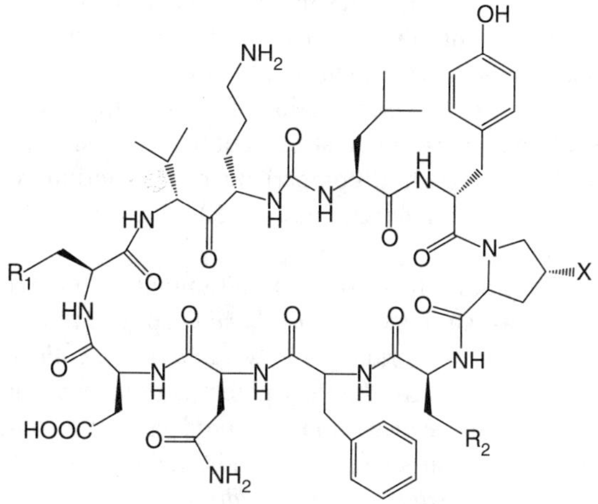

18.25a: Loloatin A R₁= [4-hydroxyphenyl] R₂= [phenyl] X=H

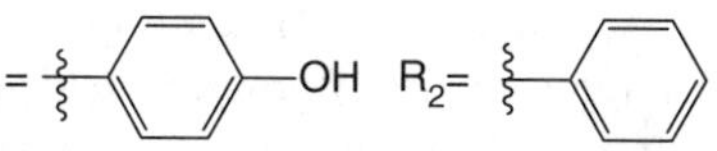

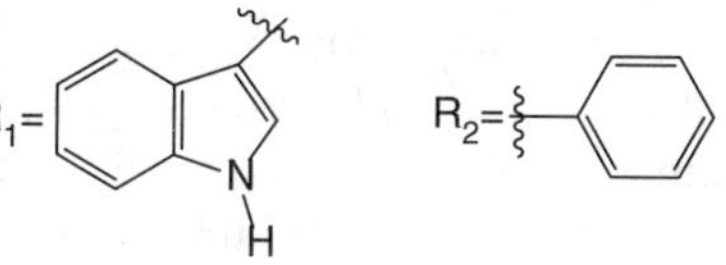

18.25b: Loloatin B R₁= [indol-3-yl] R₂= [phenyl] X=H

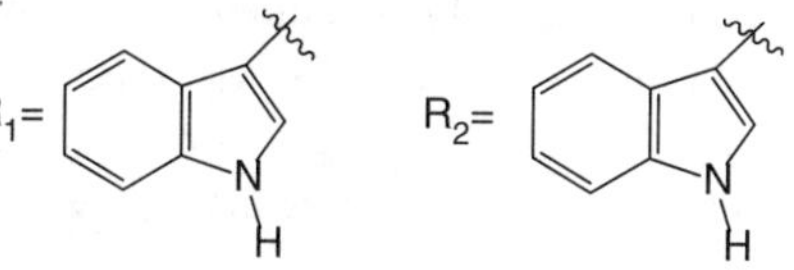

18.25c: Loloatin C R₁= [indol-3-yl] R₂= [indol-3-yl] X=H

18.25d: Loloatin D R₁= [indol-3-yl] R₂= [phenyl] X=OH

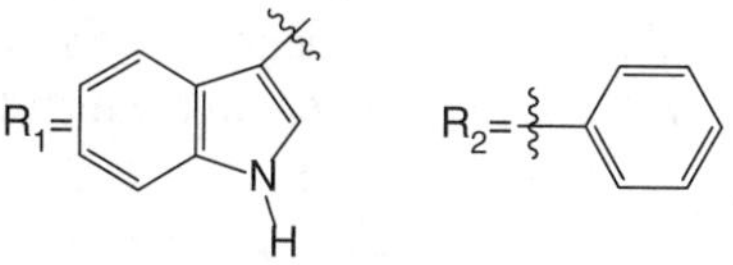

features with the tyrocidines, the latter have never been reported to demonstrate Gram-negative antibacterial activity.

Shigemori et al.[38] isolated a new cyclic alkaloid, alteramide A (Structure 18.26), produced by an *Alteromonas* species isolated from the marine sponge *Halichondria okadai* collected near Nagai, Kanagawa, Japan. Alteramide A is a macrocyclic lactam containing dienone and dienoyltetramic acid functionalities that can undergo a [4 + 4] cycloaddition to generate a hexacyclic derivative. Alteramide A exhibited cytotoxicity against murine leukemia P388 cells, murine lymphoma L1210 cells, and human epidermoid carcinoma KB cells. Related macrocylic lactams, such as ikaruga-mycin and discodermolide, have been previously isolated from the terrestrial *Streptomyces phae-ochromogenes* var. *ikuruganesis* and the sponge *Discodermia dissoluta*, respectively, but not from symbiotic bacteria. The isolation of alteramide may provide insights into the metabolic origin of discodermolide.

Investigations of microorganisms associated with the Antarctic sponge *Isodictya setifera* led to the isolation of a strain of *Pseudomonas aeruginosa* that exhibited Gram-positive antibacterial activity.[39] Fractionation of the culture broth identified a new diketopiperazine, *cyclo*-(L-proline-L-methionine) (Structure 18.27), five known diketopiperzines, and two known phenazine alka-loids. Investigations of the sponge revealed that neither metabolite was present. This suggests that either the bacterium only produces these secondary metabolites under certain environmental or seasonal conditions, or that the bacterium is present in low abundance in the sponge. Oclarit et al.[40] isolated a halophytic marine bacterium, *Janthinobacterium* sp., from an unidentified species of the Japanese sponge *Adocia*. Fermentation of this culture produced the compound *o*-aminophenol (Structure 18.28), which exhibited strong antibacterial activity against *Staphylococcus aureus* and *Bacillus subtilis*. However, a comparison of the antimicrobial activity produced by the bacterium to the antimicrobial activity associated with extracts of the host sponge revealed that the two were not related.

In the course of screening for antitumor and immunosuppressive agents, Acebal et al.,[41] from Pharma Mar, isolated two strains of *Agrobacterium* sp. from the tunicates *Ecteinascidia tubinata* collected in the mangroves of Florida and *Polycitonidae* sp. collected on the Turkish coast. *Ecteinascidia turbinata* is the source for the very potent antitumor agent ecteinascidin. Fermentation of these two strains in enriched seawater medium produced lipid-soluble products with potent antitumor activity, sesbanimide A (Structure 18.29) and C. Of interest is that sesbanimides were originally isolated from the seeds of *Sesbania drummondii* and *Sesbania punicea*, leguminous plants. This isolation from a microbial source confirms their correct metabolic origin and allows for the further investigation of sesbanimide A, which has immunosuppressive activity. The same group also investigated the same marine tunicate, *Ecteinascida turbinata*, collected in Formentera Island, Spain.[42] Isolation and fermentation of the microorganisms associated with this tunicate revealed that an *Agrobacterium* sp. produced a new cytotoxic compound. The compound was obtained from the bacterial cells by solvent extraction and purified by silica gel chromatogoraphy. Structural elucidation identified the compound as a new thiazole alkaloid substance called agrochelin and its acetyl derivative. Agrochelin (Strucure 18.30) is structurally related to yersiniabactin and yersiniophore produced by *Yersinia enterocolitica* and micacocidins A, B, C, isolated from a *Pseudomonas* sp. The IC_{50} value of agrochelin to P-388 cells was 0.053 μM, which was similar to the value obtained against mouse lymphoid, human lung colon carcinomas, and human melanoma *in vitro* cell lines. The cytotoxic activities of the acetyl derivative were substantially reduced compared with those of agrochelin.

Molluscs have the potential to be particularly rich sources of microorganisms since they prefer to feed on the top milliliter of the surface sediment and carry out a high degree of food sorting in the mantle.[43] It is thought that 97% of the sediment in the mantle is rejected as pseudofeces leaving

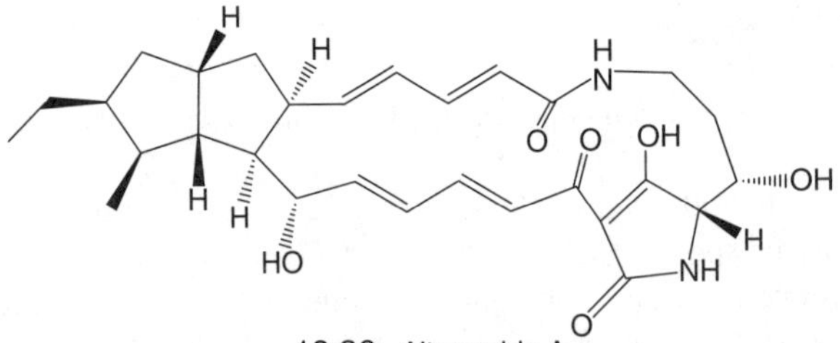

18.26: Alteramide A

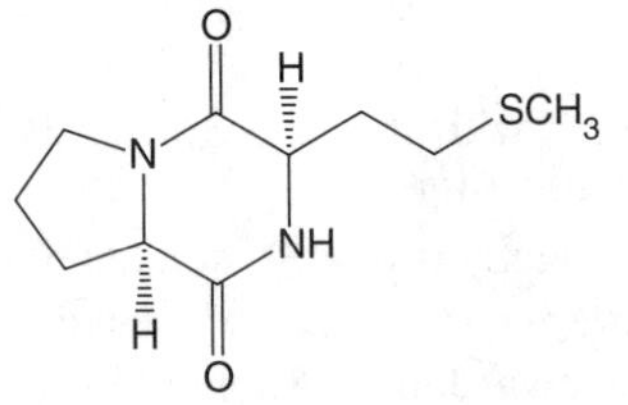

18.27: *Cyclo*-(L-proline-L-methionine)

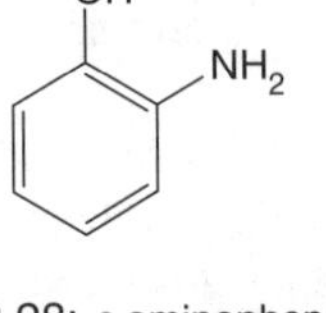

18.28: *o*-aminophenol 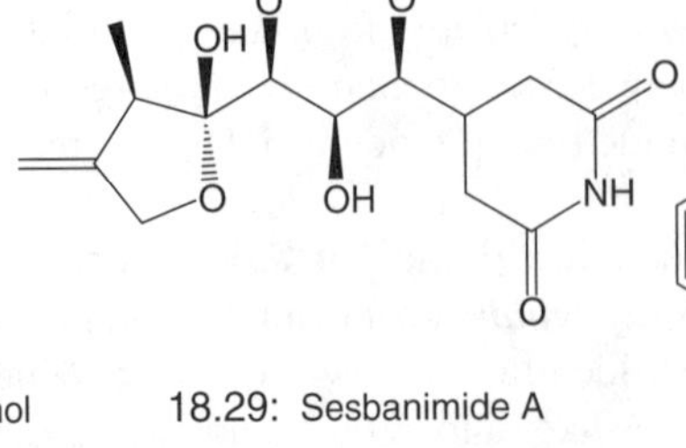   18.29: Sesbanimide A 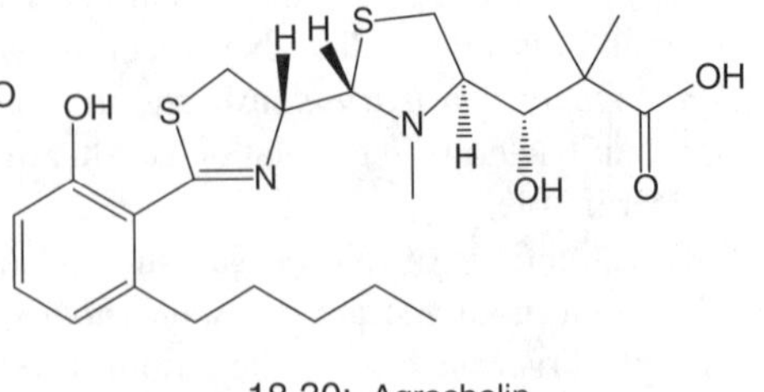   18.30: Agrochelin an enriched microbial population as a source of food.[43] Because of this sorting process, and because of microbial digestion, microorganisms are more abundant in the stomach contents of these deposit-feeders. Kobayashi et al.[44] isolated a *Flavobacterium* sp. from the marine bivalve, *Cristaria plicata*, that was collected at Ishikari Bay, Hokkaido, Japan. The culture was grown statically in enriched seawater medium and then extracted with organic solvent to yield two new sulfonolipids, flavocristamide A (Structure 18.31) and its 3,4-dihydro-analog flavocristamide B. Both flavocristamides exhibited inhibitory activity against a eukaryotic DNA replication enzyme, DNA polymerase α (IC_{50} ~ 15–20 µg/mL). Since these compounds are related to ceramide, it appears that the sulfonate group is important for the activity, since ceramide has been reported to have no effect on DNA polymerase α. In another study, a marine bacterium, *Bacillus cereus*, was isolated from the surface of a toxic snail collected in Izu Penisula, Japan.[45] Fermentation extracts produced from this culture were discovered to exhibit potent cytotoxicity activity. Two compounds, cereulide (Structure 18.32a) and homocereulide (Structure 18.32b), were isolated from the lipid extract that showed potent activity against P388 and Colon 26 tumor cell lines with an IC_{50} of 1 and 35 pg/mL, respectively. The *B. cereus* strain was never proven to be the source of the poisoning produced by the snail.

Actinomycetes have also been isolated from the relatively nutrient-rich sufaces of invertebrates and seaweeds.[6] Fenical et al.[46] isolated an unidentified Actinomycete from the surface inoculum of the Carribean brown alga *Lobophora variegata* that produced two new macrolides, lobophorins A and B (e.g., Structure 18.33). Despite their structural relationship to kijanimicin and to several related antibiotic macrolide glycosides such as tetrocarcins and chlorothricin, lobophorins A and B did not exhibit significant antibiotic properties. However, they did show potent antiinflammatory activities in the phorbol-myristate-acetate-induced mouse ear edema model. At the normal testing dose, lobophorin A and B reduced edema by 86% and 84%, respectively, and lobophorin B demonstrated *in vivo* activity when administered intraperitonly in mice.

Fenical's group[47] has also investigated actinomycetes living on the surface of marine invertebrates. They isolated a streptomycete from the surface of a gorgonian coral (*Pacifigorgia* sp.) collected from the Gulf of California, Mexico. When fermented in marine media, the isolate produced several metabolites, including the 20-hydroxy derivative of oligomycin A, the 5-deoxy derivative of enterocin, and the octalactins A and B. The octalactins belong to a new structure class, which are C_{19} ketones possessing rare eight-membered ring lactone functionalities. Octalactin A (18.34) demonstrates potent *in vitro* cyctoxicity against B16-F10 murine melanoma and HCT-116 human carcinoma cell lines. The surface of a tropical jellyfish, *Cassiopeia xamachana*, yielded a *Streptomyces* sp. that produced two new bicyclic peptides, salinamides A (Structure 18.35) and B, which have novel depsipeptide backbones.[48] The salinamides A and B exhibit activity against all Gram-positive microorganisms tested. Interestingly, salinimide A and B also demonstrate potent anti-inflamatory effects (84% inhibition at 50 µg/ear) in the inhibition of phorbol-ester-induced edema in the mouse ear model. Recently three additional minor peptides were isolated from this culture. Salinamides C and E are monocyclic depsipeptides that are likely methylated byproducts of salinamide A biosynthetic intermediates, and salinamide D contains a D-valine residue in place of the D-isoleucine moiety in salinamide A.[49]

Research at Pharma Mar resulted in the isolation of a new species of *Micromonospora* from an unidentified marine soft coral collected off the coast of Mozambique.[50] This culture produced a novel depsipeptide designated PM-93135 when fermented in a nutrient-rich medium. PM-93135 exhibited antibacterial activity against *S. aureus*, *B. subtilis*, and *Microcoous luteus* and inhibited RNA synthesis in P388 cells with an IC_{50} of 0.008 µg/mL. In addition, this compound demonstrated significant antitumor activity against P388, human lung carcinoma A-549, human colon carcinoma HT-29, and human melanoma MEL-28 with IC_{50} of 0.0002, 0.002, 0.01, and 0.0025 µg/mL, respectively. Another study involved the isolation of a *Streptomyces* sp. from the surface of a mollusc collected in Kanagawa Prefecture, Japan.[51] This culture, when fermented, produced a novel cytotoxic agent that induced apoptotic cell death. The compound, aburatubolactam C (Structure 18.36),

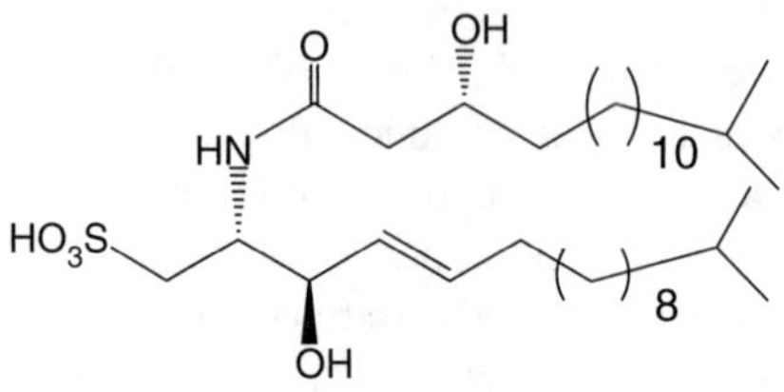

18.31: Flavocristamide A

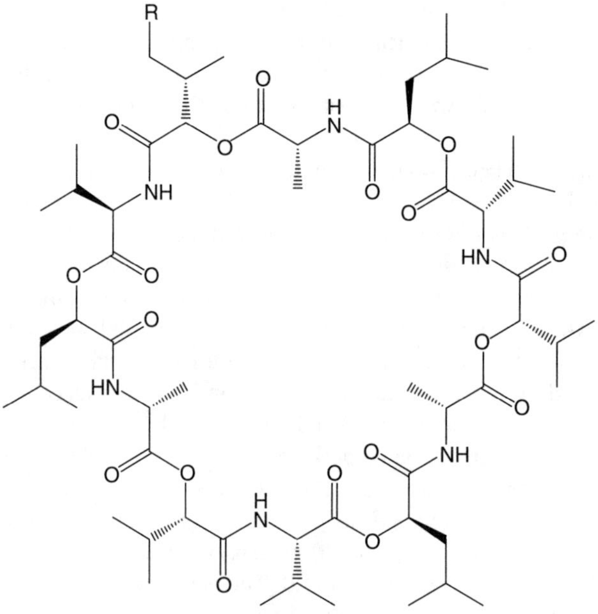

18.32a: Cerulide, R=H
18.32b: Homocerulide, R=CH$_3$

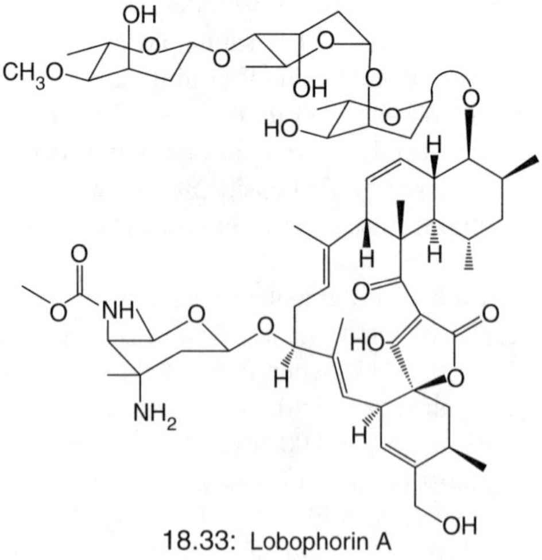

18.33: Lobophorin A

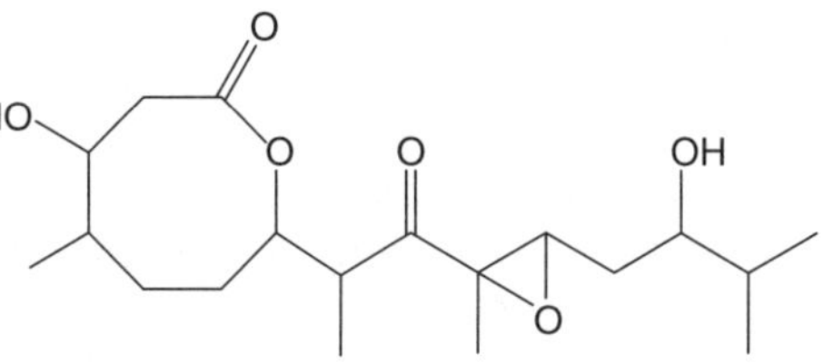

18.34: Octalactin A

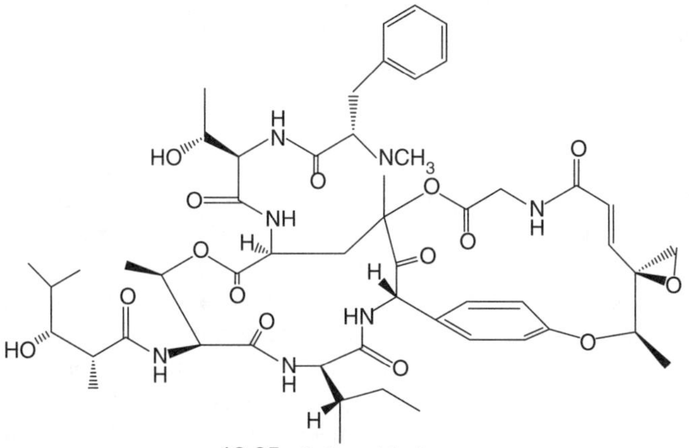

18.35: Salinamide A

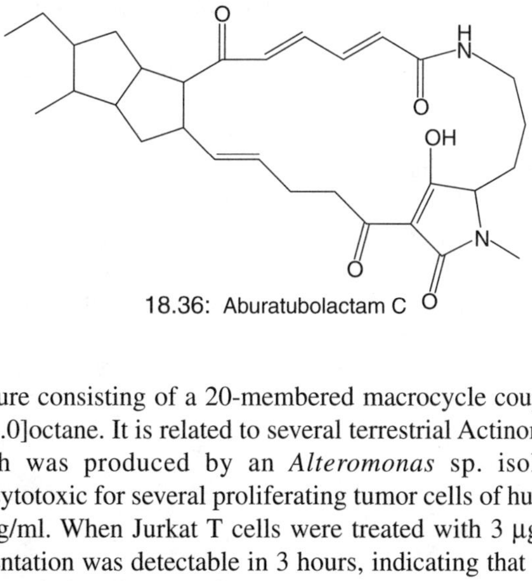

18.36: Aburatubolactam C

has a novel lactam structure consisting of a 20-membered macrocycle coupled with a unique acyl tetramine and bicyclo[3.3.0]octane. It is related to several terrestrial Actinomycete products as well as alteramide A, which was produced by an *Alteromonas* sp. isolated from a sponge. Aburatubolactam C was cytotoxic for several proliferating tumor cells of human and murine origins with IC_{50} of 0.3 to 5.8 µg/ml. When Jurkat T cells were treated with 3 µg/mL of the compound, the apototic DNA fragmentation was detectable in 3 hours, indicating that the effect of aburatubo-lactam C was attributable to induced apoptosis.

IV. SYMBIOTIC MARINE MICROORGANISMS

Even though symbiosis is a widespread phenomenon and an important agent of evolution, little is known about symbiotic relationships between prokaryotic and eukaryotic taxa.[52] The endobiotic environment is comprised of cells of microorganisms as well as the tissues of plants and animals that serve as hosts to a wide spectrum of microbial forms in a variety of relationships. These relationships include mutualism (organisms of different species live together for the benefit of both), parasitism (only the parasite derives nourishment from the host, but does not necessarily cause disease), and pathogenesis (the pathogen can cause disease in the host). Mutualistic symbiosis and disease are very important factors that affect the ecology of both microorganisms and macroorganisms in the sea. Many of these associations began as incidental interactions that later developed into a relationship of obligate mutualistic symbioses. This range of interdependence includes a wide spectrum of bacterial/host adaptations, including the biosynthesis and maintenance of unique secondary metabolites. Among the first conclusive reports of marine symbiosis are relationships between chemoautotrophic bacteria and marine invertebrates in deep-sea hydrothermal vents. The hydrothernal vents emit sulfide, which, in turn, provides the necessary energy and reducing power for chemoautotrophic bacteria. Many of these bacteria live in symbiotic asssociations with macro-invertebrates living in close proximity to these vents, sometimes directly providing their sole source of nutrition. Furthermore, some of the invertebrates living there either contain a greatly reduced digestive tract or have no digestive tract at all, forcing them to rely completely upon these endo-symbionts for their survival.[52]

Bacterial symbionts that are chemoheterotrophic have been described and isolated in pure culture. Fenical et al.[53] studied the resistance of the estuarine shrimp *Palaemon macrodactylus* to pathogens and observed that their eggs harbored bacterial epibionts. Upon removal of the bacterial epibionts, rapid infestation of the pathogenic fungi *Lagenidium callinectes* occurred. Further investigation revealed that a penicillin-sensitive *Alteromonas* species could be consistently isolated from the healthy embryos. This bacterial strain was fermented and found to produce a potent antifungal agent, 2,3-indolinedione, also known as istatin. Bacteria-free embryos could become disease-free if they were either reinoculated with the bacteria or treated with istatin. Interestingly, istatin had been known for many years as a synthetic intermediate in the production of indigo dyes but not as an antifungal agent. Gil-Turnes et al.[54] discovered a similar relationship after investigating the eggs of the American lobster *Homarus americanus*. The eggs were found to be colonized by an unidentified unicellular bacterium that produced the phenolic compound tyrosol (2-*p*-hydroxyphenol ethanol). The bacterium produces this phenol in quantities which are sufficient to control pathogenic microorganisms.

Some of the best examples of chemistry derived from marine microbial symbionts are those microorganisms responsible for producing many of the marine toxins that pose human health hazards. One of these, surugatoxin (Structure 18.37), which specifically blocks nicotinic receptors and is a causative agent of shellfish poisoning in Japan, was initially isolated from the gut of the Japanese Ivory Shell snail *Babylonia japonica*.[55] Because of the seasonal outbreaks of this toxin and the ability of the snails to become toxin-free after being introduced to a new environment, it was proposed that a microorganism may be the source of this poison. Subsequently, neosurugatoxin and prosurugatoxin, precursors of surugatoxin, were isolated from a Gram-positive *Coryneform* sp. obtained from the mid-gut of *B. japonica*.

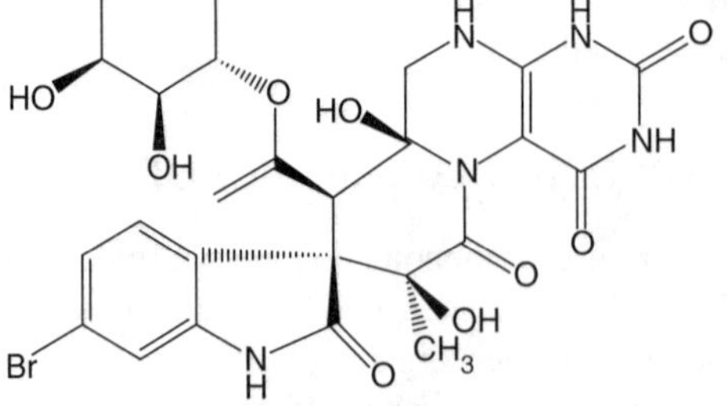

18.37: Surugatoxin

Marine sessile invertebrates are considered a particularly rich source of novel metabolites and symbiotic microorganisms.[2] Sponges, for example, may have up to 40% of their cellular volume occupied by associated bacteria. The surfaces, tissues, and internal spaces of such invertebrates provide a variety of marine microhabitats and harbor tremendous potential for diverse microorganisms. Bacteria associated with marine invertebrates experience a wide range of environmental parameters including variable pH, nutrient availability, and surface texture. Collectively, these factors may select for novel biosynthetic pathways for survival. Jackson and Buss[56] have suggested that many cryptic marine invertebrates have evolved species–species allelochemical effects against competitors. Such allelochemicals are thought to play an important role in the ecology of many algae and benthic invertebrates. Studies of sponges and holothurians have suggested that fish predation and grazing play an important role in selecting for toxicity in coral reef invertebrates. Crypticity may have evolved as a means of protection for nontoxic shell-less invertebrates, while toxic species may live both unexposed and exposed to fish predation and grazing.[56] These chemical defenses, which may be attributed to the associated symbiotic microorganisms can be quite important mediators of predation, competition, or epizoic colonization.[2] However, identification of the symbiotic microorganism(s) potentially responsible for the production of a specific defensive metabolite can be difficult. Often, only circumstantial evidence obtained from the marine invertebrate is presented to indicate that the metabolite(s) originated from a symbiotic microorganism. Such evidence usually consists of extremely low or variable metabolite yields, isolation of identical metabolites from different sources, or similarity in compound structure to metabolites already described from another microorganism.[57] In retrospect, failure to isolate the desired metabolites from microbial symbionts is not surprising. It is well known from studies of terrestrial microorganisms and now marine microorganisms, that, to be productive, the symbionts may require host-specific growth conditions, and that often these physiological parameters remain unknown.[57]

Sponges have provided more natural products with unprecedented molecular structure and bioactivities than any other phylum of marine invertebrates. In many poriferans, complex microbial communities have been described that contribute to the lives of their hosts.[58,59] In addition to providing a primary food source, microorganisms can also process waste products, transfer nutrients, produce reef-like structures, or produce secondary metabolites.[58,59] However, despite the numerous associations documented between sponges and microbial symbionts, most microorganisms remain uncultured or unculturable.[57] This remains a major challenge, and new methods of isolation that include a consortia of microorganisms in combination with growth conditions that mimic the *in situ* environment must be developed. In the 1970s through 1980s, descriptions of symbioses relied on microscopy to demonstrate the presence of specific symbionts or chemical measurements of nutrient transfer. Recently, molecular phylogenetic surveys have proven the existence of these uncultivated marine microorganisms. One example is the discovery and preliminary characterization of a marine archaeon that inhabits the tissues of a temperate water sponge.[60] The microorganism, *Cenarchaeum symbiosum* (Phylum: Crenarchaeota), inhabits a single species of sponge, *Axinella*, and grows well at temperatures of 10°C, which is over 60°C below the growth temperature optimum of any cultivated archeon species. *In situ* hybridization studies concluded which microorganism in the sponge was archael and allowed localization of the symbiont. The high abundance of a single, crenarchaeal phylotype in every specimen of *Axinella* sponge examined and the presence of active cell division in laboratory-maintained sponges over long time periods strongly suggest that this partnership is a true symbiosis.

The first definitive example of a sponge natural product being derived from its associated microorganism was demonstrated by Oclarit and co-workers.[61] They isolated a *Vibrio* species from the homogenate of the marine sponge *Hyatella* sp. collected along the coast of Oshima Island, Miyazaki, Japan. When this bacterium was cultured on marine agar, it was found to produce the compound andrimid (Structure 18.38a). This same bioactive compound was also found in the sponge extract, suggesting that the active component may be synthesized by the associated microorganism. Adrimid had been previously isolated from the culture of an *Enterobacter* sp. that is an intracellular symbiont of the Brown Planthopper *Nilaparvata lugens* and was found to exhibit potent activity against *Xanthomonas campestris* pv. *oryzae*. Needham et al.[62] also isolated andrimid and the moiramides A–C (e.g., Structure 18.38b), but from a marine *Pseudomonas fluorescens* isolated from an unidentified Alaskan tunicate. Andrimid and moiramide B both potently inhibited the growth of methicillin resistant *S. aureus*. Due to the diversity of the microorganisms producing this toxin, one can speculate that the production of this compound can be encoded by genes transferable on a plasmid.

Numerous examples exist in which structurally related or identical compounds have been reported from taxonomically distinct marine invertebrates or from animals collected in certain geographic locations.[63] The microbial-type structures of compounds isolated from marine sponges and ascidians lends support to the hypothesis that many marine invertebrate natural products may be of microbial origin. One of the most striking examples is the isolation of a new endiyne antitumor antibiotic called namenamicin, (Structure 18.39) isolated from the didemnid ascidian *Polysyncraton lithostrotum*.[64] Members of this class of compounds were first isolated from terrestrial actinomycetes in the genera *Micromonospora* and *Actinomadura*, which produce calicheamicin γ (Structure 18.40) and esperimicin, respectively.[65] Additional examples include the isoquinoline alkaloids such as ecteinascidin (Structure 18.41) and renieramycin (Structure 18.42). These compounds have been isolated from the ascidian *Ecteinascidia tubinata* and the sponge *Reiniera sp.*, with the terrestrial counterpart saframycin B (Structure 18.43), isolated from *Streptomyces lavendulae*.[66]

Faulkner et al.[67] have proposed that secondary metabolites are biosynthesized within the cells they are localized and have demonstrated that these metabolites are cellularly located within symbiotic microorganisms. Due to the failure of culturing symbiotic microorganisms, cellular localization studies have become a model of evaluating invertebrate–microbial symbioses and the production of microbially derived natural products. The chemically productive sponge *Theonella swinhoei* is well known for harboring a diverse series of biologically active secondary metabolites, many of which resemble microbial products. Bewley and Faulkner[68] analyzed the origin of these natural products by conducting separate chemical isolations from the various cell populations isolated from the intact sponge. The cells were dissociated with a "juicer" and purified by differential centrifugation. Four distinct populations were recovered: sponge cells, heterotrophic unicellular bacteria, filamentous bacteria, and from the ectosome, unicellular cyanobacteria. These were separately extracted and chemically analyzed. The macrolide swinholide A (Structure 18.44) was found to occur only in the heterotrophic unicellular bacteria, while the bicyclic peptide theopalauamide (Structure 18.45) was localized in the filamentous bacteria. These results were unexpected since both these metabolites have structural precedent as cyanobacterial products. Another unexpected result was that no major secondary metabolites were found in the extracts of either the sponge or the cyanobacteria. This work has raised as many questions as it has answered. Contrary to earlier speculation, the identification of the microorganisms that produce swinholide A and theopalauamide remains unknown. One could postulate that the filamentous bacteria have evolved from a cyanobacterial origin or that the compounds are synthesized from the cyanobacteria and have diffused to other regions of the sponge for various self-protective mechanisms. This work clearly characterizes the complexity of the ecology of marine invertebrates and their associated microorganisms and exemplifies how little we understand about the chemical ecology of marine symbionts.

Another example of a natural product produced via a symbiont results from the work of Haygood et al.[69] Using the bryozoan *Bugula neritina* as a model system, their laboratory is investigating whether the biosynthetic source of the potent bryostatins (Structure 18.46) is a marine microorganism. The bryostatins are an important family of cytotoxic macrolides based on the bryopyran ring system, and bryostatin 1 is currently in Phase II clinical trials for the treatment of leukemias, lymphomas, melanoma, and solid tumors. An important advantage of this system is that the microbial community of *B. neritina* consists of a single bacterial symbiont and few other bacteria, in contrast to the complex microbial populations of most sponges. This symbiont is a rod-shaped Gram-negative bacterium localized in the larvae of *B. neritina*. Bryostatin is also found in the larvae

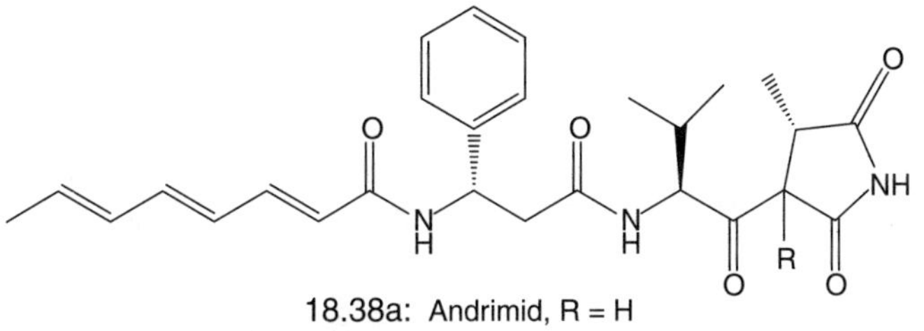

18.38a: Andrimid, R = H
18.38b: Moiramid C, R = β-OH

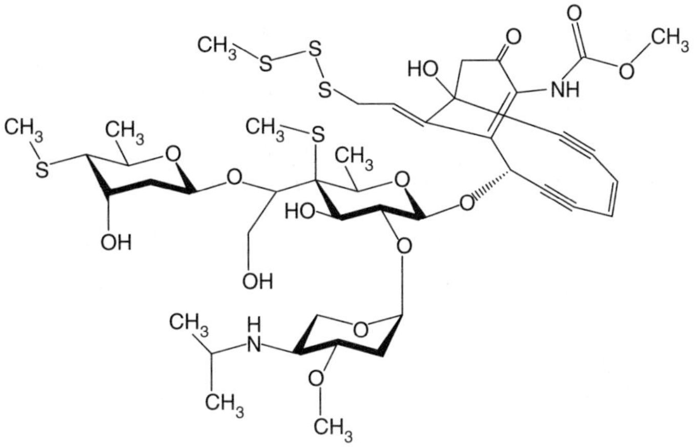

18.39: Namenamicin

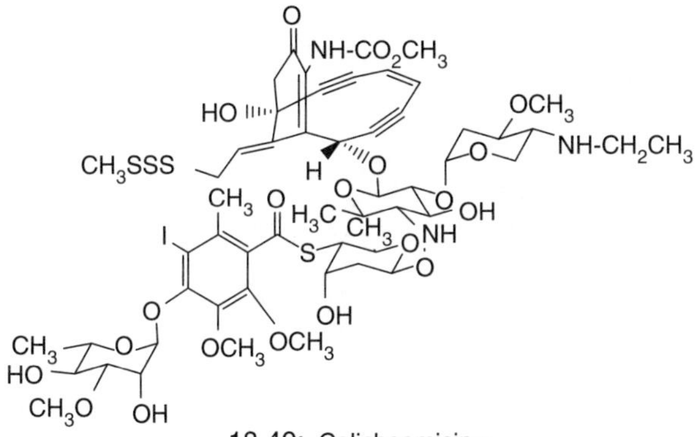

18.40: Calicheamicin γ

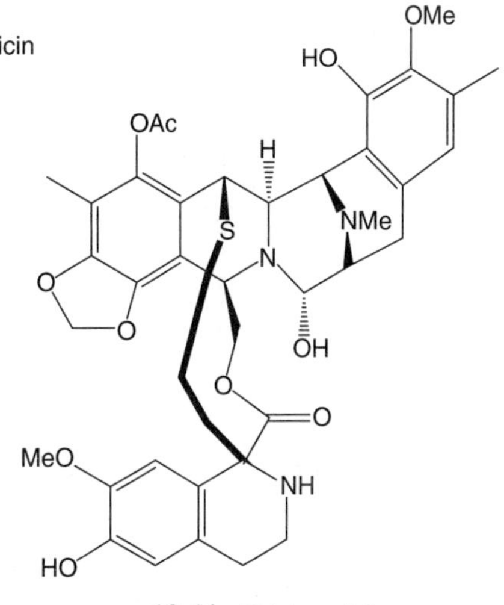

18.41: Ecteinascidin

where it may be used for chemical defense. Symbiont ribosomal sequences have been obtained, and *in situ* hybridizations have demonstrated that there are approximately 2500 symbionts per larva. In addition, analyses of sequence from larvae isolated from seven different west coast and one Atlantic coast population of *B. neritina*, shows that the same symbiont is present in all cases. Phylogenetic analysis revealed that the symbiont is a new genus of γ-proteobacterium and has been named "*Candidatus endobugula sertula*." Unfortunately, the symbiont has not been viable in culture, which may indicate it is a true symbiont. However, bryostatin is a complex polyketide. The biosynthesis of complex polyketides is well understood and the molecular biology and cloning of these pathways is a very active area of research.[70] It may be quite feasible to clone the biosynthetic genes of bryostatin and solve the problem of supply without cultivating the microorganism.

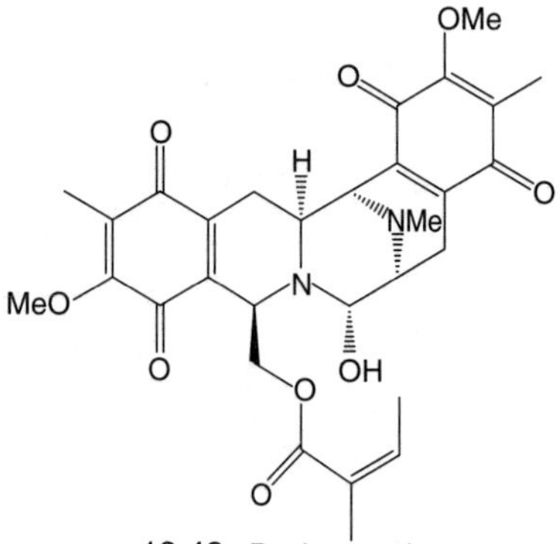

18.42: Renieramycin

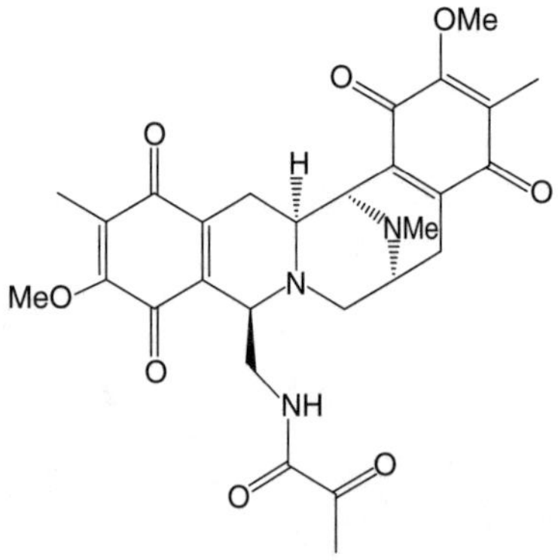

18.43: Saframycin B

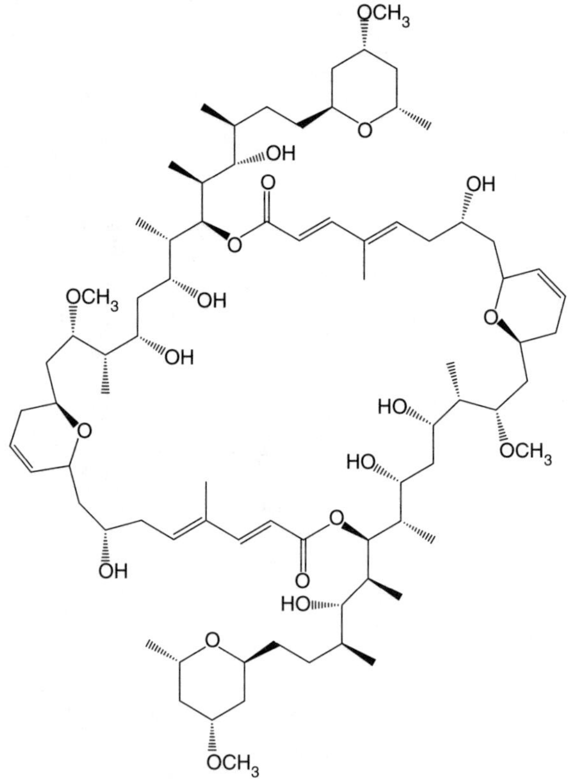

18.44: Swinholide

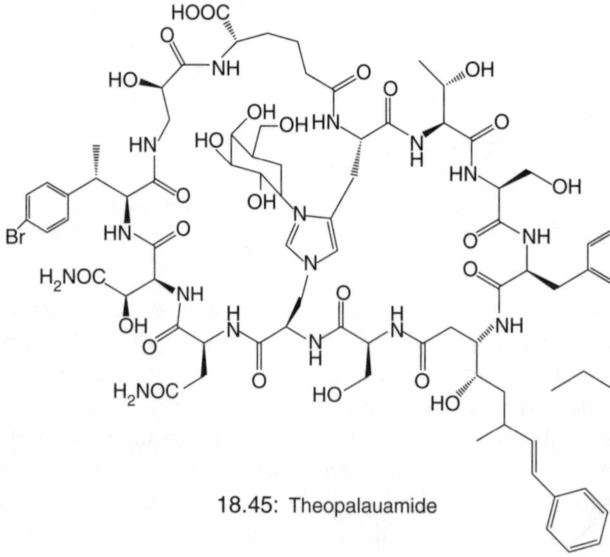

18.45: Theopalauamide

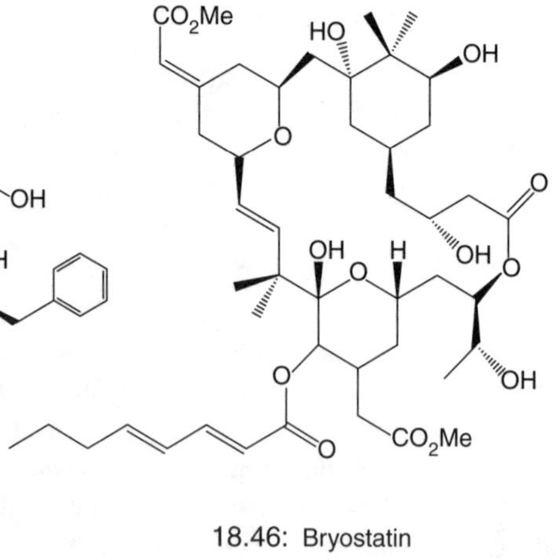

18.46: Bryostatin

V. SUMMARY

This chapter has presented examples that illustrate the diversity of microorganisms living in the sea and the plethora of natural products that has been derived from them. The structural and biosynthetic diversity of these microbial compounds coupled with their pharmacological activities is truly striking, and, clearly, the potential for discovering new natural products remains quite promising. However, most of the microorganisms and chemical entities that have been discovered are the result of traditional protocols. The majority of the eubacteria described have been isolated on conventional marine agar and cultivated in simple marine fermentation media, either in shake flasks or on solid agar surfaces. These physiological parameters are quite limited and do not mimic the natural marine environment. The difficulties encountered in the isolation and cultivation of microbial symbionts reveal the profound lack of understanding of these complex associations. Research efforts must focus on developing new isolation and fermentation methods that include a consortium of microorganisms in combination with growth conditions that are found in the original marine environment. Studies must also be initiated that focus on the effects of host metabolites upon the distribution of marine bacteria and the production of natural products. Clearly, exploration of the marine environment is just beginning, and the potential for discovering new microorganisms and novel bioactive compounds remains an exciting field of research.

REFERENCES

1. Colwell, R.R., World Bank discussion papers, in *Marine Biotechnology and Developing Countries*, Zilinskas, R.A. and Lundin, C.G., Eds., World Bank Press, Washington, D.C., 1993, IX.
2. Fenical, W., and Jensen, P.R., Pharmaceutical and bioactive natural products, in *Marine Biotechnology*, Attaway, D.A. and Zaborsky, O.R., Eds., Plenum Press, New York, 1993, 419.
3. F. Pietra, Secondary metabolites from marine microorganisms: bacteria, protozoa, algae, and fungi. Achievements and prospects, *Nat. Prod. Rep*, 14, 453, 1997.
4. Gerwick, W.H. and Sitachitta, N., Nitrogen-containing metabolites from marine bacteria, *Alkaloids*, 53, 239, 2000.
5. Sieburth, J.M., *Sea Microbes*, Oxford University Press, New York, 1979.
6. Weyland, H. and Heimke, E., Actinomycetes in the marine environment, in *The Biology of Actinomycetes '88. Proceedings of the 7th International Symposium on the Biology of Actinomycetes*, Okami, Y., Beppu, T., and Ogawara, H., Eds., Japan Science Socity, Japan, 1988, 294.
7. Jensen, P.R., Dwight, R., and Fenical, W., Distribution of actinomycetes in near-shore tropical marine sediments, *Appl. Environ. Microbiol.*, 57, 1102, 1991.
8. Okazaki, T., Kitahara, T., and Okami, Y., Studies on marine microorganisms. IV. A new antibiotic SS-228Y produced by *Chainia* isolated from shallow sea mud, *J. Antibiot.*, 28, 176, 1975.
9. Moran, M.A., Rutherford, L.T., and Hodson, R.E., Evidence for indigenous *Streptomyces* populations in a marine environment determined with a 16S rRNA probe, *Appl. Environ. Microbiol.*, 61, 3695, 1995.
10. Okami, Y., Hotta, K., Yoshida, M., Ikeda, D., Kondo, S., and Umezawa, H., New aminoglycoside antibiotics, istamycins A and B, *J. Antibiot.*, 32, 964, 1979.
11. Pathirana, C., Tapiolas, D.M., Jensen, P.R., Dwight, R., and Fenical, W., Structure determination of maduralide: a new 24-membered ring macrolide glycoside produced by a marine bacterium (Actinomycetales), *Tetrahedron Lett.*, 32, 2323, 1991.
12. Takahashi, A., Kurosawa, S., Ikeda, D., Okami, Y., and Takeuchi, T., Altemicidin, a new acaricidal an antitumor substance. I. Taxonomy, fermentation, isolation, and physico-chemical and biological properties, *J. Antibiot.*, 42, 1556, 1989.
13. Biabani, M. A. F., Baake, M., Lovisetto, B., Laatsch, H., Helmke, E., and Weyland, H., Anthranilamides: new antimicroalgal active substances from a marine *Streptomyces* sp., *J. Antibiot.*, 51, 333, 1998.
14. Jiang, Z., Jensen, P. R., and Fenical, W., Actinoflavoside, a novel flavonoid-like glycoside produced by a marine bacterium of the genus *Streptomyces*, *Tetrahedron Lett.*, 38, 5065,1997.

15. Bernan, V. S., Greenstein, M., and Maiese, W., Marine microorganisms as a source of new natural products, in *Advances in Applied Microbiology*, Neidleman, S. L. and Laskin, A. I., Eds., Academic Press, San Diego, CA, 1997, 57.

16. Parthirana, C., Jensen, P. R., Dwight, R., and Fenical, W., Rare phenazine L-quinovose esters from a marine actinomycete, *J. Org. Chem.*, 57, 740, 1992.

17. Pusecker, K., Laatsch, H., Helmke, E., and Weyland, H., Dihydrophencomycin methyl ester, a new phenazine derivative from a marine Streptomycete, *J. Antibiot.*, 50, 479, 1997.

18. Sitachitta, N., Gadepalli, M., and Davidson, B. S., New α-pyrone-containing metabolites from a marine-derived actinomycete, *Tetrahedron Lett.*, 52, 8073, 1996.

19. Pathirana, C., Jensen, P. R., and Fenical, W., Marinone and debromomarinone: antibiotic sesquiterpenoid naphthoquinones of a new structure class from a marine bacterium, *Tetrahedron Lett.*, 33, 7663, 1992.

20. Hardt, I., Jensen, P.R., and Fenical, W., Neomarinone, and new cytotoxic marinone derivatives, produced by a marine filamentous bacterium (actinomycetales), *Tetrahedron Lett.*, 41, 2073, 2000.

21. Renner, M.K., Shen, Y., Cheng, X., Jensen, P. R., Frankmoelle, W. F., Kauffman, C.A., Fenical, W., Lobkovsky, E., and Clardy, J., Cyclomarins A–C, new antiinflammatory cyclic peptides produced by a marine bacterium (*Streptomyces* sp.), *J. Am. Chem. Soc.*, 121, 11273, 1999.

22. Cheng, X. C., Jensen, P. R. and Fenical, W., Luisols A and B, new aromatic tetraols produced by an estuarine marine bacterium of the genus *Streptomyces* (Actinomycetales), *J. Nat. Prod.*, 62, 608, 1999.

23. Cheng, X. C., Jensen, P. R., and Fenical, W., Arenaric acid, a new pentacyclic polyether produced by a marine bacterium (Actinomycetales), *J. Nat. Prod.*, 62, 605, 1999.

24. Canedo, L. M., Fernandez, J. L., Faz, J. P., Acebal, C., de la Calle, F., Garcia Fravalos, D. G., and Garcia de Quesada, T., PM-94128, a new isocoumarin antitumor agent produced by a marine bacterium, *J. Antibiot.*, 50, 175, 1997.

25. Kameyama, T., Takahashi, A., Kurasawa, S., Ishizuka, M., and Okami, Y., Biscuberin, a new siderophore sensitizing tumor cells to macrophage-mediated cytolysis. I. Taxonomy of the producing organism, isolation, and biological properties, *J. Antibiot.*, 40, 1664,1987.

26. Gustafson, K., Roman, M., and Fenical, W., The macrolactins, a novel class of antiviral and cytotxic macrolides from a deep-sea marine bacterium, *J. Am. Chem. Soc.*, 111, 7519, 1989.

27. Davidson, B. S. and Schumacher, R.W., Isolation and synthesis of caprolactins A and B, new caprolactams from a marine bacterium, *Tetrahedron*, 49, 6569, 1993.

28. Schumacher, R. W., Montenegro, D. A., Davidson, B. S., and Bernan, V. S., γ-Indomycinone, a new pluramycin metabolite from a deep-sea derived actinomycete, *J. Nat. Prod.*, 58, 613, 1995.

29. Imada, C. and Okami, Y., Characteristics of marine actinomycete isolated from a deep-sea sediment and production of β-glucosidase inhibitor, *J. Mar. Biotechnol.*, 2, 109, 1995.

30. Do, H. K., Kogure, K., and Simidu, U., Identification of deep-sea sediment bacteria which produce tetrodotoxin, *Appl. Environ. Microbiol.*, 56, 1162, 1990.

31. Trischman, J. A., Jensen, P. R., and Fenical, W., Halobacillin: a cytotoxic acylpeptide of the iturin class produced by a marine *Bacillus*, *Tetrahedron Lett.*, 35, 557, 1994.

32. Fletcher, M. and Marshall, K. C., Are solid surfaces of ecological significance to aquatic bacteria?, *Adv. Microbiol. Ecol.*, 6, 199, 1982.

33. Umezawa, H., Okami, Y., Kurasawa, S., Ohnuki, T., and Ishizuka, M., Marinactin, antitumor polysaccharide produced by a marine bacterium, *J. Antibiot.*, 36, 471, 1983.

34. Yoshikawa, K., Takedera, T., Adachi, K., Nishijima, M., and Sano, H., Korormicin, a novel antibiotic specifically active against marine Gram-negative bacteria, produced by a marine bacterium, *J. Antibiot.*, 50, 949, 1997.

35. Imamura, N., Nishijima, M., Takadera, T., Adachi, K., Sakai, M. and Sano, H., New anticancer antibioitics pelagiomicins produced by a new marine bacterium *Pelagiobacter variabilis*, *J. Antibiot.*, 50, 8, 1997.

36. Gerard, J., Lloyd, R., Barsby, T., Hadan, P., Kelly, M. T., and Andersen, R. J., Massetolides A–H, antimycobacterial cyclic depsipeptides produced by two Psedomonads isolated from marine habitats, *J. Nat. Prod.*, 60, 223, 1997.

37. Gerhard, J. M., Haden, P., Kelly, M. T., and Andersen, R. J., Loloatins A–D, cyclic decapeptide antibiotics produced in culture by a tropical marine bacterium, *J. Nat. Prod.*, 62, 80, 1999.

38. Shigemori, H., Bae, M. A., Yazawa, K., Sasaki, T., and Kobayashi, J., Altermide A, a new tetracyclic alkaloid from a bacterium *Altermonas* sp. associated with the marine sponge *Halichondria okadai*, *J. Org. Chem.*, 57, 4317, 1992.

39. Jayatilake, G. S., Thornton, M. P., Leonard, A. C., Grimwade, J. E., and Baker, B. J., Metabolites from an Antarctic sponge-associated bacterium, *Pseudomonas aeruginosa*, *J. Nat. Prod.*, 59, 293, 1996.

40. Oclarit, J. M., Shinji, O., Kazuo, K., Yukiho, Y., and Susumu, I., Production of an antibacterial agent, *o*-aminophenol, by a bacterium isolated from the marine sponge, *Adocia* sp., *Fish Sci.*, 60, 559, 1994.

41. Acebal, C., Alcazar, R., Canedo, L. M., Rodriguez, P., Romero, F., and Puentes, J. L. F., Two marine agrobacterium producers of sesbanimide antibiotics, *J. Antibiot.*, 51, 64, 1998.

42. Acebal, C., Canedo, L. M., Puentes, J. L. F., Baz, J. P., and Romero, F., Agrochelin, a new cytotoxic antibiotic from a marine *Agrobacterium*, *J. Antibiot.*, 52, 983, 1999.

43. Hylleberg, J. and Gallucci, V. F., Selectivity in feeding by the deposit-feeding bivalve *Macoma nasuta.*, *Mar. Biol.*, 32, 167, 1975.

44. Kobayashi, J., Mikami, S. M., Shigemore, H., Takao, T., Shimonishi, Y., Izuta, S., and Yoshida, S., Flavocristamides A and B, new DNA polymerase α inhibitors from a marine bacterium *Flavobacterium* sp., *Tetrahedron*, 51, 10487, 1995.

45. Wang, G., Kuramoto, M., Yamada, K., Yazawa, K., and Uemura, D., Homocereulide, an extremely potent cytotoxic depsipeptide from the marine bacterium *Bacillus cereus*, *Chem. Lett.*, 791, 1995.

46. Jiang, Z., Jensen, P., and Fenical, W., Lobophorins A and B, new antiinflammatory macrolides produced by a tropical marine bacterium, *Biol. Med. Chem. Lett.*, 9, 2003, 1999.

47. Tapiolas, D. M., Roman, M., Fenical, W., Sout, T. J., and Clardy, J., Octalactins A and B, cytotoxic eight-membered ring lactones from a marine bacterium, *Streptomyces* sp., *J. Am. Chem. Soc.*, 113, 4682, 1991.

48. Trischman J., Tapiolas, D. M., Fenical, W., Dwight, R., and Jensen, P. R., Salinamide A and B antiinflammatory depsipeptides from a marine streptomycete, *J. Am. Chem. Soc.*, 116, 757, 1994.

49. Moore, B. S., Trischman, J. A., Seng, D., Kho, D., Jensen, P. R., and Fenical, W., Salinamides, antiinflammatory depsipeptides from a marine streptomycete, *J. Org. Chem.*, 64, 1145, 1999.

50. Baz, J. P., Millan, F. R., DeQuesada, T. G., and Gravalos, D. G., New Thiodepsipeptide Isolated from a Marine Actinomycete, International Patent WO 95/27730, 1995.

51. Bae, M., Yamada, K., Uemura, D., Seu, J., and Kim, Y. H., Aburatubolactam C, a novel apoptosis-inducing sunstance produced by marine *Streptomyces* sp. SCRC A-20, *J. Microbiol. Biotechnol.*, 8, 455, 1998.

52. Distal, D. L., Lane, D. J., Olsen, G. J., Giovanni, S. J., Pace, B., Pace, N. R., Stahl, D. A., and Felbeck, H., Sulfur-oxidizing bacterial endosymbionts: analysis of phylogeny and specificity by 16S rRNA sequences, *J. Bacteriol.*, 170, 2506, 1988.

53. Gil-Turnes, M. S., Hay, M. E., and Fenical, W., Symbiotic marine bacteria chemically defend crustacean embryos from a pathogenic fungus, *Science*, 246, 117, 1989.

54. Gil-Turnes, M. S. and Fenical, W., Embryos of *Homarus americanus* are protected by epibiotic bacteria, *Biol. Bull.* 182, 105, 1992.

55. Kosuge, T., Tsuji, K., Harai, K., and Fukuyama, T., First evidence of toxin production by bacteria in a marine organism, *Chem. Pharm. Bull.*, 33, 3059, 1985.

56. Buss, L. W. and Jackson, J. B. C., Competitive networks; nontransitive competitive relationships in cryptic coral reef environments, *Am. Nat.*, 113, 223, 1975.

57. Jensen, P. R. and Fenical, W., Strategies for the discovery of secondary metabolites from marine bacteria, ecological perspectives, *Annu. Rev. Microbiol.*, 48, 559, 1994.

58. Schumann-Kindel, G., Bergbauer, M., Manz, W., Szewzyk, U., and Reitner, J., Aerobic and anaerobic microorganisms in modern sponges: a possible relationship to fossilization-processes, *Facies*, 36, 268, 1997.

59. Brunton, F. R. and Dixon, O. A., Siliceous sponge-microbe biotic associations and their recurrence through the Phanerozoic as reef mound constructors, *Palaios*, 9, 370, 1994.

60. Preston, C. M., Wu, K. Y., Molinski, T. F., and Delong, E. F., A psychrophilic crenarchaeon inhabits a marine sponge: *Cenarchaeum symbiosum* gen. Nov., sp. nov., *Proc. Natl. Acad. Sci.*, 93, 6241, 1996.

61. Oclarit, J. M., Okada, H., Ohta, S., Kaminura, K., Yamaoka, Y., Lizuka, T., Miyashiro, S., and Ikegami, S., Anti-bacillus substance in the marine sponge, *Hyatella* species, produced by an associated *Vibrio* species bacterium, *Microbios*, 78, 7, 1994.

62. Needham, J., Andersen, R. J., and Kelly, M. T., Andrimid and Moiramides A–C, metabolites produced in culture by a marine isolate of the bacterium *Pseudomonas fluorescens*: structure elucidation and biosynthesis, *Tetrahedron Lett.*, 37, 1327, 1994.

63. Faulkner, D. J., He, H. Y., Unson, M. D., and Bewley, C. A., New metabolites from marine sponges: are symbionts important?, *Gazz. Chim. Ital.*, 123, 301, 1993.

64. McDonald, L. A., Capson, T. L., Kirshnamurthy, G., Ding, W. D., Ellestad, G. A., Bernan, V. S., Maiese, W. M., Lasota, P., Discafani, C., Kramer, R. A., and Ireland, C. M., Namenamicin, a new enediyne antitumor antibiotic from the marine ascidian *Polysyncraton lithostrotum*, *J. Am. Chem. Soc.*, 118, 10898, 1996.

65. Doyle, T.W. and Borders, D. B., Enediyne antitumor antibiotics, in *Enediyne Antibiotics as Antitumor Agents*, Borders, D. B. and Doyle, T. W., Eds., Marcel Dekker, New York, 1995, chap. 1.

66. Moore, B. S., Biosynthesis of marine natural products: microorganisms and macroalgae, *Nat. Prod. Rep.*, 16, 653, 1999.

67. Bewley, C. A., Holland, N. A., and Faulkner, D. J., Two classes of metabolites from *Theonella swinhoei* are localized in distinct populations of bacterial symbionts, *Experientia*, 52, 71, 1996.

68. Bewley, C. A. and Faulkner, D. J., Lithistid sponges, star performers or hosts to the stars?, *Angew. Chem. Int. Ed. Engl.*, 37, 2162, 1998.

69. Haygood, M. G., Schmidt, E. W., Davidson, S. K., and Faulkner, D. J., Microbial symbionts of the marine invertebrates: opportunities for microbial biotechnology, *Mol. Microbiol. Biotechnol.*, 1, 33, 1999.

70. Katz, L. and Donadio, S., Polyketide synthesis: prospects for hybrid antibiotics, *Annu. Rev. Microbiol.*, 47, 875, 1993.

Index

A